Teucrium Species: Biology and Applications

Milan Stanković
Editor

Teucrium Species: Biology and Applications

Springer

Editor
Milan Stanković
Department of Biology and Ecology, Faculty of Science
University of Kragujevac
Kragujevac, Serbia

ISBN 978-3-030-52158-5 ISBN 978-3-030-52159-2 (eBook)
https://doi.org/10.1007/978-3-030-52159-2

This Springer imprint is published by the registered company Springer Nature Switzerland AG
The registered company address is: Gewerbestrasse 11, 6330 Cham, Switzerland

Foreword

Nature is a major source of our current medicines, and many (semi)synthetic medicines have been developed from natural products. These compounds have been found by studying traditional medicines and by screening extracts from various organisms, particularly plants and microorganisms. In fact, about 25% of all western drugs are from plants. Microorganisms are particularly the source of antibiotics and antitumor medicines. In the meantime, for most ailments, good medicines are available; however, in the field of antibiotics and parasitic diseases, the resistance of microbes and parasites against available medicine is causing a major problem. New drugs are urgently needed, drugs that particularly should be available for the poor in developing countries at a reasonable price. As it concerns drugs cure a patient, the use is only for short periods. Consequently, the market is small if compared to, for example, an antihypertension drug that a patient takes every day for many years. Therefore, it is difficult to earn back the high costs for developing a new drug. The big pharma is thus not investing effort in developing such drugs, but if good leads are available, they may develop these further, as we can learn from the Nobel Prize of 2015. That prize shows the interest from the medical field for nature as a source of medicine, either by bioprospecting or by studying traditional medicines.

There is thus a unique opportunity for governmental and academic research institutes to look for novel leads from Nature for infectious and parasitic diseases. Particularly in Asia, with a well- documented traditional use of medicinal plants to treat patients, there is host of projects aiming at evidence-based traditional

medicines and new leads for drug development. With the enormous biodiversity, with plants only having a rather small number of species (estimations run from 250,000 to 350,000) if compared with the largest group that of the insects (some 30 million species), there are ample materials to choose from. Concerning plants, I wrote in the year 2000 that only about 15% were studied phytochemically and 6% for one or more biological activities. At that time, the Dictionary of Natural Products had 139,000 entries. At present, the Dictionary of Natural Products claims to have now more than 300,000 entries (September 2019), which means an average increase of about 8000 compounds per year. That is quite a lot, but when we assume that each biological species can make one unique compound, there should be 10–100 million natural products, that means compared with the number of known ones, there must be many more compounds waiting to be discovered. Obviously, we will not be out of work for the coming years. But at the same time, it is clear that we must make choices for the goals of our research. The choices are made on the basis of previously reported results and the needs of the society for novel medicines and other sustainable products. Also, we will have to choose what approach to use, for example, to focus on evidence-based traditional uses or at random screening of extracts, which plant species or genera should be prioritized, and what type of activity does one focus on. Many questions and few answers, only a thorough study of the literature can help us make choices.

Publications like the present book are very helpful in making our choices, as one can learn about what type of activity is already shown and if that is of sufficient interest to warrant further studies. What kind of compounds may one expect, are they novel, and do they look like good candidates for further development from hit to lead, or is it already clear from other studies that the compounds are not promising. The editor have done an excellent job in compiling all the information and making this book a good navigator for further studies, as all aspects of the plants from the genus *Teucrium* are dealt with in this book. It will be an excellent guide for making new discoveries.

Professor emeritus, Institute of Biology
Leiden University,
Leiden, The Netherlands

Robert Verpoorte

Preface

Species of the genus *Teucrium* (Lamiaceae) are widespread in different climates as well as on different habitat types especially in the Mediterranean region. With more than 300 species, *Teucrium* is one of the largest genera of the Lamiaceae family. Some species have been used in folk medicine since Antiquity—the name *Teucrium* refers to the Achaean hero Teucros (Teucer), a great archer during the Trojan War. Species of this genus are commonly known as germanders. The species of genus *Teucrium* are mainly perennial herbaceous plants, shrubs, or subshrubs, while only a small number are annual species.

Known medicinal species belonging to the genus *Teucrium* are used as sources of natural therapy and beneficial bioactive compounds. Also, some *Teucrium* species are useful as spices, for the preparation of teas and bitter drinks. The species of the genus *Teucrium* are very rich in a variety of secondary metabolites with very significant biological activities. For medicinal purposes, these species are used in the treatment of digestive and respiratory disorders, abscesses, gout, and conjunctivitis as well as in the stimulation of fat and cellulite decomposition; they also possess anti-inflammatory, antioxidative, anticancer, antimicrobial, antidiabetic, and anthelminthic properties. However, their most significant therapeutic effect is the elimination of some digestive and respiratory problems.

Teucrium species are an interesting object of research in the numerous scientific projects from various aspects of science, with multiple applications. Numerous studies on the biological activity and phytochemical composition of plant extracts and essential oils, as well as biotechnological applications of *Teucrium* species, are based on ethnobotany as a primary source. All of these influenced the need for a book publication that will integrate such aspects. On that basis, the book *Teucrium Species: Biology and Applications* includes 15 systematically grouped chapters which highlight recent advances in exploring the unique features of *Teucrium* species. The chapters describe the systematics, morphology, ecology, biogeography, ethnobotany, and phytochemistry—secondary metabolites diversity; genotoxic, antioxidant, antibacterial, antifungal, antiviral, anticancer, anticholinesterase, and antidiabetic and anti-inflammatory activity of *Teucrium* secondary metabolites; as well as applications in biotechnology, food industry, and pharmacy, including

current challenges and further perspectives. Excessive use of some medicinal plants of this genus can cause liver inflammation. This phenomenon and precaution are described in detail in a separate chapter in order to promote the safe use of certain species.

A book about *Teucrium* species will have several advantages as a literary material. First of all, this book is a comprehensive overview of the scientific information on the current achievements of research and their application of *Teucrium* species, as well as a rich source of the up-to-date bibliography of all scientific and practical fields related to *Teucrium* genus. The book will also provide the current list of the species of *Teucrium* genus, as well as color photographs of many species. Scientists in the field of biology, biotechnology, and pharmacy will have all the information and current literature in one place about *Teucrium* species from all aspects of science. The book would be initial literature review for beginners in the investigations of *Teucrium* species. Science enthusiasts and practitioners involved in medicinal plant applications will have the opportunity to get to know everything about the *Teucrium* species used in everyday life in Ethnomedicine for the treatment of various diseases and as the sources of additives as well as preservatives. The book can be used as a course literature at the university level. It is suitable for PhD students who focus on research on *Teucrium* species or other types of plant biological activity and applications.

Kragujevac, Serbia Milan Stanković

Acknowledgments

The book *Teucrium Species: Biology and Applications* includes 15 chapters written by 32 scientists in the field of Plant Biology and Ecology, Systematics, Morphology, Physiology, as well as Chemistry, Genetics, Microbiology, Molecular Biology, Pharmacognosy, Biotechnology, Food Sciences, Medicine and Pharmacy. Many of them are leading scientists in their field, my colleagues from the University, associates from foreign institutions, as well as laboratory colleagues from my research team working on the investigations of *Teucrium* species. I am highly grateful to all the chapter authors for their professional contributions and productive collaboration.

I am grateful to my colleague Prof. Dr. Teresa Navarro del Aguila (University of Malaga, Spain) for the expertise, effort, and valuable time devoted, as well as to other colleagues Prof. Dr. Trinidad Ruiz Tellez, Dr. Jose Blanco Salas (University of Extremadura, Spain) and Dr. Abhay Mishra (HNBG University, India) for friendly support.

I am thankful to the Springer's editorial team, especially to Mamta Kapila and Melanie van Overbeek, Joseph Daniel, and Anitta Camilya for valuable cooperation and guidance during initiating and editing process, as well as many other individuals acknowledged throughout the book.

I would like to specially thank Professor Emeritus Dr. Robert Verpoorte (Leiden University, The Netherlands) for supporting the book concept and idea in his Foreword to this book.

Book Editor
Prof. Dr. Milan Stanković

Contents

About the Editor

Milan Stanković (1984) is an Associate Professor of Plant Science in the Department of Biology and Ecology, Faculty of Science, University of Kragujevac, Republic of Serbia, and Head of Department of Biology and Ecology. He earned his BSc in Biology from the Faculty of Science and Mathematics at the University of Niš, in 2007. Dr. Stanković began his scientific and teaching career as an Assistant in the Department of Biology and Ecology, Faculty of Science, University of Kragujevac (2008). He acquired PhD degree in Plant Science (2012) from the same University with the thesis "Biological activity of secondary metabolites of *Teucrium* L. species from Serbian flora." He completed Postdoctoral Research at the Université François-Rabelais de Tours, France. In the Faculty of Science, he was appointed as an Assistant Professor (2013) as well as Associate Professor (2019) and taught several BSc, MSc, and PhD courses on Plant Science as well as supervised thesis and dissertations in this field. His current research is focused on the biology, ecology, and applications of *Teucrium* species. Dr. Stanković is (co-)author of over 300 references including articles in peer-reviewed journals, edited books, book chapters, conference papers, meeting abstracts, etc. He is an editor, editorial board member, and reviewer of several scientific journals. Currently, he works as an Associate Editor of *Plants* (2012-) and *Rangeland Ecology and Management* (2015–). He was a member of the scientific and organizing committee of several conferences, congresses,

and symposia in the fields of Biology and Plant Sciences. During his university career, he was a member of various scientific and professional associations such as the Society for Experimental Biology—Italy, Society for Medicinal Plants and Natural Product Research, Société Botanique de France, Organization for the Phyto-Taxonomic Investigation of the Mediterranean Area, and Serbian Biological Society. Currently he is a member of the State Commission for Expertise in the publication of school textbooks in Biology. In scientific collaboration, Dr. Stanković has published co-authored papers with colleagues from 23 various international universities from 17 countries. During his scientific career, he participated in the realization of over 30 scientific projects and studies.

List of Contributors

Christina Barda Department of Pharmacognosy and Chemistry of Natural Products, Faculty of Pharmacy, National and Kapodistrian University of Athens, Athens, Greece

Nada Bezić Department of Biology, Faculty of Science, University of Split, Split, Croatia

Lucia Caputo Department of Pharmacy, University of Salerno, Fisciano, Italy

Raffaele Coppola Department of Agricultural, Environmental and Food Sciences, University of Molise, Campobasso, Italy

Danijela Cvetković Department of Biology and Ecology, Faculty of Science, University of Kragujevac, Kragujevac, Serbia

Vincenzo De Feo Department of Pharmacy, University of Salerno, Fisciano, Italy

Laura De Martino Department of Pharmacy, University of Salerno, Fisciano, Italy

Valerija Dunkić Department of Biology, Faculty of Science, University of Split, Split, Croatia

Florinda Fratianni Institute of Food Sciences, National Research Council, Avellino, Italy

Maria-Eleni Grafakou Department of Pharmacognosy and Chemistry of Natural Products, Faculty of Pharmacy, National and Kapodistrian University of Athens, Athens, Greece

Darko Grujičić Department of Biology and Ecology, Faculty of Science, University of Kragujevac, Kragujevac, Serbia

Dragana Jakovljević Department of Biology and Ecology, Faculty of Science, University of Kragujevac, Kragujevac, Serbia

Snežana Jarić Department of Ecology, Institute for Biological Research "Siniša Stanković" - National Institute of the Republic of Serbia, University of Belgrade, Belgrade, Serbia

Irina Karadjova Department of Analytical Chemistry, Faculty of Chemistry and Pharmacy, University of Sofia, Sofia, Bulgaria

Dragoslav Marinković Serbian Academy of Sciences and Arts, Belgrade, Serbia

Nevena Mihailović Department of Chemistry, Faculty of Science, University of Kragujevac, Kragujevac, Serbia

Vladimir Mihailović Department of Chemistry, Faculty of Science, University of Kragujevac, Kragujevac, Serbia

Olivera Milošević-Djordjević Department of Biology and Ecology, Faculty of Science, University of Kragujevac, Kragujevac, Serbia

Department of Genetics, Faculty of Medical Sciences, University of Kragujevac, Kragujevac, Serbia

Milena Milutinović Department of Biology and Ecology, Faculty of Science, University of Kragujevac, Kragujevac, Serbia

Miroslava Mitrović Department of Ecology, Institute for Biological Research "Siniša Stanković"- National Institute of the Republic of Serbia, University of Belgrade, Belgrade, Serbia

Teresa Navarro Department of Botany and Plant Physiology, University of Malaga, Malaga, Spain

Marija Nazlić Department of Biology, Faculty of Science, University of Split, Split, Croatia

Filomena Nazzaro Institute of Food Sciences, National Research Council, Avellino, Italy

Taner Özcan Department of Biology Education, Necatibey Education Faculty, Balıkesir University, Balıkesir, Turkey

Dolja Pavlova Department of Botany, Faculty of Biology, University of Sofia, Sofia, Bulgaria

Pavle Pavlović Department of Ecology, Institute for Biological Research "Siniša Stanković"- National Institute of the Republic of Serbia, University of Belgrade, Belgrade, Serbia

Helen Skaltsa Department of Pharmacognosy and Chemistry of Natural Products, Faculty of Pharmacy, National and Kapodistrian University of Athens, Athens, Greece

Jelena S. Katanić Stanković Department of Chemistry, Faculty of Science, University of Kragujevac, Kragujevac, Serbia

Milan Stanković Department of Biology and Ecology, Faculty of Science, University of Kragujevac, Kragujevac, Serbia

Olgica Stefanović Department of Biology and Ecology, Faculty of Science, University of Kragujevac, Kragujevac, Serbia

Elma Vuko Department of Biology, Faculty of Science, University of Split, Split, Croatia

Nenad Zlatić Department of Biology and Ecology, Faculty of Science, University of Kragujevac, Kragujevac, Serbia

Chapter 1
Systematics and Biogeography of the Genus *Teucrium* (Lamiaceae)

Teresa Navarro

Abstract *Teucrium* L. is the second-largest genus of subfamily Ajugoideae with a subcosmopolitan distribution and 434 taxa (Govaerts RA, Paton A, Harvey Y, Navarro T, Del Rosario Garcia Pena M, World checklist of Lamiaceae. The Board of Trustees of the Royal Botanic Gardens Kew, Kew. www.kew.org/wcsp/, 2013). The Mediterranean region and surrounding floristic areas are the main center of diversity with around 90% of the total *Teucrium* species in the world. We describe the systematic of *Teucrium* following Bentham (Labiatarum genera et species. Ridgeway Sons, London, 1833), Boissier (Flora Orientalis IV. H. Georg, Geneva/Basel, 1879) and Kästner (Übersicht zur systematischen gliederung der gattung *Teucrium* L. Biocosme Mésogéen 6:63–78, 1989) with the recognition of the independent status of the section *Montana* Lazaro Ibiza (Navarro, *Teucrium* L. In: Castroviejo S et al (eds) Flora Iberica, VerbenaceaeLabiataeCallitrichaceae, vol 7. Real Jardín Botánico, CSIC, Madrid, pp 30–166, 2010). 341 herbarium specimens of 97 *Teucrium* taxa from throughout the world were studied, three taxa of its phylogenetic related genera *Spartothamnella* Briq., *Oncinocalyx* F. Muell., and *Teucridium* F. Hook, and two taxa of its segregated genera *Rubiteucris* Kudô and *Leucosceptrum* Smith. Based on the species biogeographical distribution and the main discriminant systematic characters, almost five major biogeographic and taxonomic species groups can differentiate. The group of species with 2-lipped corolla and zygomorphic calyx (sections *Pycnobotrys* Benth., *Stachyobotrys* Benth., *Scorodonia* (Hill) Schreb. and *Teucriopsis* Benth.) which provide a clear example of the Eurasian pattern of radiation from C Asia and China. The group of species with 1-lipped corolla and subactinomorphic calyx (sections *Polium* (Mill.) Schreb., *Chamaedrys* (Mill.) Schreb., *Montana* Lazaro Ibiza, *Isotriodon* Boiss., *Scordium* (Mill.) Schreb., and subsection *Pumila*), which represent the Mediterranean pattern of radiation, with a high degree of recent diversification. The typical Mediterranean species with 1-lipped corolla and actinomorphic calyx, consistent with the bulk of species of section *Teucrium* (Kästner, Übersicht zur systematischen gliederung der gattung *Teucrium* L. Biocosme Mésogéen 6:63–78, 1989). The Australian and S America 1-lipped corolla species with actinomorphic calyx, (*Teucrium* subsection

T. Navarro (✉)
Department of Botany and Plant Physiology, University of Malaga, Malaga, Spain
e-mail: tnavarro@uma.es

M. Stanković (ed.), *Teucrium Species: Biology and Applications*,
https://doi.org/10.1007/978-3-030-52159-2_1

Cretica (pro part)) (Kästner, Übersicht zur systematischen gliederung der gattung *Teucrium* L. Biocosme Mésogéen 6:63–78, 1989). Finally, the species with unclear 1-lipped corolla and actinomorphic calyx (Australian and Cape Region species belong to section *Teucrium*). The section *Teucrium* is the most heterogeneous and distinctive group within *Teucrium* with three different geographical distribution in Australian, N America and Mexico *Teucrium* while as other Ajugoideae genera closed to *Teucrium* such as Schnabelia Hand.-Mazz. All the sections occurs in the Mediterranean area, NW Africa, N Sahara Desert (Morocco, Algeria, Tunisia and Libya), N Egypt to the Ethiopia mountains, SW Arabian mountains, the Sinai Peninsula and through the arid regions up through W Asia and Europe. Section *Teucriopsis* is exclusive of Canary and Madeira islands, section *Pycnobotrys* distributed only in SE Asia and section *Leucosceptrum* Smith., exclusive from India.

Keywords *Teucrium* · *Teucrium* sections · Systematics · Biogeography · *Spartothamnella* · *Oncinocalyx* · *Teucridium* · *Rubiteucris* · Lamiaceae

1.1 Introduction

Teucrium L. is a polymorphic and widespread genus of Lamiaceae family including about 434 taxa (Appendix I) currently recognised species (Navarro and El Oualidi 2000a; Govaerts et al. 2013). They are shrubs, subshrubs and perennial herbs (rarely annuals or biennials) (Appendix II) with a subcosmopolitan distribution occurring mainly in the Mediterranean region and in the temperate parts of Asia (King 1988; Hedge 1992; Navarro and El Oualidi 2000a; Harley et al. 2004; Navarro 2010). *Teucrium* has an unusual distribution in the Lamiaceae family, only 7% of its species are present in the southern hemisphere (Australia, New Zealand, Cape Region and Argentina), while 93% are present in the north (Hedge and Miller 1977; Navarro and El Oualidi 2000a). The Mediterranean region, without doubt represents the major area of distribution, since it is represented by more of the 90% of the total of species in the world (Navarro and El Oualidi 2000a; Blanca et al. 2017) being the only genus of the subfamily Teucrioideae in the area (Cantino et al. 1992).

Teucrium has been divided into several sections mainly based on calyx shape, inflorescence structure and plant habitat. Bentham (1833) made the most relevant taxonomic classification with additions of Boisser (1879), and the most recent is whose of Kästner (1989). *Teucrium* is a paraphyletic genus (Salmaki et al. 2016) based on the placement of *Spartothamnella*, *Oncinocalyx* and *Teucridium* within it. All the sections, following Bentham (1833), occurs in the Mediterranean area, NW Africa, N Sahara Desert (Morocco, Algeria, Tunisia and Libya), N Egypt to the Ethiopia mountains, SW Arabian mountains, the Sinai Peninsula and through the arid regions up through W Asia and Europe. *Teucridium* is exclusive of Canary and Madeira islands, section *Pycnobotrys* distributed only in SE Asia and section *Leucosceptrum* exclusive from India.

The most recent hypothesis on evolutionary relationships within *Teucrium* have been proposed based on phylogenetic and chromosomatic studies, Salmaki et al. (2016) and Massoud et al. (2018) respectively. These authors proposed two main hypotheses, one of them considering the *Teucrium* section *Teucrium* (*Teucrium* core clade), as an old lineage within the genus as well as the Australian and S African lineages. The second and most speculative, consider the probable origin of the genus was the Mediterranean region, likely with the first aneuploidy originated in the Old World and dispersed in the New World (Massoud et al. 2018).

Due to the size of the taxa under study, a sample of 341 specimens of 97 *Teucrium* taxa from throughout the world have been studied in addition to the previous biosystematic revisions (Navarro and El Oualid 2000a; Navarro 1995, 2010). Species was taken to represent the total taxonomic variation and geographical distribution of *Teucrium*. For geographical regions where no infrageneric divisions of *Teucrium* have hitherto been recognised, representative species were chosen from the regional Floras. The monotypic genera *Oncinocalyx* and *Teucridium* were both examined. In *Sparthotamnella*, *Leucosceptrum* and *Rubiteucris* one member from each genera was chosen (Appendix III). The studied specimens are housed at: (Spain) MA, MAF, BC, SEV, MGC, GDAC, VAL, ABH, JACA; (Portugal) LISU; (United Kingdom) KEW, EDI; (Australia) NE, CANB; (New Zealand) CHR; (Italy) CAG; (France) MPU (for herbarium abbreviations see Holmgren et al. 1990).

In this chapter, we will discuss the systematic and biogeography of *Teucrium* focusing in the relationships within *Teucrium* infrageneric taxa and between *Teucrium* and its close related phylogenetic genera *Spartothamnella*, *Oncinocalyx* and *Teucridium*.

1.2 Systematic of *Teucrium* Genus and Its Infrageneric Delimitation Taxa

Teucrium has a corolla usually 1-lipped or 2-lipped, 5-lobed (with the anterior lobe much larger than the others), concave, most rarely unequal 5-lobed in the upper half; lobes slightly spreading, 4 posterior ± similar, (anterior lobe larger than the others). *Spartothamnella*, *Oncinocalyx* and *Teucridium* are the closest taxonomically related genera to *Teucrium* (Harley et al. 2004) and forming a phylogenetic clade (Salmaki et al. 2016). *Spartothamnella* is an endemic Australian genus that contains three species: *Spartothamnella teucriiflora* (F. Muell.) Moldenke, *S. juncea* Briq. Engl. and Prantl and *S. puberula* (F. Muell.) Maiden and Betchei. The main morphological difference between *Spartothamnella* and *Teucrium* is that the former has drupaceous fruit. *Teucridium*, is an endemic New Zealand genus represented by *T. parvifolium* F. Hook, which differs from *Teucrium* mainly in ovary form, unlobed or lobed up to a quarter part of its length in *Teucridium* and usually lobed from a quarter to half its length in *Teucrium*. *Oncinocalyx,* also an endemic genus from E Australia (represented by *Oncinocalyx betchei* F. Muell), differs from *Teucrium* in the conspicuously hooked calyx lobes. The pollen morphology

supports the phylogenetic relationships between these genera and *Teucrium* (Abu Asab and Cantino 1993), forming the *Teucriina* clade (Cantino et al. 1997), based on the operculate and verrucate pollen. The phylogenetic analysis based on DNA sequence data has confirmed the affinities between *Teucrium* (throughout *T. fruticans* L.), *Teucridium* and *Oncinocalyx* (Wagstaff and Olmstead 1997; Wagstaff et al. 1998; Lindquist and Albert 2002) and *Spartothamnella* (Steane et al. 2004). *Teucridium, Oncinocalyx, Spartothamnella* and *Teucrium fruticans* L. form a morphologically recognised species group. They are erect or straggly shrubs with many branches with entire leaves, often-axillary cymes, actinomorphic calyx, and fruit in a schizocarp of four big haired reticulate-ridged nutlets.

The delimitation of infrageneric taxa in *Teucrium* is difficult because of the great variation in most morphological characters in many of the species. Bentam's (1833) system is the most traditional classification based on floral features that divides the genus into nine sections: *Leucosceptrum, Teucropsis, Teucris, Pycnobotrys, Stachyobotrys, Scorodonia, Scordium, Chamaedrys* and *Polium*. Bentham noted that the most distinct section is *Teucris*, recognising three distinct species (*Teucrium fruticans, T. corymbosum* R. Br. and *T. pseudochamaepitys* L.) which show different ways in this section. *Scorodonia, Pycnobotrys, Stachyobotrys, Leucosceptrum* and *Teucropsis* would, when considered together, form a well characterised group based on the broad upper lip of the calyx, which is usually but not always accompanied by less reticulate achenia. As well, section *Teucropsis* is similar to section *Scorodonia* in calyx morphology. Boissier (1879) described two new sections, *Spinularia* from species included by Bentham in *Scordium* and *Isotriodon* for the Asian species unknown to Bentham. Briquet (1895-1897) retained Bentham's circumscription but segregated *Leucosceptrum*, at present included in the subfamily Lamioideae (Harley et al. 2004). Lazaro Ibiza (1896) described two new sections, *Montana* and *Pumila*, for the Iberian species segregated from section *Polium*, based on leaf morphology. Rouy (1909) created a new monospecific section; section *Botrys* based on *Teucrium botrys*, a species included by Bentham in *Scordium*. Kudô (1929) segregated *Kinostemon* based on *Teucrium* section *Pleurobotrys*. *Kinostemon* included three species: *Kinostemon pernyi* (Franch.) Kudô, *K. bidentatum* (Hemsl.) Kudô and *K. ornatum* (Hemsl.) Kudô. Kästner (1989) and Harley et al. (2004) retained all *Kinostemon* within *Teucrium*. Cohen (1956) described the subsection *Rotundifolia* under *Polium*, based on the *Teucrium rotundifolium* species group. Rivas Martinez (1974) treated *Pumilum* (*Pumila*) as a subsection under *Polium*. Valdés Bermejo and Sánchez-Crespo (1978) recognised three subsections under *Polium*; *Polium, Pumilum* and *Rotundifolia*. Puech (1978) described the subsection *Simplicipilosa* for a *Polium* species group with simple hairs, endemic to the Iberian Peninsula. The most commonly used infrasectional arrangement of *Polium* is the one proposed by Valdés Bermejo and Sánchez-Crespo (1978), with an addition made by Puech (1980). As a result, *Polium* is one of the largest (50% of the total species, following Gobaerts et al. 2013) and most complex groups within *Teucrium*. Finally, Kästner (1989) produced the first substantial revision of the genus based on corolla, inflorescence and leaf characters. However, the taxonomic sampling of section *Polium* was poor, and many critical species have not been investigated. Kästner included *Polium* as a subsection under *Chamaedrys*, *Spinularia* and *Botrys* under *Scordium*,

Pycnobotrys under *Isotriodon* and *Stachyobotrys* under *Scorodonia*. This author suggested new subsections; *Cretica* under *Teucrium*, *Canadensia* under *Scorodonia* and *Marum* under *Chamaedrys* and retained *Rubiteucris* palmata as *Teucrium palmatum* in the section *Pycnobotrys*. However, pollen morphology (Abu-Asab and Cantino 1993) and phylogenetic research segregate this genus.

The infrageneric taxonomic relationships within *Teucrium* have been investigated by different authors, karyology by Valdés Bermejo and Sánchez-Crespo (1978), growth form (Kästner 1985), chemical analyses (Harbone et al. 1986) and nutlet morphology (Marin et al. 1994) show *Teucrium* as the most distinctive section. The results of the study on trichome morphology (Manzanares et al. 1983; Navarro and El Oualidi 2000b) concluded that sections *Teucrium* and *Polium* are two well-differentiated groups. Navarro and El Oualidi (1999) investigated floral features and concluded that *Teucrium* and *Scorodonia* are the most distinct sections within *Teucrium*. A molecular study based on ITS analyses (El Oualidi et al. 1999) shows that *Polium* can be differentiated from the rest of the sections. The results obtained from pollen morphology (Abu Asab and Cantino 1993; Diez et al. 1993; Navarro et al. 2004) support Bentham's delimitation and agree with the distinctiveness of the sections *Teucrium* and *Spinularia* and establishing new taxonomic relationships between *Chamaedrys* and *Montanum*. Juan et al. (2004) studied seed amino acids and showed that sections *Scordium*, *Botrys* and *Spinularia* are different.

1.3 Systematic and Biogeographic Relationships Within *Teucrium* Infrageneric Taxa and Close Related Genera

Taxonomic relationships are established between *Oncinocalyx*, *Teucridium*, *Oncinocalyx*, *Spartothamnella* and the Austral *Teucrium* (section *Teucrium*) based on the no clearly lipped corolla. These relationships have been proven with results of phylogenetic analysis (Wagstaff and Olmstead 1997; Wagstaff et al. 1998; Lindquist and Albert 2002; Steane et al. 2004; Salmaki et al. 2016). Among them, *Spartothamnella* is the most morphologically close genus to *Teucrium*, differing substantially only based on it drupaceous fruit.

The primitive forms of subfamily Ajugoideae, such as section *Teucrium*, may have appeared in Australia (Chengyih and Hsiwen 1982). Section *Teucrium* is a basal group within *Teucrium* (Marin et al. 1994; Navarro and El Oualidi 1999; Navarro et al. 2004). This section has widespread floral characters in the structurally simple states, including the actinomorphic calyx and unclearly lipped corollas (Austral species of section *Teucrium*) present in geologically ancient areas. *Teucrium* is also the most worldwide spread section found in regions with a Mediterranean climate. As a result of climatic change (Pliocene/Pleistocene), closed forests started to open up and arid areas were established in Australia (Mummenhoff et al. 1992; Cox and Moore 1993) and in the Mediterranean (Quezel 2000). Therefore, it is not surprising to find two different centres of diversification of this section in Australia and in the Mediterranean region.

Section *Teucrium* can be separated from the rest of the sections mainly based on the actinomorphic ventral calyx (Bentham 1833; Kästner 1989; Marin et al. 1994; Navarro and El Oualidi 1999, 2000a). Three distinct groups can be recognized on the basis of corolla conformation. The first includes the 1-lipped species with spurred corollas, an annulus of hairs in the base, adpressed simple hairs in the stems and big nutlets. These are the much branched divaricate evergreen shrubs, including *Teucrium fruticans*, *T. brevifolium* Schreb., *T. malenconianum* and *T. chardonianum* Maire and Wilczek which form the most typically Mediterranean *Teucrium* with convergent traits of woody plants that belong to pre-Mediterranean lineages (Herrera 1987, 1992; Specht 1988; Verdú 2000; Verdú et al. 2003). The second is composed of the Australian *Teucrium racemosum* group characterized by their longer pedicellate flowers, 1-lipped non- spurred corollas with an annulus of hairs in the middle part of the tube. This group includes *Teucrium racemosum* R.Br., *T. corymbosum* R.Br. and *T. cubense* Jacq., and possibly, *T. integrifolium* Benth., *T. depressum* Small and *T. laciniatum* Torr. The third and most distinctive group is composed of the S African species *T. africanum* Thunb, *T. capense* Thunb., and *T. fililobum* F. Muell. ex Benth. Section *Teucriopsis* is discriminate from the rest of the sections according to previous morphological studies (Bentham 1833; Kästner 1989; Marin et al. 1994; Navarro and El Oualidi 2000a).

Trichome morphology (Bini-Maleci and Servettaz 1991; Servettaz et al. 1992) support the systematic affinities between the sections *Chamaedrys* and *Isotriodon* (Kästner 1985, 1986), both distributed in open forest, shrublands and rocky slopes and fissures in arid and sub-humid regions, mainly in the E Mediterranean area. Nevertheless, species of *Isotriodon* section are weel differenciated to be herbaceous often with a broad upper lip to the calyx and fruit in a schizocarp of four smooth or slightly reticulate coherent nutlets. An annulus of hairs in the calyx tube, a derived character (Abu Asab and Cantino 1993), also unites the species of this section.

The independent status of section *Stachyobotrys* is supported by pollen morphology (Abu Asab and Cantino 1993; Navarro et al. 2004) by chemical studies (Harbone et al. 1986) and floral features (Kästner 1989; Navarro and El Oualidi 1999). The Irano–Turanian and Mediterranean species of this section (*Teucrium lamiifolium* D'Urv., *T. hircanicum* L., *T. bracteatum* Desf. and *T. collincola* Greuter and Burdet) have a gibbous calyx, which is absent from the American species (*Teucrium canadense* L., and *T. vesicarium* Mill.). The Moroccan subshrubs, *Teucrium collincola*, seem to be the most different as regards hair type, corolla conformation and reticulate nutlets (Navarro and El Oualidi 1999). The N American and Mexican *Teucrium canadense* differs from the Mediterranean species on the basis of its tubular calyx, curved hairs and slightly reticulate nutlets, while the temperate S American *Teucrium vesicarium* differs in its rare sacciform calyx and reticulate-rugose nutlets.

The Mediterranean species of section *Scorodonia* with 2-lipped (1/4) calyx with a broad upper lip, 2-lipped corolla with straight stamens, the two last characters being the autopomorphies relative to *Teucrium* (Abu Asab and Cantino 1993). The two Chinese species, *Teucrium pernyi* (*Kinostemon pernyi*) and *T. bidentatum* (*K. bidentatum*), both with zygomorphic calyx (1/2/2), broad upper lip and slightly 2-lipped corollas, the five SE Asian species; *Teucrium veronicoides* Maxim., *T.*

stoloniferum Roxb., *T. viscidum* Blume and *T. japonicum* Will., with slightly 2-lipped corollas and zygomorphic calyx (3/2). *T. plectranthoides* Gamble appears to be separate group inside this section based on its 2-lipped calyx. The two SE Asian species, *Teucrium quadrifarium* Buchanan-Hamilton ex D. Don and *T. argutum* R. Br., and the Iberian *T. salviastrum* Schreb., all with 2-lipped corollas, calyx 1/2/2 with broad upper lip, are considered by Bentham (1833) as species belonging to the section *Scorodonia*. This author noted the strong affinity between *Teucrium quadrifarium* and *T. argutum* and the doubtful position of *T. salviastrum*.

Section *Polium* subsection *Polium* and *Simplicipilosa* were clearly different from the rest of the sections because they are the only group with sessile flowers and leaves, short corolla tube with a tuft of hairs inside. Close affinities between them have been reported (Puech 1978; Navarro and Rosúa 1988; Navarro and El Oualidi 1999, 2000b).

Species of section *Montana* can be differenciated based on the calyx, stem hair type, erect in the *Teucrium rotundifolium* group, and curved in *T. montanum* group. Systematic relationships between sections *Montana* and *Chamaedrys* are shown based on nutlet ornamentation (Marin et al. 1994) and trichome morphology (Navarro and El Oualidi 1999). Affinities between sections *Teucrium*, *Chamaedrys* and *Montana* were shown based on nutlet morphology (Marin et al. 1994), and trichome morphology (Navarro and El Oualidi 2000b).

The relationships between the 1-lipped taxa with subactinomorphic calyx, as occurs in the sections *Chamaedrys*, *Montana* and *Pumila* was shown by Kästner (1989), El Oualidi et al. (1999), Navarro and El Oualidi (2000b), Navarro et al. (2004) and between sections *Scordium* and *Polium* (on the basis of pollen morphology) by Diez et al. (1993), Navarro et al. (2004) and seed amino acids (Juan et al. 2004).

1.4 Biogeography of the Infrageneric *Teucrium* Taxa

Section *Teucrium* is distributes in the Mediterranean region, Europe, Africa, Australia, Asia and America (Fig. 1.1). Most species are perennials or short-lived herbs found in the E Mediterranean. They are branched shrubs like *Teucrium fruticans* looking like the relict elements from the old tropical Mediterranean flora. Section *Teucriopsis* comprises the semi-sclerophyllous shrub: *Teucrium heterophyllum* L'Hér from Canary Islands, and two perennial herbs; *T. abutiloides* L'Hér and *T. betonicum* L'Hér from Madeira Islands. It is an endemic section from the Macaronesian region distributed in forest, open habitats and rocky slopes. Section *Stachyobotrys* is distributed in America, Australia, Mediterranean region and Asia. The species are the perennials or short-lived herbs and half-shrubs. This section is distributed mainly on shady walls, rocky slopes and sandy areas in the semi-arid and sub-humid climate regions. It is represented by only two species in the W Mediterranean area, *Teucrium bracteatum* Desf. (Morocco and Spain) and *T. collincola* Greuter & Burdet endemic from Morocco. Section *Scorodonia* is distributed

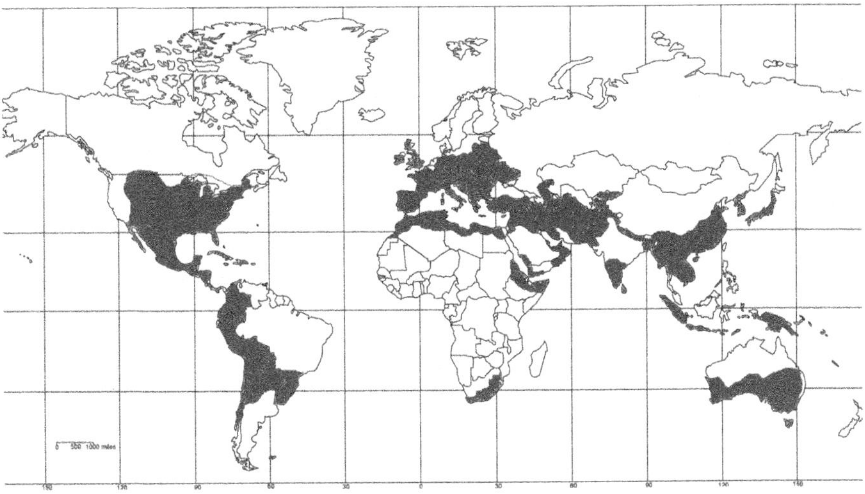

Fig. 1.1 Distribution map of *Teucrium* genus

in the Mediterranean, Europe and Asia but mainly in the humid climate of the W Mediterranean. This section includes half-shrubs and perennial herbs with asexual reproduction. Section *Scordium* is distributed in the Mediterranean, Europe and Asia. This section includes the stoloniferous erect and prostrate short-lived herbs. This is a widespread section in the meadows of the humid habitats. The most important species is the Saharo–Sindian *Teucrium scordium* L. Section *Scordium* subsection *Spinularia* is distributed in the Mediterranean, Europe and Africa and includes erect, prostrate and spinescent annual and short-lived herbs. This section is distributed mainly in the semi-arid and sub-humid regions of the W Mediterranean region (NW Africa). Section *Isotriodon* includes half-shrub species restricted to cliffs, rocky slopes and fissures in the Mediterranean semi-arid regions. Section *Chamaedrys* mainly occurring in the Mediterranean region, Europe, Africa and Asia. They are small half-shrubs and perennial herbs with asexual reproduction. The most important species is *Teucrium chamaedrys* L., distributed through the Mediterranean region. It is the only section of the genus found in the W Mediterrancan islands. Section *Polium* includes a difficult systematic group of vary diverse Mediterranean widespread species with adaptative radiations by polyploidy. The species of this section are common in exposed and disturbed areas of the Mediterranean area, N Africa (Morocco, Algeria and Tunisia). *Polium* is the most diversified section into the genus due its adaptative radiation in the Mediterranean region (Cohen 1956; Puech 1984; Navarro 1995, 2010; Navarro and El Oualidi 1999, 2000a, b; El Oualidi et al. 2002).

Section *Polium* comprises the subsection *Polium* that includes erect, prostrate, cushion-like, half-shrubs or scrubs and perennial herbs. The most important species are *T. polium* L. and *Teucrium capitatum* L., occur throughout the Mediterranean and Irano–Turanian regions. 70% of species are found in the W Mediterranean area

(SE Spain and N Morocco). Subsection *Simplicilpilosa* includes often endemic species from SE Spain, Ethiopia, and SE Arabian Peninsula and subsection *Pumila* is endemic to the Iberian Peninsula. Section *Montanum* is a diverse section throughout the Mediterranean area, NE Africa and E Mediterranean (Turkey, Greece and Cyprus), Ethiopia, E of the Arabia peninsula and W Asia mainly in from rocky slopes, fissures and cliffs.

1.5 Conclusions

Based on the biogeographical distribution of the species of *Teucrium* and the main discriminant systematic characters in the genus, such as corolla and calyx lobes structure (zygomorphy) as well as their indumentum type, almost five major biogeographic and taxonomic species groups can be differentiate. The group of the perennial herbs with 2-lipped corolla and zygomorphic calyx (sections *Pycnobotrys*, *Stachyobotrys*, *Scorodonia* and *Teucropsis*) which provide a clear example of the Eurasian pattern of radiation from C Asia and China. A group integrate for the subshrubs and half-shrubs with 1-lipped corolla and mainly subactinomorphic calyx (sections *Polium*, *Chamaedrys*, *Montana*, *Isotriodon*, *Pumila* and *Scordium*), that clearly represent the Mediterranean pattern of radiation, with a high degree of recent diversification (Hedge 1986). A third group composed by the typical Mediterranean shrubs with 1-lipped corolla and actinomorphic calyx, consistent with the bulk of species of section *Teucrium* subsection *Fruticantia* (pro part.) (Kästner 1989). A fourth group that includes the Australian and S America subshrubs and perennial herbs with 1-lipped corolla and actinomorphic calyx, which seem to be the most distinctive group (*Teucrium racemosum* group), which corresponds to the section *Teucrium* subsection *Cretica* (pro part) (Kästner 1989). This last group confirms the connection between Australasia and S American (through *Teucrium cubense*), the sole pattern of radiation in Lamiaceae in the southern hemisphere (Hedge 1992). Finally, the group formed of the perennial herbs and subshrubs with unclear 1-lipped corolla and actinomorphic calyx (Australian and Cape Region species belong to section *Teucrium*). The section *Teucrium* is the most heterogeneous and distinctive group within *Teucrium* with three different geographical distribution in Australian, N America and Mexico *Teucrium* while as other Ajugoideae genera closed to *Teucrium* such as *Schnabelia* and *Pseudocaryopteris*.

Teucrium genus is widespread in the Mediterranean, Macaronesian, Irano–Turanian and Saharo–Arabian regions with 434 taxa (Govaerts et al. 2013) in the world. The genus shows two centres of richness located in the E and W region of the Mediterranean Sea. This fact is exemplified by the geographical distribution of sections *Stachyobotrys*, *Isotriodon* and section *Polium* subsection *Simplicipilosa*. However, other sections, such as *Scordium*, have an homogeneous distribution in all the areas but are mainly represented by Saharo–Sindian species. The sections *Teucrium*, *Chamaedrys*, *Isotriodon* and *Stachyobotrys* mainly occur in the E Mediterranean and Irano–Turanian region. *Polium*, *Scorodonia* and *Spinularia* are

predominantly in the W Mediterranean area. The E Mediterranean region, the S of the Arabian Peninsula and NE Africa are centres of floristic diversity of the section *Polium*. *Teucriopsis* is restricted to the Macaronesian region. Part of the floristic richness of *Teucrium* genus is the high number of endemic species in the Mediterranean region (63% of total species in the world). The main part of them (c. 50%) in the Iberian Peninsula and Morocco and a high number (c. 18%) in Turkey and Greece.

Stachyobotrys (through *Teucrium canadense* and *T. inflatum*), *Pycnobotrys* (through *Teucrium vesicarium*) are the only taxonomic groups with amphiatlantic relationships, in which the basalmost members of these are largely Mediterranean and C Asian in origin. This pattern of distribution can be explained based on the Boreotropics hypothesis (Wolfe 1975). In this sense, *Teucrium* species have a distinct distribution in the Old and in the New World, among arid regions of S Africa, N America, N Africa and Asia (Mabberly 1997). All these sections can be considered as a remnant of widespread early Tertiary flora. Finally, section *Polium* is an example that reflects a rapid radiation and speciation with a high number of species in the Mediterranean region (Navarro and El Oualidi 2000a; El Oualidi et al. 2002).

Acknowledgement The author thanks the Dr. Gabriel Blanca Lopez (Granada University, Spain) for the assignment of the photographs that constitute the figures of this chapter. The author is also grateful to the directors and curators of the studied herbaria and Dr. Philip Cantino (Ohio University, USA) for the valuable suggestions in preparing the manuscript.

Appendices

Appendix I: List of Teucrium *Infrageneric Taxa in the World*

Teucrium L., taxa list of the accepted names from WCSP World Checklist of Selected Plant (WCSP) (familieshttp://wcsp.science.kew.org/qsearch.do). Govaerts RA, Paton A, Harvey Y, Navarro T, Del Rosario Garcia Pena M (2013) World Checklist of Lamiaceae. Kew. The Board of Trustees of the Royal Botanic Gardens, Kew.www.kew.org/wcsp/

***Teucrium* L., Sp. Pl.: 562 (1753).**

Teucrium abolhayatensis Ranjbar & Mahmoudi, Novon 25: 304 (2017).
Teucrium abutiloides L'Hér., Stirp. Nov.: 84 (1788).
Teucrium africanum Thunb., Prodr. Pl. Cap.: 95 (1800).
Teucrium afrum (Emb. & Maire) Pau & Font Quer in P.Font i Quer, Iter Marocc. 1927: n.° 510 (1928).

Teucrium afrum subsp. *afrum*.
Teucrium afrum subsp. *rhiphaeum* (Font Quer & Pau) Castrov. & Bayon, Anales Jard. Bot. Madrid 47: 513 (1989 publ. 1990).
Teucrium afrum subsp. *rubriflorum* (Font Quer & Pau) Castrov. & Bayon, Anales Jard. Bot. Madrid 47: 513 (1989 publ. 1990).

Teucrium aladagense Vural & H.Duman, Turkish J. Bot. 39: 319 (2015).
Teucrium albicaule Toelken, J. Adelaide Bot. Gard. 7: 296 (1985).
Teucrium albidum Munby, Bull. Soc. Bot. France 2: 286 (1855).
Teucrium alborubrum Hemsl., J. Linn. Soc., Bot. 26: 311 (1890).
Teucrium × alexeenkoanum Juz. in V.L.Komarov, Fl. URSS 20: 506 (1954).
Teucrium algarbiense (Cout.) Cout., Esboco Fl. Lenh. Portug., ed. 2: 262 (1936).
Teucrium alopecurus de Noé, Bull. Soc. Bot. France 2: 585 (1855).
Teucrium alpestre Sm. in J.Sibthorp & J.E.Smith, Fl. Graec. Prodr. 1: 395 (1809).
Teucrium × alvarezii Alcaraz, Sánchez-Gómez, De la Torre & S.Ríos, Datos Veg. Murcia: 144 (1991).
Teucrium alyssifolium Stapf, Denkschr. Kaiserl. Akad. Wiss., Wien. Math.-Naturwiss. Kl. 50: 104 (1885).
Teucrium amplexicaule Benth., Labiat. Gen. Spec.: 687 (1835).
Teucrium andrusi Post, Bull. Herb. Boissier 5: 758 (1897).
Teucrium angustissimum Schreb., Pl. Verticill. Unilab. Gen. Sp.: 49 (1774).
Teucrium anlungense C.Y.Wu & S.Chow, Acta Phytotax. Sin. 10: 338 (1965).
Teucrium annandalei Mukerjee, Rec. Bot. Surv. India 14: 219 (1940).
Teucrium antiatlanticum (Maire) Sauvage & Vindt, Bull. Soc. Sci. Nat. Maroc 35: 286 (1956).
Teucrium antilibanoticum Mouterde, Saussurea 4: 25 (1973).
Teucrium antitauricum Ekim, Notes Roy. Bot. Gard. Edinburgh 38: 58 (1980).
Teucrium apollinis Maire & Weiller, Bull. Soc. Hist. Nat. Afrique N. 30: 86 (1939).
Teucrium aragonense Loscos & J.Pardo in H.M. Willkomm (ed.), Ser. Inconf. Pl. Aragon.: 85 (1863).
Teucrium arduinoi L., Mant. Pl. 1: 81 (1767).
Teucrium argutum R.Br., Prodr. Fl. Nov. Holland.: 504 (1810).
Teucrium aristatum Pérez Lara, Anales Soc. Esp. Hist. Nat. 18: 90 (1889).
Teucrium aroanium Orph. ex Boiss., Diagn. Pl. Orient., ser. 2, 4: 55 (1859).
Teucrium asiaticum L., Mant. Pl. 1: 80 (1767).
Teucrium atratum Pomel, Nouv. Mat. Fl. Atl.: 304 (1874).
Teucrium aureiforme Pomel, Nouv. Mat. Fl. Atl.: 113 (1874).
Teucrium aureocandidum Andr., Ind. Horti Bot. Univ. Budapest 5: 18 (1941).
Teucrium aureum Schreb., Pl. Verticill. Unilab. Gen. Sp.: 43 (1774).

Teucrium aureum subsp. *aureum*.
Teucrium aureum subsp. *turdetanum* Devesa & Valdés Berm., Anales Jard. Bot. Madrid 41: 88 (1984).

Teucrium* × *badiae Sennen, Bol. Soc. Aragonesa Ci. Nat. 11: 230 (1912).
Teucrium balearicum (Coss. ex Pau) Castrov. & Bayon, Anales Jard. Bot. Madrid 47: 508 (1989 publ. 1990).
Teucrium balfourii Vierh., Denkschr. Kaiserl. Akad. Wiss., Wien. Math.-Naturwiss. Kl. 71: 436 (1907).
Teucrium balthazaris Sennen, Diagn. Nouv.: 95 (1936).
Teucrium baokangensis C.L.Xiang, Taxon 67: 390 (2018).
Teucrium barbarum Jahand. & Maire, Bull. Soc. Hist. Nat. Afrique N. 19: 85 (1928).
Teucrium barbeyanum Asch. & Taub. ex E.A.Durand & Barratte, Fl. Libyc. Prodr.: 191 (1910).
Teucrium* × *bergadense Sennen, Bol. Soc. Aragonesa Ci. Nat. 11: 214 (1912).
Teucrium betchei (F.Muell.) Kattari & Salmaki, Taxon 65: 818 (2016).
Teucrium betonicum L'Hér., Stirp. Nov.: 83 (1788).
Teucrium bicolor Sm. in A.Rees, Cycl. 35: n.° 25 (1817).
Teucrium bicoloreum Pau ex Vicioso, Bol. Soc. Esp. Hist. Nat. 16: 142 (1916).
Teucrium bidentatum Hemsl., J. Linn. Soc., Bot. 26: 312 (1890).
Teucrium bogoutdinovae Melnikov, Novosti Sist. Vyssh. Rast. 45: 81 (2014).
Teucrium botrys L., Sp. Pl.: 562 (1753).
Teucrium brachyandrum Puech, Naturalia Monspel., Sér. Bot. 21: 209 (1970 publ. 1971).
Teucrium bracteatum Desf., Fl. Atlant. 2: 7 (1798).
Teucrium brevifolium Schreb., Pl. Verticill. Unilab. Gen. Sp.: 27 (1774).
Teucrium* × *bubanii Sennen, Bol. Soc. Aragonesa Ci. Nat. 11: 229 (1912).
Teucrium bullatum Coss. & Balansa, Bull. Soc. Bot. France 20: 260 (1873).
Teucrium burmanicum Mukerjee, Notes Roy. Bot. Gard. Edinburgh 19: 306 (1938).
Teucrium buxifolium Schreb., Pl. Verticill. Unilab. Gen. Sp.: 42 (1774).
Teucrium campanulatum L., Sp. Pl.: 562 (1753).
Teucrium canadense L., Sp. Pl.: 564 (1753).

Teucrium canadense var. *canadense*.
Teucrium canadense var. *hypoleucum* Griseb., Cat. Pl. Cub.: 214 (1866).
Teucrium canadense var. *occidentale* (A.Gray) E.M.McClint. & Epling, Brittonia 5: 500 (1946).
Teucrium canadense var. *virginicum* (L.) Eaton, Man. Bot., ed. 5: 415 (1829).

Teucrium canum Fisch. & C.A.Mey., Index Seminum (LE, Petropolitanus) 1: 40 (1835).
Teucrium capitatum L., Sp. Pl.: 566 (1753).

Teucrium capitatum subsp. *capitatum*.
Teucrium capitatum subsp. *gracillimum* (Rouy) Valdés Berm., Acta Bot. Malac. 4: 40 (1978 publ. 1979).
Teucrium capitatum subsp. *majoricum* (Rouy) Nyman, Consp. Fl. Eur., Suppl. 2: 246 (1889).

Teucrium* × *carmelitanum Roselló, P.P.Ferrer, A.Guillén, Gómez Nav., Peris & E. Laguna, Flora Montiber. 55: 102 (2013).

Teucrium carolipaui Vicioso ex Pau, Bol. Soc. Ibér. Ci. Nat. 20: 185 (1921 publ. 1922).

Teucrium carolipaui subsp. *carolipaui*.

Teucrium carolipaui subsp. *fontqueri* (Sennen) Rivas Mart., Anales Inst. Bot. Cavanilles 31: 88 (1974).

Teucrium carthaginense Lange, Vidensk. Meddel. Naturhist. Foren. Kjøbenhavn 1881: 97 (1882).

Teucrium* × *carvalhoae A.F.Carrillo, A.Hern., Coy, Güemes & Sánchez-Gómez, Acta Bot. Malac. 22: 221 (1997).

Teucrium caucasigenum Melnikov, Novosti Sist. Vyssh. Rast. 45: 72 (2014).

Teucrium cavanillesianum Font Quer & Jeronimo, Anales Jard. Bot. Madrid 6(2): 488 (1946).

Teucrium cavernarum P.H.Davis, Kew Bull. 6: 113 (1951).

Teucrium chamaedrys L., Sp. Pl.: 565 (1753).

Teucrium chamaedrys subsp. *albarracinii* (Pau) Rech.f., Bot. Arch. 42: 385 (1941).

Teucrium chamaedrys subsp. *algeriense* Rech.f., Bot. Arch. 42: 364 (1941).

Teucrium chamaedrys subsp. *chamaedrys*.

Teucrium chamaedrys subsp. *germanicum* (F.Herm.) Rech.f., Bot. Arch. 42: 379 (1941).

Teucrium chamaedrys subsp. *gracile* (Batt.) Rech.f., Bot. Arch. 42: 389 (1941).

Teucrium chamaedrys subsp. *lydium* O.Schwarz, Repert. Spec. Nov. Regni Veg. 36: 132 (1934).

Teucrium chamaedrys var. *multinodum* Bordz., Mat. Fl. Kavk. 4(3): 57 (1916).

Teucrium chamaedrys subsp. *nuchense* (K.Koch) Rech.f., Bot. Arch. 42: 370 (1941).

Teucrium chamaedrys subsp. *olympicum* Rech.f., Bot. Arch. 42: 368 (1941).

Teucrium chamaedrys subsp. *pectinatum* Rech.f., Bot. Arch. 42: 366 (1941).

Teucrium chamaedrys subsp. *pinnatifidum* (Sennen) Rech.f., Bot. Arch. 42: 383 (1941).

Teucrium chamaedrys subsp. *sinuatum* (Celak.) Rech.f., Bot. Arch. 42: 378 (1941).

Teucrium chamaedrys subsp. *syspirense* (K.Koch) Rech.f., Ann. Naturhist. Mus. Wien 51: 427 (1941).

Teucrium chamaedrys subsp. *tauricola* Rech.f., Bot. Arch. 42: 376 (1941).

Teucrium chamaedrys subsp. *trapezunticum* Rech.f., Bot. Arch. 42: 369 (1941).

Teucrium chardonianum Maire & Wilczek, Bull. Soc. Hist. Nat. Afrique N. 26: 130 (1935).
Teucrium charidemi Sandwith, Cavanillesia 3: 38 (1930).
Teucrium chasmophyticum Rech.f., Pl. Syst. Evol. 134: 287 (1980).
Teucrium chlorocephalum Celak., Bot. Centralbl. 14: 186 (1883).
Teucrium chlorostachyum Pau & Font Quer in P.Font i Quer, Iter Marocc. 1929: n.° 370 (1930).

Teucrium chlorostachyum subsp. *chlorostachyum*.
Teucrium chlorostachyum subsp. *melillense* (Sennen ex Maire) Ouqlidi, Mathez & T.Navarro, Fl. Medit. 7: 23 (1997).

Teucrium chowii Y.H.Tong & N.H.Xia, Phytotaxa 349: 99 (2018).
Teucrium chrysotrichum Lange, Vidensk. Meddel. Naturhist. Foren. Kjøbenhavn 1881: 96 (1882).
Teucrium cincinnatum Maire, Bull. Soc. Sci. Nat. Maroc 13: 272 (1933 publ. 1934).
Teucrium clementiae Ryding, Edinburgh J. Bot. 55: 210 (1998).
Teucrium coahuilanum B.L.Turner, Lundellia 8: 7 (2005).
Teucrium × coeleste Schreb., Pl. Verticill. Unilab. Gen. Sp.: 49 (1774).
Teucrium compactum Clemente ex Lag., Gen. Sp. Pl.: 17 (1816).
Teucrium coniortodes Boiss. & Blanche in P.E.Boissier, Fl. Orient. 4: 818 (1879).
Teucrium × conquense M.B.Crespo & Mateo, Fl. Medit. 1: 197 (1991).

Teucrium × conquense nothosubsp. *conquense*.
Teucrium × conquense nothosubsp. *siyasense* A.F.Carrillo & Sánchez Gómez, Acta Bot. Malac. 21: 285 (1996).

Teucrium × contejeanii Giraudias, Bull. Soc. Études Sci. Angers 1888: 16 (1888).
Teucrium corymbiferum Desf., Fl. Atlant. 2: 8 (1798).
Teucrium corymbosum R.Br., Prodr. Fl. Nov. Holland.: 504 (1810).
Teucrium cossonii D.Wood, Bot. J. Linn. Soc. 65: 261 (1972).
Teucrium creticum L., Sp. Pl.: 563 (1753).
Teucrium cubense Jacq., Enum. Syst. Pl.: 25 (1760).
Teucrium cuneifolium Sm. in J.Sibthorp & J.E.Smith, Fl. Graec. Prodr. 1: 395 (1809).
Teucrium cyprium Boiss., Diagn. Pl. Orient. 5: 43 (1844).
Teucrium cyrenaicum (Maire & Weiller) Brullo & Furnari, Webbia 34: 167 (1979).
Teucrium davaeanum Coss., Bull. Soc. Bot. France 36: 102 (1889).
Teucrium dealianum Emb. & Maire, Bull. Soc. Sci. Nat. Maroc 15: 171 (1935).
Teucrium decaisnei C.Presl, Abh. Königl. Böhm. Ges. Wiss., ser. 5, 3: 530 (1845).
Teucrium decipiens Coss. & Balansa, Bull. Soc. Bot. France 20: 258 (1873).
Teucrium demnatense Coss. ex Batt., Contr. Fl. Atl.: 74 (1919).
Teucrium disjunctum K.R.Thiele & K.A.Sheph., Nuytsia 28: 139 (2017).

Teucrium divaricatum Sieber ex Heldr., Exsicc. (Autogr. Herb. Graec. Norm.) 1856: n.° 290 (1857).

Teucrium divaricatum subsp. *athoum* (Hausskn.) Bornm., Mitth. Thüring. Bot. Vereins, n.f., 38: 57 (1929).
Teucrium divaricatum subsp. *canescens* (Celak.) Holmboe, Bergens Mus. Skr., ser. 2, 1(2): 151 (1914).
Teucrium divaricatum subsp. *divaricatum*.
Teucrium divaricatum subsp. *graecum* (Celak.) Bornm., Mitth. Thüring. Bot. Vereins, n.f., 38: 57 (1929).

Teucrium × djebalicum Font Quer, Cavanillesia 7: 76 (1935).

Teucrium doumerguei Sennen, Exsicc. (Pl. Esp.) 1933: n.° 8874 (1934).
Teucrium ducellieri Batt., Bull. Soc. Hist. Nat. Afrique N. 8: 71 (1917).
Teucrium dumulosum (Rech.f.) Brullo & Guarino, Fl. Medit. 10: 275 (2000).

Teucrium dunense Sennen, Exsicc. (Pl. Esp.) 1925: n.° 5378 (1925).

Teucrium dunense subsp. *dunense*.
Teucrium dunense subsp. *sublittoralis* P.P.Ferrer, R.Roselló, E.Laguna, Gómez Nav., A.Guillén & J.B, Flora Montiber. 60: 79 (2015).

Teucrium eburneum Thulin, Edinburgh J. Bot. 48: 337 (1991).
Teucrium edetanum M.B.Crespo, Mateo & T.Navarro, Acta Bot. Malac. 19: 205 (1994).
Teucrium × eloualidii Sánchez-Gómez & T.Navarro, Anales Jard. Bot. Madrid 57: 167 (1999).
Teucrium embergeri (Sauvage & Vindt) El Oualidi, T.Navarro & A.Martin, Acta Bot. Malac. 22: 198 (1997).
Teucrium eremaeum Diels, Bot. Jahrb. Syst. 35: 530 (1905).
Teucrium eriocephalum Willk., Linnaea 25: 58 (1852).

Teucrium eriocephalum subsp. *almeriense* (C.E.Hubb. & Sandwith) T.Navarro & Rosua, Candollea 43: 181 (1988).
Teucrium eriocephalum subsp. *eriocephalum*.
Teucrium eriocephalum subsp. *lutescens* (Coincy) S.Puech, Rech. Teucrium Bassin Medit. Occid.: 67 (1976).
Teucrium eriocephalum subsp. *serranum* (Pau) T.Navarro & Rosua, Candollea 43: 180 (1988).

Teucrium × estevei Alcaraz, Sánchez-Gómez & J.S.Carrión, Lazaroa 9: 26 (1986 publ. 1988).
Teucrium eximium O.Schwartz, Mitt. Inst. Allg. Bot. Hamburg 10: 219 (1939).
Teucrium expassum Pau, Not. Bot. Fl. Españ. 2: 14 (1889).

Teucrium expassum subsp. *expassum*.
Teucrium expassum subsp. *neilense* Mateo & M.B.Crespo, Flora Montiber. 59: 92 (2015).

Teucrium faurei Maire, Bull. Soc. Hist. Nat. Afrique N. 20: 198 (1929).
Teucrium fililobum F.Muell. ex Benth., Fl. Austral. 5: 134 (1870).
Teucrium flavum L., Sp. Pl.: 565 (1753).

Teucrium flavum subsp. *flavum*.
Teucrium flavum subsp. *glaucum* (Jord. & Fourr.) Ronniger, Verh. K. K. Zool.-Bot. Ges. Wien 68: 234 (1918).
Teucrium flavum subsp. *gymnocalyx* Rech.f., Bot. Arch. 42: 397 (1941).
Teucrium flavum subsp. *hellenicum* Rech.f., Bot. Arch. 42: 397 (1941).

Teucrium fragile Boiss., Elench. Pl. Nov.: 77 (1838).
Teucrium franchetianum Rouy & Coincy in A.H.C.de Coincy, Ecl. Pl. Hisp.: 20 (1893).
Teucrium francisci-werneri Rech.f., Phyton (Horn) 1: 207 (1949).
Teucrium francoi M.Seq., Capelo, J.C.Costa & R.Jardim, Bot. J. Linn. Soc. 156: 643 (2008).
Teucrium freynii E.Rev. ex Willk., Suppl. Prodr. Fl. Hispan.: 159 (1893).
Teucrium fruticans L., Sp. Pl.: 563 (1753).
Teucrium gabrieliae Bornm., Beih. Bot. Centralbl. 59B: 311 (1939).
Teucrium gattefossei Emb., Bull. Soc. Sci. Nat. Maroc 13: 299 (1933 publ. 1934).
Teucrium glandulosum Kellogg, Proc. Calif. Acad. Sci. 2: 23 (1863).
Teucrium gnaphalodes L'Hér., Stirp. Nov.: 84 (1788).
Teucrium × gnaphaureum M.B.Crespo & Mateo, Fl. Medit. 1: 197 (1991).
Teucrium goetzei Gürke, Bot. Jahrb. Syst. 30: 391 (1901).
Teucrium gracile Barbey & Fors.-Major, Karpathos: 127 (1895).
Teucrium grandifolium R.A.Clement, Edinburgh J. Bot. 50: 37 (1993).
Teucrium grandiusculum F.Muell. & Tate, Trans. & Proc. Roy. Soc. South Australia 13: 108 (1890).

Teucrium grandiusculum subsp. *grandiusculum*.
Teucrium grandiusculum subsp. *pilosum* Toelken, J. Adelaide Bot. Gard. 7: 299 (1985).

Teucrium grisebachii Hieron. ex Epling, Rev. Mus. La Plata, Secc. Bot. 2: 91 (1946).
Teucrium grosii Pau in P.Font i Quer, Iter Marocc. 1927: n.° 515 (1928).
Teucrium × guarae-requenae P.P.Ferrer, E.Laguna, Gómez Nav., Roselló & Peris, Collect. Bot. (Barcelona) 31: 30 (2012).
Teucrium × guemesii J.F.Jiménez, A.F.Carrillo, M.A.Carrión, Acta Bot. Malac. 24: 205 (1999).
Teucrium gypsophilum Emb. & Maire, Pl. Rif Nov. 1: 10 (1927).
Teucrium haenseleri Boiss., Elench. Pl. Nov.: 79 (1838).

Teucrium halacsyanum Heldr., Oesterr. Bot. Z. 29: 241 (1879).
Teucrium haradjanii Briq. ex Rech.f., Svensk Bot. Tidskr. 43: 44 (1949).
Teucrium helichrysoides (Diels) Greuter & Burdet, Willdenowia 15: 79 (1985).
Teucrium heterophyllum L'Hér., Stirp. Nov.: 84 (1788).

Teucrium heterophyllum subsp. *brevipilosum* Gaisberg, Willdenowia 30: 267 (2000).
Teucrium heterophyllum subsp. *heterophyllum*.
Teucrium heterophyllum subsp. *hierrense* Gaisberg, Willdenowia 30: 367 (2000).

Teucrium heterotrichum Briq. ex Rech.f., Svensk Bot. Tidskr. 43: 43 (1949).
Teucrium heynei V.S.Kumar & Chakrab., Phytotaxa 243: 197 (2016).
Teucrium hieronymi Sennen, Diagn. Nouv.: 150 (1936).
Teucrium hifacense Pau, Bol. Soc. Aragonesa Ci. Nat. 1: 30 (1902).
Teucrium hijazicum Hedge & R.A.King, Notes Roy. Bot. Gard. Edinburgh 45: 30 (1988).
Teucrium hircanicum L., Syst. Nat. ed. 10, 2: 1095 (1759).
Teucrium homotrichum (Font Quer) Rivas Mart., Opusc. Bot. Pharm. Complut. 3: 88 (1986).
Teucrium huotii Emb. & Maire, Pl. Rif Nov. 1: 10 (1927).
Teucrium integrifolium Benth., Fl. Austral. 5: 133 (1870).
Teucrium intricatum Lange, Vidensk. Meddel. Naturhist. Foren. Kjøbenhavn 1863: 21 (1863).
Teucrium japonicum Houtt., Nat. Hist. 2(9): 282 (1778).

Teucrium japonicum var. *japonicum*.
Teucrium japonicum var. *microphyllum* C.Y.Wu & S.Chow, Acta Phytotax. Sin. 10: 334 (1965).
Teucrium japonicum var. *tsungmingense* C.Y.Wu & S.Chow, Acta Phytotax. Sin. 10: 334 (1965).

Teucrium joannis (Sauvage & Vindt) El Oualidi, T.Navarro & A.Martin, Acta Bot. Malac. 22: 198 (1997).
Teucrium jolyi Mathez & Sauvage, Trav. Inst. Sci. Chérifien, Sér. Gén. 3: 181 (1975).
Teucrium jordanicum (Danin) Faried, Nordic J. Bot. 33: 390 (2015).

Teucrium jordanicum var. *jordanicum*.
Teucrium jordanicum var. *sinaicum* (Danin) Faried, Nordic J. Bot. 33: 391 (2015).

Teucrium junceum (A.Cunn. ex Walp.) Kattari & Heubl, Taxon 65: 818 (2016).
Teucrium kabylicum Batt. in J.A.Battandier & L.C.Trabut, Fl. Algérie Tunisie: 269 (1905).
Teucrium karpasiticum Hadjik. & Hand, Willdenowia 38: 112 (2008).
Teucrium kotschyanum Poech, Enum. Pl. Cypr.: 24 (1842).
Teucrium kraussii Codd, Bothalia 12: 179 (1977).

Teucrium krymense Juz., Bot. Mater. Gerb. Bot. Inst. Komarova Akad. Nauk S.S.S.R. 14: 19 (1951).
Teucrium kyreniae (P.H.Davis) Hadjik. & Hand, Willdenowia 38: 117 (2008).
Teucrium labiosum C.Y.Wu & S.Chow, Acta Phytotax. Sin. 10: 342 (1965).
Teucrium laciniatum Torr., Ann. Lyceum Nat. Hist. New York 2: 231 (1827).
Teucrium lamiifolium d'Urv., Mém. Soc. Linn. Paris 1: 320 (1822).
Teucrium lanigerum Lag., Gen. Sp. Pl.: 17 (1816).
Teucrium laxum D.Don, Prodr. Fl. Nepal.: 109 (1825).
Teucrium leonis Sennen, Diagn. Nouv.: 35 (1936).
Teucrium lepicephalum Pau, Bol. Soc. Aragonesa Ci. Nat. 3: 286 (1904).
Teucrium leucocladum Boiss., Diagn. Pl. Orient. 5: 44 (1844).

Teucrium leucocladum var. *glandulosum* Danin, Willdenowia 27: 169 (1997).
Teucrium leucocladum var. *leucocladum*.

Teucrium leucophyllum Montbret & Aucher ex Benth., Ann. Sci. Nat., Bot., sér. 2, 6: 55 (1836).
Teucrium libanitis Schreb., Pl. Verticill. Unilab. Gen. Sp.: 48 (1774).
Teucrium lini-vaccarii Pamp., Bull. Soc. Bot. Ital. 1914: 17 (1914).
Teucrium lucidum L., Syst. Nat. ed. 10, 2: 1095 (1759).
Teucrium lusitanicum Schreb., Pl. Verticill. Unilab. Gen. Sp.: 47 (1774).

Teucrium lusitanicum subsp. *aureiforme* (Rouy) Valdés Berm., Acta Bot. Malac. 4: 44 (1978 publ. 1979).
Teucrium lusitanicum subsp. *clementiae* T.Navarro, El Oualidi, Arn.Martin & S.Puech, Acta Bot. Gallica 145: 64 (1998).
Teucrium lusitanicum subsp. *lusitanicum*.

Teucrium luteum (Mill.) Degen, Fl. Veleb. 2: 587 (1937).

Teucrium luteum subsp. *flavovirens* (Batt.) Greuter & Burdet, Willdenowia 15: 79 (1985).
Teucrium luteum subsp. *gabesianum* (S.Puech) Greuter, Willdenowia 15: 423 (1986).
Teucrium luteum subsp. *luteum*.

Teucrium macrophyllum (C.Y.Wu & S.Chow) J.H.Zheng, Fl. Hubeiensis 3: 447 (2002).
Teucrium macrum Boiss. & Hausskn. in P.E.Boissier, Fl. Orient. 4: 810 (1879).
Teucrium × maestracense M.B.Crespo & Mateo, Fl. Medit. 1: 200 (1991).
Teucrium maghrebinum Greuter & Burdet, Willdenowia 15: 80 (1985).
Teucrium × mailhoi Giraudias, Bull. Soc. Études Sci. Angers 1890: 56 (1890).

Teucrium × mailhoi nothosubsp. *mailhoi*.
Teucrium × mailhoi nothosubsp. *orientalis* I.Soriano & Aymerich, Collect. Bot. (Barcelona) 30: 103 (2011).

Teucrium malenconianum Maire, Bull. Soc. Hist. Nat. Afrique N. 25: 316 (1934).
Teucrium manghuaense Y.Z.Sun ex S.Chow, Acta Phytotax. Sin. 10: 339 (1965).

Teucrium manghuaense var. *angustum* C.Y.Wu & S.Chow, Acta Phytotax. Sin. 10: 340 (1965).
Teucrium manghuaense var. *manghuaense*.

Teucrium marum L., Sp. Pl.: 564 (1753).

Teucrium marum subsp. *marum*.

Teucrium mascatense Boiss., Diagn. Pl. Orient. 5: 44 (1844).
Teucrium massiliense L., Sp. Pl. ed. 2: 789 (1763).
Teucrium × mateoi Solanas, M.B.Crespo & De la Torre, Anales Biol., Fac. Biol., Univ. Murcia 19: 80 (1993).
Teucrium maximowiczii Prob., Sosud. Rast. Sovet. Dal'nego Vostoka 7: 301 (1995).
Teucrium melissoides Boiss. & Hausskn. in P.E.Boissier, Fl. Orient. 4: 813 (1879).
Teucrium mesanidum (Litard. & Maire) Sauvage & Vindt, Bull. Soc. Sci. Nat. Maroc 35: 289 (1956).
Teucrium micranthum B.J.Conn, Telopea 9: 803 (2002).
Teucrium microphyllum Desf., Ann. Mus. Hist. Nat. 10: 300 (1807).
Teucrium micropodioides Rouy, Naturaleza (Madrid) 2: 16 (1882).
Teucrium mideltense (Batt.) Humbert, Bull. Soc. Hist. Nat. Afrique N. 19: 240 (1928).
Teucrium miragestorum Gómez Nav., Roselló, P.P.Ferrer & Peris, Sabuco 9: 45 (2013).
Teucrium mitecum Tattou & El Oualidi, Acta Bot. Malac. 18: 154 (1993).
Teucrium moleromesae Sánchez-Gómez, T.Navarro, J.F.Jiménez, J.B.Vera, Mota & del Río, Phytotaxa 151: 59 (2013).
Teucrium montanum L., Sp. Pl.: 565 (1753).

Teucrium montanum subsp. *helianthemoides* (Adamovic) Baden in A.Strid & Kit Tan (eds.), Mount. Fl. Greece 2: 74 (1991).
Teucrium montanum subsp. *montanum*.

Teucrium montbretii Benth., Ann. Sci. Nat., Bot., sér. 2, 6: 56 (1836).

Teucrium montbretii subsp. *heliotropiifolium* (Barbey) P.H.Davis, Notes Roy. Bot. Gard. Edinburgh 21: 138 (1953).
Teucrium montbretii subsp. *judaicum* P.H.Davis, Kew Bull. 6: 117 (1951).
Teucrium montbretii subsp. *libanoticum* P.H.Davis, Kew Bull. 6: 117 (1951).
Teucrium montbretii subsp. *montbretii*.
Teucrium montbretii subsp. *pamphylicum* P.H.Davis, Kew Bull. 6: 117 (1951).
Teucrium montbretii subsp. *yildirimlii* Dinç & Dogu, Biologia (Bratislava) 67(4): 669 (2012).

Teucrium × motae Lahora & Sánchez-Gómez, Acta Bot. Malac. 35: 205 (2010).
Teucrium × mugronense P.P.Ferrer, Roselló, Gómez Nav. & Guara, Sabuco 8: 74 (2011).
Teucrium muletii Roselló, P.P.Ferrer, E.Laguna, Gómez Nav., A.Guillén & Peris, Flora Montiber. 54: 137 (2013).
Teucrium multicaule Montbret & Aucher ex Benth., Ann. Sci. Nat., Bot., sér. 2, 6: 54 (1836).

Teucrium multicaule subsp. *multicaule*.
Teucrium multicaule subsp. *planifolium* Post, Fl. Syria, ed. 2: 404 (1832).

Teucrium murcicum Sennen, Bull. Soc. Iber. Ci. Nat. 30(13): 328 (1931).
Teucrium musimonum Humbert ex Maire, Bull. Soc. Hist. Nat. Afrique N. 15: 232 (1924).
Teucrium myriocladum Diels, Bot. Jahrb. Syst. 35: 530 (1905).
Teucrium nanum C.Y.Wu & S.Chow, Acta Phytotax. Sin. 10: 337 (1965).
Teucrium × navarroi Sánchez-Gómez, Güemes, A.F.Carrillo, Coy & A.Hern., Acta Bot. Malac. 21: 285 (1996).
Teucrium novorossicum Melnikov, Novosti Sist. Vyssh. Rast. 45: 76 (2014).
Teucrium nudicaule Hook., Bot. Misc. 2: 235 (1831).
Teucrium nummularifolium Baker, Bull. Misc. Inform. Kew 1895: 185 (1895).
Teucrium odontites Boiss. & Balansa in P.E.Boissier, Diagn. Pl. Orient., ser. 2, 4: 57 (1859).
Teucrium oliverianum Ging. ex Benth., Labiat. Gen. Spec.: 668 (1835).
Teucrium omeiense Y.Z.Sun, Acta Phytotax. Sin. 10: 340 (1965).

Teucrium omeiense var. *cyanophyllum* C.Y.Wu & S.Chow, Acta Phytotax. Sin. 10: 341 (1965).
Teucrium omeiense var. *omeiense*.

Teucrium orientale L., Sp. Pl.: 562 (1753).

Teucrium orientale var. *glabrescens* Hausskn. ex Bornm., Verh. K. K. Zool.-Bot. Ges. Wien 48: 624 (1898).
Teucrium orientale subsp. *gloeotrichum* Rech.f., Fl. Iranica 150: 41 (1982).
Teucrium orientale var. *orientale*.
Teucrium orientale var. *puberulens* Ekim, Notes Roy. Bot. Gard. Edinburgh 38: 58 (1980).
Teucrium orientale subsp. *taylorii* (Boiss.) Rech.f., Fl. Iranica 150: 41 (1982).

Teucrium ornatum Hemsl., J. Linn. Soc., Bot. 26: 313 (1890).
Teucrium oxylepis Font Quer, Mem. Mus. Ci. Nat. Barcelona, Ser. Bot. 1(2): 9 (1924).
Teucrium ozturkii A.P.Khokhr., Byull. Glavn. Bot. Sada 175: 54 (1997).
Teucrium paederotoides Boiss., Fl. Orient. 4: 1268 (1879).
Teucrium pampaninii C.Du, Checkl. Shaanxi Vasc. Pl.: 297 (2016).
Teucrium parviflorum Schreb., Pl. Verticill. Unilab. Gen. Sp.: 31 (1774).

Teucrium parvifolium (Hook.f.) Kattari & Salmaki, Taxon 65: 818 (2016).
Teucrium pernyi Franch., Nouv. Arch. Mus. Hist. Nat., sér. 2, 6: 125 (1883).
Teucrium persicum Boiss., Diagn. Pl. Orient. 5: 42 (1844).
Teucrium pestalozzae Boiss., Diagn. Pl. Orient. 12: 90 (1853).
Teucrium petelotii Doan ex Suddee & A.J.Paton, Kew Bull. 61: 620 (2007).
Teucrium pilbaranum B.J.Conn, Telopea 8: 299 (1999).
Teucrium plectranthoides Gamble, Fl. Madras: 1158 (1924).
Teucrium polioides Ryding, Edinburgh J. Bot. 55: 214 (1998).
Teucrium polium L., Sp. Pl.: 566 (1753).

Teucrium polium subsp. *aurasiacum* (Maire) Greuter & Burdet, Willdenowia 15: 80 (1985).
Teucrium polium subsp. *chevalieri* Maire, Mém. Soc. Hist. Nat. Afrique N. 3: 266 (1933).
Teucrium polium subsp. *clapae* S.Puech, Naturalia Monspel., Sér. Bot. 21: 209 (1970 publ. 1971).
Teucrium polium subsp. *polium.*
Teucrium polium subsp. *purpurascens* (Benth.) S.Puech, Rech. Teucrium Bassin Medit. Occid.: 94 (1976).

Teucrium popovii R.A.King, Notes Roy. Bot. Gard. Edinburgh 45: 28 (1988).
Teucrium* × *portusmagni Sánchez-Gómez, A.F.Carrillo, A.Hern. & T.Navarro, Acta Bot. Malac. 24: 205 (1999).
Teucrium procerum Boiss. & Blanche in P.E.Boissier, Diagn. Pl. Orient., ser. 2, 4: 56 (1859).
Teucrium proctorii L.O.Williams, Fieldiana, Bot. 34: 114 (1972).
Teucrium pruinosum Boiss., Fl. Orient. 4: 808 (1879).
Teucrium pseudaroanium Parolly, Erdag & Nordt, Willdenowia 37: 252 (2007).
Teucrium* × *pseudoaragonense M.B.Crespo & Mateo, Fl. Medit. 1: 200 (1991).
Teucrium pseudochamaepitys L., Sp. Pl.: 562 (1753).
Teucrium pseudoscorodonia Desf., Fl. Atlant. 2: 5 (1798).
Teucrium* × *pseudothymifolium Sánchez-Gómez, Güemes & A.F.Carrillo, Acta Bot. Malac. 21: 283 (1996).
Teucrium puberulum (F.Muell.) Kattari & Bräuchler, Taxon 65: 818 (2016).
Teucrium pugionifolium Pau, Anales Soc. Esp. Hist. Nat. 1897: 199 (1897).
Teucrium* × *pujolii Sennen, Bol. Soc. Aragonesa Ci. Nat. 11: 214 (1912).
Teucrium pumilum Loefl. ex L., Cent. Pl. I: 15 (1755).
Teucrium pyrenaicum L., Sp. Pl.: 566 (1753).

Teucrium pyrenaicum subsp. *guarensis* P.Monts., Anales Jard. Bot. Madrid 37: 625 (1980 publ. 1981).
Teucrium pyrenaicum subsp. *pyrenaicum.*

Teucrium quadrifarium Buch.-Ham. ex D.Don, Prodr. Fl. Nepal.: 108 (1825).
Teucrium racemosum R.Br., Prodr. Fl. Nov. Holland.: 504 (1810).

Teucrium radicans Bonnet & Barratte, Explor. Sci. Tunisie, Ill. Bot.: t. 14, f. 1–8 (1895).
Teucrium ramaswamii M.B.Viswan. & Manik., Nordic J. Bot. 27: 86 (2009).
Teucrium ramosissimum Desf., Fl. Atlant. 2: 4 (1798).
Teucrium reidii Toelken & D.Dean Cunn., J. Adelaide Bot. Gard. 22: 97 (2008).
Teucrium resupinatum Desf., Fl. Atlant. 2: 4 (1798).
Teucrium rhodocalyx O.Schwartz, Mitt. Inst. Allg. Bot. Hamburg 10: 218 (1939).
Teucrium rifanum (Maire & Sennen) Maire & Sennen, Bull. Soc. Hist. Nat. Afrique N. 23: 260 (1932).
Teucrium rigidum Benth. in A.P.de Candolle, Prodr. 12: 578 (1848).
Teucrium* × *rigualii De la Torre & Alcaraz, Acta Bot. Malac. 17: 138 (1992).
Teucrium* × *riosii De la Torre & Alcaraz, Acta Bot. Malac. 17: 136 (1992).
Teucrium rivas-martinezii Alcaraz, Garre, Mart.Parras & Peinado, Willdenowia 14: 135 (1984).
Teucrium* × *riverae De la Torre & Alcaraz, Acta Bot. Malac. 17: 137 (1992).
Teucrium rixanense Ruíz Torre & Ruíz Cast., Bol. Estaçión Centr. Ecol. Madrid 3(6): 31 (1974).
Teucrium* × *robledoi De la Torre & Alcaraz, Acta Bot. Malac. 17: 136 (1992).

Teucrium × *robledoi* nothosubsp. *robledoi*.
Teucrium × *robledoi* nothosubsp. *serralaligae* Mateo & M.B.Crespo, Flora Montiber. 59: 92 (2015).

Teucrium ronnigeri Sennen, Bol. Soc. Ibér. Ci. Nat. 30(13): 47 (1931).

Teucrium ronnigeri subsp. *lagunae* (Roselló, Peris & Stübing) M.B.Crespo & P.P.Ferrer, Fl. Montiberica 42: 77 (2009).
Teucrium ronnigeri subsp. *ronnigeri*.

Teucrium rotundifolium Schreb., Pl. Verticill. Unilab. Gen. Sp.: 42 (1774).

Teucrium rotundifolium subsp. *rotundifolium*.
Teucrium rotundifolium subsp. *transatlanticum* Emb. ex Greuter, Burdet & G. Long, Med-Checkl. 3: xxix (1986).

Teucrium rouyanum Coste & Soulié, Bull. Soc. Bot. France 44: 114 (1897).
Teucrium royleanum Wall. ex Benth. in N.Wallich, Pl. Asiat. Rar. 1: 58 (1830).
Teucrium* × *rubrovirens Pau in P.Font i Quer, Iter Marocc. 1927: n.° 516 (1928).
Teucrium rupestre Coss. & Balansa, Bull. Soc. Bot. France 20: 259 (1873).
Teucrium* × *sagarrae Font Quer, Ill. Fl. Occid.: 7 (1926).
Teucrium salaminium Hadjik. & Hand, Candollea 66: 342 (2011).
Teucrium salviastrum Schreb., Pl. Verticill. Unilab. Gen. Sp.: 38 (1774).
Teucrium sandrasicum O.Schwarz, J. Roy. Hort. Soc. 74: 115 (1949).
Teucrium sanguisorbifolium (Pau & Font Quer) Dobignard, J. Bot. Soc. Bot. France 46–47: 62 (2009).

Teucrium santae Quézel & Simonn. ex Greuter & Burdet, Willdenowia 15: 81 (1985).
Teucrium sauvagei Le Houér., Bull. Soc. Bot. France 107: 101 (1960).
Teucrium scabrum Suddee & A.J.Paton, Kew Bull. 63: 675 (2008 publ. 2009).
Teucrium schoenenbergeri Nabli, Bull. Soc. Bot. France, Lett. Bot. 137: 314 (1990).
Teucrium scordium L., Sp. Pl.: 565 (1753).

Teucrium scordium subsp. *glabrescens* (Murata) Rech.f., Fl. Iranica 150: 35 (1982).
Teucrium scordium subsp. *scordioides* (Schreb.) Arcang., Comp. Fl. Ital.: 559 (1882).
Teucrium scordium subsp. *scordium.*
Teucrium scordium subsp. *serratum* (Benth.) Rech.f., Fl. Iranica 150: 34 (1982).

Teucrium × scorolepis Pajarón & A.Molina, Fontqueria 27: 163 (1989).
Teucrium serpylloides Maire & Weiller, Bull. Soc. Hist. Nat. Afrique N. 31: 36 (1940).
Teucrium sessiliflorum Benth. in A.P.de Candolle, Prodr. 12: 580 (1848).
Teucrium shanicum Mukerjee, Notes Roy. Bot. Gard. Edinburgh 19: 306 (1938).
Teucrium siculum (Raf.) Guss., Fl. Sicul. Syn. 2: 837 (1845).
Teucrium similatum T.Navarro & Rosua, Candollea 45: 583 (1990).
Teucrium simplex Vaniot, Bull. Acad. Int. Géogr. Bot. 14: 186 (1904).
Teucrium sirnakense Özcan & Dirmenci, Turkish J. Bot. 39: 312 (2015).
Teucrium socinianum Boiss., Fl. Orient. 4: 818 (1879).
Teucrium socotranum Vierh., Denkschr. Kaiserl. Akad. Wiss., Wien. Math.-Naturwiss. Kl. 71: 436 (1907).
Teucrium somalense Ryding, Edinburgh J. Bot. 55: 217 (1998).
Teucrium spinosum L., Sp. Pl.: 566 (1753).
Teucrium stachyophyllum P.H.Davis, Notes Roy. Bot. Gard. Edinburgh 21: 48 (1951).
Teucrium stocksianum Boiss., Diagn. Pl. Orient., ser. 2, 4: 58 (1859).

Teucrium stocksianum subsp. *incanum* (Hedge & Lamond) Rech.f., Fl. Iranica 150: 29 (1982).
Teucrium stocksianum subsp. *patulum* (Hedge & Lamond) Rech.f., Fl. Iranica 150: 29 (1982).
Teucrium stocksianum subsp. *stenophyllum* R.A.King, Notes Roy. Bot. Gard. Edinburgh 45: 32 (1988).
Teucrium stocksianum subsp. *stocksianum.*

Teucrium subspinosum Pourr. ex Willd., Enum. Pl.: 596 (1809).
Teucrium taiwanianum T.H.Hsieh & T.C.Huang, Taiwania 41: 86 (1996).
Teucrium tananicum Maire, Bull. Soc. Hist. Nat. Afrique N. 23: 210 (1932).
Teucrium teinense Kudô, J. Coll. Sci. Imp. Univ. Tokyo 43: 6 (1921).
Teucrium terciae (Sánchez-Gómez, M.A.Carrión & A.Hern.) Sánchez-Gómez, M.A.Carrión & A.Hern., Anales Biol., Fac. Biol., Univ. Murcia 25: 112 (2003).

Teucrium teresianum Blanca, Cueto & J.Fuentes, Nordic J. Bot. 35: 14 (2017 publ. 2016).
Teucrium teucriiflorum (F.Muell.) Kattari & Salmaki, Taxon 65: 818 (2016).
Teucrium thieleanum B.J.Conn, Telopea 11: 135 (2006).
Teucrium thymifolium Schreb., Pl. Verticill. Unilab. Gen. Sp.: 50 (1774).
Teucrium thymoides Pomel, Nouv. Mat. Fl. Atl.: 112 (1874).
Teucrium townsendii Vasey & Rose, Proc. U. S. Natl. Mus. 12: 146 (1890).
Teucrium trifidum Retz., Observ. Bot. 1: 21 (1779).
Teucrium tsinlingense C.Y.Wu & S.Chow, Acta Phytotax. Sin. 10: 334 (1965).

Teucrium tsinlingense var. *porphyreum* C.Y.Wu & S.Chow, Acta Phytotax. Sin. 10: 335 (1965).
Teucrium tsinlingense var. *tsinlingense*.

Teucrium × turianum M.B.Crespo, P.P.Ferrer, Roselló, M.A.Alonso, Juan & E. Laguna, Flora Montiber. 58: 70 (2014).
Teucrium turredanum Losa & Rivas Goday, Anales Inst. Bot. Cavanilles 25: 204 (1968).
Teucrium ussuriense Kom., Izv. Bot. Sada Akad. Nauk S.S.S.R. 30: 208 (1931 publ. 1932).
Teucrium veronicoides Maxim., Bull. Acad. Imp. Sci. Saint-Pétersbourg, sér. 3, 23: 388 (1877).
Teucrium vesicarium Mill., Gard. Dict. ed. 8: n.° 17 (1768).
Teucrium vincentinum Rouy, Naturaliste 4(3): 20 (1882).
Teucrium viscidum Blume, Bijdr. Fl. Ned. Ind.: 827 (1826).

Teucrium viscidum var. *leiocalyx* C.Y.Wu & S.Chow, Acta Phytotax. Sin. 10: 332 (1965).
Teucrium viscidum var. *longibracteatum* C.Y.Wu & S.Chow, Acta Phytotax. Sin. 10: 332 (1965).
Teucrium viscidum var. *macrostephanum* C.Y.Wu & S.Chow, Acta Phytotax. Sin. 10: 333 (1965).
Teucrium viscidum var. *miquelianum* (Maxim.) H.Hara, Bot. Mag. (Tokyo) 51: 145 (1937).
Teucrium viscidum var. *nepetoides* (H.Lév.) C.Y.Wu & S.Chow, Acta Phytotax. Sin. 10: 331 (1965).
Teucrium viscidum var. *viscidum*.

Teucrium wattii Prain, J. Asiat. Soc. Bengal, Pt. 2, Nat. Hist. 59: 317 (1890 publ. 1891).
Teucrium webbianum Boiss., Elench. Pl. Nov.: 78 (1838).
Teucrium werneri Emb., Bull. Soc. Sci. Nat. Maroc 15: 185 (1935).
Teucrium wightii Hook.f., Fl. Brit. India 4: 701 (1885).
Teucrium yemense Deflers, Voy. Yemen: 190 (1889).
Teucrium zaianum Emb. & Maire, Pl. Marocc. Nov. 2: 7 (1929).
Teucrium zanonii Pamp., Nuovo Giorn. Bot. Ital., n.s., 24: 153 (1917).

Appendix II: Photographs of Selected Teucrium *Species on Natural Habitats*

T. aureum Schreb.

T. carolipaui Vicioso ex Pau

T. carthaginense Lange

T. chrysotrichum Lange

T. compactum Clemente ex Lag.

T. dunense Sennen

T. freynii E. Rev. ex Willk.

T. gnaphalodes L'Hér.

T. haenseleri Boiss.

T. subspinosum Pourr. ex Willd.

T. intricatum Lange

T. moleromesae Sánchez-Gómez, T. Navarro, et al.

Teucrium pumilum Loefl. ex L.

Teucrium rotundifolium Schreb.

T. resupinatum Desf., Fl.

T. thymifolium Schreb.

T. barbarum Jahand. & Maire

T. bicoloreum Pau ex Vicioso

T. capitatum L.

T. eriocephalum Willk.

T. haenseleri Boiss.

T. hieronymi Sennen

T. pseudochamaepitys L.

T. similatum T. Navarro & Rosua

T. turredanum Losa & R. G.

Appendix III: List of Herbarium Specimens of Teucrium *L.*, Oncinocalyx *F. Muell.*, Sparthotamnella *Briq.*, Teucridium *F. Hook,* Rubiteucris *palmata Kudô and* Leucosceptrum canum *Sm., with Localities and Voucher. The Infrageneric Classification Follows Navarro (2010)*

Taxa of *Oncinocalyx* F. Muell., *Sparthotamnella* Briq., *Teucridium* F. Hook, *Rubiteucris palmata* Kudô and *Leucosceptrum canum* Sm.

***Oncinocalyx betchei*.** AUSTRALIA: New South Wales, Gunnedak, vi-1909, Boorman, KEW; Curlevis, M. Martha, xi-1969, Johnson & Briggs, 2991, KEW; Deriah, Barra trail, ii-1986, Mackay 806, NE 051374; Tamworth, xii-1985, Hosking, NE 047136; Deriah, Myrtle, x-1916, Mackay, NE 043201A.

***Spartothamnella teucriiflora*.** AUSTRALIA: Alice, vii-1962, Nelson 351, KEW; Uluru, v-1988, Lazarides & Palmer 45, KEW; Buró Plain, vii-1962, Nelson 351, KEW; Uluru, iv-1988, Lazarides & Palmer 045, KEW.

***Teucridium parviflorum*.** NEW ZEALAND: Rangitikei, Taihapa, Mataroa, iii-1991, Ogle, CHR 471440; Whakatane, CHR 335270. Hangaroa, Tiniroto, xii-1967, Druce 179777, CHR 179777; Taihape, ii-1979, Druce, CHR 286453; Kawhatau, Taihape, CHR 286453; Ikawetea, 1-1975, Druce, CHR 275092; Gwavas, Hawkés bay, CHR 282979; Manonui, CHR 23800; Tawhaiti, i-1920, CHR 297158; Gwavas Bush, i-1976, Druce, CHR 282979; Manuni, iii-1949, Wood, CHR23800.

***Rubiteucris palmata*.** NEPAL: Marsyandi Valley, x-1978, Davis, KEW; Sihkim, 1867, H.Hookerianum 1867, KEW.

***Leucosceptrum canum*.** NEPAL: Sinduwa, 1963, Hara et al. 6300990, KEW; Kyapra, ix-1989, keke 342, KEW; Katmandú, ii-1966, Schilling 736, KEW; Sadon, i-1958, Mc Kee 6073, KEW.

Section *Teucrium*

***T. fruticans*.** MOROCCO: Targuist, vi-1993, Jury 11301, MA 577056; Xauen, iv-1926, Font Quer, MA 99122; Igrouka, MA 595799; Targuist, MA 577056.

***T. campanulatum*.** SPAIN: Zaragoza, Bujaraloz, BCF 32764. ITALIA: Sicilia, Catania, vi-2000, Güemes, VAL 119660.

***T. aristatum*.** FRANCE: Arles, vi-2004, Yavercovski, MGC 63560. SPAIN: Sevilla, Carmona, v-1967, Mayor & Ladero, MA 278365; Huelva, Hinojos, v-2002, Talavera, MGC 51882.

***T. creticum*.** CYPRUS: Larnaca, iv-1979, McClintock 2967, EDI 178048; Lefkara, vii-1940, Davis 1917, EDI 178047; Larnaca, Cape Kiti, EDI 2967; Larnaka, MA 495430; Montes Troodos, MA 418783; Larnaca, Cape Kiti, iv-1974, EDI 178048; Lefkara, vii-1940, EDI 178047.

***T. pruinosum*.** TURKEY: Yesilhisar to Urgüp, vii-1980, Nydegger 15485, VAL 129347.

***T. pseudoscorodonia*.** MOROCCO: Tizzi Iffni, MA 9839; Melilla, MA 98937; Nador, MA 562366.

T. parviflorum. Iran (Persia); Kurdistan, Hoseynabad, Sanandaj, v-1971, Rechinger, MA 417834.

T. brevifolium. CRETA: Akroteri, iv-1884, Reverchon 205, MA 9907, MA 99072.

T. malenconianum. MOROCCO: High Atlas, Imilchil, vii-1997, Jury 17728, MA 615566; Gorges du Dades, vi-1982, FC 6966, MA 633607.

T. pseudochamaepitys. SPAIN: Madrid, Aranjuez, MA 436850; Aranjuez, vi-1958, Guinea, MA 436850; Madrid, Algodor, v-1925, Caballero & Gonzalez, MA 436863.

Section *Teucrium* (species with no clearly lipped corolla)

T. africanum. S AFRICA: Grahamstown, xii-1980, Germishuizen 7212, KEW; Fort Beaufort, xii-1902, Brink 713, KEW; Riversdale, xi-1970, Taylor7833, KEW.

T. capense. S AFRICA: Grahanstown, x-1970, Brink 154, KEW; Transnaal, 1894, Transvaalensis 5780, KEW; Transvaal, iii-1958, Vauden 514, KEW; Rustenburg, 1904, Nation, KEW; Wolmaranstad, xii-1973, Hanekon 2185, KEW; Pretoria, Plaas Rooibos, 1977, Liebenberg 8804, KEW.

T. fililobum. AUSTRALIA: x-1909, H. Morrison, KEW; Norseman, xi-1962, Aplin 1851, KEW; Norseman, x-1904, Campbell, KEW.

Section *Teucrium* (*T. racemosum* group)

T. corymbosum. AUSTRALIA: Flinders, Partacoona, ix-1969, Kuchel 2827, MA 456302; Victoria, Mueller, MA 256743; Partacuona, MA 456302; Partacoona, ix-1969, Kuchel 2827, MA 456302.

T. cubense. MEXICO: Mexico, Tampico, i-1910, Palmer 13, KEW; San Luis Potosi, xi-1975, Peterson 471, KEW; Ejido Palmas, Martinez 1204, KEW; Castaños, 1936, Wynd, KEW. ARGENTINA: Corrientes, Mercedes, ix-1997, Ferruci & al. 1291, MA 622439. CUBA: La Habana, iv-1991, Herrera, MA 256749.

T. racemosum. AUSTRALIA: Borung, iii-1903, Reader, EDI 178056; v-1802, EDI 178055; Murray, xi-1968, Cole 131, MA 4563303; Melbourne, Borung, EDI 178058; Bryant creek Reserve, Morgan, ii-1968, Cole 131, MA 456303; Victoria, You-yango, MA 256850; Victoria, iii-1903, EDI 178050.

Section *Teucriopsis* Benth.

T. heterophyllum. SPAIN: Tenerife, Malpais, ii-1946, Sventenius, VAL 138501; Tenerife, Anaga, MGC 8272.

T. betonicum. PORTUGAL: Madeira, Encominada, MA 655341; S. Jorge valley, MA 522318; Mirodoiro do Porto, MA 27852; Madeira, San Jorge, vi-1966, Sjögren, 1178, MA 522318; Madeira, Encomiada, vi-2000, Velayos 9745, MA 655341.

Section *Montana* Lazaro Ibiza

T. pyrenaicum. SPAIN: Huesca, vii-1976, Cabezudo, MGC 37970; Cantabria, 1983, Casaseca, MA 256848; Huesca, Boltaña, vii-1987, Villar et al., MA 515728.

T. rotundifolium. MOROCCO: Alto Atlas, Oukaimeden, 1999, Güemes, VAL 41211. SPAIN: Valencia, Villalonga, 1980, Mateo, VAL 800654. Alicante, Benissa, 1998, Marin Camos, VAL 105784; Jaén, Los ladrones, 1977, Muñoz-Garmendia, MA 395212; Murcia, La Molata, 1923, Cuatrecasas, MA 97822.

T. musimonum. MOROCCO: Alto Atlas, Tizi-n-Ait Hamed, Güemes 1554, VAL 41104.

T. hifacense. SPAIN: Alicante, Pego, vi-1997, Signes & Soler 7130, MA 590236; Alicante, Altea, vi-1962, Rigual, MA 374228; Alicante, Benisa, iv-1993, Soler, MA 548713.

T. montanum. GREECE: Olympos, 1983, Aguilella, VAL 7311. SPAIN: Navarra, Lokiz, vii-1989, Alejandre, MA 485511; Aramendia, vii-1987, Urrutia, MA 477842; Lokiz, vii-1987, Uribe-Echevarria, MA 489242.

T. thymifolium. SPAIN: Cuenca, v-1975, G. López, MA 433643; Cuenca, Hoz de Tragaviros, vii-1978, G. López, MA 433630.

T. buxifolium. SPAIN: Alicante, Villena, vii-1955, Rigual, MA 375540; Albacete, rio Segura, v-1987, Izuzguiza 654A, MA 434376; Alicante, Serrella, vi-1984, Mateo & Figuerola, MA 434317. Alicante, Pego, 1997, Signer, MA 590256; Alicante, Altea, Rigual, MA 374228.

T. rivasii. SPAIN: Alicante, Monford del Cid, 1997, Serra, ABH 33976; Alicante, Novelda, Rigual, MA 375675; Alicante, Raspeig, v-1962, Rigual, MA 373637; Alicante, Monforte, v-1954, Rigual, MA 375546.

T. rivas-martinezii. SPAIN: Albacete, Yeste, MGC 45032.

T. compactum. SPAIN: Almería, Nijar, vi-1986, T. Navarro, MGC 39598.

T. grosii. MOROCCO: Alhucemas, 1999, Cabezudo, MGC 28014; Alhucemas, v-1989, Diaz et al., BC 164369.

T. alopecurus. TUNICIA: Gabes, Djebel Tebaga, 1992, Molero & Vicens, VAL 95/5441.

T ducellieri. MOROCCO: Midelt, iv-133, Wilczek, MA 97959; MA 97960; Haute Moulouya, iv-1925, Jahandiez 110, MA 9796; Midelt, 1933, Wilczek, MA 97959.

T. alpestre. CRETA: Nomos, v-2002, Vitek 02104, MA 710216.

Section *Chamaedrys* (Mill.) Schreb.

T. chamaedrys. SPAIN: León, Beberino, vii-1988, Luceño 2553, MA 489290; Soria, Cameros, viii-1991, Alejandre, MA 534072; Segovia, Fuentidueña, MA 317831. GREECE: Olimpos, viii-1983, Aguililla, VAL 41896. MOROCCO: Aknoul, vi-1995, Boratynski & Romo, 8787/5, BC 165411.

T. lucidum. FRANCE: Castellan, vii-1966, Berger, MA 581566; Toudon, vii-1994, Alziaz, MA 564241; Basses Alpes, vii-1886, Reverchon, MA 986335(2); Basses Alpes, vii-1886, Annot, MA 98635.

T. flavum. SPAIN: Valencia, Oliva, VAL 51220; Alicante, Fleix, VAL 48140. Valencia, Oliva, xi-1984, Mateo, VAL 51220; Alicante, Gallinera, Mateo, VAL 83820; Alicante, Gallinera, v-1979, Mansanet & Mateo, MA 462875. ITALIA: Siracusa, Sortino, VAL 119809.

T. fragile. SPAIN: Málaga, Alcaucín, MGC 57153; Granada, Almuñecar, MGC 57152; Granada, Sierra Tejeda, MGC 57151; Málaga, Sierra Tejeda, MGC 57150.

T. intricatum. SPAIN: Almería, Vicar, MA 542131; Felix, MA 501580.

T. pugionifolium. SPAIN: Valencia, Salvacañete, iv-1988, Mateo & Crespo 060, VAL 72107.

T. webbianum. SPAIN: Albacete, Sierra Ayna, vii-1971, Rivas Godoy & Borja, MA 558023; Ciudad Real, vii-1934, Alcaraz, MA 98676; Granada, Sierra Nevada, vii-1976, Castroviejo & Valdes Bermejo 778, MA 435011.; Jaén, Sierra de Mágina, vi-1926, Cuatrecasas, BC 47385; Jaén, Albanchez, vi-1925, Cuatrecasas, BC 47383.

Section *Chamaedrys* Subsection *Marum* Kästner

T. subspinosum. SPAIN: Menorca, Cap de Favaritx, MA 435671. Menorca, Binincola, vi-1980; Valdes-Bermejo 5611, MAF 118439; Binincola, Menorca, vi-1980, Castroviejo & al., MA 377857. Menorca, Ciutadella, JACA 1253; Menorca, Mercadal, JACA 444392; Cabrera, Canal Llarg, MA 98859; Cabrera, Coll Roig, MA 383095; ITALIA: Cerdeña, Cagliari, Limbara, CAG; Cerdeña. Isla S. Prieto, Porto Banda, CAG (Herbarium Horti Botanici Caralitani).

T. balearicum. SPAIN: Mallorca, Galatzó, MA 98854; Mallorca, Valldemosa, MA 98682; Mallorca, Soller, MA98856; Menorca, Monte Toro, MA 435647; Mallorca, Formentor, iv-1934, Rivas Goday, MAF 83869; Mallorca, Soller, vi-1879, Crespi, MAF 33134; Cabrera, Coll Roig, v-1948, Palau Ferrer, MAF 33135; Mallorca, Galatzo, vii-1934, Gros, MAF 33133.

T. marum. ITALIA: Livorno, VAL 025593; Corcega, Calvi, La Revellata, VAL 25594; Corse, Bavella, vii-1929, Litardiere, MAF 60255; Corse, Calvi, vi-1989, Lambinon 89145, MAF 145468; Livorno, vi-1988, Moggi & al. 14427, MAF 145467.

T. divaricatum. SIRIA: Liban, Ain-Halesoun, vi-1934, Bertschinger, MA 470817. PALESTINA: Jerusalem, Meyers 4025, MA 98679. GREECE: Athenes, MPU; Karpathos, Rhodas, Samos, MPU. LIBANO: Libano, Vallée de Maydouché, MPU.

Section *Polium* (Mill.) Schreb.

T. aureum. SPAIN: Tarragona, Montsant, vi-1954, MA 98498; Barcelona, Montserrat, vii-1914, Caballero, MA 96328; Querol, vi-1999, Castroviejo et al.,15065, MA 628840. ITALIA: Sicilia, Ragusa, vi-2000, Güemes 3398, VAL 119567. FRANCE: Aude, vii-1944, Rechinger 458, MA 648449.

T. capitatum subsp. *capitatum*. SPAIN: Jaén, Jodar, x-1925, Cuatrecasas, BC 47262; Benidorm, v-1993, Verdú, ABH 11809; Alicante, Benitaxell, v-1992, Barber, ABH 9746. MOROCCO: Quebdana, vii-1939, Sennen, BC 138165.

T. homotrichum. SPAIN: Alicante, Sierra Mariola, 1949, Borja, MA 186190; Santa Pola, Muñoz Garmendia, MA 375840; Valencia, Pinet, 1975, Mateo, MA 484323; Valencia, Játiva, 1943, Bellot, MA 383032; Alicante, Serralla, 1981, Solanas, MA 557663.

T. dunense. SPAIN: Mallorca, El Arenal, vi-1919, Bianor, MA 98059; Valencia, Albufera, vi-1913, Sennen 7408, MA 98065; Mallorca, Font Santa, vi-1998, C. Navarro et al. 2115, MA 618607; Mallorca, El Arenal, vi-1919, Bianor & Sennen, MA 467631; Mallorca, Soller, 1919, Bianor, MA 98059; Mallorca, El Arenal, MA 98061.

T. polium. GREECE: Parnasos, viii-1983, Aguililla, VAL 07305. YUGOSLAVIA: Skopje, viii-1983, Aguililla, VAL 07312. ISRAEL: Nahal Qetalar, vi-1989, Heller, VAL 25590. ITALIA: Toscana, Moggi, VAL 04511. SPAIN: Lleida, Noguera, vii-1992, Aedo 2385, MA 511612; Huesca, Sopeira, vii-1992, Aedo 2343, MA 514397.

T. flavovirens. MOROCCO: Alto Atlas, Ait-Youly Bou Thrarar, Güemes 1511, VAL 41076; Ouarzazate, v-1985, Blanché et al. 1985, BC 165403; Ksar es Souk, vi-1985, Blanché et al., 1985, BC 165518.

T. murcicum. SPAIN: Alicante, San Miguel de Salinas, 1996, Crespo, ABH 33189; Alicante, Rabate, 1996, Crespo, ABH 33190; Mallorca, Sant Esteve, vi-1985, Benedi, MA 501870.

T. algarbiense. PORTUGAL: Tavira, Santa Maria, v-1969, Correia, Instituto Superior Agronomia de Lisboa; Margen del rio Chança, LISU 32527; Tavira, v-1907, Mendez, LISU 32524; Tavira, Mendez, LISU 32529.

T. angustissimum. SPAIN: Valencia, Requena, VAL 99910.

T. gnaphalodes. SPAIN: Madrid, Arajuez, iv-1950, Rivas Goday, MA 256782; Jaén, Sierra de Cabra, vi-1925, Cuatrecasas, BC 47364.

T. aragonense. SPAIN: Teruel, Valdealgorfa, v-1984, VAL 64593; Zaragoza, Caspe, VAL 129337; Zaragoza, v-1973, Castroviejo & Valdés Bermejo, MA 434365; Zaragoza, Mequinenza, v-1976, Segura Zubizarreta 3379, MA 256695.

T. carthaginenese. SPAIN: Murcia, Cartagena, iv-1998, Aran, 89-18-35, BC 631743; Murcia, Portman, iii-2000, Fuertes, BC 75145.

T. capitatum subsp. *majoricum*. SPAIN: Ibiza, San Rafael, v-1918, Gros, MA 98221; Mallorca, Cala Fuent, xi-1947, Palau Ferrer, MA 98223; Mallorca, Bellver, vi-1955, Palau Ferrer 763, MA 167651; Mallorca, Burguesa, vi-1988, Morales et al. 1719, MA 617911.

T. similatum. SPAIN: Almería, Sierra de Filabres, vii-1989, T. Navarro, MGC 44958; Cádiz, Grazalema, vi-1984, Aparicio et al, BC 806325; Cádiz, Sierra del Pinar, vii-1984, Martin & Silvestre, BC 806328.

T. cossonii. SPAIN: Ibiza, LLadalt, v-1920, Font Quer, MA 98451; Mallorca, Puig Mayor, vi-1980, Castroviejo et al. 1935, MA 377938; Ibiza, Santa Agnes, vi-1979, FC 2905, MA 256744; Mallorca, Soller, MA 98081.

T. rouyanum. FRANCE: Aveyron, Tourmemire, vi-1906, VAL 141047.

T. ***mideltense***. MOROCCO: Alto Atlas, Ighil M Goun, vii-1997, Güemes, VAL 41068; Oukaimeden, 1999, Güemes, VAL 41211.

Section *Polium* Subsection *Simplicipilosa* S. Puech

T. ***lanigerum***. SPAIN: Almeria, Garrucha, vii-1950, Jerónimo, BC 146340; Almería, San José, vi-1983, T. Navarro, GDAC 23078; Almería, Mojacar, ii-1985, T. Navarro & Rosúa, GDAC 23080.

T. ***charidemi***. SPAIN: Almería, Cabo de Gata, vii-1950, Losa, BC 145725; Almería, Cabo de Gata, vii-1948, Jerónimo, BC 146339.

T. ***chrysotrichium***. SPAIN: Málaga, Coín, v-1994, Navas et al., MGC 39572; Málaga, Alhaurin Perez Gómez, MGC 15748.

T. ***haenseleri***. SPAIN: Toledo, Orgaz, vii-1993, Aran, VAL 93346; Ciudad Real, Sierra Madrona, vi-1988, Sánchez Mata, VAL 77640; Málaga, Cerro Muriano, vii-1922, Gros, BC 802646.

Section *Polium* Subsection *Pumila* (Lazaro Ibiza) Rivas-Mart.

T. ***pumilum***. SPAIN: Madrid, Chinchón, vi-1988, Aran, VAL 75540; Madrid, Arganda, vii-1984, VAL 47191; Arganda, vii-1981, Aguilella, VAL 72120.

T. ***carolipaui***. SPAIN: Almería, Los Gallardos, vii-1983, Zubizarreta, VAL 11718; Alicante, La Nuncia, Mateo, VAL 46625; Villajoyosa, vi-1935, Martinez, VAL 72887.

T. ***libanitis***. SPAIN: Murcia, Jumilla, viii-1959, Borja, MA 179404.

T. ***turredanum***. SPAIN: Almería, Mojacar, ii-1985, Rosúa & T. Navarro, GDAC 22776; Almería, rio de Aguas, xii-1983, T. Navarro & Guirado, GDAC 22778; Turre, vii-1963, Rivas Goday, MAF 102670.

Section *Scordium* (Mill.) Rchb.

T. ***scordium***. SPAIN: Huelva, Doñana, vi-1977, Castroviejo et al. 2256, MA 434926; Cádiz, Algodonales, viii-1978, Silvestre, MA 284365; Málaga, Cartama, vii-1888, Reverchon, MA 98796; Ciudad Real, vii-1980, Castroviejo et al. 1976, MA 256874.

Section *Scorodonia* (Hill) Schreb.

T. ***canadense***. USA: Florida, St. Andrew bay, vi-1959, Dress & Read 7661, EDI 178059; Indiana, Riverside, viii-1958, Irvine 706, EDI 178060; Nebraska, Lancaster, vii-1988, Nieto-Feliner 2058GN, MA 472409; Chicago, Bennett,

Especially in some species, the stamens and stigma are greatly exserted from corolla (De Martino et al. 2010).

The members of the genus *Teucrium* are shrubs, subshrubs and mostly perennials, rarely annual or biennial. Leaves opposite, sessile or petiolate, simple, margines entire, toothed or lobed; inflorescences vary from simple to composites; calices vary from radial to bilabiate, tubular, campanulate or tubular-campanulate, 5-lobed, mostly 10 veins are also obvious, or sometimes the main veins are more obvious, calyx lobes differ from species to species, and all of them are same or sometimes different, the other 4 or different types of lobes, gibbouse or not at the base; corollas in very different colors, mostly short or almost equal to or longer than the calices, very rarely bilabiate and mostly one lipped (lower lip middle lobe), 5-lobed, the lower lobe is much larger than the others, usually inward; stamens 4, which of two short and two long filaments, exserted from corollas, obviously mostly longer than calices, filaments flat or curved, thecas separated; pistils with two carpels and four lobes, stigma lobes equal or almost equal, curved or not at the apex; nutlets obovate to almost orbicular, reticulated-wrinkled, with short eglandular and sessile glands or glabrous, endosperms usually available (Boissier 1879; Briquet 1895; McClintock and Epling 1946; Ekim 1982; Harley et al. 2004; Melnikov 2014; Özcan 2015).

Morphological variation can be seen in some *Teucrium* species according to the literature. Gaisberg (2000) divided *Teucrium heterophllyum* into three subspecies on the basis of their calyx lengths and the size of the branched, non-glandular trichomes on the calices and leaves. Trichome size, branched or unbranched trichomes, and the state of glandular or eglandular trichomes are used as very significant characters in many paper, too (Bentham 1833; Tutin et al. 1972; Kästner 1978; Manzanares et al. 1983; Antunes and Sevinate-Pinto 1991; Bini Maleci and Servettaz 1991; Servettaz et al. 1992; Navarro and El Oualidi 2000; Vural et al. 2015). *Teucrium orientale* L. (var. *glabrescens*, var. *orientale* and var. *puberulens*) is another species and its varieties can be easily distinguished by their trichome types and size on the adaxial and abaxial leaf surfaces. *Teucrium orientale* var. *orientale* has dense and elongated eglandular trichomes on the other hand *Teucrium orientale* var. *glabrescens* has sparse and shorter eglandular trichomes. All the three examined varieties have subsessile glandular trichomes (peltate) on the abaxial side of their leaves but only *Teucrium orientale* var. *orientale* has also adaxial side (Ecevit-Genc et al. 2015). Most of the *Teucrium* species have unbrunched trichomes on their leaves. *Teucrium polium* (Navarro and El Oualidi 2000; Lakušić et al. 2010; Eshratifar et al. 2011), *Teucrium capitatum* L. (Antunes et al. 2004) are two well known species with the branched trichomes on their leaves.

Most of the species of *Teucrium* grow mainly up on limestone cliffs, but some of endemics live on serpentine cliffs and some the other species live on basalt, volcanic tuff or marble. Section *Isotriodon* species grow only on rocky crevices and they are saxicolous plants, on the other hand most of the sect. *Scordium* members grow on wetlands. So, the species that live closest to the water are in this section (Stanković 2012).

The aim of this chapter is a general overview of the vegetative morphological characteristics of *Teucrium* species. The general characteristics of the life form,

stem, leaf structure, as well as micromorphology will be presented. Comparative morphological differentiation of *Teucrium* sections will be discussed. Each sub-chapter will show accompanying micrographs, drawings and photographs of morphological details, as well as species on natural habitat.

2.2 Sectional Classification

The calyx shape and the inflorescence structure are used to identify the sections of *Teucrium* genus. The epidermal properties and trichome types of the nutlets show apparent differences among *Teucrium* sections. Kästner (1989) produced the first substantial revision of the sectional divisions by including corolla and vegetative characters. The detailed sectional classification was given in Table 2.1. The morphology of the leaf trichomes of *Teucrium* species with differentiation into thin and thick-walled trichomes provide valuable taxonomic distinguishing at the species level (Navarro and El Oualidi 2000; Eshratifar et al. 2011).

In addition to calyx shapes and trichome types, crenation of the leaves margins and attenuate leaves are very important for the classification of *Teucrium* sections. Leaves of the sect. *Teucrium* members entire or the crenation is very undetectable at first glance to deeply dissected, and their leaves are attenuate. This section is very interesting with crenation range among species. Leaves sessile or subsessile. Also, leaves are retrorse at the abaxial surface. Villous indumentum can be seen on different aerial parts of the plants. Some species of this section especially *Teucrium creticum* and *T. fruticans* are strongly fruticose, some of them are shrubs and the others are dwarf shrublets. The species are mostly rhizomatous. *Chamaedrys*, *Polium*, *Scordium*, *Stachyobotrys*, *Scorodonia*, *Spinularia* and *Isotriodon* sections have an obvious crenation on their leaf margins. Sect. *Chamaedrys* has obviously lobulate or dentate leaves, and the leaves are attenuate or cuneate at base and mostly subsessile or lower ones shortly petiolate. Leaves are glabrous to dense and sometimes retrorse-crispate hairy. This section members are generally small shrubs or suffrutescent. Especially *Teucrium chamaedrys* is rhizomatous and it can be found as heaps in the field trips. Stem lengths of the species are generally not as tall as the members of section *Teucrium*. Sect. *Scordium* members are obviously stoloniferous herbs and mostly live near or on wetlands. Leaves are toothed not crenated. Sect. *Polium* is distinguished with its terete and villous adpressed crispate hairy stems. Their leaves are usually tomentose and oblong to narrowly obovate or linear, obtuse, crenate to the base or middle, flat or revolute-margined. *Teucrium montanum* L. has not branched trichomes and dense indumenta on its aerial parts. All the species of the section *Polium* have globose heads consist of their flowers. *Isotriodon*, *Stachyobotrys* and *Scorodonia* sections are easily seperated with their calyx structure in the genus. With this properties they have some characteristic vegetative morphology. Section *Isotriodon* members are the most saxatile species of this genus. *Isotriodon* species are growing only the E. Mediterranean Region and Turkey has the most endemic species belonging to this section. This section members are

Table 2.1 Morphological differentiation of the *Teucrium* sections

Section	*Teucrium*	*Teucriopsis*	*Chamaedrys*	*Montanum*	*Polium*	*Scordium*	*Stachyobotrys*	*Scorodonia*	*Isotriodon*	*Spinularia*
Life form	Perennial	Perennial	Perennial	Perennial	Perennial	Perennial	Biennial or perennial	Perennial	Perennial	Annual
Stem	Shrubs, sub-shrubs or stoloniferous	Shrubs to 2–(3) m, short-hairy to tomentose	Dwarf shrubs, woody at base	Dwarf, woody and rotund at the base, erect to ascendance, sparsely hairy	Erect, woody at base, rotund to quadrangular, tomentose	Stoloniferous or erect, densely villous or velutinous	Shrubs, woody at base, hirsute or villous, with short eglandular trichomes to longer multicellular trichomes	Erect, conspicuously quadrangular	Dwarf, fragile, suffruticose, rotund to quadrangular	With spiny branches, hirsute
Leaf	Orbicular-ovate, oblanceolate, linear, obtuse, margin mostly entire to deeply divided, revolute at beneath, sparsely hairy	Ovate-lanceolate, obtuse, margin entire to bluntly crenate in the upper part, tomentose	Lobate or dentate-crenate	Lanceolate, elliptic to obovate, margin entire or unobvious dentate, revolute at beneath, sparsely glandular and eglandular hairy above, denser beneath	Sessile and simple, margin entire to deeply toothed-serrate, oblong-linear, spatulate, strongly revolute beneath, tomentose	Ovate to ovate-lanceolate, oblong, oblong-ovate, conspicuously toothed, crenate, subcordate at base, not petiolate to subpetiolate, villous-tomentose	Dentate-crenate, cuneate, truncate or cordate-subcordate at the base, short eglandular trichomes to long eglandular trichomes	Lanceolate to ovate, slightly rounded or blunt at the base, sessile or petiolate, tomentose	Ovate to ovate-lanceolate, slightly acute at apex, lower ones crenate, cuneate, subcordate to truncate at base, upper ones crenate to entire	Linear to elliptic-oblanceolate, obviously dentate, not rotund beneath, sparsely hairy above and denser beneath
Bract	Orbicular-ovate, Oblanceolate, linear, lanceolate	Leaf-like, obtuse	Leaf-like, dentate	Leaf-like, entire at margin	Bracts linear-spatulate	Leaf-like, rarely entire	Linear-subulate, shorter to longer than calyx	Ovate to ovate-lanceolate, sessile	Linear to oblanceolate, acute at apex, attenuate at base, pedicellate	Oblanceolate to ovate, entire or rarely crenated at the margin

obviously fragile and dwarf suffrutescent. Leaves are entire to dentate and subsessile to petiolate. Stems are mostly terete like sect. *Polium* and slender. Section *Stachyobotrys* members have the longest crenate-dentate leaves with the members of the section *Scorodonia*, and they can be around 80 mm length. Leaves have subsessile glandular trichomes but any capitate or clavate glandular trichomes could not be seen on abaxial or adaxial sides. Petioles of *Stachyobotrys* species are distinctive in *Teucrium* genus with around 35 mm length. Stems are villous or pubescent with 1–3 celled capitate glandular trichomes above. Petioles of section *Scorodonia* members are shorter than *Stachyobotrys* species, or subsessile. The colour of general view of the species are yellowish-greenish. Plants slightly rhizomatous like *Stachyobotrys* species, softly hirsute below, crisped-pubescent above. Retrorsing of leaves can also be seen in the sections *Polium*, *Chamaedrys*, and *Isotriodon* members (Tutin et al. 1972).

2.3 Vegetative Morphology

As mentioned before, most of the taxa belonging to different sections are perennial herbs from 20 cm to 2 m tall. Also, some members in the sect. *Stachyobotrys* are biennial, and some species from the sect. *Teucrium* are annual. On the other hand, the sect. *Spinularia* has totally annual species. When we exclude the members of *Spinularia*, *Teucrium* species are at least woody at base. But some members of the sect. *Teucrium* are woody (for instance *Teucrium creticum*, *T. fruticans* etc.). Section *Teucrium* is shrubby or subshrubby, *Teucriopsis* is shrubby to 2–(3) m, *Chamaedrys* is dwarf shrubby and woody at base, *Montanum* is dwarf, woody and terete at the base, *Polium* is woody at base, terete to quadrangular, *Scordium* is stoloniferous and rhizomatous, *Stachyobotrys* is shrubby and woody at base, *Scorodonia* is shrubby, *Isotriodon* is dwarf, fragile, suffruticose, rotund to quadrangular. So, the members of sections *Teucrium, Teucriopsis, Chamaedrys, Scordium, Stachyobotrys* and *Scorodonia* are mostly phanerophytes and the members of sections *Polium, Isotriodon* and *Spinularia* are mostly chamaephytes. The stems are frequently square in cross section, but this is not a strict characteristic of all *Teucrium* species, as well as whole family (Fig. 2.1). Several *Teucrium* species feature stolons or rhizomes and can spread vegetatively (Tutin et al. 1972).

Leaf morphology of *Teucrium* species vary considerably from species to species of plant, depending largely on their life form, although some species produce more than one type of leaf. The shape and size of the leaves are somewhat variable. Variability in shape and size, apart from the individual sections and species, is also observed on the same individual. The leaves of the upper and lower branches may be of different shapes and sizes, and the well lower and upper leaves of the same branch may vary to some extent, as well as leaves of the main stem and lateral branches. Leaves diameter can vary from 0.5 to 10 cm long and 0.5 to 5 cm across. The opposite arrangement of the leaves and branches on the stem is a feature of the whole family – leaf attachments are paired at each node and decussate. *Teucrium*

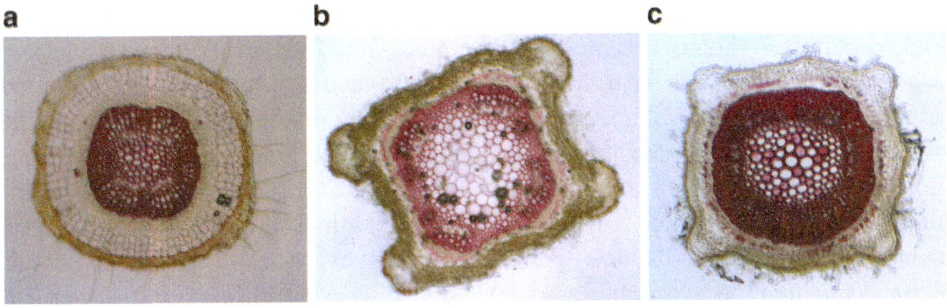

Fig. 2.1 Cross-section of *T. ekimii* (**a**), *T. sandrasicum* (**b**) and *T. divaricatum* (**c**) stem

species have a simple leaf blade, while some species have 1- to 2-pinatisect leaves. Leaf shape is characterized by most major types such as lanceolate, oblanceolate, ovate, obovate, triangular-ovate, ovate to orbicular, oblong, linear, linear-lanceolate, cuneate, cordate, truncate, eliptic, obtuse, acute, truncate, as well as different pinnatifid (Fig. 2.2). Leaf margins are very rarely entire, while the most common margin types are crenate, dentate, serrate and sinuate. The lamina is also characterized by different types of base and apex, while the surface may be glabrous, glaucous or pubescent with several degrees of hairiness, as well as differently expressed venation (Tutin et al. 1972).

Sect. *Teucrium* is characteristic in the genus with its leaf shapes. We can separate the section into four groups according to the leaf characters. For example, *Teucrium creticum* and *T. sandrasicum* have linear leaves with slightly toothed or entire margins; *Teucrium alyssifolium* and *T. pseudaroanium* have elliptic and entire leaves; *Teucrium multicaule* and *T. pestalozzae* have entire, bipartite and tripartite leaves on the same plants and *Teucrium orientale* and *T. parviflorum* have 2–3-pinnatipartite leaves. General characters of the leaves of the species belonging to section *Teucrium* are these followings: orbicular-ovate, oblanceolate, linear, obtuse, margin mostly entire to deeply divided, rotund at beneath, sparsely hairy. The leaves are mostly subsessile or rarely very short petiolate. Sect. *Teucriopsis* is characteristic with ovate-lanceolate and obtuse leaves, their margin entire to bluntly crenate in the upper part, and the leaves are obviously tomentose. Leaves of sect. *Chamaedrys* are oblong or obovate-oblong, cuneate at base, usually crenate-dentate or shortly lobulate, lobes entire or toothed. Obovate-oblong and dentate leaves are very significant to distinct the species belonging to this section. Leaves of sect. *Montanum* are lanceolate, elliptic to obovate, margin of the leaves are entire or unobvious dentate, rotund at beneath. On the other hand, the leaves of sect. *Polium* are sessile and simple, their margins are entire to deeply toothed-serrate, oblong-linear, spatulate, strongly rotund beneath, tomentose. Tomentose and spatulate leaves are distinguishing characters between *Polium* and *Montanum* sections. Sections *Scordium, Scorodonia* and *Stachyobotrys* have the longest leaves in the genus *Teucrium*. *Scordium* members have ovate to ovate-lanceolate, oblong, oblong-ovate leaves with conspicuously toothed, crenate, subcordate at base, subsessile to short petiolate, and villous or tomentose with thin or thick elongated trichomes. Leaves of sect. *Stachyobotrys* are more toothed-crenated than *Scordium* species, cuneate, truncate

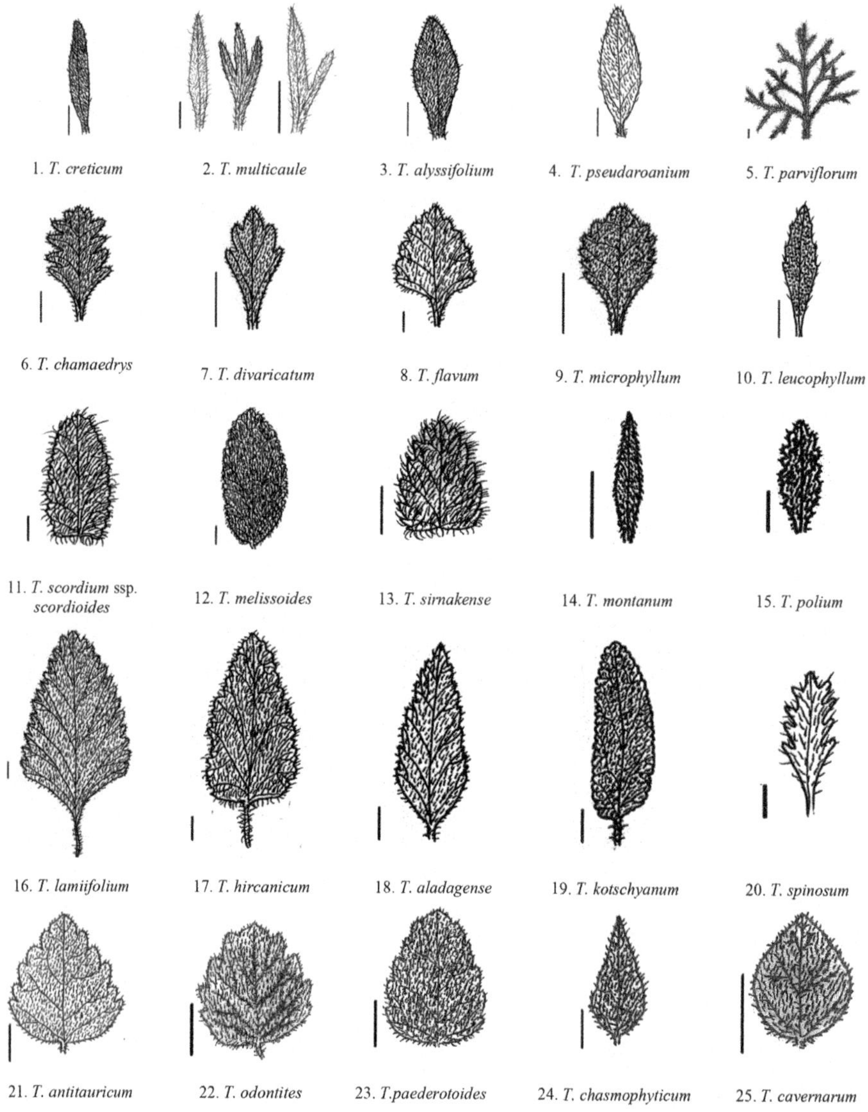

Fig. 2.2 Leaf morphology of selected *Teucrium* species 1–5: sect. *Teucrium*; 6–10: sect. *Chamaedrys*; 11–13: sect. *Scordium*; 14: sect. *Montanum*; 15: sect. *Polium*; 16–18: sect. *Stachyobotrys*; 19: sect. *Scorodonia*; 20: sect. *Spinularia*; 21–25 (scale bar 1 cm)

or cordate-subcordate at the base with short to elongated eglandular trichomes. We can see lanceolate to ovate leaves in the sect. *Scorodonia* and that means their leaves are narrower than the section *Stachyobotrys* and *Scordium* members, their leaves are slightly rounded or blunt at the base, sessile or petiolate and tomentose. The saxatile species of the genus (sect. *Isotriodon*) have ovate to ovate-lanceolate leaves, their leaves are slightly acute at apex, lower ones crenate, cuneate, subcordate to truncate

at base, upper ones crenate to entire, glabrous or slightly hairy. And the annual *Spinularia* members have linear to elliptic-oblanceolate leaves with obviously toothed, their leaves are not rotund beneath. Sect. *Chamaedrys*, *Isotriodon* and *Stachyobotrys* are not branched or branched at base, the rest of the members have very branched stems from the middle or upper sides (Tutin et al. 1972).

2.4 Micromorphology

Trichome absence or presence and also their types have a significant role in the infrageneric classification of the genus *Teucrium*, and they are the most helpful characters to separate the species. There are many studies on *Teucrium* morphology and micromorphology (Antunes and Sevinato-Pinto 1991; Bini Maleci and Servettaz 1991; Servettaz et al. 1992; Marin et al. 1994; Navarro and El Oualidi 2000; Dönmez 2006; Parolly and Eren 2007; Jurišić Grubešic et al. 2007; Dinç et al. 2008, 2009, 2011a, b; Dönmez et al. 2010; Lakušić et al. 2010; Eshratifar et al. 2011; Kremer et al. 2012; Doğu et al. 2013; Özcan 2013; Ecevit-Genç et al. 2015, 2017, 2018; Özcan et al. 2015; Vural et al. 2015; Dinç and Doğu 2016).

El Oualidi and Puech (1993) provided the basic information for the members of *Polium* section based on the structure of the calyx teeth, the structure and degree of tessellation of the leaves from the bottom, and the internal and external indumentum of the corolla. Navarro and El Oualidi (2000) studied the trichome morphology of 56 Mediterranean species of *Teucrium* members. In their study, the indumenta of leaf, calyx, corolla, and nutlet were investigated. Twelve of the 25 trichome types were specific in this study. The indumentum structures on these parts are very distinctive for *Teucrium* sections. The trichome types are basically divided into four parts: simple capitate glandular, branched capitate glandular, peltate glandular and eglandular. El Beyrouthy et al. (2009) reported that corolla indumentum was very distinctive to separate sections. Eshratifar et al. (2011) studied the micromorphological characteristics of nutlets and leaves of some members of the genus *Teucrium*, and it was stated that although pubescence had obvious differences between different species, it was similar in the same section. In another study, the differences between glandular and eglandular trichomes of *Teucrium polium* leaves were examined in winter and summer months. In this study, it was revealed that the trichome structures, the volatile oils, and antioxidant activities changed periodically and it was clearly emphasized that the dimensions of this species differed in winter and summer months (Bosabalidis and Kofidis 2002). Gaisberg (2000) made a revision of *Teucrium heterophyllum* species, and as a result of this study, he discovered two new subspecies and presented morphological differences of subspecies of this species. Trichome morphology was very distinctive for the subspecies.

Lakušić et al. (2010) examined the adaptations of 5 *Teucrium* species due to their habitat differences. In this study, it was shown that indumentum was an important factor in adaptation to xeric environments. *Teucrium scordium* was investigated morpho-anatomically in this study, and according to their findings, glandular

(peltate and capitate), and eglandular (unicellular or multicellular unbranched) trichomes were found on the leaves of this species. But Ecevit-Genç et al. (2018) did not observe any unicellular trichomes on the leaves of *Teucrium scordium* ssp. *scordium*. Jurišić Grubešić et al. (2007) investigated the trichomes of the leaves and stem of some *Teucrium* taxa from Italy. According to their findings, peltate, capitate glands and eglandular trichomes were found on *Teucrium montanum* leaves. Members of the sections *Polium* and *Montanum* are different from the other *Teucrium* sections with their inflorescences are arranged as a globose head. Also according to micromorphological studies, section *Polium* is also easily distinguished from all the other sections with the presence of branched eglandular trichomes on their leaves (*Teucrium polium*) (Ecevit-Genç et al. 2017).

Thin and thick-walled trichomes can be used to the delimitation of the species and could be regarded as a valid taxonomic character for the classification of *Teucrium* genus (Navarro and El Oualidi 2000; Eshratifar et al. 2011; Ecevit-Genç et al. 2015, 2017). Size, shape and density of glandular and eglandular trichomes are very significant for morphological differentiation of *Teucrium* species (Fig. 2.3).

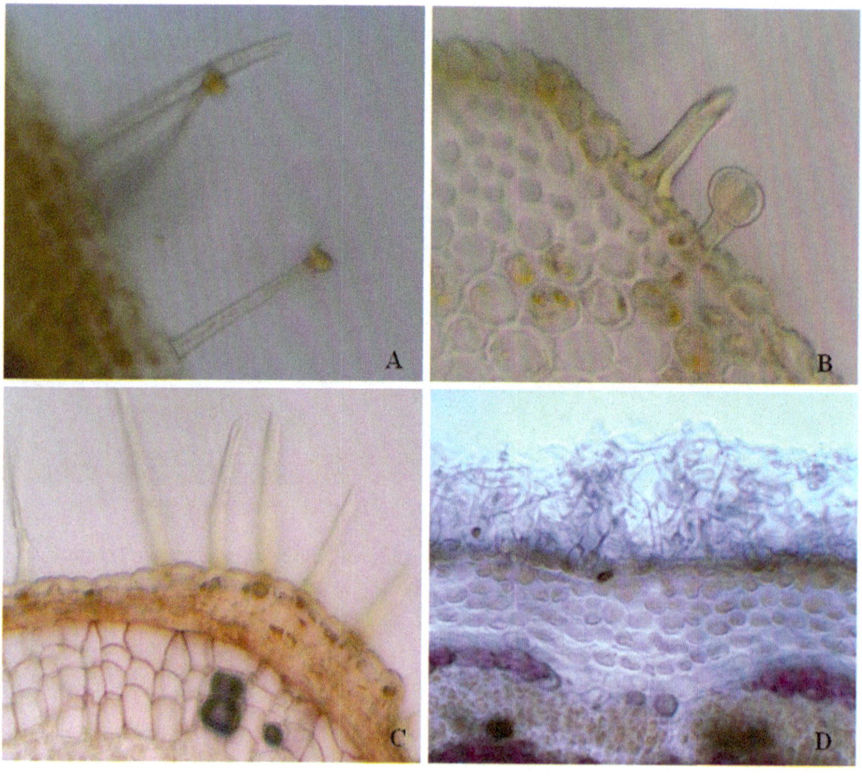

Fig. 2.3 General trichome types seen in transverse-sections of the stems: (**a**) *T. scordium* subsp. *scordioides* (sect. *Scordium*), (**b**) *T. chamaedrys* subsp. *chamaedrys* (sect. *Chamaedrys*), (**c**) *T. ekimii* and (**d**) *T. alyssifolium* (sect. *Teucrium*)

Fig. 2.4 General habitus of selected *Teucrium* species with different indumentum structure

The indument is also determined to some extent by the ecological conditions of the habitats. Figure 2.4 shows habitus of the six *Teucrium* species with different indumentum structure.

2.5 Conclusions

The genus *Teucrium* is one of the largest genera of Lamiaceae family. Certain morphological traits of this genus are characteristic of the whole family, but besides them, *Teucrium* species have many specificities regarding morphological differentiation. *Teucrium* species are perennial but a small group of species is annual. Most are shrubs, subshrubs, woody at base, rarely whole herbaceous plants. Some *Teucrium* species have stolons or rhizomes. Among the most common life forms are phanerophytes and chamephytes. The stems are frequently square in cross section. Leaves of *Teucrium* species are opposite, petiolate, sessile or subseeile, simple, with entire or toothed margin, lobed or pinnatifid. The lamina is also characterized by different types of base and apex, while the lamina surface may be glabrous, glaucous or pubescent with several degrees of hairiness, as well as differently expressed venation. The species of the genus *Teucrium* is characterized by a diverse indumentum structure composed of branched or unbranched eglandular as well as glandular trichomes. Trichome characteristics are a significant taxonomic character for the *Teucrium* genus. In addition to calyx shapes and trichome types, crenation of the leaves margins and attenuate leaves are very important for the classification of *Teucrium* sections.

Acknowledgments The author would like to thank Dr. Tuncay Dirmenci and Dr. Milan Stanković for his valuable comments and material.

References

Antunes T, Sevinate-Pinto I (1991) Glandular trichomes of *Teucrium scorodonia* L. morphology and histochemistry. Flora 185:65–70

Antunes T, Sevinate-Pinto I, Barroso JG, Cavaleiro C, Salgueiro LG (2004) Micromorphology of trichomes and composition of essential oil of *Teucrium capitatum*. Flavour Fragr J 19:336–340

Bentham G (1833) Labiatarum genera et species. Ridgeway Sons, London

Bini Maleci L, Servettaz O (1991) Morphology and distribution of trichomes in Italian species of *Teucrium* sect. *Chamaedrys* (Labiatae) – a taxonomical evaluation. Plant Syst Evol 174:83–91

Boissier PE (1879) Flora orientalis IV. H. Georg, Geneva/Basel

Bosabalidis AM, Kofidis G (2002) Comparative effects of drought stress on leaf anatomy of two olive cultivars. Plant Sci 163:375–379

Briquet J (1895–1897) Lamiaceae. In: Engler AK, Prantl K (eds) Die Natürlichen Pflanzenfamilien, vol 4(3A). Wilhelm Engelmann, Leipzig, pp 183–375

De Martino L, Formisano C, Mancini E, De Feo V, Piozzi F, Rigano D, Senatore F (2010) Chemical composition and phytotoxic effects of essential oils from four *Teucrium* species. Nat Prod Commun 5:1969–1976

Dinç M, Doğu S (2016) *Teucrium pruinosum* var. *aksarayense* var. *nov.* (Lamiaceae) from Central Anatolia, Turkey. Mod Phytomorphol 9:13–17

Dinç M, Duran A, Pınar NM, Öztürk M (2008) Anatomy, palynology and nutlet micromorphology of Turkish endemic *Teucrium sandrasicum* (Lamiaceae). Biologia 63:637–641

Dinç M, Doğu S, Bilgili B, Duran A (2009) Comparative anatomical and micromorphological studies on *Teucrium creticum* and *Teucrium orientale* var. *orientale* (sect. *Teucrium*, Lamiaceae). Nord J Bot 27:251–256

Dinç M, Doğu S, Bağcı Y (2011a) Taxonomic reinstatement of *Teucrium andrusi* from *T. paederotoides* based on morphological and anatomical evidences. Nord J Bot 29:148–158

Dinç M, Doğu S, Doğru Koca A, Kaya B (2011b) Anatomical and nutlet differentiation between *Teucrium montanum* and *T. polium* from Turkey. Biologia 66:448–453

Doğu S, Dinç M, Kaya A, Demirci B (2013) Taxonomic status of the subspecies of *Teucrium lamiifolium* in Turkey: reevaluation based on macro-and micro-morphology, anatomy and chemistry. Nord J Bot 31:198–207

Dönmez AA (2006) *Teucrium chasmophyticum* Rech. f. (Lamiaceae): a new record for the flora of Turkey. Turk J Bot 30:317–320

Dönmez AA, Mutlu B, Özçelik AD (2010) *Teucrium melissoides* Boiss. and Hausskn. ex Boiss. (Lamiaceae): a new record for Flora of Turkey. Hacet J Biol Chem 38:291–294

Ecevit-Genç G, Özcan T, Dirmenci T (2015) Micromorphological characters on nutlet and leaf indumentum of *Teucrium* sect. *Teucrium* (Lamiaceae) in Turkey. Turk J Bot 39:439–448

Ecevit-Genç G, Özcan T, Dirmenci T (2017) Nutlet and leaf micromorphology in some Turkish species of *Teucrium* L. (Lamiaceae). Phytotaxa 312:71–82

Ecevit-Genç G, Özcan T, Dirmenci T (2018) Leaf indumentum in some Turkish species of *Teucrium* (Lamiaceae). Istanbul J Pharm 48:6–11

Ekim T (1982) *Teucrium* L. In: Davis PH (ed) Flora of Turkey and the East Aegean Islands VII. Edinburgh University Press, Edinburgh, pp 53–75

El Beyrouthy M, Arnold-Apostolides N, Dupont F (2009) Trichomes morphology of six Lebanese species of *Stachys* (Lamiaceae). Flora Medit 19:129–139

El Oualidi J, Puech S (1993) Quelques marqueurs morphologiques des *Teucrium* section *Polium* du Maroc. Valeurs diagnostiques ä differents niveaux d'integration. Acta Bot Malacitana 18:163–173

Eshratifar M, Attar F, Mahdigholi K (2011) Micromorphological studies on nutlet and leaf indumentum of genus *Teucrium* L. (Lamiaceae) in Iran. Turk J Bot 35:25–35

Gaisber Von M (2000) A revision of *Teucrium heterophyllum* L'Hér. (Lamiaceae) with two new subspecies of the Canary Islands. Willdenowia 30:263–271

Govaerts RA, Paton A, Harvey Y, Navarro T, Del Rosario Garcia Pena M (2013) World checklist of Lamiaceae. The Board of Trustees of the Royal Botanic Gardens, Kew, Kew. www.kew.org/wcsp/

Harley RM, Atkins S, Budantsev A, Cantino PD, Conn B, Grayer R, Harley MM, Kok R, Krestovskaja T, Morales R, Paton AJ, Ryding O, Upson T (2004) Lamiaceae. In: Kadereit JW (ed) The families and genera of vascular plants, vol 7. Springer, Heidelberg, pp 167–276

Jurišić Grubešić R, Vladimir-Knežević S, Kremer D, Kalodera Z, Vuković J (2007) Trichome micromorphology in *Teucrium* (Lamiaceae) species growing in Croatia. Biologia 62:148–156

Kästner A (1978) Beiträge zur Wuchsformenanalyse und systematischen Gliederung von *Teucrium* L. I. Die Infloreszenzen und Blüten. Flora 167:485–514

Kästner A (1989) Ubersicht zur systematischen gliederung der gattung *Teucrium* L. Biocosme Mésogéen 6:63–77

Kremer D, Stabentheiner E, Jurisic Grubešić R, Oberlander A, Vladimir-Knežević S, Kosalec I, Ballian D (2012) A morphological and chemotaxonomic study of *Teucrium arduini* L. in Croatia, and Bosnia and Herzegovina. Plant Biosyst 146:402–412

Lakušić B, Stevanović B, Jančić R, Lakušić D (2010) Morpho-anatomical differentiation of the Balkan populations of the species *Teucrium flavum* L. (Lamiaceae). Flora 205:633–646

Manzanares P, Gómez-Campo C, Tortosa ME (1983) Estudios sobre el indumento de las especies ibéricas y baleáricas del género *Teucrium* L. (Lamiaceae). An Jard Bot Madr 40:93–106

Marin DP, Petković B, Duletić S (1994) Nutlet sculpturing of selected *Teucrium* species (Lamiaceae): a character of taxonomic significance. Plant Syst Evol 192:199–214

McClintock E, Epling C (1946) A revision of *Teucrium* in the new world, with observations on its variation, geographical distribution and history. Brittonia 5:491–510

Melnikov D (2014) The system of the genus *Teucrium* L. (Lamiaceae). Novit Syst Plant Vasc 45:63–69

Meusel H, Jager E, Rauschert S, Weinert E (1978) Vergleichende Chorologie der Zentraleuropaischen Flora II. Gustav Fischer, Jena

Navarro T, El Oualidi J (2000) Trichome morphology in *Teucrium* L. (Labiatae) a taxonomic review. An Jard Bot Madr 57:277–297

Özcan T (2013) Presence of *Teucrium microphyllum* in Turkey: morpho-anatomical, karyological and ecological studies. Biodicon 6:79–87

Özcan T (2015) The revision of the genus *Teucrium* L. (Lamiaceae). Dissertation, Balıkesir University

Özcan T, Dirmenci T, Coşkun F, Akciçek E, Güner O (2015) A new species of *Teucrium* sect. *Scordium* (Lamiaceae) from SE of Turkey. Turk J Bot 39:310–317

Parolly G, Eren O (2007) Contributions to the flora of Turkey II. Willdenowia 37:243–271

Servettaz O, Bini Maleci L, Pinetti A (1992) Micromorphological and phytochemical characters of *Teucrium marum* and *T. subspinosum* (Labiatae) from Sardinia and Balearic Islands. Plant Syst Evol 179:129–139

Stanković M (2012) Biological effects of secondary metabolites of *Teucrium* species of Serbian flora. Dissertation, University of Kragujevac

Tutin TG, Heywood VH, Burges NA, Moore DM, Valentine DH, Walters SM, Webb DA (eds) (1972) Flora Europaea III. Cambridge University Press, Cambridge

Vural M, Duman H, Dirmenci T, Özcan T (2015) A new species of *Teucrium* sect. *Stachyobotrys* (Lamiaceae) from the south of Turkey. Turk J Bot 30:318–324

Chapter 3
Morphological Characteristics of *Teucrium* Species: Generative Morphology

Teresa Navarro

Abstract The generative morphology related with corolla and calyx construction, pollen, inflorescence type and nutlets are of great importance in *Teucrium* L. These characters are the basis for the classic systematic and recent classifications. *Teucrium* has a corolla usually 1-lipped or 2-lipped (bilabiate). *Spartothamnella* Briq., *Oncinocalyx* F. Muell., and *Teucridium* F. Hook, closely related to *Teucrium*, present a corolla 5-lobed in the upper half, lobes slighty spreading, 4 posterior ± similar, anterior lobe larger than the others and not concave. The calyx in these genera is actinomorphic. This is an exceptional corolla conformation in *Teucrium* and only found in these species. The main difference between *Spartothamnella* and *Teucrium* is that the former has drupaceous fruit and between *Teucrium* and *Teucridium* is the ovary form. *Oncinocalyx* differs from *Teucrium* in the conspicuously hooked calyx lobes. *Teucrium* Sections *Pycnobotrys* Benth., *Stachyobotrys* Benth., *Scorodonia* (Hill) Schreb., and *Teucriopsis* Benth., have weakly 2-lipped corolla and zygomorphic calyx. Species of section *Teucrium* have 1-lipped spurred corolla and actinomorphic calyx except for the Australian species with non-spurred corollas. *Teucrium* show a great diversity in floral phenotypic diversity and adaptive significance to pollination and dispersal. Functional redundancy is found in the gullet-shaped corollas with nectar guides and nototribic pollination of the sections *Chamaedrys* (Mill.) Kästner, *Isotriodon* Boiss., subsection *Scordium* (Mill.) Kästner, *Montana* Lazaro Ibiza and subsection *Pumila* (Lazaro Ibiza) Rivas Mart. *Teucrium* is the only genus within the sub family Ajugoideae with two derived pollen characteristics, granulate or verrucate sculpturing and operculate colpi and with a considerable variation in pollen sculpturing. Nototribic pollination of the gullet-shaped corollas with closed gullet mechanism (by a palate of hairs associated to a slight pouch in the corolla tube) are exclusive of the section *Teucrium*. The sternotribic pollination of the brush-shaped corollas closed by a tuft of hair in the entrance of the corolla tube (base of the anterior lobe), are specific of the ginodioecious species of section *Polium* (Mill.) Schreb. Resupinate flag-shaped corollas with sternotribic pollination

T. Navarro (✉)
Department of Botany and Plant Physiology, University of Malaga, Malaga, Spain
e-mail: tnavarro@uma.es

M. Stanković (ed.), *Teucrium Species: Biology and Applications*,
https://doi.org/10.1007/978-3-030-52159-2_3

is present in subsection *Spinularia* (Boiss.) Kästner and *Stachyobotrys*. The main dispersal modes in section *Chamaedrys, Isotriodon* and *Polium* is the semachory. Sections *Teucrium* and *Teucriopsis* are wind-dispersal by pogonospores, subsection *Spinularia*, section *Scorodonia* and *Teucrium betonicum* and *Teucrium abutiloides* L'Hér from section *Teucriopsis* are ballospores.

Keywords *Teucrium* · *Spartothamnella* · *Oncinocalyx* · *Teucridium* · Corolla and calyx construction · Inflorescence · Nutlets · Pollen · Pollination · Dispersal · Lamiaceae

3.1 Introduction

The "bilabiate blossom" characterizing the Lamiaceae is a floral construction for nototribic (dorsal) pollen deposition (Claßen-Bockhoff 2007). The floral construction is a syndrome of adaptive characters from generative structures needed for pollen transfer that have evolved under phylogenetic developmental and environmental constraints. The floral construction allows understanding of the ecological and morphogenetic constraints of flower evolution in Lamiaceae (Claßen-Bockhoff 2007) and it is particularly interesting in *Teucrium* because most of the species have one-lipped flowers except for the section *Pycnobotrys, Stachyobotrys, Scorodonia* and *Teucriopsis* with weakly bilabiate (two-lipped) flowers. The *Teucrium* phylogenetic related genera, such as *Spartothamnella* Briq., *Oncinocalyx* F. Muell., and *Teucridium* F. Hook, do not have one-lipped or two-lipped (bilabiate) flowers; they own flowers with corolla of five-lobed, four posterior ± similar, and anterior lobe larger.

Some studies have also investigated the variation of floral construction in different flower morphs related to the breeding system and the evolution of separate sexes in gynodioecius *Teucrium* species (Puech 1978; El Oualidi 1987; Alados et al. 1997; Rodríguez-Riaño and Dafni 2007) and *Teucridium* genus (Merrett 2005).

Teucrium and its related genera show a great phenotypic diversity and adaptive significance in the morphological generative characters related to breeding system, pollination and other functional generative traits such as dispersal. *Teucrium* is melittophilous and pollinated mainly by bees (van der Pijl 1972) and wasps (Petanidou 1996), it is a good example of a sympetal, tubular, hermaphrodite and gynodioecious, proterandrous flower. In particular, this flower type is interesting in *Teucrium*, because it has mainly one-lipped corollas with bilateral symmetry, slightly campanulate tube and the lower lip with five asymmetrical lobes; this corolla type also occurs in related genera of the Ajugoideae. The lower lip acts as a landing platform and provides an easier access to the nectar, but in some sections, a barrier in the tube may still act as a selective device. The variations in the corolla construction, size, lobes symmetry, specialized features of lower lip, resupination of the tube are the result of a selection by pollinators, which leads an adjustment in floral forms to efficient pollen transfer and fruit set. The most common *Teucrium*

calyx is five-lobed and persistent, except when falls with the diaspores inside and plays a part in seed dispersal (van der Pjil 1972; Bouman and Meeuse 1992; Paton 1992; Navarro et al. 2009). The variation in the calyx symmetry, morphology of the tube and teeth while as functioning of the pedicel, express the adaptation to a certain kind of seed dispersal mechanisms and a protection of immature nutlets (Navarro and El Oualidi 1999). The sculpturing and size of the nutlets have an important role in dispersal mechanisms (Bouman and Meeuse 1992). Most of the species show a flowering peak at the end of spring but may also extend their flowering season into the summer and autumn (Navarro and Cabezudo 1995).

Teucrium pollen grains are trizonocolpate, isopolar and radiosymmetric. They are circular to circular-lobulate in polar view and sinuaperturate; in equatorial view they are elliptic and usually subprolate to prolate. Pollen surface is minutely perforated with granules, less often with verrucae and with a supractectal reticulum or with an intermediate sculpturing composed of short, discontinuous supratectal ridges (Navarro et al. 2004). *Teucrium* is the only genus within the sub family Ajugoideae in the Mediterranean region with two derived pollen characteristics, granulate or verrucate sculpturing and operculate colpi. Abu Asab and Cantino (1993) pointed to the considerable variation in pollen sculpturing within *Teucrium*, distinguishing three sculpturing types: the verrucate sculpturing (bearing supratectal wart-like projections), which is considered as a derived pollen character in Lamiaceae, the suprareticulate exine (supratectal ridges forming a reticulate pattern), a plesiomorphic condition within the subfamily Nepetoideae (Wagstaff 1992), and an intermediate sculpturing type transition from the verrucate to the suprareticulate composed of short discontinuous supratectal ridges. The first type is considered as the norm in *Teucrium*, while the last two are less common. Nabli (1970, 1976), El Oualidi (1987), Diez et al. (1993), Abu-Asab and Cantino (1993) and Navarro et al. (2004) described the pollen morphology of many species of section *Polium*. The pollen morphology supports the phylogenetic relationships between these genera and *Teucrium* (Abu Asab and Cantino 1993), forming the *Teucriina* clade (Cantino et al. 1997), based on the operculate and verrucate pollen.

The generative characters related with corolla and calyx construction, pollen, inflorescence type and nutlet morphology are of great importance in *Teucrium*. These characters are the basis for the classic and recent systematic classifications, the great useful for delimiting and structuring the most problematic taxonomic groups. In order to examine the floral phenotypic diversity, a sample of 341 specimens of 97 *Teucrium* taxa and its close related phylogenetic genera *Spartothamnella*, *Oncinocalyx* and *Teucridium* from throughout the world have been studied in addition to whose examined in previous works (Navarro and El Oualid 1999; Navarro 1995, 2010). This study based on the field, herbarium specimens and bibliographic references of the regional Floras. Flowers and nutlets were collected in the field or from authorized Herbarium specimens (see Appendix in Chap. 1). The main floral features, such as the inflorescence type, corolla size, corolla colour, flower sexuality and main flowering season were measures or observed. The flower form related to the pollination based on Faegry and van der Pjil (1979). The primary dispersal mode follows van der Pijl (1972), Zohary (1937) and Bouman and Meeuse (1992) and the

diaspore type is described in accordance with Dansereau and Lems (1957) and Jenny (1995). The sectional delimitation of *Teucrium* is according to Bentham (1833), Boissier (1879) and Navarro (2010). The examined specimens are listed in the Appendix included in the Chap. 1.

The aim of this study is described the phenotypic diversity and adaptive significance of the *Teucrium* flower construction to carry out an overview of the diversity of the generative characters, especially floral features related to pollination and dispersal mechanisms, providing a synthesis of the predominant generative strategies and their infrageneric diversity.

3.2 Floral Phenotypic Diversity and Adaptive Functional Significance in *Teucrium* and Related Genera

Teucrium has a corolla usually 1-lipped or 2-lipped, 5-lobed (with the anterior lobe much larger than the others), concave, occasionally unequal 5-lobed in the upper half; lobes slightly spreading, 4 posterior ± similar (anterior lobe larger than the others). *Oncinocalyx betchei, Spartothamnella teucriiflora* and *Teucridium parvifolium* present a corolla 5-lobed in the upper half, lobes slighty spreading, 4 posterior ± similar, anterior lobe larger than the others and not concave, with an annulus of hairs at the middle part of insertion of stamens in the corolla tube (Table 3.1). The calyx in these genera is actinomorphic. This is an exceptional corolla conformation in *Teucrium* and only found in these species. The main difference between *Spartothamnella* and *Teucrium* is that the former has drupaceous fruit and between *Teucrium* and *Teucridium* is the ovary form, unlobed or lobed up to a quarter part of its length in *Teucridium* and usually lobed from a quarter to half its length in *Teucrium*. *Oncinocalyx* differs from *Teucrium* in the conspicuously hooked calyx lobes. Variation in corolla construction is show in Table 3.1 and corolla characters related with pollination in Table 3.2. A schematic drawing of the general conformation of the *Teucrium* flower is shown in Fig. 3.1.

Teucrium and its close related genera also differs from the other members of Lamiaceae based on non-gynobasic style. The ovary is bicarpellary, syncarpous, 4-lobed, 4-locular, with one ovule in each cell attached to an axile placenta, the style terminal, exserted, filiform, with 2 stigmatic lobes at the top. Fruit dry, schizocarpic, splitting into 4 mericarps (nutlets), enclosed in the persistent calyx, except for the drupaceous fruit in *Spartothamnella*.

The basic inflorescence is a raceme derived from a tyrse of verticillasters by floral reduction, originating contracted verticillasters, grouping into heads (Navarro and Cabezudo 1995). The infrageneric variation from axillary flowers or lax verticillasters (2–8 flowers) to simple dense terminal or branched raceme of contracted verticillasters is show in Table 3.1.

The calyx is tubular or campanulate, urceolate (gibbous) in plate or obconical, ventral, dorsiventral or dorsally pedicellated, 10-nerved, zygomorphic (bilabiate),

Table 3.1 Floral construction of the infrageneric taxa of *Teucrium* and its related genera *Oncinocalyx, Spartothamnella* and *Teucridium*

Teucrium infrageneric taxa	Broad upper calyx lip	Calyx construction	Corolla construction	Internal indumentum of corolla	Calyx pedicellum insertion	Corolla tube short apically broadly
Teucrium (Mediterranean area)	Absent	Actinomorphic Calyx	1-lipped, llp. not forming angle with the tube	Annulus of hairs at the base of the ct.	Ventral	Present
Teucrium (*T. racemosum* group)	Absent	Actinomorphic Calyx	1-lipped, llp. not forming angle with the tube	Annulus of hairs at the middle part of the ct.	Ventral	Present
Teucrium (Austral areas)	Absent	Actinomorphic Calyx	5-lobed,4 posterior ±similar, anterior lobe larger	Annulus of hairs at the middle part of the ct.	Ventral	Present
Teucriopsis	Absent	2-lipped 1/4	2-lipped 2/3, llp. emarginated or two lobed,	Without indumentum	Dorsi-ventral	Absent
Scorodonia	Present	2-lipped 1/4	2-lipped 2/3, ul. developed	Without indumentum	Dorsi-ventral	Absent
Pycnobotrys	Present	2-lipped(1/4)/ zygomorphic (1/2/2)/ Sub-actinomorphic (3/2)	Sligth 2-lipped (pseudo-2-lipped)	Without indumentum	Dorsi-ventral	Absent
Stachyobotrys	Present	Zygomorphic (1/2/2)	Sligth 2-lipped (pseudo-2-lipped)	Without indumentum	Dorsal	Absent
Isotriodon	Absent	Zygomorphic (1/2/2)	1-lipped, llp forming angle with the ct.	Hairs at the base of ml. forming two guides	Dorsi-ventral	Absent
Chamaedrys subsection *Chamaedrys*	Absent	Sub-actinomorphic (3/2)	1-lipped, llp forming angle with the ct.	Hairs at the base of ml. forming two guides	Dorsi-ventral	Absent
Chamaedrys subsection *Marum*	Absent	Sub-actinomorphic (3/2)	1-lipped, llp forming angle with the ct.	Hairs at the base of ml. forming two guides	Dorsi-ventral	Absent

(continued)

Table 3.1 (continued)

Teucrium infrageneric taxa	Broad upper calyx lip	Calyx construction	Corolla construction	Internal indumentum of corolla	Calyx pedicellum insertion	Corolla tube short apically broadly
Montana	Absent	Sub-actinomorphic (3/2)	1-lipped, llp forming angle with the tube,	Hairs at the base of ml. forming two guides	Dorsi-ventral	Absent
Polium subsection *Polium*	Absent	Sub-actinomorphic (3/2)	1-lipped, llp not forming angle with the ct.	Hairs at the base of the ml. forming a tuf	Dorsi-ventral	Present
Polium subsection *Pumila*	Absent	Sub-actinomorphic (3/2)	1-lipped, llp forming angle with the ct.	Hairs at the base of ml. forming two guides	Dorsi-ventral	Absent
Polium subsection *Simplicipilosa*	Absent	Sub-actinomorphic (3/2)	1-lipped, llp not forming angle with the ct.	Hairs at the base of the ml. forming a tuf	Dorsi-ventral	Present
Scordium	Absent	Sub-actinomorphic (3/2)	1-lipped, llp forming angle with the ct.	Hairs at the base of ml. forming two guides	Dorsi-ventral	Absent
Scordium subsection *Botrys*	Absent	2-lipped (1/4)	1-lipped, llp forming angle with the ct.	Hairs at the base of ml. forming two guides	Dorsal	Absent
Scordium subsection *Spinularia*	Absent	2-lipped (1/4)	1-lipped, llp forming angle with the ct.	Hairs at the base of ml. forming two guides/ without indumentum	Dorsal	Absent
Oncinocalyx genus	Absent	Actinomorphic Calyx	5-lobed,4 posterior ±similar, anterior lobe larger	Annulus of hairs at the middle part of the ct.	Ventral	Present
Spartothamnella genus	Absent	Actinomorphic Calyx	5-lobed,4 posterior ±similar, anterior lobe larger	Annulus of hairs at the middle part of the ct.	Ventral	Present

(continued)

Table 3.1 (continued)

Teucrium infrageneric taxa	Broad upper calyx lip	Calyx construction	Corolla construction	Internal indumentum of corolla	Calyx pedicellum insertion	Corolla tube short apically broadly
Teucridium genus	Absent	Actinomorphic Calyx	5-lobed,4 posterior ±similar, anterior lobe larger	Annulus of hairs at the middle part of the ct.	Ventral	Present

ct corolla tube, *ml* middle lobe, *llp* latero-posterior corolla lobes

sub-actinomorphic or actinomorphic, tube within glabrous or pilose at throat, the 5 teeth, equal or unequal, some times mucronate to hooked at end or cucullate (Table 3.1).

The infrageneric adaptive functional significance of floral features to pollination and dispersal is summarized below. Very detailed illustrations of the inflorescence type, corolla and clayx construction and nutlets morphology can be consulted in Navarro (2010).

The corolla of section *Teucrium* is one-lipped, gullet-shaped and slightly personate (with a slight pouch) and shortened tubuled, lilac or white with closed gullet mechanisms by a palate of hairs; the lateral and latero-posterior lobes are developed. Flowers are axillary in the leaves or in lax verticillasters (Fig. 3.2) and the flowering occurs in spring or multi-seasonal floration in winter-spring-summer. In this group, the most important factor for attracting pollinators is presumably the strong visual signal of their large flowers described by Shmida and Dafni (1989) as flowers engaged in a "discovery advertisement" (Fig. 3.2). The campanulate actinomorphic calyx does not have a closed mechanism to protect the immature seed. They are wind-dispersal species with a big hairy nutlet, ancestral character in the genus (Marin et al. 1994; Navarro and El Oualidi 2000b). The diaspore type is pogonospore. Section *Teucrium* is a basal group within *Teucrium* (Marin et al. 1994; Navarro and El Oualidi 2000a; Navarro et al. 2004; Salmaki et al. 2016). This section has widespread floral characters in the structurally simple states, including the actinomorphic calyx and unclearly lipped corollas (Austral species of section *Teucrium*) such as *Spartothamnella, Oncinocalyx* and *Teucridium* present in geologically ancient areas.

Section *Teucriopsis* comprise endemic species from Macaronesian region. The flowers are in verticillasters or in lax axillary cymes in the upper leaves. The corolla blue, orange or cream is weakly bilabiate, wide and shortly tubular; the calyx is campanulate, actinomorphic, some with an annulus of hairs inside; the nutlets are large and haired.

Species from section *Scorodonia* have one-lipped gullet-shaped corollas with long and narrow tube and a lower lip and latero-posterior lobes well-developed without nectar guides; the calyx is bilabiate with an annulus of hairs inside. Flowers

Table 3.2 Main inflorescence type, corolla size (mm), corolla colour, main flowering season and pollinators in the infrageneric taxa of *Teucrium* L

Teucrium infrageneric taxa	Inflorescence	Corolla size (mm) Length of tube and lower lip	Corolla colour	Main flowering season	Pollinators
Teucrium	Axillary flowers or lax verticillasters	(10)12.5–14.5(22)	Lilac/white	End winter-spring-summer (III–VI)	Large solitary bees
Teucriopsis	Axillary flowers or lax verticillasters	(7.5)9.5–11.5(13.5)	Blue/orange/white	Summer (V–VI)	Large solitary bees
Chamaedrys	Lax verticillasters/sub-spirally racemes	(9.5)11.5–12.5(14.5)	Purple/pink/yellow	Summer (V–VI)	Small solitary bees, flies
Montana	Sub-spirally raceme	(8)10.5–12.5(17)	Cream/white-pink/purple/yellow	Summer (V–VII)	Solitary bees, ants, flies
Scorodonia	Verticillasters	(5.5)7–8(11)	Cream/pink	End spring-summer (IV–VI)	Long-tongued butterflies, bees
Scordium	Axillary flowers/verticillasters (2–8) flowered	7.5–8.5	Cream/pink	Summer (V–VII)	Small solitary bees
Stachyobotrys	Dense Verticillasters like spike	7.5–9.5	Cream/cream-yellow	End spring-summer (IV–VI)	Small solitary bees, flies
Scordium subsection *Spinularia*	Verticillasters (2–4) flowered	(6.8)8.5–9(15.5)	Cream/cream-pink	Summer (V–VIII)	Solitary bees
Isotriodon	Sub-spirally racemes	(5.5)8–10(14.5)	Cream/pink-purple/yellow	Spring (IV–V)	Solitary bees, flies
Polium subsection *Polium*	Simple dense terminal or branched raceme	(3)3.5–4.5(6.5)	White/cream-pink/yellow	End spring-end summer (IV–VIII)	Small solitary bees, wasps, ants, flies
Polium subsection *Simplicipilosa*	Simple dense terminal or branched raceme	(3)3.5–4.5(6.5)	White/cream	Summer (VI–VII)	Small solitary bees, wasps, ants, flies

(continued)

Table 3.2 (continued)

Teucrium infrageneric taxa	Inflorescence	Corolla size (mm) Length of tube and lower lip	Corolla colour	Main flowering season	Pollinators
Polium subsection *Pumila*	Simple dense terminal or branched raceme	5.5–6	White/ purple	Summer (V–VII)	Solitary bees, long-tongued butterflies, ants

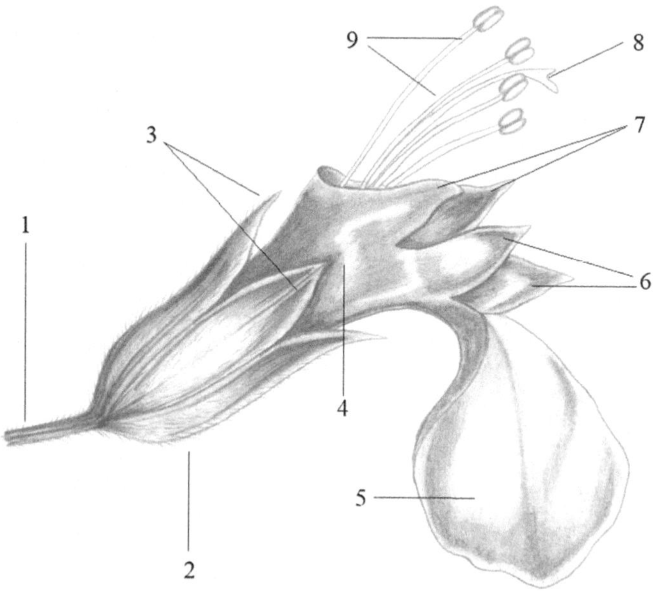

Fig. 3.1 Schematic drawing of the general conformation of the *Teucrium* flower. 1 – Flower pedicel; 2 – Calyx; 3 – Calyx lobes; 4 – Corolla tube; 5 – Anterior corolla lobe; 6 – Lateral corolla lobes; 7 – Latero-posterior corolla lobes; 8 – Style; 9 – Stamines

are disposed in verticillasters and the main flowering season is the summer Fig. 3.3. They are wind-ballistic or/and rain-ballistic species with glabrous and smooth nutlets. These features are in accordance with the ones described by Paton (1992) in *Scutellaria*.

Section *Isotriodon* includes species with infloreseence in a sub-spiral raceme; the calyx is tubular-campanulate, sub-actinomorphic, gibbous at the base, ±pedicellate with the upper tooth shorty ovate, the lower teeth narrower and acute; the corolla has long tube with well-developed lobes and two nectar guides of hairs on the lower lip; the nutlets have glandular hairs. In the species of section *Chamaedrys*, the flowers are placed in lax verticillasters or in sub-spirally lax racemes; the corolla has

Fig. 3.2 *T. fruticans* L. (section *Teucrium*)

Fig. 3.3 *T. oxylepis* Font Quer (section *Scorodonia*)

long latero-posterior lobes, acute and haired and the lower lip has two clear nectar guidelines of hairs; the calyx is tubular-campanulate and actinomorphic (Fig. 3.4) with an annulus of hairs inside and the nutlets are haired.

The greatest convergence of floral features is show in the species of section *Isotriodon*, the rocky species of section *Chamaedrys* and the rocky species of section *Montana* indicating a higher incidence of functional redundancy. The corolla is one-lipped, gullet-shaped and with regular tube (tube >5 mm), with two clear nectar guides in the lower lip, without closed gullet mechanisms and with well-developed

Fig. 3.4 *T. webbianum* Boiss. (section *Chamaedrys*)

Fig. 3.5 *T. charidemi* Sandwith (section *Polium* subsection *Simplicipilosa*)

and specialized latero-posterior lobes acting as guiding visitors. Flowers are purple, white or cream, rarely yellow, flowering mainly at the end spring and summer. They are wind dispersal species with campanulate, sub-actinomorphic calyx generally without annulus of hairs inside. They are also wind-ballistic species, or seed-dispersal inside the calyx. In this last case, the calyx is easily detached and generally dry and papery. In some species such as *Teucrium charidemi* Sandwith (Fig. 3.5), *T. compactum* Clemente ex Lag., and *T. yemense* Deflers, the inflated calyx ensure long distance dispersion by the wind, this type is found by Zohary (1937) and

Fig. 3.6 *T. scordium* L. (section *Scordium*)

Navarro et al. (2009) in desert species. This group of species shows a highly convergence exemplified by the lax inflorescence in sub-spirally simple racemes.

Species from the section *Scordium* have one-lipped, gullet-shaped corollas with nectar guides. Flowers are placed in verticillasters (2–8) flowered axillary in upper leaves; the calyx is tubular-campanulate, and sub-actinomorphic; the corolla has two nectar guides on lower lip and the nutlets are glandular haired (Fig. 3.6).

The species of section *Scordium* subsection *Spinularia* have corolla one-lipped, gullet and flag-shaped with a long and narrow tube, resupinate in some species. The strongly zygomorphic, gibbous calyx have an annulus of hairs inside. They are wind-ballistic species and possibly epizoochores, since most of the species show spinescent calyx teeth. The flowers (2–4) placed in axillary verticillasters in the upper leaves; the nutlets are small and glabrous. The species flowering mainly in spring and summer (Fig. 3.7).

In the section *Stachyobotrys,* the flowers are in dense verticillasters like spikes; the lateral lobes of the corolla are reduced or absent, in some cases, the tube is resupinate; the calyx is campanulate, bilabiate often-gibbous al the base with spinescent teeth, upper tooth weakly broad, 2 lower teeth lanceolate and the nutlets have glandular hairs. Some species from section *Stachyobotrys* such as *Teucrium collincola* Greuter & Burdet show a particular resupinate corolla without nectar guides and with a well-developed lower lip, the latero-posterior lobes are almost totally reduced and the lateral lobes are very short. The calyx is strongly zygomophic and campanulate. They are rain-ballistic or/and wind-ballistic species, some of them show the long mucronated calyx teeth, perhaps related to the epizoochory.

Species belonging to section *Polium* subsections *Polium* and *Simplicipilosa* are ginodioecious species (Fig. 3.8) with one-lipped brush-shaped corollas shortened tubuled with closed gullet mechanisms by a tuft of hairs to exclude inefficient

Fig. 3.7 *T. spinosum* L. (section *Scordium* subsection *Spinularia*)

Fig. 3.8 *T. teresianum* Blanca, Cueto & J. Fuentes (section *Montana*)

visitors and limit pollinator diversity. Flowers are compacted in the dense racemes forming the branched inflorescence (Figs. 3.8 and 3.9). Flowering mainly in spring, the ginodioecious system imposes limitations in pollination and rewards partitioning between females, generally pink and with nectar reward flowers and hermaphrodite flowers (white or cream). The calyx is sub-actinomorphic. They are wind-dispersal species and the calyx falls with the nutlet inside. Some species from arid regions have cucullate calyx teeth that close the calyx tube and protect the nutlets. In other species like *Teucrium charidemi* (Fig. 3.5) or *T. musimonum* Humbert, the dense vermiform hairs cover the calyx outside to facilitate the dispersion by the air that pushes them hovering over the ground.

Fig. 3.9 *T. lusitanicum* Schreb. (section *Polium*)

3.3 Corolla Construction and Pollination Adaptive Significance

Teucrium exhibit a great diversity in flower construction. Table 3.1 show the main infrageneric generative traits in *Teucrium* related to the pollination. *Teucrium* has with white, cream, pink and purple flowers. The orange and blue flowers are exclusive from section *Teucriopsis* and the lilac from section *Teucrium*. The most common flower construction related to pollination in *Teucrium* are the gullet-shaped corollas with nototribic pollen deposition; this type occurs in all sections of the genus except in section *Polium* subsection *Polium*. The flag-shaped corollas with sternotribic pollen deposition are rare and exclusive of the species with resupinate corollas from subsections *Spinularia* (Fig. 3.7) and section *Stachyobotrys*. In these cases, the gullet-shaped corollas are reversed and function like a flag-shaped and there may be a simple resupination through torsion of the corolla tube, and the lower lip only acts as an attraction and the lateral and latero-posterior lobes are reduced or absent (Fig. 3.7). The brush-shaped corollas with diffuse pollen deposition are exclusive of the ginodioecious flowers of the section *Polium*. The bilateral symmetric flowers such as those of the section *Teucrium* induce a more specific behavior to ensure that stigma and anthers are touched effectively (Fig. 3.2). Sternotribic and nototribic pollen deposition occur in this last section.

Teucrium species have corollas closed to ineffective insect visitors. We observed two closed gullet mechanisms related to pollination: (1) A barrier, which offers resistance and demands precision on the part of the pollinator. This barrier formed by a palate of hairs in the region of the common basal part of the latero-posterior lobes and stamine insertion, which is associated to a slight pouch (Navarro and El Oualidi 1999). When the insect poses on the lower lip, its force makes the lip bend down and the tube closed with an annulus of hair. This mechanism is specific of the species from the section *Teucrium*. (2) A barrier to exclude the inefficient visitors,

formed by an annulus of hairs from the base of the stamine insertion to the basal part of the anterior lobe. This basal part is cover densely by hair forming a tuft, which also serves as a foothold for the pollinator visitors. This mechanism is exclusive of the unspecialized ginodioecious species of section *Polium* subsections *Polium* and *Simplicipilosa* and section *Montana* (Navarro and El Oualidi 1999) and *Teucridium* genus (Merrett 2005).

The gullet-shaped corollas without closed gullet mechanism show the guiding of visitors such as the nectar guides formed by two clear rows of hairs in the adaxial side of the lower lip extending to the entrance of the corolla tube. These guidings are common in the species of the sections *Chamaedrys, lsotriodon, Scordium, Montana* and *Polium*. The calyx does not play an important role in the pollination syndromes because it does not have a known semaphyll function, but has glandular hairs emitting scent adding to the attraction for visitors.

3.4 Calyx Construction, Nutlet Morphology and Dispersal Adaptive Significance

Table 3.3 show the main infrageneric generative traits in *Teucrium* related to the dispersal. The predominant primary dispersal mode in *Teucrium* is the semachory and ballautochory. Wind-dispersal or wind-ballistic species (semachores or ballospores) are found in *Teucrium heterophyllum* L'Her from section *Teucriopsis* and in the species of the sections *Teucrium, Chamaedrys, Scordium, Isotriodon* and *Polium*. Ballautochory, wind-ballistic or rain-ballistic species (ombrochory) found in the section *Scorodonia, Stachyobotrys*, subsection *Spinularia* and in *Teucrium abutiloides* and *T. betonicum* from the section *Teucriopsis*. Epizoochory is rare in the genus, found in some species with spinescent calyx from subsection *Spinularia*. Meteo-anemochory only found in the section *Polium*.

The sculpturing and size of the nutlets (diaspore) have an important role in dispersal mechanisms (Bouman and Meeuse, 1992). Pericarp surface of the nutlets in *Teucrium* are variously ornamented (Marin et al. 1994); smooth (section *Scorodonia*); haired (section *Teucrium, Teucriopsis* and *Chamaedrys*) (Fig. 3.10); with glands (sections *Chatnaedrys, Isotriodon* and *Montana*) and very reticulated (section *Polium* subsection *Polium*) to maximize the dispersion by the wind. The variation in the type of indument of the nutlets are described in Navarro and El Oualidi (2000b).

The main diaspore type present in *Teucrium* are the semaspores associated to the species with actinomorphic or sub-actinomorphic, pedicelled calyx. This dispersal mode characterized *Thymus* (Bouman and Meeuse 1992). Ballospores are associated to species with bilabiate or strongly zygomorphic calyx and pogonospores (diaspores largest than 2 mm) with the actinomorphic and campanulated calyx.

The calyx morphology vary in relation to the protection of the immature nutlets by two type of system: (1) an annulus of hairs inside the calyx tube, considering its presence as a derived state character (Abu-Asab and Cantino 1993) frequent in other

Table 3.3 Main dispersal mode, diaspore type and diaspore size (mm) in the infrageneric taxa of *Teucrium* L

Teucrium infrageneric taxa	Dispersal mode	Diaspore type	Diaspore size (mm) length × width
Teucrium	Semachory wind-dispersal	Pogonospore	(1.5)2 × 3(3.5)
Teucriopsis	Semachory wind-dispersal	Pogonospore	(1.5)1.8 × 2.5
Chamaedrys	Semachory wind-dispersal	Semaspore/ pogonospore	(0.7)1.5 × 1.8(2.5)
Montana	Ballautochory semachory wind dispersal	Semaspore/calyx with seed inside	(0.5)1.2 × 1.3(2.5)
Scorodonia	Ballautochory semachory wind-dispersal	Ballospore	(1)1.3 × 1.5(1.8)
Scordium	Ballautochory semachory wind-dispersal	Semaspore	(0.8)1 × 1.3
Stachyobotrys	Ballautochory epizoochory	Ballospore	(1)1.3 × 1.8
Scordium subsection *Spinularia*	Ballautochory epizoochory	Semaspore/ ballospore/ sclerospore	(0.1)1 × 1 × (1.5)
Isotriodon	Ballautochory/ semachory wind-dispersal	Semaspore	(0.7)1.3 × 1.5(2.5)
Polium subsection *Polium*	Meteo-anemochory wind-dispersal	Calyx with seed inside	(0.8)1.3 × 1.5(2.5)
Polium subsection *Simplicipilosa*	Meteo-anemochory wind-dispersal	Calyx with seed inside	(0.6)1.2 × 1.5(1.7)
Polium subsection *Pumila*	Meteo-anemochory	Calyx with seed inside	(0.6)0.8 × 1(1.5)

Lamiaceae genera such as *Marrubium, Origanum, Sideritis, Ocimum* and *Scutellaria*; (2) the mature cucullate teeth closing completely the calyx tube (El Oualidi and Puech 1993). The last is exclusive of some species of section *Polium* subsection *Polium* as *Teucrium lusitanicum* Schreb., or (3) a gibbous in the base of the calyx like *Teucrium botrys* L., covered inside by long hairs, making difficult the exit the nutlets, this mechanisms is exclusive of the subsection *Spinularia.*

The pedicel of the calyx and the upper tooth in strongly zygomorphic calyces plays a role in the dispersal mechanism of *Teucrium.* The pedicel is present in all *Teucrium* flowers except in the section *Polium* subsection *Polium* and some rocky and arid species from section *Montana.* The pedicel is not rigid and acts in the ballistic mechanisms contracting itself to eject the mature nutlets (subsections

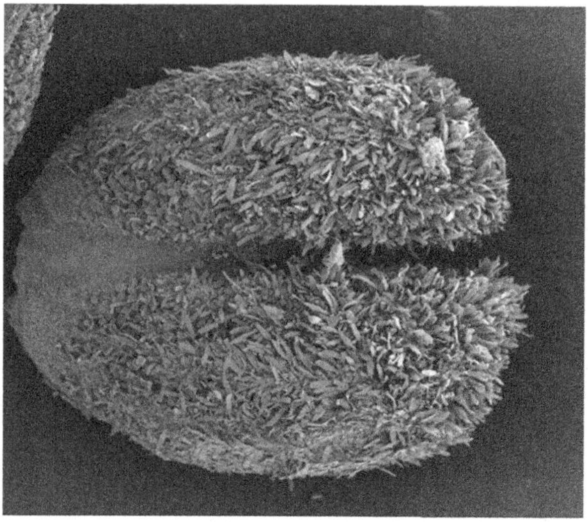

Fig. 3.10 Haired nutlet of *T. racemosum* R.Br. (section *Teucrium*)

Spinularia, section *Scorodonia,* some species from section *Stachyobotrys* and in *Teucrium abutiloides* and *T. betonicum* from section *Teucriopsis*). The broad and more or less horizontally upper lobe tooth of the calyx reinforces this ballistic action.

3.5 Conclusions

The most important discriminant generative characters within *Teucrium* are the corolla and calyx construction. Sections *Pycnobotrys, Stachyobotrys, Scorodonia* and *Teucriopsis* have species with wrakly 2-lipped corolla and zygomorphic calyx (2-lipped calyx that in the majority of cases present an upper broad lip). Sections *Polium, Chamaedrys, Montana, Isotriodon, Pumila* and *Scordium* have 1-lipped corolla and mainly sub-actinomorphic calyx. Section *Polium* is separate into two groups: species of the subsection *Simplicipilosa* with 1-lipped corolla, tube apically broading and a tuft of hairs inside, and species of subsection *Pumila,* with 1-lipped corollas and latero-posterior lobes forming an angle with the corolla tube, which has two guides of hairs at the base of the anterior corolla lobe. Section *Teucrium* is the most heterogeneous and distinctive group within *Teucrium.* Species of section *Teucrium* subsection *Fruticantia* (Kästner 1989) have 1-lipped corolla and actinomorphic calyx. Species of section *Teucrium* subsection *Cretica* (Kästner 1989) have 1-lipped spurred corolla with an annulus of hairs in the base and actinomorphic calyx and the Australian species of section *Teucrium* are characterized by their longer pedicellate flowers, 1-lipped non-spurred corollas with an annulus of hairs in the middle part of the tube.

Salmaki Y, Stefan K, Günther H, Christian B (2016) Phylogeny of non-monophyletic *Teucrium* (Lamiaceae: Ajugoideae): implications for character evolution and taxonomy. Taxon 65:805–822
Shmida A, Dafni A (1989) Blooming strategies, flower size and advertising in the "lily-grou" geophytes in Israel. Herbetia 45:11–123
van der Pijll L (1972) Functional consideration and observation and the flowers of some Lamiaceae. Blumea 20:93–103
Wagstaff SJ (1992) A phylogenetic interpretation of pollen morphology in tribe Mentehae (Lamiaceae). In: Harley RM, Reynolds T (eds) Advances in Labiate science. Royal Botanic Gardens, Kew/London, pp 113–125
Zohary M (1937) Die verbreitungsokologischen Verhaltnisse der Planzen Palastinas. Bot Centralbl Beih 56:1–155

Chapter 4
Ecology of *Teucrium* Species: Habitat Related Metal Content Dynamics

Dolja Pavlova, Irina Karadjova, Milan Stanković, and Nenad Zlatić

Abstract The chapter reviews the available data about the effect of habitat related metal content on *Teucrium chamaedrys* and *T. montanum* (Lamiaceae). The study was focused on element concentrations in plant and soil samples, both on metalliferous and non-metalliferous soils. Metal concentrations varied depending on species and habitat type. The levels of elements in plant tissues from non-metalliferous localities were always lower, compared to those from metalliferous (serpentine) ones. None of the species could not hyperaccumulate metals although the metal concentration in some of them exceeded the range, which is naturally found in plants. Depending on the nickel accumulation, both analyzed species are classified as excluders. The level of tolerance was related to the amount of metals and their bioavailability in the soil. The metal concentrations for the toxic elements were above the permissible limits for the toxic elements, in both species. The populations of the studied species demonstrated some adaptations to the serpentine habitats related to their secondary metabolites and its morphology, which is known as serpentinomorphoses. As a result of the heavy metal profiles of the soils, significantly higher values and differences in the quantity of secondary metabolites were recorded in plant populations growing on serpentines compared to non-serpentine ones. *Teucrium chamaedrys* and *T. montanum* populations on metalliferous habitats, possess morphological differences in contrast of populations on non-metalliferous habitats.

Keywords *Teucrium* · Ecology · Distribution · Metal content · Soil · *Teucrium chamaedrys* · *Teucrium montanum*

D. Pavlova (✉)
Department of Botany, Faculty of Biology, University of Sofia, Sofia, Bulgaria
e-mail: pavlova@biofac.uni-sofia.bg

I. Karadjova
Department of Analytical Chemistry, Faculty of Chemistry and Pharmacy, University of Sofia, Sofia, Bulgaria

M. Stanković · N. Zlatić
Department of Biology and Ecology, Faculty of Science, University of Kragujevac, Kragujevac, Serbia

M. Stanković (ed.), *Teucrium Species: Biology and Applications*,
https://doi.org/10.1007/978-3-030-52159-2_4

Abbreviations

Al	Aluminium
B	Boron
BAF	Bioaccumulation factor
Ca	Calcium
Cd	Cadmium
CEC	Cation exchange capacity
Co	Cobalt
CO2	Carbon dioxide
Cr	Chromium
Cu	Copper
Fe	Iron
GR	Glutathione reductase
GSH	Reducted glutathione
H	Hydrogen
H_2O_2	Hydrogen peroxide
Hg	Mercury
K	Potassium
Mg	Magnesium
Mn	Manganese
N	Nitrogen
Na	Sodium
Ni	Nickel
P	Phosphorus
Pb	Lead
ROS	Reactive oxygen species
Zn	Zinc

4.1 Introduction

The genus *Teucrium* L. is the one of the largest in Lamiaceae family. It is widely distributed in Europe, Asia, Africa and Australia and have two centers of richness located in the east and western regions of the Mediterranean area. The genus *Teucrium* comprises approximately 300 taxa of herbs and shrubs grouped in 9 sections according to Bentham (1833) with additions proposed by Boissier (1879) (Navarro and El Oualidi 2000). The Mediterranean area represents the major area of distribution of the genus, since it comprises about 90% of the total species in the world (Navarro and El Oualidi 2000; Navarro 2010).

Teucrium chamaedrys L. (wall germander) is the most important species of sect. *Chamaedrys* (Mill.) Schreb., distributed throughout the Mediterranean, mainly in the Balkan peninsula and Turkey. The species *Teucrium montanum* L. (mountain

germander) is a representative of sect. *Polium* (Mill.) Schreb., subsect. *Rotundifolia* Berm. & Sanchez Crespo, distributed in South and East Europe, the west Mediterranean area and in the mountains of the south Arabian peninsula (Navarro and El Oualidi 2000).

Reichinger (1941) considered *Teucrium chamaedrys* as the most variable taxon in the *Teucrium* genus and described 15 subspecies, most of them locally distributed. These subspecies were not initially included in Flora Europaea (Tutin et al. 1972) but, it was later added by Greuter et al. (1986), and some of them in local floras (Ekim 1982; Peev 1989). This perennial plant occurs in open forests, shrublands and rocky slopes in semi-arid, arid and sub-humid climates (Navarro and El Oualidi 2000). In the Balkan countries, the plant inhabits rocky limestone areas, dry mountain meadows and pastures, edges of sparse oak and pine forests up to 1000 m in Serbia and up to 1500 m in Bulgaria, demonstrating a high degree of morphological diversity (Diklić 1974; Peev 1989; Markova 1992). It was also reported for the serpentine territories in Bulgaria, Serbia, Albania, Turkey (Ekim 1982; Shallari et al. 1998; Pavlova et al. 2003; Pavlova 2010). *Teucrium montanum* L. is a perennial, shrub-like plant with half ligneous branches, up to 25 cm high and inhabits thermophilic calcareous rocks, dry mountain meadows and edges of forests in Europe and Anatoly (Diklić 1974; Zlatić et al. 2017).

A large number of known medicinal species, belonging to the genus *Teucrium*, are used in folk medicine and pharmacy, in food industry as spices and for bitter beverages. The most popular species in the flora of Europe and Asia (*Teucrium chamaedrys*, *T. montanum* and *T. polium*) have a history of traditional use in the Balkan countries for herbal teas and basic medicinal healing treatments (Ivancheva and Stancheva 2001; Obratov-Petković et al. 2006, 2008; Evstatieva et al. 2007; Redzić 2010). For pharmaceutical purposes aboveground plant parts (stems, leaves, flowers) are collected from natural populations and from cultivated plants (Nikolov 2006). The mentioned plants are included in the list of medicinal plants permitted for export, and annually hundreds of tons are exported (Evstatieva et al. 2007). They are used in the treatment of digestive and respiratory disorders, abscesses, gout, conjunctivitis, in the stimulation of fat and cellulite decomposition. These species possess anti-inflammatory, antioxidative, antimicrobial, antidiabetic, and antihelmintic effects (Harborne et al. 1986; Stanković et al. 2011a, 2012; Grubešić et al. 2012; Milošević-Djordjević et al. 2013; Vlase et al. 2014; Zdraveva et al. 2018).

According to Obratov-Petković et al. (2006), the distribution and the degree of presence of medicinal plants are directly correlated to the edaphic factors. Moreover, the quantity of active substances in plant tissues depends of many ecological factors that affect the vegetative plant organs (Lombini et al. 1998). The properties of soil generally depended on the combined effects of climate, biological activity, topography and the mineralogical composition of the parent rock and play a central role in the distribution and ecology of plant species and their associated biota (Jenny 1980; Whitea and Claxtonb 2004). Usually, extreme edaphic conditions, like limestone, gypsum, dolomite, granite, guano deposits, salt marshes, and even mine tailings, provide ideal settings for examining the role of the edaphic factor in the distribution, ecology and natural selection of plants (Kruckeberg 1969; Rajakaruna 2004; Rajakaruna and Boyd 2008).

Fig. 4.1 *T. chamaedrys* L. on serpentine (left) and calcareous habitat (right). (Photo D. Pavlova and M. Stanković)

Fig. 4.2 *T. montanum* L. on serpentine (left) and calcareous habitat (right). (Photo M. Stanković and N. Zlatić)

The species *Teucrium chamaedrys* (Fig. 4.1) and *T. montanum* (Fig. 4.2) demonstrate a wide ecological tolerance as they were found in open grasslands, oak woodlands, pastures, rocky calcareous and siliceous terrains, and on serpentine outcrops. They occurred either on soils that are naturally metalliferous (e.g. serpentine soils) or in territories which secondarily have been enriched with metals (industrial plants emitting metals, mine tailings, etc.). As the serpentines and their soils are characterized by toxic quantities of heavy metals (particularly Ni and Cr), low Ca/Mg ratio, drought, and wide temperature fluctuations (Kruckeberg 1984, 1992; Brooks 1987; Brady et al. 2005), they probably have the most extreme habitat conditions for these species. Most populations of *Teucrium chamaedrys* and *T. montanum* grow on non-metalliferous soils in their range and are quite sensitive to metal toxicity. There is a strong selection favoring the metal-tolerant genotypes, when plants colonize metalliferous soils, causing a rapid increase in the average tolerance of the local populations, which leads to the establishment of metal-tolerant "races" or ecotypes (Antonovics et al. 1971; Pollard et al. 2002). It could be concluded that their adaptive mechanisms includes exclusion of metals (either by restricting them to the roots

or through the absence of any uptake mechanisms), compartmentalization of metals in various organs, or toxicity tolerance as it was summarized by Kay et al. (2011) for plants growing on such soils. The level of tolerance is related to the amount of metals in the soil and their bioavailability. The metal bioavailability in the system soil-plant could be an indicator of potential risk for environment and human health, especially for medicinal plants.

This chapter will describe the influence of soil metal concentration on the therapeutically important medicinal plants – *Teucrium chamaedrys* and *T. montanum* (Lamiaceae) throughout their range in the Mediterranean region. Additionally, this chapter will describe the components of edaphic factors and their role, through metal concentrations in different types of soils. Related to soil analysis, metal accumulation by *Teucrium chamaedrys* and *T. montanum* will be presented as well as plant-soil correlations with bioaccumulation factor of selected plant species. Furthermore, this chapter will include the morphological characteristics of presented species related to the substrate and the quantity of secondary metabolites as one of the mechanisms of species adaptation to the metalliferous soil.

4.2 Soils and Their Role in the Distribution of *Teucrium* Species

4.2.1 Components of the Edaphic Factor and Its Role

The edaphic factor as a combination of numerous physical and chemical properties of the substrate significantly determines plant species distribution, diversity and composition. Very significant soil factor is combination and concentration of mineral nutrients available in the soil. Bioavailability is the proportion of total metals or metal species that are available for incorporation into biota (bioaccumulation). Among the chemical elements two classes are considered essential for plants: macronutrients and micronutrients. Macronutrients like K and Mg are required in high quantities as they form the basis of crucial cellular components like proteins and nucleic acids. Micronutrients like Zn, Mn, Cu, Fe, Ni etc. often required as cofactors for enzyme activity are needed in small amounts and, as expected, in excess they become toxic. The highly toxic elements like Cd, Pb, Hg usually can not be distinguished by the nutrient uptake system and enter into the plant causing reduced uptake of the essential nutrients and significantly reduced plant growth and quality. Plant uptake of trace elements is dependent on (1) movement of elements from the soil to the plant root, (2) elements crossing the membrane of epidermal cells of the root, (3) transport of elements from the epidermal cells to the xylem, in which a solution of elements is transported from roots to shoots, and (4) possible mobilization, from leaves to storage tissues used as food (seeds, tubers, and fruits).

Many factors can affect the efficiency of chemical elements biouptake mainly from the soil through the plant roots. Among these are factors connected with

specific plant biouptake mechanism and root system. Micro- and macronutrient homeostasis in plants is realized by various strategies for mobilization and uptake which differ for different plants species. Metal ions reach plant tissue through ion exchange, passive diffusion or through a membrane transporters localized in the root. Various defense mechanism has been developed by plants against toxic elements: extracellular immobilization near root region (using root exudates or mycorrhizal association), intracellular detoxification by the formation of complexes between metal ions and peptides – phytochelatins, or complexes with organic acids or aminoacids. On the other hand, metals tolerant for some plants are associated with increased synthesis of phytochelatins proline or antioxidant enzymes.

In addition to the factors connected with specific plant biouptake mechanism and root system, very important are factors connected with soil properties, edaphic factors and chemical elements bioavailability. Acid soils have contributed to a great extent to the solubilization of metals. It is clear that with increasing pH the solubility of Fe, Cr, Ni and Co in serpentine soils decreases. The order of increasing solubility is Fe, Cr, Ni, Co (Brooks 1987). According to Blake and Goulding (2002) Pb is not mobilized until $pH < 4.5$. Zinc is readily soluble relative to the other metals in soils and available in acid light mineral soils. It is considered that the Zn fraction associated with Fe and Mn oxides is likely to be the most available to plants. The greatest amounts of adsorbed Cu have always been found for Fe and Mn oxides, amorphous Fe and Al hydroxides, and clays (Kabata-Pendias 2011). The solubilization of Cd and its transformation into an available ionic form, which is preferentially absorbed by plant roots depend on soil pH and other soil factors (Sharma and Dubey 2006). In acid soils the mobility and availability of Cd is much higher than in non-calcareous, neutral and slightly alkaline soils. Acidification to pH 4 mobilizes 60–90% of the total soil Cd, but it is adsorbed on ion exchange surfaces and complexed with soil organic matter (Obratov-Petković et al. 2006). The rapid increase in the Cd concentration at the 0.5 mg kg^{-1} level (in straw) is considered to be related to a break of the physiological barrier controlling a metabolic absorption of this metal (Kabata-Pendias 2011).

In most cases above separation of factors is quite operational as far as plant can to some extend regulate chemical elements biouptake as a respond to the changes in the soil properties and chemical elements bioavailability. The rhizosphere is a zone at the root-soil interface controlled by plant root and root released metabolites or exudates where additionally mutually interacting physical, chemical and biological processes take place. The health of both the plant and the soil associated is dependent on this rhizosphere region and its biochemical reactions.

Evidently, equilibria between different chemical elements in soil solid phase (mineral and organic/soil solution/plants rhizosphere) is extremely complex depending on various biotic and abiotic factors. Even more, reliable determination of bioavailable concentrations of nutrients and evaluation of their behavior is almost impossible. Although various operationally defined analytical procedures have been proposed and applied, the results obtained for chemical elements, their distribution between soil fractions and their chemical species are far from real situation in a complex soil-plants system. From such point of view the investigations on soils with

unusual features (extreme pH, nutrient imbalances, limited depth, etc.) accepted as a strong selective force shaping plant evolution are very important lightening in a specific manner soil-plant connection and plant response to soil edaphic factors.

Together, these mechanisms allow plants to maximize their nutrient acquisition abilities while protecting against the accumulation of excess nutrients, which can be toxic to the plant. It is clear that the ability of plants to utilize such mechanisms exerts significant influence on plant community structure, soil ecology, ecosystem health, and biodiversity.

4.2.2 Metal Content in Serpentine Soil

Serpentine soil (formed after serpentine rocks (ultramafic rock consists of magnesium-iron silicate minerals, such as olivine and pyroxene) weathering under different climatic conditions) is a naturally occurring model system ideal for the study of the physiological responses of plants to edaphic factors. Serpentine soils are characterized with (i) low Ca content combined with elevated concentrations of Mg, which creates an unusually low Ca:Mg ratio; (ii) low concentrations of several of the macronutrients essential for plant growth, especially N, P, and K; (iii) high concentrations of toxic elements like Ni, Cr, Co and less toxic Mn. The parent material from which serpentine soil originates is highly variable leading to variations within them with respect to the absolute concentrations of Ca and Mg and the presence and concentration of toxic elements. Within serpentine ecosystems, additional abiotic stresses are also often present, including drought, low nutrient cycling rates, and shallow soil depth. This suite of abiotic factors and the sparse vegetation observed on serpentine soils has been described as the "serpentine syndrome" (Jenny 1980). It is a physiologically challenging environment for most plant species driving the development of unique physiological adaptations and high levels of endemism (Brooks 1987; Kruckeberg 1992).

Various hypotheses (reviewed by Brady et al. 2005; Kazakou et al. 2008), including tolerance to low Ca:Mg ratio, avoidance of Mg toxicity, an increased requirement for Mg, and tolerance/exclusion or hyperaccumulation of toxic elements have been discussed for plants that do grow on serpentine soil. Tolerances and adaptations range from those at the cellular level to those apparent to the naked eye. High concentrations of Ni are tolerated in some serpentine plants by exclusion, reduced transfer of Ni from root to shoot, or hyperaccumulation. Other plants adapt to a low Ca environment, by selectively up taking this nutrient rather than Mg.

Although physical-chemical features of serpentine soils where *Teucrium chamaedrys* and *T. montanum* are distributed vary considerably from site to site (Pavlova 2009; Pavlova and Karadjova 2013; Obratov-Petković et al. 2008; Branković et al. 2017; Zlatić et al. 2017) and within a site (Pavlova and Karadjova 2013), species populations are often found in open, steep landscapes with shallow and rocky soils with reduced moisture. Sparse plant cover on serpentines promotes elevated soil

temperature and erosion (Kruckeberg 1992). Each of these factors poses an additional stress to plant life.

4.2.3 Metal Content in Calcareous Soil

Calcareous parent materials are the most widespread type of sedimentary rock in the world. They are made of calcium carbonate and originally divided into chemical, over gammon and rocks of organogenic origin. Usually, they contain primers of other metals such as Mn, Fe, Mg, clays, organic matter, sand, etc. Soils formed on the calcareous geological substrate contain calcium carbonate in free form (Ewald 2003). Calcareous soils are characterized with (i) high Ca content; (ii) high concentrations of several macronutrients essential for plant growth, especially K, P, and N; (iii) low concentrations of toxic elements like Co, Cr, and Ni, which is opposite to serpentine soils (Zlatić et al. 2017). This type of soil is characteristic for dry and semi-arid areas, as well as humid and moderately humid regions, especially in those areas where the parent material is rich in $CaCO_3$. At the chemical level, the presence of $CaCO_3$ determines the alkaline reaction in calcareous soils and affects the availability of certain metals, such as N, P, K, Mg, Zn, Cu, and Fe, to plants. The carbonates presented in the soil contribute to a alkaline pH value, ranging between 7.5 and 8.5. In addition to the chemical influence of $CaCO_3$ in calcareous soils, it also affects the physical characteristics of the soil (Lambers et al. 2008).

Plant species that inhabits limestone habitats are not exposed to the negative impact of heavy metals in the substrate. Plants that inhabit such habitats are adapted to other ecological factors by specific physiological mechanisms (Zlatić et al. 2017). These species are exposed to different stress conditions in opposed to the serpentine ones. The alkaline pH value of the calcareous soil, the deficiency of water in the substrate, the high temperature and erosion, are factors that plants needs to be adapt (Lambers et al. 2008).

4.3 Plant Life on Selected Edaphic Conditions – *Teucrium chamaedrys*

4.3.1 Metal Accumulation by Teucrium chamaedrys

The physiological response of plants to the stress provided by the elevated metal concentrations in the soil can be related to the different ways to cope with metals in the soil and to their genotypic response. Plants tolerate elevated metal concentrations on serpentine soils either by a constitutive trait present in all members of the species growing either on serpentine or non-serpentine substrate or by an adaptive mechanism present only in tolerant ecotypes (Antonovics et al. 1971; Kazakou et al.

2008). *Teucrium chamaedrys* absorbs a number of elements from soil both essential (Cu, Zn, Fe, Ni, Mn) and non-essential (Cr, Co, Pb) for its biological functions. The availability of metals in a soil-plant system depends on a number of factors which include pH of the soil, soil organic matter content, cationic exchange capacity as well as plant species, stage of development, and others (Farago 1994; Kassim and Rahim 2014). The species is widely distributed without preferences to any type of rock and should be considered as a "bodenvag" species, which are those widely distributed (Fig. 4.1) in serpentine and non-serpentine habitats (Kruckeberg 1992). The significant ecological differences between populations are related to the differences in edaphic conditions of the serpentine sites from the non-serpentine sites. According to Obratov-Petković et al. (2008) there are also differences in the heavy metal uptake between the individuals of the same species, which is first of all the consequence of climatic conditions and moisture regime.

The mechanism by which a serpentine-tolerant plant copes with elevated levels of Mg and relatively insufficient quantities of Ca in the soil is considered its most defining character (Brooks 1987; Brady et al. 2005; Kazakou et al. 2008). The ratios Ca:Mg are generally always above 1 in plants. However, serpentine plants are better able to maintain a greater than 1 ratio despite minimal levels of Ca found in the soils. They either have very efficient Ca uptake systems or ability to exclude Mg despite high concentrations in the soil. The low Ca:Mg ratio in serpentine soils has elicited a wide range of adaptive responses based on either ion exclusion at the root/soil interface, selective translocation of Ca from root to shoot, sequestration of Mg in the vacuole, or internal mechanisms of tolerance (Kay et al. 2011). Calcium concentrations found in tissues of *Teucrium chamaedrys* are between 2369 and 6345 mg kg^{-1} in serpentine plant populations in Bulgaria (Pavlova 2009; Pavlova and Karadjova 2012) and between 324 and 4872.2 mg kg^{-1} in plant populations from Serbia (Obratov-Petković et al. 2008; Branković et al. 2017; Zlatić et al. 2017). Similar to previous findings (Karataglis et al. 1982; Brooks 1987; Roberts and Proctor 1992; Kay et al. 2011; Pavlova and Karadjova 2013, Branković et al. 2017), while the amounts of Ca in the serpentine soils were small, the amounts of Ca taken up by the plant were higher. Serpentine species populations are much more tolerant to low Ca levels and elevated concentrations of Mg in the soil. Calcium concentrations in *Teucrium chamaedrys* populations are in all cases lower than in Ni hyperaccumulating plants distributed in the same areas such as *Alyssum* and *Thlaspi* species. The remarkable Ca uptake ability is considered an important feature in Ni-hyperaccumulator physiology (Broadhurst et al. 2004; Chaney et al. 2007).

Calcium is one of the elements contributing to the inhibition of the heavy-metal toxicity, and it is possible that plant takes up Ca to compensate the toxic action of different toxic metals (Brooks 1987; Lombini et al. 1998; Brady et al. 2005). Studying populations of *Buxus sempervirens* L. (Common Box) in Greece, Karataglis et al. (1982) suggested that the plant developed a mechanism to permit an excess soil Ca to be taken by plant.

Several species adapted to serpentine environments have higher external and internal Mg requirements than their non-serpentine relatives and variety of responses shown suggests that the physiological basis for tolerating low Ca:Mg may involve

more than one mechanism in a given species (Brady et al. 2005; Kay et al. 2011). Normally, the high Mg of serpentine soils is reflected in high Mg concentrations in plant tissues, but this result might also indicate an unusual Mg and low Ca requirement (Proctor and Woodell 1975). The mean Mg concentrations for *Teucrium chamaedrys* in the range 0.13–0.35% and 0.11–0.19%, respectively for serpentine and non-serpentine sites, were reported by Pavlova and Karadjova (2012). The mean Mg concentrations for the species from Serbia were almost the same for serpentine and non-serpentine (calcareous) populations as reported by Zlatić et al. (2017).

Brooks and Yang (1984) found the concentration of Mg in plant tissue to be inversely proportional to the concentrations of other nutrients: B, Fe, Co, Mn, P, and Na. Their findings showed that these relationships were caused by high Mg levels in the soil; high Mg concentration is probably the major cause of infertility of serpentine soils and hence of the development of specialized serpentine floras adapted to this unfavorable edaphic condition (Kazakou et al. 2008). These data clearly suggest that the uptake of Mg comes at a cost to the plant as the uptake of other elemental nutrients is forfeited. More complex is this interaction in Ni-hyperaccumulator plants because Ni is also involved. Some publications demonstrated that in the presence of Ni internal Ca and Mg concentrations counteract Ni toxicity or in any case enhance Ni tolerance (Gabbrielli and Pandolfini 1984), or decrease Ca uptake and increase uptake of Mg (Kazakou et al. 2008). It is likely, therefore, that the Ca/Mg quotient of the soil solution has a strong influence on Ni absorption, translocation and accumulation.

The accumulation of Ni, Cr, and Co in the above-ground plant parts for all serpentine populations of *Teucrium chamaedrys* is higher than in non-serpentine ones which was proved by a number of studies (Shellari et al. 1998; Jurišić et al. 2001; Obratov-Petković et al. 2008; Pavlova 2009; Pavlova and Karadjova 2012, 2013; Branković et al. 2017; Zlatić et al. 2017). Metal concentrations also vary at population level in relation to soil properties, mainly soil metal availability. This species demonstrates, as many other, specific strategies of adaptation to elevated levels of toxic metals in soil probably result of mineral element imbalance characteristic of serpentine soils (Ater et al. 2000) or due to inter-population variation in parts of the uptake and translocation processes (Adamidis et al. 2014). Nickel concentrations measured in *Teucrium chamaedrys* serpentine populations in Bulgaria (Pavlova 2009; Pavlova and Karadjova 2012, 2013) are lower than the data of Shallari et al. (1998) for Albanian serpentine populations, and higher from data provided for this medicinal plant from Serbia (Obratov-Petković et al. 2008). It was shown that different Ni concentrations in serpentine populations correlate with the bioavailable fraction of trace elements in the soils (Pavlova and Karadjova 2013; Zlatić et al. 2017). The concentrations of Ni even higher for all serpentine populations of *Teucrium chamaedrys* were not exceptional. The amounts of Ni in *Teucrium chamaedrys* were low compared to the data for other plants growing on serpentines (Karataglis et al. 1982; Vergnano Gambi et al. 1982; Konstantinou and Babalonas 1996; Bani et al. 2013; Sawidis et al. 2014), but higher than levels normally considered to be toxic to most plants (Kabata-Pendias and Pendias 1984). The

concentrations of Ni in plants from some of the investigated sites were higher than 60 mg kg^{-1} considered to be the threshold of physiological evidence of toxicity in plants of serpentine habitats (Kazakou et al. 2008, 2010). Concentrations of Ni above 100 mg kg^{-1} were documented for some serpentine populations of *Teucrium chamaedrys* in Bulgaria and Serbia (Pavlova and Karadjova 2012; Branković et al. 2017; Zlatić et al. 2017). Values from 145.7 mg kg^{-1} Ni found in *Teucrium chamaedrys* growing on serpentine in Goč Mt. (Serbia) give reason this species to be considered as accumulating plant of Ni (Branković et al. 2017). Despite of high Ni values documented they are rarely found in species populations and can be considered exception for the species. Only a few species appear to show extreme variations in metal uptake, even when confined to metalliferous soils. This behavior may be a reflection of widely varying metal availability caused by variations in pH or other soil properties (Reeves 2017). The first record for Ni accumulation above 1000 mg kg^{-1}, threshold for Ni hyperaccumulation, reported for the species of *Teucrium* were those of Hossain (2007) after Kassim and Rahim (2014) who studied metal accumulation in *Teucrium polium* populations in Iran. The values 8140 mg kg^{-1} in leaves and 2300 mg kg^{-1} in stem are reported for serpentine populations of the species. Later on, Ni concentrations between 9678 and 14.110 mg kg^{-1} in above ground parts of *Teucrium polium* grown in serpentine soils in Turkey were reported by Yaman (2014). The same species is included in the list of accumulating species that belong to the Italian flora (Bazan and Galizia 2018), altogether with Pb accumulator *Teucrium flavum* subsp. *glaucum* and Hg accumulator *Teucrium scorodonia*.

The populations of *Teucrium chamaerdrys* growing on non-serpentine soil types also differ in their ability to accumulate Ni and in most cases reported values are below 1 mg kg^{-1} (Obratov-Petković et al. 2008; Pavlova 2009; Pavlova and Karadjova 2012). An exception is the data provided for a calcareous site from Kopaonik Mt. (Zlatić et al. 2017) where a value of 9.65 mg kg^{-1} was measured. In all cases Ni concentrations were below the levels considered threshold for Ni toxicity.

The Ellenberg's indicator values used to describe the ecological relationship amongst accumulating plants, climate and soil conditions demonstrate that most of the phytoextractors naturally grow in neutral to basic soils (Bazan and Galizia 2018). The experiments with two *Alyssum* species (including *A. murale*) showed that increasing soil pH through liming increased Ni uptake when plants are grown on Ni-rich non-serpentine soils. However, liming an serpentine soil decreased Ni uptake (Kukier et al. 2004).

The phytotoxic Ni concentrations range widely among native species populations and cultivars of *Teucrium chamaedrys* growing on different substrates (Jurišić et al. 2001; Bazan and Galizia 2018). The higher concentrations were found in cultivated plants, while in *Teucrium montanum* the highest quantities of Ni and other metal ions (Cr and Cu) were found in wild species populations. This indicates that abilities of handling metal cations are not specific for serpentine populations, but can also be found in non-serpentine populations (Vicić et al. 2013a). Hence, Ni toxicity did not appear to be a universal feature of serpentine soils, but was

dependent on other variables, including Mg concentration in particular; the influence of Ni sometimes even varied within a site (Kazakou et al. 2008).

In *Teucrium chamaedrys* from serpentine sites Ni is taken up in higher quantities than Cr, and the Ni/Cr ratio in plant tissues is always much higher than in the soil. Most data from Bulgarian and Serbian serpentines demonstrated low Cr concentrations in populations of *Teucrium chamaedrys* (Obratov-Petković et al. 2008; Pavlova 2009; Pavlova and Karadjova 2012). These results corroborate the data of Brooks (1987) that serpentine plants contain only trace quantities of Cr. The phytoextraction of Cr from serpentine plant populations in Goč Mt. (Serbia) was above the threshold of 50 mg kg^{-1} and according to Branković et al. (2017) it may be considered as accumulation. However, even on serpentine soils, Cr concentrations > 50 mg kg^{-1} are so uncommon that this may be used as an indicator of soil contamination (Reeves 2006). Low mobility of this metal in the soil also serves to possible dust contamination. Several native plants (from different families) from areas of serpentine or chromite deposits were found to accumulate Cr as much as 0.3% or 3.4% (Kabata-Pendias 2011).

Although, most soils contain significant amounts of Cr, its availability to plants is highly limited and controlled mainly by the soluble Cr contents of the soils. Cr bioaccessibility is a function of soil type and retention time and not easily translocated within plants, thus it is concentrated mainly in roots (Kabata-Pendias 2011). Data provided for some *Teucrium* species (Pavlova and Alexandrov 2003; Vicić et al. 2013a) confirmed this conclusion. There is a great difference in the accumulation of Cr by shoots and roots of various plants and in vegetable crops the ratio shoot/root varies widely from 0.005 to 0.027 (Kabata-Pendias 2011).

Besides the elevated concentrations for Ni and Cr measured for *Teucrium chamaedrys* from serpentine soils such high concentrations were reported also for Fe (Obratov-Petković et al. 2008; Pavlova and Karadjova 2012, 2013; Branković et al. 2017; Zlatić et al. 2017). The quantity of Fe in the tissues of the species growing on serpentines varied in wide range from extremely high unusual concentrations (1781.9 mg kg^{-1}) to almost typical contents for the species from Serbia (Obratov-Petković et al. 2008; Branković et al. 2017, Zlatić et al. 2017) and in range 538–2740 mg kg^{-1} for the species from Bulgaria (Pavlova and Karadjova 2012). The same high concentrations were also reported for different species growing on metalliferous soils (Cornara et al. 2007; Dudić et al. 2007; Golubović and Blagojević 2013; Franco et al. 2013, Sawidis et al. 2014). Values > 1000 mg kg^{-1} often indicate contamination by serpentine soil or dust, not easily removed by simple washing procedures (Reeves et al. 1999). However, higher Fe concentration seems to be characteristic for the serpentine flora of the Balkans and more attention should be given on this phenomenon (Babalonas et al. 1984). Such behavior of some serpentine plants might be explained with their ability to acidify the rhizosphere and to increase the solubility of ferric ions and their reduction to even more soluble ferrous ion. However, where Fe is easily soluble, plants may take up a very large amount of Fe and this is clearly shown by vegetation grown in soils derived from serpentine and excessive Fe uptake can produce toxic effects in plants. The variation among plants in their ability to absorb Fe is not always consistent and is affected by abiotic

factors – soil, climate, and biotic conditions like the stages of plant growth and the ability to accumulate Fe (Kabata-Pendias and Pendias 1984; Kabata-Pendias 2011). In most cases *Teucrium chamaedrys* growing on serpentines demonstrates abnormal levels of Fe (> 200 mg kg^{-1}) in tissues according to the Element Concentration Cadastre in Ecosystems presented at the 25th General Assembly of the International Union of Biological Sciences (Lieth and Markert 1988). The accumulation of the metal in plant organs is different depending on the species, for example Sawidis et al. (2014) mentioned that Fe in the root was more in tree, shrub and cultivated plant samples but in non-cultivated and aquatic plants the leaves showed the highest Fe values. Also, Fe concentrations in roots of *Teucrium montanum* and *T. polium* were higher than in the leaves, stems and flowers (Pavlova and Alexandrov 2003; Kassim and Rahim 2014).

The concentrations of Co in *Teucrium chamaedrys* were lower than Cr in samples suggesting that the plant adapted to serpentine soil did not accumulate Co (Wallace et al. 1982). The concentrations for Co seldom reached 10 mg kg^{-1} in plants on normal soils and even on serpentine soil this level is not often exceeded. Co concentrations in *Teucrium chamaedrys* were in range 0.9–9.4 mg kg^{-1} for serpentine populations and 0.07–0.4 mg kg^{-1} in non-serpentine populations in Bulgaria (Pavlova and Karadjova 2012). Data from serpentines in Serbia (Branković et al. 2017) also fall in this range.

The concentrations of essential elements Mn, Cu and Zn measured in *Teucrium chamaedrys* from Bulgaria and Serbia (Obratov-Petković et al. 2008; Pavlova 2009; Pavlova and Karadjova 2012, 2013; Branković et al. 2017; Zlatić et al. 2017) were within the ranges considered normal for plants both from serpentine and non-serpentine sites (Kabata-Pendias and Pendias 1984; Reeves 2006). Usually, serpentine soils are not rich in Cu, Zn, and Pb, and typical concentrations of these elements in *Teucrium chamaedrys* tissues are below the limits for toxicity. However, results for Zn and Pb concentrations were higher in species populations from Bulgaria in comparison with the data from Serbia (Obratov-Petković et al. 2008; Branković et al. 2017) and from Albania (Shallari et al. 1998) result of possible contamination from recent polymetal mining activities and aerosol deposition due to heavy road traffic near some sites. Also, the quantity of Zn in plants changes with the growing season and often shows an increase throughout the season (Antonovics et al. 1971).

Cadmium is also not part of the "serpentine syndrome" and in most serpentine soils demonstrates relatively lower levels close to typical ranges for unpolluted regions. The concentrations of this element in tissues of *Teucrium chamaedrys* reported from some serpentine sites in Bulgaria were between 0.2 and 1.43 mg kg^{-1} (Pavlova and Karadjova 2012), while in Serbia Cd was below the toxic levels (Branković et al. 2017). The degree to which plants are able to take up Cd depends on its concentration in the soil and its bioavailability modulated by the presence of organic matter, pH, redox potential, temperature and concentration of other elements (Di Toppi and Gabbrielli 1999). Cd and Pb as well, are considered to be effectively absorbed by the root and leaf systems showing a great difference in the ability of various species to accumulate them (Kabata-Pendias and Pendias 1984). In general, Cd concentration in plant tissues is in the order:

root > stem > leaves > grains, although possible absorption directly from the atmosphere can change it (Zhang et al. 2006). Cadmium is variable within species populations of *Teucrium chamaedrys* and the variability of Cd accumulation within populations can be of genetic origin or due to local environmental factors (Reeves 2006). The absorption and accumulation of Cd in plants is also dependent on the growth stage and metabolic activity (Zhang et al. 2006). Based on the difference in absorption, transport and accumulation of heavy metals, *Teucrium chamaedrys* can be classified as an excluder.

Over the diverse range of sites, sampling times of the species and climatic conditions involved in the field collections, it is not appropriate to try to draw conclusions about the detailed interactions of the major and trace elements.

4.3.2 Plant-Soil Correlations

Correlations between elements in the aerial plant parts and elements in the rhizosphere serpentine soils were calculated, taking into account all studied soils and plants (*Teucrium chamaedrys*, *Teucrium polium* and *T. montanum*) as average values for Bulgaria (Pavlova and Karadjova 2012, 2013). Positive correlation coefficients were found for Ni, Cr, Fe, Co, and Cu concentrations in plants with Ca in the soil. Significant negative correlations were found between Mg in plants and Mg in soil confirming plant ability to exclude Mg from soil. However it should be also emphasized that the correlations between measured elemental concentrations in plant populations and their rhizosphere soils are species specific (Pavlova and Karadjova 2013). Highly significant correlations at level $p < 0.01$ were observed between Fe, Ni, Cr, and Co in *Teucrium chamaedrys* samples and Fe in the soils for Bulgarian serpentine sites. However, it should be mention that all these correlation patterns are for total element content in soils, which most probably will be changed if their bioavailable fractions have been considered.

4.3.3 Bioaccumulation Factors

The uptake of heavy metals depends on their concentration in the soil solution and the rate of the transfer from the solid phase into soil solution for replenishment of the heavy metals taken up by the plant roots (Kashem and Singh 2002). The bioaccumulation factor (BAF) calculated as a ratio between the concentrations of elements in plants and in the respective soils provides information or potential bioavailability modes of absorption and accumulation of specific elements in plant tissues. For most cases in *Teucrium chamaedrys* BAF has a value below 1 (Pavlova and Karadjova 2013; Zlatić et al. 2017). BAF values higher than 1 were reported for K both from serpentine and calcareous populations of *Teucrium montanum* and *T. chamaedrys* by Zlatić et al. (2017), whereas BAF values higher than 1 were

calculated for Ca from serpentine populations of the same species. The bioaccumulation trends in the species are the following: Zn > Cu > Cd > Pb > Ni > Co > Mn > Cr > Fe. It is worth mentioning that the bioaccumulation trend follows the bioavailability of trace elements. Higher BAFs could be expected for elements like Zn, Cu, Ni, and Pb. Extractable fraction for these elements is reported to be between 2% and 8%. BAFs for elements like Fe and Cr with extractable fraction below 0.05% are much lower (Pavlova and Karadjova 2013). Almost the same is the bioaccumulation trend for the species shown by Zlatić et al. (2017) and this fact confirms the specific adaptation strategy of the species to the metals. Zinc and Cu are the metals with the highest biouptake abilities, while Cr and Fe have the lowest. Zn is regarded highly mobile from some authors, while others consider Zn to have intermediate mobility. Soluble forms of Zn are readily available to plants and uptake of Zn was reported to be linear with concentration in soils and in nutrient solution (Kabata-Pendias 2011). The mechanism of Cu uptake is different from the uptake of Zn (Antonovics et al. 1971). In the above ground plant parts of *Teucrium chamaedrys,* both from serpentine and non-serpentine soils, Cu uptake stays low and almost constant at low levels of soil Cu. The plant reaction to Cu in the soil suggests that the mechanism of exclusion is available. Plant Cu concentrations are controlled within a remarkably narrow range and extractable Cu was about 2% (Pavlova and Karadjova 2013). The bioavailability of Pb in *Teucrium chamaedrys* above-ground plant parts determined through extractable fraction in 0.43 M acetic acid was about 3%, higher compared to extractable Cu (Pavlova and Karadjova 2013). It is considered that plants are able to take up Pb from soils to a limited extent even under contaminated soil conditions (Kabata-Pendias 2011).

Ni is considered to be a major contributor to the serpentine factor because of the very low availability of Cr and the relatively low abundance of Co compared to Ni (Brooks 1987). The uptake of Ni in *Teucrium chamaedrys* populations was demonstrated to be variable and extractable Ni was about 5% (Pavlova and Karadjova 2013). Nickel is usually easily extracted from soils by plants and its contents of plants are functions of Ni forms in soils. Both plant and pedological factors affect these processes and the most pronounced factor is soil pH (Kabata-Pendias 2011). Although only 1% of the total Ni in soils is available to plants, this far exceeds the percentage of available Cr (Brooks 1987). The low solubility of Cr is reflected in the very slight uptake of this element by plants. In the case of *Teucrium chamaedrys* Cr demonstrates very low extractable fraction, below 0.05%, similarly to Fe. According to Brooks (1987) Fe in serpentine soils is very insoluble and typically only about 0.01% of the total Fe is extracted. It is possible that *Teucrium chamaedrys* strictly selects elements and the absorbed amount and this mechanism enables survival. Thus, in addition to the internal factors of control of mineral content in plant such as genetic specificity, the properties of the root system and leaves, ontogenetic development of the species, external factors should also be considered as very important (Rašić et al. 2006). It is known that the availability of some heavy metals decreases with rising pH of the soil, organic matter and clay content (Mengel and Kirkby 1987; Obratov-Petković et al. 2006; Kabata-Pendias 2011). Among the external factors soil acidity has a strong influence on ion mobility and their uptake.

Weak acidic to weak alkaline conditions of the rhizosphere soil favor strong binding of toxic element in the soil, on one hand, and optimal bioavailability of nutrients as they are essential elements (Rašić et al. 2006).

4.3.4 Species Adaptations to Metalliferous Soils: Quantity of Secondary Metabolites

Secondary metabolites play a major role in the environmental adaptation of plants. The concentrations of various secondary plant products are strongly dependent on the growing conditions and their accumulation occurs in plants subjected to stresses including various elicitors or signal molecules (Nasim and Dhir 2010; Akula and Ravishankar 2011). Heavy metals, as well as other abiotic stress factors (temperature, salinity, water, radiation, etc.), can influence biosynthetic pathways of secondary metabolites (Mahajan and Tuteja 2005) and suppress or stimulate their production. The impact of heavy metals on plant secondary metabolite production is in relation to their concentration and exposure time (Berni et al. 2018). The plant secondary metabolites play numerous roles in plant protection under different biotic and abiotic influences (Ibrahim et al. 2017). Drought resistance is attributed to flavonoids or other phenolic compounds which may also help plants to tolerate soils that are rich in toxic metals (Treutter 2006). It is confirmed that due to the presence of heavy metals certain plants increase the synthesis of phenolics (Michalak 2006; Zlatić et al. 2017; Zdraveva et al. 2018).

As a result of the specific heavy metal profiles of the soils, differences in the quantity of secondary metabolites in *Teucrium chamaedris* were found. In all serpentine plant populations significantly higher values of secondary metabolites were reported compared to non-serpentine ones (Stanković et al. 2010; Zlatić et al. 2017; Zdraveva et al. 2018). The serpentine populations of *Teucrium chamaedrys* are enriched in phenolic compounds similarly to other medicinal plants growing on serpentine substrates (Stanković et al. 2011a, b, 2015; Veličković et al. 2014) and they could be considered as appropriate sources of natural antioxidants. The total amount of phenolic compounds and flavonoids both in extracts and plant material (above-ground parts) of *Teucrium chamaedrys* populations from serpentines were higher compared to non-serpentine (calcareous) populations according to Zlatić et al. (2017). The data presented for Bulgaria (Zdraveva et al. 2018) also confirmed higher values of these compounds in serpentine populations. The content of phenols and flavonoids in plant extracts of *Teucrium chamaedrys* depends on the type of extracts and polarity of solvents used for extractions (Stanković et al. 2010).

The concentrations of phenols and flavonoids vary in the plant body of the species. Stanković et al. (2010) showed the highest concentrations of phenols and flavonoids in leaf extracts in water, acetone and petroleum ether, smaller in flower extracts, and lowest in stem extracts. The large number of glandular hairs, rich in phenolic components, is considered as a reason for higher concentrations of phenols

found in some stem extracts of *Teucrium chamaedrys* var. *glanduliferum* from Serbia (Stanković et al. 2010). The concentrations of phenols and flavonoids in different plant parts compared to those of the whole plant are also different.

The phenolic compounds contribute to the antioxidant potential of plants by neutralizing free radicals and preventing decomposition of hydroperoxides into free radicals (Jain et al. 2014). The high concentration of phenolics, as well as the properties of these molecules and the positive correlation found between the concentration of phenolic compounds and antioxidant activity of different plant extracts of *Teucrium chamaedrys*, indicated that these compounds contributed to the strong antioxidant activity (Stanković et al. 2010). The highest antiradical effect has been also detected in purified methanol *Teucrium chamaedrys* leaf and root extracts from Italy (Pacifico et al. 2009) but the methanol extracts collected from serpentine *Teucrium chamaedrys* populations possessed stronger antioxidant capacity. According to Stanković et al. (2010) both methanol and acetone extracts of *Teucrium chamaedrys* have high concentration of total phenols and flavonoids, which is in correlation with the intense antioxidant activity of these extracts and have a crucial role in the adaptation of this plant to the adverse effects of heavy metals.

The plant extracts of *Teucrium chamaedrys* populations demonstrated variation in the concentrations of tannins as well (Zdraveva et al. 2018). The tannin content was higher in serpentine samples and it was comparable to data for Croatia reported by Grubešić et al. (2012). The amounts of total polyphenols and tannins was found to vary also among the *Teucrium* species (Grubešić et al. 2012; Maleš et al. 2015) and their quantity was higher in native than in cultivated samples in coherence with the environmental factors (temperature, humidity, light intensity, water, minerals and CO_2). The content of total polyphenols and tannins in the samples of *Teucrium chamaedrys* and other studied species reported by Maleš et al. (2015) is significantly lower in comparison to data presented by Grubešić et al. (2012) because of the different method used for quantitative analysis. However, polyphenols represent most abundant and widely distributed antioxidants that have beneficial health promoting effects and confirm that *Teucrium* species are good source of polyphenols (Maleš et al. 2015).

Exposure to heavy metals leads to accumulation of harmful reactive oxygen species (ROS). Reactive oxygen species (ROS), generated during heavy metal stress, may cause lipid peroxidation that stimulates formation of highly active signaling compounds capable of triggering production of bioactive compounds (secondary metabolites) that enhance the medicinal value of the plant (Nasim and Dhir 2010; Berni et al. 2018).

Phenolic compounds and other secondary metabolites have numerous pharmacological properties and thus any environmental condition that affects either the quantity or composition of phytochemical compounds may potentially influence the efficacy of the medicinal plant product. There is limited information on specific physiological responses of medicinal plants to heavy metals in soils and the resulting changes in the efficacy of the plant (Ibrahim et al. 2017).

4.3.5 Species Adaptations to Metalliferous Soils: Morphology

Plants growing on metalliferous soils often possess morphologies distinct from their relatives growing on normal soils. Soil nutrients, moisture stress, photoperiodism, temperature effects, photosynthesis and respiration, all have elicited racial differentiation in a variety of seed plants (Kruckeberg 1969). Intraspecific variation leading to local adaptation (i.e. ecotypic differentiation) has been frequently cited (Štepankova 1996, 1997; Rajaharuna and Bohm 1999; O'Dell and Rajakaruna 2011) for species found on and off serpentine soils. The specific serpentinomorphoses are described and summarized by many authors (Pichi-Sermolli 1948; Krause 1958; Ritter-Studnička 1968). The morphological characteristics of serpentine plants are three main groups: xeromorphic foliage, reduction in stature, and increase in root system (Kruckeberg 1984). Several of these traits have been shown to be genetically determined (Westerbergh 1994).

Although *Teucrium chamaedrys* is widely distributed in serpentine and non-serpentine habitats it has high abundance and potential preference for alkaline soils (Obratov-Petković et al. 2006). The growth habit of *Teucrium chamaedrys* differs between serpentine and non-serpentine sites as it was documented for many species (Brooks 1987; Vergnano Gambi 1992; Westerbergh and Saura 1992; Mayer and Soltis 1994; Štepankova 1996, 1997; Westerbergh and Rune 1996; Bratteler et al. 2006; Wright et al. 2006; Boyd et al. 2009; Salmerorn-Sarnchez et al. 2018). The plants are typically adapted to dry soils and are of smaller stature than non-serpentine relatives. *Teucrium chamaedrys* populations are dense and well developed in mesophilous habitats on calcareous and siliceous terrains compared to serpentine populations. The serpentine populations are characterized by their reduced size – lower stems, smaller leaves, and shorter internodes compared to the non-serpentine populations (Pavlova 2009). The variation is higher for the vegetative features and not clearly expressed for the generative ones. The positive correlations within the floral and vegetative trait groups do not indicate functional relationships among traits. Confirming Conner and Sterling (1996) it was concluded that correlations between both groups of traits were due at least in part to overall size relationship and reflect common developmental pathways.

Studying ecobiology of *Teucrium chamaedrys* in some regions of Serbia Obratov-Petković et al. (2006) noted that populations of *Teucrium chamaedrys* growing on serpentines were less compact than those growing on limestone; the plants growing on serpentine were smaller than the same species on limestone; the leaves of plants on serpentine were smaller, covered with more hairs and many of them were grey-green in comparison with the plants on limestone. The leaf indumentum has adaptive value for species because the hairs allow to gain a higher rate of carbon under arid condition than the leaf acquire without hairs to avoid potentially lethal condition of soil and loss of water which allows the plant to extend its growth for a longer period into the drought (Azmat et al. 2009). Important characteristic for the individuals growing on serpentines is denser indumentum and longer simple or glandular hairs covering the stems, leaves, and bracts which is related to the dry habitat

conditions provided by the serpentine. For most serpentine populations of *Teucrium chamaedrys* are reported 3–5 celled glandular hairs densely covered plant parts – leaves, stem and bracts (Pavlova 2009). Such kind of hairs is not mentioned by Bini-Maleci and Servettaz (1991), Navarro and El Oualidi (2000) or Grubešić et al. (2007) for this species and can be considered specific for the serpentine populations. Because hairs can play an important role in various aspects of plant physiology and ecology in *Teucrium*, further investigations are needed to demonstrate the ecotypic differentiation in *Teucrium chamaedrys* found on serpentine and non-serpentine soils (Pavlova 2009).

4.4 Plant Life on Selected Edaphic Conditions – *Teucrium montanum*

4.4.1 Metal Accumulation by Teucrium montanum

Plant species are exposed to different environmental factors. Among them, the influence of geological substrate and soil was considered as the most important. The effects of the edaphic factor originate from the abiotic and biotic components of the soil as a dynamic system. The process of pedogenesis is a complex of physical and chemical process determined by the type of geological substrate, climatic conditions, living organisms, etc. Edaphic factors in plants affect the appearance of various morpho-anatomical and ecophysiological properties in the function of adaptation. Differentiated groups of species was adapted to a wider range of edaphic conditions variation, as well as groups of species specialized for a substrate making special edaphic ecotypes. Due to the development of complex adaptation mechanisms to a certain type of substrate, the distribution of some ecological groups was completely determined by the type of substrate, while for some groups, there is a certain degree of tolerance in terms of variation of edaphic conditions.

Plant species, that are widespread on habitats with a certain type of substrate, developed the ability to tolerate habitats that are characterized by completely different edaphic conditions. Soil formed on the calcareous geological substrate contained calcium carbonate in free form, which determined the alkaline reaction and limited availability of certain elements to plants, such as N, P, K, Mg, Zn, Cu, and Fe. The soil formed on the serpentine geological substrate represents a special substrate, with extreme physical and chemical characteristics. The disturbed mineral regime, the disturbed Ca:Mg ratio, the lack of essential nutrients Ca, P, N, K, increased concentrations of heavy metals, such as Mn, Ni, Cr and the variation in pH from acidic to strong alkaline, are one of the most significant characteristics of serpentine soils (Zlatić et al. 2017).

Teucrium montanum L. (Lamiaceae) is a facultative serpentinophyte (Fig. 4.2) and it is an ideal model organism for comparative analysis of ecological differentiation relative to the substrate such as serpentine and calcareous substrate (Fig. 4.3).

Fig. 4.3 Serpentine (left) and calcareous (right) habitat of *T. montanum* L. (Photo M. Stanković)

Teucrium montanum absorbs a large number of elements from soil, essential (Cu, Fe, Mn, Ni, Zn) and non-essential (Cr, Co, Pb) for its biological functions. Many authors state that the most stressful component of the plant was disturbed Ca:Mg ratio, that affects the establishment of viable populations of *Teucrium montanum* species on serpentine habitats. The Ca:Mg ratio is generally in values above 1 in plants, while the plants present on the serpentine could have values higher than 1, regardless of the minimum concentration of Ca in the soil. It has been confirmed that Ca deficiency had an influence on root elongation and shoots growth (Marschner 1995). Since Ca and Mg are available to the plants through the roots, plants need a special mechanisms to maintain the adequate Ca:Mg ratio in the species that grow on serpentine. These mechanisms allow selective Ca and Mg exclusion, despite high concentrations in the soil, as well as the translocation of Ca and Mg from the root into the above-ground plant parts (Vicić et al. 2013b).

In the study of *Teucrium montanum* populations on the serpentine localities in Serbia and Bulgaria, similar amounts of accumulated Ca and Mg was found in the above-ground plant parts. Plant populations sampled from serpentine soil accumulated higher quantities of Ca than the population sampled from calcareous soil (Pavlova and Karadjova 2012). Juranović-Cindrić et al. (2013) described that the population of *Teucrium montanum* accumulated between 30 and 40 times more Mg than Ca, which indicated that the species process a high ability to absorb Mg in habitats where the concentration of Mg is lower than the concentration of Ca.

Calcium concentrations found in tissues of *Teucrium montanum* were between 1356.5 and 2237.7 mg kg^{-1} in serpentine plant populations from Serbia (Zlatić et al. 2017), and between 2980 and 4798 mg kg^{-1} in plant populations from Bulgaria (Pavlova and Karadjova 2012). The results in presented research indicated that the amount of Ca, in soil samples and in samples of plants on the calcareous substrate, is higher in relation to the quantity in samples from the serpentine substrate. The increased amount of Ca in the soil affects the amount of Ca in the plants (Zlatić et al. 2017). Calcium is involved in 0.1–2.0% of dry matter of plants. The normal concentrations of this element in the above-ground parts of the plants were 5 g kg^{-1} (Shallari et al. 1998). The content of Ca in plants growing on serpentine soil is lower

than 0.8% (Kataeva et al. 2004). Concentration of calcium was higher in the above-ground parts of plants compared to the substrate (serpentine or calcareous), although the serpentine species are better adapted to low Ca concentrations. The plants, which have normal growth, the Ca:Mg ratio is higher than 1. The serpentine species are adapted to low Ca concentrations relative to Mg, since Mg prevents the absorption of Ca (Brooks 1987). Calcium is an essential macronutrient which participates in the construction of the cell wall, provides normal transport through membrane proteins and acts as a cofactor for many enzymes (Marschner 1995). Differences in concentration of Ca in soil and plant material from serpentine habitats could be due to the presence of organic matter and the effects of various ecological factors presented on a specific type of ecosystem (Zlatić et al. 2017).

The amount of Mg in plant samples of *Teucrium montanum* was in the range from 1476.6 to 3651.4 mg kg^{-1} (Zlatić et al. 2017), and in the range from 1932 to 4626 mg kg^{-1} in plant populations from Bulgaria (Pavlova and Karadjova 2012). The results of the study indicated that the amount of Mg in the plant samples on the serpentine substrate (Goč and Kamenica) was higher than in the samples from the calcareous substrate (Kopaonik and Durmitor) (Zlatić et al. 2017). Increased concentrations of Mg in serpentine plants population indicated their ability to adapt and survive under stressful conditions. The differences in quantity between Mg and Ca affects the appearance of undeveloped vegetation on serpentine soils. Plants could grow successfully, if the ratio of the total amount of Ca:Mg in the soil is at least in ratio 1:1 (Brooks 1987). Magnesium is macronutrient that plants used for their growth and development in large quantities (Karley and White 2009). Also, this element is located at the center of the chlorophyll molecule and had a significant function for many enzymes and chelating nucleotide formations (Shaul 2002). Lack of Mg in plants could affects the increase in the activity of antioxidant mechanisms in order to neutralize harmful effects (Cakmak and Kirkby 2008). The obtained results of the comparative analysis showed that the samples of soil and plant material from the serpentine geological substrate had a higher concentration of Mg from the calcareous substrate. Certain amounts of Mg are required by plant nutrition, and studies showed that plants could metabolically control the absorption of Mg from the soil. The differences in the concentration of Mg between serpentine localities could be attributed to the different influence of pedological processes and ecological factors that were dominate on the investigated habitats (Zlatić et al. 2017).

The amount of K in plant samples of *Teucrium montanum* was in the range from 6221.4 to 12338.6 mg kg^{-1} (Zlatić et al. 2017). The results of this study indicated that the amount of K in soil and plant samples from calcareous substrate, was higher in relation to the quantity in samples from the serpentine substrate. The increased amount of K in the soil influenced the amount of K in the plants (Zlatić et al. 2017). Potassium is one of the main nutrient, which affect the growth and development of plants. It is the most important element in plant nutrition of all alkali elements found in the soil (K, Ca, Mg, and Na) (Haby et al. 1990). Due to the disturbed water regime, plants require higher concentrations of K for adaptation to stress caused by photo-oxidative damage. Potassium had an important role in the protection of plants, which are presented in arid habitats (Sen Gupta et al. 1989).

The accumulation of Ni, Cr, and Co in the above-ground plant parts for all serpentine populations of *Teucrium montanum* were higher than in non-serpentine ones (Pavlova and Karadjova 2012; Branković et al. 2017; Zlatić et al. 2017). Element concentrations vary between the populations in relation to soil properties. Nickel quantity in plant samples for *Teucrium montanum* was in the range from 9.8 to 108.6 mg kg^{-1} (Zlatić et al. 2017). The results of this study indicated that the amount of Ni, in soil samples and in samples of plants on the serpentine substrate, was higher than the quantity in samples from the calcareous substrate, as well as that the increased amount of Ni in the soil influenced the amount of Ni in the plants. In the plant species from serpentine, the absorption of certain nutrients had been changed because of stress induced by the presence of Ni. Certain concentrations of this element in plant tissue inhibited the process of photosynthesis and transpiration. In natural conditions, the toxicity of Ni is in conjunction with serpentine soils which possess great amounts of this metal (Kabata-Pendias 2011). Nickel is very mobile in the soil and showed the ability to cross from one soil layer to another. Fe and Mg oxides could accumulate large amounts of Ni. After this accumulation, Ni is available to plants. Influence on the bioavailability of Ni in the soil depended of the presence of organic matter, clay fractions and pH values (Kastori 1993). It has been showed that the soils formed on the serpentine substrate had a higher concentration of Ni than those on limestone (Zlatić et al. 2017). The composition of the geological substrate plays an important role in the presence of Ni in the soil. The difference between localities with the same geological substrate could be attributed to the various environmental factors, that affect the certain habitat, as well as the chemical processes in the upper layers of the soil.

The amount of Cr in plant samples of *Teucrium montanum* was in the range from 4.15 to 45.3 mg kg^{-1} (Zlatić et al. 2017) and between 0.3 and 28 mg kg^{-1} in plant populations from Bulgaria (Pavlova and Karadjova 2012). The results of presented research indicated that the amount of Cr, in soil samples and in samples of plants on the serpentine substrate, was higher than the amount in samples from the calcareous substrate on the territory of Bulgaria and Serbia. The increased quantity of Cr in the soil affects the quantity of Cr in plants (Zlatić et al. 2017). Chromium content in plants is affected by the amount of soluble Cr in the soil. In the mentioned studies, there is no data based on the important role of Cr in plant metabolism. However, there are data which described the positive influence of Cr on plant growth. The most accessible for plants is Cr^{6+}, which in certain soils is unstable and whose distribution depends on the characteristics of the soil, its texture, and pH value. In some plant species from the serpentine region, the Cr concentration varies from 0.3% to 3.4% (Kabata-Pendias 2011).

The amount of Fe in the plant samples of *Teucrium montanum* was in the range from 351.9 to 2673.1 mg kg^{-1} (Zlatić et al. 2017) and between 148.5 and 1490 mg kg^{-1} in plant populations from Bulgaria (Pavlova and Karadjova 2012). In the presented studies the amount of Fe, in soil samples and in samples of plants on the serpentine substrate, was higher than the quantity in samples from the calcareous substrate. Iron is a reactive element and it is similar like Co and Ni. Studies have confirmed that the reactivity of Fe depends on the pH value and the oxidative degree

of the Fe compound (Kabata-Pendias 2011). The distribution of the Fe compound in the soil can be useful in determination the type and characteristics of the soil (Zonn 1982). The concentration of Fe in the soil solution at usual pH value varies between 30 and 550μg l^{-1}. In very acidic soils, it could be over concentration of 2000μg l^{-1}. Therefore, in calcareous soil types, with pH from 7 to 7.8, concentration of Fe ranged from 100 to 200μg l^{-1}. In sandy substrates, whose pH varies from 2.5 to 4.5, concentration of Fe ranged between 1000 and 2223μg l^{-1}. The solubility level of Fe is minimal in alkaline soils. Acidic soils are rich in soluble inorganic Fe in relation to soil from neutral and limestone substrates. Reduced quantities of Fe are characteristic for dry areas (Kabata-Pendias 2011). Although the Fe compounds in the soil are mobile, there is a tendency for the formation of mobile organic compounds and chelates. These compounds caused the distribution of Fe in the soil layers, as well as the transition of ion Fe from one layer to another that plants absorb through the root system. The quantity of Fe, that plants could absorb, depends on the changing conditions of the soil and climate, and the stage of plant development. However, when Fe is soluble, plants may absorb a higher amount (Kabata-Pendias 2011). This was clearly demonstrated in plant samples from serpentine habitats where the Fe concentration in the tested grasses were between 2127 and 3580 mg kg^{-1} (Johnston and Proctor 1977). Fe oxides bind to Mn compounds and this is the reason of why Fe is more represented in the serpentine substrate (Kabata-Pendias 2011). Certain amounts of Fe are required by plants in the nutrition. Plants regulate the concentration of the same in their tissues through metabolism. The concentration of Fe affects the composition of the parent material, pedogenetic processes, the presence of organic matter and other factors. The difference in the concentration of Fe between similar sites could be explained through different environmental factors, which are dominated in the specific habitat.

The concentrations of essential elements, Mn, Cu and Zn measured in *Teucrium montanum* from Serbia and Bulgaria (Pavlova and Karadjova 2012; Zlatić et al. 2017) were within the ranges considered normal for plants from serpentine and non-serpentine sites (Kabata-Pendias 2011). Usually, serpentine soils are not rich in Cu, Zn, and Pb. The amount of Mn in the plant samples of *Teucrium montanum* was in the range from 22.16 to 78.3 mg kg^{-1} (Zlatić et al. 2017) and between 16 and 99 mg kg^{-1} in plant populations from Bulgaria (Pavlova and Karadjova 2012). The results of the study indicated that the amount of Mn, in soil samples and in samples of plants on the serpentine substrate, was higher than the quantity in samples from the calcareous substrate. By reducing the amount of Mn in the soil, the amount of Mn increases in the above-ground parts of the plants. High levels of Mn are observed in soils that are present on serpentine soil, in soil rich in Fe and organic matter, as well as in arid and semi-arid regions of the earth (Kabata-Pendias 2011). Certain quantities of Mn are required in plant nutrition, and studies showed that plants could metabolically control the absorption of Mn from the soil (Skinner et al. 2005). Numerous studies showed the relationship between the amount of Mn and Fe represented in the soil. Manganese was significantly more present in areas where Fe oxides are present (Kabata-Pendias 2011).

The amount of Cu in plant samples of *Teucrium montanum* was in the range from 1.68 to 8.35 mg kg^{-1} (Zlatić et al. 2017) and between 5.7 and 15.4 mg kg^{-1} in plant populations from Bulgaria (Pavlova and Karadjova 2012). The quantity of Cu, in the soil samples and in the samples of plants on the serpentine substrate (Goč and Kamenica), was higher than the quantity in the samples from the calcareous substrate (Kopaonik and Durmitor). With reducing the amount of Cu in the soil, the amount of Cu in plants grows. Plants sampled on serpentine soil had a higher concentration of Cu than those on limestone. However, minor deviations were observed, which is in accordance with the above data on the bioavailability of Cu (Zlatić et al. 2017). The concentration of Cu was influenced by the presence of organic matter, pedogenetic processes, and the composition of the parent material.

According to the results presented by the study from Zlatić et al. (2017), the quantity of Zn in the above-ground plant parts of the species *Teucrium montanum* parts varied from 14.79 to 39.75 mg kg^{-1} in populations from Serbia, and between 29 and 70 mg kg^{-1} in plant populations from Bulgaria (Pavlova and Karadjova 2012). The results indicated that the quantity of Zn, in soil samples and in samples of plants from the calcareous substrate, was higher than the quantity in samples from the serpentine substrate, and that the increased amount of Zn in the soil influenced the amount of Zn in the plants. The composition of the parent material, the structure of the soil, the presence of organic matter, the ecological factors, the pH value are one of the factors that influenced the amount of Zn in the soil. The concentration of Zn in plant tissues varies between 15 and 150 mg kg^{-1}, while maximum values were up to 300 mg kg^{-1} (Brunetti et al. 2009). In the acidic environment, Zn was very soluble (Kastori 1993). In previous studies, it has been found that the lack of Zn disrupted the absorption of Fe (Graham et al. 1987). Tolerant plant species had ability to reduce the effect of increased concentrations of Zn. Mechanisms of defense are reflected in the adaptation of metabolic reactions in the plant organism. Plants accumulate Zn in some parts of the cell and tissue (Kabata-Pendias 2011).

The amount of Pb in plant samples of *Teucrium montanum* was between the detection limits to 6.02 mg kg^{-1} (Zlatić et al. 2017), and between 0.6 and 21.7 mg kg^{-1} in plant populations from Bulgaria (Pavlova and Karadjova 2012). The results of the authors suggested that the amount of Pb in the soil samples from the calcareous substrate was higher than the quantity in the samples from the serpentine substrate, while the opposite results were showed in the plant samples. Lead is the least mobile among heavy metals in the soil, although its sorption was lower than Zn and Cu (Vega et al. 2007). The Pb concentrations in soils formed on the calcareous substrate are always higher than those formed on the serpentine (Kabata-Pendias 2011). Lead behaves like Ca, therefore, it had less in plants on calcareous soil. The obtained results were in accordance with numerous studies related to the absorption of Pb by plants present on the serpentine substrate (Reeves et al. 2007; Kabata-Pendias 2011).

Based on the results presented in the studies conducted by Pavlova and Karadjova (2012) and Zlatić et al. (2017), it can be concluded that *Teucrium montanum*, which accumulate a higher amounts of heavy metals (Ni, Fe, Mn, Cu), is not a hyperaccumulator of these elements. The investigated species possess physiological mechanisms that allow adaptation of various stress conditions, caused by increased

concentrations of heavy metals in the soil, as well as specific conditions of the habitat on the serpentine substrate.

4.4.2 Plant-Soil Correlation

Correlations between elements in the above-plant parts of *Teucrium montanum* and elements in the serpentine soils from the territory of Serbia were calculated as average values (Zlatić et al. 2017). Significant positive correlations at level $p < 0.01$ were observed between Ca and Zn in *Teucrium montanum* samples and Ca in the soils for Serbian serpentine sites. Negative correlation coefficients were found for Ni and Mg plant samples and Ca in the serpentine soils. Also, the synergistic effect between elements Fe, Ni, Cr, Mn in serpentine soils were found. It has been established that the contents of Mn and Cu in the soil and those of Mn and Fe in the plants correlate. The studies confirmed the antagonistic and synergistic effects of Mn in the process of absorption of Pb and Cd; antagonistic effects of Mn in the process of N, Na and K absorption and the substantial impact of Zn on the low absorption of Mn (Kabata-Pendias 2011). Other scientists confirmed the existence of correlations between certain metals (Yan et al. 2012; Gonneau et al. 2014).

4.4.3 Bioaccumulation Factors

Concentration and the absorption of elements in plant tissues depended on the concentration in the substrate and the ability of the plant to absorb certain amounts of the same. Absorption of certain macro- and microelements is regulated by special mechanisms in plants. Bioaccumulation factor (BAF) is the ratio between the total content of metals in the plants and the content of metals in the soils, and it is used to determine the total amount of metal that the plant accumulated from the soil through the root system into organs (Pandy and Tripathi 2010). The BAF results for *Teucrium montanum* on calcareous and serpentine habitats in most cases had values below 1 (Zlatić et al. 2017). BAF values higher than 1 were observed for K from calcareous and serpentine populations of *Teucrium montanum* while values higher than 1 for Ca were observed for populations of the species of serpentine habitat (Zlatić et al. 2017). The bioaccumulation trends in the species are as follow: Zn > C u > Cr > Mg > Mn > Ni > Pb > Fe.

Factors affecting the mobility, accessibility, and dynamics of soil elements for plants were the pH value, the content of organic matter and clay in the soil, the mechanical composition and humidity of the soil, the content and the presence of hydrated Al and Fe oxides. Elements, such as Zn and Cd, are very mobile in the soil and could be easily accumulated by plants. Elements such as Cu, Mn, Ni, are moderately mobile in the soil and could be rapidly absorbed by plants, while Pb is

relatively strongly related to soil particulates and could not be transported to the above-ground parts of plants (Kabata-Pendias 2011).

Zinc is a very mobile in most types of soil. Zinc has an essential metabolic function in the plants, and in the soil solution, it is in the form of free ions or ionic complexes. Plants absorb Zn mainly from the soil in the form of divalent cation (Zn^{2+}). It has been showed that plants absorb Zn in the form of hydrated zinc (Zn^{2+}), in the form of ionic complexes and Zn organic chelates (Kabata-Pendias 2011). The copper is quite immobile in the soil and its mobility was influenced by the organic matter of the soil, the dissolved organic matter, the pH and the content of Cu in the soil. The bioavailability of the soluble Cu forms depended on the molecular weight of the Cu complex, and on their amounts in the soil. Copper is an essential element for plants. Soil pH value play an important role in the availability of Cu and its toxicity to plants. The plants can absorb small quantities of Cu, mainly in the form of Cu^{2+} ions and chelates. Copper concentration is influenced by its concentration and the presence of other ions in soil such as heavy metals, like Zn, Mn, Fe (Kastori 1993). Chromium is a crucial element for plants, which stimulate the growth and development of some plant species. Higher concentrations of Cr showed a toxic effect on plants. The effect of Cr in plants depends on its concentration in soil, but also on the chemical and physical properties of the soil (Kastori 1993). Plants could absorb Mg in the form of Mg^{+2} from the soil solution. The availability of Mg is affected by the parent material, climatic factors, as well as the accumulation capacity of the soil. In the conditions of the acidic soil reaction, the availability of Mg was reduced by its competence with Al or Mn, while the presence of carbonates and increased concentrations of Ca, K and Na could reduce the absorption of Mg in alkaline conditions (Sigel and Sigel 1990). Manganese could be present in soils in several forms (Kabata-Pendias 2011). The plants absorb Mn in acidic soil where it is in the form of Mn^{2+} ions. Absorption of Mn is metabolically controlled while higher concentrations of this element depend on the level of Mn concentration in the soil. The accumulation of Ni had a significant influence on the content of other elements in the substrate, while it had no effect on the absorption and metabolism of Fe (Kastori 1993). Dynamics of Pb in the soil is influenced by numerous factors, like pH value and the presence of organic matter. Lead is mainly occurred in the soil, in the form of Pb^{2+}. Plants poorly can absorb and move inorganic forms of Pb into the above-ground parts, while the adsorption and transport of organic forms are relatively strong. Lead have very low bioavailability, so the roots have a great ability to accumulate Pb, which is one aspect of protecting the above-ground parts of plants. The content of easily soluble Fe fractions is limited compared to the total content in the soil. Iron is a slow-moving element in the soil and showed the tendency to form organic complexes and chelates. Iron ions are not absorbed directly from the plant, but from the complexes with organic compounds. The high level of Fe compound, its precipitation on carbonates and the competence of other cations with Fe^{2+}, are responsible for the low absorption of Fe (Kabata-Pendias 2011).

4.4.4 Species Adaptations to Metalliferous Soils: Quantity of Secondary Metabolites

Secondary metabolites are compounds that plays a role in the interaction of plants with the environment and have an influence on plant adaptation on different abiotic and biotic stresses (Kliebenstein and Osbourn 2012). Biosynthetic pathways of secondary metabolites are to some extent influenced by environmental factors such as the presence of heavy metals in the substrate, altitude, water regime, radiation, etc. (Endt et al. 2002; Politycka and Adamska 2003). The quantity of secondary metabolites is different in plant parts and depends on the development phase, season and several ecological factors (Filippini et al. 2010). Secondary metabolites participate in the process of adaptation of plant organisms to the ecological environmental conditions, and their quantity varies in plant organs depending on the abiotic and biotic factors (Ramakrishna and Ravishankar 2011). The presence of heavy metals in the substrate caused increased synthesis of secondary metabolites. It has been determined experimentally that certain plants increase the synthesis of phenolic compounds due to the presence of heavy metals. Due to the increase in the number of phenolic compounds, the activity of enzymes involved in their biosynthesis is also enhanced (Michalak 2006).

Phenolic compounds represent a large group of secondary metabolites, which have one or more aromatic rings with one or more hydroxyl groups in the structure. Within the group of phenolic compounds, two subgroups are distinguished: flavonoids and phenolic acids (Quideau et al. 2011). Flavonoids are present in all plant organs, while their amount varies according to the ecological and genetic factors (Stanković et al. 2010, 2012). The role of phenolic acids in plants is various and it reflected in the process of absorption of necessary materials from the substrate and the adaptation of the plant to abiotic and biotic stresses (Mandal et al. 2010). The particularly important role of phenolic compounds is the antioxidant activity. Phenolic compounds act as an reducing agents, donors of hydrogen, and have the properties of metal healing. The antioxidant activity of these compounds is primarily the result of their role as a donor of hydrogen, after which there are less reactive phenoxyl radicals (Quideau et al. 2011).

Zlatić et al. (2017) indicated that the total amount of phenolic compounds in plant extracts of *Teucrium montanum* was in the range from 143.42 to 190.2 mg GA g^{-1} of extract, while the quantity of flavonoids ranges from 46.5 to 54.19 mg Ru g^{-1} of the extract. For *Teucrium montanum*, the concentration of phenolic compounds and flavonoids was higher in extracts from sample originated from serpentine sites than in samples from the calcareous substrate. Also, authors indicated that the total amount of phenolic compounds in plant extracts of *Teucrium montanum* was in the range from 37.11 to 66.52 mg GA g^{-1} of the extract, while the quantity of flavonoids was in range from 15.38 to 20.9 mg Ru g^{-1} of the extract. The concentration of phenolic compounds as well as flavonoids of *Teucrium montanum* was higher in extracts from samples originated from the habitat with serpentine geological substrate compared to samples from the limestone substrate (Zlatić et al. 2017).

There are differences in the quantitative composition of secondary metabolites of plants that were sampled on calcareous and serpentine substrates (Pavlova 2009). The presence of heavy metals in the soil caused increase of secondary metabolites synthesis. It had been determined that certain plants enhance the synthesis of phenolic compounds due to the presence of heavy metals (Michalak 2006; Stanković et al. 2011b).

The increase of phenolic compounds concentration in the plant extracts from *Teucrium montanum* occurred in the response of plants to stress conditions, caused by heavy metals (Lavid et al. 2001). The total amount of phenolic compounds was higher due to the increased concentration of heavy metals (Hamid et al. 2010). The antioxidant system of *Aeluropus littoralis*, under the influence of stress caused by heavy metals, goes through biochemical changes, which are reflected in increase the concentration of phenolic compounds involved in chelation (Rastgoo and Almzadeh 2011). Rusak et al. (2005) indicated that flavonoids have multiple protective functions, such as antioxidant activity. Phenols and flavonoids can be oxidized by peroxidase and react in H_2O_2 phenolic systems, which regulate the input of heavy metals in plants (Michalak 2006). Some heavy metals can cause a decrease in volume of reducted glutathione (GSH) and insufficient activity of antioxidant enzymes, especially glutathione reductase (GR) (Schützendübel and Polle 2002). Based on this knowledge, alternative antioxidants, such as flavonoids, could be produced by plants. Flavonoids build complexes with heavy metals, making them the main systems in the adaptation of plants to stress conditions caused by increasing the concentrations of heavy metals (Michalak 2006; Korkina 2007).

The qualitative and quantitative analysis of the composition of *Teucrium montanum* secondary metabolites confirmed the presence of gentisic acid which is the most represented phenolic acid in the extracts. Also, the phenolic acids, that are present in lower concentrations, are chlorogenic, coumaric, syringic, gallic, vanillin, caffeic, and ferulic acid (Tumbas et al. 2004; Čanadanović-Brunet et al. 2006). A qualitative analysis of the essential oil showed that the main components of the volatile compounds are: β-cadenine, β-selinene, α-calacorene, β-caryophyllene, β-pinene and germacrene (Vuković et al. 2008; Bežić et al. 2011).

The phenolics contribute to the antioxidant potential of plants by neutralization of free radicals (Stanković et al. 2010). Populations of plants sampled from the serpentine localities are differentiated by the increased quantity of total phenolic compounds, as well as intense antioxidant activity in relation to plant populations sampled from calcareous localities. Higher concentration of total phenolic compounds are in correlation with the intense antioxidant activity of *Teucrium montanum* extracts and have a crucial role in the adaptation of this plant to habitats with the higher concentration of heavy metals. Differences in the concentration of secondary metabolites and expressed antioxidant activity could be attributed to a different type of habitat as well as the abiotic factors on which the populations are represented. There are differences in the quantity of phenolic compounds and antioxidant activity among plant populations from serpentine soil. This difference was induced by a different amount of metal at a specific locality, a difference in thermal

and water regime, as well as the altitude at which the locality is located (Zlatić et al. 2017).

4.4.5 Species Adaptations to Metalliferous Soils: Morphology

Serpentine soils represent a special substrate for the development of flora and vegetation (Brooks 1987). Plants, which are present on the serpentine soil, possesses numerous structural and functional adaptations, which are different from the plants represented on other types of soil. Serpentine plants are characterized by "serpentine syndrome", which is a result of the disturbed relationship between Mg and Ca, heavy metal toxicity and other pedological physicochemical characteristics (Kruckeberg 1984; Brooks 1987). Additionally, to the low water potential of the soil, the plants on the serpentine have great tolerance to the increase of concentration of heavy metals from the substrate (Proctor 1999). Numerous studies have showed that populations represented in different serpentine soils have similar characteristics (Proctor and Woodell 1975; Kruckeberg 1984; Proctor and Nagy 1992). Facultative serpentinophytes showed a "serpentine syndrome" such as a complex of morpho-functional adaptations that differentiate the plant populations present on serpentine soils. These adaptations include smooth-surface leaves with waxy coatings, plants with high developed roots, plagiotropism and nanism (Kruckeberg 1984; Proctor and Woodel 1975; Brady et al. 2005).

The habitats of *Teucrium montanum* are calcareous rocks, pine forests on calcareous soil, pastures on serpentine substrate. The species occupy different type of habitats from 30 to 2000 m above sea level. Species is represented in the communities of deciduous, mixed and coniferous forests, in the communities of serpentine rocky and continental calcareous pastures, gorge and canyons, in the communities of the submediterranean region, the communities of the xerophilous type, and the slopes of the mountainous and mountain regions. In the territory of Serbia, the species *Teucrium montanum* is the most represented on the calcareous and serpentine substrate, rocky meadows and dry pastures around the pine forests (Stanković and Zlatić 2019).

Previous morpho-anatomical studies of *Teucrium montanum* from different habitats indicated that species possess phenotypic plasticity manifested through more or less xeromorphic properties. Studies of *Teucrium montanum* from calcareous and serpentine localities showed physiological and ecological differences between populations, presented on the serpentine and calcareous substrate, with specific features in the morpho-anatomical structure of serpentine plants designated as serpentinomorphoses (Stevanović and Stevanović 1985). Morpho-anatomic differences were observed in plants from different populations as a response to the adaptive abilities of the species to certain environmental conditions. In plant populations from calcareous habitats, xeromorphic characteristics were observed, which are in accordance with the general habitat conditions determined by the hygrothermal soil regime.

Teucrium montanum on these habitats is compact and densely branched bushes. The leaves are small, covered with hairs and moderately curved to the abaxial side. At the cross-section, the leaves are thicker with a distinct cuticle compared to the serpentine population. They possess a large number of vascular bundles and a more developed mesophyll. On the serpentine habitats, *Teucrium montanum* is represented in the bushy habitus form with low density. On the shoots, there are fewer leaves than on the plants from calcareous habitats. Leaves are smaller and ticker in relation to leaves from calcareous populations, while the edges of the leaves are strongly curved to the abaxial side. At the cross-section, the leaves are thin with a well-developed cuticle. Cells of mesophyll are smaller with poorly expressed intercellular space. In plants with serpentine habitats, xeromorphic characteristics are strongly expressed in connection with the serpentine substrate and the effect of physiological drought (Stevanović and Stevanović 1985).

4.5 Conclusions

This chapter is a review of the ecological, physical, chemical, morphological and geographical distribution of two selected species of the genus *Teucrium*. *Teucrium chamaedrys* and *T. montanum* demonstrate a wide ecological tolerance as they could be found in different habitats, like rocky calcareous and serpentine terrains, open grasslands, pastures, and pine forests. Both species showed the ability to absorb a large number of elements from soil both essential (Cu, Fe, Ni, Mn, Zn) and non-essential (Co, Cr, Pb) for its biological functions. Plant populations from serpentine habitats possess a higher content of heavy metals, like Mn, Ni, Cr, than calcareous one. Among the selected species *Teucrium montanum* can accumulate more metals (Cr, Fe, Ni) compared to *Teucrium chamaedrys*. Based on the difference between the ability of absorption and accumulation of heavy metals, described plant species could not be classified as metal-hyperaccumulating plants. *Teucrium chamaedrys* and *T. montanum* are classified as excluders, containing relatively low metal concentrations in their aerial parts, despite the high element concentrations in the soil. The presence of heavy metals and other environmental conditions from serpentine possess significantly higher values of secondary metabolites in plant populations, compared to non-serpentine ones. The significant morpho-anatomic differences between populations are related to the differences in edaphic conditions of the serpentine and non-serpentine sites. The serpentine populations are characterized by their reduced size – lower stems, smaller leaves, and shorter internodes, compared to the non-serpentine populations. Selected plant species collected from naturally metalliferous (serpentine) soils should be avoided because of the increased concentration of heavy metals in plant tissues. Traditional medicinal products, which contains heavy metals in concentrations hazardous to human health. Usage of these plants should be limited. Plants collected from serpentine soils require strict control. Medicinal herbs should be gathered and cultivated only from non-metalliferous sites.

References

Adamidis GC, Aloupi M, Kazakou E, Dimitrakopoulos PG (2014) Intra-specific variation in Ni tolerance, accumulation and translocation patterns in the Ni-hyperaccumulator *Alyssum lesbiacum*. Chemosphere 95:496–502

Akula R, Ravishankar GA (2011) Influence of abiotic stress signals on secondary metabolites in plants. Plant Signal Behav 6:1720–1731

Antonovics J, Bradshaw AD, Turner RG (1971) Heavy metal tolerance in plants. Adv Ecol Res 7:1–85

Ater M, Lefebvre C, Gruber W, Meerts P (2000) A phytogeochemical survey of the flora of ultramafic and adjacent normal soils in North Morocco. Plant Soil 218:127–135

Azmat R, Haider S, Nasreen H, Aziz F, Riaz M (2009) A variable alternative mechanism in adapting the plants to heavy metal environment. Pak J Bot 41:2729–2738

Babalonas D, Karataglis S, Kabassakalis V (1984) The ecology of plant populations growing on serpentine soils. III. Some plant species from North Greece in relation to the serpentine problem. Phyton 24:225–238

Bani A, Imeri A, Echevarria G, Pavlova D, Reeves R, Morel J-L, Sulçe S (2013) Nickel hyperaccumulation in the serpentine flora of Albania. Fresenius Environ Bull 22:1792–1801

Bazan G, Galizia G (2018) Geographical and ecological outline of metal(loid) accumulating plants in Italian vascular flora. Ecocycles 4:47–64

Bentham G (1833) Labiatarum genera et species. Ridgeway Sons, London

Berni R, Luyckx M, Xu X, Legay S, Sergeant K, Hausman J-F, Lutts S, Cai G, Guerriero G (2018) Reactive oxygen species and heavy metal stress in plants: impact on the cell wall and secondary metabolism. Environ Exp Bot 161:98–106

Bezić N, Vuko E, Dunkić V, Ruščić M, Blažević I, Burčul F (2011) Antiphytoviral activity of sesquiterpene-rich essential oils from four Croatian *Teucrium* species. Molecules 16:8119–8129

Bini-Maleci L, Servettaz O (1991) Morphology and distribution of trichomes in Italian species of *Teucrium* sect. *Chamaedrys* (Labiatae): a taxonomical evaluation. Plant Syst Evol 174:83–91

Blake L, Goulding KWT (2002) Effects of atmospheric deposition, soil pH and acidification on heavy metal content in soil and vegetation of semi-natural ecosystems at Rothamsted experimental station. Plant Soil 240:235–251

Boissier PE (1879) Lamiaceae. In: Flora Orientalis 4. apud H. Georg, Geneva/Basel, pp 805–822

Boyd R, Wall M, Santos S, Davis M (2009) Variation of morphology and elemental concentrations in the California nickel hyperaccumulator *Streptanthus polygaloides* (Brassicareae), soil and biota of serpentine: a world view. Northeast Nat 16:21–39

Brady K, Kruckeberg A, Bradshaw H (2005) Evolutionary ecology of plant adaptation to serpentine soils. Ann Rev Ecol Evol S 36:243–266

Branković S, Cupara S, Glisić R, Djelić G, Grbović F, Kojicić K, Milovanović O (2017) Phytoaccumulation in plants of mountain Goč in Serbia. Studia Universitatis "Vasile Goldiş", Seria Ştiinţele Vieţii 27:196–201

Bratteler M, Baltisberger M, Widmer A (2006) QTL analysis of intraspecific differences between two *Silene vulgaris* ecotypes. Ann Bot 98:411–419

Broadhurst CL, Chaney RL, Angle JS, Maugel TK, Erbe EF, Murphy CA (2004) Simultaneous hyperaccumulation of nickel, manganese and calcium in *Alyssum* leaf trichomes. Environ Sci Technol 38:5797–5802

Brooks RR (1987) Serpentine and its vegetation: a multidisciplinary approach (ecology, phytogeography and physiology). Dioscorides Press, Portland

Brooks RR, Yang XH (1984) Elemental levels and relationships in the endemic serpentine flora of the central Dyke, Zimbabwe and their significance as controlling factors for the flora. Taxon 33:392–399

Brunetti G, Soler-Rovira P, Farrag K, Senesi N (2009) Tolerance and accumulation of heavy metals by wild plant species grown in contaminated soils in Apulia region – Southern Italy. Plant Soil 318:285–298

Cakmak I, Kirkby EA (2008) Role of magnesium in carbon partitioning and alleviating photooxidative damage. Physiol Plant 133:692–704

Čanadanović-Brunet J, Djilas MS, Ćetković SG, Tumbas TV, Mandić IA, Čanadanović MV (2006) Antioxidant activities of different *Teucrium montanum* L. extracts. Int J Food Sci Technol 41:667–673

Chaney RL, Angle JS, Broadhurst CL, Peters CA, Tappero RV, Sparks DL (2007) Improved understanding of hyperaccumulation yields commercial phytoextraction and phytomining technologies. J Environ Qual 36:1429–1443

Conner JK, Sterling A (1996) Selection for independence of floral and vegetative traits: evidence from correlation patterns in five species. Can J Bot 74:642–644

Cornara L, Roccotiello E, Minganti V, Drava G, De Pellegrini R, Mariotti MG (2007) Level of trace elements in plants growing on serpentine and metalliferous soils. J Plant Nutr Soil Sci 170:781–787

Di Toppi LS, Gabrrielli R (1999) Response to cadmium in higher plants. Environ Exp Bot 41:105–130

Diklić N (1974) Teucrium. In: Jakovljević S (ed) Flore de la Republique Socialiste de Serbie VI, 1st edn. Serbian Academy of Sciences and Arts, Belgrade, pp 349–357

Dudić B, Rakić T, Sinzar J, Atanacković V, Stevanović B (2007) Differences of metal concentrations and morpho-anatomical adaptations between obligate and facultative serpentinophytes from Western Serbia. Arch Biol Sci 59:341–349

Ekim T (1982) *Teucrium* L. In: Davis PH (ed) Flora of Turkey and the East Aegean Islands VII. Edinburgh University Press, Edinburgh, pp 53–75

Endt DV, Kijne JW, Memelink J (2002) Transcription factors controlling plant secondary metabolism: what regulates the regulators? Phytochemistry 61:107–114

Evstatieva L, Hardalova R, Stoyanova K (2007) Medicinal plants in Bulgaria: diversity, legislation, conservation and trade. Phytol Balcan 13:415–427

Ewald J (2003) The calcareous riddle: why are there so many calciphilous species in the Central European flora. Folia Geobot 38:357–366

Farago ME (1994) Plants and chemical elements biochemistry, uptake, tolerance and toxicity. Wiley-VCH, Weinheim

Filippini R, Piovan A, Borsarini A, Caniato R (2010) Study of dynamic accumulation of secondary metabolites in three subspecies of *Hypericum perforatum*. Fitoterapia 81:115–119

Franco A, Rufo L, Zuluaga J, Fuente V (2013) Metal uptake and distribution in cultured seedlings of *Nerium oleander* L. (Apocynaceae) from Rio Tinto (Huelva, Spain). Biol Trace Elem Res 155:82–92

Gabbrielli R, Pandolfini T (1984) Effect of Mg^{+2} and Ca^{+2} on the response to nickel toxicity in a serpentine and nickel accumulating species. Physiol Plant 62:540–544

Golubović T, Blagojević B (2013) Concentration of heavy metals in medicinal plants in Serbia – potential health risk. Paper presented at the International science conference "Reporting for sustainability", Becici, Montenegro, 7–10 May 2013

Gonneau C, Genevois N, Frérot H, Sirguey C, Sterckeman T (2014) Variation of trace metal accumulation, major nutrient uptake and growth parameters and their correlations in 22 populations of *Noccaea caerulescens*. Plant Soil 384:271–287

Graham RD, Welch RM, Grunes DL, Cary EE, Norvel WA (1987) Effects of zinc deficiency on the accumulation of boron and other mineral nutrients in barley. Soil Sci Soc Am J 51:652–657

Greuter W, Burdet HM, Long G (1986) Med-checklist III, Conservatoire et Jardin Botaniques ville de Genève, Genève

Grubešić RJ, Vladimir-Knezevic S, Kremer D, Kalodera Z, Vukovic J (2007) Trichome micromorphology in *Teucrium* (Lamiaceae) species growing in Croatia. Biologia (Bratislava) 62:148–156

Grubešić R, Kremer D, Vladimir-Knežević S, Vuković Rodríguez J (2012) Analysis of polyphenols, phytosterols, and bitter principles in *Teucrium* L. species. Cent Eur J Biol 7:542–550

Haby VA, Russelle M, Skogley P, Earl O (1990) Soil testing for potassium, calcium and magnesium. In: Westerman RL (ed) Soil testing and plant analysis, 3rd edn. Soil Science Society of America, Madison, pp 181–228

Hamid N, Bukhari N, Jawaid F (2010) Physiological responses of *Phaseolus vulgaris* to different lead concentrations. Pak J Bot 42:239–246

Harborne JB, Tomás-Barberán FA, Williams CA, Gil MI (1986) A study of flavonoids from European *Teucrium* species. Phytochemistry 25:2811–2816

Hossain MR (2007) Flora of some heavy hyperaccumulators recorded in Penjwin and Mawat in Iraqi Kurdistan. Dissertation, Sulaimani University Iraq

Ibrahim MH, Kong YC, Mohd Zain NA (2017) Effect of cadmium and copper exposure on growth, secondary metabolites and antioxidant activity in the medicinal plant sambung nyawa (*Gynura procumbens* (Lour.) Merr). Molecules 22:1–16

Ivancheva S, Stancheva B (2001) Ethnobotany in Bulgaria. In: Ozhatay N (ed) Proceedings of the 2nd Balkan Botanical Congress, Istanbul

Jain A, Ranade R, Pritam P, Joshi N, Vavilala SL, Jain A (2014) A comparative study of antioxidant activity, total phenolic and flavonoid contents in different parts of *Helicteres isora* L. AJLS 2:292–302

Jenny H (1980) The soil resource: origin and behaviour. Springer, New York

Johnston WR, Proctor J (1977) Metal concentrations in plants and soils from two British serpentine sites. Plant Soil 46:275–278

Juranović Cindrić I, Zeiner M, Glamuzina E, Stingeder G (2013) Elemental characterisation of the medical herbs *Salvia officinalis* L. and *Teucrium montanum* L. grown in Croatia. Microchem J 107:185–189

Jurišić R, Kalodera Z, Grgić J, Grgić Z, Ćavar S (2001) Determination of copper, cobalt and nickel in *Teucrium* species growing in Croatia. Acta Pharma 51:75–80

Kabata-Pendias A (2011) Trace elements in soils and plants, 4th edn. Taylor & Francis Group, London

Kabata-Pendias A, Pendias H (1984) Trace elements in soils and plants. Taylor & Francis Group, London

Karataglis S, Babalonas D, Kabasakalis B (1982) The ecology of plant populations growing on serpentine soils. II Ca/Mg ratio and the Cr, Fe, Ni, Co concentrations as development factors of *Buxus sempervirens* L. Phyton (Austria) 22:317–327

Karley AJ, White PJ (2009) Moving cationic minerals to edible tissues: potassium, magnesium, calcium. Curr Opin Plant Biol 12:291–298

Kashem MA, Singh BR (2002) The effect of fertilizer additions on the solubility and plant-availability of Cd, Ni and Zn in soil. Nutr Cycl Agroecosyst 62:287–296

Kassim J, Rahim B (2014) Some heavy metals content in plants grown on serpentinitic soil from Penjwin and Mawat area at Kurdistan region – Iraq. Int J Agric Sci Nat Res 1:31–39

Kastori R (1993) Heavy metals and pesticides in soil – heavy metals and pesticides in the soil of Vojvodina (in Serbian). Faculty of Agriculture, Novi Sad

Kataeva MN, Alexeeva-Popova NV, Drozdova IV, Beljaeva AI (2004) Chemical composition of soils and plant species in the Polar Urals as influence by rock type. Geoderma 122:257–268

Kay K, Ward K, Watt L, Schrmske D (2011) Plant speciation. In: Harisson S, Rajakaruna N (eds) Serpentine: the evolution and ecology of a model system. University of California Press, Berkely/Los Angeles/London, pp 71–95

Kazakou E, Dimitrakopoulos PG, Baker AJM, Reeves RD, Troumbis AY (2008) Hypotheses, mechanisms and trade-off s of tolerance and adaptation to serpentine soils: from species to ecosystem level. Biol Rev 83:495–508

Kazakou E, Adamidis G, Baker A, Reeves R, Gogino M, Dimitrakopoulos P (2010) Species adaptation in serpentine soils in Lesbos Island (Greece): metal hyperaccumulation and tolerance. Plant Soil 332:369–385

Kliebenstein DJ, Osbourn A (2012) Making new molecules – evolution of pathways for novel metabolites in plants. Curr Opin Plant Biol 15:415–423

Konstantinou M, Babalonas D (1996) Metal uptake by Caryophyllaceae species from metalliferous soils in northern Greece. Plant Syst Evol 203:1–10

Korkina LG (2007) Phenylpropanoids as naturally occurring antioxidants: from plant defense to human health. Cell Mol Biol 53:15–25
Krause W (1958) Andere Bodenspezialisten. In: Michael G (ed) Handbuch der Pflanzenphysiologie, vol 4. Springer, Berlin, pp 758–806
Kruckeberg AR (1969) The implications of ecology for plant systematics. Taxon 18:92–120
Kruckeberg AR (1984) California serpentines: flora, vegetation, geology, soils, and management problems. University of California Press, Berkeley
Kruckeberg AR (1992) Plant life of western North American ultramafics. In: Roberts B, Proctor J (eds) The ecology of areas with serpentinized rocks: a world view. Kluwer Academic Publishers, Dordrecht, pp 31–73
Kukier U, Peters CA, Chaney RL, Angle JS, Roseberg RJ (2004) The effect of pH on metal accumulation in two *Alyssum* species. J Environ Qual 33:2090–2102
Lambers H, Stuart Chapin F III, Pons TL (2008) Plant physiological ecology, 2nd edn. Springer, New York
Lavid N, Schwartz A, Yarden O, Tel-Or E (2001) The involvement of polyphenols and peroxidase activities in heavy metal accumulation by epidermal glands of the waterlily (Nymphaeaceae). Planta Med 212:323–331
Lieth H, Markert B (1988) The establishment of element concentration cadasters for ecosystems (ECCE) in the different vegetation zones of the earth. Biol Int 16:7–11
Lombini A, Dinelli E, Ferrari C, Simoni A (1998) Plant-soil relationships in the serpentinite screes of Mt Prinzera (Northern Apennines, Italy). J Geochem Explor 64:19–33
Mahajan S, Tuteja N (2005) Cold, salinity and drought stresses: an overview. Arch Biochem Biophys 444:139–158
Maleš Ž, Pilepić KH, Bojić M, Tatalović Z (2015) Determination of the content of total polyphenols, non-tannin polyphenols and tannins in five species of the genus *Teucrium* L. Period Biol 117:453–455
Mandal MS, Chakraborty D, Dey S (2010) Phenolic acids act as signaling molecules in plant-microbe symbioses. Plant Signal Behav 5:359–368
Markova M (1992) *Teucrium* L. In: Kozuharov S (ed) Opredelitel na vishite rastenia v Balgaria (in Bulgarian). Nauka i Izkustvo, Sofia, pp 491–492
Marschner H (1995) Mineral nutrition of higher plants, 2nd edn. Academic, Waltham
Mayer MS, Soltis PS (1994) The evolution of endemics: a chloroplasts DNA phylogeny of the *Streptanthus glandulosus* complex (Cruciferae). Syst Bot 19:537–574
Mengel K, Kirkby EA (1987) Principles of plant nutrition, 4th edn. International Potash Institute, Worblaufen-Bern
Michalak A (2006) Phenolic compounds and their antioxidant activity in plants growing under heavy metal stress. Pol J Environ Stud 15:523–530
Milošević-Djordjević O, Stošić I, Stanković M, Grujičić D (2013) Comparative study of genotoxicity and antimutagenicity of methanolic extracts from *Teucrium chamaedrys* and *Teucrium montanum* in human lymphocytes using micronucleus assay. Cytotechnology 65:863–869
Nasim SA, Dhir B (2010) Heavy metals alter the potency of medicinal plants. Rev Environ Contam Toxicol 203:139–149
Navarro T (2010) *Teucrium* L. In: Castroviejo S et al (eds) Flora Iberica, VerbenaceaeLabiataeCallitrichaceae, vol 7. Real Jardín Botánico, CSIC, Madrid, pp 30–166
Navarro T, El Oualidi J (2000) Synopsis of *Teucrium* L. (Labiatae) in the Medditerranean region and surrounding areas. Flora Medit 10:349–363
Nikolov S (2006) Encyclopedia of medicinal plants in Bulgaria (in Bulgarian). Publishing House Trud, Sofia
O'Dell RE, Rajakaruna N (2011) Intraspecific variation, adaptation, and evolution. In: Harrison S, Rajakaruna N (eds) Serpentine: evolution and ecology in a model system. University of California Press, Berkeley, pp 97–137
Obratov-Petković D, Popović I, Belanović S, Kadović R (2006) Ecobiological study of medicinal plants in some regions of Serbia. Plant Soil Environ 52:459–467

Obratov-Petković D, Bjedov I, Belanović S (2008) The relationship between heavy metal contents and bedrock in some species of genus *Teucrium* L. in Serbia. In: Ruzichkova G (ed) Proceedings of the 5th conference on medicinal and aromatic plants of Southeast European countries, Brno

Pacifico S, D'Abrosca B, Pascarella MT, Letizia M, Uzzo P, Piscopo V, Fiorentino A (2009) Antioxidant efficacy of iridoid and phenylethanoid glycosides from the medicinal plant *Teucrium chamaedris* in cell-free systems. Bioorg Med Chem 17:6173–6179

Pandy P, Tripathi K (2010) Bioaccumulation of heavy metal in soil and different plant parts of *Albizia procera* (Roxb.) seedling. Bioscan 5:263–266

Pavlova D (2009) Morphological variation in *Teucrium chamaedrys* from serpentine and non-serpentine populations, soil and biota of serpentine: a world view. Northeast Nat 16:39–55

Pavlova D (2010) A survey of the serpentine flora in the West frontier Bulgarian Mountains (Vlahina and Ogražden). Phytol Balcan 16:97–107

Pavlova D, Alexandrov S (2003) Metal uptake in some plants growing on serpentine areas in the Eastern Rhodopes Mountains (Bulgaria). Ot Sist Bot Dirgisi 10:13–31

Pavlova D, Karadjova I (2012) Chemical analysis of *Teucrium* species (Lamiaceae) growing on serpentine soils in Bulgaria. J Plant Nutr Soil Sci 175:891–899

Pavlova D, Karadjova I (2013) Toxic elements profiles in selected medicinal plants growing on serpentines in Bulgaria. Biol Trace Elem Res 156:288–297

Pavlova D, Kozuharova E, Dimitrov D (2003) A floristic catalogue of the serpentine areas in the Eastern Rhodope Mountains (Bulgaria). Polish Bot J 48:21–41

Peev D (1989) *Teucrium* L. In: Velchev V, Kuzmanov B (eds) Flora Reipublicae Popularis Bulgaricae, vol 9. Bulgarian Academy of Sciences, Sofia, pp 241–249

Pichi-Sermolli R (1948) Flore e vegetazione delle serpentine e delle alti ofioliti dell'alta valle del Tevere (Toscana). Webbia 6(1):380

Politycka B, Adamska D (2003) Release of phenolic compounds from apple residues decomposing in soil and the influence of temperature on their degradation. Pol J Environ Stud 12:95–98

Pollard AJ, Powell KD, Harper FA, Smith JAC (2002) The genetic basis of metal hyperaccumulation in plants. Crit Rev Plant Sci 21:539–566

Proctor J (1999) Toxins, nutrient shortages and droughts: the serpentine challenge. Trees 14:334–335

Proctor J, Nagy L (1992) Ultramafic rocks and their vegetation: an overview. In: Baker AJM, Proctor J, Reeves RD (eds) The vegetation of ultramafic (serpentine) soils. Intercept Limited, Andover, pp 469–494

Proctor J, Woodell SRJ (1975) The ecology of serpentine soils. Adv Ecol Res 9:256–366

Quideau S, Deffieux D, Douat-Casassus C, Pouysegu L (2011) Plant polyphenols: chemical properties, biological activities, and synthesis. Angew Chem Int Ed 50:586–621

Rajaharuna N, Bohm B (1999) The edaphic factor and patterns of variation in *Lasthenia californica* (Asteraceae). Am J Bot 86:1576–1596

Rajakaruna N (2004) The edaphic factor in the origin of plant species. Int Geol Rev 46:471–478

Rajakaruna N, Boyd R (2008) Edaphic factor. In: Jørgensen SE, Fath BD (eds) General ecology. Encyclopedia of ecology, vol 2. Elsevier, Amsterdam, pp 1201–1207

Ramakrishna A, Ravishankar GA (2011) Influence of biotic stress signaling on secondary metabolites in plants. Plant Signal Behav 6:1720–1731

Rašić S, Dogo S, Slanković L (2006) Inorganic analysis of herbal drugs. Part II. Plant and soil analysis – diverse bioavailability and uptake of essential and toxic elements. J Serb Chem Soc 71:1095–1105

Rastgoo L, Alemzadeh A (2011) Biochemical responses of gouan (*Aeluropus littoralis*) to heavy metals stress. Aust J Crop Sci 5:375–383

Redzić S (2010) Wild medicinal plants and their usage in traditional human therapy (Southern Bosnia and Herzegovina, W. Balkan). J Med Plant Res 4:1003–1027

Reeves R (2006) Hyperaccumulation of trace elements by plants. In: Morel JL, Echevarria G, Goncharova N (eds) Phytoremediation of metal-contaminated soils, NATO science series, vol 68. Springer, Dordrecht

Reeves RD (2017) A global database for plants that hyperaccumulate metal and metalloid trace elements. New Phytol 218:407–411
Reeves RD, Baker AJM, Borhidi A, Berazain R (1999) Nickel hyperaccumulation in the serpentine flora of Cuba. Ann Bot 83:29–38
Reeves RD, Baker AJM, Becquer T, Echevarria G, Miranda ZJG (2007) The flora and biogeochemistry of the ultramafic soils of Goiás state. Brazil. Plant Soil 293:107–119
Reichinger HK (1941) Monographische stiudie über *Teucrium* Sect. *Chamaedrys*. Bot Arch 42:335–420
Ritter-Studnička H (1968) Die serpentinomorphosen der flora Bosniens. Bot Jahrb 88:443–465
Roberts BA, Proctor J (1992) The ecology of areas with serpentinized rocks: a world view. Kluwer Academic Publishers, Dordrecht
Rusak G, Gutzeit H, Ludwig-Müller J (2005) Structurally related flavonoids with antioxidative properties differentially affect cell cycle progression and apoptosis of human acute leukemia cells. Nutr Res 25:143–155
Salmerón-Sarnchez E, Fuertes-Aguilar J, Španiel S, Perrez-García FJ, Merlo E, Garrido-Becerra JA, Mota J (2018) Plant evolution in alkaline magnesium-rich soils: a phylogenetic study of the Mediterranean genus *Hormathophylla* (Cruciferae: Alysseae) based on nuclear and plastid sequences. PLoS One 13(12):e0208307. https://doi.org/10.1371/journal.pone.0208307
Sawidis T, Halley J, Llupo S, Bellos D, Veros D, Symeonidis L (2014) Nickel and iron concentrations in plants from mining area Pogradec, Albania. Environ Eng Manag J 13:861–871
Schützendübel A, Polle A (2002) Plant responses to abiotic stresses: heavy metal-induced oxidative stress and protection by mycorrhization. J Exp Bot 53:1351–1365
Sen Gupta A, Berkowitz GA, Pier PA (1989) Maintenance of photosynthesis at low leaf water potential in wheat. Plant Physiol 89:1358–1365
Shallari S, Schwartz C, Hasko A, Morel J-L (1998) Heavy metal in soils and plants of serpentine and industrial sites of Albania. Sci Total Environ 209:133–142
Sharma P, Dubey R (2006) Cadmium uptake and its toxicity in higher plants. In: Khan NA, Samiullah (eds) Cadmium toxicity and tolerance in plants. Alpha Science International Ltd, Oxford, pp 63–86
Shaul O (2002) Magnesium transport and function in plants: the tip of the iceberg. Biometals 15:309–323
Sigel H, Sigel A (1990) Compendium on magnesium and its role in biology, nutrition, and physiology, 1st edn. CRC Press, New York
Skinner WM, Martin RR, Naftel SJ, Macfie S, Séquin CF (2005) Multi-technique studies of the distribution of metals between the soil, rhizosphere and roots of *Populus tremuloides* growing in forest soil. In: Abstract of the 8th international conference on the biogeochemistry of trace elements, Adelaide
Stanković M, Zlatić N (2019) Ethnobotany of *Teucrium* species. In: Martinez JL, Muñoz-Acevedo A, Rai M (eds) Ethnobotany: local knowledge and tradition. CRC Press, Taylor & Francis Group Ltd, Oxford, pp 214–231
Stanković M, Topuzović M, Solujić S, Mihajlović V (2010) Antioxidant activity and concentration of phenols and flavonoids in the whole plant and plant parts of *Teucrium chamaerdys* L. var. *glanduliferum* Haussk. J Med Plant Res 4:2092–2098
Stanković M, Curcić M, Zizić J, Topuzović M, Solujić S, Marković S (2011a) *Teucrium* plant species as natural sources of novel anticancer compounds: antiproliferative, proapoptotic and antioxidant properties. Int J Mol Sci 12:4190–4205
Stanković M, Nicifirović N, Topuzović M, Solujić S (2011b) Total phenolic content, flavonoid concentrations and antioxidant activity of the whole plant and plant parts extracts from *Teucrium montanum* L. var. *montanum*, f. *supinum* (L.) Reichenb. Biotechnol Biotechnol Equip 25:2222–2227
Stanković M, Stefanović O, Čomić L, Topuzović M, Radojević I, Solujić S (2012) Antimicrobial activity, total phenolic content and flavonoid concentrations of *Teucrium* species. Cent Eur J Biol 7:664–671

Stanković SM, Petrović M, Godjevać D, Dajić-Stevanović Z (2015) Screening inland halophytes from the Central Balkan for their antioxidant activity in relation to total phenolic compounds and flavonoids: are there any prospective medicinal plants? J Arid Environ 120:26–32

Štepankova J (1996) Karyological variation in the group of *Myosotis alpestris* (Boraginaceae). Folia Geobot Phytotaxon 31:251–262

Štepankova J (1997) The effect of serpentine on morphological variation in the *Galium pumilum* group (Rubiaceae). Thaiszia – J Bot 7:29–40

Stevanović B, Stevanović V (1985) Morpho-anatomical characteristics of the species *Teucrium montanum* L. from different habitats. Bot Serb 9:73–88

Treutter D (2006) Significance of flavonoids in plant resistance: a review. Environ Chem Lett 4:147–157

Tumbas TV, Mandić IA, Ćetković SG, Djilas MS, Čanadanović-Brunet MJ (2004) *HPLC* analysis of phenolic acids in Mountain germander (*Teucrium montanum* L.) extracts. Acta Period Technol 35:265–273

Tutin TG, Heywood VH, Burges NA, Moore DM, Valentine DH, Walters SM, Webb DA (eds) (1972) Flobra Europaea III. Cambridge University Press, Cambridge

Vega FA, Covelo EF, Vazques JJ, Abdrade L (2007) Influence of mineral and organic components on copper, lead and zinc sorption by acid soils. J Environ Sci Health A 42:2167–2173

Veličković MJ, Dimitrijević SD, Mitić SS, Mitić NM, Kostić AD (2014) The determination of the phenolic composition, antioxidative activity and heavy metals in the extracts of *Calendula officinalis* L. Adv Technol 3:46–51

Vergnano Gambi O (1992) The distribution and ecology of the vegetation of ultramafic soils in Italy. In: Roberts B, Proctor J (eds) The ecology of areas with serpentinized rocks. A world view. Kluwer Academic Publishers, Dordrecht, pp 217–247

Vergnano Gambi O, Gabbrielli R, Pancaro L (1982) Nickel, chromium, and cobalt in plants from Italian serpentine areas. Acta Oecol 3:291–306

Vicić D, Polavder S, Stojiljkovic M, Jurišic B, Bojat N (2013a) Content and allocation of nickel, chromium, cobalt, copper and zinc in *Teucrium montanum* L. from serpentine habitats in Serbia. Acta Agric Serb 18:101–110

Vicić D, Stoiljković M, Polavder S (2013b) Tissue Mg:Ca ratio in *Teucrium montanum* L. from serpentine soils of Serbia. Soil Plant 62:31–38

Vlase L, Benedec D, Hanganu D, Damian G, Csillag I (2014) Evaluation of antioxidant and antimicrobial activities and phenolic profile for *Hyssopus officinalis*, *Ocimum basilicum* and *Teucrium chamaedrys*. Molecules 19:5490–5507

Vuković N, Milošević T, Sukdolak S, Solujić S (2008) The chemical composition of the essential oil and the antibacterial activities of the essential oil and methanol extract of *Teucrium montanum*. J Serb Chem Soc 73:299–305

Wallace A, Jones M, Alexander GV (1982) Mineral composition of native wood plants growing on a serpentine soil in California. Soil Sci 134:42–44

Westerbergh A (1994) Serpentine and non-serpentine *Silene dioica* plants do not differ in nickel tolerance. Plant Soil 167:297–303

Westerbergh A, Rune O (1996) Genetic relationship among *Silene dioica* (Caryophyllaceae) populations on and off serpentine: a review. Symb Bot Ups 46:277–284

Westerbergh A, Saura A (1992) The effect of serpentine on the population structure of *Silene dioica* (Caryophyllaceae). Evolution 46:1537–1548

Whitea PA, Claxtonb LD (2004) Mutagens in contaminated soil: a review. Mutat Res 567:227–345

Wright J, Stanton M, Scherson R (2006) Local adaptation to serpentine and non-serpentine soils in *Collinsia sparsiflora*. Evol Ecol Res 8:1–21

Yaman M (2014) *Teucrium* as a novel discovered hyperaccumulator for the phytoextraction of Ni-contaminated soils. Ekol Derg 23:81–89

Yan X, Zhang F, Zeng C, Zhang M, Devcota PL, Yao T (2012) Relationship between heavy metal concentrations in soils and grasses of roadside farmland in Nepal. Int J Environ Res Public Health 9:3209–3226

Zdraveva P, Pavlova D, Krasteva I, Pencheva I (2018) Phytochemical analysis on populations of *Teucrium chamaedrys* from serpentine sites in Bulgaria. CR Acad Bulg Sci 71:185–192
Zhang G, Wu F, Wei K, Dong Q, Dai F, Chen F, Yang J (2006) Cadmium stress in higher plants. In: Samiullah N (ed) Cadmium toxicity and tolerance in plants. Alpha Science International, Oxford, pp 87–103
Zlatić N, Stanković M, Simić Z (2017) Secondary metabolites and metal content dynamics in *Teucrium montanum* L. and *Teucrium chamaedrys* L. from habitats with serpentine and calcareous substrate. Environ Monit Assess 189:110
Zonn SV (1982) Iron in soils. Publishing House, Nauka

Chapter 5
Ethnobotanical Features of *Teucrium* Species

Snežana Jarić, Miroslava Mitrović, and Pavle Pavlović

Abstract Species of the *Teucrium* genus have been used in ethnopharmacology for centuries, helping to treat many pathophysiological conditions, such as diabetes, gastrointestinal disorders, rheumatism, inflammations, and tuberculosis. They can also be used as a diuretic, antipyretic, tonic, diaphoretic, analgesic and antihyperlipidemic. This study provides a comprehensive analysis of the traditional use of *Teucrium* species based on research conducted up to now in different parts of the world. With this aim, 72 ethnobotanical studies were analyzed and a total of 20 *Teucrium* species were identified for this analysis. Nineteen of these species are used in human ethnomedicine, 3 in veterinary ethnomedicine and 4 for other purposes. *Teucrium chamaedrys*, *T. polium* and *T. montanum* are mentioned in the largest number of studies and have the widest range of uses. The aerial parts of the plant are those that are most frequently used, with infusions (77.8%) and decoctions (48.6%) the most common methods of preparation. The use of *Teucrium* species for treating abdominal problems is mentioned in 56 of the studies and in 21 studies for treating disorders heart and blood vessels. Treating respiratory problems is referenced in 17 studies and problems in the functioning of the endocrine glands in 16 studies. The medicinal properties of *Teucrium* species can be ascribed to their chemical composition, specifically essential oils, phenolic acids, flavonoids, and other secondary metabolites. However, the long-term use of some of the preparations can have negative side effects, such as hepatotoxicity or gradual deterioration in neuromuscular coordination.

Keywords Ethnobotanical research · *Teucrium* species · *T. chamaedrys* · *T. polium* · *T. montanum*

S. Jarić (✉) · M. Mitrović · P. Pavlović
Department of Ecology, Institute for Biological Research "Siniša Stanković" - National Institute of the Republic of Serbia, University of Belgrade, Belgrade, Serbia
e-mail: nena2000@ibiss.bg.ac.rs

M. Stanković (ed.), *Teucrium Species: Biology and Applications*,
https://doi.org/10.1007/978-3-030-52159-2_5

5.1 Introduction

Medicinal plants and products made from them have been used in ethnomedicine all across the world since time immemorial. Numerous plant species are central to beliefs, methods and institutions for diagnosing and treating diseases and preventing them. Hence, botanical folk knowledge plays a crucial role in the ethnomedicine of every civilisation (Nedelcheva and Draganov 2014). Indeed, the most useful starting point for developing new pharmaceuticals is provided by the array of botanical sources, particularly metabolites and their biological activities, in addition to the various practices developed by ethnic groups over the centuries in terms of the preparation and application of herbal remedies (Jarić et al. 2007). According to estimations provided by Uritu et al. (2018), up to 70,000 plant species are used for ethnomedicine worldwide, while according to data from the World Health Organization (WHO), approximately 80% of the world's population still relies mainly on plant-based drugs (Bahmani et al. 2014). The reasons for the wide use of medicinal plants in ethnomedicine across the world include their relative ease of availability and the fact that they are quite a cost effective alternative when it comes to daily health care and self-medication; in light of this, many people, particularly those in poorer rural areas or when economic crisis hits, depend on them as home remedies for health problems (Leonti 2011). Their therapeutic value is based primarily on the link between the chemical structure of the active substances they contain and their pharmacodynamic effects on the body (Rafieian-Kopaei 2012; Jarić et al. 2014, 2015a).

Medicinal and aromatic plants come from a whole variety of plant families that often have common active ingredients (due to their biosynthetic pathways being similar). One, highly significant example is the plant family Labiatae (Lamiaceae), which includes many species that contain essential oils (Máthé 2015; Jarić et al. 2015b) and, as such, have biological and medical applications. Their high concentrations of biologically active substances mean they contribute significantly to both traditional and modern medicine (phytotherapy). This family includes aromatic herbs such as thyme, mint, oregano, basil, sage, savory, rosemary, self-heal, hyssop, and lemon balm, while other species have more limited uses (Bekut et al. 2018).

This chapter provides a comprehensive analysis of the traditional uses of *Teucrium* species from the ethnobotanical and ethnomedicine points of view, based on ethnobotanical research undertaken in different parts of the world. With this aim, 72 ethnobotanical studies were analysed and 20 species of the *Teucrium* genus were identified for the purposes of this analysis.

5.2 Botanical Characterization and Distribution of *Teucrium* Species

Teucrium L. (germander) is a large, polymorphic, cosmopolitan genus, belonging to the family Lamiaceae, within the subfamily Ajugoideae. This family has a cosmopolitan distribution with about 236 genera and contains between 6900 (Heywood et al. 2007) and 7200 species (Harley et al. 2004), although the World Checklist lists 7534 (https://wcsp.science.kew.org/qsearch.do). The genus *Teucrium* comprises more than 300 species, which are widely found in Europe, North Africa, and temperate parts of Asia, but 96% are distributed in the Mediterranean region (Tutin et al. 1972; Meusel et al. 1978; Navarro and El Oualidi 2000). About 195 taxa are present at the level of species and subspecies, 83 of which are on the Pyrenean peninsula, 72 in the northwestern part of Northern Africa (Algeria and Morocco), 61 in Asia Minor, 41 on the Balkan Peninsula, and 23 on the Apennine Peninsula (Greuter et al. 1986). In terms of European flora, the genus *Teucrium* has been divided into seven sections with 49 species (Tutin et al. 1972), while a few are spread throughout South America, mountainous tropical Northeast and South Africa, and Australia (Wielgorskaya 1995).

Species of the genus *Teucrium* are mostly perennial herbs, shrubs or subshrubs, and more rarely, annual herbaceous plants (e.g. *T. botrys*) (Tutin et al. 1972). Research into micromorphological characteristics, especially trichomes and pollen, nutlet morphology and vegetative anatomy has shown that these have taxonomic value for sectional and interspecific classification in this genus (Oybak-Dönmez and Inceoğlu 1988; Navarro and El Oualidi 2000; Jurišić Grubešić et al. 2007; Dinç et al. 2008, 2009; Eshratifar et al. 2011; Lakušić and Lakušić 2014).

5.3 Ethnobotanical and Ethnomedicinal Research into *Teucrium* Species

5.3.1 *A Historical Review of the Use of* Teucrium *Species in Ethnomedicine*

For over two millennia, *Teucrium* species have been used as medicinal herbs, with many still used in folk medicine today (Milošević-Djordjević et al. 2013). Used to treat coughs and asthma since ancient Greek times (Menichini et al. 2009), there are many other records of their use in ethnomedicine. Pharmaceutical manuscripts from Persia, dating from the period between the ninth and eighteenth centuries AD, note the use of the aerial part of *T. montanum* as a decoction to treat headaches (Zarshenas et al. 2013). Moreover, in medieval written documents, i.e. in lists of *materia medica* used by the Jewish community of Medieval (eleventh to fourteenth centuries) Cairo, found at the Taylor–Schechter Genizah collection, Cambridge, the species

Teucrium capitatum L. is mentioned twice, with its leaves and stems used in the treatment of eye diseases, stomach ailments and colic (Lev 2007). Similarly, the botanical *materia medica* of the *Iatrosophikon* – A collection of prescriptions from a monastery in Cyprus, which dates back to the Ottoman period (1571–1878), includes *Teucrium micropodioides* Rouy as an exotic taxon originating from the New World, used to treat catarrh and the common cold (Lardos 2006). In their work on the use of medicinal herbs for the treatment of rheumatic disorders in the sixteenth and seventeenth centuries, Adams et al. (2009) stated that important European herbals (Fuchs 1543; Bock 1577; Matthiolus 1590; Tabernaemontanus 1687) mention the ethnomedicinal use of *Teucrium chamaedrys* L. Namely, its aerial parts were boiled in wine and this concoction was taken for 60 days, 3 h before meals on an empty stomach. Any patient seeking to use this as a remedy for gout of the feet or hip pains also had to avoid sour and salty foods.

5.3.2 *The Use of* Teucrium *Species in Human Ethnomedicine*

Teucrium species are generally aromatic, ornamental plants and also a valuable source of pollen, while many are used in folk medicine and pharmacy. Numerous phytochemical and pharmacological studies have confirmed their medicinal and biological properties, such as hypoglycemic (Gharaibeh et al. 1988; Baluchnejadmojarad et al. 2005), hypolipidemic (Rasekh et al. 2001), hepatoprotective (Shtukmaster et al. 2010), antipyretic (Autore et al. 1984), anti-inflammatory and antiulcerogenic (Sghaier et al. 2011a), anticarcinogenic (Sghaier et al. 2011b; Stanković et al. 2011a), and antimicrobial activities (Vuković et al. 2007). These species are very rich in phenolic compounds with very strong biological activity (Yin et al. 2009; Hasani-Ranjbar et al. 2010).

Knowledge on the use of *Teucrium* species, which are used by locals in various geographical regions, has been well documented in numerous ethnobotanical studies. In this chapter, 72 ethnobotanical studies from across the world were analysed and 20 species of the genus *Teucrium* were identified for their medicinal properties. Nineteen species are used in human ethnomedicine (Table 5.1) and 3 in veterinary ethnomedicine (Table 5.2), while 4 are also used for other purposes (Table 5.3). It should be noted that some of them have multiple uses, while *Teucrium scorodonia* is only used in veterinary medicine.

Our analysis of these studies revealed that the people of Turkey, Iran, Bosnia and Herzegovina, Italy, Pakistan, Serbia and Palestine know most about *Teucrium* species and use them most frequently (Fig. 5.1). It was also determined that the species *Teucrium chamaedrys* was the most frequently mentioned of all the species of the genus *Teucrium*. Details on its use were found in 34 (47.2%) ethnobotanical studies; 13 (18.15%) of these were conducted in different parts of Turkey, 5 (7%) in Italy and the same number in Bosnia and Herzegovina. The species *Teucrium polium* is used in 23 regions of the world (31.9%), but is most familiar to and most frequently used by the people of Iran (6 studies, 26.1%) and Turkey (5 studies, 21.8%).

Table 5.1 The use of *Teucrium* species in traditional medicine

Teucrium species	Locality	Part used	Forms of preparation and method of administration	Traditional uses	References
T. arduini	Bosnia and Herzegovina	Aerial part	**I:** Infusion	Stomach diseases	Redžić (2007)
T. arduini	Bosnia and Herzegovina (Central, Southern and Western)	Aerial part	**I:** Infusion	Gastrointestinal ailments	Šarić-Kundalić et al. (2010a)
T. arduini	Bosnia and Herzegovina (Eastern, Northern and Northeastern)	Aerial part	**I:** Infusion	Diarrhoea	Šarić-Kundalić et al. (2011)
T. capitatum	Palestine (West Bank)	Leaf	**E:** Decoction (about 50 g leaves are boiled in 100 ml water and the affected area is bathed with this decoction once a day)	Psoriasis	Shawahna and Jaradat (2017)
T. capitatum subsp. *capitatum*	Lebanon (Mount Hermon)	Aerial part	**I:** Infusion	Diabetes, insomnia and neurological disorders, abdominal cramps	Baydoun et al. (2015)
T. chamaedrys	Turkey (Nigde, Aladaglar)	Aerial part	**I:** Infusion	Abdominal pain, haemorrhoids, diabetes, a painkiller, stomach pains	Özdemir and Alpınar (2015)
T. chamaedrys	Turkey (Afyonkarahisar, Inner-West Anatolia)	Leaf, flower	**I:** Infusion	A painkiller, stomach problems and haemorrhoids	Ari et al. (2015)
T. chamaedrys	Turkey (East Anatolia)	Herb/aerial part	**I:** Infusion, decoction	Toothache, kidney pains, a stomachic, indigestion, heart disease	Altundag and Ozturk (2011)
T. chamaedrys	Turkey (Edremit Gulf, Balıkesir Province)	Flowering branches	**I:** Infusion (drink one cupful twice a day for a week)	Abdominal pains, kidney stones	Polat and Satil (2012)
T. chamaedrys	Turkey (Bozyazı district of Mersin)	Aerial part	**I:** Decoction (gargle 1 glassful and spit, once a day for 3–5 days)	Mouth sores	Sargin (2015)

(continued)

Table 5.1 (continued)

Teucrium species	Locality	Part used	Forms of preparation and method of administration	Traditional uses	References
T. chamaedrys	Turkey (Hatay Province, Antakya)	Aerial part	**I:** Decoction	An antidiabetic	Güzel et al. (2015)
T. chamaedrys	Turkey (Central Anatolia)	Herb/aerial part	**I:** Infusion (as tea)	Stomachaches	Sezik et al. (2001)
T. chamaedrys	Italy (Rotonda, Pollino National Park)	Flowering tops	**I:** Decoction	A febrifuge	Di Sanzo et al. (2013)
T. chamaedrys	Italy (Mundimitar/ Montemitro)	Aerial part	**I:** Decoction	An antimalarial (in the past), hypertension	di Tizio et al. (2012)
T. chamaedrys	Italy (Dolomiti Lucane)	Aerial part	**I:** Decoction	An antimalarial	Pieroni et al. (2004)
T. chamaedrys	Italy (North-western Ligurian Alps)	Leaf	**I:** Decoction	A hypotensive, abdominal pain	Cornara et al. (2014)
T. chamaedrys	Italy (Lucania region)	Aerial part	–	An antimalarial	Pieroni and Quave (2005)
T. chamaedrys	Serbia (Kosovo, Albanian Alps)	Aerial part	**I:** Infusion	An antihaemorrhoidal	Mustafa et al. (2012)
		Whole plant		An antidiabetic	
T. chamaedrys	Serbia (Kopaonik)	Aerial part	**I:** Infusion (tea)	Gastrointestinal ailments	Jarić et al. (2007)
T. chamaedrys	Serbia (Suva planina)	Aerial part	**I:** Infusion (tea)	A digestive, stomach problems	Jarić et al. (2015a)
T. chamaedrys	Serbia (Zlatibor)	Leaf	**I:** Infusion	Digestive complaints, diarrhoea	Šavikin et al. (2013)
T. chamaedrys	Bosnia and Herzegovina (Central, Southern and Western)	Aerial part	**I:** Infusion (tea)	Spasms (a mixture of *Agrimonia* sp., *Frangula* sp., *Melissa* sp., *Mentha* sp., and *T. montanum*), diarrhoea (a mixture of *Agrimonia* sp., *Matricaria* sp., *Rosa* sp., and *Tilia* sp.)	Šarić-Kundalić et al. (2010a)

T. chamaedrys	Bosnia and Herzegovina (Eastern, Northern and North-eastern)	Aerial part	**I:** Infusion (tea)	Anaemia, digestive ailments	Šarić-Kundalić et al. (2011)
T. chamaedrys	Bosnia and Herzegovina (Pristine Village of Prokoško Lake on Mt. Vranica)	Aerial part	**I:** Infusion (tea)	Spasms	Šarić-Kundalić et al. (2010b)
T. chamaedrys	Bosnia and Herzegovina (Mt. Javor)	Aerial part	**I:** Infusion	Digestive complaints, diarrhoea	Savić et al. (2019)
T. chamaedrys	Bosnia and Herzegovina	Aerial part	**I:** Fresh juice, infusion	Diarrhoea	Redžić (2007)
T. chamaedrys	Spain (Balearic Islands, Eastern Mallorca, Mediterranean Sea)	Flowering tops	**I:** Drops	Earache	Carrió and Vallès (2012)
T. chamaedrys	Macedonia (Southeastern)	Leaf	**E:** Compress from fresh leaves on the neck area	An aphrodisiac (anti-aphrodisiac), 'anti sex crusader'	Nedelcheva et al. (2017)
		Aerial part	**I:** Infusion (tea)	Childlessness	
T. chamaedrys	Greece (Epirus, Zagori)	–	–	A haemostatic, haemorrhoids, a tonic and stimulant, rheumatism and arthritis	Vokou et al. (1993)
T. chamaedrys	Albania (Theth, a village in the Northern Albanian Alps)	Aerial part	**I:** Infusion *Aerial part dried and sold to the city markets.*	Stomachaches, fevers	Pieroni (2008)
T. chamaedrys	Israel (the Golan Heights and the West Bank region)	Leaf	**I:** Decoction (a standard decoction is prepared from 50 g leaves and stems and taken orally, 100 cc, three times a day)	Stomach and intestinal pain and inflammation, lack of appetite, jaundice	Said et al. (2002)
T. chamaedrys	Iran (North Khorasan Province)	Aerial part	–	An anti-inflammatory, an aperient, an astringent, a carminative, a diaphoretic, a digestive, a diuretic, a stimulant	Mashayekhan et al. (2015)

(continued)

Table 5.1 (continued)

Teucrium species	Locality	Part used	Forms of preparation and method of administration	Traditional uses	References
T. chamaedrys	Palestine (West Bank)	–	–	Digestive disorders	Ali-Shtayeh et al. (2000)
T. chamaedrys subsp. *chamaedrys*	Turkey (Honaz Mountain National Park, Middle Aegean Region of Turkey)	Aerial part	**I:** Decoction (use 2–3 cupfuls daily to relieve pain)	An analgesic	Kargıoğlu et al. (2010)
T. chamaedrys subsp. *chamaedrys*	Turkey (Ulukışla)	Aerial part	**E:** Decoction (bath; take a bath; take a sitz bath twice a day)	Haemorrhoids, itching	Paksoy et al. (2016)
T. chamaedrys subsp. *lydium*	Turkey (Bayramiç, Çanakkale)	Aerial part	**I:** Infusion	Eczema	Bulut and Tuzlacı (2015)
T. chamaedrys subsp. *sinuatum*	Turkey (Hakkari- Geçitli)	Aerial part	**I:** Decoction, infusion (drink one glassful of the tea from the plant twice a day)	Gastric pain, poisoning, rheumatism	Kaval et al. (2014)
T. chamaedrys subsp. *sinuatum*	Turkey (Bingöl, Solhan)	Aerial part	**I:** Infusion (drink one teacupful after meals)	Insomnia, colds and influenza, a sedative	Polat et al. (2013)
T. chamaedrys subsp. *sinuatum*	Turkey (Sivrice, Elazığ)	Aerial part	**E:** Infusion (compress)	An antispasmodic, haemorrhoids	Cakilcioglu and Turkoglu (2010)
T. chardonianum	Western Sahara (Moroccan occupied territories)	Flowering tops (dried and triturated)	**E:** Infusion	Mixed with water and applied to hair to perfume it and stimulate hair growth	Volpato et al. (2012)
T. cubense	Mexico (Nuevo León, Bustamante)	Whole plant	**I/E:** Infusion (as a tea to drink and for bathing)	Fevers	Estrada-Castillón et al. (2018)
		Leaf	**I:** Infusion (as a tea to be drunk for several weeks)	Kidney disorders	

T. divaricatum subsp. *canescens*	Cyprus	Aerial part (in flower)	**I/E:** Infusion (taken orally but also together with inhalation of the steam)	A stomachic, fevers and common colds	Arnold (1985)
			E: Infusion	Wound healing (cicatrisant)	
T. divaricatum subsp. *divaricatum*	Turkey (Mugla, Marmaris)	Aerial part	**I:** Infusion	Coughs, a tonic for the eyes, stomachaches, urinary diseases	Gürdal and Kültür (2013)
T. fruticans	Morocco (middle region of Oum *Rbai*)	Aerial part	**I:** Infusion	Influenza (in the treatment of respiratory diseases)	Fatiha et al. (2017)
T. fruticans	Spain	–	–	Cardiovascular problems, mental/nervous disorders	Gonzalez-Tejero et al. (2008)
T. lusitanicum	Spain (Granada province)	Aerial part	**I:** Decoction	Fevers	Benítez et al. (2010)
T. micropodioides	Cyprus (Paphos and Larnaca)	–	–	A digestive, mental/nervous disorders	Gonzalez-Tejero et al. (2008)
T. montanum	Bosnia and Herzegovina (Mt. Javor)	Aerial part	**I:** Infusion	Digestive complaints	Savić et al. (2019)
T. montanum	Bosnia and Herzegovina (Eastern, Northern and North-eastern)	Aerial part	**I:** –	Gastrointestinal ailments, digestive ailments, rheumatism, arthritis, hangovers, biliary tract purification, lung cancer	Šarić-Kundalić et al. (2011)
T. montanum	Bosnia and Herzegovina (Pristine Village of Prokoško Lake on Mt. Vranica)	Aerial part	**I:** Infusion/tea	Spasms, blood purification	Šarić-Kundalić et al. (2010b)
			E: Balm	Rheumatism	
T. montanum	Bosnia and Herzegovina (Lukomir)	–	**I:** –	Diabetes	Ferrier et al. (2014)
T. montanum	Bosnia and Herzegovina	Aerial part	**I:** Infusion	Liver and stomach diseases	Redžić (2007)

(continued)

Table 5.1 (continued)

Teucrium species	Locality	Part used	Forms of preparation and method of administration	Traditional uses	References
T. montanum	Serbia (Zlatibor)	Aerial part	**I:** Infusion	Digestive complaints	Šavikin et al. (2013)
T. montanum	Serbia (Mt. Suva planina)	Aerial part	**I:** Infusion	Disorders of the abdominal organs, 'male' headaches, a tonic (tea), improving appetite, an antipyretic ('bitter tea' – blend: mountain germander, wall germander, yarrow, sage, pennyroyal and oregano)	Jarić et al. (2015a)
			E: Infusion: 'bitter tea'	Tuberculosis – a bath soak: add yellow chamomile (*Anthemis tinctoria* L.) to the 'bitter tea' or set the plants mentioned above alight for the patient to inhale the 'smoke' **a combination of internal and external use is recommended*	
T. montanum	Serbia (Rtanj)	Aerial part	**I:** Infusion	Strengthening the immune system	Zlatković et al. (2014)
T. montanum	Croatia (Dubrovnik coast)	Leaf, flower	**I:** Infusion/tea, brandy	–	Dolina and Łuczaj (2014)
T. montanum	Montenegro (Mt. Prokletije)	Aerial part	**I:** Infusion/tea	Cirrhosis	Menković et al. (2011)
T. montanum	Iran = Persia (pharmaceutical manuscripts from ninth to eighteenth century AD)	Aerial part	**I:** Decoction	Headaches	Zarshenas et al. (2013)
T. oliverianum	Saudi Arabia (Al-Rass province)	Whole plant	–	Diabetes	El-Ghazali et al. (2010)
T. orientale	Iran (Hormozgan province)	Leaf, flower	**I:** Decoction	Hoarseness	Safa et al. (2013)

T. orientale subsp. *orientale*	Lebanon (Mount Hermon)	Aerial part	**I:** Infusion	Fevers	Baydoun et al. (2015)
			E: Infusion	Wounds and skin injuries	
T. parviflorum	Turkey (East Anatolia)	Aerial part	**I:** Decoction	An antihemorrhoidal	Altundag and Ozturk (2011)
T. polium	Turkey (East Anatolia)	Aerial part	**I:** Fresh, decoction	Stomach problems, diarrhoea, an antihemorrhoidal, internal diseases, diabetes, an analgesic, an anti-inflammatory, an oedema, stomachaches, a digestive, an orexigenic, a carminative, tuberculosis, abdominal pain	Altundag and Ozturk (2011)
			E: Direct application	Sunstroke, a haemostatic	
T. polium	Turkey (Midyat)	Aerial part	**I:** Infusion, decoction	Stomachaches	Akgul et al. (2018)
T. polium	Turkey (Honaz Mountain National Park, Middle Aegean Region of Turkey)	Aerial part	**I:** Decoction (use 2–3 cupfuls daily to relieve abdominal pain)	Stomach diseases	Kargıoğlu et al. (2010)
			I: Decoction (use 2–3 cupfuls daily)	Rheumatic diseases, alleviating spasms and cramps	
T. polium	Turkey (Manisa, Turgutlu)	Aerial part	**I:** Infusion (one glassful of the infusion after meals)	Haemorrhoids	Bulut and Tuzlaci (2013)
			I: Crushed with honey (eaten before breakfast)	Haemorrhoids	
			I: Decoction	Diabetes	
T. polium	Turkey (Bayramiç, Çanakkale)	Aerial part	**I:** Infusion (before breakfast)	Haemorrhoids, eczema	Bulut and Tuzlacı (2015)
T. polium	Iran (Turkmen Sahra)	Aerial part	**I:** Decoction	Digestive problems, a stomachic, liver disorders	Ghorbani (2005)
T. polium	Iran (Mt. Hezar, Southeastern Iran)	Aerial part	**I:** Hydrodistillation	Stomachaches, a carminative	Rajaei and Mohamadi (2012)

(continued)

Table 5.1 (continued)

Teucrium species	Locality	Part used	Forms of preparation and method of administration	Traditional uses	References
T. polium	Iran (Arjan, Parishan protected area in Fars Province)	Leaf, flower	**I:** Infusion	Regulating blood lipids and sugars, diabetes, a scent	Dolatkhahi et al. (2014)
T. polium	Iran (Kohghiluyeh va Boyer Ahmad province)	Aerial part	**I:** Decoction, infusion, cooked aerial part	Menstruation disorders, toothache, body and joint pains, abortions, gynaecological infections, a carminative	Mosaddegh et al. (2012)
T. polium	Iran (Hormozgan province)	Flower, leaf, seed (fresh)	**I:** Decoction	Stomachaches, abdominal pain, flatulence, diarrhoea, regulating blood pressure, menstruation in women who have just given birth, measles, eye pain, headaches	Safa et al. (2013)
			E: Powder	Scorpion stings, snake bites, wound healing	
T. polium	Iran (Kurd tribe, Dehloran and Abdanan districts, Ilam province)	Leaf, flower	**I:** –	An antiseptic for gastric problems, a breath freshener	Ghasemi Pirbalouti et al. (2013)
T. polium	Albania (Albanians of Lepushe, Northern Albanian Alps)	Aerial part	**I:** Decoction	A digestive	Pieroni et al. (2005)
T. polium	Albania	–	**I:** –	A digestive	Gonzalez-Tejero et al. (2008)
T. polium	Jordan (the Ajloun Heights region)	Aerial part	**I:** Infusion	An antispasmodic, flatulence, an antidiabetic, kidney stones	Aburjai et al. (2007)
T. polium	Jordan (Al-Mafraq region)	Leaf	**I:** Fresh, soaked, cooled and taken orally whenever needed	An anti-inflammatory for the stomach and intestines	Al-Quran (2014)
T. polium	Morocco (Seksaoua Region)	Aerial part	**I:** Decoction	Abdominal pain	Sbai-Jouilil et al. (2017)

T. polium	Israel (the Golan Heights and the West Bank region)	Leaf	**I:** Decoction (a standard decoction is prepared from 50 g plant material and taken orally, 100 cc, three times a day)	Kidney and liver diseases, diabetes, stomach and intestinal pain and inflammation	Said et al. (2002)
T. polium	Algeria	–	**I:** –	A digestive	Gonzalez-Tejero et al. (2008)
T. polium	Bosnia and Herzegovina	Aerial part	**I:** Infusion	Stomach diseases	Redžić (2007)
T. polium	Palestine	Leaf, aerial part	**I:** Decoction	Diabetes	Ali-Shtayeh et al. (2012)
T. polium	Palestine	Leaf	**I:** –	Local treatment of stomach and intestinal inflammation, an antispasmodic, an anthelmintic, smallpox	Jaradat (2005)
T. polium	Libya (Wadi Alkuf, Al-Jabal Al-Akhder)	–	–	Diabetes, gastritis, thyroiditis, anaemia, common colds, hypertension, kidney stones, arthritis, herpes, hair loss, rheumatism, scabies	El-Mokasabi (2014)
T. pruinosum	Lebanon (Mount Hermon)	Whole plant	**I:** Infusion	Gastrointestinal disorders	Baydoun et al. (2015)
			E: Infusion	Wounds, fevers	
			I: Steam	Colds	
T. royleanum	Pakistan (Dir Lower, Talash Valley)	Aerial part	**I:** Decoction	Fevers, an antiseptic and stimulant, a vermifuge (anthelmintic)	Khan et al. (2018a)
T. scordium	Bosnia and Herzegovina	Aerial part	**I:** Infusion	Diarrhoea	Redžić (2007)
T. scordium	Bosnia and Herzegovina (Central, Southern and Western)	Aerial part	**I:** Infusion	Gastrointestinal ailments	Šarić-Kundalić et al. (2010a)
T. scordium	Bosnia and Herzegovina (Eastern, Northern and Northeastern)	Aerial part	**I:** –	Diarrhoea	Šarić-Kundalić et al. (2011)

(continued)

Table 5.1 (continued)

Teucrium species	Locality	Part used	Forms of preparation and method of administration	Traditional uses	References
T. stocksianum	Pakistan (Kurram agency, lower Kurram)	Leaf, flower	**I:** Infusion	A blood purifier, an antipyretic, malaria, weight loss/obesity, an antidiabetic	Hussain et al. (2018)
T. stocksianum	Pakistan (Khyber Pakhtunkhwa, Lower Dir District, Tehsil Khall)	Leaf	**I:** Decoction	Diabetes	Irfan et al. (2018)
T. stocksianum	Pakistan (Bahadur Khel, Karak District, Khyber Pakhtunkhwa)	Seeds, leaf	**I/E:** Powder	Arthritis, coughs, a blood purifier, asthma, pneumonia, jaundice, diarrhoea	Khan et al. (2018b)
T. stocksianum	Pakistan (Madyan Valley in the Swat district)	Leaf	–	An expectorant	Ahmad et al. (2013)
T. stocksianum	Iran (Hormozgan province)	Leaf	**I:** Decoction, powder (when fresh)	Stomachaches, abdominal pain, flatulence, toxicity, emesis, stomach acidification, regulating blood pressure, lowering lipids, aiding the recovery of women who have recently given birth	Safa et al. (2013)
T. trifidum	South Africa (Southeastern Karoo)	Aerial part	**I:** Infusion	Colds, back pain, bladder problems in women, influenza	Van Wyk et al. (2008)

Table 5.2 The use of *Teucrium* species in veterinary medicine

Teucrium species	Locality	Part used	Forms of preparation and method of administration	Traditional uses	References
T. chamaedrys	Macedonia (Southeastern)	Aerial part	**E:** Compress with honey	For foot-and-mouth disease in cloven-hoofed animals	Nedelcheva et al. (2017)
T. scorodonia	Canada (British Columbia)	–	**I:** Tincture	Mastitis in cattle (cows)/Woodsage (*T. scorodonia*) tincture is infused into the udder	Lans et al. (2007)
T. trifidum	South Africa (Southeastern Karoo)	Aerial part	**I:** Infusion	Fever in sheep and cattle	Van Wyk et al. (2008)

Table 5.3 Uses of *Teucrium* species for other purposes

Teucrium species	Locality	Part used	Forms of preparation and method of administration	Traditional uses	References
T. chamaedrys	Italy (Campania, National Park of Cilento and Vallo di Diano)	Aerial part	**I:** *T. chamaedrys* with the aerial part of *Urtica dioica* L., *Gallium verum* L. and *Veronica chamaedrys* L. is chopped and put in white vinegar. This mixture is used in the preparation of cheese.	For human nutrition	Di Novella et al. (2013)
T. chamaedrys	Italy (Rotonda, Pollino National Park)	Whole plant	**I:** Infusion	As an appetizer	Di Sanzo et al. (2013)
T. polium	Egypt	–	**I:** –	For human nutrition	Gonzalez-Tejero et al. (2008)
T. cubense	Mexico (Nuevo León, Bustamante)	Whole plant	**E:** Rub the whole body with the dry plant while praying	Fright	Estrada-Castillón et al. (2018)
			Ornamental	Planted in private gardens	
T. chardonianum	Western Sahara (Moroccan occupied territories)	Flowering tops (dried and triturated)	– (In its dry state)	Burnt inside a tent to perfume the air	Volpato et al. (2012)

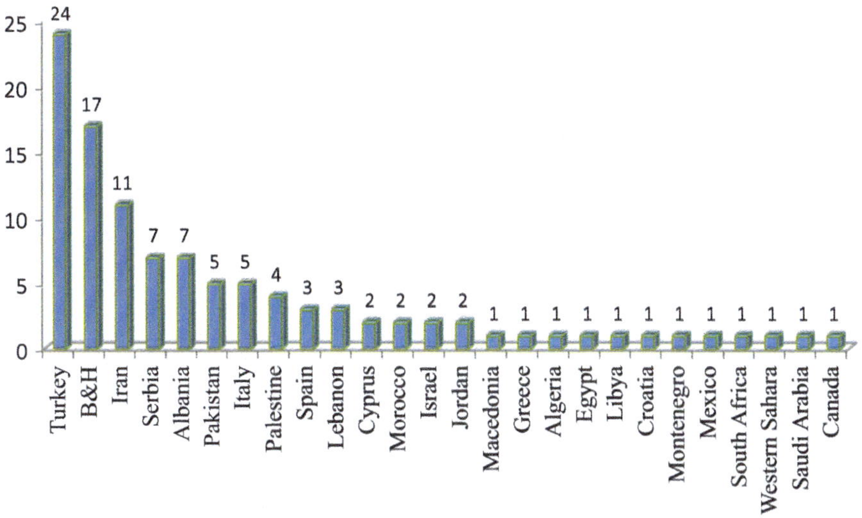

Fig. 5.1 The number of analysed studies in the different countries where *Teucrium* species are used

Furthermore, the species *Teucrium montanum* also features significantly in ethnomedicine, as is confirmed by 11 (15.3%) ethnobotanical studies. It is used most in the mountain regions of Bosnia and Herzegovina (5 studies, 45.4%) and Serbia (4 studies, 27.3%). In addition, the use of *Teucrium stocksianum* was recorded in 5 (6.94%) ethnobotanical studies conducted in Pakistan and Iran (Fig. 5.2).

Research showed that infusions (77.8%) and decoctions (48.6%) were the dominant methods for preparing and using *Teucrium* species; however, these plants were also found to be used in the form of fresh juice, drops, brandy, tincture, balm or by applying the fresh aerial parts wrapped either whole or chopped and mixed with honey in a gauze (Table 5.1). According to the available data, their internal use was the dominant form (85%), while their use externally (for rheumatism, haemorrhoids, psoriasis, wound healing, and sunstroke, as an antidote to the snake bites and scorpion stings, and as a haemostatic) was significantly lower (12%). Combined internal/external use was mentioned in 3% of the studies analysed.

Teucrium species were used most often to treat various gastrointestinal disorders (56 of the analysed studies), followed by problems related to the functioning of the heart (e.g. hypertension) and blood vessels (e.g. haemorrhoids) (21 studies), and respiratory tract problems (colds, lung cancer, tuberculosis, asthma, pneumonia, influenza, etc., 17 studies). Their use in the treatment of problems related to endocrine function disorders (diabetes, thyroiditis) was mentioned in 16 ethnobotanical studies, while 14 studies noted their being used to treat problems connected to muscle and skeletal disorders and skin diseases (Fig. 5.3). In addition, *Teucrium* species were also used to treat problems concerning the excretory organs (7 studies), fevers (7 studies), as a stimulant and tonic (6 studies), to treat inflammations (5 studies), gynaecological ailments (4 studies), malaria (4 studies), mental disorders (3) and

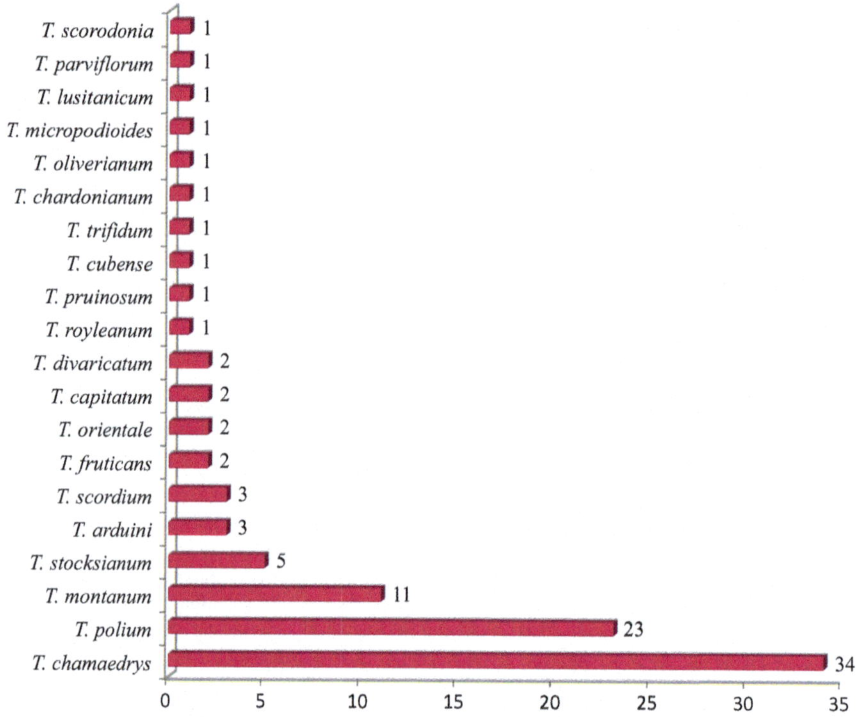

Fig. 5.2 The total number of ethnobotanical studies in which *Teucrium* species are cited

headaches (3 studies). Twenty-four studies noted health problems and diseases that were only mentioned once or twice (Other, Fig. 5.3).

5.3.3 *The Use of* Teucrium *Species in Veterinary Ethnomedicine*

Our research established that three *Teucrium* species, mentioned in three ethnobotanical studies, are used in veterinary ethnomedicine (Table 5.2). Namely, in Southeastern Macedonia, Nedelcheva et al. (2017) recorded the use of *Teucrium chamaedrys* in the treatment of foot-and-mouth disease in cloven-hoofed animals. The South African endemic species *Teucrium trifidum* (Codd 1977) is used to treat fevers in sheep and cattle (Van Wyk et al. 2008). In the British Columbia region (Canada), tincture of *Teucrium scorodonia* is used to treat mastitis in cattle, particularly cows (Lans et al. 2007). In line with this, laboratory research by Djilas et al. (2006) found that ethyl acetate, chloroform and n-butanol extracts of *Teucrium montanum* exhibit a wide range of inhibitory activities against Gram (+) and Gram (−) bacteria, representing mid-level validity for mastitis.

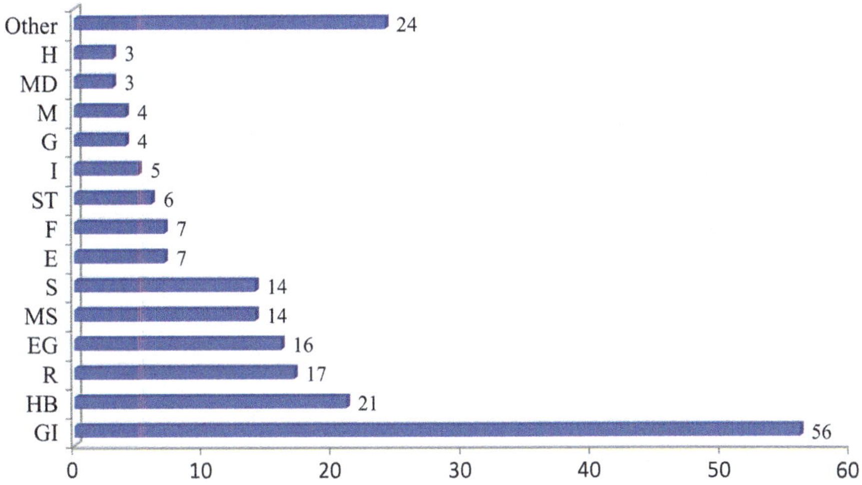

Fig. 5.3 The frequency of the use of *Teucrium* species in human therapy based on the analysed studies: **GI** – gastrointestinal disorders; **HB** – heart and blood vessels; **R** – respiratory diseases; **EG** – diseases of the endocrine glands; **MS** – muscular and skeletal problems; **S** – skin diseases; **E** – excretory organ problems; **F** – fever; **ST** – stimulant and tonic; **I** – inflammation; **G** – gynaecological ailments; **M** – malaria; **MD** – mental disorders; **H** – headaches; **Other** – toothache, earache, poisoning, lack of appetite, an aphrodisiac, oedema, sunstroke, weight loss, an analgesic, etc

5.3.4 The Use of Teucrium *Species for Other Purposes*

Besides the traditional uses for treating a variety of health problems and complaints in people and animals, *Teucrium* species are also used for other purposes. Namely, they are important alimentary plants with some used to prepare flavoured wines, herbal teas, bitters and liqueurs, while people use infusions of leaves and flowers to flavour beer in some regions (Maccioni et al. 2007). Furthermore, in studies by Saroglou et al. (2007), Ulubelen et al. (2000) and Özkan et al. (2007), this genus was found to be important in the food industry with many species serving as natural preservatives due to their antimicrobial, antioxidant and antifungal properties. Hence, in ethnobotanical research conducted in two national parks in Italy, *Teucrium chamaedrys* is mentioned as being used in cheese preparation (Di Novella et al. 2013) and as an appetizer (Di Sanzo et al. 2013) (Table 5.3). *Teucrium polium* is also used in Egypt in human nutrition (Gonzalez-Tejero et al. 2008). Estrada-Castillón et al. (2018) mentioned *Teucrium* cubense as an ornamental plant in Mexico and its use in rituals, while *Teucrium chardonianum* is used as an air freshener in the Western Sahara region (Volpato et al. 2012).

5.4 The Medicinal Properties of *Teucrium* Species

As with other Lamiaceae, the aerial organs of *Teucrium* species are covered by an indumentum of glandular and non-glandular trichomes in which essential oils are secreted. The essential oil content in these species is high and sesquiterpenes predominate in it (Cavalerio et al. 2002; Kucuk et al. 2006; Hachicha et al. 2007; Saroglou et al. 2007). In addition to sesquiterpenes, the essential oil of *Teucrium* species is a great source of neoclerodane diterpenes, in fact one of the richest, with over 220 diterpenes described so far (Piozzi et al. 2005).

The yield of essential oil in *Teucrium* species varies from species to species, but ranges from 0.5% to 1.5%, while the proportion of the main chemical constituents (primarily monoterpene/sesquiterpene hydrocarbons and oxygenated sesquiterpenes) also varies dramatically (Kovačević et al. 2001; Saroglou et al. 2007). Some of the differences in the chemical composition of the essential oil are believed to be linked to different subspecies and/or the plants' geographical origin and a range of environmental factors (Menichini et al. 2009).

Used widely due to their antioxidant and antidiabetic properties, the essential oils and volatile constituents extracted from *Teucrium* species also have a part to play in the prevention and treatment of a whole variety of human disorders, ailments and diseases, including cancer, cardiovascular problems, such as atherosclerosis and thrombosis, and bacterial and viral infections (Leyel 1984; Bruneton 1995; Edris 2007). They are also of interest ecologically, being used as antifeedants, inhibiting attack by different species of insects, due to the properties of their components, especially the diterpenes (Piozzi et al. 2005).

The chemical composition of the essential oil largely determines the medicinal properties of a plant and hence its uses. In this chapter, analysis of 72 ethnobotanical studies showed that the following species were used most, and as such, were the most popular among the human population: *Teucrium chamaedrys, T. polium* and *T. montanum*.

5.4.1 Teucrium chamaedrys *(Wall Germander)*

A perennial herbaceous plant with a half-ligneous and shrub-like low stem, *Teucrium chamaedrys* (section *Chamaedrys*) grows up to 30 cm high. It is found in rocky limestone areas, dry mountain meadows and pastures, and on the edges of sparse oak and pine forests, up to 1000 m above sea level in Central Europe, the Mediterranean region and Western Asia. Flowering and harvesting takes place between July and September, while fruiting occurs from August to September (Tutin et al. 1972; Diklić 1974; Fig. 5.4). *Teucrium chamaedrys* has a very wide range of uses in traditional medicine, particularly in its centres of distribution, as is confirmed by 34 of the analysed ethnobotanical studies. It is mainly the aerial parts that are used, most frequently when the plant is flowering, as then it has the greatest

Fig. 5.4 *Teucrium chamaedrys* L. (Photo P. Lazarević)

concentrations of active substances. It is prepared in the form of an infusion, decoction, and fresh juice, while it is usually used internally, and more rarely externally (as a compress).

It is used as a painkiller and to treat stomach pains, gastric pains, indigestion, digestive complaints, haemorrhoids, toothache, kidney pains, kidney stones, heart disease, hypertension, mouth sores, poisoning, rheumatism, insomnia, colds and flu, itching, eczema, malaria, diarrhoea, spasms, anaemia, earache, childlessness, rheumatism and arthritis. Furthermore, it is used as a stomachic, antidiabetic, analgesic, sedative, antispasmodic, febrifuge, aphrodisiac, haemostatic, tonic, and stimulant, as well as an appetizer (Table 5.1). In addition, it is used in veterinary ethnomedicine for foot-and-mouth disease in cloven-hoofed animals (Table 5.2). Nencini et al. (2014) stated that it is used as Portland Powder in traditional medicine in England for the treatment of rheumatism and gout.

The chemistry of *Teucrium chamaedris* has been thoroughly researched due to its proven in vitro antioxidant activity (Kadifkova-Panovska et al. 2005) and also as there have been several cases of hepatotoxicity arising from the use of this species (Kouzi et al. 1994; Stickel et al. 2000; Perez Alvarez et al. 2001). Namely, it is very rich in phenolic compounds, exhibiting very strong biological activity and antioxidative effects (Ozgen et al. 2006; Gursoy and Tepe 2009; Stanković et al. 2010). In a study by Prescott et al. (2011), teucrioside was identified as the main active ingredient of the plant, which was shown in preclinical research to be effective in inhibiting calcineurin, meaning it could aid in the reduction of inflammatory states.

5.4.2 Teucrium polium *(Felty Germander)*

Teucrium polium (section *Polium*) is a perennial, aromatic plant, 20–50 cm high, appearing from June to August, and found abundantly in Southwestern Asia, Europe (the Mediterranean region), and North Africa (Diklić 1974; Djabou et al. 2012; Fig. 5.5). It grows on well-drained land, hillsides, sands, in stony mountains, and in sunny regions with a semiarid and arid bio-climate. The aerial parts of this plant are characterised by a pleasant aroma and a bitter taste (Barceloux 2008).

Numerous ethnobotanical studies (23) have highlighted the major ethnomedicinal importance of this species, mainly in the treatment of abdominal pain (digestive problems, stomach and intestinal pain and inflammation, and diarrhoea), kidney and liver diseases, kidney stones, diabetes, haemorrhoids, oedema, rheumatism, tuberculosis, and sunstroke. Moreover, it is used as an analgesic, an anti-inflammatory, an orexigenic, a carminative, a haemostatic, a digestive, an antispasmodic, and in human nutrition (Table 5.1). The medicinal properties of *Teucrium polium* for treating most of these health problems have been confirmed by laboratory research, which justifies its use in traditional medicine. Likewise, numerous in vivo and in vitro studies have confirmed the various biological activities of this species. These include anti-inflammatory and antirheumatic (Tariq et al. 1989), hypoglycemic (Kasabri et al. 2011), antipyretic and antibacterial (Autore et al. 1984), antioxidant and antimicrobial (Ilhami et al. 2003; Balmekki et al. 2013), hypolipidemic (Rasekh et al. 2001), and antihypertensive properties (Suleiman et al. 1988), as well as its benefits for treating stomach disorders (Malakov and Papanov 1983; Aqel et al. 1990), improving mental performance (Perry et al. 1996), and aiding weight loss (Gharaibeh et al. 1988).

Research by Kadifkova-Panovska et al. (2005) showed that different *Teucrium polium* extracts exhibit significant free radical scavenging activity, hydroxyl radical

Fig. 5.5 *Teucrium polium* L. (Photo M. Stanković)

scavenging, and antioxidant activity in vitro. These antioxidative properties of *Teucrium polium* are the result of the presence of flavonoids (rutin, apigenin, apigenin-4, 7-dimethylether, cirsimaritin, cirsiliol, luteolin, etc.) (Sharififar et al. 2009), which has been confirmed by in vitro and in vivo studies (Djeridane et al. 2006; Hasani et al. 2007). In his research, Rajabalian (2008) demonstrated that the methanol extract of *Teucrium polium* could potentially be used as an effective and safe chemo-sensitizer agent in cancer chemotherapy. In addition, laboratory research confirmed that the ethanol extract also exhibited potent antibacterial activity against Gram (+) and Gram (−) microorganisms, while the aqueous extract of *Teucrium polium* inhibits the growth of *Saccharomyces cerevisiae* and *Yarrowia lipolytica* (Autore et al. 1984; Rojas et al. 1992; Aggelis et al. 1998; Essawi and Srour 2000; Darabpour et al. 2010). Due to the presence of sterols and flavonoids, *Teucrium polium* exhibits anti-inflammatory properties (Tariq et al. 1989), while its analgesic properties are the result of the presence of essential oils (Abdollahi et al. 2003). The antispasmodic activity of *Teucrium polium* can be put down to the high level of sesquiterpene alcohols (Kamel and Sandra 1994). In experimental conditions, it was established that the aqueous extract of the aerial parts of this species brought about a significant reduction in serum triglycerides and cholesterol in hyperlipidemic rats (Rasekh et al. 2001). Moreover, Movahedi et al. (2014) demonstrated that a decoction of *Teucrium polium* can protect liver cells against hepatocellular carcinoma in carcinogenesis-induced animal models. In one study, burn wounds healed more quickly through the topical application of *Teucrium polium* extract (Ansari et al. 2013). Similarly, the anti-inflammatory effects of this plant were demonstrated through it inhibiting carrageenan-induced inflammation in an animal study and its methanol and ethanol extracts have also been shown to be effective on veterinary pathogens (Darabpour et al. 2010).

5.4.3 **Teucrium montanum** *(Mountain Germander)*

The species *Teucrium montanum* (section *Polium*) is a perennial, shrub-like plant that is native to the Mediterranean region of Europe and North Africa, and to the Middle East (Tutin et al. 1972; Diklić 1974; Lakušić and Lakušić 2014; Fig. 5.6). It has half-ligneous branches, grows up to 25 cm high and can be found on thermophilic limestone and serpentine rocks, in dry mountain meadows and on the edges of forests. It flowers between June and September.

According to the ethnomedicinal data available, *Teucrium montanum* is used for digestive disorders, biliary tract purification, liver (cirrhosis) and stomach diseases, diabetes, spasms and blood purification, lung cancer, tuberculosis, rheumatism, arthritis, and headaches (in some regions, for "male" headaches), to improve appetite, as an antipyretic, to strengthen the immune system, as a hangover cure, and as a tonic (tea) (Table 5.1). The dominant method of preparation is an infusion, while in some areas it is used in the form of a decoction or balm. In Bosnia and Herzegovina and Serbia, *Teucrium montanum* is one of the most popular plants in traditional

Fig. 5.6 *Teucrium montanum* L. (Photo M. Stanković)

medicine and according to local people's beliefs "heals every disease" (Redžić 2010; Jarić et al. 2015a).

Phytochemical studies have shown that *Teucrium montanum* is very rich in phenolic compounds, exhibiting very strong biological activity and antioxidative effects (Čanadanović-Brunet et al. 2006; Stanković et al. 2011b), which justifies its use in traditional medicine. Furthermore, data promoting the use of *Teucrium montanum* in ethnomedicine confirms the free-radical scavenging activity of flavonoids and phenolic acids in extracts made from different solvents (Djilas et al. 2006).

5.5 Potentially Toxic Effects of *Teucrium* Species

The use of *Teucrium* species in traditional medicine spanning several centuries points unequivocally to its medicinal properties. However, there have been instances when the long-term use of preparations made from these plants can have unwanted side effects. An illustrative example of the toxic effects of one species of the genus *Teucrium* was reported by a 33-year-old woman who drank *Teucrium chamaedrys* tea every day for 2 weeks. It was found that she had symptoms of acute icteric hepatitis and all other causes of acute hepatitis were ruled out. When she stopped drinking the tea, the patient made a clinical recovery and her serum bilirubin, aminotransferase and alkaline phosphatase levels returned to normal within 9 weeks (Ural et al. 2011). This case indicates that *Teucrium chamaedrys* can cause acute icteric hepatitis, which can clinically mimic acute viral hepatitis. Similar hepatotoxicity was observed during the use of preparations made from other species of the *Teucrium* genus (Chitturi and Farrell 2008; Kotsiou and Tesseromatis 2017) with most patients presenting with very high liver aminotransferases or developing jaundice. Hepatitis and intrahepatic cholestatic liver disease were the etiology of liver

damage in these cases. *Teucrium* species are rich in neoclerodane diterpenoids, which may be the cause of hepatotoxicity (Sundaresan et al. 2006). Therefore, those suffering from hepatic abnormalities should be careful when using *Teucrium* species (Perez Alvarez et al. 2001). Likewise, taking high doses of *Teucrium* species or using them over a longer period of time might cause the gradual deterioration in neuromuscular coordination, as is supported by histopathological and biochemical evidence (Tanira et al. 1996, 1997).

5.6 Conclusions

A comprehensive analysis of the traditional use of *Teucrium* species from the ethnobotanical and ethnomedicinal points of view took in 72 ethnobotanical studies from different parts of the world. It revealed a wide range of uses for different species of this genus, but for the purposes of this chapter the traditional use of 20 species of the genus *Teucrium* has been presented. Nineteen of these species are used in ethnomedicine and 3 in veterinary ethnomedicine, while 4 are used for other purposes. It has been established that these species are most frequently used traditionally in the treatment of abdominal problems, followed by problems related to the functioning of the heart, blood vessels, and endocrine glands, muscular and skeletal problems, and skin diseases. Those species which stand out for their wide range of uses are *Teucrium chamaedrys*, *T. polium* and *T. montanum*.

The medicinal properties of *Teucrium* species and their favourable impact on numerous ailments is undeniable, as has been confirmed by many studies. On the one hand, this justifies their use in ethnomedicine; on the other hand, it leads to the need for caution, particularly in terms of dosage and treatment duration and the overall health of the patient must be taken into consideration. Namely, histopathological and biochemical studies have revealed that when *Teucrium* species are taken in high doses or over a long period of time, they might cause progressive deterioration in neuromuscular coordination and hepatic damage, which may or may not be irreversible. Therefore, it is necessary to undertake further research aimed at isolating and characterizing the constituents of *Teucrium* species so as to obtain suitable drugs.

Acknowledgments This manuscript was supported by the Ministry of Education, Science and Technological Development of the Republic of Serbia, Grants 451-03-68/2020-14/200007.

References

Abdollahi M, Karimpour H, Monsef-Esfehani HR (2003) Antinociceptive effects of *Teucrium polium* L. total extract and essential oil in mouse writhing test. Pharmacol Res 48:31–35

Aburjai T, Hudaib M, Tayyem R, Yousef M, Qishawi M (2007) Ethnopharmacological survey of medicinal herbs in Jordan, the Ajloun Heights region. J Ethnopharmacol 110:294–304

Adams M, Berset C, Kessler M, Hamburger M (2009) Medicinal herbs for the treatment of rheumatic disorders – a survey of European herbals from the 16th and 17th century. J Ethnopharmacol 121:343–359

Aggelis G, Athanassopoulos N, Paliogianni A, Komaitis M (1998) Effect of a *Teucrium polium* L. extract on the growth and fatty acid composition of *Saccharomyces cerevisiae* and *Yarrowia lipolytica*. Antonie Van Leeuwenhoek 73:195–198

Ahmad N, Anwar S, Fazal H, Abbasi BH (2013) Medicinal plants used in indigenous therapy by people of Madyan Valley in district Swat, Pakistan. Int J Med Arom Plant 3:47–54

Akgul A, Akgul A, Senol SG, Yildirim H, Secmen O, Dogan Y (2018) An ethnobotanical study in Midyat (Turkey), a city on the silk road where cultures meet. J Ethnobiol Ethnomed 14:12. https://doi.org/10.1186/s13002-017-0201-8

Ali-Shtayeh MS, Yaniv Z, Mahajna J (2000) Ethnobotanical survey in the Palestinian area: a classification of the healing potential of medicinal plants. J Ethnopharmacol 73:221–232

Ali-Shtayeha MS, Jamousa RM, Jamousa RM (2012) Complementary and alternative medicine use amongst Palestinian diabetic patients. Complement Ther Clin Pract 18:16–21

Al-Quran S (2014) Used Ethnobotany of medicinal plants by inhabitants of Al-Mafraq, Jordan. Arnaldoa 21:119–126

Altundag E, Ozturk M (2011) Ethnomedicinal studies on the plant resources of east Anatolia, Turkey. Procedia Soc Behav Sci 19:756–777

Ansari R, Sahinfard N, Namjou A, Rafieian M, Shirzad H, Rafieian-Kopaei M (2013) Ameliorative property of *Teucrium polium* on second degree burn. J Herbmed Pharmacol 2:9–11

Aqel M, Gharaibeh M, Salhab A (1990) The calcium antagonistic effect of the volatile oil of *Teucrium polium*. Pharm Biol 28:201–207

Ari S, Temel M, Kargıoğlu M, Konuk (2015) Ethnobotanical survey of plants used in Afyonkarahisar – Turkey. J Ethnobiol Ethnomed 11:84. https://doi.org/10.1186/s13002-015-0067-6

Arnold N (1985) Contribution á la connaissance ethnobotanique et mèdicinale de la flore de Chypre, vol 3. Universitè Renè Descartes de Paris, Paris, pp 1203–1210

Autore G, Capasso F, De Fusco R, Fasulo MP, Lembo M, Mascolo N, Menghini A (1984) Antipyretic and antibacterial actions of *Teucrium polium* (L.). Pharmacol Res Commun 16:21–29

Bahmani M, Shirzad H, Majlesi M, Shahinfard N, Rafieian-Kopaei M (2014) A review study on analgesic applications of Iranian medicinal plants. Asian Pac J 7:S43–S53

Balmekki N, Bendimerad N, Bekhechim C, Fernandez X (2013) Chemical analysis and antimicrobial activity of *Teucrium polium* L. essential oil from Western Algeria. J Med Plant Res 7:897–902

Baluchnejadmojarad T, Roghani M, Roghani-Dehkordi F (2005) Antinociceptive effects of *Teucrium polium* leaf extract in the diabetic rat formalin test. J Ethnopharmacol 97:207–210

Barceloux DG (2008) Medical toxicology of natural substances: foods, fungi, medicinal herbs, plants, & venomous animals. Wiley, Hoboken

Baydoun S, Lamis C, Helena D, Nelly A (2015) Ethnopharmacological survey of medicinal plants used in traditional medicine by the communities of Mount Hermon, Lebanon. J Ethnopharmacol 173:139–156

Bekut M, Brkić S, Kladar N, Dragović G, Gavarić N, Božin B (2018) Potential of selected Lamiaceae plants in anti(retro)viral therapy. Pharmacol Res 133:301–314

Benítez G, González-Tejero MR, Molero-Mesa J (2010) Pharmaceutical ethnobotany in the western part of Granada province (southern Spain): Ethnopharmacological synthesis. J Ethnopharmacol 129:87–105

Bruneton J (1995) Pharmacognosy, phytochemistry, medicinal plants. Lavoisier Publishing, Paris

Bulut G, Tuzlaci E (2013) An ethnobotanical study of medicinal plants in Turgutlu (Manisa – Turkey). J Ethnopharmacol 149:633–647

Bulut G, Tuzlaci E (2015) An ethnobotanical study of medicinal plants in Bayramiç (ÇanakkaleTurkey). Marmara Pharm J 19:268–282

Cakilcioglu U, Turkoglu I (2010) An ethnobotanical survey of medicinal plants in Sivrice (Elazığ-Turkey). J Ethnopharmacol 132:165–175

Čanadanović-Brunet JM, Djilas SM, Ćetković GS, Tumbas VT, Mandić AI, Čanadanović VM (2006) Antioxidant activities of different *Teucrium montanum* L. extracts. Int J Food Sci Technol 41:667–673

Carrió E, Vallès J (2012) Ethnobotany of medicinal plants used in Eastern Mallorca (Balearic Islands, Mediterranean Sea). J Ethnopharmacol 141:1021–1040

Cavaleiro C, Salgueiro LR, Antunes T, Sevinate-Pinto I, Barroso JG (2002) Composition of the essential oil and micromorphology of trichomes of *Teucrium salviastrum*, an endemic species from Portugal. Flavour Fragr J 17:287–291

Chitturi S, Farrell GC (2008) Hepatotoxic slimming aids and other herbal hepatotoxins. J Gastroenterol Hepatol 23:366–373

Codd LE (1977) The South African species of *Teucrium* (Lamiaceae). Bothalia 12:177–179

Cornara L, La Rocca A, Terrizzano L, Dente F, Mariotti MG (2014) Ethnobotanical and phytomedical knowledge in the North-Western Ligurian Alps. J Ethnopharmacol 155:463–484

Darabpour E, Motamedi H, Nejad SMS (2010) Antimicrobial properties of *Teucrium polium* against some clinical pathogens. Asian Pac J Trop Med 3:124–127

Di Novella R, Di Novella N, De Martino L, Mancini E, De Feo V (2013) Traditional plant use in the National park of Cilento and Vallo di Diano, Campania, Southern, Italy. J Ethnopharmacol 145:328–342

Di Sanzo P, De Martino L, Mancini E, De Feo V (2013) Medicinal and useful plants in the tradition of Rotonda, Pollino National Park, Southern Italy. J Ethnobiol Ethnomed 9:19. https://doi.org/10.1186/1746-4269-9-19

di Tizio A, Łuczaj Ł, Quave C, Redžić S, Pieroni A (2012) Traditional food and herbal uses of wild plants in the ancient South-Slavic diaspora of Mundimitar/Montemitro (Southern Italy). J Ethnobiol Ethnomed 8:21. https://doi.org/10.1186/1746-4269-8-21

Diklić N (1974) Teucrium. In: Jakovljević S (ed) Flore de la Republique Socialiste de Serbie VI, 1st edn. Serbian Academy of Sciences and Arts, Belgrade, pp 349–357

Dinç M, Duran A, Pinar M, Őztürk M (2008) Anatomy, palynology and nutlet micromorphology of Turkish endemic *Teucrium sandrasicum* (Lamiaceae). Biologia 63:637–641

Dinç M, Doğu S, Bilgili B, Duran A (2009) Comparative anatomical and micromorphological studies on *Teucrium creticum* and *Teucrium orientale* var. *orientale* (T. sect. *Teucrium*, Lamiaceae). Nord J Bot 27:251–256

Djabou N, Muselli A, Alkali H, Dib ME, Tabti B, Varesi L, Costa J (2012) Chemical and genetic diversity of two Mediterranean subspecies of *Teucrium polium* L. Phytochemistry 83:51–62

Djeridane A, Yousfi M, Nadjemi B, Boutassouna D, Stocker P, Vidal N (2006) Antioxidant activity of some Algerian medicinal plants extracts containing phenolic compounds. Food Chem 97:654–660

Djilas SM, Markov SL, Cvetković DD, Čanadanović-Brunet JM, Ćetković GS, Tumbas VT (2006) Antimicrobial and free radical scavenging activities of *Teucrium montanum*. Fitoterapia 77:401–403

Dolatkhahi M, Dolatkhahi A, Nejad JB (2014) Ethnobotanical study of medicinal plants used in Arjan – Parishan protected area in Fars province of Iran. Avicenna J Phytomed 4:402–412

Dolina K, Łuczaj Ł (2014) Wild food plants used on the Dubrovnik coast (south-eastern Croatia). Acta Soc Bot Pol 83:175–181

Edris AE (2007) Pharmaceutical and therapeutic potentials of essential oils and their individual volatile constituents: a review. Phytother Res 21:308–323

El-Ghazali GE, Al-Khalifa KS, Saleem GA, Abdallah EM (2010) Traditional medicinal plants indigenous to Al-Rass province, Saudi Arabia. J Med Plant Res 4:2680–2683

El-Mokasabi FM (2014) Floristic composition and traditional uses of plant species at Wadi Alkuf, Al-Jabal Al-Akhder, Libya. Am Eurasian J Agric Environ Sci 14:685–697

Eshratifar M, Attar F, Mahdigholi K (2011) Micromorphological studies on nutlet and leaf indumentum of genus *Teucrium* L. (Lamiaceae) in Iran. Turk J Bot 35:25–35

Essawi T, Srour M (2000) Screening of some Palestinian medicinal plants for antibacterial activity. J Ethnopharmacol 70:343–349

Estrada-Castillón E, Villarreal-Quintanilla JA, Rodríguez-Salinas MM, Encinas-Domínguez JA, González-Rodríguez H, Figueroa GR, Arévalo JR (2018) Ethnobotanical survey of useful species in Bustamante, Nuevo León, Mexico. Hum Ecol 46:117–132

Fatiha BA, Ouafae B, Souad S, Fatima EH, Jamila D, Allal D, Lahcen Z (2017) Ethnobotany study of medicinal plants used in the treatment of respiratory diseases in the middle region of Oum Rbai. Int J Environ Agric Biotechnol 2:1460–1468

Ferrier J, Šačiragić L, Chen ECH, Trakić S, Saleem A, Alikadić E, Cuerrier A, Balick MJ, Arnason JT, Redžić S (2014) Ways the Lukomir highlanders of Bosnia and Herzegovina treat diabetes. In: Pieroni A, Quave CL (eds) Ethnobotany and biocultural diversities in the Balkans. Springer, NewYork/Heidelberg/Dordrecht/London, pp 13–27

Gharaibeh MN, Elayan HH, Salhab AS (1988) Hypoglycaemic effects of *Teucrium polium*. J Ethnopharmacol 24:93–99

Ghasemi Pirbalouti A, Momeni M, Bahmani M (2013) Ethnobotanical study of medicinal plants used by Kurd tribe in Dehloran and Abdanan districts, Ilam province, Iran. Afr J Tradit Complement Altern Med 10:368–385

Ghorbani A (2005) Studies on pharmaceutical ethnobotany in the region of Turkmen Sahra, north of Iran (Part 1): general results. J Ethnopharmacol 102:58–68

Gonzalez-Tejero MR, Casares-Porcel M, Sanchez-Rojas CP, Ramiro-Gutierrez JM, Molero-Mesa J, Pieroni A, Giusti ME, Censorii E, de Pasquale C, Della A, Paraskeva-Hadijchambi D, Hadjichambis A, Houmani Z, El-Demerdash M, El-Zayatf M, Hmamouchi M, ElJohrig S (2008) Medicinal plants in the Mediterranean area: synthesis of the results of the project Rubia. J Ethnopharmacol 116:341–357

Greuter W, Burdet HM, Long G (1986) Med-checklist 3. Trust of OPTIMA, Geneva

Gürdal B, Kültür Ş (2013) An ethnobotanical study of medicinal plants in Marmaris (Mugla, Turkey). J Ethnopharmacol 146:113–126

Gursoy N, Tepe B (2009) Determination of the antimicrobial and antioxidative properties and total phenolics of two “endemic” Lamiaceae species from Turkey: *Ballota rotundifolia* L. and *Teucrium chamaedrys* C. Koch. Plant Foods Hum Nutr 64:135–140

Güzel Y, Güzel M, Miski M (2015) Ethnobotany of medicinal plants used in Antakya: a multicultural district in Hatay Province of Turkey. J Ethnopharmacol 174:118–152

Hachicha SF, Skanji T, Barrek S, Ghrabi ZG, Zarrok H (2007) Chemical composition *Teucrium alopecurus* essential oil from Tunusia. Flavour Fragr J 22:101–104

Harley RM, Atkins S, Budantsev A, Cantino PH, Conn B, Grayer R, Harley MM, Kok R, Krestovskaja T, Morales A, Paton AJ, Ryding O, Upson T (2004) Labiatae. In: Kadereit JW (vol ed), Kubitzki K (ed) The families and genera of vascular plants VII. Springer, Berlin/Heidelberg, pp 167–275

Hasani P, Yasa N, Vosough-Ghanbari S, Mohammadirad A, Dehghan G, Abdollahi M (2007) In vivo antioxidant potential of *Teucrium polium*, as compared to alpha-tocopherol. Acta Pharma 57:123–129

Hasani-Ranjbar S, Nayebi N, Larijani B, Abdollahi M (2010) A systematic review of the efficacy and safety of *Teucrium* species; from anti-oxidant to anti-diabetic effects. Int J Pharmacol 6:315–325

Heywood VH, Brummitt RK, Seberg O, Culham A (2007) Flowering plant families of the world. Firefly Books, Ontario

Hussain W, Ullah M, Dastagir G, Badshah L (2018) Quantitative ethnobotanical appraisal of medicinal plants used by inhabitants of lower Kurram, Kurram agency, Pakistan. Avicenna J Phytomed 8:313–329

Ilhami G, Metin U, Munir O, Suktru B, Irfan K (2003) Antioxidant and antimicrobial activities of *Teucrium polium* L. J Food Technol 1:9–16

Irfan M, Nabeela I, Kamil M, Khan NA, Khan H, Khalil S, Ullah S, Shah M, Jan G, Murad W (2018) Ethnomedicinal plants uses of tehsil Khall, district Lower Dir, Khyber Pakhtunkhwa, Pakistan. Int J Biosci 13:219–229

Jaradat NA (2005) Ethnopharmacological survey of natural products in Palestine. An-Najah Univ J Res Nat Sci 19:13–67

Jarić S, Popović Z, Mačukanović-Jocić M, Djurdjević L, Mijatović M, Karadzić B, Mitrović M, Pavlović P (2007) An ethnobotanical study on the usage of wild medicinal herbs from Kopaonik Mountain (central Serbia). J Ethnopharmacol 111:160–175

Jarić S, Mitrović M, Pavlović P (2014) An ethnobotanical and ethnomedicinal study on the use of wild medicinal plants in rural areas of Serbia. In: Pieroni A, Quave CL (eds) Ethnobotany and biocultural diversities in the Balkans. Springer, New York/Heidelberg/Dordrecht/London, pp 87–112

Jarić S, Mačukanović-Jocić M, Djurdjević L, Mitrović M, Kostić O, Karadžić B, Pavlović P (2015a) An ethnobotanical survey of traditionally used plants on Suva planina mountain (south-eastern Serbia). J Ethnopharmacol 175:93–108

Jarić S, Mitrović M, Pavlović P (2015b) Review of ethnobotanical, phytochemical, and pharmacological study of *Thymus serpyllum* L. Evid Based Complement Alternat Med 2015:101978. https://doi.org/10.1155/2015/101978

Jurišić Grubešić R, Vladimir-Knežević S, Kremer D, Kaloera Z, Vuković J (2007) Trichome micromorphology in *Teucrium* (Lamiaceae) species growing in Croatia. Biologia 62:148–156

Kadifkova-Panovska T, Kulevanova S, Stefova M (2005) In vitro antioxidant activity of some *Teucrium* species (Lamiaceae). Acta Pharma 55:207–214

Kamel A, Sandra P (1994) Gas chromatography – mass spectrometry analysis of the volatile oils of two *Teucrium polium* varieties. Biochem Syst Ecol 22:529–532

Kargıoğlu M, Cenkci S, Serteser A, Konuk M, Vural G (2010) Traditional uses of wild plants in the middle Aegean region of Turkey. Hum Ecol 38:429–450

Kasabri V, Afifi FU, Hamdan I (2011) In vitro and in vivo acute antihyperglycemic effects of five selected indigenous plants from Jordan used in traditional medicine. J Ethnopharmacol 133:888–896

Kaval I, Behçet L, Cakilcioglu U (2014) Ethnobotanical study on medicinal plants in Geçitli and its surrounding (Hakkari-Turkey). J Ethnopharmacol 155:171–184

Khan MT, Ahmad L, Rashid W (2018a) Ethnobotanical documentation of traditional knowledge about medicinal plants used by indigenous people in the Talash Valley of Dir Lower, northern Pakistan. J Intercult Ethnopharmacol 7:8–24

Khan S, Jan G, Bibi H, Murad W, Ullah K, Ihsanullah (2018b) An ethnomedicinal survey of plants used in traditional medicine in arid and semi-arid zone of Bahadur Khel, District Karak, Khyber Pakhtunkhwa, Pakistan. Asian J Pharmacogn 2:41–44

Kotsiou A, Tesseromatis C (2017) Hepatotoxicity of herbal medicinal products. J Med Plant Stud 5:80–88

Kouzi SA, Mc Murtry RJ, Nelson SD (1994) Hepatotoxicity of germander (*Teucrium chamaedrys* L.) and one of its constituent neoclerodane diterpenes teucrin A in the mouse. Chem Res Toxicol 7:850–856

Kovačević N, Lakušić B, Ristić M (2001) Composition of the essential oils of seven *Teucrium* species from Serbia and Montenegro. J Essent Oil Res 13:163–165

Kucuk M, Gulec C, Yasar A, Ucuncu O, Yayli N, Coskuncelebi K, Terzioglu S, Yayli N (2006) Chemical composition and antimicrobial activities of the essential oils of *Teucrium chamaedrys* subsp. *chamaedrys*, *T. orientale* var. *puberulens* and *Teucrium chamaedrys* subsp. *lydium*. Pharm Biol 44:592–598

Lakušić B, Lakušić D (2014) Morpho-anatomical differentiation of the species *Teucrium montanum* (Lamiaceae) in the Central Balkan Peninsula. Bot Serb 38:109–120

Lans C, Turner N, Khan T, Brauer G, Boepple W (2007) Ethnoveterinary medicines used for ruminants in British Columbia, Canada. J Ethnobiol Ethnomed 3:11. https://doi.org/10.1186/1746-4269-3-11

Lardos A (2006) The botanical *materia medica* of the *Iatrosophikon* – a collection of prescriptions from a monastery in Cyprus. J Ethnopharmacol 104:387–406

Leonti M (2011) The future is written: impact of scripts on the congnition, selection, knowledge and transmission of medicinal plant use and its implications for ethnobotany and ethnopharmacology. J Ethnopharmacol 134:542–555

Lev E (2007) Drugs held and sold by pharmacists of the Jewish community of medieval (11-14th centuries) Cairo according to lists of *materia medica* found at the Taylor–Schechter Genizah collection, Cambridge. J Ethnopharmacol 110:275–293

Leyel CF (1984) A modern herbal by Mrs M. Grieve. Penguin Books, Harrnondsworlh

Maccioni S, Baldini R, Tebano M, Cioni PL, Flamini G (2007) Essential oil of *Teucrium scorodonia* L. ssp. *scorodonia* from Italy. Food Chem 104:1393–1395

Malakov PY, Papanov GY (1983) Furanoid diterpenes from *Teucrium polium*. Phytochemistry 22:2791–2793

Mashayekhan A, Pourmajidian MR, Jalilvand H, Gholami MR, Teimouri MS (2015) Ethnobotanical survey of herbal remedies traditionally used in North Khorasan Province of Iran. Med Aromat Plant 4:3. https://doi.org/10.4172/2167-0412.1000192

Máthé A (ed) (2015) Medicinal and aromatic plants of the world: scientific, production, commercial and utilization aspects. Springer, Dordrecht

Menichini F, Conforti F, Rigano D, Formisano C, Piozzi F, Senatore F (2009) Phytochemical composition, anti-inflammatory and antitumour activities of four *Teucrium* essential oils from Greece. Food Chem 115:679–686

Menković N, Šavikin K, Tasić S, Zdunić G, Stešević D, Milosavljević S, Vincek D (2011) Ethnobotanical study on traditional uses of wild medicinal plants in Prokletije Mountains Montenegro. J Ethnopharmacol 133:97–107

Meusel H, Jäger E, Rauschert S, Weinert E (1978) Vergleichende chorologie der Zentraleuropäischen fora II. Gustav Fischer, Jena

Milošević-Djordjević O, Stošić I, Stanković M, Grujičić D (2013) Comparative study of genotoxicity and antimutagenicity of methanolic extracts from *Teucrium chamaedrys* and *Teucrium montanum* in human lymphocytes using micronucleus assay. Cytotechnology 65:863–869

Mosaddegh M, Naghibi F, Moazzeni H, Pirani A, Esmaeili S (2012) Ethnobotanical survey of herbal remedies traditionally used in Kohghiluyeh va Boyer Ahmad province of Iran. J Ethnopharmacol 141:80–95

Movahedi A, Basir R, Rahmat A, Charaffedine M, Othman F (2014) Remarkable anticancer activity of Teucrium polium on hepatocellular carcinogenic rats. Evid Based Complement Alternat Med 2014:726724. https://doi.org/10.1155/2014/726724

Mustafa B, Hajdari A, Krasniqi F, Hoxha E, Ademi H, Quave CL, Pieroni A (2012) Medical ethnobotany of the Albanian Alps in Kosovo. J Ethnobiol Ethnomed 8:6. https://doi.org/10.1186/1746-4269-8-6

Navarro T, El Oualidi J (2000) Synopsis of *Teucrium* L. (Labiatae) in the Mediterranean region and surrounding areas. Fl Medit 10:349–363

Nedelcheva A, Draganov S (2014) Bulgarian medical ethnobotany: the power of plants in pragmatic and poetic frames in ethnobotany and biocultural diversities in the Balkans. In: Pieroni A, Quave CL (eds) Ethnobotany and biocultural diversities in the Balkans. Springer, New York/Heidelberg/Dordrecht/London, pp 45–65

Nedelcheva A, Pieroni A, Dogan Y (2017) Folk food and medicinal botanical knowledge among the last remaining Yörüks of the Balkans. Acta Soc Bot Pol 86:3522. https://doi.org/10.5586/asbp.3522

Nencini C, Galluzzi P, Pippi F, Menchiari A, Micheli L (2014) Hepatotoxicity of *Teucrium chamaedrys* L. decoction: role of difference in the harvesting area and preparation method. Indian J Pharmacol 46:181–184

Oybak-Dönmez E, Inceoğlu Ö (1988) Pollen morphology of some *Teucrium* L. (Labiatae) species. Commun Fac Sci Univ Ank Sér C 6:133–146

Özdemir E, Alpınar K (2015) An ethnobotanical survey of medicinal plants in western part of central Taurus Mountains: Aladaglar (Nigde – Turkey). J Ethnopharmacol 166:53–65

Ozgen U, Mavi A, Terzi Z, Yildirim A, Coskun M, Houghton PJ (2006) Antioxidant properties of some medicinal Lamiaceae (Labiatae) species. Pharm Biol 44:107–112

Özkan G, Kuleasan H, Çelik S, Göktürk RS, Ünal O (2007) Screening of Turkish endemic *Teucrium montbretii* subsp. *pamphylicum* extracts for antioxidant and antibacterial activities. Food Control 18:509–512

Paksoy MY, Selvi S, Savranc A (2016) Ethnopharmacological survey of medicinal plants in Ulukıs, la (Nigde-Turkey). J Herb Med 2016:1–7

Perez Alvarez JC, Saez-Royuela F, Pena EG, Morante AL, Oses AV, Lorente JLM (2001) Acute hepatitis due to ingestion of *Teucrium chamaedrys* infusions. Gastroenterol Hepatol 24:240–243

Perry N, Court G, Bidet N, Court J, Perry EK (1996) European herbs with cholinergic activities: potential in dementia therapy. Int J Geriatr Psychiatry 11:1063–1069

Pieroni A (2008) Local plant resources in the ethnobotany of Theth, a village in the Northern Albanian Alps. Genet Resour Crop Evol 8:1197–1214

Pieroni A, Quave CL (2005) Traditional pharmacopoeias and medicines among Albanians and Italians in southern Italy: a comparison. J Ethnopharmacol 101:258–270

Pieroni A, Quave CL, Santoro RF (2004) Folk pharmaceutical knowledge in the territory of the Dolomiti Lucane, inland southern Italy. J Ethnopharmacol 95:373–384

Pieroni A, Dibra B, Grishaj G, Grishaj I, Maçai SG (2005) Traditional phytotherapy of the Albanians of Lepushe, Northern Albanian Alps. Fitoterapia 76:379–399

Piozzi F, Bruno M, Rosselli S, Maggio A (2005) Advances on the chemistry of furano-diterpenoids from *Teucrium* genus. Heterocycles 65:1221–1234

Polat R, Satıl F (2012) An ethnobotanical survey of medicinal plants in Edremit Gulf (Balıkesir – Turkey). J Ethnopharmacol 139:626–641

Polat R, Cakilcioglu U, Satıl F (2013) Traditional uses of medicinal plants in Solhan (Bingöl – Turkey). J Ethnopharmacol 148:951–963

Prescott TAK, Veitch NC, Simmonds MSJ (2011) Direct inhibition of calcineurin by caffeoyl phenylethanoid glycosides from *Teucrium chamaedrys* and *Nepeta cataria*. J Ethnopharmacol 137:1306–1310

Rafieian-Kopaei M (2012) Medicinal plants and the human needs. J Herbmed Pharmacol 1:1–2

Rajabalian S (2008) Methanolic extract of *Teucrium polium* L. potentiates the cytotoxic and apoptotic effects of anticancer drugs of vincristine, vinblastine and doxorubicin against a panel of cancerous cell lines. Exp Oncol 30:133–138

Rajaei P, Mohamadi N (2012) Ethnobotanical study of medicinal plants of Hezar Mountain Allocated in South East of Iran. Iran J Pharm Res 11:1153–1167

Rasekh HR, Khoshnood-Mansourkhani MJ, Kamalinejad M (2001) Hypolipidemic effects of *Teucrium polium* in rats. Fitoterapia 72:937–939

Redžić S (2007) The ecological aspect of ethnobotany and ethnopharmacology of population in Bosnia and Herzegovina. Coll Antropol 31:869–890

Redžić S (2010) Wild medicinal plants and their usage in traditional human therapy (Southern Bosnia and Herzegovina, W. Balkan). J Med Plant Res 4:1003–1027

Rojas A, Hernandez L, Pereda-Miranda R, Mata R (1992) Screening for antimicrobial activity of crude drug extracts and pure natural products from Mexican medicinal plants. J Ethnopharmacol 35:275–283

Safa O, Soltanipoor MA, Rastegar S, Kazemi M, Nourbakhsh Dehkordi K, Ghannadi A (2013) An ethnobotanical survey on Hormozgan province, Iran. Avicenna J Phytomed 3:64–81

Said O, Khalil K, Fulder S, Azaizeh H (2002) Ethnopharmacological survey of medicinal herbs in Israel, the Golan Heights and the West Bank region. J Ethnopharmacol 83:251–265

Sargin SA (2015) Ethnobotanical survey of medicinal plants in Bozyazı district of Mersin, Turkey. J Ethnopharmacol 173:105–126

Šarić-Kundalić B, Dobeš C, Klatte-Asselmeyer V, Saukel J (2010a) Ethnobotanical study on medicinal use of wild and cultivated plants in middle, south and west Bosnia and Herzegovina. J Ethnopharmacol 131:33–55

Šarić-Kundalić B, Fritz E, Dobeš C, Saukel J (2010b) Traditional medicine in the Pristine Village of Prokoško Lake on Vranica Mountain, Bosnia and Herzegovina. Sci Pharm 78:275–290

Šarić-Kundalić B, Dobeš C, Klatte-Asselmeyer V, Saukel J (2011) Ethnobotanical survey of traditionally used plants in human therapy of east, north and north-east Bosnia and Herzegovina. J Ethnopharmacol 133:1051–1076

Saroglou V, Arfan M, Shabir A, Hadjipavlou-Litina D, Skaltsa H (2007) Composition and antioxidant activity of the essential oil of *Teucrium royleanum* Wall. ex Benth growing in Pakistan. Flavour Fragr J 22:154–157

Savić J, Mačukanović-Jocić M, Jarić S (2019) Medical ethnobotany on the Javor Mountain (Bosnia and Herzegovina). Eur J Intern Med 27:52–64

Šavikin K, Zdunić G, Menković N, Živković J, Cujić N, Tereščenko M, Bigović D (2013) Ethnobotanical study on traditional use of medicinal plants in south-western Serbia, Zlatibor district. J Ethnopharmacol 146:803–810

Sbai-Jouilil H, Fadli A, Zidane L (2017) Survey of ethnomedicinal plants used for the treatment of gastrointestinal disorders in Seksaoua region (western high Moroccan Atlas). Ann Res Rev Biol 16:1–9

Sezik E, Yeşilada E, Honda G, Takaishi Y, Takeda Y, Tanaka T (2001) Traditional medicine in Turkey X. Folk medicine in Central Anatolia. J Ethnopharmacol 75:95–115

Sghaier MB, Harizi H, Louhichi T, Krifa M, Ghedira K, Chekir-Ghedira L (2011a) Anti- inflammatory and antiulcerogenic activities of leaf extracts and sesquiterpene from *Teucrium ramosissimum* (Lamiaceae). Immunopharmacol Immunotoxicol 33:656–662

Sghaier MB, Skandrani I, Nasr N, Franca MG, Chekir-Ghedira L, Ghedira K (2011b) Flavonoids and sesquiterpenes from *Tecurium ramosissimum* promote antiproliferation of human cancer cells and enhance antioxidant activity: a structure-activity relationship study. Environ Toxicol Pharmacol 32:336–348

Sharififar F, Dehghn-Nudeh G, Mirtajaldini M (2009) Major flavonoids with antioxidant activity from *Teucrium polium* L. Food Chem 112:885–888

Shawahna R, Jaradat NA (2017) Ethnopharmacological survey of medicinal plants used by patients with psoriasis in the West Bank of Palestine. BMC Complement Altern Med 17:4. https://doi.org/10.1186/s12906-016-1503-4

Shtukmaster A, Ljubuncic P, Bomzon A (2010) The effect of an aqueous extract of *Teucrium polium* on glutathione homeostasis in vitro: a possible mechanism of its hepatoprotectant action. Adv Pharmacol Sci 2010:938324. https://doi.org/10.1155/2010/938324

Stanković MS, Topuzović M, Solujić S, Mihailović V (2010) Antioxidant activity and concentration of phenols and flavonoids in the whole plant and plant parts of *Teucrium chamaedrys* L. var. *glanduliferum* Haussk. J Med Plant Res 4:2092–2098

Stanković M, Ćurčić M, Žižić J, Topuzović M, Solujić S, Marković S (2011a) *Teucrium* plant species as natural sources of novel anticancer compouds: antiproliferative, proapoptotic and antioxidant properties. Int J Mol Sci 12:4190–4205

Stanković M, Nićiforović N, Topuzović M, Solujić S (2011b) Total phenolic content, flavonoid concentrations and antioxidant activity, of the whole plant and plant parts extracts from *Teucrium montanum* L. var. *montanum*, f. *supinum* (L.) Reichenb. Biotechnol Biotechnol Equip 25:2222–2227

Stickel F, Egerer G, Seitz HK (2000) Hepatotoxicity of botanicals. Public Health Nutr 3:113–124

Suleiman MS, Abdul-Ghani AS, Al-Khalil S, Amin R (1988) Effect of *Teucrium polium* boiled leaf extract on intestinal motility and blood pressure. J Ethnopharmacol 22:111–116

Sundaresan PR, Slavoff SA, Grundel E, White KD, Mazzola E, Koblenz D, Rader JI (2006) Isolation and characterization of selected germander diterpenoids from authenticated *Teucrium chamaedrys* and *T. canadense* by HPLC, HPLC-mS and NMR. Phytochem Anal 17:243–250

Tanira MOM, Wasfi IA, Homsi MA, Bashir AK (1996) Neuromuscular and microvascular changes associated with chronic administration of an extract of *Teucrium stocksianum* in mice. J Pharm Pharmacol 48:1098–1102

Tanira MOM, Ali BH, Bashir AK, El-Sabban FF, Al Homsi M (1997) Neuromuscular and microvascular changes associated with chronic administration of an extract of *Teucrium stocksianum* in mice. J Pharm Pharmacol 49:301–304

Tariq M, Ageel AM, al-Yahya MA, Mossa JS, al-Said MS (1989) Antiinflammatory activity of *T. polium*. Int J Tissue React 11:185–188

Tutin TG, Heywood VH, Burges NA, Moore DM, Valentine DH, Walters SM, Webb DA (eds) (1972) Flora Europaea III. Cambridge University Press, Cambridge

Ulubelen A, Topcu G, Sonmez U (2000) Chemical and biological evaluation of genus *Teucrium*. Stud Nat Prod Chem 23:591–648

Ural O, Satilmiş Ö, Ura G, Dikici N (2011) A case: acute hepatitis associated with herbal (*Teucrium chamaedrys*) ingestion. Turk Hij Den Biyol Derg 68:135–138

Uritu CM, Mihai CT, Stanciu GD, Dodi G, Alexa-Stratulat T, Luca A, Leon-Constantin MM, Stefanescu R, Bild V, Melnic S, Tamba BI (2018) Medicinal plants of the family Lamiaceae in pain therapy: a review. Pain Res Manag 2018:7801543. https://doi.org/10.1155/2018/7801543

Van Wyk BE, De Wet H, van Heerden FR (2008) An ethnobotanical survey of medicinal plants in the southeastern Karoo, South Africa. S Afr J Bot 74:696–704

Vokou D, Katradi K, Kokkini S (1993) Ethnobotanical survey of Zagori (Epirus, Greece), a renowned centre of folk medicine in the past. J Ethnopharmacol 39:187–196

Volpato G, Kourková P, Zelený V (2012) Healing war wounds and perfuming exile: the use of vegetal, animal, and mineral products for perfumes, cosmetics, and skin healing among Sahrawi refugees of Western Sahara. J Ethnobiol Ethnomed 8:49

Vuković N, Milošević T, Sukdolak S, Solujić S (2007) Antimicrobial activities of essential oil and methanol extract of *Teucrium montanum*. Evid Based Complement Alternat Med 4:17–20

Wielgorskaya T (1995) Dictionary of generic names of seed plants. Columbia University Press, New York

World Checklist of Selected Plant Families. https://wcsp.science.kew.org/qsearch.do

Yin G, Zeng H, He M, Wang M (2009) Extraction of *Teucrium manghuaense* and evaluation of the bioactivity of its extract. Int J Mol Sci 10:4330–4341

Zarshenas MM, Zargaran A, Muller J, Mohagheghzadeh A (2013) Nasal drug delivery in traditional persian medicine. Jundishapur J Nat Pharm Prod 8:144–148

Zlatković B, Bogosavljević S, Radivojević A, Pavlović M (2014) Traditional use of the native medicinal plant resource of Mt. Rtanj (Eastern Serbia): Ethnobotanical evaluation and comparison. J Ethnopharmacol 151:704–713

Chapter 6
Phenolic Compounds Diversity of *Teucrium* Species

Vladimir Mihailović, Jelena S. Katanić Stanković, and Nevena Mihailović

Abstract A main phytochemical property of *Teucrium* species, apart from essential oils production, is the presence of phenolic compounds in their composition. The aim of this chapter was to review the main publications on the genus *Teucrium*, focusing on studies of phenolic compounds identification. A total of 78 different phenolic compounds have been reported in *Teucrium* species, such as phenolic acids, phenylethanoids, and flavonoids. Current literature describes the chemical richness of flavonoid metabolites present in *Teucrium* genus, making these compounds the most diverse group of phenolics identified in *Teucrium* plants and chemotaxonomically relevant for species belonging to this genus. The information presented in this chapter also describes all the known phenolic acids and phenylethanoids of *Teucrium* species as reported in the literature. Also, available data about the quantification of different phenolic compounds in these plants will be described and compared. Based on the published results, biosynthetic pathways for every class of phenolic compounds present in these species will also be provided. In addition, the methods for extraction of phenolic compounds from *Teucrium* species, as reported in the literature, will be compared and discussed. Taking into account that the genus *Teucrium* includes more than 300 species and that phenolic composition has been investigated for only 48 of them so far, this chapter aims to survey current literature, improve the knowledge and provide sufficient data for future research of phenolic compounds of *Teucrium* genus.

Keywords Phenolic acids · Phenylethanoids · Flavonoids · Biosynthesis of phenolic compounds

---

V. Mihailović (✉) · J. S. Katanić Stanković · N. Mihailović
Department of Chemistry, Faculty of Science, University of Kragujevac, Kragujevac, Serbia
e-mail: vladimir.mihailovic@pmf.kg.ac.rs

M. Stanković (ed.), *Teucrium Species: Biology and Applications*,
https://doi.org/10.1007/978-3-030-52159-2_6

Abbreviations

CHI	Chalcone isomerase
CHS	Chalcone synthase
DAHP	3-deoxy-D-arabino-heptulosonic acid 7-phosphate
DOPA	3,4-dihydroxyphenylalanine
EPSP	5-enolpyruvylshikimic acid 3-phosphate
F3H	Flavanone 3′-hydroxylase
FHT	Flavanone 3-hydroxylase
FLS	Flavonol synthase
FNS	Flavone synthase
GAE	Gallic acid equivalents
HPLC-DAD	High-performance liquid chromatography with diode array detector
HPLC-MS	High-performance liquid chromatography with mass spectrometry
HPLC-PDA	High-performance liquid chromatography with photodiode array detector
LC/DAD/ESI-MS^n	Liquid chromatography with diode array detector coupled to an electrospray ionization mass detector
PABA	*p*-aminobenzoic acid
TPC	Total phenolic content
UPLC–MS/MS	Ultra performance liquid chromatography-tandem mass spectrometry

6.1 Introduction

Phenolic compounds represent a widespread group of plant phytochemicals having a wide range of structures and functions (Rispail et al. 2005). They are considered to be the most important and numerous groups of secondary plant metabolites, which is ubiquitous in the plant kingdom. The term phenols include a very large and diverse group of chemical compounds which have one or more hydroxyl groups directly attached to the aromatic ring. Polyphenols are compounds with more than one phenolic hydroxyl group in its structure, attached to one or more benzene rings (Vermerris and Nicholson 2006).

Since polyphenols include a large group of compounds, with over 8000 characterized structures, there are several ways of their classification. These compounds are generally classified according to the number of phenolic rings they contain in their structure (phenolic acids, stilbenes, flavonoids, lignans, and tannins). The most accepted classification by Harborne and Simonds (1964) is based on the number of carbon atoms in the molecule. According to this classification phenolic compounds are classified to: simple phenols (C_6), phenolic acids – derivatives of benzoic acid (C_6-C_1), acetophenones and phenylacetic acids (C_6-C_2), cinnamic acids (C_6-C_3),

coumarins (C_6-C_3), flavonoids (C_6-C_3-C_6), biflavonoids (C_{30}), benzophenones (C_6-C_1-C_6), xanthones and stilbenes (C_6-C_2-C_6), quinones (C_6, C_{10}, C_{14}), tannins, lignans, and lignins (polymers). According to another type of classification, phenolic compounds are divided into simple phenols, phenylpropanoids, flavonoids, tannins, and quinones (Vermerris and Nicholson 2006).

Simple phenols generally include substituted phenols monomeric components (C_6) and acids that make up some plant tissues, such as lignins. Phenolic acids, classified as simple phenols, are divided into two subgroups: hydroxybenzoic acids and hydroxyacetic acids. Phenolic acids are rarely found free in plant species, and usually are present in the form of esters with various compounds (glucose, organic acids, flavonoids, terpenes, etc.). Well-known representatives of phenolic acids are gallic acid, salicylic acid, vanillic acid, protocatechuic acid, etc. (Brielmann et al. 2006).

The phenylpropanoids represent phenolic compounds consisting of three-carbon side chain attached to a phenol basis (C_6-C_3 skeleton). This class of phenolic compounds includes the hydroxycoumarins, phenylpropenes, and lignans. The most common phenylpropenes are hydroxycinnamates, including the caffeic, *p*-coumaric, ferulic, and sinapic acid. These are found in plants as conjugates with tartaric acid or quinic acid. Chlorogenic acids, principally 3-*O*-, 4-*O*- and 5-*O*-caffeoylquinic acids, found in high amount in coffee beans together with caffeic acid, are the main hydroxycinnamates of dietary significance (Crozier et al. 2009).

Flavonoids are the most extensive group of phenolic compounds found in plants. They represent a family of polyphenols consisting of 15 carbon atoms (Fig. 6.1) that

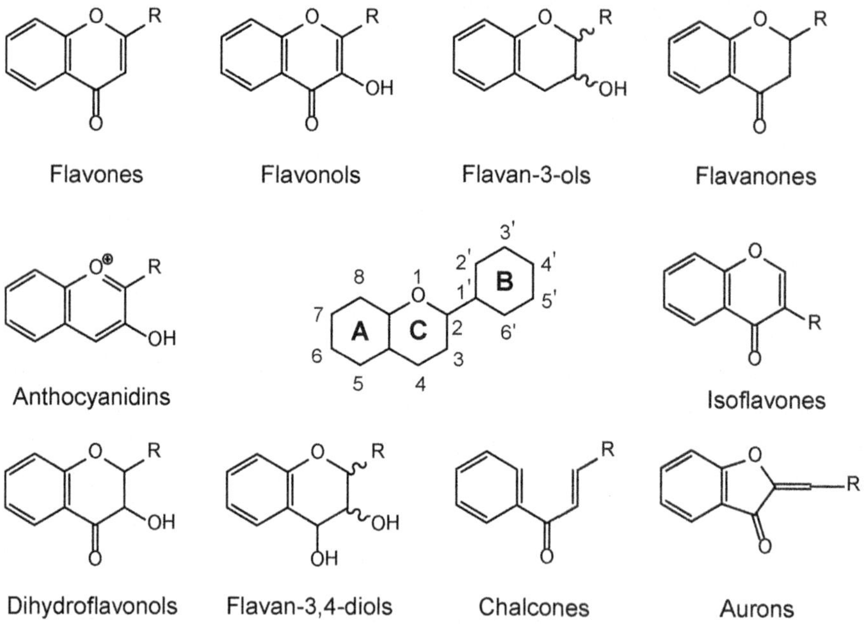

Fig. 6.1 Classes of structurally different flavonoids ($R = C_6H_5$)

are chemically defined to have a basic structure of two aromatic rings (A and B) linked via three carbon atoms that usually form an oxygenated heterocycle (C ring). This structure is most commonly described as C_6-C_3-C_6 (Ignat et al. 2011). The C ring is the key factor to define the different groups of flavonoids by providing different chemical structures containing various functional groups (hydroxy, methoxy, and glycosidic groups) and forming polymers (Fraga and Oteiza 2011). Flavonoids are usually divided into 6 main subclasses: flavonols, flavones, flavanols (flavan-3-ol), flavanones, anthocyanidins, and isoflavones (Fig. 6.1). The less represented subclasses include dihydroflavonols, flavan-3,4-diols, chalcones, dihydrochalcones, and aurons (Ignat et al. 2011).

Phenolic compounds are synthesized during the normal plant growth, but also as a response to various ecological conditions, such as, among others, stress and UV radiation (Naczk and Shahidi 2004). The synthesis of phenolic compounds in plants is performed by the biosynthetic pathway of the shikimic acid and is associated with the metabolism of carbohydrates (Isabelle et al. 2010; Steffen et al. 2003). Distribution and accumulation of phenolic compounds within the plant are not uniform but rather determined by the solubility of phenolic compounds. In that sense, water-soluble phenols are located in cellular vacuoles and other organelles, while lignans are concentrated mainly in the subepidermal and epidermal layers as the basic structural elements of the cell wall of plants, contributing to its mechanical resistance and defense against microorganisms (Naczk and Shahidi 2004). In addition, the presence of phenolic compounds in certain parts of the plant is conditioned by the stage of growth and development of the plant (Haminiuk et al. 2012).

Phenolic compounds include signaling molecules, pigments, and aromas that can attract or repel, as well as compounds that can protect plants against insects, fungus, bacteria, and viruses. The most important and most studied role of phenolic compounds of plant origin, which has an impact on human health, is their antioxidant activity. Oxidative stress can cause a range of degenerative diseases in humans, such as cancer, multiple sclerosis, autoimmune diseases, Parkinson's disease, etc. Antioxidants present in plants, e.g., phenolic acids, flavonoids, anthocyanins, and tannins, are often associated with positive effects of plants on human health (Haminiuk et al. 2012; Theriault et al. 2006).

The aim of this chapter is a comprehensive and comparative review of the phenolic compounds which are, in addition to volatile terpenoids, the most characteristic for the genus *Teucrium*. It will also review available data about the quantification of the phenolic compounds in these plants and biosynthetic pathways for every class of phenolic compounds present in these species. On the evidence so far, the methods for extraction of phenolic compounds from *Teucrium* species will be compared and discussed.

6.2 Phenolic Compounds of *Teucrium* Species

Species belonging to the genus *Teucrium* have extensively been the subject of phytochemical studies. These plants biosynthesize various compounds, predominantly phenolics, iridoid glycosides, and volatile terpenoids, found in the essential oils (De Marino et al. 2012; Frezza et al. 2019). Through phytochemical studies of *Teucrium* species, three groups of phenolic compounds, hydroxycinnamic acid derivatives, phenylethanoid glycosides, and flavonoid glycosides have been reported as the most abundant phenolics in these species. Previous phytochemical studies identified phenylethanoids as the major group of polyphenols in some *Teucrium* species (Lin et al. 2009; Mitreski et al. 2014), while specific flavonoids were reported as chemotaxonomic markers for the classification of the various species of this genus (Mitreski et al. 2014; Harborne et al. 1986). Numerous reports have been published on total phenolic content of species from the genus *Teucrium*. The results showed high percentage of total phenolics in these plants ranging from 7 to 93 mg GAE/g dry herb (Mitreski et al. 2014; Šamec et al. 2010; Zlatić et al. 2017), while total phenolic contents of examined *Teucrium* extracts were in the range from 53 to 255 mg GAE/g dry extract (Ben Sghaier et al. 2011; Khled-Khoudja et al. 2014a, b; Vlase et al. 2014; Belarbi et al. 2017; Nastić et al. 2018a, b; Farahmandfar et al. 2019) depending on the growing location, examined plant part or applied extraction procedures. This chapter will focus on major groups of phenolic compounds in the genus *Teucrium* – phenolic acids, phenylethanoids, and flavonoids.

6.2.1 Phenolic Acids

The genus *Teucrium* is characterized by the presence of diverse groups of phenolic compounds. Phenolic acids represent significant, but not so extensive group of compounds and they can be associated in the two classes: derivatives of benzoic acid (Fig. 6.2) and derivatives of cinnamic acid (Fig. 6.3) (Manach et al. 2004).

A comprehensive review of literature data on the phytochemical composition of plant species from genus *Teucrium* revealed the presence of several simple phenolic acids, derivatives of benzoic acid, in a few tested species (Table 6.1). The

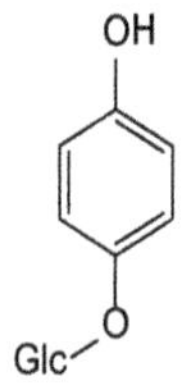

(1) *β*-Arbutin

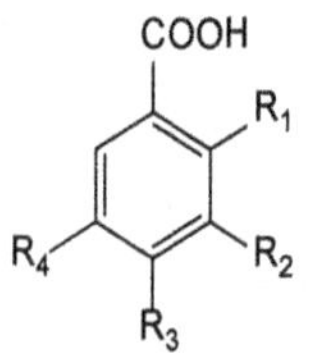

		R_1	R_2	R_3	R_4
(2)	Salicylic acid	OH	H	H	H
(3)	4-Hydroxybenzoic acid	H	H	OH	H
(4)	Protocatechuic acid	H	OH	OH	H
(5)	Gentisic acid	OH	H	H	OH
(6)	Gallic acid	H	OH	OH	OH
(7)	Vanillic acid	H	OCH_3	OH	H
(8)	Syringic acid	H	OCH_3	OH	OCH_3

Fig. 6.2 Benzoic acid derivatives identified in *Teucrium* species

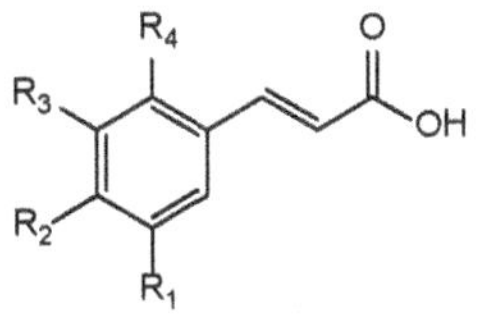

		R_1	R_2	R_3	R_4
(9)	*p*-Coumaric acid	H	OH	H	H
(10)	Caffeic acid	H	OH	OH	H
(11)	6-Hydroxy caffeic acid	OH	OH	H	OH
(12)	Ferulic acid	H	OH	OCH_3	H
(13)	Sinapic acid	OCH_3	OH	OCH_3	H

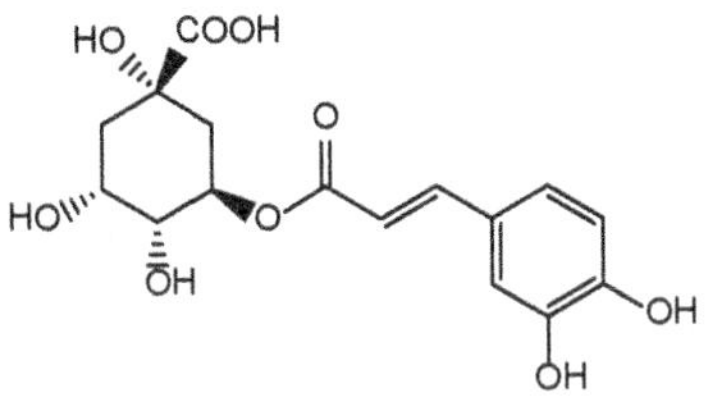

(14) Chlorogenic acid

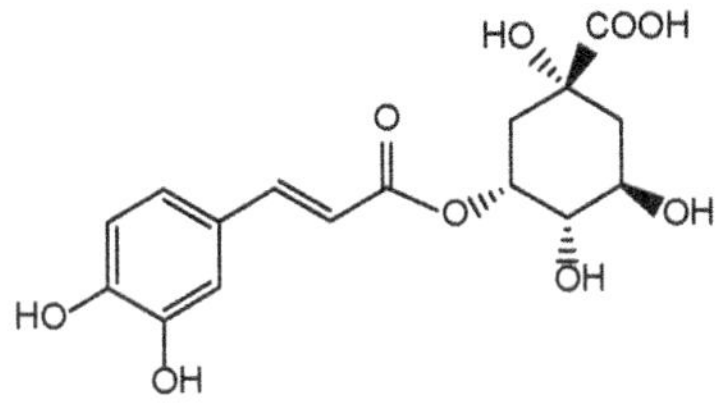

(15) Neochlorogenic acid

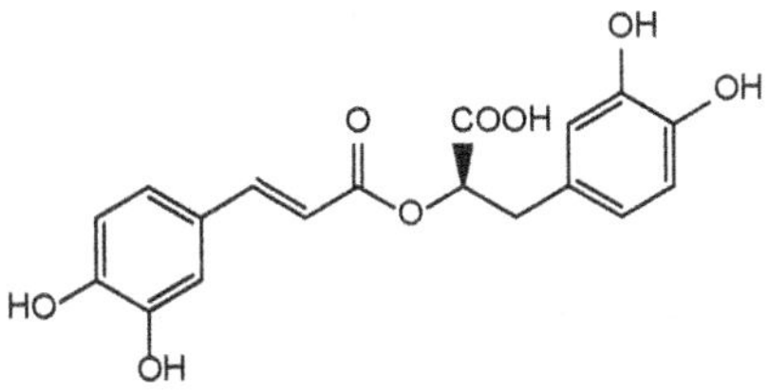

(16) Rosmarinic acid

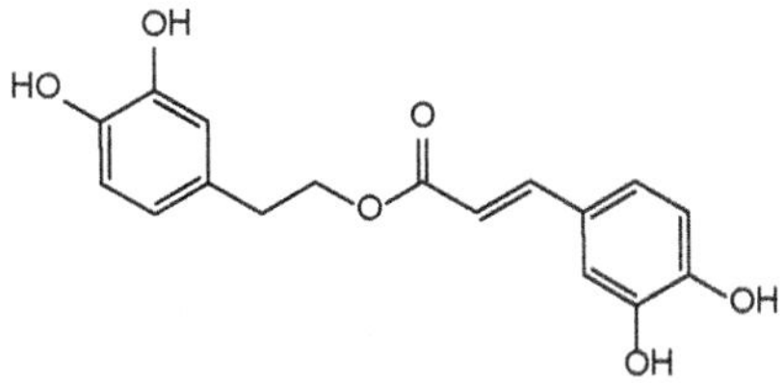

(17) Teucrol

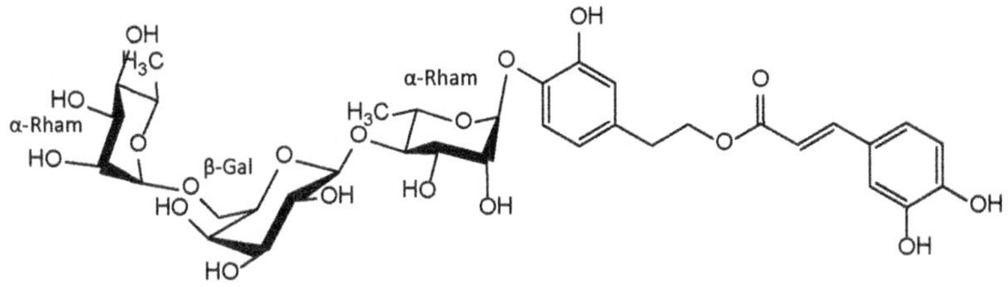

(18) Teucroside

Fig. 6.3 Hydroxycinnamic acids identified in *Teucrium* species

phytochemical composition of extracts or infusions made from *Teucrium* leaves or flowers was most frequently estimated using HPLC-DAD, HPLC-PDA, HPLC-MS, and UPLC–MS/MS analyses. Several scientific reports showed the presence of a total of eight compounds (Fig. 6.2) that can be classified as simple phenolic acids with the core structure of benzoic acid.

One of the phenolic acids derivatives (Fig. 6.2), *β*-arbutin (**1**), was identified in ethanolic extract of *Teucrium chamaedrys* aerial parts, collected in Italy. This glycosylated hydroquinone was defined by Frezza et al. (2018) as a new compound for the whole genus since it is a quite uncommon metabolite for the whole Lamiaceae family. Salicylic (**2**) and 4-hydroxybenzoic acids (**3**) were found only in *Teucrium arduini* flower and leaf extracts so far, while protocatechuic (**4**) and gentisic acids

Table 6.1 Benzoic acid derivatives identified in *Teucrium* species

No.	Compounds reported	*Teucrium* species	References
1.	*β*-Arbutin	*T. chamaedrys*	Frezza et al. (2018)
2.	Salicylic acid	*T. arduini*	Šamec et al. (2010)
3.	4-Hydroxybenzoic acid	*T. arduini*	Šamec et al. (2010)
4.	Protocatechuic acid	*T. chamaedrys*	Nastić et al. (2018a)
		T. montanum	Tumbas et al. (2004) and Nastić et al. (2018b)
		T. arduini	Šamec et al. (2010)
5.	Gentisic acid	*T. pseudoscorodonia*	Vlase et al. (2014)
		T. montanum	Tumbas et al. (2004)
		T. arduini	Šamec et al. (2010)
6.	Gallic acid	*T. polium*	Milošević-Djordjević et al. (2018)
		T. chamaedrys	Nastić et al. (2018a)
		T. montanum	Tumbas et al. (2004) and Nastić et al. (2018b)
7.	Vanillic acid	*T. polium, T. scordium*	Milošević-Djordjević et al. (2018)
		T. arduini	Šamec et al. (2010)
		T. chamaedrys	Nastić et al. (2018a)
		T. montanum	Tumbas et al. (2004) and Nastić et al. (2018b)
8.	Syringic acid	*T. pseudoscorodonia*	Belarbi et al. (2017)
		T. montanum	Tumbas et al. (2004)
		T. arduini	Šamec et al. (2010)

(**5**) were present, besides *Teucrium arduini,* also in *Teucrium montanum*. Moreover, protocatechuic acid was also identified in *Teucrium chamaedrys*, whilst gentisic acid was detected in *Teucrium pseudoscorodonia*. Gallic acid (**6**), as one of the most abundant phenolic acids in plant kingdom, was found in several *Teucrium* species, such as *Teucrium polium*, *T. chamaedrys*, and *T. montanum* in high quantities, particularly in subcritical water extracts of *T. chamaedrys* and *T. montanum* (217 and 345 mg/100 g dry extract, respectively). Generally, in eight plant species where benzoic acids were analyzed, the presence of vanillic acid (**7**) was observed in five of them (in *T. arduini*, *T. scordium, T. polium*, *T. chamaedrys*, and *T. montanum*). Literature sources revealed the presence of one more benzoic acid derivative, syringic acid (**8**) in *T. pseudoscorodonia*, *T. arduini*, and *T. montanum* extracts.

Besides simple phenolic acids, ten representatives of hydroxycinnamic acids or phenylpropanoids and their derivatives (Fig. 6.3) were identified in a notably higher number of *Teucrium* plant species, even 13 of them (Table 6.2). The hydroxycinnamic acids are significantly more represented in plants than the hydroxybenzoic acids and they are not often found in free form, but mostly as glycosylated derivatives or esters of shikimic, quinic, or tartaric acids (Manach et al. 2004). The presence of main secondary metabolites (Fig. 6.3) in plants obtained in the process of

Table 6.2 Hydroxycinnamic acids identified in *Teucrium* species

No.	Compounds reported	*Teucrium* species	References
9.	*p*-Coumaric acid	*T. chamaedrys*	Vlase et al. (2014) and Stanković (2012)
		T. montanum	Tumbas et al. (2004) and Stanković (2012)
		T. arduini	Šamec et al. (2010) and Stanković (2012)
		T. pseudoscorodonia	Belarbi et al. (2017)
		T. polium, T. scordium	Milošević-Djordjević et al. (2018) and Stanković (2012)
		T. botrys	Stanković (2012)
10.	Caffeic acid	*T. polium*	Proestos et al. (2006), Tepe et al. (2011), Milošević-Djordjević et al. (2018) and Stanković (2012)
		T. pseudoscorodonia	Belarbi et al. (2017)
		T. hyrcanicum	Oganesyan (2005)
		T. arduini	Šamec et al. (2010) and Stanković (2012)
		T. chamaedrys	Lin et al. (2009), Nastić et al. (2018a), Mitreski et al. (2014) and Stanković (2012)
		T. montanum	Tumbas et al. (2004), Nastić et al. (2018b), Mitreski et al. (2014) and Stanković (2012)
		T. scordium	Mitreski et al. (2014) and Stanković (2012)
		T. botrys	Stanković (2012)
11.	6-Hydroxy caffeic acid	*T. chamaedrys, T. flavum, T. fruticans, T. scorodonia*	Pedersen (2000)
12.	Ferulic acid	*T. polium*	Proestos et al. (2006) and Stanković (2012)
		T. pseudoscorodonia	Belarbi et al. (2017)
		T. chamaedrys	Nastić et al. (2018a) and Stanković (2012)
		T. arduini	Šamec et al. (2010), Kremer et al. (2013) and Stanković (2012)
		T. montanum	Tumbas et al. (2004), Nastić et al. (2018b) and Stanković (2012)
		T. scordium, T. botrys	Stanković (2012)
13.	Sinapic acid	*T. pseudoscorodonia*	Belarbi et al. (2017)
		T. scordium	Milošević-Djordjević et al. (2018)
		T. chamaedrys	Nastić et al. (2018a)
		T. arduini	Šamec et al. (2010)

(continued)

Table 6.2 (continued)

No.	Compounds reported	*Teucrium* species	References
14.	Chlorogenic acid	*T. canadense*	Lin et al. (2009)
		T. flavum, T. fruticans, T. marum	Pedersen (2000)
		T. polium	Milošević-Djordjević et al. (2018) and Stanković (2012)
		T. chamaedrys	Vlase et al. (2014), Nastić et al. (2018a) and Stanković (2012)
		T. arduini, T. montanum, T. botrys, T. scordium	Stanković (2012)
15.	Neochlorogenic acid	*T. chamaedrys*	Lin et al. (2009) and Mitreski et al. (2014)
		T. polium, T. montanum, T. scordium	Mitreski et al. (2014)
16.	Rosmarinic acid	*T. scorodonia*	Pedersen (2000)
		T. chamaedrys	Vlase et al. (2014) and Stanković (2012)
		T. arduini	Kremer et al. (2013) and Stanković (2012)
		T. pseudoscorodonia	Belarbi et al. (2017)
		T. montanum, T. polium, T. scordium, T. botrys	Stanković (2012)
17.	Teucrol	*T. pilosum* (syn. *T. japonicum* var. *pilosum*)	El-Mousallamy et al. (2000)
18.	Teucroside	*T. pilosum* (syn. *T. japonicum* var. *pilosum*)	El-Mousallamy et al. (2000)

phenylpropanoid biosynthesis, *p*-coumaric (**9**), caffeic (**10**), and ferulic (**12**) acids were confirmed in *T. chamaedrys*, *T. montanum*, *T. arduini*, *T. pseudoscorodonia*, *T. polium*, *T. scordium*, and *T. botrys* by different research groups. Chlorogenic acid (**14**) was identified in ten species while sinapic acid (**13**) was present in four *Teucrium* species. *p*-Coumaric acid was greatly present in *T. arduini* from Croatia, especially in flower infusions, ranging from 3.4 to 10.8 μg/g of dry weight (Šamec et al. 2010). On the other hand, much higher content of *p*-coumaric acid was found in the *Teucrium* extracts, particularly in *T. montanum* (1794 μg/g dry extract) (Tumbas et al. 2004).

Some of *Teucrium* species were able to synthesize rosmarinic acid (**16**), another very important ester of caffeic acid and 3, 4-dihydroxyphenyllactic acid. Moreover, El-Mousallamy et al. (2000) reported the presence of two derivatives of rosmarinic acid, a decarboxyrosmarinic acid or teucrol (**17**) and its' 4′-*O*-triglycoside teucroside (**18**) in *T. pilosum*.

6.2.2 Phenylethanoids and Phenylethanoid Glycosides

Besides phenolic acids, including phenylpropanoids, another very important class of phenolic secondary metabolites identified in *Teucrium* spp. are phenylethanoids and their glycosides. Phenylethanoid glycosides are secondary metabolites widely distributed in plants and they often appear together with iridoids in several plant families. Phenylethanoid glycosides are characterized by a hydroxyphenylethyl moiety (C_6-C_2 structure) to which a sugar molecule, mainly *β*-glucopyranose, is linked through glycosidic bond formation (Fu et al. 2008). Additional monosaccharides (i.e. rhamnose, apiose, galactose, xylose, etc.) and phenylpropanoid (C_6-C_3) residues are usually attached to the glucose center of the molecule (Alipieva et al. 2014). The synonyms caffeic acid glycoside esters or caffeoyl phenylethanoid glycosides are frequently in use instead of phenylpropanoid glycosides, because the most common phenylpropanoid residue in these molecules is caffeic acid (Luca et al. 2019). Verbascoside (**21**), also known as acteoside, the most common phenylethanoid glycoside in plants, has been identified in more than 200 plants (Schaluer et al. 2004). According to the previous studies, phenylethanoid glycosides have been the subject of many scientific researches in the past few decades. They possess numerous biological activities and pharmacological properties of some of them have been confirmed in certain clinical trials (Jimenez and Riguera 1994).

The phenylethanoids and phenylethanoid glycosides reported from the *Teucrium* genus are shown in Fig. 6.4. A total of 16 different phenylethanoids and phenylethanoid glycosides, listed in Table 6.3, have been recorded from 13 species belonging to *Teucrium* genus. The most common phenylethanoid glycoside reported from 12 species was verbascoside (**21**). Other most abundant phenylethanoid glycosides of *Teucrium* species are forsythoside B (**28**) reported in 4 species, poliumoside (**29**), and samioside (**31**) reported in 3 species listed in Table 6.3. Other phenylethanoids identified in *Teucrium* species were found only in one or two species.

All phenylethanoid glycosides identified in *Teucrium* genus contain caffeic acid as phenylpropanoid residue in these molecules, except leucoseptoside A (**23**) found in *T. polium*, and alyssonoside (**24**) found in *T. polium* (Mitreski et al. 2014) and *T. chamaedrys* (Mitreski et al. 2014; Frezza et al. 2018) which contain ferulic acid residue. Considering the number of monosaccharide moieties in identified phenylethanoids, phenylethanoid triglycosides are the most distributed class of these compounds in *Teucrium* genus. Among 16 reported phenylethanoids, 5 phenylethanoid diglycosides, verbascoside (**21**), 6-hydroxyacteoside (**22**), leucoseptoside A (**23**), isoverbascoside (**33**), and forsythoside A (**34**) were found in *Teucrium* species. Only two phenylethanoid aglycons, tyrosol (**1**) and 3,4-dihydroxy-*β*-phenylethanol (**2**), were found in species from this genus.

T. polium, *T. montanum*, and *T. chamaedrys* are the most studied plant species from *Teucrium* genus with the highest number of identified phenylethanoids. A total of 11 distinct phenylethanoid glycosides were reported in *T. polium* species, while 7 and 5 different phenylethanoids were identified in *T. montanum* and *T. chamaedrys*, respectively. 3,4-Dihydroxy-*β*-phenylethanol (**2**) was found only in

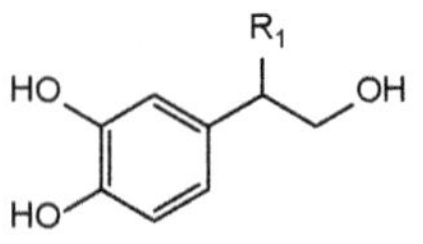

(19) Tyrosol R_1 = H
(20) 3,4-Dihydroxy-β-phenylethanol R_1 = OH

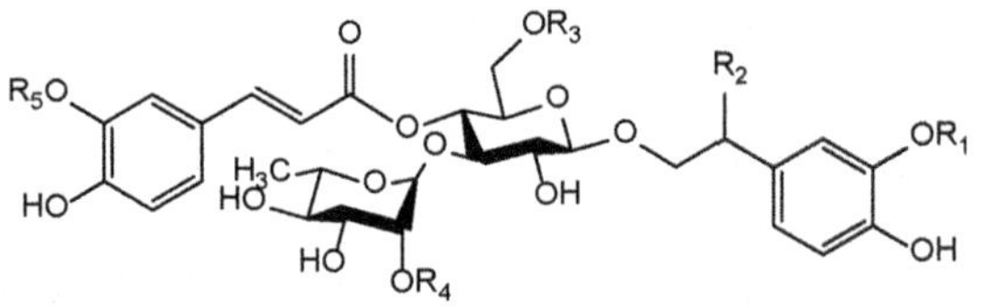

		R_1	R_2	R_3	R_4	R_5
(21)	Acteoside (Verbascoside)	H	H	H	H	H
(22)	6-Hydroxyacteoside	H	OH	H	H	H
(23)	Leucoseptoside A	H	H	H	H	CH_3
(24)	Alyssonoside	H	H	Apiose	H	CH_3
(25)	Caerulescenoside	H	H	H	Glucose	H
(26)	Echinacoside	H	H	Glucose	H	H
(27)	Castanoside A	CH_3	H	Glucose	H	H
(28)	Forsythoside B	H	H	Apiose	H	H
(29)	Poliumoside	H	H	Rhamnose	H	H
(30)	Poliumoside B	H	H	Rhamnose	Apiose	H
(31)	Samioside	H	H	H	Apiose	H
(32)	Teucreoside	H	H	H	Xylose	H

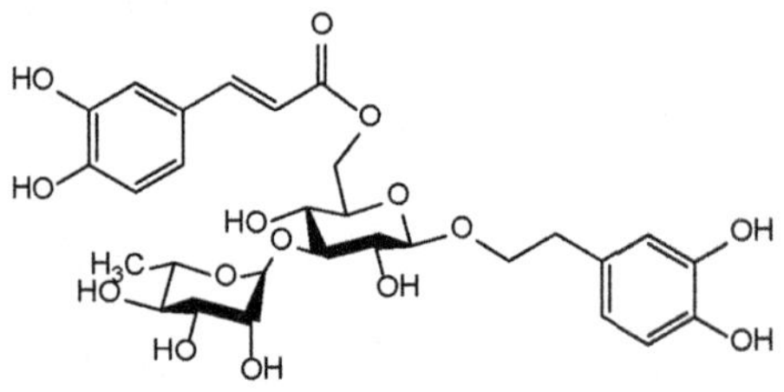

(33) Isoverbascoside

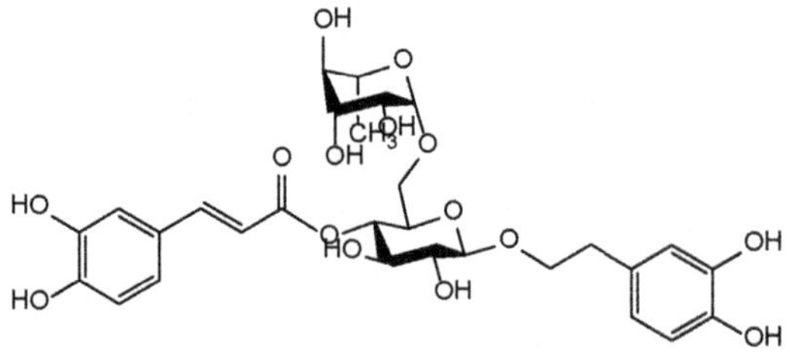

(34) Forsythoside A

Fig. 6.4 Phenylethanoids and phenylethanoid glycosides identified in *Teucrium* species

Table 6.3 Phenylethanoids and phenylethanoid glycosides identified in *Teucrium* species

No.	Compounds reported	*Teucrium* species	References
19.	Tyrosol	*T. polium*	Proestos et al. (2006)
20.	3,4-Dihydroxy-*β*-phenylethanol	*T. hyrcanicum*	Oganesyan (2005)
21.	Acteoside (Verbascoside)	*T. capitatum, T. flavum, T. fruticans, T. marum, T. scorodonia*	Pedersen (2000)
		T. polium	Pedersen (2000), Tepe et al. (2011) and Mitreski et al. (2014)
		T. chamaedrys	Frezza et al. (2018), Pedersen (2000), Mitreski et al. (2014) and Lin et al. (2009)
		T. montanum, T. scordium	Mitreski et al. (2014)
		T. polium var. *gnaphalodes*	Boghrati et al. (2016)
		T. canadense	Lin et al. (2009)
		T. belion (syn. *T. dunense*)	Andary et al. (1985)
22.	6-Hydroxyacteoside	*T. scorodonia*	Pedersen (2000)
23.	Leucoseptoside A	*T. polium*	Mitreski et al. (2014)
24.	Alyssonoside	*T. polium*	Mitreski et al. (2014)
		T. chamaedrys	Frezza et al. (2018) and Mitreski et al. (2014)
25.	Caerulescenoside	*T. montanum*	Mitreski et al. (2014)
26.	Echinacoside	*T. polium, T. montanum*	Mitreski et al. (2014)
		T. belion (syn. *T. dunense*)	Andary et al. (1985)
27.	Castanoside A	*T. montanum*	Mitreski et al. (2014)
28.	Forsythoside B	*T. chamaedrys*	Frezza et al. (2018) and Mitreski et al. (2014)
		T. polium , T. montanum, T. scordium	Mitreski et al. (2014)
29.	Poliumoside	*T. polium*	De Marino et al. (2012) and Venditti et al. (2017)
		T. polium var. *gnaphalodes*	Boghrati et al. (2016)
		T. chamaedrys	Mitreski et al. (2014)
		T. belion (syn. *T. dunense*)	Andary et al. (1985)
30.	Poliumoside B	*T. polium*	De Marino et al. (2012) and Mitreski et al. (2014)
31.	Samioside	*T. chamaedrys*	Frezza et al. (2018)
		T. polium, T. montanum	Mitreski et al. (2014)
32.	Teucreoside	*T. polium*	Tepe et al. (2011)
		T. chamaedrys	Lin et al. (2009) and Mitreski et al. (2014)
33.	Isoverbascoside	*T. canadense*	Lin et al. (2009)
34.	Forsythoside A	*T. polium, T. scordium*	Mitreski et al. (2014)
		T. belion (syn. *T. dunense*)	Andary et al. (1985)

T. hyrcanicum, while verbascoside (**21**) was the only identified phenylethanoid glycoside in *T. capitatum, T. flavum, T. fruticans,* and *T. marum.*

Only a few studies reported phenylethanoids contents in *Teucrium* species. According to the results of Mitreski et al. (2014), the verbascoside concentrations were 6.7 for *T. scordium*, 4.5 for *T. chamaedrys*, 2.8 for *T. polium,* and 2.0 mg/g dry herb for *T. montanum*. The same study showed that forsythoside B (**28**) was the most abundant phenylethanoid glycoside in these species, except *T. polium*, in the concentration range of 39.0 mg/g dry herb for *T. chamaedrys* to 10.2 mg/g dry herb for *T. montanum*, while samioside (**31**) was the dominant phenylethanoid glycoside for *Teucrium polium* in the concentration of 11.5 mg/g dry herb (Mitreski et al. 2014). Lin et al. (2009) showed that different *T. canadense* samples contain verbascoside in amounts from 0.23% to 9.66% of dry plant weight (**21**).

6.2.3 Flavonoids

Flavonoids represent the largest group of phenolic compounds found in *Teucrium* species, and totally 44 compounds belonging to this subclass of phenolics were identified in plants from this genus (Table 6.4). Four subclasses of flavonoids were detected in *Teucrium* species: flavones, flavonols, flavone-3-ols, and flavanones. Flavones are the major subclass of flavonoids recorded in these plants, represented by apigenin and luteolin derivatives (Figs. 6.5 and 6.6). As can be seen in Table 6.4, a total of 34 flavones have been reported from a total of 45 *Teucrium* species so far. Luteolin (**49**), identified in 42 *Teucrium* species, is the most common flavone within this genus. Also, luteolin 7-*O*-glucoside (**50**) and luteolin 7-*O*-rutinoside (**52**) are widely distributed flavone glycosides in these species found in 35 and 38 *Teucrium* plants, respectively (Table 6.4). On the other hand, apigenin (**35**) was identified only in 5 species, while its glycoside derivatives, such as apigenin 7-*O*-glucoside (**36**) and apigenin 7-*O*-rutinoside (**38**), were found in more than 20 different *Teucrium* plants. According to the studies dealing with flavonoids analysis of this genus, salvigenin (**43**), cirsimaritin (**44**) (Fig. 6.5), cirsiliol (**61**), and cirsilineol (**62**) (Fig. 6.6) were also reported as common flavone aglycones. Some of *Teucrium* species are able to synthesize flavone *C*-glycosides; vicenin-2 (**48**), as a common flavone *C*-glycoside found in these species, was observed in thirty of them. Harborne et al. (1986) reported the presence of two flavone *C*-glycosides, isoscutellarein 7-allosylglucoside (**37**) and hypolaetin 7-allosylglucoside (Fig. 6.5) in *T. chamaedrys* and *T. webbianum.*

As shown in Table 6.4, the most abundant flavonols reported from *Teucrium* species are quercetin glycosides, isoquercitrin (**69**) and rutin (**70**) (Fig. 6.7) which were identified in 13 and 14 *Teucrium* plants, respectively. The presence of quercetin (**68**) and myricetin (**75**) was observed in 6 *Teucrium* species. Flavan-3-ols (Fig. 6.7), catechin (**76**) and epicatechin (**77**), were reported by Nastić et al. (2018b) only for *T. chamaedrys* and *T. montanum*, while Milošević-Djordjević et al. (2018) confirmed the presence of catechin in *T. polium* and *T. scordium*. Flavanone glycoside

Table 6.4 Flavonoids identified in *Teucrium* species

No.	Compounds reported	*Teucrium* species	References
Flavones and flavone glycosides			
35.	Apigenin	*T. polium*	Sharififar et al. (2009), Milošević-Djordjević et al. (2018), Mitreski et al. (2014), Harborne et al. (1986), Carreiras et al. (1989) and Venditti et al. (2017)
		T. chamaedrys, T. montanum, T. scordium	Mitreski et al. (2014)
		T. pilosum (syn. *T. japonicum* var. *pilosum*)	Hawas et al. (2008)
36.	Apigenin 7-*O*-glucoside	*T. arduini, T. scorodonia, T. massiliense, T. asiaticum, T. spinosum, T. resupinatum, T. botrys, T. webbianum, T. divaricatum, T. villosum, T. flavum, T. intrincatum, T. fragile, T. marum, T. subspinosum, T. pyrenaicum, T. rotundifolium, T. buxifolium, T. freynii, T. thymifolium, T. cossoni, T. libanitis, T. cartaginensis, T. pumilum, T. carolipaui, T. capitatum, T. pii-fontii, T. gnaphalodes, T. eriocephalum, T. charidemi*	Harborne et al. (1986)
		T. polium	Kawashty et al. (1999) and Mitreski et al. (2014)
		T. chamaedrys, T. scordium	Harborne et al. (1986) and Mitreski et al. (2014)
		T. leucocladum	Kawashty et al. (1999)
37.	Apigenin 7-*O*-glucuronide	*T. polium*	Tepe et al. (2011)
		T. chamaedrys, T. canadense	Lin et al. (2009)
38.	Apigenin 7-*O*-rutinoside (Isorhoifolin)	*T. massiliense, T. asiaticum, T. resupinatum, T. webbianum, T. divaricatum, T. villosum, T. intrincatum, T. fragile, T. marum, T. subspinosum, T. buxifolium, T. freynii, T. cossoni, T. libanitis, T. cartaginensis, T. pumilum, T. carolipaui, T. pii-fontii*	Harborne et al. (1986)
		T. chamaedrys	Harborne et al. (1986) and Mitreski et al. (2014)
		T. scordium	Mitreski et al. (2014)
		T. polium	Venditti et al. (2017)
		T. polium var. *gnaphalodes*	Boghrati et al. (2016)

39.	Apigenin 5-galloylglucoside	*T. polium, T. leucocladum*	Kawashty et al. (1999)
40.	Apigenin *p*-coumaroylglucoside	*T. polium*	Mitreski et al. (2014)
41.	4′,7-Dimethoxy apigenin	*T. polium*	Sharififar et al. (2009) and Verykokidou-Vitsaropoulou and Vajias (1986)
42.	7-Methyl scutellarein	*T. hyrcanicum*	Oganesyan (2005)
43.	Salvigenin	*T. massiliense, T. divaricatum, T. villosum, T. subspinosum, T. pyrenaicum, T. rotundifolium, T. buxifolium, T. thymifolium, T. libanitis, T. cartaginensis, T. aragonense, T. pumilum, T. carolipaui, T. aureum, T. capitatum, T. pii-fontii, T. gnaphalodes, T. eriocephalum, T. haenseleri*	Harborne et al. (1986)
		T. polium	Rizk et al. (1986)
		T. pilosum (syn. *T. japonicum* var. *pilosum*)	Hawas et al. (2008)
44.	Cirsimaritin	*T. brevifolium, T. pseudochamaepitys, T. arduini, T. scorodonia, T. massiliense, T. asiaticum, T. spinosum, T. webbianum, T. divaricatum, T. villosum, T. intrincatum, T. fragile, T. marum, T. subspinosum, T. compactum, T. pyrenaicum, T. rotundifolium, T. buxifolium, T. freynii, T. thymifolium, T. cossoni, T. libanitis, T. cartaginensis, T. aragonense, T. pumilum, T. carolipaui, T. aureum, T. capitatum, T. pii-fontii, T. gnaphalodes, T. eriocephalum, T. haenseleri, T. charidemi*	Harborne et al. (1986)
		T. polium	Verykokidou-Vitsaropoulou and Vajias (1986), Mitreski et al. (2014), Harborne et al. (1986) and Venditti et al. (2017)
		T. chamaedrys, T. montanum	Harborne et al. (1986) and Mitreski et al. (2014)
		T. pilosum (syn. *T. japonicum* var. *pilosum*)	Hawas et al. (2008)
45.	Cirsimaritin 4′-glucoside	*T. arduini*	Harborne et al. (1986)
46.	5-hydroxy-6,7,3′,4′-tetramethoxyflavon	*T. fruticans, T. brevifolium, T. pseudochamaepitys, T. arduini, T. salviastrum, T. scordium, T. divaricatum, T. villosum, T. flavum, T. fragile, T. marum, T. subspinosum, T. compactum*	Harborne et al. (1986)

(continued)

Table 6.4 (continued)

No.	Compounds reported	*Teucrium* species	References
47.	3′,6-Dimethoxy apigenin	*T. polium*	Sharififar et al. (2009)
48.	Vicenin-2	*T. polium*	Harborne et al. (1986) and Kawashty et al. (1999)
		T. leucocladum	Kawashty et al. (1999)
		T. fruticans, T. brevifolium, T. pseudochamaepitys, T. arduini, T. massiliense, T. asiaticum, T. chamaedrys, T. webbianum, T. divaricatum, T. villosum, T. intrincatum, T. fragile, T. compactum, T. pyrenaicum, T. buxifolium, T. freynii, T. montanum, T. cossoni, T. libanitis, T. cartaginensis, T. aragonense, T. pumilum, T. carolipaui, T. aureum, T. capitatum, T. pii-fontii, T. haenseleri, T. charidemi	Harborne et al. (1986)
		T. pilosum (syn. *T. japonicum* var. *pilosum*)	Hawas et al. (2008)
49.	Luteolin	*T. fruticans, T. brevifolium, T. pseudochamaepitys, T. arduini, T. scorodonia, T. massiliense, T. salviastrum, T. asiaticum, T. scordium, T. spinosum, T. resupinatum, T. botrys, T. divaricatum, T. villosum, T. flavum, T. intrincatum, T. fragile, T. marum, T. subspinosum, T. pyrenaicum, T. rotundifolium, T. buxifolium, T. freynii, T. thymifolium, T. cossoni, T. libanitis, T. cartaginensis, T. aragonense, T. pumilum, T. carolipaui, T. aureum, T. capitatum, T. pii-fontii, T. gnaphalodes, T. eriocephalum, T. haenseleri, T. charidemi*	Harborne et al. (1986)
		T. polium	Harborne et al. (1986), Proestos et al. (2006), Tepe et al. (2011), Milošević-Djordjević et al. (2018) and Mitreski et al. (2014)
		T. chamaedrys	Harborne et al. (1986), Lin et al. (2009) and Vlase et al. (2014)
		T. hyrcanicum	Oganesyan (2005)
		T. montanum	Harborne et al. (1986) and Mitreski et al. (2014)
		T. pilosum (syn. *T. japonicum* var. *pilosum*)	Hawas et al. (2008)

(continued)

50.	Luteolin 7-*O*-glucoside	*T. arduini, T. scorodonia, T. massiliense, T. salviastrum, T. asiaticum, T. spinosum, T. resupinatum, T. botrys, T. divaricatum, T. villosum, T. flavum, T. intrincatum, T. fragile, T. marum, T. subspinosum, T. pyrenaicum, T. rotundifolium, T. buxifolium, T. freynii, T. thymifolium, T. cossoni, T. libanitis, T. cartaginensis, T. pumilum, T. carolipaui, T. aureum, T. capitatum, T. gnaphalodes, T. eriocephalum, T. haenseleri, T. charidemi*	Harborne et al. (1986)
		T. polium	Harborne et al. (1986), Tepe et al. (2011), De Marino et al. (2012), Mitreski et al. (2014) and Kawashty et al. (1999)
		T. chamaedrys	Harborne et al. (1986), Lin et al. (2009) and Mitreski et al. (2014)
		T. leucocladum	Kawashty et al. (1999)
		T. montanum, T. scordium	Harborne et al. (1986) and Mitreski et al. (2014)
51.	Luteolin 4′-*O*-glucoside	*T. polium*	De Marino et al. (2012)
52.	Luteolin 7-*O*-rutinoside	*T. fruticans, T. brevifolium, T. pseudochamaepitys, T. arduini, T. scorodonia, T. massiliense, T. salviastrum, T. asiaticum, T. spinosum, T. resupinatum, T. botrys, T. divaricatum, T. villosum, T. intrincatum, T. fragile, T. marum, T. subspinosum, T. pyrenaicum, T. rotundifolium, T. buxifolium, T. freynii, T. thymifolium, T. cossoni, T. libanitis, T. cartaginensis, T. aragonense, T. pumilum, T. carolipaui, T. aureum, T. capitatum, T. gnaphalodes, T. eriocephalum, T. haenseleri, T. charidemi*	Harborne et al. (1986)
		T. polium	Harborne et al. (1986), Tepe et al. (2011), De Marino et al. (2012) and Mitreski et al. (2014)
		T. chamaedrys	Harborne et al. (1986), Lin et al. (2009) and Mitreski et al. (2014)
		T. montanum, T. scordium	Harborne et al. (1986) and Mitreski et al. (2014)

(continued)

Table 6.4 (continued)

No.	Compounds reported	*Teucrium* species	References
53.	Luteolin 7-*O*-neohesperidoside	*T. polium*	De Marino et al. (2012)
54.	Luteolin 7-sambubioside	*T. scordium, T. spinosum, T. resupinatum, T. botrys, T. chamaedrys, T. flavum, T. compactum, T. rotundifolium, T. polium, T. pii-fontii*	Harborne et al. (1986)
55.	Diosmetin	*T. polium*	Tepe et al. (2011), Mitreski et al. (2014) and Venditti et al. (2017)
		T. chamaedrys	Lin et al. (2009)
		T. scordium	Mitreski et al. (2014)
		T. pilosum (syn. *T. japonicum* var. *pilosum*)	Hawas et al. (2008)
56.	Diosmetin 7-*O*-rutinoside	*T. spinosum, T. resupinatum, T. gnaphalodes, T. eriocephalum, T. haenseleri, T. charidemi*	Harborne et al. (1986)
		T. polium	Harborne et al. (1986) and Mitreski et al. (2014)
		T. montanum, T. scordium	Mitreski et al. (2014)
		T. chamaedrys	Lin et al. (2009) and Mitreski et al. (2014)
57.	Diosmetin *p*-coumaroylglucoside	*T. polium*	Mitreski et al. (2014)
58.	6-Hydroxyluteolin	*T. polium*	Harborne et al. (1986)
59.	6-*OH*-Luteolin 7-glucoside	*T. arduini, T. scorodonia, T. massiliense, T. asiaticum, T. flavum, T. rotundifolium, T. buxifolium, T. thymifolium, T. libanitis, T. cartaginensis, T. aragonense, T. pumilum, T. carolipaui, T. polium, T. aureum, T. capitatum, T. gnaphalodes,T. eriocephalum*	Harborne et al. (1986)
60.	6-*OH*-Luteolin 7-rhamnoside	*T. scorodonia, T. massiliense, T. buxifolium, T. thymifolium, T. libanitis, T. cartaginensis, T. aragonense, T. carolipaui, T. gnaphalodes,T. eriocephalum, T. haenseleri, T. charidemi*	Harborne et al. (1986)

61.	Cirsiliol	*T. fruticans, T. brevifolium, T. pseudochamaepitys, T. arduini, T. scorodonia, T. massiliense, T. salviastrum, T. asiaticum, T. spinosum, T. botrys, T. webbianum, T. divaricatum, T. villosum, T. flavum, T. fragile, T. marum, T. subspinosum, T. compactum, T. pyrenaicum, T. rotundifolium, T. buxifolium, T. freynii, T. thymifolium, T. cossoni, T. libanitis, T. cartaginensis, T. aragonense, T. pumilum, T. carolipaui, T. aureum, T. capitatum, T. pii-fontii, T. gnaphalodes, T. eriocephalum, T. haenseleri*	Harborne et al. (1986)
		T. polium	Rizk et al. (1986), Verykokidou-Vitsaropoulou and Vajias (1986), Harborne et al. (1986), Carreiras et al. (1989) and Mitreski et al. (2014)
		T. chamaedrys, T. montanum, T. scordium	Harborne et al. (1986) and Mitreski et al. (2014)
62.	Cirsilineol	*T. brevifolium, T. pseudochamaepitys, T. massiliense, T. scordium, T. resupinatum, T. divaricatum, T. villosum, T. marum, T. subspinosum, T. compactum, T. pyrenaicum, T. rotundifolium, T. buxifolium, T. freynii, T. montanum, T. thymifolium, T. libanitis, T. cartaginensis, T. aragonense, T. pumilum, T. carolipaui, T. aureum, T. capitatum, T. pii-fontii, T. gnaphalodes, T. eriocephalum, T. haenseleri, T. charidemi*	Harborne et al. (1986)
		T. polium	Harborne et al. (1986) and Venditti et al. (2017)
63.	Pedalitin	*T. hyrcanicum*	Oganesyan (2005)
64.	Eupatorin	*T. polium*	Verykokidou-Vitsaropoulou and Vajias (1986)
65.	Desmethoxycentaureidin	*T. pilosum* (syn. *T. japonicum* var. *pilosum*)	Hawas et al. (2008)
66.	6-Methoxyluteolin-7,3′,4′-trimethylether	*T. pilosum* (syn. *T. japonicum* var. *pilosum*)	Hawas et al. (2008)
67.	Isoscutellarein 7-allosylglucoside	*T. chamaedrys, T. webbianum*	Harborne et al. (1986)

(continued)

Table 6.4 (continued)

No.	Compounds reported	*Teucrium* species	References
68.	Hypolaetin 7-allosylglucoside	*T. chamaedrys, T. webbianum*	Harborne et al. (1986)
Flavonols and flavonol glycosides			
69.	Quercetin	*T. polium*	Stanković (2012) and Milošević-Djordjević et al. (2018)
		T. arduini	Stanković (2012) and Kremer et al. (2013)
		T. chamaedrys, T. montanum, T. botrys, T. scordium subsp. *scordioides*	Stanković (2012)
70.	Isoquercitrin	*T. chamaedrys*	Vlase et al. (2014)
		T. fruticans, T. brevifolium, T. pseudochamaepitys, T. botrys, T. compactum, T. montanum, T. aragonense, T. polium, T. aureum, T. capitatum, T. pii-fontii, T. haenseleri	Harborne et al. (1986)
71.	Rutin	*T. chamaedrys*	Stanković (2012), Vlase et al. (2014) and Zdraveva et al. (2018)
		T. pseudoscorodonia	Belarbi et al. (2017)
		T. montanum	Harborne et al. (1986), Stanković (2012) and Nastić et al. (2018b)
		T. scordium	Milošević-Djordjević et al. (2018)
		T. polium	Sharififar et al. (2009), Stanković (2012) and Milošević-Djordjević et al. (2018)
		T. arduini, T. scordium subsp. *scordioides*	Stanković (2012)
		T. botrys	Harborne et al. (1986) and Stanković (2012)
		T. fruticans, T. brevifolium, T. pseudochamaepitys, T. compactum, T. aragonense, T. haenseleri	Harborne et al. (1986)
72.	Quercitrin	*T. chamaedrys*	Vlase et al. (2014) and Mitreski et al. (2014)
		T. montanum	Mitreski et al. (2014)
73.	Isorhanmetin 3-glucoside	*T. fruticans, T. brevifolium, T. pseudochamaepitys, T. compactum, T. aragonense*	Harborne et al. (1986)

74.	Isorhamnetin 3-rutinoside	*T. fruticans, T. brevifolium, T. pseudochamaepitys, T. compactum, T. aragonense*	Harborne et al. (1986)
75.	Myricetin	*T. polium*	Stanković (2012) and Milošević-Djordjević et al. (2018)
		T. chamaedrys, T. arduini, T. montanum, T. botrys, T. scordium subsp. *scordioides*	Stanković (2012)
Flavan-3-ols			
76.	Catechin	*T. polium, T. scordium*	Milošević-Djordjević et al. (2018)
		T. chamaedrys	Nastić et al. (2018a)
		T. montanum	Nastić et al. (2018b)
77.	Epicatechin	*T. chamaedrys*	Nastić et al. (2018a)
		T. montanum	Nastić et al. (2018b)
Flavanone glycoside			
78.	Naringin	*T. montanum*	Nastić et al. (2018b)

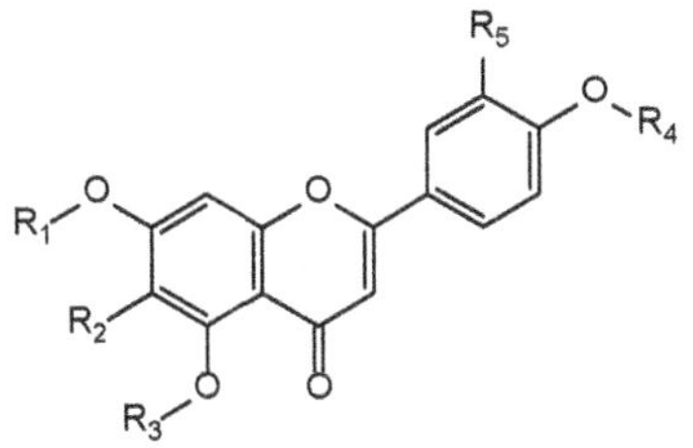

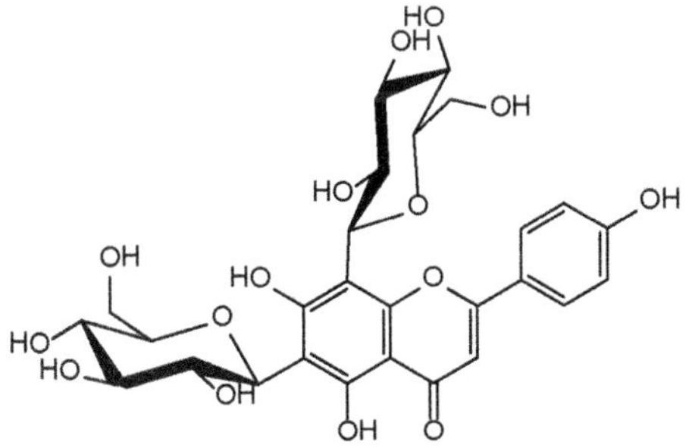

(48) Vicenin-2

	R_1	R_2	R_3	R_4	R_5
(35) Apigenin	H	H	H	H	H
(36) Apigenin 7-*O*-glucoside	Glucose	H	H	H	H
(37) Apigenin 7-*O*-glucuronide	Glucuronic acid	H	H	H	H
(38) Apigenin 7-*O*-rutinoside (Isorhoifolin)	Rutinose	H	H	H	H
(39) Apigenin 5-galloylglucoside	H	H	Glucose-galloyl	H	
(40) Apigenin *p*-coumaroyl glucoside	Glucose-*p*-coumaroyl	H	H	H	
(41) 4',7-Dimethoxy apigenin	CH_3	H	H	CH_3	H
(42) 7-Methyl scutellarein	CH_3	OH	H	H	H
(43) Salvigenin	CH_3	OCH_3	H	CH_3	H
(44) Cirsimaritin	CH_3	OCH_3	H	H	H
(45) Cirsimaritin 4'-glucoside	CH_3	OCH_3	H	Glucose	H
(46) 5-Hydroxy-6,7,3',4'-tetramethoxyflavon	CH_3	OCH_3	H	CH_3	OCH_3
(47) 3',6-Dimethoxy apigenin	H	OCH_3	H	H	OCH_3

Fig. 6.5 Flavones (apigenin and its derivatives) identified in *Teucrium* species

(Fig. 6.7), naringin (**78**), was currently reported (Nastić et al. 2018b) only for *T. montanum.*

T. polium, *T. chamaedrys*, and *T. montanum* are the most studied plants from *Teucrium* genus with the highest number of identified flavonoids. A total of 30 distinct flavonoids have been recorded in *T. polium* so far, while 22 and 17 flavonoids have been reported in *T. chamaedrys* and *T. montanum*, respectively. Apigenin 5-galloylglucoside (**39**), apigenin *p*-coumaroylglucoside (**40**), 4′,7-dimethoxy apigenin (**41**), 3′,6-dimethoxy apigenin (**47**), luteolin 4′-*O*-glucoside (**51**), luteolin 7-*O*-neohesperidoside (**53**), diosmetin *p*-coumaroylglucoside (**57**), 6-hydroxyluteolin (**58**), and eupatorin (**64**) were found only in *Teucrium polium*. 7-Methyl scutellarein (**42**) and pedalitin (**63**) were reported only in *T. hyrcanicum*, while cirsimaritin 4′-glucoside (**45**) was identified only in *T. arduini* so far.

Although there are numerous literature data about flavonoids composition of *Teucrium* plants, publications on their quantitative analysis are relatively rare. Mitreski et al. (2014) showed that flavonoid glycosides represent about 7% of total

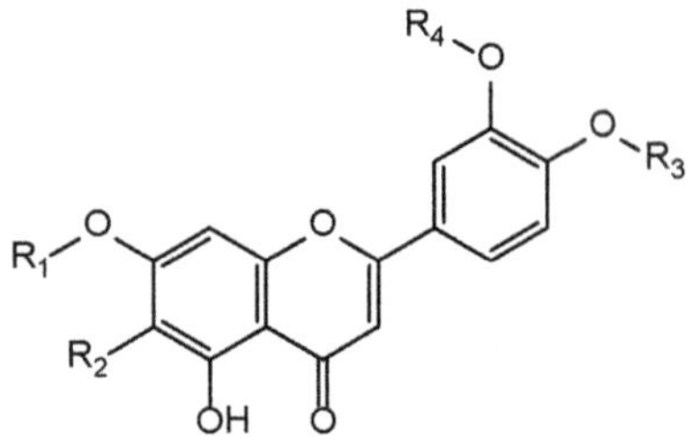

		R_1	R_2	R_3	R_4
(49)	Luteolin	H	H	H	H
(50)	Luteolin 7-*O*-glucoside	Glucose	H	H	H
(51)	Luteolin 4'-*O*-glucoside	H	H	Glucose	H
(52)	Luteolin 7-*O*-rutinoside	Rutinose	H	H	H
(53)	Luteolin 7-*O*-neohesperidoside	Neohesperidose	H	H	H
(54)	Luteolin 7-sambubioside	Sambubiose	H	H	H
(55)	Diosmetin	H	H	CH_3	H
(56)	Diosmetin 7-*O*-rutinoside	Rutinose	H	CH_3	H
(57)	Diosmetin *p*-coumaroylglucoside	Glucose-*p*-coumaroyl	H	CH_3	H
(58)	6-Hydroxyluteolin	H	OH	H	H
(59)	6-*OH*-Luteolin 7-glucoside	Glucose	OH	H	H
(60)	6-*OH*-Luteolin 7-rhamnoside	Rhamnose	OH	H	H
(61)	Cirsiliol	CH_3	OCH_3	H	H
(62)	Cirsilineol	CH_3	OCH_3	H	CH_3
(63)	Pedalitin	CH_3	OH	H	H
(64)	Eupatorin	CH_3	OCH_3	CH_3	H
(65)	Desmethoxycentaureidin	H	OCH_3	CH_3	H
(66)	6-Methoxyluteolin-7,3′,4′-trimethylether	CH_3	OCH_3	CH_3	CH_3

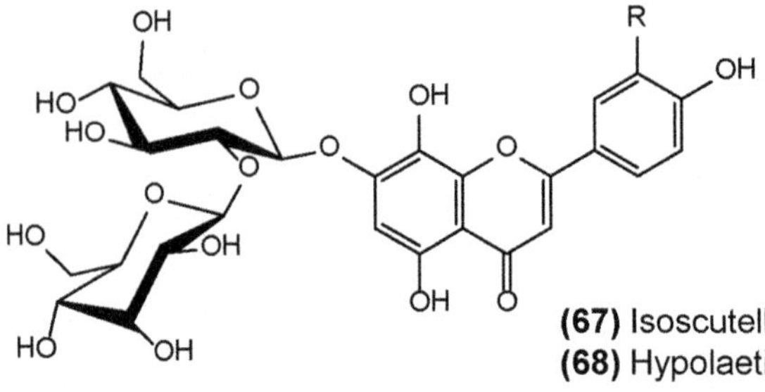

(67) Isoscutellarein 7-allosylglucoside R = H
(68) Hypolaetin 7-allosylglucoside R = OH

Fig. 6.6 Flavones (luteolin and its derivatives) identified in *Teucrium* species

phenolic content detected in *T. montanum* and *T. chamaedrys* and about 30% in *T. polium* and *T. scordium*. In the same study, rutin, luteolin, and apigenin 7-*O*-glycosides were found in the highest concentration among all identified flavonoids in *T. polium, T. chamaedrys, T. montanum*, and *T. scordium* extracts examined using LC/DAD/ESI-MS^n. According to the results of Proestos et al. (2006), *T. polium* contained 0.48 mg of quercetin in 100 g of dry plant. Vlase et al. (2014) quantified

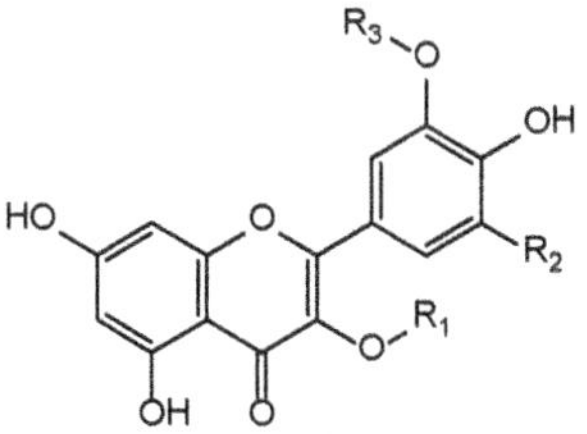

		R_1	R_2	R_3
(69)	Quercetin	H	H	H
(70)	Isoquercitrin	Glucose	H	H
(71)	Rutin	Rutinose	H	H
(72)	Quercitrin	Rhamnose	H	H
(73)	Isorhanmetin 3-glucoside	Glucose	H	CH_3
(74)	Isorhamnetin 3-rutinoside	Rutinose	H	CH_3
(75)	Myricetin	H	OH	H

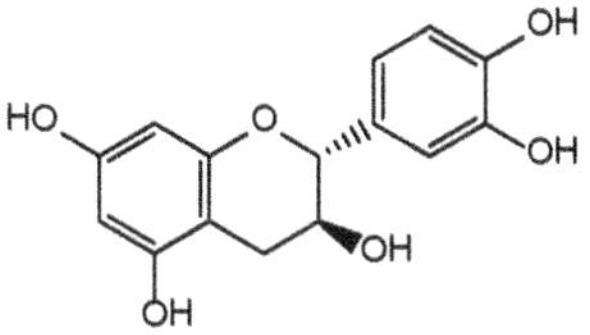

(76) Catechin

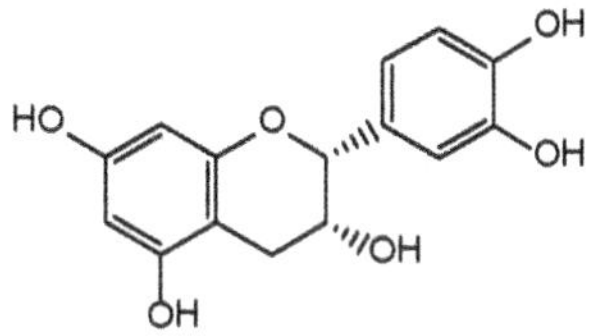

(77) Epicatechin

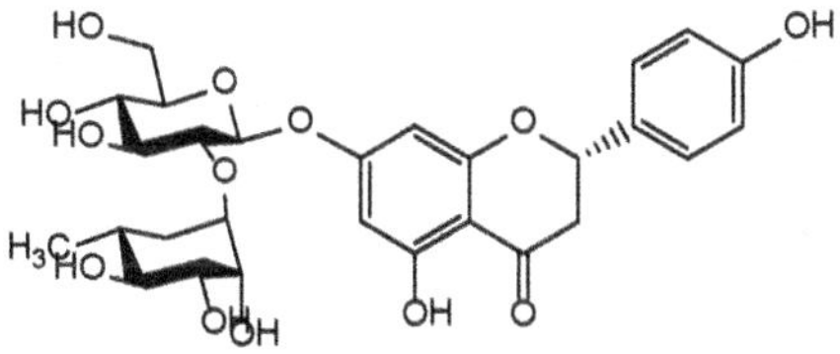

(78) Naringin

Fig. 6.7 Flavonols, flavan-3-ols and flavanones identified in *Teucrium* species

isoquercitrin, rutin, quercitrin, and luteolin in *T. chamaedrys* extract in the concentrations of 524.8, 85.42, 18.52, and 20.42 μg/g plant material, respectively, while catechin and epicatechin in subcritical water extract of *T. chamaedrys* were found in amounts of 73.7 and 54.2 mg/100 g extract, respectively (Nastić et al. 2018a). Subcritical water extraction of *T. montanum* showed that this extract contained epicatechin, rutin, and naringin in the concentrations of 120, 125, and 996 mg/100 g dry extract, respectively (Nastić et al. 2018b). The results of qualitative and quantitative analyses of flavonoids in *Teucrium* plants suggest that flavonoids represent the most diverse group of phenolic compounds present in these plant species, while their concentrations were lower compared to phenylethanoid glycosides.

6.3 Biosynthesis of Plant Phenolic Compounds

The shikimate pathway serves for the biosynthesis of the aromatic amino acids tryptophan, phenylalanine, and tyrosine. This pathway is common to plants and microorganisms, but not to animals, and because of that, the aromatic amino acids are placed in a group of the essential amino acids for humans (Dewick 2009; Vermerris and Nicholson 2006). The shikimate pathway got its name after shikimic acid, as the main intermediate compound in this pathway. Shikimic acid was named after

shikimino-ki or shikimi, the Japanese name for the fruits of star anise (*Illicum anisatum* L., Illicaceae), from which it was first isolated in 1885, many years before its role in plant metabolism was elucidated (Dewick 2009; Samuelsson and Bohlin 2015). Precursors for the formation of aromatic amino acids are phosphoenolpyruvate, from the glycolytic pathway, and erythrose 4-phosphate, from the pentose phosphate cycle (Fig. 6.8). Briefly, these two compounds condense to form firstly 3-deoxy-D-arabino-heptulosonic acid 7-phosphate (DAHP), catalyzed by DAHP synthase. Thereafter, the elimination of phosphate from DAHP and an intramolecular cyclisation occur to form 3-dehydroquinic acid, which can be reduced to quinic acid. The side reactions on 3-dehydroquinic acid can lead to the formation of some simple phenolic acids of C_6-C_1 skeleton type (protocatechuic and gallic acids). The removal of water in 3-dehydroquinic acid leads to 3-dehydroshikimic acid, from which shikimate is formed by the action of shikimate dehydrogenase. Shikimic acid is then converted with a series of enzymatic reactions to 5-enolpyruvylshikimic acid 3-phosphate (EPSP), which is transformed by chorismate synthase to chorismic acid. Chorismic acid itself represents a key precursor and branch point for two biosynthetic pathways, one is biosynthesis of tryptophan (L-Trp) and another is biosynthesis of phenylalanine (L-Phe) and tyrosine (L-Tyr) via formation of prephenic acid. The pathways to the formation of aromatic amino acids L-Phe and L-Tyr may vary in different organisms, where usually more than one route may be present in a particular species (Dewick 2009). Besides, chorismate is the starting point for biosynthesis of many other simple phenolic compounds (C_6-C_1), e.g., 4-hydroxybenzoic acid, *p*-aminobenzoic acid (PABA), and salicylic acid.

The synthesized aromatic amino acids are the precursors of the important class of phenolic compounds, the phenylpropanoids (C_6-C_3), as well as several other classes of phenolic compounds (flavonoids, monolignols, hydroxycinnamic acids, sinapoyl esters, coumarins, and stilbenes). After deamination of L-Phe and L-Tyr in plants and some microorganisms, cinnamic acid and 4-coumaric acid are formed, respectively. Thereupon, a sequence of hydroxylation and methylation reactions lead to the formation of metabolites called hydroxycinnamic acids (caffeic, ferulic, 5-hydroxyferulic, and sinapic acids, Fig. 6.8). *p*-Coumaric acid (4-coumaric acid) can be enzymatically converted to *p*-coumaroyl Coenzyme A which ends the general phenylpropanoid pathway, as further reactions lead to the biosynthesis of specific classes of compounds (Vermerris and Nicholson 2006). The mutual interaction of some hydroxycinnamic acids may lead to the formation of many other plant metabolites. For example, in the reaction of caffeoyl-CoA and quinic acid, chlorogenic acid is formed. Also, rosmarinic acid can be synthesized from L-Tyr and 4-coumaroyl-CoA via a range of enzymatic reactions. Gallic acid, besides direct synthesis in shikimate pathway, can be formed via a wide range of metabolites, such as phenylalanine, cinnamic, *p*-coumaric, caffeic, and protocatechuic acids (Dewick 2009; Vermerris and Nicholson 2006).

Starting molecule in flavonoids biosynthetic pathway, as the most abundant group of phenolic compounds in *Teucrium* species, is 4-hydroxycinnamoyl-CoA (*p*-coumaroyl-CoA) which is extended with three molecules of malonyl-CoA (Fig. 6.9). The resulting product of this reaction, catalyzed by the enzyme chalcone

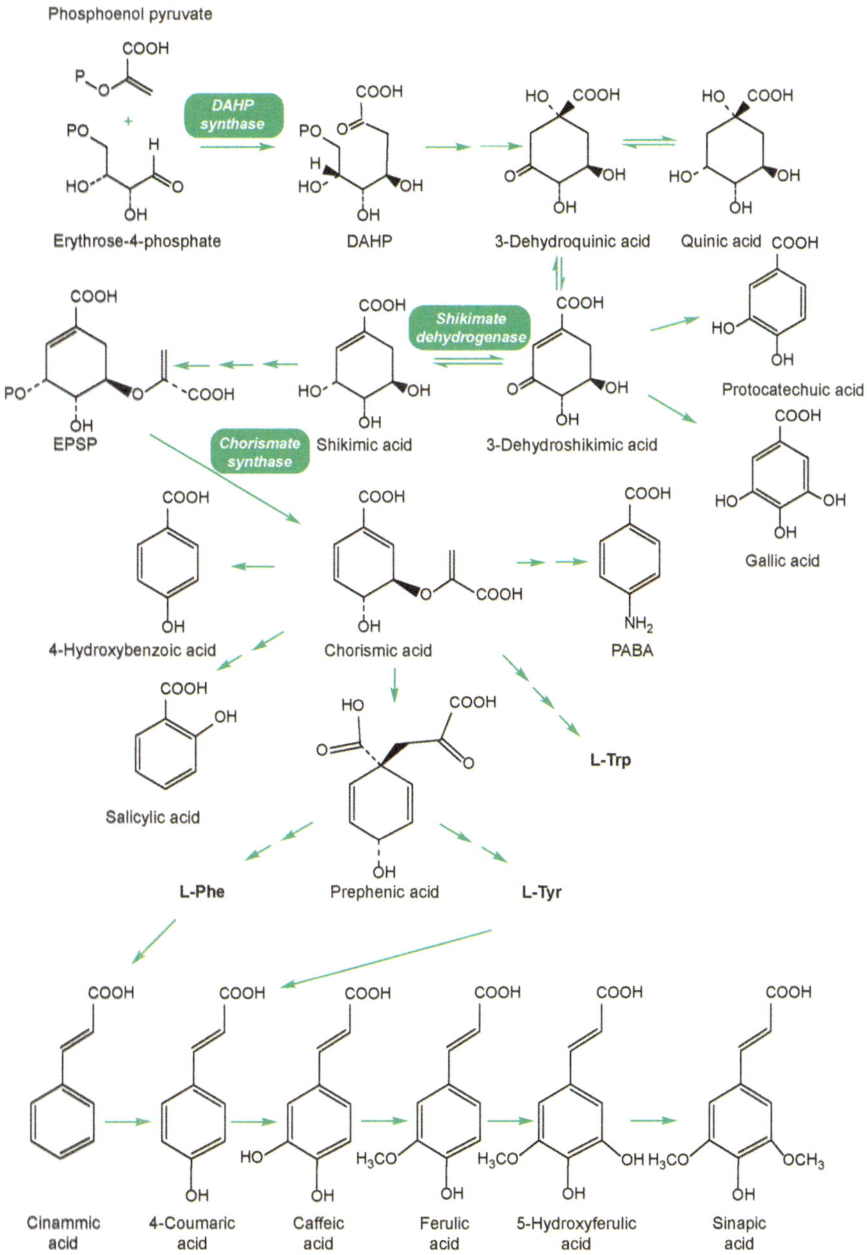

Fig. 6.8 Biosynthesis of shikimic acid with routes of aromatic amino acids (L-Trp, L-Phe, and L-Tyr) and phenylpropanoids synthesis

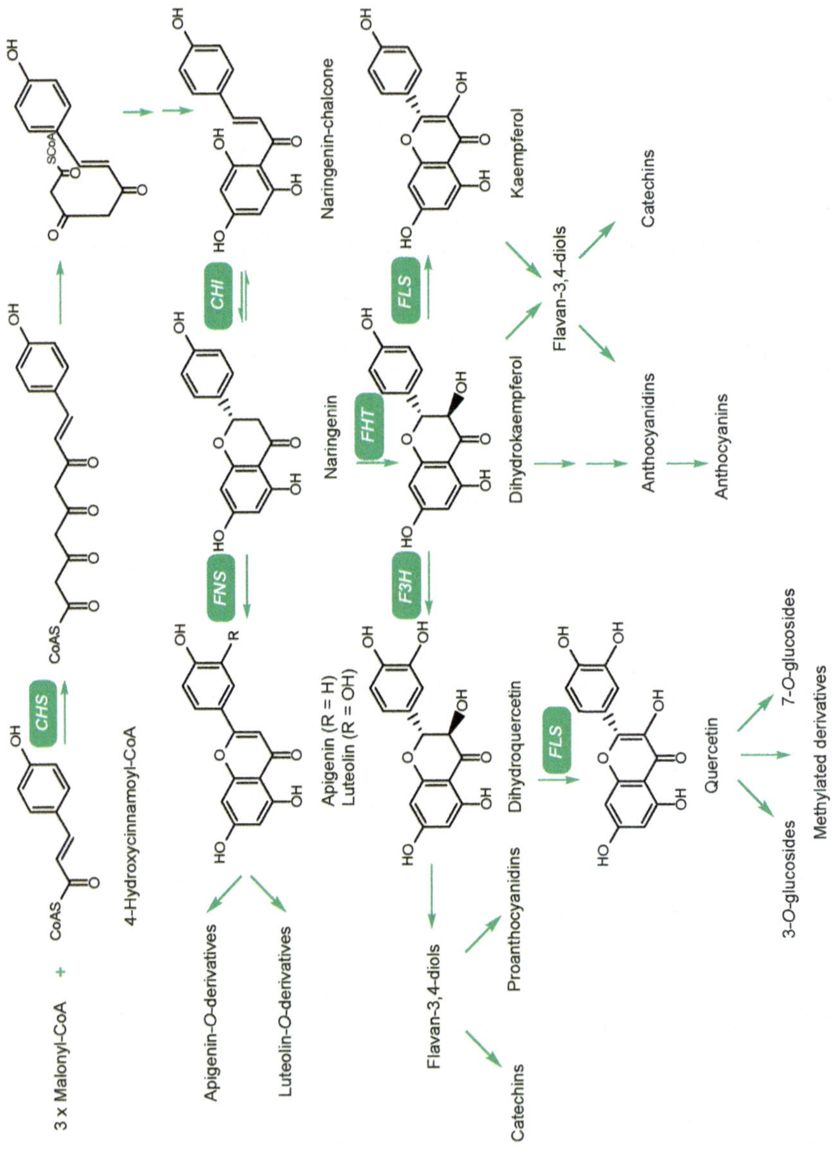

Fig. 6.9 Biosynthetic pathways for production of flavonoids. The enzymes: *CHS* chalcone synthase, *CHI* chalcone isomerase, *FNS* flavone synthase, *FHT* flavanone 3-hydroxylase, *F3H* flavanone 3′-hydroxylase, *FLS* flavonol synthase

synthase (CHS), is initially in a polyketide form, but it can be folded to generate aromatic rings. The key compound for further synthesis of flavonoids is naringenin-chalcone (4,2′,4′,6′ tetrahydroxychalcone) which is transformed by chalcone isomerase (CHI) via closing the pyrane ring to generate a flavanone naringenin. Naringenin-chalcone and chalcones generally are considered as precursors and can undergo a number of reactions to form different classes of polyphenolic metabolites, such as aurones (Dewick 1998, 2009).

Naringenin is subsequently converted by flavanone 3-hydroxylase (FHT) to the dihydrokaempferol (dihydroflavanonol). Besides that, the action of flavone synthase (FNS) on naringenin leads to the formation of flavones, apigenin (**35**) or luteolin (**49**), from which a wide range of metabolites identified in *Teucrium* species can be derived. Dihydrokaempferol can be converted by flavonol synthase (FLS) to the flavonol, kaempferol, but on the other hand, action of flavonoid 3′-hydroxylase (F3H) generates another dihydroflavonol, dihydroquercetin. This compound can be converted by FLS to its corresponding flavonol, quercetin (**69**), and thereafter to many quercetin derivatives. Kaempferol can also be enzymatically hydroxylated to quercetin (Dewick 1998). The most common form of flavonoids in plants is glycosylated for storage or transport, where the most often sugar attached to an OH group in the flavonoid is glucose (but also galactose, xylose, arabinose or rhamnose can be found) (Samuelsson and Bohlin 2015). Extension of the formed molecule with another sugar molecule takes place by addition to the first one, such as naringin (**78**, naringenin-7-hesperidoside), a compound that contains two sugar molecules, rhamnose and glucose. But, there is a possibility that two sugars can be attached to the core of the flavonoid structure, like in vicenin-2 (**48**).

As shown in Fig. 6.9, flavanones, particularly naringenin, can give rise to many different forms of their basic skeleton (C_6-C_3-C_6), leading to the formation of diverse groups of secondary metabolites in plants, e.g., flavones, flavonols, flavan-3,4-diols, catechins, anthocyanidins, proanthocyanidins, and anthocyanins.

One of the most studied biosynthetic pathways of phenylethanoid glycosides, another very important class of phenolic secondary metabolites identified in *Teucrium* spp., is a synthesis of acteoside (**21**, verbascoside) in plants, but it still remains not elucidated in detail. Tentative mechanism of biosynthesis, proposed by Alipieva et al. (2014) and Saimaru and Orihara (2010), is shown in Fig. 6.10. The acteoside molecule is comprised of phenylethanoid (C_6-C_2) and caffeic acid (C_6-C_3) moieties linked *via* α-rhamnopyranosyl-β-glucopyranose to form the disaccharide caffeoyl ester. It has been suggested that the biosynthesis of acteoside begins with the generation of aromatic acids, phenylalanine and tyrosine, in the pathway of shikimic acid. Thereafter, from L-Tyr can be synthesized the hydroxytyrosol moiety in acetoside, and caffeic acid moiety can be derived from L-Phe in phenylpropanoid pathway. Monohydroxyl compounds synthesized in this process, such as tyramine, tyrosol, and salidroside, can be effectively incorporated into acteoside, as well as dopamine and DOPA (Alipieva et al. 2014).

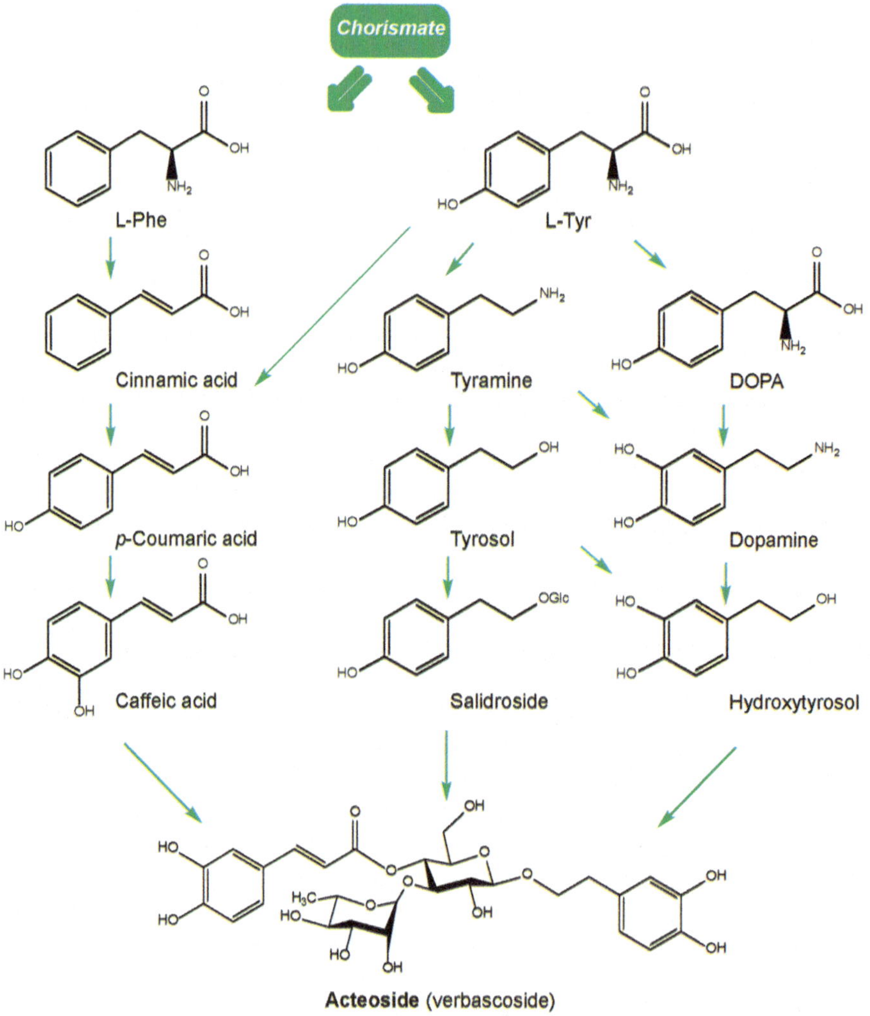

Fig. 6.10 Biosynthetic pathway of acteoside (tentative, adopted and modified from Alipieva et al. (2014) and Saimaru and Orihara (2010))

6.4 Extraction of Phenolic Compounds from *Teucrium* Species

There are a number of extraction methods that can be applied for the extraction of phenolic compounds from plant matrices, and generally can be divided into traditional and modern techniques. Traditional methods include maceration, maceration with mixing, and Soxhlet extraction. These methods are time-consuming, often giving low extraction yields, and also not suitable from the environmental protection

point of view due to the large volumes of used organic solvents. Nowadays, modern techniques for the extraction of bioactive compounds, such as ultrasonic extraction, microwave extraction, sub- and supercritical extraction, solid-phase microextraction, and accelerated solvent extraction, are increasingly used (Khoddami et al. 2013). Each of these techniques has its advantages and disadvantages, and the main goal of the chosen method is to achieve the maximum extraction yield of compounds of interest and to avoid their chemical modification (Ćujić et al. 2016).

A number of studies have shown that chosen extraction method can alter the total phenolic content (TPC) and antioxidant activity (Chan et al. 2007; Sikora et al. 2008; Yeh et al. 2014). The choice of the solvent for extraction is one of the main factors that have an influence on the phenolic compounds content. Phenolic compounds present in plant matrices have different characteristics and polarities, so they may not be soluble in a particular solvent (Turkmen et al. 2006). Many studies have shown that aqueous mixtures containing methanol, ethanol, acetone, and ethyl acetate are the most suitable solvents for the extraction of phenolic compounds in plant material. Ethanol is an appropriate solvent for polyphenol extraction. Methanol has been found to be more efficient for extraction of lower molecular weight phenolic compounds, while an application of aqueous acetone is appropriate for higher molecular weight flavonols extraction (Dai and Mumper 2010).

A wide range of extraction methods were applied for the extraction of phenolic compounds from different species of *Teucrium* genus, including traditional methods, such as maceration or Soxhlet's extraction, or modern techniques, such as ultrasonic-assisted and subcritical extraction. Farahmandfar et al. (2019) applied maceration and ultrasonic-assisted extraction with various solvents in their study, in order to analyze TPC in the aerial part of *T. polium*. The highest extraction yield was obtained by ultrasonic-assisted extraction using ethanol:water (50:50) solvent mixture. This is in agreement with the determined TPC, which was the highest (65.74 mg GAE/g extract) by applying the above-mentioned extraction technique and solvent. Considering the classic, maceration methods, this solvent mixture (ethanol:water = 50:50) also proved to be the most appropriate for TPC extraction from *T. polium* aerial part (60.50 mg GAE/g extract). No difference was found between ethanol:water (50:50) maceration and ethanol ultrasonic-assisted extraction, while ethanol maceration had intermediate content of TPC (55.29 mg GAE/g extract). Water maceration and water ultrasonic extraction showed the lowest values for TPC (47.42 and 49.60 mg GAE/g extract, respectively), which may be attributed to the presence of more nonphenol compounds (polar) in water extracts than in other extracts. Ben Sghaier et al. (2011) compared extraction yields and total phenolics, flavonoids, and tannins content in *T. ramosissimum* water (boiling), methanol and chloroform (Soxhlet's extraction) plant extracts. The aqueous extract showed the highest extraction yield (21.6%), followed by methanol extract (16.78%) and chloroform extract (6.53%). The results also showed that aqueous extract was more enriched in total phenolic compounds than the methanol and chloroform extract with TPC of 121.66 μg GAE/mg extract. However, the results of total flavonoid and tannin contents showed to be reversed considering aqueous and methanol *T. ramosissimum* extracts; methanol extract was significantly more enriched in

flavonoid and tannin compounds. The content of flavonoids and tannins in *T. ramosissimum* chloroform extract was negligible compared to methanol and aqueous extracts. Stanković et al. (2010) examined the influence of water and different organic solvents (methanol, acetone, ethyl acetate, and petroleum ether) on extraction yields and TPC in the whole plant and plant parts of *T. chamaerdys* using Soxhlet extraction. The results showed that the highest extraction yield was obtained using polar solvents, such as water and methanol. TPC in the examined extracts ranged from 15.98 to 208.17 mg GAE/g extract, and the highest TPC was found in leaf methanol extract, followed by methanol extract of the whole plant (169.50 mg GAE/g extract), while all ethyl acetate and petroleum ether extracts contained considerably lower amounts of phenolic compounds. Phenolic content of *T. chamaerdys* was determined by applying different extraction techniques and solvents in a number of studies. Considering the obtained results, it can be concluded that the highest TPC (243.65 mg GAE/g plant material) was determined in *T. chamaerdys* ethanol extract (70% EtOH at 60 °C) (Vlase et al. 2014). Also, a high amount of total phenols (176.74 mg GAE/g dry extract) in *T. chamaerdys* was obtained by applying environmentally-friendly extraction technique, such as subcritical water extraction on the temperature of 160 °C for 30 min. The application of such methods eliminates the use of organic solvents, contributing to safety, quality, and applicability of plant extracts (Nastić et al. 2018a). Nastić et al. (2018b) also applied a subcritical water technique for the phenolic compounds extraction of *T. montanum* species and examined the influence of extraction pressure and temperature on TPC. Extracts obtained at a temperature of 160 °C and 10 bar showed the highest TPC. Under optimal extraction conditions, the extraction yield of *T. montanum* was 42.63%, while TPC was 174.61 mg GAE/g dry extract.

According to literature data, it can be concluded that modern extraction techniques, such as ultrasonic-assisted extraction, using aqueous organic solvents (methanol and ethanol) and subcritical water extraction are the most effective methods for the extraction of phenolic compounds in plant material for the most of *Teucrium* species.

6.5 Conclusions

Combining the known chemical data about the phenolic composition of genus *Teucrium*, 78 different phenolic compounds have been reported from 48 species. A total of 44 different flavonoids, including flavonoid aglycons, and their *O*-glycoside and *C*-glycoside forms, have been identified in plants from this genus so far, representing the largest class of phenolic compounds of *Teucrium* species. Flavones represent the most abundant subclass of flavonoids in this genus with a total of 34 compounds reported so far. Luteolin is the most common flavone identified in 42 *Teucrium* species. Among flavone derivatives, luteolin and apigenin glycosides are also frequently found in these species, suggesting that these compounds may be chemotaxonomic markers for genus *Teucrium*.

Teucrium genus is also characterized by the presence of phenylethanoids and derivatives of benzoic and hydroxycinnamic acids. We have recorded a total of 16 different phenylethanoids and 15 phenolic acids (7 of them belonging to the group of hydroxybenzoic acids and 8 to the hydroxycinnamic acids group). The hydroxybenzoic acids have been rarely investigated in *Teucrium* species and their presence was confirmed only in 6 species, while hydroxycinnamic acids and phenylethanoids have been reported in 13 species. The most common phenylethanoid is verbascoside, found in 12 investigated plants from *Teucrium* genus. According to the quantitative analysis of phenolic compounds in *Teucrium* plants, phenylethanoids are present in higher concentration compared with flavonoids and phenolic acids.

The genus *Teucrium* includes more than 300 species, but an analysis of phenolic compounds was reported only for 48 of them so far. Taking into account the limited number of investigated *Teucrium* species, the diversity of phenolic compounds in these species is probably significantly higher. Considering the number of unexplored *Teucrium* species and biological importance of phenolic compounds identified in plants from this genus, many possibilities for the further investigation of the chemical composition and application of neglected plants of the *Teucrium* genus are opened.

Acknowledgments This work was supported by the Ministry of Education, Science and Technological Development of the Republic of Serbia (Agreements No. 451-03-68/2020-14/200122 and No. 451-03-68/2020-14/200378).

References

Alipieva K, Korkina L, Orhan IE, Georgiev MI (2014) Verbascoside – a review of its occurrence, (bio)synthesis and pharmacological significance. Biotechnol Adv 32:1065–1076

Andary C, Wylde R, Heitz A, Rascol JP, Roussel JL, Laffite C (1985) Poliumoside, a caffeic glycoside ester from *Teucrium belion*. Phytochemistry 24:362–364

Belarbi K, Atik-Bekkara F, El Haci IA, Bensaid I, Beddou F, Bekhechi C (2017) In vitro antioxidant activity and phytochemical analysis of *Teucrium pseudo-Scorodonia* Desf. Collected from Algeria. Orient Pharm Exp Med 17:151–160

Ben Sghaier M, Boubaker J, Skandrani I, Bouhlel I, Limem I, Ghedira K, Chekir-Ghedira L (2011) Antimutagenic, antigenotoxic and antioxidant activities of phenolic-enriched extracts from *Teucrium ramosissimum*: combination with their phytochemical composition. Environ Toxicol Pharmacol 31:220–232

Boghrati Z, Naseri M, Rezaie M, Pham N, Quinn RJ, Tayarani-Najaran Z, Iranshahi M (2016) Tyrosinase inhibitory properties of phenylpropanoid glycosides and flavonoids from *Teucrium polium* L. var. *gnaphalodes*. Iran J Basic Med Sci 19:804–811

Brielmann HL, Setzer WN, Kaufman PB, Kirakosyan A, Cseke LJ (2006) Phytochemicals: the chemical components of plants. In: Cseke LJ, Kirakosyan A, Kaufman PB, Warber SL, Duke JA, Brielmann HL (eds) Natural products from plants. CRC Press, Taylor & Francis Group, Boca Raton, pp 1–49

Carreiras MC, Rodríguez B, Piozzi F, Savona G, Torres MR, Perales A (1989) A chlorine-containing and two 17β-neo-clerodane diterpenoids from *Teucrium polium* subsp. *vincentinum*. Phytochemistry 28:1453–1461

Chan EWC, Lim Y, Omar M (2007) Antioxidant and antibacterial activity of leaves of *Etlingera* species (Zingiberaceae) in Peninsular Malaysia. Food Chem 104:1586–1593

Crozier A, Jaganath IB, Clifford MN (2009) Dietary phenolics: chemistry, bioavailability and effects on health. Nat Prod Rep 26:1001–1043

Ćujić N, Šavikin K, Janković T, Pljevljakušić D, Zdunić G, Ibrić S (2016) Optimization of polyphenols extraction from dried chokeberry using maceration as traditional technique. Food Chem 194:135–142

Dai J, Mumper RJ (2010) Plant phenolics: extraction, analysis and their antioxidant and anticancer properties. Molecules 15:7313–7352

De Marino S, Festa C, Zollo F, Incollingo F, Raimo G, Evangelista G, Iorizzi M (2012) Antioxidant activity of phenolic and phenylethanoid glycosides from *Teucrium polium* L. Food Chem 133:21–28

Dewick PM (1998) The biosynthesis of shikimate metabolites. Nat Prod Rep 1:17–58

Dewick PM (2009) Medicinal natural products – a biosynthetic approach, 3rd edn. Wiley, Southern Gate/Chichester

El-Mousallamy AMD, Hawas UW, Hussein SAM (2000) Teucrol, a decarboxyrosmarinic acid and its 4′-O-triglycoside, teucroside from *Teucrium pilosum*. Phytochemistry 55:927–931

Farahmandfar R, Asnaashari M, Bakhshandeh T (2019) Influence of ultrasound-assist and classical extractions on total phenolic, tannin, flavonoids, tocopherol and antioxidant characteristics of *Teucrium polium* aerial parts. J Food Meas Charact 13:1357–1363. https://doi.org/10.1007/s11694-019-00051-5

Fraga CG, Oteiza PI (2011) Dietary flavonoids: role of (−)-epicatechin and related procyanidins in cell signaling. Free Radic Biol Med 51:813–823

Frezza C, Venditti A, Matrone G, Serafini I, Foddai S, Bianco A, Serafini M (2018) Iridoid glycosides and polyphenolic compounds from *Teucrium chamaedrys* L. Nat Prod Res 32:1583–1589

Frezza C, Venditti A, Serafini M, Bianco A (2019) Phytochemistry, chemotaxonomy, ethnopharmacology, and nutraceutics of Lamiaceae. In: Atta-ur R (ed) Studies in natural products chemistry, vol 62. Elsevier, Amsterdam, pp 125–178

Fu G, Pang H, Wong Y (2008) Naturally occurring phenylethanoid glycosides: potential leads for new therapeutics. Curr Med Chem 15:2592–2613

Haminiuk CWI, Maciel GM, Plata-Oviedo MSV, Peralta RM (2012) Phenolic compounds in fruits – an overview. Int J Food Sci Technol 47:2023–2044

Harborne JB, Simmonds NW (1964) The natural distribution of the phenolic aglycones. In: Harborne JB (ed) Biochemistry of phenolic compounds. Academic, Waltham, pp 77–127

Harborne JB, Tomás-Barberán FA, Williams CA, Gil MI (1986) A chemotaxonomic study of flavonoids from European *Teucrium* species. Phytochemistry 25:2811–2816

Hawas UW, Sharaf M, El-Ansari MA (2008) Phytochemical studies and biological activity of *Teucrium pilosum*. Curr Chem Biol 2:174–177

Ignat I, Volf I, Popa VI (2011) A critical review of methods for characterisation of polyphenolic compounds in fruits and vegetables. Food Chem 126:1821–1835

Isabelle M, Lee BL, Lim MT, Koh WP, Huang D, Ong CN (2010) Antioxidant activity and profiles of common vegetables in Singapore. Food Chem 123:77–84

Jimenez C, Riguera R (1994) Phenylethanoid glycosides in plants: structure and biological activity. Nat Prod Rep 11:591–606

Kawashty SA, Gamal El-Din EM, Saleh NAM (1999) The flavonoid chemosystematics of two *Teucrium* species from Southern Sinai, Egypt. Biochem Syst Ecol 27:657–660

Khled-Khoudja N, Boulekbache-Makhlouf L, Madani K (2014a) Antioxidant capacity of crude extracts and their solvent fractions of selected Algerian Lamiaceae. Ind Crop Prod 52:177–182

Khled-Khoudja N, Boulekbache-Makhlouf L, Madani K (2014b) Phytochemical screening of antioxidant and antibacterial activities of methanolic extracts of some Lamiaceae. Ind Crop Prod 61:41–48

Khoddami A, Wilkes MA, Roberts TH (2013) Techniques for analysis of plant phenolic compounds. Molecules 18:2328–2375

Kremer D, Kosir I, Kosalec I, Končić M, Potočnik T, Čerenak A, Bezić N, Srečec S, Dunkić V (2013) Investigation of chemical compounds, antioxidant and antimicrobial properties of *Teucrium arduini* L. (Lamiaceae). Curr Drug Targets 14:1006–1014

Lin L-Z, Harnly JM, Upton R (2009) Comparison of the phenolic component profiles of skullcap (*Scutellaria lateriflora*) and germander (*Teucrium canadense* and *T. chamaedrys*), a potentially hepatotoxic adulterant. Phytochem Anal 20:298–306

Luca SV, Miron A, Ignatova S, Skalicka-Wozniak K (2019) An overview of the two-phase solvent systems used in the countercurrent separation of phenylethanoid glycosides and iridoids and their biological relevance. Phytochem Rev 18:377–403. https://doi.org/10.1007/s11101-019-09599-y

Manach C, Scalbert A, Morand C, Rémésy C, Jiménez L (2004) Polyphenols: food sources and bioavailability. Am J Clin Nutr 79:727–747

Milošević-Djordjević O, Radović Jakovljević M, Marković A, Stanković M, Ćirić A, Marinković D, Grujičić D (2018) Polyphenolic contents of *Teucrium polium* L. and *Teucrium scordium* L. associated with their protective effects against MMC-induced chromosomal damage in cultured human peripheral blood lymphocytes. Turk J Biol 42:152–162

Mitreski I, Petreska Stanoeva J, Stefova M, Stefkov G, Kulevanova S (2014) Polyphenols in representative *Teucrium* species in the flora of R. Macedonia: LC/DAD/ESI-MS^n profile and content. Nat Prod Commun 9:175–180

Naczk M, Shahidi F (2004) Extraction and analysis of phenolics in food. J Chromatogr A 1054:95–111

Nastić N, Švarc-Gajića J, Delerue-Matos C, Barroso MF, Soares C, Moreira MM, Morais S, Mašković P, Gaurina Srček V, Slivac I, Radošević K, Radojković M (2018a) Subcritical water extraction as an environmentally-friendly technique to recover bioactive compounds from traditional Serbian medicinal plants. Ind Crop Prod 111:579–589

Nastić N, Švarc-Gajića J, Delerue-Matos C, Morais S, Barroso MF, Soares C, Moreira MM (2018b) Subcritical water extraction of antioxidants from mountain germander (*Teucrium montanum* L.). J Supercrit Fluids 138:200–206

Oganesyan GB (2005) Minor phenolic compounds from *Teucrium hyrcanicum*. Chem Nat Compd 41:228–229

Pedersen JA (2000) Distribution and taxonomic implications of some phenolics in the family Lamiaceae determined by ESR spectroscopy. Biochem Syst Ecol 28:229–253

Proestos C, Sereli D, Komaitis M (2006) Determination of phenolic compounds in aromatic plants by RP-HPLC and GC-MS. Food Chem 95:44–52

Rispail N, Morris P, Webb K (2005) Phenolic compounds: extraction and analysis. In: Marquez A (ed) Lotus Japonicus handbook. Springer, Berlin, pp 349–354

Rizk AM, Hammouda FM, Rimpler H, Kamel A (1986) Iridoids and flavonoids of *Teucrium* polium herb. Planta Med 2:87–88

Saimaru H, Orihara Y (2010) Biosynthesis of acteoside in cultured cells of *Olea europaea*. J Nat Med 64:139–145

Šamec D, Gruz J, Strnad M, Kremer D, Kosalec I, Grubesić RJ, Karlović K, Lucic A, Piljac-Zegarac J (2010) Antioxidant and antimicrobial properties of *Teucrium arduini* L. (Lamiaceae) flower and leaf infusions (*Teucrium arduini* L. antioxidant capacity). Food Chem Toxicol 48:113–119

Samuelsson G, Bohlin L (2015) Drugs of natural origin – a treatise of pharmacognosy, 7th edn. Swedish Pharmaceutical Society, Stockholm

Schlauer J, Budzianowski J, Kukulczanka K, Ratajczak L (2004) Acteoside and related phenylethanoid glycosides in *Byblis liniflora* Salisb. plants propagated in vitro and its systematic significance. Acta Soc Bot Pol 73:9–15

Sharififar F, Dehghn-Nudeh G, Mirtajaldini M (2009) Major flavonoids with antioxidant activity from *Teucrium polium* L. Food Chem 112:885–888

Sikora E, Cieslik E, Leszcznska T, Filipiak-Florkiewicz A, Pisulewski PM (2008) The antioxidant activity of selected cruciferous vegetables subjected to aquathermal processing. Food Chem 107:55–59

Stanković M (2012) Biological effects of secondary metabolites of *Teucrium* species of Serbian flora. Dissertation, University of Kragujevac

Stanković MS, Topuzović M, Solujić S, Mihailović V (2010) Antioxidant activity and concentration of phenols and flavonoids in the whole plant and plant parts of *Teucrium chamaerdys* L. var. *glanduliferum* Haussk. J Med Plant Res 4:2092–2098

Steffen LM, Jacobs JDR, Stevens J, Shahar E, Carithers T, Folsom AR (2003) Associations of whole-grain, refined-grain, and fruit and vegetable consumption with risks of all-cause mortality and incident coronary artery disease and ischemic stroke: the Atherosclerosis Risk in Communities (ARIC) Study. Am J Clin Nutr 78:383–390

Tepe B, Degerli S, Arslan S, Malatyali E, Sarikurkcu C (2011) Determination of chemical profile, antioxidant, DNA damage protection and antiamoebic activities of *Teucrium polium* and *Stachys iberica*. Fitoterapia 82:237–246

Theriault M, Caillet S, Kermasha S, Lacroix M (2006) Antioxidant, antiradical and antimutagenic activities of phenolic compounds present in maple products. Food Chem 98:490–501

Tumbas V, Mandić A, Ćetković G, Đilas S, Čanadanović-Brunet J (2004) HPLC analysis of phenolic acids in mountain germander (*Teucrium montanum* L) extracts. Acta Period Technol 35:265–273

Turkmen N, Sari F, Velioglu YS (2006) Effects of extraction solvents on concentration and antioxidant activity of black and black mate tea polyphenols determined by ferrous tartrate and Folin-Ciocalteu methods. Food Chem 99:835–841

Venditti A, Frezza C, Zadeh SMM, Foddai S, Serafini M, Bianco A (2017) Secondary metabolites from *Teucrium polium* L. collected in Southern Iran. AJMAP 3:108–123

Vermerris W, Nicholson R (2006) Phenolic compounds biochemistry. Springer, Dordrecht

Verykokidou-Vitsaropoulou E, Vajias C (1986) Methylated flavones from *Teucrium polium*. Planta Med 5:401–402

Vlase L, Benedec D, Hanganu D, Damian G, Csillag I, Sevastre B, Mot AC, Silaghi-Dumitrescu R, Tilea I (2014) Evaluation of antioxidant and antimicrobial activities and phenolic profile for *Hyssopus officinalis*, *Ocimum basilicum* and *Teucrium chamaedrys*. Molecules 19:5490–5507

Yeh H-YU, Chuang C-H, Chen H-C, Wan C-J, Chen T-L, Lin L-Y (2014) Bioactive components analysis of two various gingers (*Zingiber officinale* Roscoe) and antioxidant effect of ginger extracts. LWT Food Sci Technol 55:329–334

Zdraveva P, Pavlova D, Krasteva I, Pencheva I (2018) Phytochemical analysis on populations of *Teucrium chamaedrys* from serpentine sites in Bulgaria. C R Acad Bulg Sci 71:185–192. https://doi.org/10.1136/bjsm.2008.051193

Zlatić NM, Stanković MS, Simić ZS (2017) Secondary metabolites and metal content dynamics in *Teucrium montanum* L. and *Teucrium chamaedrys* L. from habitats with serpentine and calcareous substrate. Environ Monit Assess 189:110. https://doi.org/10.1007/s10661-017-5831-8

Chapter 7
Essential Oils Diversity of *Teucrium* Species

Laura De Martino, Raffaele Coppola, Vincenzo De Feo, Lucia Caputo, Florinda Fratianni, and Filomena Nazzaro

Abstract The Mediterranean flora is characterized by an abundance of aromatic plants. These plants make a significant contribution to the Mediterranean ecological communities both in terms of species numbers and biomass. In this area, *Teucrium* is a polymorphic and cosmopolitan genus of perennial plants, the largest of the Lamiaceae family, which comprises more than 300 species, of which almost 50 are native to Europe. The members of this genus are commonly known as germanders: they are bitter, astringent and antirheumatic herbs, utilized as antispasmodic agents for gastric ulcer and intestinal inflammation, as diuretic, antiseptic, antipyretic and antihelmintic agents. Moreover, they were also used in food preparation. *Teucrium* species are rich in essential oils and the presence of various biologically active monoterpenoid and sesquiterpenoid with several biological activities has been reported. This review summarizes the chemical features of the essential oils of *Teucrium* species reported in literature: it provides an overview on their compositions with particular reference to the more abundant components and to the reported biological activities. *Teucrium polium*, *T. chamaedrys*, *T. flavum* and *T. capitatum* are the most studied taxa concerning the volatile oil composition. Moreover, it is possible to distinguish different types of essential oils, characterized by the prevalence of a specific class of compounds: in particular, we recognized volatile oils rich in caryophyllene and caryophyllene derivatives, α- and β-pinene, germacrene B and D, α-, γ-, δ-cadinene and/or cadinol, limonene, and bisabolol/bisabolene. This richness in chemodiversity confers to the essential oils from *Teucrium* species distinctive biological properties and applications, often confirming the traditional and folk uses of this genus. This review testifies the high great variability of *Teucrium* essential oils and shows that the possible factors responsible for the chemical polymor-

L. De Martino · V. De Feo (✉) · L. Caputo
Department of Pharmacy, University of Salerno, Fisciano, Italy
e-mail: defeo@unisa.it

R. Coppola
Department of Agricultural, Environmental and Food Sciences, University of Molise, Campobasso, Italy

F. Fratianni · F. Nazzaro
Institute of Food Sciences, National Research Council, Avellino, Italy

M. Stanković (ed.), *Teucrium Species: Biology and Applications*,
https://doi.org/10.1007/978-3-030-52159-2_7

phism of essential oils and consequently for their bioactivity need further investigation.

Keywords *Teucrium* · Essential oil · Caryophyllene · α-pinene · β-pinene · Germacrene D · Limonene · Bisabolol · Biological activity

Abbreviations

AChE	Acetylcholinesterase
AD	Alzheimer's disease
BChE	Butyrylcholinesterase
DPPH	2,2-Diphenyl-1-picrylhydrazyl
EOs	Essential oils
ICAM	Intercellular adhesion molecule
MIC	Minimal inhibitory concentration
MRSA	Methicillin-resistant *S. aureus*
NO	Nitric oxide
TNF-α	Tumor necrosis factor-α
VCAM	Vascular cell adhesion molecule

7.1 Introduction

The use of herbal alternative medicinal therapeutics has increased the interest of herbalists and pharmacologists over the past decade. Historically, plants have provided excellent sources of inspiration for new effective medications and, as herbal medicines, they contribute since ancient time to the human health and well-being. Although natural plants medications for over 50 years became less common in evolved countries, however, they are still used in the developing countries and poor section populations. Essential oils (EOs) and extracts from plants, herbs, and spices have long been used for culinary and medicinal purposes throughout the world (Bakari et al. 2015). In particular, essential oils are extracted from aromatic plants, generally found in temperate to warm locations such as the Mediterranean and tropical countries, where they represent an important part of the traditional pharmacopeias. EOs are secondary metabolites, deriving from a specific biosynthetic pathway (Fig. 7.1), characterized by a strong odor. They are composed principally by terpenoids, especially monoterpenes (C_{10}), and sesquiterpenes (C_{15}), although diterpenes (C_{20}) may also be present along with a variety of other molecules (De Martino et al. 2015). EOs are liquid, volatile, limpid, colored, and soluble in lipids and in organic solvents with a density lower than water. They can be present in all plant organs and generally are stored in secretory cells, cavities, canals, epidermal cells, or glandular

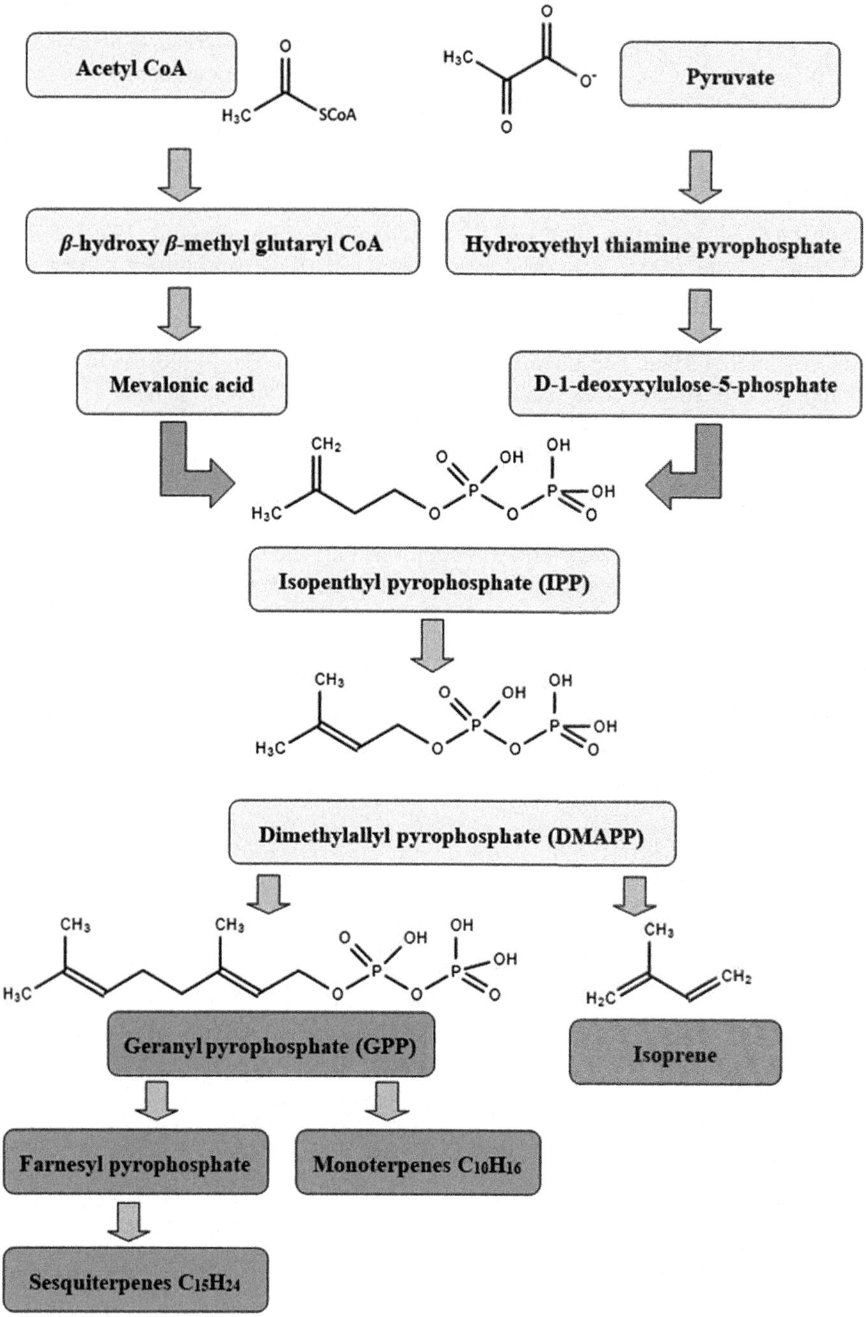

Fig. 7.1 Biosynthetic pathway of essential oils

trichomes. Known for their bactericidal, virucidal, fungicidal, and medicinal properties and for their fragrance, EOs are used also in the preservation of foods, in perfumery and as antimicrobial, analgesic, sedative, anti-inflammatory, spasmolytic, and locally anaesthetic remedies (Bakkali et al. 2008).

The Mediterranean flora is characterized by an abundance of aromatic plants. Such plants make a significant contribution to the Mediterranean ecological communities both in terms of species numbers and biomass (Vokou 1992): apart from their medical properties, plants of the genera *Thymus* (Adzet et al. 1977), *Salvia* (Ulubelen et al. 2002), *Teucrium* (Stankovic et al. 2010) and others, represent a source of substances that might benefit the food industry or agriculture or be used for the human health and well-being (De Martino et al. 2015). The multiple roles of the essential oils and their main components make them natural substances of great importance in several fields such as physiological function of growth, ecological function, development (Wink 2003), resistance against diseases and insects (Gershenzon and Dudareva 2007). They also possess antimicrobial, antiviral, antimycotic, antioxigenic, antiparasitic and insecticidal properties (Bishop 1995; Lamiri et al. 2001; Juglal et al. 2002; Moon et al. 2006; Michaelakis et al. 2007).

This chapter is focused on the essential oils extracted from different species of *Teucrium* genus; specifically, this review has the aims of: (1) describing the chemical composition and the high variability of *Teucrium* essential oils; (2) highlighting the existence of chemical variability these *Teucrium* essential oils oils, differentiated according to the dominant components; (3) making a compendium of the biological activities of the essential oils; (4) testifying the great variability of *Teucrium* essential oils, as influenced by several endogenous and exogenous factors. The chapter would a comprehensive resource on these aspects of the plant.

7.2 The Genus *Teucrium*

Teucrium is a polymorphic and cosmopolitan genus of perennial plants, the largest of the Lamiaceae family in the Mediterranean area, which comprises more than 300 species (Djabou et al. 2012), of which almost 50 are known in Europe (Menichini et al. 2009). In fact, it is a genus of temperate regions, with relatively few representatives in the tropics and the greatest concentration of species in the Mediterranean basin and Central Europe. Some of these species are annual herbs, others are shrubs or subshrubs (Djabou et al. 2011).

The name *Teucrium* derives from Teucer, a son of Talomon, king of Salamis, who was the first to use these plants for curative purposes. In fact, *Teucrium* species have been known for a long time for their medicinal properties, and they are commonly used as diuretic, antiseptic, antipyretic and antispasmodic agents for the treatment of digestive and pulmonary disorders (Lo Presti et al. 2010; Djabou et al. 2011). Members of this genus are commonly known as germanders. Most of these species are bitter, astringent and antirheumatic herbs. Moreover, *Teucrium* species were also alimentary plants and are currently used in the preparation of flavoured wines,

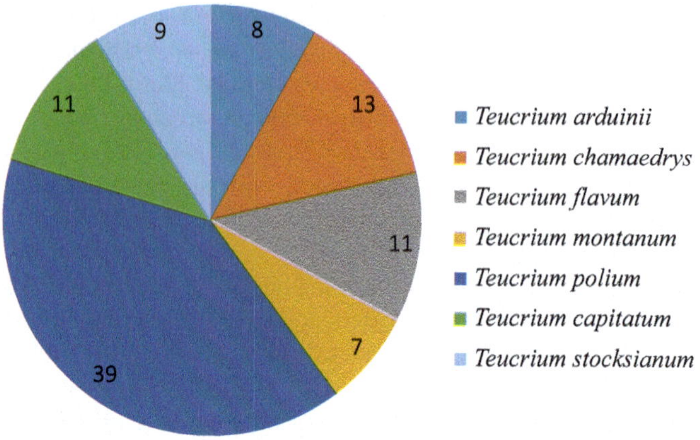

Fig. 7.2 The most studied *Teucrium* species

herbal teas, bitters and liqueurs, as well as aerial part infusions are used for flavoring beer (Maccioni et al. 2007). Some species (notably *Teucrium* fruticans L.) are valued as ornamental plants (Djabou et al. 2011).

From a taxonomical point of view, the genus *Teucrium* has been divided into nine sections identifiable through the calyx shape and the inflorescence structure (Navarro and El Oualidi 2000).

An unusual feature of *Teucrium* genus, if compared with other members of Lamiaceae, is that its flowers completely lack the upper lip of the corolla. Moreover, as in other Lamiaceae, the aerial organs of *Teucrium* species are covered by an indumentum of glandular and non-glandular trichomes (De Martino et al. 2010). *Teucrium* species are rich in essential oils and various active monoterpenoid/sesquiterpenoid have been reported in them, possessing several biological activities. This genus has drawn attention of phytochemists and pharmacognosists and the scientific literature reports a consirable number of papers: in Fig. 7.2, *Teucrium* taxa with more bibliographic reference concerning essential oil composition, are reported.

Above all, *Teucrium polium* is the species with the highest number of references: its common names are polium, pouliot de montagne, germandée tomenteuse (French); germander, mountain germander, felty germander, cat thyme, hulwort (English); camendrio di montagna, polio, polio primo, timo bianco (Italian); poleigamander, berggamander (German) (Jaradat 2015). This species grows wildly in sunny regions of semiarid and arid in bioclimates, and is capable to grow also on hillsides, sands and in the stony mountains. It is a 10–35 cm high shrub, with a white or yellow corolla contained in a globular small inflorescence. The leaves are white, tomentose on both sides; the fruits are brown and all the plant aerial parts have aromatic pleasant odor and bitter taste (Barceloux 2008). In the traditional herbal medicine, decoctions of *Teucrium polium* have been used for treatment of inflammations, gastrointestinal tract disorders, rheumatism, and diabetes. A tea from the plant is used in traditional Iranian medicine for treating many diseases such as indigestion, abdominal pain, Type 2 diabetes and common cold (Bahramikia and Yazdanparast

2012); the same formulation is used in the Palestinian folk medicine for the treatment of gastrointestinal inflammations, as well as an antispasmodic and an anthelmintic and for the treatment of claustrophobia (Jaradat 2005).

Teucrium chamaedrys L. is widely distributed in Azerbaijan, Golestan, Khorasan, Lurestan, Mazandaran and Yazd provinces of Iran (Zargari 1993): the plant has been used medicinally since ancient Greek times, with similar applications to other *Teucrium* species; in Iranian herbal medicine, it is used as a diuretic, antiseptic, antipyretic and antihelmintic agent (Djabou et al. 2011). *Teucrium flavum* L. is a 60 cm evergreen perennial shrub, with triangular-ovate leaves and yellow flowers, which appear from May to August (Jeanmonod and Gamisans 2007; Lo Presti et al. 2010). The plant can be found in the cracks of lime rocks up to 1000 m (Lo Presti et al. 2010). Previously classified in *Chamaedrys* (Miller) Schreber section, successively, on the basis of investigation of flowers and inflorescences, *Teucrium flavum* was included in section *Polium* Schreber (Lakusic et al. 2006). *Teucrium capitatum* L. [syn. *T. polium* L. ssp. *capitatum* (L.) Arcang.] is widespread diffused in the dry and stony places of Greece and almost in all Mediterranean countries: it is a perennial, prostrate species with white to bright pink flowers, crenate leaves and ramified indumentum (De Martino et al. 2010). This plant is used in folk medicine to treat diabetes and intestinal troubles (Boulila et al. 2008). The Iranian *Teucrium stocksianum* Boiss. is a polymorphous perennial plant distributed just in southern regions of Iran, 10–35 cm high and with a very pleasant aromatic odor (Bakhtiari and Asgarpanah 2015). The purple globular inflorescences appears from April to May; the fruits are dark brown with a latticed surface. This perennial aromatic herb is extensively exploited as a medicinal plant and is used in folk medicine to treat diarrhea, cough, jaundice and abdominal pain (Rahim et al. 2012). *Teucrium arduini* L. (Arduini's germander) is an endemic Illyric Balcan species, distributed in the mountains along the Adriatic coast of Croatia, Bosnia and Herzegovina, Montenegro, Serbia and Northern Albania. It is a 10–40 cm dwarf shrub, semi-woody, branchy, erect or ascending; it grows on calcareous rocky slopes up to 1600 m a.s.l. Whitish flowers form simple, very dense, up to 16 cm long inflorescences (De Martino et al. 2010). *Teucrium montanum* L. is a grass crop which has long been used both as an herbal medicine and a nourishing food. It is a widely used as a diuretic, stomachic, analgesic and antispasmodic agent. Moreover, it possess antibacterial, antifungal, anti-inflammatory and antioxidative activities (Vukovic et al. 2008).

7.3 Essential Oils of *Teucrium* Species

In literature, the essential oils of 93 different species and or subspecies are reported in more than 200 papers. It is possible to distinguish different types of essential oils or essential oils characterized by a prevalence of a specific class of compounds. The majority of *Teucrium* essential oils can be divided on the basis of predominant components: caryophyllene and caryophyllene derivatives (Table 7.1), α- and β-pinene

Table 7.1 *Teucrium* taxa with predominance of caryophyllenes

Teucrium species	Percentage (%)	References
T. alyssifolium	16.87	Semiz et al. (2016)
T. apollinis	22.4	Elabbara et al. (2014)
T. arduinii	32.9	Kremer et al. (2012)
	35.2	Kremer et al. (2013)
	13.7–50.7	Kremer et al. (2015)
	34.5	Dunkic et al. (2011)
	17.7	De Martino et al. (2010)
	24.5	Kovacevic et al. (2001)
T. botrys	20.4	Kovacevic et al. (2001)
T. capitatum	15.1	De Martino et al. (2010)
T. cavernarum	47.0	Kaya et al. (2013)
T. chamaedrys	33.9	Djabou et al. (2013a)
	7.1–52.0	Bezic et al. (2011)
	27.0–29.4	Muselli et al. (2009)
	26.9	Kovacevic et al. (2001)
T. chamaedrys ssp. *syspirense*	28.0	Kaya et al. (2009)
	23.0	Sajjadi and Shookohinia (2010)
T. chamaedrys ssp. *trapezunticum*	25.0	Kaya et al. (2009)
T. creticum	24.6	Valentini et al. (1997)
T. divaricatum	36.2	Formisano et al. (2010)
T. flavum	32.5	Hammami et al. (2015b)
	22.0	Formisano et al. (2012)
	23.1	Bezic et al. (2011)
	20.1	Menichini et al. (2009)
	30.7	Baher and Mirza (2003)
T. flavum ssp. *hellenicum*	22.2	Bellomaria et al. (1998)
T. fruticans (fruiting tops)	22.0	Flamini et al. (2001)
T. marum	1.3–39.4	Djabou et al. (2013b)
	15.1	Djabou et al. (2013a)
	27.4	Ricci et al. (2005)
T. montbretii ssp. *heliotropiifolium*	17.0	De Martino et al. (2010)
T. multicaule	16.5	Polat et al. (2010)
T. orientale var. *orientale*	42.8	Javidnia and Miri (2003)
	29.3	Kucukbay et al. (2011)
T. orientale var. *lydium*	19.7	Kucuk et al. (2006)
T. orientale var. *puberulens*	38.0	Kucukbay et al. (2011)
	21.7	Kucuk et al. (2006)
T. parviflorum	27.4	Bagci et al. (2011)
T. pestalozzae	27.6	Baser et al. (1997)
T. pilosum	21.52	Chen et al. (2010)

(continued)

Table 7.1 (continued)

Teucrium species	Percentage (%)	References
T. polium	29.0	Raei et al. (2014)
	7.1–52.0	Bezic et al. (2011)
	100.0	Al-Qudah et al. (2011)
T. quadrifarium	25.0	Tandon and Mittal (2018)
	38.3	Mohan et al. (2010)
T. scordium	31.4	Morteza-Semnani et al. (2007)
T. scorodonia ssp. *scorodonia*	20.5	Djabou et al. (2012)
	26.2–41.4	Makowczynska et al. (2016)
T. scorodonia ssp. *baeticum*	39.7	Djabou et al. (2012)
T. stocksianum	37.0	Shah et al. (2015)
T. turredanum	15.6–32.6	Blazquez et al. (2003)
T. yemense	23.4–31.3	Awadh Ali et al. (2017)

(Table 7.2), germacrene B and D (Table 7.3), α-, γ-, δ-cadinene and or cadinol (Table 7.4), limonene (Table 7.5), bisabolol/bisabolene (Table 7.6) (Fig. 7.3).

Thirty *Teucrium* taxa have been reported for an essential oil rich in caryophyllene and caryophyllene derivatives (Table 7.1). Thirteen of these taxa are reported only for this characteristic chemical characteristic: *Teucrium alyssifolium* Stapf, *T. apollinis* Maire & Weiller, *T. botrys* L., *T. cavernarum* P. H. Davis, *T. chamaedrys* ssp. *syspirense* (C. Koch) Rech. F., *T. chamaedrys* ssp. *trapezunticum* Rech. F., *T. creticum* L., *T. marum* L., *T. multicaule* Montbret & Aucher ex Benth., *T. pestalozzae* Boiss., *T. scorodonia* ssp. *baeticum* (Boiss. & Reut.) Tutin and *T. turredanum* Losa & Rivas Goday. Moreover, *Teucrium flavum* L. ssp. *hellenicum* Reich. F. essential oil showed almost identical amounts of caryophyllene (22.2%) and germacrene D (22.3%).

Seventeen taxa possess essential oils with a predominance of α- and β-pinene (Table 7.2). For nine of these species [*Teucrium algarbiense* (Cout.) Cout., *T. gnaphalodes* L'Hér., *T. libanitis* Schreb., *T. lusitanicum* Schreb., *T. melissoides* Boiss. et Hausskn. ex Boiss., *T. polium* L. ssp. *gabesianum, T. polium* L. ssp. *valentinum* and *T. stocksianum*], the literature reports only this chemical feature. Seventeen taxa possess essential oils with a predominance of germacrenes (Table 7.3). In literature, for the species *Teucrium antiatlanticum* (Maire) Sauvage & Vindt, *T. bidentatum* Hemsl., *T. labiosum* C. Y. Wu & S. Chow, *T. paederotoides* Boiss. and *T. zanonii* Pamp., only essential oils with germacrene predominance have been reported. Sixteen *Teucrium* taxa have been reported for essential oils with a predominance of cadinenes (Table 7.4). For *Teucrium carolipaui* Vicioso ex Pau, *T. maghrebinum* Greuter & Burdet, *T. persicum* Boiss. and *T. polium* ssp. *aurasiacum* (Maire) Greuter & Burdet only essential oils with cadinenes predominance have been reported. Five taxa (Table 7.5) have been reported for their limonene-rich essential oil. Among these taxa, *Teucrium flavum* L. ssp. *glaucum* (Jord. & Fourr.) Runninger, *T. polium* L. ssp. *geyrii* Maire and *T. polium* L. ssp. *aurasiacum* are known only for this chemical characteristic. For five taxa (Table 7.6) an essential oil rich in bisabolol/

Table 7.2 *Teucrium* taxa with predominance of pinenes

Teucrium species	Percentage (%)	References
T. algarbiense	18.5	Cavaleiro et al. (2004)
T. capitatum	18.0	Antunes et al. (2004)
	24.1	Djabou et al. (2013a)
	36.0	Cozzani et al. (2005)
	36.1	Mitic et al. (2012)
T. flavum	29.0	Stanic et al. (1993)
	36.2–39.6	Djabou et al. (2011)
	40.0	Djabou et al. (2013a)
T. fruticans (flowering phase)	21.0	Flamini et al. (2001)
T. gnaphalodes	27.5	Perez-Alonso et al. (1993)
T. haenseleri	42.0–54.0	Gaspar et al. (1997)
T. libanitis	9.9–21.2	Blazquez et al. (2003)
T. lusitanicum	3.3–20.4	Cavaleiro et al. (2004)
T. melissoides	44.1	Ahmadi et al. (2002)
T. polium	29.7	Bakari et al. (2015)
	23.3	Keykavousi et al. (2016)
	29.4–36.1	Lograda et al. (2014)
	19.61	Moghtader (2009)
	8.9	Nikpour et al. (2018)
	25.8	Purnavab et al. (2015)
	28.3	Sabzeghabaie and Asgarpanah (2016)
	42.8	Rowshan and Najafian (2012)
	14.43	Stanciu et al. (2006)
	19.8	Kovacevic et al. (2001)
	30.0	Cakir et al. (1998)
	11.94	Sarer and Konuklugil (1987)
T. polium ssp. *gabesianum*	49.29	Ben Othman et al. (2017)
T. polium ssp. *valentinum*	27.5	Perez-Alonso et al. (1993)
T. sandrasicum	55.67	Ali et al. (2013)
T. scordium	27.7	Kovacevic et al. (2001)
T. stocksianum	50.76	Mojab et al. (2003)
T. stocksianum ssp. *gabrielae*	36.0	Sonboli et al. (2013)
T. zanonii	14.13	Abdelshafeek et al. (2010)

bisabolene is reported. *Teucrium gracile* Barbey & Fors.-Major is known only for this characteristic. Oct-1-en-3-ol is reported as the main constituent of the essential oils of *Teucrium abutiloides* L'Hér. and *T. betonicum* L'Hér. in which the amonuts are 20% and 24%, respectively (Barroso et al. 1993). The EO of *Teucrium lepicephalum* Pau contains sabinene in percentage between 47.54% and 57.43% (Perez et al. 2000), *Teucrium leucocladum* Boiss. has an essential oil with a predominance of patchouli alcol (31.24%) (El-Shazly and Hussein 2004). The EO of *Teucrium* orientale ssp. *glabrescens* (Hausskn. ex Bornm.) Rech.f. shows a predominance of β-(26.9%) and α-cubebene (9.0%) (Oezek et al. 2012).

Table 7.3 *Teucrium* taxa with predominance of germacrenes

Teucrium species	Percentage (%)	References
T. antiatlanticum	13.2	Kartah et al. (2015)
T. antitauricum	28.2	Baser et al. (1999)
T. arduinii	17.0	Vukovic et al. (2011)
	57.8	Blazevic et al. (1992)
T. bidentatum	32.55	Chen et al. (2010)
T. chamaedrys	32.1	Bagci et al. (2010)
	16.7	Kucuk et al. (2006)
	16.5	Morteza-Semnani et al. (2005)
T. flavum ssp. *hellenicum*	22.3	Bellomaria et al. (1998)
T. fruticans (fruiting tops)	24.0	Flamini et al. (2001)
T. labiosum	34.08	Chen et al. (2010)
T. montanum	31.0	Radulovic et al. (2012)
	17.2	Bezic et al. (2011)
	15.0	Kovacevic et al. (2001)
T. orientale var. *orientale*	24.6	Oezek et al. (2012)
T. orientale var. *puberulens*	33.4	Oezek et al. (2012)
T. paederotoides	20.8	Kaya et al. (2013)
T. polium	12.5–13.8	Bendif et al. (2018)
	8.7–19.5	Lianopoulou et al. (2014)
	18.26	Mahmoudi and Nosratpour (2013)
	25.81	Belmekki et al. (2013)
	31.0	Radulovic et al. (2012)
T. polium ssp. *capitatum*	31.8	Mitic et al. (2012)
T. quadrifarium	8.8	Liu et al. (2016)
T. royleanum	28.9	Mohan et al. (2010)
T. sandrasicum	27.9	Baser et al. (1997)

Teucrium pseudochamaepitys L. has been reported for its essential oil with a predominance of hexadecanoic acid (26.1%) (Hammami et al. 2015a); the EO of *Teucrium scorodonia* ssp. *baeticum* presents N-formylmorpholine (25.07%) and morpholine, 4-acetyl- (17.61%) as the main components (Tahoum et al. 2017); *Teucrium salviastrum* Schreb. essential oil has in prevalence β-farnesene (26.1–29.3%) followed by E-caryophyllene (19.1–24.1%) (Cavaleiro et al. 2002). For *Teucrium scordium* L. ssp. *scordioides* (Schreb.) Arcang. essential oil, menthofuran (11.9%) is reported as the main component (Radulovic et al. 2012). *Teucrium massiliense* L. showed 6-methyl-3-heptyl acetate as the main compound with percentage between 19.1% (Djabou et al. 2013a) and 23.83% (Djabou et al. 2010). However, the same plant gave an essential oil rich in 3,7-dimethyloctan-2-one (15.2%) and bu-2-methylbutyrate (12.1%). Some papers reports no quantitative data on essential oil compositions of *Teucrium africanum* Thunb. (Ruiters et al. 2016), *T. brevifolium* Schreb. (Menichini et al. 2009), *T. pruinosum* Boiss. (Jaradat et al. 2018), *T. sauvagei* Le Houér. (Salah et al. 2006) and *T. trifidum* Retz. (Ruiters

Table 7.4 *Teucrium* taxa with predominance of cadinenes

Teucrium species	Percentage (%)	References
T. alopecurus	13.4	Hachicha et al. (2007a)
T. atratum	45.8	Kabouche et al. (2006)
T. capitatum	31.6	Antunes et al. (2004)
	46.2	Khani and Heydarian (2014)
	18.3	Kerbouche et al. (2015)
T. carolipaui	21.5	Pala-Paul et al. (2001)
	13.12–29.32	Perez et al. (2000)
T. divaricatum	27. 5–33.9	Tzakou et al. (1997)
T. heterophyllum	26.0–32.0	Barroso et al. (1996)
T. hircanicum	13.4	Rahimi et al. (2018)
T. maghrebinum	22.2	Formisano et al. (2009)
	21.0	De Martino et al. (2010)
T. montanum	17.2	Vukovic et al. (2007)
	17.2	Vukovic et al. (2008)
T. persicum	12.9	Javidnia et al. (2007)
	23.2	Masoudi et al. (2009)
	18.9	Miri et al. (2012)
T. polium	13.6	Vokou and Bessiere (1985)
	13.82	Guetat and Al-Ghamdi (2014)
T. polium ssp. *aurasiacum*	46.8	Kabouche et al. (2007)
T. ramosissimum	23.7–42.5	Ghazouani et al. (2016)
	37.33	Ghazouani et al. (2017)
	29.9	Hachicha et al. (2007b)
T. stocksianum	24.4–28.4	Al-Yousuf et al. (2002)
	12.7	Hisham et al. (2006a)
	12.9	Shah et al. (2012)
	37.6	Bakhtiari and Asgarpanah (2015)
T. yemense	34.9	Awadh Ali et al. (2008)

Table 7.5 *Teucrium* taxa with predominance of limonene

Teucrium species	Percentage (%)	References
T. flavum ssp. *glaucum*	21.1–31.8	Djabou et al. (2011)
	27.4	Djabou et al. (2013b)
T. lusitanicum ssp. *aureiformis*	15.9	Velasco-Negueruela and Perez-Alonso (1990)
T. mascatense	17.2	Oraimi et al. (2012)
T. polium	37.70	Vahdani et al. (2011)
T. polium ssp. *geyrii*	11.18	Roukia et al. (2013)
T. polium ssp. *aurasiacum*	34.72	Bencheikh et al. (2015)

Table 7.6 *Teucrium* taxa with predominance of bisabolol/bisabolene

Teucrium species	Percentage (%)	References
T. alopecurus	16.16	Guesmi et al. (2017)
T. flavum	35.0	Kovacevic et al. (2001)
	31.5–48.2	Lo Presti et al. (2010)
T. gracile	1.9–30.7	Arnold et al. (1998)
T. heterophyllum	20.8	Velasco-Negueruela et al. (1999)
T. polium	32.3	Sadeghi et al. (2014)

et al. 2016). For other taxa, only the classes of main components are reported: *Teucrium polium* L. var. *album* and *T. polium* var. *pilosum* (L.) Arcang. (Kamel and Sandra 1994) as well as *T. viscidum* Blume (Wei et al. 2010).

Teucrium polium essential oil is mainly represented by a pinenes type: in fact, in 12 different papers α- and/or β-pinene are reported as the most abundant components; in five papers, germacrene D and/or germacrene B are reported as the main components, while caryophyllene and derivatives predominate in other four oils. Probably, the origin of the plants can influence the essential oil composition. Yugoslavian (Kovacevic et al. 2001) and Iranian germander (Eikani et al. 1999) were mainly represented by germacrene D rich essential oil (11.9% and 23.6–13.2% respectively), while the major components of a Turkish EO were β- (18%), α-pinene (12%) and β-caryophyllene (18%) (Cakir et al. 1998). On the other hand, Jordanian germander EOs contain 8-cedren-13-ol (24.8%), β-caryophllene (8.7%), sabinene (5.2%) and germacrene D (6.8%) (Aburjai et al. 2006). The phytochemical screening of the Saudi Arabia *Teucrium polium* volatile oil identified the presence of ten isoprenoid compounds, including alcohols as linalool, cedrenol, cedrol, guaiol and terpine4-ol and the hydrocarbons, β-pinene, γ-, δ-cadinene, limonene and α-phellandrene (Hassan et al. 1979).

Thirteen studies on the EOs of *Teucrium chamaedrys* reported the presence of caryophyllene and germacrene as principal components. Moreover, Zhou et al. (2004) indicated the presence of the diterpenoid teucrin A in the oil. *Teucrium flavum* essential oils have been subject of some studies. The volatile oils appeared qualitatively quite similar and only quantitative differences occurred in the relative amounts of components. These differences were presumably due to environmental conditions and they might be explained by ambiguous taxonomic determination of the taxa studied (Djabou et al. 2011). The essential oils of *Teucrium capitatum* is variously characterized: in fact, a part from the relative aboundance of pinenes, some literature data report the predominance of oxygenated compound, as cadinol (Table 7.4), carvacrol (Menichini et al. 2009) and endo borneol (El Amri et al. 2017). Probably, the reported differences also depend on the part of the plant studied, its stage of growth as well as the nature of the soil and of the environment. In any case, a chemical polymorphism is present in this plant (El Amri et al. 2017). Ten chemotypes are reported for *Teucrium arduini*, based on the essential oil profile and molecular analyses: one with prevalence of caryophyllenes and the others with predominance of germacrene, although such a basis requires further clarification

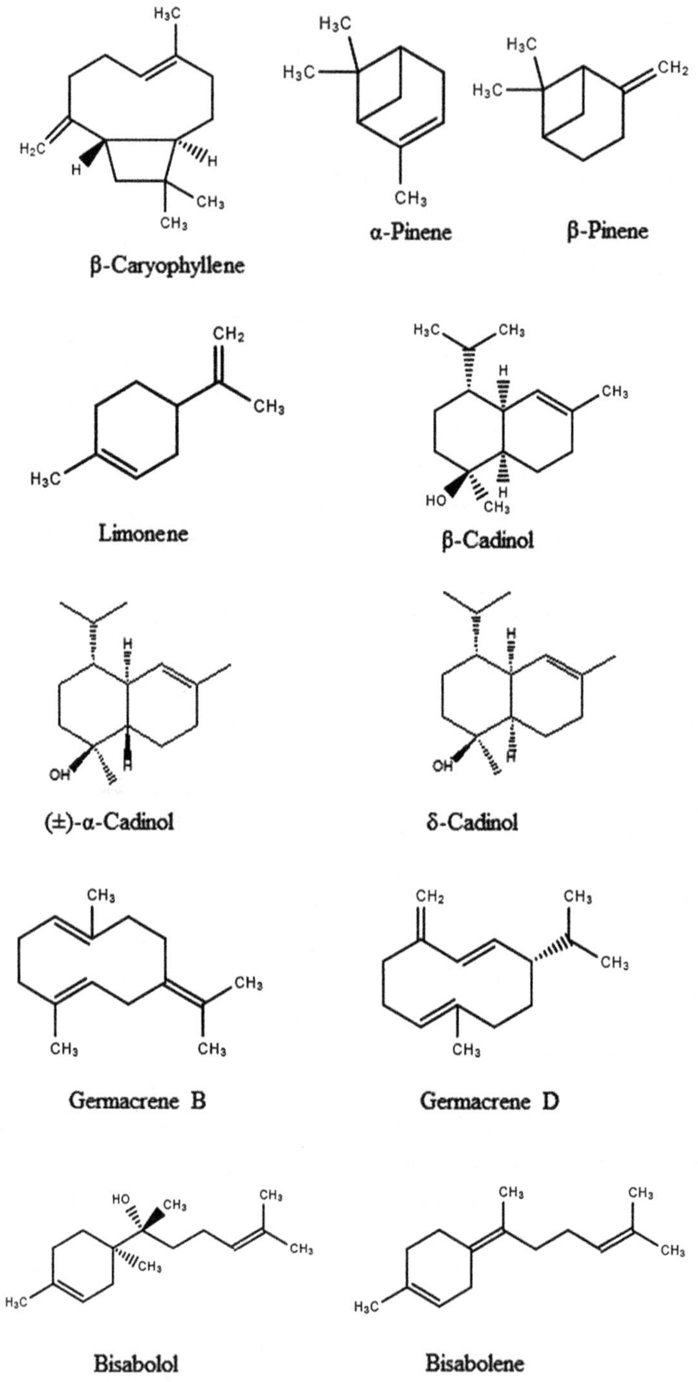

Fig. 7.3 The main components of *Teucrium* essential oils

(Kremer et al. 2015). *Teucrium montanum* EOs showed the presence of sesquiterpene hydrocarbons as characteristic compounds and monoterpenes as minor constituents, with an high prevalence of germacrene and cadinene (Vukovic et al. 2007, 2008; Bezic et al. 2011; Radulovic et al. 2012). The essential oils of *Teucrium stocksianum* are mainly constituted by cadinene and/or cadinol (Al-Yousuf et al. 2002; Hisham et al. 2006a; Shah et al. 2012). Similar results were obtained also by Bakhtiari and Asgarpanah (2015), who found α-cadinol (37.6%), α-pinene (9.7%) and caryophyllene oxide (4.9%) as the main components of the fruits.

On the basis of the available literature, the most part of *Teucrium* essential oils are characterized by the presence of sesquiterpenes. In this feature, caryophyllenes constitute the most common principal class of compounds in combination with germacrenes and, in minor extent, with cadinenes. In some taxa, caryophyllenes are combined with pinenes. Only in few taxa monoterpenes predominate, with pinenes and limonenes as main classes. A generalization about the chemical expression of *Teucrium* essential oils need certainly of more studies: there is no homogeinity in chemical characterization of EOs and some species (for example *Teucrium polium, T. capitatum, T. chamaedrys* and *T. flavum*) have been more extensively studied. These *Teucrium* accessions showed a high degree of variability in essential oil composition: in fact, *Teucrium polium* appear the most variable species, with 6 different types of essential oils. Also *Teucrium flavum* and *T. capitatum* show essential oils with 5 and 4 different types, respectively. The high degree of chemical variability, a part on intrinsic and extrinsic factors linked to plant and environment, appears in part attributable to different methodological approaches used in the studies.

7.4 Biological Activity of the Genus *Teucrium*

Generally, most, if not all, countries of the Mediterranean area have a consolidated experience in the utilization and in the study *Teucrium*. At present, due to the concerns about the harmful consequences and side effects showed by many drugs, an in-depth study on the biological properties of natural substances is considered with increasingly strong interest. Thus, research concerning also *Teucrium* species is not an exception. Many biological properties of the genus *Teucrium* are well known and exploited in folk medicine. Such properties are ascribable both to the hydroalcholic extracts and to the essential oils giving rise from most of the plant.

Based on the study of the chemical composition of several EOs derived from *Teucrium*, it is now established that the biological properties, with a positive effect on health, are essentially due to the presence of some molecules, such as sesquiterpenes, of which the essential oil from many species of *Teucrium* is rich (Bezic et al. 2011). The main effects due to the *Teucrium* EOs can be listed as below indicated. Naturally, not all species of this genus exhibit the same properties, which depend also on the genetic factors, as well as on the environmental conditions of growth, climate, and geography, as therefore demonstrated for many other species of plants. Furthermore, for the same organ and the same plant, the extraction methods, the

area of origin as well as the operating conditions can affect the chemical composition of the oil or extract, subsequently their biological activities. In general, however, *Teucrium* species are reported for their antibacterial, antioxidant, antidiabetic, antielmintic, antirheumatic, and anticancer effects (Miri et al. 2012; Mahmoudi and Nosratpour 2013; Guesmi et al. 2018).

7.4.1 Antibacterial Activity

The antimicrobial activity has been evaluated for different species of *Teucrium*. The essential oils of three African endemic species, *Teucrium africanum, T. kraussii* Codd and *T. trifidum* have been assayed to determine their minimal inhibitory concentration (MIC) against *Bacillus cereus*, *Escherichia coli*, *Klebsiella pneumoniae*, *Moraxella catarrhalis*, *Pseudomonas aeruginosa*, *Staphylococcus aureus* and *Streptococcus pyogenes* (Ruiters et al. 2016); *Teucrium africanum* showed antibacterial activity against *Escherichia coli* and *Streptococcus pyogenes*. The essential oil from croatian *Teucrium arduini* showed antimicrobial activity against *Staphylococcus aureus*, *Enterococcus faecalis*, *Escherichia coli*, *Pseudomonas aeruginosa*, *Candida albicans*, *Microsporum gypseum* and *Aspergillus brasiliensis*, with MIC values ranging from 6.25 to 37.50 mg/ml. Fungal species were susceptible with MIC values from 7.81 to 25.00 mg/ml (Kremer et al. 2012). The same species was tested by Vukovic et al. (2011) against *Klebsiella pneumonia*e with a MIC of 6.25 μl/ml. Moreover, with a MIC value of 25 μl/ml, the oil showed a good antibacterial activity against *Staphylococcus aureus* (ATCC 25923) and *S. aureus* (FSB 30), as well as against *Enterobacter cloacae* (FSB 22) and *Proteus mirabilis* (FSB 34). *Teucrium polium* ssp. *aurasiacum* was included in a comparative study of antimicrobial activity of five Lamiaceae essential oils from Algeria (Kabouche et al. 2005). Three populations of *Teucrium polium* from the same region of eastern Algeria were evaluated to determine the potential antibacterial activity: the essential oil of the Beni Aziz population has a high antibacterial activity against *Escherichia coli*, *Staphylococcus aureus* and the yeast *Saccharomyces cerevisiae*. The population of Boutaleb showed significant activity against *Bacillus cereus* and no action against *Staphylococcus aureus* and the yeast *Saccharomyces cerevisiae* (Lograda et al. 2014).

Belmekki et al. (2013) and Fertout-Mouri et al. (2017) studied the antibacterial activity of the essential oil of *Teucrium polium* from Western Algeria: the oil showed moderate inhibitory effects on *Bacillus cereus*, *Enterococcus faecalis*, *Escherichia coli* and *Staphylococcus aureus*. This latter recorded the highest sensitivity level (100%); *Escherichia coli*, *Proteus mirabilis* and *Pseudomonas aeruginosa* exhibits a resistance (33.33%) towards the tested oil. Moreover, Sevindik et al. (2016) highlighted that the antimicrobial effects of essential oil from *Teucrium polium* growing in North Anatolian region were stronger against *Pseudomonas aeruginosa* and methicillin-resistant *Staphylococcus aureus* (MRSA), whereas *Escherichia coli* Q157:H7 showed the smallest inhibition zone (11 mm). Considering the alarming spread of antibiotic resistance to many bacterial pathogens, interest in natural plant

products as alternative antibacterial agents has increased: thus, Raei et al. (2014) demonstrated that *Teucrium polium* essential oil may have a potential for use against multidrug resistant organisms such as clinical isolates of *Klebsiella pneumoniae*. An Iranian *Teucrium polium* essential oil was active against *Streptococcus pneumoniae* and *Staphylococcus aureus* (Vahdani et al. 2011).

The essential oils of *Teucrium polium* ssp. *gabesianum* (Ben Othman et al. 2017) and *Teucrium polium* L. ssp. *aurasianum* (Bencheikh et al. 2015) were assayed for a possible antimicrobial activity. The essential oil of the first subspecies was most effective against *Proteus mirabilis*, *Staphylococcus aureus* and *Citrobacter freundei* and against some fungi as *Microsporum canis*, *Scopulariopsis brevicaulis* and *Trichophyton rubrum*; the oil of the second one was active against *Staphylococcus aureus*. The EOs of *Teucrium chamaedrys* L. ssp. *chamaedrys*, *T. orientale* L. var. *puberulens*, and *T. chamaedrys* L. ssp. *lydium* were tested for their antimicrobial activity, showing moderate antibacterial activity against Gram-positive and Gram-negative bacteria, but no antifungal activity against two yeast-like fungi (Kucuk et al. 2006). Food spoilage is an enormous economic problem worldwide: the research is focus on discovery of new antimicrobial agents and biological additives from plants. Mahmoudi et al. (2015) investigated the growth and survival of *Salmonella typhimurium* and *Lactobacillus casei* at a 7-day interval during storage (28 days) by addition of *Teucrium polium*, that showed the best growth inhibition of *Salmonella typhimurium* at different doses tested. *Teucrium polium* essential oil was evaluated also against some phytobacteria: the oil did not show antibacterial activity against *Pseudomonas aeruginosa* and *Pectobacterium carotovorum*; instead it showed antibacterial activity against *Pantoea agglomerans*, *Brenneria nigrifluens*, *Streptomyces scabies* and *Ralstonia solanacearum* (Purnavab et al. 2015).

The antibacterial activity of Corsican *Teucrium marum*, *T. massiliense*, *T. chamaedrys*, *T. scorodonia* L. ssp. *scorodonia*, *T. capitatum*, *T. flavum* ssp. *glaucum* and *T. flavum* L. ssp. *flavum* was evaluated against the foodborne pathogens, *Campylobacter jejuni* and *Listeria innocua* and four toxi-infectious bacteria (*Staphylococcus aureus*, *Staphylococcus epidermidis*, *Enterobacter aerogenes* wild type (CIP 60.86) and the multi-drug resistant bacterium *E. aerogenes* (EAEP289) (Djabou et al. 2013a). Almost all *Teucrium* essential oils (with exception of *Teucrium scorodonia* ssp. *scorodonia* and *T. chamaedrys*) inhibited the bacterial growth: *Campylobacter jejuni* was found to be extremely sensitive to all the Corsican *Teucrium* oils. suggesting that Corsican *Teucrium* essential oils have the potential to be used as food preservatives and to prevent the growth of nosocomial bacteria (Djabou et al. 2013a). The essential oil of *Teucrium capitatum* growing in Morocco has been evaluated for its potential antimicrobial activity against the bacterial strain *Staphylococcus aureus*, two yeasts (*Candida albicans* and *C. glabrata*), the mold *Aspergillus niger*, and 5 dermatophytes: *Microsporum canis*, *M. gypseum*, *Trichophyton rubrum*, *T. mentagrophytes* and *Epidermophyton floccosum*. The results showed a large antimicrobial activity of this essential oil (El Amri et al. 2017). *Teucrium marum* essential oil was reported by Ricci et al. (2005) for its antimicrobial activity against four phytopathogenic fungi: *Rhyzoctonia solani* resulted the most sensitive microorganism.

The essential oil of *Teucrium hircanicum* L. exhibited a dose-dependent antimicrobial activity against *Bacillus subtilis*, *Staphylococcus aureus*, *Salmonella typhi*, *Pseudomonas aeruginosa* and *Aspergillus niger* but it did not showed antimicrobial activity against *Escherichia coli* and *Candida albicans* (Morteza-Semnani et al. 2011). The essential oil of *Teucrium leucocladum* was tested for its bacteriostatic and antifungal activities, showing broad and potent activity against *Pseudomonas aeruginosa*, *Bacillus subtilis* and *Candida albicans* (El-Shazly and Hussein 2004). The antimicrobial activity of the essential oil of *Teucrium mascatense* Boiss. from Oman was evaluated against a panel of 17 bacterial and six fungal strains by the disk diffusion method: the oil inhibited the growth of all test organisms at various levels as reported by Hisham et al. (2006b). Vukovic et al. (2007, 2008) studied the antibacterial and antifungal activities of the essential oil of *Teucrium montanum* against several bacterial and fungal strains: the highest activities were obtained specifically against *Klebsiella pneumoniae*, *Bacillus subtilis*, *B. mycoides* and *Enterobacter cloacae*.

Thoppil et al. (2001) studied the antimicrobial activity of the essential oil of *Teucrium plectranthoides* Gamble against seven bacteria and eight fungi using the filter paper disk diffusion method. The growth of almost all microorganisms was inhibited suggesting the possible use of this essential oil as a source of potential microbicides. With the same method, the antimicrobial activities of the essential oil from *Teucrium polium* L. ssp. *geyrii* was evaluated (Roukia et al. 2013) against *Staphylococcus aureus* ATCC 27923, *Escherichia coli* ATCC 25922, *Pseudomonas aeruginosa* ATCC 25853 and *Candida albicans*, showing mainly an efficient fungicidal activity. *Teucrium stocksianum* essential oil exhibited profound antibacterial potential against both Gram-positive and Gram-negative bacterial strains as reported by Shah et al. (2016) and Hisham et al. (2006a). Finally, two different populations of *Teucrium yemense* Deflers were evaluated for their antimicrobial activity: one of these essential oils showed antimicrobial activity against *Staphylococcus aureus*, *Bacillus cereus*, *Aspergillus niger* and *Botrytis cinerea* (Awadh Ali et al. 2017).

7.4.2 Cytotoxic Effects

Anticancer effects have been attributed to different species of *Teucrium*. The hydrophobic and hydrophilic fractions of the essential oil of *Teucrium alopecurus* Noë, for instance, were studied in the context of cancer prevention and therapy. In particular, the molecular mechanism involved was elucidated, with the aim to determine the effectiveness of such fractions against cancer, through their capability to trigger apoptosis and suppress the tumorigenesis, using as a model human colon cancer cells. The EO acted clearly as free radical scavenger and inhibited the growth and proliferation of several cancer cells, in a time and concentration dependent way. The EO was also capable to induce the apoptotic process, by acting on caspase-3, -8 and -9, as well as on poly-adenosine diphosphate ribose polymerase. In addition, it was able to suppress the gene expression that is involved in the different steps (from

survival to proliferation, from the invasion to angiogenesis until metastatic events occurring in human colon cancer cells), without inducing toxicity. Phenolics and sesquiterpenes (in particular δ-limonene, α-bisabol, humulene, thymol and (+)-epi-bicyclosesquiphellandrene) might be also capable to alleviate the inflammatory process (Guesmi et al. 2018). In vitro, through the Ames test, the EO of *Teucrium ramosissimum* Desf. showed antimutagenic effects against sodium azide, aflatoxin B1, benzo[α]pyrene, and 4-nitro-o-phenylenediamine by the bacterial reverse mutation assay in *Salmonella typhimurium* TA98, TA100, and TA1535, with and without exogenous metabolic activation (Sghaier et al. 2010). The EO of *Teucrium pruinosum* from Palestine exhibited in vitro cytotoxic activity against HeLa cervical adenocarcinoma cells. This EO showed also potential antioxidant activity and COX-1 and COX-2 inhibitory effects, with IC_{50} values of 0.25 μg/ml and 0.5 μg/ml, respectively, demonstrating also potential anti-inflammatory properties (Jaradat et al. 2018).

The essential oil of *Teucrium alopecurus* demonstrated the capability to exert in vitro anti-inflammatory and anti-cancer activity by a combination of several actions, such as the block of IκBα kinase, downregulation of NF-κB activation, enhancement of apoptosis and destruction of NF-κB-regulated gene expression (Guesmi et al. 2017). Moreover, terpenes and an hydrolate from this EO oil triggered apoptotic events dependent on caspases activation and PARP cleavage in human colon cancer cells through decreased protein expressions (Guesmi et al. 2018). The essential oil of *Teucrium yemense* demonstrated positive effects against HT-29 human colorectal adenocarcinoma cell lines, with an IC_{50} of 43.7 μg/ml. Coherent with this, this oil was active against both MCF-7 and MDA-MB-231 human breast adenocarcinoma cells (Awadh Ali et al. 2017).

7.4.3 Anti-cholinesterase Activity

Alzheimer's disease (AD) represents actually the fifth leading age-related neurological disorders in the world. AD is characterized by several symptoms, such as cognitive dysfunction, memory loss, psychological changes, and disability, which increase day by day, due to an oxidative stress-induced scarceness in cholinergic neurotransmission, the accumulation of amyloid plaques, and neurofibrillary tangles in the brain (Reitz 2012; Ayaz et al. 2015). In this context, acetylcholinesterase (AChE) and butyrylcholinesterase (BChE) inhibitor compounds, such as donepezil, galanthamine, rivastigmine, tacrine, etc., represent, at present, the majority of drugs approved for clinical use in the symptomatic AD relief. Nevertheless, EOs target several pathways and thus provide better healing.

Members of the Lamiaceae family and their phytochemicals are known for pharmacological and some clinical effects relevant to dementia. Among Lamiaceae plants, especially *Salvia*, *Rosmarinus* and *Melissa* species, have neuroprotective properties known for many years, besides some other Lamiaceae plants, such as *Teucrium polium*. These plants have been used since ancient times in Turkey for memory enhancement (Perry and Howes 2011; Tuncer 1978). The EOs of some

species of *Teucrium* showed an anticholinesterase activity, thus a putative capability to bind to the acetylcholinesterase enzyme, inhibiting its activity. For this reason, as cholinesterase could lead to an increase in the acetylcholine that acts at the nervous level with consequent bradycardia, hypotension, bronchoconstriction and prolonged muscle contraction, the EO with anticholinesterase activity could be used in the pharmacological field (treatment of Alzheimer's disease, or as antidotes for anticholinergic poisoning). *Teucrium polium* EO, extracted from plants collected in Algeria, exhibit such property, in particular against butyrylcholinesterase, resulting also more effective than *Thymus algeriensis* Boiss. & Reut. (Bendjabeur et al. 2018).

7.4.4 Antinociceptive Activity

Pain is a sensorial modality that applies nature from a protective point of view. However, it often can cause discomfort. At present, opiates and non-steroidal anti-inflammatory drugs are the most diffused analgesic drugs used in the current pharmacology to mitigate the pain: however, they cannot be so often used, due to their side effects. Thus, in last decades, the search for new analgesic compounds was one of the main aims for scientists and pharmaceutical companies (Mattison et al. 1998). Therefore, it is well known that herbs and medicinal plants represent a significant font of substances capable to exert potential therapeutic effects (Farnsworth 1989; Blumenthal 2000). Thus, the research and study of plant species traditionally used to mitigate or eliminate the pain could be seen as a logical research strategy, in search for new analgesic drugs (Rang et al. 1998). The genus *Teucrium* can supply, both in the form of aqueous extract (Zendehdel et al. 2011) and in form of essential oils, products with several interesting properties. The EO of *Teucrium polium* is the most known and studied as analgesic. The anti-visceral pain properties discovered for this EO might be comparable to those exhibited by hyoscine and indomethacin. This suggests a good place for it in antispasmodic therapies in humans. The presence of flavonoids and sterols might be also responsible for the anti-inflammatory activity of this plant (Abdollahi et al. 2003). The antispasmodic activity exhibited by *Teucrium polium* is widely demonstrated through the study of plants collected in different countries of the Mediterranean/Middle East. *Teucrium polium* collected in South Arabia, for instance, exhibited a powerful antispasmodic activity and the presence of β-pinene, limonene, α-phellandrene, γ- and δ-cadinenes, and of some alcohols such as linalool, terpinen-4-ol, cedrol, cedrenol and guaiol, which could contribute, further than the alleviation of the pain, also to the other well-known properties provided by the species (Hassan et al. 1979).

Few *Teucrium* EOs were studied for their anti-nociceptive activity. The essential oil of *Teucrium stocksianum* from Pakistan, showed outstanding anti-nociceptive activity. It was observed that, in an animal model, the decrease of writhes occurred using from 20 to 80 mg/kg (b.w), with the maximum of writhes inhibition noted at a concentration of 80 mg/kg (b.w). This oil contains mainly α-pinene, myrcene, limonene, linalool and caryophyllene oxide, all of them with antinociceptive effects.

The high potency of the EO of *Teucrium stocksianum,* as compared to diclofenac used as positive control, might be due to the synergistic effect of some of its components (Shah et al. 2012).

7.4.5 Anti-inflammatory Activity

Free radical species and nitric oxide (NO) or its derivatives represent key denominators in carcinogenesis, further to have biological significance in viral and bacterial infections. Treatment with chemical carcinogens, such as carbon tetrachloride and heterocyclic amines, also generated superoxide. Furthermore, most experimental solid tumours have elevated levels of iNOS in the tumoral tissue, and NO thus generated facilitates vascular permeability, which quickens nutritional supply to the tumor tissue and hence supports the rapid tumour growth (Maeda and Akaike 1998). Inflammation can be considered as a normal protective response induced by tissue injury or infection and functions against microorganisms and non-self cells, when these invade the body, and as response to remove dead or damaged host cells (Stevenson and Hurst 2007; Miguel 2010). In the inflammatory response, we find an increase of permeability of endothelial lining cells and influxes of blood leukocytes into the interstitium, oxidative burst, with release of cytokines [interleukins and tumor necrosis factor-α (TNF-α)]. At the same time, it occurs also an induction of the activity of several enzymes (in particular oxygenases, nitric oxide synthases, peroxidases) and of the arachidonic acid metabolism. In addition, the expression of cellular adhesion molecules, such as intercellular adhesion molecule (ICAM) and vascular cell adhesion molecule (VCAM) can be observed (Gomes et al. 2008). Further than the capability of some essential oils to scavenge free radicals, there is also evidence that some of them possess anti-inflammatory activity. Under this aspect, several essential oils, including those obtained from some species of *Teucrium,* can show beneficial effect by inhibiting the production of inflammatory mediators in macrophages.

The incubation of cell such as RAW 264.7 cells with essential oils obtained from *Teucrium* spp. can positively cause a significant inhibitory effect on the LPS-induced nitrite production, with IC_{50} values lower than those of the positive control, indomethacin. Among the essential oils recovered by the *Teucrium* spp., the oil with the most effective inhibitory action against LPS-induced NO production could be that obtained from *Teucrium brevifolium,* which, as reported by Menichini et al. (2009), gave an IC_{50} value of 7.1 μg/ml, followed by *Teucrium montbretii* ssp. *heliotropiifolium* (Berbey) P. H. Davis and *T. capitatum,* which IC_{50} values resulted of 16.5 and 29.4 μg/ml, respectively. *T. flavum* EO shows a weaker activity (IC_{50} = 41.4 μg/ml). Such activities are noteworthy if compared with those of other essential oils. For instance, the essential oil from the *Farfugium japonicum* (L.) Kitam. (Asteraceae) produced a 25% of inhibition of NO production at a concentration of 100 μg/ml (Kim et al. 2008); on the other hand, the EO from *Cinnamomum osmophloeum*

Kaneh. (Lauraceae) can inhibit by 100% the NO production when used at a concentration of 50 μg/ml (Tung et al. 2008).

A survey of literature reveals that some *Teucrium* EOs, such as that of *Teucrium buxifolium* Schreb., have potent anti-inflammatory activity against both the acute and the delayed phases of inflammation (Fernández Puntero et al. 1997). Some essential oils, containing major constituents similar to those in the oil of *Teucrium brevifolium,* have shown anti-inflammatory activity. For example, the EO of *Garcinia brasiliensis* Mart. (Clusiaceae), which contains spathulenol (8.7%) as a major constituent (Martins et al. 2008), exhibited anti-inflammatory activity, in the rat-paw oedema model induced by carrageenan. The highest anti-inflammatory activities exhibited by *Teucrium capitatum* and *T. montbretii* ssp. *heliotropiifolium* could be probably due to their major components, caryophyllene and carvacrol, which showed anti-inflammatory activity in previous studies (Sosa et al. 2005; Passos et al. 2007).

7.4.6 Antioxidant and Hypoglycemic Activities

Generally, the species of *Teucrium* exhibit a good antioxidant activity, which can be a support for other biological properties. Several *in vitro* studies demonstrated the antioxidant potentiality of *Teucrium polium* with a lot of the assays usually performed in this context, starting from the DPPH test to metal chelating activity until the evaluation of the reducing power (Abdollahi et al. 2003). *Teucrium polium* EO even demonstrated the same potency of tocopherol (Couladis et al. 2003).

Hypoglycemic activity has been reported – in addition for the flavonoids – also for the volatile oils of different species of *Teucrium.* Traditionally, especially in the Mediterranean countries, some *Teucrium* species, such as *Teucrium polium* and *T. mascatense* are used for their hypoglycemic activities by the native inhabitants and recommended by the herbalists (Afifi et al. 2005; Rehman et al. 2016). However, the EO from *Teucrium polium* ssp. *polium* Briq., and its major terpene components, showed strong antioxidant, anti-inflammatory and antimicrobial activities (Kerbouche et al. 2015).

7.4.7 Allelopathic and Biocide Effects

Allelopathic effects gives rise from the release, by the plant, of certain compounds, known as allelochemicals, which are biosynthesized in the plant as secondary metabolites (essenatial oils, polyphenols, coumarins, terpenoids, etc.) and may be present in all tissues including pollens (Whittaker and Feeny 1971). The isolation and identification of EOs and/or components with allelopathic, anthelminthic and in general biocide activity is of great interest, due to the increasing request, in food, health and agriculture, of substances/compounds of natural origin that can be used

with safety, are no toxic and have a great spectrum of action without induce resistance (Ahmad et al. 2011; De Martino et al. 2015).

The EO of *Teucrium royleanum* Wall. ex Benth. has been evaluated as allelopathic agent (Ahmad et al. 2011). De Martino et al. (2010) studied the chemical composition and the phytotoxic effect of four *Teucrium* EOs. All oils, although rich in sesquiterpenes (50.0–61.9%) differed for composition, being caryophyllene and caryophyllene oxide the main components of *Teucrium arduini*, germacrene D, δ-cadinene and γ-cadinene predominated in *Teucrium maghrebinum*. Carvacrol and caryophyllene were abundant in *Teucrium capitatum*, while carvacrol, caryophyllene oxide and caryophyllene were the most main components in *Teucrium montbretii* ssp. *heliotropiifolium*. The germination of radish and garden cress was less sensitive to the four essential oils. The radicle elongation, above all of radish, was significantly inhibited by all oils, in particular by that of *Teucrium arduini*. Among the main components of the oils, monoterpenes resulted the more active compounds. Some of the sesquiterpenes isolated by the essential oil of *Teucrium ramosissimum*, in particular homalomenol C, 4β-hydroxy-11,12,13-trinor-5-eudesmen-1,7-dione, oxo-T-cadinol and 1β,4β,6β-trihydroxyeudesmane exhibit a significant *in vitro* antiplasmodial activity against *Plasmodium falciparum*, with IC_{50} values ranging from 1.2 to 5.0 μg/ml, without any cytotoxic effect upon the human diploid lung cell line MRC-5 (Sathiyamoorthy et al. 1999; Henchiri et al. 2009).

7.5 Conclusions

This review testifies the great variability of the essential oils produced by different species of *Teucrium*, as influenced by several endogenous and exogenous factors and by the possible presence of chemotypes. The essential oils constitute the basis for most of the uses of *Teucrium* species in traditional medicine and alimentation.

Teucrium polium, *T. chamaedrys*, *T. flavum* and *T. capitatum* are the more studied taxa concerning the essential oil composition. The literature data permit to categorize the essential oils in some groups, following the prevalence of some components, even if the majority of the essential oils contain a prevalence of sesquiterpenic compounds. Caryophyllene and caryophyllene derivatives, α- and β-pinene, germacrene B and D, α-, γ-, δ-cadinene and or cadinol, limonene, bisabolol/bisabolene have been identified as the marker components in *Teucrium* essential oils. However, *Teucrium* volatile oils are charaterized by a high degree of chemical polymorphism and often the same species possessed essential oils with very different chemical profiles. This inhomogeneity could reflect differences in genetic structure, environmental effects, and/or a combination of both factors. Moreover, in most cases, also the seasonal influence or climatic factors, such as altitude, annual rainfall, winter cold stress, summer precipitation, summer drought stress, evapotranspiration and humidity could contribute to explain the complex essential oil variability.

These differences constitute an asset both in terms of biodiversity and in terms of practical use of *Teucrium* species, primarily for their biological properties. In fact,

the available literature reports data about antimicrobial, antioxidant, antinflammatory, cytotoxic, nociceptive and allelopathic activity of the essential oils, often confirming the traditional uses of *Teucrium* species.

References

Abdelshafeek KA, Elgattar AA, Zarkoon AH, Alwahash MA, Shahat AA (2010) Investigation of the volatile oils, lipid constituents and biological activity of *Ballota andreuzziana*, *Teucrium zanonii* and *Verbena tenuisecta* in Libya. Asian Pac J Trop Med 3:594–601

Abdollahi M, Karimpour H, Monsef-Esfehani HM (2003) Antinociceptive effects of *Teucrium polium* L. total extract and essential oil in mouse writhing test. Pharmacol Res 48:31–35

Aburjai T, Hudaib M, Cavrini V (2006) Composition of the essential oil from Jordanian germander (*Teucrium polium* L.). J Essent Oil Res 18:97–99

Adzet T, Granger R, Passet J, San Martin R (1977) Chemical polymorphism in the genus *Thymus*: taxonomic importance. Biochem Syst Ecol 5:269–272

Afifi FU, Al-Khalidi B, Khalil E (2005) Studies on the *in vivo* hypoglycemic activities of two medicinal plants used in the treatment of diabetes in Jordanian traditional medicine following intranasal administration. J Ethnopharmacol 100:314–318

Ahmad S, Arfan M, Khan AL, Ullah R, Hussain J, Muhammad Z, Khan R, Khan N, Watanabe KN (2011) Allelopathy of *Teucrium royleanum* Wall. ex Benth. from Pakistan. J Med Plant Res 5:765–772

Ahmadi L, Mirza M, Shahmir F (2002) Essential oil of *Teucrium melissoides* Boiss. et Hausskn. ex Boiss. J Essent Oil Res 14:355–356

Ali C, Ahmet E, Herken EN, Idris A (2013) Phytochemistry and biological activities of terpene-rich essential oil (TREo) from *Teucrium sandrasicum*. Res J BioTechnol 8:60–63

Al-Qudah TS, Shibli RA, Alali FQ (2011) *In vitro* propagation and secondary metabolites production in wild germander (*Teucrium polium* L.). *In Vitro* Cell Dev Biol Plant 47:496–505

Al-Yousuf MH, Bashir AK, Dobos A, Veres K, Nagy G, Mathe I, Blunden G (2002) The composition of the essential oil of *Teucrium stocksianum* from the United Arab Emirates. J Essent Oil Res 14:47–48

Antunes T, Sevinate-Pinto I, Barroso JG, Cavaleiro C, Salgueiro LR (2004) Micromorphology of trichomes and composition of essential oil of *Teucrium capitatum*. Flavour Fragr J 19:336–340

Arnold N, Bellomaria B, Valentini G, Arnold HJ (1998) Study of the chemical composition of essential oil of *Teucrium gracile* W. Barbey and Major endemic to Karpath. Riv Ital EPPOS 26:41–46

Awadh Ali NA, Wurster M, Arnold N, Lindequist U, Wessjohann L (2008) Chemical composition of the essential oil of *Teucrium yemense* Deflers. Rec Nat Prod 2:25–32

Awadh Ali NA, Chhetri BK, Dosoky NS, Shari K, Al-Fahad AJA, Wessjohann L, Setzer WN (2017) Antimicrobial, antioxidant, and cytotoxic activities of *Ocimum forskolei* and *Teucrium yemense* (Lamiaceae) essential oils. Medicine 4:E17

Ayaz M, Junaid M, Ullah F, Sadiq A, Khan MA, Ahmad W, Shah MR, Imran M, Ahmad S (2015) Comparative chemical profiling, cholinesterase inhibitions and anti-radicals properties of essential oils from *Polygonum hydropiper* L: a preliminary anti-Alzheimer's study. Lipids Health Dis 14:141

Bagci E, Yazgyn A, Hayta S, Cakilcioglu U (2010) Composition of the essential oil of *Teucrium chamaedrys* L. (Lamiaceae) from Turkey. J Med Plant Res 4:2588–2590

Bagci E, Hayta S, Yazgin A, Dogan G (2011) Composition of the essential oil of *Teucrium parviflorum* L. (Lamiaceae) from Turkey. J Med Plant Res 5:3457–3460

Baher ZF, Mirza M (2003) Volatile constituents of *Teucrium flavum* L. from Iran. J Essent Oil Res 15:106–107

Bahramikia S, Yazdanparast R (2012) Phytochemistry and medicinal properties of *Teucrium polium* L. (Lamiaceae). Phytother Res 26:1581–1593

Bakari S, Ncir M, Felhi S, Hajlaoui H, Saoudi M, Gharsallah N, Kadri A (2015) Chemical composition and in vitro evaluation of total phenolic, flavonoid, and antioxidant properties of essential oil and solvent extract from the aerial parts of *Teucrium polium* grown in Tunisia. Food Sci Biotechnol 24:1943–1949

Bakhtiari M, Asgarpanah J (2015) Volatile constituents of *Teucrium stocksianum* Boiss. fruits from South of Iran. J Essent Oil Bear Plants 18:1174–1179

Bakkali F, Averbeck S, Averbeck D, Idaomar M (2008) Biological effects of essential oils – a review. Food Chem Toxicol 46:446–475

Barceloux DG (2008) Medical toxicology of natural substances: foods, fungi, medicinal herbs, plants, & venomous animals. Wiley, New York

Barroso JG, Pedro LG, Figueiredo AC, Antunes T, Sevinate-Pinto I, Scheffer JJC (1993) The essential oils of two endemic *Teucrium* species from Madeira: *T. abutiloides* L'Hér. and *T. betonicum* L'Hér. Flavour Fragr J 8:277–280

Barroso JG, Figueiredo AC, Pedro LG, Antunes T, Sevinate-Pinto I, Fontinha SS, Scheffer JJC (1996) Composition of the essential oil of *Teucrium heterophyllum* L'Hér. grown on Madeira. Flavour Fragr J 11:129–132

Baser KHC, Demircakmak B, Duman H (1997) Composition of the essential oils of three *Teucrium* species from Turkey. J Essent Oil Res 9:545–549

Baser KHC, Demirci B, Duman H, Aytac Z (1999) Composition of the essential oil of *Teucrium antitauricum* T. Ekim. J Essent Oil Res 11:61–62

Bellomaria B, Arnold N, Valentini G (1998) Essential oil of *Teucrium flavum* subsp. *hellenicum* from Greece. J Essent Oil Res 10:131–133

Belmekki N, Bendimerad N, Bekhechi C, Fernandez X (2013) Chemical analysis and antimicrobial activity of *Teucrium polium* L. essential oil from Western Algeria. J Med Plant Res 7:897–902

Ben Othman M, Bel Hadj Salah-Fatnassi K, Ncibi S, Elaissi A, Zourgui L (2017) Antimicrobial activity of essential oil and aqueous and ethanol extracts of *Teucrium polium* L. subsp. *gabesianum* (L.H.) from Tunisia. Physiol Mol Biol Plants 23:723–729

Bencheikh SE, Ladjel S, Goudjil MB, Mehani M, Zighmi S (2015) Chemical analysis, antimicrobial and antioxidant activity of the essential oil of *Teucrium polium* ssp. *aurasiacum*, Labiatae. Pharma Chem 7:308–314

Bendif H, Lazali M, Souilah N, Miara MD, Kazernaviciute R, Baranauskiene R, Venskutonis PR, Maggi F (2018) Supercritical CO_2 extracts and essential oils from *Teucrium polium* L. growing in Algeria: chemical composition and antioxidant activity. J Essent Oil Res 30:488–497

Bendjabeur S, Benchabane O, Bensouici C, Hazzit M, Baaliouamer A, Bitam A (2018) Antioxidant and anticholinesterase activity of essential oils and ethanol extracts of *Thymus algeriensis* and *Teucrium polium* from Algeria. J Food Meas Charact 12:2278–2288

Bezic N, Vuko E, Dunkic V, Ruscic M, Blazevic I, Burcul F (2011) Antiphytoviral activity of sesquiterpene-rich essential oils from four Croatian *Teucrium* species. Molecules 16:8119–8129

Bishop CD (1995) Antiviral activity of the essential oil of *Melaluca alternifolia* (Maiden & Betche) Cheel. (tea tree) against tobacco mosaic virus. J Essent Oil Res 7:641–644

Blazevic N, Kalodera Z, Petricic J, Plazibat M (1992) Essential oil content and composition of *Teucrium arduini* L. J Essent Oil Res 4:223–225

Blazquez MA, Perez I, Boira H (2003) Essential oil analysis of *Teucrium libanitis* and *T. turredanum* by GC and GC-MS. Flavour Fragr J 18:497–501

Blumenthal M (2000) Herbal medicines. Integrative Medicine Communications, Austin 419–423

Boulila A, Béjaoui A, Messaoud C, Boussaid M (2008) Variation of volatiles in Tunisian populations of *Teucrium polium* L. (Lamiaceae). Chem Biodivers 5:1389–1400

Cakir A, Duru M, Emin Harmandar M, Ciriminna R, Passannanti S (1998) Volatile constituents of *Teucrium polium* L. from Turkey. J Essent Oil Res 10:113–115

Cavaleiro C, Salgueiro LR, Antunes T, Sevinate-Pinto I, Barroso JG (2002) Composition of the essential oil and micromorphology of trichomes of *Teucrium salviastrum*, an endemic species from Portugal. Flavour Fragr J 17:287–291

Cavaleiro C, Salgueiro LR, Miguel MG, Proenca da Cunha A (2004) Analysis by gas chromatography-mass spectrometry of the volatile components of *Teucrium lusitanicum* and *Teucrium algarbiensis*. J Chromatogr A 1033:187–190

Chen Q, Zhang Q, Yang Z, Zhao C, Ye C (2010) Component analysis of volatile oil from *Teucrium labiosum, T. bidentatum* and *T. pilosum* by SPME/GC-MS. Zhongguo Yaofang 21:1013–1016

Couladis M, Tzakou O, Verykokidou E, Harvala C (2003) Screening of some Greek aromatic plants for antioxidant activity. Phytother Res 17:194–195

Cozzani S, Muselli A, Desjobert J-M, Bernardini A-F, Tomi F, Casanova J (2005) Chemical composition of essential oil of *Teucrium polium* subsp. *capitatum* (L.) from Corsica. Flavour Fragr J 20:436–441

De Martino L, Formisano C, Mancini E, De Feo V, Piozzi F, Rigano D, Senatore F (2010) Chemical composition and phytotoxic effects of essential oil's from four *Teucrium* species. Nat Prod Commun 5:1969–1976

De Martino L, Nazzaro F, Mancini E, De Feo V (2015) Essential oils from Mediterranean aromatic plants. In: Preedy VR, Eatson RR (eds) The Mediterranean diet, an evidence-based approach. Elsevier, Amsterdam, pp 649–661

Djabou N, Paolini J, Desjobert J-M, Allali H, Baldovini N, Costa J, Muselli A (2010) Qualitative and quantitative analysis of volatile components of *Teucrium massiliense* L. – identification of 6-methyl-3-heptyl acetate as a new natural product. Flavour Fragr J 25:475–487

Djabou N, Battesti M-J, Allali H, Desjobert J-M, Varesi L, Costa J, Muselli A (2011) Chemical and genetic differentiation of Corsican subspecies of *Teucrium flavum* L. Phytochemistry 72:1390–1399

Djabou N, Muselli A, Allali H, Dib MEA, Tabti B, Varesi L, Costa J (2012) Chemical and genetic diversity of two Mediterranean subspecies of *Teucrium polium* L. Phytochemistry 83:51–62

Djabou N, Lorenzi V, Guinoiseau E, Andreani S, Giuliani M-C, Desjobert J-M, Bolla J-M, Costa J, Berti L, Luciani A, Muselli A (2013a) Phytochemical composition of Corsican *Teucrium* essential oils and antibacterial activity against foodborne or toxi-infectious pathogens. Food Control 30:354–363

Djabou N, Andreani S, Varesi L, Tomi F, Costa J, Muselli A (2013b) Analysis of the volatile fraction of *Teucrium marum* L. Flavour Fragr J 28:14–24

Dunkic V, Bezic N, Vuko E (2011) Antiphytoviral activity of essential oil from endemic species *Teucrium arduini*. Nat Prod Commun 6:1385–1388

Eikani MH, Goodarznia I, Mirza M (1999) Comparison between the essential oil and supercritical carbon dioxide extract of *Teucrium polium* L. J Essent Oil Res 11:470–472

El Amri J, El Badaoui K, Haloui Z (2017) The chemical composition and the antimicrobial properties of the essential oil extracted from the leaves of *Teucrium capitatum* L. Asian J Pharm Clin Res 10:112–115

Elabbara FA, Habel AM, Bozkeh NMA, El-Tuonsi ATM, Awin TM (2014) The essential oil composition of *Teucrium apollinis* (Lamiaceae) from Libya. Chem Sin 5:26–27

El-Shazly AM, Hussein KT (2004) Chemical analysis and biological activities of the essential oil of *Teucrium leucocladum* Boiss. (Lamiaceae). Biochem Syst Ecol 32:665–674

Farnsworth NR (1989) Screening plants for new medicines. In: Wilson EO (ed) Biodiversity, part II. National Academy Press, Washington, pp 83–97

Fernández Puntero B, Iglesias Peinado I, Villar del Fresno AM (1997) Anti-inflammatory and antiulcer activity of *Teucrium buxifolium*. J Ethnopharmacol 55:93–98

Fertout-Mouri N, Latreche A, Mehdadi Z, Toumi-Benali F, Khaled MB (2017) Chemical composition and antibacterial activity of the essential oil of *Teucrium polium* L. of Tessala Mount (Western Algeria). Phytothérapie 15:346–353

Flamini G, Cioni PL, Morelli I, Maccioni S, Monti G (2001) Composition of the essential oil of *Teucrium fruticans* L. from the Maremma Regional Park (Tuscany, Italy). Flavour Fragr J 16:367–369

Formisano C, Rigano D, Senatore F, Al-Hillo MRY, Piozzi F, Rosselli S (2009) Analysis of essential oil from *Teucrium maghrebinum* Greuter et Burdet growing wild in Algeria. Nat Prod Commun 4:411–414

Formisano C, Napolitano F, Rigano D, Arnold Apostolides N, Piozzi F, Senatore F (2010) Essential oil composition of *Teucrium divaricatum* Sieb. ssp. *villosum* (Celak.) Rech. fil. growing wild in Lebanon. J Med Food 13:1281–1285

Formisano C, Rigano D, Senatore F, Bruno M, Maggio A, Piozzi F (2012) Chemical composition of the essential oil of *Teucrium flavum* ssp. *flavum* from Zakynthos, Greece. Rec Nat Prod 6:306–310

Gaspar H, Brito P, Fernando MS, de La Torre MC, Rodriguez B, Barroso JG, Figueiredo AC (1997) Composition of the essential oil of *Teucrium haenseleri* Boiss. Flavour Fragr J 12:355–357

Gershenzon J, Dudareva N (2007) The function of terpene natural products in the natural world. Nat Chem Biol 3:408–414

Ghazouani N, Abderrabba M, Bouajila J (2016) *Teucrium ramosissimum* (Lamiaceae): volatile composition, seasonal variation, and pharmaceutical activity. Anal Lett 49:1258–1271

Ghazouani N, Sifaoui I, Bachrouch O, Abderrabba M, Piñero JE, Lorenzo-Morales J (2017) Essential oil composition and anti *Acanthamoeba* studies of *Teucrium ramosissimum*. Exp Parasitol 183:207–211

Gomes A, Fernandes E, Lima JLFC, Mira L, Corvo ML (2008) Molecular mechanisms of anti-inflammatory activity mediated by flavonoids. Curr Med Chem 15:1586–1605

Guesmi F, Prasad S, Tyagi AK, Landoulsi A (2017) Antinflammatory and anticancer effects of terpenes from oily fractions of *Teucruim alopecurus*, blocker of IκBα kinase, through downregulation of NF-κB activation, potentiation of apoptosis and suppression of NF-κB-regulated gene expression. Biomed Pharmacother 95:1876–1885

Guesmi F, Tyagi AK, Prasad S, Landoulsi A (2018) Terpenes from essential oils and hydrolate of *Teucrium alopecurus* triggered apoptotic events dependent on caspases activation and PARP cleavage in human colon cancer cells through decreased protein expressions. Oncotarget 9:32305–32320

Guetat A, Al-Ghamdi FA (2014) Analysis of the essential oil of the germander (*Teucrium polium* L.) aerial parts from the northern region of Saudi Arabia. Int J Appl Biol Pharm 5:128–135

Hachicha SF, Skanji T, Barrek S, Zarrouk H, Ghrabi ZG (2007a) Chemical composition of *Teucrium alopecurus* essential oil from Tunisia. J Essent Oil Res 19:413–415

Hachicha SF, Skanji T, Barrek S, Ghrabi ZG, Zarrouk H (2007b) Composition of the essential oil of *Teucrium ramosissimum* Desf. (Lamiaceae) from Tunisia. Flavour Fragr J 22:101–104

Hammami S, Jmii H, El Mokni R, Khmiri A, Faidi K, Dhaouadi H, El Aouni MH, Aouni M, Joshi RK (2015a) Essential oil composition, antioxidant, cytotoxic and antiviral activities of *Teucrium pseudochamaepitys* growing spontaneously in Tunisia. Molecules 20:20426–20433

Hammami S, El Mokni R, Faidi K, Falconieri D, Piras A, Procedda S, Mighri Z, El Aouni MH (2015b) Chemical composition and antioxidant activity of essential oil from aerial parts of *Teucrium flavum* L. subsp. *flavum* growing spontaneously in Tunisia. Nat Prod Res 29:2336–2340

Hassan MMA, Muhtadi FJ, Al-Badr AA (1979) GLC-mass spectrometry of *Teucrium polium* oil. J Pharm Sci 68:800–801

Henchiri H, Bodo B, Deville A, Dubost L, Zourgui L, Raies A, Grellier P, Mambu L (2009) Sesquiterpenoids from *Teucrium ramosissimum*. Phytochemistry 70:1435–1441

Hisham A, Pathare N, Al-Saidi S (2006a) The composition and antimicrobial activity of the essential oil of *Teucrium stocksianum* subsp. *stocksianum* leaf from Oman. Nat Prod Commun 1:195–199

Hisham A, Pathare N, Al-Saidi S, Al-Salmi A (2006b) The composition and antimicrobial activity of leaf essential oil of *Teucrium mascatenses* Boiss. from Oman. J Essent Oil Res 18:465–468

Jaradat N (2005) Ethnopharmacological survey of natural products in Palestine. An-Najah Univ J Res – A (Nat Sci) 19:14–67

Jaradat NA (2015) Review of the taxonomy, ethnobotany, phytochemistry, phytotherapy and phytotoxicity of germander plant (*Teucrium polium* L.). Asian J Pharm Clin Res 8:13–19

Jaradat N, Al-lahham S, Abualhasan MN, Bakri A, Zaide H, Hammad J, Hussein F, Issa L, Mousa A, Speih R (2018) Chemical constituents, antioxidant, cyclooxygenase inhibitor, and cytotoxic activities of *Teucrium pruinosum* Boiss. essential oil. Biomed Res Int 2018:4034689/1–4034689/9. https://doi.org/10.1155/2018/4034689

Javidnia K, Miri R (2003) Composition of the essential oil of *Teucrium orientale* L. ssp. *orientale* from Iran. J Essent Oil Res 15:118–119

Javidnia K, Miri R, Khosravi AR (2007) Composition of the essential oil of *Teucrium persicum* Boiss. from Iran. J Essent Oil Res 19:430–432

Jeanmonod D, Gamisans J (2007) Flora Corsica. Edisud, Aix-en-Provence, pp 638–640

Juglal S, Govinden R, Odhav B (2002) Spice oils for the control of co-occurring mycotoxinproducing fungi. J Food Prot 65:683–687

Kabouche Z, Boutaghane N, Laggoune S, Kabouche A, Ait-Kaki Z, Benlabed K (2005) Comparative antibacterial activity of five Lamiaceae essential oils from Algeria. Int J Aromather 15:129–133

Kabouche A, Touafek O, Nacer A, Kabouche Z, Bruneau C (2006) Volatile oil constituents of *Teucrium atratum* pomel from Algeria. J Essent Oil Res 18:175–177

Kabouche A, Kabouche Z, Ghannadi A, Sajjadi SE (2007) Analysis of the essential oil of *Teucrium polium* ssp. *aurasiacum* from Algeria. J Essent Oil Res 19:44–46

Kamel A, Sandra P (1994) Gas chromatography-mass spectrometry analysis of the volatile oils of two *Teucrium polium* varieties. Biochem Syst Ecol 22:529–532

Kartah B, Harhar H, Elmonfalouti H, Gharby S, Guillaume D, Charrouf Z (2015) Chemical composition of the essential oil of *Teucrium antiatlanticum* (Lamiaceae). Pharma Chem 7:23–25

Kaya A, Demirci B, Baser KHC (2009) Compositions of essential oils and trichomes of *Teucrium chamaedrys* L. subsp. *trapezunticum* Rech. Fil. and subsp. *syspirense* (C. Koch) Rech. Fil. Chem Biodivers 6:96–104

Kaya A, Demirci B, Dinc M, Dogu S, Baser KHC (2013) Compositions of the essential oils of *Teucrium cavernarum* and *Teucrium paederotoides*, two endemic species from Turkey. J Essent Oil Bear Plants 16:588–594

Kerbouche L, Hazzit M, Ferhat M-A, Baaliouame A, Miguel MG (2015) Biological activities of essential oils and ethanol extracts of *Teucrium polium* subsp. *capitatum* (L.) Briq. and *Origanum floribundum* Munby. J Essent Oil Bear Plants 18:1197–1208

Keykavousi M, Tarzi BG, Mahmoudi R, Bakhoda H, Kabudari A, Mahalleh SFRP (2016) Study of antibacterial effects of *Teucrium polium* essential oil on *Bacillus cereus* in cultural laboratory and commercial soup. Carpath J Food Sci Technol 8:193–201

Khani A, Heydarian M (2014) Fumigant and repellent properties of sesquiterpene-rich essential oil from *Teucrium polium* subsp. *capitatum* (L.). Asian Pac J Trop Med 7:956–961

Kim JY, Oh TH, Kim BJ, Kim SS, Lee NH, Hyun CG (2008) Chemical composition and anti-inflammatory effects of essential oil from *Farfugium japonicum* flower. J Oleo Sci 57:623–628

Kovacevic NN, Lakusic BS, Ristic MS (2001) Composition of the essential oils of seven *Teucrium* species from Serbia and Montenegro. J Essent Oil Res 13:163–165

Kremer D, Mueller Dragojevic I, Dunkic V, Vitali D, Stabentheiner E, Oberlaender A, Bezic N, Kosalec I (2012) Chemical traits and antimicrobial activity of endemic *Teucrium arduini* L. from Mt Biokovo (Croatia). Cent Eur J Biol 7:941–947

Kremer D, Kosir IJ, Kosalec I, Koncic MZ, Potocnik T, Cerenak A, Bezic N, Srecec S, Dunkic V (2013) Investigation of chemical compounds, antioxidant and antimicrobial properties of *Teucrium arduini* L. (Lamiaceae). Curr Drug Targets 14:1006–1014

Kremer D, Bolaric S, Ballian D, Bogunic F, Stesevic D, Karlovic K, Kosalec I, Vokurka A, Vukovic Rodriguez J, Randic M, Bezic N, Dunkic V (2015) Morphological, genetic and phytochemical variation of the endemic *Teucrium arduini* L. (Lamiaceae). Phytochemistry 116:111–119

Kucuk M, Gulec C, Yasar A, Ucuncu O, Yayli N, Coskuncelebi K, Terzioglu S, Yayli N (2006) Chemical composition and antimicrobial activities of the essential oils of *Teucrium chamaedrys* subsp. *chamaedrys*, *T. orientale* var. *puberulens*, and *T. chamaedrys* subsp. *lydium*. Pharm Biol 44:592–599

Kucukbay FZ, Yildiz B, Kuyumcu E, Gunal S (2011) Chemical composition and antimicrobial activities of the essential oils of *Teucrium orientale* var. *orientale* and *Teucrium orientale* var. *puberulens*. Chem Nat Compd 47:833–836

Lakusic B, Lakusic D, Jancic R, Stevanovic B (2006) Morpho-anatomical differentiation of the Balkan populations of the species *Teucrium flavum* L. (Lamiaceae). Flora 201:108–119

Lamiri A, Lhaloui S, Benjilali B, Berrada M (2001) Insecticidal effects of essential oils against Hessian fly, *Mayetiola destructor* (Say). Field Crop Res 71:9–15

Lianopoulou V, Bosabalidis AM, Patakas A, Lazari D, Panteris E (2014) Effects of chilling stress on leaf morphology, anatomy, ultrastructure, gas exchange, and essential oils in the seasonally dimorphic plant *Teucrium polium* (Lamiaceae). Acta Physiol Plant 36:2271–2281

Liu XC, Liu SL, Liu ZL (2016) GC-MS Analysis of the essential oil and insecticidal activity of *Teucrium quadrifarium* Buch.-Ham. (Lamiaceae) aerial parts against *Liposcelis bostrychophila*. J Essent Oil Bear Plants 19:1794–1800

Lo Presti M, Crupi ML, Costa R, Dugo G, Mondello L, Ragusa S, Santi L (2010) Seasonal variations of *Teucrium flavum* L. essential oil. J Essent Oil Res 22:211–216

Lograda T, Ramdani M, Chalard P, Figueredo G, Deghar A (2014) Chemical analysis and antimicrobial activity of *Teucrium polium* L. essential oil from Eastern Algeria. Am J Adv Drug Deliv 2:697–710

Maccioni S, Baldini R, Tebano M, Cioni PL (2007) Essential oil of *Teucrium scorodonia* L. ssp. *scorodonia* from Italy. Food Chem 104:1393–1395

Maeda H, Akaike T (1998) Nitric oxide and oxygen radicals in infection, inflammation, and cancer. Biochemistry 63:854–1007

Mahmoudi R, Nosratpour S (2013) *Teucrium polium* L. essential oil: phytochemical component and antioxidant properties. Int Food Res J 20:1697–1701

Mahmoudi R, Zare P, Nosratpour S (2015) Application of *Teucrium polium* essential oil and *Lactobacillus casei* in yoghurt. J Essent Oil Bear Plants 18:477–481

Makowczynska J, Sliwinska E, Kalemba D, Piatczak E, Wysokinska H (2016) In vitro propagation, DNA content and essential oil composition of *Teucrium scorodonia* L. ssp. *scorodonia*. Plant Cell Tissue Organ Cult 127:1–13

Martins FT, Doriguetto AC, de Souza TC, de Souza KRD, Dos Santos MH, Moreira MEC, Barbosa LCA (2008) Composition, and anti-inflammatory and antioxidant activities of the volatile oil from the fruit peel of *Garcinia brasiliensis*. Chem Biodivers 5:251–258

Masoudi S, Aghajani Z, Rustaiyan A, Feizbakhsh A, Kakhky AM (2009) Volatile constituents of *Teucrium persicum* Boiss., *Thymus caucasicus* Willd. ex Ronniger subsp. *grossheimii* (Ronniger) Jalas. and *Marrubium crassidens* Boiss. three Labiatae herbs growing wild in Iran. J Essent Oil Res 21:5–7

Mattison N, Trimple AG, Lasagna I (1998) New drug development in the United States, 1963 through 1984. Clin Pharmacol Ther 43:290–301

Menichini F, Conforti F, Rigano D, Formisano C, Piozzi F, Senatore F (2009) Phytochemical composition, anti-inflammatory and antitumour activities of four *Teucrium* essential oils from Greece. Food Chem 115:679–686

Michaelakis A, Theotokatos SA, Koliopoulos G, Chorianopoulos NG (2007) Essential oils of *Satureja* species: insecticidal effects on *Culex pipens* larvae (Diptera: Culicidae). Molecules 12:2567–2578

Miguel MG (2010) Antioxidant and anti-inflammatory activities of essential oils: a short review. Molecules 15:9252–9287

Miri A, Monsef-Esfahani HR, Amini M, Amanzadeh Y, Hadjiakhoondi A, Hajiaghaee R, Ebrahimi A (2012) Comparative chemical composition and antioxidant properties of the essential oils and aromatic water from *Teucrium persicum* Boiss. Iran J Pharm Res 11:573–581

Mitic V, Jovanovic O, Stankov-Jovanovic V, Zlatkovic B, Stojanovic G (2012) Analysis of the essential oil of *Teucrium polium* ssp. *capitatum* from the Balkan Peninsula. Nat Prod Commun 7:83–86

Moghtader M (2009) Chemical composition of the essential oil of *Teucrium polium* L. from Iran. Am-Eurasian J Sustain Agric 5:843–846

Mohan L, Pant CC, Melkani AB, Dev V (2010) Terpenoid composition of the essential oils of *Teucrium royleanum* and *T. quadrifarium*. Nat Prod Commun 5:939–942

Mojab F, Javidnia K, Yazdani D, Rustaiyan A (2003) Essential oil of the aerial parts of *Teucrium stocksianum* Boiss. subsp. *stocksianum* (Lamiaceae) from Iran. J Med Plant 2:49–54

Moon T, Wilkinson JM, Cavanagh HMA (2006) Antiparasitic activity of two *Lavandula* essential oils against *Giardia duodenalis*, *Trichomonas vaginalis* and *Hexamita inflata*. Parasitol Res 99:722–728

Morteza-Semnani K, Akbarzadeh M, Rostami B (2005) The essential oil composition of *Teucrium chamaedrys* L. from Iran. Flavour Fragr J 20:544–546

Morteza-Semnani K, Saeedi M, Akbarzadeh M (2007) Essential oil composition of *Teucrium scordium* L. Acta Pharma 57:499–504

Morteza-Semnani K, Saeedi M, Akbarzadeh M (2011) Chemical composition and antimicrobial activity of essential oil of *Teucrium hyrcanicum* L. J Essent Oil Bear Plants 14:770–775

Muselli A, Desjobert J-M, Paolini J, Bernardini A-F, Costa J, Rosa A, Dessi MA (2009) Chemical composition of the essential oils of *Teucrium chamaedrys* L. from Corsica and Sardinia. J Essent Oil Res 21:138–143

Navarro T, El Oualidi J (2000) Sinopsis of *Teucrium* L. (Lamiaceae) in the Mediterranean region and surrounding areas. Flora Medit 10:349–363

Nikpour H, Mousavi M, Asadollahzadeh H (2018) Qualitative and quantitative analysis of *Teucrium polium* essential oil components by GC-MS coupled with MCR and PARAFAC methods. Phytochem Anal 29:590–600

Oezek G, Oezek T, Dinc M, Dogu S, Baser KHC (2012) Chemical diversity of volatiles of *Teucrium orientale* L. var. *orientale*, var. *puberulens*, and var. *glabrescens* determined by simultaneous GC-FID and GC/MS techniques. Chem Biodivers 9:1144–1154

Oraimi MMAL, Al-Sabahi JN, Weli AM, Selim D, Akhtar MS, Al-Riyami Q, Al-Khanjari S (2012) Chemical composition of essential oil of *Tecurium mascatenses* Boiss. grown in Oman. J Essent Oil Bear Plants 15:822–826

Pala-Paul J, Perez-Alonso MJ, Velasco-Negueruela A, Garcia-Jimenez N, Jimenez Garcia R, Vargas L (2001) Composition of the essential oil of *Teucrium carolipaui* Pau grown in Spain. J Essent Oil Res 13:452–453

Passos GF, Fernandes ES, da Cunha FM, Ferreira J, Pianowski LF, Campos MM, Calixto JB (2007) Anti-inflammatory and anti-allergic properties of the essential oil and active compounds from *Cordia verbenacea*. J Ethnopharmacol 110:323–333

Perez I, Blazquez MA, Boira H (2000) Chemotaxonomic value of the essential oil compounds in species of *Teucrium pumilum* aggregate. Phytochemistry 55:397–401

Perez-Alonso MJ, Velasco-Negueruela A, Lopez-Saez JA (1993) The essential oils of two Iberian *Teucrium* species. J Essent Oil Res 5:397–402

Perry E, Howes M-JR (2011) Medicinal plants and dementia therapy: herbal hopes for brain aging? CNS Neurosci Ther 17:683–698

Polat T, Ozer H, Ozturk E, Cakir A, Kandemir A, Demir Y (2010) Chemical composition of the essential oil of *Teucrium multicaule* Montbret et Aucher ex Bentham from Turkey. J Essent Oil Res 22:443–445

Purnavab S, Ketabchi S, Rowshan V (2015) Chemical composition and antibacterial activity of methanolic extract and essential oil of Iranian *Teucrium polium* against some of phytobacteria. Nat Prod Res 29:1376–1379

Radulovic N, Dekic M, Joksovic M, Vukicevic R (2012) Chemotaxonomy of Serbian *Teucrium* species inferred from essential oil chemical composition: the case of *Teucrium scordium* L. ssp. *scordioides*. Chem Biodivers 9:106–122

Raei F, Ashoori N, Eftekhar F, Yousefzadi M (2014) Chemical composition and antibacterial activity of *Teucrium polium* essential oil against urinary isolates of *Klebsiella pneumoniae*. J Essent Oil Res 26:65–69

Rahim G, Qureshi R, Gulfraz M, Arshad M, Rahim S (2012) Preliminary phytochemical screening and ethnomedicinal uses of *Teucrium stocksianum* from Malakand Division. J Med Plant Res 6:704–707

Rahimi MA, Nazeri V, Andi SA, Sefidkon F (2018) Variation in essential oil composition of *Teucrium hircanicum* L. from Iran-A rich source of (E)-α-bergamotene. Nat Prod Res 33:1–6

Rang HP, Dale MM, Ritter JM (1998) Pharmacology. Churchill Livingston, New York, pp 614–616

Rehman NU, Al-Sahai JMS, Hussain H, Khan AL, Gilani SA, Abbas G, Hussain J, Al-Sabahi J, Al-Harrasi A (2016) Phytochemical and pharmacological investigations of *Teucrium mascatense*. Int J Phytomed 8:567–579

Reitz C (2012) Alzheimer's disease and the amyloid cascade hypothesis: a critical review. Int J Alzheimers Dis 2012:369808. https://doi.org/10.1155/2012/369808

Ricci D, Fraternale D, Giamperi L, Bucchini A, Epifano F, Burini G, Curini M (2005) Chemical composition, antimicrobial and antioxidant activity of the essential oil of *Teucrium marum* (Lamiaceae). J Ethnopharmacol 98:195–200

Roukia H, Mahfoud HM, Didi OHM (2013) Chemical composition and antioxidant and antimicrobial activities of the essential oil from *Teucrium polium geyrii* (Labiatae). J Med Plant Res 7:1506–1510

Rowshan V, Najafian S (2012) Comparison of volatile compounds in *Teucrium polium* L. by headspace and hydrodistillation techniques. Int J Appl Biol Pharm 3:151–157

Ruiters AK, Tilney PM, Van Vuuren SF, Viljoen AM, Kamatou GPP, Van Wyk B-E (2016) The anatomy, ethnobotany, antimicrobial activity and essential oil composition of southern African species of *Teucrium* (Lamiaceae). S Afr J Bot 102:175–185

Sabzeghabaie A, Asgarpanah J (2016) Essential oil composition of *Teucrium polium* L. fruits. J Essent Oil Res 28:77–80

Sadeghi H, Jamalpoor S, Shirzadi MH (2014) Variability in essential oil of *Teucrium polium* L. of different latitudinal populations. Ind Crop Prod 54:130–134

Sajjadi SE, Shookohinia Y (2010) Composition of the essential oil of *Teucrium chamaedrys* L. subsp. *syspirense* (C. Koch) Rech. Fil. growing wild in Iran. J Essent Oil Bear Plants 13:175–180

Salah KBH, Mahjoub MA, Chaumont JP, Michel L, Millet-Clerc J, Chraeif I, Ammar S, Mighri Z, Aouni M (2006) Chemical composition and in vitro antifungal and antioxidant activity of the essential oil and methanolic extract of *Teucrium sauvagei* Le Houerou. Nat Prod Res 20:1089–1097

Sarer E, Konuklugil B (1987) Investigation of the volatile oil of *Teucrium polium* L. Doga: Turk Tip ve. Eczacilik Dergisi 11:317–325

Sathiyamoorthy P, Lugasi-Evgi H, Schlesinger P, Kedar I, Gopas J, Pollack Y, Golan-Goldhirsh A (1999) Screening for cytotoxic and antimalarial activities in desert plants of the Negev and Bedouin market plant products. Pharm Biol 37:188–195

Semiz G, Celik G, Gonen E, Semiz A (2016) Essential oil composition, antioxidant activity and phenolic content of endemic *Teucrium alyssifolium* Stapf (Lamiaceae). Nat Prod Res 30:2225–2229

Sevindik E, Abaci ZT, Yamaner C, Ayvaz M (2016) Determination of the chemical composition and antimicrobial activity of the essential oils of *Teucrium polium* and *Achillea millefolium* grown under North Anatolian ecological conditions. Biotechnol Biotechnol Equip 30:375–380

Sghaier MB, Boubaker J, Neffati A, Limem I, Skandrani I, Bhouri W, Bouhlel I, Kilani S, Chekir-Ghedira L, Ghedira K (2010) Antimutagenic and antioxidant potentials of *Teucrium ramosissimum* essential oil. Chem Biodivers 7:1754–1763

Shah SMM, Ullah F, Shah SMH, Zahoor M, Sadiq A (2012) Analysis of chemical constituents and antinociceptive potential of essential oil of *Teucrium stocksianum* Boiss. collected from the North West of Pakistan. BMC Complement Altern Med 12:244–249

Shah SMM, Ali Khan F, Ali M, Shah SMS, Afridi MS, Ullah N (2015) Molecular profiling of the essential oil of *Teucrium stocksianum* collected at three different stages. Am-Eurasian J Toxicol Sci 7:279–285

Shah SMM, Khan S, Shah SMH, Shah SM (2016) Saifoor antimicrobial and cytotoxic potential of the essential oil of *Teucrium stocksianum* Boiss. Am-Eurasian J Toxicol Sci 8:42–46

Sonboli A, Bahadori MB, Dehghan H, Aarabi L, Savehdroudi P, Nekuei M, Pournaghi N, Mirzania F (2013) Chemotaxonomic importance of the essential-oil composition in two subspecies of *Teucrium stocksianum* Boiss. from Iran. Chem Biodivers 10:687–694

Sosa S, Altinier G, Politi M, Braca A, Morelli I, Della Loggia R (2005) Extracts and constituents of *Lavandula multifida* with topical anti-inflammatory activity. Phytomedicine 12:271–277

Stanciu G, Mititelu M, Popescu M (2006) The GC-MS characterization of the volatile oil from *Teucrium polium* L. An Univ "Ovidius" Constanta, S Chim 17:119–122

Stanic G, Petricic J, Blazevic N, Plazibat M (1993) Essential oil of *Teucrium flavum* L. from Croatia. J Essent Oil Res 5:625–627

Stankovic MS, Topuzovic M, Solujic S, Mihailovic V (2010) Antioxidant activity and concentration of phenols and flavonoids in the whole plant and plant parts of *Teucrium chamaedrys* L. var. *glanduliferum* Haussk. J Med Plant Res 4:2092–2098

Stevenson DE, Hurst RD (2007) Polyphenolic phytochemicals-just antioxidants or much more? A review. Cell Mol Life Sci 64:2900–2916

Tahoum N, Arakrak A, Bakkali M, Jatunov SS, Franconetti AG, Guillen MG, Escribano FC, Laglaoui A (2017) Chemical composition of essential oil of *Teucrium pseudoscorodonia* subsp. *baeticum*, endemic plant of north of Morocco. Int J Adv Res 5:914–919

Tandon S, Mittal AK (2018) Insecticidal and growth inhibitory activity of essential oils of *Boenninghausenia albiflora* and *Teucrium quadrifarium* against *Spilarctia obliqua*. Biochem Syst Ecol 81:70–73

Thoppil JE, Minija J, Tajo A, Deena MJ (2001) Antimicrobial activity of *Teucrium plectranthoides* Gamble essential oil. J Nat Remedies 1:155–157

Tuncer H (1978) Utilization of Wild Plants as Medicine, vol. II. Ministry of Food and Agriculture, Atak Printhouse, Ankara-Turkey

Tung YT, Chua MT, Wang SY, Chang ST (2008) Anti-inflammation activities of essential oil and its constituents from indigenous cinnamon (*Cinnamomum osmophloeum*) twigs. Bioresour Technol 99:3908–3913

Tzakou O, Roussis V, Loukis A, Harvala C, Galati EM, Germanò MP (1997) Essential oil analysis of *Teucrium divaricatum* Heldr. ssp. *divaricatum* growing in Greece. Flavour Fragr J 12:113–115

Ulubelen A, Birman H, Oksuz S, Topcu G, Kolak U, Barla Demirkoz A, Voelter W (2002) Cardioactive diterpenes from the roots of *Salvia eriophora*. Planta Med 68:818–821

Vahdani M, Faridi P, Zarshenas MM, Javadpour S, Abolhassanzadeh Z, Moradi N, Bakzadeh Z, Karmostaji A, Mohagheghzadeh A, Ghasemi Y (2011) Major compounds and antimicrobial activity of essential oils from five Iranian endemic medicinal plants. Pharm J 3:48–53

Valentini G, Bellomaria B, Arnold N (1997) Essential oil of *Teucrium creticum* L. from Cyprus. J Essent Oil Res 9:649–652

Velasco-Negueruela A, Perez-Alonso MJ (1990) Essential oils of germanders endemic to Spain. I. *Teucrium lusitanicum* subsp. *aureiformis*. An Bromatol 41:241–248

Velasco-Negueruela A, Perez-Alonso MJ, Pala-Paul J, Ramos-Vazquez P, De Paz Perez PL, Vallejo Garcia MC (1999) Composition of the essential oil of *Teucrium heterophyllum* L'Hér. grown in the Canary Islands. J Essent Oil Res 11:553–555

Vokou D (1992) The allelophatic potential of aromatic shrubs in phryganic (east Mediterranean) ecosystems. In: Rizvi SJH, Rizvi V (eds) Allelopathy: basic and applied aspects. Springer, Dordrecht, pp 303–320

Vokou D, Bessiere JM (1985) Volatile constituents of *Teucrium polium*. J Nat Prod 48:498–499

Vukovic N, Milosevic T, Sukdolak S, Solujic S (2007) Antimicrobial activities of essential oil and methanol extract of *Teucrium montanum*. Evid Based Complement Alternat Med 4:17–20

Vukovic N, Milosevic T, Sukdolak S, Solujic S (2008) The chemical composition of the essential oil and the antibacterial activities of the essential oil and methanol extract of *Teucrium montanum*. J Serb Chem Soc 73:299–305

Vukovic N, Sukdolak S, Solujic S, Mihailovic V, Mladenovic M, Stojanovic J, Stankovic MS (2011) Chemical composition and antimicrobial activity of *Teucrium arduini* essential oil and cirsimarin from Montenegro. J Med Plant Res 5:1244–1250

Wei Z, Zhen H, Lu H, Mei L, Wei H (2010) GC-MS analysis of the chemical components of essential oils in *Teucrium viscidum*. Chin J Exp Trad Med Formul 16:91–92

Whittaker RH, Feeny PP (1971) Allelochemics: chemical interactions between species. Science 171:757–770

Wink M (2003) Evolution of secondary metabolites from an ecological and molecular phylogenetic perspective. Phytochemistry 64:3–19

Zargari A (1993) Medicinal plants, vol 4. Tehran University Publications, Tehran, pp 127–130

Zendehdel M, Taati M, Jadidoleslami M, Bashiri Dezfouli A (2011) Evaluation of pharmacological mechanisms of antinociceptive effect of *Teucrium polium* on visceral pain in mice. Iran J Vet Res 12:292–297

Zhou S, Koh H-L, Gao Y, Gong Z, Lee EJD (2004) Herbal bioactivation: the good, the bad and the ugly. Life Sci 74:935–968

Chapter 8
Secondary Metabolites of *Teucrium* Species with Toxic Effects

Maria-Eleni Grafakou, Christina Barda, and Helen Skaltsa

Abstract The genus *Teucrium* L. has a long-term use in folk medicine, however there are some implications for human health related to the consumption of germander-containing products. Numerous cases of hepatotoxicity have been reported, mostly in Europe, where the plants were used as herbal products for weight loss and still are used in folklore medicine to treat hypercholesteraemia and diabetes. During the last decades several cases of germander toxicity including chronic hepatitis as well as acute cytolytic hepatitis and a case of death have been reported. The most well known toxic *Teucrium* species are *Teucrium chamaedrys*, *T. polium* and *T. capitatum*, the use of which as herbal medicines and dietary supplements should be strictly controlled. The neo-clerodane diterpenoids occurring in *Teucrium* species have been the subject of intensive studies, since they have been correlated to the toxic effects of the genus. Studies in mice showed that the hepatotoxicicity of the genus is being induced by this type of diterpenes, which are transformed into their reactive metabolites by P450. Teucrin A and teuchamaedrin A have been subjected to numerous studies in order to reveal the mechanism of action. This chapter aims to summarize the diterpenes isolated from *Teucrium* spp., as well as the hepatotoxicity of some of them.

Keywords *Teucrium* · Toxicity · Hepatotoxicity · Diterpenes

Abbreviations

CEFS	Council of Europe Committee of Experts on Flavouring Substances
FDA	Food and Drug Administration
NFG	Nerve growth factor
PP	Pyrophosphate
UV	Ultraviolet

M.-E. Grafakou · C. Barda · H. Skaltsa (✉)
Department of Pharmacognosy and Chemistry of Natural Products, Faculty of Pharmacy, National and Kapodistrian University of Athens, Athens, Greece
e-mail: skaltsa@pharm.uoa.gr

M. Stanković (ed.), *Teucrium Species: Biology and Applications*,
https://doi.org/10.1007/978-3-030-52159-2_8

8.1 Introduction

Natural products are structurally diverse compounds, deriving from natural sources, mainly plants, in the form of primary and secondary metabolites (Dewick 1999). Secondary metabolites differ from primary metabolites in having a districted distribution throughout the plant kingdom (Anulika et al. 2016). Plants and other living organisms have uniquely evoluted to biosynthesize secondary metabolites, compounds which unlike primary metabolites are generally not essential for growth, development or reproduction of the organism (Theis and Lerdau 2003). However, secondary metabolites play significant roles in defence mechanisms against insects, pretadors and microorganisms, as attractants for pollinators and seed-dispersing animals, allelopathic agents in natural habitats, barriers to abiotic stressors such as UV, salty or drought environmental conditions (Wink 2009; Theis and Lerdau 2003; Pichersky and Gang 2000). Because of structure similarity to humans metabolites, secondary metabolites from plants exert biological activities of a wide range, thus plants have been used for a variety of internal and external ailments since ancient times (Dias et al. 2012). Many secondary metabolites, however, act as toxins, which is explained by their major role in defence mechanisms (Wink 2009). In western civilization, there is an increasing use of herbal medicines, because of their supposed safety in contrast to chemical drugs, however, using herbal remedies may pose health hazards (De Smet 2002). Plant derived products should not be assumed safe; safety is evaluated either in the case of herbal medicinal products or isolated compounds. There are even reports from poisoning associated with widely used medicinal plants, which are usually due to misidentification of the plants or incorrectly preparation and administration (Nasri Hamid Shirzad 2013; Karimi et al. 2015).

The genus *Teucrium* L. (germander) of the Lamiaceae family, comprises about 300 species, distributed worldwide, mainly in the Mediterranean area (Tutin et al. 1972; Greuter et al. 1986). Members of the genus have a long term use in traditional medicine and the first records of the medicinal use of *Teucrium* date back to Greek mythology. Probably the name *Teucrium* comes from Teukros (Teucer), the best Achaean archer in the Trojan War (son of Telamon, the king of the island Salamina); his step-brother Aias healed using this plant the wounds of Teukros caused by Hector (Carnoy 1959). Later on, Aias was killed by Hector. At his return to Salamina, Teukros was kicked out by his father, because he did not defend his brothers' death. The bitter taste and the toxicity of the plant make allusion to the Teukros' misfortune. According to Dioscorides, the plant as an oxycrate beverage, or combined with figs, helps the spleen; it has also anticonvulsant, antitussive, diuretic, abortive and emmenagogue activities. Also, the poultices embedded in germander decoctions are antidotes against venomous bites; combined with honey it could be used to treat old ulcers and with oil in ocular discomforts (Berendes 1902; Wellmann 1958); The medicinal uses of the genus include the treatment of digestive and respiratory disorders, tuberculosis, jaundice, rheumatics, diabetes, kidney diseases, skin

inflammations and for weight loss (Stanković et al. 2011; Pacifico et al. 2012; Stanković and Zlatić 2019).

Teucrium species despite being popular in folk medicine, have been also associated to potential toxicity, especially hepatotoxicity which has been attributed to their content of furano-*neo*-clerodane diterpenes (Ulubelen et al. 2000). As a result, the aim of the present study is to summarize the secondary metabolites of the genus *Teucrium* with toxic effects. A selection of the relevant data was made through a search using the keyword "Teucrium" and "toxicity", "hepatotoxicity", "diterpenes" in PubMed, Scopus, Google Scholar and Science Direct databases. Only papers in English language are included.

8.2 Diterpenes from the Genus *Teucrium*

Phytochemical investigations have revealed a plethora of bioactive secondary metabolites produced by members of the genus *Teucrium*, including monoterpenes, iridoids, sesquiterpenes, diterpenes, sterols, saponins, polyphenolic compounds, flavonoids, tannins, coumarins, glycosides, alkaloids, and essential oils (Djabou et al. 1974; Piozzi et al. 1998; Hao et al. 2013; Fatima 2016) However, the most abundant compounds are diterpenes, especially *neo*-clerodane type diterpenes; so far, more than 230 compounds with this skeleton have been isolated from the aerial part of the taxa (Table 8.1; Piozzi et al. 1998, 2005), while this group of secondary metabolites has been described as chemotaxonomic markers of the genus *Teucrium* (Ulubulen et al. 2000). Moreover, some rearranged *neo*-clerodane diterpenes, as well as some abietane type diterpenes have been isolated, mainly from the roots of *Teucrium* taxa (Ulubulen et al. 2000). *Neo*-clerodanes have been also isolated from other plant sources; in addition to the Lamiaceae family, those diterpenes have been reported from plants of Asteraceae, Salicaceae and Menispermaceae families (Li et al. 2016). This group of compounds have captured the attention of the scientific community, as they can exhibit a wide range of pharmacological activities, such as antitumor, NGF-potentiating, antibiotic, anti-ulcer, hypoglycaemic, hypolipidaemic, antiplatelet; however, insect antifeedant activity and opioid receptor agonist effects have been assumed to be the most significant activities (Li et al. 2016).

8.2.1 Chemistry of Diterpenes in Teucrium *spp*

The following section summarizes the diterpenes isolated from *Teucrium* spp. Our study aims to outline the up-to-date investigated *Teucrium* species, characterized by the presence of diterpenes (Table 8.1).

Table 8.1 Literature Survey of ditepenes from *Teucrium* spp.

Teucrium species	Isolated diterpenes	References
T. abutiloides	12-epiteupolin II, montanin C, teubutilin A, teubutilin B	Torre et al. (1990)
T. alpestre	3-acetylteumicropin, teumicropin, teupyrenone, deacetylteupyrenon	Piozzi et al. (1997)
T. asiaticum	Auropolin, teuftin	Camps et al. (1987)
	Auropolin, teucrasiatin	Rodríguez et al. (1996)
	Auropolin, teucrasiolide, teucrasiatin	Rodríguez et al. (1997)
T. barbeyanum	Teucrin A, teucrin F, teucrin G	Bruno et al. (1985)
T. betonicum	19-acetylgnaphalin (teucrin H_3), 6β-hydroxyteuscordin, 6α-hydroxyteuscordin, teucrin E, teucrin H_2, teucvin, teubetonin	Gaspar et al. (1995)
T. bicolor	Montanin C, 12-epi-teucvin, teucvin, teucrin H2, teupolin I, 12-epi- teupolin I, (12R)-epi- teuscordon	Labbe et al. (1989)
T. botrys	19-deacetylteuscorodol, 6β-hydroxyteuscordin, montanin D, teubotrin, teucvidin, teuchamaedrin C	Torre et al. (1986)
T. brevifolium	Teubrevin C, teubrevin D, teubrevin E, teubrevin F, teubrevin G, teubrevin H, teubrevins I	Rodríguez et al. (1995)
	Teubrevin A, teubrevin B	Rodríguez et al. (1994)
T. buxifolium	19-acetylteulepicin, 19-acetylgnaphalin	Savona et al. (1986)
T. canadense	18-acetylmontanin D, 12-epiteupolin II, isoteuflin, teucvidin, teuflin, teucvin, teuscorodal, (12R)-teupolin I	Bruno et al. (1989)
T. capitatum	19-acetylgnaphalin, lolin, teucapitatin	Marquez et al. (1981)
T. chamaedrys	2(*S*)-15,16-epoxy-19-hydroxy-*neo*-cleroda-13(16),14-dien-18,6α:20,12-diolide	Bedir et al. (2003)
	Dihydroteugin, syspirensin A, teuchamaedryn D, teucrin A, teucroxide	Elmastas et al. (2016)
	Chamaedryoside A, chamaedryoside B, chamaedryoside C	Fiorentino et al. (2009)
	Teucrin F, teucrin G	Bruno et al. (1985)
	Teucrin A	Kouzi et al. (1994)
	Dihydroteugin, 6α-hydroxyteuscordin, teuchamaedrin C	Papanov and Malakov (1985)
T. chartaginense subsp. *homotrichum*	19-acetylgnaphalin, eriocephalin	Bruno et al. (1985)
T. cosoonii	Teucossiin A, teucossiin B, montanin H	Alcazar et al. (1992)

(continued)

Table 8.1 (continued)

Teucrium species	Isolated diterpenes	References
T. cubense	Eugarzasadone	Dominguez et al. (1974)
T. cuneifolium	Deacetylteupyrenone	Piozzi et al. (1997)
T. divaricatum subsp. *canescens*	2-deoxychamaedroxide, 6β-hydroxy-teuscordin, montanin D, dihydroteugin, teuflin, teucrin H_2, teuflidin, teucrin A, teucrin F, teucrin G	Bruno et al. (1987)
T. divaricatum subsp. *divaricatum* syn. *T. divaricatum*	Teucrin A	Rodríguez et al. (1996)
T. divaricatum subsp. *villosum*	Hydroxyteuscordin, dihydroteugin, montanin D, teuflin, teuscordinone, teuflidin, teucrin A, 6β- teugin	Piozzi et al. (1997)
	Villosin A, villosin B, villosin C, imbricatolic acid, dehydroabietic acid, teuvincenone B, teucrin A	Ulubelen et al. (1994)
T. flavum subsp. *glaucum*	Teuflin, teuflavin, teutlavoside	Savona et al. (1984)
T. flavum subsp. *hellenicum*	Teucvidin, 12-epiteucvidin, teuflin	Piozzi et al. (1997)
T. fruticans	Fruticolone, isofruticolone, 8β-hydroxyfruticolone, 6-acetylteucjaponin B, 7β-hydroxyfruticolone, 11-hydroxyfruticolone, deacetylfruticolone, 6-acetyl-10-hydroxyteucjaponin B	Coll and Tandrón (2004)
	Difuranofruticol, deoxyfruticolone, 10-hydroxyteucjaponin B, 7,8-didehydrofruticolone	Coll and Tandrón (2005)
	Teufruintins A–G, fruticolone, isofruticolone, 6α-hydroxyfruticolone, 6-acetyl-10-hydroxyteucjaponin B, 6-acetyl-teucjaponin B, 8β-hydroxyfruticolone, teucretol, 11-hydroxyfruticolone	Lv et al. (2015)
	Fruticolide, fruticolone, isofruticolone, 8β-hydroxyfruticolone	Bruno et al. (1991)
	Teuvincenone F, teuvincenone G, teuvincenone E	Sexmero Cuadrado et al. (1992)
T. gnaphalode	Gnaphalin, 19-acetylgnaphalin	Savona et al. (1979)
T. gracile	Teugracilin A, 3-*O*-deacetylteugracilin A, teugracilin B, teugracilin C, teumicropodin	Bruno et al. (1991)
	19- acetylteulepicin, teugracilin D, teugracilin E	Bruno et al. (1992a)
T. hyrcanicum	Teucrin H_1, teucrin H_2, teucrin H_3, teucrin H_4	Gács-Baitz et al. (1978)
T. intricatum	Teucvin	Bruno et al. (1985)
T. japonicum	Teuponin	Zhi-da et al. (1991)

(continued)

Table 8.1 (continued)

Teucrium species	Isolated diterpenes	References
T. polium subsp. *vincentinum*	Teuvincentin A, teuvincentin B, teuvincentins C, 19-acetylgnaphalin, eriocephalin, isoeriocephalin, 3-deacetyl-20-epi-teulanigin	Carreiras et al. (1990)
	Euvincenone A, euvincenone B, euvincenone C, euvincenone D	Torre et al. (1990)
T. pyrenaicum	Teupyrin A, teupyrin B, teutlin, teucrin H_2, 6*α*-hydroxyteuscordin	Fernández et al. (b)
T. tomentosum	Euctomin, along with 6-*O*-acetyl-teucmicropodin, teugracilin A, 6*α*-hydroxyteuscordin, 6*α*-acetoxyteuscordin	Aravind et al. (2010)
	Teuctosin, teuflin, teucrin H_2, 6*β*-hydroxyteuscordin, 6*β*-acetylteuscordin, montanin D	Kumari et al. (2003)
	3*β*-acetoxy-4*α*,18:15,16-diepoxy-6*β*, 12-dihydroxyneoclerođa-13(16),14-dien-19,20-olide	Kumari et al. (2003)
T. racemosum	Teuracemin, eutrifidin, 4*α*,18- epoxytafricanin A, 20-oxo-teufiavin	Rodríguez et al. (1995)
T. rivas-martinezii	19-acetylgnaphalin, 19-acetylteulepicin	Rodríguez et al. (1996)
T. royleanum	Royleanumin	Ahmad et al. (2014)
T. salviastrum	Teusalvin A, teusalvin B, teusalvin C, teusalvin D, teusalvin E, teusalvin F, teucvidin, teucroxide	Torre et al. (1986)
T. sandrasicum	Sandrasin A, 6-deacetylsandrasin A, sandrasin B	Topcu et al. (1996)
T. scordium	Teucrin F, teucrin G, 6*α*-hydroxyteuscordin	Bruno et al. (1985)
	Teucrin E, 6*α*-hydroxyteuscordin, 6*β*-hydroxyteuscordin, teuscordinone, teugin, dihydroteugin, teucjaponin B acetate, teucroxide, 2,3-Dehydroteucrin E, 2*β*-6*α*-Dihydroxyteuscordin, 2*β*-Hydroxyteuscordinone, 6,20 Bisdeacetylteupyreinidin, 6-deacetylteupyreinidin	Jakupovic et al. (1985)
	Teuscordinon	Papanov et al. (1981)
	2-keto-19-hydroxyteuscordin, teucrin E, teucrin H_4	Papanov and Malakov (1985)
T. scorodonia subsp. *euganeum* syn. *T. siculum*	Teuilin	Bruno et al. (1985)
T. scorodonia	Teuflin, teuscorolide, teuscorodin, teupolin I, teuscorodal, teuscorodol, teuscorodonin, 2*α*-hydroxyteuscorolide	Bruno et al. (1985)

(continued)

Table 8.1 (continued)

Teucrium species	Isolated diterpenes	References
T. viscidum	Teuvisone, biteuvisones A, biteuvisones B, 2α,11,12-trihydroxy-7β,20-epoxy-8,11,13-abietatriene	Gao et al. (2015)
	19-acetyl-teuspinin, 6α-acetoxyteuscordin, deoxychamaedroxide, isoteuflin, kinalborin C, teucvisin A, teucvisin B, teucvisin C, teucvisins D, teucvisin E, teucvin, teucvidin, teuflin, teuspinin, 2- teuscordal	Lv et al. (2014)
T. viscidum var. *miquelianum*	Teucvin, teuflin (6-epiteucvin), teucvidin	Node et al. (1981)
T. yemense	Fatimanol A, fatimanol B, fatimanol C, fatimanol D, fatimanol E, fatimanone, teulepicephin	Nur-E-Alam et al. (b)
	3-*O*-deacetylteugracilin, teugracilin B	Nur-E-Alam et al. (2017a)
	6β-*O*-acetyl-3β-hydroxyteucroxylepin, teucryemin, 19-*O*-acetylteucryemin, teucryeminone	Sattar et al. (1995)

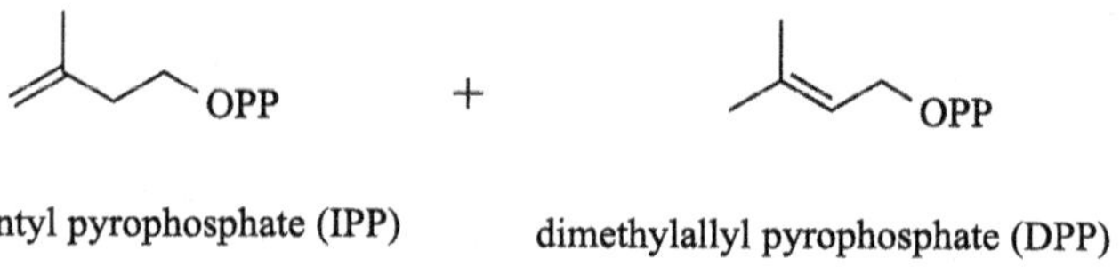

Fig. 8.1 Building blocks of terpenoids

8.2.2 Biosynthesis of Diterpenes

Terpenes or isoprenoids are a large and structural diverse class of secondary metabolites formed through the mevalonate pathway or the deoxyxylulose phosphate pathway. Biosynthetically, monoterpenes, sesquiterpenes, diterpenes, sesterterpenes, triterpenes, derived from C5 isoprene units commonly joined in a head to tail orientation and they are classified according to the number of isoprene units. The active building blocks biochemically are the diphosphate esters dimethylallyl diphosphate and isopentenyl diphosphate. The precursor of diterpenes (C20) is the geranylgeranyl diphosphate (Figs. 8.1 and 8.2) and it is further modified by cyclization reactions, involving numerous of parallel steps to yield the mentioned natural compounds (Table 8.1). Their large structural diversity is attributed to their different biosynthetic pathways, and for this reason the diterpenes are classified by their biosynthesis (Bruneton 1993). The biosynthetic pathway of cyclic diterpenes comprises two different types of cyclization A and B. In the first case, its characteristic conversion into cebrenes or casbanes. From the cebrene, a group of macrocyclic diterpenes is initiated. The casbane, leads to the formation of skeletons such as tigliane, daphnane, etc. The second type (B) of cyclization is characteristic of the upper terpenes. This point of biosynthesis is critical because it can lead either to the normal order (n-) or to the enantiomer type (*ent*-) of labdane skeleton. N-labdane derivatives are believed to be derived from the labdadienyl PP, while the antipodes

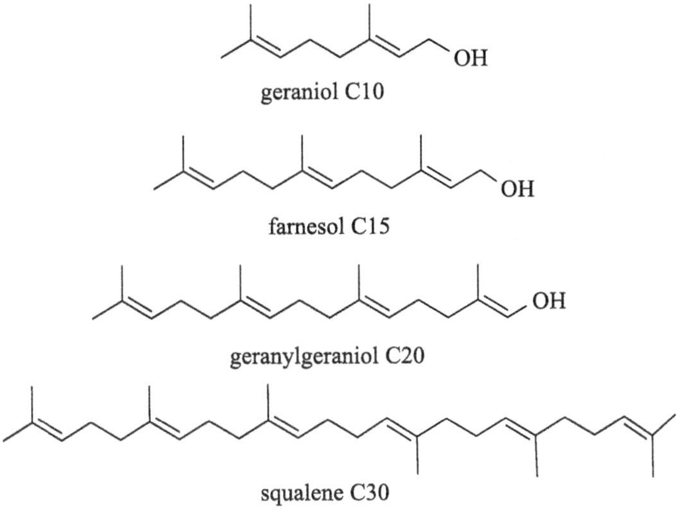

Fig. 8.2 Precursors of terpenes

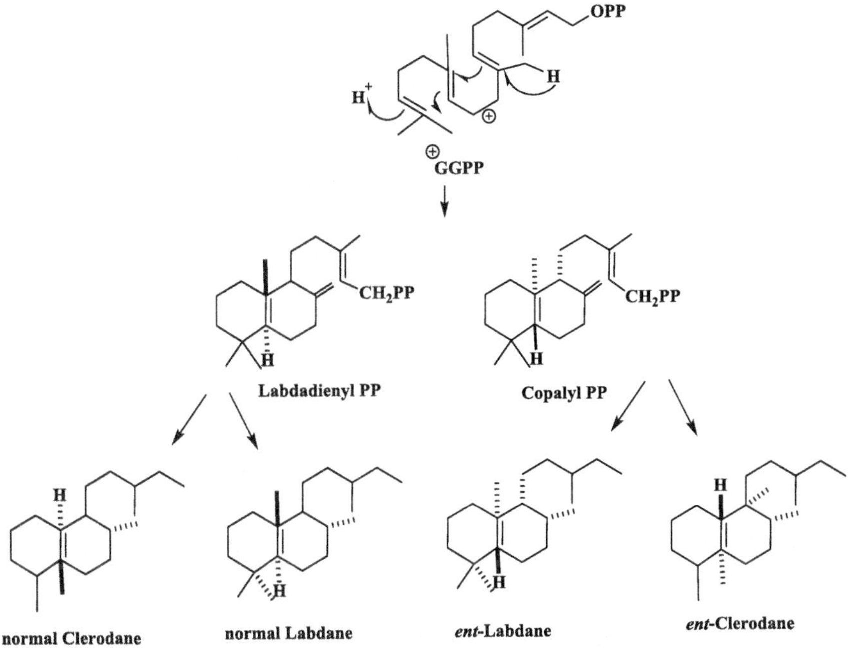

Fig. 8.3 Biosynthetic pathway of clerodane and labdane diterpenoids

from its enantiomer, copalyl diphosphate. Through the above, we can understand the basic principles in the biosynthesis of labdanes and clerodanes which appear to be closely related (Fig. 8.3) (Merritt and Ley 1992; Mac Milan and Beale 1999).

8.3 Hepatotoxicity Related to *Teucrium* spp.

In modern civilization, the use of herbal remedies is becoming an increasingly popular form of alternative medicine because of their assumed safety, but a survey on the literature reveals the concerns related to their health risks. One such health hazard, is hepatoxicity; liver injury has been associated with herbal medicines and especially with terpenoid-containing dietary supplements (Chitturi and Farrell 2008; Hasani-Ranjbar et al. 2010). Wild germander (*Teucrium chamaedrys*) was one from the first herbal products to be related with cases of acute liver failure, attributed to its content of diterpene derivatives.

In 1986, an advertisement promoted germander tea bags and capsules for weight loss and mild diarrhoea in France, while germander was also consumed as a constituent of aromatized wines and liquors (Stickel and Shouval 2015). The use of high amounts of germander, previously regarded as safe, led to an epidemic of cytolytic hepatitis (De Berardinis et al. 2000; Stickel et al. 2000). Numerous cases of hepatic injury were reported (Larrey et al. 1992; De Berardinis et al. 2000), mainly acute cytolytic hepatitis, as well as some reports of chronic hepatitis. Most patients recovered after discontinuation of the herbal remedy except for those with cirrhosis, and a case of death, because of fulminant hepatic necrosis (Mostefa-Kara et al. 1992). The daily intake was ranging between 600 and 1620 mg of germander daily (Larrey et al. 1992), and the developed hepatitis (2 months after permanent administration) was described by jaundice and increased amino-transferase levels and serum bilirubin levels, but in some cases of chronic course of liver disease also fibrosis and cirrhosis have been described. More women consumed germander products, in an attempt of losing weight, which reflects the higher levels of toxicity in this gender (Stickel et al. 2000).

As a result of these indices, all herbal preparations of *Teucrium chamaedrys* were banned in France in 1992, and no further cases were recorded in this country. In Italy 1996, the sale of germander products was also prohibited, while more *Teucrium* spp. (*Teucrium marum*, *T. montanum*, *T. polium*, *T. scordium* and *T. scorodonia*) are included in "Plants that contain substances that are definitely toxic to humans" (Carratù et al. 2010). Similar prohibitions had been applied also in 1997 in Belgium for products containing both *Teucrium chamaedrys* and *T. polium* (Gori et al. 2011). FDA permits *Teucrium chamaedrys* in alcoholic beverages only, which is in accordance with the recommendation of the CEFS that the use of germander products including teucrin A should not extended beyond flavoured beverages (European Commission 2003).

Despite those restrictions, two further cases were reported in Canada in 1996 (Laliberté and Villeneuve 1996) and in more recent years, more cases were reported, such as two in Spain in 2001 (Pérez Alvarez et al. 2001), and four more in Italy, in 2011 and 2014 (Gori et al. 2011; Nencini et al. 2014).

Moreover, hepatotoxicity has also been related with other members of the genus *Teucrium*. In 1995 the first case of liver transplantation for severe acute liver injury after treatment for 10 days with *Teucrium polium* for hypercholesterolaemia was reported in Greece (Mattéi et al. 1995). Regarding *Teucrium polium* (golden germander) more cases have been reported; acute cholestatic hepatitis has been reported in two cases in Greece, after the use of the plant as hypoglycaemic or hypolipidaemic agent (Polymeros et al. 2002; Mazokopakis et al. 2004). Furthermore, the use of *Teucrium polium* products for the same reasons led to three other cases of hepatitis in Greece (Starakis et al. 2006; Savvidou et al. 2007).

Teucrium capitatum is also related to hepatotoxicity. Dourakis et al. (2002) reported a patient with acute icteric hepatocellular necrosis in Greece, who used *Teucrium capitatum* because of hypercholesterolaemia and hyperglycaemia. Moreover, *Teucrium viscidum* has been linked to hepatotoxicity. Poon et al. (2008) reported a case of acute hepatotoxicty from a patient in China who used a preparation of *Teucrium viscidum* for low back pain.

8.3.1 Mechanism of Action

At the time of the first records of hepatotoxicity related to *Teucrium* spp. and the resulting restrictions to the use of germander containing products, studies relevant to the phytochemical profiles of the *Teucrium* plants have not been conducted yet, thus the chemical constituents which were responsible for the hepatotoxicity remained unknown (Stickel et al. 2000). The next step was the evaluation of wild germander hepatotoxicity in animal studies, both in vivo and in vitro (Kouzi et al. 1994; Loeper et al. 1994; Lekehal et al. 1996; Fau et al. 1997). Except from *Teucrium chamaedris*, also *Teucrium polium* has been evaluated in animal models. More specifically, studies in diabetic rats proved the hypoglycaemic properties of *Teucrium polium*, but also revealed hepatotoxic effects (Zal et al. 2001; Shahraki et al. 2007). In addition, Mehdinia et al. (2013) reported decreased body weight and changes in liver parameters in experiments with petroleum ether extracts of *Teucrium polium* in mice. Regarding *Teucrium stocknaum,* a toxicological investigation in rats did not revealed significant hepatotoxicity rather than neurotoxicity (Tanira et al. 1996)

Regarding the mechanism of action, studies in mice revealed that the hepatotoxic constituents of germander are the furano-*neo*-clerodane diterpenes, which are transformed into their reactive metabolites by P450, especially CYP3A (Kouzi et al. 1994; Loeper et al. 1994). Kouzi et al. (1994) showed that the exodiation of the furan ring of those diterpenes through the CYP3A-metabolism is required for the hepatotoxicity of teucrin A, a major diterpene from the aerial parts of *Teucrium chamaedrys*. Further studies in isolated rat hepatocytes showed that teucrin A and teuchamaedrin.

A (Fig. 8.4) are being metabolized in electrophilic derivatives, which in turn cause apoptosis through decreasing of cellular thiols and [Ca^+]-involving mechanisms (Lekehal et al. 1996; Fau et al. 1997). Covalent modification of proteins

Fig. 8.4 Stuctures teucrin A (1) and teuchamaedrin A (2)

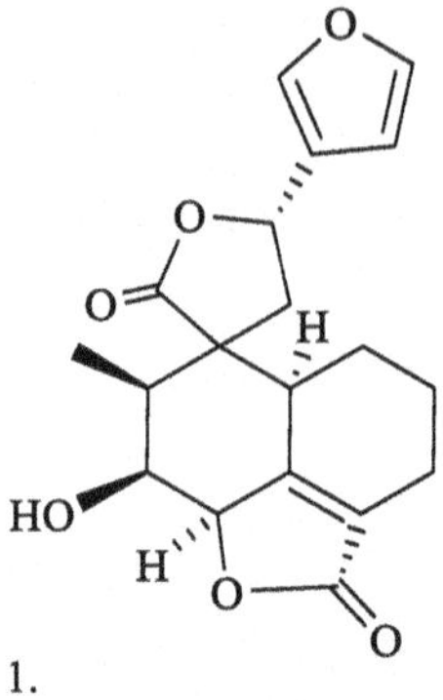

1.

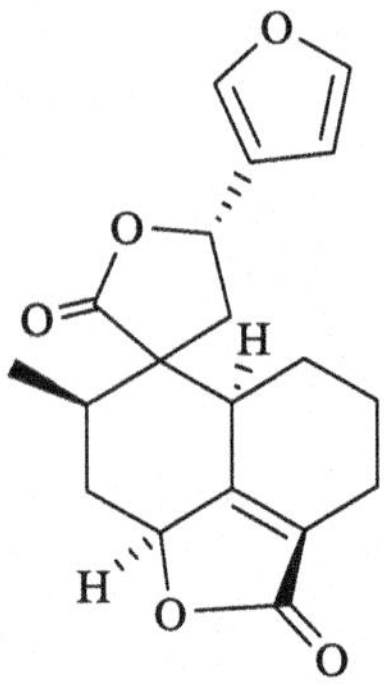

2.

(mitochondrial, endoplasmatic, cytosolic, peroxisomal and secreted proteins) has been linked to the hepatotoxicity of the reactive metabolites of teucrin A (Druckova et al. 2007). Furthermore, these reactive metabolites may cause hepatotoxicity through an immunoallergic reaction, as anti-microsomal epoxide hydrolase autoantibodies have been found in the sera of patients who have been long term using germander products (De Berardinis et al. 2000; Larrey and Faure 2011).

8.3.2 Hepatoprotective Activity, Cell Cultures and Possible Protection from Total Polyphenolics

A survey on the literature reveals some controversial studies. Panovska et al. (2007) evaluated the hepatoprotective activity of ethyl acetate extract of *Teucrium polium* against carbon tetrachloride induced liver damage in rats, and reported significant liver regeneration after 7 days injection of *Teucrium polium* extract. These results are in accordance with some similars studies (Al-Kubaisy 2013; Rahmouni et al. 2019). Forouzandeh et al. (2013) further evaluated the hepatoprotective activity of this plant, and reported significant inhibition of acetaminophen-induced hepatotoxicity in mice. Similar results were reported for another member of the genus, *Teucrium polium-geyrii*, attributed to its phenolic content (Baali et al. 2016). Moreover, a safety data review (De Vincenzi et al. 2003), reports that teucrin A from *Teucrium chamaedrys* showed no mutagenic effects and no hepatotoxicity, as the changes in body weights of the rats in comparison to the absence of other morphological findings can be regarded as an adaptative metabolic effect.

Teucrium chamaedrys contains substances that are both hepatoprotective and hepatotoxic, saponines and *neo*-clerodane diterpenoids, respectively (Stickel et al. 2000). A study revealed that cell cultures from leaf explants of *Teucrium chamaedrys* (both solid and liquid) continue to biosynthesize phenylethanoid glycosides, especially teucrioside, but not *neo*-clerodanes (Antognoni et al. 2012). More recently, two cases of acute hepatotoxicity induced by the use of *Teucrium chamaedrys* were evaluated in relation to the diterpene concentration and differences in the

preparation of the traditional family decoction, which had been used without toxic effects. The study showed the patients used a decoction, in which the extraction time was much higher, resulting in high levels of teucrin A, low antioxidant activity and total polyphenolics content in comparison to the traditional family preparation (Nencini et al. 2014).

8.4 Conclusions

The current review reveals that *Teucrium* genus contain several toxic species, the most well known are *Teucrium chamaedrys*, *T. polium* and *T. capitatum*, responsible for several cases of hepatotoxicity in S. Europe, where the plants were used as herbal products for weight loss and still are used in folklore medicine to treat hypercholesteraemia and diabetes. The use of these plants as herbal medicines and dietary supplements should be strictly controlled.

The plants have been reported to contain a plethora of secondary metabolites, belonging to terpenes, phenolics, alkaloids and essential oils. The *neo*-clerodane diterpenoids occurring in *Teucrium* species have been the subject of intensive studies, since they have been correlated to the toxic effects of the genus.

References

Ahmad S, Ulllah Abd Elsalam RM, Fouad H, Bibi A, Tariq Jan M, Shad AA, Arfan M (2014) One new royleanumoate from *Teucrium royleanum* Wall. Ex Benth. Sci World J 2014:581629. https://doi.org/10.1155/2014/581629

Alcazar R, de la Torre MC, Rodríguez B, Bruno M, Piozzi F, Savona G, Arnold NA (1992) Neo-clerodane diterpenoids from three species of *Teucrium*. Phytochemistry 31:3957–39661

Al-Kubaisy K (2013) Hepatoprotective activity of *Teucrium polium* against CCl_4 induced hepatoxicity in male albino rats. Int J Indig Med Plant 46:1206–1210

Al-Yahya MA, El-Feraly FS, Chuck Dunbar D, Muhammad I (2002) Neo-clerodane diterpenoids from *Teucrium oliverianum* and structure revision of teucrolin E. Phytochemistry 59:409–414

Antognoni F, Iannello C, Mandrone M, Scognamiglio M, Fiorentino A, Giovannini PP, Poli F (2012) Elicited *Teucrium chamaedrys* cell cultures produce high amounts of teucrioside, but not the hepatotoxic neo-clerodane diterpenoids. Phytochemistry 81:50–59

Anulika NP, Ignatius EO, Raymond ES, Osasere O, Hilda A (2016) The chemistry of natural product: plant secondary metabolites. Int J Technol Enhanc Emerg Eng Res 4:1–7

Aravind S, Balachandran J, Ramanujam Ganesh M, Krishna Kumari GN (2010) Further antifeedant neo-clerodanes from *Teucrium tomentosum*. Nat Prod Res 24:7–12

Baali N, Belloum Z, Baali S, Chabi B, Pessemesse L, Fouret G, Ameddah S, Benayache F, Benayache S, Feillet-Coudray C, Cabello G, Wrutniak-Cabello C (2016) Protective activity of total polyphenols from *Genista quadriflora* munby and *Teucrium polium* geyrii maire in acetaminophen-induced hepatotoxicity in rats. Nutrients 8:193. https://doi.org/10.3390/nu8040193

Bedir E, Tasdemira D, Calis I, Zerbeb O, Sticherb O (1999) Neo-clerodane diterpenoids from *Teucrium polium*. Phytochemistry 51:921–925

Bedir E, Manyam R, Khan IA (2003) Neo-clerodane diterpenoids and phenylethanoid glycosides from *Teucrium chamaedrys* L. Phytochemistry 63:977–983

Berendes J (1902) Des Pedianos Dioskurides aus Anazarbos Arzneimittellehre. In: Sandig M (ed) Übersetzt und Mit Erklärungen Versehen. Enke Verlag (reprint 1970) Wiesbaden

Bianco A, Ramunno A, Serrilli AM, Lo Castro M, Ballero M, Serafini M (2004) Phytochemical characters of *Teucrium marum* from Sardinia: an endemic plant. Nat Prod Res 18:557–564

Boneva IM, Malakov PY, Papanov GY (1988) 12-Epiteupolin II, a neo-clerodane diterpenoid from *Teucrium lamiifolium*. Phytochemistry 27:295–297

Bruno M, Dominguez G, Lourenffo A, Piozzi F, Rodríguez B, Savona G, de la Torre MC, Arnold NA (1991) Neo-clerodane diterpenoids from Teucrium gracile. Phytochemistry 30:3693–3697

Bruno M, Alcázar R, de la Torre MC, Piozzi F, Rodríguez B, Savona G, Perales A, Arnold NA (1992a) Neo- and seco-neo-clerodane diterpenoids from *Teucrium gracile* and *T. Fruticans*. Phytochemistry 31(10):3531–3534

Bruno M, Piozzi F, Rodríguez B, Savona G, de la Torre MC, Servettaz O (1992b) Neo-clerodane diterpenes from *Teucrium* species. Phytochemistry 31(12):4366–4367

Bruneton J (1993) Pharmacognosie, phytochimie, plantes medicinales, 2nd edn. Technique et Documentation-Lavoisier, Paris

Bruno M, Piozzi F, Rodríguez B, Savona G, Servettaz O (1985) Ent-clerodane diterpenoids from six further species of *Teucrium*. Phytochemistry 24:2597–2599

Bruno M, Piozzi F, Savona G, Maria C, Rodríguez B, de la Torre MC, Servettaz O (1987) 2-Deoxychamaedroxide, A neo-clerodane diterpenoid from *Teucrium divaricatum*. Phytochemistry 26:2859–2861

Bruno M, Piozzi F, Savona G, Maria C, Rodríguez B (1989) Neo-clerodane diterpenoids from *Teucrium canadense*. Phytochemistry 28:3539–3541

Bruno M, Dominguez G, Lourenffo A, Piozzi F, Rodríguez B, Savona G, de la Torre MC, Arnold NA (1991) Neo-clerodane diterpenoids from *Teucrium gracile*. Phytochemistry 30:3693–3697

Bruno M, Bondı ML, Rosselli S, Piozzi F, Al-Hillo MRY, Lamara K, Ladjel S (2000) Neoclerodane diterpenoids from *Teucrium maghrebinum*. J Nat Prod 63:1029–1031

Bruno M, Bondì ML, Rosselli S, Maggio A, Piozzi F, Arnold NA (2002) Neoclerodane diterpenoids from *Teucrium montbretii* subsp. *libanoticum* and their absolute configuration. J Nat Prod 65:142–146

Bruno M, Maggio AM, Piozzi F, Puech S, Rosselli S, Simmonds MSJ (2003) Neoclerodane diterpenoids from *Teucrium polium* subsp. *polium* and their antifeedant activity. Biochem Syst Ecol 31:1051–1056

Bruno M, Rosselli S, Maggio A, Piozzi F, Scaglioni L, Arnold NA, Simmonds MSJ (2004) Neoclerodanes from *Teucrium orientale*. Phytochemisty 31:4366–4367

Camps F, Coll J, Dargallo O, Rius J, Miravitlles C (1987) Clerodane diterpenoids from *Teucrium* and *Ajuga* plants. Phytochemistry 26:1475–1479

Carnoy A (1959) Dictionnaire étymologique des noms grecs de plantes. In: Bibliothèque du Muséon, vol 46. Publications Universitaires, Louvain

Carratù B, Federici E, Gallo FR, Geraci A, Guidotti M, Multari G, Palazzino G, Sanzini E (2010) Plants and parts of plants used in food supplements: an approach to their safety assessment. Ann Ist Super Sanita 46:370–388

Carreiras MC, Rodríguez B, de la Terre MC, Perales A, Torres MR, Savona G, Piozzi F (1990) Rearranged abietane diterpenoids from the root of *Teucrium polium* subsp. *vincetinum*. Tetrahedron 46:847–860

Castro A, Moco S, Coll J, Vervoort J (2010) LC-MS-SPE-NMR for the isolation and characterization of neo-clerodane diterpenoids from *Teucrium luteum* subsp. *flavovirens*. J Nat Prod 73:962–965

Chitturi S, Farrell GC (2008) Hepatotoxic slimming aids and other herbal hepatotoxins. J Gastroenterol Hepatol 23:366–373

Coll J, Tandrón Y (2004) Neo-clerodane diterpenes from *Teucrium fruticans*. Phytochemistry 65:387–392

Coll J, Tandrón Y (2005) Isolation and structure elucidation of three neo-clerodane diterpenes from *Teucrium fruticans* L. (Labiatae). Phytochemistry 66:2298–2303
De Berardinis V, Moulis C, Maurice M, Beaune P, Pessayre D, Pompon D, Loeper J (2000) Human microsomal epoxide hydrolase is the target of germander-induced autoantibodies on the surface of human hepatocytes. Mol Pharmacol 58:542–551
De Smet PAGM (2002) Herbal Remedies. N Engl J Med 347:2046–2056
de Torre L, Rodríguez B, Plozzi F, Savona G, Brunot M, Carreiras MC (1990) Neo-clerodane from diterpenoids *Teucrium abutiloides*. Phytochemistry 29:579–584
De Vincenzi M, Maialetti F, Silano M (2003) Constituents of aromatic plants: Teucrin A. Fitoterapia 74:746–749
Dewick PM (1999) The biosynthesis of C5–C25 terpenoid compounds. Nat Prod Rep 16:97–130
Dias DA, Urban S, Roessner U (2012) A historical overview of natural products in drug discovery. Meta 2:303–336
Djabou N, Lorenzi V, Guinoiseau E, Andreani S, Giuliani MC, Dominguez XA, Merijania NA, Gonzalez BI (1974) Terpenoids from *Teucrium cubense*. Phytochemistry 13:754–755
Dourakis SP, Papanikolaou IS, Tzemanakis EN, Hadziyannis SJ (2002) Acute hepatitis associated with herb (*Teucrium capitatum* L.) administration. Eur J Gastroenterol Hepatol 14:693–695
Druckova A, Mernaugh RL, Ham AJL, Marnett LJ (2007) Identification of the protein targets of the reactive metabolite of teucrin A in vivo in the rat. Chem Res Toxicol 20:1393–1408
Dominguez XA, Merijanian A, Gonzalez B (1974) Terpenoids of *Teucrium cubense*. Phytochemistry 13:754–755
Elmastas M, Erenler R, Isnac B, Aksit H, Sen O, Genc N, Demirtas I (2016) Isolation and identification of a new neo-clerodane diterpenoid from *Teucrium chamaedrys* L. Nat Prod Res 30:299–304
Fatima N (2016) A review on *Teucrium oliveranum*, a plant found abundantly in Saudi Arabia. Sci Int (Lahore) 28:1229–1231
Fau D, Lekehal M, Farrell G, Moreau A, Moulis C, Feldmann G, Haouzi D, Pessayre D (1997) Diterpenoids from germander, an herbal medicine, induce apoptosis in isolated rat hepatocytes. Gastroenterology 113:1334–1346
Fernández P, Rodríguez B, Savona G, Piozz F (1986a) Neo-clerodane diterpenoids from *Teucrium polium* subsp. *capitatum*. Phytochemistry 25:181–181
Fernández P, Rodríguez B, Villegast JA, Perales A, Savona G, Piozzi F, Bruno M (1986b) Neo-clerodane diterpenoids from *Teucrium pyrenaicum* and *T. subspinosum*. Phytochemistry 25:1405–1409
Fernandez-Gadea F, Rodríguez B, Savona G, Piozzi F (1984) Isoeriocephalin and 20-deacetyleriocephalin, neoclerodane diterpenoids from *Teucrium lanigerum*. Phytochemistry 23:1113–1181
Fiorentino A, D'Abrosca B, Ricci A, Pacifico S, Piccolella S, Monaco P (2009) Structure determination of chamaedryosides A-C, threenovel nor-neo-clerodane glucosides from *Teucrium chamaedrys*, by NMR spectroscopy. Magn Reson Chem 47:1007–1012
Fiorentino A, D'Abrosca B, Pacifico S, Scognamiglio M, D'Angelo G, Monaco P (2010) Abeo-Abietanes from *Teucrium polium* roots as protective factors against oxidative stress. Bioorg Med Chem 18:8530–8536
Fiorentino A, D'Abrosca B, Pacifico S, Scognamiglio M, D'Angelo G, Gallicchio M, Chambery A, Monaco P (2011) Structure elucidation and hepatotoxicity evaluation against HepG2 human cells of neo-clerodane diterpenes from *Teucrium polium* L. Phytochemistry 72:2037–2044
Fontana G, Paternostro MP, Savona G, Rodríguez B, de la Torre MC (1998) Neoclerodane diterpenoids from *Teucrium massiliense*. J Nat Prod 61:1242–1247
Forouzandeh H, Azemi ME, Rashidi I, Goudarzi M Kalantari H (2013) Study of the protective effect of *Teucrium polium* L. extract on acetaminophen-induced hepatotoxicity in mice. Iran J Pharm Res 12:123–129
Gács-Baitz E, Radics L, Oganessian GB, Mnatsakanian VA (1978) Teucrins H1-H4, novel clerodane-type diterpenes from *Teucrium hyrcanicum*. Phytochemistry 17:1967–1973

Gao C, Han L, Zheng D, Jin H, Gai C, Wang J, Zhang H, Zhang L, Fu H (2015) Dimeric abietane diterpenoids and sesquiterpenoid lactones from *Teucrium viscidum*. J Nat Prod 78:630–638

Gaspar H, Palma F, De la Torre M, Rodríguez B, Perales A (1995) A rearranged homo-neo-clerodane diterpenoid from *Teucrium betonicum*. Tetrahedron 51:2363–2368

Gori L, Galluzzi P, Mascherini V, Gallo E, Lapi F, Menniti-Ippolito F, Raschetti R, Mugelli A, Vannacci A, Firenzuoli F (2011) Two contemporary cases of hepatitis associated with *Teucrium chamaedrys* L. decoction use. Case reports and review of literature. Basic Clin Pharmacol Toxicol 109:521–526

Greuter W, Burdet HM, Long G (1986) Med-Checklist Conservatoire et Jardin Botanique de la Ville de Genève. Genéve 3:366–380

Hao X, Zhang Z, Zhan G, Xue Y, Luo Z, Yao G, Zhang Y (2013) Chemical constituents from *Teucrium viscidum*. Biochem Syst Ecol 51:78–82

Hasani-Ranjbar S, Nayebi N, Larijani B, Abdollahi M (2010) A systematic review of the efficacy and safety of *Teucrium* species; from anti-oxidant to anti-diabetic effects. Int J Pharmacol 6:315–325

Jakupovic J, Baruah RN, Bohlmann F, Quack W (1985) New clerodane derivatives from *Teucrium scordium*. Planta Med 51:341–342

Karimi A, Majlesi M, Rafieian-Kopaei M (2015) Herbal versus synthetic drugs; beliefs and facts. J Nephropharmacol 4:27–30

Kisiel W, Piozzi F, Grzybek J (1995) Terpenoids from *Teucrium montanus* subsp. *pannonicum*. Planta Med 61:191–192

Kouzi SA, McMurtry RJ, Nelson SD (1994) Hepatotoxicity of germander (*Teucrium chamaedrys* L.) and one of its constituent neoclerodane diterpenes Teucrin A in the mouse. Chem Res Toxicol 7:850–856

Kumari GNK, Aravind S, Balachandran J, Ganesh MR, Soundarya Devi S, Rajan SS, Malathi R, Ravikumar K (2003) Antifeedant neo-clerodanes from *Teucrium tomentosum* Heyne. (Labiatae). Phytochemistry 64:1119–1123

Labbe C, Polanco MI, Castillo M (1989) 12-Epi-Teuscordonin and other neoclerodanes from *Teucrium bicolor*. J Nat Prod 52:871–874

Ladjel S, Laamara K, Al-Hillo MRY, Pays M (1994) Neo-clerodane diterpenoids from *Teucrium polium* ssp. *aurasianum*. Phytochemistry 37:1663–1666

Laliberté L, Villeneuve JP (1996) Hepatitis after the use of germander, a herbal remedy. Can Med Assoc J 154:1689–1692

Larrey D, Faure S (2011) Herbal medicine hepatotoxicity: a new step with development of specific biomarkers. J Hepatol 54:599–601

Larrey D, Vial T, Pauwels A, Castot A, Biour M, David M, Michel H (1992) Hepatitis after germander (*Teucrium chamaedrys*) administration: another instance of herbal medicine hepatotoxicity. Ann Intern Med 117:129–132

Lekehal M, Pessayre D, Lereau JM, Moulis C, Fourasté I, Fau D (1996) Hepatotoxicity of the herbal medicine germander: metabolic activation of its furano diterpenoids by cytochrome P450 3A depletes cytoskeleton- associated protein thiols and forms plasma membrane blebs in rat hepatocytes. Hepatology 24:212–218

Li R, Morris-Natschkeb SK, Lee KH (2016) Clerodane diterpenes: sources, structures, and biological activities. Nat Prod Rep 33:1166–1226

Loeper J, Descatoire V, Letteron P, Moulis C, Degott C, Dansette P, Fau D, Pessayre D (1994) Hepatotoxicity of germander in mice. Gastroenterology 106:464–472

Lv H, Luo J, Zhu M, Shan S, Kong L (2014) Teucrisins A–E, five new neo-Clerodane diterpenes from *Teucrium viscidum*. Chem Pharm Bull 62:472–476

Lv HW, Luo JG, Zhu MD, Zhao HJ, Kong LY (2015) Neo-Clerodane diterpenoids from the aerial parts of *Teucrium fruticans* cultivated in China. Phytochemistry 119:26–31

Mac Milan J, Beale ME (1999) Diterpene biosynthesis. In: Barton SD, Nakanishi K, Meth-Cohn (eds) Isoprenoids including cartenoids and steroids. Comprehensive natural products chemistry, vol 2. Elsevier, Amsterdam, pp 217–243

Malakov PY, Papanov GY (1983) Furanoid diterpenes from *Teucrium polium*. Phytochemistry 22:2791–2793

Malakov PY, Papanov GY, Ziesche J (1982) Teupolin III, a furanoid diterpene from *Teucrium polium*. Phytochemistry 21:2597–2598

Malakov PY, Papanov GY, Boneva IM (1992) Neo-clerodane diterpenoids from *Teucrium montanum*. Phytochemistry 31:4029–4030

Malakov PY, Papanov GY, Boneva IM, de la Torre MC, Rodríguez B (1993) Teulamioside, a neo-clerodane glucoside from *Teucrium lamiifolium*. Phytochemistry 34:1095–1098

Marquez C, Rabanal RM, Valverde S, Eguren L, Perales A, Fayos J (1981) Diterpenes from *Teucrium capitatum* L. x-ray crystal and molecular structure of lolin. Tetrahedron Lett 22:2823–2826

Mattéi A, Rucay P, Samuel D, Feray C, Reynes M, Bismuth H (1995) Liver transplantation for severe acute liver failure after herbal medicine (*Teucrium polium*) administration. J Hepatol 22:597. https://doi.org/10.1016/0168-8278(95)80458-7

Mazokopakis E, Lazaridou S, Tzardi M, Mixaki J, Diamantis I, Ganotakis E (2004) Acute cholestatic hepatitis caused by *Teucrium polium* L. Phytomedicine 11:83–84

Mehdinia Z, Eftekhar-Vaghefi R, Mehrabani M, Nabipoor F, Mehdinia N, nematolahi Mahani N (2013) The effect of petroleum ether fraction of *Teucrium polium* extract on laboratory mouse liver. J Kerman Univ Med Sci 20:279–291

Merritt AT, Ley SV (1992) Clerodane diterpenoids. Nat Prod Rep 9:243–287

Mostefa-Kara N, Pauwels A, Pines A, Biour M, Levy VG (1992) Fatal hepatitis after herbal tea. Lancet 340:674. https://doi.org/10.1016/0140-6736(92)92209-X

Nasri Hamid Shirzad H (2013) Toxicity and safety of medicinal plants. J HerbMed Pharmacol 2:21–22

Nencini C, Galluzzi P, Pippi F, Menchiari A, Micheli L (2014) Hepatotoxicity of *Teucrium chamaedrys* L. decoction: role of difference in the harvesting area and preparation method. Indian J Pharm 46:181–184

Node M, Sai M, Fujita E (1981) Isolation of the diterpenoid teuflin (6-epiteucvin) from *Teucrium viscidum* var. *miquelianum*. Phytochemistry 20:757–760

Nur-e-Alam M, Kanthasamy G, Yousaf M, Alqahtani AS, Ghabbour HA, Al-Rehaily AJ (2017a) Spectroscopy and crystal structures of natural stereoisomers of neoclerodane diterpenoids from *Teucrium yemense* of Saudi medicinal plant. J Mol Struct 1147:616–621

Nur-E-Alam M, Yousaf M, Ahmed S, Al-Sheddi ES, Parveen I, Fazakerley DM, Bari A, Ghabbour HA, Threadgill MD, Whatley KCL, Hoffmann KF, Al-Rehaily AJ (2017b) Neoclerodane diterpenoids from Reehal Fatima, *Teucrium yemense*. J Nat Prod 80:1900–1908

Pacifico S, D'Abrosca B, Scognamiglio M, D'Angelo G, Gallicchio M, Galasso S, Monaco P, Fiorentino A (2012) NMR-based metabolic profiling and in vitro antioxidant and hepatotoxic assessment of partially purified fractions from Golden germander (*Teucrium polium* L.) methanolic extract. Food Chem 135:1957–1967

Panovska TK, Kulevanova S, Gjorgoski I, Bogdanova M, Petrushevska G (2007) Hepatoprotective effect of the ethyl acetate extract of *Teucrium polium* L. against carbontetrachloride-induced hepatic injury in rats. Acta Pharma 57:241–248

Papanov GY, Malakov PY (1985) 2-keto-19-hydroxyteuscordin, a neo-clerodane diterpene from *Teucrium scordium*. Phytochemistry 24:297–299

Papanov GY, Malalov P, Bohlmann F (1981) Teuscordinon, a furanoid diterpene from *Teucrium scordium*. Phytochemistry 20:170–171

Pérez Alvarez J, Sáez-Royuela F, Gento Peña E, López Morante A, Velasco Osés A, Martín Lorente J (2001) Acute hepatitis due to ingestion of *Teucrium chamaedrys* infusions. Gastroenterol Hepatol 24:240–243

Pichersky E, Gang DR (2000) Genetics and biochemistry of secondary metabolites in plants: an evolutionary perspective. Trends Plant Sci 5:439–445

Piozzi F, Bruno M, Ciriminna R, Fazio C, Vassallo N, Arnold NA, De La Torre MC, Rodríguez B (1997) Putative hepatotoxic neoclerodane diterpenoids from *Teucrium* species. Planta Med 63:483–484

Piozzi F, Bruno M, Rosselli S (1998) Further furoclerodanes from *Teucrium* species. Heterocycles 48:2185–2203

Piozzi F, Bruno M, Rosselli S, Maggio A (2005) Advances on the chemistry of furano-diterpenoids from *Teucrium* genus. Heterocycles 65:1221–1234

Polymeros D, Kamberoglou D, Tzias V (2002) Acute cholestatic hepatitis caused by *Teucrium polium* (Golden Germander) with transient appearance of antimitochondrial antibody. J Clin Gastroenterol 34:100–101

Poon WT, Chau TL, Lai CK, Tse KY, Chan YC, Leung KS, Chan YW (2008) Hepatitis induced by *Teucrium viscidum*. Clin Toxicol 46:819–822

Rahmouni F, Daoud S, Rebai T (2019) *Teucrium polium* attenuates carbon tetrachloride-induced toxicity in the male reproductive system of rats. Andrologia 51:1–8

Rodríguez B, de la Torre M, Bruno M, Fazio C, Piozzi F, Savona G, Perales A, Arnold NA (1994) Rearranged neo-Clerodane diterpenoids from *Teucrium brevifolium*. Tetrahedron 50:2289–2296

Rodríguez B, de la Torre MC, Jimeno ML, Bruno M, Fazio C, Piozzi F, Savona G, Perales A (1995) Rearranged neo-clerodane diterpenoids from *Teucrium brevifolium* and their biogenetic pathway. Tetrahedron 51:837–848

Rodríguez B, de la Torre MC, Bruno M, Piozzi F, Vassallo N, Ciriminna R, Servettaz O (1996) Neo-clerodane diterpenoids from three species of *Teucrium*. Phytochemistry 43:435–438

Rodríguez B, de la Torre MC, Bruno M, Piozzi F, Vassallo N, Ciriminna R, Servettaz O (1997) A neo-clerodane diterpenoid from *Teucrium asiaticum*. Phytochemistry 45:383–385

Rodríguez B, Robledo A, Pascual-Villalobos MJ (2009) Rearranged abietane diterpenoids from the root of *Teucrium lanigerum*. Biochem Syst Ecol 37:76–79

Sattar EA, Mossa JS, Muhammad I, El-Feraly FS (1995) Neo-clerodane diterpenoids from *Teucrium yemense*. Phytochemistry 40:1737–1741

Savona G, Paternostro M, Piozzi F (1979) New furanoid diterpenes from *Teucrium gnaphalodes* L'hep. Tetrahedron Lett 4:379–582

Savona G, Piozzi F, Servettaz O, Maria C, Rodríguez B, Gadea FF, Martin-Loma M (1984) A neo-clerodane glucoside and neo-clerodane diterpenoids from *Teucrium flavum* subsp. *glaucum*. Phytochemistry 23:843–848

Savona G, Piozzi F, Servettaz O, Rodríguez B, Hueso-Rodriguez JA, De La Torre MC (1986) Neo-clerodane diterpenoids from *Teucrium lepicephalum* and *Teucrium buxifolium*. Phytochemistry 25:2569–2572

Savvidou S, Goulis J, Giavazis I, Patsiaoura K, Hytiroglou P, Arvanitakis C (2007) Herb-induced hepatitis by *Teucrium polium* L.: report of two cases and review of the literature. Eur J Gastroenterol Hepatol 19:507–511

Sexmero Cuadrado MJ, de la Torre MC, Rodríguez B, Bruno M, Piozzi F, Savona G (1991) Neo-clerodane diterpenoids from *Teucrium oxylepls* subsp. *marianum*. Phytochemistry 30:4079–4082

Sexmero Cuadrado MJ, Bruno M, de la Torre MC, Piozzin F, Savona G, Rodríguez B (1992) Rearranged abietane diterpenoids from the root of two *Teucrium* species. Phytochemistry 31:1697–1701

Shahraki MR, Arab MR, Mirimokaddam E, Palan MJ (2007) The effect of *Teucrium polium* (Calpoureh) on liver function, serum lipids and glucose in diabetic male rats. Iran Biomed J 11:65–68

Stanković SM, Zlatić N (2019) Ethnobotany of *Teucrium* species. In: Martinez LJ, Muñoz-Acevedo A, Rai M (eds) Ethnobotany: local knowledge and traditions. CRC Press, Boca Raton, pp 214–231

Stanković MS, Curcic MG, Zizic JB, Topuzovic MD, Solujic SR, Markovic SD (2011) *Teucrium* plant species as natural sources of novel anticancer compounds: Antiproliferative, proapoptotic and antioxidant properties. Int J Mol Sci 12:4190–4205

Starakis I, Siagris D, Leonidou L, Mazokopakis E, Tsamandas A, Karatza C (2006) Hepatitis caused by the herbal remedy *Teucrium polium* L. Eur J Gastroenterol Hepatol 18:681–683

Stickel F, Shouval D (2015) Hepatotoxicity of herbal and dietary supplements: an update. Arch Toxicol 89:851–865

Stickel F, Egerer G, Seitz HK (2000) Hepatotoxicity of botanicals. Public Health Nutr 3:113–124

Tanira MOM, Wasfi IA, Al Homsi M, Bashir AK (1996) Toxicological effects of *Teucrium stocksianum* after acute and chronic administration in rats. J Pharm Pharmacol 48:1098–1102

Theis N, Lerdau M (2003) The evolution of function in plant secondary metabolites. Int J Plant Sci 164(S3):S93–S102. https://doi.org/10.1086/374190

Topcu G, Eriş C, Che CT, Ulubelen A (1996) C-10 oxygenated neo-clerodane diterpenes from *Teucrium sandrasicum*. Phytochemistry 42:775–778

Torre de la M, Fernandez-Gadea F, Michavila A, Rodríguez B, Piozzi F, Savona G (1986) Neo-clerodane diterpenoids from *Teucrium botrys*. Phytochemistry 25:2385–2387

Torre de la M, Rodríguez B, Bruno M, Savona G, Piozzi F, Servettaz O (1988) Neo-clerodane diterpenoids from *Teucrium micropodioldes*. Phytochemistry 27:213–216

Tutin TG, Heywood VH, Burges NA, Moore DM, Valentine DH, Walters SM, Webb DA (1972) Flora Europaea III. Cambridge University Press, Cambridge

Ulubelen A, Topcu G, Olcal S (1994) Rearranged abietane diterpenes from *Teucrium divaricatum* subsp. *villosum*. Phytochemistry 37:1371–1375

Ulubelen A, Topcu G, Sönmez U (2000) Chemical and biological evaluation of genus *Teucrium*. Stud Nat Prod Chem 23:591–648

Venditti A, Frezza C, Trancanella E, Zadeh SMM, Foddai S, Sciubba F, Delfini M, Serafini M, Bianco A (2017) A new natural neo-clerodane from *Teucrium polium* L. collected in Northern Iran. Ind Crop Prod 97:632–638

Wellmann M (1958) Pedanii Dioscuridis Anazarbei De Materia Medica. Berlin Libri III, 97

Wink M (2009) Mode of action and toxicology of plant toxins and poisonous plants. Julius-Kuhn-Archiv 421:93–112

Zal F, Vasei M, Rasti M, Vessal M (2001) Hepatotoxicity associated with hypoglycemic effects of *Teucrium polium* in diabetic rats. Arch Iran Med 4:188–192

Zhi-da MIN, Ning XIE, Pei Z, Shou-xun Z, Chong-shu W, Qi-tai Z (1991) A neo-clerodane diterpene from *Teucrium japonicum*. Phytochemistry 30:4175–4177

Chapter 9
Genotoxic Activity of Secondary Metabolites of *Teucrium* Species

Darko Grujičić, Dragoslav Marinković, and Olivera Milošević-Djordjević

Abstract *Teucrium* species have been used in traditional medicine for the treatment of different diseases and for this reason, it is very important to know their effect on the human genome. In this chapter are presented literature data of genotoxic potential of secondary metabolites (phenolic acids and flavonoids) and different extracts from *Teucrium* species using different valid methods for the detection of genotoxicity. The presented results of investigating the genotoxic effect of methanolic extracts from *Teucrium chamaedrys, T. montanum, T. scordium, T. arduini* and *T. flavum*, methanolic, ethanolic, and aqueous extracts from *Teucrium polium*, as well as methanolic, ethanolic and chloroform extracts from *Teucrium ramosissimum* using micronucleus assay in human peripheral blood lymphocytes (PBLs), Comet assay in PBLs, Ames test/*Salmonella typhimurium*. It is known that the biological activities of the herbal extracts associated with its phytochemical composition. Phytochemical analyses of examined extracts of *Teucrium* species showed that they were rich in polyphenolic compounds and could be considered responsible for antimutagenic properties of the plants. Presented results have shown that the investigated of the *Teucrium* species possess the protective properties against antineoplastic drug-induced genetic damage in a different type of cells. There are two possible genoprotective effects, antimutagenic/antioxidative (*Teucrium polium* and *T. ramosissimum*) and proapoptotic (*Teucrium chamaedrys, T. montanum, T. scordium, T. arduini* and *T. flavum*). The investigated *Teucrium* species showed a genotoxic effect in high concentrations, while in combination with known mutagenic in the same concentrations had a protective effect. Presented investigations from dif-

D. Grujičić
Department of Biology and Ecology, Faculty of Science, University of Kragujevac, Kragujevac, Serbia

D. Marinković
Serbian Academy of Sciences and Arts, Belgrade, Serbia

O. Milošević-Djordjević (✉)
Department of Biology and Ecology, Faculty of Science, University of Kragujevac, Kragujevac, Serbia

Department of Genetics, Faculty of Medical Sciences, University of Kragujevac, Kragujevac, Serbia
e-mail: olivera.djordjevic@pmf.kg.ac.rs

M. Stanković (ed.), *Teucrium Species: Biology and Applications*,
https://doi.org/10.1007/978-3-030-52159-2_9

ferent genotoxicity studies have showed that secondary metabolites have a genoprotective effect on different types of cells (PBLs, mice bone marrow cells, PCEs; Chinese hamster ovary cells, CHO; bacteria *Salmonella typhimurium*). Moreover, present results clearly showed that tested extracts *Teucrium* species had a protective effect on the genotoxicity of different agents (mitomycin C, MMC; N-methyl-N-nitro-N-nitrosoguanidine, MNNG; benzo[a] pyrene, B[a]P; aflatoxin B1, AFB1; hydrogen peroxide, H_2O_2) in a different type of cells in vitro and, therefore, they can be a potential source of new pharmaceutics that will have a significant protective effect on induced genetic damage by chemotherapeutics in healthy cells.

Keywords Genotoxic activity · Secondary metabolites · *Teucrium* species · Methanolic extracts · Flavonoids · Phenolic acids

Abbreviations

BN	Binucleated Cells
CA	Chromosome aberrations
CBMN	Cytokinesis Block Micronucleus Assay
CHO cells	Chinese hamster ovary cells
DNA	Deoxyribonucleic acid
DPPH	2,2-diphenyl-1- picrylhydrazyl radical
EMS	Ethyl methanesulfonate
H_2O_2	Hydrogen peroxide
HPLC	High-Performance Liquid Chromatography
MMC	Mitomycin C
MN	Micronuclei
NDI	Nuclear Division Index
PBLs	Peripheral blood lymphocytes
PCEs	Polychromatic erythrocytes
SD	Standard deviation

9.1 Introduction

Plants have been used for centuries in the treatment of many human diseases. According to data from the World Health Organization (WHO), about 70–80% of the world's population in developing countries relies on plants for their primary health care (Chan 2003). Medicinal plants can be considered the main source of new chemical substances with potential therapeutic effects (Blumenthal 2000; Celik 2012). It is known that the specific parts of the herb (aerial parts, root, leaves, fruit,

flowers, and seeds) are already being used in form of suitable preparations (tablets, teas, extracts, creams or tinctures) in therapeutic purposes (Saad et al. 2017).

Teucrium L. (Lamiaceae) is a large, polymorphic, and cosmopolitan genus of perennial plants, the largest member of the family Lamiaceae distributed in Europe, North Africa, and the temperate parts of Asia, but mainly in the Mediterranean area, including more than 300 species (Tutin et al. 1972). In the flora of Europe, the genus *Teucrium* has been divided into seven sections with 49 species.

The preservation of the genetic integrity of human populations is very important, and any disturbance can either induce or contribute to, the onset of disease. It is known that are the genotoxic agents involved in the pathogenesis of several chronic degenerative diseases, including neurodegenerative and cardiovascular disorders, diabetes, chronic inflammation, cancer and aging (Lopez-Romero et al. 2018). Genotoxic agents present in the environment and occupational are numerous, and they are divided into physical, chemical and biological, and according to their origin, they can be natural and synthetical. The mutagenic effect they produce can be on somatic or germ cells are showed in Fig. 9.1.

Precisely for these reasons that a relatively new scientific discipline, genotoxicology, has been created, the primary goal of which is the detection of genotoxic agents, which ability to induce genetic damage regardless of the level of life organization, from bacteria to human cells (Izquierdo-Vega et al. 2017). Over the last 30 years, different methodologies, strategies and approaches have been developed to asses chemicals that could demonstrate genotoxic and/or carcinogenic effects, in vitro or in vivo, on somatic or germ cells (Sponchiado et al. 2016; Lopez-Romero et al. 2018). Accordingly, at what level the changes in genetic material are detected,

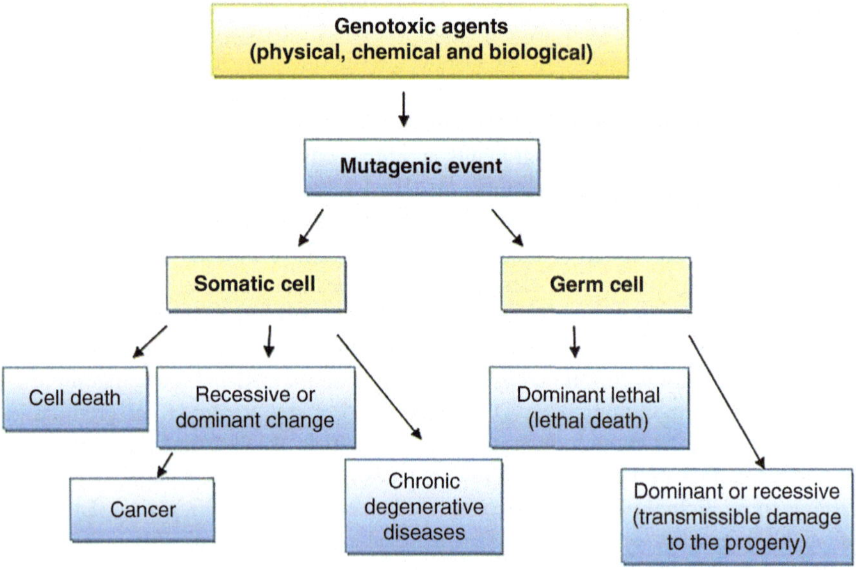

Fig. 9.1 Effect of genotoxic agents on mutagenic event in somatic and germ cells

all tests are divided into (1) tests detect changes in the level of DNA molecules (sister chromatid exchanges, SCE; single cell gel electrophoresis SCGE, or Comet assay); (2) gene mutation detection tests (Ames test, bacteria *Salmonella typhimurium*; SLRL test; *Drosophila melanogaster*); and (3) tests for detection chromosomal damage (chromosomal aberrations, CA; micronucleus test, MN).

The CA, SCEs, and MN assay in PBLs, as well as Comet assay, are used as the most useful assays to detect the potential genotoxicity of different agents (Joksić 1990; Celik 2012). They have been considered to be markers of early biological effects of carcinogen exposure (Liou et al. 2002). Numerous studies are showed that different drugs presented genotoxic effect in human peripheral blood lymphocytes (PBLs) in vivo (Milošević-Djordjević et al. 2003; Dimitrijević et al. 2006; Grujičić et al. 2007, 2008) and in vitro (Grujičić et al. 2009; Milošević-Djordjević et al. 2010a, b); as well as environmental contamination (Milošević-Djordjević et al. 2004, 2005, 2007). Traditionally, short term tests (STTs) for evaluating the genotoxic potential of hazardous chemicals were introduced and modified decades ago. STTs include the Ames test, in vivo cytogenetics tests, and the MN assay (Ren et al. 2017).

Each test has its advantages and disadvantages, but, overall, it is sensitive methods that are rapid, simple, and able to evaluate the genotoxic effect, provide an opportunity for assessment antigenotoxic effect in the same cells. The compounds that reduce the DNA damage caused by genotoxic agents called antigenotoxic agents (De Flora and Ferguson 2005; Ferguson et al. 2005). Numerous studies have shown the antigenotoxic potential of different plants (Behravan et al. 2011; Andrade et al. 2016; Lopez-Romero et al. 2018), mushroom (Tubić et al. 2019; Kosanić et al. 2019), and lichens (Turkez et al. 2012; Grujičić et al. 2014; Kosanić et al. 2016).

The detection of novel bioactive phytocompounds in counteracting mutagenic and carcinogenic effects is now gaining credence. In recent decades, numerous studies have been carried out in the last decades in order to identify compounds that might protect humans against DNA damage and its consequences. There are continual efforts all over the world to explore the rich biodiversity of edible (fruits, vegetables) as well as medicinal plants and other edible non-toxic plants in pursuit of the most effective phyto-antimutagens (Bhattacharya 2011; Lopez-Romero et al. 2018). In general, the antimutagens have been classified as desmutagens and bio-antimutagens. According to Kada and Shimoi (1987) antimutagenic agents are divided into two large groups. The first group is desmutagens, agents that can inactivate the mutagenic agents extensively before they reach the DNA and prevent the formation of premutation lesions. The second group belongs to bio-antimutagens, who reduce the effect of mutagens by acting inside the cell and participating in the suppression of mutations after DNA damage. Prevent treatment of premutation lesion in mutations, modulating replication and repair of DNA molecules (Fig. 9.2).

Intensive research over the last few decades has shown the beneficial protective effects of medicinal plant extracts on different cell cultures both in vitro (Ljubuncic et al. 2006; Leskovac et al. 2007; Milošević-Djordjević et al. 2013, 2018) and

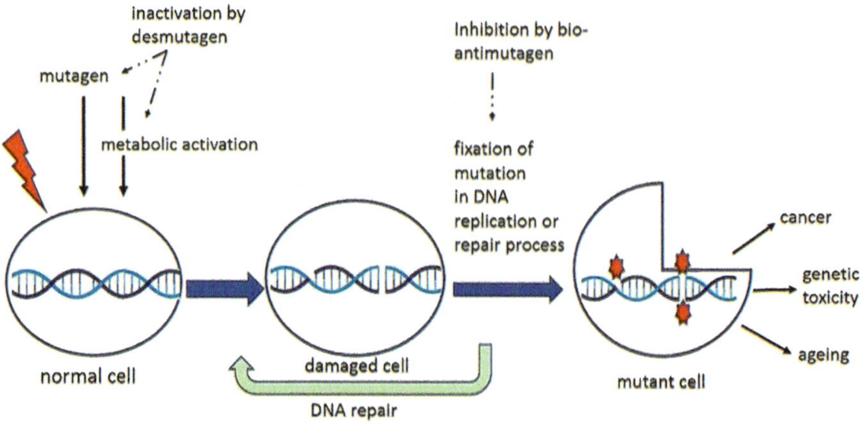

Fig. 9.2 Schematic representation showing the process of antimutagenesis

in vivo conditions (Damasceno et al. 2016; Rahmouni et al. 2017). Numerous studies were showed that species of the *Teucrium* genus have significant biological effects such as antimicrobial, antioxidant, cytotoxic and anti-cancerogenic properties. In Table 9.1 presented some of the biological effects and application in the traditional medicine of *Teucrium* species.

Genotoxicity assessment is an indispensable component in the safety assessment, aiming to prevent certain substances from affecting human health. Concerning the plant use, especially medicinal, long historical use of many practices of traditional medicine has indicated the safety and efficacy of traditional medicine. On the other side, some of studies have revealed that some plants frequently used in traditional medicine are potentially genotoxic (Marques et al. 2003; Ananthi et al. 2010; Mello-Reis et al. 2011; Regner et al. 2011; Shin et al. 2011). Therefore, it is extremely important to assess genotoxicity of plant extracts or secondary metabolites in order to verification their mutagenic potential since medicinal plants are widely used in traditional medicine and as a resource for the development of new drugs (Mello-Reis et al. 2011). However, to provide its efficacy more scientific research is needed. Moreover, in order to evaluate the genotoxicity risk of medicinal plants, it is necessary to review the literature and, as well, explore the genotoxicity assays most used and, then, compare these to the current legal requirements (Šavikin et al. 2013; Harutyunyan et al. 2019).

In relation to the described biological effects, the genotoxic properties of *Teucrium* species have been slightly studied. Accordingly, the purpose of this chapter is to collect available data on previous genotoxicity studies of *Teucrium* species. This chapter unifies scientific knowledge currently and to answer some questions concerning the genotoxic and antimutagenic properties of the investigated *Teucrium* species. In this chapter, we are showing the genotoxic and antimutagenic effect of different extracts of so far tested species *Teucrium* species.

Table 9.1 Biological effects and application in the traditional medicine of analysed *Teucrium* species

Teucrium species	Application in traditional medicine	References	Biological effects	References
T. chamaedrys	Gastrointestinal disorders: diarrhea, metabolic disorders, diuretic	Redžić (2007) and Jarić (2007)	Anti-inflammatory Antirheumatoid Antioxidative	Redžić (2007) and Jarić (2007)
T. montanum	Gastrointestinal disorders	Redžić (2007) and Jarić (2007)	Anti-inflammatory Antibacterial Antioxidative	Redžić (2007) and Jarić (2007)
T. polium	Gastrointestinal disorders, inflammation, eczema, urinary tract inflammation, diabetes, rheumatic disease mental performance	Everest and Ozturk (2005), Abu-Irmaileh and Afifi (2003), Said et al. (2002) and Perry et al. (1996)	Anti-inflammatory Antirheumatic Antimicrobial Antihypertensive Hypolipidemic Hypoglicemic Antioxidative	Balmekki et al. (2013), Kasabri et al. (2011), Kadifkova-Panovska et al. (2005), Rasekh et al. (2001), Tariq et al. (1989) and Suleiman et al. (1988)
T. scordium	Gastrointestinal disorders: diarrhea, fever, hemorrhoids, and intestinal parasites, bronchial ailments	Morteza-Semnani et al. (2007)	Anti-inflammatory	Morteza-Semnani et al. (2007)
T. arduini	Gastrointestinal disorders	Redžić (2007) and Jarić (2007)	Antimicrobial Antiproliferative Antiphytoviral Antioxidative	Kremer et al. (2013), Stanković et al. (2012), Dunkić et al. (2011), Šamec et al. (2010) and Vuković et al. (2010)
T. flavum	Heal skin wounds gastrointestinal disorders	Dall'Acqua et al. (2008), Bellomaria et al. (1998) and Calatayud et al. (1998)	Anti-inflammatory Antidiabetic Hypotensive Antioxidant	Dall'Acqua et al. (2008), Bellomaria et al. (1998), Calatayud et al. (1998) and Barrachina et al. (1995)

9.2 Evaluation of Genotoxic and Antimutagenic Activities of Medicinal Plants

9.2.1 PBLs as a Model System in Genetic Toxicology

Studies on cultured cells have shown that the PBLs is an extremely sensitive indicator in vitro and in vivo induced chromosome structural changes (Lerda et al. 2005; Milošević-Djordjević et al. 2013; Turkez et al. 2015). The PBLs are the best material for determination of cytogenetic effects (Eroglu et al. 2011; Grujičić et al. 2016; Milošević-Djordjević et al. 2018). PBLs have proved to be exceptional markers of exposure to a variety of mutagenic agents, since they represent long living cells able to accumulate various types of cytogenetic damages for years. Via the blood circulation lymphocytes are distributed through the entire organism. Thus, the chromosomal aberrations are not just the indicators of the state of the lymphocytes in the circulation but can also be induced in any part of the organism (Lou et al. 2007; Holland et al. 2011: Fortin et al. 2015).

9.2.2 MN Assay in Genetic Toxicology

For screening chromosomal damage in PBLs, MN assay is the most frequently used assay (Fenech 2000; Fenech and Bonassi 2011). Otherwise, Countryman and Heddle (1976) are the time were the first who detected MN in these cells investigating chromosome aberrations. MN test is one of the valid tests used as a quick check for detection of acute disorders caused by chemical substances in cells that have undergone cell division. Following the frequency of MN gives us insight into the level of chromosomal aberrations in mutagen-exposed cells (Speit et al. 2012; Kirsch-Volders et al. 2014; Ghosh and Godderis 2016). MN originates mainly from either acentric chromosome fragments or whole chromosomes. Acentric chromosome fragments originate from unrepaired DNA strand breaks or misrepaired DNA strand breaks. These acentric fragments often occur in conjunction with forming dicentric chromosomes. MN may also arise from dicentric chromosome fragments during anaphase. Malsegregation of whole chromosomes may be induced if the chemical either causes centromere/kinetochore malfunction or impairs the mitotic spindle or centrosome (Fenech et al. 2011; IAEA 2011; Kirsch-Volders et al. 2011; Luzhna et al. 2013; Fortin et al. 2015; Guo et al. 2019).

MNs are formed spontaneously or are induced, as a cellular response to the effects of genotoxic agent. For years, the presence of MN in PBLs has been used as biological dosimetry at exposure to ionizing radiation (Joksić 1990; Vral et al. 2011; Ryu et al. 2014, 2016), as well as to chemical genotoxic agents in working and living environment (Huang et al. 2009). Several studies have shown that air pollution can contribute to the increase of MN frequency in adults, as well as in children (Neri et al. 2006; Milošević-Djordjević et al. 2007; Hallare et al. 2008; Feretti et al. 2014).

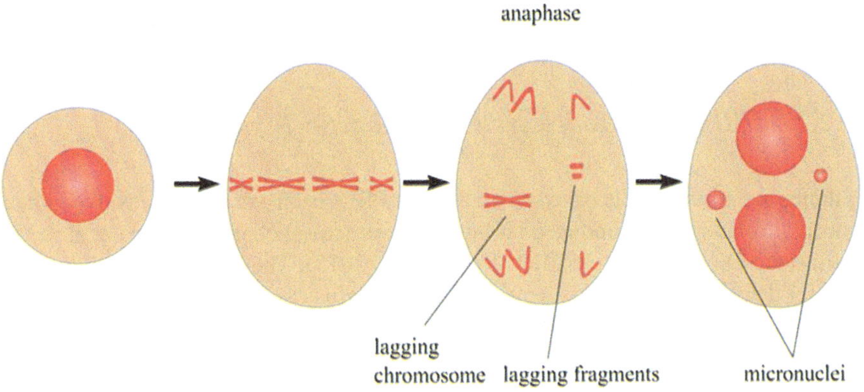

Fig. 9.3 Mechanisms of micronuclei formation during cell division

The correlation between exposure to various inorganic chemicals, among which heavy metals were the most prominent, and the MN frequency has also been determined, mostly in workplace settings (Bonassi et al. 2005). Research has shown that MN frequency can also be influenced by dietary, genetic, and some lifestyle factors (Bonassi et al. 2011; Fenech and Bonassi 2011; Kadioglu et al. 2012). High MN frequency in PBLs indicates a high level of CAs, and, therefore, MN is an important biomarker for the risk assessment and prediction of cardiovascular disease (Andreassi and Botto 2003), and different cancers (Bonassi et al. 2007; Milošević-Djordjević et al. 2010a, b, 2011a, b; Vrndić et al. 2013; Stošić et al. 2014; Pardini et al. 2017). MN is a small additional nucleus which is located near the main nucleus in the cytoplasm (Fig. 9.3). Its diameter is usually between 1/16 and 1/3 of the diameter of the main nucleus. They are mostly oval or round, but they can also be pyramidal, hemispheric, elliptical, cylindrical or very rarely irregular. MN adopts the same texture as the main nucleus. Generally, there is only one MN in a cell, but occasionally there can be two. Some cells can contain triple or more MN, but very rarely (Fenech et al. 2003; Fenech 2007).

One of the most commonly used cytogenetic assays in the genotoxic assessment of different agents is the cytokinesis-block micronucleus (CBMN) assay in cultivated human PBLs and is simultaneously used as a method in human biomonitoring studies (Decordier et al. 2009; Speit et al. 2011). Fenech and Morley have developed the CBMN assay in human lymphocytes (Fenech and Morley 1985). It is one of the most commonly used methods for measuring structural and numerical chromosomal changes in human cells in vitro and in vivo (Fenech 2007; Fenech et al. 2011; OECD Library 2010; Nersesyan et al. 2016; Gajski et al. 2018). The advantage of CBMN assay is its speed and simplicity, as well as the possibility of MN frequency analysis in binucleated (BN) cells. Using this test, thousands of lymphoblasts can be analyzed on a single-sample basis, which provides a more objective viewpoint as far as the damage to the genetic material is concerned. CBMN assay ensures simultaneous information on chromosomal breakage, chromosomal

rearrangement, and gene amplification. MN test has a great advantage over other cytogenetic tests because it offers tools for a damage assessment at the level of function and mitotic spindle integrity (Mateuca et al. 2006).

The analysis of CBMN assay is based on the cultivation of PBLs and it allows the identification of cells that have undergone only one division in culture. These results were achieved with cytochalasin B. It was added 44 h after the beginning of incubation, before the first mitosis, to block the division of cytoplasmic cytokinesis without inhibiting nuclear division. This is the mechanism of the formation of BN cells. Cytochalasin B may help to establish the difference among mononuclear cells (those that did not divide in vivo) or those that have completed one or more nuclear divisions. Generally, MN has been used as a biomarker of genome instability, chromosomal damages and cancer risk, integrating acquired mutations and genetic susceptibility towards mutations (Holland et al. 2008; Iarmarcovai et al. 2008; Vrndić et al. 2013). In recent years, the CBMN assay has evolved into a "cytome" assay (CBMN Cyt) of chromosomal instability in which more nuclear anomalies and many more cells in the sample can be included in the analysis and categorized to provide a more comprehensive view of DNA damage, cytostasis, and cell death. The key components of the CBMN Cyt assay are nuclear buds (NBUDs), nucleoplasmic bridges (NPBs) and apoptotic and necrotic cells (Fenech 2006, 2007).

In addition to chromosomal aberrations, the second cytogenetic parameter that can be detected by the CBMN assay is the nuclear division index (NDI), as a significant indicator proliferative status of the cells. NDI is calculated according to the formula proposed by Eastmond and Tucker (1989) as follows: $NDI = ((1 \times M1) + (2 \times M2) + (3 \times M3) + (4 \times M4))/N$, where M1 to M4 represents the number of cells with one to four nuclei and N is the total number of cells scored (Fig. 9.4). The NDI is the parameter that shows the number of cell cycles through which the treated cells have passed in culture (Surralles et al. 1995; Fenech 2000). Having determined NDI value we can also establish the cytostatic and cytotoxic effect in the lymphocytes that can be the level of mutagenic response (Fenech 2007). The numerous authors proposed NDI as a screening marker for different cancers (El-Zein et al. 2011; Ionescu et al. 2011) because is NDI marker of cell proliferation in cultures and is considered a measure of general toxicity (Speit 2013; Ionescu et al. 2015). Cells with greater chromosomal damage will either die before cell division or may be less able to enter in this phase (Santos-Mello et al. 2004; Minozzo et al. 2004). Therefore,

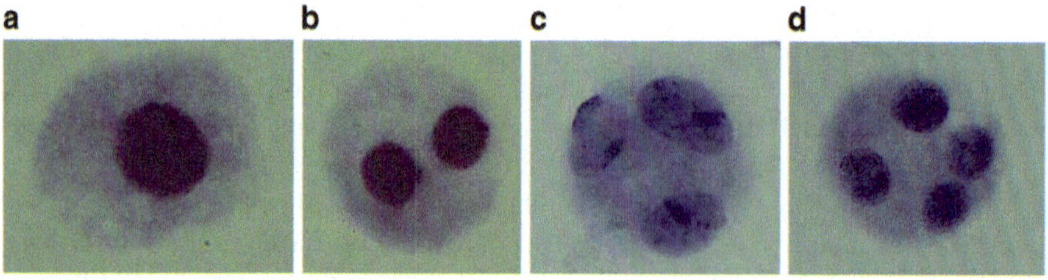

Fig. 9.4 Representative image of cells for calculating NDI: (**a**) mononucleated cell; (**b**) binucleated cell; (**c**) trinucleated cell; (**d**) tetranucleated cell

more a cell will accumulate genetic alterations, less it will be able to divide, and NDI will be lower (Ionescu et al. 2015).

CBMN assay is widely used to determine the mutagenic effect of both physical (Das and Karuppasamy 2009) and chemical agents (OECD Library 2010; Kirsch-Volders et al. 2014). According to a number of authors, this method is used as an alternative to fixed metaphase analysis. It is particularly suitable for population monitoring studies, when fast and accurate information, whether a substance is genotoxic or not, is wanted. For that reason, there has been a growing interest in the use of the MN assay to detect genotoxic agents. The latest research showed that the MN assay as a test is most frequently used in the evaluation of genotoxic risks of medicinal plants (Sponchiado et al. 2016).

9.3 Phytochemical Characteristics of the Examined *Teucrium* Species

Phenolic compounds are products of secondary plant metabolism and occur during normal development of plants, but also in stress conditions, for example during UV radiation, infection, after injury, etc. In the past research of secondary plant metabolites, more than 8000 phenolic compounds have been identified (Bravo 1998; Ferguson 2001; Ghasemzadeh and Ghasemzadeh 2011; Kenndey 2014). These compounds are present in all plant organs, leaves, flowers, tree, root, fruits and seeds of plants (Stanković 2012). There are different classifications of phenolic compounds with more than 15 classes different in chemical structure: from simple molecules to high molecular weight polymers. The most common polyphenolic compounds are divided into phenolic acids, flavonoids, stilbene, and lignans.

Phenolic acids are hydroxybenzoic acid and hydroxycinnamic acid derivatives. The most prominent benzoic acid derivatives are gallic acid, protocatechuic acid, vanillic acid, and the most commonly derivatives of cinnamic acid are the *p*-coumaric, caffeic and ferulic acid. Flavonoids are a class of polyphenol secondary metabolites. They are presented broadly in plants and diets and have various bioactive effects (antiviral, anti-inflammatory, cardioprotective, antidiabetic, antiaging, etc.). Moreover, flavonoids, shown to have anticancer activity notably. This was confirmed through numerous in vitro and in vivo studies (Havsteen 2002; Esmaeili et al. 2014; Hammami et al. 2015; Veeramuthu et al. 2017).

Flavonoids are frequently found as glycosylated or esterified forms, consisting of C6—C3—C6 rings, namely rings A and B linked by three-carbon-ring C. Due to their poor bioavailability, however, information about associations between structure and biological fate is limited. This review, therefore, attempts to bring some order into relationships between structure, activity as well as pharmacokinetics of bioactive flavonoids (Wang et al. 2018). Flavonoids are a group of natural compounds with variable phenolic structures and are found in plants. In 1930 a new substance was isolated from oranges. At that time, it was believed to be a member

of a new class of vitamins and was designated as vitamin P (Batra and Sharma 2013). Later it became clear that this substance was a flavonoid (rutin) and till now more than 4000 varieties of flavonoids have been identified (Middleton 1998; Kumar and Pandey 2013). They can be divided into a variety of classes such as flavones (apigenin, luteolin, and chrysin), flavonols (quercetin, kaempferol, and myricetin), flavanones (naringenin, and hesperetin), and others (Veeramuthu et al. 2017).

The extracts of the investigated *Teucrium* species had a very rich content of secondary metabolites such as phenolic compounds (Stanković 2012; Milošević-Djordjević et al. 2018). The most important polyphenolic compounds in plants are phenolic acid and flavonoids (Petti and Scully 2009; Khoddami et al., 2013). They play an important role in reducing the risk of cardiovascular disease, as antioxidants, and exhibited a wide variety of other biological activities, including hepatoprotective and antibacterial activities (Tiwari 2001). Due to their importance in plants and human health, it would be useful to know the concentration of the polyphenolic compounds and biological activities that could indicate their potentials as therapeutic agents (Benedec et al. 2013).

Results of qualitative analysis of methanolic extracts of examined *Teucrium* species are presented in Table 9.2. The analyzes were performed by High-performance liquid chromatography – HPLC (Stanković 2012; Milošević-Djordjević et al. 2018; Grujičić et al. 2020). Presented results showed that the most prevalent phenolic acids were chlorogenic acid (for *Teucrium chamaedrys* and *T. polium*), caffeic acid (for *Teucrium arduini* and *T. flavum*) and rosmarinic acid (for *Teucrium montanum*).

Table 9.2 The of polyphenolic contents of methanolic extract from analyzed *Teucrium* species

Teucrium species	Phenolic acids ↑ ↓		Flavonoids ↑ ↓		References
T. chamaedrys	Chlorogenic acid	Caffeic acid *p*-coumaric acid	Rutin	Myricetin Kaemferol	Stanković (2012)
T. montanum	Rosmarinic acid	Caffeic acid *p*-coumaric acid	Rutin	Kaempferol Myricetin	
T. polium	Chlorogenic acid *p*-coumaric acid	Caffeic acid Vanillic acid	Catechin Apigenin	Luteolin Quercetin	Milošević-Djordjević et al. (2018)
T. scordium	Vanillic acid *p*-coumaric acid	Sinapic acid	Catechin	Rutin	
T. arduini	Caffeic acid Ferulic acid	Chlorogenic acid Vanillic acid	Quercetin	Epicatechin Rutin	Grujičić et al. (2020)
T. flavum	Caffeic acid *p*-coumaric acid	Vanilic acid Chlorogenic acid	Quercetin	Epicatechin Catechin	

P-coumaric acid was present in *Teucrium polium*, *T. scordium* and *T. flavum* extracts, as a second most abundant phenolic acid. Analyzing the content of flavonoids, present results showed that *Teucrium chamedrys* and *T. montanum* extracts contained the highest content of rutin, while the lowest contents were measured for myricetin and kaempferol. The extracts of *Teucrium polium* and *T. scordium* had the highest content of catechin and apigenin, while the lowest contents were measured for luteolin and rutin. Both extracts of *Teucrium arduini* and *T. flavum* contained the highest content of quercetin, and the lowest contents were measured for epicatechin. The second most abundant flavonoid in the extract of *Teucrium polium* was apigenin.

9.4 Genotoxic Activity of *Teucrium* Extracts in PBLs

The results of the genotoxic activity of methanolic extracts of examined *Teucrium* species on PBLs of healthy donors are shown in Table 9.3 (Milošević-Djordjević et al. 2013, 2018; Grujičić et al. 2020). The results indicated that the extract of *Teucrium chamaedrys* did not show genotoxic effect in all tested concentrations, while both extracts of *Teucrium montanum* and *T. polium* only in the highest concentration (1000 μg/ml) significantly induced chromosomal damage. On the other hand, *Teucrium scordium* showed genotoxic effect in the two highest tested concentrations (500 and 1000 μg/ml). Also, the high concentrations (250–1000 μg/ml) extracts of *Teucrium arduini* and *T. flavum* have a genotoxic effect. In relation to cellular kinetics, present results have show that, in separately treatment PBLs, only extract of *Teucrium scordium* significantly decreased NDI values in two highest tested concentrations (500 and 1000 μg/ml), while others plant extracts did not change NDI values in comparation to negative control (untreated) cells.

All plant extracts, except *Teucrium chameaedrys*, in a separately treatment, with an increased concentration, showed a genotoxic effect. *Teucrium arduini* and *T. flavum* showed a genotoxic effect at concentrations from 250 to 1000 μg/ml, while at the lowest test concentration of extracts (125 μg/ml) none of the six analyzed plants showed a genotoxic effect. None of the plants influence on cells kinetics, except *Teucrium scordium*, which in higher concentrations (500 and 1000 μg/ml) significantly decreased NDI. These results may be explained by high levels of flavonoids contained in higher concentrations of the extract which can possibly contribute to the pro-mutagenic effect, as pro-oxidants in generating free radicals that damage DNA. Stopper et al. (2005), Silva et al. (2000) and Boos and Stopper (2000) showed that flavonoids may cause genetic damage in a variety of prokaryotic and eukaryotic systems. Also, numerous studies provide evidence for dual biological activities of flavonoids, which are showed that in high concentrations can act both protective and mutagenic (Bhattachar 2011; Kunwar and Priyadarsini 2011; Baldim et al. 2017). The possible mechanism has shown in Table 9.4.

The anti-mutagenic/pro-mutagenic and antioxidant/pro-oxidant activities largely depend on the concentration used (Labieniec et al. 2003; Labieniec and Gabryelak 2003; Stopper et al. 2005), as well as the physiological parameters and

Table 9.3 The results of genotoxic activities of different methanolic extracts concentrations from *Teucrium* species in vitro

Teucrium species	Concentration (μg/ml)	MN in comparison to negative controls	NDI in comparison to negative controls	Effect	References
T. chamaedrys	125	No sig. difference	No sig. difference	No genotoxic	Milošević-Djordjević et al. (2013)
	250				
	500				
	1000				
T. montanum	125	No sig. difference	No sig. difference	No genotoxic	
	250				
	500				
	1000	Sig. ↑ MN*		Genotoxic	
T. polium	125	No sig. difference in MN	No sig. difference	No genotoxic	Milošević-Djordjević et al. (2018)
	250				
	500				
	1000	Sig. ↑ MN*		Genotoxic	
T. scordium	125	No sig. difference	No sig. difference	No genotoxic	
	250				
	500	Sig. ↑ MN*	Sig. ↓ NDI*	Genotoxic	
	1000				
T. arduini	125	No sig. difference	No sig. difference	No genotoxic	Grujičić et al. (2020)
	250	Sig. ↑ MN*		Genotoxic	
	500				
	1000				
T. flavum	125	No sig. difference	No sig. difference	No genotoxic	
	250	Sig. ↑ MN*		Genotoxic	
	500				
	1000				

*significant difference in comparison to negative controls; *sig.* significant

physiological conditions in the body (Skibola and Smith 2000; Noel et al. 2006). The genotoxic damage caused by flavonoids at higher concentrations may be due to DNA intercalation and inhibition of DNA topoisomerase II and inhibition of key enzymes involved in hormone metabolism (Skibola and Smith 2000; Sharma 2013). Some phenolic acids which otherwise as well-known antioxidants are, possible can be pro-oxidants, depending on concentration, and thus being a significant cause of DNA damage. Accordingly, Erdem et al. (2012) demonstrated that a high dose of the vanillic acid itself had genotoxic effects on DNA in PBLs in vitro, and when used at an appropriately low dose this acid could prevent oxidative DNA damage. Similar results were obtained by Bhat et al. (2007) that caffeic acid induced prooxidant DNA breakage in PBLs.

Table 9.4 The possible mechanism of the protective/mutagenic effect of high concentrations of flavonoides in vitro

Mutagenic effect	Protective effect
(1) Pro-oxidant effect, generating free radicals 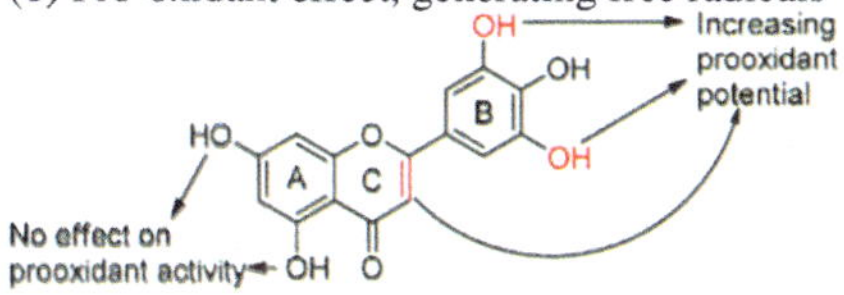 Structure–prooxidant activity relationship in flavonoids (Eghbaliferiz and Iranshahi 2016) (2) DNA interaction and inhibition of DNA topoisomerase II, enzyme who play importance role in the mitotic chromosome segregation after DNA replication (Skibola and Smith 2000)	(1) Genoprotective: (a) Inhibition of genotoxic effects- antioxidant effect and scavenging of free radicals 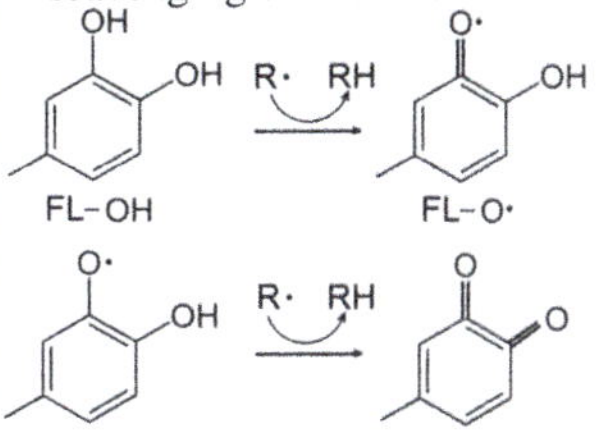 Scavenging of ROS (R*) by flavonoids (Fl-OH) (Gupta et al. 2010) (b) proapoptotic effect for cells with collected chromosome damage (Milošević-Djordjević et al. 2013, 2018)
(3) Inhibition of key enzymes involved in hormone metabolisms, and alter the activity other key metabolizing enzymes (Sharma 2013)	(2) Modulation of signal transduction pathways (3) Induce antioxidant enzymes and detoxifying enzymes (Bahramikia and Yazdanparast 2012)
(4) Unrepaired or misrepaired oxidative DNA strand breaks and mutation (Skibola and Smith 2000)	(4) As blocking agents thus preventing the metabolic activation of promutagens (Daradka and Alshibly 2012)

Of all flavonoids, perhaps the greatest scientific interest was caused by quercetin. It is known that the prooxidant property of quercetin is important as it is related to the generation of mutagenic quinone-type metabolites. Mutagenic property of quercetin has been demonstrated in bacteria (Czeczot et al. 1990) as well as in mammalian cell culture (Silva et al. 2000), including human DNA (Duthie et al. 1997). Mazumdar et al. (2011) showed that quercetin induced MN and CAs in polychromatic erythrocytes in bone marrow cells in vivo. The findings same authors suggest that quercetin may result in genomic instability in the tested dose range and significant reduction in MMC induced genotoxicity in the highest dose tested. However, numerous studies have shown that the genotoxicity of one compound present in the extract can be alleviated by the other compound from extract when they are found together since the extracts act as a mixture of different compounds (De Marino et al. 2014; Andrade et al. 2016).

9.5 Protective Effect of Plant Extracts Against Mitomycin C (MMC) Induced Chromosome Damage in PBLs

To determine the protective effects of plant extracts against MMC-induced chromosomal damages in PBLs, the cell cultures were treated with the tested concentrations of extracts in combination with MMC (mitomycin C). MMC is a natural antitumor antibiotic drug used in clinical chemotherapy regimens in various carcinomas (Verweij and Pinedo 1990; Sontakke and Fulzele 2009). However, MMC induces chromosomal aberrations as a non-cell-cycle-specific agent, therefore showing strong genotoxic effects in different cell lines (Ho and Schärer 2010; Vasquez 2010). During metabolism, the MMC can generate reactive oxygen species (ROS) such as superoxide radical anions, hydrogen peroxide and hydroxyl radicals (Albertini et al. 2003; Ortega-Gutierrez et al. 2009). ROS induce numerous lesions in DNA including deletions, base modifications and single and double-strand breakages (Cadet et al. 1994; Tiwari 2001).

It is known that free radicals involve the MN formation (Vijayalaxmi et al. 1999) and MMC generates MN-induced genotoxic damage in animal models (Hayashi et al. 1992; Grisolia 2002). Thus, for example, MMC induced genotoxic effect which can be detected by different methods such as CA, SCEs (Krishnaja and Sharma 2008) and MN frequency (Fauth et al. 2000) in the cultures of PBLs. In the studies from Milošević-Djordjević et al. (2013, 2018) it was observed that MMC induced micronuclei in PBLs in comparison to untreated PBLs. These chromosomal damages were, to some extent, the result of the activity of free radicals formed during the metabolism of MMC. Similar results have been obtained in other studies that have shown that MMC caused genotoxicity in PBLs (Sontakke and Fulzele 2009; Erdem et al. 2012; Turkez et al. 2012, 2015). Therefore, MMC is often used as a positive control in genotoxicology studies (Siddique et al. 2008; Santovito et al. 2011; Akyil et al. 2015; Turkez et al. 2015).

The effects of methanolic extracts of *Teucrium* species on MN frequency and NDI in cultured PBLs treated with MMC showed in Table 9.5 (Milošević-Djordjević et al. 2013, 2018; Grujičić et al. 2020). All the tested concentrations of extracts of *Teucrium chamaedrys, T. polium, T. arduini* and *T. scordium* (except on the lowest, 125 μg/ml) significantly reduced MMC-induced MN in PBLs in a dose-dependent manner. The extract of *Teucrium montanum* significantly reduced MMC-induced MN frequency only in lowest concentration, at 125 μg/ml, however, with increasing concentration protective effect of extract decreased. The extract *Teucrium flavum* in lower concentrations (125–500 μg/ml) evident no significance effect on decrease MMC-induced MN frequency, while at the highest concentration (at 1000 μg/ml) it significantly reduces MN frequency, and expressed protective effect on PBLs. Cell proliferation was evaluated by NDI, which indicates the average number of the cell cycle in culture. Many authors showed that significant decrease of NDI in the cultures of PBLs treated with various agents in vitro was associated with the activation of apoptosis mechanism (Hess et al. 1999; Abou-Eisha et al. 2004). Present results showed that high concentrations of extracts *Teucrium chamaedrys*, *T. scordium*,

Table 9.5 Genotoxic/antigenotoxic effects different concentrations of methanolic extracts from *Teucrium* species on PBLs in vitro

Teucrium species	Concentrations extract (μg/ml) + MMC (0.5 μg/ml)	MN in comparison to positive controls	NDI in comparison to positive controls	Effect against MMC	References
T. chamaedrys	125 + MMC	Sig. ↓ MN*	No sig. difference	Protective	Milošević-Djordjević et al. (2013)
	250 + MMC				
	500 + MMC		Sig. ↓ NDI*		
	1000 + MMC				
T. montanum	125 + MMC	Sig. ↓ MN*	Significant ↓ NDI*	Protective	
	250 + MMC	No sig. difference			
	500 + MMC				
	1000 + MMC				
T. polium	125 + MMC	Sig. ↓ MN*	No sig. difference	Protective	Milošević-Djordjević et al. (2018)
	250 + MMC				
	500 + MMC				
	1000 + MMC				
T. scordium	125 + MMC	No sig. difference	No sig. difference	Protective	
	250 + MMC	Sig. ↓ MN*			
	500 + MMC				
	1000 + MMC		Sig. ↓ NDI*		
T. arduini	125 + MMC	Sig. ↓ MN*	No sig. difference	Protective	Grujičić et al. (2020)
	250 + MMC				
	500 + MMC				
	1000 + MMC		Sig. ↓ NDI*		
T. flavum	125 + MMC	No sig. difference	No sig. difference	Protective	
	250 + MMC				
	500 + MMC				
	1000 + MMC	Sig. ↓ MN*	Sig. ↓ NDI*		

*significant difference in comparison to positive controls; *sig.* significant

T. arduini and *T. flavum*, in combined treatment with MMC, a significant decrease of NDI in PBLs. The extract of *Teucrium montanum* significant reduced NDI in all tested concentrations, while extract of *Teucrium polium* non reduced NDI not even in one tested concentration.

Researches Milošević-Djordjević et al. (2013, 2018) and Grujičić et al. (2020) showed that the methanolic extracts of the genus *Teucrium* clearly indicate a protective effect against MMC-induced MN. Presented results showed that methanolic extracts of *Teucrium chamaedrys*, *T. polium* and *T. arduini* in all concentrations, reduced MMC-induced genetic damage, reducing the frequency of MN in PBLs. The most of all tested plants (*Teucrium chamaedrys*, *T. scordium*, *T arduini* and *T. flavum*) are at high concentrations affecting on cellular proliferation because NDI reduced, can be explained as a proapoptotic effect these plant extracts. It is known

that a significant decrease in NDI indicates that cells with the greater chromosomal damage did not survive in culture and were entered a process of apoptosis and that they were as such eliminated (Abou-Eisha et al. 2004; Minozzo et al. 2004; Milošević-Djordjević et al. 2011a). The methanolic extract of *Teucrium montanum* is it significantly reduced the MMC-induced MN frequency at the lowest tested concentration but is in all concentrations reduces NDI in MMC treated PBLs, which also indicates its proapoptotic effect. The methanolic extract of *Teucrium polium* significantly reduced MN in all concentrations but without the reduction of NDI, which indicating the antimutagenic effect this extract. All the concentrations of *Teucrium scordium* (except the lowest one) significantly decreased the genotoxic effect of MMC. However, the strongest protective activity was observed in the highest tested concentrations.

This plant extract reduced significantly NDI at the highest concentration. The reduced proliferation of cells indicates and the proapoptotic mechanism of the action of the plant extract and the introduction of cells with significant genetic damage in apoptosis. The similar effect was showed the methanolic extract of *Teucrium flavum* which is reduced MN frequency in PBLs only in the highest concentration, and reduced significantly NDI only at the highest concentration, indicating a pro-apoptotic effect. Therefore, we think that MN is a possible mechanism for introduction to cell death – apoptosis. Similar results were obtained from Hess et al. (1999) and Abou-Eisha et al. (2004) who studied NDI values in cultured PBLs treated with various agents. In Fig. 9.5 were presented BN cells after MMC treatment PBLs, while in Fig. 9.6 were presented BN cells after combined treatment plant extracts and MMC. The antimutagenic effect detected in tested plants may be attributed to actions of secondary metabolites who presence in the extract. It is known that secondary metabolites to have antioxidant properties are known to have inhibitory effects on the genotoxic actions of several known mutagens (Ferguson 2001; Matkowski and Wołniak 2005; Bhattachar 2011). The active ingredients, secondary metabolites of the extract have a role of collecting free radicals, thus alleviating the mutagenic effect of MMC. Many polyphenols in plant extracts have ideal structure

a **b** **c**

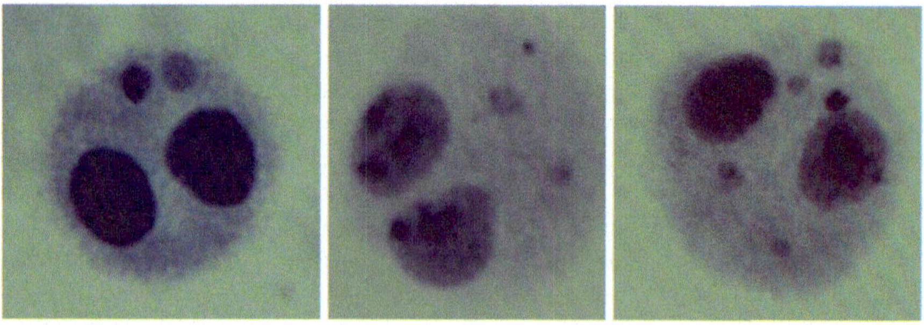

Fig. 9.5 The binucleated (BN) cells after MMC treatment: (**a**) BN with 2 MN; (**b**) BN with 3MN; (**c**) BN with 5MN

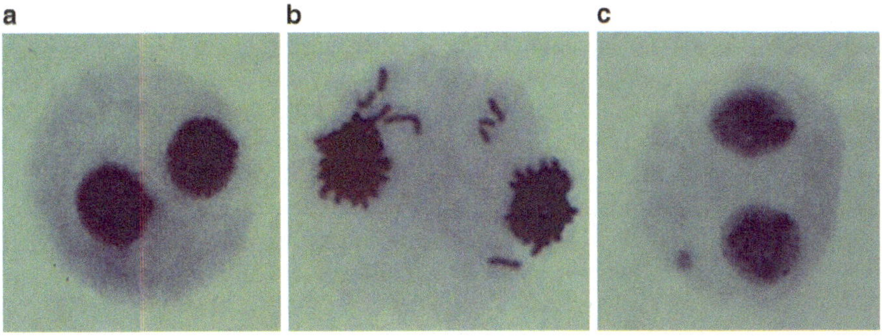

Fig. 9.6 The binucleated (BN) cells after treatment plant extracts of *Teucrium* species against MMC-induced MN (**a**) without MN (normal BN cell), (**b**) BN cell with MNs in development; (**c**) BN cell with an MN

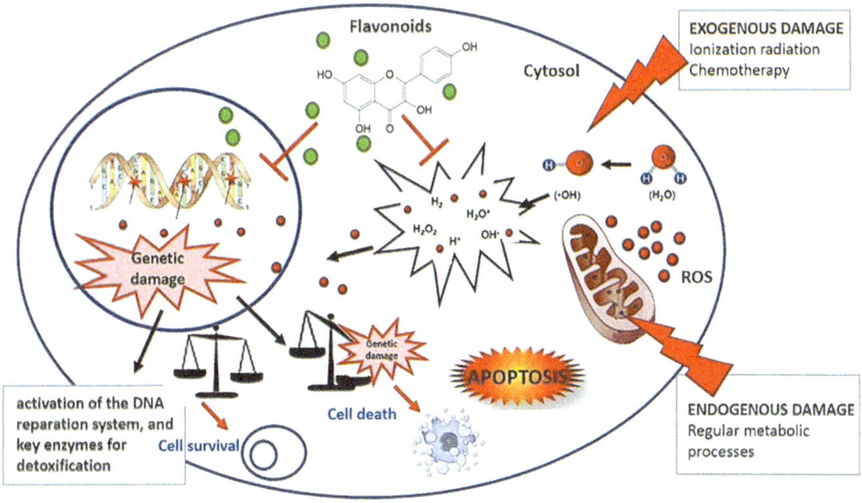

Fig. 9.7 Protective effect of plant secondary metabolites in human cell

and redox properties for free radical scavenging and can chelate transition metal ions thus preventing the formation of ROS (Montoro et al. 2004).

On the other hand, numerous studies have demonstrated that plant-derived natural compounds exhibit protective activities against genotoxicity caused by oxidative stress (Glei and Poll-Zobel 2006; Plazar et al. 2008; Behravan et al. 2011; Mazumdar et al. 2011). The primarily, the antimutagenic effect can be attributed to flavonoids (Calomme et al. 1996), tannins (Baratto et al. 2003) and total polyphenols (Ben Ammar et al. 2008), which cause by radical scavenging effects (Edenharder and Grunhage 2003) (Fig. 9.7).

The phenolic compounds and many flavonoids were reported to have the capacity to scavenge mutagens or free radicals and the ability to neutralize reactive

oxygen species (Rice-Evans et al. 1997; Saxena et al. 2012). Flavonoids and phenolic acids are known to be beneficial to human health and disease prevention thanks to their strong antioxidant activity, capacity to scavenge free radicals. Numerous studies demonstrated that plant extracts possess significant antioxidative properties (Sghaier et al. 2011), as well as that, exist a positive correlation between antioxidative and antimutagenic properties of tested plant extracts. Also, exist a positive correlation between total polyphenolic content, total flavonoid content, antimutagenic, antigenotoxic, and antioxidant activities (Ben Sghaier et al. 2011a, b).

9.6 Protective Effect of Phenolic Acids

The protective effects of the methanolic extracts of *Teucrium* species can be explained by the presence of phytochemical compounds. Phytochemical analysis showed that the extracts were rich in polyphenolics. From phenolic acids presence chlorogenic, rosmarinic, vanillic, caffeic and ferulic acid, and of flavonoids: rutin, catechin, apigenin, quercetin. In Table 9.6 presence review of some literature data of the genoprotective effect of phenolic acids, using application different genotoxicology tests and on different cell types.

The results Cinkilic et al. (2013) showed that chlorogenic and quinic acids had a protective effect in vitro radiation-induced DNA damage in PBLs from two healthy human donors, using the comet assay. The comet assay is a well-known test system used to analyzed DNA damaging agents in vitro and in vivo (Singh et al. 1988). Chlorogenic acid is an important plant polyphenol that is widely distributed in the leaves and fruits of dicotyledonous plants, such as coffee beans. In the investigations from Milošević-Djordjević et al. (2013, 2018) chlorogenic acid most presented in *Teucrium chamaedrys* and *T. polium* methanolic extracts.

Most phytochemicals act by inhibiting the binding of the free radicals to DNA (Lahouel and Fillastre 2004) although the antioxidant properties of chlorogenic acid are primarily attributed to its capacity to donate hydrogen atoms of the phenolic ring to free radicals (Farah et al. 2008). In a few studies, antioxidant properties of chlorogenic acid were shown using superoxide anion and 2,2-diphenyl-1- picrylhydrazyl (DPPH) radical scavenging method (Chen and Ho 1997; Sato et al. 2011). Chlorogenic and quinic acids have a vicinal hydroxyl group on an aromatic residue, which might be the cause of antimutagenic, anticarcinogenic and antioxidant activities in vitro, allowing the compound to scavenge ROS (Sato et al. 2011). Consequently, the decrease in the DNA damage in human lymphocytes might be attributed to the proposed antioxidant mechanisms of quinic and chlorogenic acid.

Furtado et al. (2008) showed that treatments with different concentrations of rosmarinic acid combined with doxorubicin a significantly reduced the frequency of micronuclei compared to Swiss mice treated with doxorubicin only. Although the mechanism the protective effect of rosmarinic acid are not completely understood, the antioxidant activity of rosmarinic acid might explain its effect against doxorubicin mutagenicity. In the next research, authors Furtado et al. (2010) demonstrated

Table 9.6 Literature data of protective effect of phenolic acids using different tests and cell types

The phenolyc acid	Effect	Genotoxicology test	Model system	References
Chlorogenic acid 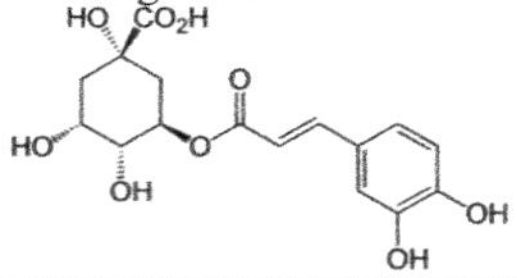	Radioprotective, antimutagenic (decreased the DNA damage induced by X-ray irradiation)	Comet assay	PBLs	Cinkilic et al. (2013)
Rosmarinic acid 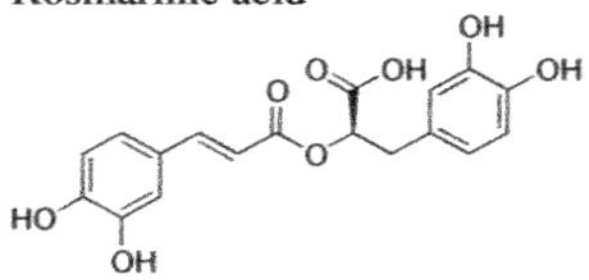	Antimutagenic, protective effect chemotherapeutic agent doxorubicin	MN test	PCEs of Swiss mice	Furtado et al. (2008)
	Antimutagenic, protective effect chemotherapeutic agent doxorubicin	MN assay, Comet test	V79 cells (CHO cells)	Furtado et al. (2010)
	Protective effect against ethanol-induced DNA damage in mice	MN assay, Comet test	Mice bone marrow	De Oliveira et al. (2012)
Vanilic acid 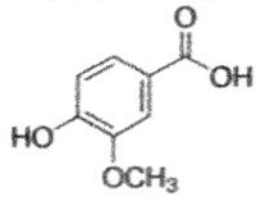	Antimutagenic, (decreased DNA damage induced H_2O_2)	CBMN assay, Comet test	PBLs	Taner et al. (2016)
	Chemoprotective, antimutagenic against MMC induced genomic damage in human lymphocytes at low dose	CBMN assay, Comet test	PBLs	Erdem et al. (2012)
	Antimutagenic	Ames test	*Salmonella typhimurium*	Shaughnesy et al. (2001)
	Chemoprotective, antimutagenic against MMC	MN test	Mice bone marrow cells	Inouye et al. (1988)
	Reduced DNA damage induced EMS and H_2O_2	CA	V79 cells (CHO cells)	Tamai et al. (1992)

The phenolyc acid	Effect	Genotoxicology test	Model system	References
Caffeic acid 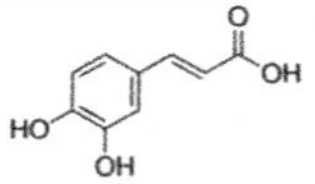	Radioprotective, antimutagenic against radiation induced DNA damage	CBMN assay, CA, Comet test	PBLs	Devipriya et al. (2008a)
Ferulic acid 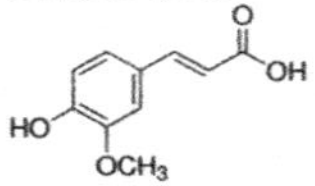	Prevention of radiation-induced MN and dicentric aberrations in human lymphocytes	CA, CBMN assay	PBLs	Prasad et al. (2006)

H_2O_2 hydrogen peroxide, *EMS* ethyl methanesulfonate, *PBLs* peripheral blood lymphocytes, *PCEs* polychromatic erythrocytes, *CA* chromosome aberrations, *CHO cells* Chinese hamster ovary cells

that rosmarinic acid showed the same, antimutagenic effect in V79 cells. For the evaluate they were used the MN assay and comet test. The results of De Oliveira et al. (2012) showed that rosmarinic acid decreases the levels of DNA damage induced by ethanol probably due to it exerts antioxidant activity.

Numerous in vitro and in vivo studies have shown that phenolic acids exhibit powerful effects on the biological response by scavenging free radical. Vanillic acid found in vanilla is natural plant phenolic acid which are secondary aromatic plant products that are presumed to possess many physiological and pharmacological functions. According to Milošević-Djordjević et al. (2018) vanillic acid is the most presented in *Teucrium scordium* methanolic extract. Vanillic acid is an oxidized form of vanillin and exhibits more free radical scavenging activity then vanillin (Sasaki et al. 1990). Taner et al. (2016) showed that vanillic acid indicated antioxidant activity, as well as did not have cytotoxic and genotoxic effects alone at the tested concentrations in compared with the controls. Some author demonstrated that vanillic acid decreased DNA damage induced by hydrogen peroxide in PBLs. Vanillic acid shows an antigenotoxic effect on MMC-induced genetic damage in healthy PBLs measured by the CBMN and alkaline comet assays. Vanilic acid could prevent oxidative damage to DNA and chromosomes when used at an appropriately low dose. Despite the antioxidant and antigenotoxic capabilities of vanillic acid, care should be taken with regard to its concentration in the daily diet (Erdem et al. 2012). Earlier studies also confirmed the protective role of vanillic acid against MMC, EMS and hydrogen peroxide in different animal models (Inouye et al. 1988; Tamai et al. 1992).

Caffeic acid (3,4-dihydroxicinnamic acid) is a naturally occurring polyphenol widely distributed in plant-derived materials such as fructus, vegetables, tea, wine, olive oil, coffee bean (Challis and Bartlett 1975). Of all polyphenolic acids, coffee acid was present in the largest amount in *Teucrium flavum* and *T. arduini* methanolic extracts (Grujičić et al. 2020). The investigations (Prasad et al. 2006; Devipriya et al. 2008a) were showed that this polyphenol compound has protective properties in PBLs against radiation-induced genetic damage (MN, dicentric aberration, comet attributes).

9.7 Protective Effect of Flavonoides

In recent years, there has been a growing academic and industrial interest in the health benefits of flavonoids (Chebil et al. 2007; Resende et al. 2012; Kozłowska and Szostak-Węgierek 2014). In Table 9.7 gives an overview of some literature data of the genoprotective effect of flavonoids, using application different genotoxicology tests (CBMN assay, CA, Comet test, Ames test) on different cell types (PBLs, polychromatic erythrocytes- PCEs, V79 cell, *Salmonella typhimurium*).

Rutin is most abundant flavonoid in *Teucrium chamaedrys* and *T. montanum* methanolic extracts (Stanković 2012), while quercetin mostly abundant in *Teucrium arduini* and *T. flavum* methanolic extracts. According to our knowledge, there are

Table 9.7 Literature data of protective effect of secondary metabolites, flavonoides using different tests and cell types

The flavonoides	Effect	Genotoxicology test	Model system	References
Rutin 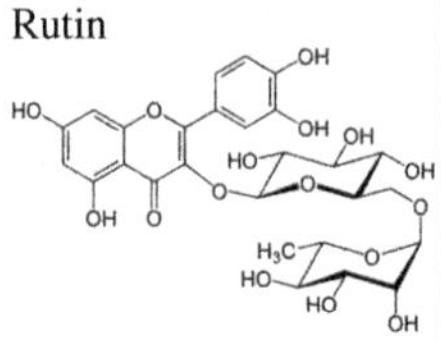	Radioprotective, antimutagenic against radiation induced damage	CBMN assay	PBLs	Patil et al. (2017)
	Radioprotective, antimutagenic against radiation induced damage	Comet assay	PBLs	
	Antimutagenic, (decreased DNA damage induced H_2O_2)	CA	PBLs	Boligon et al. (2012)
	Chemoprotective, antimutagenic against MMC	Comet assay	PBLs	Undeğer et al. (2004)
Catchin 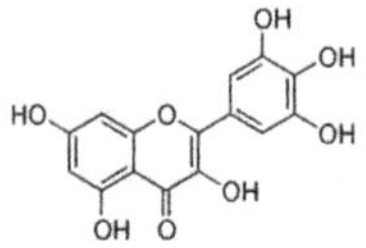	Antigenotoxic potential catechin hydrate against cadmium toxicity	DNA fragmentation assay	PBLs	Alshatwi et al. (2014)
Apigenin 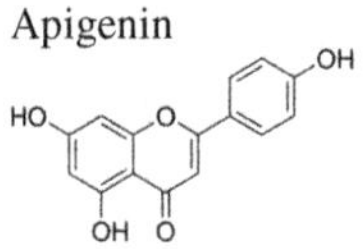	Chemoprotective, antimutagenic against ethinylestradiol	CA, SCEs	PBLs	Siddique et al. (2010)
	Antimutagenic, (decreased DNA damage induced H_2O_2)	CBMN assay SCE	PBLs	Siddique and Afzal (2009a)
	Chemoprotective, antimutagenic against MMC	MN test	Mice bone marrow cells	Siddique and Afzal (2009b)
	Chemoprotective, antimutagenic against MMC and cyclophosphamide (anti-cancerous drug)	CA, SCE	Mice bone marrow cells	Siddique et al. (2008)
	Radioprotective, except in highest concentration	CBMN assay	PBLs	Rithidech et al. (2005)

(continued)

Table 9.7 (continued)

The flavonoides	Effect	Genotoxicology test	Model system	References
Quercetin 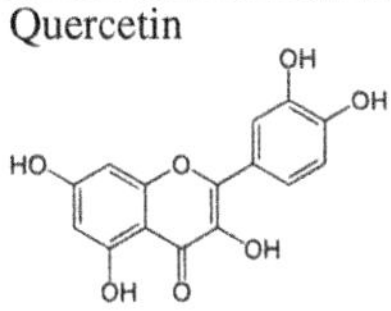	Radioprotective, antimutagenic	CBMN assay	PBLs	Patil et al. (2017)
	Radioprotective, antimutagenic against radiation induced DNA damage	Comet assay	PBLs	
	Chemoprotective, antimutagenic against MMC	MN test	Mice bone marrow cells	Mazumdar et al. (2011)
	Chemoprotective, antimutagenic against MMC and idarubicin induced DNA strand breaks	DNA strand breakage assay	pBR322 plasmid DNA	Çelik and Arinç (2010)
	Radioprotective, antimutagenic against radiation induced DNA damage	CBMN assay Comet test	PBLs *in vitro*	Devipriya et al. (2008b)
	Chemoprotective, antimutagenic, against t-butylhydroperoxide	Ames	*Salmonella typhimurium* TA102	Geetha et al. (2005)
	Chemoprotective, antimutagenic against MMC	Comet assay	PBLs	Undeğer et al. (2004)

CA chromosomal aberrations, *SCEs* sister chromatid exchange

few studies which determined the antigenotoxic and radioprotective effects of rutin and quercetin on PBLs (Undeğer et al. 2004; Boligon et al. 2012). Study Patil et al. (2017) demonstrates the protective effect of rutin and quercetin against radiation-induced DNA damage in PBLs. The authors explained that these results may be partly attributed to scavenging of radiation-induced free radicals and also by the inhibition of radiation-induced oxidative stress.

Methanolic extract of *Teucrium polium* and *T. scordium* are characterized by a high presence of flavonoids catechin (Milošević-Djordjević et al. 2018). Catechin is a polyphenolic flavonoid that has been isolated from a variety of natural source, including tea leaves, grape seeds, and the wood and bark of trees such as acacia and mahogany. Catechin is a more potent antioxidant then ascorbate or tocopherol in certain in vitro assays of lipid peroxidation (Alshatwi 2010). Catechins (flavan-3-ol) are typical natural phenols. Catechin hydrate, also known as taxofilin (Makena et al. 2009), a type of flavonoid that can be found mostly in the Siberian larch (*Larix sibirica*) and silymar in extract from milk thistle seeds (Hoffman 2003). Some studies indicate the antigenotoxic potential of catechin hydrate against cadmium toxicity in PBLs (Alshatwi et al. 2014).

The second most abundant flavonoid in the extract of *Teucrium polium* is apigenin. Some studies have shown that apigenin is potent in reducing the genotoxic

damage inducing by MMC, thereby reducing the chances of developing secondary tumors during the therapy (Siddique et al. 2008). Similarly, Siddique and Afzal (2009a, b) indicated a protective role of apigenin against the genotoxicity of H_2O_2 and MMC both on PBLs and mouse bone marrow cells. Also, Siddique et al. (2010) showed a clear reduction in genotoxic damage in PBLs induced by ethinylestradiol observed with increasing doses of apigenin, suggesting a protective role for apigenin during ethynylestradiol therapy. On the other hand, apigenin had protective effects on radiation-induced chromosome damage in human lymphocytes (Rithidech et al. 2005; Sharma 2013). Quercetin, a flavonol group of plant flavonoid, has generated immense interest because of its potential antioxidant, antiproliferative, chemoprotective, anti-inflammatory and gene expression modulating properties (Mazumdar et al. 2011). Quercetin as the most common flavonoid is found in the methanolic extracts of *Teucrium arduini* and *T. flavum* (Grujičić et al. 2020).

The findings Mazumdar et al. (2011) suggest that quercetin may result in genomic instability in the tested dose range and a significant reduction in MMC-induced genotoxicity in the highest dose tested. Pre-treatment with quercetin significantly reduced the frequency of MMC-induced MN as well as CA. These effects of quercetin should be taken into account when assessing the potential use of quercetin as a therapeutic agent. The same authors point out that further studies are required to elucidate the possible interaction of quercetin with DNA as well as with other DNA damaging agents like MMC in vivo. On the other hand, Çelik and Arinç (2010) concluded that quercetin is highly effective in reducing the DNA damage caused by the antitumor agents, idarubicin, and MMC, following bioactivation by P450 reductase. Devipriya et al. (2008b) showed that quercetin significantly decreased radiation-induced genetic damage and biochemical changes in PBLs. The important mechanisms offered by quercetin in protecting lymphocytes from radiation is free radical scavenging property, DNA binding potential, increasing the antioxidant status and antiperoxidative potential.

Although some studies have shown that quercetin in Ames test is regarded as mutagenic (Gaspar et al. 1993), later in vitro studies indicate that quercetin is protective against genotoxicants and regarded as antimutagenic, as well as that in vitro free radical scavenging activity of quercetin correlated well with the antimutagenic activity (Geetha et al. 2005). In 1999, the International Agency for Research on Cancer (IARC) concluded that quercetin is not classified carcinogenic to humans. In the U.S. and Europe, supplements of quercetin are commercially available, and beneficial effects of quercetin supplements were reported in clinical trials (Okamoto 2005).

9.8 Protective Effect of Secondary Metabolites of *Teucrium* Species

However, although numerous studies have confirmed the antimutagenic effect of phenolic components of secondary metabolites from herbal extracts, it should be noted that the extracts are a mixture of different compounds. Since a plant extract is the mixture of natural compounds, their different biological activities (antioxidant, antimicrobial, antigenotoxic and anticancer) are not only the result of different activities of individual components and may also be the result of their interactions, possibly producing different effects on the overall activity of extracts. Therefore, numerous authors suggested that the activity of the extracts may be the result of synergistic effects of several compounds (Romero-Jimenez et al. 2005; Daradka and Alshibly 2012; Kosanić et al. 2016; Milošević-Djordjević et al. 2013, 2018). On the other hand, it is well documented that often a mixture of plant polyphenols present in nature, by their synergistic collaboration acts better than its individual components separately (Rice-Evans et al. 1997; Havsteen 2002). As well it is quite possible that uptake of complex plant-derived mixture modulates the genotoxicity of one flavonoid by the antigenotoxic capacity of another flavonoid and vice-versa (Di Virgilio et al. 2004).

Table 9.8 gives an overview of some literature data of genoprotective effect of different *Teucrium* species extracts using application different genotoxicology tests and on different cell types. According to our knowledge, there are very few scientific studies who investigated of genotoxic properties of the extracts genus *Teucrium*. Except of the investigate by Milošević-Djordjević et al. (2018), genotoxic effect of *Teucrium polium* extract, also, has been investigated by Khader et al. (2007, 2010). The extract of *Teucrium polium*, had protective, antimutagenic effect against MMC-induced genetic damage, and this extract had no effect on proliferation PBLs in vitro. These results are consistent with previous results from Milošević-Djordjević et al. (2013) which showed that both methanolic extracts of *Teucrium chamaedrys* and *T. montanum* administered alone in PBLs in vitro, did not significantly affect the NDI in the concentrations tested.

Numerous studies have shown that extracts of *Teucrium polium* have significant effects on cell proliferation in different cell types. Khader et al. (2007) demonstrated that aqueous extract of *Teucrium polium* significantly reduced the mitotic indices in primary rat hepatocyte cultures after simultaneous treatment with mutagen *N*-methyl-*N*-nitro-*N*-nitrosoguanidine (MNNG), indicating that the extract enhances MNNG-mediated cytotoxicity. In a later study, Khader et al. (2010) showed that the ethanolic extract of *Teucrium polium* had the antimutagenic effect on primary rat hepatocyte cultures before, during and after the treatment with MNNG. Also, investigations demonstrated that the extract of *Teucrium polium* potentiates the cytotoxic and apoptotic effects of anticancer drugs (vincristine, vinblastine, and doxorubicin) against a panel of cancerous cell lines (Rajabalian 2008).

Similar results were obtained on another plant species as a *Teucrium ramosissimum*. Sghaier et al. (2016) showed that methanolic extract of *Teucrium*

Table 9.8 Protective effect of different extracts of *Teucrium* species

Teucrium species	Extract	Protective effects	Test/model system	References
T. chamaedrys	Methanolic	Proapoptotic against MMC	MN assay/PBLs in vitro	Milošević-Djordjević et al. (2013)
T. montanum	Methanolic	Proapoptotic against MMC	MN assay/PBLs in vitro	
T. polium	Methanolic	Antimutagenic against MMC (free radical scavenging activities)	MN assay/PBLs in vitro	Milošević-Djordjević et al. (2018)
	Aqueous	Reduced the mitotic index	Primary rat hepatocyte in vitro	Khader et al. (2007)
	Ethanolic	Antimutagenic (free radical scavenging activities), during and after treatment with MNNG	Primary rat hepatocyte in vitro	Khader et al. (2010)
T. scordium	Methanolic	Proapoptotic (at highest concentration)	MN assay/PBLs in vitro after treatment with MMC	Milošević-Djordjević et al. (2018)
T. arduini	Methanolic	Proapoptotic	MN assay/PBLs in vitro after treatment with MMC	Grujičić et al. (2020)
T. flavum	Methanolic	Proapoptotic	MN assay/PBLs in vitro after treatment with MMC	Grujičić et al. (2020)
T. ramosissimum	Methanolic	Antimutagenic, against hydrogen peroxide (H_2O_2)	Comet assay/human lymphocytes	Sghaier et al. (2016)
	Methanolic, ethanolic, chloroform	Antimutagenic, against AFB1, nitrofurantoin	Ames test/*Salmonella typhimurium*	Ben Sghaier et al. (2011a)
	Methanolic ethanolic chloroform	Antimutagenic, against AFB1, nitrofurantoin	SOS chromotest/*Escherichia coli* PQ37	Ben Sghaier et al. (2011b)
	Total flavonoids, ethyl acetate, petroleum ether	Antimutagenic, against SA, AFB1, B[a]P, NOPD	Ames test/*Salmonella typhimurium*	Sghaier et al. (2011)
	Essential oil	Antimutagenic, against SA, AFB1, B[a]P, NOPD	Ames test/*Salmonella typhimurium*	Ben Sghaier et al. (2010)

MNNG mutagen *N*-methyl-*N*-nitro-*N*-nitrosoguanidine, *SA* sodium azide, *AFB1* aflatoxin B 1, *B[a]P* benzo[a] pyrene, *NOPD* 4-nitro-*o*-phenylenediamine

ramosissimum had antigenotoxic properties against genotoxicity induced by the direct-acting mutagen, hydrogen peroxide (H_2O_2), using comet assay. The same authors have opinions that tested *Teucrium ramosissimum* extract possess potent antioxidant, antiproliferative and antigenotoxic activities, which could be derived from compounds such as flavonoids and polyphenols. The studies Ben Sghaier et al. (2011a, b) has demonstrated that *Teucrium ramosissimum* extracts (methanolic, ethanolic, and chloroform) possess antimutagenic activity against all the tested genotoxicants (aflatoxin B1, benzo[a]pyrene, NOPD -4-nitro-o-phenylenediamine and sodium azide) in the *Salmonella* assay systems and SOS chromotest in the *Escherichia coli*. Antimutagenic activity of the tested extracts may be ascribed to flavonoids, tannins and total polyphenols (Calomme et al. 1996; Baratto et al. 2003; Ben Ammar et al. 2008), at least partly to their antioxidative properties but we cannot exclude other additionally mechanisms. *Teucrium ramosissimum* extracts could give rise to antimicrobial, anti-inflammatory agents and could be promising candidates for further studies designed to obtain more evidence on their components with potential chemopreventive activity. The results Sghaier et al. (2011) demonstrated that total oligomers flavonoids (TOF), ethyl acetate (EA) and petroleum ether (PE) extracts from aerial parts of *Teucrium ramosissimum* was assessed using Ames *Salmonella* tester strains TA98, TA100 and TA1535 with and without metabolic activation (S9). In addition, all extracts showed important free radical scavenging activity toward the radical's DPPH and ABTS except the PE extract. Same authors suspected an eventual correlation between antiradical and antimutagenic effects of *Teucrium ramosissimum* extracts. Antioxidant potential expressed by the different extracts may provide a common mechanism for inhibiting the genotoxicity of both direct and indirect tested mutagens. In fact, phenolic compounds are ubiquitous in fruit, vegetables, and nuts, several of which have been reported to be inhibitors of chemical carcinogenesis (Fiala et al. 1985; Stich 1991). The study Ben Sghaier et al. (2010) showed that *Teucrium ramosissimum* essential oil possesses antimutagenic effects against SA, AFB1, B[a]P, and NOPD. Finally, the antimutagenic and antioxidant properties of the essential oil from *Teucrium ramosissimum* and its constituents could make it a promising candidate for future applications in chemoprevention and therapy. Antigenotoxic plant extracts can counter or prevent the adverse effects caused by DNA-damaging chemicals. Indeed, it has been shown that plant-derived polyphenolic compounds can exert an antigenotoxic activity (Zhao et al. 1999; Bhatia et al. 2001).

The previous researches (Milošević-Djordjević et al. 2013, 2018; Grujičić et al. 2020) have shown that investigated *Teucrium* species possess the protective properties against MMC-induced genetic damage in PBLs. In our opinion, two genoprotective effects are possible: (1) antimutagenic, antioxidant effect, and (2) proapoptotic effect. The type of genoprotective effect depends on the influence of the extract on cell kinetics in the culture of PBLs. We think that the extracts do not affect cellular kinetics (NDI is not changed), but they reduce MMC-induced MN frequency. Those extracts have an antimutagenic, antioxidant effect that manifest as free radical scavenging activities. In this way, the plant extract relieves the harmful mutagenic effect of antitumor drug on healthy human cells (PBLs). If the extract influence on PBLs

is such that NDI is decreased and that the cell kinetics slows down, with, of course, additionally reduction of MMC-induced MN frequency, this extract shows a genoprotective, proapoptotic effect. This means that apoptotic mechanisms are activated for cells that have suffered significant damage to the genetic material and such cells are eliminated.

Presented observations indicate that these genoprotective properties can belong to antioxidative type, as in methanolic extracts of *Teucrium polium* and *T. ramosissimum*, or proapoptotic type, as noted in all other tested extracts *Teucrium* species (*Teucrium chamaedrys, T. montanum, T. scordium, T. arduini* and *T. flavum*). That means that except *Teucrium polium* and *T. ramosissimum*, all the others tested extracts have proapoptotic properties that can be responsible for their protective effects against MMC-induced genetic damage. These finds are in agreement with reports of other authors who pointed out that antimutagenic properties of plant extracts depend on the phytochemical composition and the content of secondary metabolites. The different genoprotective effects of tested *Teucrium* extracts on MMC-induced mutagenicity can be explained with differences in the phytochemical content of the polyphenols present. Phytochemical analyses of methanolic extract of *Teucrium* species showed that the extracts were rich in polyphenolic compounds (Stanković 2012), and could be considered responsible for antimutagenic properties of the plant. In support of this, the fact is that many studies on the different cell lines (normal and cancer) suggest that polyphenols, separately and in combination with chemotherapeutic agents induced apoptosis (Rajabalian 2008; Ćurčić et al. 2012). It is an important mechanism for the maintenance of cellular homeostasis in both in chemoprevention and chemotherapy of cancer. Studies on cultured cells have shown that the human PBLs is a sensitive indicator both in vitro and in vivo induced chromosome damage (Lerda et al. 2005). In fact, the MN assay provides a measure of both chromosome breakage and chromosome loss or nondisjunction in clastogenic and aneugenic events, respectively (Karaman et al. 2009). An increase of the frequency of chromosomal aberrations in PBLs is in associated with an increased overall risk of cancer (Bonassi et al. 2000; Natarajan 2002; Rossner et al. 2005; Boffetta et al. 2007).

The antimutagenic activity of compounds derived from plants may be due to a variety of mechanisms, such as inhibition of genotoxic effect, signal transduction modulation, antioxidant activity, and scavenging of free radicals (Yen and Chen 1995; Marnewick et al. 2000; Fragoso et al. 2008). The recent studies have suggested the use of medicinal plants as antimutagenic agents in the prevention of genotoxic effects of different chemotherapeutic agents (Verschaeve et al. 2004; Sibanda and Okoh 2007; Kumar et al. 2012; Vlastos et al. 2013). Since the use of chemotherapeutic drugs can cause the emergence of secondary carcinoma (Vlastos et al. 2013), which results from genetic damage in healthy cells, the reduction of the risk of this phenomenon is of great importance (Unal et al. 2013). Natural antioxidant compounds are important agents for potential use in chemoprevention and/or during or after cancer chemotherapy, showed extensive studies, both in vitro and in vivo (Çelik and Arinç 2010; Veeramuthu et al. 2017; Zyad et al. 2018).

The recent research has shown the dual activity of flavonoids presenting both antioxidative and prooxidant activities revealed that the existence of a balance between these two features could be important to the development of adequate therapeutic strategies (Baldim et al. 2017). Additionally, presented results have shown that it is necessary to be careful when using these extracts because plants except *Teucrium chamaedrys* have shown a genotoxic effect in high concentrations, while they in combination with MMC in the same concentrations, all plants, had a protective effect.

9.9 Conclusions

An overview of the literature data suggests that *Teucrium* extracts have a significant protective effect against various genotoxic agents (MMC, AFB1, B[a]P, NOPD), alleviating their effect, either as potent antioxidants or introducing cells into the apoptosis process. Presented results showed that secondary metabolites, as well as extracts from *Teucrium* species, have a genoprotective effect on different types of cells. The different genoprotective effects of tested *Teucrium* extracts can be explained with differences in the phytochemical content of the polyphenols present. Generally, presented results clearly showed that tested extracts *Teucrium* species, as well as secondary metabolites, had a protective effect on the genotoxicity of the different agents in different type cells in vitro and, therefore, they can be a potential source of new pharmaceutics that will have a significant protective effect on induced genetic damage by chemotherapeutics in healthy cells.

Acknowledgments This work was supported by the Ministry of Education, Science and Technological Development of the Republic of Serbia (Grants III41010 and ON175103). The authors are grateful to the volunteers for providing blood samples.

References

Abou-Eisha A, Marcos R, Creus A (2004) Genotoxicity studies on the antimicrobial drug sulfamethoxazole in cultured human lymphocytes. Mutat Res 14:51–56

Abu-Irmaileh BE, Afifi FU (2003) Herbal medicine in Jordan with special emphasis on commonly used herbs. J Ethnopharmacol 89:193–197

Akyıl D, Özkara A, Erdoğmuş SF, Eren Y, Konuk M, Sağlam E (2015) Micronucleus assay in human lymphocytes after exposure to alloxydim sodium herbicide in vitro. Cytotechnology 67:1059–1066

Albertini A, Bolognese A, Guerra M, Lavecchia A, Macciantelli D, Marcaccio M, Novellio E, Paolucci F (2003) Antitumor agents 4. Characterization of free radicals produced during reduction of the antitumor drug 5H-pyridophenoxazin-5-one: an EPR study. Biochemistry 42:11924–11931

Alshatwi AA (2010) Catechin hydrate suppresses MCF-7 proliferation through TP53/Caspase-mediated apoptosis. J Exp Clin Cancer Res 17:167. https://doi.org/10.1186/1756-9966-29-167

Alshatwi AA, Hasan TN, Alqahtani AM, Syed NS, Shafi GS, Al-Assaf AH, Al-Khalifa AS (2014) Delineating the anti-cytotoxic and anti-genotoxic potentials of catehin hydrate against cadmium toxicity in human peripheral blood lymphocytes. Environ Toxicol Pharmacol 38:653–662

Ananthi R, Chandra N, Santhiya ST, Ramesh A (2010) Genotoxic and antigenotoxic effects of *Hemidesmus indicus* R.Br. root extract in cultured lymphocytes. J Ethnopharmacol 3:558–560

Andrade AF, Alves JM, Corrêa MB, Cunha WR, Veneziani RC, Tavares DC (2016) In vitro cytotoxicity, genotoxicity and antigenotoxicity assessment of *Solanum lycocarpum* hydroalcoholic extract. Pharm Biol 54:2786–2790

Andreassi MG, Botto N (2003) Genetics instability, DNA damage and atherosclerosis. Cell Cycle 2:224–227

Bahramikia S, Yazdanparast R (2012) Phytochemistry and medicinal properties of *Teucrium polium* L. (Lamiaceae). Phytother Res 26:1581–1593

Baldim JL, de Alcântara BGV, Domingos ODS, Soares MG, Caldas IS, Novaes RD, Oliveira TB, Lago JHG, Chagas-Paula DA (2017) The correlation between chemical structures and antioxidant, prooxidant, and antitrypanosomatid properties of flavonoids. Oxidative Med Cell Longev 2017:3789856. https://doi.org/10.1155/2017/3789856

Balmekki N, Bendimerad N, Bekhechim C, Fernandez X (2013) Chemical analysis and antimicrobial activity of *Teucrium polium* L. essential oil from Western Algeria. J Med Plant Res 7:897–902

Baratto MC, Tattini M, Galardi C, Pinelli P, Romani A, Visioli F, Basosi R, Pongi R (2003) Antioxidant activity of galloyl quinic derivates isolated from *P. lentiscus* leaves. Free Radic Res 37:405–412

Barrachina MD, Bello R, Martinez-Cuesta MA, Esplugues J, Primo-Yúfera E (1995) Anti-inflammatory activity and effects on isolated smooth muscle of extract from different *Teucrium* species. Phytother Res 9:368–371

Batra P, Sharma AK (2013) Anti-cancer potential of flavonoids: recent trends and future perspectives. Biotech 3:439–459

Behravan J, Mosafa F, Soudmand N, Taghiabadi E, Razavi BM, Karimi G (2011) Protective effects of aqueous and ethanolic extracts of *Portulaca oleracea* L. aerial parts on H_2O_2-induced DNA damage in lymphocytes by comet assay. J Acupunct Meridian Stud 4:193–197

Bellomaria B, Arnold N, Valentini G (1998) Essential oil of *Teucrium flavum* subsp. *hellenicum* from Greece. J Essent Oil Res 10:131–133

Ben Ammar R, Kilani S, Bouhlel I, Ezzi L, Skandrani I, Boubaker J, Ben Sghaier M, Naffeti A, Mahmoud A, Chekir-Ghedira L, Ghedira K (2008) Antiproliferative antioxidant and antimutagenic activities of flavonoid-enriched extracts from (Tunisian) *Rhamnus alaternus* L. Combination with the phytochemical composition. Drug Chem Toxicol 31:61–80

Ben Sghaier M, Boubaker J, Neffati A, Limem I, Skandrani I, Bhouri W, Bouhlel I, Kilani S, Chekir-Ghedira L, Ghedira K (2010) Antimutagenic and antioxidant potentials of *Teucrium ramosissimum* essential oil. Chem Biodivers 7:1754–1763

Ben Sghaier M, Bhouri W, Neffati A, Boubaker J, Skandrani I, Bouhlel I, Kilani S, Chekir- Ghedira L, Ghedira K (2011a) Chemical investigation of different crude extracts from *Teucrium ramosissimum* leaves: correlation with their antigenotoxic and antioxidant properties. Food Chem Toxicol 49:191–201

Ben Sghaier M, Boubaker J, Skandrani I, Bouhlel I, Limem I, Ghedira K, Chekir-Ghedira L (2011b) Antimutagenic, antigenotoxic and antioxidant activities of phenolic enriched extracts from *Teucrium ramosissimum*: combination with their phytochemical composition. Environ Toxicol Pharmacol 31:220–232

Benedec D, Vlase L, Oniga I, Mot AC, Damian G, Hanganu D, Duma M, Silaghi-Dumitrescu R (2013) Polyphenolic composition, antioxidant and antibacterial activities for two Romanian subspecies of *Achillea distans* Waldst. et Kit. ex Willd. Molecules 18:8725–8739

Bhat SH, Azmi AS, Hadi SM (2007) Prooxidant DNA breakage induced by caffeic acid in human peripheral lymphocytes: involvement of endogenous copper and a putative mechanism for anticancer properties. Toxicol Appl Pharmacol 218:249–255

Bhatia N, Agarwal C, Agarwal R (2001) Differential responses of skin cancer-chemopreventive agents silibinin, quercetin, and epigallocatechin 3-gallate on mitogenic signaling and cell cycle regulators in human epidermoid carcinoma A431 cells. Nutr Cancer 39:292–299

Bhattachar S (2011) Natural antimutagens: a review. Res J Med Plant 5:116–126

Blumenthal M (2000) Herbal medicine: expanded commission and monographs. Integrat Med Commun, Newton

Boffetta P, van der Hel O, Norppa H, Fabianova E, Fucic A, Gundy S, Lazutka J, Cebulska-Wasilewska A, Puskailerova D, Znaor A, Kelecsenyi Z, Kurtinaitis J, Rachtan J, Forni A, Vermeulen R, Bonassi S (2007) Chromosomal aberrations and cancer risk: results of a cohort study from Central Europe. Am J Epidemiol 165:36–43

Boligon AA, Sagrillo MR, Machado LF, de Souza Filho O, Machado MM, da Cruz IB, Athayde ML (2012) Protective effects of extracts and flavonoids isolated from *Scutia buxifolia* Reissek against chromosome damage in human lymphocytes exposed to hydrogen peroxide. Molecules 14:5757–5769

Bonassi S, Hagmar L, Strömberg U, Montagud AH, Tinnerberg H, Forni A, Heikkilä P, Wanders S, Wilhardt P, Hansteen IL, Knudsen LE, Norppa H (2000) Chromosomal aberrations in lymphocytes predict human cancer independently of exposure to carcinogens. European study group on cytogenetic biomarkers and health. Cancer Res 60:1619–1625

Bonassi S, Ugolini D, Kirsch-Volders M, Stromberg U, Vermeulen R, Tucker JD (2005) Human population studies with cytogenetic biomarkers: review of the literature and future prospectives. Environ Mol Mutagen 45:258–270

Bonassi S, Znaor A, Ceppi M, Lando C, Chang WP, Holland N, Kirsch-Volders M, Zeiger E, Ban S, Barale S, Bigatti MP, Bolognessi C, Cabulska-Wasilewska A, Fabianova E, Fucic A, Hagmar L, Joksic G, Martelli A, Migliore L, Mirkova E, Scarfi MR, Zijno A, Norppa H, Fenech M (2007) An increased micronucleus frequency in peripheral blood lymphocytes predicts the risk of cancer in humans. Carcinogenesis 28:625–631

Bonassi S, Coskun E, Ceppi M, Lando C, Bolognesi C, Burgaz S, Holland N, Kirsh-Volders M, Knasmueller S, Zeiger E, Carnesoltas D, Cavallo D, da Silva J, de Andrade VM, Demircigil GC, Odio AD, Donmez-Altuntas H, Gattas G, Giri A, Giri S, Gómez-Meda B, Gómez-Arroyo S, Hadjidekova V, Haveric A, Kamboj M, Kurteshi K, Martino-Roth MG, Montoya RM, Nersesyan A, Pastor-Benito S, Salvadori DMF, Shaposhnikova A, Stopper H, Thomas P, Torres-Bugarín O, Yadav AS, González G, Fenech M (2011) The HUman MicroNucleus project on eXfoLiated buccal cells (HUMNXL): the role of life-style, host factors, occupational exposures, health status, and assay protocol. Mutat Res 728:88–97

Boos G, Stopper H (2000) Genotoxicity of several clinically used topoisomerase II inhibitors. Toxicol Lett 116:7–16

Bravo L (1998) Polyphenols: chemistry, dietary sources, metabolism and nutritional significance. Nutr Rev 56:317–333

Cadet J, Ravanat JL, Buchko GW, Yeo HC, Ames BN (1994) Singlet oxygen DNA damage: chromatographic and mass spectrometric analysis of damage products. Methods Enzymol 234:79–88

Calatayud S, Bello R, Beltran B, Primo-Yúfera E, Esplugues J (1998) Cardiovascular effects of the methanol and dichloromethanol extracts from *Teucrium flavum* L. Phytother Res 12:68–69

Calomme M, Pieters L, Vlietink A, Berghe DV (1996) Inhibition of bacterial mutagenesis flavonoids. Planta Med 92:222–226

Celik TA (2012) Potential genotoxic and cytotoxic effects of plant extracts. In: Bhattacharya A (ed) A compendium of essays on alternative therapy, 1st edn. IntechOpen, Croatia, pp 233–250

Çelik H, Arinç E (2010) Evaluation of the protective effects of quercetin, rutin, resveratrol, naringenin and trolox against idarubicin-induced DNA damage. J Pharm Pharm Sci 13:231–241

Challis BC, Bartlett CD (1975) Possible cocarcinogenic effects of coffee constituents. Nature 254:532–533

Chan K (2003) Some aspects of toxic contaminants in herbal medicines. Chemosphere 52:1361–1371

Chebil L, Anthoni J, Humeau C, Gerardin C, Engasser JM, Ghoul M (2007) Enzymatic acylation of flavonoids: effect of the nature of the substrate, origin of lipase, and operating conditions on conversion yield and regioselectivity. J Agric Food Chem 55:9496–9502
Chen JH, Ho CT (1997) Antioxidant activities of caffeic acid and its related hydroxycinnamic acid compounds. J Agric Food Chem 45:2374–2378
Cinkilic N, Cetintas SK, Zorlu T, Vatan O, Yilmaz D, Cavas T, Tunc S, Ozkan L, Bilaloglu R (2013) Radioprotection by two phenolic compounds: chlorogenic and quinic acid, on X-ray induced DNA damage in human lymphocytes in vitro. Food Chem Toxicol 53:359–363
Coutryman PI, Heddle JA (1976) The product of micronuclei from chromosome aberration in irradiated cultures of human lymphocyte. Mutat Res 41:321–332
Ćurčić MG, Stanković MS, Mrkalić EM, Matović ZD, Banković DD, Cvetković DM, Djačić DS, Marković SD (2012) Antiproliferative and proapoptotic activities of methanolic extracts from *Ligustrum vulgare* L. as an individual treatment and in combination with palladium complex. Int J Mol Sci 13:2521–2543
Czeczot H, Tudek B, Kusztelak J, Szymczyk T, Dobrowolska B, Glinkowska G, Malinowski J, Strzelecka H (1990) Isolation and studies of the mutagenic activity in the Ames test of flavonoids naturally occurring in medical herbs. Mutat Res 240:209–216
Dall'Acqua S, Cervellati R, Loi MC, Innocenti G (2008) Evaluation of in vitro antioxidant properties of some traditional Sardinian medicinal plants: investigation of the high antioxidant capacity of *Rubus ulmifolius*. Food Chem 106:745–749
Damasceno JL, Oliveira PF, Miranda MA, Leandro LF, Acésio NO, Ozelin SD, Bastos JK, Tavares DC (2016) Protective effects of *Solanum cernuum* extract against chromosomal and genomic damage induced by methyl methanesulfonate in Swiss mice. Biomed Pharmacother 83:1111–1115
Daradka HM, Alshibly NMY (2012) Effect of *Artemisia alba* L. extract against ethinylestradiol induced genotoxic damage in cultured human lymphocytes. Afr J Biotechnol 11:15246–15250
Das B, Karuppasamy CV (2009) Spontaneous frequency of micronuclei among the newborns from high level natural radiation areas of Kerala in the southwest coast of India. Int J Radiat Biol 85:272–280
De Flora S, Ferguson LR (2005) Overview of mechanisms of cancer chemopreventive agents. Mutat Res 591:8–15
De Marino S, Festa C, Zollo F, Nini A, Antenucci L, Raimo G, Iorizzi M (2014) Antioxidant activity and chemical components as potential anticancer agents in the olive leaf (*Olea europaea* L. cv Leccino) decoction. Anti Cancer Agents Med Chem 14:1376–1385
De Oliveira NC, Sarmento MS, Nunes EA, Porto CM, Rosa DP, Bona SR, Rodrigues G, Marroni NP, Pereira P, Picada JN, Ferraz AB, Thiesen FV, Da Silva J (2012) Rosmarinic acid as a protective agent against genotoxicity of ethanol in mice. Food Chem Toxicol 50:1208–1214
Decordier I, Papine A, Plas G, Roesems S, Vande Loock K, Moreno-Palomo J, Cemeli E, Anderson D, Fucic A, Marcos R, Soussaline F, Kirsch-Volders M (2009) Automated image analysis of cytokinesis-blocked micronuclei: an adapted protocol and a validated scoring procedure for biomonitoring. Mutagenesis 24:85–93
Devipriya N, Sudheer AR, Menon VP (2008a) Caffeic acid protects human peripheral blood lymphocytes against gamma radiation-induced cellular damage. J Biochem Mol Toxicol 22:175–186
Devipriya N, Sudheer AR, Srinivasan M, Menon VP (2008b) Quercetin ameliorates gamma radiation-induced DNA damage and biochemical changes in human peripheral blood lymphocytes. Mutat Res 654:1–7
Di Virgilio AL, Iwami K, Wätjen W, Kahl R, Degen GH (2004) Genotoxicity of the isoflavones genistein, daidzein and equol in V79 cells. Toxicol Lett 15:151–162
Dimitrijević A, Milošević-Djordjević O, Gujičić D, Arsenijević S (2006) Micronucleus frequency in women with *Chlamidia trachomatis* infection before and after therapy. Mutat Res 608:43–48
Dunkić V, Bezić N, Vuko E (2011) Antiphytoviral activity of essential oil from endemic species *Teucrium arduini* L. Nat Prod Commun 6:1385–1388

Duthie SJ, Johnson W, Dobson VL (1997) The effect of dietary flavonoids on DNA damage (strand breaks and oxidised pyrimidines) and growth in human cells. Mutat Res 390:141–151

Eastmond DA, Turcker JD (1989) Identification of aneuploidy inducing agents using cytokinesis-blocked human lymphocytes and anti-kinetochore antibody. Environ Mol Mutagen 13:34–43

Edenharder R, Grünhage D (2003) Free radical scavenging abilities of flavonoids as mechanism of protection against mutagenicity induced by *tert*-butyl hydroperoxide or cumene hydroperoxide in *Salmonella typhimurium* TA102. Mutat Res 9:1–18

Eghbaliferiz S, Iranshahi M (2016) Prooxidant activity of polyphenols, flavonoids, anthocyanins and carotenoids: updated review of mechanisms and catalyzing metals. Phytother Res 30:1379–1391

El-Zein R, Vral A, Etzel CJ (2011) Cytokinesis blocked micronucleus assay and cancer risk assessment. Mutagenesis 26:101–106

Erdem MG, Cinkilic N, Vatan O, Yilmaz D, Bagdas D, Bilaloglu R (2012) Genotoxic and anti-genotoxic effects of vanillic acid against mitomycin C-induced genomic damage in human lymphocytes in vitro. Asian Pac J Cancer Prev 13:4993–4998

Eroglu HE, Koca I, Yıldırım I (2011) In vitro cytotoxic potential of newly synthesized furo[3,2-c] pyran-4-one derivatives in cultured human lymphocytes. Cytotechnology 63:407–413

Esmaeili S, Hamzeloo-Moghadam M, Ghaffari S, Mosaddegh M (2014) Cytotoxic activity screening of some medicinal plants from south of Iran. Res J Pharmacogn 1:19–25

Everest A, Ozturk E (2005) Focusing on the ethnobotanical uses of plants in Mersin and Adana provinces (Turkey). J Ethnobiol Ethnomed 6:1–6

Farah A, Monteiro M, Donangelo CM, Lafay S (2008) Chlorogenic acids from green coffee extract are highly bioavailable in humans. J Nutr 138:2309–2315

Fauth E, Scherthan H, Zankl H (2000) Chromosome painting reveals specific patterns of chromosome occurrence in MMC and diethyl stilbestrol- induced micronuclei. Mutagenesis 15:459–467

Fenech M (2000) The in vitro micronucleus technique. Mutat Res 455:81–95

Fenech M (2006) Cytokinesis-block micronucleus assay evolves into a "cytome" assay of chromosomal instability, mitotic dysfunction and cell death. Mutat Res 600:58–66

Fenech M (2007) Cytokinesis block micronucleus cytome assay. Nat Protoc 2:1084–1104

Fenech M, Bonassi S (2011) The effect of age, gender, diet and lifestyle on DNA damage measured using micronucleus frequency in human peripheral blood lymphocytes. Mutagenesis 26:43–49

Fenech M, Morley AA (1985) Measurement of micronuclei in lymphocytes. Mutat Res 147:29–36

Fenech M, Chang WP, Kirsch-Volders M, Holland N, Bonassi S, Zeiger E (2003) Human MicronNucleus project. HUMN project: detailed description of the scoring criteria for the cytokinesis-block micronucleus assay using isolated human lymphocyte cultures. Mutat Res 534:65–75

Fenech M, Kirsch-Volders M, Natarajan AT, Surralles J, Crott JW, Parry J, Norppa H, Eastmond DA, Tucker JD, Thomas P (2011) Molecular mechanisms of micronucleus, nucleoplasmic bridge and nuclear bud formation in mammalian and human cells. Mutagenesis 26:125–132

Feretti D, Ceretti E, De Donno A, Moretti M, Carducci A, Bonetta S, Marrese MR, Bonetti A, Covolo L, Bagordo F, Villarini M, Verani M, Schilirò T, Limina RM, Grassi T, Monarca S, Casini B, Carraro E, Zani C, Mazzoleni G, Levaggi R, Gelatti U, the MAPEC_LIFE Study Group (2014) Monitoring air pollution effects on children for supporting public health policy: the protocol of the prospective cohort MAPEC study. BMJ Open 4(9):e006096. https://doi.org/10.1136/bmjopen-2014-006096

Ferguson LR (2001) Role of plant polyphenols in genomic stability. Mutat Res 475:89–111

Ferguson LR, Bronzetti G, De Flora S (2005) Mechanistic approaches to chemoprevention of mutation and cancer. Mutat Res 591:3–7

Fiala ES, Reddy BS, Weisburger JH (1985) Naturally occurring anticarcinogenic substances in foodstuffs. Annu Rev Nutr 5:295–321

Fortin F, Bonvalot Y, Pham TCV, Ouellet N, Ayotte P, Viau C, Lemieux N (2015) Biomarkers of genotoxicity measured in human lymphocytes exposed to benzo[a]pyrene: aneugenic effect, and involvement multiple primary DNA lesions. Integr Pharm Toxicol Genotoxicol 1:21–32

Fragoso V, do Nascimento NC, Moura DJ, e Silva AC, Richter MF, Saffi J, Fett-Neto AG (2008) Antioxidant and antimutagenic properties of the monoterpene indole alkaloid psychollatine and the crude foliar extract of *Psychotria umbellata* Vell. Toxicol in Vitro 22:559–566

Furtado MA, de Almeida LC, Furtado RA, Cunha WR, Tavares DC (2008) Antimutagenicity of rosmarinic acid in Swiss mice evaluated by the micronucleus assay. Mutat Res 657:150–154

Furtado RA, Araújo FRR, Resende FA, Cunha WR, Tavares DC (2010) Protective effect of rosmarinic acido on V79 cells evaluated by the micronucleus and comet assays. J Appl Toxicol 30:254–259

Gajski G, Gerić M, Oreščanin V, Garaj-Vrhovac V (2018) Cytokinesis-block micronucleus cytome assay parameters in peripheral blood lymphocytes of the general population: contribution of age, sex, seasonal variations and lifestyle factors. Ecotoxicol Environ Saf 148:561–570

Gaspar J, Laires A, Monteiro M, Laureano O, Ramos E, Rueff J (1993) Quercetin and the mutagenicity of wines. Mutagenesis 8:51–55

Geetha T, Malhotra V, Chopra K, Kaur IP (2005) Antimutagenic and antioxidant/prooxidant activity of quercetin. Indian J Exp Biol 43:61–67

Ghasemzadeh A, Ghasemzadeh N (2011) Flavonoids and phenolic acids: role and biochemical activity in plants and human. J Med Plant Res 5:6697–6703

Ghosh M, Godderis L (2016) Genotoxicity of ethylene oxide: a review of micronucleus assay results in human population. Mutat Res 770:84–91

Glei M, Pool-Zobel BL (2006) The main catechin of green tea, (-)-epigallocatechin-3-gallate (EGCG), reduces bleomycin-induced DNA damage in human leucocytes. Toxicol in Vitro 20:295–300

Grisolia CK (2002) A comparison between mouse and fish micronucleus test using cyclophosphamide, mitomycin C and various pesticides. Mutat Res 518:145–150

Grujičić D, Milošević-Djordjević O, Arsenijević S, Marinković D (2007) The effect of combined therapy with ritodrine, erithromycin and verapamil on the frequency of micronuclei in peripheral blood lymphocytes of pregnant women. Clin Exp Med 7:11–15

Grujičić D, Milošević-Djordjević O, Arsenijević S, Marinković D (2008) Treatment of pregnant women with a betamimetic and verapamil increases the micronuclei frequency in umbilical cord blood lymphocytes. Tohoku J Exp Med 215:363–371

Grujičić D, Stošić I, Milošević-Djordjević O (2009) The antibiotic erythromycin did not affect micronucleus frequency in human PHA-stimulated lymphocytes. Arch Biol Sci 61:179–185

Grujičić D, Stošić I, Kosanić M, Stanojković T, Ranković B, Milošević-Djordjević O (2014) Evaluation of in vitro antioxidant, antimicrobial, genotoxic and anticancer activities of lichen *Cetraria islandica*. Cytotechnology 66:803–813

Grujičić D, Radović M, Arsenijević S, Milošević-Djordjević O (2016) Cytogenetic biomarkers in detection of genotoxic effects of gestagens in peripheral blood lymphocytes in vitro and in vivo. Eur J Med Genet 59:624–633

Grujičić D, Marković A, TubićVukajlović J, Stanković M, Radović Jakovljević M, Ćirić A, Djordjević K, Planojević N, Milutinović M, Milošević-Djordjević O (2020) Genotoxic and cytotoxic properties of two medical plants (*Teucrium arduini* L. and *Teucrium flavum* L.) in relation to their polyphenolic contents. Mutat Res Genet Toxicol Environ Mutagen 852:503168. https://doi.org/10.1016/j.mrgentox.2020.503168

Guo X, Ni J, Liang Z, Xue J, Fenech MF, Wang X (2019) The molecular origins and pathophysiological consequences of micronuclei: new insights into an age-old problem. Mutat Res 779:1–35

Gupta VK, Kumria R, Garg M, Gupta M (2010) Recent updates on free radicals scavenging flavonoids: an overview. Asian J Plant Sci 9:108–117

Hallare AV, Gervasio MK, Gervasio PL, Acacio-Claro PJ (2008) Monitoring genotoxicity among gasoline station attendants and traffic enforcers in the City of Manila using the micronucleus assay with exfoliated epithelial cells. Environ Monit Assess 156:331–341

Hammami S, Jmii H, El Mokni R, Khmiri A, Faidi K, Dhaouadi H, El Aouni MH, Aouni M, Joshi RK (2015) Essential oil composition, antioxidant, cytotoxic and antiviral activities of *Teucrium pseudochamaepitys* growing spontaneously in Tunisia. Molecules 16:20426–20433

Harutyunyan K, Balayan K, Tadevosyan G, Hayrapetyan M, Ruzanna Musayelyan R, Grigoryan R, Khondkaryan L, Sarkisyan N, Babayan N (2019) Genotoxic potential of selected medicinal plant extracts in human whole blood cultures. J Herbmed Pharmacol 8:160–162

Havsteen BH (2002) The biochemistry and medical significance of the flavonoids. Pharmacol Ther 96:67–202

Hayashi M, Kodama Y, Awogi T, Suzuki T, Asita AO, Sofuni T (1992) The micronucleus assay using peripheral blood reticulocytes from mitomycin C- and cyclophosphamide-treated rats. Mutat Res 278:209–213

Hess DA, Sisson ME, Suria H, Wijsman J, Puvanesasingham R, Madrenas J, Rieder NJ (1999) Cytotoxicity of sulfonamide reactive metabolites: apoptosis and selective toxicity of $CD8^+$ cells by the hydroxylamine of sulfamethoxazole. FASEB J 13:1688–1698

Ho TV, Schärer OD (2010) Translesion DNA synthesis polymerases in DNA interstrand crosslink repair. Environ Mol Mutagen 51:552–566

Hoffmann D (2003) Medical herbalism: the science and practice of herbal medicine. Healing Art Press, Vermont

Holland N, Bolognesi S, Kirsch-Volders M, Bonassi S, Zeiger E, Knasmueller S, Fenech M (2008) The micronucleus assay in human buccal cells as a tool for biomonitoring DNA damage: the HUMN project perspective on current status and knowledge gaps. Mutat Res 659:93–108

Holland N, Fucic A, Merlo DF, Sram R, Kirsch-Volders M (2011) Micronuclei in neonates and children: effects of environmental, genetic, demographic and disease variables. Mutagenesis 26:51–56

Huang P, Huang B, Weng H, Nakayama K, Morimoto K (2009) Effect of lifestyle on micronuclei frequency in human lymphocytes in Japanese hard-metal workers. Prev Med 48:383–388

IAEA (2011) Cytogenetic dosimetry: applications in preparedness for and response to radiation emergencies. In: EPR-biodosimetry 2011. International Atomic Energy Agency, Vienna

Iarmarcovai G, Ceppi M, Botta A, Orsiere T, Bonassi S (2008) Micronuclei frequency in peripheral blood lymphocytes of cancer patients: a meta-analysis. Mutat Res 659:274–283

Inouye T, Sasaki YF, Imanishi H, Watanebe M, Ohta T, Shirasu Y (1988) Suppression of mitomycin C-induced micronuclei in mouse bone marrow cells by post-treatment with vanillin. Mutat Res 202:93–95

Ionescu EM, Ciocirlan M, Becheanu G, Nicolaie T, Ditescu C, Teiusanu A, Gologan S, Arbanas T, Diculescu M (2011) Nuclear division index may predict neoplastic colorectal lesions. Maedica 6:173–178

Ionescu EM, Nicolaie T, Ionescu MA, Becheanu G, Andrei F, Diculescu M, Ciocirlan M (2015) Predictive cytogenetic biomarkers for colorectal neoplasia in medium risk patients. J Med Life 8:398–403

Izquierdo-Vega JA, Morales-González JA, Sánchez Gutiérrez M, Betanzos-Cabrera G, Sosa-Delgado SM, Sumaya-Martínez MT, Morales-González Á, Paniagua-Pérez R, Madrigal-Bujaidar E, Madrigal-Santillán E (2017) Evidence of some natural products with antigenotoxic effects. Part 1: fruits and polysaccharides. Nutrients 9(2):102. https://doi.org/10.3390/nu9020102

Jarić S, Popović Z, Mačukanović-Jocić M, Djurdjevič L, Mijatović M, Karadžić B, Mitrović M, Pavlović P (2007) An ethnobotanical study on the usage of wild medicinal herbs from Kopaonik Mountain (Central Serbia). J Ethnopharmacol 111:160–175

Joksić G (1990) Comparative study of the incidence of chromosomal aberrations and MN in human lymphocytes during internal radionuclide contamination. Dissertation, University of Belgrade

Kada T, Shimoi K (1987) Desmutagens and bio-antimutagens -their modes of action. BioEssays 7:113–116

Kadifkova-Panovska T, Kulevanova S, Stefova M (2005) In vitro antioxidant activity of some *Teucrium species* (Lamiaceae). Acta Farm 55:207–214

Kadioglu E, Kocabas NA, Demircigil GC, Coskun E, Ozcagli E, Durmaz E, Karahalil B, Burgaz S, Sardas S (2012) Assessment of individual susceptibility to baseline DNA and cytogenetic damage in a healthy Turkish population: evaluation with lifestyle factors. Genet Test Mol Biomarkers 16:1157–1164

Karaman A, Kadı M, Kara F (2009) Sister chromatid exchange and micronucleus studies in patients with Behçet's disease. J Cutan Pathol 36:831–837

Kasabri V, Afifi FU, Hamdan I (2011) In vitro and in vivo acute antihyperglycemic effects of five selected indigenous plants from Jordan used in traditional medicine. J Ethnopharmacol 133:888–896

Kennedy DO (2014) Polyphenols and the human brain: plant "secondary metabolite" ecologic roles and endogenous signaling functions drive benefits. Adv Nutr 5:515–533

Khader M, Eckl PM, Bresgen N (2007) Effects of aqueous extracts of medicinal plants on MNNG-treated rat hepatocytes in primary cultures. J Ethnopharmacol 112:199–202

Khader M, Bresgen N, Eckl PM (2010) Antimutagenic effects of ethanolic extracts from selected Palestinian medicinal plants. J Ethnopharmacol 127:319–324

Khoddami A, Wilkes MA, Roberts TH (2013) Techniques for analysis of plant phenolic compounds. Molecules 18:2328–2375

Kirsch-Volders M, Plas G, Elhajouji A, Lukamowicz M, Gonzalez L, Vande Loock K, Decordier I (2011) The in vitro MN assay in 2011: origin and fate, biological significance, protocols, high throughput methodologies and toxicological relevance. Arch Toxicol 85:873–899

Kirsch-Volders M, Bonassi S, Knasmueller S, Holland N, Bolognesi C, Fenech M (2014) Commentary: critical questions, misconceptions and a road map for improving the use of the lymphocyte cytokinesis-block micronucleus assay for in vivo biomonitoring of human exposure to genotoxic chemicals – a HUMN project perspective. Mutat Res 759:49–58

Kosanić M, Ranković B, Stanojković T, Stošić I, Grujičić D, Milošević-Djordjević O (2016) *Lasallia pustulata* lichen as possible natural antigenotoxic, antioxidant, antimicrobial and anticancer agent. Cytotechnology 68:999–1008

Kosanić M, Ranković B, Stanojković T, Radović-Jakovljević M, Ćirić A, Grujičić D, Milošević-Djordjević O (2019) *Craterellus cornucopioides* edible mushroom as source of biologically active compounds. Nat Prod Commun 2019:1–6

Kozłowska A, Szostak-Węgierek D (2014) Flavonoids – food sources and health benefits. Rocz Panstw Zakl Hig 65:79–85

Kremer D, Košir IJ, Kosalec I, Zovko Končić M, Potočnik T, Čerenak A, Bezić N, Srečec S, Dunkić V (2013) Investigation of chemical compounds, antioxidant and antimicrobial properties of *Teucrium arduini* L. (Lamiaceae). Curr Drug Targets 14:1006–1014

Krishnaja AP, Sharma NK (2008) Variability in cytogenetic adaptive response of cultured human lymphocytes to mitomycin C, bleomicyn, quinacrine dihydrochloride, Co60 gamma-rays and hypothermia. Mutagenesis 23:77–86

Kumar S, Pandey AK (2013) Chemistry and biological activities of flavonoids: an overview. Sci World J 2013:162750. https://doi.org/10.1155/2013/162750

Kumar RV, Venkatrajireddy G, Bikshapathi T, Reddy MK (2012) Antioxidant-the maximum expressed activity among 63 medicinal plants. J Phyto Pharmacol 1:1–13

Kunwar A, Priyadarsini KI (2011) Free radicals, oxidative stress and importance of antioxidants in human health. J Med Allied Sci 1:53–60

Labieniec M, Gabryelak T (2003) Effects of tannins on Chinese hamster cell line B14. Mutat Res 539:127–135

Labieniec M, Gabryelak T, Falcioni G (2003) Antioxidant and pro-oxidant effects of tannins in digestive cells of the freshwater mussel *Unio tumidus*. Mutat Res 539:19–28

Lahouel M, Fillastre JP (2004) Role of flavonoids in the prevention of hematotoxicity due to chemotherapeutic agents. Haema 7:313–320

Lerda D, Biaggi Bistoni M, Peralta N, Ychari S, Vazquez M, Bosio G (2005) Fumonisins in foods from Cordoba (Argentina), presence and genotoxicity. Food Chem Toxicol 43:691–698

Leskovac A, Joksic G, Jankovic T, Savikin K, Menkovic N (2007) Radioprotective properties of the phytochemically characterized extracts of *Crataegus monogyna*, *Cornus mas* and *Gentianella austriaca* on human lymphocytes in vitro. Planta Med 73:1169–1175

Liou SH, Chen YH, Loh CH, Yang T, Wu TN, Chen CJ, Hsieh LL (2002) The association between frequencies of mitomycin C-induced sister chromatid exchange and cancer risk in arseniasis. Toxicol Lett 28:237–423

Ljubuncic P, Dakwar S, Portnaya I, Cogan U, Azaizeh H, Bomzon A (2006) Aqueous extracts of *Teucrium polium* possess remarkable antioxidant activity in vitro. Evid Based Complement Alternat Med 3:329–338

López-Romero D, Izquierdo-Vega JA, Morales-González JA, Madrigal-Bujaidar E, Chamorro-Cevallos G, Sánchez-Gutiérrez M, Betanzos-Cabrera G, Alvarez-Gonzalez I, Morales-González Á, Madrigal-Santillán E (2018) Evidence of some natural products with antigenotoxic effects. Part 2: plants, vegetables, and natural resin. Nutrients 10:1954. https://doi.org/10.3390/nu10121954

Lou J, He J, Zheng W, Jin L, Chen Z, Chen S, Lin Y, Xu S (2007) Investigation the genetic instability in the peripheral lymphocytes of 36 untreated lung cancer patients with comet assay and micronucleus assay. Mutat Res 617:104–110

Luzhna L, Kathiria P, Kovalchuk O (2013) Micronuclei in genotoxicity assessment: from genetics to epigenetics and beyond. Front Genet 4:131. https://doi.org/10.3389/fgene.2013.00131

Makena PS, Pierce SC, Chung KT, Sinclair SE (2009) Comparative mutagenic effects of structurally similar flavonoids quercetin and taxifolin on tester strains *Salmonella typhimurium* TA102 and *Escherichia coli* WP-2 uvrA. Environ Mol Mutagen 50:451–459

Marnewick JL, Gelderblom WC, Joubert E (2000) An investigation on the antimutagenic properties of South African herbal teas. Mutat Res 20:157–166

Marques RC, de Medeiros SR, Dias Cda S, Barbosa-Filho JM, Agnez-Lima LF (2003) Evaluation of the mutagenic potential of yangambin and of the hydroalcoholic extract of *Ocotea duckei* by the Ames test. Mutat Res 20:117–120

Mateuca R, Lombaert N, Aka PV, Decodier I, Kirsch-Volders M (2006) Chromosomal changes: induction, detection methods and applicability in human biomonitoring. Biochemist 88:1515–1531

Matkowski A, Wołniak D (2005) Plant phenolic metabolites as the free radical scavengers and mutagenesis inhibitors. BMC Plant Biol 5:S23. https://doi.org/10.1186/1471-2229-5-S1-S23

Mazumdar M, Giri S, Giri A (2011) Role of quercetin on mitomycin C induced genotoxicity: analysis of micronucleus and chromosome aberrations in vivo. Mutat Res 3:147–152

Melo-Reis PR, Bezerra LS, Vale MA, Canhête RF, Chen-Chen L (2011) Assessment of the mutagenic and antimutagenic activity of *Synadenium umbellatum* Pax latex by micronucleus test in mice. Braz J Biol 71:169–174

Middleton EJ (1998) Effect of plant flavonoids on immune and inflammatory cell function. Adv Exp Med Biol 439:175–182

Milošević-Djordjević O, Grujičić D, Marinković D, Arsenijević S, Banković S (2003) Effects of various doses of gestagens on micronuclei frequency in human peripheral blood lymphocytes of pregnant women. Hum Reprod 18:433–436

Milošević-Djordjevic O, Grujičić D, Arsenijević S, Marinković D (2004) The frequency of micronuclei among newborns from Kragujevac, Central Serbia, after NATO bombing in the Spring of 1991. Russ J Ecol 35:426–430

Milošević-Djordjević O, Grujičić D, Arsenijević S, Marinković D (2005) Monitoring of lymphocyte micronuclei among newborns from Kragujevac in central Serbia before and after environmental contamination. Tohoku J Exp Med 205:1–9

Milošević-Djordjević O, Grujičić D, Arsenijević S, Brkić M, Ugrinović S, Marinković D (2007) Micronuclei in cord blood lymphocytes as a biomarker of transplacental exposure to environmental pollutants. Tohoku J Exp Med 213:231–239

Milošević-Djordjević O, Grujičić D, Vasković Ž, Marinković D (2010a) High micronuclei frequency in peripheral blood lymphocytes of untreated cancer patients irrespective of gender, smoking and cancer site. Tohoku J Exp Med 220:115–120

Milošević-Djordjević O, Grujičić D, Joksić G, Marinković D (2010b) In vitro evaluation of the genotoxicity of ritodrine and verapamil in human lymphocytes. Hum Exp Toxicol 30:398–405

Milošević-Djordjević O, Stošić I, Vučković M, Grujičić D, Marinković D (2011a) Baseline and therapy-induced chromosome damages in peripheral blood lymphocytes of breast cancer patients assessed by the micronucleus assay. J BUON 16:437–443

Milošević-Djordjević O, Stošić I, Grujičić D, Banković D, Arsenijević S (2011b) Cervical precancerous lesions – chromosomal instability in peripheral blood lymphocytes in relation to lesion stage, age and smoking habits. Acta Obstret Gynecol Scand 90:1082–1087

Milošević-Djordjević O, Stošić I, Stanković M, Grujičić D (2013) Comparative study of genotoxicity and antimutagenicity of methanolic extracts from *Teucrium chamaedrys* and *Teucrium montanum* in human lymphocytes using micronucleus assay. Cytotechnology 65:863–869

Milošević-Djordjević O, Radović Jakovljević M, Marković A, Stanković M, Ćirić A, Marinković D, Grujičić D (2018) Polyphenolic contents of *Teucrium polium* L. and *Teucrium scordium* L. associated with their protective effects against MMC-induced chromosomal damage in cultured human peripheral blood lymphocytes. Turk J Biol 42:152–162

Minozzo R, Deimling LI, Petrucci Gigante L (2004) Micronuclei in peripheral blood lymphocytes of workers exposed to lead. Mutat Res 565:53–60

Montoro P, Braca A, Pizza C, De Tommasi N (2004) Structure-antioxidant activity relationships of flavonoids isolated from different plant species. Food Chem 92:349–355

Morteza-Semnani K, Saeedi M, Akbarzadeh M (2007) Essential oil composition of *Teucrium scordium* L. Acta Pharma 57:499–504

Natarajan AT (2002) Chromosome aberrations: past, present and future. Mutat Res 25:3–16

Neri M, Ugolini D, Bonassi S, Fucic A, Holland N, Knudsen LE, Sram RJ, Ceppi M, Bocchini V, Merlo DF (2006) Children's exposure to environmental pollutants and biomarkers of genetic damage. II. Results of a comprehensive literature search and meta-analysis. Mutat Res 612:14–39

Nersesyan A, Fenech M, Bolognesi C, Mišík M, Setayesh T, Wultsch G, Bonassi S, Thomas P, Knasmüller S (2016) Use of the lymphocyte cytokinesis-block micronucleus assay in occupational biomonitoring of genome damage caused by in vivo exposure to chemical genotoxins: past, present and future. Mutat Res 770:1–11

Noel S, Kasinathan M, Rath SK (2006) Evaluation of apigenin using in vitro cytochalasin blocked micronucleus assay. Toxicol in Vitro 20:1168–1172

OECD Library, Test no. 487: in vitro mammalian cell micronucleus test. OECD guidelines for the testing of chemicals, section 4. Health effects, 2010, https://doi.org/10.1787/20745788

Okamoto T (2005) Safety of quercetin for clinical application. Int J Mol Med 16:275–278

Ortega-Gutierrez S, Lopez-Vicente M, Lostale F, Fuentes-Broto L, Martinez-Ballarin E, Garcia JJ (2009) Protective effects of melatonin against mitomycin C-induced genotoxic damage in peripheral blood of rats. J Biomed Biotechnol 2009:791432. https://doi.org/10.1155/2009/791432

Pardini B, Viberti C, Naccarati A, Allione A, Oderda M, Critelli R, Preto M, Zijno A, Cucchiarale G, Gontero P, Vineis P, Sacerdote C, Matullo G (2017) Increased micronucleus frequency in peripheral blood lymphocytes predicts the risk of bladder cancer. Br J Cancer 116:202–210

Patil SL, Swaroop K, Kakde N, Somashekarappa HM (2017) In vitro protective effect of rutin and quercetin against radiation-induced genetic damage in human lymphocytes. Indian J Nucl Med 32:289–295

Perry N, Court G, Bidet N, Court J, Perry EK (1996) European herbs with cholinergic activities: potential in dementia therapy. Int J Geriatr Psychiatry 11:1063–1069

Petti S, Scully C (2009) Polyphenols, oral health and disease: a review. J Dent 37:413–423

Plazar J, Filipic M, Groothuis GM (2008) Antigenotoxic effect of Xanthohumol in rat liver slices. Toxicol in Vitro 22:318–327

Prasad NR, Srinivasan M, Pugalendi KV, Menon VP (2006) Protective effect of ferulic acid on gamma-radiation-induced micronuclei, dicentric aberration and lipid peroxidation in human lymphocytes. Mutat Res 28:129–134

Rahmouni F, Hamdaoui L, Badraoui R, Rebai T (2017) Protective effects of *Teucrium polium* aqueous extract and ascorbic acid on hematological and some biochemical parameters against carbon tetrachloride (CCL_4) induced toxicity in rats. Biomed Pharmacother 91:43–48

Rajabalian S (2008) Methanolic extract of *Teucrium polium* L. potentiates the cytotoxic and apoptotic effects of anticancer drug of vincristine, vinblastine and doxorubicin against a panel of cancerous cell lines. Exp Oncol 30:133–138

Rasekh HR, Khoshnood-Mansourkhani MJ, Kamalinejad M (2001) Hypolipidemic effects of *Teucrium polium* in rats. Fitoterapia 72:937–939

Redžić SS (2007) The ecological aspect of ethnobotany and ethnopharmacology of population in Bosnia and Herzegovina. Coll Antropol 31:869–890

Regner GG, Gianesini J, Von Borowski RG, Silveira F, Semedo JG, Ferraz Ade B, Wiilland E, Von Poser G, Allgayer M, Picada JN, Pereira P (2011) Toxicological evaluation of Pterocaulon polystachyum extract: a medicinal plant with antifungal activity. Environ Toxicol Pharmacol 31:242–249

Ren N, Atyah M, Chen WY, Zhou CH (2017) The various aspects of genetic and epigenetic toxicology: testing methods and clinical applications. J Transl Med 15:110. https://doi.org/10.1186/s12967-017-1218-4

Resende FA, Vilegas W, Dos Santos LC, Varanda EA (2012) Mutagenicity of flavonoids assayed by bacterial reverse mutation (Ames) test. Molecules 17:5255–5268

Rice-Evans CA, Miller NJ, Paganga G (1997) Antioxidant properties of phenolic compounds. Trends Plant Sci 2:152–159

Rithidech KN, Tungjai M, Whorton EB (2005) Protective effect of apigenin on radiation-induced chromosomal damage in human lymphocytes. Mutat Res 1:96–104

Romero-Jimenez M, Campos-Sanchez J, Analla M, Munoz-Serrano A, Alonso-Moraga A (2005) Genotoxicity and anti-genotoxicity of some traditional medicinal herbs. Mutat Res 585:147–155

Rossner P, Boffetta P, Ceppi M, Bonassi S, Smerhovsky Z, Landa K, Juzova D, Srám RJ (2005) Chromosomal aberrations in lymphocytes of healthy subjects and risk of cancer. Environ Health Perspect 113:517–520

Ryu TH, Kim JH, Kim JK (2014) Radiation exposure dose in human blood lymphocytes as assessed by the CBMN assay. J Ecol Environ 37:195–200

Ryu TH, Kim JH, Kim JK (2016) Chromosomal aberrations in human peripheral blood lymphocytes after exposure to ionizing radiation. Genome Integr 7:5. https://doi.org/10.4103/2041-9414.197172

Saad B, Zaid H, Shanak S, Kadan S (2017) Anti-diabetes and anti-obesity medicinal plants and phytochemicals – safety, efficacy and action mechanisms. Springer, Basel, pp 21–55

Said O, Khalil K, Fulder S, Azaizeh H (2002) Ethnopharmacological survey of medicinal herbs in Israel, the Golan Heights and the West Bank region. J Ethnopharmacol 83:251–265

Šamec D, Gruz J, Strnad M, Kremer D, Kosalec I, Grubešić RJ, Karlović K, Lucic A, Piljac-Žegarac J (2010) Antioxidant and antimicrobial properties of *Teucrium arduini* L. (Lamiaceae) flower and leaf infusions (*Teucrium arduini* L. antioxidant capacity). Food Chem Toxicol 48:113–119

Santos-Mello R, Kwan D, Norman A (2004) Chromosome aberrations and T-cell survival in human lymphocytes. Radiat Res 60:482–488

Santovito A, Cervella P, Delpero M (2011) In vitro aneugenic effects of the fungicide thiabendazole evaluated in human lymphocytes by the micronucleus assay. Arch Toxicol 85:689–693

Sasaki YF, Ohta T, Imanishi H, Watanabe M, Matsumoto K, Kato T, Shirasu Y (1990) Suppressing effects of vanillin, cinnamaldehyde, and anisaldehyde on chromosome aberrations induced by x-rats in mice. Mutat Res 243:299–302

Sato Y, Itagaki S, Kurokawa T, Ogura J, Kobayashi M, Hirano T, Sugawara M, Iseki K (2011) In vitro and in vivo antioxidant properties of chlorogenic acid and caffeic acid. Int J Pharm 403:136–138

Šavikin K, Zdunić G, Menković N, Živković J, Ćujić N, Tereščenko M, Bigović D (2013) Ethnobotanical study on traditional use of medicinal plants in South-Western Serbia, Zlatibor district. J Ethnopharmacol 19:803–810

Saxena M, Saxena J, Pradhan A (2012) Flavonoids and phenolic acids as antioxidants in plants and human health. Int J Pharm Sci Rev Res 16:130–134

Sghaier MB, Bhouri W, Bouhlel I, Skandrani I, Boubaker J, Ghekir-Ghedira L, Ghedira K (2011) Inhibitory effect of *Teucrium ramosissimum* extracts on aflatoxin B 1, benzo[a]pyrene, 4-nitro-o-phenylenediamine and sodium azide induced mutagenicity: correlation with antioxidant activity. S Afr J Bot 77:730–740

Sghaier MB, Ismail MB, Bouhlel I, Ghedira K, Chekir-Ghedira L (2016) Leaf extracts from *Teucrium ramosissimum* protect against DNA damage in human lymphoblast cell K562 and enhance antioxidant, antigenotoxic and antiproliferative activity. Environ Toxicol Pharmacol 44:44–52

Sharma NK (2013) Modulation of radiation-induced and mitomycin C-induced chromosome damage by apigenin in human lymphocytes in vitro. J Radiat Res 54:789–797

Shaughnessy DT, Setzer RW, DeMarini DM (2001) The antimutagenic effect of vanillin and cinnamaldehyde on spontaneous mutation in *Salmonella* TA104 is due to a reduction in mutations at GC but not AT sites. Mutat Res 480–481:55–69

Shin IS, Seo CS, Ha HK, Lee MY, Huang DS, Huh JI, Shin HK (2011) Genotoxicity assessment of Pyungwi-san (PWS), a traditional herbal prescription. J Ethnopharmacol 27:696–703

Sibanda T, Okoh AI (2007) The challenges of overcoming antibiotic resistance: plant extracts as potential sources of antimicrobial and resistance modifying agents. Afr J Biotechnol 6:2886–2896

Siddique YH, Afzal M (2009a) Protective effect of apigenin against hydrogen peroxide induced genotoxic damage on cultured human peripheral blood lymphocytes. J Appl Biomed 7:35–43

Siddique YH, Afzal M (2009b) Antigenotoxic effect of apigenin against mitomycin C induced genotoxic damage in mice bone marrow cells. Food Chem Toxicol 47:536–539

Siddique YH, Beg T, Afzal M (2008) Antigenotoxic effect of apigenin against anti-cancerous drugs. Toxicol in Vitro 22:625–631

Siddique YH, Ara G, Beg T, Afzal M (2010) Anticlastogenic effect of apigenin in human lymphocytes treated with ethinnylestradiol. Fitoterapia 81:590–594

Silva ID, Gaspar J, da Costa GG, Rodrigues AS, Laires A, Rueff J (2000) Chemical features of flavonols affecting their genotoxicity. Potential implications in their use as therapeutical agents. Chem Biol Interact 124:29–51

Singh NP, McCoy MT, Tice RR, Schneider EL (1988) A simple technique for quantitation of low levels of DNA damage in individual cells. Exp Cell Res 175:184–191

Skibola CF, Smith MT (2000) Potential health impacts of excessive flavonoid intake. Free Radic Biol Med 29:375–383

Sontakke YA, Fulzele RR (2009) Cytogenetic study on genotoxicity of antitumor-antibiotic Mitomycin C. Biomed Res 20:40–44

Speit G (2013) Does the recommended lymphocyte cytokinesis-block micronucleus assay for human biomonitoring actually detect DNA damage induced by occupational and environmental exposure to genotoxic chemicals? Mutagenesis 28:375–380

Speit G, Zeller J, Neuss S (2011) The in vivo or ex vivo origin of micronuclei measured in human biomonitoring studies. Mutagenesis 26:107–110

Speit G, Linsenmeyer R, Schütz P, Kuehner S (2012) Insensitivity of the in vitro cytokinesis-block micronucleus assay with human lymphocytes for the detection of DNA damage present at the start of the cell culture. Mutagenesis 27:743–747

Sponchiado G, Adam ML, Silva CD, Soley BS, de Mello-Sampayo C, Cabrini DA, Correr CJ, Otuki MF (2016) Quantitative genotoxicity assays for analysis of medicinal plants: a systematic review. J Ethnopharmacol 178:289–296

Stanković M (2012) Biological effects of secondary metabolites of *Teucrium* species of Serbian flora. Dissertation, University of Kragujevac

Stanković M, Stefanović O, Čomić LJ, Topuzović M, Radojević I, Solujić S (2012) Antimicrobial activity, total phenolic content and flavonoid concentrations of *Teucrium* species. Cent Eur J Biol 7:664–671

Stich HF (1991) The beneficial and hazardous effects of simple phenolic compounds. Mutat Res 259:307–324

Stopper H, Schmitt E, Kobras K (2005) Genotoxicity of phytoestrogens. Mutat Res 574:139–155

Stošić I, Grujičić D, Arsenijević S, Brkić M, Miloševic-Djordjević O (2014) Glutathione S-transferase T1 and M1 polymorphisms and risk of uterine cervical lesions in women from central Serbia. Asian Pac J Cancer Prev 15:3201–3205

Suleiman MS, Abdul-Ghani AS, Al-Khalil S, Amin R (1988) Effect of *Teucrium polium* boiled leaf extract on intestinal motility and blood pressure. J Ethnopharmacol 22:111–116

Surralles J, Xamena N, Creus A, Catalan J, Norppa H, Marcos R (1995) Induction of micronuclei by five pyrethroid insecticides in whole-blood and isolated human lymphocytes cultures. Mutat Res 341:169–184

Tamai K, Tezuka H, Kuroda Y (1992) Direct modifications by vanillin in cytotoxicity and genetic changes induced by EMS and H_2O_2 in cultured Chinese hamster cells. Mutat Res 268:231–237

Taner G, Özkan Vardar D, Aydin S, Aytaç Z, Başaran A, Başaran N (2016) Use of in vitro assays to assess the potential cytotoxic, genotoxic and antigenotoxic effects of vanillic and cinnamic acid. Drug Chem Toxicol 40:183–190

Tariq M, Ageel AM, al-Yahya MA, Mossa JS, al-Said MS (1989) Anti-inflammatory activity of *Teucrium polium*. Int J Tissue React 11:185–188

Tiwari AK (2001) Imbalance in antioxidant defence and human diseases: multiple approach of natural antioxidants therapy. Curr Sci 81:1179–1187

Tubić J, Grujičić D, Radović Jakovljević M, Ranković B, Kosanić M, Stanojković T, Ćirić A, Milošević-Djordjević O (2019) Investigation of biological activities and secondary metabolites of *Hydnum repandum* acetone extract. Farmacia 67:174–183

Turkez H, Aydin E, Aslan A (2012) *Xanthoria elegans* (Link) (lichen) extract counteracts DNA damage and oxidative stress of mitomycin C in human lymphocytes. Cytotechnology 64:679–686

Turkez H, Aydın E, Geyikoglu F, Cetin D (2015) Genotoxic and oxidative damage potentials in human lymphocytes after exposure to terpinolene in vitro. Cytotechnology 67:409–418

Tutin TG, Heywood VH, Burges NA, Moore DM, Valentine DH, Walters SM, Webb DA (eds) (1972) Flora Europaea III. Cambridge University Press, Cambridge

Unal F, Taner G, Yuzbasioglu D, Yilmaz S (2013) Antigenotoxic effect of lipoic acid against mitomycin-C in human lymphocyte cultures. Cytotechnology 65:553–565

Undeğer U, Aydin S, Başaran AA, Başaran N (2004) The modulating effects of quercetin and rutin on the mitomycin C induced DNA damage. Toxicol Lett 15:143–149

Vasquez KM (2010) Targeting and processing of site-specific DNA interstrand crosslinks. Environ Mol Mutagen 51:527–539

Veeramuthu D, Raja WRTR, Al-Dhabi NA, Savarimuthu I (2017) Flavonoids: anticancer properties. In: Justino J (ed) Flavonoids: from biosynthesis to human health, 1st edn. IntechOpen, Croatia, pp 287–303

Verschaeve L, Kestens V, Taylor JL, Elgorashi EE, Maes A, Van Puyvelde L, De Kimpe N, Van Staden J (2004) Investigation of the antimutagenic effects of selected South African medicinal plant extracts. Toxicol in Vitro 18:29–35

Verweij J, Pinedo HM (1990) Mitomycin C: mechanism of action, usefulness and limitations. Anti-Cancer Drugs 1:5–13

Vijayalaxmi MML, Reiter RJ, Herman TS (1999) Melatonin and protection from genetic damage in blood and bone marrow: whole-blood irradiation studies in mice. J Pineal Res 27:221–225

Vlastos D, Mademtzoglou D, Drosopoulou E, Efthimiou I, Chartomatsidou T, Pandelidou C, Asryrakaki M, Chalatsi E, Mavragani-Tsipidou P (2013) Evaluation of the genotoxic and anti-genotoxic effects of Chios mastis water by the in vitro micronucleus test on human lymphocytes and the in vivo wing somatic test on *Drosophila*. PLoS One 8(7):e69494. https://doi.org/10.1371/journal.pone.0069494

Vral A, Fenech M, Thierens H (2011) The micronucleus assay as a biological dosimeter of in vivo ionising radiation exposure. Mutagenesis 26:11–17

Vrndić O, Milošević-Djordjević O, Mijatović-Teodorović LJ, Jeremić M, Stošić I, Grujičić D, Živančević-Simonović S (2013) Correlation between micronuclei frequency in peripheral blood lymphocytes and retention of 131-I in thyroid cancer patients. Tohoku J Exp Med 229:115–124

Vuković N, Sukodlak S, Solujić S, Mihailović V, Mladenović M, Stojanović J, Stanković SM (2010) Chemical composition and antimicrobial activity of *Teucrium arduini* essential oil and cirsimarin from Montenegro. J Med Plant Res 5:1244–1250

Wang T, Li Q, Bi K (2018) Bioactive flavonoids in medicinal plants: structure, activity and biological fate. Asia J Pharm Sci 13:12–23

Yen GC, Chen HY (1995) Antioxidant activity of various tea extracts in relation to their antimutagenicity. J Agric Food Chem 43:27–32

Zhao J, Wang J, Chen Y, Agarwal R (1999) Anti-tumor-promoting activity of a polyphenolic fraction isolated from grape seeds in the mouse skin two-stage initiation-promotion protocol and identification of procyanidin B5-3′-gallate as the most effective antioxidant constituent. Carcinogenesis 20:1737–1745

Zyad A, Leouifoudi I, Tilaoui M, Mouse HA, Khouchani M, Jaafari A (2018) Natural products as cytotoxic agents in chemotherapy against cancer. In: Celik TA (ed) Cytotoxicity, 1st edn. IntechOpen, Croatia, pp 65–88

Chapter 10
Antioxidant Activity of Secondary Metabolites of *Teucrium* Species

Milan Stanković

Abstract *Teucrium* species are characterized by valuable secondary metabolites content and long traditional ethnopharmacological use. Bearing in mind that the therapeutic benefit of medicinal plants is often based on their antioxidant activity, this chapter is a comprehensive review of the antioxidant activity of medicinal and potentially medicinal species of the genus *Teucrium*, as well as the secondary metabolites content and diversity including methods for antioxidant activity determination. Analysis of the current literature indicates that antioxidant activity has been tested for over 30 species of this genus using different methodological approaches. *Tucrium polium* is the most tested species followed by *Teucrium chamaedrys* and *Tucrium montanum*. Plant extracts are often used for antioxidant activity analysis followed by essential oils and isolated chemical compounds. The most commonly used methods include DPPH – 2,2-diphenyl-1-picrylhydrazyl (DPPH) radical scavenging capacity, Ferric Reducing Antioxidant Power (FRAP), β-carotene/linoleic acid assay, 2,2′-azino-bis-(3-ethylbenzothiazoline-6-sulphonic acid) radical scavenging capacity (ABTS) etc. *Teucrium* species is characterized by the presence of monoterpenes and sesquiterpenes, as well as flavonoids, phenylethanoids and phenolic acids. The significant correlation of the antioxidant activity and phenolic content was determined in many studies. The antioxidant potential of some *Teucrium* species has been confirmed by in vivo testing. Variability during vegetative period, habitat-related variability as well as variability between plant parts were determined for the antioxidant activity and content of the active compounds of some *Teucrium* species. Numerous studies confirm that species of the genus *Teucrium* possess secondary metabolites with significant antioxidant effects.

Keywords *Teucrium* species · Antioxidant activity · Plant extracts · Phenolics · Flavonoids · Essential oil · Antioxidant assays

M. Stanković (✉)
Department of Biology and Ecology, Faculty of Science, University of Kragujevac, Kragujevac, Serbia
e-mail: mstankovic@kg.ac.rs

M. Stanković (ed.), *Teucrium Species: Biology and Applications*,
https://doi.org/10.1007/978-3-030-52159-2_10

Abbreviations

ABTS	2,2′-azino-bis-(3-ethylbenzothiazoline-6-sulphonic acid) radical scavenging capacity
BDE	Bond dissociation energy
BR	Briggs-Rauscher oscillating reaction
CUPRAC	Cupric Reducing Antioxidant Capacity assay
DPPH	2,2-diphenyl-1-picrylhydrazyl (DPPH) radical scavenging capacity
EPR	Electron Paramagnetic Resonance radicals detection
FRAP	Ferric Reducing Antioxidant Power
HAPX	Hemoglobin Ascorbate Peroxidase activity inhibition
HAT	Hydrogen atom transfer
HRSA	Hydroxyl Radical Scavenging Activity
IP	Ionization potential
NBT/riboflavine	Nitroblue Tetrazolium riboflavine assay
ORAC	Oxygen Radical Absorbance Capacity
OSI	Oxidative Stability Index
RP	Reducing Power
SET	Single electron transfer
SOD	Superoxide Dismutase Activity assay
SRSA	Superoxide Radical Scavenging Activity
TAC	Total Antioxidant Capacity
TAP	Total Antioxidant Power
TBA	Thiobarbituric Acid Assay
TEAC	Trolox Equivalent Antioxidant Capacity
X/XOD	Xanthine/Xanthine Oxidase assay
XOA	Xanthine Oxidase Activity assay

10.1 Introduction

Genus *Teucrium* (germander) is a member of the Lamiaceae family. It includes more than 300 species and together with genera *Salvia*, *Stachys*, *Plectranthus*, and *Hyptis* makes the group of the most numerous genera of this plant family. The species of genus *Teucrium* are mainly perennial herbaceous plants, shrubs or subshrubs, while only a small number are annual. The species of this genus are present in moderate climate regions, mostly in Mediterranean and upper Asia. In Europe the most species are found in the south, southwest and southeast areas, thus making this continent characterised by the greatest number of species of this genus. In Asia, the greatest number of species are found in the southwest part, while on the territory of Africa it is present in the far north areas, and only a small number is in the far south areas. In America the areal of genus *Teucrium* covers the south areas of North

America and the southwest areas of South America. In Australia, these species are found in the southern parts as well as on some surrounding islands (Stanković 2012).

The species of genus *Teucrium* are important plant group for research in biology and ecology, as well as in the domains of applied disciplines such as biotechnology, pharmacy and food and drink industries. One of significant aspects of for development of the application of *Teucrium* species is their ethno botanic use which dates thousands of years ago. The species of genus *Teucrium* are very popular in traditional medicine from the very beginnings of plant use in the treatment of diseases. The species of this genus were used in the treatments of tuberculosis, scurvy, jaundice, rheumatism and diseases of digestive tract. Therapeutic products, prepared from the parts of these plants above the ground are characterised by bitter taste, whereby their positive effect on digestive tract is found. High contents of aromatic compounds with antioxidant and antimicrobial activity contributes to their effect in treatment of many infectious diseases. Due to their bitter aromatic compounds, the species of genus *Teucrium* are used in cookery as spices and additives in preparation of soft and alcoholic drinks. Because of the presence of the several toxic neoclerodane diterpenes application of some *Teucrium* species should be controlled and careful (Keršek 2006; Starakis et al. 2006; Savvidou et al. 2007; Stanković 2012).

Due to characteristic qualitative and quantitative contents of secondary metabolites in the group of flavonoids, phenolic acids and mono and sesquiterpenes, the species of *Teucrium* genus are significant sources of antioxidant, antimicrobial and anticancer natural components, and bearers of many therapeutic properties. For that reason, plant extracts, essential oils and isolated components are extensively analysed for their biological activity. The literature data show that plant extracts, essential oils and isolated secondary metabolites of the genus *Teucrium* are largely analysed due to their antioxidant activity. Detailed review of literature showed that the research results of the antioxidant activity of more than 30 species were published in more than a 80 scientific papers. The greatest number of data for the antioxidant activity of the species of this genus are found for *Teucrium polium* (felty germander), *Teucrium chamaedrys* (wall germander) and *Teucrium montanum* (mountain germander). The chapter shows the review of literature about antioxidant activity of secondary metabolites of the *Teucrium* species. In introductory parts general information on secondary metabolites as bearers of antioxidant role is presented together with the mechanisms of their effect. It is followed by the table of plant species of genus *Teucrium,* analysed for antioxidant activity of their extracts, essential oils and isolated components, together with the methodological approach to the investigation.

10.2 Plant Secondary Metabolites

Besides primary, secondary metabolism also occurs in plant cells, and its products, i.e. secondary metabolites are not of essential importance for the plant, but are most often the result of plant adaptation to environmental conditions. The term

"secondary" was first used by A Kossel in 1891; it designated metabolic pathways that resulted in products that were not essential, but were involved in the reaction of a plant organism to environmental factors (Edreva et al. 2008). These substances play a role in ecological interaction of plants; they show regulatory effect and protect plants from the attack of predators, parasites and other negative factors, and also play a role in the processes of pollination and seed distribution by animals attracting. Their role in protective response to unfavourable abiotic influences (radiation, temperature changes, disturbances in water regime etc.) is significant. They appeared during evolution of plants; mutually, they differ in chemical structure and quantity in various groups of taxon, thus they can be characteristic of only one species or taxon of higher rank. Quantitative and qualitative composition of secondary metabolites changes during a vegetation period, season or a day, depending on the state of a plant, the presence or absence of stress etc. (Oh et al. 2009).

Metabolic pathways of secondary biomolecules are extremely complex. The enzymes that are included in metabolic pathways of secondary metabolites are mainly under the influence of a great number of ecological factors, while the pathways of regulation of these enzymes are a significant domain of scientific research today, since many products of secondary metabolism in plants are extremely important, especially for pharmaceutical and food industry (Zwenger and Basu 2008; Quideau et al. 2011). According to their chemical composition, plant secondary metabolites are divided into two groups. The first group includes molecules without nitrogen atoms – phenolic compounds and terpenoids. The second group includes secondary metabolites with nitrogen atoms in their molecules, i.e. alkaloids (Wink 2004; Roberts 2007). Besides a great number of biological effects in plants, secondary metabolites that are isolated from plants show strong biological effects. Plant secondary metabolites are characterised by wide use in pharmaceutical industry, medicine, agricultural and food production. A wide spectrum of their application is related to structural properties of their molecules, which shows their antioxidant, antimicrobial and anticancer activity, as well as many therapeutic effects (Quideau et al. 2011). Due to varied chemical structure, secondary metabolites produce effects at all levels of organisation in biological systems, which is the reason of their intensive application. Their pharmacological effects are very important, thus they are used in therapy and prevention of many pathological states (Fraga 2010).

10.2.1 Biological Effects of Plant Secondary Metabolites

In addition to their significant biological role in the interactions of plants in an ecosystem, secondary metabolites that are isolated from a plant organism show biological activity in both in vitro and in vivo conditions. The mechanism of their activity in the given biological system involves mainly stimulation or inhibition of enzymatic activity and other metabolic processes where their therapeutic benefit is expressed (Korkina 2007). Plants have mechanisms to defend themselves from occurrence of oxidative stress, as well as mechanisms to reduce negative impact of

oxidative stress. Antioxidant activity of phenolic compounds is based on their chemical structure, which allows them to be donors of hydrogen atoms and electrons. In addition to the stated processes, there are a few significant mechanisms of antioxidant activity of phenolic compounds and other secondary metabolites in plant cells and tissues (Blokhina et al. 2003; Michalak 2006). Besides important and necessary biological effects in the process of the plant response to negative impact of environmental factors, i.e. stress conditions, isolated plant secondary metabolites possess different biological effects (Quideau et al. 2011). The biological activity of plant secondary metabolites is based on their capability to react with numerous regulatory molecules and other cellular and subcellular structures, thus having positive or negative impact on a great number of metabolic processes. The effect of secondary metabolites is most often manifested through the control of enzyme reactions, control of hormone activity, gene expression, cellular transport, cellular division, immune response and many other physiological processes. Acting by stimulatory or inhibitory mechanisms, the secondary metabolites show antioxidant, antimicrobial, antiproliferative, apoptotic, anti-inflammatory, antihypertensive and many other activities (Briskin 2000).

10.2.2 Antioxidant Activity of Plant Secondary Metabolites

Free radicals are molecules, ions or atoms of carbon, oxygen, nitrogen or sulphur with unpaired electrons that make them very reactive. Free radicals occur during a great number of metabolic processes in an organism under the influence of negative effects of physical and chemical environmental factors. They most often occur during reactions such as homolytic cleavage within a molecule, oxidoreduction reactions, thermolysis, radiolysis, transmission of one electron from transient ion of metal to organic molecule, the effect of ozone, nitrogen (IV) – oxide and singlet oxygen as well as during numerous enzymatic processes in the organism (Sen et al. 2010). The most common and most reactive free radical and non-radical forms are oxygen reactive forms that can appear as superoxide anion radicals, hydroxyl radicals, hydroperoxyl radicals, singlet oxygen, carbon dioxide and carbon monoxide radicals (Halliwel and Whiteman 2004). The mechanism of negative effect of free radical forms in biological systems is based on their reaction with biomolecules such as nucleic acids, proteins, lipids and enzymes, thus causing their damage and inhibiting their primary biological role in cells. The increased production of free radicals in a human organism intensifies its ageing process and causes a great number of pathological conditions such as neurodegenerative changes, cancer genesis, cardiovascular disorders and inflammatory processes (Valko et al. 2006).

Harmful effects induced by free radicals are prevented by the actions of antioxidant substances which inhibit their activity. Antioxidants are substances which partly inhibit or completely prevent oxidation of substrates exposed to the oxidation process. The mechanism of effect of antioxidant substances is based on their role of a donor of electron or hydrogen atom, i.e. their scavenging ability, then on their

capability to chelate metal ions (Fe^{2+}, Cu^{2+}, Zn^{2+}, Mg^{2+}), thus reducing their redox potential, destruction of hydro peroxide of lipid molecules which create non-radical forms etc. (Gupta and Sharma 2006). Antioxidant capacity of phenolic compounds depends on the number and position of hydroxyl groups. In addition, significant physical and chemical characteristics are bond dissociation energy (BDE) and ionization potential (IP). One of leading mechanisms of antioxidant activity of phenolic compounds is the role of a donor of hydrogen atom (HAT – hydrogen atom transfer) and the role of a donor of electron (SET – single electron transfer) to the molecule exposed to oxidation process (Quideau et al. 2011).

10.3 Antioxidant Activity of Secondary Metabolites from *Teucrium* Species

Numerous studies confirm that species of the genus *Teucrium* are an important plant group with a variety of secondary metabolites with antioxidant effects. A review of the literature revealed that over 30 species of this genus have been studied for the antioxidant activity using different model systems. *Teucrium* species with traditional application have been analyzed in many studies of antioxidant activity. Based on the literature review presented in Table 10.1, it was found over 80 references about antioxidant activity of the *Teucrium* species, as well as results about 34 tested *Teucrium* species such as *Teucrium alyssifolium*, *T. arduini*, *T. barbeyanum*, *T. botrys*, *T. cavernarum*, *T. chamaedrys*, *T. creticum*, *T. flavum*, *T. fruticans*, *T. hyrcanicum*, *T. marum*, *T. massiliense*, *T. montanum*, *T. montbretii*, *T. muscatense*, *T. orientale*, *T. oxylepis*, *T. parviflorum*, *T. persicum*, *T. pilosum*, *T. polium*, *T. pruinosum*, *T. pseudochamaepitys*, *T. pseudoscorodonia*, *T. ramosissimum*, *T. royleanum*, *T. salviastrum*, *T. sandrasicum*, *T. sauvagei*, *T. scordium*, *T. siculum*, *T. stocksianum*, *T. subspinosum*, and *T. yemense*. Review of current literature data about the antioxidant activity analysis of *Teucrium* species including sample types as well as methods for antioxidant activity determination were presented in Table 10.1.

The results for the antioxidant activity and amount of phenolic compounds and flavonoids of leaf, flower and stem extracts obtained by water extraction, as well as the composition of the essential oil, were determined for *Teucrium alyssifolium* (Semiz et al. 2016). The water extract of the leaves of this plant showed the highest antioxidant activity and content of phenolic compounds. The main components of the essential oil were trans-β-caryophyllene, ar-curcumene and bisabolene. The antioxidant activity of *Teucrium arduini* has been investigated using various methods. Leaf and flower infusions of *Teucrium arduini* were analyzed in order to determine phenolic content and antioxidant activity. Antioxidant activity was evaluated using the FRAP, DPPH and ABTS radical scavenging assays. The authors found significant antioxidant activity as well as a positive correlation between antioxidant activity and the amount of phenolic compounds (Šamec et al. 2010). Screening of

Table 10.1 Antioxidant activity of *Teucrium* species – review of analyzed species, sample types as well as methods for antioxidant activity determination

Teucrium species	Sample type – extract/essential oil	Method used	References
T. alyssifolium	Leaf, flower and stem water extract	DPPH	Semiz et al. (2016)
T. arduini	Infusions	FRAP, DPPH, ABTS	Šamec et al. (2010)
	Water, methanolic, acetone, ethyl acetate and petroleum ether extract	DPPH, FRAP, BR	Stanković (2012)
	Leaf, flower, stem ethanolic extract	DPPH, Fe^{2+} chelating activity, β-carotene/linoleic acid assay	Kremer et al. (2013)
	Ethanolic extracts	DPPH, TAC	Vladimir-Knežević et al. (2014)
T. barbeyanum	Petroleum ether, dichloromethane, methanolic, chloroform, ethyl acetate, butanolic and water extract	DPPH, ABTS, TEAC, FRAP	Alwahsh et al. (2015)
T. botrys	Water, methanolic, acetone, ethyl acetate and petroleum ether extract	DPPH, FRAP, BR	Stanković (2012)
	Water, methanolic, acetone, ethyl acetate and petroleum ether extract	DPPH	Stanković et al. (2014)
T. cavernarum	Methanolic extract	DPPH, ABTS, TEAC	Goger et al. (2019)
T. chamaedrys	Diethyl ether, ethyl acetate and n-butanolic extract	DPPH, HRSA, β-carotene/linoleic acid assay	Kadifkova Panovska et al. (2005)
	Water extract	DPPH, RP, TBA	Ozgen et al. (2006)
	Isolated chemical compounds	DPPH, TBA	Pacifico et al. (2009)
	Methanolic extract	DPPH, β-carotene/linoleic acid assay	Gursoy and Tepe (2009)
	Water, methanolic, acetone, ethyl acetate and petroleum ether extract	DPPH	Stanković et al. (2010a)
	Water, methanolic, acetone, ethyl acetate and petroleum ether extract	DPPH, FRAP, BR	Stanković (2012)
	Ethanolic extracts	DPPH, TAC	Vladimir-Knežević et al. (2014)
	Ethanolic extract	DPPH, TEAC, HAPX, EPR	Vlase et al. (2014)
	Methanolic-water extract	DPPH	El Guiche et al. (2015)
	Methanolic, ethanolic, acetone and ethyl acetate extract	DPPH	Stanković and Zlatić (2017)
	Methanolic extract and water infusion	DPPH	Zlatić et al. (2017)
	Methanolic extract	DPPH	Faiku et al. (2019)

(continued)

Table 10.1 (continued)

Teucrium species	Sample type – extract/essential oil	Method used	References
T. creticum	Ethanolic extract	DPPH, β-carotene/linoleic acid assay	Husein et al. (2014)
T. flavum	Essential oil	DPPH	Hammami et al. (2015b)
	Ethanolic extract	DPPH, SOD	Acquaviva et al. (2017)
	Methanolic, ethanolic, acetone and ethyl acetate extract	DPPH	Stanković and Zlatić (2017)
T. fruticans	Ethanolic extract	DPPH, SOD	Acquaviva et al. (2017)
	Methanolic, ethanolic, acetone and ethyl acetate extract	DPPH	Stanković and Zlatić (2017)
	Methanolic extract	DPPH, BR	Stanković et al. (2017)
T. hyrcanicum	Methanolic extract	DPPH, FRAP	Golfakhrabadi et al. (2015)
T. marum	Essential oil	DPPH, XOA, 5-lipoxygenase test	Ricci et al. (2005)
	Dichlormethane, acetone and ethanolic extract	DPPH, ABTS, β-carotene/linoleic acid assay	Poli et al. (2007)
T. massiliense	Essential oil	DPPH, Inhibition of lipid peroxide formation aasay	Giamperi et al. (2008)
T. montanum	Diethyl ether, ethyl acetate and n-butanolic extract	DPPH, HRSA, β-carotene/linoleic acid assay	Kadifkova Panovska et al. (2005)
	Petroleum ether, chloroform, ethyl acetate, n-butanolic and water extract	DPPH	Djilas et al. (2006)
	Water, methanolic, acetone, ethyl acetate and petroleum ether extract	DPPH	Stanković et al. (2011a)
	Water, methanolic, acetone, ethyl acetate and petroleum ether extract	DPPH, FRAP, BR	Stanković (2012)
	Ethanolic extracts	DPPH, TAC	Vladimir-Knežević et al. (2014)
	Methanolic extract and water infusion	DPPH	Zlatić et al. (2017)
	Water extract	DPPH, FRAP	Nastić et al. (2018)
T. montbretii	Methanolic extract	DPPH, Phosphomolybdenum assay	Ozkan et al. (2007)
T. muscatense	n-hexane, chloroform, ethanolic, n-buthanolic and water extract	DPPH, ABTS, TBA	Ur Rehman et al. (2017)

Teucrium species	Sample type – extract/essential oil	Method used	References
T. orientale	Petroleum ether, chloroform, acetone and methanolic extract	DPPH, Thiocyanate method	Cakir et al. (2006)
	Essential oil and methanolic extract	DPPH, β-carotene-linoleic acid assay	Amiri (2010)
T. oxylepis	Methanolic extract	DPPH, BR	Stanković et al. (2017)
T. parviflorum	Ethanolic and water extract	ABTS, DPPH, RP, SRSA, H_2O_2 scavenging capacity, Metal chelating activity	Turkoglu et al. (2010)
T. persicum	Essential oil	DPPH, FRAP, RP	Monsef-Esfahani et al. (2010)
T. pilosum	Isolated chemical compounds	DPPH, α-glucosidase inhibitory activity assay	Mou et al. (2009)
T. polium	Diethyl ether, ethyl acetate and n-butanolic extract	DPPH, HRSA, β-carotene/linoleic acid assay	Kadifkova Panovska et al. (2005)
	Water extract	β-carotene/linoleic acid assay, X/XOD, HRSA	Ljubunčić et al. (2006)
	Ethanolic extract	DPPH, TAP, TBA (*in vivo*)	Hasani et al. (2007)
	Methanolic extract	ABTS	Tarawneh et al. (2010)
	Methanolic, acetone and ethyl acetate extract	DPPH, FRAP, BR	Stanković et al. (2011b)
	Water, methanolic, acetone, ethyl acetate and petroleum ether extract	DPPH, FRAP, BR	Stanković (2012)
	Water, methanolic, acetone, ethyl acetate and petroleum ether extract	DPPH	Stanković et al. (2012)
	Methanolic extract	DPPH, FRAP	Belmekki and Bendimerad (2012)
	Methanolic, chloroform and ethyl acetate extract	XOA, SRSA, β-carotene/linoleic acid assay, Fe^{2+} chelating activity, DPPH, FRAP	Boumerfeg et al. (2012)
	Crude extracts and isolated chemical compounds	DPPH, RP, XOA, β-carotene/linoleic acid assay	De Marino et al. (2012)
	Isolated chemical compounds	DPPH, ABTS, ORAC	D'Abrosca et al. (2013)

(continued)

Table 10.1 (continued)

Teucrium species	Sample type – extract/essential oil	Method used	References
	Methanolic extract	DPPH, ABTS	Chedia et al. (2013)
	Essential oil	DPPH, β-carotene/linoleic acid assay	Mahmoudi and Nosratpour (2013)
	Ethanolic extract	DPPH, β-carotene/linoleic acid assay	Husein et al. (2014)
	Ethanolic extracts	DPPH, TAC	Vladimir-Knežević et al. (2014)
	Methanolic extract	TAC, DPPH, FRAP	Bilto et al. (2015)
	Essential oil and acetone extract	DPPH, RP, β-carotene/linoleic acid assay	Bakari et al. (2015)
	Essential oil and ethanolic extract	DPPH, RP	Kerbouche et al. (2015)
	Methanolic extract	DPPH, FRAP	Dridi et al. (2016)
	Methanolic extract	DPPH, BR	Stanković et al. (2017)
	Essential oil	DPPH, β-carotene/linoleic acid assay	Sayyad and Farahmandfar (2017)
	Water-ethanolic extract	DPPH	Fazeli-Nasab et al. (2017)
	n-hexane, dichloromethane, chloroform, ethyl acetate extract and methanolic fractions	ABTS	Dehghan and Sadani (2018)
	Essential oil, ethanolic extract	β-carotene/linoleic acid assay, FRAP, CUPRAC, ABTS, DPPH, SRSA	Bendjabeur et al. (2018)
	Methanolic, ethanolic, ethyl acetate, chloroform, hexane and water extract	DPPH, ABTS, RP	Ait Chaouche et al. (2018)
	Methanolic extract	DPPH, FRAP (*in vivo*)	Krache et al. (2018)
	Water and ethanolic extract	DPPH, β-carotene/linoleic acid assay, OSI	Farahmandfar et al. (2019)
	Methanolic, ethanolic, water and ethyl acetate extract	DPPH, FRAP, TAC	El Atki et al. (2019a)
	Essential oil	DPPH, FRAP, TAC	El Atki et al. (2019b)

Teucrium species	Sample type – extract/essential oil	Method used	References
T. pruinosum	Essential oil	DPPH	Jaradat et al. (2018)
T. pseudochamaepitys	Essential oil	DPPH	Hammami et al. (2015a)
T. pseudoscorodonia	Butanolic fraction of leaves, crude methanolic extract and ethyl acetate fraction	TAC, DPPH, FRAP, β-carotene/linoleic acid assay	Belarbi et al. (2017)
T. ramosissimum	Essential oil	X/XOD, NBT/riboflavine, DPPH	Ben Sghaier et al. (2010)
	Crude extracts	NBT/riboflavine, X/XOD	Ben Sghaier et al. (2011a)
	Crude extracts	NBT/riboflavine, DPPH, ABTS	Ben Sghaier et al. (2011b)
	Methanolic extract	CUPRAC, RP, FRAP	Ben Sghaier et al. (2016)
T. royleanum	Essential oil	DPPH	Saroglou et al. (2007)
T. salviastrum	Dichloromethane, ethanolic and water-ethanol extract; essential oil	DPPH	Cabral et al. (2010)
T. sandrasicum	Essential oil	TEAC	Ali et al. (2013)
	Water-ethanolic and water-methanolic extract	DPPH, ABTS, RP, β-carotene/linoleic acid, Phosphomolybdenum assays	Kaska et al. (2019)
T. sauvagei	Methanolic extract	DPPH, ABTS	Salah et al. (2006)
T. scordium	Water, methanolic, acetone, ethyl acetate and petroleum ether extract	DPPH, FRAP, BR	Stanković (2012)
T. siculum	Ethanolic extract	DPPH, SOD	Acquaviva et al. (2017)
T. stocksianum	Ecetone, butanolic, chloroform, ethyl acetate, ethanolic, methanolic, n-hexane and petroleum ether extract	DPPH	Rahim et al. (2013)
	Methanolic extract	DPPH	Shah et al. (2014)
	Water extract	DPPH	Shah and Shah (2015)
	Methanolic and water extract	DPPH	Uddin et al. (2016)

(continued)

Table 10.1 (continued)

Teucrium species	Sample type – extract/essential oil	Method used	References
T. subspinosum	Dichlormethane, acetone and ethanolic extract	DPPH, ABTS, β-carotene/ linoleic acid assay	Poli et al. (2007)
T. yemense	Essential oil	DPPH	Ali et al. (2017)

ABTS 2,2′-azino-bis-(3-ethylbenzothiazoline-6-sulphonic acid) radical scavenging capacity, *BR* Briggs-Rauscher oscillating reaction, *CUPRAC* Cupric Reducing Antioxidant Capacity assay, *DPPH* 2,2-diphenyl-1-picrylhydrazyl (DPPH) radical scavenging capacity, *EPR* Electron Paramagnetic Resonance radicals detection, *FRAP* Ferric Reducing Antioxidant Power, *HAPX* Hemoglobin Ascorbate Peroxidase activity inhibition, *HRSA* Hydroxyl Radical Scavenging Activity, *NBT/riboflavine* Nitroblue Tetrazolium riboflavine assay, *ORAC* Oxygen Radical Absorbance Capacity, *OSI* Oxidative Stability Index, *RP* Reducing Power, *SOD* Superoxide Dismutase Activity assay, *SRSA* Superoxide Radical Scavenging Activity, *TAC* Total Antioxidant Capacity, *TAP* Total Antioxidant Power, *TBA* Thiobarbituric Acid Assay, *TEAC* Trolox Equivalent Antioxidant Capacity, *X/XOD* Xanthine/Xanthine Oxidase assay, *XOA* Xanthine Oxidase Activity assay

antioxidant activity by analyzing extracts of different polarity from plant organs and whole herb using DPPH, FRAP and BR methods showed significant antioxidant activity of leaf methanolic extract (Stanković 2012). Antioxidant activity of *Teucrium arduini* was also investigated in a study of Kremer et al. (2013). The content of quercetin, ferulic and rosmarinic acid, as well as the antioxidant activity of ethanol leaf, flower and stem extracts using DPPH, Fe^{2+} chelating activity and *β*-carotene/linoleic acid assay were determined in this study. Further, the antioxidant activity of the ethanol extract of *Teucrium arduini* was examined by DPPH and TAC method (Vladimir-Knežević et al. 2014).

In a detailed study of the antioxidant activity of different extracts from *Teucrium barbeyanum*, ethyl acetate and buthanol extract showed significant antioxidant activity as well as highest totalphenolic content. In this study, 5-hydroxy-3,6,7,4′-tetramethoxyflavone, salvigenin, 5-hydroxy-6,7,3′,4′-tetramethoxyflavone, chrysosplenetin, cirsilineol, cirsimaritin, cirsiliol, apigenin, luteolin, methyl caffeate and 4-hydroxybenzoic acid were identified as significant chemical constituents (Alwahsh et al. 2015). The antioxidant activity of extracts of different polarity from *Teucrium* botrys were investigated by application of DPPH, FRAP and BR methods. The results showed that polar extracts of this species possess the highest antioxidant activity but moderate compared to other *Teucrium* species (Stanković 2012; Stanković et al. 2014). In the further studies, antioxidant activity and phenolic content of methanolic extract was determined for *Teucrium cavernarum*. In this experiment, DPPH, ABTS, TEAC assays were used for analysis. In qualitative study of secondary metabolites, forsythoside A, forsythoside B, luteolin-7-gucoside, cirsimaritin and cirsiliol were determined as the major compounds (Goger et al. 2019).

The antioxidant activity of *Teucrium chamaedrys* was determined for extracts of different polarity, infusions, as well as for the isolated chemical components. A significant level of antioxidant activity was determined in the study of diethyl ether, ethyl acetate and n-butanolic extract by DPPH, HRSA, *β*-carotene/linoleic acid assay (Kadifkova Panovska et al. 2005), as well as of water extract of *Teucrium chamaedrys* by DPPH, RP and TBA methods (Ozgen et al. 2006). In a study of the antioxidant activity of the isolated chemical components of *Teucrium chamaedrys*, significant activity was determined for phenylethanoid and iridoid glycosides (Pacifico et al. 2009). Remarkable antioxidant activity of this species was determined for polar fractions using DPPH, *β*-carotene/linoleic acid assays (Gursoy and Tepe 2009). In the comparative analysis of extracts with different polarity such as water, methanolic, acetone, ethyl acetate and petroleum ether extract, significant antioxidant activity correlated with the amount of phenolic compounds was determined. In these comparative studies, the variability of the amount of phenolic compounds and antioxidant activity were determined for the extracts from different plant parts of *Teucrium chamaedrys* such as leaf, flower and stem extracts, as well as for *Teucrium chamaedrys* samples at different stages of development, such as before flowering, during flowering, and after flowering. The results showed that leaf extracts obtained with polar solvents and extracts from plants during flowering period possess the highest amount of phenolic compounds and antioxidant activity (Stanković et al. 2010a; Stanković 2012; Stanković and Zlatić 2017). Further

studies analyzed the antioxidant activity and phytochemical composition of ethanolic extracts (Vladimir-Knežević et al. 2014; Vlase et al. 2014). In the tested ethanolic extracts, the presence of isoquercitrin, rutin, luteolin, quercitrin, *p*-coumaric, rosmarinic, chlorogenic and gentisic acid was determined. The antioxidant activity of ethanol extracts was confirmed by the DPPH and TAC methods (Vladimir-Knežević et al. 2014) as well as DPPH, TEAC, HAPX, EPR (Vlase et al. 2014). Antioxidant activity assay was conducted for methanolic-water (El Guiche et al. 2015) and methanolic (Faiku et al. 2019) extracts of *Teucrium chamaedrys* by DPPH method.

Within the group of the analyzed medicinal plant species, antioxidant activity was determined for *Teucrium creticum* and *Teucrium polium* using DPPH and *β*-carotene-linoleic acid assay. In addition, the amount of total phenolic compounds and flavonoids was determined, as well as a strong correlation between the amount of these compounds and free radical scavenging activity and total antioxidant activity (Husein et al. 2014). Various studies of biological activity have examined the antioxidant activity of *Teucrium flavum*. Hammami et al. 2015b identified *β*-caryophyllene and *α*-humulene as the most abundant components of essential oil and its moderate antioxidant activity. The antioxidant activity of the ethanolic extract was evaluated by application of DPPH and SOD-like methods (Acquaviva et al. 2017) as well as of the methanolic, ethanolic, acetone and ethyl acetate extract by DPPH method (Stanković and Zlatić 2017). The results showed that the extracts of this species have a significant amount of secondary metabolites with antioxidant activity. Also, *Teucrium* fruticans extracts were tested to evaluate the content of phenolic compounds as well as antioxidant and other biological activities (Acquaviva et al. 2017; Stanković and Zlatić 2017; Stanković et al. 2017). By testing methanolic, ethanolic, acetone and ethyl acetate extracts using DPPH, SOD-like and BR methods, the antioxidant activity of *Teucrium* fruticans extracts was determined. Also, methanolic extract from *Teucrium hyrcanicum* showed antioxidant activity and significant amount of flavonoids. In this study, antioxidant activity was determined using the DPPH and FRAP method (Golfakhrabadi et al. 2015).

The antioxidant activity of *Teucrium marum* has been studied by testing essential oil (Ricci et al. 2005) and extracts (Poli et al. 2007). Qualitative analysis of the essential oil revealed that the isocaryophyllene, *β*-bisabolene, *β*-sesquiphellandrene, *α*-santalene and *α*-caryophyllene were the most components. The antioxidant activity of the essential oil was evaluated using the DPPH test, 5-lipoxygenase test and XOA assay. Dichlormethane, acetone and ethanolic extract of *Teucrium marum* were analyzed by application of DPPH, ABTS, *β*-carotene/linoleic acid assay. A significant level of antioxidant activity was determined for polar *Teucrium marum* extracts. In the analysis of the essential oil chemical composition from *Teucrium massiliense*, within 35 identified compounds, 3,7-dimethyloctan-2-one, butyl 2-methylbutyrate, linalool, linalyl acetate, zingiberene, *γ*-cadinene are the most represented. In this study, the antioxidant activity of the essential oil was determined using the DPPH and lipid peroxidation methods (Giamperi et al. 2008).

The antioxidant activity of secondary metabolites of *Teucrium montanum* has been investigated in several studies. The antioxidant activity of the diethyl ether,

ethyl acetate and n-butanolic extract of this species was determined by the DPPH, HRSA, *β*-carotene/linoleic acid assays. The authors also found that the presence and quantitative composition of flavonoids were associated with high values for antioxidant activity (Kadifkova Panovska et al. 2005). Significant antioxidant activity of *Teucrium montanum* was confirmed by testing petroleum ether, chloroform, ethyl acetate, n-butanolic and water extract using DPPH method (Djilas et al. 2006). In a comparative study of the antioxidant activity of *Teucrium montanum*, water, methanolic, acetone, ethyl acetate and petroleum ether extract of the whole plant, flowers, leaves and stems were investigated using DPPH, FRAP, BR methods. In addition, antioxidant activity of the extracts obtained from *Teucrium montanum* sampled during different phenophases, such as before flowering, during flowering and after flowering were evaluated. The results of the studies indicate that the quantity of secondary metabolites as well as the antioxidant activity of the extracts depend on the plant part and the phenological phase of sample collection (Stanković et al. 2011a; Stanković 2012). Also, the variability in the amount and antioxidant activity of secondary metabolites in relation to the type of geological substrate on the habitat was examined for *Teucrium montanum*. The results of this comparative study showed that plants from harshest habitats possess a higher amount of secondary metabolites with possible adaptive significance, as well as notable antioxidant activity (Stanković et al. 2010b; Zlatić et al. 2017). Further studies analyzed the antioxidant activity of water (Vladimir-Knežević et al. 2014) and ethanolic (Nastić et al. 2018) extract of *Teucrium montanum*. The results of the phytochemical study of these extracts confirmed the presence of naringin, gallic, chlorogenic, caffeic and ferulic acid in the investigated extracts.

The total amount of phenolic compounds, flavonoids and flavonols was determined for the methanolic extract of *Teucrium montbretii*. In addition to these methods, antioxidant activity has been determined by use DPPH and phosphomolybdenum assay (Ozkan et al. 2007). In the phytochemical and pharmacological study of *Teucrium muscatense* n-hexane, chloroform, ethanolic, n-buthanolic and water extract was examined. Antioxidant activity was determined by DPPH, ABTS, TBA assays (Ur Rehman et al. 2017). GC/MS analysis of the essential oil showed higher percentage of linalool, linalyl acetate, and *β*-eudesmol. The antioxidant activity of *Teucrium* orientale was analyzed in two studies. In the first study, Chakir et al. (2006) analyzed the petroleum ether, chloroform, acetone and methanolic extract of this species using the DPPH and thiocyanate method. In this study, the authors found higher antioxidant activity as well as the amount of phenolic compounds for acetone extract during the flowering phase. In the second study, the essential oil and methanolic extract were tested using DPPH and *β*-carotene-linoleic acid assay (Amiri 2010). Forty compounds have been found in the essential oil of this species. The major constituents were linalool, caryophyllene oxide, 1,8-cineol, *β*-pinene, 3-octanol, *β*-caryophyllene and germacrene-D. High antioxidant activity was found for polar fractions of *Teucrium* orientale extracts.

In screening of phytochemical composition and biological activity of *Teucrium oxylepis*, methanolic extract was tested for the content of phenolic compounds as well as for the antioxidant activity by DPPH and BR methods (Stanković et al.

2017). Ethanolic and water extracts from leaves and flowers of *Teucrium parviflorum* were tested for antioxidant activity. Different methods were used for this study such as ABTS, DPPH, RP, SRSA, H_2O_2 scavenging capacity, metal chelating activity (Turkoglu et al. 2010.). For the essential oil of *Teucrium persicum* seasonal variation in chemical content and antioxidant activity has been studied (Monsef-Esfahani et al. 2010). Essential oil were analyzed during pre-flowering, flowering and post-flowering phenological phase. Analysis of the chemical profile of the isolated oils revealed the presence of sesquiterpene hydrocarbons, oxygenated monoterpenes and sesquiterpenes. They contained α-terpinyl acetate, α-cadinene, 1,4-cadinadiene, linalool and cadinol. For the analyzed oils, good antioxidant activity was measured by DPPF, FRAP and RP method. In the conducted research of *Teucrium pilosum* the authors are isolated chemical constituents and analyzed the antioxidant activity. The antioxidant activity was tested by the DPPH and alpha-glucosidase inhibitory activity assay for glyceryl tristearate, 2,5-dioxolanone, fernenol, stigmasta-5,22-dien-3Pol, 24-nor cholesta-5,22 (E)-dien-3beta-ol, ca-spinasterol, (3,4-dihydroxyphenyl) acrylate, 3,4-dihydroxy phenyl acrylic acid (Mou et al. 2009).

The antioxidant activity of *Teucrium polium* has been analyzed in a large number of studies. This species is one of the most tested medicinal plant from the *Teucrium* genus. In the conducted studies, different extracts, isolated components and essential oils were examined. In addition to investigating antioxidant activity, the quantitative composition of secondary metabolites was examined (Table 10.1). Significant antioxidant activity of *Teucrium polium* extracts have been reported in numerous studies. In the study of the antioxidant activity of *Teucrium polium*, extracts obtained using solvents of different polarity were investigated, such as acetone, chloroform, diethyl ether, ethanolic, ethyl acetate, hexane, methanolic, n-butanolic, petroleum ether, water and water-ethanolic extracts. Antioxidant activity has been determined using almost all known methods for antioxidant activity determination – DPPH, FRAP, β-carotene/linoleic acid assay, ABTS, BR, X/XOD, XOA, HRSA, TAP, TAC, TBA, SRSA, Fe^{2+} chelating activity, RP and OSI method (Kadifkova Panovska et al. 2005; Ljubunčić et al. 2006; Hasani et al. 2007; Tarawneh et al. 2010; Stanković et al. 2011b, 2012, 2017; Stanković 2012; Belmekki and Bendimerad 2012; Boumerfeg et al. 2012; Chedia et al. 2013; Husein et al. 2014; Vladimir-Knežević et al. 2014; Bilto et al. 2015; Dridi et al. 2016; Fazeli-Nasab et al. 2017; Ait Chaouche et al. 2018; Dehghan and Sadani 2018; Krache et al. 2018; Farahmandfar et al. 2019; El Atki et al. 2019a). Additionally, antioxidant activity of *Teucrium polium* has also been identified for isolated components of plant extracts such as flavones (D'Abrosca et al. 2013) as well as phenyletanoid and iridoid glycosides (De Marino et al. 2012). The composition and antioxidant activity of *Teucrium polium* essential oil have been analyzed in several studies. The results of these chemical analyzes indicate that spathulenol, β-pinene, β-myrcene, germacrene B, germacrene D, bicyclogermacrene, *t*-cadinol and linalool were the predominant components in *Teucrium polium* essential oil. The results also show significant antioxidant activity of the essential oil determined using various methods such as DPPH, β-carotene/linoleic acid assay, RP, FRAP, TAC and CUPRAC (Mahmoudi and Nosratpour 2013; Bakari et al. 2015; Kerbouche et al. 2015; Sayyad and

Farahmandfar 2017; Bendjabeur et al. 2018; El Atki et al. 2019b). In a comparative study of antioxidant activity, whole plant extracts and extracts from different plant parts such as leaves, flowers and stem were analyzed. The results of these tests show that *Teucrium polium* leaf extract contain the highest amount of phenolic compounds and show the significant antioxidant activity (Stanković 2012; Stanković et al. 2012). In some studies of the antioxidant activity and chemical composition of *Teucrium polium* extracts, the presence of various phenolic compounds such as luteolin, apigenin, myricetin, quercetin, kaempferol, resveratrol, coumaric, rosmarinic, chlorogenic, ferulic and caffeic acid (Kadifkova Panovska et al. 2005; Stanković 2012; Vladimir-Knežević et al. 2014).

In a study by Jaradat et al. (2018) the chemical composition of the essential oil and the biological activity were studied for *Teucrium pruinosum*. The authors found that agarospirol and caryophyllene were the most represented components of the essential oil. For the isolated essential oil, antioxidant activity was determined using the DPPH method in addition to other methods of biological activity. Also, the essential oil of *Teucrium pseudochamepitys* was analyzed in next experiment (Hammami et al. 2015a). Of 31 identifed compounds hexadecanoic acid was most abundant component followed by caryophyllene oxide, myristicin and α-cubebene. Belarbi et al. (2017) studied antioxidant activity as well as the amount of active secondary metabolites of different extracts from *Teucrium pseudoscorodonia.* For this experiment, the authors analyzed extracts and fractions of different polarity. Total phenolic, flavonoid, and condensed tannin contents followed by antioxidant assays such as TAC, DPPH, FRAP, β-carotene/linoleic acid assay were determined. The authors also showed a correlation of antioxidant activity with total phenolic and flavonoid content.

Essential oil and various extracts from *Teucrium ramosissimum* have been examined in several studies using different enzymatic and non-enzymatic methods such as X/XOD assay, NBT/riboflavine, ABTS, DPPH, CUPRAC, RP and FRAP assays. The result of conducted studies has demonstrated that *Teucrium ramosissimum* extracts possess potent antioxidant activity which could be derived from compounds such as polyphenols (Ben Sghaier et al. 2010, 2011a, b, 2016). The chemical composition of the essential oil and the antioxidant activity have also been investigated for *Teucrium royleanum* (Saroglou et al. 2007). For the essential oil of this species, sesquiterpene hydrocarbons are the most represented group of chemical compounds among which β-santalene and cis-α-bisabolene were the predominant compounds. The antioxidant activity was investigated using the DPPH method. By same method, antioxidant activity of dichloromethane, ethanolic and water-ethanol extract from *Teucrium salviastrum* was determined, as well as total phenolic content (Cabral et al. 2010). The antioxidant activity of the *Teucrium sandrasicum* was investigated for essential oil by TEAC method (Ali et al. 2013), as well as for water-ethanolic and water-methanolic extract using different methods such as DPPH, ABTS, RP, β-carotene/linoleic acid and phosphomolybdenum metal chelating assays (Kaska et al. 2019). In addition to testing for antioxidant activity, the composition of the essential oil was analyzed. The main components of the *Teucrium sandrasicum* essential oil were detected as α-pinene, β-pinene, germacrene-D and sabinene. For

the analyzed extracts, the presence of certain phenolic acids and flavonoids was examined. Also, the antioxidant activity of essential oil and the methanolic leaf extracts from *Teucrium sauvagei* was investigated by DPPH and ABTS method. Results of the analysis of the chemical composition of the essential oil show that β-eudesmol, T-cadinol, α-thujene, γ-cadinene, and sabinene were the most represented chemical components (Salah et al. 2006).

In order to analyze the antioxidant activity of *Teucrium scordium*, aqueous, water, methanolic, acetone, ethyl acetate and petroleum ether extract were tested using DPPH, FRAP and BR methods. The results showed that the methanolic extract possess the most significant level of antioxidant activity compared to the extracts obtained using other solvents (Stanković 2012). Infloresence ethanolic extract from *Teucrium siculum* was tested for antioxidant activity by DPPH and SOD-like methods. Based on phenolic content determined as well as the antioxidant activity, authors indicate important biological potential of this *Teucrium* species (Acquaviva et al. 2017). Also, *Teucrium stocksianum* extracts have been tested in various studies using the DPPH method. Acetone, butanolic, chloroform, ethyl acetate, ethanolic, water, methanolic, n-hexane and petroleum ether extract were tested in four studies (Rahim et al. 2013; Shah et al. 2014; Shah and Shah 2015; Uddin et al. 2016). Studies have shown that *Teucrium stocksianum* is characterized by significant antioxidant potential, especially extracts obtained by polar solvents. In a study of *Teucrium subspinosum*, in dichlormethane, acetone and ethanolic extract, the presence of verbascoside and arabinosyl-verbascoside has been determined. These compounds are associated with a significant level of antioxidant activity determined by DPPH, ABTS, β-carotene/linoleic acid assays (Poli et al. 2007). For *Teucrium yemense* essential oil Ali et al. (2017) identifed remarkable radical inhibition potential by DPPH method, as well as α-pinene, (E)-caryophyllene, α-humulene, δ-cadinene, caryophyllene oxide, α-cadinol and shyobunol as the major compounds.

Analysis of the presented papers on the study of the antioxidant activity of *Teucrium* species indicates that *Tucrium polium* (over 30 papers) is the most tested species followed by *Teucrium chamaedrys* (over ten papers), and *T. montanum* (over five papers), while less than five papers were found for the other analyzed *Teucrium* species (Table 10.1). The antioxidant activity of the extracts obtained by solvents of different polarity is the most common way of testing in relation to the antioxidant activity of the essential oils and isolated chemical components. The most commonly used methods include DPPH – 2,2-diphenyl-1-picrylhydrazyl (DPPH) radical scavenging capacity, Ferric Reducing Antioxidant Power (FRAP), β-carotene/linoleic acid assay, 2,2′-azino-bis-(3-ethylbenzothiazoline-6-sulphonic acid) radical scavenging capacity (ABTS) etc. In the conducted studies of the antioxidant activity of *Teucrium* species, the quantitative and qualitative composition of secondary metabolites have been analyzed in many studies. The most frequently analyzed is the amount of total phenolic compounds and flavonoids, as well as the chemical composition of essential oils (Ricci et al. 2005; Saroglou et al. 2007; Giamperi et al. 2008; Amiri 2010; Ben Sghaier et al. 2010; Cabral et al. 2010; Monsef-Esfahani et al. 2010; Mahmoudi and Nosratpour 2013; Bakari et al. 2015; Hammami et al. 2015a, b; Kerbouche et al. 2015; Ali et al. 2017; Sayyad and Farahmandfar 2017;

Bendjabeur et al. 2018; Jaradat et al. 2018; El Atki et al. 2019b) and plant extracts (Kadifkova Panovska et al. 2005; Poli et al. 2007; Stanković 2012; Kremer et al. 2013; Vladimir-Knežević et al. 2014; Vlase et al. 2014; Alwahsh et al. 2015; Goger et al. 2019; Nastić et al. 2018). Essential oil of *Teucrium* species is characterized by the presence of monoterpenes (limonene, α- and β-pinene) and sesquiterpenes (caryophyllene and its derivates, germacrene B and D, α-, γ-, δ-cadinene, cadinol, bisabolol), while phenolics of this genus is characterized by the presence of flavonoids (luteolin and apigenin), phenylethanoids (verbascoside) and phenolic acids (hydroxybenzoic and hydroxycinnamic acids). The significant correlation between the antioxidant activity and content of phenolic compounds, as well as the antioxidant potential of the presented chemical compounds in many individual studies (Quideau et al. 2011) indicates that species of the genus *Teucrium* are a source of secondary metabolites with significant antioxidant potential.

10.3.1 Correlation Between Phenolic Content and Antioxidant Activity

A significant correlation between antioxidant activities and phenolic content of *Teucrium* species was found for *Teucrium arduini* (Stanković 2012), *T. barbeyanum* (Alwahsh et al. 2015), *T. botrys* (Stanković 2012), *T. chamaedrys* (Ozgen et al. 2006; Gursoy and Tepe 2009; Stanković et al. 2010a; Stanković 2012; Zlatić et al. 2017), *T. montanum* (Stanković et al. 2011a; Stanković 2012; Zlatić et al. 2017), *T. orientale* (Amiri 2010), *T. polium* (Stanković 2012; Stanković et al. 2012; Vladimir-Knežević et al. 2014; Bakari et al. 2015; Dehghan and Sadani 2018; Krache et al. 2018; El Atki et al. 2019a) and *T. scordium* (Stanković 2012).

When examining the results of the numerous studies in various fields of science, the correlation coefficient is used as a measure for the total ratio of statistical relationships. The correlation coefficient is a parameter of summarizing the relationship between two variables in a single number. The correlation coefficient has values between −1 and + 1 and is it marked with the symbol (r). When the correlation values are close to 0, it is understood that there is no significant correlation between the variables being compared, or it is very small. When the correlation values are close to +1 it means that there is a positive relationship between the two variables. This further implies that as the value of the first variable increases, the values of the second variable increase, which is why they are in a positive correlation. When the correlation values are −1, this indicates a negative relationship between the two variables, which means that as the values of the first variable increase, the values of the second variable decrease (Schober et al. 2018).

In biological research, the correlation coefficient is used as one of the numerous statistical methods. When comparing the results of the total amount of phenolic compounds in the plant extracts and their antioxidant activity, the determination of the correlation coefficient is an essential part. Phenols are very important plant

constituents because of their scavenging ability on free radicals due to the presence of hydroxyl groups. Therefore, the phenolic content of plants may contribute directly to their antioxidant activity (Tosun et al. 2009). Numerous researchers have shown that there is a highly significant correlation between phenolics and antioxidant activity which supports the hypothesis that phenolic compounds directly contribute to the total antioxidant capacity of plant species (Tosun et al. 2009; Moein and Moein 2010; Stanković et al. 2010a; Alwahsh et al. 2015; Zlatić et al. 2017). Stanković et al. (2010a) reported a significant correlation ($r = 0.701$) between total phenolic compounds and antioxidant activity of different plant extracts of *Teucrium chamaedrys*. Bakari et al. (2015) noted a strong positive correlation between total phenols and IC_{50} of DPPH (coefficient of determination $R^2 = 0.999$) for the acetone extract of species *Teucrium polium*. Also, for the same species is determined a significant correlation between antioxidant activities and contents of polyphenols in methanol, ethanol, water, ethyl acetate, and petroleum ether extracts (Stanković et al. 2012). El Atki et al. (2019a) studied a correlation between the results of antioxidant tests and the contents of total phenolics of two *Teucrium polium* subspecies. They reported a significant and negative correlation between DPPH and total phenols ($r^2 = -0.78$). For the ferric reducing power test, the IC_{50} values were significantly and negatively correlated with total phenols ($r^2 = -0.81$). Zlatić et al. (2017) showed a significant positive correlation between total phenols and antioxidant activity in extract $r = 0.900$ ($p < 0.05$) and plant material $r = 0.957$ ($p < 0.01$) for *Teucrium chamaedrys* and *T. montanum* species sampled on limestone and serpentinite geological substrates. For six species and one subspecies of the genus *Teucrium* examined by Stanković (2012), a correlation factor ($r = 0.911$) was determined between the total phenol content and the antioxidant activity of DPPH by the 140 tested extracts. The previous research has shown that species belonging to the genus *Teucrium* can be used as good antioxidants because of the strong correlation between total phenol content and antioxidant activity.

The presented values for the correlation between total phenol content and antioxidant activity (DPPH) had different values. Since a lower IC_{50} value indicates higher antioxidant activity, it can be observed that with increasing phenol content, activity against DPPH radicals increases, i.e. the IC_{50} value decreases. The linear correlation is therefore negative and strong. Some results of correlation analysis indicate that antiradical activity is not necessarily correlated with the content of phenolic compounds. When considering the antioxidant potential of the sample, it is necessary to analyze content of all present active components. Some studies suggest that it is not always possible to correlate total phenolics and antioxidant capacity. This could be explained by the presence of different active compounds in the plant and the structural forms included in the nature of the phenolic groups. Phenolics include complex compounds with one or more hydroxyl groups (-OH), with phenylpropanoid side chains and with additional aromatic/heterocyclic ring structures, which differentiates the antioxidant activity of the tested extracts. Antioxidant activity is not simply reflective of total phenolic substances in a complex mixture. Antioxidant activity depends on the number of molecules of certain compounds and their structural forms that have potential radical scavenging

activity. This indicates that phenolic compounds are partial, if not completely, responsible for the antioxidant activity observed in the tested species of the genus *Teucrium* (Stanković 2012).

10.3.2 Methods Used for Determining the Antioxidant Activities of Teucrium *Species*

For determination of the antioxidant activity of plant extracts, essential oils as well as isolated chemical compounds from *Teucrium* species, about 30 analytical methods have been used (Table 10.1). The most commonly applied assays include DPPH – 2,2-diphenyl-1-picrylhydrazyl (DPPH) radical scavenging capacity followed by Ferric Reducing Antioxidant Power (FRAP), β-carotene/linoleic acid assay and 2,2′-azino-bis-(3-ethylbenzothiazoline-6-sulphonic acid) radical scavenging capacity (ABTS), as well as moderately applied methods such as: Total Antioxidant Capacity (TAC), Reducing Power (RP), Briggs-Rauscher oscillating reaction (BR), Hydroxyl Radical Scavenging Activity (HRSA), Trolox Equivalent Antioxidant Capacity (TEAC), Thiobarbituric Acid Assay (TBA), Nitroblue Tetrazolium riboflavine assay (NBT/riboflavine), Superoxide Dismutase Activity assay (SOD), Superoxide Radical Scavenging Activity (SRSA), Xanthine/Xanthine Oxidase assay (X/XOD), XOA – Xanthine Oxidase Activity assay (XOA), Fe^{2+} chelating activity, Cupric Reducing Antioxidant Capacity assay (CUPRAC), Phosphomolybdenum assay, Electron Paramagnetic Resonance radicals detection (EPR), Hemoglobin Ascorbate Peroxidase activity inhibition (HAPX), Oxygen Radical Absorbance Capacity (ORAC), Oxidative Stability Index (OSI), Total Antioxidant Power (TAP), 5-lipoxygenase test, α-glucosidase inhibitory activity assay, H_2O_2 scavenging capacity, Thiocyanate method and Metal chelating activity and Inhibition of lipid peroxide formation aasay.

Each listed method will always have advantages and disadvantages, which need to be taken into account in terms of complexity, sample preparation, required reagents and equipment, the chemical mechanism, the quantification method, values presentation and its relevance in biological systems. Scientific evidence indicates that DPPH is the most used assay for the analysis of the antioxidant activity (Table 10.1). The method was described by Brand-Williams et al. (1995) and adopted with suitable modifications for different samples of natural and synthetic origin. The DPPH method is based on the ability of the stable 2, 2-diphenyl-1-picrylhydrazyl (DPPH) free radical to react with different hydrogen donors as well as absorbance reduction of the radical anion DPPH• by antioxidants. Based on the chemical structure, DPPH is very stable organic nitrogen free radical. In dry form, the DPPH is a dark-purple colored powder with absorbance peak at 515–528 nm. After dissolution, dark-purple color loses its intensity, depending on the concentration. In the reduction reaction by antioxidant, the color of the solution changes to yellow, based on the concentration of the reduced molecules in the reaction mixture.

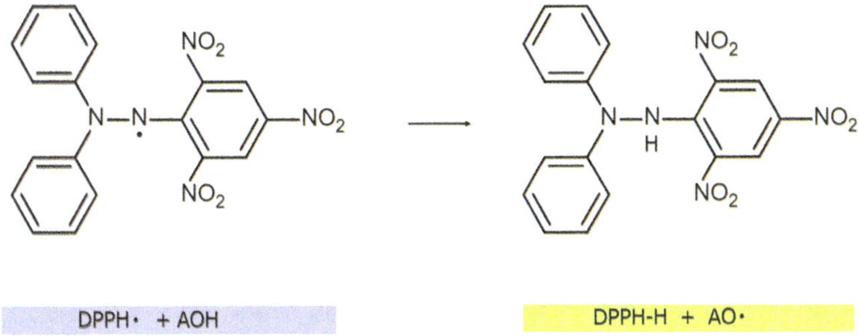

Fig. 10.1 Structure of DPPH and its reduction by an antioxidant

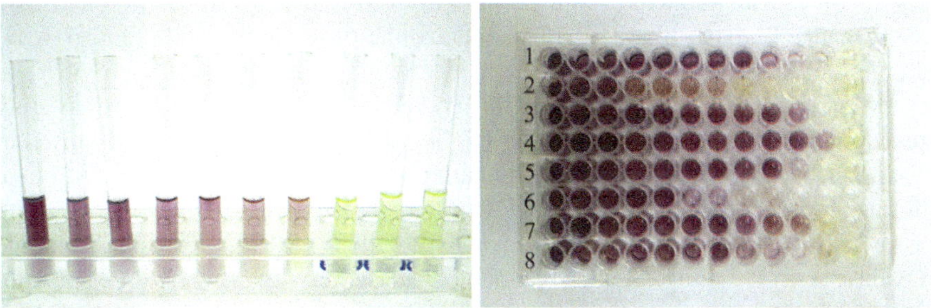

Fig. 10.2 Samples prepared for DPPH assay after reaction with reagent – sample series with increasing concentration gradient of the one plant extract for spectrophotometric absorbance reading and eight plant extract series for microplate absorbance reading

Antioxidant compounds as donors of the hydrogen atom reduce the stable 1,1-diphenyl-2-picrylhydrazine radical by transforming it into 1,1-diphenyl-2-(2,4,6-trinitrophenyl)-hydrazine (Fig. 10.1).

During this process, the purple color of the initial solution derived from the stable DPPH radical is changed to the yellow due to the presence of a newly formed 1,1-diphenyl-2-(2,4,6-trinitrophenyl)-hydrazine which resulting in a decrease in absorbance at 517 nm. This property allows visual and spectrophotometric monitoring of the reaction, i.e., the quantity of radicals at the beginning can be determined by changing of the absorbance at 517 nm or by the ECP signal of the DPPH radical. In an experimental study of the antioxidant activity of plant extracts, the intensity of color change from violet to yellow is in correlation with the antioxidative potential of the plant extract (Fig. 10.2), which can be monitored by spectrophotometric measurement. Based on this, the DPPH method is widely used to measure the ability of different substances to act as free radical scavengers or hydrogen donors, as well as to evaluate antioxidant activity of plant compounds as plant extracts or isolated metabolites (Stanković et al. 2019).

Plant extracts obtained by various solvents have been found to be the most common form of samples for testing the antioxidant activity of *Teucrium* species followed by essential oils as well as isolated chemical components. In studies of the antioxidant activity of *Teucrium* species methanolic, ethanolic, acetone, n-butanolic, chloroform, ethyl acetate, n-hexane, petroleum ether, dichlormethane, diethyl ether and water extracts were tested (Table 10.1). Essential oils of *Teucrium* species are characterized by significant chemical composition and biological activity. In previous studies, the antioxidant activity and chemical composition of essential oils have been investigated for *Teucrium flavum* (Hammami et al. 2015b), *T. marum* (Ricci et al. 2005), *T. massiliense* (Giamperi et al. 2008), *T. orientale* (Amiri 2010), *T. persicum* (Monsef-Esfahani et al. 2010), *T. polium* (Mahmoudi and Nosratpour 2013; Bakari et al. 2015; Kerbouche et al. 2015; Sayyad and Farahmandfar 2017; Bendjabeur et al. 2018; El Atki et al. 2019b), *T. pruinosum* (Jaradat et al. 2018), *T. pseudochamaepitys* (Hammami et al. 2015a), *T. ramosissimum* (Ben Sghaier et al. 2010), *T. royleanum* (Saroglou et al. 2007), *T. salviastrum* (Cabral et al. 2010) and *T. yemense* (Ali et al. 2017). In addition to investigating the chemical composition and biological activity of plant extracts and essential oils, some of the more detailed studies have assaying isolated chemical components. The antioxidant activity of isolated chemical constituents for the species of the genus *Teucrium* was determined for phenylethanoid and iridoid glycosides from *Teucrium chamaedrys* (Pacifico et al. 2009), glyceryl tristearate, 2,5-dioxolanone, fernenol, stigmasta-5,22-dien-3Pol, 24-nor cholesta-5,22 (E)-dien-3beta-ol, ca-spinasterol, (3,4-dihydroxyphenyl) acrylate, 3,4-dihydroxy phenyl acrylic acid from *Teucrium pilosum* (Mou et al. 2009) and flavones (D'Abrosca et al. 2013) as well as phenyletanoid and iridoid glycosides (De Marino et al. 2012) from *Teucrium polium*.

10.3.3 Antioxidant Activity of Traditionally Used **Teucrium** *Species*

Some species of the genus *Teucrium* are known medicinal plants with long traditional use (Fig. 10.3). Teas and other herbal remedies are used to treat certain diseases and disorders of the digestive and respiratory organs as well as skin disorders and diabetes (Stanković 2012). Investigations of biological activity as well as the phytochemical composition of many plants are based on their previous traditional use. In the studies of the *Teucrium* species conducted so far, many medicinal species of this genus has been the subject of different biological activities. The antioxidant activity of extracts, essential oils, and isolated chemical components has been tested for many medicinal species of the genus *Teucrium*, such as *T. chamaedrys* (Kadifkova Panovska et al. 2005; Ozgen et al. 2006; Pacifico et al. 2009; Gursoy and Tepe 2009; Stanković et al. 2010a; Stanković 2012; Vladimir-Knežević et al. 2014; Vlase et al. 2014; El Guiche et al. 2015; Stanković and Zlatić 2017; Zlatić et al. 2017; Faiku et al. 2019), *T. creticum* (Husein et al. 2014), *T. flavum* (Hammami et al. 2015b; Acquaviva et al. 2017; Stanković and Zlatić 2017), *T. massiliense* (Giamperi et al.

Fig. 10.3 Selected species of the genus *Teucrium* with traditional ethnomedicinal use. (Photo M. Stanković)

2008), *T. montanum* (Kadifkova Panovska et al. 2005; Djilas et al. 2006; Stanković et al. 2011a; Stanković 2012; Vladimir-Knežević et al. 2014; Zlatić et al. 2017; Nastić et al. 2018), *T. arduini* (Šamec et al. 2010; Stanković 2012; Kremer et al. 2013; Vladimir-Knežević et al. 2014), *T. polium* (Kadifkova Panovska et al. 2005; Ljubunčić et al. 2006; Hasani et al. 2007; Tarawneh et al. 2010; Stanković et al. 2011b, 2012, 2017; Stanković 2012; Belmekki and Bendimerad 2012; Boumerfeg et al. 2012; De Marino et al. 2012; D'Abrosca et al. 2013; Chedia et al. 2013;

Mahmoudi and Nosratpour 2013; Husein et al. 2014; Vladimir-Knežević et al. 2014; Bilto et al. 2015; Bakari et al. 2015; Kerbouche et al. 2015; Dridi et al. 2016; Sayyad and Farahmandfar 2017; Fazeli-Nasab et al. 2017; Dehghan and Sadani 2018; Bendjabeur et al. 2018; Ait Chaouche et al. 2018; Krache et al. 2018; Farahmandfar et al. 2019; El Atki et al. 2019a, b), *T. pruinosum* (Jaradat et al. 2018), *T. pseudochamaepitys* (Hammami et al. 2015a), *T. ramosissimum* (Ben Sghaier et al. 2010, 2011a, b, 2016), *T. sandrasicum* (Ali et al. 2013; Kaska et al. 2019), *T. stocksianum* (Rahim et al. 2013; Shah et al. 2014; Shah and Shah 2015; Uddin et al. 2016) *T. scordium* (Stanković 2012) and *T. yemense* (Ali et al. 2017). The results of the above studies of antioxidant activity of *Teucrium* medicinal species, as well as the amounts and diversity of their secondary metabolites as bioactive compounds justify their previous traditional use in the treatment of many diseases and health disorders.

10.3.4 *In Vivo Study of the Antioxidant Activity of* Teucrium *Species*

For extracts, essential oils, as well as isolated chemical components of *Teucrium* species, antioxidant activity was investigated using a model system under in vitro conditions. A review of the literature indicates that some experiments for the determination of antioxidant activity have been carried out using an in vivo mode system, whereby selected laboratory animals have been treated with bioactive substances of these plant species. Hasani et al. (2007) investigated the in vivo antioxidant potential of ethanolic extract from *Teucrium polim* and compared the obtained results with α-tocopherol as a reference antioxidant for comparison. The ethanol extract was administered to laboratory Wistar rats by intragastric intubation at various concentrations. Antioxidant activity was measured by three methods (DPPH, TAP and TBA) in serum. Results obtained by used assays indicate that *Teucrium polium* effectively inhibits oxidative stress in vivo. An in vivo study of antioxidant activity was also conducted by Krache et al. (2018) for methanolic extract of *Teucrium polium*. Adults of laboratory mice were treated with methanol extract of *Teucrium polium*, after which the blood parameters of antioxidant activity were measured using several methods. Significant antioxidant potential of *T. polium* has been demonstrated using these in vivo model systems.

10.3.5 *Variability of the Antioxidant Activity of* Teucrium *Species*

Variability during vegetative period, habitat-related variability as well as variability between plant parts were determined for the antioxidant activity and content of the active compounds of some *Teucrium* species. Dynamics of the phenolic content and

antioxidant activity for *Teucrium* orientale (Cakir et al. 2006), *Teucrium chamaedrys* and *Teucrium montanum* (Stanković 2012), as well as essential oil composition and antioxidant activity of *Teucrium persicum* (Monsef-Esfahani et al. 2010) during several successive phases such as before flowering, during flowering and after flowering has been investigated. The results of these studies showed that plants during the flowering phase contain a significant amount of bioactive compounds and exhibits the strongest antioxidant activity. Habitat-related variability has been studied for *Teucrium* species in several comparative studies. Variability in the amount and composition of secondary metabolites as well as antioxidant activity in relation to the type of geological substrate, soil mineral composition, habitat type altitude, latitude etc., were investigated for *Teucrium chamaedrys* (Stanković et al. 2011c; Stanković 2012, 2018; Zlatić et al. 2017; Faiku et al. 2019), *T. montanum* (Stanković 2011, 2012; Zlatić et al. 2017), *T. polium* (Stanković 2012, 2013) and *T. scordium* (Stanković 2012). The results of the studies indicate significant ecological variability of the investigated parameters as well as their adaptive importance to the specific habitats. Ecological conditions on the habitats with a specific geological substrate such as serpentinite, halophytic habitats as well as high altitude habitats all affect the phenolic content and diversity and antioxidant activity of the investigated *Teucrium* species. The plant part variability determination is based on a comparative analysis of the antioxidant activity of extracts obtained from different plant parts such as inflorescences, flowers, leaves, stems etc. The variability of the phenolic content and antioxidant activity in relation to plant part of *Teucrium* species was investigated for *Teucrium arduini* (Šamec et al. 2010; Stanković 2012), *T. botrys* (Stanković 2012; Stanković et al. 2014), *T. chamaedrys* (Stanković et al. 2010a; Stanković 2012), *T. flavum*, *T. fruticans* and *T. siculum* (Acquaviva et al. 2017), *T. montanum* (Stanković et al. 2010b, 2011a; Stanković 2012), *T. polium* (Ljubunčić et al. 2006; Stanković 2012; Stanković et al. 2012) and *T. scordium* (Stanković 2012). The obtained results showed that the leaves and flower parts are the richest in active compounds and that they exhibit significant antioxidant activity compared to the other plant parts.

10.4 Conclusions

Based on the long traditional application and the significant content of secondary metabolites, the antioxidant activity of *Teucrium* species has been analyzed in numerous studies using various methodological approaches. It was found over 80 references about antioxidant activity of *Teucrium* species, as well as results for 34 tested *Teucrium* species. *Tucrium polium* (over 30 papers) is the most tested species followed by *Teucrium chamaedrys* (over ten papers), and *T. montanum* (over five papers), while less than five papers were found for the other analyzed *Teucrium* species.

For determination of the antioxidant activity of *Teucrium* species, about 30 analytical methods have been used. The most commonly applied methodological

Briskin DP (2000) Medicinal plants and phytomedicines, linking plant biochemistry and physiology to human health. Plant Physiol 124:507–514

Cabral C, Francisco V, Cruz M, Lopes M, Salgueiro L, Sales F, Batista M (2010) Potential antioxidant and anti-inflammatory properties in *Teucrium salviastrum* Schreb. Planta Med 76:237

Cakir A, Mavi A, Kazaz C, Yildirim A, Kufrevioglu OI (2006) Antioxidant activities of the extracts and components of *Teucrium orientale* L. var. *orientale*. Turk J Chem 30:483–494

Chedia A, Ghazghazi H, Brahim H, Abderrazak M (2013) Secondary metabolite, antioxidant and antibacterial activities of *Teucrium polium* L. methanolic extract. Int J Agron Plant Prod 4:1790–1797

D'Abrosca B, Pacifico S, Scognamiglio M, D'Angelo G, Galasso S, Monaco P, Fiorentino A (2013) A new acylated flavone glycoside with antioxidant and radical scavenging activities from *Teucrium polium* leaves. Nat Prod Res 27:356–363

De Marino S, Festa C, Zollo F, Incollingo F, Raimo G, Evangelista V, Iorizzi M (2012) Antioxidant activity of phenolic and phenylethanoid glycosides from *Teucrium polium* L. Food Chem 133:21–28

Dehghan H, Sadani S (2018) Antioxidant activity and phenolic content of the fractions from some Iranian antidiabetic plants. ASAG 2:73–77

Djilas SM, Markov SL, Cvetković DD, Čanadanović-Brunet MJ, Ćetković SG, Tumbas TV (2006) Antimicrobial and free radical scavenging activities of *Teucrium montanum*. Fitoterapia 77:401–403

Dridi A, Hadef Y, Bouloudani L (2016) Determination of total phenol, flavonoid, antioxidant and antimicrobial activity of methanolic extract of *Teucrium polium* L. in Algerian east. IJPPR 8:1566–1570

Edreva A, Velikova V, Tonsev T, Dagnon S, Gurel A, Aktas L, Gesheva E (2008) Metabolites: diversity of functions and mechanisms. Gen Appl Plant Physiol 34:67–78

El Atki Y, Aouam I, El Kamari F, Taroq A, Lyoussi B, Taleb M, Abdellaoui A (2019a) Total phenolic and flavonoid contents and antioxidant activities of extracts from *Teucrium polium* growing wild in Morocco. Mater Today Proce 13:777–783

El Atki Y, Aouam I, El Kamari F, Taroq A, Lyoussi B, Oumokhtar B, Abdellaoui A (2019b) Phytochemistry, antioxidant and antibacterial activities of two Moroccan *Teucrium polium* L. subspecies: preventive approach against nosocomial infections. Arab J Chem 13(2):3866–3874. https://doi.org/10.1016/j.arabjc.2019.04.001

El Guiche R, Tahrouch S, Amri O, El Mehrach K, Hatimie A (2015) Antioxidant activity and total phenolic and flavonoid contents of 30 medicinal and aromatic plants located in the south of Morocco. IJNTR 1:7–11

Faiku F, Buqaj L, Haziri A (2019) Phytochemicals and antioxidant study of *Teucrium chamaedrys* (L.) plant. Agric For 65:137–145

Farahmandfar R, Asnaashari M, Bakhshandeh T (2019) Influence of ultrasound-assist and classical extractions on total phenolic, tannin, flavonoids, tocopherol and antioxidant characteristics of *Teucrium polium* aerial parts. J Food Meas Charact 13:1357–1363

Fazeli-Nasab B, Rahnama M, Mazarei A (2017) Correlation between antioxidant activity and antibacterial activity of nine medicinal plant extracts. Mazandaran Univ Med Sci 27:63–78

Fraga GC (ed) (2010) Plant phenolics and human health: biochemistry, nutrition, and pharmacology. Wiley, Hoboken

Giamperi L, Bucchini A, Fraternale D, Cara P, Ricci D, Epifano F, Genovese S, Curini M (2008) Chemical composition and antioxidant activity of the essential oil of *Teucrium massiliense* L. J Essent Oil Res 20:446–449

Goger F, Kaya A, Dinc M, Dogu S (2019) Phenolic compounds determination and antioxidant activity of *Teucrium cavernarum*. Eskişehir Tech Univ J Sci Technol C 8(2):229–237

Golfakhrabadi F, Yousefbeyk F, Mirnezami T, Laghaei P, Hajimahmoodi M, Khanavi M (2015) Antioxidant and antiacetylcholinesterase activity of *Teucrium hyrcanicum*. Pharm Res 7:15–19

Gupta VK, Sharma SK (2006) Plants as natural antioxidants. Nat Prod Rad 5(4):326–334

Shah SMM, Sadiq A, Shah SMH, Ullah F (2014) Antioxidant, total phenolic contents and antinociceptive potential of *Teucrium stocksianum* methanolic extract in different animal models. BMC Complement Altern Med 14:181

Stanković SM (2011) Ecological study of *Teucrium montanum* L. – population, phenological and plant part variability of secondary metabolites concentration. In: Abstracts of the international botanical congress "Botanikertagung 2011", Berlin, Germany, 18–22 September 2011

Stanković M (2012) Biological effects of secondary metabolites of *Teucrium* species of Serbian flora. Dissertation, University of Kragujevac

Stanković SM (2013) Inter-population variability in secondary metabolites content of *Teucrium polium* L. from the localities in the Balkan Peninsula. In: Abstracts of the 4th Croatian botanical symposium with international participation, Split, Croatia, 27–29 September 2013

Stanković M (2018) Interpopulation variability of *Teucrium chamaedrys* L. (Lamiaceae) phenolic compounds. Paper presented at the 2nd congress of biologists, Serbia, 25–30 September 2018

Stanković M, Zlatić N (2017) The total quantity of phenolic compounds and antioxidant activity of the selected species of the genus *Teucrium*. In: Proceedings of the 22th conference about biotechnology with international participation, University of Kragujevac, Serbia, 10–11 March 2017

Stanković SM, Topuzović M, Solujić S, Mihailović V (2010a) Antioxidant activity and concentration of phenols and flavonoids in the whole plant and plant parts of *Teucrium chamaedrys* L. var. *glanduliferum* Haussk. J Med Plant Res 4:2092–2098

Stanković SM, Topuzović M, Solujić S, Pavlović D, Marković A, Nićiforović N, Mihailović V (2010b) Total phenol and flavonoid content of extracts from different plant parts of *Teucrium montanum* L. 58th international congress and annual meeting of the Society for Medicinal Plant and Natural Product Research as well as the 7th Tannin conference offered as presymposium, Berlin, Germany. Planta Med 76(12):1261–1262

Stanković SM, Nićiforović N, Topuzović M, Solujić S (2011a) Total phenolic content, flavonoid concentrations and antioxidant activity, of the whole plant and plant parts extracts from *Teucrium montanum* L. var. *montanum*, f. *supinum* (L.) Reichenb. Biotechnol Biotechnol Equip 25:2222–2227

Stanković SM, Jukić M, Burčul F, Miloš M, Politeo O, Carev I (2011b) Biological effects and phenolic content of felty germander (*Teucrium polium* L. subsp. *polium*). 59th international congress and annual meeting of the Society for Medicinal Plant and Natural Product Research, Antalya, Turkey. Planta Med 77(12):1397

Stanković SM, Vassilev K, Stanković NM, Milošević T, Topuzović M, Marković A, Solujić S (2011c) Inter-population variation in phenolic content of *Teucrium chamaedrys* L. from the localities in the Balkan Peninsula. 59th international congress and annual meeting of the Society for Medicinal Plant and Natural Product Research, Antalya, Turkey. Planta Med 77(12):1374

Stanković SM, Nićiforović N, Mihailović V, Topuzović M, Solujić S (2012) Antioxidant activity, total phenolic content and flavonoid concentrations of different plant parts of *Teucrium polium* L. subsp. *polium*. Acta Soc Bot Pol 81:117–122

Stanković SM, Jakovljević D, Topuzović M, Zlatković B (2014) Antioxidant activity and contents of phenolics and flavonoids in the whole plant and plant parts of *Teucrium botrys* L. Oxid Commun 37:522–532

Stanković SM, Stojanović-Radić Z, Blanco-Salas J, Vázquez-Pardo MF, Ruiz-Téllez T (2017) Screening of selected species from Spanish flora as a source of bioactive substances. Ind Crop Prod 95:493–501

Stanković M, Jakovljević D, Stojadinov M, Stevanović ZD (2019) Halophyte species as a source of secondary metabolites with antioxidant activity. In: Hasanuzzaman M, Nahar K, Öztürk M (eds) Ecophysiology, abiotic stress responses and utilization of halophytes. Springer, Singapore

Starakis I, Siagris D, Leonidou L, Mazokopakis E, Tsamandas A, Karatza C (2006) Hepatitis caused by the herbal remedy *Teucrium polium* L. Eur J Gastroenterol Hepatol 18(6):681–683

Tarawneh KA, Irshaid F, Jaran AS, Ezealarab M, Khleifat KM (2010) Evaluation of antibacterial and antioxidant activities of methanolic extracts of some medicinal plants in northern part of Jordan. J Biol Sci 10:325–332

Tosun M, Ercisli S, Sengul M, Ozer H, Polat T (2009) Antioxidant properties and total phenolic content of eight *Salvia* species from Turkey. Biol Res 41:175–181

Turkoglu S, Celik S, Turkoglu D, Cakilcioglu U, Bahsi M (2010) Determination of the antioxidant properties of ethanol and water extracts from different parts of *Teucrium parviflorum* Schreber. Afr J Biotechnol 9:6797–6805

Uddin G, Rauf A, Siddiqui BS, Khan H, Barkatullah UR (2016) Antinociceptive, antioxidant and phytochemical studies of Pakistani medicinal plants. Pak J Pharm Sci 29:929–933

Ur Rehman N, Al-Sahai JMS, Hussain H, Khan AL, Gilani AS, Abbas G, Hussain J, Sabahi NJ, Al-Harrasi A (2017) Phytochemical and pharmacological investigation of *Teucrium muscatense*. Int J Phytomed 8:567–579

Valko M, Rhodes CJ, Moncol J, Izakovic M, Mazur M (2006) Free radicals, metals and antioxidants in oxidative stress-induced cancer. Chem Biol Interact 160:1–40

Vladimir-Knežević S, Blažeković B, Kindl M, Vladić J, Lower-Nedza AD, Brantner AH (2014) Acetylcholinesterase inhibitory, antioxidant and phytochemical properties of selected medicinal plants of the Lamiaceae family. Molecules 19:767–782

Vlase L, Benedec D, Hanganu D, Damian G, Csillag I, Sevastre B, Mot AC, Silaghi-Dumitrescu R, Tilea I (2014) Evaluation of antioxidant and antimicrobial activities and phenolic profile for *Hyssopus officinalis, Ocimum basilicum* and *Teucrium chamaedrys*. Molecules 19:5490–5550

Wink M (2004) Phytochemical diversity of secondary metabolites. Encyclopedia of plant and crop science. Marcel Dekker, New York

Zlatić MN, Stanković SM, Simić SZ (2017) Secondary metabolites and metal content dynamics in *Teucrium montanum* L. and *Teucrium chamaedrys* L. from habitats with serpentine and calcareous substrate. Environ Monit Assess 189:110

Zwenger S, Basu C (2008) Plant terpenoids: applications and future potentials. Biotechnol Mol Biol Rev 3:1–7

Chapter 11
Antiviral Activity of Secondary Metabolites of *Teucrium* Species

Elma Vuko, Nada Bezić, Marija Nazlić, and Valerija Dunkić

Abstract Unefficiency of chemical treatments and recent plant protection that is focused on natural products, have pointed the importance of finding natural substances that can help in a control of plant virus diseases. Current knowledge about the antiviral effects of essential oils, although limited, indicates the potential of these secondary metabolites to control or at least reduce the spread of viral infection. Essential oils isolated form *Teucrium* species showed antiviral activity in tobacco mosaic virus (TMV) and cucumber mosaic virus (CMV) infected plants. The aim of this chapter is to present findings dealing with antiphytoviral potential of selected species of genus *Teucrium* and to improve the knowledge about possible application of essential oils for the prevention of diseases caused by plant viruses. Essential oil of *Teucrium arduini* applied on the leaves of local host plants, significantly reduced the number of lesions on both TMV and CMV infected plants. Aside from *Teucrium arduini*, four additional *Teucrium* species (*T. montanum, T. polium, T. chamaedrys* and *T. flavum*) are also source of bioactive molecules with antiphytoviral activity against CMV infection. This promising biological activity of *Teucrium* species opens new area of research that could help in a control of virus diseases of plants.

Keywords *Teucrium* · Essential oil · Antiphytoviral activity · CMV · TMV · β-caryophyllene · Caryophyllene oxide

Abbreviations

CMV	Cucumber mosaic virus
EO	Essential oil
TMV	Tobacco mosaic virus

E. Vuko · N. Bezić · M. Nazlić · V. Dunkić (✉)
Department of Biology, Faculty of Science, University of Split, Split, Croatia
e-mail: dunkic@pmfst.hr

M. Stanković (ed.), *Teucrium Species: Biology and Applications*,
https://doi.org/10.1007/978-3-030-52159-2_11

11.1 Introduction

The aromatic plants produce a diverse array of organic compounds, which are known as secondary metabolites. Secondary metabolites are often referred as specialized metabolites since it is often impossible to distinguish primary and secondary metabolic reactions. Essential oils are secondary metabolites that are important for life and survival plants for numerous reasons, such as physiological function of growth, ecological function and development, and resistance against diseases and insects (Wink 2003; Gershenzon and Dudareva 2007). Xerophyte plants produce essential oils in response to the specific conditions of life, such as hypersensitivity reaction to the temperature cross flow and water stress (Kremer et al. 2012). Numerous aromatic species are rich in strongly-bioactive phenolic compounds. Since ancient times, aromatic plants have been used as herbal formulations in diverse forms, like tinctures, teas, powders and poultices, for their growing interest as alternative therapies for the prevention of various diseases (Bhuwan et al. 2015). Biological effects of essential oils are often attributed to prooxidant effects at the cellular level. Thus, they are widely used for their bactericidal, virucidal, fungicidal, antiparasitic, and insecticidal properties in pharmaceutical, sanitary, cosmetic, agricultural, and food industries (Bakkali et al. 2008; Poma et al. 2018). In addition, many of them show a valuable potential as anticancer therapeutic agents (Russo et al. 2015).

The antimicrobial activity of the essential oils constituents has already been described by many authors (Nostro et al. 2007; Ćavar et al. 2008; Vazirian et al. 2019; Sojić et al. 2019). Clinical observations reported antibacterial, antifungal, and antiviral activities of *Melaleuca alternifolia* tea tree oil whereas in vitro experiments ascribed most of biological properties to complex interaction among different components (Brun et al. 2019). Drevinskas et al. (2019) described novel chromatographic data segmentation method which demonstrates the capability of finding what volatile substances are responsible for antiviral and cytotoxic effects in the medicinal plant extracts. Their findings suggested that β-cis-caryophyllene, linalool, and eucalyptol possess antiviral activity, while thujones do not, and α-thujone, β-thujone, cis-*p*-menthan-3-one, and estragole show cytotoxic effects. Antiviral activity of essential oils highlight their potential use in fresh produce sanitation (Battistini et al. 2019). Described antiviral activity of different plant extracts demonstrate that secondary plant metabolites could be new natural antiviral agents. For example, *Helichrysum arenarium* and *H. armenium* showed notable antiviral activity against both Herpes simplex virus Type-1 and Parainfluenza-3 virus (Kutluk et al. 2018), *Osmunda regalis* essential oil showed antiviral activity against Coxsackievirus B4 (Bouazzi et al. 2018), *Teucrium chamaedrys* extracts against herpes simplex virus type 2 (Todorov et al. 2015), essential oil of *Salvia fruticosa* has a strong activity against *Herpes simplex* virus (Sivropoulou et al. 1997) and the activity of savory's essential oils against HIV has been documented (Yamasaki et al. 1998).

Significant variations of essential oil components have been observed between different species, but also within populations or within individual species, with implications for plant–pathogen interactions. The activity of essential oils in plant

defense can be direct, through toxicity, or some volatile organic compounds act as specific cues to attract parasitoids (Giamakis et al. 2001; Padovan et al. 2013). Many individual compounds act as antifungal, antibacterial or allelopathic agents, or for priming of systemic defences in both the host and neighbouring plants (Zulak and Bohlmann 2010; Sharifi-Rad et al. 2017). Although various substances of natural and synthetic origin have been assessed for their antiphytoviral activity, only limited number of scientific publications have revealed the antiphytoviral activity of essential oils. In this sense, essential oil of *Melaleuca alternifolia* (Maiden & Betche) Cheel reduced lesion number in TMV-infected plants (Bishop 1995), while *Plectranthus tenuiflorus* essential oil inhibited infection with Tobacco Necrosis Virus (TNV) (Othman and Shoman 2004). Essential oil of *Satureja montana* L. ssp. *variegata* (Host) P.W. Ball as well as its dominant phenol compounds thymol and carvacrol applied simultaneously with the infecting virus, reduced the number of local lesions on both tobacco mosaic virus and cucumber mosaic virus infected plants (Dunkić et al. 2010). The essential oils of *Eryngium* species reduced the number of viral lesions in plants infected with satellite associated cucumber mosaic virus (Dunkić et al. 2013). Plants treated with essential oil of *Helichrysum italicum* showed delay in the development and strong reduction of local lesions in the early stage of infection with tobacco mosaic virus (Bezić et al. 2016). Essential oil from *Micromeria graeca* caused delay in the development and reduction of the number of lesions in the plants infected with satellite associated cucumber mosaic virus (Vuko et al. 2012).

Results dealing with antiphytoviral activity indicate the role of these plant metabolites in response to viral infection. Qualitative differences in essential oil composition of different plant species do not exhibit a uniform response that would explain the mechanism of antiphytoviral activity. Overall conclusion based on our previous results is that sesquiterpene-rich oils can reduce viral infection in local host plants (Dunkić et al. 2010, 2011; Bezić et al. 2011; Vuko et al. 2012). It is known that synthesis of certain *Arabidopsis thaliana* sesquiterpenes, including β-caryophyllene, is induced by the phytohormones gibberelline and jasmonate. Thus, that the production of sesquiterpenes is obviously integrated within the gibereline and jasmonate signalling pathways (Hong et al. 2012).

Many authors described the different biological activities and medical properties of *Teucrium* species as a traditional plants (Khazaei et al. 2018; Vujanović et al. 2019). GC-MS analysis showed sesquiterpenes as abundant components in the oil composition of selected *Teucrium* species, indicating its possible antiphytoviral activity (Dunkić et al. 2011). Thus, we focused on this promising biological activity of *Teucrium* that is unsufficiently explored and consequently opens new area of research that could help in a control of diseases caused by plant viruses. The aim of this chapter is to present findings dealing with antiphytoviral potential of selected species of genus *Teucrium* and to improve the knowledge about possible application of essential oils as antiphytoviral agents. Findings presented in this chapter indicate that sesquiterpene-rich essential oils of *Teucrium* species are potent inhibitors of viral infection and natural substances with possible role in the control of plant virus diseases.

11.2 Tobacco Mosaic Virus and Cucumber Mosaic Virus as a Worldwide Occurring Disease Agents

Tobacco mosaic virus (TMV) is a positive-sense single stranded RNA virus belonging to genus Tobamovirus, family Virgaviridae. TMV causes mottling and discoloration of leaves of an infected plants, especially tobacco and other members of the Solanaceae. It was the first infectious agent to be recognised as being distinct from bacteria and thus the first to bear the designation "virus". It can infect over 350 different plant species. TMV can multiply only inside a living cell but it can survive in a dormant state in dead tissue. Opposite to TMV, most other viruses die when the plant tissue dies. TMV is one of the most stable viruses that retains ability to infect plants for years after the infected plant part died. Infection by TMV causes serious losses on many crops including ornamental plants, tomatoes, peppers and many others. Although the virus almost never kills plants, infection lowers the quality and quantity of the crop. As opposed to fungicidal chemicals used to control fungal diseases, there are no efficient chemical treatments that protect plant parts from virus infection. Additionally, there are no known chemical treatments used under field conditions that eliminate viral infections of plants. Therefore, plants infected by viruses remain so. Thus, control of TMV is primarily focused on reducing and eliminating sources of the virus as the most important practice in controlling tobacco mosaic virus infection.

Cucumber mosaic virus (CMV) is the type member of the genus *Cucumovirus* in the family *Bromoviridae*. This plant pathogenic virus has a worldwide distribution and the widest host range among plant viruses infecting about 1000 plant species. As a worldwide occurring disease agent, it has been reported in many crops, for example in tomatoes, peppers, cucurbits and cucumbers inducing diverse symptoms (Galliteli 2000). As described above, chemicals can not cure a plant of this virus, or of any other virus infection. CMV is a small virus with icosahedral shape and a tripartite genome and two subgenomic RNAs (Shi et al. 1997; Kwon and Chung 2000). Some strains of CMV encapsidate satellite RNA (satRNA) that is completely dependent on the CMV for its replication and spread. Accompanied by their helper virus, different variants of satRNA can induce severe symptoms on infected plants (Garciá-Arenal and Palukaitis 1999).

Recent plant protection that is focused on natural products emphasizes the importance of finding natural substances that can help in a control of plant virus diseases. Current knowledge about the antiviral effects of essential oils, although limited, indicates the potential of these secondary metabolites to control or at least reduce the spread of viral infections. Accordingly, we present bioactivity of selected *Teucrium* species that showed inhibition in development of local lesions in tobacco mosaic virus and cucumber mosaic virus infected plants.

11.3 Antiphytoviral Activity of *Teucrium arduini*

Family Lamiaceae includes about 3500 species spread all over the world, especially in the Mediterranean area. Plants in the family Lamiaceae are a very important source of essential oils and other biologically active molecules. Genus *Teucrium* (Lamiaceae), includes 300 species widespread all around the world, chiefly in the Northern Hemisphere. In the flora of Europe the genus *Teucrium* includes 49 species, of which 13 are widespread in the Croatian flora (Trinajstić 2008; Nikolić 2011). *Teucrium arduini* is Mediterranean endemic plant with restricted range in the Western Balkans, distributed along the Adriatic Coast from the Istra Peninsula in Croatia in the north to Albania in the south (Tutin et al. 1972; Šilić 1990). Representatives of *Teucrium* genus have been used for more than 2000 years as medicinal herbs (Ulubelen et al. 2000; Bedir et al. 2003). Previous chemical research on *Teucrium* species revealed the presence of volatile oils, tannins, terpenoids (mostly diterpenoids), flavonoids and steroid compounds (Šamec et al. 2010). *Teucrium* species were found to possess a wide range of biological activities such as antioxidant, antimicrobial, antifungal, larvicidal, antispasmodic, antinociceptive, antiinflammatory, antiulcer, hypoglycemic, antiacetylcholinesterase and hepatoprotective activities (Vuković et al. 2007; Menichini et al. 2009), some of which are attributed to the presence of the essential oils. Despite a wide spectrum of biological activities of essential oils and secondary metabolites in general, there is only little information available about the effects of these compounds on viruses or viral infection in plant systems. Thus, all results showing antiphytoviral potential of secondary plant metabolites are valuable resource for further research and possible development of a new natural antiviral compounds. Antiphytoviral activity of *Teucrium arduini* essential oil against Tobacco mosaic virus (TMV) and Cucumber mosaic virus (CMV) infection indicate that plants of this genus could be used as a natural source of bioactive molecules for control of plant virus diseases.

Our results showed that essential oil of *Teucrium arduini* is characterized by a high concentration of sesquiterpene hydrocarbons (43.8%) with β-caryophyllene (19.9%) being the major compound, followed by oxygenated sesquiterpenes (19.6%) with caryophyllene-oxide (14.6%) as the main fraction. Essential oil, β-caryophyllene and caryophyllene oxide applied as a spray solution on the leaves of local host plants prior to virus inoculation, significantly reduced the number of local lesions. Treatment with essential oil reduced lesion number on both TMV and CMV infected plants. The percentage of inhibition was 25.7% on plants infected with TMV, and 21.9% on CMV-infected plants (Table 11.1) (Dunkić et al. 2011).

The main oil components, sesquiterpene β-caryophyllene and oxigenated sesquiterpene caryophyllene oxide, applied individually to plants prior to virus inoculation, showed significant antiviral activity against CMV, but weak activity against TMV. β-caryophyllene and caryophyllene oxide inhibited lesions on the leaves of CMV-infected plants with percentage 30.8% and 36.9%, respectively (Table 11.2). Antiviral efficiency of both components was increased in relation to the essential oil. These sesquiterpene-essential oil components are probably responsible for

Table 11.1 *Teucrium arduini* essential oil, *β*-caryophyllene and caryophyllene oxide activity on TMV and CMV infection (Dunkić et al. 2011)

	TMV			CMV		
	Mean of L.L ± SEM (treated)	Mean of L.L ± SEM (control)	% of inhibition	Mean of L.L ± SEM (treated)	Mean of L.L ± SEM (control)	% of inhibition
T. arduini essential oil	26 ± 1.3*	35 ± 1.3*	25.7	25 ± 1.1*	32 ± 0.9*	21.9
β-caryophyllene	52.6 ± 1.8	56.9 ± 2.1	7.6	13.7 ± 1.6*	19.8 ± 2.1*	30.8
Caryophyllene-oxide	53.5 ± 2.1	56.9 ± 2.1	5.9	12.5 ± 1.4*	19.8 ± 2.1*	36.9

Mean of L.L the mean number of local lesions, *SEM* Standard Error Mean. * Significance reduction in disease compared with control ($p \leq 0.05$)

Table 11.2 Antiviral activity of *Teucrium flavum*, *T. chamaedris*, *T. polium* and *T. montanum* essential oils on cucumber mosaic virus (Bezić et al. 2011)

Species	Mean of L.L ± SEM	% of inhibition
Control plants	7.0 ± 0.5	/
T. polium	4.1 ± 0.3*	41.4
T. flavum	5.4 ± 0.3*	22.9
T. montanum	3.9 ± 0.4*	44.3
T. chamaedrys	5.2 ± 0.4*	25.7

Mean of L.L the mean number of local lesions, *SEM* Standard Error Mean. * Significance reduction in disease compared with control ($p \leq 0.05$)

antiviral activity of *Teucrium arduini* essential oil against CMV infection. Inhibition of TMV infection could be explained with antiviral activity of other oil components or their synergistic effect. This is supported by the fact that different extracts of many plants were tested for antiviral activity and it was explained due to their content of diterpenoids and essential oils (Romero et al. 1989; Bishop 1995).

Teucrium arduini essential oil and dominant components of the oil reduce infection of plants with CMV and TMV by inducing resistance response in the host plants. Main oil components *β*-caryophyllene and caryophyllene oxide are responsible for described activity, at least in the reduction of CMV infection. The possibility of other mechanisms as well as antiviral activity of other essential oil compounds, remains as a questions for further research. Answering to above questions will improve our understanding of antiviral activity of *Teucrium arduini* essential oil.

11.4 Antiphytoviral Activity of *Teucrium montanum*, *T. polium*, *T. chamaedrys* and *T. flavum*

Evaluation of antiphytoviral effects of the volatiles of additional four *Teucrium* species on the development of local lesions in CMV infected plants indicated that, aside from *Teucrium arduini*, other *Teucrium* species are also source of bioactive molecules with antiphytoviral activity. A comparison of the mean number of lesions on the essential oil-treated local host plants with the corresponding control plants showed activity of *Teucrium* essential oils in reduction of CMV infection. *Teucrium montanum* essential oil showed the strongest antiviral effect with percentage of inhibition 44.3%, followed by *T. polium* with 41.4%, *T. chamaedrys* with 25.7% and *T. flavum* with 22.9% (Table 11.2) (Bezić et al. 2011). The chemical composition of essential oils of above *Teucrium* species is characterised with sesqiuiterpenes *β*-caryophyllene and germacrene D, that are represented in relatively high percentages. Since we above discussed about antiphytoviral potential of sesquiterpene-rich oils, activity of these four *Teucrium* species against CMV infection confirms that these aromatic species are valuable source of bioactive molecules. With the exception of *Teucrium montanum*, the percentage of *β*-caryophyllene in the essential oil correlates with its antiviral activity. Essential oil of *Teucrium montanum* showed the strongest reduction of local lesions, despite the fact that *β*-caryophyllene is not the most abundant component in the *Teucrium montanum* essential oil. The most abundant component in the oil of *Teucrium montanum* is germacrene D, followed by *β*-pinene, *β*-caryophyllene and limonene. These components or their synergistic effect could be responsible for strongest antiphytoviral effect of *Teucrium montanum* essential oil compared to other three oils. *β*-caryophyllene is the main component in the essential oil composition of *Teucrium polium, T. chamaedrys* and *T. flavum* and the percentage of *β*-caryophyllene in the essential oil correlates with antiviral activity of these oils. Comparing the percentages of inhibition, essential oils of these four *Teucrium* species showed stronger activity against CMV infection than above described *Teucrium arduini*. In addition, *Teucrium montanum* and *T. polium* essential oil showed significantly stronger antiviral activity against CMV than *Satureja montana* (Dunkić et al. 2010). Our conclusion is that sesquiterpene-rich essential oils of *Teucrium* species are potent inhibitors of CMV infection and natural substances with possible role in the control of plant virus diseases.

11.5 Conclusions

Although no single answer is expected to explain the antiphytoviral activity of essential oils, further gathering of information will improve the existing knowledge about antiphytoviral activity of secondary plant metabolites as well as our understanding of the role of essential oils in plant-virus interactions. In this sense, species of the genus *Teucrium* are worth exploring, since several *Teucrium* species showed

potent activity against CMV and TMV infection. Antiviral activity of the *Teucrium* essential as well as the antiviral activity of the main oil components, indicated the role of β-caryophyllene and caryophyllene oxide in the activity of the essential oil in the plant defence against viruses. Other components of essential oil, such as germacrene D as abundant compound or additional, lower represented components, could also be responsible and sinergistically contribute to effectiveness of essential oil in reducing plant virus infection. Understanding the defence response mechanisms at the molecular level will improve our understanding of the plant defence responses and enhance our knowledge about the antiphytovial potential of essential oils and their use in the prevention of viral plant diseases. The above presents several directions for future research that should aim to clarify the mechanism of antiviral activity of essential oils.

References

Bakkali F, Averbeck S, Averbeck D, Idaomar M (2008) Biological effects of essential oils. A review. Food Chem Toxicol 46:446–475

Battistini R, Rossini I, Ercolini C, Goria M, Callipo MR, Maurella C, Pavoni E, Serracca L (2019) Antiviral activity of essential oils against hepatitis A virus in soft fruits. Food Environ Virol 11:90–95

Bedir E, Manyam R, Khan IA (2003) Neo-clerodane diterpenoids and phenylethanoid glycosides from *Teucrium chamaedrys* L. Phytochemistry 63:977–983

Bezić N, Vuko E, Dunkić V, Ruščić M, Blažević I, Burčul F (2011) Antiphytoviral activity of sesquiterpene-rich essential oils from four Croatian *Teucrium* species. Molecules 16:8119–8129

Bezić N, Vuko E, Ruščić M, Dunkić V (2016) *Helichrysum italicum* (Roth) G. Don – essential oil composition and activity on tobacco mosaic virus infection. J Plant Physiol Pathol 4:1–3

Bhuwan KC, Nasser A, Awadh A, William NS (2015) A survey of chemical compositions and biological activities of Yemeni aromatic medicinal plants. Medicine 2:67–92

Bishop CD (1995) Antiviral activity of the essential oil of *Melaluca alternifolia* (Maiden & Betche) Cheel (tea tree) against tobacco mosaic virus. J Essent Oil Res 7:641–644

Bouazzi S, Jmii H, El Mokni R, Faidi K, Falconieri D, Piras A, Jaidane H, Porcedda S, Hammami S (2018) Cytotoxic and antiviral activities of the essential oils from Tunisian Fern *Osmunda regalis*. S Afr J Bot 118:52–57

Brun P, Bernabe G, Filippini R, Piovan A (2019) In vitro antimicrobial activities of commercially available tea tree (*Melaleuca alternifolia*) essential oils. Curr Microbiol 76:108–116

Ćavar S, Maksimović M, Šolić ME, Jerković-Mujkić A, Bešta R (2008) Chemical composition and antioxidant and antimicrobial activity of two *Satureja* essential oils. Food Chem 111:648–653

Drevinskas T, Maruska A, Telksnys L, Hjerten S, Stankevicius M, Lelesius R, Mickiene R, Karpovaite A, Salomskas A, Tiso N (2019) Chromatographic data segmentation method: a hybrid analytical approach for the investigation of antiviral substances in medicinal plant extracts. Anal Chem 91:1080–1088

Dunkić V, Bezić N, Vuko E, Cukrov D (2010) Antiphytoviral activity of *Satureja montana* L. ssp. *variegata* (Host) P. W. Ball essential oil and phenol compounds on CMV and TMV. Molecules 15:6713–6721

Dunkić V, Bezić N, Vuko E (2011) Antiphytoviral activity of essential oil from endemic species *Teucrium arduini* L. Nat Prod Commun 6:1385–1388

Dunkić V, Vuko E, Bezić N, Kremer D, Ruščić M (2013) Composition and antiviral activity of the essential oils of *Eryngium alpinum* and *E. amethystinum*. Chem Biodivers 10:1894–1902

Gallitelli D (2000) The ecology of cucumber mosaic virus and sustainable agriculture. Virus Res 71:9–21

Garciá-Arenal F, Palukaitis P (1999) Satellites and defective viral RNAs. In: Vogt PK, Jackson AO (eds) Current topics in microbiology and immunology. Springer, Berlin/Heidelberg/New York, pp 37–63

Gershenzon J, Dudareva N (2007) The function of terpene natural products in the natural world. Nat Chem Biol 3:408–414

Giamakis A, Kretsi O, Chinou I, Spyropoulos CG (2001) *Eucalyptus camaldulensis*: volatiles from immature flowers and high production of 1,8-cineole and beta-pinene by in vitro cultures. Phytochemistry 58:351–355

Hong GJ, Xue XY, Mao YB, Wang LJ, Chen XY (2012) *Arabidopsis* MYC2 interacts with DELLA proteins in regulating sesquiterpene synthase gene expression. Plant Cell 24:2635–2648

Khazaei M, Nematollahi-Mahani SN, Mokhtari T, Sheikhbahaei F (2018) Review on *Teucrium polium* biological activities and medical characteristics against different pathologic situations. J Contemp Med Sci 4:1–6

Kremer D, Dragojević MI, Stabentheiner E, Vitali D, Kopričanec M, Ruščić M, Kosalec I, Bezić N, Dunkić V (2012) Phytochemical and micromorphological traits of endemic *Micromeria pseudocroatica* Šilić (Lamiaceae). Nat Prod Commun 7:1667–1670

Kutluk I, Aslan M, Orhan IE, Ozcelik B (2018) Antibacterial, antifungal and antiviral bioactivities of selected *Helichrysum* species. S Afr J Bot 119:252–257

Kwon CS, Chung WI (2000) Differential roles of the 5′ untranslated regions of cucumber mosaic virus RNAs 1, 2, 3 and 4 in translational competition. Virus Res 66:175–185

Menichini F, Conforti F, Rigano D, Formisano C, Piozzi F, Senatore F (2009) Phytochemical composition, anti-inflammatory and antitumour activities of four *Teucrium* essential oils from Greece. Food Chem 115:679–686

Nikolić T (ed) (2011) Flora croatica database – Department of Botany, Faculty of Science, University of Zagreb, Croatia. http://hirc.botanic.hr/fcd. Accessed 3 Apr 2019

Nostro A, Roccaro AS, Bisignano G, Marino A, Cannatelli MA, Pizzimenti FC, Cioni PL, Procopio F, Blanco AR (2007) Effects of oregano, carvacrol and thymol on *Staphylococcus aureus* and *Staphylococcu epidermidis* biofilms. J Med Microbiol 56:519–523

Othman BA, Shoman SA (2004) Antiphytoviral activity of the *Plectranthus tenuiflorus* on some important viruses. Int J Agric Biol 6:844–849

Padovan A, Keszei A, Foley WJ, Kulheim C (2013) Differences in gene expression within a striking phenotypic mosaic *Eucalyptus* tree that varies in susceptibility to herbivory. BMC Plant Biol 13(1):29

Poma P, Labbozzetta M, Notarbartolo M, Bruno M, Maggio A, Rosselli S, Sajeva M, Zito P (2018) Chemical composition, in vitro antitumor and pro-oxidant activities of *Glandora rosmarinifolia* (Boraginaceae) essential oil. PLoS One 13:e0196947. https://doi.org/10.1371/journal.pone.0196947

Romero E, Tateo F, Debiaggi M (1989) Antiviral activity of *Rosmarinus officinalis* L. extract. Mitt Geb Lebensmittelunters Hyg 80:113–119

Russo R, Corasaniti MT, Bagetta G, Morrone LA (2015) Exploitation of cytotoxicity of some essential oils for translation in cancer therapy. Evid Based Complement Alternat Med 2015:397821. https://doi.org/10.1155/2015/397821

Šamec D, Gruz J, Strnad M, Kremer D, Kosalec I, Jurišić Grubešić R, Karlović K, Lucić A, Piljac-Žegarac J (2010) Antioxidant and antimicrobial properties of *Teucrium arduini* L. (Lamiaceae) flower and leaf infusions (*Teucrium arduini* L. antioxidant capacity). Food Chem Toxicol 48:113–119

Sharifi-Rad J, Sureda A, Tenore GC, Daglia M, Sharifi-Rad M, Valussi M, Tundis R, Sharifi-Rad M, Loizzo MR, Ademiluyi AO, Sharifi-Rad R, Ayatollahi SA, Iriti M (2017) Biological activities of essential oils: from plant chemoecology to traditional healing systems. Molecules 22:70. https://doi.org/10.3390/molecules22010070

Shi BJ, Ding SW, Symons RH (1997) In vivo expression of an overlapping gene encoded by the cucumoviruses. J Gen Virol 78:237–241
Šilić Č (1990) Endemic plants. Svjetlost, Sarajevo
Sivropoulou A, Nikolaou C, Papanikolaou E, Kokkini S, Lanaras T, Arsenakis M (1997) Antimicrobial, cytotoxic and antiviral activities of *Salvia fruticosa* essential oil. J Agric Food Chem 45:3197–3201
Sojić B, Pavlić B, Tomović V, Ikonić P, Zeković Z, Kocić-Tanackov S, Durović S, Skaljac S, Jokanović M, Ivić M (2019) Essential oil versus supercritical fluid extracts of winter savory (*Satureja montana* L.) – assessment of the oxidative, microbiological and sensory quality of fresh pork sausages. Food Chem 28:280–286
Todorov D, Pavlova D, Hinkov A, Shishkova K, Dragolova D, Kapchina-Toteva V, Shishkov S (2015) Effect of *Teucrium chamaedrys* l. extracts on herpes simplex virus type 2. Cr Acad Bulg Sci 68(12):1519–1526
Trinajstić I (2008) Plant communities of Croatia. Academy of Forestry Sciences, Zagreb
Tutin TG, Heywood VH, Burges NA, Moore DM, Valentine DH, Walters SM, Webb DA (1972) Flora Europaea III. Cambridge University Press, Cambridge
Ulubelen A, Topcu G, Sönmez Ü (2000) Chemical and biological evaluation of genus *Teucrium*. In: Rahman A (ed) Bioactive natural products. Studies in natural products chemistry, vol 23. Elsevier, Amsterdam, pp 591–648
Vazirian M, Hamidian K, Noorollah M, Manayi A, Samadi N (2019) Enhancement of antibiotic activity and reversal of resistance in clinically isolated methicillin-resistant *Staphylococcus aureus* by *Trachyspermum ammi* essential oil. Res J Pharmacogn 6:1–10
Vujanović M, Zengin G, Durović S, Mašković P, Cvetanović A, Radojković M (2019) Biological activity of extracts of traditional wild medicinal plants from the Balkan Peninsula. S Afr J Bot 120:213–218
Vuko E, Dunkić V, Bezić N, Ruščić M, Kremer D (2012) Chemical composition and antiphytoviral activity of essential oil of *Micromeria graeca*. Nat Prod Commun 7:1227–1230
Vuković N, Milošević T, Sukdolak S, Solujić S (2007) Antimicrobial activities of essential oil and methanol extract of *Teucrium montanum*. Evid Based Complement Alternat Med 4:17–20
Wink M (2003) Evolution of secondary metabolites from an ecological and molecular phylogenetic perspective. Phytochemistry 64:3–19
Yamasaki K, Nakano M, Kawahata T, Mori H, Otake T, Ueba N, Oishi I, Inami R, Yamane M, Nakamura M, Murata H, Nakanishi T (1998) Anti-HIV-1 activity of herbs in Labiatae. Biol Pharm Bull 21:829–833
Zulak KG, Bohlmann J (2010) Terpenoid biosynthesis and specialized vascular cells of conifer defense. J Integr Plant Biol 52:86–97

Chapter 12
Antibacterial and Antifungal Activity of Secondary Metabolites of *Teucrium* Species

Olgica Stefanović

Abstract *Teucrium* species are known for their medicinal properties and have exhibited different biological activities including broad-spectrum antimicrobial activity. Considering that *Teucrium* species produce various bioactive compounds (phenolic acids, flavonoids, saponines, alkaloids, monoterpenes, neo-clerodane diterpenes, sesquiterpenes, essential oils) they could be an important source of new antimicrobial compounds. The expanding of knowledge on the antimicrobial plant compounds has opened wide opportunities for their application in medicine, pharmacy, and food industry. Accordingly, the aim of this chapter was to collect and summarize the results of antimicrobial (antibacterial and antifungal) studies. The results for 44 *Teucrium* species were processed including both the activity of different types of plant extracts and essential oils. Antimicrobial properties were based on in vitro determination of zones of growth inhibition and minimal inhibitory concentrations, using diffusion and dilution method. *Teucrium* species established broad-spectrum antimicrobial activity. Generally, the species were more active against Gram-positive than Gram-negative bacteria and fungi. The essential oils were more potent than plant extracts. The following *Teucrium* species exhibited a promising antimicrobial activity: plant extracts of *Teucrium flavum*, *T. fruticans*, *T. siculum*, *T. yemense*, *T. sokotranum*, *T. persicum* and *T. scordium*, especially against Gram-positive bacteria, and essential oils from *Teucrium orientale*, *T. africanum*, *T. ramosissimum*, *T. mascatence*, *T. yemense*, *T. massiliense* and *T. scordonia*. *Teucrium polium* was one of the most tested *Teucrium* species and has exhibited pronounced activity. The activity was observed against important pathogenic bacteria (*Staphylococcus* sp., *Bacillus* sp., *Enterococcus* sp., *Streptococcus* sp., *Escherichia coli*, *Pseudomonas aeruginosa*, *Klebsiella pneumoniae*, *Salmonella* sp.) and fungi (*Candida* sp., *Trychophyton* sp.). In the last 20 years, the significant database of antimicrobial properties of *Teucrium* species is formed. However, the mechanisms of activity are still poorly explored. Only with exact knowledge of these mecha-

O. Stefanović (✉)
Department of Biology and Ecology, Faculty of Science, University of Kragujevac, Kragujevac, Serbia
e-mail: olgica.stefanovic@pmf.kg.ac.rs

M. Stanković (ed.), *Teucrium Species: Biology and Applications*,
https://doi.org/10.1007/978-3-030-52159-2_12

nisms, it will be possible to develop a new generation of standardized, effective biopreparations. Future studies on bioavailability, pharmacodynamics, and mechanisms of action will contribute to the development of new *Teucrium* antimicrobial agents.

Keywords *Teucrium* · Plant extracts · Essential oils · Bacteria · Fungi · Antibacterial effects · Antifungal effects

Abbreviations

AcOH	Acetone
BuOH	Buthanol
CFU	Colony forming unit
$CHCl_3$	Chloroform
conc.	Concentration
DIZ	Diameter of inhibition zone
DMC	Dihlormethane
EtAc	Ethyl acetate
EtOH	Ethanol
G^-	Gram-negative bacteria
G^+	Gram-positive bacteria
Hex.	Hexane
MDR	Multi-drug resistant
MeOH	Methanol
MIC	Minimal inhibitory concentrations
MRSA	Methicillin-resistant *Staphylococcus aureus*
RPMI	Roswell Park Memorial Institute
TLC	Thin layer chromatography

12.1 Introduction

Various medicinal plants were, for a long time, used for infectious diseases healing. Medicinal plants synthesized different volatile and non-volatile compounds which exhibit inhibitory effects on growth and division of pathogenic microorganisms. Essential oils, balsams, resins of many aromatic plants (exp. *Salvia officinalis*, *Matricaria chamomilla*, *Eucalyptus globulus*, *Syzygium aromaticum*, *Myrtus communis*) are used as remedies for oral hygiene, tooth decay and gingivitis, wound healing. The essential oils of *Thymus vulgaris*, *T. serphyllum*, *Origanum vulgare*, *Ocimum basilicum* content thymol and carvacrol, compounds which interrupt the growth of microorganisms. *Melaleuca alternifolia* essential oil is a common

therapeutic agent to treat acne and other skin infectious problems. The well-known aromatic plants with the antimicrobial property are *Allium sativum*, *A. cepa* and other species of *Allium* genus which by hydrolysis of glycosides realized volatile sulfur-containing bioactive compounds. Antimicrobial activity possesses and numerous nonvolatile plant's compounds efficient for wound healing, infection, and inflammation of mucous membrane of the digestive tract, digestive disorders, and diarrhea. Proanthocyanidins from *Vaccinium macrocarpon* are used to treat urinary tract infections. Moreover, the plant- and spice-based antimicrobial compounds derived from *Cinnamomum* sp., *Syzygium aromaticum*, *Sinapsis alba*, *Laurus nobilis*, *Carum carvi*, *Coriandrum sativum*, *Cuminum cyminum*, *Origanum vulgare*, *Rosmarinus officinalis* and *Salvia officinalis* are used as natural preservatives for food preservation and food safety (Stefanović et al. 2009, 2012; Tajkarimi et al. 2010; Saleem et al. 2010; Ličina et al. 2013; Antolak and Kregiel 2017; Ionescu 2018).

The healing properties of species from genus *Teucrium* (family Lamiaceae) are well-known. In traditional medicine, the most common uses of these plants are as a diuretic, diaphoretic, tonic, analgesic, antipyretic, antirheumatic and antiseptic agent. In the Serbian herbal medicine, *Teucrium* species (*Teucrium chamaedrys*, *T. montanum*, *T. polium* and *T. scordium*) are used as a tonic, antirheumatic, digestion stimulant as well as an antiseptic agent (Sarić 1989). In recent years, numerous studies on *Teucrium* species bioactivities including antispasmodic, anti-inflammatory, antihypertensive, hypoglycemic, anticancer, antimicrobial and antioxidant, were reported. The expanding of knowledge on the antimicrobial plant compounds has opened wide opportunities for their application in both the medical and cosmetic purposes, and, also, in the food industry for storing and preserving food products. Accordingly, the goal of this chapter was an up-to-date literature review on antimicrobial activity of *Teucrium* species summarizing the results and selecting the most potent species for further potential application.

12.2 Bioactive Plant Compounds

Medicinal and prophylactic properties of medicinal plants are attributed to the synthesized bioactive organic compounds, the products of secondary metabolism of plants. The one of widely accepted classification of secondary metabolites is into three large classes: phenolic compounds (8000 types), terpenes (25,000 types), and alkaloids (12,000 types) (Antolak and Kregiel 2017).

The phenolic compounds are one of the largest groups of secondary metabolites that have exhibited antimicrobial activity. Important subclasses in this group of compounds include phenols, phenolic acids, quinones, flavonoids, tannins, and coumarins. Phenols are a class of chemical compounds consisting of a hydroxyl functional group (-OH) attached to an aromatic phenolic group. The site(s) and the number of hydroxyl groups on the phenol group are thought to be related to their

relative toxicity (Cowan 1999). Quinones have aromatic rings with two ketone substitutions. Quinones have the potential to form an irreversible complex with nucleophilic amino acids in proteins (Cowan 1999). Flavonoids are also hydroxylated phenolic substances but occur as a C6-C3 unit linked to an aromatic ring. They are classified according to their biosynthetic origin into the following classes: flavones, flavonols, flavanones, flavanols, anthocyanidins, isoflavones (Cowan 1999; Cushnie and Lamb 2011). Hydrolyzable and condensed tannins are found in almost every plant part: bark, wood, leaves, fruits, and roots. Hydrolyzable tannins are polyesters of gallic acid (or its derivatives), and the sugar. The condensed tannins are formed by the coupling of flavan-3-ol or flavan-3,4-diol monomers. Coumarins are benzo-α-pyrones and could be categorized as simple coumarins and cyclic coumarins (furanocoumarins and pyranocoumarins) (Ojala 2001).

Terpenes are organic compounds built up from isoprene subunits. According to the number of isoprene subunits, there are monoterpenes (C10), sesquiterpenes (C15), diterpenes (C20), triterpenes (C30), tetraterpenes (C40) and polyterpenes. Monoterpenes, diterpenes, and sesquiterpenes are the primary constituents of the essential oils (Kovačević 2004).

Alkaloids, one of the earliest isolated bioactive compounds from plants, are divided into the three classes: true alkaloids – heterocyclic nitrogen compounds derived from amino acids, protoalkaloidis – derived from amino acids, but not contain nitrogen in a heterocyclic ring and pseudoalkaloids – not derived from amino acids (Kovačević 2004).

12.3 Mechanisms of Antimicrobial Activity of Plant Compounds

The antimicrobial efficiency of plant compounds depends on several factors: (i) characteristics of target microorganism (the type, genus, species, strain), (ii) characteristics of plant material (botanical source, composition of the bioactive compounds as well as time of harvesting, stage of development or method of extraction) and (iii) chemical properties (hydrophilicity, lipophilicity, concentration, pH value).

The antibacterial activity occurs through several mechanisms: (i) disruption of membrane structure and function (including the efflux system), (ii) interruption of DNA/RNA synthesis and function, (iii) interference with intermediary metabolism, (iv) induction of coagulation of cytoplasmic constituents (Cowan 1999; Kovačević 2004; Buzzini et al. 2008; Cushnie and Lamb 2011; Daglia 2012; Radulović et al. 2013; Coppo and Marchese 2014). The mechanisms of antibacterial action for the particular class of bioactive compounds are summarized in Table 12.1.

In addition, plant compounds are being able to suppress bacterial toxin production by reducing the expression of major virulence genes. Selected plant extracts inhibit the production of cholera toxin by *Vibrio cholerae*, reduce the production of *Staphylococcus aureus* α-hemolysin, enterotoxins and toxic shock syndrome toxin

Table 12.1 Mechanisms of antibacterial activity of plant compounds

Class of plant compounds	Mechanism of action
Phenols	Changing the permeability of cell membrane and causing the leakage of cellular content, interfere with membrane proteins resulting in structure disrupting
Flavonoids	Cell membrane disruption, inhibition of nucleic acid synthesis (caused by topoisomerase inhibition), inhibition of energy metabolism (caused by NADH-cytochrome c reductase or ATP synthase inhibition), disturbance of cell wall and cell membrane synthesis
Quinones	Formation of irreversible complex with surface-exposed adhesins, cell wall polypeptides, and membrane-bound enzymes
Proanthocyanidins	Destabilization of cell membrane, inhibition of extracellular microbial enzymes, direct actions on microbial metabolism, deprivation of the substrates required for microbial growth
Gallotannins	Inactivation of membrane-bound proteins, strong affinity for iron
Coumarins	Decrease cell respiration
Essential oils	Cell membrane disruption
Alkaloids	Intercalate with DNA, inhibition of enzymes (esterase, DNA- polymerase, RNA-polymerase) or cell respiration

1, reduce the production of vero toxin and inactivate Shiga toxins (Upadhyay et al. 2014). Moreover, it was observed that plant extracts and essential oils inhibit bacterial biofilm formation, motility, attachment, and cell communication (Huber et al. 2003; Borges et al. 2015; Stefanović et al. 2015; Muruzović et al. 2016; Silva et al. 2016). Biofilm is defined as a structured community of bacterial cells enclosed in a self-produced polymeric matrix and adherent to an inert or living surface (Hall-Stoodley et al. 2004). The ability to form biofilm is recognized as an additional virulence factor which contributes to the long-lasting persistence of bacterium. In biofilms, bacteria show great tolerance toward antibacterial agents and host immune defenses what makes difficulties in control and eradication of bacterial infections (Høiby et al. 2010).

Studies investigating the mechanism of antifungal activity showed that plant compounds (essential oils, plant extracts) demonstrate several negative effects on fungal cell: (i) cause changes in cell membrane function or structure and, also, in the membranes of organelles (ii) inhibit key enzymes in normal physiological metabolism (respiration and energy production), (iii) inhibit the synthesis of the main polymers of the fungal cell wall, (iv) induce differential expression of critical genes including those involved in the cellular drug response, oxidation-reduction processes, pathogenesis, and the cellular starvation response (Carson et al. 2006; Li et al. 2013, 2014, 2016).

12.4 Extraction of Plant Compounds

Medicinal plants may have been specifically cultivated or collected from the wild. After harvesting, plant material should be cleaned, sorted, and then subjected to extraction. Fresh or dried plant material can be used. Extraction and isolation of the bioactive compounds from the plant tissues can be performed by using different procedures and methods (solvent extraction, distillation, supercritical fluid extraction). The appropriated extraction procedure used depends on the physical and chemical properties of the plant material, texture, water content in plant material, as well as the nature and kind of the compound to be isolated.

Solvent extraction is a process of separation of active compounds from plant material using different solvents. During extraction, solvents diffuse into the plant tissues and solubilise compounds with similar polarity. After the extraction process has finished, solvents have been evaporated, so that an extract is a concentrated mixture of plant active compounds. Successful extraction is largely dependent on the type of solvent used in the extraction procedure. The desirable properties of extracting solvents are: selectivity, large extraction capacity, no interference with plant components, no toxicity, and high evaporation level. The most often tested extracts are: water extract as a sample of extract that primarily used in traditional medicine and extracts from organic solvents such as methanol, ethanol as well as ethyl acetate, acetone, chloroform, dichlormethane (Ncube et al. 2008).

Distillation is a process of separation of volatile active compounds from plant material by physical process of evaporation, treating plant material with steam or boiling. After cooling and condensation of vaporized compounds, the water distillate is formed. Due to essential oils are insoluble in the water and because of specific gravity, essential oils are separated at the surface or top of the water distillate and further collected. Essential oils are extracted by steam distillation or hydrodistillation (Tongnuanchan and Benjakul 2014).

Supercritical fluid extraction is a diffusion-based process of separation of plant active compounds using supercritical fluids as the extracting solvent. Supercritical fluids are formed by heating of gases above their critical temperatures or by compressing of liquids above their critical pressure. Carbon dioxide is the most used supercritical fluid. Supercritical fluids have dissolving properties that depend on their density and can be controlled with temperature and pressure. In relation to solvent extraction, supercritical fluid extraction has some advantages such as low extraction temperature, minimal degradation of the active substances, selectively dissolving of compounds, and faster extraction (Janaćković et al. 2017).

12.5 Antimicrobial Susceptibility Methods

Antimicrobial (antibacterial and antifungal) activity of plant extracts, essential oils, and pure compounds might be analysed using several methods: diffusion, dilution and bioautography method (Fig. 12.1). Diffusion and dilution method are frequently

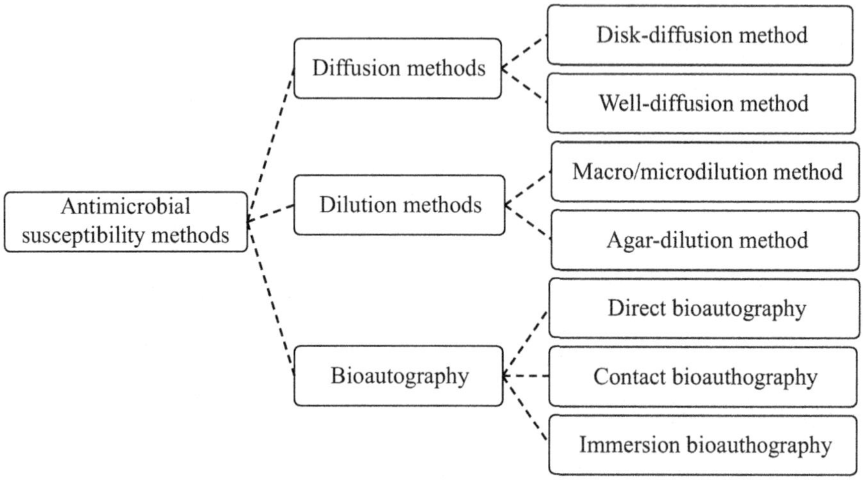

Fig. 12.1 Classification of antimicrobial susceptibility methods

used. These methods are standard antimicrobial susceptibility testing methods used for antibiotics/antifungal drugs recommended by the Clinical and Laboratory Standards Institute, USA.

Diffusion method is a qualitative test which, according to a size of a diameter of the zone of growth inhibition, shows the intensity of antimicrobial effect of the tested compound and allows classification of microorganisms as susceptible or resistant. The tested compound has applied on disk or into a hole in agar medium inoculated with a microorganism to be tested. During incubation, the compound diffuses into the medium inhibiting the growth of a microorganism. The reference medium for testing is Mueller-Hinton agar for bacteria and Mueller-Hinton agar supplemented with 2% glucose for fungi. The density of inoculum should be equivalent to 0.5 McFarland standard (1–2 × 10^8 CFU/ml). The results are recorded as zones of growth inhibition and when the diameter of the inhibition zone is wider, the antimicrobial activity of the compounds is higher (Fig. 12.2). The rate of diffusion of the tested compounds through the agar is dependent on solubility properties and the molecular weight of the tested compound. Hence, this method is not suitable to natural antimicrobial compounds that are scarcely soluble or insoluble in water and thus their hydrophobic nature prevents uniform diffusion through the agar medium (Klancnik et al. 2010).

Dilution (macrodilution and microdilution) method is more appropriate when investigating the activity of plant products. It can be carried out by using test tubes (macrodilution) or 96-wells microtiter plates (microdilution). The serial twofold dilutions of a tested compound are prepared in a liquid medium, Mueller-Hinton broth for bacteria and RPMI 1640 broth for fungi. The density of final inoculum is 5 × 10^5 CFU/ml for bacteria and 5 × 10^2 – 2.5 × 10^3 cell/ml for yeasts. This method is quantitative, the activity of a compound or a plant extract is determined as minimal inhibitory concentration (MIC). MIC is defined as the lowest concentration of

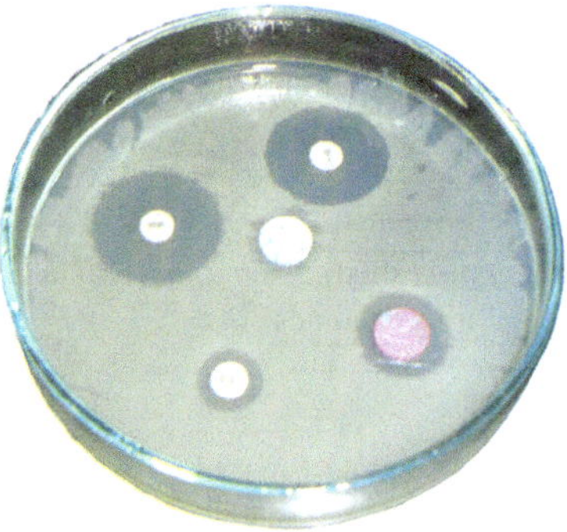

Fig. 12.2 Diffusion method – the zones of growth inhibition around disks indicate the activity of tested compound

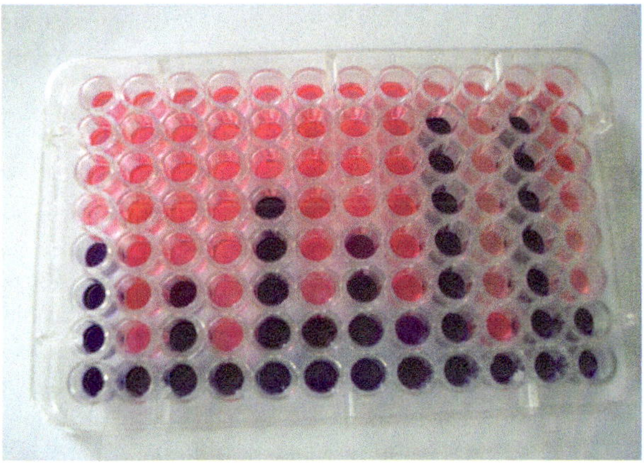

Fig. 12.3 Microdilution method – the change of colour of indicator (resazurine) from blue to pink indicates the growth of bacteria

tested compound able to inhibit the visible growth of microorganisms. Microbial growth (growth of bacteria and yeasts) could be assessed either visually by grading turbidity or spectrophotometrically by measuring optical density. Moreover, easy MIC detection and increased credibility of this method allow using of growth indicators which changing the colour indicate the growth of microorganisms (Cos et al. 2006; Fig. 12.3).

Once more method used for determination of MIC is agar dilution method. In this method, dilutions of a plant extract or a compound are prepared by adding into a precise volume of molten agar medium which then poured into sterile petri plates. Each agar plate containing a different concentration of tested compound. MIC is recorded as the lowest concentration of the compound that completely inhibits the growth, no colony observed.

Bioautography rely on fractionalisation of the extracts on its different components simplifies the identification of compounds with antimicrobial potency. Bioautography may be direct, when microorganisms grow directly on a TLC plate, then contact, when the active compound is transferred from a TLC plate to inoculated agar, and immersion when the inoculated agar medium is spilled over a TLC plate. Clear (white) zones on the TLC plate indicate antimicrobial activity of the extracts and their components (Rahalison et al. 1991).

12.6 Antimicrobial Activity of *Teucrium* Species

Antibacterial and antifungal activity of *Teucrium* species has been extensively investigated. In the following sections, the results of the literature search of databases Scopus, PubMed, Google Scholar for period 1999–2019 (the last accessed was on April 1, 2019) were presented. According to published data, 16% of species of the genus *Teucrium* (total of 44 species) were screened for antimicrobial testing. The genus *Teucrium* includes 268 species. The species from different regions (Serbia, Croatia, Italy, Romania, Turkey, Pakistan, Jordan Iran, Tunisia, Algeria, Mexico) were investigated, so the results were summarised and compared.

Evidently, bioactive compounds from *Teucrium* species were frequently isolated in a form of plant extracts. Plant extracts were prepared from fresh or dried plant material using conventional extraction methods (Soxhlet extraction, maceration, decoction) and most recent supercritical fluid extraction. The most commonly used extracting solvents were methanol, ethanol, and water. The more detailed information on plant parts used or extraction procedure and solvents used are presented in Table 12.2.

The essential oils were isolated by hydrodistillation or steam distillation using a Clevenger-type apparatus. A total of 20 *Teucrium* species essential oils were analysed and screened for antimicrobial activity (Table 12.3). For the most part, antimicrobial activity was determined using diffusion or dilution method. However, activity has also been demonstrated using time-kill assays. Antimicrobial activity studies were carried out on numerous bacterial species (Gram-positive and Gram-negative strains; sensitive and resistant, pathogens and opportunistic pathogens, clinical isolates and standard strains) and fungal species (yeasts, dermatophytes, and filamentous fungi, human and phytopathogens, clinical isolates and standard strains).

Teucrium species exhibited broad-spectrum antimicrobial activity (Fig. 12.4). Mostly, antimicrobial properties of *Teucrium* species were based on their activity against planktonic bacteria (bacteria in suspensions). In most environments, bacteria form biofilms, surface-attached communities. Many chronic infections (urinary tract infections, middle-ear infections, tooth decay and paradontosis, prostheses-associated infection) or problems with contamination in food processing industry are being connected with biofilms (Costerton et al. 1999; Wood et al. 2011). So, in recent years, the investigation of new active plant-derived antibiofilm agents is

Table 12.2 List of different plant extracts, plant part used, tested microorganisms, and results of *Teucrium* species antimicrobial activity

Teucrium species (part used)	Type of extract (extraction method)	Tested microorganisms	DIZ (tested concentration) range, MIC range		References
			Antibacterial activity	Antifungal activity	
T. polium (flowers) Iran	Ethanol extract (Maceration)	489 isolates of *Staphylococcus aureus*	DIZ (30 mg/ml) 10–11 mm	Not tested	Mansouri (1999)
T. polium (aerial parts) Turkey	Ethanol extract (Soxhlet)	*Listeria monocytogenes*, *L. ivanovii*, *L. innocua*, *L. murrayi*	DIZ (133 mg/ml) 10–11 mm MIC 12.5–50μg/ml	Not tested	Altanlar et al. (2006)
T. polium (aerial parts) Jordan	Ethanol extract (Maceration)	*S. aureus*, *Pseudomonas aeruginosa*, *Candida albicans*	DIZ (80 ppm) 14–15 mm MIC 80 ppm	Not active	Khalil et al. (2009)
T. polium (aerial parts) Jordan	Methanol extract (Soxhlet)	*Escherichia coli* ATCC 25922, *P. aeruginosa* ATCC 27853, *Proteus mirabilis* ATCC 426, *Enterobacter cloaceae* ATCC 29004, *S. aureus* ATCC 25923	DIZ (10 mg/ml) 11–26 mm MIC 1.2–2.4 mg/ml	Not tested	Tarawneh et al. (2010)
T. polium (aerial parts) Iran	Ethanol, methanol extracts (Maceration)	*Brucella melitensis*	DIZ_{EtOH} (100–400μg/ml) 9–12 mm DIZ_{MeOH} (100–400μg/ml) 8–11 mm	Not tested	Motamedi et al. (2010)
T. polium (aerial parts) Iran	Ethanol, methanol extracts (Maceration)	*Bacillus pumilis*, *B. anthracis*, *B. licheniformis*, *B. cereus*, *S. aureus*, *S. epidermidis*, *E. coli*, *Yersinia enterocolitica*, *S. typhy*, *P. mirabilis*, *Bordetella bronchispetica*, *Streptococcus pyogenes*	DIZ_{EtOH} (400 mg/ml) 9–18 mm DIZ_{MeOH} (600 mg/ml) 9–26 mm	Not tested	Darabpour et al. (2010)

T. polium (aerial parts) Egypt	Ethanol, ethyl acetate, chloroform, n-hexane extracts (Maceration)	*Trichophyton rubrum*, *T. tonsurans*, *C. albicans*, *Chrysosporium tropicum*, *Paecilomyces lilacinus*, *P. variotii*, *Scopulariopsis bervicaulis*	Not tested	MIC_{EtOH} 75– > 150 mg/ml MIC_{EtAc} 75– > 150 mg/ml MIC_{CHCl3} 25–125 mg/ml $MIC_{Hex.}$ 25– > 150 mg/ml	Hashem (2011)
T. polium (aerial parts) Turkey	Water extract (Maceration)	*Klebsiella pneumoniae*, *Haemophilus influenzae* ATCC 49766, *P. aeruginosa* ATCC 10145, *Acinetobacter baumannii*, *Streptococcus pneumonia* ATCC 19615, *S. pyogenes* ATCC 13615, *S. aureus* ATCC 25923, *S. epidermidis* ATCC 12228, *Mycobacterium tuberculosis*, *C. albicans* ATCC 10231, *C. parapsilosis* ATCC 22019, *C. krusei*, *C. tropicalis*	MIC 64μg/ml (G^+ bacteria) MIC 16–32μg/ml (G^- bacteria)	MIC 8–64μg/ml	Deliorman Orhan et al. (2012)
T. polium (aerial parts) Serbia	Methanol, acetone, ethyl acetate extracts (Maceration)	*S. aureus* ATCC 25923, *S. aureus*, *E. coli* ATCC 25922, *E. coli*, *P. aeruginosa*, *C. albicans* ATCC 10231, *C. albicans*, *A. niger*	MIC_{MeOH} 0.3–20 mg/ml MIC_{AcOH} 0.3–20 mg/ml MIC_{EtAC} 0.3, 10 mg/ml (*S. aureus* ATCC 25923, *P. aeruginosa*)	MIC_{MeOH} > 20 mg/ml MIC_{AcOH} > 20 mg/ml MIC_{EtAC} > 10 mg/ml	Stanković et al. (2012)
T. polium (aerial parts) Iran	Ethanol extract (Maceration)	*S. typhi*, *P. aeruginosa*, *B. cereus*, *L. monocytogenes*, *S. aureus*	DIZ (50 mg/ml) 6–21 mm MIC 6.25–50 mg/ml	Not tested	Mirzaei et al. (2013)
T. polium (aerial parts) Palestine	Ethanol extract (Maceration)	*S. aureus* ATCC 29213, *B. subtilis* ATCC 6633, *M. luteus* ATCC 10240, *E. coli* ATCC 10536, *P. aeruginosa*, *K. pneumoniae*, *C. albicans* ATCC 10231, *A. niger* ATCC 16404	DIZ (25 mg/ml) 8.7–18.3 mm	DIZ (25 mg/ml) 19.7 mm (*C. albicans*)	Qabaha (2013)

(continued)

Table 12.2 (continued)

Teucrium species (part used)	Type of extract (extraction method)	Tested microorganisms	DIZ (tested concentration) range, MIC range		References
			Antibacterial activity	Antifungal activity	
T. polium (leaves) Algeria	Methanol extract (Maceration)	*E. coli* ATCC 25922, *S. aureus* ATCC 6538	DIZ (2 mg/ml) 14–15 mm MIC 13 mg/ml (*S. aureus*) MIC 13 mg/ml (*E. coli*)	Not tested	Khaled-Khodja et al. (2014)
T. polium (aerial parts) Algeria	Methanol extract (Maceration)	*S. aureus* ATCC 25923, *E. coli* ATCC 25922, *S. typhimurium* ATCC 10428, *P. aeruginosa* ATCC 27853, *E. faecalis* ATCC 29212, *K. pneumoniae* ATCC 700603, *B. subtilis* ATCC 7033	DIZ (600 mg/ml) 11.5–26 mm MIC 3.125–50 mg/ml	Not tested	Dridi et al. (2016)
T. polium subsp. *gabesianum* (aerial parts) Tunisia	Ethanol, water extracts (Maceration)	*E. coli* ATCC 25922, *E. faecalis* ATCC 29212, *S. aureus* ATCC 25923, *P. aeruginosa* ATCC 27853, *C. freundeii*, *P. mirabilis*, *C. albicans*, *Cryptococcus neoformans*, *Trichophyton rubrum*, *T. soudanense*, *Microsporum canis*, *A. fumigatus*, *Scopulariopsis brevicaulis*	MIC_{EtOH} 0.156–0.625 mg/ml MIC_{H2O} 0.156–0.625 mg/ml	MIC_{EtOH} 0.625– > 1 mg/ml MIC_{H2O} > 1 mg/ml	Ben Othman et al. (2017)
T. polium (aerial parts) Pakistan	Ethanol, ethyl acetate, dichloromethane acetone, hexane extracts (Maceration)	*E. coli* ATCC 15224, *B. subtilis* ATCC 6663, *S. aureus* ATCC 29213, *S. flexneri* ATCC 14028, *A. flavus* ATCC 32611, *A. niger*, *C. albicans* ATCC 2091, *T. longifusus*	EtOH (1 mg/ml) 37–73% of inhibition EtAc (1 mg/ml) 24–56% DCM (1 mg/ml) 25–37% AcOH (1 mg/ml) 30–31% Hex. (1 mg/ml) 10–35%	EtOH (1 mg/ml) 40–70% of inhibition EtAc (1 mg/ml) 10–65% DCM (1 mg/ml) 27–58% AcOH (1 mg/ml) 15–25% Hex. (1 mg/ml) 25–30%	Ali et al. (2018)

T. polium (leaves, flowers, stem) Turkey	Ethanol, acetone, ether extracts (Maceration)	*B. subtilis* NRRL 558, *B. cereus* NRRL 3711, *S. aureus* NRRL 767, *S. epidermidis* NRRL 4377, *S. lutea* NRRL 4370, *E. coli* NRRL 3008, *E. aerogenes* NRRL 3567, *S. typhimurium* NRRL 4420, *Shigella* sp., *C. albicans* NRRL 27077, *Pichia membranifaciens* NRRL 2026, *S. cerevisiae* NRRL 2034, *Shizosaccharomyces pombe* NRRL 12796, *Zygosaccharomyces rouxii* NRRL 229	DIZ (1 mg/ml) 8–14 mm (G^+ bacteria) DIZ (1 mg/ml) 8–9 mm (G^- bacteria)	DIZ (1 mg/ml) 8–18 mm	Kunduhoglu et al. (2011)
T. chamaedrys subsp. *chamaedrys* (flowers) Turkey			DIZ (1 mg/ml) 8–22 mm (G^+ bacteria) DIZ (1 mg/ml) 8–13 mm (G^- bacteria)	DIZ (1 mg/ml) 8 mm	
T. chamaedrys Turkey	Polar and non-polar fractions of methanol extract (Soxhlet)	*S. pneumoniae*, *B. cereus*, *Acinetobacter lwoffii* ATCC 19002, *E. coli*, *K. pneumoniae*, *Clostridium perfringens*, *C. albicans*, *C. krusei* ATCC 6258	MIC_{polar} > 72 mg/ml $MIC_{non\text{-}polar}$ 9– > 72 mg/ml	MIC_{polar} > 72 mg/ml $MIC_{non\text{-}polar}$ 18–36 mg/ml	Gursoy and Tepe (2009)
T. chamaedrys (aerial parts) Turkey	Water, ethanol, ethyl acetate, hexane extracts (Maceration)	*B. cereus*, *S. aureus* ATCC 6538P, *E. coli* O157:H7, *K. pneumoniae*, *C. albicans* ATCC 10239, *S. aureus* ATCC 49444, *L. monocytogenes* ATCC 13076, *E. coli* ATCC 25922, *S. typhymurium* ATCC 14028, *C. albicans* ATCC 10231	MIC_{H2O} 12.5–150 mg/ml MIC_{EtOH} 400 mg/ml (G^+ bacteria) MIC_{EtAC} 3.12–12.5 mg/ml $MIC_{Hex.}$ – not active MIC (*S. aureus*) 3.12 mg/ml	MIC_{EtAC} 400 mg/ml	Ceyhan et al. (2012)
T. chamaedrys (aerial parts) Serbia	Methanol, acetone, ethyl acetate extracts (Maceration)	*S. aureus* ATCC 25923, *E. coli* ATCC 25922, *S. aureus*, *E. coli*, *P. aeruginosa*, *C. albicans* ATCC 10231, *C. albicans*, *A. niger*	MIC_{MeOH} 0.078–5 mg/ml MIC_{AcOH} 5– > 20 mg/ml MIC_{EtAC} 0.6– > 10 mg/ml	MIC_{MeOH} > 20 mg/ml MIC_{AcOH} 10 mg/ml (*A. niger*) MIC_{EtAC} 10 mg/ml (*A. niger*)	Stanković et al. (2012)

(continued)

Table 12.2 (continued)

Teucrium species (part used)	Type of extract (extraction method)	Tested microorganisms	DIZ (tested concentration) range, MIC range		References
			Antibacterial activity	Antifungal activity	
T. chamaedrys (aerial parts) Romania	Ethanol extract (Maceration)	*S. aureus* ATCC 49444, *L. monocytogenes* ATCC 13076, *E. coli* ATCC 25922, *S. typhimurium* ATCC 14028, *C. albicans* ATCC 10231	DIZ 11–20 mm	DIZ 22 mm	Vlase et al. (2014)
T. chamaedrys (aerial parts) Serbia	Water extract (Supercritical water extraction)	*S. aureus* ATCC 25923, *E. coli* ATCC 25922, *B. subtilis* ATCC 6633, *P. vulgaris* ATCC 13315, *P. mirabilis* ATCC 14153, *K. pneumoniae* ATCC 13883, *C. albicans* ATCC 10231, *A. niger* ATCC 16404	MIC 78.125–312.5μg/ml	MIC 78.125–156.25μg/ml	Nastić et al. (2018)
T. montanum (aerial parts) Serbia	Petroleum ether, chloroform, ethyl acetate, n-butanol (Maceration)	*P. aeruginosa*, *E. coli*, *S. aureus*, *S. lutea*, *Bacillus* sp.	DIZ_{Ether} (10 mg/ml) 0 mm, MIC_{Ether} 10– > 10 mg/ml DIZ_{CHCl3} (10 mg/ml) 15–18 mm, MIC_{CHCl3} 5–10 mg/ml DIZ_{EtAC} (10 mg/ml) 16–19.8 mm, MIC_{EtAC} 1– > 10 mg/ml $DIZ_{Butanol}$ (10 mg/ml) 16–20 mm, $MIC_{Butanol}$ 1– > 10 mg/ml	Not tested	Djilas et al. (2006)

T. montanum (aerial parts) Serbia	Methanol, acetone, ethyl acetate extracts (Maceration)	*S. aureus* ATCC 25923, *E. coli* ATCC 25922, *S. aureus*, *E. coli*, *P. aeruginosa*, *C. albicans* ATCC 10231, *C. albicans*, *A. niger*	MIC_{MeOH} 0.15–5 mg/ml MIC_{AcOH} 0.15–20 mg/ml MIC_{EtAC} 0.3, 10 mg/ml (*S. aureus* ATCC, *P. aeruginosa*)	MIC_{MeOH} > 20 mg/ml MIC_{AcOH} 20 mg/ml (*A. niger*) MIC_{EtAC} > 20 mg/ml	Stanković et al. (2012)
T. arduini (leaves, flowers) Croatia	Infusions	*B. subtilis* NCTC 8236, *S. aureus* ATCC 6538, *E. coli* ATCC 10535, *P. aeruginosa* ATCC 27853, *C. albicans* ATCC 10231, *A. niger* ATCC 16404	MIC_{leaves} 25–50 mg/ml (*B. subtilis*) MIC_{leaves} 1.56–4.16 mg/ml (*S. aureus*) $MIC_{flowers}$ 16.66 mg/ml (*S. aureus*)	Not active	Šamec et al. (2010)
T. arduini (aerial parts) Serbia	Methanol, acetone, ethyl acetate extracts (Maceration)	*S. aureus* ATCC 25923, *E. coli* ATCC 25922, *S. aureus*, *E. coli*, *P. aeruginosa*, *C. albicans* ATCC 10231, *C. albicans*, *A. niger*	MIC_{MeOH} 0.15–10 mg/ml MIC_{AcOH} 0.6– > 20 mg/ml MIC_{EtAC} 0.6– > 20 mg/ml	MIC_{MeOH} > 20 mg/ml MIC_{AcOH} > 20 mg/ml MIC_{EtAC} > 20 mg/ml	Stanković et al. (2012)
T. arduini (leaves, flowers, stem) Croatia	Ethanol extract (Ultrasonication)	*S. aureus* ATCC 6538, *E. faecalis* ATCC 21212, *E. coli* ATCC 10535, *P. aeruginosa* ATCC 27853, *C. albicans* ATCC 10231, *A. brasiliensis* ATCC 16404, *Microsporum gypseum*	MIC_{leaves} 2–4 mg/ml $MIC_{flowers}$ 2– > 4 mg/ml MIC_{stem} > 4 mg/ml	MIC_{leaves} 2– > 4 mg/ml $MIC_{flowers}$ 1– > 4 mg/ml MIC_{stem} 2– > 4 mg/ml	Kremer et al. (2013)

(continued)

Table 12.2 (continued)

Teucrium species (part used)	Type of extract (extraction method)	Tested microorganisms	DIZ (tested concentration) range, MIC range		References
			Antibacterial activity	Antifungal activity	
T. scordium subsp. *scordioides* (aerial parts) Serbia	Methanol, dichlormetan, cyclohexane extracts (Maceration)	*S. aureus* ATCC 25923, *S. epidermidis* ATCC 12228, *M. luteus* ATCC 10240, *E. faecalis* ATCC 29212, *B. subtilis* ATCC 6633BB, *B. cereus* ATCC 11778, *E. coli* ATCC 25922, *P. aeruginosa* ATCC 27853, *K. pneumoniae* NCIMB 9111, *C. albicans* ATCC 10259	DIZ_{MeOH} (100 mg/ml) 6.5 mm (*P. aeruginosa*) DIZ_{DCM} (100 mg/ml) 6–11 mm (*P. aeruginosa, B. subtilis*) $DIZ_{Hex.}$ (100 mg/ml) 6.5–11 mm	Not active	Kundaković et al. (2011)
T. scordium subsp. *scordium* Serbia	Methanol, acetone, ethyl acetate extracts (Maceration)	*S. aureus* ATCC 25923, *E. coli* ATCC 25922, *S. aureus*, *E. coli*, *P. aeruginosa*, *C. albicans* ATCC 10231, *C. albicans*, *A. niger*	MIC_{MeOH} 0.3–20 mg/ml MIC_{AcOH} 0.6– > 20 mg/ml MIC_{EtAC} 0.6, 10 mg/ml (*S. aureus* ATCC, *P. aeruginosa*)	MIC_{MeOH} > 20 mg/ml MIC_{AcOH} > 20 mg/ml MIC_{EtAC} 5–10 mg/ml	Stanković et al. (2012)
T. scordium subsp. *scordioides* (aerial parts) Serbia			MIC_{MeOH} 0.3–20 mg/ml MIC_{AcOH} 0.6 mg/ml (*S. aureus* ATCC) MIC_{EtAC} 0.6, 10 mg/ml (*S. aureus* ATCC, *P. aeruginosa*)	MIC_{MeOH} > 20 mg/ml MIC_{AcOH} > 20 mg/ml MIC_{EtAC} > 10 mg/ml	
T. stocksianum (whole plant) Pakistan	Methanol extract and sub-fractions: n-hexane, chloroform, ethyl acetate	*E. coli*, *S. aureus*, *S. typhi*, *S. flexneri*, *B. subtilis*, *A. niger*, *A. flavus*, *A. fumigatus*, *Fusarium solani*	DIZ_{MeOH} (1 mg/ml) 12–24 mm $DIZ_{Hex.}$ (1 mg/ml) 12–26 mm DIZ_{CHCl3} (1 mg/ml) 14–26 mm DIZ_{EtAC} (1 mg/ml) 21–26 mm	MIC_{MeOH} 2–3 mg/ml $MIC_{Hex.}$ 2.5–4 mg/ml MIC_{CHCl3} 2.5–3 mg/ml MIC_{EtAC} 2–3 mg/ml	Shah et al. (2015a, b)

T. africanum (leaves, flowers, stem) South Africa	Water, methanol, dichloromethane – methanol extracts (Maceration)	*B. cereus* ATCC 11778, *E. coli* ATCC 8739, *K. pneumoniae* ATCC 13883, *Moraxella catarrhalis* ATCC 23246, *P. aeruginosa* ATCC 27858, *S. aureus* ATCC 25923, *S. pyogenes* ATCC 8668	MIC 1– > 8 mg/ml (G^+ bacteria) MIC 0.125– > 8 mg/ml (G^- bacteria)	Not tested	Ruiters et al. (2016)
T. kraussii (leaves, flowers, stem) South Africa			MIC 0.8–6.6 mg/ml (G^+ bacteria) MIC 1–8 mg/ml (G^- bacteria)	Not tested	
T. trifidum (leaves, flowers, stem) South Africa			MIC 1– > 8 mg/ml (G^+ bacteria) MIC 1– > 8 mg/ml (G^- bacteria)	Not tested	
T. montbretii subsp. *pamphylicum* (aerial parts) Turkey	Methanol extract (Soxhlet)	*E. coli*, *S. typhi*, *Yersinia enterocolitica*, *S. aureus*, *S. pneumoniae*, *E. aerogenes*, *L. monocytogenes*, *Lactobacillus reuteri*, *L. acidophilus*, *Micrococcus luteus*	DIZ (10% extract) 13.5–22.5 mm	Not tested	Özkan et al. (2007)
T. flavum (flowers) Italy	Ethanol (80%) extract (Maceration)	32 clinical isolates and 8 standard strains of bacteria	MIC 1024–16,384µg/ml (G^+ bacteria) MIC 8192– > 16,384µg/ml (G^- bacteria)	Not tested	Acquaviva et al. (2018)
T. fruticans (flowers) Italy			MIC 256–8192µg/ml (G^+ bacteria) MIC 2048–8192µg/ml (G^- bacteria)	Not tested	
T. siculum (flowers) Italy			MIC 256–8192µg/ml (G^+ bacteria) MIC 1024–8192µg/ml (G^- bacteria)	Not tested	

(continued)

Table 12.2 (continued)

Teucrium species (part used)	Type of extract (extraction method)	Tested microorganisms	DIZ (tested concentration) range, MIC range		References
			Antibacterial activity	Antifungal activity	
T. oliverianum (aerial parts) Saudi Arabia	Methanol (80%) extract (Maceration)	*K. pneumoniae*, *P. vulgaris*, *P. aeruginosa*, *Serratia marcescens*, *B. cereus*, *M. luteus*, *M. roseus*, *S. aureus*, *A. flavus*, *A. ochraceus*, *C. albicans*, *F. moniliforme*	DIZ (10 mg/ml) 9 mm (*B. cereus*, *K. pneumoniae*)	Not active	Shahat et al. (2017)
T. orientale var. *glabrescens* Turkey	Water extract (Maceration)	*E. coli* ATCC 25922, *Y. pseudotuberculosis* ATCC 911, *P. aeruginosa* ATCC 43288, *S. aureus* ATCC 25923, *E. faecalis* ATCC 29212, *B. cereus* 709 Roma; *M. smegmatis* ATCC 607, *C. albicans* ATCC 60193, *S. cerevisiae*	DIZ 6–16 mm	Not active	Yildirmiş et al. (2017)
T. cubense (aerial parts) Mexico	Water extract (Maceration)	*S. aureus*, *S. haemolyticus*, *E. faecalis*, *P. aeruginosa*, *E. coli*, *Acinetobacter lwoffii*, *A. baumanii*, *Burkholderia cepacin*, *C. albicans*, *C. tropicalis*, *T. belgeii*	DIZ (375µg/disk) 22 mm (*A. lwoffii*) DIZ (750µg/disk) 10 mm (*P. aeruginosa*)	DIZ (750µg/disk) 12 mm (*C. albicans*)	Jacobo-Salcedo et al. (2011)
T. yemense (aerial parts) Yemen	Methanol, water extracts (Maceration)	*S. aureus* ATCC 6538, *B. subtilis* ATCC 6059, *M. flavus*, *E. coli* ATCC 11229, *P. aeruginosa* ATCC 27853, *C. maltose*, multiresistant strains – *S. epidermidis*, *S. haemolyticus*, *S. aureus*	DIZ_{MeOH} (4 mg/ml) 9–26 mm MIC_{MeOH} 250–500µg/ml DIZ_{H2O} (4 mg/ml) 10–16 mm MIC_{H2O} > 1000µg/ml	Not active	Mothana et al. (2009a, b)
T. sokotranum (aerial parts) (Soqotra Island) Yemen			DIZ_{MeOH} (4 mg/ml) 10–22 mm MIC_{MeOH} 500–1000µg/ml DIZ_{H2O} (4 mg/ml) 10–22 mm MIC_{H2O} 250–1000µg/ml	Not active	

T. persicum (aerial parts) Iran	Methanol extract (Maceration)	*E. coli* PTCC 1338, *B. subtilis* PTCC 1023, *S. aureus* PTCC 1112, *S. epidermis* PTCC 1114, *P. aeruginosa* PTCC 1074, *S. typhi* PTCC 1693, *K. pneumonia* PTCC 1031, *A. niger* PTCC 5010, C. *albicans* PTCC 5027	MIC 0.5–2 mg/ml	MIC 2 mg/ml	Javidnia et al. (2009)
T. leucocladum (aerial parts) Egypt	n-Hexane-ether extract (Maceration)	*E. coli*, *P. aeruginosa*, *S. aureus*, *B. subtilis*, *C. albicans*	DIZ (1 mg/ml) 16–19 mm	DIZ (1 mg/ml) 15 mm	El-Shazly and Hussein (2004)
T. botrys (aerial parts) Serbia	Methanol, acetone, ethyl acetate extracts (Maceration)	*S. aureus* ATCC 25923, *E. coli* ATCC 25922, *S. aureus*, *E. coli*, *P. aeruginosa*, *C. albicans* ATCC 10231, *C. albicans*, *A. niger*	MIC_{MeOH} 1.25 mg/ml (*S. aureus* ATCC) MIC_{AcOH} 2.5 mg/ml (*S. aureus* ATCC) MIC_{EtAC} > 10 mg/ml	MIC_{MeOH} > 20 mg/ml MIC_{AcOH} 10 mg/ml (*A. niger*) MIC_{EtAC} 5–10 mg/ml	Stanković et al. (2012)
T. bicolor (aerial parts) Mexico	Methanol extract (Maceration)	*C. albicans*, *C. parapsilosis*, *C. tropicalis*, *C. krusei*, *C. glabrata*, *M. tuberculosis*	MIC 125μg/ml (*M. tuberculosis*)	MIC 31.2–125μg/ml	Garza et al. (2017)
T. sauvagei (leaves) Tunisia	Methanol extract (Maceration)	Dermatophytes	Not tested	Active against dermatophytes	Salah et al. (2006)
T. royleanum (whole plant) Pakistan	Methanol extract and sub-fractions: n-butanol, chloroform, ethyl acetate (Maceration)	*E. coli*, *B. subtilis*, *K. pneumoniae*, *S. flexneri*, *S. aureus*, *S. typhi*, *P. aeruginosa*, *T. longifusus*, *C. albicans*, *C. glaberata*, *A. flavus*, *M. canis*, *F. solani*	DIZ_{MeOH} (3 mg/ml) 9–16 mm DIZ_{BuOH} (3 mg/ml) 5–6 mm DIZ_{CHCl3} (3 mg/ml) 4–24 mm DIZ_{EtAC} (3 mg/ml) 14–25 mm	MeOH (24 mg/ml) 40–50% of inhibition $CHCl_3$ (24 mg/ml) 53–87% EtAc (24 mg/ml) 40–71% BuOH (24 mg/ml) 70% (*T. longifusus*)	Ahmad et al. (2008)

Table 12.3 List of major oils' compounds, tested microorganisms, and results of *Teucrium* species essential oils antimicrobial activity

Teucrium species	Major oil's compounds	Tested microorganisms	DIZ (tested concentration) range, MIC range		References
			Antibacterial activity	Antifungal activity	
T. polium Iran	Limonene (37.70%), 2,4 di-tetra-butylphenol (10.81%) p-cymene (8.20%)	*Staphylococcus aureus*, *S. epidermidis*, *Streptococcus pneumoniae*, *Klebsiella pneumoniae*, *Escherichia coli*	DIZ (10µl) 7–17 mm MIC 5–50µl	Not tested	Vahdani et al. (2011)
T. polium Algeria	Germacrene, bicyclogermacrene, β-pinene, carvacrol	*S. aureus* ATCC 25923, *Enterococcus feacalis* ATCC 29212, *Bacillus cereus* ATCC 11778, *E. coli* ATCC 25922, *Pseudomonas aeruginosa* ATCC 27853, *Aspergillus flavus*, *Fusarium oxysporum*, *Rhizopus stolonifer*	DIZ (15µl) 9–16 mm MIC 3–5µl/ml	2–10µl/ml – 6.58–25.9% of inhibition	Belmekki et al. (2013)
T. polium subsp. *capitatum* Italy (Corsica Island)	α-pinene (24.71%), α-thujene (8.1%), terpinen-4-ol (6.6%), limenone (5.2%)	*S. aureus*, *S. epidermidis*, *Listeria innocua*, *Campylobacter jejuni*, *Enterobacter aerogenes*, *E. aerogenes* (MDR)	DIZ (15µl) 10–44 mm MIC 0.2–12.5 mg/ml	Not tested	Djabou et al. (2013)
T. polium Iran	β-caryophyllene (29%), farnesene (13%), β-pinene (11%), germacrene D (6.5%), α-pinene (5.5%)	15 clinical isolates of *K. pneumoniae*	DIZ 14–28.5 mm MIC 0.62–1.25 mg/ml	Not tested	Raei et al. (2014)
T. polium Iran	α-pinene (25.79%), myrcene (12.50%)	*P. aeruginosa*, *Pantoea agglomerans*, *Brenneria nigrifluens*, *Rhizobium radiobacter*, *Rhizobium vitis*, *Streptomyces scabies*, *Ralstonia solanacearum*, *Xanthomonas campestris*, *Pectobacterium cartovorum*	Active against – *R. solanacearum*, *P. agglomerans*, *B. nigrifluens*, *S. scabies*	Not tested	Purnavab et al. (2015)

T. polium subsp. *capitatum* Algeria	t-cadinol (18.3%), germacrene D (15.3%), β-pinene (10.5%)	*S. aureus* ATCC 6538, *B. subtilis* ATCC 9372, *P. aeruginosa* ATCC 9027, *K. pneumoniae* ATCC 4352, *Candida albicans* ATCC 24433	DIZ (20µl) 14.7–35.3 mm	DIZ (20µl) 13.7 mm	Kerbouche et al. (2015)
T. polium Morocco	Not analysed	*S. aureus* ATCC 29213, *E. coli* ATCC 25922, *P. aeruginosa* ATCC 27853, *C. albicans* ATCC 10231, *A. brasiliensis* ATCC 16404	DIZ (10µl) 12 mm (*S. aureus*) MIC 1.3µl/ml (*S. aureus*)	DIZ (80µl) 10.33 mm (*C. albicans*)	Boukhira et al. (2016)
T. polium Turkey	(Z)-β-farnesene (15.49%), β-phellandrene (10.77%), α-farnesene (10.71%)	*S. aureus* (MRSA), *S. aureus* ATCC 6538, *P. aeruginosa*, *E. coli* O157:H7, *B. cereus*	DIZ (5µl) 11–15 mm	Not tested	Sevindik et al. (2016)
T. polium subsp. *gabesianum* Tunisia	β-pinene (35.9%), α-pinene (13.32%), α-thujene (8.46%)	*E. coli* ATCC 25922, *E. faecalis* ATCC 29212, *S. aureus* ATCC 25923, *P. aeruginosa* ATCC 27853, *Citrobacter freundeii*, *Proteus mirabilis C. albicans*, *Cryptococcus neoformans*, *Trichophyton rubrum*, *T. soudanense*, *Microsporum canis*, *A. fumigatus*, *Scopulariopsis brevicaulis*	MIC 0.078–0.156 mg/ml	MIC 0.062– > 1 mg/ml	Ben Othman et al. (2017)
T. marum subsp. *marum* Italy (Sardinia Island)	Isocaryophyllene (20.24%), β-bisabolene (14.73%), β-sesquiphellandrene (11.27%), α-santalene (10.97%), dolichodial (9.38%), α-caryophyllene (7.18%)	*F. oxysporum*, *Botrytis cinerea*, *Rhyzoctonia solani*, *Alternaria solani*	Not tested	MIC 250–3800µg/ml	Ricci et al. (2005)
T. marum Italy (Corsica Island)	Caryophyllene oxide (9.8%), (E)-α-bergamotene (8.2%), β-bisabolene (7.5%)	*S. aureus*, *S. epidermidis*, *L. innocua*, *C. jejuni*, *E. aerogenes*, *E. aerogenes* (MDR)	DIZ (15µl) 8–40 mm MIC 0.4–1 mg/ml	Not tested	Djabou et al. (2013)

(continued)

Table 12.3 (continued)

Teucrium species	Major oil's compounds	Tested microorganisms	DIZ (tested concentration) range, MIC range		References
			Antibacterial activity	Antifungal activity	
T. arduini Montenegro	Germacrene D (16.98%), β-caryophyllene (14.98%), β-burbonene (5.59%), α-amorphene (4.68%), linalool (7.05%), α-terpinolene (5.25%), 1-octene-3-ol (4.69%)	*B. subtilis*, *E. cloaceae*, *E. faecalis* ATCC 29212, *E. coli* ATCC 25922, *K. pneumoniae*, *M. lysodeikticus* ATCC 4698, *P. mirabilis*, *S. aureus* ATCC 25923, *S. aureus*, *C. albicans* ATCC 10259	MIC 6.25–50µl/ml	MIC 50µl/ml	Vuković et al. (2011)
T. arduini Croatia	β-caryophyllene (32.9%), germacrene D (16.4%), borneol (5.4%)	*S. aureus* ATCC 6538, *E. faecalis* ATCC 21212, *E. coli* ATCC 10535, *P. aeruginosa* ATCC 27853, *C. albicans* ATCC 10231, *M. gypseum*, *A. brasiliensis*	MIC 6.25–37.5 mg/ml	MIC 7.81–25 mg/ml	Kremer et al. (2012)
T. chamaedrys Italy (Corsica Island)	(E)-α-caryophyllene (33.9%), germacrene D (18.5%), α-humulene (7.5%), δ-cadinene (4.6%)	*S. aureus*, *S. epidermidis*, *L. innocua*, *C. jejuni*, *E. aerogenes*, *E. aerogenes* (MDR)	DIZ (15µl) 6–22 mm MIC 1– > 50 mg/ml	Not tested	Djabou et al. (2013)
T. chamaedrys subsp. *chamaedrys* Turkey	Germacrene D (16.7%), α-pinene (15.8%), β-caryophyllene (11.8%), β-pinene (8.9%), β-myrcene (4.1%)	*E. coli* ATCC 35218, *K. pneumoniae* ATCC 13883, *Yersinia pseudotuberculosis* ATCC 911, *Serratia marcescens* ATCC 13880, *E. faecalis* ATCC 29212, *S. aureus* ATCC 25923, *B. subtilis* ATCC 6633, *C. albicans* ATCC 60193, *C. tropicalis* ATCC 13803	DIZ (1000µg/ml) 5–15 mm	Not active	Küçük et al. (2006)
T. chamaedrys subsp. *lydium* Turkey	β-caryophyllene (19.7%), α-pinene (12.5%), germacrene D (9.3%), β-pinene (6.6%), caryophyllene oxide (6.1%)		DIZ (1000µg/ml) 5–15 mm	Not active	
T. orientale var. *puberulens* Turkey	β-caryophyllene (21.7%), 2-methyl cumarone (20.0%), germacrene D (10.6%), α-humulene (4.8%), δ-cadinene (4.1%)		DIZ (1000µg/ml) 5–15 mm	Not active	

T. orientale var. *orientale* Turkey	*β*-caryophyllene (15.3–19.0%), germacrene D (14.2–12.8%), caryophyllene oxide (14.0–19.0%)	*S. aureus* ATCC 29213, *E. faecalis* ATCC 29212, *E. coli* ATCC 25922, *P. aeruginosa* ATCC 27853, *C. albicans*, *C. tropicalis*	MIC 100–400μg/ml	MIC 25–50μg/ml	Kucukbay et al. (2011)
Teucrium orientale var. *puberulens* Turkey	*β*-caryophyllene (15.3–19.0%), germacrene D (14.2–12.8%), caryophyllene oxide (14.0–19.0%)		MIC 50–400μg/ml	MIC 12.5–25μg/ml	
T. massiliense Italy (Corsica Island)	6-methyl-3-heptyl acetate (19.1%), 3-octanyl acetate (7.0%), pulegone (6.9%), germacrene D (6.1%)	*S. aureus*, *S. epidermidis*, *L. innocua*, *C. jejuni*, *E. aerogenes*, *E. aerogenes* (MDR)	DIZ (15μl) 6–41 mm MIC 0.8–6 mg/ml	Not tested	Djabou et al. (2013)
T. scorodonia subsp. *scorodonia* Italy (Corsica Island)	(E)-*α*-caryophyllene (21.1%), germacrene B (8.3%), *α*-humulene (6.9%), germacrene D (6.7%), *α*-cubebene (6.2%)		DIZ (15μl) 6–26 mm MIC 0.2– > 50 mg/ml	Not tested	
T. flavum subsp. *glaucum* Italy (Corsica Island)	Limenone (27.4%), *α*-pinene (12.2%), (Z)-*α*-ocimene (6.0%), (E) phytol (4.5%)		DIZ (15 μl) 16–39 mm MIC 0.2–1.5 mg/ml	Not tested	
T. africanum South Africa	*β*-cubebene (20.5%), *α*-cubebene (29.3%)	*B. cereus* ATCC 11778, *E. coli* ATCC 8739, *K. pneumoniae* ATCC 13883, *Moraxella catarrhalis* ATCC 23246, *P. aeruginosa* ATCC 27858, *S. aureus* ATCC 25923, *S. pyogenes* ATCC 8668	MIC 0.16– > 8 mg/ml	Not tested	Ruiters et al. (2016)
T. kraussii South Africa	Not analysed		MIC 4 mg/ml (*S. pyogenes*)	Not tested	
T. trifidum South Africa	*β*-cubebene (31.1%), *α*-cubebene (11.4%), *β*-caryophyllene (7.7%)		MIC 2– > 8 mg/ml	Not tested	
T. leucocladum Egypt	Patchouli alcohol (31.24%), *β*-pinene (12.66%), *α*-pinene (10.99%), *α*-cadinol (19.27%), viridiflorol (5.36%), myrcene (5.35%)	*E. coli*, *P. aeruginosa*, *S. aureus*, *B. subtilis*, *C. albicans*	DIZ (20 mg/ml) 12–30 mm	DIZ (20 mg/ml) 26 mm	El-Shazly et al. (2004)

(continued)

Table 12.3 (continued)

Teucrium species	Major oil's compounds	Tested microorganisms	DIZ (tested concentration) range, MIC range		References
			Antibacterial activity	Antifungal activity	
T. stocksianum subsp. *stocksianum* Oman	α-cadinol (7.6%), β-selinene (6.4%), trans-verbenol (5.9%), caryophyllene oxide (5.7%), α-phellandren-8-ol (5.0%), verbenone (5.0%), δ-cadinene (5.1%)	*S. aureus* NCTC 6571, *E. coli* NCTC 10418, *P. aeruginosa* NCTC 10662, *S. aureus*, *S. albus*, *S. epidermidis*, *S. mitis*, *S. sanguis*, *M. luteus*, *B. subtilis*, *B. cereus*, *E. coli*, *E. aerogenes*, *K. pneumoniae*, *S. typhy*, *P. vulgaris*, *P. aeruginosa*, *C. albicans*, *Sacharomyces cerevisiae*, *R. stolonifer*, *Penicillium notatum*, *F. oxysporum*	DIZ (2 mg/disk) 9–18.5 mm MIC 4.5–5 mg/ml (G^+ bacteria) MIC 6.5–11 mg/ml (G^- bacteria)	DIZ (2 mg/disk) 3.5–6 mm	Hisham et al. (2006a, b)
T. mascatense Oman	Linalool (27.8%), linalyl acetate (12.6%), β-eudesmol (10.1%), α-bergamotene (5.0%)		DIZ (0.2 mg/disk) 3.5–17.5 mm MIC 1.5–2.5 mg/ml (G^+ bacteria) MIC 6–8.5 mg/ml (G^- bacteria)	DIZ (1.6 mg/disk) 3.5–6.5 mm	
T. hyrcanicum Iran	(E)-β-farnesene (34.1%)	*B. subtilis*, *S. aureus*, *S. typhi*, *P. aeruginosa*, *E. coli*, *A. niger*, *C. albicans*	Active against *B. subtilis*, *S. aureus*, *S. typhi*, *P. aeruginosa*	Active against *A. niger*	Morteza-Semnani et al. (2011)
T. ramosissimum Tunisia	β-eudesmol (44.52%), caryophyllene oxide (9.36%), α-thujene (5.51%), sabinene (4.71%), *t*-cadinol (3.9%)	*S. aureus* ATCC 25923, *E. faecalis* ATCC 29212, *E. coli* ATCC 25922, *S. enteridis* ATCC 13076, *S. typhymurium* NRRL B- 4420	MIC 0.24–0.36 mg/ml	Not tested	Ben Sghaier et al. (2007)
T. yemense Yemen	α-pinene (6.6%), (E)-caryophyllene (19.1%) α-humulene (6.4%), δ-cadinene (6.5%), caryophyllene oxide (4.3%), α-cadinol (9.5%), shyobunol (4.6%)	*E. coli* ATCC 10536, *P. aeruginosa* ATCC 25619, *S. aureus* ATCC 29737, *B. cereus*, *A. niger* ATCC 16888, *B. cinerea* ATCC 126943, *C. albicans* ATCC 90028	MIC 156–1250μg/ml	MIC 313–1250μg/ml	Ali et al. (2017)

T. sauvagei Tunisia	β-eudesmol, *t*-cadinol, α-thujene, gamma-cadinene, sabinene	Dermatophytes	Not tested	Active against dermatophytes	Salah et al. (2006)
T. divaricatum subsp. *villosum* Lebanon	(E)-caryophyllene (30.1%), caryophyllene oxide (6.1%)	G^+ bacteria, G^- bacteria	Active against G^+ bacteria	Not tested	Farmisano et al. (2010)
T. plectranthoides India	Not analysed	*B. megaterium, B. subtilis, E. coli, S. aureus, P. aeruginosa, P. vulgaris, A. niger, Xanthomonas campestris, A. parasiticus, Rhizopus oryzae, Rhizoctonia oryzae-sativae, Colletotrichum musae, F. solani, C. albicans, Alternaria brassicicola*	DIZ 20–38 mm	DIZ 16–27 mm	Thoppil et al. (2001)

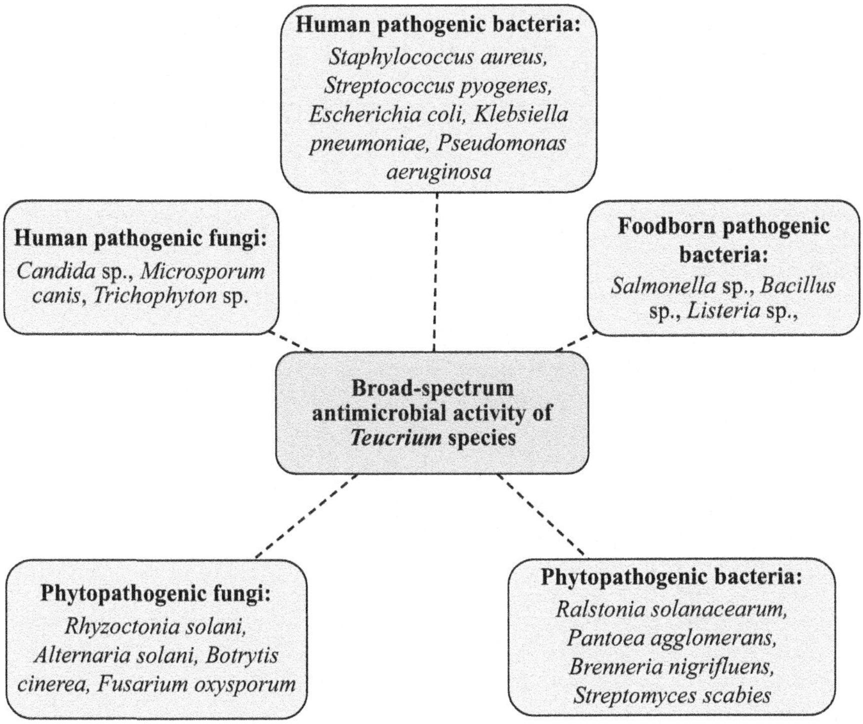

Fig. 12.4 The most susceptible bacteria and fungi to *Teucrium* species

necessary. So far, only, Elmasri et al. (2014, 2015) have evaluated antibiofilm activity of *Teucrium* species. The authors have noticed that several sesquiterpenes and flavonoids from *Teucrium polium* were effective in prevention of *Staphylococcus aureus* biofilm formation.

12.6.1 *Antimicrobial Activity of* Teucrium *Species Plant Extracts*

Generally, three levels of investigation of antimicrobial activity of plant extracts are established: (i) testing of crude extract, (ii) testing of sub-fractions, and (iii) testing of pure compounds. According to the literature search, crude extracts of *Teucrium* species were frequently researched. Several studies were focused on testing of sub-fractions of extracts, but no studies on testing of isolated pure compounds. Also, the mechanism of antimicrobial action of extracts from *Teucrium* species has not been determined yet.

The MIC values are used as cut-off points for classification of the antimicrobial potential of plant extracts. The one of that classification was defined by Tamokou

et al. (2017), as follows: MIC below 100μg/ml – highly active, 100 ≤ MIC ≤ 512μg/ml – significantly active, 512 < MIC ≤ 2048μg/ml – moderately active, MIC > 2048μg/ml – low active, MIC > 10 mg/ml – not active. The stated cut-off points were applied for interpretation of antimicrobial activity of extracts from *Teucrium* species. Summary results of antimicrobial activity are presented in Table 12.2. In the following paragraphs, the most significant results were discussed.

Teucrium polium was one of the most tested *Teucrium* species and plant materials, collected from different regions, have been screened for antimicrobial activity. Extracts from *Teucrium polium* originated from Tunisia and Turkey showed the most promising antimicrobial activity in relation to commonly used Gram-positive bacteria (*Staphylococcus aureus, S. epidermidis, Bacillus subtilis, B. cereus, Micrococcus luteus, Enterococcus faecalis*), Gram-negative bacteria (*Escherichia coli, Proteus mirabilis, Pseudomonas aeruginosa, Klebsiella pneumoniae, Salmonella typhi, Enterobacter cloaceae*) and fungi (*Candida albicans, Aspergillus niger, A. fumigatus, Trychophyton rubrum*). The MIC values of ethanol and water extract from Tunisian *Teucrium polium* were 0.156–0.625 mg/ml against bacteria and 0.625 mg/ml against fungal species *Microsporum canis* (Ben Othman et al. 2017). Moreover, Turkish *Teucrium polium* extract, in a form of decoction, showed better activity with MIC range of 16–64μg/ml for bacteria and 8–64μg/ml for fungi (Deliroman Orhan et al. 2012). The methanol extract from Jordanian plant showed moderate antibacterial activity, the best activity was noticed against *Escherichia coli* (MIC of 1.2 mg/ml) (Tarawneh et al. 2010). In addition, the ethanol extract of plant material from the same region were active on *Staphylococcus aureus* and *Pseudomonas aeruginosa* at MIC of 80 ppm (Khalil et al. 2009). On the other side, organic solvent extracts, tested in other studies, exhibited low antimicrobial activity or they were not active, MIC values were 3.125–50 mg/ml for bacteria and up to ˃ 150 mg/ml for fungi (Hashem 2011; Stanković et al. 2012; Mirzei et al. 2013; Khaled-Khodja et al. 2014; Dridi et al. 2016). Using time-kill assay, the ethanol extract at concentration of 0.3 mg/ml has no effect on the growth of *Escherichia coli* O157 NCTC 1290 (Mashreghi and Niknia 2012). In the study of Mansouri (1999), *Teucrium polium* flower ethanol extract had very low antistaphylococcal activity. Against 489 isolates of *Staphylococcus aureus*, at concentration of 30 mg/ml, it showed activity against 0.61% of tested isolates. The antilisterial activity of ethanol extract was noticed, the inhibitory zones of 10–11 mm (at conc. 133 mg/ml) and MIC values 12.5–50μg/ml were detected (Altanlar et al. 2006). Against pathogenic bacteria, *Brucella melitensis*, the extracts of *Teucrium polium* showed moderate activity compared to other medicinal plants tested in the same study (Motamedi et al. 2010). Furthermore, the synergistic activity of *Teucrium polium* extract and antibiotics was investigated. The methanol extract in combination with antibiotics (amoxicillin, chloramphenicol, neomycin, doxycycline, clarithromycin, cephalexin, and nalidixic acid) enhanced the activity of antibiotics against the resistant *Escherichia coli* strain (Darwish and Aburjai 2010).

Studies on *Teucrium chamaedrys* extracts indicated their similar antimicrobial activity even that plant material was collected in the different regions of the world.

In general, low to moderate antimicrobial activity was noticed and activity was higher against Gram-positive bacteria then Gram-negative bacteria and fungi (Table 12.2). In the recent paper (Nastić et al. 2018), the authors have reported the activity of *Teucrium chamaedrys* extract obtained by supercritical water extraction. If we compare the results with the results of the study (Stanković et al. 2012) where plant material was extracted by maceration, the higher activity of supercritical water extract was obtained. The plant materials were from the same region (Serbia). In *Teucrium chamaedrys* supercritical water extract several phenolic compounds were identified: vanillin acid, caffeic acid, epicatechin, ferulic acid, sinapic acid, gallic acid, protocatechuic acid, catechin, and chlorogenic acid. In addition, in ethanol extract of *Teucrium chamaedrys* from Romania the following phenolic compounds, luteolin, p-coumaric acid, isoquercitrin, rutin, quercitrin, were isolated (Vlase et al. 2014).

Teucrium montanum is one of the most popular medicinal plants in Serbia. Two independent research groups investigated the antimicrobial activity of this plant and they observed that the extracts (methanol, acetone, ethyl acetate, butanol) were active against *Staphylococcus aureus* strains (MIC of 0.15–1 mg/ml) (Djilas et al. 2006; Stanković et al. 2012). In addition, the methanol, acetone, ethyl acetate extracts were not active against tested fungi (Stanković et al. 2012).

Teucrium arduini is an endemic Illyric Balkan species distributed in Croatia, Bosnia and Herzegovina, Montenegro, Serbia and northern Albania. Šamec et al. (2010) have conducted a detailed study including six different localities in Croatia. The leaf and flower infusions were tested. The lowest MIC values were found against *Staphylococcus aureus* ATCC 6538 (MIC of 1.56 and 4.16 mg/ml), depending on the location from which plant material (leaves) was collected. *Teucrium arduini* flower and leaf infusions did not exhibit activity against Gram-negative bacterial species, or against fungal species tested in this study. Except for water extract in form of infusion, antimicrobial activity of organic solvent extracts from different location (Croatia and Serbia) were tested and promising activity against Gram-positive bacteria (MIC of 0.15–2 mg/ml) was noticed (Stanković et al. 2012; Kremer et al. 2013). In Croatian samples of *Teucrium arduini* several phenolic compounds were identified: protocatechuic acid, 4-hydroxybenzoic acid, salicylic acid, gentisic acid, ferulic acid, vanillic acid, caffeic acid, syringic acid, sinapic acid, 4-coumaric acid, rosmarinic acid and quercetin (Kremer et al. 2013).

Two subspecies of *Teucrium scordium* (*T. scordium* subsp. *scordioides* and *T. scordium* subsp. *scordium*) have been subjected to antimicrobial testing of two research groups from Serbia. The plant material was collected from different localities and organic solvent extracts were prepared. The promising activity was noticed against *Staphylococcus aureus* ATCC 25922 (MIC of 0.3–0.6 mg/ml) (Kundaković et al. 2011; Stanković et al. 2012).

Numerous studies of antimicrobial activity of different *Teucrium* endemic species, rare species or species characteristic for a particular region were, also, performed. According to detected MIC values, extracts of *Teucrium flavum*, *T. fruticans*, *T. siculum*, *T. yemense*, *T. sokotranum* and *T. persicum* exhibited pronounced activity against tested Gram-positive bacteria, MIC range was 250–2000µg/ml

(Table 12.2). *Teucrium bicolor* methanol extract was active against *Mycobacterium tuberculosis* (MIC of 125µg/ml) and *Candida* species (MIC of 31.2–125µg/ml) (Garza et al. 2017). Three endemic species (*Teucrium africanum, T. kraussii* and *T. trifidum*) from southern Africa were active against some important pathogens; *Teucrium africanum* against *Escherichia coli* with MIC value of 0.13 mg/ml. *Teucrium kraussii* against *Streptococcus pyogenes* with MIC value of 0.8 mg/ml and *Teucrium trifidum* showed activity against *Pseudomonas aeruginosa* with MIC value of 0.5 mg/ml (Ruiters et al. 2016). Oppositely, extracts of *Teucrium oliverianum, T. orientale, T. royleanum* and *T. botrys* showed low activity (Table 12.2).

12.6.2 Antimicrobial Activity of Teucrium *Species Essential Oils*

Essential oils are recognized as the most active products of plant metabolism. They present the mixture of bioactive compounds including monoterpenes (hydrocarbons and oxygenated derivatives), sesquiterpenes (hydrocarbons and oxygenated derivatives) and aliphatic compounds (alkanes, alkenes, ketones, aldehydes, esters, and alcohols). Generally, essential oils have strong antimicrobial properties and could be used as food preservatives or therapeutic antimicrobial agents. The level of *Teucrium* essential oils bioactivity was determined according to cut-off points, it was estimated that MIC values lower than 2 mg/ml indicate significant activity of essential oils (Van Vuuren and Viljoen 2006). Hence, most of the essential oils from *Teucrium* species exhibited significant activity against Gram-positive bacteria (*Staphylococcus* sp., *Streptococcus* sp., *Bacillus* sp.) and Gram-negative bacteria (*Escherichia coli, Pseudomonas aeruginosa, Klebsiella pneumoniae, Salmonella* sp.) and fungi (*Candida* sp.). Essential oils were isolated from *Teucrium* plants originated from different regions, especially from the Mediterranean basin. It was obvious that the chemical composition and the amounts of main constituents (mono- and sesquiterpene hydrocarbons and oxygenated sesquiterpenes) differed notably and showing intraspecific variations in the chemical composition of the essential oils according to environment and/or genetic parameters. The essential oils' main compounds as well as detailed results of antimicrobial activity were summarized in Table 12.3. In the following sections, the most significant results were discussed.

Antibacterial activity of *Teucrium polium* essential oil was investigated in several studies. The oil in the quantity of 5–20µl inhibit the growth of *Staphylococcus aureus* and *Bacillus subtilis* (Vahdani et al. 2011; Belmekki et al. 2013; Djabou et al. 2013; Kerbouche et al. 2015; Boukhira et al. 2016; Sevindik et al. 2016). Moreover, *Teucrium polium* essential oil extracted from plant material collected in North Anatolia (Turkey) showed promising activity against *Pseudomonas aeruginosa* and methicillin-resistant *Staphylococcus aureus* (Sevindik et al. 2016). In addition, essential oil of *Teucrium polium* originated from Iran was active against clinical isolates of *Klebsiella pneumoniae* (Raei et al. 2014). Interestingly, *Teucrium*

polium and *T. marum* essential oils were active against some phytopathogenic bacteria and fungi (Ricci et al. 2005; Purnavab et al. 2015). Essential oil of *Teucrium marum* inhibits the growth of *Rhyzoctonia solani* at concentration of 250μg/ml. Furthermore, *Teucrium orientale*, *T. africanum*, *T. ramosissimum*, *T. mascatense* and *T. yemense* essential oils showed excellent activity against tested bacteria (Table 12.3). As promising antibacterial agents can be *Teucrium africanum* essential oil against *Streptococcus pyogenes* (MIC of 0.16 mg/ml), *Teucrium ramosissimum* against *Salmonella enterica*, *S. typhimurium*, *Escherichia coli* (MIC of 0.28–0.24 mg/ml), *Teucrium mascatense* and *T. yemense* against *Staphylococcus aureus* and *Bacillus subtilis* (MIC of 156μg/ml). Multi-drug resistant strain of *Enterobacter aerogenes* was susceptible to *Teucrium massiliense* and *T. scordonia* essential oil (MIC of 0.2–0.4 mg/ml) (Djabou et al. 2013). On the other side, *Teucrium stocksianum*, *T. kraussii*, *T. trifidum*, *T. leucoclada*, *T. chamedrys* and *T. arduini* exhibited low antibacterial activity (Table 12.3).

Antifungal activity of *Teucrium* essential oils was, especially, evaluated against *Candida* species and dermatophytes (Table 12.3). Essential oils of *Teucrium orientale* subsp. *orientale* and *T. orientale* subsp. *puberulens* could be promising anti-*Candida* agents. They were active at concentrations range of 12.5–50μg/ml (Kucukbay et al. 2011). In addition, *Teucrium polium* and *T. sauvagei* were active against tested dermatophytes (Salah et al. 2006; Ben Othman et al. 2017).

Essential oils obtained from *Teucrium* species possess significant broad-spectrum antimicrobial activity, however, there is limited information available about their mechanisms of action. Better understanding the mechanisms of action will give more detailed information about their potency and potential application as antimicrobial agents.

12.7 Conclusion

This chapter provides a detailed review of the antimicrobial activity of *Teucrium* species, their active constituents, and their potential as sources of antibacterial and antifungal agents. The relevant literature summary showed that *Teucrium* species exhibited a diverse range of antimicrobial properties. The promising activity and potential application in control of bacteria and fungi possess the following species: plant extracts of *Teucrium polium*, *T. flavum*, *T. fruticans*, *T. siculum*, *T. yemense*, *T. sokotranum*, *T. persicum*, *T. scordium*, and essential oils from *Teucrium polium*, *T. orientale*, *T. africanum*, *T. ramosissimum*, *T. mascatence*, *T. yemense*, *T. massiliense*, *T. scordonia*. They were active against important pathogenic bacteria (*Staphylococcus* sp., *Bacillus* sp., *Enterococcus* sp., *Streptococcus* sp., *Escherichia coli*, *Pseudomonas aeruginosa*, *Klebsiella pneumoniae*, *Salmonella* sp.) and fungi (*Candida* sp., *Trychophyton* sp.). However, these results are conducted on the basis of in vitro studies. The future studies on mechanisms of action of plant extracts, essential oils, and pure active compounds will contribute to the development of new *Teucrium* antimicrobial agents. Furthermore, in vivo testing of activity, toxicity and

bioavailability will determine their actual relevance for treatment of infectious diseases or used as food preservatives.

Acknowledgments This work was supported by the Ministry of Education, Science and Technological Development of the Republic of Serbia (Grants OI173032 and III41010).

References

Acquaviva R, Genovese C, Amodeo A, Tomasello B, Malfa G, Sorrenti V, Tempera G, Addamo AP, Ragusa S, Rosa T, Menichini F, Di Giacomo C (2018) Biological activities of *Teucrium flavum* L., *Teucrium fruticans* L., and *Teucrium siculum* rafin crude extracts. Plant Biosyst 152:720–727

Ahmad B, Mukaram Shah SM, Bashir S, Begum H (2008) Antibacterial and antifungal activities of *Teucrium royleanum* (Labiatea). J Enzyme Inhib Med Chem 23:136–139

Ali NAA, Chhetri BK, Dosoky NS, Shari K, Al-Fahad AJA, Wessjohann L, Setzer WN (2017) Antimicrobial, antioxidant, and cytotoxic activities of *Ocimum forskolei* and *Teucrium yemense* (Lamiaceae) essential oils. Fortschr Med 4:17. https://doi.org/10.3390/medicines4020017

Ali F, Jan AK, Khan NM, Ali R, Mukhtiar M, Khan S, Khan SA, Aziz R (2018) Selective biological activities and phytochemical profiling of two wild plant species, *Teucrium polium* and *Capsicum annum* from Sheringal, Pakistan. Chiang Mai J Sci 45:881–887

Altanlar N, Saltan Çitoğlu G, Yılmaz BS (2006) Antilisterial activity of some plants used in folk medicine. J Pharm Biol 44:91–94

Antolak H, Kregiel D (2017) Food preservatives from plants. In: Karunaratne DN, Pamunuwa G (eds) Food additives. IntechOpen, Croatia, pp 45–87

Belmekki N, Bendimerad N, Bekhechi C, Fernandez X (2013) Chemical analysis and antimicrobial activity of *Teucrium polium* L. essential oil from Western Algeria. J Med Plant Res 7:897–902

Ben Othman M, Salah-Fatnassi KBH, Ncibi S, Elaissi A, Zourgui L (2017) Antimicrobial activity of essential oil and aqueous and ethanol extracts of *Teucrium polium* L. subsp. *gabesianum* (LH) from Tunisia. Physiol Mol Biol Plants 23:723–729

Ben Sghaier M, Chraief I, Skandrani I, Bouhlel I, Boubaker J, Kilani S, Neffati A, Mahmoud A, Hammami M, Chekir-Ghedira L, Ghedira K (2007) Chemical composition and antimicrobial activity of the essential oil of *Teucrium ramosissimum* (Lamiaceae). Chem Biodivers 4:1480–1486

Borges AJ, Saavedra M, Simoes M (2015) Insights on antimicrobial resistance, biofilms and the use of phytochemicals as new antimicrobial agents. Curr Med Chem 22:2590–2614

Boukhira S, Balouiri M, Bousta F, Moularat S, Taleb MS, Bousta D (2016) Antimicrobial activities of essential oil of five plant species from Morocco against some microbial strains. Int J Pharm Phytochem Res 8:1901–1906

Buzzini P, Arapitsas P, Goretti M, Branda E, Turchetti B, Pinelli P, Ieri F, Romani A (2008) Antimicrobial and antiviral activity of hydrolysable tannins. Mini-Rev Med Chem 8:1179–1187

Carson CF, Hammer KA, Riley TV (2006) *Melaleuca alternifolia* (tea tree) oil: a review of antimicrobial and other medicinal properties. Clin Microbiol Rev 19:50–62

Ceyhan N, Keskin D, Uğur A (2012) Antimicrobial activities of different extracts of eight plant species from four different family against some pathogenic microorganisms. J Food Agric Environ 10:193–197

Coppo E, Marchese A (2014) Antibacterial activity of polyphenols. Curr Pharm Biotechnol 15:380–390

Cos P, Vlietinck AJ, Berghe DV, Maes L (2006) Anti-infective potential of natural products: how to develop a stronger in vitro "proof-of-concept". J Ethnopharmacol 106:290–302

Costerton JW, Stewart PS, Greenberg EP (1999) Bacterial biofilm: a common cause of persistent infections. Science 284:1318–1322
Cowan MC (1999) Plant products as antimicrobial agents. Clin Microbiol Rev 12:564–582
Cushnie T, Lamb AJ (2011) Recent advances in understanding the antibacterial properties of flavonoids. Int J Antimicrob Agents 38:99–107
Daglia M (2012) Polyphenols as antimicrobial agents. Curr Opin Biotechnol 23:174–181
Darabpour E, Motamedi H, Nejad SM (2010) Antimicrobial properties of *Teucrium polium* against some clinical pathogens. Asian Pac J Trop Med 3:124–127
Darwish RM, Aburjai TA (2010) Effect of ethnomedicinal plants used in folklore medicine in Jordan as antibiotic resistant inhibitors on *Escherichia coli*. BMC Complement Altern Med 10:9. https://doi.org/10.1186/1472-6882-10-9
Deliroman Orhan D, Özçelik B, Hoşbaş S, Vural M (2012) Assessment of antioxidant, antibacterial, antimycobacterial, and antifungal activities of some plants used as folk remedies in Turkey against dermatophytes and yeast-like fungi. Turk J Biol 36:672–686
Djabou N, Lorenzi V, Guinoiseau E, Andreani S, Giuliani MC, Desjobert JM, Bolla JM, Costa J, Berti L, Luciani A, Muselli A (2013) Phytochemical composition of Corsican *Teucrium* essential oils and antibacterial activity against foodborne or toxi-infectious pathogens. Food Control 30:354–363
Djilas SM, Markov SL, Cvetković DD, Čanadanović-Brunet JM, Ćetković GS, Tumbas VT (2006) Antimicrobial and free radical scavenging activities of *Teucrium montanum*. Fitoterapia 77:401–403
Dridi A, Hadef Y, Bouloudani L (2016) Determination of total phenol, flavonoid, antioxidant and antimicrobial activity of methanolic extract of *Teucrium polium* L. Algerian East. Int J Pharmacogn Phytochem Res 8:1566–1570
Elmasri WA, Hegazy M-EF, Aziz M, Koksal E, Amor W, Mechref Y, Hamood AN, Cordes DB, Paré PW (2014) Biofilm blocking sesquiterpenes from *Teucrium polium*. Phytochemistry 103:107–113
Elmasri WA, Yang T, Tran P, Hegazy M-EF, Hamood AN, Mechref Y, Pare PW (2015) *Teucrium polium* phenylethanol and iridoid glycoside characterization and flavonoid inhibition of biofilm-forming *Staphylococcus aureus*. J Nat Prod 78:2–9
El-Shazly AM, Hussein KT (2004) Chemical analysis and biological activities of the essential oil of *Teucrium leucocladum* Boiss. (Lamiaceae). Biochem Syst Ecol 32:665–674
Formisano C, Napolitano F, Rigano D, Arnold NA, Piozzi F, Senatore F (2010) Essential oil composition of *Teucrium divaricatum* Sieb. ssp. *villosum* (Celak.) Rech. fil. growing wild in Lebanon. J Med Food 13:1281–1285
Garza BA, Arroyo JL, González GG, González EG, de Torres NW, Aranda RS (2017) Anti-fungal and anti-mycobacterial activity of plants of Nuevo Leon, Mexico. Pak J Pharm Sci 30:17–21
Gursoy N, Tepe B (2009) Determination of the antimicrobial and antioxidative properties and total phenolics of two "endemic" Lamiaceae species from Turkey: *Ballota rotundifolia* L. and *Teucrium chamaedrys* C. Koch. Plant Foods Hum Nutr 64:135–140
Hall-Stoodley L, Costerton JW, Stoodley P (2004) Bacterial biofilms: from the natural environment to infectious diseases. Nat Rev Microbiol 2:95–108
Hashem M (2011) Antifungal properties of crude extracts of five Egyptian medicinal plants against dermatophytes and emerging fungi. Mycopathologia 172:37–46
Hisham A, Pathare N, Al-Saidi S (2006a) The composition and antimicrobial activity of the essential oil of *Teucrium stocksianum* subsp. *stocksianum* leaf from Oman. Nat Prod Commun 1:195–199
Hisham A, Pathare N, Al-Saidi S, Al-Salmi A (2006b) The composition and antimicrobial activity of leaf essential oil of *Teucrium mascatenses* Boiss. from Oman. J Essent Oil Res 18:465–468
Høiby N, Bjarnsholt T, Givskov M, Molin S, Ciofu O (2010) Antibiotic resistance of bacterial biofilms. Int J Antimicrob Agents 35:322–332
Huber B, Eberl L, Feucht W, Polster J (2003) Influence of polyphenols on bacterial biofilm formation and quorum-sensing. Z Naturforsch C 58c:879–884

Ionescu MI (2018) Are herbal products an alternative to antibiotics. In: Kirmusaoğlu S (ed) Bacterial pathogenesis and antibacterial control. IntechOpen, Croatia, pp 3–23

Jacobo-Salcedo MD, Alonso-Castro AJ, Salazar-Olivo LA, Carranza-Alvarez C, González-Espíndola LÁ, Domínguez F, Maciel-Torres SP, García-Lujan C, González-Martínez MD, Gómez-Sánchez M, Estrada-Castillón E (2011) Antimicrobial and cytotoxic effects of Mexican medicinal plants. Nat Prod Commun 6:1925–1928

Janaćković P, Rajčević N, Gavrilović M (2017) Phytochemical practicum (in Serbian). University of Belgrade, Belgrade

Javidnia K, Miri R, Assadollahi M, Gholami M, Ghaderi M (2009) Screening of selected plants growing in Iran for antimicrobial activity. Iran J Sci Technol (Sci) 33:329–333

Kerbouche L, Hazzit M, Ferhat MA, Baaliouamer A, Miguel MG (2015) Biological activities of essential oils and ethanol extracts of *Teucrium polium* subsp. *capitatum* (L.) Briq. and *Origanum floribundum* Munby. J Essent Oil Bear Plants 18:1197–1208

Khaled-Khodja N, Boulekbache-Makhlouf L, Madani K (2014) Phytochemical screening of antioxidant and antibacterial activities of methanolic extracts of some Lamiaceae. Ind Crop Prod 61:41–48

Khalil A, Dababneh BF, Al-Gabbiesh AH (2009) Antimicrobial activity against pathogenic microorganisms by extracts from herbal Jordanian plants. J Food Agric Environ 7:103–106

Klancnik A, Piskernik S, Jersek B, Mozina SS (2010) Evaluation of diffusion and dilution methods to determine the antibacterial activity of plant extracts. J Microbiol Methods 81:121–126

Kovačević N (2004) Basics of pharmacognosy (in Serbian). Serbian School Book, Belgrade

Kremer D, Dragojević Müller I, Dunkić V, Vitali D, Stabentheiner E, Oberländer A, Bezić N, Kosalec I (2012) Chemical traits and antimicrobial activity of endemic *Teucrium arduini* L. from Mt Biokovo (Croatia). Cent Eur J Biol 7:941–947

Kremer D, Joze Kosir I, Kosalec I, Zovko Koncic M, Potocnik T, Cerenak A, Bezic N, Srecec S, Dunkic V (2013) Investigation of chemical compounds, antioxidant and antimicrobial properties of *Teucrium arduini* L. (Lamiaceae). Curr Drug Targets 14:1006–1014

Küçük M, Gülec C, Yaşar A, Üçüncü O, Yaylı N, Coşkunçelebi K, Terzioğlu S, Yaylı N (2006) Chemical composition and antimicrobial activities of the essential oils of *Teucrium chamaedrys*. subsp. *chamaedrys*., *T. orientale*. var. *puberulens*., and *T. chamaedrys*. subsp. *lydium*. Pharm Biol 44:592–599

Kucukbay ZF, Yildiz B, Kuyumcu E, Gunal S (2011) Chemical composition and antimicrobial activities of the essential oils of *Teucrium orientale* var. *orientale* and *Teucrium orientale* var. *puberulens*. Chem Nat Compd 47:833–836

Kundaković T, Milenković M, Topić A, Stanojković T, Juranić Z, Lakušić B (2011) Cytotoxicity and antimicrobial activity of *Teucrium scordium* L. (Lamiaceae) extracts. Afr J Microbiol Res 5:2692–2696

Kunduhoglu B, Pilatin S, Caliskan F (2011) Antimicrobial screening of some medicinal plants collected from Eskisehir, Turkey. Fresenius Environ Bull 20:945–952

Li W-R, Shi Q-S, Ouyang Y-S, Chen Y-B, Duan S-S (2013) Antifungal effects of citronella oil against *Aspergillus niger* ATCC 16404. Appl Microbiol Biotechnol 97:7483–7492

Li W-R, Shi Q-S, Liang Q, Huang X-M, Chen Y-B (2014) Antifungal effect and mechanism of garlic oil on *Penicillium funiculosum*. Appl Microbiol Biotechnol 98:8337–8346

Li W-R, Shi Q-S, Dai H-Q, Liang Q, Xie X-B, Huang X-M, Zhao G-Z, Zhang L-X (2016) Antifungal activity, kinetics and molecular mechanism of action of garlic oil against *Candida albicans*. Sci Rep 6:22805. https://doi.org/10.1038/srep22805

Ličina BZ, Stefanović OD, Vasić SM, Radojević ID, Dekić MS, Čomić LJR (2013) Biological activities of the extracts from wild growing *Origanum vulgare* L. Food Control 33:498–504

Mansouri S (1999) Inhibition of *Staphylococcus aureus* mediated by extracts from Iranian plants. J Pharm B 37:375–377

Mashreghi M, Niknia S (2012) The effect of *Peganum harmala* and *Teucrium polium* alcoholic extracts on growth of *Escherichia coli* O157. Jundishapur J Microbiol 5:511–515

Mirzaei A, Toori MA, Mirzaei N, Shirazi RG (2013) Antioxidant, antimicrobial and antimutogenic potential of 4 Iranian medicinal plants. Life Sci J 10:1085–1091

Morteza-Semnani K, Saeedi M, Akbarzadeh M (2011) Chemical composition and antimicrobial activity of essential oil of *Teucrium hyrcanicum* L. J Essent Oil Bear Plants 14:770–775

Motamedi H, Darabpour E, Gholipour M, Nejad SMS (2010) *In vitro* assay for the anti-Brucella activity of medicinal plants against tetracycline-resistant *Brucella melitensis*. J Zhejiang Univ Sci B 11:506–511

Mothana RA, Gruenert R, Bednarski PJ, Lindequist U (2009a) Evaluation of the *in vitro* anticancer, antimicrobial and antioxidant activities of some Yemeni plants used in folk medicine. Pharmazie 64:260–268

Mothana RA, Lindequist U, Gruenert R, Bednarski PJ (2009b) Studies of the *in vitro* anticancer, antimicrobial and antioxidant potentials of selected Yemeni medicinal plants from the island Soqotra. BMC Complement Altern Med 9:7. https://doi.org/10.1186/1472-6882-9-7

Muruzović MZ, Mladenović KG, Stefanović OD, Vasić SM, Čomić LR (2016) Extracts of *Agrimonia eupatoria* L. as sources of biologically active compounds and evaluation of their antioxidant, antimicrobial, and antibiofilm activities. J Food Drug Anal 24:539–547

Nastić N, Švarc-Gajić J, Delerue-Matos C, Barroso MF, Soares C, Moreira MM, Morais S, Mašković P, Srček VG, Slivac I, Radošević K, Radojković M (2018) Subcritical water extraction as an environmentally-friendly technique to recover bioactive compounds from traditional Serbian medicinal plants. Ind Crop Prod 111:579–589

Ncube NS, Afolayan AJ, Okoh AI (2008) Assessment techniques of antimicrobial properties of natural compounds of plant origin: current methods and future trends. Afr J Biotechnol 7:1797–1806

Ojala T (2001) Biological screening of plant coumarins. Dissertation, University of Helsinki

Özkan G, Kuleaşan H, Çelik S, Göktürk RS, Ünal O (2007) Screening of Turkish endemic *Teucrium montbretii* subsp. *pamphylicum* extracts for antioxidant and antibacterial activities. Food Control 18:509–512

Purnavab S, Ketabchi S, Rowshan V (2015) Chemical composition and antibacterial activity of methanolic extract and essential oil of Iranian *Teucrium polium* against some of phytobacteria. Nat Prod Res 29:1376–1379

Qabaha KI (2013) Antimicrobial and free radical scavenging activities of five Palestinian medicinal plants. Afr J Tradit Complement Altern Med 10:101–108

Radulović NS, Blagojević PD, Stojanović-Radić ZZ, Stojanović NM (2013) Antimicrobial plant metabolites: structural diversity and mechanism of action. Curr Med Chem 20:932–952

Raei F, Ashoori N, Eftekhar F, Yousefzadi M (2014) Chemical composition and antibacterial activity of *Teucrium polium* essential oil against urinary isolates of *Klebsiella pneumoniae*. J Essent Oil Res 26:65–69

Rahalison L, Hamburger M, Hostettmann K, Manod M, Frenk E (1991) A bioautographic agar overlay method for the detection of antifungal compounds from higher plants. Phytochem Anal 2:199–203

Ricci D, Fraternale D, Giamperi L, Bucchini A, Epifano F, Burini G, Curini M (2005) Chemical composition, antimicrobial and antioxidant activity of the essential oil of *Teucrium marum* (Lamiaceae). J Ethnopharmacol 98:195–200

Ruiters AK, Tilney PM, Van Vuuren SF, Viljoen AM, Kamatou GP, Van Wyk BE (2016) The anatomy, ethnobotany, antimicrobial activity and essential oil composition of southern African species of *Teucrium* (Lamiaceae). S Afr J Bot 102:175–185

Salah KB, Mahjoub MA, Chaumont JP, Michel L, Millet-Clerc J, Chraeif I, Ammar S, Mighri Z, Aouni M (2006) Chemical composition and *in vitro* antifungal and antioxidant activity of the essential oil and methanolic extract of *Teucrium sauvagei* Le Houerou. Nat Prod Res 20:1089–1097

Saleem M, Nazir M, Shaiq Ali M, Hussain H, Lee YS, Riaz N, Jabbar A (2010) Antimicrobial natural products: an update on future antibiotic drug candidates. Nat Prod Rep 27:238–254

Šamec D, Gruz J, Strnad M, Kremer D, Kosalec I, Grubešić RJ, Karlović K, Lucic A, Piljac-Žegarac J (2010) Antioxidant and antimicrobial properties of *Teucrium arduini* L. (Lamiaceae) flower and leaf infusions. Food Chem Toxicol 48:113–119

Sarić M (1989) Medicinal plants of Serbia (in Serbian). Serbian Academy of Science and Arts, Belgrade, p 640

Sevindik E, Abacı ZT, Yamaner C, Ayvaz M (2016) Determination of the chemical composition and antimicrobial activity of the essential oils of *Teucrium polium* and *Achillea millefolium* grown under North Anatolian ecological conditions. Biotechnol Biotechnol Equip 30:375–380

Shah SM, Ayaz M, Khan AU, Ullah F, Farhan, Shah AU, Iqbal H, Hussain S (2015a) 1,1-Dipheny 1,2-picrylhydrazyl free radical scavenging, bactericidal, fungicidal and leishmanicidal properties of *Teucrium stocksianum*. Toxicol Ind Health 31:1037–1043

Shah S, Sadiq A, Gul F (2015b) Antibacterial potential of methanolic extracts and sub-fractions of *Teucrium stocksianum* Bioss collected from Malakand division Pakistan. Pharmacol Online 1:8–12

Shahat AA, Mahmoud EA, Al-Mishari AA, Alsaid MS (2017) Antimicrobial activities of some Saudi Arabian herbal plants. Afr J Tradit Complement Altern Med 14:161–165

Silva LN, Zimmer KR, Macedó AJ, Trentin DS (2016) Plant products targeting bacterial virulence factors. Chem Rev 116:9162–9236

Stanković M, Stefanović O, Čomić LJ, Topuzović M, Radojević I, Solujić S (2012) Antimicrobial activity, total phenolic content and flavonoid concentrations of *Teucrium* species. Cent Eur J Biol 7:664–671

Stefanović O, Čomić LJ, Stanojević D (2009) Inhibitory effect of *Torilis anthriscus* on growth of microorganisms. Cent Eur J Biol 4:493–498

Stefanović O, Stanojević D, Čomić LJ (2012) Synergistic antibacterial activity of *Salvia officinalis* and *Cichorium intybus* extracts and antibiotics. Acta Pol Pharm 69:457–463

Stefanović OD, Tešić JD, Čomić LR (2015) *Melilotus albus* and *Dorycnium herbaceum* extracts as source of phenolic compounds and their antimicrobial, antibiofilm, and antioxidant potentials. J Food Drug Anal 23:417–424

Tajkarimi MM, Ibrahim SA, Cliver DO (2010) Antimicrobial herb and spice compounds in food. Food Control 21:1199–1218

Tamokou JDD, Mbaveng AT, Kuete V (2017) Antimicrobial activities of African medicinal spices and vegetables. In: Kuete V (ed) Medicinal spices and vegetables from Africa: therapeutic potential against metabolic, inflammatory, infectious and systemic diseases. Academic, Waltham, pp 207–237

Tarawneh KA, Irshaid F, Jaran AS, Ezealarab M, Khleifat KM (2010) Evaluation of antibacterial and antioxidant activities of methanolic extracts of some medicinal plants in northern part of Jordan. J Biol Sci 10:325–332

Thoppil JE, Minija J, Tajo A, Deena MJ (2001) Antimicrobial activity of *Teucrium plectranthoides* Gamble essential oil. J Nat Rem 1:155–157

Tongnuanchan P, Benjakul S (2014) Essential oils: extraction, bioactivities, and their uses for food preservation. J Food Sci 79:1231–1249

Upadhyay A, Upadhyaya I, Kollanoor-Johny A, Venkitanarayanan K (2014) Combating pathogenic microorganisms using plant-derived antimicrobials: a minireview of the mechanistic basis. Biomed Res Int 2014:761741. https://doi.org/10.1155/2014/761741

Vahdani M, Faridi P, Zarshenas MM, Javadpour S, Abolhassanzadeh Z, Moradi N, Bakzadeh Z, Karmostaji A, Mohagheghzadeh A, Ghasemi Y (2011) Major compounds and antimicrobial activity of essential oils from five Iranian endemic medicinal plants. Pharmacogn J 3:48–53

Van Vuuren SF, Viijoen AM (2006) A comparative investigation of the antimicrobial properties of indigenous South Africa aromatic plants with popular commercially available essential oils. J Essent Oil Res 18:66–71

Vlase L, Benedec D, Hanganu D, Damian G, Csillag I, Sevastre B, Mot AC, Silaghi-Dumitrescu R, Tilea I (2014) Evaluation of antioxidant and antimicrobial activities and phenolic profile for *Hyssopus officinalis*, *Ocimum basilicum* and *Teucrium chamaedrys*. Molecules 19:5490–5507

Vuković N, Sukdolak S, Solujić S, Mihailović V, Mladenović M, Stojanović J, Stanković M (2011) Chemical composition and antimicrobial activity of *Teucrium arduini* essential oil and cirsimarin from Montenegro. J Med Plant Res 5:1244–1250
Wood TK, Hong SH, Ma Q (2011) Engineering biofilm formation and dispersal. Trends Biotehnol 29:87–94
Yildirmiş S, Aliyazicioglu R, Emre Eyupoglu O, Ozgen U, Alpay Karaoglu S (2017) Biological activity and characterization of volatile compounds of *Teucrium orientale* var. *glabrescens* by SPME and GC-FID/MS. J Food Biochem 41:e12284. https://doi.org/10.1111/jfbc.12284

Chapter 13
Anticancer Activity of Secondary Metabolites of *Teucrium* Species

Milena Milutinović and Danijela Cvetković

Abstract Cancer is one of the life-threatening diseases and currently leading human health problem. Conventional use of chemotherapeutics in anticancer therapy is faced with some problems, such as their being harmful for the organism and development of the cancer cell resistance to the current therapeutic agents. There is a great need to discover and develop new anticancer drugs, more effective and less toxic for healthy cells, such as plant-derived compounds. There is a variety of experimental and epidemiological evidence that plant extracts and secondary metabolites from them achieve anticancer activity. Many of the species from the genus *Teucrium* have been used extensively in the traditional medicine as medicinal plants for treatment of numerous diseases, due to a variety of their therapeutic properties. Bearing in mind the advantages of utilization of plant products as anticancer agents, we revised recent literature and searched bibliographic databases about anticancer properties of *Teucrium* species, with an aim to give survey of their investigation and emphasize the need for further research in order to improve the anticancer treatment. This paper presents an overview of the current *in vitro* and *in vivo* data that support their anticancer potential. The extracts of *Teucrium* species and their secondary metabolites showed antiproliferative effects on multiple cancer cell lines, multiple signaling pathways, and altered the expression of genes and proteins involved in cancer development, cell cycle, apoptosis, angiogenesis, metastasis. The comparative review of results may contribute to better understanding of possible mechanisms of action and stimulate further examination for application of *Teucrium* species as adjuvant chemotherapeutics.

Keywords Apoptosis · Angiogenesis · Cancer · Cytotoxicity · Invasion · Migration · ROS · *Teucrium* species

M. Milutinović (✉) · D. Cvetković
Department of Biology and Ecology, Faculty of Science, University of Kragujevac, Kragujevac, Serbia
e-mail: milena.milutinovic@pmf.kg.ac.rs

M. Stanković (ed.), *Teucrium Species: Biology and Applications*,
https://doi.org/10.1007/978-3-030-52159-2_13

Abbreviations

AO/EB	Acridine Orange/Ethidium Bromide
Bcl-2	B-cell lymphoma 2
Bid	BH3 interacting-domain death agonist
Cdks	Cyclins dependent kinases
CKIs	Cdk inhibitors
CLL	Chronic Lymphocytic Leukemia
CXCL-12	CXC Motif Chemokine Ligand 12
CXCR-4	CXC Receptor 4
DISC	Death-Inducing-Signaling Complex
ECM	Extracellular Matrix
EMT	Mesenchymal-epithelial Transition
FGF	Fibroblast Growth Factor
GSH	Glutathione
HIF-1α	Hypoxia Inducible Factor-1α
ICAM-1	Intracellular Adhesion Molecule
iNOS	Inducible Nitric Oxide Synthase
MMP	Matrix metalloproteinase
MTT	3-[4,5-dimethylthiazol-2-yl]-2,5-diphenyltetrazolium bromide
NCI	National Cancer Institute
$O_2^{\cdot -}$	Superoxide anion radical
PARP	Poly ADB Ribose Polymerase
PI	Propidium Iodide
RNS	Reactive Nitrogen Species
ROS	Reactive Oxygen Species
Smac/DIABLO	Second mitochondria-derived activator of caspase/Direct Inhibitor of Apoptosis-Binding protein
STAT3	Signal Transducer and Activator of Transcription 3
tBid	Truncated Bid
TNF	Tumor Necrosis Factor
VCAM1	Vascular Cell Adhesion Molecule 1
VEGF	Vascular Endothelial Growth Factor
XIAP	Inhibitor of caspase function proteins
γ-GCS	γ-glutamyl cysteine synthetase

13.1 Introduction

Cancer can be explained as uncontrolled cell growth, due to disorders in the basic cell processes (Cooper 2000; Ma and Yu 2006). It begins with changes in genetic material, i.e. mutations, but also may be the result of epigenetic interactions, mainly in protooncogenes, tumor suppressor, DNA repair genes and genes which regulate

apoptosis. Changes in those genes are the most common causes of cancer development, together with impact of hereditary factors (Levine and Puzio-Kuter 2010; Ngo et al. 2015). Cancer cells are characterized by multiple molecular, biochemical, morphological and behavioral characteristics. Some of them are dysregulation and abnormality in cell cycle and division, production of own growth signals, evasion of apoptosis and contact inhibition, changes in communication with surrounding cells, altered metabolism, angiogenic, invasive and metastatic potential, etc. (Hanahan and Weinberg 2000).

Cancer is one of the leading health problems, with growing incidence in last years (Siegel et al. 2019). There are many strategies and attempts to overcome and treat this disease with very heterogeneous symptoms, depending on its primary and/or secondary location. Numerous *in vitro* and *in vivo* investigations have been carried out in order to find an effective anticancer treatment strategy and potential way to overcome this disease (Cvetković et al. 2017; Zeng 2018). The main approach to cancer treatment is chemotherapy, which is desirable to be fully effective and safe. However, currently used drugs have many side effects, thus affecting normal cells and developing drug resistance in cancer cells. In search for new drugs the most important goal is their high potential of drugs against cancer cells, together with the safe targeted therapy and avoidance of harmful effects after the treatment (Artemov et al. 2015; Zugazagoitia et al. 2016).

13.1.1 Plant-Derived Anticancer Compounds

Plants have been traditionally used for cancer prevention and they are a valuable source of potentially active substances in the anticancer therapy. The literature data indicate that consumption of fruits and vegetables, teas from medicinal plants, various tinctures and plant extracts, as well as numerous compounds isolated from plants have impact on chemo-prevention and/or chemo-therapy of cancer (Kotecha et al. 2016; Roy et al. 2017).

Primary positive effects of plants are related to their chemo-preventive role in the stage of cancer development. The preventive role of plants includes their antioxidative activity and reduction of free radicals, protection of biomolecules by preventing carcinogens to bind on targeting size through modulation of carcinogen-activation and carcinogen-detoxification enzymes, thus disabling their interaction with DNA, RNA or proteins. In this way, they protect biomolecules from harmful effects of cancer causative agents, like chemical carcinogen, radiation, infective agents, thus inhibiting the DNA damage (Desai et al. 2008; Kotecha et al. 2016).

An important role of plants and plant-derived compounds is their ability to inhibit initiation, promotion and progression of cancer, due to targeting key molecules involved in cancer cell signaling. The most important mechanisms of their anticancer success are stimulation and modulation of immune system, modulation of redox status, inhibition of growth and survival of malignant transformed cells, modulation of inner signal transmission paths, induction of apoptosis, suppression

of promotion and progression of cancer, and antiangiogenic, antiinvasive antimigratory potentials (Demain and Vaishnav 2011; Newman and Cragg 2016; Blowman et al. 2018). Besides their own anticancer properties, naturally derived compounds, due their additive or synergistic effects, can be used in combination with chemotherapeutics, thus leading to increased efficacy of therapy. They can increase cell response sensitivity to induction of apoptosis or inhibition of proliferation and decrease harmful effects of chemotherapy. Their potential is based on selective cytotoxic activity on cancer cells, as well as low toxicity on normal cells (Greenwell and Rahman 2015; Milutinović et al. 2015a).

Based on the mentioned properties of plants, the investigation of plant extracts as the mixture of secondary metabolites or their individual consideration is one of the most important aspects in discovering new compounds with possible anticancer activity. It has been the focus of many studies (Ćurčić et al. 2012a; Stanković et al. 2012; Milutinović et al. 2015b; Alimpić et al. 2015, 2017a). This actual field of research started with discovery of vinblastine and vincristine from *Catharanthus roseus*, paclitaxel from *Taxus brevifolia*, followed by campotechin and podophyllotoxins and many other discovered natural products or their derivatives (Prakash et al. 2013; Newman and Cragg 2016; Thomford et al. 2018). Currently, many of these drugs are involved in advanced clinical trial and clinically used as anticancer agents (Newman and Cragg 2004; Butler 2008; Seca and Pinto 2018). Among currently used chemotherapeutics in the treatment of cancer, the secondary metabolites from plant and their semi-synthetic derivates take a significant place. Many of them, mainly phenolic compounds, which include flavonoids (flavanols, isoflavones, flavanones, anthocyanins), stilbenes and phenolic acid, terpenoids and alkaloids have been identified as potent anticancer agents. Some of them have been used directly or after chemical modifications in different strategies of cancer therapy, with more pronounced activity and lower toxicity compared to synthetic drugs (Fabricant and Fransworth 2001; Rayan et al. 2017; Shin et al. 2018; Seca and Pinto 2018).

13.2 *Teucrium* Species and Their Anticancer Properties

The genus *Teucrium* consists of diverse species with variety of habitats and medicinal use. Due to their positive effects on human health, many were traditionally known as medicinal and aromatic plants. They were used for diverse applications, mainly in pharmaceutical industry and treatments of many pathological conditions (Bahramikia and Yazdanparast 2012). Because of their bitter taste, some were used for improvement of digestion as tea or tonic, then in treatment of gastric pain, ulceration and generally, digestive tract (Sarkhail 2011; Sghaier et al. 2012). Diverse therapeutic and biological activities of *Teucrium* species have been reported such as antibacterial, antioxidant, antifungal, antidiabetic, antispasmodic, antioxidant, antiinflammatory, antiulcer, hypoglycemic, anti-acetylcholinesterase, and hepatoprotective (Ulubelen et al. 2000; Ljubunčić et al. 2006; Jaradat 2015; Purnavab et al. 2015; Othman et al. 2017; Khazaei et al. 2018). Various groups of bioactive

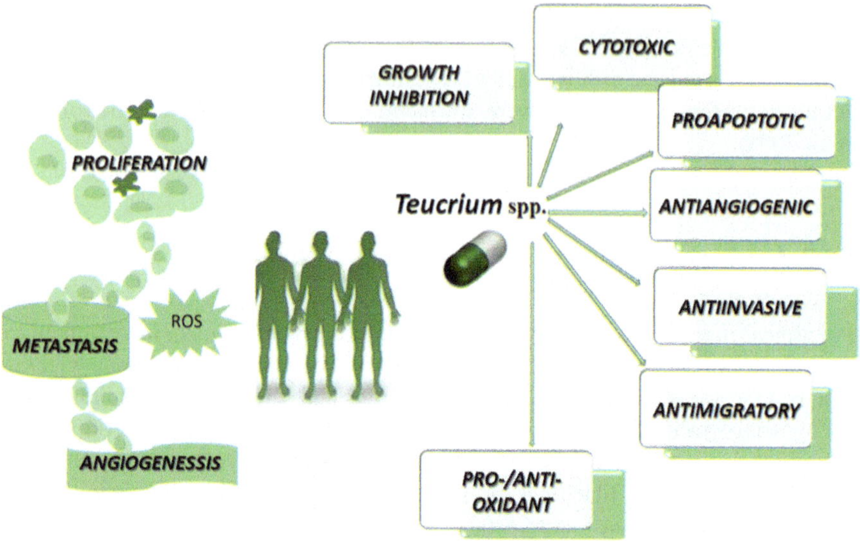

Fig. 13.1 The most important mechanisms of anticancer activity of extracts, essential oils and secondary metabolites from *Teucrium* species. *Teucrium* species have impact on multiple signaling pathways, genes and/or proteins involved in cancer development, cell cycle, apoptosis, redox status, migration, angiogenesis, etc

molecules were isolated from *Teucrium* species, but terpenoids and flavonoids were the most potent for achievement of its pharmacological and therapeutic effects (Bahramikia and Yazdanparast 2012). Besides, many other compounds were detected and isolated, such as phenolics, flavonoids, iridoids, sterols, tannins, saponins, terpenoids, alkaloids and glycosides of many others (Bedir et al. 1999; Bahramikia and Yazdanparast 2012; Elmasri et al. 2015, 2016; Fatima 2016; Milutinović et al. 2019). The observed anticancer properties of *Teucrium* species, and their various mechanisms of anticancer actions (Fig. 13.1), were the consequence of chemical contents and presence the high concentration and wide range of secondary metabolites. There were many reports that confirmed positive correlation between qualitative and/or quantitative content of phenolic compounds in plants and their biological activity (Stanković et al. 2012; Alimpić et al. 2017b; Duletić-Laušević et al. 2018).

In the focus of investigation are the extracts from *Teucrium* species obtained by different solvents as the mixture of total chemical constituents. Plant extracts consist of various secondary metabolites with different chemical structures, so they may produce diverse therapeutic effects in comparison to an individual compound. Their anticancer properties are not only the result of activities of individual components, but also of their interactions, since they act in a synergistic, additive or antagonistic manner (Freeman et al. 2010). Crude extracts were the most investigated, however, several phytochemicals obtained from different species from the genus *Teucrium* showed anticancer properties (Stanković et al. 2011; Shah et al. 2014; Ali

et al. 2017). The essential oils from the genus were also evaluated (Guesmi et al. 2018), because essential oils were reported as potent anticancer agents *in vitro* and *in vivo* models against different cell lines (Blowman et al. 2018). Some of biological activities of *Teucrium* species were attributed to components of essential oils (Ali et al. 2008).

13.2.1 Antiproliferative and Cytotoxic Activity

Preclinical investigations of anticancer potential of plant extracts include them *in vitro* testing on various human cancer cell lines. There are different human cell line panels, used for development of anticancer drugs and screening for potential activity by using proliferation assays. One of the panels, with 60 tumor cell lines, was recommended by NCI (Shoemaker 2006). This panel was required to include the most common tumors, like lungs, prostate, colorectal, stomach and liver in the population of men and breast, colorectal, lungs, cervix and stomach in women (Torre et al. 2015). In the discovery of new cancer drug, it is important to select candidates for phase II and III trials. Besides investigation in cytotoxicity in phase I, it is important to examine the mechanism of their action (Zhou et al. 2000) and evaluation *in vivo* model systems as a xenograft model, hollow fiber assay, genetically engineered mouse model, etc. (Kinghorn et al. 2009; Kumar et al. 2016).

Table 13.1 shows the effects of tested extracts and essential oils from various *Teucrium* species on different cell lines, according to our knowledge and available literature data. The available results about cytotoxic activity, expressed as the IC_{50} values, the concentration which inhibits 50% of cell growth.

Many *Teucrium* species have been investigated for possible antiproliferative activity and inhibition of cancer cell growth. According to results presented in the Table 13.1 it is evident that *Teucrium polium* is the most investigated; the literature data are available for some others as well, while many species in this genus have not been investigated yet. In the genus *Teucrium,* the most active secondary metabolites against tumor cells *in vitro* and *in vivo* are diterpenoids and flavonoids. The presence of diterpenoids was confirmed by many authors (Bruno et al. 1987, 2002; Avula et al. 2003). Neo-clerodane diterpenoids, Teucrins A, B, C, D, E and F were isolated from *Teucrium oliveranum* (de la Torre et al. 1991; Al-Yahia et al. 2002) and characterized as potential anticancer agents (Islam 2017). *Teucrium* diterpenoids were also reported against P 388 lymphocytic leukemia in mice (Nagao et al. 1982). Saponins from *Teucrium stocksianum* were recommended for future investigation of anticancer activity (Shah et al. 2014). Constituents from *Teucrium chamaedrys* were also specified, and the flavonoids, various glycosides, saponins, numerous neo-clerodane diterpenoids and the most dominant phenylethanoid glycosides (Forsythoside B, Samioside, Teucroside and Teucreoside) were found (Popa and Reinbold 1972; Savona et al. 1982; Piozzi et al. 1997; Milutinović et al. 2019). The content of essential oils from some *Teucrium* species was also evaluated; they contained high concentration of compounds with confirmed cytotoxic activity on some

Table 13.1 IC_{50} values (μg/ml) of *Teucrium* species in different cell lines, available in the literature data

Teucrium species	Type of extract	Experimental method	Time of exposure	Cell line	Effects IC_{50} (μg/ml)	References
T. persicum	M	MTT	48	Jurkat – T cell leukemia	> 200	Amirghofran et al. (2010)
				K562 – Chronic myelogenous leukemia	> 200	
				Raji – Burkitt's lymphoma	> 200	
				Fen – Bladder carcinoma	196	
				HeLa – Cervix epitheloid carcinoma	69	
T. persicum	M	MTT	48	PC-3 – Prostate cancer	142	Tafrihi et al. (2014)
				SW480 – Colon adenocarcinoma	79	
				T47D – Ductal epithelial breast tumor	50	
				NIH-3 T3 – Fibroblast	143	
T. sandrasicum	M	MTT	72	HeLa – Cervix epitheloid carcinoma	513	Artun et al. (2016)
				Vero – Normal kidney cell	593	
T. ramosissimum	M	MTT	48	K562 – Chronic myelogenous leukemia	150	Sghaier et al. (2012)
	CH				200	
	TOF				550	
	EA				600	
	AQ				800	

(continued)

Table 13.1 (continued)

Teucrium species	Type of extract	Experimental method	Time of exposure	Cell line	Effects IC_{50} (μg/ml)	References
T. mascatense	HE	ABA	24	MCAS – Ovarian mucinous cystadenocarcinoma	15.24	Said et al. (2017)
				MDA-MB-231 – Breast adenocarcinoma	61.08	
	CH			MCAS – Ovarian mucinous cystadenocarcinoma	> 100	
				MDA-MB-231 – Breast adenocarcinoma	33.80	
	EA			MCAS – Ovarian mucinous cystadenocarcinoma and MDA MB-231 – Breast adenocarcinoma	> 100	
	AQ					
T. mascatense	M	MTT	72	MCF-10A – Normal epithelial cell line	45.83	Panicker et al. (2019)
				MCF-7 – Breast adenocarcinoma	227.00	
				MDA-MB-231 – Breast adenocarcinoma	232.80	
				HeLa – Cervix epithelioid carcinoma	196.40	
T. brevifolium	EO	MTT	48	CACO-2 – Colorectal adenocarcinoma	164	Mencini et al. (2009)
				C32 – Amelanotic melanoma	> 200	
				COR-L23 – Large lung cell carcinoma	80.7	
				RAW246-7 – Murine monocytic macrophage	> 100	
T. oliverianum	E	MTT	24	MCF-7 – Breast adenocarcinoma	> 1000	Ali et al. (2014)
T. orientale	E	MTT	48	PC-3 – Prostate cancer cells	> 300	Asadi-Samani et al. (2018)

(continued)

Table 13.1 (continued)

Teucrium species	Type of extract	Experimental method	Time of exposure	Cell line	Effects IC_{50} (μg/ml)	References
T. yemense				Du-145 – Prostate cancer cell	> 300	
	EO	XTT	72	HT-29 – Colorectal adenocarcinoma	43.7	Ali et al. (2017)
		MTT	48	MCF-7 – Breast adenocarcinoma	24.4	
				MDA-MB-231 – Breast adenocarcinoma	59.9	
T. arduini	M	MTT	72	HeLa – Cervix epithelioid carcinoma	137.51	Stanković et al. (2015)
				Fem-X – Malignant melanoma	> 200	
				K562 – Chronic myelogenous leukemia	113.38	
				MDA-MB-361 – Breast adenocarcinoma	~ 200	
T. arduini	M	MTT	24	HCT-116 – Colorectal carcinoma	114.16	Stanković et al. (2011)
			72		0.37	
T. botrys	M	MTT	72	HeLa – Cervix epithelioid carcinoma	164.23	Stanković et al. (2015)
				Fem-X – Malignant melanoma	~ 200	
				K562 – Chronic myelogenous leukemia	98.78	
				MDA-MB-361 – Breast adenocarcinoma	~ 200	
T. botrys	M	MTT	24	HCT-116 – Colorectal carcinoma	116.38	Stanković et al. (2011)
			72		183.15	
T. flavum	EO	MTT	48	CACO-2 – Colorectal adenocarcinoma	> 200	Mencini et al. (2009)
				C32 – Amelanotic melanoma	> 200	
				COR-L23 – Large lung cell carcinoma	104.2	
				RAW246-7 – Murine monocytic macrophage	> 100	

(continued)

Table 13.1 (continued)

Teucrium species	Type of extract	Experimental method	Time of exposure	Cell line	Effects IC_{50} (μg/ml)	References
T. montberietii	EO	MTT	48	CACO-2 – Colorectal adenocarcinoma	92.2	Mencini et al. (2009)
				C32 – Amelanotic melanoma	135	
				COR-L23 – Large lung cell carcinoma	143	
				RAW246-7 – Murine monocytic macrophage	> 100	
T. scordioides	M	MTT	72	HeLa – Cervix epithelioid carcinoma	139.96	Stanković et al. (2015)
				MDA-MB-361 – Breast adenocarcinoma	196.5	
				Fem-X – Malignant melanoma	> 200	
				K562 – Chronic myelogenous leukemia	96.63	
T. scordium subsp. *scordioides*	M	MTT	24	HCT-116 – Colorectal carcinoma	143.46	Stanković et al. (2011)
			72		72.83	
T. scordium	M	MTT	24	HCT-116 – Colorectal carcinoma	17.04	Stanković et al. (2011)
			72		59.02	
T. scordium	M	MTT	72	HeLa – Cervix epithelioid carcinoma	144.82	Stanković et al. (2015)
				Fem-X – Malignant melanoma	~ 200	
				K562 – Chronic myelogenous leukemia	102.48	
				MDA-MB-361 – Breast adenocarcinoma	~ 200	
T. scordium	CX	KBR	72	MDA-MB-361 – Breast adenocarcinoma	130.33	Kundaković et al. (2011)
	DM				189.89	
	M				> 500	
	CX			MDA-MB-453 – Breast carcinoma	367.28	
	DM				131.01	
	M				371.55	

(continued)

Table 13.1 (continued)

Teucrium species	Type of extract	Experimental method	Time of exposure	Cell line	Effects IC_{50} (μg/ml)	References
T. chamaedrys	M	MTT	72	HeLa – Cervix epithelioid carcinoma	146.47	Stanković et al. (2015)
				Fem-X – Malignant melanoma	190.16	
				K562 – Chronic myelogenous leukemia	102.71	
				MDA-MB-361 – Breast adenocarcinoma	188.28	
T. chamaedrys	M	MTT	24	HCT-116 – Colorectal carcinoma	< 50	Stanković et al. (2011)
			72		190.07	
T. chamaedrys	M	MTT	24	SW480 – Colon adenocarcinoma	53.27	Milutinović et al. (2019)
	EA				126.89	
	A				163.44	
	M		72		20.87	
	EA				63.57	
	A				36.13	
T. montanum	M	MTT	72	HeLa – Cervix epithelioid carcinoma	152.34	Stanković et al. (2015)
				Fem-X – Malignant melanoma	196.44	
				K562 – Chronic myelogenous leukemia	99.15	
				MDA-MB-361 – Breast adenocarcinoma	> 200	
T. montanum	M	MTT	24	HCT-116 – Colorectal carcinoma	< 50	Stanković et al. (2011)
			72		75.73	
T. montanum	M	MTT	24	SW480 – Colon adenocarcinoma	135.79	Nikodijević et al. (2016)
			72		372.69	
			24	MDA-MB-231 – Breast adenocarcinoma	199.32	
			72		174.39	
T. polium	AQ	MTT	72	REYF-1 – Glioblastoma multiforme	> 1000	Eskandary et al. (2007)
	M				95	
T. polium	M	MTT	24	HT-29 – Colorectal adenocarcinoma	4	Khodaei et al. (2018)
			48		3	
			72		6	

(continued)

Table 13.1 (continued)

Teucrium species	Type of extract	Experimental method	Time of exposure	Cell line	Effects IC_{50} (μg/ml)	References
T. polium	M	MTT	72	MCF-7 – Breast adenocarcinoma	118.5	Abu-rish et al. (2016)
				T47D – Ductal epithelial breast tumor	98.5	
				CACO-2 – Colorectal adenocarcinoma	102.2	
				HRT18 – Rectum adenocarcinoma	89.78	
				A375.S2 – Malignant melanoma	61.1	
				WM1361A – Malignant melanoma	99.64	
T. polium	M	MTT	24	HCT-116 – Colorectal carcinoma	77.83	Stanković et al. (2011)
			72		253.39	

(continued)

Table 13.1 (continued)

Teucrium species	Type of extract	Experimental method	Time of exposure	Cell line	Effects IC_{50} (μg/ml)	References
T. polium	90% E	ABA	48	MCF-7 – Breast adenocarcinoma	184	Alzeer et al. (2014)
				B16F10 – Malignant melanoma	803	
				HeLa – Cervix epithelioid carcinoma	420	
	80% M			MCF-7 – Breast adenocarcinoma	104	
				B16F10 – Malignant melanoma	426	
				HeLa – Cervix epithelioid carcinoma	460	
	A			MCF-7 – Breast adenocarcinoma	140	
				B16F10 – Malignant melanoma	129	
				HeLa – Cervix epithelioid carcinoma	173	
	5% AA			MCF-7 – Breast adenocarcinoma	360	
				B16F10 – Malignant melanoma	> 1000	
				HeLa – Cervix epithelioid carcinoma	> 1000	
	AV			MCF-7 – Breast adenocarcinoma	400	
				B16F10 – Malignant melanoma	> 1000	
				HeLa – Cervix epithelioid carcinoma	> 1000	
	GV			MCF-7 – Breast adenocarcinoma	360	
				B16F10 – Malignant melanoma	> 1000	
				HeLa – Cervix epithelioid carcinoma	> 1000	
	AQ			MCF-7 – Breast adenocarcinoma	650	
				B16F10 – Malignant melanoma	> 1000	
				HeLa – Cervix epithelioid carcinoma	> 1000	

(continued)

Table 13.1 (continued)

Teucrium species	Type of extract	Experimental method	Time of exposure	Cell line	Effects IC_{50} (µg/ml)	References
T. polium (seeds)	M	MTT	72	MCF-7 – Breast adenocarcinoma	20.2	El-Naggar et al. (2018)
				HepG-2 – Hepatocellular carcinoma	143.1	
T. polium	M	MTT	24	Skmel -3 – Malignant melanoma	83	Rajabalian (2008)
				SW480 – Colorectal adenocarcinoma	139	
				MCF-7 – Breast adenocarcinoma	174	
				EJ – Bladder carcinoma	108	
				Saos-2 – Osteoblastoma	109	
				A431 – Epidermoid carcinoma	93	
				KB – Oral cavity epidermal cells	174	
T. polium	M	MTT	24	SW480 – Colorectal adenocarcinoma	242.93	Nikodijević et al. (2016)
			72		292.13	
			24	MDA-MB-231 – Breast adenocarcinoma	429.37	
			72		118.26	
T. polium	E	MTT	24	HUVEC – Umbilical vein endothelial cells	98.7	Sheikhbahaei et al. (2018)
			72		~ 100	
T. polium spp. *capitatum*	EO	MTT	48	CACO-2 – Colorectal adenocarcinoma	52.7	Menichini et al. (2009)
				C32 – Amelanotic melanoma	91.2	
				COR-L23 – Large lung cell carcinoma	104	
				RAW246-7 – Murine monocytic macrophage	> 100	
T. polium	E	MTT	48	A549 – Lung adenocarcinoma	90	Nematollahi-Mahani et al. (2007)
				BT20 – Breast ductal carcinoma	106	
				MCF-7 – Breast adenocarcinoma	140	
				PC12 – Mouse pheochromocytoma	120	

(continued)

Table 13.1 (continued)

Teucrium species	Type of extract	Experimental method	Time of exposure	Cell line	Effects IC_{50} (μg/ml)	References
T. polium	M	MTT	72	HeLa – Cervix epithelioid carcinoma	148.02	Stanković et al. (2015)
				Fem-X – Malignant melanoma	199.79	
				K562 – Chronic myelogenous leukemia	116.75	
				MDA-MB-361 – Breast adenocarcinoma	~ 200	

MTT cell viability assay, *TBS* Trypan Blue Staining, *KBR* Kenacid Blue R assay, *ABA* Alamar Blue assay, *M* methanolic, *E* ethanolic, *A* acetone, *AQ* aqueous, *HE* hexane, *CH* chloroform, *EA* ethyl-acetate, *TOF* total flavonoid enriched extract, *AA* acetic acid, *AV* apple vinegar, *GV* grape vinegar, *CX* cyclohexane extract, *DM* dichloromethane extract, *EO* essential oil

cancer cell lines, like (*E*)-caryophyllene, *α*-humulene and *α*-cadinol on MCF-7, HT-29 and other cell lines (Jun et al. 2011; Sylvestre et al. 2007). Additionally, many other compounds were detected in *Teucrium yamense* essential oil, such as *α*-pinene, *β*-pinene, linalool, 3-octanol, *α*-cadinol, caryophyllene oxide, 8-cedren-13-ol, and (*E*)-*β*-farnesene (Ali et al. 2017).

The criteria of cytotoxic activity for crude extracts was recommended by NCI (Suffnes and Pezuto 1990). According this criterion, the extract which inhibited 50% of cell growth, in the concentration less than or approximately 30 μg/ml was considered highly cytotoxic and presented promising anticancer potential. The results obtained by many authors (Table 13.1) indicated that some of the observed IC_{50} values for the most investigated *Teucrium polium* on MCF-7 cells (El-Naggar et al. 2018) and HT-29 cells (Khodaei et al. 2018) were very close to this criterion. Among other *Teucrium* species, a significant cytotoxic activity was observed for *Teucrium yamense* essential oil on breast MCF-7 and HT-29 colon cancer cells (Ali et al. 2017), *Teucrium arduini*, *T. chamaedrys*, *T. scordium* and *T. montanum* on HCT-116 cells (Stanković et al. 2011), *Teucrium chamaedrys* on SW480 cells (Milutinović et al. 2019) and *Teucrium mascateus* on MCF-10A cells (Panicker et al. 2019). Variety of human tumor cell lines obtained from different origins have been selected for investigation of anticancer activity of *Teucrium* species. The presented IC_{50} values indicate that *Teucrium* species induced higher or lower cytotoxicity, depending on the origin and nature of cancer cell lines. A significant cytotoxic activity has been noticed in cells of colorectal carcinomas, thus indicating that *Teucrium* species show organ-specific activity, which is in correlation with traditional use of this plant in the treatment of digestive disorders.

In addition to the results presented in Table 13.1, some other studies indicate significant inhibition of cancer cells growth induced by *Teucrium* species, where IC_{50} values were not calculated. So, aqueous extract of *Teucrium polium* inhibited

growth of hepatocellular carcinoma Hep2 cells, (Abdulrazzaq 2017), and lung cancer cells H322 and A549, 48 h after treatment (Haidara et al. 2011). Also, the methanol extract of *Teucrium polium* inhibited growth of human leukemia cell lines MOLT-4 and JMV-13 and chronic lymphocytic leukemia CLC, mainly in concentrations higher than 100 μg/ml, 24 and 48 h after treatment (Zarić et al. 2019). Dominant secondary metabolites, as the bioactive compounds in *Teucrium polium* were flavonoids like rutin and apigenin and neo-clerodane diterpenoids (Piozzi et al. 1997; Kadifkova et al. 2005). Chemically identified and isolated saponins were also present and showed cytotoxic activity on several cell lines (Elmasri et al. 2015). Additionally, many other compounds were identified in *Teucrium polium* extract with potential anticancer properties such as tannins, terpenoids, saponins, flavonoids, sterols, β-caryophyllene, monoterpenes, diterpenoids and many others (Hassan et al. 1979; Vokou and Bessiere 1985; Khazaei et al. 2018). *Teucrium polium* had the high level of phenolics, including flavonoids, anthocyanins and saponins which were involved in cytotoxic activity on MCF-7 and HepG-2 cells (El-Naggar et al. 2018). *Teucrium sandrasicum* was reported to inhibit cell proliferation of HeLa, MCF-7, and L929 cells (Tarhan et al. 2016). Hydrophobic and hydrophilic fraction of essential oils from *Teucrium alopecurus* inhibited cell proliferation of multiple myeloma U266 cells, pancreatic carcinoma Panc28, chronic myelogenous leukemia KBM-5, tonque squamosus carcinoma SCC4, breast cancer MCF-7 and colorectal cancer HCT-116 cells, 1, 3 and 5 days after treatment. Breast cancer MCF-7 cells and Panc28 cells were the most sensitive, while treatments induced moderate sensitivity over KBM-5 and HCT-116 cells (Guesmi et al. 2018). Chemical composition of essential oil from *Teucrium alopecurus* was evaluated, which confirmed that sesquiterpenes were detected in predominant percentages, like δ-cadinene, α-humulene, nerolidyl acetate and others (Hachicha et al. 2007). The cytotoxic activity correlated with the presence of these sesquiterpene compounds, which inhibited proliferation of several cell lines (El-Hadri et al. 2010; Sharma et al. 2013).

The potential of *Teucrium* species to inhibit cancer cell growth was examined with additional methods. Thus, authors report that methanol extract from *Teucrium polium* reduces colony forming efficacy in BT20, A549, MCF-7 and PC12 cell lines (Nematollahi-Mahani et al. 2007) and clonogenic potential of HCT-116 and KBM cells, 9 days after treatment (Guesmi et al. 2018). The same extract induced cell growth inhibition and decreased colony formation in Saos-2, Skmel-3 and EJ cell lines (Rajabalian 2008). The inhibition of REYF-1 cell growth, treated by *Teucrium polium* methanol extract was evaluated by soft agar clonogenic assay. The study reported its clonogenic efficiency of 70.77 ± 5.27% (Eskandary et al. 2007).

The cell cycle is regulated by control mechanisms which provide the cycle progression in a desirable direction. Cyclins, Cdks and CKIs have an important role in the cell cycle regulation (Fisher et al. 2012). The changes in cyclin levels lead to activation of cyclin-Cdk complex, which affects its own target proteins and induces specific cellular events (Lim and Kaldis 2013). Some of *Teucrium* species achieve antiproliferative effects by deregulation of cell cycle progression, like *Teucrium polium* in lung cancer cell lines H322 and A549. The increased number of treated

cells, stained by PI were in Sub-G1 phase of the cell cycle, detected by Flow cytometry analysis, 48 h after treatment (Haidara et al. 2011). Fractions of essential oil from *Teucrium polium* also exert an effect on the *c-myc* gene with role in cell cycle progression. Their expression was reduced in treated HCT-116, MiaPaca-2, Panc28, KBM-5, SCC4, MCF-7, U266, HL-60, RAW 264.7 cells compared to nontreated controls, thus leading to growth arrest (Guesmi et al. 2018). The essential oil from *Teucrium alopecurus* inhibited cell cycle progression in G2/M phase after 24 h. The identified compound from EO called thymol was reported to induce cell cycle arrest at the G0/g1 phase (Yin et al. 2012). In HCT-116 cells treated by hydrophobic and hydrophilic fractions of *Teucrium alopecurus* Cyclin D1 and *p21*, cell cycle regulator protein decreased, thus contributing to reduced cell proliferation (Guesmi et al. 2018). These results showed that fractions of *Teucrium alopecurus* inhibited STAT-3 phosphorylation and induced repression of STAT-3 activation as regulator of diverse cellular processes, whose inhibition led to reduced proliferation and induction of apoptosis (Sherry et al. 2009).

According to the presented IC_{50} values, time dependence was not always observed. In some cases, the extension of exposure time did not show the increased activity. On the contrary, after longer time cells recovered and adapted to treatments. This was noticed in *Teucrium polium* treatment of HT-29 cells, where the IC_{50} value for 72 h was higher compared to 24 and 48 h (Khodaei et al. 2018), and in HCT-116 cells (Stanković et al. 2011), SW480 cells (Nikodijević et al. 2016) and HUVEC (Sheikhbahaei et al. 2018), after 72 compared to 24 h. Similarly, the time dependence for cytotoxicity was not observed for *Teucrium bortys*, *T. scordium*, *T. chamaedrys* on HCT-116, and for *Teucrium montanum* in HCT-116 and SW480 cells (Stanković et al. 2011; Nikodijević et al. 2016).

Some data indicate cell selectivity of *Teucrium* species, apropos their cytotoxic activity on cancer cells, while they did not influence normal cells as the healthy control. *Teucrium mascatense* was not cytotoxic on normal fibroblasts (Said et al. 2017), *Teucrium alopecurus* on HUVECs cells (Guesmi et al. 2018), *Teucrium polium* on peripheral blood mononuclear cells (PBMCs) isolated from 16 healthy peoples (Zarić et al. 2019), *Teucrium persicum* on adherent fibroblasts (Tafrihi et al. 2014) and *Teucrium chamaedrys* on normal HaCaT keratinocytes (Milutinović et al. 2019). This selectivity and absence of effect on normal cells are significant results for further evaluation of these plants related to anticancer activity.

Differences in effects are observed for various extracts depending on the solvent used in extraction procedure. Dissolving of natural compounds with anticancer activities depends on their structure, therefore the selection of solvent is the first step in natural product processing. Methanol and ethanol extracts, able to extract polar and non-polar compounds were mainly investigated. Many authors reported that methanol was the most potent solvent for extraction of secondary metabolites with anticancer activity (Dai and Mumper 2010; Gupta et al. 2012; Mitrović et al. 2011; Milutinović et al. 2019). In addition, hot and cold water, acetone and ethyl-acetate extract were also commonly used solvents. Lipophilic compounds were most often extracted by petrol and chloroform. Healthy safe apple and grape vinegar of *Teucrium polium* were investigated as well. In the study of Alzeer et al. (2014),

acetone was the most potent organic solvent for extraction of secondary metabolites from *Teucrium polium* with antiproliferative activity on MCF-7, B16F10 and Hela cell lines, while the coconut water was not recommended for natural compound extraction.

13.2.2 Proapoptotic Activity

Apoptosis is natural highly regulated cell mechanism of programmed cell death, essential for development and maintaining of tissue homeostasis. Different conditions, extra- and intra-cellular signals may induce apoptosis. Besides physiological condition, apoptosis is involved in molecular basis of many apoptosis-related diseases, such as cancer (Kroemer et al. 2007; Fuchs and Steller 2011). The dysregulation and reduced rate of apoptosis contribute to pathogenesis of cancer (Elmore 2007; Fulda 2009). In addition, apoptosis is relevant targeting mechanism in anticancer treatment. Targeting of apoptosis is effective by modulation of its signaling pathways, especially since cancer cells develop multiple mechanisms to avoid apoptosis. The direct targeting of apoptosis could offer new possibilities for improvement of cancer therapy (Pfeffer and Singh 2018).

Among the investigated anticancer properties of natural compounds, the potential to modulate apoptosis is one of the most significant, as a desirable way of induced cytotoxicity in cancer cells. Many studies suggest that secondary metabolites delivered from plants, separately or in combination with chemotherapeutic drugs promote apoptosis in different cell lines (Milutinović et al. 2015a; Wang et al. 2018). Their ability to interfere and regulate key-proteins on gene and/or protein level provide novel opportunities for development of cancer drugs (Hassan et al. 2014; Wang et al. 2018).

Apoptosis is characterized by common, well-known morphological changes and biochemical events (Fig. 13.2). These characteristics, like DNK fragmentation and condensation, membrane blebbing, loss of adhesion to extracellular matrix, size reduction, and activation of the most important caspases have been used for apoptosis detection (Kalinichenko and Matveeva 2008; Filipović et al. 2014).

There are two major pathways for apoptosis induction, the receptor-mediated (extrinsic) and the mitochondrial-mediated (intrinsic) pathways. Both ways result in activation of protein family of enzymes, named caspases, which are initiator or effector molecules in apoptosis signaling (Green and Llambi 2015; Pfeffer and Singh 2018). The involvement of the death responsible receptors, such as FAS and TNF receptor superfamily, include formation of the protein complex named DISC, with appropriate adapter proteins resulting from activation of caspase-8 and/or -10. Once-activated caspase-8 down-regulates cascade of caspases, including effector caspase-3, or alternatively, cleaves Bid into tBid (Pobezinskaya and Liu 2012). Bid is the protein of the Bcl-2 protein family, which is translocated as tBid to mitochondria after cleavage by caspase-8, where it stimulates permeabilization of mitochondrial membrane (Kroemer et al. 2007). The intrinsic pathway includes mitochondria

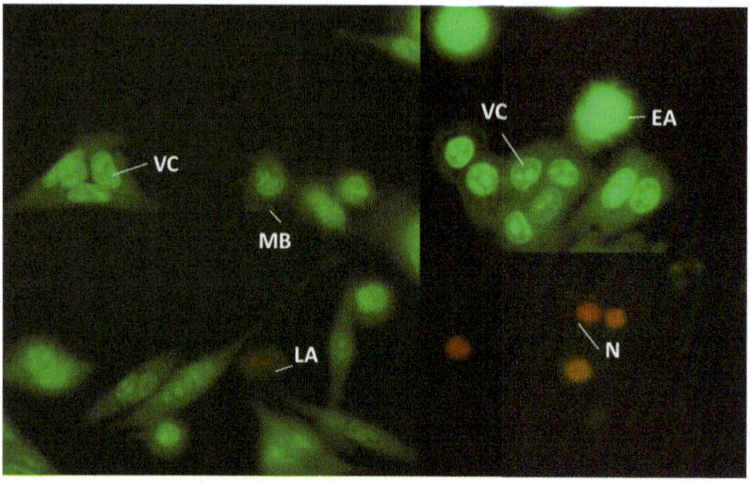

Fig. 13.2 Morphological features of viable, apoptotic and necrotic cells detected by AO/EB assay. *VC* viable cells, *EA* early apoptosis, *MB* membrane blebbing, *LA* late apoptosis, *N* necrosis

and mitochondrial proteins. It is regulated by Bcl-2 family of proteins, which controls and modulates the release of apoptotic factors. There are different structures and functionally distinct proteins within the family, divided into anti-apoptotic (Bcl-2, Bcl-xL, Bcl-W; MCL-1) and pro-apoptotic (Bax, Bak, Bim, Bid, Puma and Noxa) members (Yip and Reed 2008; Hata et al. 2015). They regulate permeability of outer mitochondrial membrane, releasing cytochrome *c* into the cytosol, formation of the complex named apoptosome which causes the activation of the initiator caspase-9, followed by activation of effector caspase-3 (Elmore 2007). The crucial event may be also releasing of Smac/DIABLO, with ability for bind to endogenous XIAPs (Lopez and Tait 2015). Contrary to apoptosis as regulated and controlled cell death, necrosis is passive, uncontrolled process which includes the release of inflammatory cellular substances, inflammation and effects on the surrounding cells (Fink and Cookson 2005).

13.2.3 Teucrium-*Targeting Molecules and Key Proteins Involved in Apoptosis Process*

Some mechanisms related to how apoptosis pathways are targeted by extracts of *Teucrium* species and how apoptosis can be utilized for cancer therapy are presented. There are many extracts from *Teucrium* species or their secondary metabolites that target various molecules in both intrinsic and extrinsic pathways (Nikodijević et al. 2016; Milutinović et al. 2019; Panicker et al. 2019; Zarić et al. 2019). A significant strategy for cancer treatment is increase and stimulation of proapoptotic and/or reduction and inhibition of antiapoptotic molecules (Wang

et al. 2018). There are many surveys on proapoptotic activity of extracts from *Teucrium* species detected by different methods, based on staining, fluorescence observation, morphology, DNK labeling, etc. (Rajabalian 2008; Stanković et al. 2011; Sheikhbahaei et al. 2018). The changes on the membrane of apoptotic cells include exposition of phosphatidylserine on the cell surface during apoptosis, aimed at recognition by phagocytes. This may be detectable by Annexin V/PI staining on Flow cytometer, as has been detected in *Teucrium polium* treated HT-29 cells compared to nontreated control (Khodaei et al. 2018). The induced cell death was due to increased mitochondrial membrane permeability. The extract of *Teucrium polium* also induces proapoptotic activity in human lung cancer H322 and A549 cell lines (Haidara et al. 2011). According to morphological observation by DAPI staining, it was observed that *Teucrium polium* methanol extract induced apoptosis in Skmel-3, Saos-2, MCF-7, A431, SW480 and EJ cells (Rajabalian 2008). Methanol extracts of *Teucrium chamaedrys, T. montanum, T. scordium. T. scordium* spp. *scordioides, T. botrys* and *T. arduini* from Serbian flora predominantly induced proapoptotic activity, according to their cytotoxic effects on HCT-116 cells and dose and time dependence (Stanković et al. 2011). Treatments with extracts of these *Teucrium* species caused morphological changes characteristic for early and late apoptosis, observed by AO/EB method. Methanol, ethyl-acetate and acetone extracts from *Teucrium chamaedrys* also induce apoptosis in SW480 cells 24 and 72 h after treatment, detected by AO/EB methods (Milutinović et al. 2019). In this study apoptosis was dominantly induced via activation of Fas receptors and caspase-8 in the receptor-mediated pathway, but increased caspase-9 activity indicate some role and mitochondria in induced apoptosis. Apoptotic morphological changes and increased number of early and late apoptosis were observed in the treatment on HUVECs cells by *Teucrium polium* extract, detected by AO/EB assay after 72 h, which was achieved by increased proapoptotic *Bax* and decreased *Bcl-2* gene expression (Sheikhbahaei et al. 2018). *Teucrium persicum* achieved the antiproliferative activity on SW480 and T47D cells by induction of apoptosis and detected by Flow cytometry and DNK fragmentation experiment (Tafrihi et al. 2014). One of the main events in apoptosis was activation of various caspases and downstream effector protein PARP (Ba and Garg 2011; Milutinović et al. 2015a). The extracts from *Teucrium montanum* and *T. polium* showed a significant proapoptotic activity in SW480 and MDA-MB-231 (Nikodijević et al. 2016). In this study, apoptosis was induced via activation of extinct pathway, by increased Fas receptor expression and caspase-8 activity and increased caspase-9 activity, thus suggesting an inclusion of mitochondrial pathway in MDA-MB-231 cells. The death of SW480 cells occurred in caspase-independent way. Panicker et al. (2019) reported that dichloromethane fraction of *Teucrium mascatense* induced caspase dependent apoptosis in MCF-7 cells by cleaved caspase-7 and PARP protein, as well as activation of caspase-8 and -9. This treatment induced mainly early apoptosis 24 h after treatment. Cytotoxic activity of *Teucrium polium* methanol extracts was induced by apoptosis in CLL lymphocytes isolated from 28 patients (Zarić et al. 2019). This extract induced Bax translocation and decreased the level of cellular Bcl-2 protein, where the ratio of Bax:Bcl-2 proteins was in favor of Bax that caused release of cytochrome *c* to

cytosol and activation of caspase-3. Tarhan et al. (2016) also reported that hydroethanolic and ethyl-acetate extract of *Teucrium sandrasicum* induced apoptosis in HeLa and MCF-7 cells, by affecting mitochondrial membrane permeability and activating caspase-9, while only ethyl-acetate extract activated caspase-3 in HeLa cells.

Some studies report the necrotic activity of *Teucrium* species. This may be related to the results obtained about hepatotoxicity of some constituents of *Teucrium* (Chitturi and Farrell 2008; Lin et al. 2009). Hepatotoxicity of *Teucrium chamaedrys* was investigated and discussed in many studies (Forouzandeh et al. 2013; Nencini et al. 2014). *Teucrium polium* also induced hepatotoxicity in *Swiss albino* mice after prolonged herb administration (Krache et al. 2015). Several neo-clerodanes diterpenes were isolated from leaf extracts of *Teucrium polium* and evaluated for their hepatotoxicity, for which low toxicity was confirmed to hepatoblastoma cancer HepG2 cells (Fiorentino et al. 2011). Thus *Teucrium ramosissimum* extract inhibited K562 cells proliferation, but DNA ladder profile suggested that cell death was not caused by apoptotic mechanism (Sghaier et al. 2012). Other authors reported that extract of *Teucrium polium* induced necrosis by increased ROS production. High amount of ROS induced necrosis of HT-29 cells, through instinct pathway and modifications of Sirt3 as an important member of Sirtuin family of mitochondrial regulatory proteins (Khodaei et al. 2018). The impact on mitochondria and induction of oxidative stress led to cell death in cells of colorectal cancer HT-29, mainly by necrosis, too.

The greatest number of studies report that apoptosis was mainly induced by polyphenols (Sharma et al. 2018). However, antitumor properties of essential oils and their bioactive constituents were well examined in many repots (Blowman et al. 2018). Essential oil from *Teucrium alopecurus* was also investigated for its pro-apoptotic potential (Guesmi et al. 2018). It induced apoptosis in a time- and concentration dependence in HCT-116, U266, SCC4, Panc28, KBM5 and MCF-7 cells. It was shown that essential oil from *Teucrium alopecurus* triggered the increase of cleaved PARP expression, activation of initiator caspases-8 and -9, and consequently caspase-3 in HCT-116 cells. Downstream expression levels of Bcl-2 genes were also observed.

13.2.4 Impact on Redox Status in Cancer Cells

Oxygen and nitrogen derived species such as superoxide radical, singlet oxygen, hydrogen peroxide, hydroxyl radical, nitric oxide, peroxynitrite, and others have been related to the etiology of many human diseases, including cancer (Waris and Ashan 2006). Overgeneration of ROS and RNS in cells induce oxidative and nitrosative stress, often with modifications of biomolecules – proteins, lipids, DNA and RNA as the consequence. Their damages can lead to activation and changes in the key proteins necessary for initiation and progression of carcinogenesis. Overexpressed levels of ROS are associated with various human cancers (England

and Cotter 2005; Gibellini et al. 2010). Redox regulation has an impact on basic cellular processes such as proliferation, regulation of various signaling pathways in the cell and many transcription factors, apoptosis, ageing, etc. (Mates et al. 2008; Gibellini et al. 2010). It is very important to understand and examine the ROS and RNS as important mediators in the cancer cells, which offer various opportunities for pharmacological intervention (Waris and Ashan 2006).

There are numerous attempts, experiments *in vitro* and *in vivo* to reduce ROS and RNS generation by exogenous antioxidants (Marković et al. 2011; Di Meo et al. 2016). Plants are the most studied and accessible source of exogenous antioxidants, rich with numerous secondary metabolites with antioxidant properties (Kasote et al. 2015). Phenol antioxidants directly reduce reactive ROS and RNS intermediaries, and in this way, they may stop the chain of oxidative propagation via their scavenging. The antioxidative activity of plant phenolics depend of their structure, number and position of substituents (hydroxyl and methoxy groups). Some natural antioxidants act as prooxidants in certain conditions, such as in presence of transition metals (Sakihama et al. 2002; Ullah et al. 2013; Ćurčić et al. 2014). Prooxidative and antioxidative potential also depends on oxidative potential inside the target cell. Cancer cells contains higher level of endogenous oxidative stress and changes oxidative status compared to normal cell. Thus, cancer cells can be more sensitive to phenolics in comparison to normal cells, which leads to additional increase in production of reactive oxygen intermediaries and induction of apoptosis. So, the phenolics-induced ROS generation contribute to their cytotoxicity and proapoptotic activity (Milutinović et al. 2015a). The role of natural polyphenol in cancer therapies is widely and intensively discussed in the current literature data (Mileo and Micadei 2015).

Teucrium species consist of many secondary metabolites with antioxidative activity (Alwahsh et al. 2015). Ardestani et al. (2008) showed that *Teucrium polium* processed antioxidative activity via presented quercetin in the extract that protected lipid and proteins from oxidation in pancreatic tissue in the rats. *Teucrium* species *in vitro* condition, measured by DPPH and many other assays show antioxidative activity (Vlase et al. 2014; Golfakhrabadi et al. 2015; Özer et al. 2018). However, in cancer cell lines they may behave and act differently. Beside the ability to scavenge free radicals and enhance antioxidative defenses by stimulation of synthesis and activity of antioxidative enzymes, some constituents from the *Teucrium* species can stimulate ROS production. In this way, stimulated oxidative stress in the cancer cells showed to be effective for their removal (Vallejo et al. 2017). Measured by Flow cytometry, the oily fraction of *Teucrium alopecurus* stimulated ROS generation in HCT-116 colon cancer cells, while its hydrophilic fraction induced nonsignificant amount of ROS (Guesmi et al. 2018). Khodaei et al. (2018) showed that *Teucrium polium* extracts increased ROS levels in HT-29 cells 6 and 12 h after the treatment. This hyperproduction of ROS was responsible for anticancer properties via mitochondria alternation, increased Sirt3 activity that correlated with ROS overproduction and induction of necrosis. The other authors suggested that plant extracts, including the *Teucrium* species induced oxidative stress by enhanced ROS production, followed by loss of cell function and induction of apoptosis, mainly by

mediated mitochondrial permeability and cytochrome *c* release associated with caspase activation (Stanković et al. 2011; Milutinović et al. 2015a; Zhang et al. 2016). It seems that the concentration of produced ROS is responsible for cell choice, i.e. whether the cells will die with apoptosis or necrosis. The methanol extract of *Teucrium chamaedrys* induced apoptosis partially by increasing of $O^{2.-}$ level and activation of caspase 9 in SW480 cells (Milutinović et al. 2019). In HCT-116 cells seven different species *Teucrium polium*, *T. chamaedrys*, *T. montanum*, *T. scordium*, *T. scordium* spp. *scordioides*, *T. arduini* and *T. botrys*, also induced prooxidant effects and increased $O^{2.-}$ level, which was probably responsible for their observed proapoptotic activity (Stanković et al. 2011). Higher production of $O^{2.-}$ was observed after 24 h in comparison to longer exposure time (72 h), when some of them acted as antioxidants, which correlated with cell recovery, adaptation to treatments and lower cytotoxity of extracts. Similarly, Živanović et al. (2016) reported that *Teucrium polium* and *T. montanum* intensified $O^{2.-}$ concentrations in treated SW480 colon and MDA-MB-231 breast cancer cells, thus indicating prooxidant effects of these species after 24 h; after 72 h they induced antioxidant activity in some concentrations. The reason for observed activities may lie in chemical structure and properties of *Teucrium* constituents, which reduce and/or oxidize substrates depending on the environmental conditions, the cell endogenous protective system, and adaptation to oxidative stress. The level of NO, as an important cell signaling molecule involved in many processes was also examined in some studies. Thus, literature data shows that extract of *Teucrium polium* significantly reduced NO production in HUVECs (Sheikhbahaei et al. 2018), as well as *Teucrium polium* and *T. montanum* in SW480 and MDA-MB-231 cells (Živanović et al. 2016) and several *Teucrium* species in HT-116 cells (Stanković et al. 2011). The reports of other authors confirmed the investigated effects of plant extracts on NO in cancer cell lines; they noted that many plant phenolics had scavenger activity for RNS such as NO and peroxynitrites (Ćurčić et al. 2014; Jeong and Jeong 2010; Cvetković et al. 2019). These findings indicate that plant constituents, associated with endogenous generation of NO may be related to synthesis by reduced or increased expression of iNOS (Ćurčić et al. 2014; Cvetković et al. 2019). The effects of *Teucrium* species on redox status in cancer cells require examination of antioxidant defense system, as the protective system that may be insufficient or attenuated in cancer cells. Živanović et al. (2016) reported that GSH, as the major component of these systems decreased in SW480 cells treated by *Teucrium polium* and *T. montanum*, as the consequence of induced oxidative stress. On the contrary, *Teucrium polium* increased the level of reduced GSH in HepG2 cells, as well as glutathione peroxidase and glutathione reductase activities (Shtukmaster et al. 2010). The results of these authors indicate that hepatoprotective activity of *Teucrium polium* was due to enhanced intracellular GSH level. Also, *Teucrium polium* ethyl-acetate extract exhibited preventive and antioxidative properties in K562 erythroleukemic cells, in which oxidative stress was pre-induced by 2-deoxy-d-ribose (Yazdanparast and Ardestani 2009). This antioxidative activity was mainly due to enhanced GSH content and its increased synthesis by γ-GCS.

13.3 Antiinvasive, Antiangiogenic and Antimigratory Potential

Metastatic potential is the ability of malignant cells to separate from primary cancer site, invade through ECM of surrounding tissues, blood and lymphoid vessels and migrate to other tissues and organs, thus forming secondary cancers (Friedl and Wolf 2010; Gupton and Gertler 2007). These features of cancer cells are one of the main problems in cancer therapy (Friedl and Wolf 2010). Invasive potential is the ability of cancer cells to have an effect on basement membrane and modulate the ECM. The initial transformation of malignant cells shows changes in shape and polarity of cells that acquire invasive phenotype. The changes occur through reorganization of cell cytoskeleton and impaired extracellular bounding and adherence. β-Catenin is a component of cell adherens junction, whereby its intracellular and intranuclear localization promotes cancer initiation and progression (Pećina-Šlaus 2003). *E*-Cadherin is one of the proteins required for the cell adherents and connectivity into the epithelial tissues (Jeanes et al. 2008). One of crucial events and initial steps of invasion and metastasis is the destruction of *E*-Cadherin/β-Catechin complex, which takes part in the important process called EMT (Fan et al. 2012; Tafrihi and Nakhaei Sistani 2017). The downregulation of *E*-Cadherin results in loss of epithelial integrity, leading to EMT. In cancer patients, the therapy takes the direction of reverse EMT. The cells with invasive potential have the ability to overproduce proteolytic enzymes – MMPs which degrade basement membrane and ECM and thus facilitate the penetration of malignant cells into the surrounding tissues (Gupton and Gertler 2007; Friedl and Wolf 2010; Munson et al. 2013; Weinberg 2014).

Angiogenesis, that creates new from the pre-existing blood vessels, has the important role in cancer progression. In poorly differentiated cancers the vascular network cannot support intensive cell proliferation. Thus "*angiogenic switch*" is initiated in the cancer due to rapid reduction of pO_2, lack of nutrients and other factors, after which numerous factors induce production of chemotaxis and movement of endothelial cells towards the tumor mass (Folkman 2007). In hypoxic condition *HIF-1α* is stabilized; it activates expression of genes with role in angiogenesis. The increase of VEGF expression leads to increased vascular permeability, while FGF allows growth of endothelial cells. HIF-1α, together with VEGF, thus inducing increased expression of MMPs, especially MMP-9 (van Hinsbergh and Koolwijk 2008; Cvetković et al. 2017). The factors that are released in microenvironment activate tumor-associated macrophages which produce angiogenic factors such as VEGF and MMPs and additionally contribute to angiogenesis (Guo et al. 2013). Under the influence of VEGF, stromal fibroblasts in tumor tissues produce CXCL-12, bonded to CXCR-4 and other cancer cell receptors, which initiate formation of the new blood vessels and mobilization of progenitor cells from bone marrow (Zheng et al. 2007; Chavakis et al. 2008).

Plant-derived products have been widely examined as antiinvasive, antiangiogenic and antimigratory agents, and great number of strategies was devised aimed

at the prevention and treatment of cancer (Šeklić et al. 2016; Lu et sar. 2016; Cvetković et al. 2019). The reason of their therapeutic potential lies in their ability to modulate these processes through their impact on different molecular signaling, limited toxicity and availability (Klongkumnuankarn et al. 2015). Since the cancer invasion and metastases lead to death of cancer patients, drugs with antiinvasive, antiangiogenic and antimigratory potential are significantly challenging for scientific research. In this regard some of *Teucrium* species were evaluated.

The results of wound healing assay, which measures the movement of cell population, indicate the antiinvasive activity of *Teucrium persicum*, where the extract induces decrease in wound healing PC-3 cell sheets. The gelatin zymography assay also indicates its antiinvasive potential due to the reduced level of MMP-2 and -9 for approximately 50–60% in PC-3 cells, measured 48 h after treatment (Tafrihi et al. 2014). The extract of *Teucrium persicum* enhanced *E*-Cadherin and *β*-Catenin expression and membrane localization, thus inducing the formation of different phenotype and reverse EMT in less than 20% of the treated PC-3 cell population. *Teucrium persicum* changed morphology of invasive and poorly differentiated PC-3 cells, which lost their mesenchymal phenotype under the treatment and transformed into mesenchymal epithelial phenotype, with closely attached epithelial-like cells (Tafrihi et al. 2014). Kandouz et al. (2010), report that the extract from *Teucrium polium* inhibited invasion and migration of prostate PC-3 and DU-145 carcinoma cell lines. It inhibited the invasion and motility of human prostate cancer cells via the restoration of *E*-Cadherin/Catenin complex. Other authors also reported that *Teucrium polium* induced differentiation of epithelial phenotype, EMC, and reduced cell invasion and motility of H322 and A549 lung cell lines (Haidara et al. 2011). The mechanism of these actions included re-localization of *E*-Cadherin and *β*-Catenin phosphorylation due to Src dephosphorylation. Tafrihi and Nakhaei Sistani (2017) reported that *Teucrium persicum* and *T. polium* downregulated MMPs, *E*-Cadherin and inhibited EMT and metastases. *Teucrium polium* showed ability to inhibit micro-vessel formation in concentrations of 50 and 25 μg/ml, on EA.hy926 human endothelial cell line, which suggested its antiinvasive potential (Abdallah et al. 2018). Essential oil from *Teucrium alopecurus* also has an impact on crucial component involved in invasion and metastasis due to reduction of CXCR-4 level, ICAM-1, VCAM-1 and MMP-9 expression in Panc28, Miapaca, and HL60 cancer cells (Guesmi et al. 2018). The VEGF and ICAM-1 are also inhibited in HCT-116 cells (Guesmi et al. 2018).

The examination of antimigratory potential of methanol extract from *Teucrium montanum* on SW480 and MDA-MB-231 cells showed some opposite results considering the used concentrations. The concentration of 1 μg/ml reduced cell migration, while the 10 and 50 μg/ml had the promigratory potential on both cell lines, monitored in the real time on RTCA xCELLigence system (Živanović et al. 2016). In the same study, *Teucrium polium* caused antimigratory effects on both SW480 and MDA-MB-231 cells, in treatment periods longer than 12 h. Sheikhbahaei et al. (2018) also reported that *Teucrium polium* had antimigratory activity on HUVECs cells, measured by wound healing (scratch method).

13.4 *Teucrium* Species in Combined Treatments

Many studies propose combined therapy that includes two or more therapeutic agents. There are attempts for application of chemo-therapeutics together with plant extracts or plant-derived compounds. Some secondary metabolites from plants can reduce side effects of chemotherapy and radiotherapy, when applied in cotreatment (Chung et al. 2004; Mokhtari et al. 2017; Zhang et al. 2018). A combined therapy may achieve many benefits compared to administration of drugs alone. Primarily, it can increase cytotoxicity by synergistic or additive interactions between combined drugs. A higher cell sensitivity in combined treatments is also observed, as well as enhanced efficacy or more potent cytotoxicity, due to targeting on different pathways, induction of apoptosis rather than necrosis, reduction of cancer cell drug resistance, etc. (Khdair et al. 2010; Ćurčić et al. 2012b; Milutinović et al. 2015a). One of the main benefits is prevention of healthy cells from toxic effects (Blagosklonny 2005). There are some statistical tests and calculations of interactions, such as *CalcuSyn*, combination index analysis, *CI index* (Rajabalian 2008; Milutinović et al. 2015a).

Among *Teucrium* species only the *Teucrium polium* was evaluated for anticancer testing in combined therapy. Sheikhbahaei et al. (2018) evaluated the effects of ethanol extract of *Teucrium polium* in combination with *tranilast* on HUVECs cells. The results showed synergistically enhanced cytotoxic activity in combined treatment, as well as increased apoptosis due to changed ratio between Bax and Bcl-2 gene expression, observed 72 h after treatment. This combined treatment also showed the ability to reduce NO concentration, as well as antimigratory potential of HUVECs cells (Sheikhbahaei et al. 2018). Other authors also investigated *Teucrium polium* in combination with other drugs (Rajbalian 2008). The combined treatments of *Teucrium polium* with *vinblastine*, *vincristine* and *doxorubicine* also induced increased proapoptotic activity compared to single treatments on MCF-7, A431, SW480, Skmel3, KB and EJ cell lines, without synergistic effects on normal fibroblasts. The calculated CI index indicated synergistic interaction of these drugs with methanolic extract of *Teucrium polium*.

13.5 *In Vivo* Anticancer Evaluation of *Teucrium* Species

Although *Teucrium* species, as medicinal plants are widely used for therapy of different disorders and extensively investigated in preclinical *in vitro* experiments, there are insufficient data about their anticancer properties based on *in vivo* studies. Movahedi et al. (2014) investigated protective role of *Teucrium polium* on liver cells against hepatocellular carcinoma on animal model system, in which cancerogenesis was induced. According to the investigated biochemical parameters, the results suggest that *Teucrium polium* may have had protective role for liver cells in 40 male rats, 8 ± 1 weeks old. Similar investigation was conducted by Shahat et al. (2016),

who reported data about *Teucrium oliverianum* anticancer potential in chemically-induced hepatocellular carcinoma in rats. These results indicate hepatoprotective, antiproliferative activity and antiangiogenic potential of *Teucrium oliverianum*. The antioxidative activity of *Teucrium polium* im adult male Wistar rats was also observed and comparable with α-tocopherol, as positive control for antioxidant activity (Hasani et al. 2007). Considering that increased concentrations of ROS and RNS and oxidative damage are in the basis of almost all kinds of cancers, the antioxidative role of *Teucrium polium* may have a protective effect (Liou and Storz 2010). Guesmi et al. (2018) reported anticancer properties of essential oil from *Teucrium alopecurus in vitro*, but also showed its nontoxic effects and no damage on histological review between treated and nontreated groups of mice, as the experimental model.

13.6 Conclusions

There are reports of many authors about anticancer properties of different species from the genus *Teurium*, that influence multistep cancerogenseis. Various preclinical findings and results of several *in vitro* and *in vivo* studies report preventive and pronounced anticancer properties of different *Teucrium* species, where the most investigated was *Teucrium polium*. In conclusion, after the reviewed literature data on anticancer properties, the genus *Teucrium* is a promising source for new anticancer plant-derived compounds, that exhibit anticancer activity through antiproliferative activity, activation of the apoptotic pathways, modulation of apoptotic key molecules and redox status in the cancer cells, with antiinvasive, antimetastatic and antiangiogenic activities. The most investigated effect was antiproliferative activity on variety of cell lines *in vitro*; there are fewer data about their possible mechanism of action, impact on migration, angiogenesis and invasion. Therefore, more data about these activities on greater number of model systems, especially on *in vivo* models are desirable. Also, given the diversity of the genus *Teucrium*, it is necessary to examine other species of this genus that have not been examined yet. Additional preclinical studies and clinical trials are certainly still required in order to elucidate the full spectrum of anticancer properties of the *Teucrium* species alone, or in synergistic combination with others for their commercial utilization and future perspective in drug design.

Acknowledgments This work was supported by the Ministry of Education, Science and Technological Development of the Republic of Serbia (Grant III41010).

References

Abdallah Q, Al-Deeb I, Bader A, Hamam F, Saleh K, Abdulamajid A (2018) Anti-angiogenic activity of Middle East medicinal plants of the Lamiaceae family. Mol Med Rep 18:2441–2448

Abdulrazzaq IH (2017) The Comparison effect of nickel (II) and cadmium (II) complexes with aqueous extract of *Tecrium polium* L (Ja,adah) plant on hepatocellular carcinoma cell line HeP2. Baghdad Sci J 14:371–378

Abu-rish YE, Kasabri V, Hudaib MM, Mashalla SH, AlAlawi LH, Tawaha K, Mohammad MK, Mohamed YS, Bustanji Y (2016) Evaluation of antiproliferative activity of some traditional anticancer herbal remedies from Jordan. Trop J Pharm Res 15:469–474

Ali NAA, Wurster M, Arnold N, Lindequist U, Wessjohann L (2008) Chemical composition of the essential oil of *Teucrium yemense* Deflers. Rec Nat Prod 2:25–32

Ali MA, Abul Farah M, Al-Hemaid FM, Abou-Tarboush FM (2014) In vitro cytotoxicity screening of wild plant extracts from Saudi Arabia on human breast adenocarcinoma cells. Genet Mol Res 13:3981–3990

Ali NAA, Chhetri BK, Dosoky NS, Shari K, Al-Fahad AJA, Wessjohann L, Setzer WN (2017) Antimicrobial, antioxidant, and cytotoxic activities of *Ocimum forskolei* and *Teucrium yemense* (Lamiaceae) essential oils. Medicine 4(2):pii: E17. https://doi.org/10.3390/medicines4020017

Alimpić A, Pljevljakušić D, Savikin K, Knezević A, Ćurčić M, Veličković D, Stević T, Petrović G, Matevski V, Vukojević J, Marković S, Marin P, Duletić-Laušević S (2015) Composition and biological effects of *Salvia ringens* (Lamiaceae) essential oil and extracts. Ind Crop Prod 76:702–709

Alimpić A, Knežević A, Milutinović M, Stević T, Šavikin K, Stajić M, Marković S, Marin P, Matevski V, Duletić-Laušević S (2017a) Biological activities and chemical composition of *Salvia amplexicaulis* Lam. extracts. Ind Crop Prod 105:1–9

Alimpić A, Knežević A, Šavikin K, Ćurčić M, Veličković D, Stević T, Matevski V, Stajić M, Marković S, Marin PD, Duletić-Laušević S (2017b) Composition and biological activities of different extracts of *Salvia jurisicii*, a rare and endemic Macedonian species. Plant Biosyst 151:1002–1011

Alwahsh MA, Khairuddean M, Chong WK (2015) Chemical constituents and antioxidant activity of *Teucrium barbeyanum*. Rec Nat Prod 9:159–163

Al-Yahya MA, El-Feraly FS, Dunbar DC, Muhammad I (2002) Neo-Clerodane diterpenoids from *Teucrium oliverianum* and structure revision of teucrolin E. Phytochemistry 59:409–414

Alzeer J, Vummidi BR, Arafeh R, Rimawi W, Saleem H, Luedtke NW (2014) The influence of extraction solvents on the anticancer activities of Palestinian medicinal plants. J Med Plant Res 8:408–415

Amirghofran Z, Zand F, Javidnia K, Miri K (2010) The cytotoxic activity of various herbals against different tumor cells: an in vitro study. Iran Red Crescent Med J 12:260–265

Ardestani A, Yazdanparast R, Jamshidi S (2008) Therapeutic effects of *Teucrium polium* extract on oxidative stress in pancreas of streptozotocin-induced diabetic rats. J Med Food 11:525–532

Artemov A, Aliper A, Korzinkin M, Lezhnina K, Jellen L, Zhukov N, Roumiantsev S, Gaifullin N, Zhavoronkov A, Borisov N, Buzdin A (2015) A method for predicting target drug efficiency in cancer based on the analysis of signaling pathway activation. Oncotarget 6(30):29347–29356

Artun FT, Karagoz A, Ozcan G, Melikoglu G, Anil S, Kultur S, Sutlupinar N (2016) In vitro anticancer and cytotoxic activities of some plant extracts on HeLa and Vero cell lines. JBUON 21:720–725

Asadi-Samani M, Rafieian-Kopaei M, Lorigooini Z, Shirzad H (2018) A screening of growth inhibitory activity of Iranian medicinal plants on prostate cancer cell lines. Biomedicine 8:16–21

Avula B, Manyam RB, Bedir E, Khan IA (2003) HPLC analysis of neo-clerodane diterpenoids from *Teucrium chamaedrys*. Pharmazie 58:494–496

Ba X, Garg NJ (2011) Signaling mechanism of Poly(ADP-Ribose) polymerase-1 (PARP-1) in inflammatory diseases. Am J Pathol 178:946–955

Bahramikia S, Yazdanparast R (2012) Phytochemistry and medicinal properties of *Teucrium polium* L. (Lamiaceae). Phytother Res 26:1581–1593

Bedir E, Tasdemir D, Alis IC, Zerbe O (1999) Neo-clerodane diterpenoids from *Teucrium polium*. Phytochemistry 51:921–925

Blagosklonny MV (2005) Overcoming limitations of natural anticancer drugs by combining with artificial agents. Trends Pharmacol Sci 26:77–81

Blowman K, Magalhães M, Lemos MFL, Cabral C, Pires IM (2018) Anticancer properties of essential oils and other natural products. Evid Based Complement Alternat Med 2018:3149362. https://doi.org/10.1155/2018/3149362

Bruno M, Piazz F, Savona G, Rodriguez B, De Latorre MC, Servettaz O (1987) 2-deoxy chamaedroxide, a neo-clerodane diterpenoid from *Teucrium divaricatum*. Phytochemistry 26:2859–2861

Bruno M, Bondi ML, Rosselli S, Maggio A, Piozzi F, Arnold NA (2002) Neoclerodane diterpenoids from *Teucrium montbretii* subsp. *libanoticum* and their absolute configuration. J Nat Prod 65:142–146

Butler MS (2008) Natural products to drugs: natural product-derived compounds in clinical trials. Nat Prod Rep 25:475–516

Chavakis E, Carmona G, Urbich C, Göttig S, Henschler R, Penninger JM, Zeiher AM, Chavakis T, Dimmeler S (2008) Phosphatidylinositol-3-kinase-gamma is integral to homing functions of progenitor cells. Circ Res 102:942–949

Chitturi S, Farrell GC (2008) Hepatotoxic slimming aids and other herbal hepatotoxins. J Gastroenterol Hepatol 23:366–373

Chung VQ, Tattersall M, Cheung HT (2004) Interactions of a herbal combination that inhibits growth of prostate cancer cells. Cancer Chemother Pharmacol 53:384–390

Cooper GM (2000) The cell: a molecular approach, 2nd edn. Sinauer Associates, Sunderland

Ćurčić M, Stanković M, Radojević I, Stefanović O, Lj Č, Topuzović M, Đačić D, Marković S (2012a) Biological effects, total phenolic content and flavonoid concentrations of fragrant yellow onion (*Allium flavum* L.). Med Chem 8:46–51

Ćurčić M, Stanković M, Mrkalić E, Matović Z, Banković D, Cvetković D, Đačić D, Marković S (2012b) Antiproliferative and proapoptotic activities of methanolic extracts from *Ligustrum vulgare* L. as an individual treatment and in combination with palladium complex. Int J Mol Sci 13:2521–2534

Ćurčić M, Stanković M, Cvetković D, Topuzović M, Marković S (2014) *Ligustrum vulgare* L.: in vitro free radical scavenging activity and pro-oxidant properties in human colon cancer cell lines. Dig J Nanomater Biostruct 9:1689–1697

Cvetković D, Živanović M, Milutinović M, Djukić T, Radović M, Cvetković A, Filipović ND, Zdravković ND (2017) Real-time monitoring of cytotoxic effects of electroporation on breast and colon cancer cell lines. Bioelectrochemistry 113:85–94

Cvetković DM, Jovankić JV, Milutinović MG, Nikodijević DD, Grbović FJ, Ćirić AR, Topuzović MD, Marković SD (2019) The anti-invasive activity of *Robinia pseudoacacia* L. and *Amorpha fruticosa* L. on breast cancer MDA-MB-231 cell line. Biologia 74:915–928

Dai J, Mumper JR (2010) Plant phenolics: extraction, analysis and their antioxidant and anticancer properties. Molecules 15:7313–7352

de la Torre MC, Bruno M, Piozzi F, Savona G, Rodriguez B, Omar AA (1991) Teucrolivins D–F, neo-clerodane derivatives from *Teucrium oliverianum*. Phytochemistry 30(5):1603–1606

Demain AL, Vaishnev P (2011) Natural products for cancer chemotherapy. Microb Biotechnol 4:687–699

Desai AG, Qazi GN, Ganju RK, El-Tamer M, Singh J, Saxena AK, Bedi YS, Taneja SC, Bhat SK (2008) Medicinal plants and cancer chemoprevention. Curr Drug Metab 9:581–591

Di Meo S, Reed TT, Venditti P, Victor M (2016) Role of ROS and RNS sources in physiological and pathological conditions. Oxidative Med Cell Longev 2016:1245049. https://doi.org/10.1155/2016/1245049

Duletić-Laušević S, Alimpić A, Šavkin K, Knežević A, Milutinović M, Stevć T, Vukojević J, Marković S, Marin P (2018) Composition and biological activities of Lybian *Salvia fruticosa* Mill. and *S. lanigera* Poir. extracts. S Afr J Bot 117:101–109

El-Hadri A, Del Río MG, Sanz J, Coloma AG, Idaomar M, Ozonas BR, González JB, Reus MS (2010) Cytotoxic activity of α-humulene and transcaryophyllene from *Salvia officinalis* in animal and human tumor cells. An R Acad Nac Farm 76:343–356

Elmasri WA, Hegazy MF, Mechref Y, Paré PW (2015) Cytotoxic saponin poliusaposides from *Teucrium polium*. RSC Adv 5:27126–27133

Elmasri WA, Hegazy MEF, Mechref Y, Paré PW (2016) Structure-antioxidant and anti-tumor activity of *Teucrium polium* phytochemicals. Phytochem Lett 15:81–87

Elmore S (2007) Apoptosis: a review of programmed cell death. Toxicol Pathol 35:495–516

El-Naggar SA, Germoush MO, Abdel-Farid IB, Elgebaly HA, Alkazendar AA (2018) Phytochemical analysis and anticancer screening of some indigenous plants grown in Saudi Arab. J Can Biomed Res 1:19–27

England K, Cotter TG (2005) Direct oxidative modification of signaling proteins in mammalian cells and their effects on apoptosis. Redox Rep 10:237–245

Eskandary H, Rajabalian S, Yazdi T, Eskandari E, Fatehi K, Ganjooei NA (2007) Evaluation of cytotoxic effect of *Teucrium polium* on a new glioblastoma multiforme cell line (REYF-1) using MTT and soft agar clonogenic assays. Int J Pharmacol 3:435–437

Fabricant DS, Fransworth NR (2001) The value of plants used in traditional medicine for drug discovery. Environ Health Perspect 109:69–75

Fan L, Wang H, Xia X, Rao Y, Ma X, Ma D, Wu P, Chen G (2012) Loss of *E*-Cadherin promotes prostate cancer metastasis via upregulation of metastasis-associated gene 1 expression. Oncol Lett 4:1225–1233

Fatima N (2016) A review on *Teucrium oliveranum*, a plant found abundantly in Saudi Arabia. Sci Int (Lahore) 28:1229–1231

Filipović N, Djukić T, Radović M, Cvetković D, Ćurčić M, Marković S, Peulić A, Jeremić B (2014) Electromagnetic field investigation on different cancer cell lines. Cancer Cell Int 14:84. https://doi.org/10.1186/s12935-014-0084-x

Fink SL, Cookson BT (2005) Apoptosis, pyroptosis, and necrosis: mechanistic description of dead and dying eukaryotic cells. Infect Immun 73:1907–1916

Fiorentino A, D'Abrosca B, Pacifico S, Scognamiglio M, D'Angelo G, Gallicchio M, Chambery A, Monaco P (2011) Structure elucidation and hepatotoxicity evaluation against HepG2 human cells of neo-clerodane diterpenes from *Teucrium polium* L. Phytochemistry 72:2037–2044

Fisher D, Krasinska L, Coudreuse D, Novak B (2012) Phosphorylation network dynamics in the control of cell cycle transitions. J Cell Sci 125:4703–4711

Folkman J (2007) Angiogenesis: an organizing principle for drug discovery? Nat Rev Drug Discov 6:273–286

Forouzandeha H, Azemib ME, Rashidic I (2013) Study of the protective effect of *Teucrium polium* L. extract on acetaminophen-induced hepatotoxicity in mice. Iran J Pharm Res 12:123–129

Freeman BL, Eggett DL, Parker TL (2010) Synergistic and antagonistic interactions of phenolic compounds found in navel oranges. J Food Sci 75:C570-6. https://doi.org/10.1111/j.1750-3841.2010.01717.x

Friedl P, Wolf K (2010) Plasticity of cell migration: a multiscale tuning model. J Cell Biol 188:11–19

Fuchs Y1, Steller H (2011) Programmed cell death in animal development and disease. Cell 147:742–758

Fulda S (2009) Tumor resistance to apoptosis. Int J Cancer 124:511–515

Gibellini L, Pinti M, Nasi M, De Biasi S, Roat E, Bertoncelli L, Cossarizza A (2010) Interfering with ROS metabolism in cancer cells: the potential role of quercetin. Cancers (Basel) 2:1288–1311

Golfakhrabadi F, Yousefbeyk F, Mirnezami T, Laghaei P, Hajimahmoodi M, Khanavi M (2015) Antioxidant and antiacetylcholinesterase activity of *Teucrium hyrcanicum*. Pharm Res 7:15–19

Green DR, Llambi F (2015) Cell death signaling. Cold Spring Harb Perspect Biol 7:a006080. https://doi.org/10.1101/cshperspect.a006080

Greenwell M, Rahman PKSM (2015) Medicinal plants: their use in anticancer treatment. Int J Pharm Sci Res 6:4103–4112

Guesmi F, Tyagi AK, Prasad S, Landoulsi A (2018) Terpenes from essential oils and hydrolate of *Teucrium alopecurus* triggered apoptotic events dependent on caspases activation and PARP cleavage in human colon cancer cells through decreased protein expressions. Oncotarget 9:32305–32320

Guo C, Buranych A, Sarkar D, Fisher PB, Wang X-Y (2013) The role of tumor-associated macrophages in tumor vascularization. Vasc Cell 5:20. https://doi.org/10.1186/2045-824X-5-20

Gupta A, Naraniwal M, Kothari V (2012) Modern extraction methods for preparation of bioactive plant extracts. Int J Appl Nat Sci 1(1):8–26

Gupton SL, Gertler FB (2007) Filopodia: the fingers that do the walking. Sci STKE 2007(400):re5. https://doi.org/10.1126/stke.4002007re5

Hachicha S, Skanji T, Barrek S, Zarrouk H, Ghrabi ZG (2007) Chemical composition of *Teucrium alopecurus* essential oil from Tunisia. J Essent Oil Res 19:413–415

Haidara K, Alachkar A, Al Moustafa A (2011) *Teucrium polium* plant extract provokes significant cell death in human lung cancer cells. Health 3:366–369

Hanahan D, Weimberg RA (2000) The hallmarks of cancer. Cell 100:57–70

Hasani P, Yasa N, Vosough-Ghanbari S, Mohammadirad A, Dehghan G, Abdollahi M (2007) In vivo antioxidant potential of *Teucrium polium*, as compared to α-tocopherol. Acta Pharma 57:123–129

Hassan M, Muhtadi F, Al-Badr A (1979) GLC-mass spectrometry of *Teucrium polium* oil. J Pharm Sci 68:800–801

Hassan M, Watari H, AbuAlmaaty A, Ohba Y, Sakuragi N (2014) Apoptosis and molecular targeting therapy in cancer. Biomed Res Int 2014:150845. https://doi.org/10.1155/2014/150845

Hata AN, Engelman JA, Faber AC (2015) The BCL-2 family: key mediators of the apoptotic response to targeted anti-cancer therapeutics. Cancer Discov 5(5):475–487

Islam MT (2017) Diterpenes and their derivatives as potential anticancer agents. Phytother Res 31(5):691–712

Jaradat NA (2015) Review of the taxonomy, ethnobotany, phytochemistry, phytotherapy and phytotoxicity of Gramander plant (*Teucrium polium*). Asian J Pharm Clin Res 8(2):13–19

Jeanes A, Gottardi CJ, Yap AS (2008) Cadherins and cancer: how does Cadherin dysfunction promote tumor progression? Oncogene 27(55):6920–6929

Jeong JB, Jeong HJ (2010) Rheosmin, a naturally occurring phenolic compound inhibits LPS-induced iNOS and COX-2 expression in RAW264.7 cells by blocking NF-kappaB activation pathway. Food Chem Toxicol 48:2148–2153

Jun NJ, Mosaddik A, Moon JY, Jang KC, Lee DS, Ahn KS, Cho SK (2011) Cytotoxic activity of β-caryophyllene oxide isolated from *Jeju guava* (*Psidium cattleianum* Sabine) leaf. Rec Nat Prod 5:242–246

Kadifkova PT, Kulevanova S, Stefova M (2005) In vitro antioxidant activity of some *Teucrium* species (Lamiaceae). Acta Pharma 55:207–214

Kalinichenko SG, Matveeva NY (2008) Morphological characteristics of apoptosis and its significance in neurogenesis. Neurosci Behav Physiol 38:333–344

Kandouz M, Alachkar A, Zhang L, Dekhil H, Chehna F, Yasmeen A, Al Moustafa AE (2010) *Teucrium polium* plant extract inhibits cell invasion and motility of human prostate cancer cells via the restoration of the *E*-Cadherin/Catenin complex. J Ethnopharmacol 129:410–415

Kasote DM, Katyare SS, Hegde MV, Bae H (2015) Significance of antioxidant potential of plants and its relevance to therapeutic applications. Int J Biol Sci 11:982–991

Khazaei M, Nematollahi-Mahani NS, Mokhtari T, Sheikhbahaei F (2018) Review on *Teucrium polium* biological activities and medical characteristics against different pathologic situations. J Contemp Med Sci 4:1–6

Khdair A, Chen D, Patil Y, Ma L, Dou QP, Shekhar MP, Panyam J (2010) Nanoparticle-mediated combination chemotherapy and photodynamic therapy overcomes tumor drug resistance. J Control Release 141:137–144

Khodaei F, Ahmadi K, Kiyani H, Hashemitabar M, Rezaei M (2018) Mitochondrial effects of *Teucrium polium* and *Prosopis farcta* extracts in colorectal cancer cells. Asian Pac J Cancer Prev 19:103–109

Kinghorn AD, Chin Y-W, Swanson SM (2009) Discovery of natural product anticancer agents from biodiverse organisms. Curr Opin Drug Discov Devel 12(2):189–196

Klongkumnuankarn P, Busaranon K, Chanvorachote P, Sritularak B, Jongbunprasert V, Likhitwitayawuid K (2015) Cytotoxic and antimigratory activities of phenolic compounds from *Dendrobium brymerianum*. Evid Based Complement Alternat Med 2015:350410. https://doi.org/10.1155/2015/350410

Kotecha R, Takami A, Espinoza JL (2016) Dietary phytochemicals and cancer chemoprevention: a review of the clinical evidence. Oncotarget 7(32):52517–52529

Krache I, Boussoualmi N, Charef N, Hayat Trabsa H, Ouhida S, Benbacha F, Daamouchei ZEY, Benzidane N, Baghiani A, Khennouf S, Arrar L (2015) Evaluation of acute and chronic toxic effects of Algerian germander in Swiss albino mice. J Appl Pharm Sci 5:27–32

Kroemer G, Galluzzi L, Brenner C (2007) Mitochondrial membrane permeabilization in cell death. Physiol Rev 87:99–163

Kumar S, Bajaj S, Bodla RB (2016) Preclinical screening methods in cancer. Indian J Pharm 48(5):481–486

Kundaković T, Milenković M, Topić A, Stanojković T, Juranić Z, Lakušić B (2011) Cytotoxicity and antimicrobial activity of *Teucrium scordium* L. (Lamiaceae) extracts. Afr J Microbiol Res 5(19):2950–2954

Levine AJ, Puzio-Kuter AM (2010) The control of the metabolic switch in cancers by oncogenes and tumor suppressor genes. Science 330:1340–1344

Lim S, Kaldis P (2013) Cdks, cyclins and CKIs: roles beyond cell cycle regulation. Development 140(15):3079–3093

Lin LZ, Harnly JM, Upton R (2009) Comparison of the phenolic component profiles of skullcap (*Scutellaria lateriflora*) and germander (*Teucrium canadense* and *T. chamaedrys*), a potentially hepatotoxic adulterant. Phytochem Anal 20:298–306

Liou GY, Storz P (2010) Reactive oxygen species in cancer. Free Radic Res 44(5):479–496

Ljubunčić P, Dakwar S, Portnaya I, Cogan U, Azaizeh H, Bomzon A (2006) Aqueous extracts of *Teucrium polium* possess remarkable antioxidant activity in vitro. Evid Based Complement Alternat Med 3:329–338

Lopez J, Tait SWG (2015) Mitochondrial apoptosis: killing cancer using the enemy within. Br J Cancer 112:957–962

Lu K, Bhat M, Basu S (2016) Plants and their active compounds: natural molecules to target angiogenesis. Angiogenesis 19(3):287–295

Ma X, Yu H (2006) Global burden of cancer. Yale J Biol Med 79:85–94

Marković SD, Žižić JB, Djačić DS, Obradović AD, Ćurčić MG, Cvetković DM, Đorđević NZ, Ognjanović BŠ, Štajn ŠA (2011) Alteration of oxidative stress parameters in red blood cells of rats after chronic in vivo treatment with cisplatin and selenium. Arch Biol Sci 63:991–999

Mates JM, Segura JA, Alonso FJ, Marquez J (2008) Intracellular redox status and oxidative stress: implications for cell proliferation, apoptosis, and carcinogenesis. Arch Toxicol 82:273–299

Menichini F, Conforti F, Rigano D, Formisano C, Piozzi F, Senatore F (2009) Phytochemical composition, anti-inflammatory and antitumor activities of four *Teucrium* essential oils from Greece. Food Chem 115:679–686

Mileo AM, Miccadei S (2015) Polyphenols as modulator of oxidative stress in cancer disease: new therapeutic strategies. Oxidative Med Cell Longev 2016:6475624. https://doi.org/10.1155/2016/6475624

Milutinović M, Stanković M, Cvetković D, Maksimović V, Šmit B, Pavlović R, Marković S (2015a) The molecular mechanisms of apoptosis induced by *Allium flavum* L. and synergistic effects with new-synthesized Pd(II) complex on colon cancer cells. J Food Biochem 9:238–250
Milutinović MG, Stanković MS, Cvetković DM, Topuzović MD, Mihailović VB, Marković SD (2015b) Antioxidant, antiproliferative and proapoptotic activities of leaves and seed cones from European Yew (*Taxus baccata* L.). Arch Biol Sci 67(2):525–534
Milutinović MG, Maksimović VM, Cvetković DC, Nikodijević DD, Stanković MS, Pešić MS, Marković SD (2019) Potential of *Teucrium chamaedrys* L. to modulate apoptosis and biotransformation in colorectal carcinoma cells. J Ethnopharmacol 240:1–10
Mitrović T, Stamenković S, Cvetković V, Tošić S, Stanković M, Radojević I, Stefanović O, Čomić LJ, Đačić D, Ćurčić M, Marković M (2011) Antioxidant, antimicrobial and antiproliferative activities of five lichen species. Int J Mol Sci 12(8):5428–5448
Mokhtari RB, Homayouni TS, Baluch N, Morgatskaya E, Kumar S, Das B, Yeger H (2017) Combination therapy in combating cancer. Oncotarget 8(23):38022–38043
Movahedi A, Basir R, Rahmat A, Charaffedine M, Othman F (2014) Remarkable anticancer activity of *Teucrium polium* on hepatocellular carcinogenic rats. Evid Based Complement Alternat Med 2014:726724. https://doi.org/10.1155/2014/726724
Munson J, Bonner M, Fried L, Hofmekler J, Arbiser J, Bellamkonda R (2013) Identifying new small molecule anti-invasive compounds for glioma treatment. Cell Cycle 12(14):2200–2209
Nagao Y, Ito N, Kohno T, Kuroda H, Fujita E (1982) Antitumor activity of *Rabdosia* and *Teucrium* diterpenoids against P 388 lymphocytic leukemia in mice. Chem Pharm Bull (Tokyo) 30:727–729
Nematollahi-Mahani SN, Rezazadeh-Kermani M, Mehrabani M, Nakhaee N (2007) Cytotoxic effects of *Teucrium polium* on some established cell lines. Pharm Biol 45:295–298
Nencini C, Galluzzi P, Pippi F, Menchiari A, Micheli L (2014) Hepatotoxicity of *Teucrium chamaedrys* L. decoction: role of difference in the harvesting area and preparation method. Indian J Pharm 46:181–184
Newman DJ, Cragg GM (2004) Marine natural products and related compounds in clinical and advanced preclinical trials. J Nat Prod 67:1216–1238
Newman DJ, Cragg GM (2016) Natural products as sources of new drugs from 1981 to 2014. J Nat Prod 79:629–661
Ngo DC, Ververis K, Tortorella SM, Karagiannis TC (2015) Introduction to the molecular basis of cancer metabolism and the Warburg effect. Mol Biol Rep 42:819–823
Nikodijević D, Milutinović M, Cvetković D, Stanković M, Živanović M, Marković S (2016) Effects of *Teucrium polium* L. and *Teucrium montanum* L. extracts on mechanisms of apoptosis in breast and colon cancer cells. Krag J Sci 38:147–159
Othman MB, Salah-Fatnassi KBH, Ncibi S, Elaissi A, Zourgui L (2017) Antimicrobial activity of essential oil and aqueous and ethanol extracts of *Teucrium polium* L. subsp. *gabesianum* (L.H.) from Tunisia. Physiol Mol Biol Plants 23:723–729
Özer Z, Kilic T, Carikci S, Yilmaz H (2018) Investigation of phenolic compounds and antioxidant activity of *Teucrium polium* L. decoction and infusion. Baun Fen Bil Enst Dergisi 20:212–218
Panicker NG, Balhamar SOMS, Akhlaq S, Qureshi MM, Rizvi TS, Al-Harrasi A, Hussain J, Mustafa F (2019) Identification and characterization of the caspase-mediated apoptotic activity of *Teucrium mascatense* and an isolated compound in human cancer cells. Molecules 24(5):pii: E977. https://doi.org/10.3390/molecules24050977
Pećina-Šlaus N (2003) Tumor suppressor gene E-cadherin and its role in normal and malignant cells. Cancer Cell Int 3(1):17. https://doi.org/10.1186/1475-2867-3-17
Pfeffer CM, Singh ATK (2018) Apoptosis: a target for anticancer therapy. Int J Mol Sci 19:448
Piozzi F, Bruno M, Ciriminna R, Fazio C, Vassallo N, Arnold NA, de la Torre MC, Rodriguez B (1997) Putative hepatotoxic neoclerodane diterpenoids from *Teucrium* species. Planta Med 63:483–484
Pobezinskaya YL, Liu Z (2012) The role of TRADD in death receptor signaling. Cell Cycle 11(5):871–876

Popa DP, Reimbold AM (1972) Bitter substances from *Teucrium chamaedrys*. Khim Prir Soedin 8:67–69
Prakash O, Kumar A, Kumar P, Ajeet (2013) Anticancer potential of plants and natural products: a review. Am J Pharmacol Sci 1(6):104–115
Purnavab S, Ketabchi S, Rowshan V (2015) Chemical composition and antibacterial activity of methanolic extract and essential oil of Iranian *Teucrium polium* against some of phytobacteria. Nat Prod Res 29(14):1376–1379
Rajabalian S (2008) Methanolic extract of *Teucrium polium* L potentiates the cytotoxic and apoptotic effects of anticancer drugs of vincristine, vinblastine and doxorubicin against a panel of cancerous cell lines. Exp Oncol 30:133–138
Rayan A, Raiyn J, Falah M (2017) Nature is the best source of anticancer drugs: indexing natural products for their anticancer bioactivity. PLoS One 12(11):e0187925. https://doi.org/10.1371/journal.pone.0187925
Roy A, Attre T, Bharadvaja N, Tiezzi A, Karpiński TM (2017) Anticancer agent from medicinal plants: a review. In: Tiezzi A, Karpinski TM (eds) New aspects in medicinal plants and pharmacognosy. JB Books, Poznań
Said SA, Tamimi Y, Akhtar S, Weli AM, Al-Khanjari SS, Al-Riyami QA (2017) British in vitro anticancer activity of selected medicinal plants from Oman. J Pharm Res 15(5):1–8
Sakihama Y, Cohen MF, Grace SC, Yamasaki H (2002) Plant phenolic antioxidant and prooxidant activities: phenolics-induced oxidative damage mediated by metals in plants. Toxicology 177:67–80
Sarkhail P (2011) Advances on the pharmacological use of *Teucrium* spp. (Germanders). Int J Pharmacol 7:548–549
Savona G, Garcia-Alvarez MC, Rodriguez B (1982) Dihydroteugin, a neo-clerodane diterpenoid from *Teucrium chamaedrys*. Phytochemistry 21:721–723
Seca AML, Pinto DCGA (2018) Plant secondary metabolites as anticancer agents: successes in clinical trials and therapeutic application. Int J Mol Sci 19(1):263. https://doi.org/10.3390/ijms19010263
Šeklić D, Stanković M, Milutinović M, Topuzović M, Štajn A, Marković S (2016) Cytotoxic, antimigratory, pro-and antioxidative activities of extracts from medicinal mushrooms on colon cancer cell lines. Arch Biol Sci 68:93–105
Sghaier MB, Skandrani I, Khochtali MS, Bhouri W, Ghedira K, Chekir-Ghedira L (2012) In vitro evaluation of antioxidant, cytotoxic and apoptotic activities of different extracts from the leaves of *Teucrium ramosissimum* (Lamiaceae). J Med Plant Res 6:3818–3825
Shah SMM, Sadiq A, Shah SMH, Khan S (2014) Extraction of saponins and toxicological profile of *Teucrium stocksianum* boiss extracts collected from District Swat, Pakistan. Biol Res 47:65. https://doi.org/10.1186/0717-6287-47-65
Shahat AA, Mahmoud EA, Al-Mishari AA, Alsaid MS (2016) Antimicrobial activities of some Saudi Arabian herbal plants. Afr J Tradit Complement Altern Med 13:62–70
Sharma A, Bajpai VK, Shukla S (2013) Sesquiterpenes and cytotoxicity. Nat Prod 2013:3515–3550
Sharma A, Kaur M, Katnoria JK, Nagpal AK (2018) Polyphenols in food: cancer prevention and apoptosis induction. Curr Med Chem 25:4740–4757
Sheikhbahaei F, Khazaei M, Nematollahi-Mahani SN (2018) *Teucrium polium* extract enhances the anti-angiogenesis effect of Tranilast on human umbilical vein endothelial cells. Adv Pharm Bull 8:131–139
Sherry MM, Reeves A, Wu JK, Cochran BH (2009) STAT3 is required for proliferation and maintenance of multipotency in glioblastoma stem cells. Stem Cells 27:2383–2392
Shin SA, Moon SY, Kim WY, Paek SM, Park HH, Lee CS (2018) Structure-based classification and anti-cancer effects of plant metabolites. Int J Mol Sci 19:2651. https://doi.org/10.3390/ijms19092651
Shoemaker RH (2006) The NCI 60 human tumor cell line anticancer drug screen. Nat Rev Cancer 6:813–823

Shtukmaster S, Ljubuncic P, Bomzon A (2010) The effect of an aqueous extract of *Teucrium polium* on glutathione homeostasis in vitro: a possible mechanism of its hepatoprotectant action. Adv Pharmacol Sci 2010:938324. https://doi.org/10.1155/2010/938324
Siegel RL, Miller KD, Jemal A (2019) Cancer statistics. CA Cancer J Clin 69:7–34
Stanković M, Ćurčić M, Žižić J, Topuzović M, Solujić S, Marković S (2011) *Teucrium* plant species as natural sources of novel anticancer compounds: antiproliferative, proapoptotic and antioxidant properties. Int J Mol Sci 12:4190–4205
Stanković M, Radojević I, Ćurčić M, Vasić S, Topuzović M, Čomić LJ, Marković S (2012) Evaluation of biological activities of goldmoss stonecrop (*Sedum acre* L.). Turk J Biol 36:580–588
Stanković M, Mitrović T, Matić IZ, Topuzović M, Stamenković S (2015) New values of *Teucrium* species: in vitro study of cytotoxic activities of secondary metabolites. Not Bot Hortic Agrobo 43(1):41–46
Suffness M, Pezzuto JM (1990) Assays related to cancer drug discovery. In: Hostettmann K (ed) Methods in plant biochemistry: assays for bioactivity. Academic, Waltham, pp 71–133
Sylvestre M, Pichette A, Lavoie S, Longtin A, Legault J (2007) Composition and cytotoxic activity of the leaf essential oil of *Comptonia peregrina* (L.) Coulter. Phytother Res 21:536–540
Tafrihi M, Nakhaei Sistani R (2017) *E*-cadherin/β-catenin complex: a target for anticancer and antimetastasis plants/plant-derived compounds. Nutr Cancer 69:702–722
Tafrihi M, Toosi S, Minaei T, Gohari AR, Niknam V, Arab Najafi SM (2014) Anticancer properties of *Teucrium persicum* in PC-3 prostate cancer cells. Asian Pac J Cancer Prev 15:785–791
Tarhan L, Nakipoğlu M, Kavakcıoğlu B, Tongul B, Nalbantsoy A (2016) The induction of growth inhibition and apoptosis in HeLa and MCF-7 cells by *Teucrium sandrasicum*, having effective antioxidant properties. Appl Biochem Biotechnol 178:1028–1041
Thomford NE, Senthebane DA, Rowe A, Munro D, Seele P, Maroyi A, Dzobo K (2018) Natural products for drug discovery in the 21st century: innovations for novel drug discovery. Int J Mol Sci 19(6):1578. https://doi.org/10.3390/ijms19061578
Torre LA, Bray F, Siegel RL, Ferlay J, Lortet-Tieulent J, Jemal A (2015) Global cancer statistics, 2012. CA Cancer J Clin 65:87–108
Ullah FM, Ahmed A, Khan HY, Zubair H, Sarkar F, Hadi SM (2013) The prooxidant action of dietary antioxidants leading to cellular DNA breakage and anticancer effects: implications for chemotherapeutic action against cancer. Cell Biochem Biophys 67:431–438
Ulubelen A, Topu G, Sönmez U (2000) Chemical and biological evaluation of genus *Teucrium*. Stud Nat Prod Chem 23:591–648
Vallejo MJ, Salazar L, Grijalva M (2017) Oxidative stress modulation and ROS-mediated toxicity in cancer: a review on in vitro models for plant-derived compounds. Oxidative Med Cell Longev 2017:4586068. https://doi.org/10.1155/2017/4586068
van Hinsbergh VW, Koolwijk P (2008) Endothelial sprouting and angiogenesis: matrix metalloproteinases in the lead. Cardiovasc Res 78:203–212
Vlase L, Benedec D, Hanganu D, Damian G, Csillag I, Sevastre B, Mot AC, Silaghi-Dumitrescu R, Tilea I (2014) Evaluation of antioxidant and antimicrobial activities and phenolic profile for *Hyssopus officinalis*, *Ocimum basilicum* and *Teucrium chamaedrys*. Molecules 19:5490–5507
Vokou D, Bessiere JM (1985) Volatile constituents of *Teucrium polium*. J Nat Prod 48:498–499
Wang Y, Zhong J, Bai J, Tonga R, Anc F, Jiaod P, Hea L, Zenga D, Longa E, Yana J, Yua J, Caia L (2018) The application of natural products in cancer therapy by targeting apoptosis pathways. Curr Drug Metab 19:739–749
Waris G, Ahsan H (2006) Reactive oxygen species: role in the development of cancer and various chronic conditions. J Carcinog 5:14. https://doi.org/10.1186/1477-3163-5-14
Weinberg RA (2014) The biology of cancer, 2nd edn. Garland Science, New York/London
Yazdanparast R, Ardestani AA (2009) Suppressive effect of ethyl acetate extract of *Teucrium polium* on cellular oxidative damages and apoptosis induced by 2-deoxy-D-ribose: role of de novo synthesis of glutathione. Food Chem 114:1222–1230

Yin QH, Yan FX, Zu XY, Wu YH, Wu XP, Liao MC, Deng SW, Yin LL, Zhuang YZ (2012) Anti-proliferative and pro-apoptotic effect of carvacrol on human hepatocellular carcinoma cell line HepG-2. Cytotechnology 64:43–51

Yip KW, Reed JC (2008) BCL-2 family proteins and cancer. Oncogene 77:6398–6406

Zarić M, Zarić RZ, Mitrović M, Nikolić I, Čanović P, Milosavljević Z, Jovanović D, Sekulić M, Zelen I (2019) *Teucrium polium* induces apoptosis in peripheral blood lymphocytes leukemia. Vojnosanit Pregl 2019:5. https://doi.org/10.2298/VSP181010005Z

Zeng Y (2018) Advances in mechanism and treatment strategy of cancer. Cell Mol Biol 64(6):1–3

Zhang L, Li J, Zong L, Chen X, Chen K, Jiang Z, Nan L, Li X, Li W, Shan T, Ma Q, Ma Z (2016) Reactive oxygen species and targeted therapy for pancreatic cancer. Oxidative Med Cell Longev 2016:1616781. https://doi.org/10.1155/2016/1616781

Zhang Q-Y, Wang F-X, Jia K-K, Kong L-D (2018) Natural product interventions for chemotherapy and radiotherapy-induced side effects. Front Pharmacol 9:1253. https://doi.org/10.3389/fphar.2018.01253

Zheng H, Fu G, Dai T, Huang H (2007) Migration of endothelial progenitor cells mediated by stromal cell-derived factor-1alpha/CXCR-4 via PI3K/Akt/eNOS signal transduction pathway. J Cardiovasc Pharmacol 50:274–280

Zhou B-N, Hoch JM, Johnson RK, Mattern MR, Eng W-K, Ma J, Hecht SM, Newman DJ, Kingston DGI (2000) Use of COMPARE analysis to discover new natural product drugs: isolation of camptothecin and 9-methoxycamptothecin from a new source. J Nat Prod 63:1273–1276

Živanović MN, Stojanović AZ, Cvetković DM, Milutinović MG, Stanković MS, Marković SD (2016) Effects of *Teucrium* spp. extracts on migratory potential and redox status on human colon SW-480 and breast MDA-MB-231 cells. Krag J Sci 38:161–172

Zugazagoitia J, Guedes C, Ponce S, Ferrer I, Molina-Pinelo S, Paz-Ares L (2016) Current challenges in cancer treatment. Clin Ther 38:1551–1566

Chapter 14
Anticholinesterase, Antidiabetic and Anti-inflammatory Activity of Secondary Metabolites of *Teucrium* Species

Nenad Zlatić and Milan Stanković

Abstract The chapter reviews the data about anticholinesterase, antidiabetic, and anti-inflammatory activity of the active substances from *Teucrium* species. Extracts and essential oils from the species of *Teucrium* genus possess a wide range of secondary metabolites that exhibit biological activity. Discussed examinations are focused on the determination of the qualitative and quantitative composition of the active substances from *Teucrium* species, as well as the anticholinesterasc, antidiabetic and anti-inflammatory activity. Many authors state that Alzheimer's disease, diabetes mellitus, and inflammatory diseases correlate because they are precursors of neurodegenerative diseases. Secondary metabolites isolated from plants of the genus *Teucrium* have been shown to be potential inhibitors of acetylcholinesterase. Also, examined metabolites possess antihyperglycemic activity and reduce blood glucose levels. Extracts of the species from the genus *Teucrium* exhibit anti-inflammatory activity in in vitro and in vivo conditions by inhibiting carrageenan-induced inflammation and significantly reduced serum levels of triglyceride and cholesterol. As a result, it has been shown that the species of the genus *Teucrium* are rich in compounds such as flavonoids, phenolic acids and terpenoids that directly contribute to anticholinesterase, antidiabetic and anti-inflammatory activity. Species of the genus *Teucrium* are good candidates for further examination in order to the treatment of the previously mentioned disorders.

Keywords *Teucrium* species · Acetylcholinesterase · Alzheimer's disease · Diabetes mellitus · Inflammation

---

N. Zlatić (✉) · M. Stanković
Department of Biology and Ecology, Faculty of Science, University of Kragujevac, Kragujevac, Serbia
e-mail: nenad.zlatic@pmf.kg.ac.rs

M. Stanković (ed.), *Teucrium Species: Biology and Applications*,
https://doi.org/10.1007/978-3-030-52159-2_14

Abbreviations

ACh	Acetylcholine
AChE	Acetylcholinesterase
AChI	Acetylcholine iodide
AD	Alzheimer's disease
COX	Cyclooxygenase
DTNB	5,5-dithio-bis-2-nitrobenzoic acid
FDA	Food and drug administration
HLA	Human leukocyte antigen
IC_{50}	Half maximal inhibitory concentration
iNOS	Inducible nitric oxide synthase
LDL	Low density lipoproteins
mRNA	Messenger ribonucleic acid
NO	Nitric oxide
STZ	Single dose streptozotocin
TNB	2-nitro-5-mercaptobenzoic acid

14.1 Introduction

Genus *Teucrium* L. belongs to the family Lamiaceae. Some species belonging to the genus *Teucrium* are known medicinal plants that have wide application. Medicinal species of the genus *Teucrium* are used in the treatment of rheumatism, diabetes, inflammatory and gastrointestinal diseases. The most common groups of secondary metabolites from the *Teucrium* species are flavonoids, phenolic acids, monoterpenes, diterpenes, sesquiterpenes, and others. Due to the specific quantitative and qualitative composition of secondary metabolites, plant species of the genus *Teucrium* exhibit various types of biological activity such as antioxidant, anticancer, anti-inflammatory, antidiabetic, antimicrobial and antiviral activity (Stanković 2012.

This chapter describes the main features of medicinal plants from the genus *Teucrium* on acetylcholinesterase inhibition, diabetes mellitus, and inflammatory processes. Additionally, a brief description of the methods used in the examination of anticholinesterase, antidiabetic and anti-inflammatory activity will be shown.

14.2 Acetylcholinesterase

Acetylcholinesterase (AChE) is serine hydrolysis belonging to the group of enzymes that affect different types of carboxylic esters (Čolović et al. 2013). Acetylcholine is a monomeric molecule consisting of 12 collected beta plates surrounded by 14 alpha helices and possesses a molecular weight of about 60,000. The

acetylcholinesterase molecule consists of several essential parts, peripheral anionic site, the catalytically active site and the aromatic throat (aromatic "gorge"). The peripheral anionic site plays an important role in recognizing the conformation of residues in the aromatic throat and in the catalytic site. The catalytic active site consists of an esterase and anion subunit. Esterase subunit is composed of three amino acids (Ser200, His440, and Glu327). The anionic sub spot plays a role in binding the positively charged quaternary amine of the choline group of acetylcholine (Nađpal 2017).

Acetylcholine (Fig. 14.1) is one of the essential neurotransmitters in the brain. In the central nervous system, it is found in interneurons and some cholinergic nerve fibers. Also, acetylcholine has an important role in the processes of memory and learning, as well as the maintenance of consciousness (Sisodia and Tanzi 2006; Topcu and Kusman 2014). In the peripheral nervous system, acetylcholine participates in the transmission of nerve signals in the motor plate and causes muscle contraction, slows down the pulse, enlarges blood vessels, influences the increased secretion of gastric glands, enhances peristaltic intestines (Takeda et al. 2004).

Science has not yet explained in the best way how choline is brought to the nerve endings, while acetyl groups are fed through the choline acetyltransferase enzyme derived from coenzyme A. Acetylcholine is found in presynaptic nerve endings in a state that can be immediately released or store in the form of surplus (Orhan et al. 2007; Čolović et al. 2013). Depending on whether there is a pulse, it remains unclear how acetylcholine is released from presynaptic endings. Although it is considered that calcium plays a major role in the process. Acetylcholine is released from the presynaptic neuron into the synaptic crack during transmission and binds to acetylcholine receptors (muscarinic and nicotinic) in the postsynaptic membrane, transmitting the neuron signal. The acetylcholine degradation enzyme is called acetylcholinesterase (Takeda et al. 2004; Sisodia and Tanzi 2006; Vladimir-Knežević et al. 2014). The acetylcholinesterase enzyme separates the acetylcholine to the choline an acetate group, which prevents the recurrence of muscular irritation. Released choline is transported to the presynaptic neuron and it is used for the resynthesis of acetylcholine by the activity of choline acetyltransferase enzyme (Čolović et al. 2013). Each molecule of acetylcholinesterase enzyme inhibits an average of 25,000 acetylcholine molecules per second, thereby enzyme having an appreciable catalytic activity. The choline formed in the process of decomposition of acetylcholine is recycled to recover acetylcholine. The main form of acetylcholinesterase is found in the brain and muscles. Acetylcholinesterase is encoded by AChE in humans (Čolović et al. 2013; Nađpal 2017).

A neurodegenerative disease such as Alzheimer's disease indicates a condition where neurons of the spinal cord and brain collapse (Mattson 2004). This leads to a loss of muscular coordination and cognitive dysfunction. Alzheimer's disease (AD), first described by neuropsychiatrist Alois Alzheimer in 1906, is a progressive

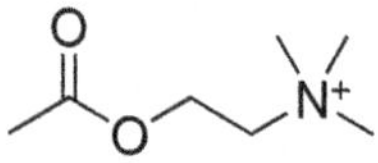

Fig. 14.1 Chemical structure of acetylcholine (ACh)

degeneration, which most often affects the elderly in developed countries (Nađpal 2017). The disease is recognized by the loss of short-term memory and disorientation (Akhondzadeh et al. 2003; Topcu and Kusman 2014). In morphological terms, Alzheimer's disease is characterized by the appearance of D-amyloid deposits and neurofibrillary loops produced by hyperphosphorylated τ-proteins (Adewusi et al. 2011). Neurochemical is characterized by loss of activity of cholinergic neurotransmitters, primarily acetylcholine (ACh) in the cerebral cortex (Orhan et al. 2007).

In people suffering from Alzheimer's disease, choline-acetyltransferase has been significantly reduced in the central nervous system. The loss of choline acetyltransferase is particularly pronounced in brain regions where the areas responsible for memory are located, i.e. in the cortex and hippocampus of the brain. Treatment of people suffering from Alzheimer's disease implied the use of acetylcholine, but such attempts to treat the disease have not shown great success (Nađpal 2017). Today's attempts to treat affected people are aimed at increasing cholinergic transmission by blocking the activity of acetylcholinesterase that degrades acetylcholine in synaptic cracks. Also, there are acetylcholinesterase inhibitors that are used in medicine but with varying degrees of efficacy in treatment (Mattson 2004).

14.2.1 Inhibition of Acetylcholinesterase

Pharmacological inhibitors of acetylcholinesterase are very important in the context of diseases where reduced neurotransmission activity mediates acetylcholine. In a disease such as Alzheimer's, there is a loss of cholinergic neurons in the brain, while some autoimmune diseases of the antibody lead to loss of nicotine acetylcholine receptors in the motor plate (Čolović et al. 2013). Alzheimer's disease is one of the most common forms of dementia, where there is a significant loss of brain mass. Despite numerous findings, it is still not known what causes Alzheimer's disease. It is considered that disease is developed by decreasing holing-acetyltransferase activity and inhibition of acetylcholinesterase enzyme is important for the treatment of the disease. The inhibition of acetylcholine increases its concentration in the synapses, which allows long-term stimulation of the receptors and the control of the disease (Soreq and Seidman 2001).

Depending on the mode of action, acetylcholine inhibitors are divided into irreversible and reversible. Reversible inhibitors are protagonists in the treatment of Alzheimer's disease symptoms. In the treatment of Alzheimer's disease, drugs that compensate for ACh deficiency, including ACh precursors, muscarinic and nicotinic receptor antagonists, and AChE inhibitors (Akhondzadeh et al. 2003). Inhibition of AChE is one of the most widely used treatment models for Alzheimer's disease (Akhondzadeh et al. 2003; Orhan et al. 2007). Some of the synthetic and natural remedies, such as galanthamine (Razadyne®, Reminyl®) donepezil (Aricept®), rivastigmine (Exelon®), and tacrine (Cognex®) are approved by the Food and Drug Administration, FDA) (Topcu and Kusman 2014; Nađpal 2017).

These drugs respond by enhancing the transmission of cholinergic neurons in the frontal region of the brain and compensating the loss of functional brain cells. Additionally, the treatment of Alzheimer's disease by drugs are used in the treatment of symptoms of neurological disorders such as Parkinson's disease, dementia with Levi's bodies and Myasthenia gravis. Irreversible acetylcholine inhibitors have a toxic activity whose function is reflected in the accumulation of acetylcholine in synaptic cracks, resulting in disorders in neurotransmission (Čolović et al. 2013). Synthetic acetylcholinesterase inhibitors exhibit side effects. Side effects are detected in the disorder of the gastrointestinal tract and hepatoxicity. There is still no answer in the science of developing acetylcholine inhibitors that manage in the brain without effect on other organs.

Commercially common inhibitors of acetylcholinesterase enzymes such as donepezil, tacrine, and rivastigmine have a damaging effect on human health, and now trying to find alternative substances of plant origin that will replace synthetic drugs. Tacrine has a short-acting time, unlike other drugs, and tablets should be taken several times a day, although it has been shown that taking this drug causes headache, muscle pain, loss of aptitude, nausea, diarrhea, etc. Rivastigmine and donepezil have similar effects because they are reversible acetylcholinesterase inhibitors, but their effect is reduced after prolonged usage (Martorana et al. 2010; Nađpal 2017).

14.2.2 Plants as Natural Sources of Anticholinesterase Agents

Nowadays, science is focused on the testing of plant extracts of certain species in order to inhibit acetylcholine and treat neurodegenerative diseases (Dastmalchi et al. 2007; Zhao et al. 2013). Galantamine is one of the most effective acetylcholinesterase inhibitors isolated from plant extracts of *Galanthus nivalis*. Galantamine does not exhibit adverse effects such as synthetic inhibitors and is used for clinical purposes. Moreover, other compounds that exhibit significant anti-acetylcholinesterase activity have been isolated (Mukherjee et al. 2007). The α-viniferin is isolated from *Caragana chamlagu* and exhibits significant anti-acetylcholinesterase activity, as well as huperzine-A which isolated from *Huperzia serrata*. Then, physostigmine isolated from the seeds of *Physostigma venenosum* is a reversible acetylcholinesterase enzyme inhibitor. Also, ursolic acid derived from commercial extract (*Origanum majorana*) exhibits significant anti-acetylcholinesterase activity (Mukherjee et al. 2007; Roseiro et al. 2012). *Ginkgo biloba* is used in traditional medicine for the prevention of degenerative brain disorders and helps maintain concentration. The chemical composition of this type of extract includes bilobalide, ginkgolide, and other compounds such as kaempferol and quercetin (Čolović et al. 2013; Nađpal 2017).

In addition to herbal extracts, there are data about using essential oils from herbs that are potent inhibitors of acetylcholinesterase enzymes due to the presence of certain compounds. Many medicinal plants are used in medicine as stimulants for

cognitive stimulation, for the treatment of depression and Alzheimer's disease. Plant species that are used in traditional medicine in the treatment of dementia belong to the genus *Salvia*, which is at the same time one of the most numerous genera belonging to the family Lamiaceae (Vladimir-Knežević et al. 2014).

14.2.3 Lamiaceae Species as a Sources of Anticholinesterase Agents

Numerous plant species belonging to different families such as Amaryllidaceae, Lamiaceae, Papaveraceae have been investigated in order to find effective acetylcholinesterase inhibitors. Some plant alkaloids, ursolic and oleic acid have been shown to be good inhibitors of acetylcholinesterase (Topcu and Kusman 2014). Plant species belonging to the Lamiaceae family are known for their medicinal properties because they possess a wide diversity of secondary metabolites that exhibit their pronounced biological activity. The biological activity of plant secondary metabolites is based on their ability to react with many regulatory molecules and other cell and subcellular structures, thereby affecting, either positively or negatively, a large number of metabolic processes. Some Lamiaceae species possess essential oils, flavonoids and hydroxycinnamic acid which are one of the main bioactive compounds. Genera belonging to the Lamiaceae family, such as *Lavandula, Origanum, Mentha, Salvia, Rosmarinus, Thymus, Teucrium*, and *Calamintha* are well known for their neuroprotective effect and usage in traditional medicine in the treatment of many diseases (Topcu and Kusman 2014).

Plants of Lamiaceae family are rich in phenolic acids, primarily rosmarinic and chlorogenic acid, and also include gentisic, caffeine, proto-catheic, vanillic, ferulic and souric acid (*Lavandula, Nepeta, Rosmarinus, Teucrium, Salvia*). Rosmarinic acid is characteristic of the some species from Lamiaceae family (Vladimir-Knežević et al. 2014). Also, they are rich in triterpene derivatives, ursans and oleans, where ursol, betulin and oleic acid are present in the highest percentage (*Salvia, Rosmarinus*) (Jäger et al. 2007). The plants of the Lamiaceae family, as well as other plant species, synthesize flavones and flavonols that, in addition to other secondary metabolites, contribute to the biological activity of these plants (Vladimir-Knežević et al. 2014). The species of the Lamiaceae family were objects of numerous pharmacological studies showing that their extracts and essential oils show a number of biological effects in vitro and in vivo conditions. The plant's essential oils belonging to the family include components such as sesquiterpenes, monoterpenes, diterpenes, aliphatic and aromatic components. The biological activity of extracts and essential oils may be significant in the treatment of neurodegenerative disorders, in particular, Alzheimer's disease (Topcu and Kusman 2014).

Vladimir-Knežević et al. (2014) examined selected Lamiaceae species to the ability to inhibit acetylcholinesterase. In their results, it has been shown that extracts of several plant species such as *Teucrium chameadrys, Thymus vulgaris, Mentha*

piperita, Salvia officinalis, Saturea montana inhibit over 75% acetylcholinesterase activity. The ethyl alcohol extract of *Salvia trilobacterium* inhibited acetylcholinesterase at 0.71 mg/ml, while the *Melissa officinalis* extract was completely inactive (Orhan and Aslan 2009). Ferreira et al. (2006) examined the extracts and essential oils of ten plants on the ability to inhibit acetylcholinesterase. They noticed that the species *Lavandula pedunculata, Mentha suaveolens* and *Melissa officinalis* show the inhibitory effect on acetylcholinesterase among the tested plant species. A high degree of acetylcholinesterase inhibition was observed by examining essential oil of *Mentha, Origanum* and *Saturea* species, while moderate and low inhibition of the activity of acetylcholinesterase enzymes was observed by examining the essential oil of plant species belonging to the genera *Ocimum, Lavandula*, and *Salvia* (Orhan et al. 2008). The results of the antiacetylcholinesterase effect of plant extracts and essential oils isolated from the species of the Lamiaceae family were presented in numerous studies (Orhan et al. 2007; Orhan and Aslan 2009; Topcu and Kusman 2014; Vladimir-Knežević et al. 2014). The studies show that plant species belonging to the Lamiaceae family are a natural source of acetylcholinesterase inhibitors and can be used in the prevention and treatment of neurodegenerative diseases (Vladimir-Knežević et al. 2014).

14.2.4 **Teucrium** *Species as a Sources of Anticholinesterase Agents*

Species of the genus *Teucrium* are known for their medicinal properties. Extracts of *Teucrium* species possess different secondary metabolites that exhibit biological activity. Secondary metabolites isolated from extracts of *Teucrium* species, have been shown to be potential inhibitors of acetylcholinesterase (Orhan and Aslan 2009; Vladimir-Knežević et al. 2014). Vladimir-Knežević et al. (2014) examined the ethanol extracts of *Teucrium* species on the ability to inhibit acetylcholinesterase using Ellman's colorimetric assay. The authors state that the extracts of the tested species such as *Teucrium arduini, T. chamaedrys, T. montanum*, and *T. polium* showed strong inhibitory activity against acetylcholinesterase. The percentage of the inhibition of acetylcholinesterase for the ethanolic extracts of these plants was obtained for tested concentrations of 0.25, 0.50 and 1 mg/ml, respectively. The same authors state that all investigated species of the genus *Teucrium* had a percentage of inhibition of acetylcholinesterase over 80% at 1 mg/ml. Among the investigated species, *Teucrium arduini* exhibits the highest ability to inhibit acetylcholinesterase enzymes over 95% at 1 mg/ml, followed by *Teucrium chamaedrys* that neutralize the enzyme near to 90% at the same concentration. *Teucrium polium* shows similar values as *Teucrium chamaedrys*, 86% at a concentration of 1 mg/ml. Among the tested species, *Teucrium montanum* shows the lowest ability of neutralizing acetylcholinesterase enzymes, about 80% at a concentration of 1 mg/ml. According to the ability to inhibit the enzyme acetylcholinesterase, the investigated plant species

have the following order *T. arduini* > *T. chamaedrys* > *T. polium* > *T. montanum* at a concentration of 1 mg/ml. At a concentration of 0.50 mg/ml *Teucrium chamaedrys* and *T. polium* exhibit the highest ability to inhibit acetylcholinesterase enzyme above 60%. *Teucrium arduini* shows slightly less inhibitory capacity than the previous two species, 58%. *Teucrium montanum*, in this case, shows the lowest ability to neutralize acetylcholinesterase enzymes, about 45% at a concentration of 0.50 mg/ml. The order of plants for a concentration of 0.50 mg/ml would be *T. chamaedrys* > *T. polium* > *T. arduini* > *T. montanum*. At a concentration of 0.25 mg/ml *Teucrium chamaedrys* and *T. arduini*, the highest inhibition ability is greater than 35%. *Teucrium montanum* has a higher ability to inhibit acetylcholinesterase enzymes 32% from *Teucrium polium* 28% at a minimum concentration of 0.25 mg/ml. The investigated plant species of the genus *Teucrium* at a concentration of 0.25 mg/ml have the following order *T. chamaedrys* > *T. arduini* > *T. montanum* > *T. polium*.

Golfakhrabadi et al. (2015) investigated the influence of the methanol extract of *Teucrium hyrcanicum* on the inhibition of acetylcholinesterase enzymes at the same time compared to donepezil as a positive control. The results showed that the IC_{50} values for the methanol extract of *Teucrium hyrcanicum* and donepezil were 2.12 mg/ml and 0.013 mg/ml, respectively. At a concentration of 1 mg/ml, the inhibitory activity of the methanol extract of the tested species was 40% (Golfakhrabadi et al. 2015). The anticholinesterase activity was investigated for the *Teucrium royleanum*, using organic solvents of different polarities (Ahmad et al. 2007). The authors state that the greatest ability to neutralize the acetylcholinesterase enzyme is observed for the ethyl acetate fraction, 83.62%. The chloroform fraction with the ability to neutralize the enzyme is 70.5%. The n-butanol fraction showed moderate activity against acetylcholinesterase 59.55%, while crude extract of *Teucrium royleanum* also showed moderate activity against acetylcholinesterase enzymes, 52.4% (Ahmad et al. 2007). A qualitative examination of *Teucrium royleanum* essential oil identified compounds such as sesquiterpene hydrocarbons, β-caryophyllene, germacrene D, alpha-humulene and linalool (Mohan et al. 2010).

Preliminary investigations of the inhibitory activity of plant extracts of species belonging to the genus *Teucrium* on the enzyme acetylcholinesterase indicate that the extracts of these plants have a strong or moderate ability to inhibit this enzyme. By examining the composition of *Teucrium* phenolic compounds from the group of phenolic acids, the most common are rosmarinic and chlorogenic acid, as well as from the group of flavonoids the most common are luteolin and apigenin. Experiments showed that rosmarinic acid contributes to the inhibition of acetylcholinesterase enzymes (Vladimir-Knežević et al. 2014). For the species of this genus, the presence of a monoterpene and sesquiterpene is characteristic (Monsef-Esfahani et al. 2010). Plants of genus *Teucrium* are rich in terpene components which are known that possess inhibitory properties of the enzyme acetylcholinesterase (Yoo and Park 2012; Vladimir-Knežević et al. 2014).

14.2.5 Methods for Determination of the Anticholinesterase Activity

One of the most prominent methods present in scientific studies of acetylcholinesterase inhibition is Ellman's method (Ellman et al. 1961; Burčul 2008). It is a spectrophotometric method based on the reaction of Ellman's reagent DTNB (5,5-dithiobis-(2-nitrobenzoic acid)) with a thiol group of substrates formed by the action of acetylcholinesterase. The reaction of DTNB with thiol groups leads to the formation of TNB (2-nitro-5-mercaptobenzoic acid) which is further hydrolyzed in the water at neutral or alkaline pH. Free TNB has a yellow color. The maximum absorption DTNB is at a wavelength $\lambda = 320$ nm, while the amount of TNB is read at a wavelength at $\lambda = 412$ nm. Reaction mixture should consist of: 100 ml acetylcholinesterase concentration 0.3 U/mg dissolved in 0.1 M phosphate buffer at an alkaline pH = 8; then 50 ml DTNB; acetylcholine iodide concentration 0.3 mM dissolved in 0.1 M phosphate buffer at neutral pH = 7; 50 ml sample (inhibitor or alcohol extract) dissolved in methanol; 850 ml 0.1 M phosphate buffer alkaline pH = 8.

In order to determine the possible non-enzymatic hydrolysis of AChI, two reference solutions should be made. The first reference solution instead of AChE was buffered (1000 ml) instead of the ethanol inhibitor (50 ml) and DTNB and AChI (50 ml). The other reference solution contains everything the same as the first reference solution only that AChI is excluded so that it contains buffer (1000 ml), ethanol, AChE, and DTNB (50 ml). Additionally, it is necessary to prepare a control solution where the 50 ml inhibitor is replaced with 50 ml methanol. It is important to note that before the experiment it is necessary to check that methanol or some other organic solvent does not interfere with AChE inhibition. The reference solutions should be incubated at room temperature for a period of 30 min. The reaction begins with the addition of 50 ml AChI as the substrate and after 6 min the absorbance is measured at a wavelength $\lambda = 412$ nm.

14.3 Diabetes Mellitus

Early studies have shown that cerebral cortex and hippocampus neurons are characterized by an inadequate response and decreased secretion of insulin effect, which indicates the emergence of neurodegenerative diseases (Craft 2007; Chiung-Chun et al. 2010). Many authors consider that insulin plays an important role in central energy metabolism and that insulin regulation disorders can influence the pathogenesis of neurodegenerative diseases, such as Alzheimer's disease. Numerous studies have shown a significant correlation between the occurrence of Alzheimer's disease and Type 2 diabetes mellitus (Craft 2007; Sims-Robinson et al. 2010).

Diabetes mellitus is a chronic metabolic disorder characterized by hyperglycemia, or high blood glucose levels. The onset of the disease is the result of inadequate

blood glucose control by insulin hormones synthesized in the pancreas (Zimmet et al. 2001). Hyperglycemia occurs due to partial or complete insulin deficiency. The onset of hyperglycemia affects the onset of long-term complications, with cardiovascular, peripheral and cerebrovascular disorders being distinguished (Alberti and Zimmet 1998).

Type 1 diabetes is characterized by the destruction of β-cells leading to an insulin deficiency. It is a condition where the pancreas does not create or creates very little insulin because the β-cells of the pancreas are destroyed by the autoimmune mechanism. Markers of immune destruction of β-cells are autoantibodies against β-cells when initially detected hyperglycemia. This disease is also associated with certain HLA genes, and the presence of certain alleles may be predominant or protective (Devendra et al. 2004).

Type 2 diabetes mellitus is predominantly insulin resistance with relative insulin deficiency and predominantly insulin secretory deficiency with insulin resistance. Type 2 diabetes accounts for 90 to 95% of the total number of diabetes cases in the world and is due to the presence of insulin resistance in skeletal muscles and liver and reduced insulin production, so there is a relative but rarely absolute insulin deficiency. Type 2 diabetes is a progressive disease where the production of insulin decreases as the disease progresses. Insulin resistance develops due to genetic defects in combination with environmental factors, primarily obesity, and physical inactivity. As the disease progresses, insulin resistance remains relatively stable, and insulin production declines progressively (Kahn 2003; Rewers 2012).

Treatment of diabetes involves a complex of factors that involve the exercise, diet, and drugs. Drugs used to treat diabetes of mainly synthetical origin and have certain negative side effects. Medicinal plants are used in the treatment of many disorders and have a significant role in the treatment of diabetes mellitus as a serious metabolic disorder. It has been shown experimentally that medicinal plants used in the treatment of diabetes do not show any negative effects. Plant species are rich in compounds such as phenolics, tannins, alkaloids and flavonoids that improve the efficiency of pancreatic tissues by increasing insulin secretion or decreasing intestinal absorption of glucose (Kooti et al. 2016).

14.3.1 Plants as Natural Sources of Antidiabetic Agents

Numerous plant species belonging to different families of Amaryllidaceae, Asteraceae, Apiaceae, Cucurbitaceae, Fabaceae, Gentianaceae, Lamiaceae, Liliaceae, Rosaceae, etc. were investigated in order to find effective substances for the purpose of treating diabetes (Moradi et al. 2018). Extracts of certain plant species have been shown to modulate the metabolic pathways such as glycogen synthesis, glycolysis, cholesterol synthesis and release of insulin (Prabhakar and Doble 2008). The extract of *Carthamus tinctorius* (Asteraceae) is used in the treatment of Type 1 and Type 2 diabetes, because the presence of flavonoids, such as quercetin and kaempferol, is determined by phytochemical analyzes, which are the causes of

hypoglycemic effects of these compounds. *Swertia punicea* (Gentianaceae) was investigated in order to determine the hypoglycaemic effect, which can be used in the treatment of diabetes because it promotes insulin resistance in the diabetic mice (Moradi et al. 2018).

The aqueous root extract of *Sarcopoterium spinosum* (Rosaceae) was examined in order to determine the anti-diabetic effect of progressive hyperglycemia in mice that are genetically diseased with diabetes. The results showed that the aqueous extract of the roots of the investigated species showed positive results in the treatment of diseased mice. Also, the aqueous extract of *Liriope spicata* (Liliaceae) positively influences the reduction of blood glucose levels and promotes glucose tolerance and resistance of insulin in the mice (Moradi et al. 2018). Ethanol extract of *Momordica charantia* (Cucurbitaceae) possess antihyperglycemic and hypoglycemic effect in normal and streptozotocin-diabetic rats. A four-week experiment conducted on Type 2 diabetes mellitus rats showed that the *Allium sativum* extract (Alliaceae) enhanced glycemic control by increased insulin secretion and enhanced insulin sensitivity, which showed the antihyperglycemic and antihyperlipidemic effect of this type of extract (Moradi et al. 2018). The study of the effects of plant extracts on the treatment of diabetes was carried out by (Huseini et al. 2006; Rao and Nammi 2006; Das et al. 2009; Abeywickrama et al. 2011; George et al. 2011; Das and Barman 2012; Rabiei et al. 2014; Jiao et al. 2017; Moradi et al. 2018).

14.3.2 Lamiaceae Species as a Sources of Antidiabetic Agents

Some plant species belonging to the Lamiaceae family are used in the treatment of diabetes due to the presence of compounds that exhibit pronounced biological activity. Genera belonging to the Lamiaceae family, such as *Ocimum, Origanum, Salvia, Teucrium, Lamium*, etc. are known for their antidiabetic activity (Patil et al. 2011; Moradi et al. 2018). *Origanum vulgare* aqueous extract affects the reduction in glucose levels of glycosylated hemoglobin, and pancreatic amylase in diabetic rats. Extract usage in of dietary in diabetes rats influenced the content of glycogen in the liver, body weight and the level of urea. Plant parts of *Salvia nemorosa* contain glycosides such as salvionosides and megastigmands. It has been shown experimentally that in diabetes mice, insulin levels are increased using extracts, which increases the activity of insulin (Sadeghzadeh et al. 2008). By isolating a luteolin compound from a plant species belonging to the genus *Perilla* it has been shown a positive effect on the development of diabetic nephropathy (Moradi et al. 2018). Flavonoid luteolin reduces creatinine levels and diabetic rats and prevents the increase in urea protein in 24 h. Aqueous and ethanolic extract of the *Ocimum tenuiflorum* resulted in decreased levels of blood glucose, fatty acids, lipid peroxides and low-density lipoproteins (Hussain et al. 2001, 2015). A tetracyclic terpenoid isolated from a hydro alcoholic extract of *Ocimum tenuiflorum* exhibited potential antidiabetic properties (Patil et al. 2011). Hannan et al. (2006) showed that ethanol,

butanol, aqueous and ethyl acetate fractions of *Ocimum tenuiflorum* stimulated insulin secretion.

14.3.3 *Teucrium Species as a Sources of Antidiabetic Agents*

Secondary metabolites isolated from extracts of *Teucrium* species are good agents against diabetes (Gharaibeh et al. 1988; Rasekh et al. 2001; Esmaeili and Yazdanparast 2004; Ardestani et al. 2008; Vahidi et al. 2010; Alamgeer et al. 2013; Dehghan et al. 2013; Sabet et al. 2013). Dehghan et al. (2013) have shown that the methanol extract of *Teucrium orientale* possesses antihyperglycemic activity. Using extract in the diabetic rat's diet it has been shown that serum glucose has decreased by 40%. Mechanisms that have influenced the decrease in serum glucose levels are strong anti-oxidant properties of *Teucrium orientale* compounds and increased peripheral glucose utilization. Studies have shown that *Teucrium orientale* is rich in flavonoids (Cakir et al. 2006) while flavonoids possess hypoglycemic properties as they increase the oxidative metabolisms of the diabetic states (Dehghan et al. 2013). Alamger et al. (2013) examined the effect of crude extract from *Teucrium stocksianum* on antidiabetic activity. The authors have shown that crude powder of tested species reduces blood glucose levels in diabetic rabbits at a dose of 250 and 500 mg/kg. Testing methanol, aqueous and ethyl acetate extract from *Teucrium stocksianum* noted that at a dose of 500 mg/kg reduce the blood glucose level in diabetes rabbits. The ethyl acetate extract showed better results than methanol and aqueous extract. Also, ethyl acetate extract produced a significant increase in the serum insulin level of diabetic rabbits.

The most noticeable *Teucrium* species in terms of antidiabetic effect is *Teucrium polium*. The *Teucrium polium* extract reduces the level of glucose and triglyceride in the serum at a dose of 4% (Vahidi et al. 2010). In vitro studies have shown that extracts of this species affect the increase in insulin levels and show a hypoglycemic effect in normal rats after only one dose. The highest hypoglycemic effect (50%) was noticed 8 h after a single administration of 125 mg/kg. Moghimi et al. (2017) state that the hypoglycemic activity of *Teucrium polium* extracts was referred to as the presence of secondary metabolites such as flavonoids, sterols, and volatile oils as the active compounds. Flavonoids have hypoglycemic effects by insulin release from the pancreas which stimulates glucose utilization.

14.3.4 *Methods for Determination of the Antidiabetic Activity*

Methods based on the study of antidiabetic activity consist of monitoring results before and after treatment in vivo and in vitro conditions. Basically, male Wistar rats are used in several groups. The first is a control group, the second is mice that are diabetic, and the third group consists of treated mice with diabetes. Type 1 diabetes

was induced by autoimmune in mice by administering low-dose intraperitoneal injections of STZ (20 mg/kg body weight), dissolved in normal saline for several days. The presence of hyperglycemia is confirmed by diabetes. Depending on the duration of the experiment, after several days of the first STZ injection, plant aqueous extract is administered in a dose of 100 mg/kg for several weeks. The level of serum glucose is monitored at the beginning, mid and end of the experiment (Sabet et al. 2013).

14.4 Inflammation

Inflammation represents the defensive reaction of the organism to harmful exogenous or endogenous factors through the immune system (Abbas et al. 2010). The release of the organism from harmful agents includes a series of changes that take place in the intercellular matrix and blood vessels. The basic function of the inflammation process is to isolate the tissue and remove damaged cells (Medzhitov 2008; Soehnlein and Lindbon 2010; Ashley et al. 2012). There are several types of inflammation. Chronic inflammation leads to necrosis and tissue fibrosis. This type of inflammation affects the development of degenerative diseases, primarily Alzheimer's disease, diabetes, cancer, atherosclerosis, etc. (Murphy and Weaver 2017). Acute inflammation occurs after a few minutes or hours after tissue injury. This is a short-term process characterized by symptoms such as redness of the tissue. Acute inflammation is a good defense mechanism in the fight against bacteria.

Symptoms that are followed by inflammatory processes include tissue redness, pain, and high temperature (Brune and Hinz 2004). The onset of an inflammatory reaction is aimed at repairing the tissue after the elimination of the infectious agent. The transition of the inflammatory process into the tissue recovery process is the interruption of the production of pro-inflammatory cytokines and the start of the production of liposomes. Lipoxins activate monocytes that remove necrotic cells and inhibit neutrophil recruitment. Also, lipoxins initiate tissue remodeling (Medzhitov 2008). The inflammatory process involves the synthesis of interleukins, prostaglandins and certain hemotoxins. Inflammation occurs by stimulation activation on the membrane that activates the hydrolysis of the phospholipid membrane by phospholipase A into arachidonic acid, which further represents the substrate for cyclooxygenase and lipooxygenase enzyme.

It has been shown that inflammatory processes are involved in the etiology of many diseases as well as Alzheimer's disease (Breitner 1996). The drugs that are used today to relieve the symptoms of inflammation are non-steroidal anti-inflammatory drugs (Beale and Block 2010; Lemke et al. 2013).

14.4.1 Plants as Natural Sources of Anti-inflammatory Agents

Nowadays, plant extracts are used to reduce inflammatory processes (Mahesh and Sathish 2008). It has been shown experimentally that medicinal plants used in the treatment of inflammatory processes do not show any side effects. Numerous plant species belonging to different families of Aristolochiaceae, Apocynaceae, Asteraceae, Lamiaceae, Plantaginaceae, Salicaceae, etc. have been investigated in order to find effective substances to reduce the symptoms of inflammation. In addition to a number of medicinal properties, the anti-inflammatory activity of plants is appreciated in traditional medicine. The anti-inflammatory activity of herbal extracts reduces carrageenan-induced edema in the rats, with the oral application of these extracts. Similar results have been shown in numerous studies of medicinal plants (Dharmasiri et al. 2003; Li et al. 2003; Ojewole 2005; Rodriguez Silva et al. 2008; Kumar et al. 2009; Chandrashekar et al. 2010; Shah and Seth 2010; Sreejith et al. 2010; Garg and Paliwal 2011; Shah et al. 2011; Vishal et al. 2014; Verma 2016).

14.4.2 Lamiaceae Species as a Sources of Anti-inflammatory Agents

The Lamiaceae family includes about 250 genera with over 7500 species. A large number of species have been widely used in pharmacy. Some plant species belonging to the Lamiaceae family have been administered against COX. The biologically active compounds present in the species of this family are the main sources of COX inhibitors (Pang et al. 1996). Plant species belonging to the Lamiaceae genera, such as *Lavandula, Glechoma, Lamium, Mentha, Marrubium, Origanum, Ocimum, Rosmarinus, Salvia, Stachys, Sideritis* and *Teucrium* are known for their anti-inflammatory activity (Mihai et al. 2018).

Species of the genus *Ocimum* are used to treat inflammation and assuage of chronic pain in the joints. The anti-inflammatory property of *Ocimum basilicum* and *O. santanum* was confirmed by Singh et al. (1996). Phenolic compounds present in the investigated plants inhibit carrageenan-induced and arachidonic acid and leukotriene-induced paw edema, possibly by blocking the enzymatic activity of both COX and lipoxygenases (Singh et al. 1996). Phenolic compounds are responsible for the anti-inflammatory properties of several medicinal plants (Pongprayoon et al. 1991; Dewhirst 1980). The species belonging to the genus *Glechoma* have the ability to elaborate long-chain unsaturated fatty acids with anti-inflammatory potentials (Kuhn et al. 1989). *Thymus vulgaris* extracts possess inhibitory roles over the nitric oxide (NO) by limiting iNOS mRNA expression, and also are used in traditional medicine for inflammatory skin disorders (Vigo et al. 2004; Alabdullatif et al. 2017). Methanolic extracts derived from species which belong to the genus *Lamium*, such as *L. purpureum* and *L. garganicum*, reducing inflammatory pain in a model of ear edema and in carrageenan-induced paw edema (Akkol et al. 2008). Studies have

shown that the hydrochloric extract of the species *Stachys* has better anti-inflammatory activity than a nonsteroidal anti-inflammatory drug, indomethacin. The extract of the investigated species was able to attenuate both early and delayed phases of carrageenan-induced inflammation over the 50–200 mg/kg dose range. The studies show that plant species belonging to family Lamiaceae is a natural source of anti-inflammatory compounds and that plants can be used in the treatment of inflammation (Mihai et al. 2018).

14.4.3 **Teucrium** ***Species as a Sources of Anti-inflammatory Agents***

Species belonging to the genus *Teucrium* possess secondary metabolites that, in addition to numerous biological activities, exhibit in vitro and in vivo anti-inflammatory activity (Tariq et al. 1989; Barrachina et al. 1995; Puntero et al. 1997; Radhakrishnan et al. 2001; Shakhanben 2001; Menichini et al. 2009; Cabral et al. 2010; Farshchi et al. 2010; Pourmotabed et al. 2010; Miri et al. 2015; Shah and Shah 2015; Rahmouni et al. 2017).

Amraei et al. (2018) examined the effect of *Teucrium polium* hydroalcoholic extract on anti-inflammatory activity. Administration of the extract significantly reduced the serum levels of triglyceride, cholesterol, and LDL-cholesterol. Additionally, the 170 mg/kg dose of extract was most effective in reducing serum levels of inflammatory markers. Authors Tariq et al. (1989) showed that a dose of 500 mg/kg of body weight of the ethanol extract of species *Teucrium polium* inhibited carrageenan-induced inflammation and reduced granuloma formation. According to a number of literary data, *Teucrium polium* contains bioactive compounds such as phenolic compounds and flavonoids. It has been shown that flavonoids reduce inflammatory processes by activating several pathways (Houshmand et al. 2015).

Pourmotabbed et al. (2010) examined the effect of aqueous extract of *Teucrium chamaedrys* on the anti-inflammatory activity, which showed that a dose-dependent inhibition of the edema was performed using 25–250 mg/kg administered 1 h before carrageenan-induced paw swelling. For extracts of *Teucrium hyrcanicum*, it has been shown to exhibit analgesic and anti-inflammatory activity in carrageenan-induced paw edema, and formalin pain tests (Farshchi et al. 2010). Extracts of *Teucrium stocksianum* significantly reduced the paw edema at a dose of 400 mg/kg in rats. However, the extract did not show any anti-inflammatory effect when administered orally in a sub-acute study using a cotton-pellet method but was effective orally in an acute study using rat paw edema (Radhakrishnan et al. 2001). Miri et al. (2015) examined aqueous extract from *Teucrium persicum* on anti-inflammatory activity. The authors have shown that aqueous extract of this species at concentrations of 100, 200 and 400 mg/kg shows an insignificant anti-inflammatory effect in the cotton pellet induced granuloma model in the mice. The authors assume that the

activity was dependent on the dose, method, or period of time. El-Ashmawy (2018) investigate the anti-inflammatory effects and the phytochemical constituents of the methanol extract from *Teucrium oliverianum* using the carrageenan-induced rat paw edema (acute and sub-acute models) and the terpentine oil-induced granuloma pouch bioassay. The methanolic extract of *Teucrium oliverianum* exhibited anti-inflammatory activity in both phases of carrageenan-induced acute edema test in a dose-dependent manner.

14.4.4 Methods for Determination of the Anti-inflammatory Activity

To investigate the anti-inflammatory effect, there are many methods of which carrageenan-induced inflammatory edema is most commonly used in the hind paw of rats. Inhibition of carrageenan-induced inflammation is one of the most suitable test procedures for monitoring anti-inflammatory agents. Carrageenan is a muco-polysaccharide isolated from Sea moss *Chondrus* in order to experimentally induce arthritis. Carrageenan causes an acute inflammation that occurs in two stages. The first or early stage occurs after 1 h of injection of carrageenan. At this stage, edema develops due to the effects of serotonin and histamine. A second or late phase occurs after 2 h of injection of carrageenan where vascular permeability is maintained by bradykinin and prostaglandins. These mediators induce pain and contribute to the inflammatory response. Studies have shown that late edema is sensitive to anti-inflammatory drugs and is used to estimate the antiphlogistic effect of natural products (El-Ashmawy 2018).

14.5 Conclusions

This chapter is a review of anticholinesterase, antidiabetic and anti-inflammatory activity of several species of the genus *Teucrium.* Extracts from *Teucrium arduini, T. chamaedrys, T. montanum*, and *T. polium*, showed strong inhibitory activity against acetylcholinesterase. Extracts of some *Teucrium* species reduces blood glucose levels in diabetic rats and rabbits. Also, the use of plant extracts significantly reduce inflammatory markers and inhibited carrageenan-induced inflammation. By examining the qualitative and quantitative composition of secondary metabolites of *Teucrium* species, the presence of active phenolic acids was observed. The most common are rosmarinic and chlorogenic acid. Also, the most common are luteolin and apigenin from the group of flavonoids. It has been shown that these biologically active compounds contribute to anticholinesterase, antidiabetic and anti-inflammatory activity.

References

Abbas A, Lichtman A, Pillai S (2010) Cellular and molecular immunology, 8th edn. Elsevier, Amsterdam

Abeywickrama KRW, Ratnasooriya WD, Amarakoon AM (2011) Oral hypoglycaemic, antihyperglycaemic and antidiabetic activities of Sri Lankan Broken Orange Pekoe Fannings (BOPF) grade black tea (*Camellia sinensis* L.) in rats. J Ethnopharmacol 135:278–286

Adewusi EA, Moodley N, Steenkamp V (2011) Antioxidant and acetylcholinesterase inhibitory activity of selected southern African medicinal plants. S Afr J Bot 77:638–644

Ahmad B, Mukarram Shah SM, Khan H, Hassan Shah SM (2007) Enzyme inhibition activities of *Teucrium royleanum*. J Enzyme Inhib Med Chem 22:730–732

Akhondzadeh S, Noroozian M, Mohammadi M, Ohadinia S, Jamshidi AH, Khani M (2003) *Salvia officinalis* extract in the treatment of patients with mild to moderate Alzheimer's disease: a double blind, randomized and placebo-controlled trial. J Clin Pharm Ther 28:53–59

Akkol EK, Yalçın FN, Kaya D, Çalıs I, Yesilada E, Ersoz T (2008) In vivo anti-inflammatory and antinociceptive actions of some *Lamium* species. J Ethnopharmacol 118:166–172

Alabdullatif M, Boujezza I, Mekni M, Taha M, Kumaran D, Yi QL, Landoulsi A, Ramirez-Arcos S (2017) Enhancing blood donor skin disinfection using natural oils. Transfusion 57:2920–2927

Alamgeer RM, Bashir S, Mushtaq MN, Khan HU, Malik MNH, Qayyum A, ur Rahaman MS (2013) Comparative hypoglycemic activity of different extracts of *Teucrium stocksianum* in diabetic rabbits. Bangladesh J Pharmacol 8:186–193

Alberti KG, Zimmet PZ (1998) Definition, diagnosis and classification of diabetes mellitus and its complications. Part 1: diagnosis and classification of diabetes mellitus provisional report of a WHO consultation. Diabet Med 15:539–553

Amraei M, Ghorbani A, Seifinejad Y, Mousavi SF, Mohamadpour M, Shirzadpour E (2018) The effect of hydroalcoholic extract of *Teucrium polium* L. on the inflammatory markers and lipid profile in hypercholesterolemic rats. J Inflamm Res 11:265–272

Ardestani A, Yazdanparast R, Jamshidi S (2008) Therapeutic effects of *Teucrium polium* extract on oxidative stress in pancreas of streptozotocin-induced diabetic rats. J Med Food 11:525–532

Ashley TN, Weil MZ, Nelson JR (2012) Inflammation: mechanisms, costs, and natural variation. Annu Rev Ecol Evol Syst 43:385–406

Barrachina MD, Bello R, Martínez-Cuesta MA, Esplugues J, Primo-Yúfera E (1995) Antiinflamatory activity and effects on isolated smooth muscle of extracts from different *Teucrium* species. Phytother Res 9:368–371

Beale JM, Block JH (2010) Wilson and Gisvold's textbook of organic medicinal and pharmaceutical chemistry, 12th edn. Lippincott Williams & Wilkins, Philadelphia

Breitner JC (1996) The role of antiinflammatory drugs in the prevention and treatment of Alzheimer's disease. Annu Rev Med 47:401–411

Brune K, Hinz B (2004) The discovery and development of antiinflamatory drugs. Arthritis Rheumatol 50:2391–2399

Burčul F (2008) The biological effects of essential oil and water extract of the plant (*Achillea millefolium* L.). MSc thesis, University of Split

Cabral C, Francisco V, Cruz M, Lopes M, Salgueiro L, Sales F, Batist M (2010) Potential antioxidant and anti-inflammatory properties in *Teucrium salviastrum* Schreb. Planta Med 76:1163–1374. https://doi.org/10.1055/s-0030-1264535

Cakir A, Mavia A, Kazaz C, Yildirim A, Kufrevioglu OI (2006) Antioxidant activities of the extracts and components of *Teucrium orientale* L. var. *orientale*. Turk J Chem 30:483–494

Chandrashekar KS, Thakur A, Prasanna KS (2010) Anti-inflammatory activity of *Moringa oleifera* stem bark extracts against carrageenen induced rat paw edema. J Chem Pharm Res 2:179–181

Chiung-Chun H, Cheng-Che L, Kuei-Sen H (2010) The role of insulin receptor signaling in synaptic plasticity and cognitive function. Chang Gung Med J 33:115–125

Čolović BM, Krstić DZ, Lazarević-Pašti TD, Bondžić MA, Vasić MV (2013) Acetylcholinesterase inhibitors: pharmacology and toxicology. Curr Neuropharmacol 11:315–335

Craft S (2007) Insulin resistance and Alzheimer's disease pathogenesis: potential mechanisms and implications for treatment. Curr Alzheimer Res 4:147–152
Das S, Barman S (2012) Antidiabetic and antihyperlipidemic effects of ethanolic extract of leaves of *Punica granatum* in alloxan-induced non-insulin-dependent diabetes mellitus albino rats. Indian J Pharm 44:219–224
Das M, Sarma BP, Khan AK, Mosihuzzaman M, Nahar N, Ali L, Bhowmik A, Begum R (2009) The antidiabetic and antilipidemic activity of aqueous extract of *Urtica dioica* L. on type 2 diabetic model rats. J Biol Sci 17:1–6
Dastmalchi K, Dorman HJD, Viorela H, Hiltunen R (2007) Plants as potential source for drug development against Alzheimer's disease. Int J Biomed Pharm Sci 1:83–104
Dehghan G, Tahmasebpour N, Hosseinpour Feizi MA, Sheikhzadeh F, Banan Khojasteh SM (2013) Hypoglycemic, antioxidant and hepato- and nephroprotective effects of *Teucrium orientale* in streptozotocin diabetic rats. Pharmacology 1:182–189
Devendra D, Liu E, Eisenbarth GS (2004) Type 1 diabetes: recent developments. BMJ 328:750–754
Dewhirst TE (1980) Structure-activity relationships for inhibition of prostaglandin cyclooxygenase by phenolic compounds. Prostaglandins 20:209–222
Dharmasiri MG, ayakody JRAC, Galhena G, Liyanage SSP, Ratnasooriya WD (2003) Anti-inflammatory and analgesic activities of mature fresh leaves of *Vitex negundo*. J Ethnopharmacol 87:199–206
El-Ashmawy IM (2018) Anti-inflammatory and phytoconstituents of *Teucrium oliverianum* Ging. Ex. Benth. Eur J Pharm Med Res 5:406–411
Ellman G, Courtney D, Valentino A, Featherstone R (1961) A new and rapid colorimetric determination of acetylcholinesterase activity. Biochem Pharmacol 7:88–95
Esmaeili MA, Yazdanparast R (2004) Hypoglycaemic effect of *Teucrium polium*: studies with rat pancreatic islets. J Ethnopharmacol (1):27–30
Farshchi A, Ghiasi G, Asl AA (2010) Antinociceptive and antiinflammatory effects of *Teucrium hyrcanicum* aqueous extract in male mice and rats. Physiol Pharmacol 14:78–84
Ferreira A, Proença C, Serralheiro MLM, Araújo MEM (2006) The in vitro screening for acetylcholinesterase inhibition and antioxidant activity of medicinal plants from Portugal. J Ethnopharmacol 108:31–37
Garg VK, Paliwal SK (2011) Anti-inflammatory activity of aqueous extract of *Cynodon dactylon*. Int J Pharmacol 7:370–375
George C, Lochner A, Huisamen B (2011) The efficacy of *Prosopis glandulosa* as antidiabetic treatment in rat models of diabetes and insulin resistance. J Ethnopharmacol 137:298–304
Gharaibeh MN, Elayan HH, Salhab AS (1988) Hypoglycemic effects of *Teucrium polium*. J Ethnopharmacol 24:93–99
Golfakhrabadi F, Yousefbeyk F, Mirnezami T, Laghaei P, Hajimahmoodi M, Khanavi M (2015) Antioxidant and antiacetylcholinesterase activity of *Teucrium hyrcanicum*. Pharm Res 7:15–19
Hannan JM, Marenah L, Ali L, Rokeya B, Flatt PR, Abdel-Wahab YH (2006) *Ocimum sanctum* leaf extracts stimulate insulin secretion from perfused pancreas, isolated islets and clonal pancreatic beta-cells. J Endocrinol 189:127–136
Houshmand G, Goudarzi M, Forouzandeh H, Nazari A, Nourollahi V (2015) Evaluation of the analgesic effects of *Teucrium* extract on rats using the formalin test. J Babol Univ Med Sci 17:33–39
Husain I, Chander R, Saxena JK, Mahdi AA, Mahdi F (2015) Antidyslipidemic effect of *Ocimum sanctum* leaf extract in streptozotocin induced diabetic rats. Indian J Clin Biochem 30:72–77
Huseini HF, Fakhrzadeh H, Larijani B, Samani AS (2006) Review of anti-diabetic medicinal plant used in traditional medicine. J Med Plants 1(S2):1–8
Hussain EHMA, Jamil K, Rao M (2001) Hypoglycaemic, hypolipidemic and antioxidant properties of tulsi (*Ocimum sanctum* Linn) on streptozotocin induced diabetes in rats. Indian J Clin Biochem 16:190–194
Jäger A, Eldeen I, Van Staden J (2007) COX-1 and -2 activity of rose hip. Phytother Res 21:1251–1252

Jiao Y, Wang X, Jiang X, Kong F, Wang S, Yan C (2017) Antidiabetic effects of *Morus alba* fruit polysaccharides on high-fat diet- and streptozotocin-induced type 2 diabetes in rats. J Ethnopharmacol 199:119–127

Kahn SE (2003) The relative contribution of insulin resistance and beta-cell dysfunction to the pathophysiology of type 2 diabetes. Diabetologia 46:3–19

Kooti W, Farokhipour M, Asadzadeh Z, Ashtary-Larky D, Asadi-Samani M (2016) The role of medicinal plants in the treatment of diabetes: a systematic review. Electron Physician 8:1832–1842

Kuhn H, Wiesner R, Alder L, Schewe T (1989) Occurrence of free and esterified lipoxygenase products in leaves of *Glechoma hederacea* L. and other Labiatae. Eur J Biochem 186:155–162

Kumar A, Panghal S, Mallapur SS, Kumar M, Ram V, Singh BK (2009) Antiinflammatory activity of *Piper longum* fruit oil. Indian J Pharm Sci 71:454–456

Lemke TL, Williams DA, Roche VF, Zito SW (2013) Foye's principles of medicinal chemistry, 7th edn. Lippincott Williams & Wilkins, Philadelphia

Li RW, David Lin G, Myers SP, Leach DN (2003) Anti-inflammatory activity of Chinese medicinal vine plants. J Ethnopharmacol 85:61–67

Mahesh B, Sathish S (2008) Antimicrobial activity of some important medicinal plant against plant and human pathogens. World J Agric Sci 4:839–843

Martorana A, Esposito Z, Koch G (2010) Beyond the cholinergic hypothesis: do current drugs work in Alzheimer's disease? CNS Neurosci Ther 16:235–245

Mattson MP (2004) Pathways towards and away from Alzheimer's disease. Nature 430:631–639

Medzhitov R (2008) Origin and physiological roles of inflammation. Nature 454:428–435

Menichini F, Conforti F, Rigano D, Formisano C, Piozzi F, Senatore F (2009) Phytochemical composition, anti-inflammatory and antitumour activities of four *Teucrium* essential oils from Greece. Food Chem 115:679–686

Miri A, Sharifi-Rad J, Tabriyian K, Nasiri AA (2015) Antinociceptive and anti-inflammatory activities of *Teucrium persicum* Boiss. extract in mice. Scientifica. https://doi.org/10.1155/2015/972827

Moghimi Z, Dehghan S, Sayadi S, Ghanbarian A (2017) Therapeutic effect and mechanism of *Teucrium polium* in diabetes mellitus: a review article. Res J Pharmacogn 4(S):73–73

Mohan L, Pant CC, Melkani AB, Dev V (2010) Terpenoid composition of the essential oils of *Teucrium royleanum* and *T. quadrifarium*. Nat Prod Commun 5:939–942

Monsef-Esfahani HR, Hajiaghaee R, Shahverdi AR, Khorramizadeh MR, Amini M (2010) Flavonoids, cinnamic acid and phenyl propanoid from aerial parts of *Scrophularia striata*. Pharm Biol 48:333–336

Moradi B, Abbaszadeh S, Shahsavari S, Alizadeh M, Beyranvand F (2018) The most useful medicinal herbs to treat diabetes. Biomed Res Ther 5:2538–2551

Mukherjee PK, Kumar V, Mal M, Houghton PJ (2007) Acetylcholinesterase inhibitors from plants. Phytomedicine 14:289–300

Murphy K, Weaver C (2017) Janeway's immunobiology, 9th edn. Garland Science, New York

Nađpal J (2017) Phytochemical screening and biological activity of extracts and traditional products from fruits of wild roses (*Rosa* L.; Rosaceae). Dissertation, University of Novi Sad

Ojewole J (2005) Antinociceptive, anti-inflammatory and antidiabetic effects of *Bryophyllum pinnatum* (Crassulaceae) leaf aqueous extract. J Ethnopharmacol 99:13–19

Orhan I, Aslan M (2009) Appraisal of scopolamine-induced antiamnesic effect in mice and in vitro antiacetylcholinesterase and antioxidant activities of some traditionally used Lamiaceae plants. J Ethnopharmacol 122:327–332

Orhan I, Kartal M, Naz Q, Ejaz A, Yilmaz G, Kan Y, Konuklugil B, Sener B, Choudhary MI (2007) Antioxidant and anticholinesterase evaluation of selected Turkish *Salvia* species. Food Chem 103:1247–1254

Orhan I, Aslan S, Kartal M, Şener B, Başer KHC (2008) Inhibitory effect of Turkish *Rosmarinus officinalis* L. on acetylcholinesterase and butyrylcholinesterase enzymes. Food Chem 108:663–668

Pang L, Las-Heras BD, Hoult JR (1996) A novel diterpenoid labdane from *Sideritis javalambrensis* inhibits eicosanoid generation from stimulated macrophages but enhances arachidonate release. Biochem Pharmacol 51:863–868

Patil R, Patil R, Ahirwar B, Ahirwar D (2011) Isolation and characterization of anti-diabetic component (bioactivity-guided fractionation) from *Ocimum sanctum* L. (Lamiaceae) aerial part. Asian Pac J Trop Med 4:278–282

Pongprayoon U, Baeckstrom P, Jacobsson U, Lindstrom M, Bohlin L (1991) Compounds inhibiting prostaglandin synthesis isolated from *Ipomoea pescaprae*. Planta Med 57:515–518

Pourmotabed A, Farschehi A, Ghiasi G, Khatabi PM (2010) Analgesic and anti-inflammatory activity of *Teucrium chamaedrys* leaves aqueous extract in male rats. Iran J Basic Med Sci 13:119–125

Prabhakar PK, Doble M (2008) A target based therapeutic approach towards diabetes mellitus using medicinal plants. Curr Diabetes Rev 4:291–308

Puntero BF, Peinado II, del Fresno AMV (1997) Anti-inflammatory and antiulcer activity of *Teucrium buxifolium*. J Ethnopharmacol 55:93–98

Rabiei Z, Rafieian-Kopaei M, Mokhtari S, Shahrani M (2014) Effect of dietary ethanolic extract of *Lavandula officinalis* on serum lipids profile in rats. Iran J Pharm Res 13:1295–1301

Radhakrishnan R, Zakaria MNM, Islam MW, Kamil M, Ismail A, Chan K, Al-Attas A (2001) Analgesic and anti-inflammatory activities of *Teucrium stocksianum*. Pharm Biol 39:455–459

Rahmouni F, Hamdaoui L, Rebai T (2017) In vivo anti-inflamatory activity of aqueus extract of *Teucrium polium* against carrageenan-induced inflammation in experimental models. Arch Physiol Biochem 123:313–321

Rao NK, Nammi S (2006) Antidiabetic and renoprotective effects of the chloroform extract of *Terminalia chebula* Retz. seeds in streptozotocin-induced diabetic rats. BMC Complement Altern Med 6(1):17. https://doi.org/10.1186/1472-6882-6-17

Rasekh HR, Khoshnood-Mansourkhani MJ, Kamalinejad M (2001) Hypolipidemic effects of *Teucrium polium* in rats. Fitoterapia 72:937–939

Rewers M (2012) The fallacy of reduction. Pediatr Diabetes 13:340–343

Rodriguez Silva D, Baroni S, Svidzinski AE, Bersani-Amado CA, Cortez DA (2008) Anti-inflammatory activity of the extract, fractions and amides from the leaves of none *Piper ovatum* Vahl (Piperaceae) none. J Ethnopharmacol 116:569–573

Roseiro LB, Rauter AP, Mourato Serralheiro ML (2012) Polyphenols as acetylcholinesterase inhibitors: structural specificity and impact on human disease. Nutr Aging 1:99–111

Sabet Z, Roghani M, Najafi M, Maghsoudi Z (2013) Antidiabetic effect of *Teucrium polium* aqueous extract in multiple low-dose streptozotocin-induced model of type 1 diabetes in rat. J Basic Clin Pathophysiol 1:34–38

Sadeghzadeh F, Eidi A, Parivar K, Mazooji A (2008) Hypoglycemic effect of alcoholic extract of *Salvia nemorosa* in normal and diabetic male rats. Res Med 32:233–238

Shah BN, Seth AK (2010) Anti-inflammatory activity of fruits of *Abelmoschus esculentus* Linn. Pharmacology online 1:208–212

Shah SMM, Shah SMH (2015) Phytochemicals, antioxidant, antinociceptive and anti-inflammatory potential of the aqueous extract of *Teucrium stocksianum* bioss. BMC Complement Altern Med. 15:351. https://doi.org/10.1186/s12906-015-0872-4

Shah BN, Seth AK, Maheshwari KM (2011) A review on medicinal plants as a source of anti-inflammatory agents. Res J Med Plants 5:101–115

Shakhanbeh J (2001) *Teucrium polium* inhibits nerve conduction and carrageenan induced inflammation in the rat skin. Turk J Med Sci 31:15–21

Sims-Robinson C, Kim B, Rosko A, Feldman EL (2010) How does diabetes accelerate Alzheimer disease pathology? Nat Rev Neurol 6:551–559

Singh S, Majumdar DK, Rehan HMS (1996) Evaluation of anti-inflammatory potential of fixed oil of *Ocimum sanctum* (Holy basil) and its possible mechanism of action. J Ethnopharmacol 54:19–26

Sisodia SS, Tanzi RE (eds) (2006) Alzheimer disease advances in genetics, molecular and cellular biology. Springer, Basel

Soehnlein O, Lindbon L (2010) Phagocyte partnership during the onset and resolution of inflammation. Nat Rev Immunol 10:427–439

Soreq H, Seidman S (2001) Acetylcholinesterase-new roles for an old actor. Nat Rev Neurosci 2:294–302

Sreejith G, Latha PG, Shine VJ, Anuja GI, Suja SR, Sini S, Shyama S, Pradeep S, Shikha P, Rajasekharan S (2010) Anti-allergic, antiinflammatory and anti-lipidperoxidant effects of *Cassia occidentalis* Linn. Indian J Exp Biol 48:494–498

Stanković M (2012) Biological effects of secondary metabolites of *Teucrium* species of Serbian flora. Dissertation, University of Kragujevac

Takeda M, Tanaka T, Cacabelos R (2004) Molecular neurobiology of Alzheimer disorder and related disorders. Karger AG, Basel

Tariq M, Ageel AM, Al-Yahya MA, Mossa JS, Al-Said MS (1989) Anti-inflammatory activity of *Teucrium polium*. Int J Tissue React 11:185–188

Topcu G, Kusman T (2014) Lamiaceae family plants as a potential anticholinesterase source in the treatment of Alzheimer's disease. Bezmialem Sci 2:1–25

Uritu CM, Mihai CT, Stanciu G-D, Dodi G, Alexa-Stratulat T, Luca A, Leon-Constantin M-M, Stefanescu R, Bild V, Melnic S, Tamba BI (2018) Medicinal plants of the family Lamiaceae in pain therapy: a review. Pain Res Manag 8:7801543. https://doi.org/10.1155/2018/7801543

Vahidi AR, Dashti-Rahmatabadi MH, Bagheri SM (2010) The effect of *Teucrium polium* boiled extract in diabetic rats. Iran J Diabetes Obes 2:27–32

Verma S (2016) Medicinal plants with anti-inflammatory activity. J Phytopharmacol 5:157–159

Vigo E, Cepeda A, Perez-Fernandez R, Gualillo O (2004) In-vitro anti-inflammatory effect of *Eucalyptus globulus* and *Thymus vulgaris*: nitric oxide inhibition in J774A.1 murine macrophages. J Pharm Pharmacol 56:257–263

Vishal V, Ganesh SN, Mukesh G, Ranjan B (2014) A review on some plants having anti-inflammatory activity. J Phytopharmacol 3:214–221

Vladimir-Knežević S, Blažeković B, Kindl M, Vladić J, Lower-Nedza AD, Brantner AH (2014) Acetylcholinesterase inhibitory, antioxidant and phytochemical properties of selected medicinal plants of the *Lamiaceae* family. Molecules 19:767–782

Yoo KY, Park SY (2012) Terpenoids as potential anti-Alzheimer's disease therapeutics. Molecules 17:3524–3538

Zhao Y, Dou J, Wu T, Aisa HA (2013) Investigating the antioxidant and acetylcholinesterase inhibition activities of *Gossypium herbaceam*. Molecules 18:951–962

Zimmet P, Alberti KGMM, Shaw J (2001) Global and societal implication of the diabetes epidemic. Nature 414:782–787

Chapter 15
Application of *Teucrium* Species: Current Challenges and Further Perspectives

Dragana Jakovljević and Milan Stanković

Abstract Increased demand for products originating from natural sources lead to novel agricultural and biotechnological techniques and create the urgent need for implementations of these techniques to appropriate response for enlarged interests. The plants from the *Teucrium* L. genus are well-known sources of natural products with wide areas of application, primarily due to the high content of secondary metabolites with significant biological activities. Utilization of *Teucrium* species in diverse areas of the industry makes them suitable candidates for various biotechnological investigations as well as further applications of obtained results. The primary objective of this chapter is to summarize the most diverse areas of application from species of the *Teucrium* genus, regarding various biotechnological procedures that maximize the exploitation but also provide sustainable conservation and rational utilization of biodiversity both under in vivo as well as under in vitro conditions. With the high content of valuable compounds and a multidisciplinary approach, *Teucrium* species are the primary target for trait manipulation and offers major opportunities for the major application of plant biotechnology.

Keywords *Teucrium* · Application · Biotechnology · Plant tissue culture · Useful compounds

D. Jakovljević (✉) · M. Stanković
Department of Biology and Ecology, Faculty of Science, University of Kragujevac, Kragujevac, Serbia
e-mail: dragana.jakovljevic@pmf.kg.ac.rs

M. Stanković (ed.), *Teucrium Species: Biology and Applications*,
https://doi.org/10.1007/978-3-030-52159-2_15

15.1 Introduction

Increase in world population and searching for novel products, together with the recognition that the human health is related to the nutritional quality of the food which is threatened under adverse climatic changes, create the urgent need for agriculture and plant biotechnologies implementations (Altman and Hasegawa 2012).

Among the members of the Lamiaceae family, genus *Teucrium* L. includes plant species which diverse areas of benefits for humans make them highly applicable traditionally as well as industrially. Traditionally, a large number of species belonging to the genus *Teucrium* have a long history of use in folk medicine and pharmacy. They are used in treatments of various disorders, mainly due to the high content of metabolites with significant biological activities. In addition, numerous studies suggest that extracts from *Teucrium* species and isolated bioactive compounds possess antioxidative, antibacterial, antidiabetic, anticancer and anti-inflammatory activities (Stankovic et al. 2011). From the industrial point of view, morphological and ecophysiological variability of the *Teucrium* species, and relatively high number of species included, emphasis the importance of plants from the genus *Teucrium* as biotechnological factories for new plant-based biomaterials.

Consumption of herbals from their natural sources is widespread and increasing, however, harvesting from the wild may lead to the loss of diversity and habitat destruction. Therefore, the use of controlled environments may successfully overcome these difficulties and could provide manipulation of phenotypic variation in bioactive compounds (Canter et al. 2005). The modern era of plant biotechnology coordinate with the ability to grow plant cells and tissues in vitro, to regenerate and clone new plants as well as to modify their genetic characteristics by traditional and molecular breeding, including genetic modification, and genome editing (Altman 2019). Having in mind diverse areas of applications of plants from the genus *Teucrium*, the main goal of this chapter is to provide information's about diverse areas of application of *Teucrium* species together with further implications of various biotechnological procedures that maximize the exploitation but also provide sustainable conservation and rational utilization of biodiversity both under in vivo as well as under in vitro conditions.

15.2 Diverse Areas of *Teucrium* Species Application

The species of the genus *Teucrium* have been popular since the very beginning of the use of herbs in the treatment of diseases. Data on healing properties can be found in Greek mythology and nowadays the species from this genus are used in the treatments of tuberculosis, scurvy, jaundice, and rheumatism. The high content of aromatic compounds with antimicrobial activity contributes to the treatment of many infectious diseases. Besides, the above-ground parts are very bitter, and due to the content of bitter aromatic compounds, plants from the genus *Teucrium* are

used as an additive in cooking as well in the preparation of beverages (Keršek 2006). Among species of the genus *Teucrium* that are presented in the Flora of the Republic of Serbia, *Teucrium chamaedrys*, *T. montanum*, *T. polium*, and *T. scordium* are most often used for the preparation of teas, as well as for other medicinal preparations. In most cases, the green above-ground parts of plants with flowers are taken, which are then dried in the shade (Stanković 2012).

Teucrium chamaedrys is overall one of the most widely used species of the *Teucrium* genus. The benefits came from teas that have a very bitter taste, and for this purpose, the tops of twigs with the flowers are used *(Teucrii herba et folium)*. This plant parts are most often used for the treatment of digestive organs. Secondary metabolites that give this plant a bitter taste stimulative affect the digestive organs and intensify secretions, alleviate the effects of gastritis and increase the secretion of gastric acid. It is also used in the treatment of various infections and immunodeficiency (Pieroni et al. 2004; Bagci et al. 2010). *Teucrium montanum* is very often used to alleviate gastric problems and to improve digestion. This plant species is also effective as an additional therapy for the treatment of infectious diseases of the respiratory tract and tuberculosis. It can also be used as a medicinal bath (Stanković 2012). *Herba teucrii montani* can be used as a tincture but also in the form of wine and brandy (Keršek 2006). Teas, tinctures, and oils prepared from above-ground parts of *Teucrium polium* are used as an auxiliary in the treatment of hypertension, diabetes, and diseases of the digestive organs as well as for the control of hypercholesterolemia (Afifi et al. 2009; Vahidi et al. 2010). *Teucrium scordium* is used to stimulate the function of organs of the digestive system, in the treatments of infected wounds, but also the treatment of plague, cough, and fever (Redžić 2007).

In spite of all the confirmed advantages obtained from these species, it is necessary to keep in mind that after long-term use of teas and other medicinal preparations originated from *Teucrium polium* and *Teucrium chamaedrys* symptoms of mild hepatotoxicity may be established. After a large number of studies conducted, it has been concluded that secondary metabolites from the group of neoclerodane diterpenes, in particular, teucrin A, cause the symptoms of damage and liver inflammation (Starakis et al. 2006; Savvidou et al. 2007). However, after the dose and time-dependent testing, it has been established that hepatotoxic effect occurs only after higher doses in the body, and for a long period, while acute treatment has no negative effect on the organism (Khleifat et al. 2002). Therefore, careful and moderate use is recommended without any further health consequences. The structure of teucrin A (secondary metabolite from the group of neoclerodane diterpenes) is presented in Fig. 15.1.

Fig. 15.1 Chemical structure of teucrin A

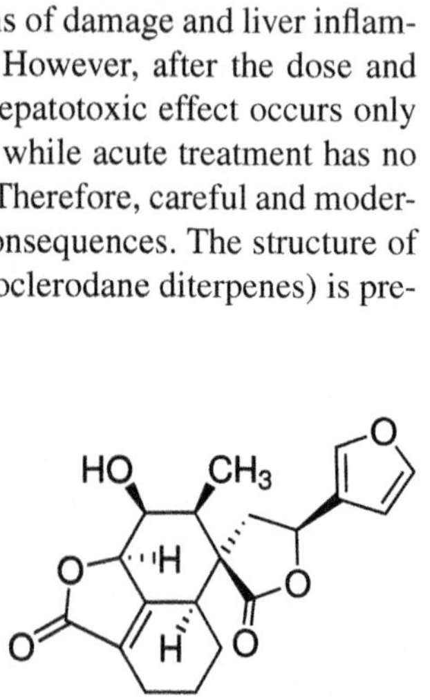

The fact that many *Teucrium* species show antibacterial, antifungal and antioxidant activities regarding them useful as natural preservative ingredients with significant importance in food industries (Bagci et al. 2010). The foodborne diseases together with food spoilage caused by a variety of microorganisms at various stages (from production to sale and distribution) are among the most important concerns for the food industry. The increase of bacterial resistance to antibiotics leads to considerable interests in investigations of the antimicrobial effects of different extracts against a wide range of bacteria to develop other classes of antimicrobials useful for the food preservation (Belmekki et al. 2013). Özkan et al. (2007) demonstrated that extracts obtained from *Teucrium montbretii* preventing the autooxidation in oils and oil-bearing foods. According to these authors, the growth of food-borne pathogens or spoilage organisms can be inhibited when high concentrations of herb extracts were applied in food products.

According to Djabou et al. (2013), the foodborne pathogen *Campylobacter jejuni* was extremely sensitive to oils of *Teucrium* species. Authors found that oils from *Teucrium marum, T. massiliense, T. chamaedrys, T. scorodonia* subsp. *scorodonia, T. polium* subsp. *capitatum, T. flavum* subsp. *glaucum* and *T. flavum* subsp. *flavum* has the potential to be used as food preservatives and to prevent the growth of nosocomial bacteria. On the other hand, López et al. (2007) showed that aerial parts of *Teucrium chamaedrys*, in the view of their antioxidant and antifungal properties, possess potential to be used in order to obtain new active products with applications in the food industry.

Sadrizadeh et al. (2018) demonstrated that the addition of *Teucrium polium* essential oils to milk products significantly increased survival of *Lactobacillus casei* after 20-day storage at 4 °C. Besides, *Teucrium polium* essential oils and probiotic fermentation reduced the population of *Escherichia coli* during the storage time and the effects were stronger in combination than using them separately. These authors recommended combining effect of *Teucrium polium* extract and probiotic due to their complementary antimicrobial effects and practically no side effects. Also, because of good appearance without any signs of spoilage organisms, it is recommended to use this method to reduce the *Escherichia coli* growth and population in the food system.

15.3 Biotechnology of *Teucrium* Species

The area under medicinal and aromatic plants cultivation is increasing worldwide due to the use of natural therapies by a growing segment of the population, and at the same time, it is necessary to optimize the technologies to maximize production and to cover the market demand (Burducea et al. 2019). Environmental stresses (drought, salinity, high temperature, nutrients) usually decrease the yield of medicinal plants and increase the content of secondary metabolites (Hosseini et al. 2018; Jakovljević et al. 2019), and therefore, the studies focusing on agronomic conservation and management of plant species become necessary, particularly due to scarce

information in literature about the cultivation of important plants (de Mesquita Arruda et al. 2018).

15.3.1 Plant Cell and Tissue Culture

Plant cell and tissue culture are among the most prominent and viable plant biotechnology techniques. Although is developed primarily for basic research on cellular and tissue differentiation and morphogenesis, tissue culture research in last decades and manipulation of cells and organs in aseptic conditions together with further growth in a medium under defined light intensity, humidity, and temperature allowed the increasing of uniformity and maintaining the same genetic characteristics of plants. In addition, the production of bioactive compounds under these conditions make additional efforts for sustainable conservation and rational utilization of biodiversity (Karuppusamy 2009; Altman and Hasegawa 2012; Dias et al. 2016). Based on the principle of cellular totipotency, plant cells and tissues may be cultured in terms of undifferentiated calli and cell suspensions, as well as organ culture (shoots and roots), and for producers, plant cell, tissue and organ culture are of great interest as an alternative for obtaining chemicals from plant species by reducing the time interval to harvest (Rodríguez-Sahagún et al. 2012).

15.3.1.1 In Vitro Seed Germination

The combination between biochemistry and biotechnology lead to significant improvement on production yield, however, initiation of in vitro culture and cultivation of some herbs has proven difficult due to low germination rates or lack of knowledge about the specific ecological requirements (Canter et al. 2005). It is demonstrated that seeds from *Teucrium* species have low germination capacities (Nadjafi et al. 2006; Luna et al. 2007; Moreira et al. 2010) and that germination characteristics of seeds in species which show lower germination using conventional techniques could be increased using in vitro methods (Pickens et al. 2003; Jakovljević et al. 2017). The effect of various pretreatments, culture conditions and storage time on in vitro germination of seeds from *T. capitatum* were examined by Papafotiou and Martini (2016). Pretreatments, such as cold stratification, scarification with sandpaper, dipping in concentrated sulfuric acid, or dipping in boiling water were tested on in vitro germinated seeds. It was demonstrated that seeds without any pretreatment germinated at lower than 10% whereas dipping in concentrated H_2SO_4 for 15 or 20 min was the most effective pretreatment. In addition to the better germination capacities, explants excised from in vitro-grown seedlings were established at higher rates compared with those collected from plants grown in a greenhouse.

15.3.1.2 Overcoming Habitat Encroachment

In addition to the low rate of seed germination which restrict massive propagation and further growth of *Teucrium* species, due to the all potential valuable properties, these plants are under various stresses (physical as well as anthropogenic) like unappropriated harvesting, overgrazing, and unsustainable development (Ahmad et al. 2002; Bouhouche and Ksiksi 2007). Because of irregular grazing and immoderate picking up for folk medicine usage, biodiversity of *Teucrium polium* decreased significantly in rapid eradication and extinction (Rad et al. 2014). Due to the selective collection for use in traditional medicine, and deforestation and habitat encroachment, wild populations of *Teucrium capitatum* are subjected to pressure (Papafotiou and Martini 2016). Recent investigations of *Teucrium stocksianum* area of cultivation revealed the need for urgent interventions due to the disappearing of this species at an alarming rate (Shinwari and Gilani 2003). Al-Qudah et al. (2011) pointed out that the biodiversity of *Teucrium polium* declined dramatically in recent years which could be attributed to habitat encroachment by urban and agricultural development, deterioration and deforestation, depletion of major water sources, as well as due to the illegal collection. Having all that in mind, the development of an in vitro protocols is of critical importance since it will provide plants that can be used for reintroduction in their natural habitats, to conserve it from being an endangered, and to maintain the germplasm (Bouhouche and Ksiksi 2007; Al-Qudah et al. 2011; Rad et al. 2014).

15.3.1.3 Micropropagation

As a large-scale clonal propagation, micropropagation is the most important (practical and economical) application of plant biotechnology (Da Silva et al. 2015) and it has been useful particularly in the case of valuable, rare or endangered plant species. This in vitro method is also an appropriate tool for multiplication of disease-free plants, faster cloning and the conservation of desired genotypes, in a very short time (Máthé et al. 2015). Plants can be regenerated either by the growth and proliferation of existing axillary and apical meristems or by the regeneration of adventitious shoots whereas adventitious buds and shoots are formed de novo and meristems are initiated from explants – leaves, petioles, hypocotyls, floral organs and roots (Iliev et al. 2010). Micropropagation was established for the *Teucrium fruticans* (Frabetti et al. 2009), *Teucrium stocksianum* (Bouhouche and Ksiksi 2007), *Teucrium capitatum* (Papafotiou and Martini 2016), *Teucrium polium* (Al-Qudah et al. 2011; Rad et al. 2014), *Teucrium scorodonia* (Makowczyńska et al. 2016). Habitus of *Teucrium fruticans* together with the complete micropropagation cycle of *Teucrium fruticans* (according to Frabetti et al. 2009) is presented in Figs. 15.2 and 15.3. Besides the fact that controlled growth system contributes to the reduction of the pressure on the natural populations, it also provides manipulation of phenotypic variation in the concentration of medicinally important compounds through the increase of uniformity and predictability of extracts (Canter et al. 2005).

Fig. 15.2 *Teucrium fruticans* L. during the flowering stage (Photo M. Stanković)

15.3.1.4 Improvement of Production Yield

The plants from the genus *Teucrium* is well-known for the production of a significant amount of biologically active secondary metabolites, however, the quantitative and qualitative characteristics of the produced metabolites are conditioned by the plant organs, as well as the physicochemical and morphological characteristics of the substrate. Additionally, the quantitative and qualitative composition of secondary metabolites of *Teucrium* species is characterized by the seasonal variations (Stanković et al. 2012a, b, 2014). Interpopulation variability in the total quantity of phenolic compounds and antioxidant activity was determined for *Teucrium camaedrys* and *T. montanum* in individuals from the different serpentine and calcareous habitats (Zlatić et al. 2017). For *Teucrium polium* essential oils it is established variability conditioned by the different latitudinal populations (Sadeghi et al. 2014). In addition to the variability in content and activity of biologically valuable secondary metabolites due to environmental conditions, there are several reports of hepatotoxicity caused by extracts obtained from *Teucrium* species (Larrey et al. 1992; Kouzi et al. 1994). According to Chitturi and Farrell (2008), *Teucrium*-associated hepatotoxicity was recognized with *Teucrium chamaedrys* (germander) and is attributed to reactive metabolites (epoxides) generated by CYP3A metabolism of its constituent neoclerodane diterpenoids (teucrin A) since the specific auto-antibody was identified from long-term drinkers of germander tea. In addition, *Teucrium capitatum* and *T. polium* was linked with the acute liver failure and acute hepatitis due to necrosis

Fig. 15.3 Complete micropropagation cycle of *Teucrium fruticans* (according to Frabetti et al. 2009): (**a**) shoot initiation from nodal segments; (**b**) and (**c**) multiple shoots regenerating on medium; (**d**) root development after 14 days of incubation; (**e**) in vitro-rooted shoot ready for transfer into soil after 24 days of incubation; (**f**) root development after the acclimatization stage; (**g**) plant acclimatized to greenhouse conditions

or occasionally massive hepatocyte necrosis (Dourakis et al. 2002; Savvidou et al. 2007).

It is well established that biotechnological techniques can be applied to improve yield and uniformity, to modify potency or toxicity, as well as to increase the production of active phytochemical compounds (Canter et al. 2005). Additionally, the advanced biotechnological tools resulted in plants with improved agricultural benefits and expanded the use of plants to produce a variety of novel biomaterials (Altman and Hasegawa 2012). In vitro developed protocols for *Teucrium* species are advantageous in terms of the speed of plant regeneration since regenerated plants were obtained within 3 months after the initiation of the culture protocol; also, it is possible to use these plants as a source of tissues for the biochemical

characterization of biologically active compounds (Bouhouche and Ksiksi 2007). According to Makowczyńska et al. (2016), it is possible to obtain approximately 300 and 400 in vitro derived *Teucrium scorodonia* plants from a single shoot or nodal explants. Additionally, these plants can be suitable for sustainable production of *β*-caryophyllene and germacrene D. Rad et al. (2014) demonstrated that amount of total flavonoids was different in tissue culture raised *Teucrium polium* plants comparing to the field-grown plants. These authors proved that the extracts of tissue culture plants contained the highest amount of flavonoid content and the greatest antioxidant activity compared to the field-grown wild plant.

Several limitation factors of plant cell biotechnology that hampers its extensive commercialization can be overcome through the elicitation which implies exogenous addition of elicitors (chemical compounds from abiotic and biotic sources) in the growth medium. This procedure stimulates stress responses in plants and leading to the enhanced synthesis and accumulation of secondary metabolites or the induction of novel secondary metabolites since induced stress activates several defense-related genes or inactivates non-defense-related genes (Naik and Al–Khayri 2016; Narayani and Srivastava 2017).

Antognoni et al. (2012) established in vitro culture of *Teucrium chamaedrys* with the possibilities to elicit the production of phenylethanoid glycosides, particularly teucrioside in the levels higher than those found in the leaves of intact plants. The absence of neoclerodane diterpenoids in in vitro culture of *Teucrium chamaedrys* makes these results particularly significant having in mind that these diterpenes may induce symptoms of hepatotoxicity.

The high content of phenylpropanoid compounds (particularly phenolics) together with significant biological and bioactive characteristics of these metabolites in *Teucrium* species are well documented (Stanković 2012; Stanković et al. 2012a, b, 2014) and is usually part of the complex defense mechanism in an adverse environment. The phenylpropanoid compounds are mainly synthetized through the shikimate pathway and occurs via a number of enzymatic reactions (Jakovljević et al. 2017). In general, enhanced biosynthesis of phenylpropanoid compounds through the phenylpropanoid biosynthetic pathway is a characteristic of adaptation to the stressful environmental conditions which is resulting in accumulation of phenolic compounds (Jakovljević et al. 2017, 2019). Different elicitors (biotic and abiotic) is used in order to increase the production of phenolic compounds. In Fig. 15.4 (according to Dias et al. 2016) are presented some of the elicitors used to increase production of phenolic compounds and pointed out parts of the biosynthetic pathways were elicitors may enhance the synthesis of these valuable metabolites.

There is considerable interest in manipulating plant biosynthetic pathways to produce secondary metabolites, but there are significant interests in the production of drug precursors, food components or pesticides (Canter et al. 2005). Generally, exogenous supply of different elicitors to different culture systems of plants create stress like cellular environment and, therefore, the elicitors not only bring yield enhancement of existing biomolecules but also found to alter/synthesize secondary metabolites qualitatively through the discovery of entirely novel compounds (Giri and Zaheer 2016). Further, extensive appraisal of elicitor use under controlled

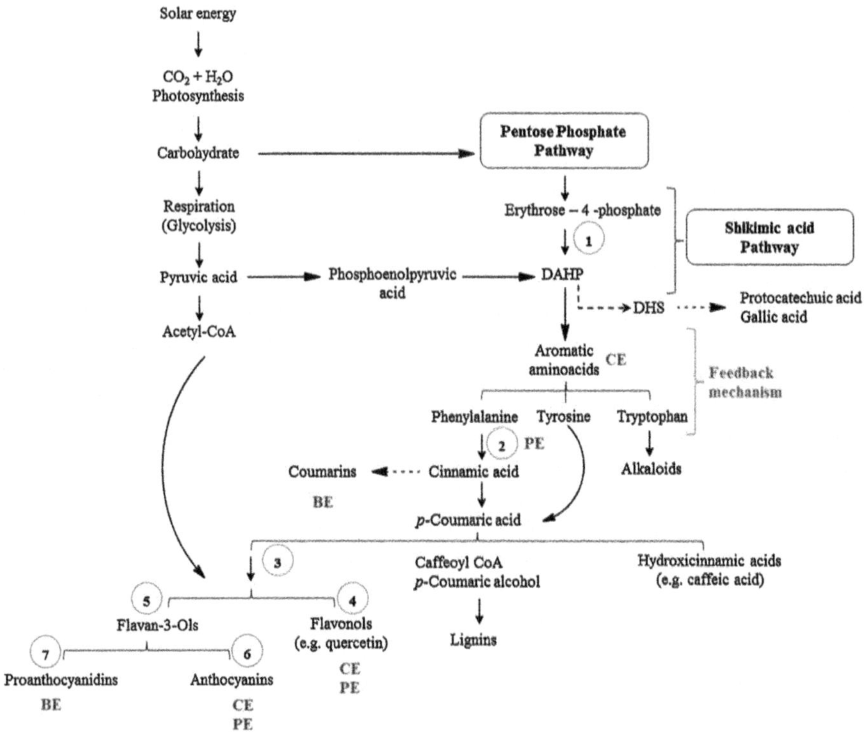

Fig. 15.4 Biosynthetic pathway of some phenolic compounds and the influence of elicitation (according to Dias et al. 2016). CO_2 carbon dioxide, H_2O water, *Acetil-CoA* Acetil-Coenzyme A, *DAHP* 3-Deoxy-O-arabino-heptulosonate phosphate, *DHS* 3-Dehydroquinate, *BE* biological elicitation, *CE* chemical elicitation, *PE* physical elicitation; Enzymes involved in the biosynthesis are marked with rounded dashed black forms: 1-DAHP synthase (3-Deoxy-O-arabino-heptulosonate phosphate synthase); 2-PAL (Phenylalanine ammonia-lyase); 3-CHS (Chalcones synthase), CHI (Chalcones isomerase), F3H (Flavanone-3-hydroxylase); 4-FLS (Flavonol synthase); 5-LAR (Leucoanthocyanidin reductase); 6-LDOX (Leucoanthocyanidin dioxygenase)

conditions will be useful for the understanding of *Teucrium* secondary metabolites synthesis, as well as for the production and in vitro enhanced production of biotechnologically applicable metabolites.

15.3.2 Hydroponic Culture

Among the soilless cultivation technologies with natural or artificial media used and with possibilities for easy environmental control, the hydroponic systems offer several additional benefits such as the ability to reuse water and nutrients, prevention of soil-borne diseases and pests, as well as higher yields and higher quality products. Therefore, demand for the hydroponic system has increased dramatically (Lommen

2007; Brentlinger 2007; Lee and Lee 2015). Hydroponic *Teucrium polium* contains flavonoids, iridoids, as well as terpenoids and phenylpropanoid glycosides with phytoestrogen-like activity (Galstyan et al. 2010). In addition to phytoestrogen-like activity, extracts from hydroponic *Teucrium polium* demonstrated neuroprotective activity following bilateral ovariectomy (Simonyan and Chavushyan 2015). Cultivation of *Teucrium polium* in open hydroponics provides the ecologically clean yield and enrichment in the content of apigenin, luteolin, verbascoside, poliumoside and teupolioside (Oganesyan et al. 1991). Simonyan and Chavushyan (2016) investigated the effects of hydroponic *Teucrium polium* on hippocampal neuronal activity and morpho-histochemistry of bilateral ovariectomized (OVX) rats. Authors concluded that *Teucrium polium* reduced OVX-induce neurodegenerative alterations in entorhinal cortex-hippocamp circuitry and facilitated neuronal survival by modulating the activity of neurotransmitters and network plasticity.

15.4 Application of *Teucrium* Useful Compounds

Although the natural products originating from plants (in particular phenylpropanoid-derived products) has a broad spectrum of applications relevant to human health, key challenges still remain to be met in the application of biotechnology for improvement in phenylpropanoid and flavonoid natural product profiles of plants and in exploitation of their full biotechnological potential (Ververidis et al. 2007). The wealth of plant-derived metabolites further emphasizes the importance of plants as biotechnological factories for new biomaterials (Altman and Hasegawa 2012).

For extracts obtained from *Teucrium* species, significant antioxidant activity has been identified, and the results suggest that there is a correlation between the intensity of the antioxidant activity and the amount of phenolic compounds. The ability of *Teucrium* spp. extracts to neutralize harmful free radicals, determined through the total antioxidant activity of secondary metabolites, was established for ethanol, aqueous, chloroform, acetone and methanol extracts, both in vitro and in vivo conditions (Gülçin et al. 2003; Kadifkova-Panovska et al. 2005; Katalinic et al. 2005; Ljubuncic et al. 2006; Hasani et al. 2007). Potential allelopathic effects of neoclerodane diterpenes from *Teucrium chamaedrys* was estimated on germination and seedling growth of *Dactylis hispanica* Roth, *Petrorhagia velutina* (Guss.) Ball et Heyw., and *Phleum subulatum* (Savi) Asch et Gr. (Fiorentino et al. 2009).

The insecticidal activity of plants from the *Teucrium* genus was confirmed, both under in vivo and in vitro conditions (Masanori et al. 2000). The insecticidal activity of essential oils from *Teucrium polium* was confirmed by treating the *Musca domestica* L. (housefly), wherein the secondary metabolites of this plant species inhibit the activity of the digestive enzymes of the house fly larvae (Bigham et al. 2010). By treating potatoes (*Solanum tuberosum* L.) with secondary metabolites *Teucrium* spp., it has been found that ericcephalin, teucrin A, teucorolide and tequin affect the increase in potato plant resistance through the effects on Colorado Potato beetle (*Leptinotarsa decemlineata* Say) larvae (Ortego et al. 1995). In addition to insecticidal, *Teucrium* spp. secondary metabolites also possess antimicrobial activity (Gülçin et al. 2003; Šamec et al. 2010).

Based on the in vivo testing of essential oils obtained from *Teucrium polium*, *T. montanum*, *T. chamaedrys*, *T. flavum* and *T. arduini* it was established that β-caryophyllene and germacrene possess significant antiphytoviral activity. Results demonstrated that essential oils from *Teucrium* plants incibit tissue damages on *Chenopodium quinoa* Willd. infected with cucumber mosaic virus (CMV) (Dunkić et al. 2011; Bezić et al. 2011).

Based on the study of the anticancer effects of *Teucrium polium*, it has been found that metabolites from the extracts of this plant can be used in the treatment of prostate cancer because they have the ability to inhibit signal pathways leading to cancer metastasis (Kandouza et al. 2010). Treating the cellular culture of multiform glioblastoma, alveolar adenocarcinoma, breast cancer and adrenal gland with the extracts of *Teucrium polium*, it has been found that secondary metabolites of this plant exhibit significant cytotoxic activity under in vitro conditions (Eskandary et al. 2007; Nematollahi-Mahani et al. 2007). By examining the effects of an aqueous *Teucrium polium* extract on the NSCLS cell line of lung cancer, it has been found that the extract exhibits strong antiproliferative activity by inhibiting the cell cycle by causing cell death (Haïdara et al. 2011).

The methanolic extract of *Teucrium polium* enhances the effect of vincristine, vinblastine, and doxorubicin used in cancer therapy and has a significant antibacterial activity (Zerroug et al. 2011). During in vivo testing of the aqueous extract from the above-ground parts of the *Teucrium polium*, strong antioxidant activity in the pancreatic tissue was associated with a hypoglycaemic effects in artificially induced diabetes in rats (Gharaibeh et al. 1988; Yazdanparast et al. 2005; Shahraki et al. 2006; Ardestani et al. 2008; Esmaeili et al. 2009; Mirghazanfari et al. 2010). By examining the effects of the aqueous and ethyl acetate extract of *Teucrium polium* in in vivo conditions on some diagnostic parameters of the liver, significant hepatoprotective activity was demonstrated (Kadifkova-Panovska et al. 2005; Shtukmaster et al. 2010). In addition, through the in vivo testing of *Teucrium polium* on artificially induced stomach ulcer, the significant antiulcer activity was established (Mehrabani et al. 2009). Based on the in vivo testing of extracts and essential oil from *Teucrium polium*, it was demonstrated that secondary metabolites from this plant species have an antinociceptive effect (Abdollahi et al. 2003). Antiinflammatory and analgesic activity was determined during in vivo testing of water extract of *Teucrium chamaedrys* and *Teucrium polium* (Tariq et al. 1989; Pourmotabbed et al. 2010).

The testing with crude plant extracts and isolation and characterization of the constituents responsible for the activity of extracts is a very common starting approach in biotechnology of plant-derived bioactive compounds. In addition, preparation of plant extracts is relatively simple, low in cost and time investment. However, it is crucial to carefully document identity and processing of the plant material, and the extraction method, since the extraction method can strongly influence the biological activity as well as the composition of the extracts (Koehn and Carter 2005; Stankovic et al. 2014; Atanasov et al. 2015). These issues need to be included in order to maximize the exploitation potential of plants from the genus *Teucrium*.

15.5 Phytoremediation

Several anthropogenic activities, including mining, smelting, the burning of fossil fuels, as well as the manufacture of fertilizers and pesticides, may lead to heavy metal contamination. The heavy metal contamination can be toxic at very low concentration to both plants and animals, and due to its potentially harmful effects, there is a high level of interests in methods which aim to eliminate the heavy metals at minimal coast. Additionally, abandoned mining areas are a significant problem due to the high and toxic concentration of heavy metals and the absence of vegetation over most of the surface. Consequentially, the application of disruptive technologies is inappropriate due to the high cost and potential impacts on the environment, the modification of landscape and soil properties. Generally, the removal of heavy metals can be performed using a variety of high cost physical, thermal, and chemical treatments (which may cause secondary pollutions) or through the biological means (Cao et al. 2009; Rascioa and Izzo 2011; Ozkan et al. 2014; Yaman 2014).

The use of plant species to clean up soil and water is called phytoremediation. The phytoremediation is emerging technology and has gained importance since the use of plant species which are hyperaccumulators is a cost-effective, promising and environmentally friendly technology to remove, extract, or inactivate metal ions in the soil using plants (Fairbrother et al. 2007; Ozkan et al. 2014). Although widespread phytoremediation application is inhibited due to the long remediation time and the production of large amounts of metal-contaminated biomass (Stals et al. 2010; Ozkan et al. 2014), phytoremediation can be considered a good solution since the application of phytostabilization (a phytotechnology that allows stabilization of contaminants in soil) and phytoextraction technique and its optimization in order to lower reclamation times can provide effective protection (Cao et al. 2009).

The bioavailability of elements in plants depends on the characteristics of the soil, climatic conditions, and the genotype and agronomic behavior of the plant (Chojnacka et al. 2005; Martínez-López et al. 2014). Further remediation success of hyperaccumulator plants depends on several factors including the transfer of the metal from the root to the above-ground parts, the growth of the plant, as well as large biomass and the high uptake capacity of the plants (Shah and Nongkynrih 2007; Yaman 2014).

According to Martínez-López et al. (2014), species *Teucrium dunense* grow in sandy soils and loose soils in wadis and dunes close to the sea under intermediate content of arsenic. The soils with intermediate content of arsenic are formed of materials from the beds of ephemeral streams, as a result of hydric erosion of calcareous materials, mica schist and mining materials. However, when there is evidence of a strong mining influence, *Teucrium dunense* also grow in the soils that contain high concentrations of arsenic.

In addition to the soils with moderate or high arsenic content, for *Teucrium* spp. there are possibilities to be used in soils containing nickel, cadmium, lead, and zinc. According to Safari Sinegani and Dastjerdi (2008) *Teucrium polium*, grown at the metalliferous site of Iran, tolerated and absorbed moderate levels of cadmium and lead. On the other hand, Yaman (2014) analyzed the roots and above ground parts of the *Teucrium polium* plants grown in serpentine and non-serpentine soils in terms of nickel (Ni) and cobalt (Co) hyperaccumulation. The author demonstrated that the ratios of Ni/Co concentrations in the roots and above-ground parts of *Teucrium polium* grown in the serpentine soil were significantly higher than the ratios for *Teucrium polium* grown in the non-serpentine soils. Also, according to obtained results, species *Teucrium polium* has been suggested for phytoextraction of Ni in contaminated soils.

In order to investigate the possibility of using native Mediterranean plants species for the application of phytostabilization and assisted phytoextraction in the remediation of abandoned mining sites, *Teucrium flavum* subsp. *glaucum* was tested by Cao et al. (2009). This species is chamaephyte that grows on carbonatic substrata and can colonize the mining waste when pedogenetic process has started. Authors showed that *Teucrium flavum* subsp. *glaucum* possess a high tolerance for lead and zinc. However, besides the great ability to accumulate lead in the aerial parts, this plant species possess low survival and biomass production, and therefore, the further application of assisted phytoextraction needs to be optimized. Having in mind that the phytoremediation can be regarded as a promising low-cost tool for the remediation of the heavy metal-contaminated soils, further investigations of *Teucrium* species in terms of their growth, phytostabilization, and phytoextraction can find diverse practical applications.

15.6 Further Perspectives

As pointed out by Canter et al. (2005), on a large scale, wild harvesting and continual selection of the wild-growing individuals, or those with the desirable traits, may lead to degradation of the wild population, and therefore, appropriate use is not possible without cultivation and conservation of wild population. In this order, different strategies may be used, including the optimization of culture conditions, selection of high producing lines, explant modifications, as well as presence or absence of different environmental conditions. On the other hand, as is pointed out previously, the modern era of plant biotechnology coordinate with the ability to grow and to regenerate plant cells and tissues in vitro, to clone new plants or to modify plant genetic characteristics by genetic modification, and genome editing (Altman 2019).

With the high content of valuable compounds, *Teucrium* species are the primary target for trait manipulation and offers major opportunities for the major application of plant biotechnology. A multidisciplinary approach involving genetic and metabolic engineering will go a long way in enhancing the production of targeted metabolites for commercial applications and direct manipulation of DNA sequences to alter gene expression in medicinal plants is an area that is ripe for expansion (Canter et al. 2005; Bindu et al. 2018). In addition to the enhancement of concentration of the desired compounds, the introducing novel biosynthetic pathways to a variety of species allowing improvement of plant nutritional or commercial value, and therefore metabolic engineering can be used to manipulate flux through both primary and secondary metabolic pathways towards products of interest (Wilson and Roberts 2014).

15.7 Conclusions

Increased in world population and increasing demand for products originating from natural sources lead to the development of novel agricultural and biotechnological techniques and implementations of these techniques to appropriate responses. With the high diversity in morphological, anatomical, phytochemical, ecophysiological and genetic characteristics, as well as in quantity and quality of valuable compounds, the species from the genus *Teucrium* have become well-known for wide areas of application. The variations in form and habitual characteristics have enabled the domestication of different *Teucrium* species and the exploitation of their horticultural potential. The growth and development under unfavorable environmental conditions, particularly in sandy soils and in soils containing heavy metals, making *Teucrium* plants possible candidates to be used in phytoextraction and phytoremediation. The high content of phenolic compounds and essential oils with significant biological activities led to the use of species of this genus both traditionally and industrially. Isolated bioactive compounds possess antioxidative, antibacterial, antifungal, genotoxic, anticancer and numerous other biological activities. To reduce time interval to harvest or to reduce the pressure on the natural populations and to provide manipulation of variation in the concentration of important compounds, *Teucrium* species may be cultured in vitro as undifferentiated calli or cell suspensions, as well as the culture of shoots and roots. Finally, a multidisciplinary approach involving genetic and metabolic engineering can be used in the manipulation of products of interest.

Acknowledgments This work was supported by the Ministry of Education, Science and Technological Development of the Republic of Serbia (Grant III41010).

References

Abdollahi M, Karimpoura H, Monsef-Esfehani RH (2003) Antinocicenptive effects of *Teucrium polium* L. total extract and essential oil in mouse writing test. Pharmacol Res 48:31–35

Afifi FU, Abu-Irmaileh BE, Al-Noubani RA (2009) Comparative analysis of the essential oils of *Teucrium polium* L. grown in different arid and semi arid habitats in Jordan. Jord J Pharm Sci 2:42–52

Ahmad H, Ahmad A, Jan MM (2002) The medicinal plants of salt range. J Biol Sci 2:175–177

Al-Qudah TS, Shibli RA, Alali FQ (2011) In vitro propagation and secondary metabolites production in wild germander (*Teucrium polium* L.). In Vitro Cell Dev Biol Plant 47:496–505

Altman A (2019) Plant tissue culture and biotechnology: perspectives in the history and prospects of the International Association of Plant Biotechnology (IAPB). In Vitro Cell Dev Biol Plant 55:590–594. https://doi.org/10.1007/s11627-019-09982-6

Altman A, Hasegawa PM (2012) Introduction to plant biotechnology 2011: basic aspects and agricultural implications. In: Altman A, Hasegawa PM (eds) Plant biotechnology and agriculture: prospects for the 21st century. Academic, Waltham

Antognoni F, Iannello C, Mandrone M, Scognamiglio M, Fiorentino A, Giovannini PP, Poli F (2012) Elicited *Teucrium chamaedrys* cell cultures produce high amounts of teucrioside, but not the hepatotoxic neo-clerodane diterpenoids. Phytochemistry 81:50–59

Ardestani A, Yazdanparast R, Jamshidi S (2008) Therapeutic effects of *Teucrium polium* extract on oxidative stress in pancreas of streptozotocin-induced diabetic rats. J Med Food 11:525–532

Atanasov AG, Waltenberger B, Pferschy-Wenzig EM, Linder T, Wawrosch C, Uhrin P et al (2015) Discovery and resupply of pharmacologically active plant-derived natural products: a review. Biotechnol Adv 33:1582–1614

Bagci E, Yazgın A, Hayta S, Cakılcıoglu U (2010) Composition of the essential oil of *Teucrium chamaedrys* L.(Lamiaceae) from Turkey. J Med Plant Res 4:2588–2590

Belmekki N, Bendimerad N, Bekhechi C (2013) Chemical analysis and antimicrobial activity of *Teucrium polium* L. essential oil from Western Algeria. J Med Plant Res 7:897–902

Bezić N, Vuko E, Dunkić V, Ruščić M, Blažević I, Burčul F (2011) Antiphytoviral activity of sesquiterpene-rich essential oils from four Croatian *Teucrium* species. Molecules 16:8119–8129

Bigham M, Hosseinina V, Nabavi B, Talebi K, Esmaeilzadeh NS (2010) Effects of essential oil from *Teucrium polium* on some digestive enzyme activities of *Musca domestica*. Entomol Res 40:37–45

Bindu KH, Mythili JB, Radhika RM (2018) Genetic engineering in medicinal and aromatic plants. In: Rout GR, Peter KV (eds) Genetic engineering of horticultural crops. Academic, Waltham, pp 249–271

Bouhouche N, Ksiksi T (2007) An efficient in vitro plant regeneration system for the medicinal plant *Teucrium stocksianum* Boiss. Plant Biotechnol Rep 1:179–184

Brentlinger D (2007) New trends in hydroponic crop production in the U. S. Acta Hortic 742:31–34

Burducea M, Zheljazkov VD, Lobiuc A, Pintilie CA, Virgolici M, Silion M, Asandulesa M, Burducea I, Zamfirache MM (2019) Biosolids application improves mineral composition and phenolic profile of basil cultivated on eroded soil. Sci Hortic 249:407–418

Canter PH, Thomas H, Ernst E (2005) Bringing medicinal plants into cultivation: opportunities and challenges for biotechnology. Trends Biotechnol 23:180–185

Cao A, Carucci A, Lai T, Bacchetta G, Casti M (2009) Use of native species and biodegradable chelating agents in the phytoremediation of abandoned mining areas. J Chem Technol Biotechnol 84:884–889

Chitturi S, Farrell GC (2008) Hepatotoxic slimming aids and other herbal hepatotoxins. J Gastroenterol Hepatol 23:366–373

Chojnacka K, Chojnacki A, Górecka H, Górecki H (2005) Bioavailability of heavy metals from polluted soils to plants. Sci Total Environ 337:175–782

Da Silva JAT, Cardoso JC, Dobránszki J, Zeng S (2015) *Dendrobium* micropropagation: a review. Plant Cell Rep 34:671–704 de Mesquita Arruda MV, Oliveira FFM, Sampaio MV, da Silva Fernandes MDS, da Silva Dias N, de Albuquerque CC, dos Santos Fernandes C (2018) Influence of nutrition and water stress in *Hyptis suaveolens*. Ind Crop Prod 125:511–519

Dias MI, Sousa MJ, Alves RC, Ferreira IC (2016) Exploring plant tissue culture to improve the production of phenolic compounds: a review. Ind Crop Prod 82:9–22

Djabou N, Lorenzi V, Guinoiseau E, Andreani S, Giuliani MC, Desjobert JM, Bolla JM, Costa J, Berti L, Luciani A, Muselli A (2013) Phytochemical composition of Corsican *Teucrium* essential oils and antibacterial activity against foodborne or toxi-infectious pathogens. Food Control 30:354–363

Dourakis SP, Papanikolaou IS, Tzemanakis EN, Hadziyannis SJ (2002) Acute hepatitis associated with herb (*Teucrium capitatum* L.) administration. Eur J Gastroenterol Hepatol 14:693–695

Dunkić V, Bezić N, Vuko E (2011) Antiphytoviral activity of essential oil from endemic species *Teucrium arduini*. Nat Prod Commun 9:1385–1388

Eskandary H, Rajabalian S, Yazdi T, Eskandari M, Fatehi K, Ganjooei AN (2007) Evaluation of cytotoxic effect of *Teucrium polium* on a new glioblastoma multiforme cell line (REYF-1) using MTT and soft agar clonogenic assays. Int J Pharmacol 3:435–437

Esmaeili AM, Zohari F, Sadeghi H (2009) Antioxidant and protective effects of major flavonoids from *Teucrium polium* on β-cell destruction in a model of streptozotocin-induced diabetes. Planta Med 75:1418–1420

Fairbrother A, Wenstel R, Sappington K, Wood W (2007) Framework for metals risk assessment. Ecotoxicol Environ Saf 68:145–227

Fiorentino A, D'Abrosca B, Esposito A, Izzo A, Pascarella MT, D'Angelo G, Monaco P (2009) Potential allelopathic effect of neo-clerodane diterpenes from *Teucrium chamaedrys* (L.) on stenomediterranean and weed cosmopolitan speciec. Biochem Syst Ecol 37:349–353

Frabetti M, Gutiérrez-Pesce P, Mendoza-de Gyves E, Rugini E (2009) Micropropagation of *Teucrium fruticans* L., an ornamental and medicinal plant. In Vitro Cell Dev Biol Plant 45:129–134

Galstyan HM, Revazova LV, Topchyan HV (2010) Digital indices and microscopic analyses of wild growing and overgrowing of *Teucrium polium* L. in hydroponic conditions. New Armenian Med J 4:104. https://doi.org/10.1186/s12906-016-1407-3

Gharaibeh MN, Elayan HH, Salhab AS (1988) Hypoglycemic effects of *Teucrium polium*. J Ethnopharmacol 24:93–99

Giri CC, Zaheer M (2016) Chemical elicitors versus secondary metabolite production in vitro using plant cell, tissue and organ cultures: recent trends and a sky eye view appraisal. Plant Cell Tissue Organ Culture 126:1–18

Gülçin İ, Uğuz M, Oktay M, Beydemir S, Küfrevioğlu IÖ (2003) Antioxidant and antimicrobial activities of *Teucrium polium* L. J Food Technol 1:9–16

Haïdara K, Alachkar A, Moustafa AA (2011) Teucrium polium plant extract provokes significant cell death in human lung cancer cells. Health 3:366–369

Hasani P, Yasa N, Vosough-Ghanbari S, Mohammadirad A, Dehghan G, Abdollahi M (2007) In vivo antioxidant potential of *Teucrium polium*, as compared to α-tocopherol. Acta Pharma 57:123–129

Hosseini MS, Samsampour D, Ebrahimi M, Abadía J, Khanahmadi M (2018) Effect of drought stress on growth parameters, osmolyte contents, antioxidant enzymes and glycyrrhizin synthesis in licorice (*Glycyrrhiza glabra* L.) grown in the field. Phytochemistry 156:124–134

Iliev I, Gajdošová A, Libiaková G, Jain SM (2010) Plant micropropagation. In: Davey M, Anthony P (eds) Plant cell culture: essential methods. Wiley, Hoboken, pp 1–23

Jakovljević D, Stanković M, Bojović B, Topuzović M (2017) Regulation of early growth and antioxidant defense mechanism of sweet basil seedlings in response to nutrition. Acta Physiol Plant 39:243. https://doi.org/10.1007/s11738-017-2548-9

Jakovljević D, Topuzović M, Stanković M (2019) Nutrient limitation as a tool for the induction of secondary metabolites with antioxidant activity in basil cultivars. Ind Crop Prod 138. https://doi.org/10.1016/j.indcrop.2019.06.025

Kadifkova-Panovska T, Kulevanova S, Stefova M (2005) In vitro antioxidant activity of some *Teucrium* species (Lamiaceae). Acta Pharma 55:207–214

Kandouza M, Alachkarb A, Zhangc L, Dekhila H, Chehnab F, Yasmeena A, Al Ala-Edin M (2010) Teucrium polium plant extract inhibits cell invasion and motility of human prostate cancer cells via the restoration of the e-cadherin/catenin complex. J Ethnopharmacol 129:410–415

Karuppusamy S (2009) A review on trends in production of secondary metabolites from higher plants by in vitro tissue, organ and cell cultures. J Med Plant Res 3:1222–1239

Katalinic V, Milos M, Kulisic T, Jukic M (2005) Screening of 70 medicinal plant extracts for antioxidant capacity and total phenols. Food Chem 94:550–557

Keršek E (2006) Medicinal herbs in wine and alcohol (in Croatian), V.B.Z., Zagreb

Khleifat K, Shakhanbeh J, Tarawneh K (2002) The chronic effects of *Teucrium polium* on some blood parameters and histopathology of liver and kidney in the rat. Turk J Biol 26:65–71

Koehn FE, Carter GT (2005) The evolving role of natural products in drug discovery. Nat Rev Drug Discov 4:206–220

Kouzi SA, McMurtry RJ, Nelson SD (1994) Hepatotoxicity of germander (*Teucrium chamaedrys* L.) and one of its constituent neoclerodane diterpenes teucrin A in the mouse. Chem Res Toxicol 7:850–856

Larrey D, Vial T, Pauwels A, Castot A, Biour M, David M, Michel H (1992) Hepatitis after germander (*Teucrium chamaedrys*) administration: another instance of herbal medicine hepatotoxicity. Ann Intern Med 117:129–132

Lee S, Lee J (2015) Beneficial bacteria and fungi in hydroponic systems: types and characteristics of hydroponic food production methods. Sci Hortic 195:206–215

Ljubuncic P, Dakwar S, Portnaya I, Cogan U, Azaizeh H, Bomzon A (2006) Aqueous extracts of *Teucrium polium* possess remarkable antioxidant activity in vitro. Evid Based Complement Alternat Med 3:329–338

Lommen WJM (2007) The canon of potato science: 27. Hydroponics. Potato Res 50. https://doi.org/10.1007/s11540-008-9053-x

López V, Akerreta S, Casanova E, García-Mina JM, Cavero RY, Calvo MI (2007) In vitro antioxidant and anti-rhizopus activities of Lamiaceae herbal extracts. Plant Foods Hum Nutr 62:151–155

Luna B, Moreno JM, Cruz A, Fernández-González F (2007) Heat-shock and seed germination of a group of Mediterranean plant species growing in a burned area: an approach based on plant functional types. Environ Exp Bot 60:324–333

Makowczyńska J, Sliwinska E, Kalemba D, Piątczak E, Wysokińska H (2016) In vitro propagation, DNA content and essential oil composition of *Teucrium scorodonia* L. ssp. *scorodonia*. Plant Cell Tiss Org 127:1–13

Martínez-López S, Martínez-Sánchez MJ, Pérez-Sirvent C, Bech J, Martínez MDCG, García-Fernandez AJ (2014) Screening of wild plants for use in the phytoremediation of mining-influenced soils containing arsenic in semiarid environments. J Soils Sediments 14:794–809

Masanori M, Sumiko K, Kiochiro K (2000) Study of flavonoids and antifeedant activity of *Teucrium* species. J Agric Food Chem 48:1888–1894

Máthé Á, Hassan F, Kader AA (2015) *In vitro* micropropagation of medicinal and aromatic plants. In: Ákos M (ed) Medicinal and aromatic plants of the world. Springer, Dordrecht, pp 305–336

Mehrabani D, Rezaee A, Azarpira N, Fattahi MR, Amini M, Tanideh N, Panjehshahin MR, Saberi-Firouzi M (2009) The healing effects of *Teucrium polium* in the repair of indomethacin-induced gastric ulcer in rats. Saudi Med J 30:494–499

Mirghazanfari MS, Keshavarz M, Nabavizadeh F, Soltani N, Kamalinejad M (2010) The effect of *Teucrium polium* L. extracts on insulin release from in situ isolated perfused rat pancreas in a newly modified isolation method: the role of Ca^{2+} and K^{+} channels. Iran Biomed J 14:178–185

Moreira B, Tormo J, Estrelles E, Pausas JG (2010) Disentangling the role of heat and smoke as germination cues in Mediterranean Basin flora. Ann Bot (Lond) 105:627–635

Nadjafi F, Bannayan M, Tabrizi L, Rastgoo M (2006) Seed germination and dormancy breaking techniques for *Ferula gummosa* and *Teucrium polium*. J Arid Environ 64:542–547

Naik PM, Al-Khayri JM (2016) Abiotic and biotic elicitors-role in secondary metabolites production through in vitro culture of medicinal plants. In: Shanker A (ed) Abiotic and biotic stress in plants-recent advances and future perspectives. Intech Open, Rijeka, pp 247–277

Narayani M, Srivastava S (2017) Elicitation: a stimulation of stress in in vitro plant cell/tissue cultures for enhancement of secondary metabolite production. Phytochem Rev 16. https://doi.org/10.1007/s11101-017-9534-0

Nematollahi-Mahani SN, Rezazadeh-Kermani M, Mehrabani M, Nakhaee N (2007) Cytotoxic effects of *Teucrium polium* on some established cell lines. Pharm Biol 45:295–298

Oganesyan GB, Galstyan AM, Mnatsakanyan VA, Shashkov AS, Agababyan PV (1991) Phenylpropanoid glycosides of *Teucrium polium*. Chem Nat Compd 27:556–559

Ortego F, Rodríguez B, Castañera P (1995) Effects of neo-clerodane diterpenes from *Teucrium* on feeding behavior of Colorado Potato beetle larvae. J Chem Ecol 21:1375–1386

Özkan G, Kuleaşan H, Çelik S, Göktürk RS, Ünal O (2007) Screening of Turkish endemic *Teucrium montbretii* subsp. *pamphylicum* extracts for antioxidant and antibacterial activities. Food Control 18:509–512

Ozkan A, Banar M, Cokaygil Z, Kulac A, Yalcin G, Taspinar K, Altay A (2014) Pyrolysis of hyperaccumulator plants used for the phytoremediation of lead contaminated soil. Ekoloji 23:51–56

Papafotiou M, Martini AN (2016) In vitro seed and clonal propagation of the Mediterranean aromatic and medicinal plant *Teucrium capitatum*. Hortic Sci 51:403–411

Pickens KA, Affolter JM, Wetzstein HY, Wolf JH (2003) Enhanced seed germination and seedling growth of *Tillandsia eizii* in vitro. Hortic Sci 38:101–104

Pieroni A, Quave CL, Santoro RF (2004) Folk pharmaceutical knowledge in the territory of the Dolomiti Lucane, inland southern Italy. J Ethnopharmacol 95:373–384

Pourmotabbed A, Farshchi A, Ghiasi G, Khatabi MP (2010) Analgesic and anti-inflammatory activity of *Teucrium chamaedrys* leaves aqueous extract in male rats. Iran J Basic Med Sci 13:119–125

Rad FA, Jafari M, Khezrinejad N, Miandoab MP (2014) An efficient plant regeneration system via direct organogenesis with in vitro flavonoid accumulation and analysis of genetic fidelity among regenerants of *Teucrium polium* L. Hortic Environ Biotechnol 55:568–577

Rascioa N, Izzo FN (2011) Heavy metal hyperaccumulating plants: how and why do they do it? And what makes them so interesting? Plant Sci 180:169–181

Redžić SS (2007) The ecological aspect of ethnobotany and ethnopharmacology of population in Bosnia and Herzegovina. Coll Antropol 31:869–890

Rodríguez-Sahagún A, Gutierrez-Lomelí M, Castellanos-Hernández O (2012) Plant cell and tissue culture as a source of secondary metabolites. In: Orhan IE (ed) Biotechnological production of plant secondary metabolites. Bentham Science Publishers, Sharjah, pp 3–20

Sadeghi H, Jamalpoor S, Shirzadi MH (2014) Variability in essential oil of *Teucrium polium* L. of different latitudinal populations. Ind Crop Prod 54:130–134

Sadrizadeh N, Khezri S, Dehghan P, Mahmoudi R (2018) Antibacterial effect of *Teucrium polium* essential oil and *Lactobacillus casei* probiotic on *Escherichia coli* O157: H7 in Kishk. Appl Food Biotechnol 5:131–140

Safari Sinegani AA, Dastjerdi FS (2008) The potential of Irankoh indigenous plant species for the phytoremediation of cadmium and lead contaminated land. Soil Sediment Contam 17:181–188

Šamec D, Gruz J, Strnad M, Kremerc D, Kosalec I, Grubešić JR, Karlović K, Lucic A, Piljac-Žegarac J (2010) Antioxidant and antimicrobial properties of *Teucrium arduini* L. (Lamiaceae) flower and leaf infusions (*Teucrium arduini* L. antioxidant capacity). Food Chem Toxicol 48:113–119

Savvidou S, Goulis J, Giavazis I, Patsiaoura K, Hytiroglou P, Arvanitakis C (2007) Herb-induced hepatitis by *Teucrium polium* L.: report of two cases and review of the literature. Eur J Gastroenterol Hepatol 19:507–511

Shah K, Nongkynrih JM (2007) Metal hyperaccumulation and bioremediation. Biologia Plantarum 51:618–634

Shahraki RM, Arab RM, Mirimokaddam E, Palan JM (2006) The effect of *Teucrium polium* (Calpoureh) on liver function, serum lipids and glucose in diabetic male rats. Iran Biomed J 11:65–68

Shinwari ZK, Gilani SS (2003) Sustainable harvest of medicinal plants of Astore, Northern Pakistan. J Ethnopharmacol 2:289–298

Shtukmaster S, Ljubuncic P, Bomzon A (2010) The effect of an aqueous extract of *Teucrium polium* on glutathione homeostasis in vitro: a possible mechanism of its hepatoprotectant action. Adv Pharmacol Sci. https://doi.org/10.1155/2010/938324

Simonyan KV, Chavushyan VA (2015) Neuroprotective activity of hydroponic *Teucrium polium* following bilateral ovariectomy. Metab Brain Dis 30:785–792

Simonyan KV, Chavushyan VA (2016) Protective effects of hydroponic *Teucrium polium* on hippocampal neurodegeneration in ovariectomized rats. BMC Complement Altern Med 16. https://doi.org/10.1186/s12906-016-1407-3

Stals M, Thijssen E, Vangronsveld J, Carleer R, Schreurs S, Yperman J (2010) Flash pyrolysis of heavy metal contaminated biomass from phytoremediation: influence of temperature, entrained flow and wood/leaves blended pyrolysis on the behaviour of heavy metals. J Anal Appl Pyrolysis 87:1–7

Stanković M (2012) Biological effects of secondary metabolites of *Teucrium* species of Serbian flora. Dissertation, University of Kragujevac

Stankovic MS, Curcic MG, Zizic JB, Topuzovic MD, Solujic SR, Markovic SD (2011) *Teucrium* plant species as natural sources of novel anticancer compounds: antiproliferative, proapoptotic and antioxidant properties. Int J Mol Sci 12:4190–4205

Stanković MS, Niciforovic N, Mihailovic V, Topuzovic M, Solujic S (2012a) Antioxidant activity, total phenolic content and flavonoid concentrations of different plant parts of *Teucrium polium* L. subsp. *polium*. Acta Soc Bot Pol 81:117–122

Stanković M, Stefanović O, Čomić L, Topuzović M, Radojević I, Solujić S (2012b) Antimicrobial activity, total phenolic content and flavonoid concentrations of *Teucrium* species. Open Life Sci 7:664–671

Stankovic MS, Jakovljevic D, Topuzovic M, Zlatkovic B (2014) Antioxidant activity and contents of phenolics and flavonoids in the whole plant and plant parts of *Teucrium botrys* L. Oxid Commun 37:522–532

Starakis I, Siagris D, Leonidou L, Mazokopakis E, Tsamandas A, Karatza C (2006) Hepatitis caused by the herbal remedy *Teucrium polium* L. Eur J Gastroenterol Hepatol 18:681–683

Tariq M, Ageel AM, al-Yahya MA, Mossa JS, al-Said MS (1989) Anti-inflammatory activity of *Teucrium polium*. Int J Tissue React 11:185–188

Vahidi AR, Dashti-Rahmatabadi MH, Bagheri SM (2010) The effects of *Tecrium polium* boiled extract in diabetic rats. Iran J Diabetes Obes 2:27–31

Ververidis F, Trantas E, Douglas C, Vollmer G, Kretzschmar G, Panopoulos N (2007) Biotechnology of flavonoids and other phenylpropanoid-derived natural products. Part I: chemical diversity, impacts on plant biology and human health. Biotechnol J 2:1214–1234

Wilson SA, Roberts SC (2014) Metabolic engineering approaches for production of biochemicals in food and medicinal plants. Curr Opin Biotechnol 26:174–182

Yaman M (2014) *Teucrium* as a novel discovered hyperaccumulator for the phytoextraction of Ni-contaminated soils. Ekoloji 23:81–89

Yazdanparast R, Ali Esmaeili M, Helan AJ (2005) Teucrium polium extract effects pancreatic function of streptozotocin diabetic rats: a histopathological examination. Iran Biomed J 9:81–85

Zerroug MM, Zouaghi M, Boumerfeg S, Baghiani A, Nicklin J, Arrar L (2011) Antibacterial activity of extracts of *Ajuga iva,* and *Teucrium polium*. Adv Environ Biol 5:491–495

Zlatić N, Stanković M, Simić Z (2017) Secondary metabolites and metal content dynamics in *Teucrium montanum* L. and *Teucrium chamaedrys* L. from habitats with serpentine and calcareous substrate. Environ Monit Assess 189:110. https://doi.org/10.1007/s10661-017-5831-8

Index

M. Stanković (ed.), *Teucrium Species: Biology and Applications*,
https://doi.org/10.1007/978-3-030-52159-2

U

V

The manufacturer's authorised representative in the EU is Springer Nature Customer Service Centre GmbH, Europaplatz 3, 69115 Heidelberg, Germany. If you have any concerns regarding our products, please contact ProductSafety@springernature.com

Printed and bound by CPI Group (UK) Ltd, Croydon, CR0 4YY
30/09/2026
02225445-0011

Lecture Notes in Nanoscale Science and Technology

Volume 38

Lecture Notes in Nanoscale Science and Technology (LNNST) aims to report latest developments in nanoscale science and technology research and teaching – quickly, informally and at a high level. Through publication, LNNST commits to serve the open communication of scientific and technological advances in the creation and use of objects at the nanometer scale, crossing the boundaries of physics, materials science, biology, chemistry, and engineering. Certainly, while historically the mysteries in each of the sciences have been very different, they have all required a relentless step-by-step pursuit to uncover the answer to a challenging scientific question, but recently many of the answers have brought questions that lie at the boundaries between the life sciences and the physical sciences and between what is fundamental and what is application. This is no accident since recent research in the physical and life sciences have each independently cut a path to the edge of their disciplines. As both paths intersect one may ask if transport of material in a cell is biology or is it physics? This intersection of curiosity makes us realize that nanoscience and technology crosses many if not all disciplines. It is this market that the proposed series of lecture notes targets.

Virat Khanna
Editor

Carbon-Based Nanocomposites for Sustainable Applications, Volume I

Fundamentals and Material Innovations

Springer

Editor
Virat Khanna
University Centre for Research and Development
Chandigarh University
Mohali, Punjab, India

ISSN 2195-2159 ISSN 2195-2167 (electronic)
Lecture Notes in Nanoscale Science and Technology
ISBN 978-3-031-95305-7 ISBN 978-3-031-95306-4 (eBook)
https://doi.org/10.1007/978-3-031-95306-4

This Springer imprint is published by the registered company Springer Nature Switzerland AG
The registered company address is: Gewerbestrasse 11, 6330 Cham, Switzerland

I dedicate this book to my elder brother, ***Late Er. Sameer Khanna****, because of whom all this is happening. His unwavering support, inspiration, and guidance have played a pivotal role in shaping my academic and professional journey. His influence continues to inspire me, and this work stands as a tribute to his lasting impact on my life.*

Preface

In the modern age of industry and research, nanotechnology has gained advancements over the past few decades. In this growth, carbon-based nanocomposites have emerged as one of the most transformative material classes due to their unique structural, electrical, and mechanical properties. All these properties have paved the way for cutting-edge applications across various industries in almost all applications like energy storage, manufacturing, automobile, biomedicals, etc.

This book, *Carbon-Based Nanocomposites for Sustainable Applications: Volume 1—Fundamentals and Material Innovations*, is a detailed resource for the fundamental principles and innovations in the synthesis and development of these advanced carbon-based materials. The volume provides a detailed coverage of fabrication techniques, structural characteristics, and tailored functionalities of various carbonaceous nanomaterials. The different chapters in this volume cover a broad spectrum of topics, starting with the fundamental principles and fabrication techniques of carbon-based nanocomposites. These chapters include various carbonaceous materials such as graphene, carbon nanotubes, carbon nanofibers, and graphitic carbon nitrides, exploring their synthesis methods, structural properties, and integration into composite materials. Various chapters also cover their applications in diverse fields, including aerospace, biomedical engineering, lightweight composites, 3D printing, stealth technology, and wearable electronics. The book also highlights their significance in energy storage, sustainable urban development, and environmental remediation. The concluding chapters of this book cover emerging challenges, scalability concerns, and the prospects of these materials, emphasizing their role in the advancement of sustainable technology.

The purpose of this edited book volume is to provide researchers, academicians, and industry professionals with an in-depth understanding of the fundamental aspects of carbon-based nanocomposites and their application potential. By bridging the gap between theoretical knowledge and practical applications, this book tries to contribute toward the development of innovative, sustainable, and high-performance materials for future technological advancements.

We extend our sincere gratitude to the contributing authors, whose expertise and dedication have shaped the content of this volume. Their valuable contributions ensure that this book remains a relevant and authoritative reference for researchers and practitioners in the field of nanomaterials and sustainable applications.

Mohali, Punjab, India Virat Khanna

Acknowledgments

The completion of this book would not have been possible without the support, encouragement, and guidance of many individuals and institutions. I would like to express my deepest gratitude to **Chandigarh University, University Centre for Research and Development (UCRD)** for providing me with an environment that fosters academic growth and research excellence. Their unwavering support has been instrumental in shaping my journey.

I sincerely thank all the contributors of the chapters in this book. Their hard work, dedication, and expertise have been instrumental in shaping this volume into a valuable resource. Additionally, I am deeply grateful to the reviewers for their insightful feedback, which has significantly enhanced the quality of the content.

I extend my heartfelt thanks to my colleagues, whose collaboration, insights, and encouragement have been invaluable throughout this process. Their expertise and dedication have greatly enriched the quality of this book.

A special note of appreciation goes to my parents, whose love, sacrifices, and unwavering belief in me have been the cornerstone of my achievements. Their support has given me the strength and motivation to pursue my passion for research and academia.

Above all, I am deeply grateful to my wife, **Anisha Tandon**, for her patience, understanding, and encouragement. Her constant support has been my driving force, allowing me to dedicate the necessary time and effort to this project. I also extend my heartfelt gratitude to my daughter, **Avishka Khanna**, whose love and joy continue to be my greatest source of motivation. Her presence in my life reminds me of the importance of perseverance and dedication.

I am grateful to Series Editors **Prof. Gang Wei, Prof. Zhiming M. Wang, Prof. Greg Salamo,** and **Prof. Stefano Bellucci**, for giving a chance in contributing to the series in the form of this edited volume.

I extend my heartfelt appreciation to **Ms. Nemul Khan**, Publishing Editor of the book series Lecture Notes in Nanoscale Science and Technology, for her unwavering support and guidance throughout this process. I am also grateful to

Ms. Tina Silva, Production Editor, and **Mr. Jeffrey Taub**, Project Coordinator at Springer, for their valuable assistance in ensuring the seamless production of this book.

To everyone who has contributed to this journey, directly or indirectly, I sincerely thank you. Your support and encouragement have made this book a reality.

Contents

About the Editor

Virat Khanna is an accomplished academic and researcher with over 15 years of experience in higher education and scientific research. He has authored more than 70 publications in high-impact, peer-reviewed journals indexed in SCI, Scopus, and Web of Science. His research contributions span nanotechnology, materials science, and sustainable development. Dr. Khanna has served as an editor for multiple books and conference proceedings with leading international publishers, including Springer, CRC Press, IGI Global, and Bentham Science. He is actively involved in scholarly publishing and currently serves on the editorial boards of *Scientific Reports* (Nature Group) and *Discover Nano* (Springer). Additionally, he holds the position of Academic Editor for the *Journal of Nanotechnology* and *Advanced Materials Science and Engineering*, both published by Wiley. As a recognised leader in his field, he serves as the Book Series Editor for *Recent Advances in Materials for Nanotechnology*, published by CRC Press (Taylor & Francis). He has also guest-edited several special issues for renowned international journals, including those published by Springer and De Gruyter. Dr. Khanna has delivered numerous invited seminars, keynote addresses, and workshops at prestigious international conferences. His contributions to research and academia have been recognised with multiple best research paper awards.

Fabrication Techniques of Carbonaceous Nanomaterial-Based Nanocomposites

Pramod Kumar Singh, M. S. Gaur, and Sergej Il'kovič

1 Introduction of Carbon-Based Nanocomposites (CNCs)

Nanocomposites are the most advanced materials obtained by combining nanotechnology and carbon science, which gives rise to nano-dimensional structures with unique properties referred as carbon-based nanocomposites (CNCs). These processes include chemical vapor deposition, sol-gel synthesis, and self-assembly which all require a great deal of control over the composition, structure, and morphology for tailoring materials based on their use. CNCs exhibit extraordinary mechanical properties, high thermal and electrical conductivity, large surface area, and unique optical inertness (transparency) bio-friendly characteristics and chemical stability that make them suitable for advanced scientific technological application [1, 2]. Graphene and carbon nanoparticles, as well as carbon nanotubes (CNTs), are a type of CNCs that have different properties such as structural regularity, large surface area open for functionalization on the basal plane, electrical conductivity, and chemical inertness; they also show biological compatibility with no toxicity during in vivo measurements. As a monolayer sp^2-hybridized carbon structure, graphene with its large π-conjugated network has shown the remarkable properties including excellent heat transfer capability; yet we are inspired to fabricate BN counterparts because of their higher thermal stability [3, 4]. This modification of graphene and CNTs is now being widely explored by researchers for diverse applications like catalyst carriers, energy storage, solar cells, absorbents, biomedical,

P. K. Singh (✉) · M. S. Gaur
Department of Physics, Hindustan College of Science and Technology, Farah, Mathura, Uttar Pradesh, India

S. Il'kovič
Faculty of Humanities and Natural Sciences, Presov University, Presov, Slovak Republic

V. Khanna (ed.), *Carbon-Based Nanocomposites for Sustainable Applications, Volume I*, Lecture Notes in Nanoscale Science and Technology 38,
https://doi.org/10.1007/978-3-031-95306-4_1

textiles type composites, and sensors [5–9]. Carbon materials are used as supports to disperse metal nanoparticles in heterogeneous catalysis, which is a significant approach for various organic transformations resulting in highly demanded and valuable products necessary for the production of pharmaceuticals and biomedical applications, agriculture, and material sciences [10–12]. Research in this area has already made remarkable progress and oriented toward the substantial success; it emerged as a significant research trend for organic reactions using carbon nanocomposites [13–16]. This chapter illustrates the advances in methodologies for preparing metal nanostructures and carbon materials and elaborates its application on catalysis of organic transformation as shown by recent breakthroughs made.

1.1 Composition of Carbon-Based Nanocomposites

Carbonaceous Nanomaterial-Based Nanocomposites (CNCs) are unique composite materials made by incorporating desired carbon nanomaterial in certain substance(s), enabling the expression of a variety of special properties and functionalities. Prominent among them are carbon nanotubes (CNTs), graphene, fullerenes, and carbon nanofibers (CNFs) because of their remarkable mechanical strength, electrical conductivity, and thermal stability [1, 9, 17–20]. The carbon nanomaterials are incorporated within matrices like polymers (such as epoxy, polystyrene), metals (aluminum, copper), and ceramics (silicon carbide, alumina), to offer flexibility along with improved mechanical properties or high temperature sustainability of the composites [7, 8, 10, 11] and metal nanoparticles (e.g., gold, silver), oxides (e.g., titanium dioxide, zinc oxide), and biomolecules (e.g., proteins, enzymes)-functional strategy to increase electrical conductivity, catalytic activity, antimicrobial effects, and biocompatibility [13, 14, 21]. Chemical modifications such as the incorporation of functional groups (carboxyl, hydroxyl) or polymer grafting help in surface modification for better compatibility and dispersion in matrix among carbonaceous nanomaterials [9, 22, 23]. This level of precision in both the selection and designing components is crucial for wide range of applications, like energy storage, catalysis, development of biomedical devices or structural materials, and so on [9, 22, 24–26]. The methods of synthesis and processing procedures employed greatly influence the properties as well as performance delivered by cellulose nanocrystals, making them an important research area in materials science[4, 12, 22].

1.2 Significance and Potential of CNCs

CNC_S have excellent mechanical properties, high electrical and thermal conductivity, and large specific surface area. However, contemporary and tradition materials occupy a larger portion of the material. This makes CNCs more desirable than conventional materials over a wide range of applications [27, 28]. The peculiarities of

2D materials have obliterated a variety of application areas, e.g., electronics as well as structural material [29–31]. Progress in CNCs is intertwined with the emergence of high-performance energy storage technologies, such as batteries and supercapacitors. The temperature gained by insulator particles, and these properties lead to increased capacity of production of stored charge transfer, resulting in great improvement in the performance [32]. Thus, they are pretty important in laying the foundation for newer generations of the storage options to prepare for the inevitable demand for sustainable and efficient balancing solutions [33]. Carbon-based nanocomposite shows potential to act as high-surface area substrates for anchoring metal nanoparticles which play a key role in heterogeneous catalytic processes. It enhances the stability and activity of catalysts, the critical components for synthesizing many of the relevant laws of practical chemicals for medicine, agriculture, or environmental protection [17, 34]. They can render catalytic processes better by making more efficient and sustainable chemical [35, 36]. CNCs nanocomposites offer superb biocompatibility and can be surface-modified for controlled bio-interfacing. This makes CNCs suitable for applications like drug delivery, imaging, biosensing, and tissue engineering. At the level of biomedical applications, their wide versatility augments precision and targeting in delivering therapies; this serves to improve patient outcomes and enrich developments made within medical research [37–39]. Owing to high-surface area and tunable surface chemistry, CNCs present an effective solution for addressing a number of environmental concerns. These properties enable very strong adsorption capabilities, which is why they are suitable to be used in water and air purification [40, 41]. Using environmental remediation technologies and integrating CNCs in those will help to treat more effectively contaminants and alleviate the serious pollution facing the environment today, so we can create much better natural environments [42, 43]. The exceptional electrical characteristics of the CNCs can develop flexible and wearable electronic circuits as well as high-speed transistors among other advanced sensors. Thereby, innovation in consumer electronics and IoT can correspondingly innovate to offer smarter devices which are more efficient and connected [44–47]. CNCs are employed in the reinforcement of polymers, metals, and ceramics leading to composites possessing far better mechanical properties. These materials have found application in aerospace, automotive, and building industries with better properties compared to those of the base matrix system through all aspects [48–51]. CNCs increase the efficiency of energy devices, reduce the utilization of precious metals in catalysis, and facilitate environmentally-friendly manufactured goods. This application has enabled various industries to move toward a form of eco-friendly practices and development of clean technology [52–54]. CNC-based research activities offer a vast potential for fundamental studies in materials science as not only new phenomena can be uncovered but also new applications can emerge. Research in this area is crucial for technology advancements, responding to the most urgent challenges worldwide and encouraging innovative leaps across several fields of science and industry [55, 56]. Figure 1 shows the schematic representation of carbon-based composites in various applications.

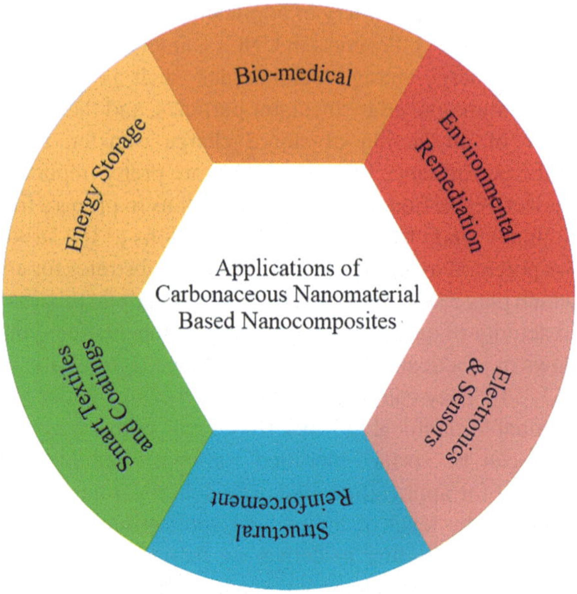

Fig. 1 Schematic representation of carbon-based composites in various applications

1.3 Scope of the Chapter

This chapter presents recent advances in the synthesis and use of CNC with approaches that improve both their mechanical properties and thermal and electrical conductivity. It describes production methods ranging from those that mix solutions to those blending melts, as well as a layer-by-layer assembly, that can all be used to create CNCs for critical applications such as the energy storage, the support of catalytic reactions, and medical use cases. Attention is also given to topics such as localized nanoparticle distribution and environmental challenges demonstrating that CNC materials can encourage growth in fields of study. Furthermore, future endeavors involve creating cost scalable methods for producing CNC materials by harnessing AI technology and eco-friendly chemistry to expand their application, in cutting edge technologies.

2 Fabrication Techniques

Carbonaceous nanomaterial-based nanocomposites, made by integrating carbon nanotubes or graphene with various matrices, exhibit enhanced properties for diverse applications. Key fabrication methods (as shown in Fig. 2) include solution

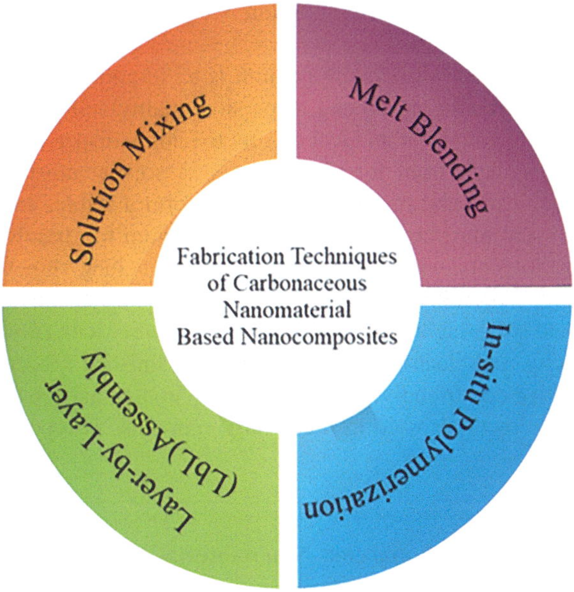

Fig. 2 Fabrication techniques of carbonaceous nanomaterial-based nanocomposite

mixing for bulk production, melt blending for mechanical enhancement, in situ polymerization for strong interfacial bonding, layer-by-layer assembly for precise control [57]. These techniques enable the creation of composites used in electronics, sensors, automotive, aerospace, energy storage, and biomedical fields, driving innovation with their tailored properties.

2.1 Solution Mixing Methods

2.1.1 Principles of Solution Mixing

It is an effective and versatile technology for producing a carbon-based nanocomposite. The process starts by dispersing carbon nanomaterials (such as carbon nanotubes or graphene) in a solvent to make them evenly dispersed. The resulting suspension is then mixed with a polymer matrix, leading to as yet enhanced material. The secret of making successful solution mixing is for nanomaterials dispersion to be put in the polymer matrix throughout. This is crucial with respect to the composite's mechanical, electrical, and thermal properties. Advances in recent years in improving dispersion through better solvent choice and optimized mixing methods have been noted in studies [19, 58, 59].

2.1.2 Selection of Solvents and Dispersion Media

The choice of solvent and dispersion medium is an important factor in preparing solutions, directly affecting at the end dispersion quality and properties of nanocomposites. It's not enough for an ideal solvent to simply disperse carbon nanomaterials and prevent them from forming clusters. A solvent must be compatible, meaning it's chemically and physically suited to form a stable, uniform polymer solution. The solvent must also be a compatible medium for dissolving polymers. Some common solvents with these characteristics include dimethylformamide (DMF), N-methyl-2-pyrrolidone (NMP), and dichlorobenzene which all have good boiling points and solvating abilities. Along with solubility with respect to the concentration, toxicity, and clean-up after the process is finished. Recent studies have surveyed alternative eco-friendly techniques for environmental protection purposes on industrial scale methods used to make large amounts, as reported in different literatures [60–62].

2.1.3 Preparation of Carbonaceous Nanomaterial-Based Nanocomposites Suspensions

In order to ensure the nanomaterial is evenly dispersed within the polymer matrix, preparing a carbon nanomaterial suspension is absolutely vital. The specific steps involved in such preparation are as follows: Generally, you just take nanomaterials and put them in the solvent you've chosen. But it all depends on how things go sonication to break up agglomerates and ensure good mixture is necessary. As an example, sonication uses high-frequency (20 kHz) ultrasonic waves to induce cavitation bubbles in the liquid. These suddenly implode, emitting energy that will split nanomaterial agglomerates [63]. Sometimes, in order to aid with the dispersion of nanomaterials, the surface modifier may need a little chemical adjustment. For instance, surfactants or functional agents are used to modify the surface properties of nanomaterials, improving dispersibility [64]. Adding a surfactant to CNTs dispersed in DMF was greatly beneficial, instead of re-aggregating and forming clumps later on as in the case without help this helped them maintain their stability better and find a place within the resulting composite material which is uniformly mixed [65].

2.1.4 Mixing Methods

To obtain uniform dispersion of carbon nanomaterials in solution mixing, following techniques and equipment are used. For these methods, it depends on the efficiency and scale of operation as well as specific material properties being mixed—one might be more appropriate than others at different times. The most widely employed techniques include the following:

2.1.4.1 Magnetic Stirring

A magnetic stir bar rotates using applied external magnetic field in the suspension and breaks the nanomaterials apart through agitation. It is suitable for low viscosity systems and commonly employed in laboratory scale preparations. Nonetheless, the method may fail to disperse strong agglomerates effectively [66].

2.1.4.2 Ultrasonication

The solvent is saturated with high-frequency sound waves, creating localized high temperatures and pressures which can destroy clusters of nanomaterials. It's good for dispersing a wide range of materials, though it's usually used together with chemical dispersing agents help to reach the higher concentrations of the solvent or other media than would otherwise be possible. The studies [67] have shown that ultrasonication is effective in making graphene/polymer composites having improved electrical conductivity.

2.1.4.3 High-Shear Mixing

High-shear mixers exert strong mechanical forces on the suspension and therefore are very suitable for blending high-viscosity systems. This technique can efficiently disperse nanomaterials within the polymer matrix, although shearing forces may eventually damage delicate substances. Recent work [68, 69] has resulted in a process that caters to industrial needs.

2.1.5 Factors Affecting Dispersion Quality

The shape, size, and surface chemistry of the carbon nanomaterials play an important role in their dispersion in the base material. As reported in literature, nanosized particle with functionalized surface generally disperses more easily as compared to the other materials [70]. Dispersion depends on the viscosity, boiling point, and polarity of the solvent. Solvents with high level of both polar and non-polar character can efficiently disperse these nanoparticles, thus providing superior dispersion due to better stability against aggregation. Uniform dispersion can only be achieved under certain mixing conditions, including the time of mixing, temperature during a manufacture scenario or scale-up context, and mechanical intensity factors such as sonication power with an ultrasound probe tip moving by shear in reciprocal mode. The optimum condition of these processes must be obtained in order to balance the good dispersion quality and good physical properties [71]. The utilization of surfactants in the polymer or host material can enhance the stability of the suspension by preventing re-aggregation of the nanomaterials [72]. Increased concentrations can lead to significant aggregation, which complicates the attainment of a homogeneous dispersion; however, increased concertation up to 10 wt.% may produce best results. The properties of the composite material can be effectively managed through precise control of its characteristics [18].

2.1.6 Advantages

- **Uniform Dispersion:** Solution mixing facilitates helps in the uniform dispersion of carbon nanomaterials = in the polymer matrix, which improves the composite prosperity.
- **Scalability:** It can be scaled up easily for industrial production, so to be used on a large scale.
- **Versatility:** Solution mixing is compatible with a variety of polymers and solvents, enabling tailored properties of the composite.
- **Low Temperature Process**: It works at almost 100 °C temperatures, minimizing the possibility of thermal degradation of the polymer matrix and the carbon nanomaterials.
- **Precise Control on Composition:** The concentration of carbon nanomaterials can be accurately adjusted that allows to finely control electrical, thermal, and mechanical properties of composite.
- **Enhanced Interaction:** Simultaneously, this process helps to facilitate interactions between the polymer and carbon nanomaterials, which can lead to significant improvements in the mechanical strength and electrical conductivity of the resulting composite.
- **Ease of Processing:** The process is fairly simple with closer equipment and steps leading to low production costs.
- **Potential for Functionalization**: Carbon nanomaterials can be functionalized prior to mixing, which means greater scope to customize the properties of the composite.
- **Customization of Nanocomposite Structure:** The ability to prepare composites in different structures (coating, film, or bulk material) provides flexibility in terms of application needs (e.g., radiation-resistant).

2.1.7 Disadvantages

- **Solvent Compatibility:** The solvent for the polymer, as well as the carbon nanomaterials, can be relatively limited. Certain solvents can be toxic, expensive, or become hard to eliminate.
- **Residual Solvent:** Since the total evaporation of the solvent is difficult, residual solvents may remain and correlate with composite performance and stability.
- **Processing Times:** The process can take time, particularly during stages where solvents need to be removed and materials need to dry, likely not optimal for fast production needs.

- **Environmental Impact:** The safety profile of the drug needs to be observed as the drug is prepared through a process involving water, which in turn can be hazardous to the ecology.
- **Aggregation:** Carbon nanomaterials have the possibility of aggregating, which will unevenly distribute the nanomaterials and will degrade the polymer composite properties.
- **Limited Material Compatibility:** As not all polymers are soluble in the same common solvents which can restrict the number of materials that can be used.
- **Complexity in functionalization:** Functionalization is possible but tends to be more complex, additional steps to functionalize and require the exact functionalization process.
- **Potential Degradation:** Certain polymers or carbon nanomaterials may degrade or lose their desirable properties in specific solvents, which limiting the selection of materials.
- **Mechanical Property Challenges:** Weak interfacial bonding or incompatibility between the polymer and carbon nanomaterials can lead to poor mechanical properties.
- **Energy Consumption:** The drying process especially if done at high temperatures or under vacuum conditions can be energy, intensive consequently increasing both the cost and environmental footprint of the production process.

2.2 Melt Blending Techniques: Introduction and Principle

One of the most widely used processes is the melt blending method, which disperses carbonaceous nanomaterial-based nanocomposites in a molten polymer matrix. Synthesis by melt compounding, it is a important process to achieve uniform dispersion and enhancement of mechanical, thermal, electrical properties of the final composites. Melt blending is the most important working mechanism which applies shear forces to disperse the nanoparticles properly in a polymer matrix to improve matrix and nanofiller interfacial interactions and minimize the filler aggregation. Usually starts by heating the polymer up to its melting point and adding CNTs or other carbon nanomaterials in the form of graphene or other carbon allotropes [73]. This is followed by process steps in which the polymer melt is subjected to high-shear conditions typically using equipment such as twin-screw extruders or internal mixers [74]. It is important to note the influence of several factors such as processing temperature, polymer viscosity, and type and content of carbon nanomaterial on melt processing efficiency. The major advantage of this method is that it can produce solvent-free composites with better properties and hence is an environmentally friendly alternative to the solution methods [75]. The unique features of carbon nanomaterials and their successful dispersion in the polymer matrix led to an enhancement of mechanical strength, thermal stability, and electrical conductivity of the corresponding nanocomposites. Besides, melt blending is considerable for tailored composite properties based on the processing parameters (temperature, shear rate and filler loading) [76]. The increase of electrical conductivity via increased filler content usually being offset by the negative effect of too high a filler loading leading to agglomeration and therefore lower mechanical performance, and therefore a fine balance of acceptable product properties. Surface modifications are essential for enhanced affinity of carbon nanomaterials with polymer matrices to enable their dispersion during melt blending [77]. Nanomaterials were reported to enhance the mechanical properties of composites with a small addition to polymer matrices [68, 78], mainly due to the improved thermal performance, and there are commonly described techniques using compatibilizers for functionalization of nanomaterials to significantly affect interfacial interactions at the stage of synthesis [79]. Melt blending is, however, an important technique to prepare carbon-based nanocomposites from the use of such materials capable of introducing diverse carbon fillers into polymer matrices resulting in a significant improvement in mechanical characteristics [78]. Details of this approach are given in the literature [80–83].

2.2.1 Sample Preparation Method

Carbon-based nanocomposites are an important class of materials which combine carbon nanomaterial like CNTs and graphene with polymer matrices, so preparation has been carried out through various techniques among that melt blending is the most commonly employed. This process is preferable for producing composites having superior mechanical, thermal, and electrical properties which are good candidate materials for application in aerospace, electronics, and automotive industry [84, 85]. The melt blending process refers to an approach that disperses carbon nanomaterials in a molten polymer matrix when the mixture becomes balance of temperature and shear condition [78]. Uniform distribution of the nanomaterials in polymer is a key to tuning the properties of composite so that it can be best utilized. The process is initiated by the selection and preparation of polymer, carbon nanomaterials. The typically used polymers are polypropylene, polyethylene, and polystyrene; the carbon nanomaterials may be multi-walled carbon nanotubes (MWCNTs) or graphene [86]. Heating a polymer up to its melting point converts it into a viscous liquid. This is a very important step, since it permits the polymer to flow and mix with carbon nanomaterials [87]. The carbon nanomaterials are added to the mixture after it has been melted. When loaded with filler, the compositions herein generally contain from 1wt.% to about 10wt.%, preferably closer to 1wt.% [88]. Twin-screw extruders or internal mixers should be employed to subject the mixture under high-shear forces. These high-shear conditions result in the disruption of carbon nanomaterials agglomerates and encourage homogeneous distribution within polymer matrix. Once the composite is enough mixed, it should be allowed to cool down which lead by solidifying into shape. These can be achieved by extrusion, injection moldering, or film casting.

2.2.2 Importance of Processing Parameter

The properties of carbon-based nanocomposites, prepared through melt blending, are significantly influenced by processing conditions such as temperature, shear rate, filler loading, and processing time. The processing temperature affects polymer viscosity and nanomaterial dispersion, with optimal ranges varying based on the materials used. Polycarbonate reinforced with carbon nanotubes, for example, is best processed within a temperature range of 200 and 250 °C where dispersion could be effectively achieved but degradation could be avoided [89, 90]. While shear rates improve the dispersion of fillable, they should be controlled to avoid breakdown of the nanomaterials. As the content level increases, agglomeration may form and subsequently can lead to weakening of the composite, limiting the enhancements in mechanical, thermal, and electrical performance. The processing time should be tailored to allow for sufficient dispersal of the polymer without causing it to degrade when using a typical polymer for 10–15 min. Fracture surface morphologies indicate that the cooling rates during quenching after blending affect both the crystallinity of the composite and its mechanical properties, such that cooling mechanisms must be suitably applied to meet compositional requirements [91–93]. Optimally tailoring these parameters is vital for engineering high-performing application-specific carbon-based nanocomposites.

2.2.3 Advantages of Melt Blending Processes

- **Uniform Dispersion:** Ensures a homogeneous distribution of nanoparticles within the polymer matrix.
- **Scalability:** Suitable for large-scale production and industrial applications.
- **Cost-Effective:** Relatively low processing costs compared to other methods.
- **Compatibility:** Compatible with various types of polymers and nanoparticles.
- **Environmental Stability:** Enhanced stability of composites in harsh environments.
- **Improved Mechanical Properties:** Significant improvements in tensile strength, elasticity, and toughness.
- **Thermal Stability:** Enhanced thermal conductivity and stability of the resulting composites.
- **Ease of Processing:** Simple and straightforward processing techniques.

- **Customizability:** Allows for tailoring of composite properties by adjusting the blend ratios and processing conditions.
- **Energy Efficiency:** Requires lower energy input compared to other synthesis methods.

2.2.4 Limitations of Melt Blending Processes

- **Agglomeration Issues:** Nanoparticle's agglomerate to form clusters and deposits also creating non-uniformity in reagent particles.
- **Processing Temperatures:** High processing temperatures may degrade the resulting composites.
- **Low Nanoparticle Loading:** Obtaining a full nanoparticle loading over a common substrate with acceptable processability.
- **Viscosity Challenges:** An increased viscosity can make processing more difficult and require the need for special equipment.
- **Interfacial Adhesion:** The ability to further improve the properties of composites depends significantly on the interfacial adhesion between the polymer matrix and the nanoparticles.
- **Complex rheology:** The melting process is complex and can cause problems during the processing.

2.3 Layer-by-Layer Assembly and Advanced Fabrication Methods

Layer-by-layer (LbL) assembly is a well-controlled process for material preparation through the stepwise deposition of components with complementary features. The functioning principle of LbL assembly is based on the interaction of oppositely charged materials on a substrate, comprising electrostatic interactions, hydrogen bonds, van der Waals forces, and covalent bonds. Because of this, layer thickness, composition and function can be precisely controlled producing complex nanostructures [94]. These usually consist of alternate immersions of a substrate in solutions of polycations and solutions of polyions followed by rinsing and drying steps. The simplest method for their preparation is LbL assembly, which is extensively used to create thin films, coatings, and functional nano-composites for drug delivery, sensors, and energy storage. For a typical LbL assembly, it employs polyelectrolytes like poly (diallyldimethylammonium chloride) (PDDA) and poly (styrene sulfonate) (PSS). This multilayer is formed by the repeated deposition of these materials, with PDDA contributing to the stability of the multilayer structure while PSS helps to stabilize the films functionally [95–100].

2.3.1 Process Steps of the Layer-by-Layer (LbL) Assembly Process

2.3.1.1 Surface Preparation

The first step of the layer-by-layer (LbL) assembly process is the surface preparation to enhance the surface charge properties and achieve strong adhesion of initial layer. The grade it reaches is determined by its cleaning methods used to eliminate contaminants such as dirt, oils, and organic residues that could interfere with layers to be added later [101]. Cleanliness methods include washing with organic solvents (e.g., ethanol, acetone, or isopropanol rather than water) that dissolve any residual organic contaminants. And lastly, oxygen or argon treatments serve as a plasma to remove any organic materials residuals and functionalize the surface increasing its hydrophilicity [102]. For materials like glass or silicon, a piranha solution that contained concentrated sulfuric acid and hydrogen peroxide was used to destroy everything organic and replace it with hydroxyl groups, resulting in lower surface tension. Following cleaning, poly(ethyleneimine) (PEI) and other kinds of polyelectrolytes can help stabilize the charge state with a positively charged surface so that the first layer is able to stick strongly [103].

2.3.1.2 Layer Deposition

After surface treatment, the substrate enters the phase of layer deposition. In this stage, both negatively charged substances are adsorbed stepwise. First, the material to be deposited is dissolved in a solvent so that it can cover a nanoparticle. The substrate is then placed into the solution. For example, if you are using negatively charged graphene oxide as an initial substance and the substrate given a negative charge, electrostatic forces materialized in a more even layer [104]. And such behavior would be followed for a long time indeed; probably, the duration tuned maybe just a few minutes but sometimes several hours. Finally, the substrate is rinsed with deionized water. This serves to remove lightly affixed material and leaves only one, even overlayer behind cutoff points for metals in a current left outside [105]. As an example, a PEI layer may be applied to a substrate with positive charge. This could then be submerged in the solution of negatively charged graphene oxide mentioned above and deionized water used for washing only wipe off any lingering material [106].

2.3.1.3 Rinsing and Drying

To ensure uniformity and strong adhesive of each layer, after each deposition stage, the substrate gets a full rinse and drying [107]. Rinsing with deionized water is done in order to eliminate any particles or polymers that are unrinsed, loosely bound to the substrate or not bound at all and to protect the layer's uniformity. Drying methods include nitrogen gas, ordinary drying by air and gentle sketching to ensure that a deposited layer is adhered securely prepared for subsequent deposition cycle. This also is critical at this stage in order to prevent cross-contamination of layers and maintain the integrity of a multilayered structure [108].

2.3.1.4 Repeating the Cycle

The final step in the layer-by-layer (LbL) assembly process is to repeat a series of immersions, adsorptions, rinses, and dryings in order for one side of the desired number may accumulate layer-by-layer, until it reaches thickness sufficient for desired functionality and utility [109]. Cycle by cycle we add monolayers to the substrate which itself must undergo layer upon thickness is built up. If a film is to have 50 layers, then the substrate will be passed through this deposition process 50 times. This very controlled and systematic approach makes it possible to create targeted ultrathin layers of composite materials, which are suited for a wide range application that require high performance, for example, electrical energy storage systems or pliable sensors [110].

2.3.2 Techniques for Depositing Layers

2.3.2.1 Dip Coating

A simple and effective method most frequently used in LbL assembly is dip coating. It was developed with the aim of immersing substrates in solutions containing the materials you want to put them on them, and then steadily withdrawing them. It brings about on substrate a uniform layer with even thickness. The quality of coating mainly depends on three factors: the solution concentration, the duration of immersion, and withdrawal speed. For example, people have used dip coating to apply layers of graphene oxide onto silicon substrates and lay carbon nanotube layers onto those, thereby to enhance the electrical conductivity mechanical strength at the end composite [111].

2.3.2.2 Spin Coating

Spin coating is a popular process in LbL assembly used to cover thin films. Inside this procedure, a small volume of the coating solution is deposited at the center of the slide, and spun at high speed. The centrifugal force scatters the solution uniformly on the coast slide as if it were all one piece creating a uniform thin film. Film thickness can be adjusted by changing the spin speed, duration of spinning and viscosity. Spin coating was used successfully to produce graphene oxide/polymer nanocomposites with uniform layers that always scored as good or better than any used, leading to stronger mechanical performance and thermal resistance [112]. Higher-resolution small-scale applications can now use this technique, but some fine-tuning of the parameters may be needed to guarantee consistent results on varying substrates [113].

2.3.2.3 Spray Coating

Spray coating simply means atomizing coating solution into small drops, which are then spread onto the substrate to form a thin layer. Highly flexible, this process can be used to coat many substrates of different sizes and shapes.

Through adjusting spray parameters like nozzle type, spray distance, and solution flow rate, the thickness and uniformity of coatings can be controlled. Spray coating has been used by scientists producing multi-walled carbon nanotube/polymer composites on flexible substrates, displaying improved electrical conductivity and flexibility does [114]. With extensive surface applications and the coating of complex substrate shapes, spray-coating offers major benefits. It is widely used in automotive and aerospace industries, as well the manufacture of flexible electronics and sensors [115].

2.3.3 Factors Influencing Layer-by-Layer (LbL)

2.3.3.1 Surface Charge Density

The density of charge is determined largely by the substrate and its adsorbed layers. It has a strong influence on both quality and stability. Higher charge density generally results in stronger electrostatic interactions and better adhesion of layers [116, 117].

2.3.3.2 pH of Solutions

The pH level will change the charge on a polyelectrolyte, with different effects upon its interaction and assembly upon substrates. To take one example, controlling the pH also determines the degree of ionization for a weak polyelectrolyte, and thus influences assembly processes [118].

2.3.3.3 Ionic Strength

Salts present in solution can screen electrostatic interactions, affecting thickness and regularity of the layers. Higher ionic strength will give thicker but less uniform layers because there's less electrostatic repulsion [119, 120].

2.3.3.4 Adsorption Time

The period of contact with each solution affects how much material is adsorbed and what the total thickness becomes. Longer adsorption times will give thicker films accordingly [121].

2.3.3.5 Rinsing Conditions

Effective rinsing process is important to remove extra or loosely bound materials, and it will ensure the stability of the layer. This process supports in achieving well defined and more stable multilayer structures [119].

2.3.3.6 Temperature

The temperature of the solution can affect the mobility and viscosity of the molecules, which influence the process of the assembly. To achieve the more uniform layer, higher temperature is required, which increases the diffusion rate of the molecules [122].

2.3.3.7 Material Compatibility

The interaction between the substrate and polyelectrolyte nanoparticles influences the adhesion and stability of the layers. The use of incompatible materials can result in poor adhesion as well as delaminated films.[116]

2.3.4 Advantages

- Precise control over film thickness and composition.
- Ability to incorporate a wide range of materials, including nanomaterials.
- Versatility in terms of substrate compatibility.
- Scalability for large-area coatings.

2.3.5 Limitation

- Time-consuming process due to multiple deposition and rinsing steps.
- Challenges in achieving uniform coatings on substrates with complex geometries.
- Sensitivity to environmental conditions, such as pH and ionic strength.
- Potential for weak interlayer adhesion if not carefully optimized.

2.4 In Situ Polymerization

In situ polymerization is one of the methods known for the synthesis of carbon nanoparticles in a nanocomposite approach whereby the introduction of carbon nanostructures such as carbon nanotubes and graphene into a polymer matrix is achieved. This is used to study nanocomposites consisting of high-density polyethylene on carbon nanotubes, observed enhanced thermal and morphological characteristics [123]. In situ polymerization occurs by polymerizing monomers in the presence of well-dispersed carbon nanoparticles ensured for proper distribution in the polymer matrix. Materials with better mechanical, thermal, and electrical properties are produced so as to be denoted as key materials for electronics, aerospace, energy storage, and biomedical devices [124]. In situ polymerization has therefore become more prudent than traditional approaches such as melt blending and solution mixing for the attainment of homogeneous dispersion of nanoparticles with improved interfacial bonding for better load transfer and functional performance of composites [125]. More so, functionalization of the nanoparticles and controlled polymerization techniques may be used for tuning the properties of the composites to adopt a favorable molecular architecture toward enhanced interfacial adhesion [126]. Undergoing refinements in research allow the in situ polymerization to embrace some of the ongoing challenges with nanoparticle agglomeration, thereby enabling future development of high-performance materials [127, 128].

2.4.1 Principle of In Situ Polymerization

In situ polymerization consists to begin the polymerization process with nanohybrid dispersed. These nanomaterials, which serve as nucleation sites, allow the polymer chains to be formed around these resulting in a stage of monomer polymerization. This method ensures the direct intimate contact at the solid/liquid interface, enhancing stress transfer between polymer and nanomaterials as well as interfacial interactions [129, 130].

2.4.2 Preparation Method

In situ polymerization is a well-established technological approach used by advanced material designers to synthesize carbon nanoparticle-based nanocomposites and takes advantage of the unique properties offered by nanostructured carbon materials including CNTs, graphene, and fullerenes. This process typically starts with the dispersion of carbon nanoparticles into a solution of monomers. The dispersion should also overcome the propensity of nanoparticle agglomeration, caused by the fact that van der Waals forces are inherently present between the nanoparticles, which can reduce the performance of a nanocomposite by the presence of weak points [131]. Methods to solve this problem include ultrasonication or surfactant usage such as SDS. In addition, chemical functionalization, such as the treatment of CNTs with nitric acid, can further improve the attachment of functional groups to CNTs, such as —OH or —COOH, significantly improving the compatibility of nanoparticles with monomer [128]. This strengthens the interfacial adhesion between nanoparticles and the polymer, which is important for load transfer and overall material performance. After a uniform dispersion is reached, polymerization starts. Monomer polymerizes around the dispersed nanoparticles embedding it into the polymer matrix via macrophysical trapping during the growth of polymer chains. The polymerization may be initiated thermally, chemically, or photochemically with ultraviolet light employing a photoinitiator like Irgacure [132]. Functionalized nanoparticles typically serve as the nucleating site and promote even growth of polymer around the nanoparticle. Common techniques include atom transfer radical polymerization (ATRP) and reversible addition fragmentation chain transfer polymerization (RAFT), which are more favored due to the control of molecular architecture and weight distribution [126, 133]. In polymerization, polymer chains embed the NPs and thus lead to composites with a homogeneous bulk composition where NPs are well encapsulated in the matrix. The encapsulation guarantees strong adhesion at the interface, resulting in improved mechanical properties and thermal stability of the final composite [134]. After polymerization, the nanocomposite is cured to solidify the matrix and to remove any residual solvents or monomers, usually by means of vacuum drying. The material obtained is subsequently characterized to ensure the structural and functional characteristics. The dispersion of nanoparticles is typically characterized using SEM and TEM, and the corresponding property improvements are shown using tensile and thermal tests [135]. In situ polymerization has emerged as an efficient integration route for fabricating polymer nanocomposites with optimized properties for use in applications including flexible electronics, aerospace parts, biomedical devices, and energy storage devices. Its capacity to disperse nanoparticles in a uniform manner and create strong polymer-nanoparticle interfaces has made it one of the cornerstones of advanced material science. Researchers keep improving this approach in light of issues like aggregation of nanoparticles and tuning reaction parameters for novel applications and material development [136, 137].

2.4.3 Advantages

- **Enhanced Dispersion:** In situ polymerization facilitates the dispersion of carbon nanoparticles, in the polymer matrix, which is far more effective than ex situ techniques in ensuring uniform dispersion, in order to achieve the material properties that are intended.
- **Strong Interfacial Bonding**: It facilitates strong interfacial adhesion; this enhances load transfer coupling and ultimately improves their mechanical properties.
- **Customizable Properties:** Depending on the monomers selected and the polymerization process controlled, properties of nanocomposites can be tailored to meet needs of individual applications.

2.4.4 Limitation

- **Agglomeration of Nanoparticles:** Carbon nanoparticles generally tend to agglomerate which makes it difficult for them to disperse uniformly. This challenge is often tackled using methods (e.g., the functionalization of nanoparticles and the use of surfactants).
- **Control of Polymerization Conditions:** To ensure the polymerization process is successful, it is essential to manage the conditions in order to avoid any reactions and to attain the specific molecular weight and structure of the polymer.
- **Compatibility Issues:** Ensuring that the polymer matrix and carbon nanoparticles are compatible is essential, for the transfer of stress and the overall performance of the nanocomposite.

3 Comparative Analysis of Fabrication Techniques

Table 1 provides the comparative analysis of different criteria various fabrication techniques of carbon nanocomposites

Table 1 Comparative analysis of different parameters for various fabrication techniques

S.No.	Criteria	Solution mixing	Melt blending	Layer-by-Layer (LbL) assembly	In situ polymerization	References
1	**Dispersion quality**	Good with proper solvent and sonication	Moderate, often requires compatibilizers	Excellent, precise control over each layer	Excellent, polymerization occurs in presence	[138]
2	**Processing temperature**	Room temperature or slightly elevated	High temperatures	Typically, room temperature	Varies depending on monomer and initiator	[139]
3	**Equipment complexity**	Simple (stirring, sonication)	Complex (extruders, internal mixers)	Simple to moderate (deposition equipment)	Moderate (polymerization setup)	[140]
4	**Cost**	Low to moderate	High due to equipment and energy costs	Low to moderate	Moderate to high	[141]
5	**Scalability**	Moderate	High	Low to moderate	High	[141]
6	**Control over structure**	Limited, dependent on mixing efficiency	Limited, dependent on shear forces	Very high, each layer can be precisely controlled	High, can tailor polymer growth around nanomaterial	[140]
7	**Suitability for thermoplastics**	Good	Excellent	Poor	Good	[142]
8	**Suitability for thermosets**	Excellent	Moderate to good	Poor to moderate	Excellent	[143]
9	**Interfacial bonding**	Moderate, enhanced with functionalized nanomaterials	Good, further improved with compatibilizers	Excellent, strong electrostatic/chemical bonds	Excellent, strong covalent bonds	[143]
10	**Property enhancement**	Mechanical, electrical, and thermal improvements	Mechanical improvements mainly	Excellent mechanical and barrier properties	Broad range: Mechanical, thermal, electrical	[144]
11	**Environmental impact**	Solvent disposal concerns	High energy consumption	Low, often water-based	Depends on initiators and monomers used	[143]
12	**Production time**	Moderate to long	Short to moderate	Long due to multi-step process	Moderate to long	[145]
13	**Material compatibility**	Wide range of polymers and nanomaterials	Mainly thermoplastics	Limited by deposition method	Wide range of polymers and nanomaterials	[146]
14	**Reproducibility**	Moderate, dependent on mixing conditions	High	High	High	[147]
15	**Application areas**	Coatings, composites, electronics	Automotive, packaging, consumer goods	Sensors, membranes, coatings	Aerospace, biomedical, electronics	[148]

4 Techniques for Enhancing Dispersion in Nanocomposites

Uniform dispersion and alignment of carbon nanoparticles in polymer matrices are essential to improve the properties of carbon-based nanocomposites. To ensure good dispersion during the preparation process, various techniques are utilized. Ultrasonication generates high-order shock waves, which separates carbon nanomaterial aggregates to enhance not only their dispersion but also their mechanical properties but should be avoided due to a very high energy demand [149]. Chemical functionalization introduced chemical groups into carbon nanomaterials that enhance interaction with the polymer matrix; curing of the thermoset resin is suspected to adversely affect dispersion, thereby affecting mechanical properties [150]. Compatibilizers are typically block copolymers or surfactants that improve the interfacial interactions between nanofillers and polymer matrix, restraining agglomeration of filler particles and leading to improved mechanical properties. Magnetic alignment aligns the magnetic carbon nanofillers inside the polymer matrix under an applied magnetic field that enhances both electrical conductivity and reinforcement performance [151–153].

5 Challenges in Preparation of Carbon-Based Nanocomposites

The development of carbon-based nanocomposites is essential due to their distinct properties and wide-ranging potential applications such as energy storage, electronics, and medicine. However, the preparation of these materials presents significant challenges. Achieving a uniform dispersion of nanoparticles within the polymer matrix is a critical technical hurdle, as agglomeration can diminish the mechanical, thermal, and electrical properties of the composite. Additionally, to preserve these properties, it is crucial to control the interface between the polymer matrix and the nanoparticles. In addition, the exact control over the fabrication parameters such as temperature, pressure, and molar concentration of precursors are also the main reasons to avoid the defects, hence precursors need to be aligned properly for carbon nanomaterials. Adopting laboratory-scale production for commercial purposes leads to more complexities like reproducibility and high cost of synthesized raw materials like high-purity carbon nanotubes or graphene, etc. Environmental and safety effects of use and disposal of nanomaterials also need to be adequately addressed. However, to overcome these challenges, novel fabrication techniques are being proposed. This characterization helps to both improve the performance of carbon-based nanocomposites at the micro or nanoscales and scale it up to practical applications through strategies such as the use of functionalized nanoparticles, additive manufacturing, and bringing in hybrid materials.

6 Conclusion: Future Research Directions and Potential Breakthroughs

Polymers formed with carbon nanostructures like carbon nanotubes, graphene, and fullerenes allow for a wide range of uses in biomedical, environmental, electronic, and energy fields. The large surface area possessed by CNCs and the regularity in their structuring provide a basis for elevating the performance of processes such as catalysis, energy storage, and pollutant removal over conventional materials. Their fabrication techniques including solution mixing, melt blending, and in situ polymerization have unique merits and demerits in terms of achieving homogenous distribution and strong interfacial adhesion within the matrices. Future work centers on efforts to develop economically viable processes, improve environmental impact, and use sophisticated techniques such as 3D printing and artificial intelligence to enhance the properties of CNCs and their performances in their intended applications. The increasing attention is focused on CNCs using the patented method of green chemistry which increases the performance of the CNC and which further advancement in its green synthesis and functionalization technologies will render them indispensable in the crafting of sustainable high-performance materials for various industrial applications. The materials developed for carbon-based nanocomposites research include the ones that withstand the existing barriers and broaden the overall horizons in the field of material science. We have been able to suggest a possibility that seems very promising a solution in this direction is within the reach is to create clean technologies for the processes of nanocomposite synthesis in the first place aimed at environmental safety. This implies the application of clean technologies and the green chemistry approach throughout the production of nanocomposites. An additional yet equally significant aspect is the synthesis of carbon-based nanocomposites and the reason why they exhibit special physical and chemical properties. Gaining more insights into the interactions at the nanoscale level offers opportunities to engineer materials with specific properties. All of these efforts are complimented by the rapidly evolving application of machine learning and artificial intelligence to the design and optimization of nanocomposites. These tools may help avoid labor-intensive trial and error approaches which are the hallmark of developing new materials by predicting desirable attributes of nanocomposites from theoretical first principles. The 3D printing deposition process of required nanostructures will be very effective in the near future owing the construction procedures which employ machine learning and AI algorithms. 3D printing is the fastest and the most accurate method for fabricating fine and monolithic structures, mass fabrication, low cost, and high designs impact.

References

1. Ambika, & Singh, P. P. (2021b). Carbon-based nanocomposites: Preparation and application in environmental pollutants removal. In M. Jawaid, A. Ahmad, N. Ismail, & M. Rafatullah (Eds.), *Environmental remediation through carbon based nano composites* (pp. 203–229). Springer Nature. https://doi.org/10.1007/978-981-15-6699-8_10
2. Sudharsana, C., Anvarsha, N., & Kalyani, P. (2024a). *Carbon-based nanocomposites: A comprehensive review of their multifunctional applications*. IntechOpen. https://doi.org/10.5772/intechopen.114402
3. Wu, Q., Yang, L., Wang, X., & Hu, Z. (2020a). Carbon-based nanocages: A new platform for advanced energy storage and conversion. *Advanced Materials, 32*(27), 1904177. https://doi.org/10.1002/adma.201904177
4. Wu, Q., Yang, L., Wang, X., & Hu, Z. (2020b). Mesostructured carbon-based nanocages: An advanced platform for energy chemistry. *Science China Chemistry, 63*(5), 665–681. https://doi.org/10.1007/s11426-020-9748-0
5. Buyuknalcaci, F. N., Polat, Y., Negawo, T. A., Döner, E., Alam, M. S., Hamouda, T., & Kilic, A. (2018). Carbon nanotube-based nanocomposites for wind turbine applications. In M. Jawaid & M. M. Khan (Eds.), *Polymer-based nanocomposites for energy and environmental applications* (pp. 635–661). Woodhead Publishing. https://doi.org/10.1016/B978-0-08-102262-7.00024-6
6. Carvalho, A. P. A., & Conte Junior, C. A. (2020). Green strategies for active food packagings: A systematic review on active properties of graphene-based nanomaterials and biodegradable polymers. *Trends in Food Science & Technology, 103*, 130–143. https://doi.org/10.1016/j.tifs.2020.07.012
7. Colusso, E., & Martucci, A. (2021). An overview of biopolymer-based nanocomposites for optics and electronics. *Journal of Materials Chemistry C, 9*(17), 5578–5593. https://doi.org/10.1039/D1TC00607J
8. Durairaj, A., Maruthapandi, M., Saravanan, A., Luong, J. H. T., & Gedanken, A. (2022). Cellulose nanocrystals (CNC)-based functional materials for supercapacitor applications. *Nanomaterials (Basel, Switzerland), 12*(11), 1828. https://doi.org/10.3390/nano12111828
9. Wu, Q., Yang, L., Wang, X., & Hu, Z. (2017). From carbon-based nanotubes to nanocages for advanced energy conversion and storage. *Accounts of Chemical Research, 50*(2), 435–444. https://doi.org/10.1021/acs.accounts.6b00541
10. Giannakoudakis, D. A., Zormpa, F. F., Margellou, A. G., Qayyum, A., Colmenares-Quintero, R. F., Len, C., Colmenares, J. C., & Triantafyllidis, K. S. (2022). Carbon-based nanocatalysts (CnCs) for biomass valorization and hazardous organics remediation. *Nanomaterials, 12*(10), 10. https://doi.org/10.3390/nano12101679
11. Rana, A., Khan, I., & Saleh, T. A. (2021). Advances in carbon nanostructures and nanocellulose as additives for efficient drilling fluids: Trends and future perspective—A review. *Energy & Fuels, 35*(9), 7319–7339. https://doi.org/10.1021/acs.energyfuels.0c04341
12. Stojanovska, E., Calisir, M. D., Ozturk, N. D., & Kilic, A. (2019). Carbon-based foams: Preparation and applications. In A. Khan, M. Jawaid, Inamuddin, & A. M. Asiri (Eds.), Nanocarbon and its composites (pp. 43–90). Woodhead Publishing. https://doi.org/10.1016/B978-0-08-102509-3.00003-1
13. Gobena, S. T., & Woldeyonnes, A. D. (2024). A review of synthesis methods, and characterization techniques of polymer nanocomposites for diverse applications. *Discover Materials, 4*(1), 52. https://doi.org/10.1007/s43939-024-00119-0
14. Norouzi, S., Barati, A., & Noroozi, R. (2019). Computational studies on mechanical properties of carbon-based nanostructures reinforced nanocomposites. *Journal of Computational Applied Mechanics, 50*(2), 413–419. https://doi.org/10.22059/jcamech.2019.293497.458
15. Rao, N., Singh, R., & Bashambu, L. (2021). Carbon-based nanomaterials: Synthesis and prospective applications. *Materials Today: Proceedings, 44*, 608–614. https://doi.org/10.1016/j.matpr.2020.10.593

16. Shalauddin, M., Basirun, W. J., Shamima, A., Hashem, A., & Rahman, M. A. (2023). Nanocellulose-based carbon nanocomposite for the electrochemical sensing application for pharmaceuticals: A review. *Malaysian NANO—An International Journal, 3*(1), 1. https://doi.org/10.22452/mnij.vol3no1.1
17. Ambika, & Singh, P. P. (2021a). Carbon nanocomposites: The potential heterogeneous catalysts for organic transformations. *Current Organic Chemistry, 25*(3), 332–350. https://doi.org/10.2174/1385272824999200401124820
18. Madhavi, V., Vijaya Bhaskar Reddy, A., & Madhavi, G. (2021). Synthesis, characterization, and properties of carbon nanocomposites and their application in wastewater treatment. In M. Jawaid, A. Ahmad, N. Ismail, & M. Rafatullah (Eds.), *Environmental remediation through carbon based nano composites* (pp. 61–83). Springer Nature. https://doi.org/10.1007/978-981-15-6699-8_4
19. Sudharsana, C., Anvarsha, N., Kalyani, P., Sudharsana, C., Anvarsha, N., & Kalyani, P. (2024b). Carbon-based nanocomposites: A comprehensive review of their multifunctional applications. In *Nanocomposites—Properties, preparations and applications*. IntechOpen. https://doi.org/10.5772/intechopen.114402
20. Yan, Q.-L., Gozin, M., Zhao, F.-Q., Cohen, A., & Pang, S.-P. (2016). Highly energetic compositions based on functionalized carbon nanomaterials. *Nanoscale, 8*(9), 4799–4851. https://doi.org/10.1039/C5NR07855E
21. Zamfirova, G. (2018). Indentation methods for the characterization of carbon-based polymer nanocomposites. In A. F. Ismail & P. S. Goh (Eds.), *Carbon-based polymer nanocomposites for environmental and energy applications* (pp. 79–111). Elsevier. https://doi.org/10.1016/B978-0-12-813574-7.00004-6
22. Khan, K., Tareen, A. K., Aslam, M., Mahmood, A., Khan, Q., Zhang, Y., Ouyang, Z., Guo, Z., & Zhang, H. (2020). Going green with batteries and supercapacitor: Two dimensional materials and their nanocomposites based energy storage applications. *Progress in Solid State Chemistry, 58*, 100254. https://doi.org/10.1016/j.progsolidstchem.2019.100254
23. Patel, K. D., Singh, R. K., & Kim, H.-W. (2019). Carbon-based nanomaterials as an emerging platform for theranostics. *Materials Horizons, 6*(3), 434–469. https://doi.org/10.1039/C8MH00966J
24. Felix, L. C., Ede, J. D., Snell, D. A., Oliveira, T. M., Martinez-Rubi, Y., Simard, B., Luong, J. H. T., & Goss, G. G. (2016). Physicochemical properties of functionalized carbon-based nanomaterials and their toxicity to fishes. *Carbon, 104*, 78–89. https://doi.org/10.1016/j.carbon.2016.03.041
25. Khan, A., Jawaid, M., Asiri, A. M., & Inamuddin. (2018). *Nanocarbon and its composites: Preparation, properties and applications*. Woodhead Publishing.
26. Mauter, M. S., & Elimelech, M. (2008). Environmental applications of carbon-based nanomaterials. *Environmental Science & Technology, 42*(16), 5843–5859. https://doi.org/10.1021/es8006904
27. Mortazavi, B., Shojaei, F., Shahrokhi, M., Azizi, M., Rabczuk, T., Shapeev, A. V., & Zhuang, X. (2020). Nanoporous C3N4, C3N5 and C3N6 nanosheets; novel strong semiconductors with low thermal conductivities and appealing optical/electronic properties. *Carbon, 167*, 40–50. https://doi.org/10.1016/j.carbon.2020.05.105
28. Paszkiewicz, S., Szymczyk, A., Janowska, I., Jedrzejewski, R., Linares, A., Ezquerra, T. A., Wagner, H., Tenne, R., & Rosłaniec, Z. (2017). Comparative study on the properties of poly(trimethylene terephthalate)-based nanocomposites containing multi-walled carbon (MWCNT) and tungsten disulfide (INT-WS2) nanotubes. *Polymers for Advanced Technologies, 28*(6), 645–657. https://doi.org/10.1002/pat.3964
29. Díez-Pascual, A. M., Luceño Sánchez, J. A., Peña Capilla, R., & García Díaz, P. (2018). Recent developments in graphene/polymer nanocomposites for application in polymer solar cells. *Polymers, 10*(2), 2. https://doi.org/10.3390/polym10020217
30. Hu, J., Zhang, S., & Tang, B. (2021). 2D filler-reinforced polymer nanocomposite dielectrics for high-k dielectric and energy storage applications. *Energy Storage Materials, 34*, 260–281. https://doi.org/10.1016/j.ensm.2020.10.001

31. Wang, H., Yuan, X., Zeng, G., Wu, Y., Liu, Y., Jiang, Q., & Gu, S. (2015). Three dimensional graphene based materials: Synthesis and applications from energy storage and conversion to electrochemical sensor and environmental remediation. *Advances in Colloid and Interface Science, 221*, 41–59. https://doi.org/10.1016/j.cis.2015.04.005
32. Li, X., Sun, X., Gao, Z., Hu, X., Ling, R., Cai, S., Zheng, C., & Hu, W. (2018b). Highly reversible and fast sodium storage boosted by improved interfacial and surface charge transfer derived from the synergistic effect of heterostructures and pseudocapacitance in SnO_2-based anodes. *Nanoscale, 10*(5), 2301–2309. https://doi.org/10.1039/C7NR07533B
33. Li, M., Deng, Q., Wang, J., Jiang, K., Hu, Z., & Chu, J. (2018a). In situ carbon encapsulation of vertical MoS_2 arrays with SnO_2 for durable high rate lithium storage: Dominant pseudocapacitive behavior. *Nanoscale, 10*(2), 741–751. https://doi.org/10.1039/C7NR07359C
34. Franco, F., Rettenmaier, C., Jeon, H. S., & Cuenya, B. R. (2020). Transition metal-based catalysts for the electrochemical CO_2 reduction: From atoms and molecules to nanostructured materials. *Chemical Society Reviews, 49*(19), 6884–6946. https://doi.org/10.1039/D0CS00835D
35. Li, X., Li, B., Zhang, Q., & Li, X. (2023). Preparation of carbon dots-metal nanoparticles nanocomposites and their application in heterogeneous catalysis. *Nano Futures, 7*(4), 042001. https://doi.org/10.1088/2399-1984/ad03b2
36. Shen, K., Chen, X., Chen, J., & Li, Y. (2016). Development of MOF-derived carbon-based nanomaterials for efficient catalysis. *ACS Catalysis, 6*(9), 5887–5903. https://doi.org/10.1021/acscatal.6b01222
37. Khorsandi, D., Jenson, S., Zarepour, A., Khosravi, A., Rabiee, N., Iravani, S., & Zarrabi, A. (2024). Catalytic and biomedical applications of nanocelluloses: A review of recent developments. *International Journal of Biological Macromolecules, 268*, 131829. https://doi.org/10.1016/j.ijbiomac.2024.131829
38. Long, W., Ouyang, H., Hu, X., Liu, M., Zhang, X., Feng, Y., & Wei, Y. (2021). State-of-art review on preparation, surface functionalization and biomedical applications of cellulose nanocrystals-based materials. *International Journal of Biological Macromolecules, 186*, 591–615. https://doi.org/10.1016/j.ijbiomac.2021.07.066
39. Sunasee, R., Hemraz, U. D., & Ckless, K. (2016). Cellulose nanocrystals: A versatile nanoplatform for emerging biomedical applications. *Expert Opinion on Drug Delivery, 13*(9), 1243–1256. https://doi.org/10.1080/17425247.2016.1182491
40. Bai, L., Liu, Y., Ding, A., Ren, N., Li, G., & Liang, H. (2019). Surface coating of UF membranes to improve antifouling properties: A comparison study between cellulose nanocrystals (CNCs) and cellulose nanofibrils (CNFs). *Chemosphere, 217*, 76–84. https://doi.org/10.1016/j.chemosphere.2018.10.219
41. Ivbanikaro, A. E., & Adeniyi, A. (2023). Cellulose nanocrystals: Source, production and application as a coagulant for water purification. *Water Practice and Technology, 18*(5), 1148–1171. https://doi.org/10.2166/wpt.2023.064
42. Akhlamadi, G., & Goharshadi, E. K. (2021). Sustainable and superhydrophobic cellulose nanocrystal-based aerogel derived from waste tissue paper as a sorbent for efficient oil/water separation. *Process Safety and Environmental Protection, 154*, 155–167. https://doi.org/10.1016/j.psep.2021.08.009
43. Jackson, J., Moallemi, A., Chiao, M., & Plackett, D. (2021). The use of surface-modified nanocrystalline cellulose integrated membranes to remove drugs from waste water and as polymers to clean oil sands tailings ponds. *Polymers, 13*(22), 22. https://doi.org/10.3390/polym13223899
44. Islam, M. S., Chen, L., Sisler, J., & Tam, K. C. (2018). Cellulose nanocrystal (CNC)–inorganic hybrid systems: Synthesis, properties and applications. *Journal of Materials Chemistry B, 6*(6), 864–883. https://doi.org/10.1039/C7TB03016A
45. Jin, X., Wang, S., Sang, C., Yue, Y., Xu, X., Mei, C., Xiao, H., Lou, Z., & Han, J. (2022). Patternable nanocellulose/Ti3C2Tx flexible films with tunable photoresponsive and electromagnetic interference shielding performances. *ACS Applied Materials & Interfaces, 14*(30), 35040–35052. https://doi.org/10.1021/acsami.2c11567

46. Pei, Z., Yu, Z., Li, M., Bai, L., Wang, W., Chen, H., Yang, H., Wei, D., & Yang, L. (2021). Self-healing and toughness cellulose nanocrystals nanocomposite hydrogels for strain-sensitive wearable flexible sensor. *International Journal of Biological Macromolecules, 179*, 324–332. https://doi.org/10.1016/j.ijbiomac.2021.03.023
47. Rdest, M., & Janas, D. (2021). Carbon nanotube wearable sensors for health diagnostics. *Sensors, 21*(17), 17. 10.3390/s21175847.
48. Esawi, A. M. K., & El Borady, M. A. (2008). Carbon nanotube-reinforced aluminium strips. *Composites Science and Technology, 68*(2), 486–492. https://doi.org/10.1016/j.compscitech.2007.06.030
49. Hiremath, N., & Bhat, G. (2017). 4—High-performance carbon nanofibers and nanotubes. In G. Bhat (Ed.), *Structure and properties of high-performance fibers* (pp. 79–109). Woodhead Publishing. https://doi.org/10.1016/B978-0-08-100550-7.00004-8
50. Say, Y., Guler, O., & Dikici, B. (2020). Carbon nanotube (CNT) reinforced magnesium matrix composites: The effect of CNT ratio on their mechanical properties and corrosion resistance. *Materials Science and Engineering: A, 798*, 139636. https://doi.org/10.1016/j.msea.2020.139636
51. Silvestre, J., Silvestre, N., & de Brito, J. (2015). An overview on the improvement of mechanical properties of ceramics nanocomposites. *Journal of Nanomaterials, 2015*(1), 106494. https://doi.org/10.1155/2015/106494
52. Meng, H., Liu, Y., Liu, H., Pei, S., Yuan, X., Li, H., & Zhang, Y. (2020). ZIF67@MFC-Derived Co/N-C@CNFs interconnected frameworks with graphitic carbon-encapsulated co nanoparticles as highly stable and efficient electrocatalysts for oxygen reduction reactions. *ACS Applied Materials & Interfaces, 12*(37), 41580–41589. https://doi.org/10.1021/acsami.0c12069
53. Song, L., Chang, J., Ma, Y., Tan, X., Xu, Y., Guo, L., Chen, Z., Zhao, T., Li, Y., Liu, Y., Zhang, Y., & Chu, W. (2020). Cobalt/nitrogen codoped carbon nanosheets derived from catkins as a high performance non-noble metal electrocatalyst for oxygen reduction reaction and hydrogen evolution reaction. *RSC Advances, 10*(71), 43248–43255. https://doi.org/10.1039/D0RA08750E
54. Zhu, Y. P., Guo, C., Zheng, Y., & Qiao, S.-Z. (2017). Surface and interface engineering of noble-metal-free electrocatalysts for efficient energy conversion processes. *Accounts of Chemical Research, 50*(4), 915–923. https://doi.org/10.1021/acs.accounts.6b00635
55. Correa-Baena, J.-P., Hippalgaonkar, K., Van Duren, J., Jaffer, S., Chandrasekhar, V. R., Stevanovic, V., Wadia, C., Guha, S., & Buonassisi, T. (2018). Accelerating materials development via automation, machine learning, and high-performance computing. *Joule, 2*(8), 1410–1420. https://doi.org/10.1016/j.joule.2018.05.009
56. Liu, Y., Zhao, T., Ju, W., & Shi, S. (2017). Materials discovery and design using machine learning. *Journal of Materiomics, 3*(3), 159–177. https://doi.org/10.1016/j.jmat.2017.08.002
57. Biglari, N., & Zare, E. N. (2024). Conjugated polymer-based composite scaffolds for tissue engineering and regenerative medicine. *Alexandria Engineering Journal, 87*, 277–299. https://doi.org/10.1016/j.aej.2023.12.041
58. Liu, C.-X., & Choi, J.-W. (2012). Improved dispersion of carbon nanotubes in polymers at high concentrations. *Nanomaterials, 2*(4), 4. https://doi.org/10.3390/nano2040329
59. Sadeghi, S., Zehtab Yazdi, A., & Sundararaj, U. (2015). Controlling short-range interactions by tuning surface chemistry in HDPE/graphene nanoribbon nanocomposites. *The Journal of Physical Chemistry B, 119*(35), 11867–11878. https://doi.org/10.1021/acs.jpcb.5b03558
60. Bahru, R., Mohamed, A. R., & Muaz, A. Z. M. F. (2018). Effect of different suspension concentrations of carbon nanotubes in dimethylformamide for electrophoretic deposition. *Materials Research Express, 5*(8), 086407. https://doi.org/10.1088/2053-1591/aad438
61. Parmentier, M., Gabriel, C. M., Guo, P., Isley, N. A., Zhou, J., & Gallou, F. (2017). Switching from organic solvents to water at an industrial scale. *Current Opinion in Green and Sustainable Chemistry, 7*, 13–17. https://doi.org/10.1016/j.cogsc.2017.06.004

62. Wei, J., Saharudin, M. S., Vo, T., & Inam, F. (2017). Dichlorobenzene: An effective solvent for epoxy/graphene nanocomposites preparation. *Royal Society Open Science, 4*(10), 170778. https://doi.org/10.1098/rsos.170778
63. Yang, K., Yi, Z., Jing, Q., Yue, R., Jiang, W., & Lin, D. (2013). Sonication-assisted dispersion of carbon nanotubes in aqueous solutions of the anionic surfactant SDBS: The role of sonication energy. *Chinese Science Bulletin, 58*(17), 2082–2090. https://doi.org/10.1007/s11434-013-5697-2
64. Jiang, L., Gao, L., & Sun, J. (2003). Production of aqueous colloidal dispersions of carbon nanotubes. *Journal of Colloid and Interface Science, 260*(1), 89–94. https://doi.org/10.1016/S0021-9797(02)00176-5
65. Pramanik, C., Gissinger, J. R., Kumar, S., & Heinz, H. (2017). Carbon nanotube dispersion in solvents and polymer solutions: Mechanisms, assembly, and preferences. *ACS Nano, 11*(12), 12805–12816. https://doi.org/10.1021/acsnano.7b07684
66. Li, P., Tang, C., Xiao, X., Jia, Y., & Chen, W. (2022). Flammable gases produced by TiO_2 nanoparticles under magnetic stirring in water. *Friction, 10*(7), 1127–1133. https://doi.org/10.1007/s40544-021-0505-5
67. Franks, R., Morefield, S., Wen, J., Liao, D., Alvarado, J., Strano, M., & Marsh, C. (2008). A study of nanomaterial dispersion in solution by wet-cell transmission electron microscopy. *Journal of Nanoscience and Nanotechnology, 8*(9), 4404–4407. https://doi.org/10.1166/jnn.2008.18351
68. Trofa, M., D'Avino, G., Fabiano, B., & Vocciante, M. (2020). Nanoparticles synthesis in wet-operating stirred media: Investigation on the grinding efficiency. *Materials, 13*(19), 19. https://doi.org/10.3390/ma13194281
69. Vocciante Marco, Trofa Marco, D., Gaetano, A., & Andrea, R. (2019). Nanoparticles synthesis in wet-operating stirred media: Preliminary investigation with dem simulations. *Chemical Engineering Transactions, 73*, 31–36. https://doi.org/10.3303/CET1973006
70. Jin, Y., Beaucage, G., Vogtt, K., Jiang, H., Kuppa, V., Kim, J., Ilavsky, J., Rackaitis, M., Mulderig, A., Rishi, K., & Narayanan, V. (2017). A pseudo-thermodynamic description of dispersion for nanocomposites. *Polymer, 129*, 32–43. https://doi.org/10.1016/j.polymer.2017.09.040
71. Staudinger, U., Krause, B., Steinbach, C., Pötschke, P., & Voit, B. (2014). Dispersability of multiwalled carbon nanotubes in polycarbonate-chloroform solutions. *Polymer, 55*(24), 6335–6344. https://doi.org/10.1016/j.polymer.2014.10.012
72. Abare, A. Y. (2023). Carbon-based nanomaterials and their properties. In T. Khan, M. Jawaid, K. A. Ahmad, & B. Singh (Eds.), *Nanomaterials: The building blocks of modern technology: Synthesis, properties and applications* (pp. 263–278). Springer Nature. https://doi.org/10.1007/978-981-99-4149-0_13
73. Oyarzabal, A., Cristiano-Tassi, A., Laredo, E., Newman, D., Bello, A., Etxeberría, A., Eguiazabal, J. I., Zubitur, M., Mugica, A., & Müller, A. J. (2017). Dielectric, mechanical and transport properties of bisphenol A polycarbonate/graphene nanocomposites prepared by melt blending. *Journal of Applied Polymer Science, 134*(13), 44654. https://doi.org/10.1002/app.44654
74. Verma, P., Saini, P., Malik, R. S., & Choudhary, V. (2015). Excellent electromagnetic interference shielding and mechanical properties of high loading carbon-nanotubes/polymer composites designed using melt recirculation equipped twin-screw extruder. *Carbon, 89*, 308–317. https://doi.org/10.1016/j.carbon.2015.03.063
75. Logakis, E., Pandis, C., Pissis, P., Pionteck, J., & Pötschke, P. (2011). Highly conducting poly(methyl methacrylate)/carbon nanotubes composites: Investigation on their thermal, dynamic-mechanical, electrical and dielectric properties. *Composites Science and Technology, 71*(6), 854–862. https://doi.org/10.1016/j.compscitech.2011.01.029
76. Rostami, A., Masoomi, M., Fayazi, M. J., & Vahdati, M. (2015). Role of multiwalled carbon nanotubes (MWCNTs) on rheological, thermal and electrical properties of PC/ABS blend. *RSC Advances, 5*(41), 32880–32890. https://doi.org/10.1039/C5RA04043D

77. Luo, J., Cerretti, G., Krause, B., Zhang, L., Otto, T., Jenschke, W., Ullrich, M., Tremel, W., Voit, B., & Pötschke, P. (2017). Polypropylene-based melt mixed composites with single-walled carbon nanotubes for thermoelectric applications: Switching from p-type to n-type by the addition of polyethylene glycol. *Polymer, 108*, 513–520. https://doi.org/10.1016/j.polymer.2016.12.019
78. Khatua, B. B., Maiti, S., Shrivastava, N. K., & Suin, S. (2013). A strategy for achieving low percolation and high electrical conductivity in melt-blended polycarbonate (PC)/multiwall carbon nanotube (MWCNT) nanocomposites: Electrical and thermo-mechanical properties. eXPRESS. *Polymer Letters, 7*(6), 505–518. https://doi.org/10.3144/expresspolymlett.2013.47
79. Jiang, F., Wu, S., Wei, Y., Zhang, L., & Hu, G. (2008). The study on the microstructures and high performances of melt blending polyurethane/multiwalled carbon nanotubes composites. *Polymers and Polymer Composites, 16*(8), 509–518. https://doi.org/10.1177/096739110801600804
80. Chaubey, A. K., Kumar, R., & Sahoo, T. (2015). Effect of mechanical mixing on the microstructure and mechanical properties of Al-based metal matrix composites reinforced with nanocrystalline Al-Ca intermetallics. *IOP Conference Series: Materials Science and Engineering, 75*(1), 012011. https://doi.org/10.1088/1757-899X/75/1/012011
81. Chen, X. L. J. P. J. (2010). Numerical study on distributive mixing performance in pin mixing section of single-screw extruder. *China Plastics, 24*(02), 109. https://doi.org/10.19491/j.issn.1001-9278.2010.02.024
82. Goharpey, F., Foudazi, R., Nazockdast, H., & Katbab, A. A. (2008). Determination of twin-screw extruder operational conditions for the preparation of thermoplastic vulcanizates on the basis of batch-mixer results. *Journal of Applied Polymer Science, 107*(6), 3840–3847. https://doi.org/10.1002/app.27401
83. Kazmer, D. O., Masato, D., Piccolo, L., Puleo, K., Krantz, J., Venoor, V., Colon, A., Limkaichong, J., Dewar, N., Babin, D., & Sayer, C. (2021). Multivariate modeling of mechanical properties for hot runner molded bioplastics and a recycled polypropylene blend. *Sustainability, 13*(14), 14. https://doi.org/10.3390/su13148102
84. Breuer, O., & Sundararaj, U. (2004). Big returns from small fibers: A review of polymer/carbon nanotube composites. *Polymer Composites, 25*(6), 630–645. https://doi.org/10.1002/pc.20058
85. Jones, W. E., Chiguma, J., Johnson, E., Pachamuthu, A., & Santos, D. (2010). Electrically and thermally conducting nanocomposites for electronic applications. *Materials, 3*(2), 2. https://doi.org/10.3390/ma3021478
86. Rane, A. V., Kanny, K., Abitha, V. K., & Thomas, S. (2018). Methods for synthesis of nanoparticles and fabrication of nanocomposites. In S. Mohan Bhagyaraj, O. S. Oluwafemi, N. Kalarikkal, & S. Thomas (Eds.), *Synthesis of inorganic nanomaterials* (pp. 121–139). Woodhead Publishing. https://doi.org/10.1016/B978-0-08-101975-7.00005-1
87. Cruz, S. M., & Viana, J. C. (2015). Melt blending and characterization of carbon nanoparticles-filled thermoplastic polyurethane elastomers. *Journal of Elastomers & Plastics, 47*(7), 647–665. https://doi.org/10.1177/0095244314534097
88. dos Anjos, E. G., Moura, N. K., Antonelli, E., Baldan, M. R., Gomes, N. A., Braga, N. F., Santos, A. P., Rezende, M. C., Pessan, L. A., & Passador, F. R. (2023). Role of adding carbon nanotubes in the electric and electromagnetic shielding behaviors of three different types of graphene in hybrid nanocomposites. *Journal of Thermoplastic Composite Materials, 36*(8), 3209–3235. https://doi.org/10.1177/08927057221124483
89. Xiang, D., Guo, J., Kumar, A., Chen, B., & Harkin-Jones, E. (2017). Effect of processing conditions on the structure, electrical and mechanical properties of melt mixed high density polyethylene/multi-walled CNT composites in compression molding. *Materials Testing, 59*(2), 136–147. https://doi.org/10.3139/120.110974
90. Zhou, S., Hrymak, A., & Kamal, M. (2016). Electrical and morphological properties of microinjection molded polystyrene/multiwalled carbon nanotubes nanocomposites. *Polymer Engineering and Science, 56*, 1182–1190. https://doi.org/10.1002/PEN.24352

91. El Sawi, I., Olivier, P. A., Demont, P., & Bougherara, H. (2012). Investigation of the effect of double-walled carbon nanotubes on the curing reaction kinetics and shear flow of an epoxy resin. *Journal of Applied Polymer Science, 126*(1), 358–366. https://doi.org/10.1002/app.36988
92. Stanciu, N.-V., Stan, F., Sandu, I.-L., Fetecau, C., & Turcanu, A.-M. (2021). Thermal, rheological, mechanical, and electrical properties of polypropylene/multi-walled carbon nanotube nanocomposites. *Polymers, 13*(2), 2. https://doi.org/10.3390/polym13020187
93. Tang, H., Chen, J.-B., Wang, Y., Xu, J.-Z., Hsiao, B. S., Zhong, G.-J., & Li, Z.-M. (2012). Shear flow and carbon nanotubes synergistically induced nonisothermal crystallization of poly(lactic acid) and its application in injection molding. *Biomacromolecules, 13*(11), 3858–3867. https://doi.org/10.1021/bm3013617
94. Kotov, N. A., Mamedov, A. A., Guldi, D. M., Prato, M., Wicksted, J., & Hirsch, A. (2004). Nanocomposites, layer-by-layer assembly. In *Encyclopedia of polymer science and technology*. John Wiley & Sons. https://doi.org/10.1002/0471440264.pst455
95. Bhalerao, U. M., Acharya, J., Halve, A. K., & Kaushik, M. P. (2014). Controlled drug delivery of antileishmanial chalcones from layer-by-layer (LbL) self-assembled PSS/PDADMAC thin films. *RSC Advances, 4*(10), 4970–4977. https://doi.org/10.1039/C3RA44611E
96. Hensel, R. C., Pereira-da-Silva, M., Riul, A., & Rodrigues, V. (2020). Dielectric permittivity and surface charge density in layer-by-layer poly(diallyldimethylammonium chloride)/poly(styrenesulfonate). *Nanostructured Films: Implications for Biosensing., 3*, 1749–1754. https://doi.org/10.1021/acsanm.9b02447
97. O'Neal, J. T., Dai, E., Zhang, Y., Clark, K. B., Wilcox, K., George, I., Ramasamy, N. E., Enriquez, D., Batys, P., Sammalkorpi, M., & Lutkenhaus, J. (2017). QCM-D investigation of swelling behavior of layer-by-layer thin films upon exposure to monovalent ions. *Langmuir: The ACS Journal of Surfaces and Colloids, 34*(3), 999–1009. https://doi.org/10.1021/acs.langmuir.7b02836
98. Vidyasagar, A., Sung, C., Gamble, R., & Lutkenhaus, J. (2012). Thermal transitions in dry and hydrated layer-by-layer assemblies exhibiting linear and exponential growth. *ACS Nano, 6*(7), 6174–6184. https://doi.org/10.1021/nn301526b
99. Wang, J., Hu, H., Wang, X., Xu, C., Zhang, M., & Shang, X. (2011). Preparation and mechanical and electrical properties of graphene nanosheets–poly(methyl methacrylate) nanocomposites via in situ suspension polymerization. *Journal of Applied Polymer Science, 122*(3), 1866–1871. https://doi.org/10.1002/app.34284
100. Zan, X., Hoagland, D. A., Wang, T., & Su, Z. (2012). Ion dispositions in polyelectrolyte multilayer films. *Macromolecules, 45*(21), 8805–8812. https://doi.org/10.1021/ma3014492
101. Andres, C. M., & Kotov, N. A. (2010). Inkjet deposition of layer-by-layer assembled films. *Journal of the American Chemical Society, 132*(41), 14496–14502. https://doi.org/10.1021/ja104735a
102. Shim, B., Podsiadlo, P., Lilly, D. G., Agarwal, A., Lee, J., Tang, Z., Ho, S., Ingle, P., Paterson, D., Lu, W., & Kotov, N. (2007). Nanostructured thin films made by dewetting method of layer-by-layer assembly. *Nano Letters, 7*(11), 3266–3273. https://doi.org/10.1021/NL071245D
103. Seo, G. H., Jeong, U., Son, H., Shin, D., & Kim, S. J. (2017). Effects of layer-by-layer assembled PEI/MWCNT surfaces on enhanced pool boiling critical heat flux. *International Journal of Heat and Mass Transfer, 109*, 564–576. https://doi.org/10.1016/J.IJHEATMASSTRANSFER.2017.02.014
104. Akagi, T., Fujiwara, T., & Akashi, M. (2012). Rapid fabrication of polylactide stereocomplex using layer-by-layer deposition by inkjet printing. *Angewandte Chemie, 51*(22), 5493–5496. https://doi.org/10.1002/anie.201201586
105. Beyer, S., Mak, W. C., & Trau, D. (2007). Reverse-phase LbLEncapsulation of highly water soluble materials by layer-by-layer polyelectrolyte self-assembly. *Langmuir, 23*(17), 8827–8832. https://doi.org/10.1021/la7011777
106. Lee, T., Min, S. H., Gu, M., Jung, Y. K., Lee, W., Lee, J. U., Seong, D. G., & Kim, B.-S. (2015). Layer-by-layer assembly for graphene-based multilayer nanocomposites: Synthesis and

applications. *Chemistry of Materials, 27*(11), 3785–3796. https://doi.org/10.1021/acs.chemmater.5b00491
107. Srivastava, S., & Kotov, N. (2008). Composite layer-by-layer (LBL) assembly with inorganic nanoparticles and nanowires. *Accounts of Chemical Research, 41*(12), 1831–1841. https://doi.org/10.1021/ar8001377
108. Zhang, X., Chen, H., & Zhang, H. (2007). Layer-by-layer assembly: From conventional to unconventional methods. *Chemical Communications, 14*, 1395–1405. https://doi.org/10.1039/B615590A
109. Xu, H., Schönhoff, M., & Zhang, X. (2012). Unconventional layer-by-layer assembly: Surface molecular imprinting and its applications. *Small, 8*(4), 517–523. https://doi.org/10.1002/smll.201101884
110. Zhao, Q., Fellinger, T.-P., Antonietti, M., & Yuan, J. (2012). Nitrogen-doped carbon capsules via poly(ionic liquid)-based layer-by-layer assembly. *Macromolecular Rapid Communications, 33*(13), 1149–1153. https://doi.org/10.1002/marc.201200020
111. Hong, J., & Kang, S. (2011). Carbon decorative coatings by dip-, spin-, and spray-assisted layer-by-layer assembly deposition. *Journal of Nanoscience and Nanotechnology, 11*(9), 7771–7776. https://doi.org/10.1166/JNN.2011.4737
112. Zhuk, I., Jariwala, F. B., Attygalle, A., Wu, Y., Libera, M., & Sukhishvili, S. (2014). Self-defensive layer-by-layer films with bacteria-triggered antibiotic release. *ACS Nano, 8*(8), 7733–7745. https://doi.org/10.1021/NN500674G
113. Hong, J., & Park, H. (2011). Fabrication and characterization of block copolymer micelle multilayer films prepared using dip-, spin- and spray-assisted layer-by-layer assembly deposition. *Colloids and Surfaces A: Physicochemical and Engineering Aspects, 381*, 7–12. https://doi.org/10.1016/J.COLSURFA.2011.02.014
114. De, S., Patel, A., & Lutkenhaus, J. (2020). Layer-by-layer assembly of polymers and anisotropic nanomaterials using spray-based approach. *Journal of Materials Research, 35*, 1163–1172. https://doi.org/10.1557/jmr.2020.44
115. Krogman, K., Lyon, K. F., & Hammond, P. (2008). Metal ion reactive thin films using spray electrostatic LbL assembly. *The Journal of Physical Chemistry B, 112*(46), 14453–14460. https://doi.org/10.1021/jp8028853
116. Ganzevles, R. A., Kosters, H., Vliet, T., Stuart, M., & Jongh, H. D. (2007). Polysaccharide charge density regulating protein adsorption to air/water interfaces by protein/polysaccharide complex formation. *The Journal of Physical Chemistry: B, 111*(45), 12969–12976. https://doi.org/10.1021/JP075441K
117. Guo, S., Zhu, X., Li, M., Shi, L., Ong, J. L. T., Jańczewski, D., & Neoh, K. (2016). Parallel control over surface charge and wettability using polyelectrolyte architecture: Effect on protein adsorption and cell adhesion. *ACS Applied Materials & Interfaces, 8*(44), 30552–30563. https://doi.org/10.1021/ACSAMI.6B09481
118. Dobrynin, A., Deshkovski, A., & Rubinstein, M. (2001). Adsorption of polyelectrolytes at oppositely charged surfaces. *Macromolecules, 34*, 3421–3436. https://doi.org/10.1021/MA0013713
119. Claesson, P., Dėdinaitė, A., & Rojas, O. (2003). Polyelectrolytes as adhesion modifiers. *Advances in Colloid and Interface Science, 104*, 53–74. https://doi.org/10.1016/S0001-8686(03)00036-8
120. Steitz, R., Jaeger, W., Klitzing, R., & v. (2001). Influence of charge density and ionic strength on the multilayer formation of strong polyelectrolytes. *Langmuir, 17*(15), 4471–4474. https://doi.org/10.1021/la010168d
121. Decher, G., Hong, J.-D., & Schmitt, J. (1992). Buildup of ultrathin multilayer films by a self-assembly process: III. Consecutively alternating adsorption of anionic and cationic polyelectrolytes on charged surfaces. *Thin Solid Films, 210–211*, 831–835. https://doi.org/10.1016/0040-6090(92)90417-A
122. Chen, Y. L., Chen, S., Frank, C., & Israelachvili, J. (1992). Molecular mechanisms and kinetics during the self-assembly of surfactant layers. *Journal of Colloid and Interface Science, 153*(1), 244–265. https://doi.org/10.1016/0021-9797(92)90316-E

123. Trujillo, M., Arnal, M. L., Müller, A. J., Laredo, E., Bredeau, S., Bonduel, D., & Dubois, P. (2007). Thermal and morphological characterization of nanocomposites prepared by in-situ polymerization of high-density polyethylene on carbon nanotubes. *Macromolecules, 40*(17), 6268–6276. https://doi.org/10.1021/ma071025m
124. Ma, Y., Chiu, P. L., Serrano, A., Ali, S. R., Chen, A. M., & He, H. (2008). The electronic role of DNA-functionalized carbon nanotubes: Efficacy for in situ polymerization of conducting polymer nanocomposites. *Journal of the American Chemical Society, 130*(25), 7921–7928. https://doi.org/10.1021/ja7112189
125. Kim, J., Hong, S. M., Kwak, S., & Seo, Y. (2009). Physical properties of nanocomposites prepared by in situpolymerization of high-density polyethylene on multiwalled carbon nanotubes. *Physical Chemistry Chemical Physics, 11*(46), 10851–10859. https://doi.org/10.1039/B913527H
126. Park, C., Ounaies, Z., Watson, K. A., Crooks, R. E., Smith, J., Lowther, S. E., Connell, J. W., Siochi, E. J., Harrison, J. S., & Clair, T. L. S. (2002). Dispersion of single wall carbon nanotubes by in situ polymerization under sonication. *Chemical Physics Letters, 364*(3), 303–308. https://doi.org/10.1016/S0009-2614(02)01326-X
127. Mao, H., & Wang, X. (2020). Use of in-situ polymerization in the preparation of graphene/polymer nanocomposites. *New Carbon Materials, 35*(4), 336–343. https://doi.org/10.1016/S1872-5805(20)60493-0
128. Sahoo, N. G., Rana, S., Cho, J. W., Li, L., & Chan, S. H. (2010). Polymer nanocomposites based on functionalized carbon nanotubes. *Progress in Polymer Science, 35*(7), 837–867. https://doi.org/10.1016/j.progpolymsci.2010.03.002
129. Adnan, M. M., Dalod, A. R. M., Balci, M. H., Glaum, J., & Einarsrud, M.-A. (2018). In situ synthesis of hybrid inorganic–polymer nanocomposites. *Polymers, 10*(10), 10. https://doi.org/10.3390/polym10101129
130. Wu, X., Takeshita, S., Tadumi, K., Dong, W., Horiuchi, S., Niino, H., Furuya, T., & Yoda, S. (2019). Preparation of noble metal/polymer nanocomposites via in situ polymerization and metal complex reduction. *Materials Chemistry and Physics, 222*, 300–308. https://doi.org/10.1016/j.matchemphys.2018.10.031
131. Abousalman-Rezvani, Z., Eskandari, P., Roghani-Mamaqani, H., & Salami-Kalajahi, M. (2020). Functionalization of carbon nanotubes by combination of controlled radical polymerization and "grafting to" method. *Advances in Colloid and Interface Science, 278*, 102126. https://doi.org/10.1016/j.cis.2020.102126
132. Roghani-Mamaqani, H., Haddadi-Asl, V., & Salami-Kalajahi, M. (2012). In situ controlled radical polymerization: A review on synthesis of well-defined nanocomposites. *Polymer Reviews, 52*(2), 142–188. https://doi.org/10.1080/15583724.2012.668153
133. Homenick, C. M., Lawson, G., & Adronov, A. (2007). Polymer grafting of carbon nanotubes using living free-radical polymerization. *Polymer Reviews, 47*(2), 265–290. https://doi.org/10.1080/15583720701271237
134. Azani, M.-R., & Hassanpour, A. (2024). High-performance polymer nanocomposites: Advanced fabrication methods and critical insights. *Journal of Polymer Research, 31*(6), 168. https://doi.org/10.1007/s10965-024-04025-0
135. Basheer, B. V., George, J. J., Siengchin, S., & Parameswaranpillai, J. (2020). Polymer grafted carbon nanotubes—Synthesis, properties, and applications: A review. *Nano-Structures & Nano-Objects, 22*, 100429. https://doi.org/10.1016/j.nanoso.2020.100429
136. Díez-Pascual, A. M. (2020). Carbon-based polymer nanocomposites for high-performance applications. *Polymers, 12*(4), 4. https://doi.org/10.3390/polym12040872
137. Ehtisham Khan, M. (2021). State-of-the-art developments in carbon-based metal nanocomposites as a catalyst: Photocatalysis. *Nanoscale Advances, 3*(7), 1887–1900. https://doi.org/10.1039/D1NA00041A
138. Bhawal, P., Ganguly, S., Das, T., Mondal, S., Nayak, L., & Das, N. (2019). A comparative study of physico-mechanical and electrical properties of polymer-carbon nanofiber in wet and melt mixing methods. *Materials Science and Engineering: B, 245*, 95–106. https://doi.org/10.1016/J.MSEB.2019.05.020

139. Weimin, Y. (2006). Preparation of TP/TLCP in situ composites by temperature difference blending. *Journal of Beijing University of Chemical Technology*. https://consensus.app/papers/preparation-tptlcp-situ-composites-temperature-weimin/424ae507f9e95b7ea7bec270a16cacf7/
140. Nair, P., George, K. E., & Jayakrishnan, N. (2017). Studies on mechanical behavior high impact polystyrene/vinyl clay nanocomposites: Comparison between in situ polymerization and melt mixing. *Polymer Composites, 38*, 68–76. https://doi.org/10.1002/PC.23561
141. Pötschke, P., Pegel, S., Claes, M., & Bonduel, D. (2008). A novel strategy to incorporate carbon nanotubes into thermoplastic matrices. *Macromolecular Rapid Communications, 29*(3), 244–251. https://doi.org/10.1002/marc.200700637
142. Chen, Z., Cui, C., Jin, C., Li, X., Zhou, Y., Shao, Y., Ma, L., Zhang, Y., & Wang, T. (2023). Tough and thermostable polybutylene terephthalate (PBT)/vitrimer blend with enhanced interfacial compatibility. *Macromolecular Rapid Communications, 44*(10), 2200972. https://doi.org/10.1002/marc.202200972
143. Yu, X., Wei, C., & Zhang, F. (2006). Studies on the mechanical and dynamic mechanical properties of thermotropic liquid crystalline polymer/unsaturated polyester/glass fiber in situ hybrid composites. *Polymers for Advanced Technologies, 17*, 534–539. https://doi.org/10.1002/PAT.741
144. Quiles-Carrillo, L., Montanes, N., Pineiro, F., Jorda-Vilaplana, A., & Torres-Giner, S. (2018). Ductility and toughness improvement of injection-molded compostable pieces of polylactide by melt blending with poly(ε-caprolactone) and thermoplastic starch. *Materials, 11*(11), 11. https://doi.org/10.3390/ma11112138
145. Zheng, Y., Ji, X., Yin, M., Shen, J., & Guo, S. (2017). Strategy for fabricating multiple-shape-memory polymeric materials via the multilayer assembly of co-continuous blends. *ACS Applied Materials & Interfaces, 9*(37), 32270–32279. https://doi.org/10.1021/acsami.7b10345
146. Carrero, A., Grieken, R., Suárez, I., & Paredes, B. (2012). Development of a new synthetic method based on in situ strategies for polyethylene/clay composites. *Journal of Applied Polymer Science, 126*, 987–997. https://doi.org/10.1002/APP.36830
147. Liow, S. S., Dou, Q., Kai, D., Karim, A., Zhang, K., Xu, F., & Loh, X. (2016). Thermogels: In situ gelling biomaterial. *ACS Biomaterials Science & Engineering, 2*(3), 295–316. https://doi.org/10.1021/ACSBIOMATERIALS.5B00515
148. Morris, B. A. (2017). Polymer blending for packaging applications. In B. A. Morris (Ed.), *The science and technology of flexible packaging* (pp. 149–177). William Andrew Publishing. https://doi.org/10.1016/B978-0-323-24273-8.00006-X
149. Yeo, E. S. Y., Mathys, G. I., Brack, N., Thostenson, E. T., & Rider, A. N. (2017). Functionalization and dispersion of carbon nanomaterials using an environmentally friendly ultrasonicated ozonolysis process. *Journal of Visualized Experiments, 123*, 55614. https://doi.org/10.3791/55614
150. Liu, P. (2005). Modifications of carbon nanotubes with polymers. *European Polymer Journal, 41*(11), 2693–2703. https://doi.org/10.1016/j.eurpolymj.2005.05.017
151. Thomassin, J.-M., Huynen, I., Jerome, R., & Detrembleur, C. (2010). Functionalized polypropylenes as efficient dispersing agents for carbon nanotubes in a polypropylene matrix; application to electromagnetic interference (EMI) absorber materials. *Polymer, 51*(1), 115–121. https://doi.org/10.1016/j.polymer.2009.11.012
152. Wu, S., Peng, S., & Wang, C. H. (2018). Multifunctional polymer nanocomposites reinforced by aligned carbon nanomaterials. *Polymers, 10*(5), 5. https://doi.org/10.3390/polym10050542
153. Xie, X.-L., Mai, Y.-W., & Zhou, X.-P. (2005). Dispersion and alignment of carbon nanotubes in polymer matrix: A review. *Materials Science and Engineering: R: Reports, 49*(4), 89–112. https://doi.org/10.1016/j.mser.2005.04.002

Graphene-Based Nanocomposites

Papiya Bhowmik, Gaurav Arora, Manoj Kumar Singh, Sanjay Mavinkere Rangappa, and Suchart Siengchin

1 Introduction to Graphene-Based Nanocomposites

The physicists Andre Geim and Konstantin Novoselov at the University of Manchester did a groundbreaking research on the separation of graphene in 2004 and brought a turning point in materials science which resulted in earning a Nobel Prize in Physics in 2010 [1, 2]. The two-dimensional graphene material sparks a wave of novelty in 2004 in the field of nanotechnology. The separation process using regular adhesive tape to take off layers of graphite raised in plain distinction to the surprising properties of graphene [3, 4]. Researchers were fixated the studies on other nanomaterials, such as carbon nanotubes, before the finding of graphene, which also have brilliant electrical and mechanical properties [5–11]. However, the exceptional properties of graphene have beaten other nanomaterials. The finding marked a major breakthrough in nanotechnology, opening up paths for the progress of graphene-based materials, including graphene-based nanocomposites.

The two-dimensional structure graphene combined with its unparalleled properties, makes it a seamless nanomaterial for creating nanocomposites [12–17]. Its ability to interact successfully with other materials at the nanoscale leads to substantial improvements in properties when combined with polymers, ceramics, and metals [12, 14, 18]. The field of nanocomposites had already been gaining thrust, but graphene's discovery represented as a catalyst, accelerating progress intensely.

P. Bhowmik
Department of Forest and Wood Technology, House M, Linnaeus University, Växjö, Sweden

G. Arora (✉) · M. K. Singh · S. M. Rangappa · S. Siengchin
Natural Composites Research Group Lab, Department of Materials and Production Engineering, The Sirindhorn International Thai-German Graduate School of Engineering (TGGS), King Mongkut's University of Technology North Bangkok (KMUTNB), Bangkok, Thailand

V. Khanna (ed.), *Carbon-Based Nanocomposites for Sustainable Applications, Volume I*, Lecture Notes in Nanoscale Science and Technology 38,
https://doi.org/10.1007/978-3-031-95306-4_2

Researchers speedily began exploring the possibility of incorporating graphene into various matrices, finding that even small quantities could significantly enhance the chemical, physical, mechanical, thermal, and electrical properties of the resulting composite material [3, 8, 12, 19–26].

Graphene-based nanocomposites are revolutionary materials formed by reinforcing graphene into matrices of polymers, ceramics, or metals. The excellent tensile strength, brilliant electrical conductivity, large surface area, and outstanding thermal conductivity make graphene a crucial candidate meant for creating nanocomposites. Graphene-based nanocomposites are transforming numerous fields with their inimitable properties as shown in Fig. 1 [27]. The influence of graphene-based nanocomposites is discussed in the following section:

(a) **Electronics and Sensors:** The electrical conductivity properties of graphene make it an ideal choice for flexible wearable electronic applications. Graphene-

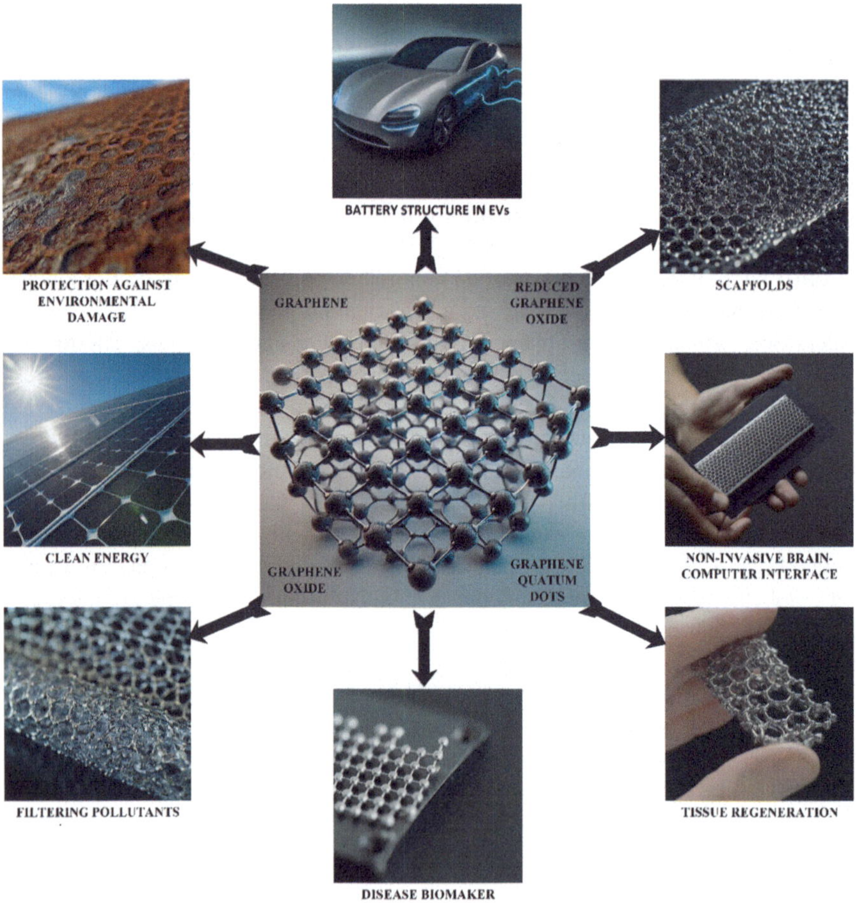

Fig. 1 Applications of graphene and its composites

based nanocomposites improve performance when used for sensors, transistors, and transparent conductive films because they outperform traditional materials such as silicon and indium tin oxide. Graphene's ability to respond to environmental shifts creates valuable opportunities for the development of advanced chemical and biological sensors [3, 26, 28–36].

(b) **Energy Storage:** The performance of batteries and supercapacitors is being enhanced through the transformational effects of graphene nanocomposites in energy storage. Graphene integration into lithium-ion battery electrodes has led to significant enhancements in battery energy density as well as faster charging times and extended lifespans. The exceptional surface area and conductivity of graphene fuel major advancements result in more powerful and sustainable energy storage solutions. [3, 13, 14, 37, 38].

(c) **Structural Materials**: Graphene's remarkable strength-to-weight ratio renders it attractive for use in high-performance structural materials. It is well understood that the integration of graphene into polymers thermoplastics markedly improves the mechanical properties of the resulting nanocomposite. These attributes make it ideal for aerospace, automotive, and civil construction. In addition, the use of grapheme nanofibers in polymers could yield composites with improved thermal stability alongside enhanced fire resistance, further extending its applications. [15, 21, 39].

(d) **Biomedical Applications**: Graphene nanocomposites are advancing in significance in a variety of fields, including drug delivery, biosensors, and so on. Graphene's biocompatibility makes it an excellent candidate for targeted drug delivery systems. Furthermore, scaffolds based on graphene are developing into effective tissue engineering materials, providing advanced methods for repairing injured tissues. [3, 12, 21, 28, 40].

2 Properties, Structure, and Synthesis of Graphene

2.1 Properties and Structure

Graphene is a two-dimensional material involving a single layer of carbon atoms organized in a honeycomb framework [17]. This form of atomic architecture is mainly the reason behind graphene's unique electrical, mechanical, and thermal properties. The singular synthetic structure of carbon gives graphene incredible strength and flexibility and makes it one of the most advanced materials in contemporary science and engineering [17].

Two-dimensional graphene is composed entirely of carbon atoms bonded to three other carbon atoms in a planar configuration, as depicted in Fig. 2. Carbon, periodic table Group 14, has four outer electrons so can make as much as four covalent bonds. In graphene, each carbon atom forms three sigma (σ) bonds with its three neighboring carbon atoms in a trigonal planar arrangement, resulting in one

Fig. 2 A three-dimensional illustration of graphene's atomic structure, depicting a single layer of carbon atoms arranged in a hexagonal lattice, with sigma (σ) bonds connecting adjacent atoms and delocalized pi (π) bonds extending above and below the plane

delocalized bonding electron being left over. The sigma bonds are formed due to sp^2 hybridization, whereby one "*s*" and two "*p*" orbitals hybridize, forming three equivalent hybrid orbitals that are 120° apart from each other, giving rise to the hexagonal structure [17].

The *p* orbitals, which aren't involved in the sigma bonds, create another set of bonds called pi (π) bonds [17]. These mobile pi (π) electrons allow graphene to conduct electricity exceptionally well. The strong sigma bonds and the delocalized pi (π) electrons give graphene its outstanding strength and flexibility.

Graphene is considered as a two-dimensional nanomaterial due to its formation of a single layer of atoms, efficiently reducing its thickness to a single atom. The carbon atoms are closely packed in a honeycomb matrix and an tidy form that is both very effective and balanced [17]. This honeycomb structure means the most even arrangement for carbon atoms in a two-dimensional system, thereby diminishing the total energy of the system. In this framework, each carbon atom is located halfway among its three adjoining carbon atoms, resulting in a hexagonal pattern of periodic units. The bond length of two adjacent carbon atoms in graphene measures 0.142 nanometers (approximately), and the lattice constant is 0.246 nanometers (approximately) [17]. Even under significant mechanical strain, graphene's atomic structure holds together remarkably well due to the robust arrangement of carbon atoms.

The atomic structure of graphene also results in another remarkable feature, i.e., its incomparable electrical conductivity compared to other nanomaterials [4, 17]. The π-cloud created by delocalized electrons separating the plane of carbon atoms into two plane [41]. The unconstrained movement of these electrons across the whole graphene sheet is not being restricted to a specific bond/atom. This electron agility is the fundamental factor that makes graphene a remarkable conductor of electricity. Certainly, graphene is a zero-bandgap semiconductor, which means its conduction and valence bands overlap which allows electrons to move freely, resulting in very little resistance to electrical flow [41]. This exceptional electronic structure has been a key factor for applications in nanoelectronics [13]. It could be integrated in high-speed transistors, sensors, and other devices that demand materials with extraordinary conductivity [42].

The atomic structure of graphene is also responsible for the extraordinary mechanical properties. The strong sp^2 carbon–carbon bonds within the honeycomb framework give graphene its extraordinary tensile strength, establishing it as one of the strongest nanomaterial [21]. Graphene can support huge stress before breaking. It has 130 GPa tensile strength which is about 100 times robust in comparison to steel [41]. Also, the two-dimensional structure permits it to be very elastic and adaptable. It can be strained by up to 20% of its original length without breaking, making it supreme for applications in flexible electronics, wearable devices, and other technologies where both strength and flexibility are vital [11, 43].

2.2 *Synthesis of Graphene*

The synthesis of graphene can be broadly divided into two main approaches, i.e., bottom-up and top-down [28]. Figure 3 provides a visual representation of these two approaches [4].

- Bottom-Up Methods: These methods start with a carbon source, acting as the precursor for graphene formation. The process involves building graphene from smaller units, like atoms or molecules.
- Top-Down Methods: In contrast, these methods begin with a larger graphite source. They involve breaking down graphite into smaller graphene sheets.

2.2.1 Chemical Vapor Deposition (CVD)

CVD is a proven method for growing graphene, and since its development, it's been the subject of extensive research and experimentation aimed at optimizing the process [8, 44–47]. The CVD growth of monolayer graphene films was achieved using polycrystalline copper (Cu) foil as a substrate [45]. Using Cu as a catalyst for graphene growth offers several benefits. It simplifies the transfer of the graphene, it's a cost-effective option, and it gives excellent control over how many layers of graphene are

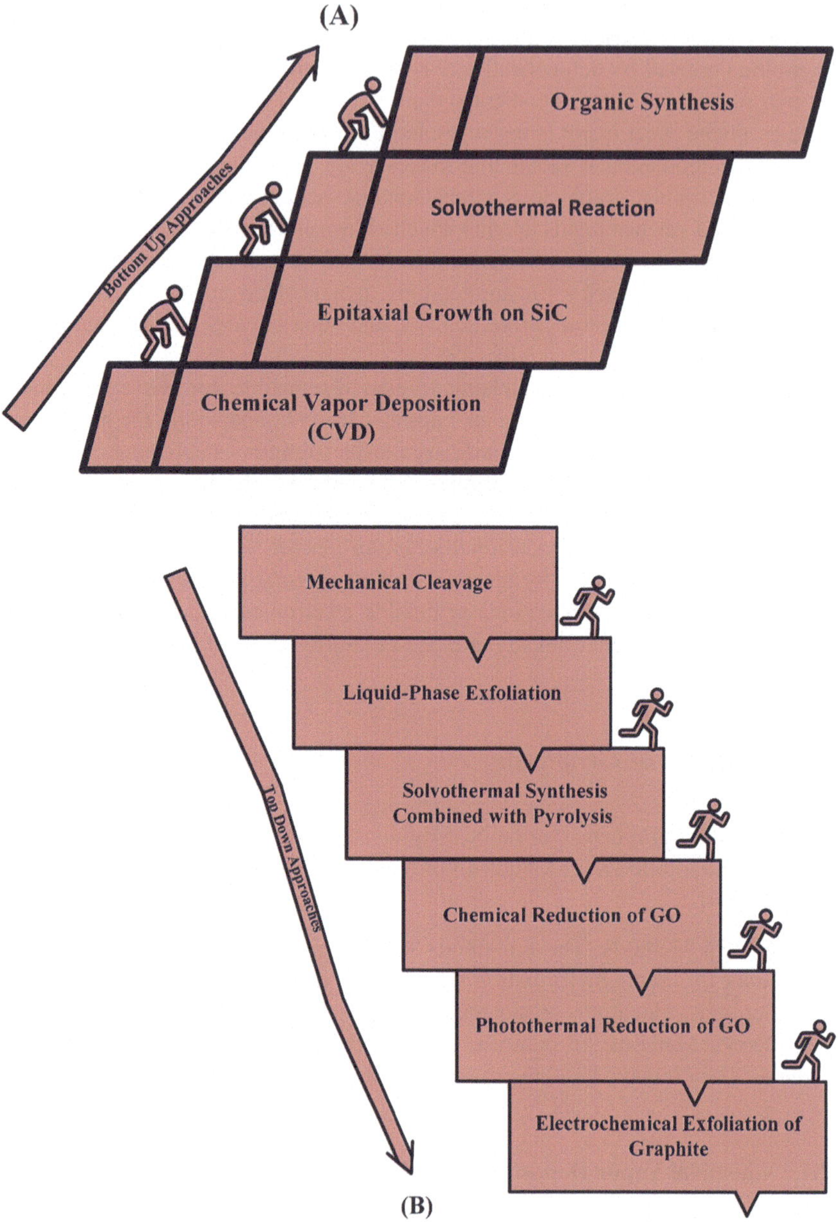

Fig. 3 The main approaches to synthesize graphene, i.e., (**a**) bottom-up and (**b**) top-down [4, 28]

formed. The growth of graphene on Cu is a surface adsorption process. This means carbon atoms from the precursor gas adsorb onto the copper surface and form a monolayer of graphene. This is mostly due to the little solvent compatibility of carbon in Cu, which confirms the formation of largely single-layer graphene, making it a flawless substrate for large-scale graphene synthesis. The stages to produce a large-scale graphene synthesis using CVD technique are illustrated in Fig. 4 [46, 47].

2.2.2 Liquid-Phase Exfoliation (LPE)

LPE is a technique to produce graphene from graphite in a liquid and then using energy to pull apart the individual graphene layers [48]. This method is extremely adaptable, valid for diverse environments and on numerous substrate categories. The ultrasound-induced forces and interactions with solvent molecules drove this exfoliation process [48]. The principle behindhand LPE includes simplifying the separation of graphene layers. In graphite, these layers are firmly bound by sturdy electrostatic forces, needing substantial mechanical energy to overcome. To diminish the energy needed, the attractive forces among the graphene layers are disordered. This is achieved by submerging the bulk graphite in a particular liquid, which is then followed by the exfoliation process. Figure 5 illustrates the main processing steps to produce the graphite [48].

2.2.3 Mechanical Exfoliation

Over the past decade, most graphene used in investigation has been produced through mechanical exfoliation [49]. In this practice, graphite samples are positioned on the gluey side of adhesive tape, pushed onto a substrate, and then peeled away, parting behind small flakes of graphene of few microns' width. While the graphene formed through this method is of outstandingly high quality, permitting researchers to study its physical and mechanical properties but it is not applied for industries with large-scale production. Figure 6 illustrates the mechanical exfoliation process [49].

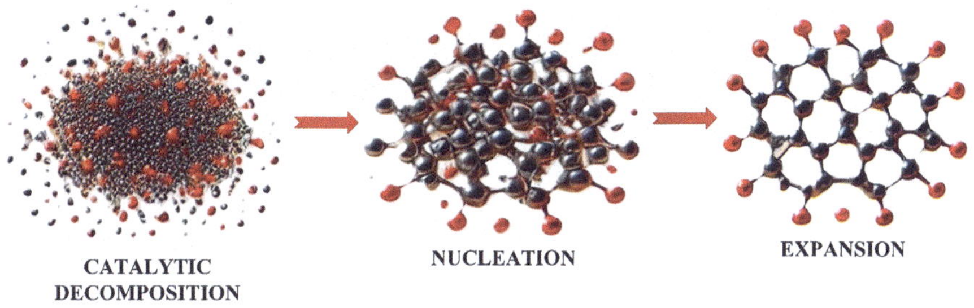

Fig. 4 CVD production technique [46]

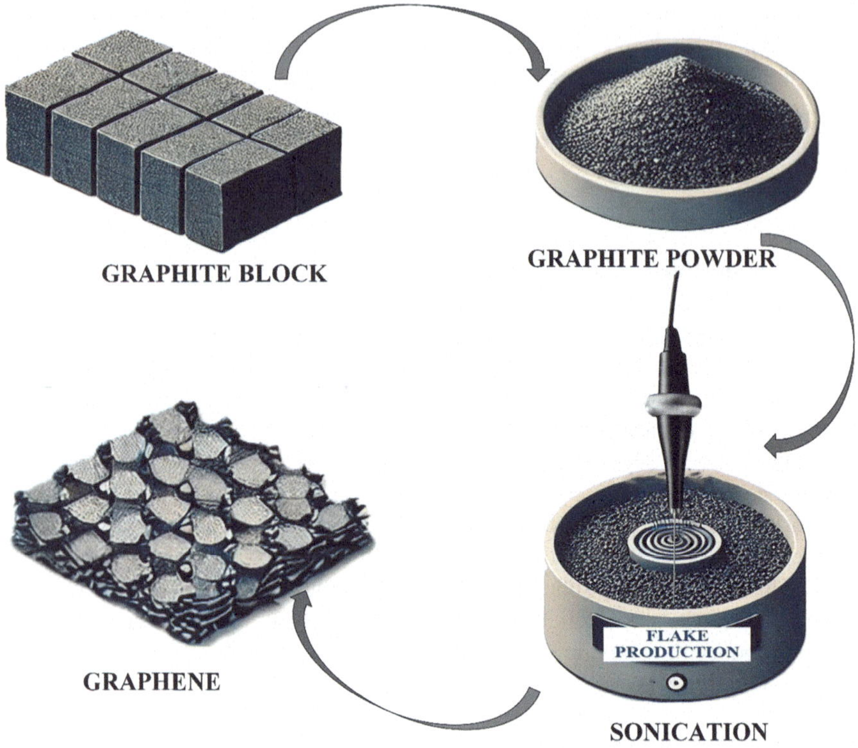

Fig. 5 A schematic of liquid exfoliation technique to produce graphene [48]

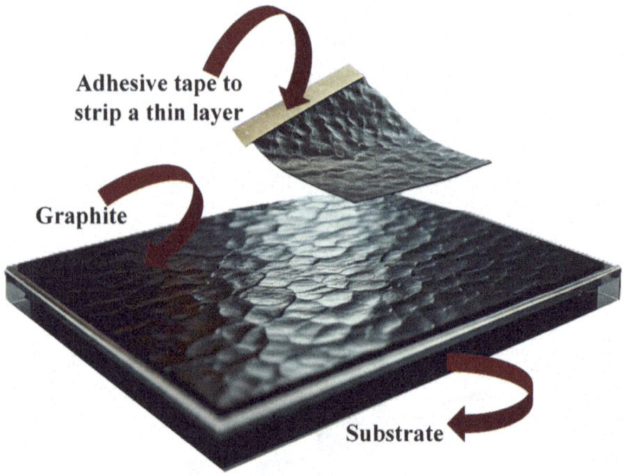

Fig. 6 A schematic of mechanical exfoliation technique to produce graphene [50]

3 Synthesis Methods for Graphene Nanocomposites

Graphene nanocomposites, which associate the incomparable properties of graphene with materials like polymers, metals, or ceramics, are produced through numerous procedures to enhance their properties. The choice of synthesis technique depends on the desired application and the materials involved. Some of the utmost commonly used procedures for manufacturing graphene nanocomposites are as follows:

3.1 In Situ Polymerization

In situ polymerization is a generally used technique for manufacturing graphene-based polymer nanocomposites, where graphene is unified into a polymer matrix throughout the polymerization process [51]. In situ polymerization encompasses the subsequent steps:

- Dispersion of Graphene: Graphene, graphene oxide (GO), or functionalized graphene is first spread into a monomer solution. To ensure the even dispersal of graphene in the monomer solution is crucial for producing a well-integrated composite. Functionalization of graphene can improve its compatibility with the polymer by familiarizing reactive groups.
- Initiation of Polymerization: The monomers' polymerization ensues in the existence of graphene, started by chemical catalysts, heat, or radiation. As the polymerization begins, the monomer units linked to the graphene sheets, producing a strong bond between the polymer chains and the graphene surface.
- Formation of the Nanocomposite: As the polymer nurtures, graphene becomes surrounded within the polymer matrix. The consequence is a nanocomposite material where the polymeric chains edge and interrelate with the graphene sheets, improving the composite's overall properties. Table 1 provides the summary of key literature that proves the advantages of in situ polymerized nanocomposites.

3.2 Solution Blending

Solution blending is an extensively used and comparatively simple process to synthesize graphene-based nanocomposites, particularly for polymer matrices [54]. This method includes the mixing of graphene or its derivatives with a polymer in a solvent, confirming uniform distribution and interaction among the graphene sheets and the polymer matrix [54]. A stepwise procedure is shown in Fig. 7.

Table 1 Summary of key literature that proves the advantages of in situ polymerized nanocomposites

Nanocomposite	Key points	References
Graphene-polydimethylsiloxane (PDMS)	1. The tensile strength and elastic modulus have increased by 54% and 36%, respectively	[29]
Nanocomposites	2. A relative permittivity of 4.216 has been obtained	
	3. The sensor prototype demonstrates exceptional sensitivity, achieving a response 45 times higher than that of pristine PDMS	
Polyamide-6 (PA6)/graphene oxide (GO) nanocomposites	1. Interfacial interaction appears to be exceptional in the composites due to the amount of functionalization on functionalized-GO sheets and the enhancement in Young's modulus at low GO content	[20]
Graphene nanosheets-filled poly-methyl methacrylate (PMMA) nanocomposites	1. The composites exhibited high electrical conductivity of 0.57 S/m, with a low percolation threshold of 0.3 wt%	[16]
	2. The elastic modulus, tensile strength, Rockwell hardness, and nanoindentation hardness reached maximum values, approximately 14%, 41%, 5%, and 40% higher than those of neat PMMA, respectively, at a low percolation threshold of 0.3 wt%	
	3. The impact strength of the PMMA polymer is significantly enhanced with the addition of 2.0 wt% nanosheets, reaching a high value of 12.89 KJ/m^2—Approximately 194% higher than that of neat PMMA	
PMMA/reduced graphene oxide nanocomposites	1. Electrical conductivity results indicate that the sheet-casting method achieves a conductivity of around 10^{-3} S/cm at an RGO loading of 2.0 wt%, making it suitable for EMI-shielding applications	[52]
	2. The 2 wt% RGO-loaded sheet-cast composite demonstrates nearly the equivalent strength as a 1 wt% in situ/PMMA bead composite, suggesting improved dispersion and stronger filler-matrix interactions in the sheet-cast composite	
Polypropylene/graphene nanosheet nanocomposites	1. In these nanocomposites, Young's modulus is enhanced by 50%, tensile strength rises by 25%, whereas elongation at break is reduced by 30%	[43]
	2. The in situ polymerization method used to synthesize these nanocomposites likely encourages the growth of polymeric chains around graphene nanosheets (GNS), which aids in their dispersion but may inhibit more effective inter-particle contact	
High-density polyethylene (HDPE)/graphene nanocomposites	1. The addition of graphene improves both the storage modulus and loss modulus of the material	[53]
	2. Graphene enhances the fire resistance properties	

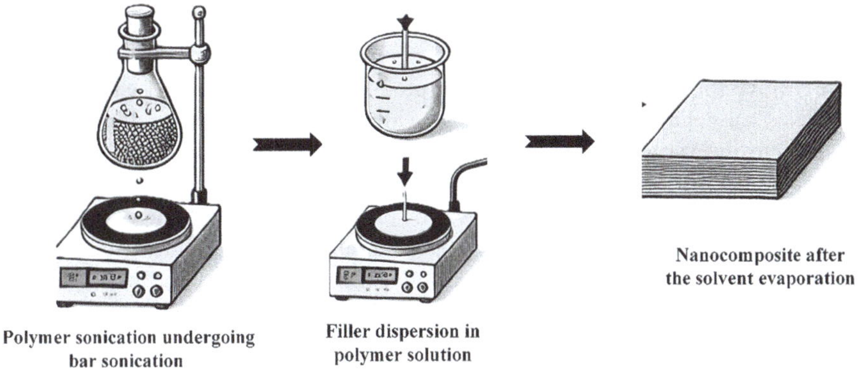

Fig. 7 Solution blending procedure to produce the nanocomposites [54]

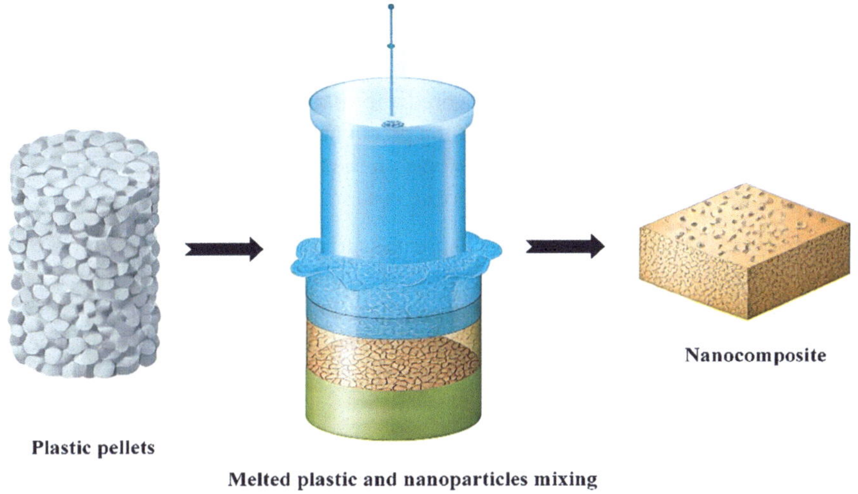

Fig. 8 Melt blending technique for producing nanocomposites [54]

3.3 Melt Blending

Melt blending is a common method for producing composites because it is efficient and compatible with existing industrial processes [54]. Figure 8 illustrates the melt blending process which involves the mixing of fillers into a matrix above the melting temperature of the matrix [54]. This method is valid for thermoplastic polymers only. The various advantages like cost-effectiveness and flexibility lead this method to use for polymer nanocomposites. But, a constraint of formulation and ideal mixing is essential to define for the polymers.

3.4 Sol-Gel Method

A highly versatile method with efficient production of graphene-based nanocomposites is sol-gel method. This method provides a robust platform to reinforce graphene into the variety of matrices [55]. Sol-gel method is a two-step method in which the sol is formed first and then gelation will be done. The sol-gel process is particularly attractive because it allows for precise control over the composition, structure, and morphology of the resulting material, making it a supreme method for fabricating graphene-based composites with tailor-made properties.

3.5 Chemical Vapor Deposition (CVD)

The most extensively effective and used method for the production of high-quality graphene is CVD for the use in graphene nanocomposites [56]. The need of high-performance applications which require large area but few layer graphene with exceptional uniformity and purity favored the discovery of CVD method. This method decomposes the carbon-containing precursor gas (methane under high temperature) for the admission of carbon atoms on the top of a substrate. The generalized process to fabricate the nanocomposite is as follows [56]:

- **Substrate Preparation**: In the CVD process, graphene is classically synthesized on metallic substrates such as copper (Cu) or nickel (Ni). Copper is often preferred due to its low carbon solubility, which limits the growth of excess graphene layers, leading to a more controlled and uniform single-layer graphene formation. The substrate is usually cleaned and prepared to ensure it is free from impurities that might affect graphene growth.
- **Introduction of Precursor Gas**: A carbon source (typically methane (CH_4) or acetylene) is introduced into the reaction cavity along with a carrier gas such as hydrogen (H_2) or argon (Ar). The chamber is heated to 800 °C to 1000 °C temperatures. Below these environments, the carbon-containing precursor gas decomposes, and the carbon atoms diffuse on top surface of the substrate.
- **Graphene Growth**: On the substrate, carbon atoms adsorb and form a hexagonal lattice, which is the basic structure of graphene. On copper substrates, the growth method is self-controlling, meaning that it naturally tends to form a single layer due to copper's low solubility for carbon. On other metals like nickel, multiple layers of graphene can form due to higher carbon solubility.
- **Cooling and Transfer**: Once the graphene has formed on the metal substrate, the system is gradually cooled down. For many applications, the graphene must be transferred from the metal substrate onto another substrate (such as polymers, ceramics, or other composites) for incorporation into nanocomposites. This transfer process generally includes coating the graphene with a polymer film, etching off the metal substrate, and subsequently transferring the graphene onto the target material.

4 Types of Graphene-Based Nanocomposites and Their Properties

Graphene-based nanocomposites signify a class of unconventional materials reinforcing graphene into a host matrix. These nanocomposites influence the unique properties of graphene, such as remarkable electrical conductivity, mechanical strength, and thermal stability, to improve the performance of parent material. Depending on the host matrix, graphene-based nanocomposites can be classified into divergent types, each with separate features and applications. The vital types include polymer-based, ceramic-based, metal-based, and cementitious graphene nanocomposites [57–61].

4.1 Polymer-Based Graphene Nanocomposites

The most commonly studied or researched nanocomposites are polymer-based graphene nanocomposites. In this type, graphene or graphene oxide or reduced graphene oxide are spread within a polymer matrix, increasing the material's properties such as electrical conductivity, mechanical strength, and thermal stability [62–69].

- Properties:
  - Mechanical properties: The reinforcement of graphene significantly improves the mechanical properties such as Young's modulus and tensile strength, and impact resistance of the polymer. This makes graphene/polymer nanocomposites perfect for structural applications [70].
  - Electrical conductivity: Graphene imparts electrical conductivity to insulating polymers, aiding the formation of conductive nanocomposites valuable in electronic and sensor applications [71, 72].
  - Thermal stability: Graphene surges the thermal conductivity and heat resistance of polymers, enabling them appropriate for high-temperature applications [64].
- Types of Polymer Matrices:
  - Thermoplastics: Polymers such as polyethylene (PE), polystyrene (PS), polypropylene (PP), and polyamide (PA6) are frequently used matrices in graphene nanocomposites [73]. These materials display discriminating mechanical properties and electrical conductivity due to graphene reinforcement.
  - Thermosets: Thermosetting polymers (like epoxy, polyurethane, etc.) are also benefitted from the uniform distribution of graphene, which improves their mechanical strength and thermal stability [74, 75].
  - Elastomers: The flexibility of elastomers like natural rubber or silicone is preserved by reinforcing graphene and boosting the mechanical properties [76, 77].

- Applications:
  - Flexible electronics and biomedical field: The improved electrical conductivity and elasticity make polymer/graphene nanocomposites perfect for flexible electronic devices and biomedical field [78–80].
  - Automotive and aerospace: Graphene-based nanocomposites are strong nanomaterials for automotive and aerospace applications with low weight and focusing on sustainability with durability [80, 81].
  - Energy storage devices: Supercapacitors and batteries requiring high electrical conductivity with mechanical stability use graphene-based nanocomposites [71, 82–85].

Table 2 summarizes the comparison of different properties of various graphene-reinforced nanocomposites.

Table 2 Comparison of different properties for various graphene-reinforced nanocomposites [70–72, 74–78, 80, 83, 85–93]

Properties	Polymer-based nanocomposites	Ceramic-based nanocomposites	Metal-based nanocomposites	Cementitious nanocomposites
Mechanical properties	Tensile strength and Young's modulus increased by up to 50% Enhanced flexibility	Improved fracture toughness, hardness, and wear resistance less brittle	Increased strength, hardness, and wear resistance while maintaining ductility	Increased compressive and tensile strength improved durability
Electrical properties	Moderate conductivity; can be made conductive for sensors and electronics	Typically insulating; graphene enhances conductivity, making it useful in sensors	High electrical conductivity; suitable for electronic and thermal management applications	Self-sensing properties due to graphene's conductivity; useful for structural health monitoring
Thermal properties	Improved thermal stability and heat dissipation suitable for high-temp environments	Significant improvement in thermal conductivity. High heat resistance	Enhanced thermal conductivity for heat dissipation, making it ideal for heat sinks	Better resistance to thermal degradation, enhancing durability
Optical properties	Good transparency in films; suitable for flexible electronics and displays	Not widely studied for optical applications, but can be used in specialized fields	Limited optical use but potentially useful for plasmonic applications	No significant optical properties, mainly used in construction

4.2 *Ceramic-Based Graphene Nanocomposites*

The integration of graphene into ceramics improves the brittle nature by increasing the electrical, mechanical, and thermal properties of ceramic-based graphene nanocomposites. Also, the toughness and flexibility improved with the addition of graphene in brittle but strong ceramic-based graphene nanocomposites.

- Properties:
  - Mechanical strength and toughness: The reinforcement of graphene into ceramics amplifies the fracture toughness, wear resistance, and hardness by avoiding the crack propagation in ceramic-based nanocomposites [86].
  - Electrical conductivity: The conductive ceramic based-nanocomposites are produced with the inclusion of graphene which is essential for electronic and sensor applications [92].
  - Thermal conductivity: Graphene increases the thermal properties of ceramics, permitting for enhanced heat dissipation, which is critical in high-temperature environments [94].
- Types of Ceramic Matrices:
  - Alumina (Al_2O_3): Alumina-based graphene nanocomposites are used in wear-resistant coatings and cutting tools [95]. The addition of graphene in alumina ceramics improves the material's toughness and mechanical properties.
  - Silicon Carbide (SiC): Graphene improves the mechanical strength and electrical conductivity of silicon carbide ceramics, which are used in cutting edge engineering applications, specifically aerospace parts and electronic devices [94].
  - Zirconia (ZrO_2): The addition of graphene in zirconia matrix is recognized due to its high strength and toughness making it suitable for biomedical and structural applications [94].
- Applications:
  - Aerospace: Ceramic-based graphene nanocomposites are used in aerospace components including thermal resistance and high strength [94].
  - Biomedical implants: Graphene increases the mechanical properties and biocompatibility of ceramic materials, making them appropriate for implants such as dental and bone prostheses [96].
  - Cutting tools: The wear resistance and toughness of ceramic-based graphene nanocomposites make them perfect for cutting tools and abrasive applications [97].

4.3 Metal-Based Graphene Nanocomposites

Metal-based graphene nanocomposites composed of a metallic matrix reinforced with graphene to improve the various properties (thermal, electrical, and mechanical) of the metal. The integration of graphene into metals can improve their hardness, strength, and conductivity while sustaining their ductility.

- Properties:
  - Mechanical properties: The toughness, robustness, and deformation of these nanocomposites increase by reinforcing the graphene which can intensify the tensile strength, hardness, and wear resistance of metals in various applications [87].
  - Electrical conductivity: The excellent electrical conductivity of graphene can improve the conductive properties of metals, making them appropriate for electronic applications [87].
  - Thermal conductivity: Graphene enhances the thermal conductivity of metal composites, allowing for more effective heat dissipation [87].
  - These composites have superior mechanical, electrical, and thermal properties compared to other graphene nanocomposites, shown in Figs. 9, 10, and 11, respectively.
- Types of Metal Matrices:
  - Aluminum (Al): Aluminum-based graphene nanocomposites offer lightweight materials with enhanced strength and conductivity making it ideal for aerospace and automotive industries [103].

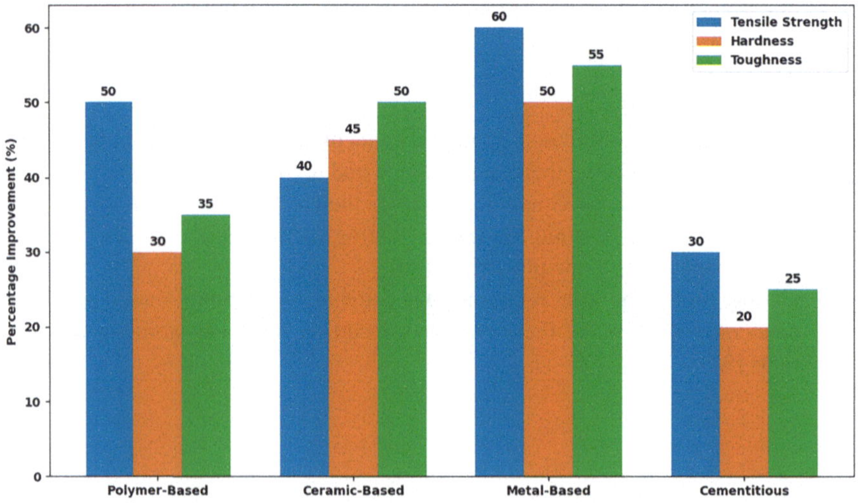

Fig. 9 Comparison of mechanical properties for various graphene nanocomposites [71, 98–101]

- Copper (Cu): Copper-based graphene nanocomposites have enhanced electrical and thermal conductivity, making them suitable for heat sinks and electronic components [102].
- Magnesium (Mg): Magnesium-based graphene nanocomposites provide lightweight materials with high strength-to-weight ratios, useful in structural applications where weight is a concern [87].

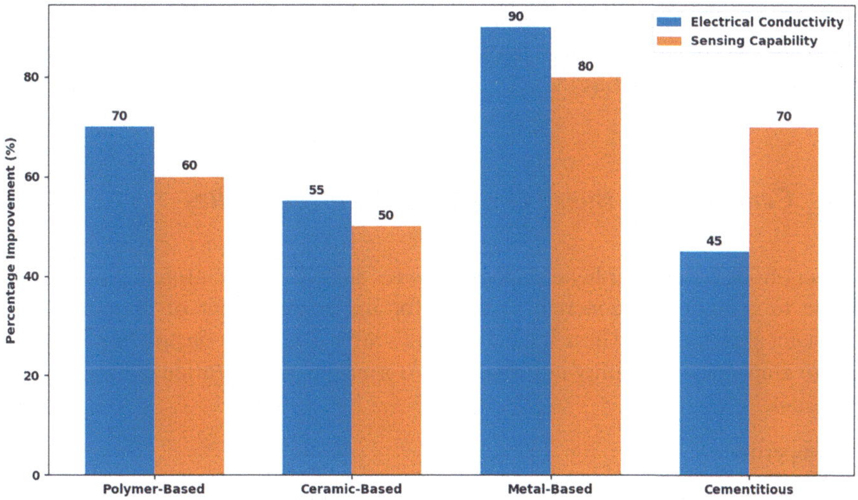

Fig. 10 Comparison of electrical properties for various graphene nanocomposites [71, 72, 77, 78, 91, 97]

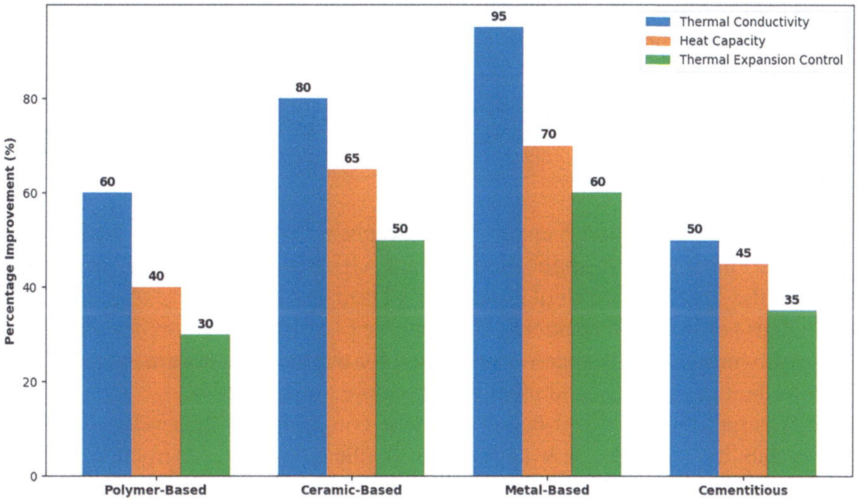

Fig. 11 Comparison of thermal properties for various graphene nanocomposites [83, 91, 94, 97, 102]

- Applications:
  - Aerospace and automotive: Metal-based graphene nanocomposites are used in structural components requiring lightweight materials with good thermal conductivity and high strength [90].
  - Electronics: The excellent electrical conductivity of metal-based graphene nanocomposites pointing toward their use in electronic circuits and conductive coatings [87, 102].
  - Thermal management: The excellent thermal conductivity of metal-based graphene nanocomposites leads their use in heat sinks and other thermal management applications [87].

4.4 Cementitious-Based Graphene Nanocomposites

Cementitious-based graphene nanocomposites include the reinforcement of graphene to cement and concrete matrices for the enhancement of durability and mechanical properties. These nanocomposites are used in the construction industry due to graphene's capability to boost the performance of conventional cementitious materials.

- Properties:
  - Mechanical strength: The strength, i.e., compressive and tensile of concrete increases by incorporating graphene, thus making it more resilient to cracking and deformation [101, 104].
  - Durability: The durability of concrete after reinforcement of graphene increasing its resistance to water and chemical penetration, which helps in preventing the degradation over time [91].
  - Self-sensing properties: The integration of graphene provides the self-sensing capabilities to these nanocomposites and thus allowing the exposure of stress and damage in concrete structures [105–107].
- Applications:
  - Smart infrastructure: Cement-based graphene nanocomposites are used in smart infrastructure applications, where self-sensing competences can monitor the structural health of buildings and bridges [105–107].
  - High-performance concrete: The improved mechanical properties and durability make these nanocomposites ideal for use in high-performance concrete for buildings, roads, and other infrastructure projects [104].
  - Waterproofing: Cement-based graphene composites with developed impermeability are used in waterproofing applications for construction materials [107].

5 Challenges in Graphene Nanocomposites

5.1 Dispersion Challenges

5.1.1 Uniform Dispersion in Matrices

One of the primary encounters in fabricating graphene nanocomposites is attaining uniform distribution of graphene within the host matrix. Graphene has a solid inclination to agglomerate due to larger surface area and van der Waals forces [108]. This agglomeration primes to poor distribution and negatively impacts the composite's various properties. A well-dispersed composite is important for creating optimum graphene and matrix interactions which result in improved performance. However, the local stress concentrations due to poor distribution of graphene in matrix result in reducing the composite's overall properties [108].

5.1.2 Methods to Improve Dispersion

- **Functionalization**: One of the method to functionalize the graphene is surface/chemical treatment which increase the interaction with the matrix but changes the intrinsic properties of graphene and creating new complications [109, 110].
- **Exfoliation Techniques**: Techniques such as sonication or high-shear mixing have been used to distribute graphene uniformly, but these methods are not always accessible or cost-effective for large-scale manufacturing [111].

5.1.3 Challenges in Different Matrices

- **Polymer Matrices**: The spreading of graphene in polymers (like epoxy, polycarbonate, and polyamide) faces challenges due to the hydrophobic nature of graphene nature and the viscosity of the polymer [111].
- **Metal Matrices**: Graphene-based metal nanocomposites require sophisticated techniques like powder metallurgy, electrochemical deposition, etc., but attaining uniform distribution of graphene remains problematic [112].

5.2 Scalability of Synthesis

Although the production of graphene using methods like CVD, mechanical exfoliation, and liquid-phase exfoliation is achievable, but using these methods for large-scale applications is still an important challenge [4, 13, 24]. Some of the crucial issues include:

- Although CVD produces high-quality graphene, it is costly, energy-intensive, and slow, making it inappropriate for mass production.
- Liquid-phase exfoliation method has possibility for scalability, but the subsequent graphene regularly has defects, dropping its efficacy in high-performance applications.
- Reduction of graphene oxide to reduced graphene oxide (rGO) is scalable but produces lower-quality graphene than exfoliation or CVD techniques.

5.2.1 Cost Considerations

Graphene's commercialization is hindered by the high costs of producing high-quality material at scale. Industries such as electronics, aerospace, and construction require graphene that meets rigorous performance standards while also being economically viable for large-scale applications. To adopt the use of graphene in nanocomposites widely in various sectors, a balance between the quality and cost-effective synthesis is essential. Although the current production methods are suitable for research and development but are still time-consuming and expensive to meet the demand at a larger scale [113].

The superior quality graphene necessity for advanced applications is characteristically met by using methods like CVD, which is known for manufacturing single-layer or few-layer graphene with brilliant electrical, mechanical, and thermal properties. However, CVD is very expensive due to the requirement for high-purity gases, high temperatures, and complex vacuum systems. The slow deposition rate and limited substrate size in CVD process further boundary production scalability. To become commercially feasible, industries require access to graphene that is not only high quality but also low cost to produce in bulk, and this has confirmed hard to achieve [4, 13, 24].

Alternatively, other synthesis methods such as liquid-phase exfoliation, reduced graphene oxide (rGO) production, and solution processing offer a more cost-effective path to graphene production, mostly for applications where pristine graphene is not an compulsion [113]. These methods are capable of producing graphene in larger quantities, but they characteristically producing graphene with lower purity and more defects, which can harmfully affect the material's properties, especially in high-performance applications like microelectronics and optical devices [114]. The trade-off among quality and scalability remains one of the crucial obstacles in the wide acceptance of graphene-based nanocomposites.

5.3 Environmental and Health Implications

5.3.1 Environmental Impact

The environmental considerations have been raised by the wide production and application of graphene-based nanocomposites. As the production of graphene includes the use of harsh chemicals, energy-intensive processes, and occasionally the release of hazardous by-products [17, 115], thus priming toward the environmental degradation if not managed appropriately.

- **Waste Generation**: The reduction of graphene oxide (GO) often involves chemical processes that can produce hazardous waste products [116].
- **Energy Usage**: Processes like CVD involve substantial energy input, contributing to the carbon footprint [117].
- **Post-Use Disposal**: If graphene is reinforced into non-biodegradable matrices, the environmental impact of graphene-based nanocomposites at the end of their lifespan is a growing concern.

5.3.2 Health Implications

The health risks which are related to exposure of graphene and its derivatives are not fully understood, but early studies have recommended that graphene particles could cause health problems when gasped or exposed to the skin. Table 3 provides potential health risks of graphene-based nanocomposites

- **Inhalation Risks**: Graphene particles, due to their infinitesimal size, can be inhaled, potentially leading to respiratory issues such as lung irritation or chronic diseases [119].
- **Toxicity**: Functionalized graphene may bearing enlarged toxicity risks due to the chemical modifications involved while offering improved properties [119].
- **Regulatory Gaps**: There are currently no standardized safety guidelines for the handling, exposure, and disposal of graphene nanomaterials or nanocomposites [118].

5.3.3 Regulatory and Safety Challenges

The lack of strict regulations and protection protocols for the manufacture, use, and disposal of graphene nanocomposites is a significant hindrance to their wider incorporation into industrial applications. As products involving graphene continue to gain traction in industries such as electronics, energy, aerospace, and construction, their potential effects on the environment and health are receiving increasing attention. The absence of uniform guidelines for the harmless handling, processing, and dumping of graphene materials postures an important challenge for both manufacturers and regulatory bodies, rising alarms about the long-term feasibility of these advanced materials.

Table 3 Potential health risks of graphene-based nanocomposites [118]

Exposure route	Potential health risk	Preventive measures
Inhalation	Respiratory issues	Proper handling, PPE (masks)
Skin contact	Irritation, toxicity	Use of protective clothing
Ingestion	Unclear but potential systemic effects	Good lab practices, safety protocols

Due to the possible risks, global standards for the safe handling and processing of graphene materials must be developed. Global regulatory bodies such as the European Union (EU), Occupational Safety and Health Administration (OSHA), and other international agencies must collaborate to develop final recommendations for facilitating the safe application of graphene nanocomposites during their life cycle, from synthesis to disposal [120]. In addition, industries need to be encouraged to employ green synthesis techniques, which minimize the application of dangerous chemicals and also lower the environmental impact of graphene production.

6 Summary

Graphene nanocomposites are a significant advancement in materials science, with properties better than their parent materials and therefore are found to be useful in various industries. Their improved mechanical, optical, electrical, and thermal properties are due to the addition of graphene, whose material is stronger, highly conductive, thermally stable, and transparent. These properties are also enhanced when graphene is dispersed homogeneously in matrices like polymers, metals, ceramics, and cementitious materials to create graphene nanocomposites for use in the aerospace sector, energy storage devices, sensors, and in building construction. The processing methods to incorporate graphene into different matrices have a very significant influence in the resultant distinctive properties of graphene-based nanocomposites.

Sophisticated methods like in situ polymerization, CVD, and solution blending have been thoroughly investigated and optimized to produce high-performance nanocomposites. Every synthesis path has a specific advantage and drawback, which adds to the overall efficiency and applicability of the final composite. In situ polymerization is one of the most common synthesis paths for polymer-based graphene nanocomposites. In this approach, graphene or graphene derivatives (e.g., graphene oxide) are included during the polymerization process, and polymer chains form around dispersed graphene sheets. This results in favorable interfacial adhesion between the matrix and the graphene, which effectively enhances the mechanical performance of the composite. The in situ method also enables dispersion, which works to reduce the stacking of graphene sheets, a frequent problem with other synthesis techniques. Despite this method, however, perfect dispersion is difficult to obtain because graphene sheets tend to restack or agglomerate, which may inhibit the electrical and thermal properties of the composite.

CVD is also a suitable technique for the synthesis of graphene films with high quality. In CVD, graphene layers are developed on metallic substrates like copper or nickel via chemical reaction among carbon-containing gases like methane. It is in much demand for preparing large-scale or few-layer graphene that can be transferred onto substrates for electronics and sensor applications. Although CVD-grown pristine and defect-free quality graphene has superior electrical and mechanical properties, the process is costly, and upscaling to industrial-level production is very difficult. Additionally, the transfer of graphene layers from the metal substrate to a material of interest may introduce defects, further complicating the synthesis process. Solution blending is another key approach to synthesizing graphene-based nanocomposites, especially for polymers. This technique entails dispersion of graphene or graphene derivatives in a solvent and then blending with a polymer solution. By evaporation or solidification of the solvent, a composite is obtained. Solution blending is the preferred technique since it is simple and scalable, and it is a viable method for large-scale production in real-world applications. Uniform dispersion of graphene throughout the matrix is still challenging to obtain. Improper dispersion may result in inhomogeneous mechanical and electrical properties, reducing the overall composite performance.

The various forms of graphene nanocomposites—like polymer-based, ceramic-based, metal-based, and cement-based graphene nanocomposites—each have their own advantages for particular applications. For instance, graphene nanocomposites based on polymers find extensive applications in the electronics, aerospace, and automotive sectors because they are lightweight, flexible, and electrically conductive [40, 121–123]. The addition of even small amounts of graphene can lead to dramatic improvements in tensile strength, flexibility, and conductivity, making them ideal for use in flexible electronics, supercapacitors, and structural components. To be specific, the mechanical strength of composites of polymers can be improved by up to 50–70%, while their electrical conductivity can be improved by 105 to 107 times depending on dispersion and integration of graphene. Ceramic nanocomposites with graphene are valued for hardness, wear resistance, and stability at high temperature and find applications in biomedical implants, heat shields, and cutting tools. Graphene is capable of toughening ceramic matrices by suppressing the crack growth in the material, the biggest limitation of traditional ceramics. The enhancement of toughness can reach 30–50% relative to standard ceramics without affecting the inherent material properties, like hardness and thermal stability.

Metal-based graphene nanocomposites take advantage of the superior mechanical and electrical properties of graphene to improve the properties of metals. Graphene is typically added to metal matrices to create materials that are stronger, lighter, and more electrically conductive. In metal-based composite materials, graphene can increase tensile strength by up to 100%, reduce the material's weight, and improve electrical conductivity by up to 50%, depending on the quality and distribution of the graphene. These hybrid materials find particular application in the aerospace, automotive, and defense industries, where lightweight but high strength materials play a significant role in fuel efficiency and overall performance. Moreover, cementitious graphene nanocomposites have tremendous advantages in the construction sector, where enhanced resistance and crack-free materials are of utmost importance. Incorporation of graphene oxide (GO) in concrete has been found to improve durability as well as compressive strength by up to 40–50% as graphene acts as a crack bridge and crack propagation preventer as a reinforcing agent. Graphene also offers novel prospects of self-heating and self-sensing concrete via its electric and thermal conductance, a feature that has the potential to transform infrastructure management by allowing real-time monitoring of structural conditions and damage detection. The various synthesis methods employed for graphene-based nanocomposites possess distinct benefits and drawbacks; however, they cumulatively facilitate the creation of materials possessing improved mechanical properties, optical transparency, thermal conductivity, and electrical conductivity. The nanocomposites are being developed for application-specific use in flexible electronics and energy storage devices, aerospace and building materials, with each use deriving unprecedented benefits from graphene. As the issues of regular dispersion and mass production are progressively resolved, the application of graphene nanocomposites in industries will be expected to grow exponentially, thereby creating new possibilities for designing next-generation materials.

In electronics, graphene nanocomposites are of great significance in the design of flexible electronics, supercapacitors, and sensors due to the superb electrical conductivity, flexibility, and thinness. For instance, graphene composites can show over 107 S/m electrical conductivity, far surpassing traditional materials, and as a result, they find application in devices like transparent electrodes and flexible touchscreens. Besides, graphene-based supercapacitors have also exhibited energy densities of 10–100 Wh/kg, thereby enabling improved energy storage capabilities over conventional capacitors. These characteristics render graphene nanocomposites highly attractive for wearable electronics, portable equipment, and future computing technologies. In aerospace and automotive applications, graphene nanocomposites are revolutionizing materials design by creating a remarkable improvement in the weight-to-strength ratio of various components. By adding graphene to polymer, metal, or ceramic matrices, manufacturers can create lightweight components that maintain or even surpass the strength and durability afforded by conventional materials. Graphene-reinforced composites, for instance, can cut the weight of aerospace components by as much as 30–40% while maintaining structural integrity, which translates to less fuel consumption and emissions for aviation and automotive applications. In the automotive industry, graphene nanocomposites have been demonstrated to increase the performance of vehicles significantly, with some reports indicating 60–80% improvement in tensile strength and as much as 50% in toughness over traditional composite materials. These qualities are of especial use in the manufacturing of lighter vehicle body structures and improved electric vehicle batteries, which can lead to increased driving range and increased overall vehicle efficiency.

In construction, graphene nanocomposites are powering the performance and durability improvement of cementitious materials. Graphene addition to cement and concrete has been shown to improve durability as well as anti-cracking ability. Research indicates the improvement of compressive strength by up to 40% due to the inclusion of graphene oxide (GO) in cement, increasing the service life of buildings and infrastructure dramatically. Further, graphene's high thermal conductivity and water resistance behavior assist in lessening the microcrack development caused by thermal changes, therefore lessening frequent maintenance and repair requirements. Moreover, graphene-reinforced concrete with self-sensing capacity possesses the potential to continuously monitor its structural health, hence improving the safety of massive infrastructure projects such as bridges and dams. Other than that, nanocomposites containing graphene are finding their way into renewable energy technologies, more so in solar cells and fuel cells. The integration of graphene in photovoltaic cells allows for an increase in electron mobility, thereby facilitating improved energy conversion efficiency. For example, graphene-integrated perovskite solar cells have demonstrated power conversion efficiencies exceeding 22%, indicating great potential for the advancement of renewable energy technologies. Likewise, graphene's potential to enhance catalytic activity in fuel cells can result in more efficient energy production, thereby reducing the use of platinum-based catalysts by as much as 80%.

Graphene nanocomposites are leading innovation in a wide range of industries with the significant improvement of material performance and enabling the development of new technologies. The synergy of high strength, good electrical conductivity, thermal stability, and light weight properties of graphene composites is offering revolutionary benefits in electronics, aerospace, automotive, construction, and energy sectors, which lead to more efficient, sustainable, and high-performance products [124]. Although they possess enormous potential, graphene nanocomposites face several challenges that must be overcome with utmost caution in order to achieve their full commercial and industrial potential. Dispersion issues are one of the significant barriers because graphene would tend to agglomerate naturally and thus not disperse uniformly in the matrix. Such an inhomogeneous dispersion degrades the mechanical, electrical, and thermal properties of the different composites and greatly hinders its functionality in real-world applications. The ability to obtain homogeneous graphene dispersion throughout the matrix—whether polymers, metals, or ceramics—is crucial to realizing the unique properties of graphene in a reproducible and predictable manner. Scalability of synthesis is another significant issue. While scale-down synthesis methods such as CVD and mechanical exfoliation produce good-quality graphene, the methods are generally expensive, time-consuming, and energy-consuming, and therefore difficult to scale up for commercial use. Also, more scalable processes like liquid-phase exfoliation result in lower-quality graphene with defects that decrease the material's properties. Affordable and scalable synthesis strategies capable of yielding the highest quality graphene on a commercial scale are still at the top of the agenda for producers and researchers.

Apart from that, the environmental and health issues of graphene nanocomposites are some of the greatest challenges. Graphene manufacturing often involves the use of harmful chemicals, energy-intensive processes, and potential generation of toxic waste, which raises the question of environmental sustainability. Moreover, the long-term health effects of graphene exposure, especially in nanoparticle form, are still not well understood. Inhalation or skin contact of graphene nanoparticles may pose health risks such as respiratory or skin irritation. The disposal at the end of the lifecycle of graphene-based nanocomposites can result in environmental contamination, especially when the composites are embedded in non-biodegradable matrices. Regulatory gaps regarding the use, production, and disposal of graphene also impede its broader use, with a lack of standard protocols on safety precautions or disposal systems. Graphene nanocomposites will witness phenomenal advancements in the times to come, with researchers actively working on overcoming these challenges. Process optimization of the manufacturing process will be a priority area. Development of green synthesis routes, i.e., eco-friendly chemical reduction processes or energy-efficient exfoliation processes, can reduce the environmental impact of graphene production. Greener processes would not only reduce the carbon footprint of the process but also restrict the formation of toxic by-products, making large-scale production more sustainable. In addition, the establishment of functionalization techniques can enhance the compatibility of graphene with different matrices, allowing easier dispersion and integration of graphene into the host materials. This can result in more homogeneous composites with improved properties for a broad spectrum of applications.

One other possible area of future research is the development of recyclable nanocomposites. By using recyclable or end-of-life decomposable materials, researchers can create more environmentally friendly composites, with a lower possibility of generating waste and pollution. Furthermore, the development of standardized safety protocols and regulatory frameworks for handling, production, and disposal of graphene-based materials will be necessary to avert health risks and enable safe large-scale application. These safety standards will also help industries adopt graphene-based technologies with greater confidence, considering that their environmental and health impacts are minimized. After these technical and regulatory challenges are surpassed, graphene nanocomposites are likely to be a critical component of the materials of tomorrow. They have the potential to transform industries like electronics, where their electrical conductivity and flexibility could enable more powerful, lighter, and more energy-efficient devices. Graphene nanocomposites could lead to breakthroughs in the energy sector with more efficient supercapacitors and batteries. In medicine, graphene's antibacterial activity and biocompatibility offer valuable potential for biomedical devices and drug delivery systems. In environmental science, graphene composites can become a key to water purification, air filtration, and sustainable construction. As these challenges are increasingly overcome, graphene nanocomposites will undoubtedly be at the forefront of transforming industries, enabling new applications, and advancing technologies worldwide.

References

1. Anjali, C., & Kizhakayil, R. N. (2024). Graphene—The material of the era BT. In M. N. Naseer, M. Ikram, A. A. Zaidi, Y. Abdul Wahab, & M. R. Johan (Eds.), *Graphene-based photocatalysts for hydrogen production and environmental remediation* (pp. 3–47). Springer Nature Switzerland. https://doi.org/10.1007/978-3-031-68464-7_1
2. Dehghani, A., Lashgari, S. M., & Ramezanzadeh, B. (2022). Introduction to graphene-based materials and their composites. In S. M. Rangappa, J. Parameswaranpillai, V. Ayyappan, M. G. Motappa, S. Siengchin, & C. Soutis (Eds.), *Innovations in graphene-based polymer*

composites (Woodhead publishing series in composites science and engineering) (pp. 1–47). Woodhead Publishing. https://doi.org/10.1016/B978-0-12-823789-2.00020-0

3. Dananjaya, V., Marimuthu, S., Yang, R., Grace, A. N., & Abeykoon, C. (2024). Synthesis, properties, applications, 3D printing and machine learning of graphene quantum dots in polymer nanocomposites. *Progress in Materials Science, 144*, 101282. https://doi.org/10.1016/j.pmatsci.2024.101282
4. Ibrahim, A., Klopocinska, A., Horvat, K., & Abdel Hamid, Z. (2021). Graphene-Based Nanocomposites: Synthesis, Mechanical Properties, and Characterizations. Polymers, 13(17), 2869. https://doi.org/10.3390/polym13172869.
5. Arora, G. (2024). Experimental analysis of blended nanocomposites processed using compression molded microwave process. *Proceedings of the Institution of Mechanical Engineers, Part N: Journal of Nanomaterials, Nanoengineering and Nanosystems*. https://doi.org/10.1177/23977914231224054
6. Arora, G., & Pathak, H. (2020). Multi-scale computational analysis of carbon-nanotube–polymer composite BT. In B. B. Biswal, B. K. Sarkar, & P. Mahanta (Eds.), *Advances in mechanical engineering* (pp. 205–216). Springer Singapore.
7. Arora, G., Singh, M. K., Pathak, H., & Zafar, S. (2021). Micro-scale analysis of HA-PLLA bio-composites: Effect of the interpenetration of voids on mechanical properties. *Materials Today Communications, 28*, 102568. https://doi.org/10.1016/j.mtcomm.2021.102568
8. Balaji, K. V., Shirvanimoghaddam, K., & Naebe, M. (2024). Multifunctional basalt fiber polymer composites enabled by carbon nanotubes and graphene. *Composites Part B: Engineering, 268*, 111070. https://doi.org/10.1016/j.compositesb.2023.111070
9. Dwivedi, K., Arora, G., & Pathak, H. (2022). Fatigue crack growth in CNT-reinforced polymer composite. Journal of micromechanics and molecular. *Physics, 7*(3–4), 173–184. https://doi.org/10.1142/S242491302241003X
10. Saeed, M., Haq, R. S. U., Ahmed, S., Siddiqui, F., & Yi, J. (2024). Recent advances in carbon nanotubes, graphene and carbon fibers-based microwave absorbers. *Journal of Alloys and Compounds, 970*, 172625. https://doi.org/10.1016/j.jallcom.2023.172625
11. Wang, J., Li, S., Yang, L., Liu, B., Xie, S., Qi, R., Zhan, Y., & Xia, H. (2024). Graphene-based hybrid fillers for rubber composites. *Molecules, 29*(5), 1009. https://doi.org/10.3390/molecules29051009
12. Kim, C. H., Lee, S. Y., Rhee, K. Y., & Park, S. J. (2024). Carbon-based composites in biomedical applications: A comprehensive review of properties, applications, and future directions. *Advanced Composites and Hybrid Materials, 7*(2), 1–22. https://doi.org/10.1007/s42114-024-00846-1
13. Rathi, V., Brajpuriya, R., Gupta, R., Parmar, K. P. S., & Kumar, A. (2024). Graphene-derived composites: A new frontier in thermoelectric energy conversion. *Energy Advances, 3*(2), 389–412. https://doi.org/10.1039/d3ya00526g
14. Rubesh Ashok Kumar, S., Vasvini Mary, D., Suganya Josephine, G. A., & Riswan Ahamed, M. A. (2024). Graphene/GO/rGO based nanocomposites: Emerging energy and environmental application—Review. *Hybrid Advances, 5*, 100168. https://doi.org/10.1016/j.hybadv.2024.100168
15. Yan, M., Lan, Y. X., & Yeh, J. M. (2024). Application of graphene and its derivatives in anticorrosion coatings: Research advances and future perspectives. *Journal of the Taiwan Institute of Chemical Engineers, 154*, 105130. https://doi.org/10.1016/j.jtice.2023.105130
16. Zhang, L. Y., & Zhang, Y. F. (2016). In situ fast polymerization of graphene nanosheets-filled poly(methyl methacrylate) nanocomposites. *Journal of Applied Polymer Science, 133*(19), 1–7. https://doi.org/10.1002/app.43423
17. Zhang, T., Gao, X., Li, J., Xiao, L., Gao, H., Zhao, F., & Ma, H. (2024a). Progress on the application of graphene-based composites toward energetic materials: A review. *Defence Technology, 31*, 95–116. https://doi.org/10.1016/j.dt.2023.01.016

18. Nagoriya, H., & Arora, G. (2024). Exploring multi-scale thermal behaviour in pin-on-disc systems for organic, metallic, ceramic, and polymer composites. *Mechanics of Advanced Composite Structures*. https://doi.org/10.22075/macs.2024.33789.1640
19. Arora, G., & Pathak, H. (2021). Mesoscale analysis of polymer-CNT composites for evaluation of Elasto-plastic and Thermo-elastic properties. In *Advanced computational methods in mechanical and materials engineering* (pp. 273–293). CRC Press. https://doi.org/10.1201/9781003202233-17
20. Barkoula, N. M., Alcock, B., Cabrera, N. O., & Peijs, T. (2008). Flame-Retardancy properties of intumescent ammonium poly(phosphate) and mineral filler magnesium hydroxide in combination with graphene. *Polymers and Polymer Composites, 16*(2), 101–113.
21. Elhaes, H., Ibrahim, A., Osman, O., & Ibrahim, M. A. (2024). Molecular modeling analysis for functionalized graphene/sodium alginate composite. *Scientific Reports, 14*(1), 1–14. https://doi.org/10.1038/s41598-024-64698-x
22. Huo, S., Guo, Y., Yang, Q., Wang, H., & Song, P. (2024). Two-dimensional nanomaterials for flame-retardant polymer composites: A mini review. *Advanced Nanocomposites, 1*(1), 240–247. https://doi.org/10.1016/j.adna.2024.07.001
23. Jin, J., Su, Y., Guo, J., Zhang, Z., Peng, B., Chen, M., & Wu, L. (2024). Flexible composite films with ultrahigh through-plane thermal conductivity yet low graphene content. *Composites Part B: Engineering, 286*, 111780. https://doi.org/10.1016/j.compositesb.2024.111780
24. Lai, S., Huang, T., Liu, P., Wang, H. B., Yang, S. C., Liu, X. H., Yang, K., Zhang, Q. L., Liu, J. H., & Hu, J. T. (2024). Progress and prospects of graphene-based materials in lithium batteries. *Rare Metals, 43*(5), 1886–1905. https://doi.org/10.1007/s12598-023-02565-w
25. Wang, H., Qian, P., Zhang, L., & Shi, H. (2022a). Graphene/polymer composite membranes for vanadium redox flow battery applications. In S. M. Rangappa, J. Parameswaranpillai, V. Ayyappan, M. G. Motappa, S. Siengchin, & C. Soutis (Eds.), *Innovations in graphene-based polymer composites* (Woodhead publishing series in composites science and engineering) (pp. 487–520). Woodhead Publishing. https://doi.org/10.1016/B978-0-12-823789-2.00014-5
26. Yang, J., Maulik, G., He, S., Nag, A., Deng, S., Afsarimanesh, N., & Gao, J. (2024). Graphene allotropes-based electrochemical sensors to detect catechol molecules. *Sensors and Actuators A: Physical, 368*, 115088. https://doi.org/10.1016/j.sna.2024.115088
27. Bhowmik, P., & Arora, G. (2022). Graphene/metal matrix nanocomposites: Synthesis, characterization, and properties. In *Metal matrix composites* (pp. 109–148). CRC Press. https://doi.org/10.1201/9781003194910-5
28. Goodrum, R., Weldekidan, H., Li, H., Mohanty, A. K., & Misra, M. (2024). Graphene-based nanostructures from green processes and their applications in biomedical sensors. *Advanced Industrial and Engineering Polymer Research, 7*(1), 37–53. https://doi.org/10.1016/j.aiepr.2023.03.001
29. Hong, H., Zhang, J., Zhu, Y., Tse, S. D., Guo, H., Lai, Y., Xi, Y., He, L., Zhu, Z., Yin, K., & Sun, L. (2024). In situ polymer-solution-processed graphene–PDMS nanocomposites for application in intracranial pressure sensors. *Nanomaterials, 14*(5), 399. https://doi.org/10.3390/nano14050399
30. Kong, M., Yang, M., Li, R., Long, Y. Z., Zhang, J., Huang, X., Cui, X., Zhang, Y., Said, Z., & Li, C. (2024). Graphene-based flexible wearable sensors: Mechanisms, challenges, and future directions. *International Journal of Advanced Manufacturing Technology, 131*(5–6), 3205–3237. https://doi.org/10.1007/s00170-023-12007-7
31. Li, N., Wang, R., Zhang, L., Zhou, J., & Jiao, T. (2022a). Graphene-based polymer composite films. In S. M. Rangappa, J. Parameswaranpillai, V. Ayyappan, M. G. Motappa, S. Siengchin, & C. Soutis (Eds.), *Innovations in graphene-based polymer composites* (Woodhead publishing series in composites science and engineering) (pp. 309–331). Woodhead Publishing. https://doi.org/10.1016/B978-0-12-823789-2.00019-4
32. Li, Z., Huang, L., Cheng, L., Guo, W., & Ye, R. (2024). Laser-induced graphene-based sensors in health monitoring: Progress, sensing mechanisms, and applications. *Small Methods, 8*(11), e2400118. https://doi.org/10.1002/smtd.202400118

33. Şimşek, B., Uygunoğlu, T., Fidan, U., Ceran, Ö. B., & Eryeşil, Ö. (2024). Graphene nanoplatelet/polypropylene pellets in cement-based composites: Mechanical, piezoresistive and microstructural properties. *Construction and Building Materials, 420*, 1–10. https://doi.org/10.1016/j.conbuildmat.2024.135549
34. Verma, A., Parashar, A., & van Duin, A. C. T. (2022a). Graphene-reinforced polymeric membranes for water desalination and gas separation/barrier applications. In S. M. Rangappa, J. Parameswaranpillai, V. Ayyappan, M. G. Motappa, S. Siengchin, & C. Soutis (Eds.), *Innovations in graphene-based polymer composites* (Woodhead publishing series in composites science and engineering) (pp. 133–165). Woodhead Publishing. https://doi.org/10.1016/B978-0-12-823789-2.00009-1
35. Verma, S., Mohanty, S., & Nayak, S. K. (2022b). Graphene-based polymer coatings. In S. M. Rangappa, J. Parameswaranpillai, V. Ayyappan, M. G. Motappa, S. Siengchin, & C. Soutis (Eds.), *Innovations in graphene-based polymer composites* (Woodhead publishing series in composites science and engineering) (pp. 121–131). Woodhead Publishing. https://doi.org/10.1016/B978-0-12-823789-2.00018-2
36. Zhang, Z., Liu, Q., Ma, H., Ke, N., Ding, J., Zhang, W., & Fan, X. (2024b). Recent advances in graphene-based pressure sensors: A review. *IEEE Sensors Journal, 24*(16), 25227–25248. https://doi.org/10.1109/JSEN.2024.3419243
37. Luo, W., Zou, M. M., Luo, L., Ma, Y., Chen, W., Hu, X., Li, Q., & Jiang, X. (2024). Efficient enhancement of photothermal conversion of polymer-coated phase change materials based on reduced graphene oxide and polyethylene glycol. *Journal of Energy Storage, 78*, 109950. https://doi.org/10.1016/j.est.2023.109950
38. Zhu, Q., Ong, P. J., Goh, S. H. A., Yeo, R. J., Wang, S., Liu, Z., & Loh, X. J. (2024). Recent advances in graphene-based phase change composites for thermal energy storage and management. *Nano Materials Science, 6*(2), 115–138. https://doi.org/10.1016/j.nanoms.2023.09.003
39. Joseph, A., Mahesh, V., Mahesh, V., & Harursampath, D. (2022). Structural analysis of graphene-based composites. In S. M. Rangappa, J. Parameswaranpillai, V. Ayyappan, M. G. Motappa, S. Siengchin, & C. Soutis (Eds.), *Innovations in graphene-based polymer composites* (Woodhead publishing series in composites science and engineering) (pp. 91–120). Woodhead Publishing. https://doi.org/10.1016/B978-0-12-823789-2.00006-6
40. Eslahi, N., Lotfi, R., Zandi, N., Mazaheri, M., Soleimani, F., & Simchi, A. (2022). Graphene-based polymer nanocomposites in biomedical applications. In S. M. Rangappa, J. Parameswaranpillai, V. Ayyappan, M. G. Motappa, S. Siengchin, & C. Soutis (Eds.), *Innovations in graphene-based polymer composites* (Woodhead publishing series in composites science and engineering) (pp. 199–245). Woodhead Publishing. https://doi.org/10.1016/B978-0-12-823789-2.00016-9
41. Rai, R., & Kumar, S. (2024). A comprehensive review on graphene: Historical development. *Properties, Applications, Challenges and Future Scope, 1*(4), 150–166. https://doi.org/10.61552/JMES.2024.04.004
42. Mural, P. K. S., Sharma, M., Madras, G., & Bose, S. (2015). A critical review on in situ reduction of graphene oxide during preparation of conducting polymeric nanocomposites. *RSC Advances, 5*(41), 32078–32087. https://doi.org/10.1039/c5ra02877a
43. Milani, M. A., González, D., Quijada, R., Basso, N. R. S., Cerrada, M. L., Azambuja, D. S., & Galland, G. B. (2013). Polypropylene/graphene nanosheet nanocomposites by in situ polymerization: Synthesis, characterization and fundamental properties. *Composites Science and Technology, 84*, 1–7. https://doi.org/10.1016/j.compscitech.2013.05.001
44. Brownson, D. A. C., & Banks, C. E. (2012). The electrochemistry of CVD graphene: Progress and prospects. *Physical Chemistry Chemical Physics, 14*(23), 8264–8281. https://doi.org/10.1039/C2CP40225D
45. Deokar, G., Avila, J., Razado-Colambo, I., Codron, J.-L., Boyaval, C., Galopin, E., Asensio, M.-C., & Vignaud, D. (2015). Towards high quality CVD graphene growth and transfer. *Carbon, 89*, 82–92. https://doi.org/10.1016/j.carbon.2015.03.017
46. Eigler, S. (2016). Graphene synthesis. In *Graphene technology* (pp. 19–61). Wiley. https://doi.org/10.1002/9783527687541.ch2

47. Plutnar, J., Pumera, M., & Sofer, Z. (2018). The chemistry of CVD graphene. *Journal of Materials Chemistry C, 6*(23), 6082–6101. https://doi.org/10.1039/C8TC00463C
48. Ciesielski, A., & Samorì, P. (2014). Graphene via sonication assisted liquid-phase exfoliation. *Chemical Society Reviews, 43*(1), 381–398. https://doi.org/10.1039/C3CS60217F
49. Yi, M., & Shen, Z. (2015). A review on mechanical exfoliation for the scalable production of graphene. *Journal of Materials Chemistry A, 3*(22), 11700–11715. https://doi.org/10.1039/C5TA00252D
50. Gao, E., Lin, S.-Z., Qin, Z., Buehler, M. J., Feng, X.-Q., & Xu, Z. (2018). Mechanical exfoliation of two-dimensional materials. *Journal of the Mechanics and Physics of Solids, 115*, 248–262. https://doi.org/10.1016/j.jmps.2018.03.014
51. Mao, H., & Wang, X. (2020). Use of in-situ polymerization in the preparation of graphene/polymer nanocomposites. *New Carbon Materials, 35*(4), 336–343. https://doi.org/10.1016/S1872-5805(20)60493-0
52. Tripathi, S. N., Saini, P., Gupta, D., & Choudhary, V. (2013). Electrical and mechanical properties of PMMA/reduced graphene oxide nanocomposites prepared via in situ polymerization. *Journal of Materials Science, 48*(18), 6223–6232. https://doi.org/10.1007/s10853-013-7420-8
53. Shehzad, F., Daud, M., & Al-Harthi, M. A. (2016). Synthesis, characterization and crystallization kinetics of nanocomposites prepared by in situ polymerization of ethylene and graphene. *Journal of Thermal Analysis and Calorimetry, 123*(2), 1501–1511. https://doi.org/10.1007/s10973-015-5087-x
54. Chen, W., Weimin, H., Li, D., Chen, S., & Dai, Z. (2018). A critical review on the development and performance of polymer/graphene nanocomposites. *Science and Engineering of Composite Materials, 25*(6), 1059–1073. https://doi.org/10.1515/secm-2017-0199
55. Innocenzi, P., Malfatti, L., Lasio, B., Pinna, A., Loche, D., Casula, M. F., Alzari, V., & Mariani, A. (2014). Sol–gel chemistry for graphene–silica nanocomposite films. *New Journal of Chemistry, 38*(8), 3777–3782. https://doi.org/10.1039/C4NJ00535J
56. Bilisik, K., & Akter, M. (2021). Graphene nanocomposites: A review on processes, properties, and applications. *Journal of Industrial Textiles, 51*(3_Suppl), 3718S–3766S. https://doi.org/10.1177/15280837211024252
57. Chang, H., & Wu, H. (2013). Graphene-based nanocomposites: Preparation, functionalization, and energy and environmental applications. *Energy & Environmental Science, 6*(12), 3483–3507. https://doi.org/10.1039/C3EE42518E
58. Ji, L., Meduri, P., Agubra, V., Xiao, X., & Alcoutlabi, M. (2016). Graphene-based nanocomposites for energy storage. *Advanced Energy Materials, 6*(16), 1502159. https://doi.org/10.1002/aenm.201502159
59. Lawal, A. T. (2019). Graphene-based nano composites and their applications: A review. *Biosensors and Bioelectronics, 141*, 111384. https://doi.org/10.1016/j.bios.2019.111384
60. Papageorgiou, D. G., Kinloch, I. A., & Young, R. J. (2017). Mechanical properties of graphene and graphene-based nanocomposites. *Progress in Materials Science, 90*, 75–127. https://doi.org/10.1016/j.pmatsci.2017.07.004
61. Zakaulla, M. (2022). Modeling and prediction of tribological properties of polyetheretherketone composite reinforced with graphene and titanium powder using artificial neural network. In S. M. Rangappa, J. Parameswaranpillai, V. Ayyappan, M. G. Motappa, S. Siengchin, & C. Soutis (Eds.), *Innovations in graphene-based polymer composites* (Woodhead publishing series in composites science and engineering) (pp. 333–352). Woodhead Publishing. https://doi.org/10.1016/B978-0-12-823789-2.00017-0
62. Abbasi, H., Antunes, M., & Velasco, J. I. (2022). Graphene polymer foams and sponges' preparation and applications. In S. M. Rangappa, J. Parameswaranpillai, V. Ayyappan, M. G. Motappa, S. Siengchin, & C. Soutis (Eds.), *Innovations in graphene-based poly-*

mer composites (Woodhead publishing series in composites science and engineering) (pp. 353–376). Woodhead Publishing. https://doi.org/10.1016/B978-0-12-823789-2.00012-1
63. Idumah, C. I., & Hassan, A. (2016). Emerging trends in graphene carbon based polymer nanocomposites and applications. *Reviews in Chemical Engineering, 32*(2), 223–264. https://doi.org/10.1515/revce-2015-0038
64. Lawal, A. T. (2020). Recent progress in graphene based polymer nanocomposites. *Cogent Chemistry, 6*(1), 1833476. https://doi.org/10.1080/23312009.2020.1833476
65. Najafi, M. D., Kowsari, E., Ramakrishna, S., & AliAkbari, R. (2022). Functionalization of graphene composites using ionic liquids and applications. In S. M. Rangappa, J. Parameswaranpillai, V. Ayyappan, M. G. Motappa, S. Siengchin, & C. Soutis (Eds.), *Innovations in graphene-based polymer composites* (Woodhead publishing series in composites science and engineering) (pp. 445–461). Woodhead Publishing. https://doi.org/10.1016/B978-0-12-823789-2.00013-3
66. Nguyen, T.-P., & Le Rendu, P. (2022). Graphene-based polymer composites for photocatalytic applications. In S. M. Rangappa, J. Parameswaranpillai, V. Ayyappan, M. G. Motappa, S. Siengchin, & C. Soutis (Eds.), *Innovations in graphene-based polymer composites* (Woodhead publishing series in composites science and engineering) (pp. 377–432). Woodhead Publishing. https://doi.org/10.1016/B978-0-12-823789-2.00008-X
67. Paszkiewicz, S., & Szymczyk, A. (2019). Graphene-based nanomaterials and their polymer nanocomposites. In P. N. Karak (Ed.), *Nanomaterials and polymer nanocomposites* (pp. 177–216). Elsevier. https://doi.org/10.1016/B978-0-12-814615-6.00006-0
68. Shah, R., Kausar, A., Muhammad, B., & Shah, S. (2015). Progression from graphene and graphene oxide to high performance polymer-based nanocomposite: A review. *Polymer-Plastics Technology and Engineering, 54*(2), 173–183. https://doi.org/10.1080/03602559.2014.955202
69. Terrones, M., Martín, O., González, M., Pozuelo, J., Serrano, B., Cabanelas, J. C., Vega-Díaz, S. M., & Baselga, J. (2011). Interphases in graphene polymer-based nanocomposites: Achievements and challenges. *Advanced Materials, 23*(44), 5302–5310. https://doi.org/10.1002/adma.201102036
70. Hu, K., Kulkarni, D. D., Choi, I., & Tsukruk, V. V. (2014). Graphene-polymer nanocomposites for structural and functional applications. *Progress in Polymer Science, 39*(11), 1934–1972. https://doi.org/10.1016/j.progpolymsci.2014.03.001
71. Chee, W. K., Lim, H. N., Huang, N. M., & Harrison, I. (2015). Nanocomposites of graphene/polymers: A review. *RSC Advances, 5*(83), 68014–68051. https://doi.org/10.1039/C5RA07989F
72. Khanam, P. N., Ponnamma, D., & AL-Madeed, M. A. (2015). In K. K. Sadasivuni, D. Ponnamma, J. Kim, & S. Thomas (Eds.), *Electrical properties of graphene polymer nanocomposites BT—Graphene-based polymer nanocomposites in electronics* (pp. 25–47). Springer International Publishing. https://doi.org/10.1007/978-3-319-13875-6_2
73. Trivedi, D. N., & Rachchh, N. V. (2022). Graphene and its application in thermoplastic polymers as nano-filler—A review. *Polymer, 240*, 124486. https://doi.org/10.1016/j.polymer.2021.124486
74. Tang, L.-C., Wan, Y.-J., Yan, D., Pei, Y.-B., Zhao, L., Li, Y.-B., Wu, L.-B., Jiang, J.-X., & Lai, G.-Q. (2013). The effect of graphene dispersion on the mechanical properties of graphene/epoxy composites. *Carbon, 60*, 16–27. https://doi.org/10.1016/j.carbon.2013.03.050
75. Xie, H., Li, C., & Wang, Q. (2024). Thermosetting polymer modified asphalts: Current status and challenges. *Polymer Reviews, 64*(2), 690–759. https://doi.org/10.1080/15583724.2023.2286706
76. Kumar, V., Lee, D.-J., & Park, S.-S. (2023). Multi-functionality prospects in functionalized and pristine graphene nanosheets reinforced silicone rubber composites: A focused review. *FlatChem, 41*, 100535. https://doi.org/10.1016/j.flatc.2023.100535

77. Zhu, Q., Wang, Z., Zeng, H., Yang, T., & Wang, X. (2021). Effects of graphene on various properties and applications of silicone rubber and silicone resin. *Composites Part A: Applied Science and Manufacturing, 142*, 106240. https://doi.org/10.1016/j.compositesa.2020.106240
78. Díez-Pascual, A. M., & Rahdar, A. (2022). Graphene-based polymer composites for flexible electronic applications. *Micromachines, 13*(7). https://doi.org/10.3390/mi13071123
79. Kholghi Eshkalak, S., Kowsari, E., & Ramakrishna, S. (2022). 3D printing of graphene-based composites and their applications in medicine and health care. In S. M. Rangappa, J. Parameswaranpillai, V. Ayyappan, M. G. Motappa, S. Siengchin, & C. Soutis (Eds.), *Innovations in graphene-based polymer composites* (Woodhead publishing series in composites science and engineering) (pp. 463–485). Woodhead Publishing. https://doi.org/10.1016/B978-0-12-823789-2.00011-X
80. Silva, M., Alves, N. M., & Paiva, M. C. (2018). Graphene-polymer nanocomposites for biomedical applications. *Polymers for Advanced Technologies, 29*(2), 687–700. https://doi.org/10.1002/pat.4164
81. Lashgari, S. M., Dehghani, A., Ramezanzadeh, B., & Tehrani, M. E. H. N. (2022). Graphene-based polymer composites in corrosion protection applications. In S. M. Rangappa, J. Parameswaranpillai, V. Ayyappan, M. G. Motappa, S. Siengchin, & C. Soutis (Eds.), *Innovations in graphene-based polymer composites* (Woodhead publishing series in composites science and engineering) (pp. 559–581). Woodhead Publishing. https://doi.org/10.1016/B978-0-12-823789-2.00021-2
82. Iqbal, M. J., Iqbal, M. Z., & Fazal, A. (2022). Graphene/polymer composite application on supercapacitors. In S. M. Rangappa, J. Parameswaranpillai, V. Ayyappan, M. G. Motappa, S. Siengchin, & C. Soutis (Eds.), *Innovations in graphene-based polymer composites* (Woodhead publishing series in composites science and engineering) (pp. 583–610). Woodhead Publishing. https://doi.org/10.1016/B978-0-12-823789-2.00022-4
83. Sun, Y., & Shi, G. (2013). Graphene/polymer composites for energy applications. *Journal of Polymer Science Part B: Polymer Physics, 51*(4), 231–253. https://doi.org/10.1002/polb.23226
84. Wang, X., Guo, W., Cai, W., & Hu, Y. (2022b). Graphene-based polymer composites for flame-retardant application. In S. M. Rangappa, J. Parameswaranpillai, V. Ayyappan, M. G. Motappa, S. Siengchin, & C. Soutis (Eds.), *Innovations in graphene-based polymer composites* (Woodhead publishing series in composites science and engineering) (pp. 61–89). Woodhead Publishing. https://doi.org/10.1016/B978-0-12-823789-2.00004-2
85. Zhang, X., & Samorì, P. (2017). Graphene/polymer nanocomposites for supercapacitors. *ChemNanoMat, 3*(6), 362–372. https://doi.org/10.1002/cnma.201700055
86. Ahmad, I., Islam, M., Subhani, T., & Zhu, Y. (2016). Toughness enhancement in graphene nanoplatelet/SiC reinforced Al_2O_3 ceramic hybrid nanocomposites. *Nanotechnology, 27*(42), 425704. https://doi.org/10.1088/0957-4484/27/42/425704
87. Baig, Z., Mamat, O., & Mustapha, M. (2018). Recent Progress on the dispersion and the strengthening effect of carbon nanotubes and graphene-reinforced metal nanocomposites: A review. *Critical Reviews in Solid State and Materials Sciences, 43*(1), 1–46. https://doi.org/10.1080/10408436.2016.1243089
88. Bidgoli, E. M.-R., & Arefi, M. (2022). Free vibration analysis of microplates reinforced with functionally graded graphene nanoplatelets. In S. M. Rangappa, J. Parameswaranpillai, V. Ayyappan, M. G. Motappa, S. Siengchin, & C. Soutis (Eds.), *Innovations in graphene-based polymer composites* (Woodhead publishing series in composites science and engineering) (pp. 521–557). Woodhead Publishing. https://doi.org/10.1016/B978-0-12-823789-2.00003-0
89. Harito, C., Zaidi, S. Z. J., Putra, B. R., Hardiansyah, A., Khalil, M., & Yuliarto, B. (2022). Synthesis of graphene polymer composites having high filler content. In S. M. Rangappa, J. Parameswaranpillai, V. Ayyappan, M. G. Motappa, S. Siengchin, & C. Soutis (Eds.), *Innovations in graphene-based polymer composites* (Woodhead publishing series in com-

posites science and engineering) (pp. 49–60). Woodhead Publishing. https://doi.org/10.1016/B978-0-12-823789-2.00002-9
90. Sabet, M. (2024). Unveiling advanced self-healing mechanisms in graphene polymer composites for next-generation applications in aerospace, automotive, and electronics. *Polymer-Plastics Technology and Materials, 63*(15), 2032–2059. https://doi.org/10.1080/25740881.2024.2365282
91. Shamsaei, E., de Souza, F. B., Yao, X., Benhelal, E., Akbari, A., & Duan, W. (2018). Graphene-based nanosheets for stronger and more durable concrete: A review. *Construction and Building Materials, 183*, 642–660. https://doi.org/10.1016/j.conbuildmat.2018.06.201
92. Wang, Q., Ramírez, C., Watts, C. S., Borrero-López, O., Ortiz, A. L., Sheldon, B. W., & Padture, N. P. (2020a). Fracture, fatigue, and sliding-wear behavior of nanocomposites of alumina and reduced graphene-oxide. *Acta Materialia, 186*, 29–39. https://doi.org/10.1016/j.actamat.2019.12.035
93. Wang, Y., Ma, R., Hu, K., Kim, S., Fang, G., Shao, Z., & Tsukruk, V. V. (2016). Dramatic enhancement of graphene oxide/silk nanocomposite membranes: Increasing toughness, strength, and Young's modulus via annealing of interfacial structures. *ACS Applied Materials & Interfaces, 8*(37), 24962–24973. https://doi.org/10.1021/acsami.6b08610
94. Ramírez, C., Belmonte, M., Miranzo, P., & Osendi, M. I. (2021). Applications of ceramic/graphene composites and hybrids. *Materials, 14*(8), 2071. https://doi.org/10.3390/ma14082071
95. Mudra, E., Shepa, I., Hrubovcakova, M., Koribanich, I., Medved, D., Kovalcikova, A., Vojtko, M., & Dusza, J. (2021). Highly wear-resistant alumina/graphene layered and fiber-reinforced composites. *Wear, 484–485*, 204026. https://doi.org/10.1016/j.wear.2021.204026
96. Xie, H., Cao, T., Rodríguez-Lozano, F. J., Luong-Van, E. K., & Rosa, V. (2017). Graphene for the development of the next-generation of biocomposites for dental and medical applications. *Dental Materials, 33*(7), 765–774. https://doi.org/10.1016/j.dental.2017.04.008
97. Nieto, A., Bisht, A., Lahiri, D., Zhang, C., & Agarwal, A. (2017). Graphene reinforced metal and ceramic matrix composites: A review. *International Materials Reviews, 62*(5), 241–302. https://doi.org/10.1080/09506608.2016.1219481
98. Gobena, S. T., & Woldeyonnes, A. D. (2024). A review of synthesis methods, and characterization techniques of polymer nanocomposites for diverse applications. *Discover Materials, 4*(1), 52. https://doi.org/10.1007/s43939-024-00119-0
99. Khalid, N., Siddiqi, H. M., & Park, O. O. K. (2021). Thermally resistant, mechanically stable, and processable triphenylamine-based hyperbranched copolyimides. *Journal of Applied Polymer Science, 138*(6), 49790. https://doi.org/10.1002/app.49790
100. Shaharuddin, M. I. (2022). Handgrip automotive prototype of polypropylene reinforced benzoyl treated kenaf and sugar palm fibers: A facile flexural strength and hardness studies. *Applied Science and Engineering Progress, 15*(2), 1–11. https://doi.org/10.14416/j.asep.2022.04.005
101. Wang, J., Dong, S., Yu, X., & Han, B. (2020b). Mechanical properties of graphene-reinforced reactive powder concrete at different strain rates. *Journal of Materials Science, 55*(8), 3369–3387. https://doi.org/10.1007/s10853-019-04246-5
102. Ali, S., Ahmad, F., Yusoff, P. S. M. M., Muhamad, N., Oñate, E., Raza, M. R., & Malik, K. (2021). A review of graphene reinforced cu matrix composites for thermal management of smart electronics. *Composites Part A: Applied Science and Manufacturing, 144*, 106357. https://doi.org/10.1016/j.compositesa.2021.106357
103. Seyed Pourmand, N., & Asgharzadeh, H. (2020). Aluminum matrix composites reinforced with graphene: A review on production, microstructure, and properties. *Critical Reviews in Solid State and Materials Sciences, 45*(4), 289–337. https://doi.org/10.1080/10408436.2019.1632792
104. Jayasooriya, D., Rajeev, P., & Sanjayan, J. (2022). Application of graphene-based nanomaterials as a reinforcement to concrete pavements. *Sustainability, 14*(18), 11282. https://doi.org/10.3390/su141811282

105. Abedi, M., Fangueiro, R., & Gomes Correia, A. (2021). A review of intrinsic self-sensing cementitious composites and prospects for their application in transport infrastructures. *Construction and Building Materials, 310*, 125139. https://doi.org/10.1016/j.conbuildmat.2021.125139
106. Jiang, X., Lu, D., Yin, B., & Leng, Z. (2024). Advancing carbon nanomaterials-engineered self-sensing cement composites for structural health monitoring: A state-of-the-art review. *Journal of Building Engineering, 87*, 109129. https://doi.org/10.1016/j.jobe.2024.109129
107. Li, W., Qu, F., Dong, W., Mishra, G., & Shah, S. P. (2022b). A comprehensive review on self-sensing graphene/cementitious composites: A pathway toward next-generation smart concrete. *Construction and Building Materials, 331*, 127284. https://doi.org/10.1016/j.conbuildmat.2022.127284
108. Saravanan, N., Rajasekar, R., Mahalakshmi, S., Sathishkumar, T. P., Sasikumar, K. S. K., & Sahoo, S. (2014). Graphene and modified graphene-based polymer nanocomposites—A review. *Journal of Reinforced Plastics and Composites, 33*(12), 1158–1170. https://doi.org/10.1177/0731684414524847
109. Ioniţă, M., Vlăsceanu, G. M., Watzlawek, A. A., Voicu, S. I., Burns, J. S., & Iovu, H. (2017). Graphene and functionalized graphene: Extraordinary prospects for nanobiocomposite materials. *Composites Part B: Engineering, 121*, 34–57. https://doi.org/10.1016/j.compositesb.2017.03.031
110. Yu, W., Sisi, L., Haiyan, Y., & Jie, L. (2020). Progress in the functional modification of graphene/graphene oxide: A review. *RSC Advances, 10*(26), 15328–15345. https://doi.org/10.1039/d0ra01068e
111. Rasyotra, A., Thakur, A., Gaykwad, B., Chakrabarty, S., Bayad, I., Parikh, J., & Jasuja, K. (2023). A review of low-cost approaches to synthesize graphene and its functional composites. *Journal of Materials Science, 58*(10), 4359–4383. https://doi.org/10.1007/s10853-023-08304-x
112. Ranjan, R., & Bajpai, V. (2021). Graphene-based metal matrix nanocomposites: Recent development and challenges. *Journal of Composite Materials, 55*(17), 2369–2413. https://doi.org/10.1177/0021998320988566
113. Deng, B., Liu, Z., & Peng, H. (2019). Toward mass production of CVD graphene films. *Advanced Materials, 31*(9), 1800996. https://doi.org/10.1002/adma.201800996
114. Paton, K. R., Varrla, E., Backes, C., Smith, R. J., Khan, U., O'Neill, A., Boland, C., Lotya, M., Istrate, O. M., King, P., Higgins, T., Barwich, S., May, P., Puczkarski, P., Ahmed, I., Moebius, M., Pettersson, H., Long, E., Coelho, J., et al. (2014). Scalable production of large quantities of defect-free few-layer graphene by shear exfoliation in liquids. *Nature Materials, 13*(6), 624–630. https://doi.org/10.1038/nmat3944
115. Arsenov, D., Beljin, J., Jović, D., Maletić, S., Borišev, M., & Borišev, I. (2023). Nanomaterials as endorsed environmental remediation tools for the next generation: Eco-safety and sustainability. *Journal of Geochemical Exploration, 253*, 107283. https://doi.org/10.1016/j.gexplo.2023.107283
116. Singh, R. K., Kumar, R., & Singh, D. P. (2016). Graphene oxide: Strategies for synthesis, reduction and frontier applications. *RSC Advances, 6*(69), 64993–65011. https://doi.org/10.1039/C6RA07626B
117. Khanna, V., Bakshi, B. R., & Lee, L. J. (2008). Carbon nanofiber production. *Journal of Industrial Ecology, 12*(3), 394–410. https://doi.org/10.1111/j.1530-9290.2008.00052.x
118. Hoseini-Ghahfarokhi, M., Mirkiani, S., Mozaffari, N., Abdolahi Sadatlu, M. A., Ghasemi, A., Abbaspour, S., Akbarian, M., Farjadian, F., & Karimi, M. (2020). Applications of graphene and graphene oxide in smart drug/gene delivery: Is the world still flat? *International Journal of Nanomedicine, 15*, 9469–9496. https://doi.org/10.2147/IJN.S265876
119. Chiticaru, E. A., & Ionita, M. (2022). Graphene toxicity and future perspectives in healthcare and biomedicine. *FlatChem, 35*, 100417. https://doi.org/10.1016/j.flatc.2022.100417

120. Sousa, S. P. B., Peixoto, T., Santos, R. M., Lopes, A., Paiva, M. D., & Marques, A. T. (2020). Health and safety concerns related to CNT and graphene products, and related composites. *Journal of Composites Science, 4*(3), 106. https://doi.org/10.3390/jcs4030106
121. Lotfi, R., Eslahi, N., Koohkhezri, M., & Moghadasi, H. (2022). 3D printing of graphene polymer composites. In S. M. Rangappa, J. Parameswaranpillai, V. Ayyappan, M. G. Motappa, S. Siengchin, & C. Soutis (Eds.), *Innovations in graphene-based polymer composites* (Woodhead publishing series in composites science and engineering) (pp. 247–281). Woodhead Publishing. https://doi.org/10.1016/B978-0-12-823789-2.00010-8
122. Verma, A., Jain, N., Parashar, A., Singh, V. K., Sanjay, M. R., & Siengchin, S. (2020). Lightweight graphene composite materials. In *Lightweight polymer composite structures* (pp. 1–20). CRC Press. https://doi.org/10.1201/9780429244087-1/LIGHTWEIGHT-GRAPHENE-COMPOSITE-MATERIALS-AKARSH-VERMA-NAMAN-JAIN-AVINASH-PARASHAR-VINAY-SINGH-SANJAY-SUCHART-SIENGCHIN
123. Verma, A., Jain, N., & Sethi, S. K. (2022c). Modeling and simulation of graphene-based composites. In S. M. Rangappa, J. Parameswaranpillai, V. Ayyappan, M. G. Motappa, S. Siengchin, & C. Soutis (Eds.), *Innovations in graphene-based polymer composites* (Woodhead publishing series in composites science and engineering) (pp. 167–198). Woodhead Publishing. https://doi.org/10.1016/B978-0-12-823789-2.00001-7
124. Bhowmik, P., Kant, R., & Singh, H. (2024). Green composites for sustainable applications. In *Modern materials and manufacturing techniques* (pp. 1–25). CRC Press. https://doi.org/10.1201/9781032703046-1/GREEN-COMPOSITES-SUSTAINABLE-APPLICATIONS-PAPIYA-BHOWMIK-RAVI-KANT-HARPREET-SINGH

Carbon Nanotubes-Based Nanocomposites: Materials System, Applications, and Processing

Satnam Singh

1 Introduction

Carbon nanotubes (CNTs) have revolutionized the field of material science and especially in the domain of nanotechnology due to their exceptional mechanical, electrical, and thermal properties [1, 2]. Since the discovery in the early 1990s, CNTs have gained a lot of traction in the materials field due to their ability of providing significant property enhancements as reinforcement in the composite materials, leading to the development of CNT-based nanocomposites [3]. These nanomaterials exhibit superior strength, lightweight characteristics, enhanced electrical conductivity, and outstanding thermal stability, making them invaluable across various industries, including aerospace, biomedical, electronics, and energy sectors [4–7]. CNTs can be broadly classified into two types, i.e., single-walled carbon nanotubes (SWCNTs) and multi-walled carbon nanotubes (MWCNTs). SWCNTs consist of a single graphene sheet rolled into a cylindrical form with diameter ranging from 0.5 to 3 nm with few micrometers of length, whereas MWCNTs have multiple concentric graphene layers 2 to >100 walls which grow vertically. The simplest form of MWCNT is DWNTs, i.e., double wall carbon nanotubes which have two graphene cylindrical walls and are more stable and stiffer than SWCNTs but have lower tensile strength [8]. Due to the presence of sp^2 carbon–carbon bonding at atomic level in CNTs, the mechanical strength is excellent, the elastic modulus varies between 1 and 2 TPa, stiffness can reach up to 1000 GPa, and the thermal

S. Singh (✉)
Multi-Disciplinary Engineering Department, The NorthCap University,
Gurgaon, Haryana, India
e-mail: satnamsingh@ncuindia.edu

V. Khanna (ed.), *Carbon-Based Nanocomposites for Sustainable Applications, Volume I*, Lecture Notes in Nanoscale Science and Technology 38,
https://doi.org/10.1007/978-3-031-95306-4_3

conductivity up to 6000 W/mK [9, 10]. Such unique structure coupled with extraordinary mechanical strength, high aspect ratio, and exceptional flexibility contributes to the prevalent applications of CNTs in the advanced nanocomposite materials. The use of CNTs as reinforcements with various material systems, including polymers, metals, and ceramics, has led to significant advancements in nanocomposite technology with numerous advanced applications. CNT-reinforced polymer composites have found applications in lightweight structural components, high strength nanocomposites, flexible and wearable electronics, and high-performance coatings [11–13]. The applications and properties of CNTs based materials depend on the matrix materials selected. For example, metal-CNT composites enhance the mechanical strength and electrical conductivity of structural materials, while ceramic-CNT composites exhibit improved thermal stability and wear resistance, making them suitable for high-temperature applications. The use of polymers reinforced CNTs contributed to flexible sensors and electronic devices and can be used as biomaterials [14, 15].

CNTs have been widely used as reinforcements in many MMCs (Metal Matrix Composites) which provide unmatchable properties compared to the conventional materials such as steels. The major factors used for the selection of the MMC materials depend on the various requirements such as reinforcement size, matrix material, desired properties, processing routes, and processing strategy as shown in Fig. 1 [16, 17].

The selection of reinforcement types and sizes depends on the functional properties to be achieved. Further, the presence of single size particles (monomodal) or dual size particles (bimodal) or more than two sizes of particles (multimodal) can affect the properties significantly. The processing strategy defines the overall cost and performance over the time [18, 19]. Bulk material processing involves more cost but provides consistence higher performance over the service period, whereas the processing in form of coatings and claddings is cost-effective but for limited working period. Several processing techniques have been developed to effectively disperse and incorporate CNTs into different matrices, with uniform distribution and strong interfacial bonding. Some of the traditional methods such as solution mixing, melt blending, and in situ polymerization [20, 21] have been widely utilized in processing of nanocomposites. In addition, some of the emerging techniques like microwave heating [22], laser processing [23], and hybrid approaches [24] have shown promise in optimizing CNT dispersion and improving composite performance. With continuous research and advancements in nanocomposite technology, CNT-based nanocomposite materials are expected to play a crucial role in future applications, ranging from next-generation electronics and energy storage to biomedicine and sustainable engineering solutions. The display of multifunctional properties and ability to be tailored at the nanoscale allows CNTs to be further explored for the development of smart and adaptive materials, marking their territory in material science and engineering.

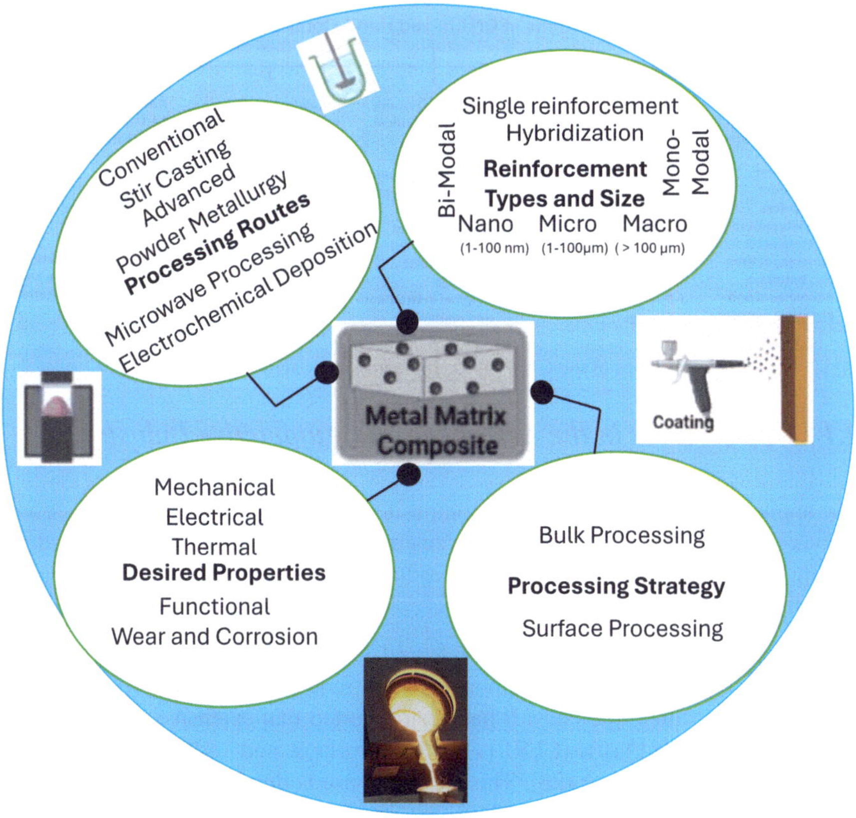

Fig. 1 Various parameters used for the selection of MMCs

2 Applications of Carbon Nanotubes-Based Nanocomposites

The superior and unparalleled properties of CNTs have enabled their integration into various nanocomposite materials, enhancing the performance, applications, and reliability across multiple industries. The major applications include electronics and sensors, nanoelectronics and transistors, aerospace, automotive, biomedical applications, energy and environmental applications, and structural and mechanical applications [25–27]. Figure 2 presents some of the key applications of the carbon nanotube-based composites.

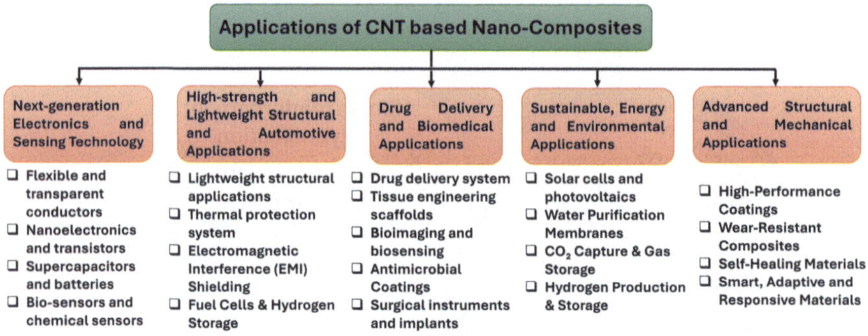

Fig. 2 Various applications of the CNT-based nanocomposites across the domains

2.1 Applications in the Aerospace and Automotive Industry

The applications of CNT-based nanocomposites are increasing in the aerospace and automotive industry. CNTs as reinforcement enhance structural strength, reduce weight, and improve thermal stability, leading to fuel efficiency and durability in aircraft and automobile components.

- **Lightweight composites**: CNT-reinforced polymer composites have replaced conventional metal-based components thus reducing the overall weight and improved the fuel efficiency. It has been reported that extensive use of Carbon Fiber Reinforced Plastic (CFRP) is ~80% of volume and ~50% of weight of the Boeing 787 Dreamliner [28]. This extensive use is due to the high strength-to-weight ratio and durability.
- **Enhanced impact resistance**: The reinforcement of CNTs in the matrix increases the mechanical properties such as toughness, making aerospace and automotive parts more resilient to mechanical stresses.
- **Thermal shielding**: CNT-reinforced composites provide excellent thermal insulation and resistance to extreme temperatures which is crucial for the advanced space applications.

2.2 Applications of CNT-Based Nanocomposites in Electronics and Conductive Materials

CNTs possess high electrical conductivity which is in the range of 6000 W/mk which makes them ideal for electronic components such as flexible displays, conductive coatings, sensors, and energy storage devices like supercapacitors and lithium-ion batteries [29, 30].

- **Flexible and wearable electronics**: The applications of CNT-based transparent electrodes allow them to be used in the bendable screens, sensors, and smart textiles.
- **Next-generation transistors**: The CNT-reinforced materials allow them to be one of the best field-effect transistors (CNT-FETs) which demonstrate superior performance as compared to silicon-based transistors, leading to ultra-fast computing. This allows more reliable computing devices to enhance computational analysis.

2.3 Applications of CNT Nanocomposites in Biomedical Applications

The properties of biocompatibility and functionalization make CNTs suitable for drug delivery, biosensors, and tissue engineering applications [31, 32]. This provides CNT-based nanocomposites a potential advantage in regenerative medicine and cancer therapy.

- **Targeted drug delivery**: The functionalized CNTs allow precise delivery of therapeutic agents to cancer cells and minimizing other side effects. This improves the drug efficiency and effectiveness with lower negative side effects.
- **Biosensors**: The CNTs based electrodes and composites enhance the sensitivity of glucose monitors, disease detection assays, and wearable health devices [33, 34].
- **Scaffold materials**: The CNT-infused hydrogels and scaffolds promote cellular growth and tissue regeneration for organ repair and are extensively used in the biomaterials [35].

2.4 Applications of the CNT-Based Nanocomposites in the Structural and Construction Materials

The CNT reinforcement enhances the mechanical properties of concrete, polymers, and various coatings, thus improving their durability and resistance to environmental conditions. These properties allow CNT-based materials to be extensively used in:

- **Self-healing materials**: CNT-based composites integrate self-repairing mechanisms that improve longevity in infrastructure and increase reliability of structures.
- **Smart coatings**: Anticorrosive CNT coatings enhance the durability of bridges, pipelines, and offshore structures under extreme conditions.

- **Earthquake-resistant buildings**: CNT-reinforced concrete offers superior load-bearing capacity and resistance to seismic activities due to better energy absorption.

2.5 *CNT Nanocomposites in Energy and Environmental Applications*

The applications of CNT-based composites are extensively found in solar cells, fuel cells, and water purification membranes [36, 37] which contribute to the sustainable energy and environmentally friendly technologies.

- **High-efficiency solar cells**: The CNTs improve the electron transport and enhance the efficiency of photovoltaic panels.
- **Hydrogen storage**: CNTs-based nanocomposites act as superior adsorbents for hydrogen, facilitating clean energy applications in various sectors.
- **Water purification membranes**: CNT-integrated filters and hybrids of MXenes and CNT can remove contaminants at the nanoscale and help in efficient desalination and wastewater treatment solutions. These properties allow CNTs to be used in the sustainable environmentally friendly technologies.

3 Material Systems in CNT-Based Nanocomposites

The incorporation of CNTs into various matrix materials has resulted in the creation of diverse nanocomposite systems, each designed for specialized applications. CNT-based nanocomposites can be broadly classified into four main types: polymer-CNT, metal-CNT, ceramic-CNT, and hybrid CNT composites. Each category possesses distinct characteristics that make it highly applicable across multiple industries, such as aerospace, automotive, electronics, biomedical, and energy sectors. Table 1 presents the different material systems of CNT and matrices along with their potential applications.

The functional properties can be imparted to the materials based on the selection of different matrix combinations. Polymer-CNT composites have emerged as an alternative to conventional materials in the material science domain due to their unmatched characteristics of lightweight nature, ease of processing, and enhanced mechanical and electrical properties. The incorporation of CNTs into polymer matrices such as epoxy, polypropylene, and polystyrene leads to the development of polymer nanocomposites which reflects superior structural performance and functional attributes. In the aerospace sector, polymer-CNT composites are highly utilized in lightweight structural components for aircraft and spacecraft with high strength-to-weight ratio. Their demand has been increased many folds, and many aircrafts use more than 50% of these nanocomposites to enhance impact

Table 1 CNT-based nanomaterial systems and their applications

Ref. No.	Material system	Matrix type	Applications
[6, 38]	Polymer-CNT composites	Epoxy, polypropylene, polystyrene	Aerospace, automotive, wearable electronics, sensors
[39, 40]	Metal-CNT composites	Aluminum, copper, titanium	Structural components, automotive parts, electrical contacts, sports equipment
[41, 42]	Ceramic-CNT composites	Alumina, silicon carbide	High-temperature applications, sensors
[43, 44]	Hybrid CNT composites	CNT-graphene, CNT-polymer blends	Energy storage, biomedical applications

resistance and fatigue performance, while lowering the overall cost and weight. Additionally, their capability to provide electromagnetic interference (EMI) shielding makes them highly suitable and valuable for safeguarding the sensitive aerospace electronic equipment. In the automotive sector, these composites contribute to the manufacturing of lightweight body panels, stronger interior components, and high-temperature bearing engine parts, leading to the overall improved fuel efficiency without compromising durability. The presence of CNTs provides superior thermal conductivity which leads to the effective heat dissipations in comparison to conventional materials. Furthermore, polymer-CNT composites play a crucial role in wearable electronics, sensor technology, and sports industry. Their flexibility and conductivity make them ideal for health monitoring devices, motion detectors, and smart textiles, and in biosensors for medical diagnostics and monitoring.

Another important class of materials is CNT-based MMC where metal matrix is used for dispersing of CNTs. Metal matrices like aluminum, copper, and titanium are commonly used for development of nanocomposites. The reinforcement of CNT significantly enhances mechanical strength, wear resistance, and electrical conductivity of such materials. In aerospace and automotive industries, aluminum-CNT composites are widely used for lightweight, high-strength structural applications, whereas titanium-CNT composites provide exceptional corrosion resistance and thermal stability at higher temperatures, making them suitable for many military and aerospace applications. These nanocomposites contribute to the development of high-temperature materials with high strength and are used in development of new generation engine components. Automotive components such as engine parts, brake systems, and chassis materials are produced from MMCs due to the superior mechanical properties of metal-CNT composites. Copper-based CNT composites exhibit enhanced electrical and thermal conductivities, making them suitable for electrical contacts, power transmission lines, and circuit boards. The improved efficiency in power distribution systems due to reduced electrical resistance makes these materials instrumental in the energy sector. Further, the lightweight and durable nature of metal-CNT composites makes them a preferred choice for manufacturing high-performance sports equipment like bicycle frames, golf clubs, and protective gear, where impact resistance coupled with lower weight is crucial.

The reinforcement of CNTs into ceramic matrices such as alumina and silicon carbide leads to the development of nanoceramic composites which have significantly improved mechanical properties. These nanomaterials while retaining their inherent properties of ceramics, such as high hardness, corrosion resistance, and thermal stability also exhibit improved strength and toughness. These composites are extensively used in the production of components which are exposed to high-temperature applications such as turbine blades, heat shields, and refractory components. The silicon carbide-CNT composites find extensive applications in the wear-resistant coatings, claddings, grinding wheels, and cutting tools due to better toughness and durability while maintaining resistance to thermal shock and mechanical wear. These CMCs (Ceramic Matrix Composites) are also employed in advanced sensor technologies, particularly in high-temperature sensors applications for industrial monitoring and aerospace applications. The high electrical conductivity of CNTs facilitates their utilization in gas sensors, strain, and pressure monitoring sensors that operate in harsh environments which make these materials effective in advanced monitoring capabilities in various industries.

Sometimes, the CNTs are mixed with other reinforcements to form the hybrid nanocomposites. CNTs are used with other nanomaterials such as graphene, polymers, or metal nanoparticles to offer multifunctional properties that enhance mechanical strength, electrical conductivity, and chemical stability. These advanced composites have found extensive applications in energy storage and biomedical applications. CNT-graphene hybrid composites are widely used in supercapacitors and lithium-ion batteries due to their excellent charge storage capacity and high electrical conductivity. These characteristics coupled with strength contribute to longer battery life and improved performance. These materials are also explored in various applications for the development of fuel cells and hydrogen storage where these materials enhance the efficiency of next-generation energy systems. In biomedical applications, hybrid CNT-polymer composites are used for controlled drug delivery systems with specific target and sustained release of therapeutic agents. CNT-graphene materials are also used for the bio-scaffolds in the tissue engineering applications while facilitating faster cell regeneration and growth. The associated antimicrobial properties of these materials make them suitable for bio-implant coatings and wound healing applications, contributing to advancements in the biomedical technology.

The various carbon nanotube-based nanocomposites represent an advanced class of materials which has wide applications spanning across multiple industries. The properties of polymer-CNT composites provide lightweight, electrically conductive solutions for aerospace, automotive, and wearable electronics, while metal-CNT composites enhance structural and electrical performance in various engineering fields. Ceramic-CNT composites deliver superior high-temperature resistance and wear properties for industrial and aerospace applications, and hybrid CNT composites combine multiple nanomaterials to drive innovations in energy storage and biomedical applications. The various advanced field of CNT-based nanocomposites are shown in Fig. 3.

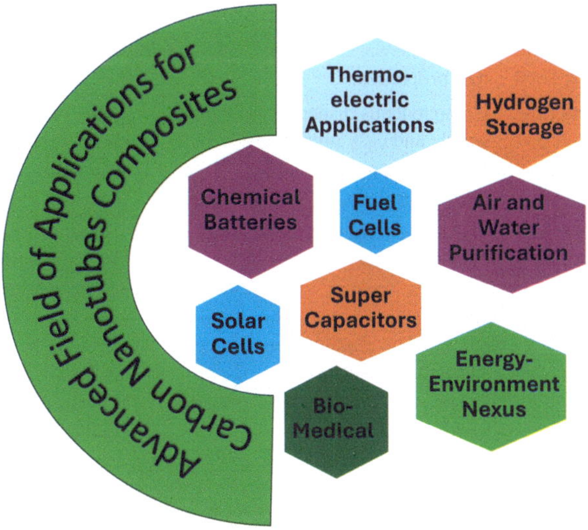

Fig. 3 Advanced fields of CNT-based nanocomposites applications

4 Processing Routes for CNT-Based Nanocomposites

There are various techniques available for the processing of CNT-based nanocomposites. The purpose of these processing routes is to incorporate CNTs into different matrices, influencing the composite's properties and performance. Some of the common processing methods include solution mixing, melt blending, in situ polymerization, electrospinning, and spark plasma sintering. Table 2 summarizes key processing routes for CNT-based nanocomposites along with their properties and applications.

4.1 Emerging Processing Routes for CNT Nanocomposites

As discussed, carbon nanotube (CNT) nanocomposites exhibit exceptional mechanical, electrical, and thermal properties, making them promising materials for advanced applications. However, the traditional processing routes are energy-intensive and involve hazardous solvents and complex procedures, thus hindering their cost-effectiveness and widespread adoption for various applications. This necessitates the exploration of new and innovative processing routes which are sustainable and environmentally friendly. Some of the key challenges for the successful integration of CNTs into a matrix material are homogeneous dispersion, strong defect free interfacial bonding, and scalability. Achieving uniform distribution of CNTs within the matrix is crucial to obtain high strength and best performance from

Table 2 Common processing routes, properties, and applications of CNT-based nanocomposites

Ref. No.	Processing method	Key properties	Applications
[45]	Solution mixing	Homogeneous dispersion of CNTs, simple and easy processing	Used for coatings, flexible electronics
[46]	Melt blending	It is cost-effective, used for thermoplastics-CNT nanocomposites	Automotive parts, packaging industries
[47]	In situ polymerization	Provides strong interfacial bonding, uniform dispersion	Used in aerospace, biomedical implants, energy storage devices and sensors
[48]	Electrospinning	Provides high aspect ratio, improves electrical conductivity, mechanical strength, and surface area	Used in electronics, sensors, filtration, biomedical scaffolds
[49]	Spark plasma sintering	Enhanced mechanical properties, rapid densification	Structural materials, conductive films

the CNT-reinforced composites. Agglomeration of CNT particles while processing can severely affect the composite's performance. Another major challenge is to obtain strong interfacial bonding between CNTs and the matrix for effective load transfer and enhanced mechanical properties. Scalability and cost-effectiveness are further challenges in promoting CNT-based nanocomposites which require the new industrial production processes while maintaining economic viability.

Recent advancements in processing techniques have led to the development of innovative methods for integrating CNTs into nanocomposites, enhancing their dispersion and overall performance. A diverse range of emerging processing techniques is being explored to address these challenges and promote sustainability and is presented in Fig. 4.

Green solvent-based processing [50] using water as a solvent with surfactants and techniques like sonication and high-shear mixing enables effective CNT dispersion while minimizing environmental impact. The use of ionic liquids offers unique properties such as negligible vapor pressure and tunable characteristics, facilitating CNT dispersion and processing with reduced environmental concerns. The use of bioinspired solvents derived from renewable resources, such as terpenes and alcohols, provides an eco-friendly alternative to traditional organic solvents. Solid state processing [51] for such materials involves mechanical milling which involves grinding CNTs with the matrix material, leading to intimate mixing and improved interfacial bonding. However, controlling potential damage to the CNTs during milling is crucial for properties optimization. Another important process is Friction Stir Processing (FSP) which can be used to incorporate CNTs into metallic matrices. It generates heat through friction, offering a potential pathway for energy-efficient processing route. The use of Additive Manufacturing (3D Printing) technologies can further help in precise manufacturing of such composite with tailored properties [52]. Various techniques such as Fused Deposition Modeling (FDM) can be used which allows the direct 3D printing of CNT-polymer composites. This can be used to produce complex shapes with precise control over CNT

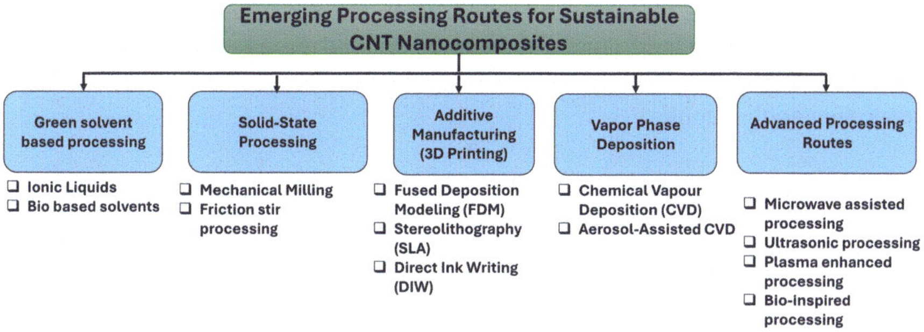

Fig. 4 Emerging processing routes for sustainable CNT-based nanocomposite

distribution and optimization of properties. Stereolithography (SLA) is another process which utilizes photopolymerization to create 3D objects, offering high resolution and the potential for incorporating functionalized CNTs to develop composites with functional properties. Further advancing the 3D printing technologies, Direct Ink Writing (DIW) technique enables the precise deposition of CNT-containing inks, providing control over the final microstructure and properties of the composite. These techniques are well utilized in electronics and sensor development. Plasma spraying is another technique which can be used to deposit CNT coatings onto substrates, thus improving their wear resistance and thermal conductivity. Some new processing techniques such as bioinspired processing are used for CNT-based nanocomposites. The application of self-assembly through biomolecules or templates to guide the assembly of CNTs can lead to highly ordered structures with enhanced properties. Enzyme-mediated reactions can be employed to functionalize CNTs and promote interfacial bonding which offers a biocompatible and environmentally friendly approach.

Chemical Vapor Deposition (CVD) allows for the direct growth of CNTs onto substrates, enabling the fabrication of integrated nanocomposite structures with quality and high order properties [53]. A lot of work on microwave-assisted processing is carried out by the researchers due to unique characteristics of microwave heating which involves rapid and uniform energy transfer, reducing processing time and energy consumption [54, 55]. It can be used for various stages, including synthesis, dispersion, and curing of CNT nanocomposites. Ultrasonic waves can be effectively used for dispersion of CNTs in various media, improving homogeneity and reducing processing times.

Microwave heating is gaining attention due to its rapid, energy-efficient processing, which promotes uniform CNT distribution and improved interfacial bonding within polymer and ceramic matrices [56, 57]. Laser processing techniques [58], including laser ablation and laser-assisted deposition, offer precise control over CNT alignment and structural modification, thus enabling the processing of high-performance composites for electronic and structural applications. Additionally, hybrid processes [59] that combine conventional and novel methods, such as electrospinning with plasma treatment or 3D printing with ultrasonic dispersion, are opening new avenues for fabricating multifunctional CNT-based composites with tailored properties. These emerging techniques hold promise for scalable, cost-effective, and high-quality production of advanced nanocomposites for next-generation applications.

5 Future of CNT-Based Nanocomposite Advanced Materials

The future of CNT-based nanocomposites is promising and getting improved with continuous advancements in material engineering and nanotechnology. Key areas of development include enhanced dispersion techniques, smart and multifunctional composites, sustainable and scalable production, and integration with emerging technologies. The horizon of materials science is ablaze with the potential of carbon nanotube (CNT)-infused nanocomposites, already marking its path in advanced engineering applications. These materials have not only shown incremental improvements, but they represent a quantum leap, promising to unlock unprecedented performance across a spectrum of critical applications. Refined surface chemistries and precision-engineered sustainable processing techniques will lead to the perfect distribution of CNTs within their host matrices, such as polymers, ceramics, or metals. This will translate into materials exhibiting a synergistic blend of unparalleled performance with improved structure-property correlations. Polymer nanocomposites are set to transcend their traditional roles, evolving into dynamic, multifunctional entities in the domain of lightweight, ultra-strong structures for aerospace, military, and electronics. The advent of self-healing polymers, reinforced by CNT networks, will usher in an era of materials that autonomously mend damage, ensuring longevity and reliability in the most demanding environments such as space applications. For ceramic and metal matrices, CNTs are the alchemists which transform the brittle nature of materials to high strength and toughness. Advanced fabrication methodologies, like microwave processing, controlled powder consolidation, and high-energy spark plasma sintering, will enable the creation of nanocomposites capable of withstanding extreme temperatures and stresses. These materials are destined to be one of the classes for high-performance systems, from turbine engines to next-generation cutting tools. Furthermore, CNT-based sensors and actuators will lead an era of precision and sensitivity. Actuators capable of nanoscale precision will drive the development of micro robotics and advanced medical devices. The future of these materials is bright and will allow for tailor-made nanocomposites which will be optimized for specific applications.

6 Conclusion

Carbon nanotube-based nanocomposites represent an advanced class of materials with vast potential across multiple industries from aerospace to structural applications to sensor technologies to wearable electronics and biomedical applications.

The research and development in material engineering, processing techniques, and functionalization strategies will further solidify their role in the next-generation technologies. This will lead to high-performance, multifunctional, tailorable and sustainable materials for global applications. Still cost and scalability are the major hindrances for CNT-based nanocomposite materials growth. New processing routes coupled with sustainability characteristics will lead to the development of variety of applications in the near future for these materials.

References

1. Choudhary, M., Sharma, A., Aravind Raj, S., Sultan, M. T. H., Hui, D., & Shah, A. U. M. (2022). Contemporary review on carbon nanotube (CNT) composites and their impact on multifarious applications. *Nanotechnology Reviews, 11*(1), 2632–2660.
2. Wu, Y., Zhao, X., Shang, Y., Chang, S., Dai, L., & Cao, A. (2021). Application-driven carbon nanotube functional materials. *ACS Nano, 15*(5), 7946–7974.
3. Nurazzi, N. M., Sabaruddin, F. A., Harussani, M. M., Kamarudin, S. H., Rayung, M., Asyraf, M. R. M., et al. (2021). Mechanical performance and applications of CNTs reinforced polymer composites—A review. *Nanomaterials, 11*(9), 2186.
4. Kausar, A., Rafique, I., & Muhammad, B. (2016). Review of applications of polymer/carbon nanotubes and epoxy/CNT composites. *Polymer-Plastics Technology and Engineering, 55*(11), 1167–1191.
5. Harussani, M. M., Sapuan, S. M., Nadeem, G., Rafin, T., & Kirubaanand, W. (2022). Recent applications of carbon-based composites in defence industry: A review. *Defence Technology, 18*(8), 1281–1300.
6. Gadtya, A. S., Mahaling, R. N., Rout, L., & Moharana, S. (2024). Carbon nanotube-based polymer nanocomposites for aerospace applications. In *Carbon nanotube-polymer nanocomposites: Synthesis, properties, and applications* (pp. 435–459). Singapore.
7. Pathania, Y., Ahluwalia, P. K., & Kapoor, P. (2024). Industrial applications of carbon nanomaterials. In *Carbon-based nanomaterials for green applications* (pp. 469–504). Wiley.
8. Gupta, N., Gupta, S. M., & Sharma, S. K. (2019). Carbon nanotubes: Synthesis, properties and engineering applications. *Carbon Letters, 29*(5), 419–447.
9. Neubauer, E., Kitzmantel, M., Hulman, M., & Angerer, P. J. C. S. (2010). Potential and challenges of metal-matrix-composites reinforced with carbon nanofibers and carbon nanotubes. *Composites Science and Technology, 70*(16), 2228–2236.
10. Bakshi, S. R., Lahiri, D., & Agarwal, A. (2010). Carbon nanotube reinforced metal matrix composites—A review. *International Materials Reviews, 55*(1), 41–64.
11. Mohd Nurazzi, N., Asyraf, M. M., Khalina, A., Abdullah, N., Sabaruddin, F. A., Kamarudin, S. H., et al. (2021). Fabrication, functionalization, and application of carbon nanotube-reinforced polymer composite: An overview. *Polymers, 13*(7), 1047.
12. Fenta, E. W., & Mebratie, B. A. (2024). Advancements in carbon nanotube-polymer composites: Enhancing properties and applications through advanced manufacturing techniques. *Heliyon, 10*(1), e36490.
13. Yim, Y. J., Yoon, Y. H., Kim, S. H., Lee, J. H., Chung, D. C., & Kim, B. J. (2025). Carbon nanotube/polymer composites for functional applications. *Polymers, 17*(1), 119.
14. Aghababaei Tafreshi, O., Saadatnia, Z., Ghaffari-Mosanenzadeh, S., Kumar, A., Salari, M., Mohseni Taromsari, S., et al. (2023). Flexible, thermally stable, and Ultralightweight polyimide-CNT aerogel composite films for energy storage applications. *ACS Applied Materials & Interfaces, 15*(43), 50360–50377.
15. Yuan, S., Zheng, Y., Chua, C. K., Yan, Q., & Zhou, K. (2018). Electrical and thermal conductivities of MWCNT/polymer composites fabricated by selective laser sintering. *Composites Part A: Applied Science and Manufacturing, 105*, 203–213.

16. Singh, K., Khanna, V., Singh, S., Bansal, S. A., Chaudhary, V., & Khosla, A. (2023). Paradigm of state-of-the-art CNT reinforced copper metal matrix composites: Processing, characterizations, and applications. *Journal of Materials Research and Technology, 24*, 8572–8605.
17. Singh, S., Gupta, D., & Jain, V. (2019). Microwave processing and characterization of nickel powder based metal matrix composite castings. *Materials Research Express, 6*(8), 0865b1.
18. Babu, A., Arora, H. S., Behera, S. N., Sharma, M., & Grewal, H. S. (2018). Towards highly durable bimodal composite claddings using microwave processing. *Surface and Coatings Technology, 349*, 655–666.
19. Mårtensson, P., Zenkert, D., & Åkermo, M. (2016). Method for the cost-efficient and weight-efficient material diversity and partitioning of a carbon fibre composite body structure. *Proceedings of the Institution of Mechanical Engineers, Part D: Journal of Automobile Engineering, 230*(1), 49–60.
20. Xie, X. L., Mai, Y. W., & Zhou, X. P. (2005). Dispersion and alignment of carbon nanotubes in polymer matrix: A review. *Materials Science and Engineering: R: Reports, 49*(4), 89–112.
21. Khan, F. S. A., Mubarak, N. M., Khalid, M., Khan, M. M., Tan, Y. H., Walvekar, R., et al. (2022). Comprehensive review on carbon nanotubes embedded in different metal and polymer matrix: Fabrications and applications. *Critical Reviews in Solid State and Materials Sciences, 47*(6), 837–864.
22. Singh, S., Gupta, D., Jain, V., & Sharma, A. K. (2015). Microwave processing of materials and applications in manufacturing industries: A review. *Materials and Manufacturing Processes, 30*(1), 1–29.
23. Gu, D., Rao, X., Dai, D., Ma, C., Xi, L., & Lin, K. (2019). Laser additive manufacturing of carbon nanotubes (CNTs) reinforced aluminum matrix nanocomposites: Processing optimization, microstructure evolution and mechanical properties. *Additive Manufacturing, 29*, 100801.
24. Sadeghi, B., Cavaliere, P., Roeen, G. A., Nosko, M., Shamanian, M., Trembošová, V., et al. (2019). Hot rolling of MWCNTs reinforced Al matrix composites produced via spark plasma sintering. *Advanced Composites and Hybrid Materials, 2*, 549–570.
25. Jafari, S. (2018). Engineering applications of carbon nanotubes. In *Carbon nanotube-reinforced polymers* (pp. 25–40). Elsevier.
26. Abubakre, O. K., Medupin, R. O., Akintunde, I. B., Jimoh, O. T., Abdulkareem, A. S., Muriana, R. A., et al. (2023). Carbon nanotube-reinforced polymer nanocomposites for sustainable biomedical applications: A review. *Journal of Science: Advanced Materials and Devices, 8*(2), 100557.
27. Dhanaselvam, P. S., Karthikeyan, B., & Vanitha, P. (2024). Principle and application of C-based nanomaterials: Graphite, graphene, and carbon nanotube (CNT). In *Handbook of emerging materials for semiconductor industry* (pp. 49–64). Singapore.
28. Giurgiutiu, V. (2022). *Stress, vibration, and wave analysis in aerospace composites: SHM and NDE applications*. Academic Press.
29. Subhedar, K. M., & Singh, B. P. (2023). Introduction to carbon nanotubes and graphene and their emerging applications. In *Emerging applications of carbon nanotubes and graphene* (pp. 1–22). CRC Press.
30. Subramanyam, S., Trabelsi, Y., Mahajan, P., Khanna, V., Thakur, A., Chahal, S., et al. (2024). Advancing sustainable energy storage with MXenes: A comprehensive analysis of fabrication techniques and applications. *Journal of Energy Storage, 101*, 114067.
31. de Almeida Barcelos, K., Garg, J., Soares, D. C. F., de Barros, A. L. B., Zhao, Y., & Alisaraie, L. (2023). Recent advances in the applications of CNT-based nanomaterials in pharmaceutical nanotechnology and biomedical engineering. *Journal of Drug Delivery Science and Technology, 87*, 104834.
32. Huang, B. (2020). Carbon nanotubes and their polymeric composites: The applications in tissue engineering. *Biomanufacturing Reviews, 5*(1), 3.
33. Erdem, Ö., Derin, E., Zeibi Shirejini, S., Sagdic, K., Yilmaz, E. G., Yildiz, S., et al. (2022). Carbon-based nanomaterials and sensing tools for wearable health monitoring devices. *Advanced Materials Technologies, 7*(3), 2100572.

34. Yuwen, T., Shu, D., Zou, H., Yang, X., Wang, S., Zhang, S., et al. (2023). Carbon nanotubes: A powerful bridge for conductivity and flexibility in electrochemical glucose sensors. *Journal of Nanobiotechnology, 21*(1), 320.
35. Ahadian, S., Yamada, S., Ramón-Azcón, J., Estili, M., Liang, X., Nakajima, K., et al. (2016). Hybrid hydrogel-aligned carbon nanotube scaffolds to enhance cardiac differentiation of embryoid bodies. *Acta Biomaterialia, 31*, 134–143.
36. Chowdhury, Z. Z., Sagadevan, S., Johan, R. B., Shah, S. T., Adebesi, A., Md, S. I., & Rafique, R. F. (2018). A review on electrochemically modified carbon nanotubes (CNTs) membrane for desalination and purification of water. *Materials Research Express, 5*(10), 102001.
37. Deshmukh, M. A., Park, S. J., Hedau, B. S., & Ha, T. J. (2021). Recent progress in solar cells based on carbon nanomaterials. *Solar Energy, 220*, 953–990.
38. Reghunadhan, A., Shajkumar, A., Abraham, J., & Nimitha, K. C. (2022). Carbon nanotube composites for aerospace applications. In *Carbon nanotubes* (pp. 259–280). Apple Academic Press.
39. Suárez, S., Reinert, L., & Mücklich, F. (2016). Carbon nanotube (CNT)-reinforced metal matrix bulk composites: Manufacturing and evaluation. In *Diamond and carbon composites and nanocomposites*. IntechOpen.
40. Firkowska, I., Boden, A., Vogt, A. M., & Reich, S. (2011). Effect of carbon nanotube surface modification on thermal properties of copper–CNT composites. *Journal of Materials Chemistry, 21*(43), 17541–17546.
41. Balázsi, K., Almansoori, A., & Balázsi, C. (2024). Superior ceramics: Graphene and carbon nanotube (CNT) reinforcements. *Ceramics, 7*(4), 1758–1778.
42. Almansoori, A., Balázsi, K., & Balázsi, C. (2024). Advances, challenges, and applications of graphene and carbon nanotube-reinforced engineering ceramics. *Nanomaterials, 14*(23), 1881.
43. Lu, Z., Foroughi, J., Wang, C., Long, H., & Wallace, G. G. (2018). Superelastic hybrid CNT/graphene fibers for wearable energy storage. *Advanced Energy Materials, 8*(8), 1702047.
44. Jyoti, J., Gupta, T. K., Singh, B. P., Sandhu, M., & Tripathi, S. K. (2022). Recent advancement in three dimensional graphene-carbon nanotubes hybrid materials for energy storage and conversion applications. *Journal of Energy Storage, 50*, 104235.
45. Ke, K., Wang, Y., Liu, X. Q., Cao, J., Luo, Y., Yang, W., et al. (2012). A comparison of melt and solution mixing on the dispersion of carbon nanotubes in a poly (vinylidene fluoride) matrix. *Composites Part B: Engineering, 43*(3), 1425–1432.
46. Albdiry, M. (2024). Effect of melt blending processing on mechanical properties of polymer nanocomposites: A review. *Polymer Bulletin, 81*(7), 5793–5821.
47. Saeed, K., Park, S. Y., Haider, S., & Baek, J. B. (2009). In situ polymerization of multiwalled carbon nanotube/nylon-6 nanocomposites and their electrospun nanofibers. *Nanoscale Research Letters, 4*, 39–46.
48. Kaur, N., Kumar, V., & Dhakate, S. R. (2016). Synthesis and characterization of multiwalled CNT–PAN based composite carbon nanofibers via electrospinning. *Springerplus, 5*, 1–7.
49. Okoro, A. M., Lephuthing, S. S., Oke, S. R., Falodun, O. E., Awotunde, M. A., & Olubambi, P. A. (2019). A review of spark plasma sintering of carbon nanotubes reinforced titanium-based nanocomposites: Fabrication, densification, and mechanical properties. *JOM, 71*, 567–584.
50. Sharifi, J., Rizvi, G., & Fayazfar, H. (2024). Sustainable 3D printing of enhanced carbon nanotube-based polymeric nanocomposites: Green solvent-based casting for eco-friendly electrochemical sensing applications. *The International Journal of Advanced Manufacturing Technology, 131*(9), 4825–4837.
51. Jain, V. K. S., Yazar, K. U., & Muthukumaran, S. (2019). Development and characterization of Al5083-CNTs/SiC composites via friction stir processing. *Journal of Alloys and Compounds, 798*, 82–92.
52. Ghoshal, S. (2017). Polymer/carbon nanotubes (CNT) nanocomposites processing using additive manufacturing (three-dimensional printing) technique: An overview. *Fibers, 5*(4), 40.
53. Ju, J., Deng, N., Zhang, D., Yan, J., Li, L., Kang, W., & Cheng, B. (2019). Facile construction of PCNF&CNT composite material by one-step simultaneous carbonization and chemical vapor deposition. *Journal of Materials Science, 54*(2), 1616–1628.

54. Singh, S., Gupta, D., & Kaushal, S. (2020). Dry sliding wear performance of Ni–SiC composites developed through an in situ microwave casting process. *Journal of Tribology, 142*(10), 101704.
55. Rangari, V. K., Bhuyan, M. S., & Jeelani, S. (2010). Microwave processing and characterization of EPON 862/CNT nanocomposites. *Materials Science and Engineering: B, 168*(1–3), 117–121.
56. Toozandehjani, M., Ostovan, F., Jamaludin, K. R., Amrin, A., Matori, K. A., & Shafiei, E. (2020). Process– microstructure– properties relationship in Al– CNTs– Al_2O_3 nanocomposites manufactured by hybrid powder metallurgy and microwave sintering process. *Transactions of Nonferrous Metals Society of China, 30*(9), 2339–2354.
57. Kaushal, S., Singh, S., Bohra, S., Gupta, D., Jain, V., & Kapoor, M. (2024). Sustainable microwave processing and surface characterization of powdered tungsten reinforced copper metal matrix (Cu-Wx) castings. *Results in Surfaces and Interfaces, 16*, 100249.
58. Alheshibri, M., Elsayed, K., Haladu, S. A., Magami, S. M., Al Baroot, A., Ercan, İ., et al. (2022). Synthesis of ag nanoparticles-decorated on CNTs/TiO_2 nanocomposite as efficient photocatalysts via nanosecond pulsed laser ablation. *Optics & Laser Technology, 155*, 108443.
59. Wu, B., Wu, Y., Zhang, M., Guo, H., Liu, T., Lin, G., & Kuang, T. (2024). Enhanced mechanical and multifunctional properties of GNPs/CNTs hybridized PLA nanocomposites by implementing dual-processing of pickering emulsion-melt blending methods. *Macromolecular Materials and Engineering, 310*, 2400306.

Carbon Nanofibre-Based Nanocomposites

Kamaljit Singh, Sachin Jaidka, and Shivinder Singh

1 Introduction

Carbon nanofibres (CNFs) are cylindrical nanostructures made primarily of graphene layers arranged in different configurations depending on the synthesis method. They are distinct from carbon nanotubes (CNTs), which are purely cylindrical in shape, while CNFs can have multiple forms like stacked cup, fishbone, or helical structures. The dimensions of CNFs, with diameters in the range of tens to hundreds of nanometres and lengths extending to several microns, enable unique properties such as a very high aspect ratio [1]. This aspect ratio is crucial for nanocomposites because it maximizes the surface area for interactions with the host matrix, ensuring effective stress transfer and enhancement of other functional properties like electrical conductivity and thermal stability [2].

Nanocomposites involve combining nanoscale reinforcements with matrices to create materials that exhibit superior mechanical, thermal, electrical, and chemical properties. These reinforcements improve performance by modifying the load transfer mechanisms, thermal conductivity pathways, or electrical percolation networks

K. Singh (✉)
Division of Research and Innovation, Department of Mechanical Engineering, Chandigarh Engineering College, Chandigarh Group of Colleges, Jhanjeri, Mohali, Punjab, India

S. Jaidka
Division of Research and Innovation, Department of Applied Sciences, Chandigarh Engineering College, Chandigarh Group of Colleges, Jhanjeri, Mohali, Punjab, India

S. Singh
Department of Mechanical Engineering, Chandigarh Engineering College, Chandigarh Group of Colleges, Jhanjeri, Mohali, Punjab, India

V. Khanna (ed.), *Carbon-Based Nanocomposites for Sustainable Applications, Volume I*, Lecture Notes in Nanoscale Science and Technology 38,
https://doi.org/10.1007/978-3-031-95306-4_4

[3, 4]. Nanocomposites find applications in industries ranging from automotive and aerospace to electronics and healthcare.

2 Synthesis and Characteristics of Carbon Nanofibres

2.1 Synthesis of Carbon Nanofibres

CNFs can be synthesized through various methods, each impacting the final structural and chemical properties of the fibres. A more detailed explanation of the synthesis method is given below:

Chemical Vapour Deposition (CVD): CVD is widely adopted due to its ability to produce high-quality CNFs with controlled morphology. The process involves heating a carbon-containing gas (e.g., methane, acetylene) at elevated temperatures (600–1200 °C), causing the gas to decompose and deposit carbon on the surface of a metal catalyst like nickel or cobalt. The carbon atoms diffuse into the metal and precipitate out, forming fibres. CVD can be tuned to produce either hollow or solid nanofibres by adjusting parameters like gas flow rates, temperature, and catalyst size [5]. CVD also allows for the incorporation of functional groups on the surface of CNFs, which can enhance the bonding between CNFs and a matrix. This is especially important when CNFs are used in polymer nanocomposites, where interfacial adhesion is critical for performance [6]. A study demonstrated the effect of varying the temperature during CVD synthesis. CNFs grown at 750 °C exhibited a higher graphitic content, thus improving their electrical conductivity. Meanwhile, those grown at 600 °C showed better mechanical properties due to the higher density of structural defects, which improved bonding with polymer matrices [7].

Electrospinning: Electrospinning is a versatile technique where a polymer solution containing carbon precursors is subjected to a high voltage to produce ultrafine fibres [8]. Afterward, the fibres undergo carbonization in an inert atmosphere, which leads to the formation of CNFs. This method allows for the production of aligned CNFs, a key feature when directional properties (such as tensile strength or conductivity) are required in the nanocomposite [9].

Templating Techniques: Templating involves the use of a scaffold or mold to direct the growth of CNFs. Porous anodized alumina is often used as a template because of its well-defined pore structure, which guides carbon deposition. Once the template is filled with carbon, it is removed by etching, leaving behind nanofibres with controlled diameters and alignment [1]. This method is particularly advantageous for applications requiring precise geometrical control.

2.2 Structural and Morphological Characteristics

The morphology and structure of CNFs are directly influenced by their synthesis method. Their graphitic arrangement determines their electrical, thermal, and mechanical properties.

Stacked-Cup Structures: These CNFs consist of conical graphene layers stacked along the fibre axis. They exhibit high mechanical strength due to the strong carbon–carbon bonds, and their structure allows for good load transfer in nanocomposites [10].

Fishbone Structures: The graphene layers in these CNFs are inclined relative to the fibre axis. This inclination increases the available surface area, making fishbone CNFs particularly effective in catalytic applications or as fillers in composite materials where large surface interactions are necessary [11].

Turbostratic Structures: These structures are more disordered compared to the other types, with randomly oriented graphene layers. This randomness decreases the electrical conductivity of the CNFs but can improve flexibility and toughness, making them suitable for applications requiring materials with high-impact resistance [12].

Surface area and porosity: CNFs typically have a surface area ranging from 100 to 500 m^2/g, depending on the synthesis conditions. The high surface area is critical for applications like catalysis, energy storage, and composite reinforcement, where the interface between the CNF and the matrix or the reaction medium dictates performance [13].

3 Carbon Nanofibre-Based Nanocomposites

CNF-based nanocomposites offer distinct advantages across various matrices: polymer, ceramic, and metal. Their unique properties allow for the enhancement of mechanical, thermal, and electrical behaviours.

3.1 Polymer-Based CNF Nanocomposites

Polymers are lightweight, flexible materials, but their inherent mechanical and thermal properties are often insufficient for demanding applications. CNFs reinforce polymers, improving tensile strength, modulus, toughness, and thermal conductivity [14, 15]. Polypropylene (PP) is a widely used thermoplastic polymer known for its low density and cost. However, its mechanical properties limit its use in high-stress environments. CNFs significantly improve the tensile modulus of PP by creating a rigid, high-strength network within the matrix [15–17]. Additionally, CNFs enhance the electrical conductivity of PP, making it suitable for electrostatic discharge (ESD) applications, where charge dissipation is critical [18]. In automotive components, CNF-reinforced PP composites are used to reduce weight while maintaining strength, which is crucial for improving fuel efficiency. Additionally, the improved electrical properties allow for safer use in electronic housings where EMI shielding is required [19, 20].

Some epoxy resins are known for their excellent adhesion, chemical resistance, and mechanical properties, making them ideal for coatings, adhesives, and composites. CNFs further improve these properties, particularly fracture toughness and thermal stability. This makes epoxy/CNF nanocomposites particularly suitable for aerospace applications where weight reduction and thermal performance are essential. The alignment of CNFs within the epoxy matrix can also be controlled to optimize properties [21, 22]. Aligned CNFs in epoxy matrices show enhanced electrical conductivity along the alignment direction, which is beneficial for developing anisotropic conductive adhesives for advanced electronic packaging.

3.2 Ceramic-Based CNF Nanocomposites

Ceramic materials, while excellent in high-temperature environments, suffer from brittleness. CNFs, when embedded in ceramic matrices, improve toughness and reduce the tendency of ceramics to undergo catastrophic failure upon fracture. Alumina (Al_2O_3) is a ceramic material with high hardness and chemical stability, widely used in cutting tools and wear-resistant coatings. However, its brittleness limits its use in high-impact environments. CNFs enhance the toughness of alumina by creating a crack-bridging mechanism. This crack-bridging prevents the rapid propagation of cracks, allowing the material to absorb more energy before failure. Alumina/CNF composites have been used in advanced aerospace components, where both heat resistance and impact toughness are required [23–26]. These materials offer a balanced combination of strength and durability at high temperatures, which is critical for turbine blades and high-wear parts.

3.3 Metal-Based CNF Nanocomposites

Metals are inherently strong and conductive, but the inclusion of CNFs can further improve their mechanical and thermal properties, leading to better performance in applications like heat sinks, structural materials, and conductive components. Aluminium is a lightweight, corrosion-resistant metal but suffers from relatively low strength compared to other metals like steel. CNFs provide reinforcement, enhancing both tensile strength and yield strength. Additionally, CNFs improve the thermal conductivity of aluminium, making it an ideal candidate for use in heat-dissipating components like heat exchangers and thermal management systems. The main challenge in producing metal/CNF composites is achieving uniform dispersion of CNFs within the matrix [26–29]. Methods like ball milling and spark plasma sintering (SPS) help address this challenge by facilitating better distribution and interfacial bonding between CNFs and the metallic matrix. Different type of CNF structure and its composite materials are shown in Table 1.

Table 1 Different types of CNF structure and processing techniques of CNF-based composites

Morphology	Composite materials	Processing technique	Attributes	References
Porous structure	Polyacrylonitrile@ activated carbon nanofibre	Electrospinning and carbonization	Mesopore and large surface area	[30]
	Polyacrylonitrile@ polyaniline nanofibres	Electrospinning and carbonization	Nitrogen abundant, free standing, better conductivity	[31]
	PAN/CoFe-PBAs/ $CaCO_3$	Electrospinning and acidic treatment	Conductive 3D architecture	[32]
Stacked up structure	PSSA-*g*-CSNF	Solution casting techniques	Facile transportation of proton	[33]
	MoS_2/CNF modified with N and O heteroatoms	Electrospinning	95% adsorption capacity for over 10 cycles towards Hg^{2+}	[34]
	PDMS/f-C@BNNS	Electrostatic spinning	Double layered thermally conductive network	[35]
Helical	Natural rubber and CNF (HCNF@NR)	Open two roll laboratory mill	Increasing the tensile force caused the helical CNFs to align in the stress direction	[36]
	C@Fe_2O_3/HCNFs	Followed two-step method	The battery electrode material exhibited prolonged battery life and high stability	[37]
	N–CNF-CNT/TiO_2	Lesser expensive electrospinning method	Enhanced electrode performance	[38]
Tubular	CNF/graphene aerogel	Chemical cross-linking, freeze-drying, and subsequent reduction	Improved mechanical behaviour	[39]
	δ-MnO_2@Fe-CNF	Two-step synthesis	Improved electrode performance	[40]
	WS_2@CNFs	Hydrothermal process	Shown considerable sensitivity towards NO_2	[41]

4 Properties of CNF-Based Nanocomposites

4.1 Mechanical Properties

The mechanical properties of CNF-based nanocomposites are significantly enhanced due to the high tensile strength and modulus of the CNFs themselves. CNFs have a tensile strength of up to 10 GPa and Young's modulus of about 600 GPa, much higher than that of most polymer or ceramic matrices [42, 43].

In several investigations, the mechanical behaviour of the nanomaterials has been broadly studied to establish the relationship among the matrices and the reinforcement materials. The addition of nanofibres can improve the mechanical properties of the composite material significantly. For example, CNFs/polyaniline aerogel-based composite material was developed using electrospinning development technique. The developed composite of CNF/PANI had demonstrated outstanding flexibility along with high strength [44]. CNFs have been grown on the graphene surface using the CVD techniques exhibited favourable results in terms of flexural strength and modulus of flexure as shown in Fig. 1a. To analyse the effect of CNFs, they were grown on the mono and several layer graphene sheets. Fe-C bonding was verified by the chemical compositional analysis through deposition of atoms of carbon on the Fe_2O_3 surface. Also, the increment in the atomic ratio of the C/O further confirmed the growth of CNFs. The nanocomposites of polyamide 6.6 revealed 14.7% and 14% increase in the flexural strength and modulus of flexure. Also, the tensile modulus for 0.5 wt.% of several layer graphene oxide hybrid composite was increased nearly by 24%. This has been a result of the intercalated structure which improves the nucleation process, thereby enhancing the mechanical properties of the composite [45].

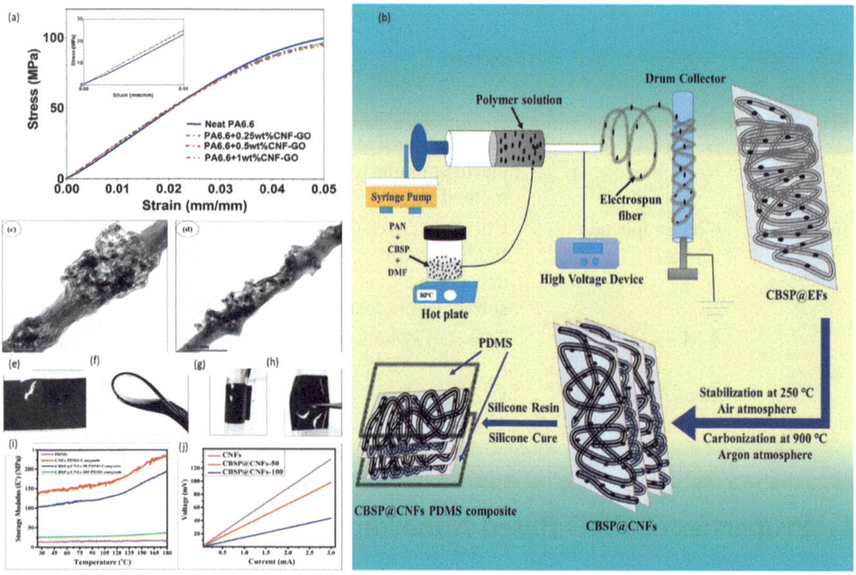

Fig. 1 (**a**) Flexural strength of CNF/PANI composite [45], (**b**) electrospinning method of synthesizing CNF base composite using polymer solution, (**c**) and (**d**) morphology of CBSP@CNF PDMS composite at different magnifications, (**e–h**) mechanical flexibility of the composite, (**i**) storage modulus for CNF composite with respect to temperature, and (**j**) comparison of electrical behaviour of different material compositions [46]

Materials to be used in electromagnetic interference shielding need to be elastic and flexible [47]. Considering the utmost importance of the application, hybrid nanocomposite of CNF incorporated with carbon black and polydimethylsiloxane was developed using electrospinning as shown in Fig. 1b. In terms of evaluating the elasticity or the stiffness of the samples, storage modulus was investigated. As shown in Fig. 1j, the modulus of storage increased as expected with the addition of CNFs. The attained values of storage modulus for CBSP@CNFs-50 were 16.1 and 175.8 while for CBSP@CNFs were 135.0 and 29.7 MPa, respectively. However, CNFs distort under loading conditions which directly impact the modulus of storage for CBSP@CNFs PDMS composites [46].

Strain sensors with high flexibility also need to deliver high sensitivity and strength. Nanocomposites of conductive polymer can fulfil these requirements in developing flexible sensors. For instance, conductive polymer based on PVDF along with CNF and ionic liquid resulted in high mechanical performance and thermal stability. The composite sample without ionic liquid exhibited tensile strength of 5 and 48 MPa modulus. While the maximum elongation depicted was ~5.8309 mm by the ionic liquid included composite sample although the microstructural images revealed pores [48]. Similarly, UHPC was incorporated with CNF for developing self-sensing sensors. The developed material was evaluated under uniaxial compression. With perpendicular orientation, the CNF composite materials sample undergo stress to maximum of 138.05 MPa for 0.0045 of corresponding strain [49].

During an investigation for improving the asphalt pavement, CNFs were added into the bitumen to obtain favourable outcomes. The observation from the study depicted the enhancement in the mechanical properties; 0.4 wt.% CNF in the mixture enhanced the fatigue behaviour of the material. Also, the resistance to the deformation was also reduced. However, exceeding this compositional level had a drastic effect on resilient modulus of the material. For better outcomes, the linear structure of carbon fibre should not be distorted after proper dispersion the mixture. The obtained results are indicative of the performance enhancement of Asphalt concrete behaviour with suitable addition of nanofibre, their composition and their properties [50]. In a different study, CNFs were grown on the graphene using CVD technique. To compare the mechanical properties, single- and multi-layered graphene were analysed after growth of CNF on their surface. The presence of Fe-C and Fe_2O_3 confirmed the growth of CNFs. The polyamide 6.6 nanocomposite produced after mixing the CNF grown multigraphene sheet (0.5 wt.% GO) sample exhibited 14.7% and 14% more flexural strength and flexural modulus in comparison to polyamide 6.6-mono graphene composite sample [45].

Epoxy functionalized carbon nanotubes (CNTs) were utilized as an intermediate layer in CNF-reinforced nanocomposite. The designed material led to synchronized development of CNF composite. The orientation of the CNT and polysulfone on the surface of CNT and epoxy resin prepregs using electrospinning was attained. As the content of CNT/epoxy composite increased, the mechanical attributes of the composite were also increased. Nearly 22% increase in the flexural modulus and 13% in flexural strength were exhibited by 20% CNT/epoxy resin composite [51].

4.2 Electrical Properties

CNFs, due to their graphitic nature, form conductive networks when properly dispersed in an insulating matrix. For example, a polymer composite that contains as little as 0.5 wt% CNFs can reach the percolation threshold, turning the insulating polymer into a conductive composite [52]. Conductive CNF-polymer composites are used in applications like antistatic coatings, where CNFs form networks that dissipate the accumulated electrical charge, thus preventing static build-up.

The electrical conductivity of the CNF/PANI@PANI (CP@PANI) composite observed in a research exhibited better results as compared to the conductivity of carbon fibre aerogel, CNFs/PANI (CP) samples as exhibited in Fig. 2a. Introduction of PANI with the aerogel enabled the conductive network. Thereby, enhancing the electrical conductivity of the CP samples as compared to CFA. Further, the coating of PANI on the surface of the aerogel producing CP@PANI further shown higher conductivity as compared to CP samples. This is ascribed to the conductive nature of the PANI that established a denser pathway for conductivity [44]. Different smart sensing materials based on the CNFs are indispensable for devising laser-composite interaction for diverse application. In this regard, electrical behaviour of CNF composite was analysed for monitoring the laser characteristics as well as degradation. The CNF sample was applied under the infrared laser that was functionalizing at 1064 nm at a frequency of 35 kHz. The pulse duration for the laser was kept at 8 ns for nearly 20 s.

Conductive polymer composites offer great potential for the development of electronic devices using 3D printing technology. Ag@CNFs with a considerable aspect ratio and remarkable 3D printing characteristic of thermoplastic polymer were designed. As shown in Fig. 2b, the electrical conductivity of the composite enhanced continuously with the increasing concentration of the CNFs in the composite [53].

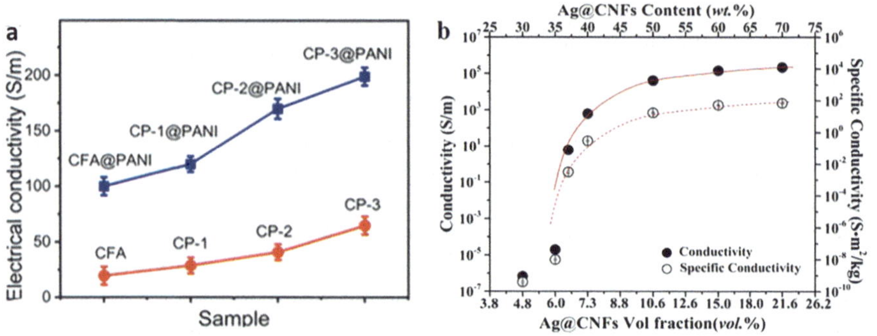

Fig. 2 (**a**) The comparison of electrical conductivity of different materials based on CNFs [44] and (**b**) enhancement of electrical conductivity of Ag@CNF with increasing volume fraction [53]

For supercapacitor, carbon black was added with the CNFs and PANI to develop freestanding electrodes. The synthesized electrode had exhibited 1062.7 Fg^{-1} of specific capacitance. Also, the retention capacity was 72.2% at 1 and 20 Ag^{-1} current densities, respectively. The presence of carbon black and PANI within the composite acted as functionalizing nanofillers for CNFs; while the wettability among the electrodes has been enhanced by the PANI. The synergistic effect of the materials in a symmetric supercapacitor enables the electrodes to reach maximum measured energy density of 54.3 $Whkg^{-1}$ [54]. Interestingly, the behaviour of CNF in comparison to graphene oxide-based CNF was compared. It was revealed that 5 wt.%GO modified CNF when synthesized using electrospinning exhibited better electrical conductivity of 0.39 S/cm which was about 2.2 times more than that of the individual fibre with 0.19 S/cm [55].

4.3 Thermal Properties

NFs' thermal conductivity, which ranges between 200 and 400 W/mK, far exceeds that of most polymers and ceramics. As a result, incorporating CNFs into a polymer or metal matrix dramatically improves the composite's thermal management capabilities. CNF-based thermal interface materials (TIMs) are used in electronics to bridge the gap between heat-producing components (like CPUs) and heat sinks, ensuring efficient heat transfer and preventing overheating [56, 57].

The thermal conductivity of CP@PANI composite has shown improved results for electromagnetic shielding application. The CFA sample exhibited a very low thermal conductivity (0.085 W/mK) owing to the higher porosity of ~82%. This is due to the fact that barrier is established by the air to the transfer of heat. The presence of of PANI also had an insignificant effect on the thermal behaviour of the material. The thermal conductivity for the CP-3@PANI sample was 0.104 W/mK Fig. 3a, due to PANI innate conductivity property [44]. The SEM and TEM images of CNF@TEGO and CNF@GO are exhibited in Fig. 3c–f. While Fig. 3g shows the Raman spectroscopy comparison of TEGO, GO and their CNF hybrid samples. After processing through CVD process, a shift in their geometry can be observed through Raman peaks. The crystalline behaviour of nanocomposites was examined using the DSC method, which also allowed for a thorough comprehension of the nucleation impact of hybrid additives. The plain PA6.6 and 1 wt.%t hybrid reinforced PA6.6 nanocomposites' DSC curves are shown in Fig. 3h. When comparing the nanocomposites specimens to the clean PA6.6 polymer, it was discovered that there was little variation in the melting temperature (Tm). These findings demonstrated that by permitting crystal growth during crystallization, graphene-based compounds functioned as efficient nucleating agents. But in the bulk matrix, we were unable to identify any discernible differences between single- and multilayer-based architectures [45].

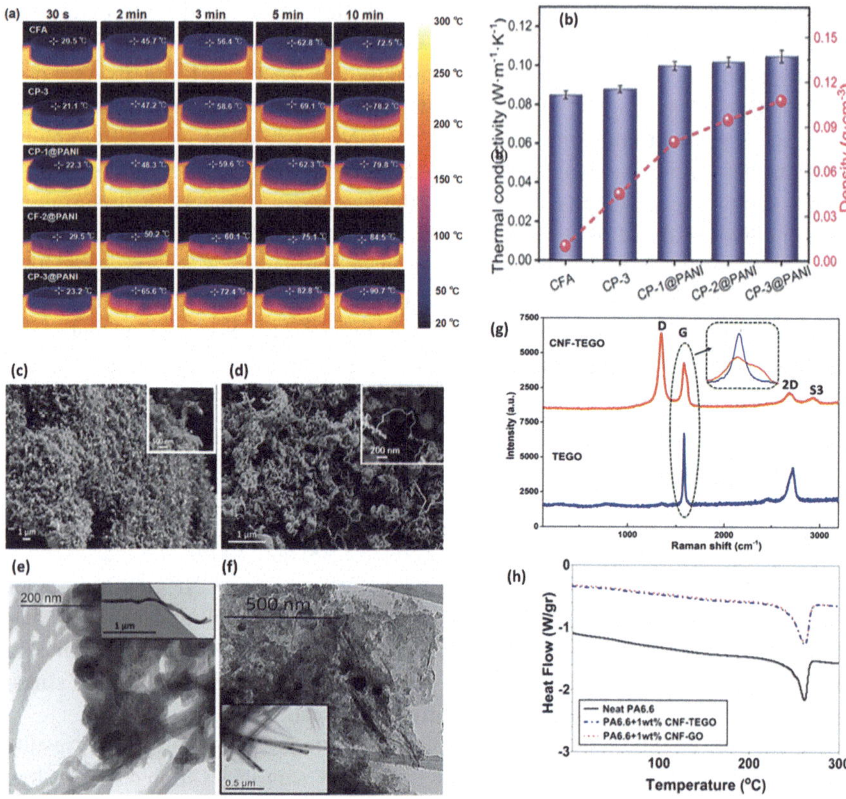

Fig. 3 (**a**) Thermograms of different material composition at 300 °C heating for 0–10 min, (**b**) thermal behaviour of these material exhibiting their utilization as insulating substances [44], (**c, d**) and (**e, f**) exhibit SEM and TEM images of CNF growth on TRGO and GO, (**g**) Raman spectroscopy of pristine TEGO and CNF grown TEGO, and DSC profile curves for different nanocomposite for 1 wt.% CNF in TEGO and GO, (**h**) DSC curves for plain PA6.6 and 1 wt.% hybrid reinforced PA6.6 nanocomposites [45]

Epoxy-based nanocomposite developed using varying weight proportions of CNFs was examined to curing degree within the dynamic and isothermal regime using the Tunnelling Atomic Force Microscopy method. Due to an efficient connection between the CNFs that are incorporated within the epoxy matrix (T20BD) to became an integral part of the cross-linked structure, the nanocomposites filled with CNF heat treated at 2500 °C (CNF2500) are distinguished by the presence of conductive paths at a relatively very low nanofibre content, increasing conductivity beyond the electrical percolation threshold (EPT). In the isothermal regime, each formulation analysis has exhibited DC values greater than 92%. In filled epoxy samples, CNF2500 achieves approximately 100% DC, improving curing efficiency. This attests to the correct establishment of cross-linked networks [58].

5 Applications of CNF-Based Nanocomposites

CNF-based nanocomposite finds application in numerous fields of science, engineering, and technology as shown in Fig. 4. The application of CNF composite ranges from energy storage to aerospace engineering and environment remediations to thermal management system.

5.1 *Structural Applications*

In aerospace and automotive industries, CNF-reinforced composites reduce weight while maintaining structural integrity, leading to fuel efficiency and improved performance.

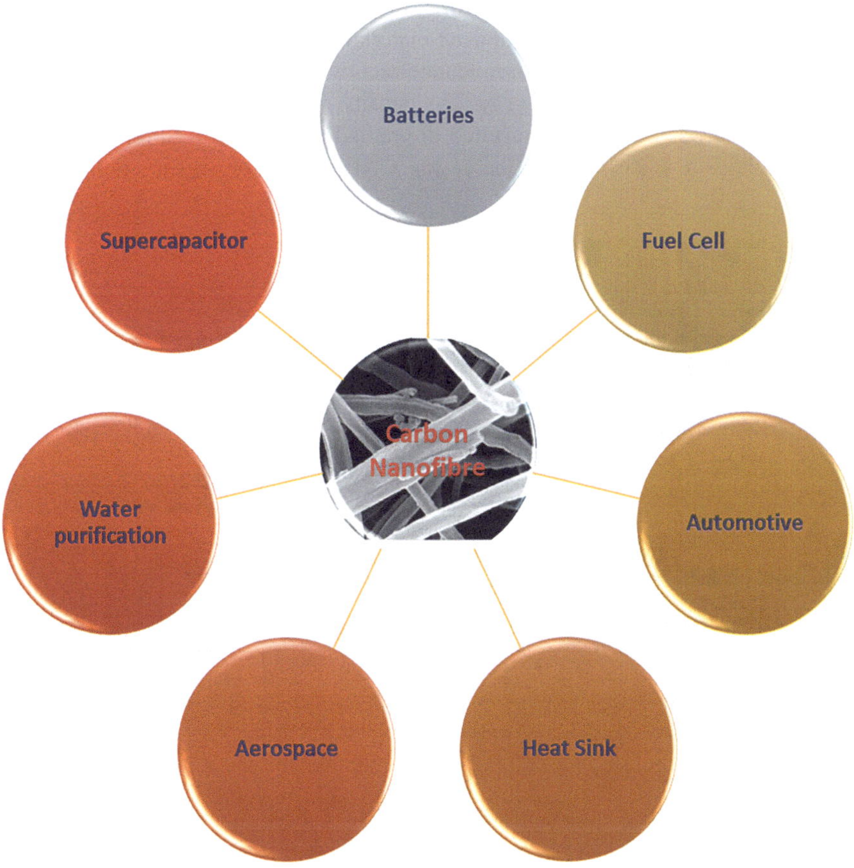

Fig. 4 Possible areas of application of CNF-based composite materials

Automotive: CNF/polypropylene composites are used in bumpers, body panels, and engine covers to reduce weight without sacrificing durability. Numerous investigations carried out recently have demonstrated the exceptional mechanical attributes of CNFs which are highly desired compared to conventionally used fibre in different sectors.

CNFs exhibit considerable tensile strength due to which they have attained huge attention as a filler material for polymer-based composites. When these fibres with high mechanical strength are incorporated within the composite material in shorter length can be adopted for fulfilling the rising demand of the automotive segment. Such composite materials can effectively cater to the strength prerequisite of both the structural and non-structural elements. In this regard, CNF with varying weight percentage from 1 to 4 was embedded into the phenolic resin. The mechanical behaviour of the developed material was evaluated and revealed that 4 wt.% CNF/phenolic resin has attained better results with tensile strength reaching 43.8 MPa and hardness of 37.8 HV [59].

Similarly, aluminium, alumina, and CNF-based powder composite were prepared using the cold spray technology for structural applications. The morphology of the individual material and composite material are shown in Fig. 5a–f. It is evident from figures that CNF has significant effect on the microstructure of the material composite built after cold spray. The density of the material increased while agglomeration decreased. This resulted in gradual enhancement in microhardness of the composite material with increasing CNF proportion in the composite as

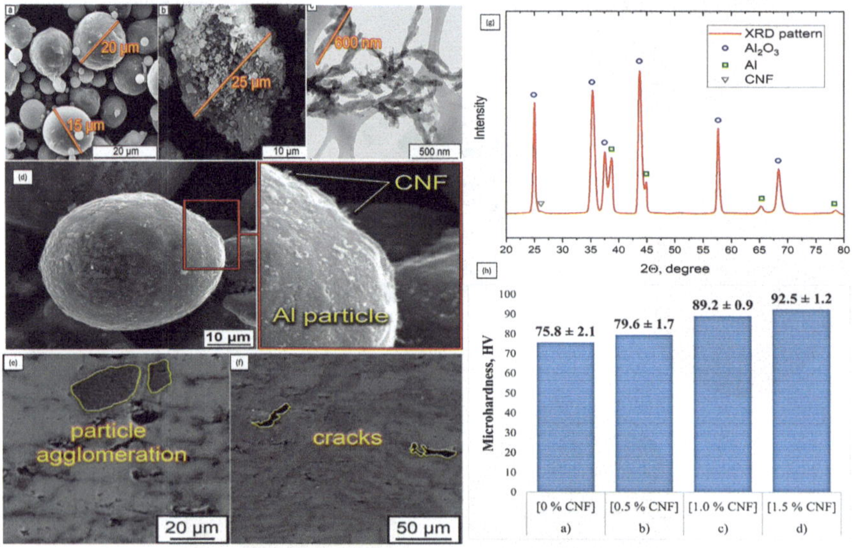

Fig. 5 Morphology of (**a**) aluminium powder, (**b**) alumina powder, (**c**) pristine CNF, (**d**) CNF dispersion on the Al powder surface after blending, (**e, f**) defects occurred within the composite material, (**g**) XRD pattern of Al-Al_2O_3—1.5% CNF sample, and (**h**) microhardness of composite samples with increasing CNF concentration in Al-Al_2O_3 matrix [60]

shown in Fig. 5h. Highest hardness occurred for 1.5 wt.% CNF composite which was ~20% higher compared to another samples. This enhancement is attributed to the dispersive mechanism of strengthening [60].

Aerospace: CNF/epoxy composites are used in fuselage panels and turbine blades, where high strength-to-weight ratios are critical. For example, CNFs were heat treated at two different temperatures, i.e., 1100 °C and 2500 °C and inserted in the epoxy resin as filler material with varying compositions. It was revealed that higher temperature treatment process reduced the defects on CNFs walls. Eventually improving the structural integrity and electrical behaviour of the nanocomposite developed in comparison to 1100 °C CNF/epoxy resin with comparatively lower enhanced storage modulus. Also, the samples with high-temperature heat treatment exhibited higher oxidation resistance compared to lower heat treatment temperature sample. CNF 2500-based nanocomposite attained 100%DC even under isothermal curing. Thereby, forming an extremely cross-linked architecture [61].

Considering the possible requirement of aerospace industry, a flame retardant has been developed recently using CNFs and epoxy resin named glycidyl polyhedral oligomeric silsesquioxane. The material developed used 0.5 wt.% of CNF and 5 wt.% of epoxy resin for improving the mechanical performance and thermal stability along with resistance to flame of resulting composite materials. The microstructural images of CNF and composite materials are exhibited in Fig. 6a–e. The analysis exhibited that for both the resin and resin/GPOSS matrix system, the storage modulus had similar results for temperature between 30 and 165 °C. While the storage modulus enhances by 700 MPa for 0.5 wt.% CNF-based composite for temperature ranging from −30 °C to 180 °C. Figure 6g shows the loss factor for the samples that were isothermally solidified [62].

5.2 Electrical and Electronic Applications

CNF/polymer composites are highly effective at shielding electronic devices from electromagnetic interference, which is crucial in high-frequency applications like communications and radar systems.

Due to the exceptional electrical, electrochemical, and mechanical properties, CNFs have a gained significant attention of the researchers. Their porous architecture, vast surface area, and numerous active sites are the reason for their application in the energy conservation and energy generation. Owing to these features, CNF-based composites are widely investigated for utilization in advanced electronic devices, supercapacitors, fuel cells, and hydrogen processing techniques.

In energy storage applications, CNF-based materials have shown significant effect in supercapacitors as electrode material due to their larger surface area and conductivity. These attributes play significant contribution in enhancing the charge storage capacity of the electrochemical devices. The interconnected porous architecture of CNF enables rapid transportation of metal ions and transfer of charges. This leads to improvement in the distinct aspects of the storage devices, such as increased power density, capacitance, and cycling stability. In particular, combining CNF with metallic oxide, polymers, or carbonaceous material like CNTs and graphene exhibits superior electrochemical performance that makes them highly suitable hybrid material for energy applications [63].

Fascinatingly, the composite materials based on CNFs are regarded as highly potential electrode material for utilization in fuel cells, including proton-exchange membrane fuel cells and solid oxide fuel cells. Their surface area and tunable properties assist in improving the fuel oxidation and oxygen reduction process. Moreover,

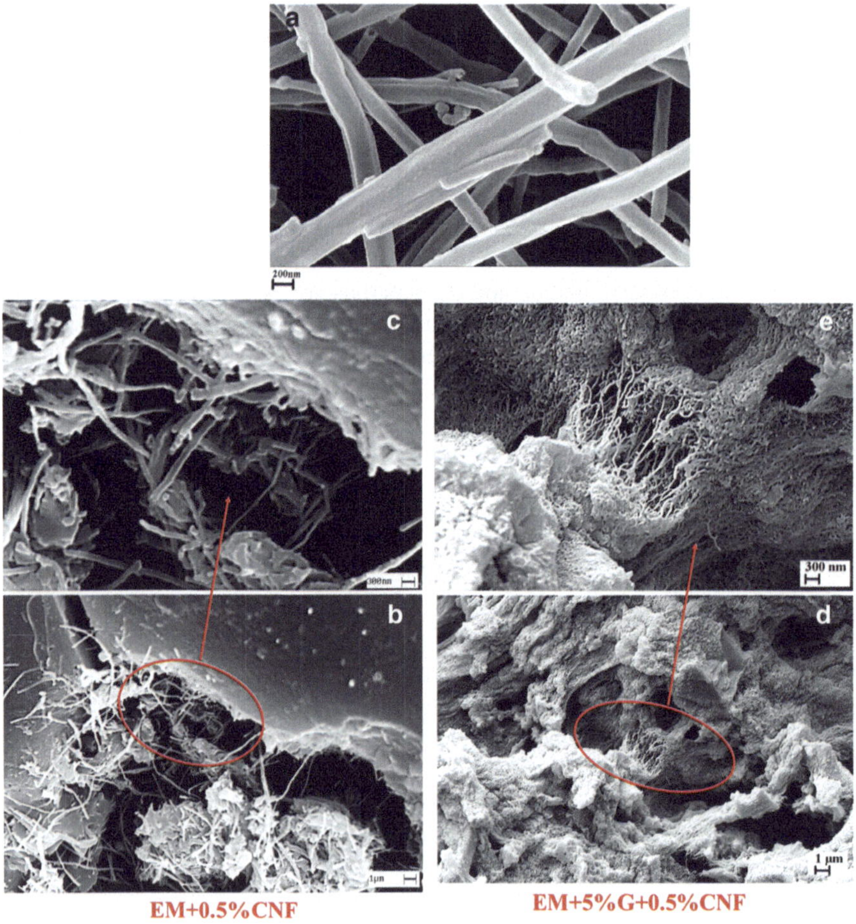

Fig. 6 SEM image exhibiting the (**a**) microstructure of pristine CNF, (**b**) the microstructure of 0.5% CNF-based composite with enlarge inset image (**c**), (**d**) the microstructure of 0.5%CNF+5% graphene and epoxy matrix composite with enlarge inset image, (**e**) absorbance rate of 0.5%CNF+5%graphene and epoxy matrix composite for different temperature range and period, (**f**) the loss factor for the isothermally solidified epoxy samples, and (**g**) the loss factor for the samples that were isothermally solidified [62]

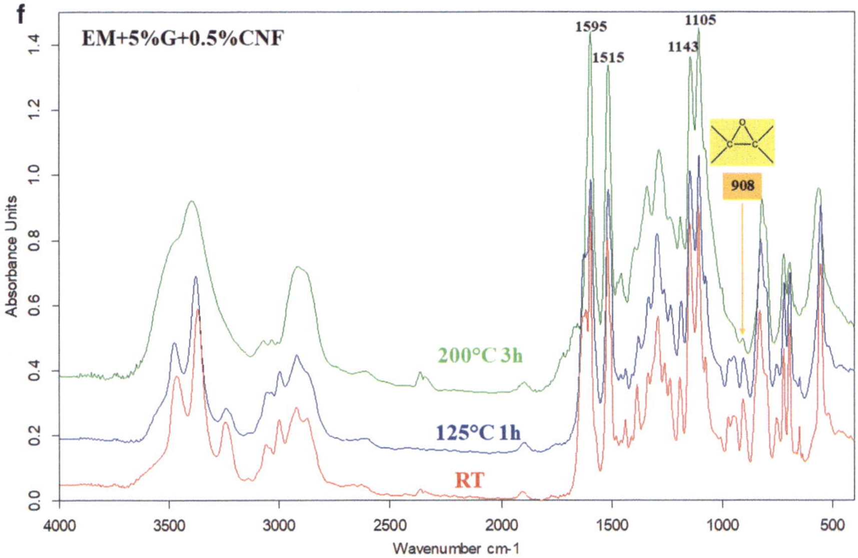

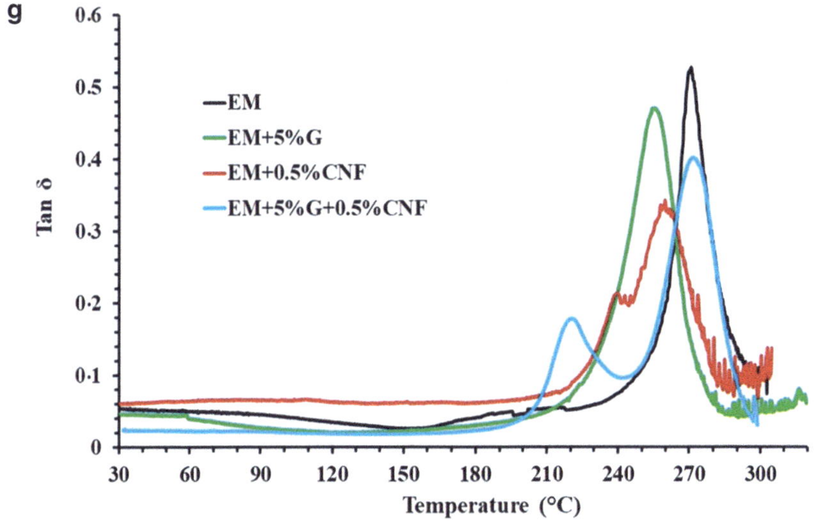

Fig. 4.6 (continued)

the incorporation of heteroatoms such as nitrogen, sulphur, or boron can further enhance their electrochemical activity,which eventually could lead to a more sustainable and cost-effective alternative to conventional material-based catalysts support [64].

Beyond fuel cells, CNF-based nanocomposites have also demonstrated remarkable possibility in batteries. CNFs have been investigated by researchers to enhance the potential of lithium ion, sodium ion, and lithium–sulphur batteries. The excellent light weight morphology, enhanced conductivity, and excellent mechanical integrity facilitate their utilization as superior electrode and cathode materials. Their successful incorporation in the battery electrodes can have substantial impact on the charging/ discharging cycles and battery life. Also, the conductive pathway of CNFs helps in reducing the degradation of the electrodes, minimizes the volume fluctuations, and enhances the ion diffusion kinetics [65].

5.3 Thermal Management Systems

Heat sinks and TIMs: CNF-based composites are increasingly used in electronics cooling systems, where they help dissipate heat from high-power devices. CNFs are known for their self-standing, light weight, and stable behaviour which makes them suitable for EMI shielding and joule heating, but their lower conductivity limits their thermal performance-based application. However, the existing techniques to improve the conductivity show trade off with other properties such as flexibility. Thus, to ensure high conductivity without sacrificing the flexibility, the researchers designed N-doped CNFs using the electro spun method followed by annealing. In nanoparticles of TiN and Co, the CNFs were successfully embedded revealing a superior Ti-N covalent bonding improving the robustness and Co enhancing the conductivity of CNF. Due to this combined effect, the material exhibited faster rate of heating and cooling (28/31 °C/s) along with better joules efficiency (1508 °CW^{-1} cm^2). Furthermore, the 78% decrease in the MXene sensor restoration time indicates the high potential of the TCCNF mat-based heater as an electrode for wearable electronics the TCCNF mat-based heater exhibits great potential as an electrode for wearable devices [66]. In another study, to improve the thermal performance of TIMs, magnetic alignment and freeze casting method were applied to develop directional porous channels using aligned CNFs with dual aligned scaffolds. The scaffolds were diffused within the silicon rubber through vacuum assistance. This maximized the CNFs axial conductivity and developed a continuous path for enhanced heat conduction. Using this approach, a thermal conductivity of 4.66 W/mK was attained for 7.7 vol% of CNF which was 1.5 times more than traditional freeze cast material composite and 25 times more than the individual silicon rubber [67].

A different application of non-structural insulation in foam concrete is vital. CNFs based carbon nanofiber-lightweight cementitious composites (CNF-LCC) was developed using micro-foam bubbles so as to enhance the physico-mechanical properties of the foam composite. The CNF-LCC was enabled to provide better thermal properties due to presence of CNFs and foam bubbles. The thermal conductivity at the beginning stage raised to 20–150°C due to evaporation of water, then decreases due to defects such as pores and cracks and increases slightly 600–800 due to radiation on foam bubbles. The ABAQUS computational simulation confirmed the thermal behaviour of the foam composite at elevated temperature. This porous structure of CNF-LCC absorbs CO_2 pressure beyond 670 °C while preventing expansion, making it a thermally stable and efficient insulation material [68].

5.4 Environmental and Energy Applications

5.4.1 Water Filtration

The complicated amalgamation of contaminants, including organic matter, chemical pesticides, heavy metallic components, and various inorganic substances that come from a variety of sectors, including textiles, pharma, and farming, makes it difficult for traditional process of separation to filter wastewater from industrial facilities.

The high surface area of CNFs allows for the adsorption of pollutants in water treatment systems. CNF-based membranes can remove heavy metals and organic contaminants efficiently. A general purification technique worthy for its effectiveness, affordability, and simplicity is the development and formulation of adsorbents for treating water using carbon nanomaterials. Different form of carbon materials such as CNFs, carbon nanotube, and graphene has shown great promise in eradicating pollutants from the wasted water.

In a comparative investigation, different materials were tested for the removal of cadmium from waste water. Activated CNFs (ACNFs), magnesium oxide (MgO), and PAN-based AGNFs based material exhibited 100% outcome compared to granular activated carbon material, which showed 97% effectiveness. In contrast to GAC, the ACNF/MgO and PAN/ACNF have demonstrated better adsorption capability [69, 70]. It has been revealed in previous examinations that MgO has high adsorption properties due to its substantial specific surface area, and its integration into ACNFs would have significantly enhanced their adsorption capacity [12]. In another study, CNFs-titanium oxide and PAN-based hybrid membrane were tested for rejecting pollutants such as Pb^{2+}, Cd^{2+}, and Cu^{2+}. Its performance was also evaluated under the varying pH, contact period, and concentration of metal ions. It was revealed that low pH reduced the adsorption rate of metal ion due to electrostatically enhanced repulsion. Higher pH increased both the precipitation and adsorption, reaching 58% for cadmium ion, ~75% for copper ions and lead ions [71].

Energy Storage: CNF-based composites are used in lithium-ion batteries and supercapacitors. CNFs' conductivity and surface area enhance the performance of electrodes, leading to higher energy storage capacity and faster charge/discharge cycles. In recent efforts by researcher for developing lithium-ion batteries with high energy and power density, improved cycle stability 3D CNF has attained huge attention as electrode material. This is due to the vast surface area, more carbon sites are highly essential for rapid electron migration, limiting the volume changes to enhance the battery performance [72]. 3D CNF can be an effective strategy to use in batteries as substrate for solving lithium metal deposition or dissolution cycle problem. To illustrate, the deposition rate of Li ions is affected by the conductivity of the CNFs. This has been found that extremely conductive CNFs within the 3D CNF matrix established a conductive pathway for movement of electron and enabled Li deposition on the outside surface of the matrix [73].

However, poor interaction among the CNF and the active materials could affect the performance of the electrode in batteries. To eradicate this issue, a study proposed synthesis of metal oxide with ultrafine structure on nitrogen enriched 3D CNF surface. The strong chemical bonding between the particles of CNFs and metal oxides resulted in rapid kinetics and reduced structural instability of the developed 3D composite material. This further led the 3D material (CO_3O@CNFs) to reach higher capacity and stabilized Li ion storage (1199 mAh/g at a current density of 200 mA/g) [74].

6 Challenges and Future Perspectives

While CNF-based nanocomposites offer many benefits, several challenges remain, particularly in terms of scalability and cost. Achieving uniform dispersion of CNFs, optimizing interfacial bonding, and developing low-cost production methods are key areas of ongoing research. One of the greatest challenges in fabricating CNF-based nanocomposites is the tendency of CNFs to agglomerate, forming bundles due to van der Waals forces. Various techniques, such as surface functionalization and the use of surfactants, have been developed to improve dispersion, but ensuring uniform distribution remains a challenge, especially in metal matrices. For CNFs to effectively transfer stress and enhance properties, strong interfacial bonding between CNFs and the matrix is essential. Surface functionalization of CNFs, such as oxidation or chemical grafting, can improve the adhesion between CNFs and polymer or metal matrices.

While CNFs offer superior properties, their production is still relatively expensive compared to traditional materials. Advances in synthesis methods, particularly large-scale production techniques, are needed to reduce costs and enable the widespread adoption of CNF-based nanocomposites.

7 Conclusion

The development of carbon nanofibre-based nanocomposites marks a notable leap forward in the field of materials science, with substantial implications for a wide range of industrial applications. Carbon nanofibres (CNFs) are renowned for their exceptional mechanical strength, electrical conductivity, and thermal stability, attributes that are significantly enhanced when they are incorporated into nanocomposites. By integrating CNFs into various matrices, such as polymers, ceramics, and metals, these nanocomposites exhibit improved performance characteristics compared to their conventional counterparts. In terms of mechanical properties, CNFs contribute to increased tensile strength, stiffness, and toughness of the composites. Their high aspect ratio ensures efficient load transfer, resulting in materials that are both stronger and more durable. For electrical applications, the inclusion of CNFs enables the creation of conductive networks within otherwise insulating matrices, which is particularly valuable for developing advanced electronic devices and electromagnetic interference shielding.

References

1. Zhou, X., Wang, Y., Gong, C., Liu, B., & Wei, G. (2020). Production, structural design, functional control, and broad applications of carbon nanofiber-based nanomaterials: A comprehensive review. *Chemical Engineering Journal, 402*, 126189.
2. Idumah, C. I., & Obele, C. M. (2021). Understanding interfacial influence on properties of polymer nanocomposites. *Surfaces and Interfaces, 22*, 100879.
3. Zhang, F., Feng, Y., & Feng, W. (2020). Three-dimensional interconnected networks for thermally conductive polymer composites: Design, preparation, properties, and mechanisms. *Materials Science and Engineering: R: Reports, 142*, 100580.
4. Tarawneh, M. A., et al. (2020). Mechanical reinforcement with enhanced electrical and heat conduction of epoxy resin by polyaniline and graphene nanoplatelets. *Nanotechnology Reviews, 9*(1), 1550–1561.
5. Fan, Z.-J., et al. (2011). Nanographene-constructed carbon nanofibers grown on graphene sheets by chemical vapor deposition: High-performance anode materials for lithium ion batteries. *ACS Nano, 5*(4), 2787–2794.
6. Sacco, L. N., & Vollebregt, S. (2023). Overview of engineering carbon nanomaterials such as carbon nanotubes (CNTs), carbon nanofibers (CNFs), graphene and nanodiamonds and other carbon allotropes inside porous anodic alumina (PAA) templates. *Nanomaterials, 13*(2), 260.
7. Wang, H., & Moore, J. J. (2012). Low temperature growth mechanisms of vertically aligned carbon nanofibers and nanotubes by radio frequency-plasma enhanced chemical vapor deposition. *Carbon N Y, 50*(3), 1235–1242.
8. Lee, J. K. Y., et al. (2018). Polymer-based composites by electrospinning: Preparation & functionalization with nanocarbons. *Progress in Polymer Science, 86*, 40–84.
9. Gan, Y. X., & Gan, J. B. (2020). Advances in manufacturing composite carbon nanofiber-based aerogels. *Journal Of Composites Science, 4*(2), 73.
10. Vera-Agullo, J., Varela-Rizo, H., Conesa, J. A., Almansa, C., Merino, C., & Martin-Gullon, I. (2007). Evidence for growth mechanism and helix-spiral cone structure of stacked-cup carbon nanofibers. *Carbon N Y, 45*(14), 2751–2758.
11. Cheng, H.-Y., Zhu, Y.-A., Sui, Z.-J., Zhou, X.-G., & Chen, D. (2012). Modeling of fishbone-type carbon nanofibers with cone-helix structures. *Carbon N Y, 50*(12), 4359–4372.
12. Monthioux, M., Noé, L., Kobylko, M., Wang, Y., Cazares-Huerta, T. C., & Pénicaud, A. (2017). Determining the structure of graphene-based flakes from their morphotype. *Carbon N Y, 115*, 128–133.
13. Darpentigny, C., Nonglaton, G., Bras, J., & Jean, B. (2020). Highly absorbent cellulose nanofibrils aerogels prepared by supercritical drying. *Carbohydrate Polymers, 229*, 115560.
14. Kargarzadeh, H., et al. (2017). Recent developments on nanocellulose reinforced polymer nanocomposites: A review. *Polymer (Guildford), 132*, 368–393.
15. Ray, S. S., & Ojijo, V. (2018). *Processing of polymer-based nanocomposites*. Springer.
16. Saba, N., & Jawaid, M. (2017). Recent advances in nanocellulose-based polymer nanocomposites. In *Cellulose-reinforced nanofibre composites* (pp. 89–112). Woodhead Publishing.

17. Quan, G., et al. (2024). Construction of cellulose nanofiber/carbon nanotube synergistic network on carbon fiber surface to enhance mechanical properties and thermal conductivity of composites. *Composites Science and Technology, 248*, 110454.
18. Nisticò, R., D'Arienzo, M., Di Credico, B., Mostoni, S., & Scotti, R. (2022). The role of inorganic fillers in electrostatic discharge composites. *Inorganics (Basel), 10*(12), 222.
19. Zhang, Y., Ruan, K., & Gu, J. (2021). Flexible sandwich-structured electromagnetic interference shielding nanocomposite films with excellent thermal conductivities. *Small, 17*(42), 2101951.
20. Khanna, V., & Bakshi, B. R. (2009). Carbon nanofiber polymer composites: Evaluation of life cycle energy use. *Environmental Science & Technology, 43*(6), 2078–2084.
21. Yusuf, J., Sapuan, S. M., Rashid, U., Ilyas, R. A., & Hassan, M. R. (2024). Thermal, mechanical, and morphological properties of oil palm cellulose nanofibril reinforced green epoxy nanocomposites. *International Journal of Biological Macromolecules, 278*, 134421.
22. Rana, S., Alagirusamy, R., & Joshi, M. (2011). Development of carbon nanofibre incorporated three phase carbon/epoxy composites with enhanced mechanical, electrical and thermal properties. *Composites. Part A, Applied Science and Manufacturing, 42*(5), 439–445.
23. Liu, H., et al. (2022). Additive manufacturing of nanocellulose/polyborosilazane derived CNFs-SiBCN ceramic metamaterials for ultra-broadband electromagnetic absorption. *Chemical Engineering Journal, 433*, 133743.
24. Dusza, J., & Šajgalík, P. (2008). Ceramic nanocomposites. *Powder Metallurgy Progress, 8*(4), 279.
25. Liu, X., et al. (2022). Nanocellulose-polysilazane single-source-precursor derived defect-rich carbon nanofibers/SiCN nanocomposites with excellent electromagnetic absorption performance. *Carbon N Y, 188*, 349–359.
26. Behera, A., Sahu, P., Mohapatra, S. K., & Ghadei, S. K. (2020). Characterization of carbon based nanofibers in nanocomposites and their applications. *Materials Today Proceedings, 33*, 5714–5719.
27. Fang, Y., Chang, H., Li, J., Li, Z., & Zhang, D. (2024). Recent advances in metal nanocomposite-based electrochemical (bio) sensors for pharmaceutical analysis. *Critical Reviews in Analytical Chemistry, 54*(6), 1680–1706.
28. Pottathara, Y. B., & Kawasaki, H. (2021). Nanocomposite of nanocellulose with metal nanoparticles for printable and flexible electronics. In *Nanocellulose based composites for electronics* (pp. 215–236). Elsevier.
29. Hansa, S. S., & Kim, T. (2022). Metal nanocomposites for energy and environmental applications. In *Metal nanocomposites for energy and environmental applications* (pp. 293–318). Springer.
30. Lee, H.-M., Kim, H.-G., An, K.-H., & Kim, B.-J. (2014). Effects of pore structures on electrochemical behaviors of polyacrylonitrile-based activated carbon nanofibers by carbon dioxide activation. *Carbon Letters, 15*(1), 71–76. https://doi.org/10.5714/CL.2014.15.1.071
31. Miao, F., Shao, C., Li, X., Wang, K., & Liu, Y. (2016). Flexible solid-state supercapacitors based on freestanding nitrogen-doped porous carbon nanofibers derived from electrospun polyacrylonitrile@polyaniline nanofibers. *Journal of Materials Chemistry A, 4*(11), 4180–4187. https://doi.org/10.1039/C6TA00015K
32. Yin, D., Liu, J., Bo, X., & Guo, L. (2020). Cobalt-iron selenides embedded in porous carbon nanofibers for simultaneous electrochemical detection of trace of hydroquinone, catechol and resorcinol. *Analytica Chimica Acta, 1093*, 35–42. https://doi.org/10.1016/j.aca.2019.09.057
33. Neeshma, M., Dhanasekaran, P., Unni, S. M., & Bhat, S. D. (2023). Short side chain perfluorosulfonic acid composite membrane with covalently grafted cup stacked carbon nanofibers for polymer electrolyte fuel cells. *Journal of Membrane Science, 668*, 121240. https://doi.org/10.1016/j.memsci.2022.121240
34. Duan, Y., et al. (2022). Embedding of ultra-dispersed MoS_2 nanosheets in N,O heteroatom-modified carbon nanofibers for improved adsorption of Hg^{2+}. *Composites Communications, 31*, 101106. https://doi.org/10.1016/j.coco.2022.101106

35. Wang, Z., Fan, L., Li, R., Xu, Y., & Fu, Q. (2022). Preparation of polymer composites with high thermal conductivity by constructing a 'double thermal conductive network' via electrostatic spinning. *Composites Communications, 36*, 101371. https://doi.org/10.1016/j.coco.2022.101371
36. Zhang, W., Liu, S., Gong, Y., Zhang, Y., Chen, J., & Fu, Q. (2021). A facile strategy for observation of helical carbon nanofiber fillers in polymer matrix under different stresses. *Polymer Testing, 93*, 106977. https://doi.org/10.1016/j.polymertesting.2020.106977
37. Jin, Y., Zhang, W., Qing, T., Chen, G., Zeng, S., & Huang, J. (2023). Constructing 3D sandwich-like carbon coated Fe_2O_3/helical carbon nanofibers composite as a superior lithium-ion batteries anode. *Journal of Electroanalytical Chemistry, 929*, 117098. https://doi.org/10.1016/j.jelechem.2022.117098
38. Thirugnanam, L., Palanisamy, M., Kaveri, S., Ramaprabhu, S., Pol, V. G., & Dutta, M. (2021). TiO_2 nanoparticle embedded nitrogen doped electrospun helical carbon nanofiber-carbon nanotube hybrid anode for lithium-ion batteries. *International Journal of Hydrogen Energy, 46*(2), 2464–2478. https://doi.org/10.1016/j.ijhydene.2020.10.149
39. Kang, S., Qiao, S., Cao, Y., Hu, Z., Yu, J., & Wang, Y. (2022). Compression strain-dependent tubular carbon nanofibers/graphene aerogel absorber with ultrabroad absorption band. *Chemical Engineering Journal, 433*, 133619. https://doi.org/10.1016/j.cej.2021.133619
40. Li, J., et al. (2021). Fe-regulated δ-MnO_2 nanosheet assembly on carbon nanofiber under acidic condition for high performance supercapacitor and capacitive deionization. *Applied Surface Science, 542*, 148715. https://doi.org/10.1016/j.apsusc.2020.148715
41. Xu, Y., et al. (2021). Edge-enriched WS_2 nanosheets on carbon nanofibers boosts NO_2 detection at room temperature. *Journal of Hazardous Materials, 411*, 125120. https://doi.org/10.1016/j.jhazmat.2021.125120
42. Hvizdoš, P., Puchý, V., Duszová, A., & Dusza, J. (2011). Carbon nanofibers reinforced ceramic matrix composites. In *Nanofibers-production, properties and functional applications*. IntechOpen.
43. Hammel, E., et al. (2004). Carbon nanofibers for composite applications. *Carbon N Y, 42*(5–6), 1153–1158.
44. Chen, M., et al. (2025). Carbon nanofiber/polyaniline composite aerogel with excellent electromagnetic interference shielding, low thermal conductivity, and extremely low heat release. *Nanomicro Letters, 17*(1), 80. https://doi.org/10.1007/s40820-024-01583-2
45. Sarac, E. C., et al. (2019). Performance comparison of CVD grown carbon nanofiber based on single- and multi-layer graphene oxides in melt-compounded PA6.6 nanocomposites. *Open Journal of Composite Materials, 09*(02), 99–123. https://doi.org/10.4236/ojcm.2019.92005
46. Sharma, G. K., & James, N. R. (2022). Carbon black incorporated carbon nanofiber based polydimethylsiloxane composite for electromagnetic interference shielding. *Carbon Trends, 8*, 100177. https://doi.org/10.1016/j.cartre.2022.100177
47. Anju, V. P., Manoj, M., Mohanan, P., & Narayanankutty, S. K. (2019). A comparative study on electromagnetic interference shielding effectiveness of carbon nanofiber and nanofibrillated cellulose composites. *Synthetic Metals, 247*, 285–297. https://doi.org/10.1016/j.synthmet.2018.12.021
48. Prasad, B., Gill, F. S., Panwar, V., & Anoop, G. (2019). Development of strain sensor using conductive poly(vinylidene fluoride) (PVDF) nanocomposite membrane reinforced with ionic liquid (IL) & carbon nanofiber (CNF). *Composites Part B: Engineering, 173*, 106990. https://doi.org/10.1016/j.compositesb.2019.106990
49. Joshi, B., et al. (2023). Development of robust ultra-high-performance carbon nanofiber aggregates (UHPCNFAs) for structural health monitoring. *Engineering Structures, 279*, 115559. https://doi.org/10.1016/j.engstruct.2022.115559
50. Jahromi, S. G. (2015). *Effect of carbon nanofiber on mechanical behavior of asphalt concrete*. http://penerbit.uthm.edu.my/
51. Li, P., et al. (2015). Synchronous effects of multiscale reinforced and toughened CFRP composites by MWNTs-EP/PSF hybrid nanofibers with preferred orientation. *Composites.*

Part A, Applied Science and Manufacturing, 68, 72–80. https://doi.org/10.1016/j.compositesa.2014.09.010

52. Zhou, Z., et al. (2020). Electrical properties of low dosage carbon nanofiber/cement composite: Percolation behavior and polarization effect. *Cement and Concrete Composites, 109*, 103539.
53. Wei, H., et al. (2019). Direct 3D printing of hybrid nanofiber-based nanocomposites for highly conductive and shape memory applications. *ACS Applied Materials & Interfaces, 11*(27), 24523–24532. https://doi.org/10.1021/acsami.9b04245
54. Sun, L., et al. (2023). Rational construction of hierarchical nanocomposites by growing dense polyaniline nanoarrays on carbon black-functionalized carbon nanofiber backbone for freestanding supercapacitor electrodes. *Journal of Energy Storage, 61*, 106738. https://doi.org/10.1016/j.est.2023.106738
55. Huang, H.-D., Guo, Z., Yang, P., Chen, P., & Wu, J. (2021). Electrical conductivity and hydrophobicity of graphene oxide-modified carbon nanofibers. *Chemical Physics Letters, 771*, 138551. https://doi.org/10.1016/j.cplett.2021.138551
56. Liao, S.-Y., et al. (2021). Flexible liquid metal/cellulose nanofiber composites film with excellent thermal reliability for highly efficient and broadband EMI shielding. *Chemical Engineering Journal, 422*, 129962.
57. Pan, C., et al. (2023). A multifunctional flexible composite film with excellent insulation flame retardancy, thermal management and solar-thermal conversion properties based on CNF-modified mica/electrospun fibrous networks structure. *Solar Energy Materials and Solar Cells, 261*, 112530.
58. Raimondo, M., Naddeo, C., & Guadagno, L. (2024). Thermal analysis and electrical performance by Tunneling Atomic Force Microscopy of carbon nanofiber-based nanocomposites. *Catalogo Prodotti Ricerca, OC5*, 35–36. Accessed: Feb. 04, 2025. [Online]. Available: https://www.iris.unisa.it/handle/11386/4859221
59. Natrayan, L., Merneedi, A., Bharathiraja, G., Kaliappan, S., Veeman, D., & Murugan, P. (2021). Processing and characterization of carbon nanofibre composites for automotive applications. *Journal of Nanomaterials, 2021*, 1–7. https://doi.org/10.1155/2021/7323885
60. Nalivaiko, A. Y., Doroshenko, V. V., Kuang, N., Ozherelkov, D. Y., Pelevin, I. A., & Gromov, A. A. (2022). Synthesis of Al–Al_2O_3–CNF composite by cold spray method: Powder preparation and synthesized objects characterization. *Nanomaterials, 12*(9), 1559. https://doi.org/10.3390/nano12091559
61. Guadagno, L., Naddeo, C., & Raimondo, M. (2023). Thermal, mechanical and electrical performance of structural epoxy resins filled with carbon nanofibers. *Journal of Thermal Analysis and Calorimetry, 148*(23), 13095–13106. https://doi.org/10.1007/s10973-023-12521-6
62. Guadagno, L., Pantelakis, S., Strohmayer, A., & Raimondo, M. (2022). High-performance properties of an aerospace epoxy resin loaded with carbon nanofibers and glycidyl polyhedral oligomeric silsesquioxane. *Aerospace, 9*(4), 222. https://doi.org/10.3390/aerospace9040222
63. Sun, C., et al. (2023). Advanced carbons nanofibers-based electrodes for flexible energy storage devices. *Advanced Functional Materials, 33*, 52. https://doi.org/10.1002/adfm.202305606
64. Yu, S., Xu, S., Khan, R., Zhao, H., & Li, C. (2024). Research developments in the application of electrospun nanofibers in direct methanol fuel cells. *Catalysis Science & Technology, 14*(4), 820–834. https://doi.org/10.1039/D3CY01509B
65. Tong, Z., Huang, L., Lei, W., Zhang, H., & Zhang, S. (2021). Carbon-containing electrospun nanofibers for lithium–sulfur battery: Current status and future directions. *Journal of Energy Chemistry, 54*, 254–273. https://doi.org/10.1016/j.jechem.2020.05.059
66. Ju, Z., Li, P., Zhao, X., Ma, J., Xu, H., & Liu, Y. (2022). Flexible TiN/co@carbon nanofiber mats for high-performance electromagnetic interference shielding and joule heating applications. *Carbon N Y, 196*, 612–620. https://doi.org/10.1016/j.carbon.2022.05.034
67. Zhan, C., Cui, W., Li, L., Quan, X., Zhang, Y., & Xiao, F. (2023). Dual-aligned carbon nanofiber scaffolds as heat conduction path to enhance thermal conductivity of polymer composites. *Composites Science and Technology, 231*, 109823. https://doi.org/10.1016/j.compscitech.2022.109823

68. Wang, S., Ng, Y. H., Tan, K. H., & Dasari, A. (2021). Thermal properties of carbon nanofibers enhanced lightweight cementitious composite under high temperature. *Construction and Building Materials, 307*, 124358. https://doi.org/10.1016/j.conbuildmat.2021.124358
69. Yusof, N. (2016). Adsorption of cadmium (ii) ions by POLYACRYLONITRILE-based activated carbon nanofibers/magnesium oxide as its adsorbents. *Malaysian Journal of Analytical Science, 20*(6), 1467–1473. https://doi.org/10.17576/mjas-2016-2006-27
70. Tian, P., et al. (2013). Synthesis of porous hierarchical MgO and its superb adsorption properties. *ACS Applied Materials & Interfaces, 5*(23), 12411–12418. https://doi.org/10.1021/am403352y
71. Kumar, P. S., Venkatesh, K., Gui, E. L., Jayaraman, S., Singh, G., & Arthanareeswaran, G. (2018). Electrospun carbon nanofibers/TiO_2-PAN hybrid membranes for effective removal of metal ions and cationic dye. *Environmental Nanotechnology, Monitoring and Management, 10*, 366–376. https://doi.org/10.1016/j.enmm.2018.08.006
72. Liu, Y., et al. (2020). Facile fabrication of Fe_3O_4 nanoparticle/carbon nanofiber aerogel from Fe-ion cross-linked cellulose nanofibrils as anode for lithium-ion battery with superhigh capacity. *Journal of Alloys and Compounds, 829*, 154541. https://doi.org/10.1016/j.jallcom.2020.154541
73. Matsuda, S. (2019). The effect of electrical conductivity on lithium metal deposition in 3D carbon nanofiber matrices. *Carbon N Y, 154*, 370–374. https://doi.org/10.1016/j.carbon.2019.07.104
74. Zhang, K., Xiong, F., Zhou, J., Mai, L., & Zhang, L. (2020). Universal construction of ultrafine metal oxides coupled in N-enriched 3D carbon nanofibers for high-performance lithium/sodium storage. *Nano Energy, 67*, 104222. https://doi.org/10.1016/j.nanoen.2019.104222

Graphitic Carbon Nitride-Based Nanocomposites

Nupur Aggarwal, Shubham Kumar Patial, Suman Singh, and Naveen Kumar

1 Introduction

In the recent decades, with the growing recognition of environmental sustainability and energy conservation, solar energy usage has attracted intensive research work [1]. Among the many techniques developed for pollutant removal and solar energy conversion, semiconductor photocatalysts have been developed as a promising one [2–4]. Nonetheless, the application of many photocatalysts has been hampered by such challenges as wide band gaps and low solar energy utilization efficiencies. Conventional photocatalysts like TiO_2, for instance, are limited to using only the portion of light that falls within the ultraviolet range and thus finally rather poorly in presence of visible light, so the search turned toward the development of efficient visible light driven (VLD) alternatives. As such, a great number of modified TiO_2 systems and alternative photocatalysts have been developed [5, 6]. However, despite

N. Aggarwal
Department of Physics, Chandigarh University, Gharuan, Punjab, India

S. K. Patial
Academy of Scientific and Innovative Research (AcSIR), Ghaziabad, Uttar Pradesh, India

Materials Science and Sensor Applications, CSIR-Central Scientific Instruments Organisation, Chandigarh, Punjab, India

School of Science, STEM College, RMIT University, Melbourne, VIC, Australia

S. Singh
Academy of Scientific and Innovative Research (AcSIR), Ghaziabad, Uttar Pradesh, India

Materials Science and Sensor Applications, CSIR-Central Scientific Instruments Organisation, Chandigarh, Punjab, India

N. Kumar (✉)
Department of Physics, Galgotias University, Greater Noida, Uttar Pradesh, India

V. Khanna (ed.), *Carbon-Based Nanocomposites for Sustainable Applications, Volume I*, Lecture Notes in Nanoscale Science and Technology 38,
https://doi.org/10.1007/978-3-031-95306-4_5

the progress made, designing abundant, stable, and easy-to-fabricate photocatalysts with high visible light activity continue to be a big challenge [7].

g-C_3N_4, a polymeric semiconductor, is gaining importance as the most promising VLD photocatalyst. Its unique heptazine-based structure and high degree of condensation impart remarkable physicochemical stability and tunable electronic structure, and it has a moderate bandgap of approximately 2.7 eV [8]. It thus becomes the best candidate to be used in photocatalytic applications under solar energy. Moreover, it is low cost and can be simply synthesized through scalable methods from low-cost precursors like cyanamide [9], urea [10], thiourea [11], melamine [12], and dicyandiamide [13]. Nonetheless, pure g-C_3N_4 has disadvantages such as rapid recombination of photogenerated electron-hole pairs, small specific surface area, and poor visible light utilization efficiency [9–13]. Numerous strategies have been explored by researchers to modify g-C_3N_4 and reduce its limitations. Its two-dimensional layered structure makes it highly suitable for hybridization with other materials. A variety of enhancement techniques have emerged in recent years coupling g-C_3N_4 with other semiconductors (e.g., TaON [14], Bi_2WO_6 [15], graphene [16]); constructing mesoporous structures, doping with metallic and non-metallic species (e.g., Fe [17], Ag [18], Au [19], S [20], B [21], P [22]) and sensitization with organic dyes [23]. Among these strategies, the formation of heterostructures was found particularly effective in the improvement of photocatalytic performance thanks to driving efficient charge separation and hindering the recombination of electron-hole pairs.

Heterostructures involving g-C_3N_4 can be formed with both visible light-responsive narrow-bandgap materials (e.g., CdS [24], Bi_2WO_3 [25], BiOI [26]) and UV-light-responsive wide-bandgap photocatalysts (e.g., TiO_2 [27], ZnO [28], $ZnWO_4$ [28]). The coupling of g-C_3N_4 with various semiconductors has progressed over the last few years toward the development of advanced nanocomposites. For instance, hybrid TiN/g-C_3N_4 materials made through magnetron sputtering showed enhanced photocatalytic properties, while g-C_3N_4/TaON composites were utilized for pollutant degradation. Other works with Au/g-C_3N_4 nanocomposites exhibited remarkable photocatalytic hydrogen evolution in metal-semiconductor junctions. This chapter highlights the recent advances in the design, synthesis, and possible applications of g-C_3N_4-based nanocomposites. Their role in such environmental and energy applications will be covered, ranging from the degradation of pollutants to hydrogen evolution and CO_2 conversion. The chapter closes with prospects, challenges, and strategies of scaling up the g-C_3N_4-based nanocomposites toward spurring still more innovation and wider acceptance into this promising area of research.

2 Synthesis of Graphitic Carbon Nitride

2.1 *Methods of Synthesis*

Graphitic carbon nitride (g-C_3N_4), a carbon-based polymeric photocatalyst, has become the focus of great interest, owing to its abundant starting materials, tunability in optical properties, and controllable morphology. This has broadened the scope of its applicability to photocatalysis, sensors, and dye-sensitized solar cells (DSSCs) [30–32]. Recently, efforts have sought the development of cost-effective and efficient syntheses for g-C_3N_4 [32]. Several approaches have been employed, such as physical vapor deposition [33], chemical vapor deposition [34], polymerization [35], and thermal condensation [36]. Other techniques, such as solvothermal, hydrothermal [37], and solid-state reactions [38], are other effective methods for g-C_3N_4 synthesis. Ultra-thin g-C_3N_4 nanosheets have been produced by mechanical exfoliation techniques that show improved features for advanced applications. Varying synthetic methods enable tailoring the structure and functionality of g-C_3N_4 to meet the demands of specific applications [39].

2.2 *Characterization Techniques*

Characterization methods form the basis for confirming very early successful synthesis of pristine and modified g-C_3N_4. Structural, compositional, and optical properties of the g-C_3N_4 were analyzed by means of different assessment techniques, as presented in Fig. 1. The carbon-to-nitrogen atomic ratio, which, for g-C_3N_4, really should be ~0.75, can basically be estimated by the elemental analysis. Usually, however, a lesser ratio might signify the presence of impurities [40]. The elemental composition and chemical states in materials could also be examined by XPS, providing insight into the structural configuration of g-C_3N_4. The high-resolution spectra for C1s and N1s would relate to the binding environment, such as the triazine ring structure [41]. For doped materials, XPS will be capable of identifying the valence state and coordination environment of the dopants. X-ray diffraction (XRD) is commonly employed to determine the crystalline structure of g-C_3N_4. It typically exhibits characteristic peaks at approximately 13.1° and 27.4°, corresponding to in-plane and interlayer stacking. Any changes in these peaks can reflect modifications or dopants in the material [42, 43]. Fourier Transform Infrared Spectroscopy (FTIR) is used to identify functional groups; for instance, peaks around 800 and 1470 cm^{-1} are associated with the triazine ring, while peaks at 1335 cm^{-1} and 1613 cm^{-1} correspond to C-N and C=N bonds. For doped g-C_3N_4, shifts in these

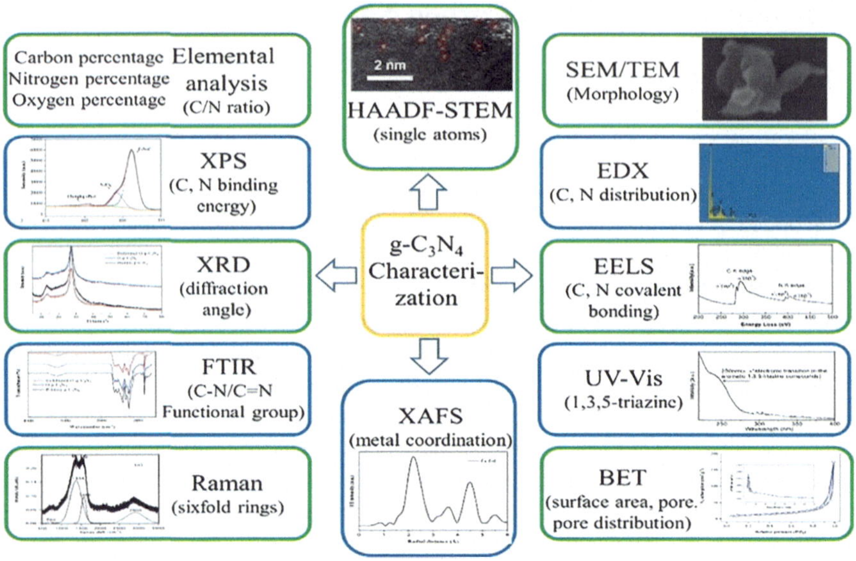

Fig. 1 The commonly used techniques for characterizing g-C_3N_4 [47]

peaks can indicate the successful modification [44, 45]. UV-Vis spectroscopy further supports structural confirmation by analyzing optical properties, with typical absorption bands between 230 and 300 nm due to $\pi \rightarrow \pi^*$ transitions in triazine rings [46].

Electron Energy Loss Spectroscopy (EELS) offers insights into the covalent bonding between carbon and nitrogen through distinct C-K-edge and N-K-edge features. These features, reflecting π^* and σ^* transitions, confirm the chemical bonding in g-C_3N_4 [45]. On the other hand, Raman spectroscopy provides structural information, with peaks at approximately 1390 and 1570 cm^{-1}, indicating defects and graphitic phases [48]. For morphological analysis, techniques such as Scanning Electron Microscopy (SEM), Transmission Electron Microscopy (TEM), and High-Resolution TEM (HRTEM) are commonly used to observe the layered structure of g-C_3N_4. However, these methods cannot distinguish g-C_3N_4 from other carbon-based materials. Elemental distribution can be analyzed using Energy-Dispersive X-ray (EDX) mapping, while extended X-ray Absorption Fine Structure (EXAFS) helps investigate the coordination environment in metal-doped samples. In cases of single-atom doping, High-Angle Annular Dark Field Scanning TEM (HAADF-STEM) can confirm the dispersion of individual dopant atoms [49]. Due to the complex nature of g-C_3N_4 and its derivatives, a single technique may not fully characterize its structure. Combining methods like FTIR, XRD, and XPS is often necessary for a more comprehensive understanding, especially in modified forms. Such multi-faceted approaches are essential to elucidate the intricate properties of g-C_3N_4-based nanocomposites.

3 Types of Graphitic Carbon Nitride-Based Nanocomposites

g-C_3N_4 has emerged as a promising material due to its excellent thermal stability, tunable electronic properties, and visible light driven photocatalytic activity. Nonetheless, in actual application, such phenomena have become limited due to rapid recombination of charge carriers and low conductivity and surface area. To overcome these disadvantages, g-C_3N_4 has been developed as nanocomposites with different materials that have drastically improved their photocatalytic, electronic, and mechanical properties.

3.1 Metal Oxide Composites

For photochemical pollution control, hydrogen production, and pollutant degradation, metal oxides like TiO_2, ZnO, SnO_2, Fe_2O_3, and CeO_2 are frequently coupled with g-C_3N_4 to promote the particle separation and photocatalytic performance [50]. To reduce the recombination of photogenerated electron-hole pairs, these oxides act efficiently as charge scavengers, thereby improving the overall photocatalytic activities. It was reported that, among all studied composites, TiO_2/g-C_3N_4 nanocomposites are the most active material due to its relatively high redox potential and proper band alignment [51].

3.2 Metal Sulfide Composites

There are many different metal sulfides such as ZnS, MoS_2, CdS, and WS_2 [52] that have been studied intensively for their superior light absorption properties. The combination of these metal sulfides with g-C_3N_4 provides an opportunity for the formation of heterojunctions that enhance charge transfer capabilities and extend light absorption into the visible spectrum. For example, CdS/g-C_3N_4 nanocomposites manifest high hydrogen production rates during photocatalytic water splitting [53]. Similarly, both WS_2 [54] and WS_2 [55], due to their two-dimensional structures, are ideal candidates for photocatalysis and electrochemical energy storage because of their high conductivities and exceptional catalyzing properties.

3.3 Polymer Composites

There exist several polymer-g-C_3N_4 nanocomposites showing potential candidature for advanced applications. Polyaniline (PANI), polypyrrole (PPy), and polyvinyl alcohol (PVA) [56] are polymers exploited in g-C_3N_4 nanocomposites in a view to improving flexibility, processability, and functional properties. These polymer composites also help improve charge mobility, surface interaction, and overall stability. Due to its very high conductivity and photocatalytic efficiencies, PANI/g-C_3N_4 has found application in supercapacitors, sensors, and organic pollutant degradation [57]. The hybrids of conducting polymers-g-C_3N_4 endow such materials with gained much charge carrier mobility and use better visible light.

3.4 Other Composite Materials

Other functional materials such as metal-organic frameworks, nitrides, and phosphides are hybridized with g-C_3N_4 (besides metal oxides and sulfides) to enhance their catalytic efficiency and electronic properties. The tunably adjustable porous surface structure makes MOFs suitable for myriad applications. They are good support materials to g-C_3N_4 due to their enhanced adsorption and catalytic capacities. In other photocatalytic and electrocatalytic applications, the phosphides CoP [58] and Ni_2P [59] were used to enhance charge transport and kinetic reaction. There is great promise coming from such hybrid materials spanning the gamut from CO_2 reduction and nitrogen fixation to biomedical applications.

3.5 Carbon Composites

Graphene, CNTs, and activated carbon, as carbon-based materials, have likewise been modified with g-C_3N_4 to enhance conductivity, surface area, and charge transport [60]. Such graphene/g-C_3N_4 composites would exhibit further increased electron transfer rates for photocatalytic systems, energy storage, and environmental remediation applications. Treating such structures in relation to supercapacitors and lithium-ion batteries would be a great benefit for CNTs/g-C_3N_4 composites owing to their structural stability and, furthermore, conductivity [61]. Their enhanced performance is mainly expressed in gas sensors and dye degradation, followed by applications in energy conversion.

4 Properties of Graphitic Carbon Nitride-Based Nanocomposites

g-C_3N_4-based nanocomposites are emerging as the novel materials with exciting characteristics, making them suitable for photocatalysis, energy storage, and environmental remediation. The structural, optical, electrical, and chemical features of

$g-C_3N_4$-based nanocomposites will ultimately define their role in different application domains.

4.1 Structural Properties

The structural features of $g-C_3N_4$ composites are determined principally by the method of synthesis, their composition, and the interaction with the incorporated material. $g-C_3N_4$ is structurally layered like graphite, with stronger in-plane covalent bonding between carbon and nitrogen atoms within the layer and weak van der Waals interactions between different layers. When it is combined with metal oxides, sulfides, polymers, or carbon-based materials, the interlayer spacing, surface area, and porosity of $g-C_3N_4$ will greatly be altered, improving charge transfer and adsorption. Metal oxide/$g-C_3N_4$ nanocomposites are usually favorable for mesoporous preparations with large surface area, which is beneficial for catalytic activity [62]. Similarly, these composites also exhibit improved mechanical stability and conductivity with the crude lattice type of graphene/$g-C_3N_4$ [63]. The aforementioned details about the fashionable structural variants of the nanocomposites greatly control their optical and electronic behavior and avail them efficient for the energy and environmental applications.

4.2 Optical Properties

Optical properties of $g-C_3N_4$-based nanocomposites are of prime importance for both photocatalysis and optoelectronic applications. $g-C_3N_4$ exhibits narrow bandgap (2.7 eV) [64], which enables it to efficiently harness visible light. However, due to fast electron-hole recombination, its overall photocatalytic activity is somewhat limited. It is due to the formation of nanocomposites that the optical properties of $g-C_3N_4$ can be tuned. The metal sulfide/$g-C_3N_4$ nanocomposites, for example, CdS/$g-C_3N_4$, exhibit enhanced visible light absorption and efficient charge separation because of their matching band structures [65]. Further down the road, doping of $g-C_3N_4$ with different elements, for example, phosphorus or sulfur, might enable tuning of the bandgap. There are also carbonaceous materials like graphene that enhance $g-C_3N_4$ to absorb more light and increase the separation of the carriers. Such concomitant merits further promote photocatalytic activity and optoelectronic performance.

4.3 *Electrical Properties*

The electrical conductivity of g-C_3N_4-based nanocomposites can be used in energy storage, sensors, and electronic devices. g-C_3N_4 displays moderate electrical conductivity but granted with highly localized states, where the electrical conduction is laden with high recombination rates of charge carriers. With the addition of conductive materials such as graphene, carbon nanotubes, and conducting polymers, electrical conductivity of g-C_3N_4 can be made to increase exponentially. Graphene/g-C_3N_4 nanocomposites have shown to increase charge transport and electron mobility in supercapacitors and batteries [66]. Likewise, metal oxide/g-C_3N_4 composites like Fe_2O_3/g-C_3N_4 yield elevated charge separation, which is a welcome boon when used in photocatalysis and electronic applications [67]. With such modifications, charge carrier recombination is decreased, the electrical conductivities are successfully raised, and in general, the device performance in energy and environmental applications is highly improved.

4.4 *Chemical Properties*

The stability, reactivity, and catalytic efficiency of g-g-C_3N_4-based nanocomposites are a result of their interesting chemical features. g-g-C_3N_4 is chemically stable at an ambient temperature and can resist oxidation and acidic or basic conditions. However, the chemical activity can be improved with functionalization or by interaction with other materials. Metal oxide/g-C_3N_4 nanocomposites usually display improved catalytic activity due to their high surface reactivity. TiO_2/g-C_3N_4 nanocomposites possess higher oxidation and reduction abilities, being fitted for water splitting and pollutant degradation [68]. Another example is polymer/g-C_3N_4 nanocomposites, which impart improved surface interaction in their application for sensing and adsorption. Besides, oxygen, phosphorus, or sulfur in the composition, as these elements will act as extra reactive sites in the process of doping, thereby allowing g-C_3N_4 to be more reactive and selective. Such chemical modifications have made g-C_3N_4-based nanocomposites valuable in areas such as photocatalysis, energy storage, and environmental remediation.

5 Applications of Graphitic Carbon Nitride-Based Nanocomposites

g-C_3N_4-based nanocomposites have attracted attention due to its remarkable physicochemical properties, including tunable bandgap, high thermal stability, and excellent photocatalytic activity. Figure 2 shows the various applications belonging to g-C_3N_4-based nanocomposites.

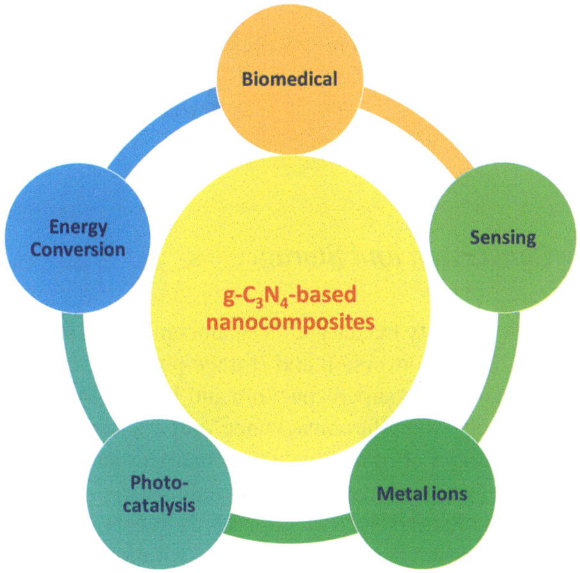

Fig. 2 Applications of g-C_3N_4-based nanocomposites

5.1 Photocatalysis

One of the most widely studied applications of g-C_3N_4-based nanocomposites is photocatalysis. g-C_3N_4 exhibits a narrow bandgap, so it is capable of absorbing visible light, but its photocatalytic efficiency is compromised due to rapid charge carrier recombination. To improve photocatalytic activity, therefore, often g-C_3N_4 composites will be prepared with such materials as metal oxides (TiO_2, ZnO), metal sulfides (CdS, MoS_2), and carbon-based materials (graphene, carbon nanotubes), giving rise to enhanced charge separation and extended light absorption. Applications include hydrogen evolution from water splitting, CO_2 reduction, and degradation of organic pollutants. TiO_2/g-C_3N_4 nanocomposites showed enhanced performance in solar-driven water splitting because of the synergistic charge transfer mechanism [69].

5.2 Environmental Remediation

The application of g-C_3N_4-based nanocomposites is seen in environmental remediation, in particular degrading pollutants and removing contaminants in the gas and liquid states, mainly due to the strong redox properties of the composites for breaking down organic pollutants like dyes, pesticides, and pharmaceutical residues. g-C_3N_4/metal oxide composites can induce high photocatalytic activity for the treatment of wastewater. In combination with metal-organic frameworks (MOFs), g-C_3N_4 optimizes adsorption and catalytic efficiencies for more effective removal of heavy metals and toxic gases [70]. Catalytic degradation involving the introduction of noble metals like Ag or Au is enhanced, well beyond any practical consideration of the composites for environmental applications [71].

5.3 *Energy Conversion and Storage*

High-charge storage capability makes g-C_3N_4 nanocomposites suitable for different nanocomposites for energy conversion and storage purposes. In particular, g-C_3N_4 is used as an active material in supercapacitors and lithium-ion batteries due to its charge storage capability. Specific capacitance and cycling stability are greatly increased with the inclusion of other conductive matrices such as graphene or carbon nanotubes. In energy conversion, g-C_3N_4-based nanocomposites have been utilized in solar cells, fuel cells, and electrocatalysis [72]. g-C_3N_4-based electrocatalysts exhibit wonderful oxygen reduction and hydrogen evolution reactions, and they are also very cost-effective alternatives to platinum-based catalysts in such responses [73]. Further, activity is enhanced for applications related to energy by hybridization with transition metal sulfides and phosphides.

5.4 *Sensing and Biosensing*

Due to their enormous surface area, chemical stability, and tunable electronic properties, g-C_3N_4-based nanocomposites provide excellent sensing features [74]. They are widely used in gas sensors, electrochemical sensors, and biosensors to detect environmental pollutants, biomolecules, and pathogens. While enhancing electron transfer, they boost electrochemical sensitivity and selectivity of sensors. g-C_3N_4-based nanocomposites have shown excellent performance in biosensing applications, from the detection of glucose, DNA, proteins, etc., [75]. Combined with metal nanoparticles like Au or Ag, this further increases their performance in the sensing area and renders them suitable for application in medical diagnostics and environmental monitoring.

6 Conclusion

Because of their manifold applications in environmental remediation and energy conversion, g-C_3N_4-based materials have attracted significant attention. The nanocomposite incorporation of g-C_3N_4 displays special properties concerning higher photocatalytic activity, large adsorption capacity, wide range of light absorption, and efficient charge carrier separation and transport. Such developments have enabled their application in the pollution degradation, hydrogen generation, CO_2 reduction, and disinfection processes. Various heterostructures have also been developed, involving metal free, semiconductor-semiconductor, and multi-composite systems, to enhance light absorption, charge separation, and catalytic efficiency. Despite considerable advances, practical deployment still faces difficulties owing to surface area, catalytic activity, and stability limitations. Numerous strategies such as doping, exfoliation, and building heterojunctions have been developed to improve the morphology modification in g-C_3N_4 to tackle these issues. However, the relationship complicating the correlations between different morphologies and/or doping strategies and the potential catalytic activity remains unclear. In particular, still needing investigation are, among others, studies about controlling doping levels, stability enhancing of doped active sites, or the synergistic effects from co-doping. They also mention that many studies have so far given attention only to the synthesis and modification of g-C_3N_4 but left open the creation of a large-scale method for producing few-layer g-C_3N_4 and their characterization. Obtaining precise parameters of g-C_3N_4 such as structure, stoichiometry, and transformation during synthesis is crucial in realizing its full potential. Furthermore, the environmental impact of modified g-C_3N_4-based nanomaterials has to be systematically assessed to ensure their secure and sustainable utilization.

References

1. Chen, C., Ma, W., & Zhao, J. (2010). Semiconductor-mediated photodegradation of pollutants under visible-light irradiation. *Chemical Society Reviews, 39*(11), 4206–4219.
2. Liu, G., et al. (2010). Titania-based photocatalysts—Crystal growth, doping and heterostructuring. *Journal of Materials Chemistry, 20*(5), 831–843.
3. Zhang, D., Li, G., & Yu, J. C. (2010). Inorganic materials for photocatalytic water disinfection. *Journal of Materials Chemistry, 20*(22), 4529–4536.
4. Hernández-Alonso, M. D., et al. (2009). Development of alternative photocatalysts to TiO_2: Challenges and opportunities. *Energy & Environmental Science, 2*(12), 1231–1257.
5. Kumar, S., et al. (2021). Metal nitrides and graphitic carbon nitrides as novel Photocatalysts for hydrogen production and environmental remediation. In S. Balakumar, V. Keller, & M. V. Shankar (Eds.), *Nanostructured materials for environmental applications* (pp. 485–519). Springer International Publishing.
6. Tang, Y.-B., et al. (2018). Preparation, characterization of nanocomposites plasmonic photocatalyst C@(Fe_3O_4–halloysite nanotubes)–ag/AgCl and its photocatalytic activity. *Desalination and Water Treatment, 110*, 144–153.
7. Zhao, Z., Sun, Y., & Dong, F. (2015). Graphitic carbon nitride based nanocomposites: A review. *Nanoscale, 7*(1), 15–37.
8. Wang, X., et al. (2009). A metal-free polymeric photocatalyst for hydrogen production from water under visible light. *Nature Materials, 8*(1), 76–80.
9. Takanabe, K., et al. (2010). Photocatalytic hydrogen evolution on dye-sensitized mesoporous carbon nitride photocatalyst with magnesium phthalocyanine. *Physical Chemistry Chemical Physics, 12*(40), 13020–13025.
10. Dong, F., et al. (2011). Efficient synthesis of polymeric g-C_3N_4 layered materials as novel efficient visible light driven photocatalysts. *Journal of Materials Chemistry, 21*(39), 15171–15174.

11. Hong, J., et al. (2012). Mesoporous carbon nitride with in situ sulfur doping for enhanced photocatalytic hydrogen evolution from water under visible light. *Journal of Materials Chemistry, 22*(30), 15006–15012.
12. Wang, Y., et al. (2012). Graphite-like C_3N_4 hybridized $ZnWO_4$ nanorods: Synthesis and its enhanced photocatalysis in visible light. *CrystEngComm, 14*(15), 5065–5070.
13. Zhang, G., et al. (2012). Polycondensation of thiourea into carbon nitride semiconductors as visible light photocatalysts. *Journal of Materials Chemistry, 22*(16), 8083–8091.
14. Yan, S. C., et al. (2010). Organic–inorganic composite photocatalyst of g-C_3N_4 and TaON with improved visible light photocatalytic activities. *Dalton Transactions, 39*(6), 1488–1491.
15. Ge, L., Han, C., & Liu, J. (2011). Novel visible light-induced g-C_3N_4/Bi_2WO_6 composite photocatalysts for efficient degradation of methyl orange. *Applied Catalysis B: Environmental, 108*, 100–107.
16. Xiang, Q., Yu, J., & Jaroniec, M. (2011). Preparation and enhanced visible-light photocatalytic H_2-production activity of graphene/C_3N_4 composites. *The Journal of Physical Chemistry C, 115*(15), 7355–7363.
17. Chen, X., et al. (2009). Fe-g-C_3N_4-catalyzed oxidation of benzene to phenol using hydrogen peroxide and visible light. *Journal of the American Chemical Society, 131*(33), 11658–11659.
18. Ge, L., et al. (2011). Enhanced visible light photocatalytic activity of novel polymeric g-C_3N_4 loaded with ag nanoparticles. *Applied Catalysis A: General, 409*, 215–222.
19. Li, X.-H., Wang, X., & Antonietti, M. (2012). Mesoporous gC 3 N 4 nanorods as multifunctional supports of ultrafine metal nanoparticles: Hydrogen generation from water and reduction of nitrophenol with tandem catalysis in one step. *Chemical Science, 3*(6), 2170–2174.
20. Chang, C., et al. (2013). Photodegradation of bisphenol a by highly stable palladium-doped mesoporous graphite carbon nitride (Pd/mpg-C_3N_4) under simulated solar light irradiation. *Applied Catalysis B: Environmental, S142–143*, 553–560.
21. Liu, G., et al. (2010). Unique electronic structure induced high Photoreactivity of Sulfur-doped graphitic C_3N_4. *Journal of the American Chemical Society, 132*(33), 11642–11648.
22. Yan, S., Li, Z., & Zou, Z. (2010). Photodegradation of rhodamine B and methyl orange over boron-doped g-C_3N_4 under visible light irradiation. *Langmuir, 26*(6), 3894–3901.
23. Yan, H., & Huang, Y. (2011). Polymer composites of carbon nitride and poly(3-hexylthiophene) to achieve enhanced hydrogen production from water under visible light. *Chemical Communications, 47*(14), 4168–4170.
24. Fu, J., et al. (2013). Novel C_3N_4–CdS composite photocatalysts with organic–inorganic heterojunctions: In situ synthesis, exceptional activity, high stability and photocatalytic mechanism. *Journal of Materials Chemistry A, 1*(9), 3083–3090.
25. Wang, Y., et al. (2012). Enhancement of photocatalytic activity of Bi_2WO_6 hybridized with graphite-like C_3N_4. *Journal of Materials Chemistry, 22*(23), 11568–11573.
26. Jiang, D., et al. (2013). Novel p–n heterojunction photocatalyst constructed by porous graphite-like C_3N_4 and nanostructured BiOI: Facile synthesis and enhanced photocatalytic activity. *Dalton Transactions, 42*(44), 15726–15734.
27. Zhou, X., et al. (2012). A carbon nitride/TiO_2 nanotube array heterojunction visible-light photocatalyst: Synthesis, characterization, and photoelectrochemical properties. *Journal of Materials Chemistry, 22*(34), 17900–17905.
28. Liu, W., et al. (2012). Facile synthesis of g-C_3N_4/ZnO composite with enhanced visible light photooxidation and photoreduction properties. *Chemical Engineering Journal, 209*, 386–393.
29. Di, Y., et al. (2010). Making metal carbon nitride heterojunctions for improved photocatalytic hydrogen evolution with visible light. *ChemCatChem, 2*(7), 834–838.
30. Dong, G., et al. (2014). A fantastic graphitic carbon nitride (g-C_3N_4) material: Electronic structure, photocatalytic and photoelectronic properties. *Journal of Photochemistry and Photobiology C: Photochemistry Reviews, 20*, 33–50.
31. Thomas, A., et al. (2008). Graphitic carbon nitride materials: Variation of structure and morphology and their use as metal-free catalysts. *Journal of Materials Chemistry, 18*(41), 4893–4908.

32. Deng, Y., & Ezyske, C. M. (2011). Sulfate radical-advanced oxidation process (SR-AOP) for simultaneous removal of refractory organic contaminants and ammonia in landfill leachate. *Water Research, 45*(18), 6189–6194.
33. Sun, X., et al. (2019). Physical vapour deposition (PVD): A method to fabricate modified gC 3 N 4 sheets. *New Journal of Chemistry, 43*(17), 6683–6687.
34. Zhang, Z., et al. (1999). Corrosion resistance studies on α-C_3N_4 thin films deposited on pure iron by plasma-enhanced chemical vapour deposition. *Journal of Materials Science Letters, 18*, 685–687.
35. Zhang, J., et al. (2009). Synthesis of a carbon nitride structure for visible-light catalysis by copolymerisation. *Angewandte Chemie, 2*(122), 451–454.
36. Yang, J., et al. (2011). Synthesis and characterisation of nitrogen-rich carbon nitride nanobelts by pyrolysis of melamine. *Applied Physics A, 105*, 161–166.
37. Hu, C., et al. (2017). Rapid synthesis of g-C_3N_4 spheres using microwave-assisted solvothermal method for enhanced photocatalytic activity. *Journal of Photochemistry and Photobiology A: Chemistry, 348*, 8–17.
38. Zhang, S., et al. (2014). Retracted article: Polymer nanodots of graphitic carbon nitride as effective fluorescent probes for detecting Fe^{3+} and Cu^{2+} ions. *Nanoscale, 6*(8), 4157–4162.
39. Sun, H., et al. (2017). An efficient exfoliation method to obtain graphitic carbon nitride nanosheets with superior visible-light photocatalytic activity. *International Journal of Hydrogen Energy, 42*(12), 7930–7937.
40. Wang, S., Liu, Y., & Wang, J. (2020). Iron and sulfur co-doped graphite carbon nitride (FeOy/sg-C_3N_4) for activating peroxymonosulfate to enhance sulfamethoxazole degradation. *Chemical Engineering Journal, 382*, 122836.
41. Yang, X., et al. (2016). Facile fabrication of acidified g-C_3N_4/g-C_3N_4 hybrids with enhanced photocatalysis performance under visible light irradiation. *Applied Catalysis B: Environmental, 193*, 22–35.
42. Ge, L. (2011). Synthesis and photocatalytic performance of novel metal-free g-C_3N_4 photocatalysts. *Materials Letters, 65*(17–18), 2652–2654.
43. Chen, X., et al. (2016). Growth of g-C_3N_4 on mesoporous TiO_2 spheres with high photocatalytic activity under visible light irradiation. *Applied Catalysis B: Environmental, 188*, 342–350.
44. Cui, Y., et al. (2012). Metal-free activation of H_2O_2 by g-C_3N_4 under visible light irradiation for the degradation of organic pollutants. *Physical Chemistry Chemical Physics, 14*(4), 1455–1462.
45. Li, X., et al. (2009). Preparation and characterisation of graphitic carbon nitride through pyrolysis of melamine. *Applied Physics A, 94*, 387–392.
46. Jiménez-Salcedo, M., Monge, M., & Tena, M. T. (2019). Photocatalytic degradation of ibuprofen in water using TiO_2/UV and g-C_3N_4/visible light: Study of intermediate degradation products by liquid chromatography coupled to high-resolution mass spectrometry. *Chemosphere, 215*, 605–618.
47. Wang, J., & Wang, S. (2022). A critical review on graphitic carbon nitride (g-C_3N_4)-based materials: Preparation, modification and environmental application. *Coordination Chemistry Reviews, 453*, 214338.
48. Anoshkin, I. V., et al. (2017). Single walled carbon nanotube quantification method employing the Raman signal intensity. *Carbon, 116*, 547–552.
49. Qian, K., et al. (2021). Single-atom Fe catalyst outperforms its homogeneous counterpart for activating peroxymonosulfate to achieve effective degradation of organic contaminants. *Environmental Science & Technology, 55*(10), 7034–7043.
50. Reenamole, G. (2023). Metal oxide nanomaterials for visible light photocatalysis. In *Emerging nanomaterials for catalysis and sensor applications* (pp. 103–130). CRC Press.
51. Zhang, L., He, X., Xiuwen, X., Liu, C., Duan, Y., Hou, L., Zhou, Q., et al. (2017). Highly active TiO_2/g-C_3N_4/G photocatalyst with extended spectral response towards selective reduction of nitrobenzene. *Applied Catalysis B: Environmental, 203*, 1–8.

52. Baral, S. C., Sasmal, D., Hupele, M., Lenka, S., & Sen, S. (2024). Recent advances in hydrogen production using sulfide-based photocatalysts. *arXiv*. Preprint arXiv:2412.05658.
53. Chen, L., Yiming, X., & Chen, B. (2019). In situ photochemical fabrication of CdS/g-C_3N_4 nanocomposites with high performance for hydrogen evolution under visible light. *Applied Catalysis B: Environmental, 256*, 117848.
54. Xu, H., Sun, L., Li, W., Gao, M., Zhou, Q., Li, P., Yang, S., & Lin, J. (2022). Facile synthesis of hierarchical g-C_3N_4@ WS2 composite as Lithium-ion battery anode. *Chemical Engineering Journal, 435*, 135129.
55. Peng, W.-c., & Li, X.-y. (2014). Synthesis of MoS_2/g-C_3N_4 as a solar light-responsive photocatalyst for organic degradation. *Catalysis Communications, 49*, 63–67.
56. Kar, D. K., Mishra, S., Mohanty, P. S., & Si, S. (2024). Two-dimensional nanomaterials-based polymer nanocomposites for catalytic and photocatalytic applications. In *Two-dimensional nanomaterials-based polymer nanocomposites: Processing, properties and applications* (pp. 541–578). Scrivener Publishing.
57. Li, P., Xu, H., Liu, F., Shi, J., Gao, X., & Li, J. (2023). Room-temperature NH_3 sensors based on polyaniline-assembled graphitic carbon nitride nanosheets. *ACS Applied Nano Materials, 6*(7), 5145–5154.
58. Guo, F., Huang, X., Chen, Z., Sun, H., & Chen, L. (2020). Prominent co-catalytic effect of CoP nanoparticles anchored on high-crystalline g-C_3N_4 nanosheets for enhanced visible-light photocatalytic degradation of tetracycline in wastewater. *Chemical Engineering Journal, 395*, 125118.
59. Liu, E., Jin, C., Chenhui, X., Fan, J., & Xiaoyun, H. (2018). Facile strategy to fabricate Ni_2P/g-C_3N_4 heterojunction with excellent photocatalytic hydrogen evolution activity. *International Journal of Hydrogen Energy, 43*(46), 21355–21364.
60. Harun, N. A., Masleeza, N. S., & Ramli, Z. A. C. (2022). Progress of g-C_3N_4 and carbon-based material composite in fuel cell application. *International Journal of Energy Research, 46*(12), 16281–16315.
61. Song, L., Kang, X., & Zhang, S. (2018). CNT/g-C_3N_4 photocatalysts with enhanced hydrogen evolution ability for water splitting based on a noncovalent interaction. *International Journal of Energy Research, 42*(4), 1649–1656.
62. Pourmadadi, M., Aghababaei, N., & Abdouss, M. (2024). Photocatalytic activation of peroxydisulfate by UV-LED through rGO/g-C_3N_4/SiO_2 nanocomposite for ciprofloxacin removal: Mineralization, toxicity, degradation pathways, and application for real matrix. *Chemosphere, 359*, 142374.
63. Akram, M. Y., Ashraf, T., Tong, L., Yin, X., Dong, H., & Huiling, L. (2024). Architecting high-performance photocatalysts: A review of modified 2D/2D graphene/g-C_3N_4 heterostructures. *Journal of Environmental Chemical Engineering, 12*, 113415.
64. Giannakopoulou, T., Papailias, I., Todorova, N., Boukos, N., Liu, Y., Jiaguo, Y., & Trapalis, C. (2017). Tailoring the energy band gap and edges' potentials of g-C_3N_4/TiO_2 composite photocatalysts for NOx removal. *Chemical Engineering Journal, 310*, 571–580.
65. Gao, H., Zhang, S., Jinzhang, X., Dou, Y., Zhou, J., & Zhou, R. (2020). Activating and optimizing activity of CdS@ g-C_3N_4 heterojunction for photocatalytic hydrogen evolution through the synergistic effect of phosphorus doping and defects. *Journal of Alloys and Compounds, 834*, 155201.
66. Sharma, M., Yadav, V., Diwakar, P., Negi, C. M. S., & Alvi, P. A. (2024). Carbon-based 2D materials: Synthesis, characterization, and their green energy applications. In *Carbon-based nanomaterials for green applications* (pp. 95–118). Wiley.
67. Bharathi, P., Chidhambaram, N., & Thirumurugan, A. (2023). Conjoining Fe_2O_3 and g-C_3N_4 semiconductors to form Fe_2O_3/g-C_3N_4 nanocomposites for photocatalytic applications: An in situ integration approach. *Journal of Materials Research, 38*(21), 4736–4746.
68. Nallusamy, S., Asaithambi, S., Pandiaraj, S., Rahaman, M., Elangovan, V., Eswaramoorthy, N., & Nandagopal, C. (2024). Cerium-modified TiO_2/g-C_3N_4 nanocomposites with synergistic

effect on enhancing visible-light photocatalytic activity employing cationic dyes. *Colloids and Surfaces A: Physicochemical and Engineering Aspects, 685*, 133175.
69. Sewnet, A., Abebe, M., Asaithambi, P., & Alemayehu, E. (2022). Visible-light-driven g-C_3N_4/TiO_2 based heterojunction nanocomposites for photocatalytic degradation of organic dyes in wastewater: A review. *Air, Soil and Water Research, 15*, 11786221221117266.
70. Sharma, S. K., Kumar, A., Sharma, G., Wang, T., Iglesias-Juez, A., & Dhiman, P. (2023). Recent advances in g-C_3N_4/metal organic frameworks heterojunctions for high-performance photocatalytic environmental remediation and energy production. *Journal of Molecular Liquids, 382*, 121890.
71. Zedan, M., Li, X., Amin, R. M., ElFeky, S. A., & Zedan, A. F. (2022). Photodeposited silver versus gold over g-C_3N_4 toward photocatalytic oxidation of organic water pollutants. *Egyptian Journal of Chemistry, 65*(132), 119–136.
72. Li, W., Sohail, M., Usama Anwar, T. A., Taha, A. G., Al-Sehemi, S. M., Al-Ghamdi, A. A., et al. (2022). Recent progress in g–C_3N_4–based materials for remarkable photocatalytic sustainable energy. *International Journal of Hydrogen Energy, 47*(49), 21067–21118.
73. Yuda, A., & Kumar, A. (2022). A review of g-C_3N_4 based catalysts for direct methanol fuel cells. *International Journal of Hydrogen Energy, 47*(5), 3371–3395.
74. Kotbi, A., Imran, M., Kaja, K., Rahaman, A., Ressami, E. M., Lejeune, M., Lakssir, B., & Jouiad, M. (2022). Graphene and g-C_3N_4-based gas sensors. *Journal of Nanotechnology, 2022*(1), 9671619.
75. Wang, Y., Zhang, R., Zhang, Z., Cao, J., & Ma, T. (2019). Host–guest recognition on 2D graphitic carbon nitride for nanosensing. *Advanced Materials Interfaces, 6*(23), 1901429.

Carbonized Biomass/Polymer Nanofibers for Sustainable Applications

Kamaljit Singh, Navjot Singh Mahal, Shivinder Singh, and Sarabjit Singh

1 Introduction

With the increasing demand for sustainable and eco-friendly materials, the development of advanced materials derived from the renewable resources has become an essential area of research and innovation. In recent years, carbonized biomass and polymer nanofibers have emerged as highly influential materials in this field. Considering the critical issue of energy and environment, transforming the biomass waste into carbon materials would assist in reducing pollution and enhance conventional carbon synthesizing techniques [1]. Several renewable biomass materials such as sawdust, wood residue, bamboo, and natural fungus are available lavishly as agricultural and forest waste. Also, their low cost makes them most suitable resources for synthesizing carbon-based materials [2]. The lignocellulosic materials such as lignin have numerous benefits over conventional materials as carbonized biomass source. Lignin, a complex aromatic polymer which is available in the plant's cells has high carbon content and superior stability under rising temperature [3]. Additionally, the chemical structure and flexibility of lignin can facilitate the production of carbon nanofibers with desired traits, such as high surface area and porosity. These attributes are highly essential for application in environmental remediation, and energy storage [4].

Besides the carbonized biomass, polymer nanofibers have also been regarded as the versatile class of materials with wide range of applications [5]. These nanofibers

K. Singh (✉) · S. Singh · S. Singh
Department of Mechanical Engineering, Advanced Centre of Research & Innovation (ACRI) Chandigarh Engineering College, CGC, Jhanjeri, Mohali, Punjab, India

N. S. Mahal
Department of Mechanical Engineering, University School of Mechanical Engineering, Rayat Bahra University, Kharar, Mohali, Punjab, India

V. Khanna (ed.), *Carbon-Based Nanocomposites for Sustainable Applications, Volume I*, Lecture Notes in Nanoscale Science and Technology 38,
https://doi.org/10.1007/978-3-031-95306-4_6

have diameters in the nanometer range and possess extremely favourable properties such as high surface area to volume ratio, porosity, and flexibility [6]. Also, they can be further modified or combined with different nanomaterials such as nanosilica to produce biopolymer nanocomposite with superior mechanical, thermal, and barrier properties [7]. Nanofibers can be developed by using adequate selection of polymers and additives along with proper processing technique. Polymers are classified as synthetic, natural, or mixed polymers that can be processed through different processing techniques such as electrospinning, template synthesis and self-assembly polymerization [8].

Fascinatingly, sustainable applications of carbonized biomass and polymer nanofibers are critical to replace conventional materials. The carbonized biomass offers cost-effective and ample alternative source for energy storage, filtration, and catalysis [9]. The polymer nanofibers having vast surface area and tunable properties can contribute to development of advanced materials for application in environment remediation, biomedical applications and energy storage systems [10, 11]. These materials improve the sustainability by utilizing the waste resources, reducing reliance on fossil-based materials and minimizing the environment impact [12]. Their remarkable physicochemical properties make them the most suitable for high-performance applications in supercapacitors, fuel cells and thermal management systems [13].

The aim of this review is to examine the importance of carbonized biomass and polymer nanofibers in sustainable applications focusing on their potential in improving energy efficiency and environmental sustainability. It also discusses the different production, characteristics, and important uses of these materials. Also, it provides insights into the challenges confronted during their processing and implementation, revealing the possible solutions for improved performance and scalability.

2 Historical Background and Previous Research

2.1 Early Developments in Carbonized Biomass Nanofibers

Over the past decades, the field of carbonized biomass nanofibers have witnessed a paradigm shift from its early conceptual stages to its current state of the art applications. The carbon-based nanofibers have attained significant attention of the researchers and industries simultaneously due to their exceptional characteristics that are required for specific applications [14]. The origin of carbon nanofiber can be tracked back to the late nineteenth century. For the first time, carbonized cotton and bamboo fibres were used as filament for light bulb design [15]. Since then the field has progressed considerably.

One of the crucial developments is related to integration of carbonaceous material like carbon nanotube into conventional ones. This breakthrough has flourished the growth of wearable technology and soft sensor research. The inclusion of carbon nanotubes (CNT) and graphene has facilitated tailoring the properties of these materials, including mechanical flexibility and enhanced thermal resistivity [16–18]. Furthermore, alternative biodegradable biomass materials such as cellulose and lignin have been explored substantially for the synthesis of carbon nanofiber [19, 20]. Figure 1 provides an idea of progression in CNF till now.

2.2 *Evolution of Polymer Nanofibers*

The field of polymer nanofibers has witnessed remarkable advancements over the years, with each decade showcased innovative breakthroughs that have dynamically transformed the realm of materials. During the early period, polymer nanofibers were fabricated using techniques such as chemical synthesis and nanolithography [21]. With further advancements, electrospinning has emerged as a widely adopted technique for continuous production of polymer nanofibers. The optimization of process parameters, such as applied voltage, distance between spinnerets and their relative angle, has significant impact on the quality of nanofibers [21].

However, in recent years, functionalization of polymer nanofibers has become popular which enables tailoring their properties to fulfil industrial needs. Several investigations depicted that incorporation of additives and chemical modifications into polymer matrix enables the nanofiber to exhibit enhanced characteristics, including improved mechanical strength and durability, responsiveness and multi-functional capabilities [22]. The usefulness of polymer nanofibers has led to their widespread adoption in a diverse range of applications, ranging from electronic and textile industries and petrochemical to environment remediation [23].

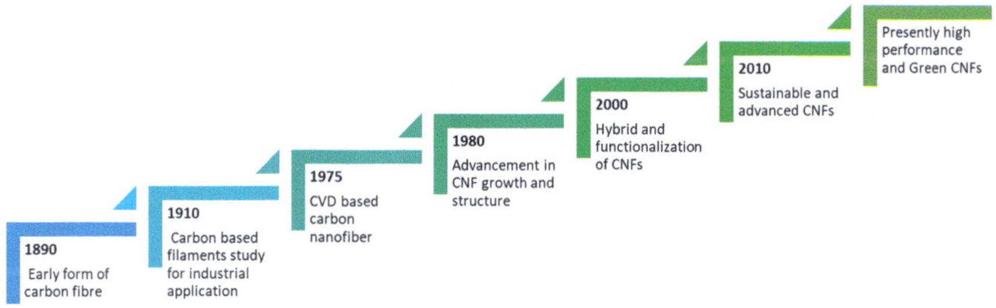

Fig. 1 Development of carbon nanofiber from concept to present stage

3 Carbon Nanofiber Synthesis Methods

Several methods have been developed for the synthesis of carbonized biomass and polymer nanofibers, enabling the formation of materials with superior performance for sustainable uses.

3.1 Chemical Vapour Deposition (CVD)

CVD has been regarded as one of the most useful technique for the fabrication of a wide range of carbon-based nanomaterials, including nanofibers, owing to its versatility and ability to control the morphological and structural behaviour of materials, precisely [24]. Scalability and industrial application are two especial appealing aspects of the CVD method, which are essential for the broad use of these nanomaterials [25]. CVD-produced CNFs are highly efficient and inexpensive. For example, using a simple two step technique highly refined and vertically aligned CNFs were developed in CVD, utilizing two different solutions, a carbon source with camphor and ACN as well as a catalytic solution with ferrocene and acetonitrile. The CNFs were produced at atmospheric pressure with temperature of 850 °C in a traditional CVD reactor. The MWCNF were coated with polymethyl methacrylate (PMMA) on the Si substrate (as shown in Fig. 2). The material was able to retain the

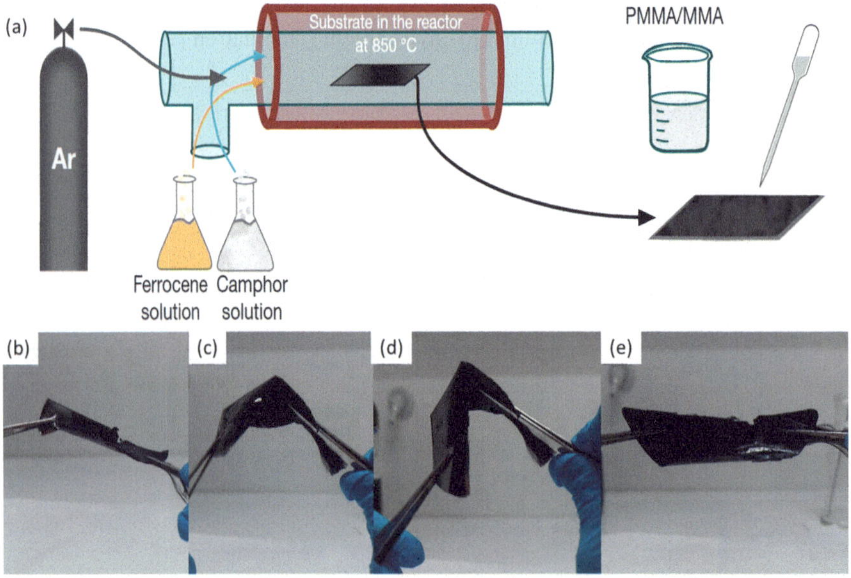

Fig. 2 Illustrates (**a**) CVD process for development of carbon nanofiber using polymeric solution, and (**b–e**) exhibits the flexibility of the developed CNF material [26]

structure and resulted in a flexible, conductive thin film suitable for flexible electrodes [26].

In an experimental study, spiral structured CNFs were developed after doping with phosphorus through CVD at 450 °C of acetylene pyrolysis under the influence of Ni and followed by annealing. The microstructural appearance of the material confirmed the spiral structure having vast surface area. Also, in terms of performance measure, the materials showed superior electrical conductivity [27]. In a different study, composite material of iron carbide and nitrogen-doped CNFs (Fe_3C/N-CNF) were effectively synthesized at a low annealing temperature using straightforward CVD technique. The whole process was simplified as the polypyrene was the nitrogen source, acetylene for carbon source and iron oxide for iron source and catalyst. Evidently, the polymerization period for the PPY precursor was critical aspect to govern the rate of CNF development for a uniform material. The obtained material was analysed for LIB electrode which revealed considerable cycling stability and enhanced performance [28].

Numerous agricultural crop residues have been examined as a precursor for activated carbon and are constantly gaining attention from research community. Carbon precursors such as carbon nanotubes that are produced from agricultural waste include significant content of cellulose, hemicellulose as well as lignin [29]. For instance, linseeds oil was utilized in CVD at 850 °C for the synthesis of carbon nanofiber. For the deposition of the CNFs an alumina substrate was selected and located in a quartz tube at the middle of split type furnace. Hydrogen was selected as carrier gas for the migration of linseeds oil vapour within the hot zone. The developed CNFs had uniformity with a mean diameter of ~533 nm and a length of ~40 μm. The electrical conductivity for the same reported was considerably better [30].

3.2 Electrospinning Techniques

Electrospinning has become a widely applied technique for the fabrication of nanofibers, in particularly carbon-based materials from polymers and bio-based polymers [31, 32]. The fibres synthesized through electrospinning technique are more diverse in comparison to other methods. They can be obtained from numerous materials including natural sources such as fungi and chitin to synthetic materials. Also, different materials can be added to create composite materials [33, 34]. For example, nanoparticles of SiO_2 were added into the solution of cellulose acetate with in the acetone/DMAc so as to develop cellulose solution for electrospinning. The addition of silicon dioxide nanoparticles has significant effect on the fibre size [35]. Further studies revealed that addition of SiO_2 has added considerable enhancement in the electrical behaviour of the CNFs. This is attributed to the improvement in porosity of CNFs [36, 37].

Interestingly, using a combination of different technologies, including electrospun, carbonization as well as cold plasma method biomass-based porous carbon nanofibers were synthesized out of the chitosan and cellulose. The hydrophobic property of the developed BCNFs was changed to hydrophilic after the plasma treatment. Also, the BCNFs exhibited improved adsorption mechanism due to synergistic effect of both the electrostatic reactions and van der Waal forces [38]. Further, Biomass tar was processed via electrospinning and utilized to create CNFs. Cobalt-nickel (CoNi) nanosheets were subsequently grown onto the CNF surfaces. This process resulted in a unique three-dimensional, open, petal-like nanofiber architecture. The electrochemical performance of the CoNi-CNF composite electrode-based supercapacitor exhibited improvement. This evident that electrospinning is simple and cost-effective method for producing material based on carbonized biomass for enhancing energy efficiency of energy devices [39].

The growing need to enhance the performance of energy storage devices, such as supercapacitors and fuel cells, is accelerating the rapid advancement of electrode materials and separators. Electrospun PAN-based carbon nanofibers, owing to their unique and tailorable structure, flexibility and surface functionalities have significantly incited the advancement and widespread usage of supercapacitors. For obtaining porous CNF from acetylated sugarcane bagasse and polyacrylonitrile (PAN), both electrospinning and carbonization techniques were applied as shown in Fig. 3. This approach provided a significant way to produce biomass-based carbon materials that can be utilized in energy storage devices [40]. For biosensing application also electrospun carbon nanofiber has been widely investigated. A flexible CNF membrane deposited with in the $NiMoO_4$ nanoparticles was produced for efficient sensing of glucose. The CNFs fabricated with electrospinning and carbonization methods with in nitrogen atmosphere resulted in excellent glucose detection capabilities [41]. Similarly, electrospun CNFs were embedded into the Co_3O_4 nanoparticle for sensing the enzyme-free glucose as a free-standing electrode in a biosensor [42].

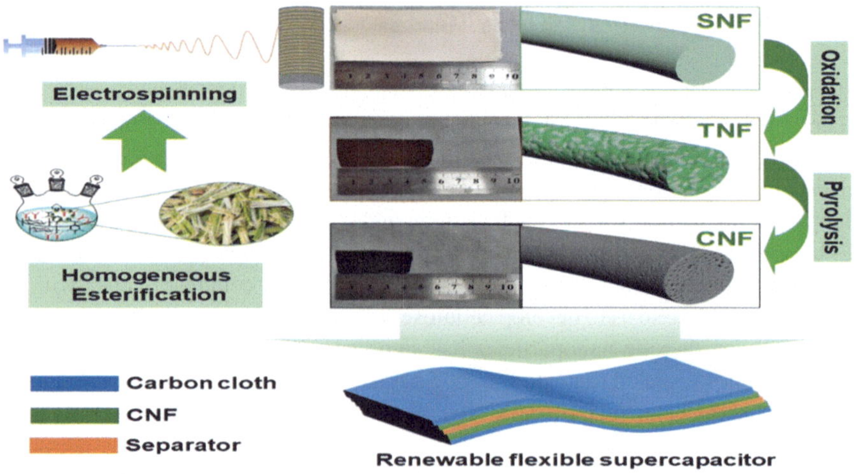

Fig. 3 Demonstrate CNF base flexible material prepared with electrospinning technique [40]

3.3 Green Synthesis

Green synthesis of carbon nanofiber uses biomass such as woods, leaves, and other low-value biomaterials as precursor. Metal catalysts employed in the production of CNPs can alternatively be substituted with green synthesis using biomass as the starting material [43]. Distinct type of CNPs, including porous structured nanofibers and nanotubes can be developed based on the reaction environment and material of the precursor [43]. For instance, using coconut coir lignocellulosic biomass, carbon nanosheets have been obtained through the combination of hydrothermal carbonization and pyrolysis methods [44]. Green synthesis methods can be further be classified as hydrothermal carbonization synthesis, pyrolysis carbonization synthesis and sol-gel synthesis.

3.4 Hydrothermal Carbonization Synthesis

Hydrothermal synthesis is regarded as an effective technique for fabricating the carbon materials from biomass in sub-critical water. Figure 4 illustrate the production system of carbon nanoparticle using solvent-based hydrothermal carbonization. In this technique, different chemical phenomenon, including hydrolysis, condensation as well as dehydration is catalysed. This leads to the synthesis of the functionalized carbon nanomaterials [45]. Different parameters such as type of biomass precursor, doping, and operating temperature can have significant effect on the microstructure and composition of the carbon material. While uniform carbon nanospheres can be extracted from the glucose at 180 °C residing for 10 h, the furfural exhibits carbon sphere structure with varying dimensions for 24 h of reaction period [46].

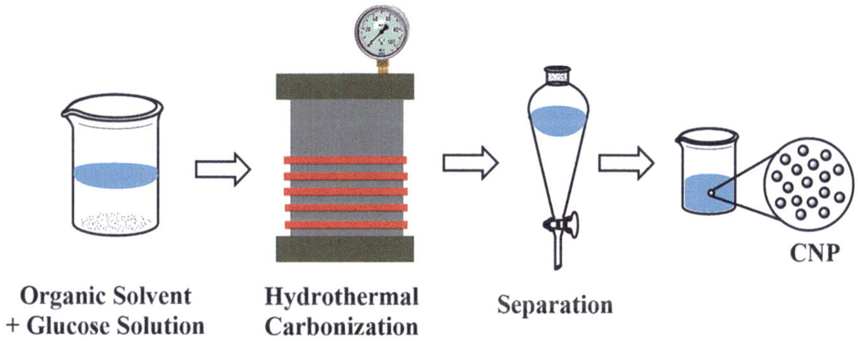

Fig. 4 Shows the different steps involved in hydrothermal carbonization synthesis of CNFs [47]

3.5 Pyrolytic Carbonization Synthesis

Pyrolytic synthesis technique has a capacity to tailor the morphology and performance of the derived material after selecting the appropriate carbonization parameters. On the other hand, hydrothermal synthesis may produce poor carbon architecture and unfavourable pores within the structure. This is the reason that pyrolytic carbonization has become more popular for deriving biomass-based carbon nanomaterial. This process facilitates utilization of high temperature processing to form large surface area and porous materials. Controlling the heating rate, the behaviour of the resulting pyrolytic material can be significantly altered. For instance, rapid rate of pyrolysis decomposes the biomass rapidly. This leads to production of more volatile matters than carbon. Conversely, slower rate of heating is enables production of more char than tar [48]. Figure 5 shows the principle of converting lignocellulosic biomass into carbon by-product using pyrolysis synthesis method.

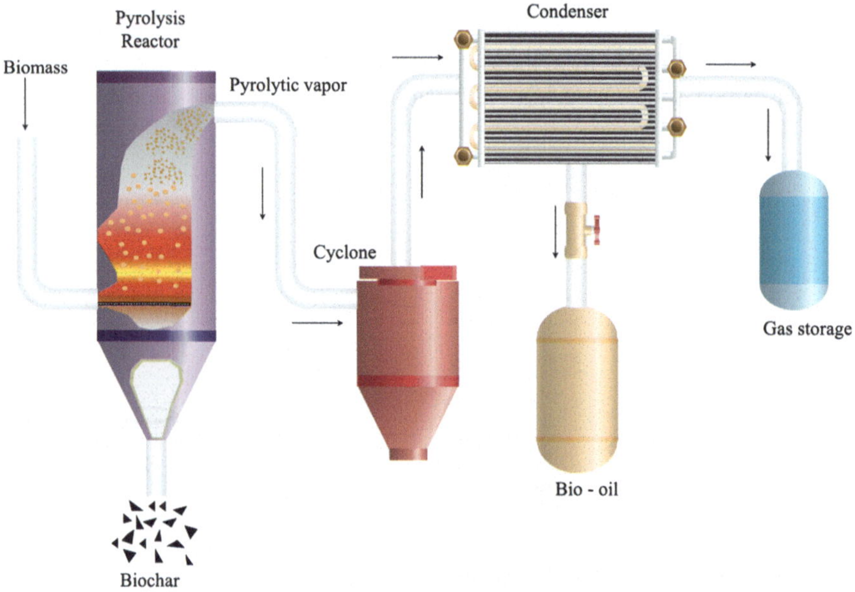

Fig. 5 Presents the pyrolytic process using high processing temperature for the formation of CNFs [49]

3.6 Sol-Gel Synthesis

The sol-gel method offers a versatile, inexpensive, and simple approach that enables the preparation of carbon substances and the fabrication of nanoparticles with controlled composition [51]. The critical steps involved in preparing carbon materials using sol-gel method are depicted in Fig. 6. To attain consistent carbonization, this method involves dispersing the biomass-derived precursors within the sol-gel matrix. Carbon nanofibers with tunable porosity and surface functionalities can be synthesized by keeping the resultant gel under high temperature conditions within an inert environment [52].

3.7 Comparison of Different Synthesis Methods

The significant merits, such as simple, economical, and scalability of green synthesis techniques over traditional methods makes them more favourable for synthesizing biomass-based carbon materials.

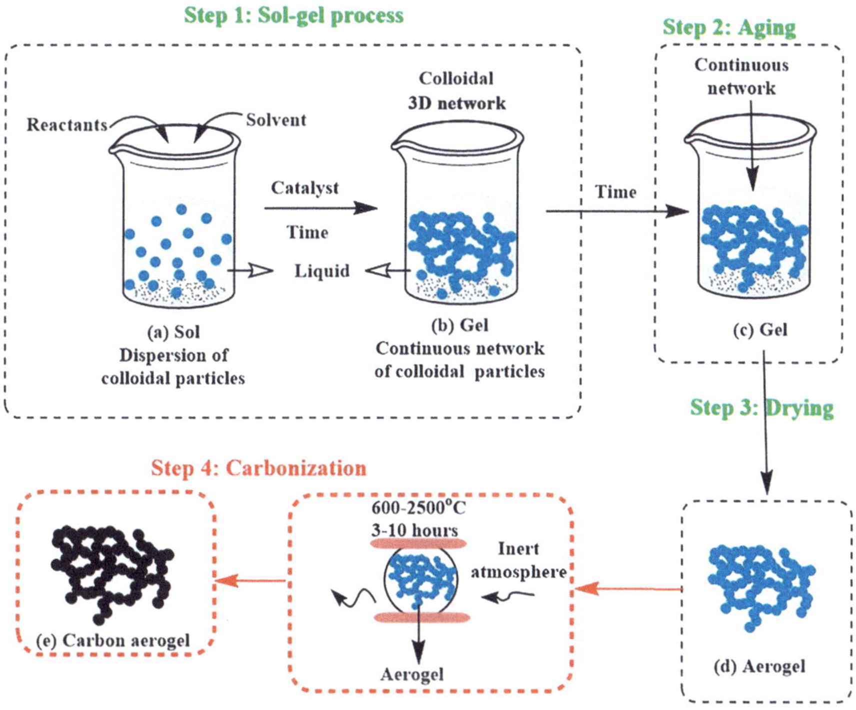

Fig. 6 Illustrates the different stages of synthesizing CNFs using Sol-Gel technique [50]

Interestingly, the hydrothermal carbonization method is a low-cost technique as it utilizes inexpensive feedstocks from agricultural waste, operates in low temperature condition and no axillary chemical agents are required. Further, the carbon yield can be easily scalable for fulfilling the industrial requirements through reaction optimization [53]. Similarly, sol-gel method is economical due to its low-temperature processing, minimal use of chemical solutions, and the use of inexpensive biomass precursors for production of carbon materials. Also, the scalability can be enhanced through optimal reaction conditions while the pyrolysis method involves higher temperature range for developing carbon content. However, due to cheaper biomass from wood residues, the resultant carbon material can have significant economic value. Additionally, the process can be scalable for increased production with minimal impact on the environment [54].

Unlike traditional methods, green methods consume minimum energy. The high energy demand, expensive equipment and complex processing are few limitations of conventional methods for large-scale applications. While the traditional methods release substantial pollutants, contribute to environmental degradation, green technologies are eco-friendlier and align towards sustainable development goals. Furthermore, traditional methods are complex and difficult to scale. Conversely, traditional methods facilitate good control over the microstructure of the nanofiber which includes diameter and alignment. The good control is highly crucial for physico-mechanical properties. Additionally, these processes produce nanomaterials with minimum impurities that could affect their performances [55].

4 Properties and Characterization

The structural and morphological characteristics of carbonized biomass/polymer nanofibers are crucial factors to assess their performance and effectiveness for various applications. In general, the CNFs produced have dimensional uniformity and smooth [56, 57]. However, depending on the processing conditions the microstructure of the CNFs may vary. Figure 7a, b depicts the scanning electron microscopy images of CNFs produced using CVD and electrospinning methods. While the microstructure of the CNF produced via CVD has non-uniformity and are exhibit random bending orientation. CNF fabricated using electrospinning has uniform dimension and smooth [58]. Additionally, incorporating different materials can also have significant effect on the morphology of the CNFs. For example, as shown in Fig. 7c, d the addition of silica nanoparticle (350 and 500 wt%) increased the porosity of the CNF [59]. Optimized nanopores can facilitate ion transport which in turn accelerates electrochemical reactions in energy storage. Besides, 3D CNFs produced using plasma enhanced CVD were coated with Si had a distinctive microstructure as shown in Fig. 7e–g. This structure which is comparable to graphitic cones has sidewalls containing graphitic edges. Due to this, the operating performance of lithium battery was enhanced notably [60].

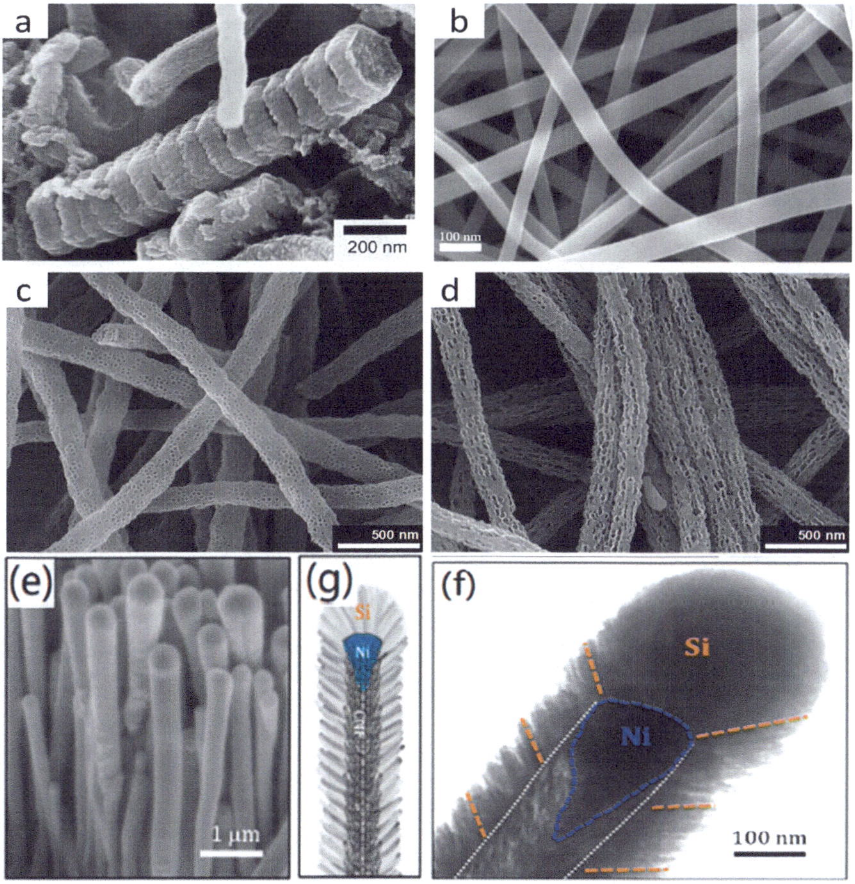

Fig. 7 show the SEM images of (**a**) non-uniform CNF produced via CVD technique and (**b**) uniform and smooth CNF fabricated with electrospinning [58], (**c**) CNF with less porosity after addition of 350 wt.% Si and (**d**) CNF with more porosity after addition of 500 wt.% Si [59], (**e–g**) CNF with coned structure [60]

The XRD patterns of carbonized biomass nanofiber typically exhibits broad diffraction peaks. These peaks indicate an amorphous or partially graphitic nature of biomass CNFs. The most common peaks, as shown in Fig. 8a, b for CNFs produced using poplar lignin, pine lignin and reed lignin as well as phosphorus doped porous CNF, appears around $2\theta = 24–27°$ and $2\theta = 42–45°$ corresponding to (002) and (001) plane of graphitic carbon [61, 62]. More sharp and intense peak suggests more graphitization, while a boarder and less intense peak indicates amorphous or disordered carbon structure. However, processing conditions, presence of doping, and synthesis using polymers can have significant effect on the XRD pattern of the carbon nanofibers [63, 64].

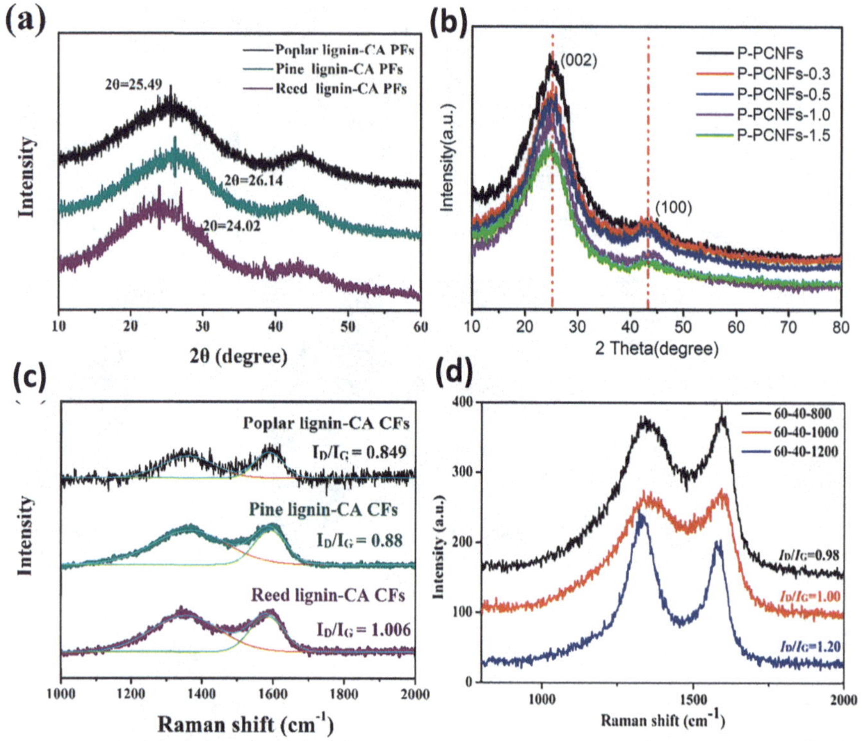

Fig. 8 Plots for (**a**) XRD profile of CNFs prepared with different lignin precursors [61], (**b**) XRD pattern for a porous type of CNF [62], compare the Raman spectra of (**c**) for lignin-based CNF material [61], and (**d**) porous CNF material [65]

The Raman spectra of carbon nanofibers usually exhibit two prominent peaks as shown in Fig. 8c, d. First is the D-band nearly at 1350 cm^{-1} representing the disorder or defect band owing to presence of defects within the carbon structure. The other is the G-band at 1580 cm^{-1} reflecting the degree of graphitization or ordered structure of carbon material.

Further, the carbon nanofiber derived from polymers such as polyacrylonitrile or polyaniline may also exhibit similar D and G bands [63, 64]. The ratio of D-band to G-band intensity is critical parameter for assessing the level of structural disorder in nanofibers.

5 Sustainable Applications in Energy Storage

Biomass/polymer CNFs have been applied and examined in supercapacitors and batteries due to their exceptional properties. They can enable rapid ion transport and enhance cyclic performance for next generation energy storage devices.

Supercapacitors have been regarded as an important technology for storage of energy owing to their capability in delivering high power density, fast charging rate and better battery cycle [66]. Usually the electrodes for the supercapacitors are designed and developed with non-renewable and costly materials such as iron oxide and iridium oxide [67, 68]. Consequently, it is essential to develop and fabricate electrode with materials which are not only abundantly available but are also cheaper. For example, Lignin with high carbon fraction is a biomass polymer which is easily available and useful due to its favourable attributes for electrodes and electrolyte materials [69, 70]. Lignin-based CNFs are extensively produced using electrospinning technology to produce fibres with small diameter and large surface area [71].

Fascinatingly, a co-spinned cellulose acetate along with lignin generated LCNF exhibited an excellent capacitance of up to 151 Fg^{-1} at a current density of 1 Ag^{-1} [72]. Similarly, electrospun alkali lignin-based carbon nanofibers with varying with varying molecular weight and phenolic hydroxyl solution were produced after pre-oxidation and drying process. The schematics of the process are shown in Fig. 9a. The developed materials were designated as F1 for higher concentrations and F2 for Lower concentrations of phenolic hydroxyl and molecular weight. The resultant LCNF, designated as KOH/F2-based F2-C-800/F2-C-800 exhibited great improvement in its electrochemical performance. In contrast to unmodified counterpart, the modified CNF had 17.1% increase in the energy density and a 17% increase in specific capacitance. Figure 9b shows the remarkable capacitance retention of 89.6% [73].

Furthermore, in a different study a binder free, porous flexible electrode of biomass-derived CNF incorporated in a nickel oxide hybridized with cobalt tetra oxide was successfully synthesized. The specific capacitance of the well-structured composite material reached maximum at 629.5 Fg^{-1}. Additionally, at 757.4 W kg^{-1} of power density; the energy density of the supercapacitor designed with this electrode reached to 31.5 Wh kg^{-1}. The synergistic combination of properties of the bimetallic oxides and the 3D network of the carbon nanofibers promoted effective interlayer ion transport [56]. For comparison of carbon materials in supercapacitor,

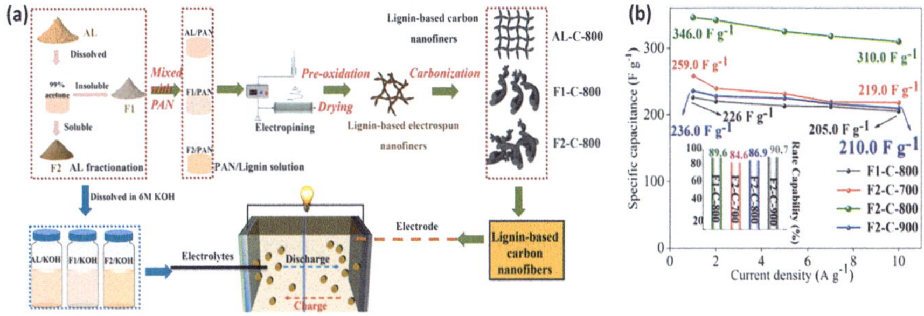

Fig. 9 (**a**) Schematics of electrospun CNF production using PAN designated as F1 and F2 and application in supercapacitor and (**b**) Plot for the specific capacitance of developed F1 and F2 materials for different current densities [73]

CNF, and porous carbon were synthesized from a single polymer precursor via one-step carbonization. Excellent supercapacitor performance was achieved by both materials due to enlarged surface area and structured graphitization. Nonetheless, CNF electrodes showed better specific capacitance of about 360 Fg^{-1} at a scan rate of 10 mV s^{-1} which is ~24% more than of porous carbon electrode (290 Fg^{-1}). This is attributed to the better interconnected architecture and nanoscale morphology of CNFs [74].

Moreover, porous carbon nanofibers (PCNFs) with a high surface area (1204 m^2 g^{-1}) were synthesized using a template technique from a high amylose starch. The PCNFs exhibited a structured micro/mesoporous framework. When used as a supercapacitor electrode, the PCNFs achieved a specific capacitance of 344 F g^{-1} at a current density of 1.0 A g^{-1}. Also, the PCNFs demonstrated 99.9% capacitance retention for over 10,000 cycles and reached an energy density of 12 Wh kg^{-1} at 125 W kg^{-1}. It is evident that HAS is a cost-effective, green material for producing high-performance carbon nanofibers for energy storage [75]. The exceptional traits of lignin have a substantial effect on the polyaniline polymerization and the subsequent generation of CNF. A study showed that in situ polymerization of polyaniline-modified lignin produced CNFs have excellent capacitance of 420 Fg^{-1} at current density of 1.0 Ag^{-1}. During its application in a supercapacitor, it showed 98.2% retention of capacitance in 10,000 cycles. Also, for 8 kW kg^{-1} Power density, the energy density of 29.4 Wh kg^{-1} was attained [76].

Another substantial mean for enhancing the supercapacitor performance using carbon-based electrode is by doping. For example, N-doped porous carbon nanofiber membranes with cross-linked structures were prepared using eccentric co-axial electrospinning followed by the carbonization. The developed membrane nanofiber exhibited high electrical conductivity (25.4 S cm^{-1}) and structural flexibility. In order to create hybrid electrodes, these membranes were decorated with the $NiCo_2O_4$. This hybridization maximizes the specific capacitance to 1474.2 F g^{-1} at 0.5 A g^{-1}. Additionally, the capacitance retention recorded was 78% for 10 A g^{-1}. These composites were developed for use as supercapacitor electrode materials. Also, the energy density of 53 Wh kg^{-1} was upheld by the asymmetric supercapacitor using these hybrid electrodes [77]. Beside this, in another study an invasive weed Parthenium hysterophorus was transformed into a highly porous activated carbon with an ultrahigh surface area of 4014 m^2g^{-1} and a large pore volume of 2.0419 cm^3 g^{-1}. The resulting carbon material exhibited a crumpled morphology and oxygen-rich surface functionalities. The material demonstrated a specific capacitance of 270 Fg^{-1} with 98.5% capacitance retention for over 30,000 cycles at 10 Ag^{-1}. Interestingly the developed material also assisted in removing the organic pollutants from water. After ten cycles, the membrane maintained 91.6% rejection indicating its potential to be a sustainable water filtration technology [78].

6 Sustainable Applications in Environmental Remediation

6.1 Detection of Heavy Metal Ions

Heavy metals in the environment endanger human health and disrupt ecological equilibrium [79]. The sensitivity, specificity, and accuracy of conventional detection techniques like atomic absorption spectrometry are constrained. The fabrication of electrochemical aptasensors has become a viable method for heavy metal detection in recent years [80]. These aptasensor-based technologies are a useful tool for on-site and real-time heavy metal detection because of their excellent sensitivity, specificity, and accuracy [80]. Carbon nanofibers are one of the materials that have demonstrated significant promise in this area. Owing to its extraordinary qualities, which include a high surface-to-volume ratio, great electrical conductivity, and remarkable chemical stability, carbon nanofibers have attracted a lot of interest from scientists and engineers developing cutting-edge environmental monitoring sensing technologies [81].

Carbon nanofibers have been shown to improve heavy metal detection in electrochemical aptasensors, even while other metal ions are interfering with each other [80]. For instance, an experimental study developed a rapid electrochemical sensor using copper oxide-dispersed graphitic carbon nanofibers (CuO-CNFs) for real-time pollutant monitoring. The sensor detected antibiotics, heavy metal ions, and pesticides in under 15 s of time period, with a defined detection ranges (1–20 ppm for one type). It also detected multiple pollutants together in tap water including cadmium (cd) and lead (pb). The sensor was highly accurate, reliable, and consistent, making it strong for finding different pollutants in water [82].

Another study also focused on identifying the Cd and Pb type heavy pollutants. Anodic stripping voltammetry (ASV) was applied in the investigation for developing an electrode made of CNF and Nafion-modified glassy carbon for the detection of heavy metals, particularly lead (Pb^{2+}) and cadmium (Cd^{2+}). When compared to an unmodified simple electrode, the electrode with modification had its detection sensitivity increased by eightfold. It obtained detection limits of 1.5 nM for cadmium and 0.9 nM for lead under suitable conditions [83]. Similarly, a modified glassy carbon electrode was fabricated to develop a highly sensitive and selective electrochemical sensor for the detection of mercury ions (Hg^{2+}). Figure 10a shows the steps for detection of Hg^{2+} ions with the developed electrode. The electrode incorporated mesoporous carbon nanofibers (MCNF) for improved conductivity, bis(indolyl)methane (BIM) for Hg^{2+} capture, and Nafion for stability. The sensor effectively identified Hg^{2+} in Yellow River water samples, showed no interference from other heavy metals, and reached a relative linear coefficient to 0.996 and sensing limit of 0.3 nM for Hg^{2+} (Fig. 10b) [84].

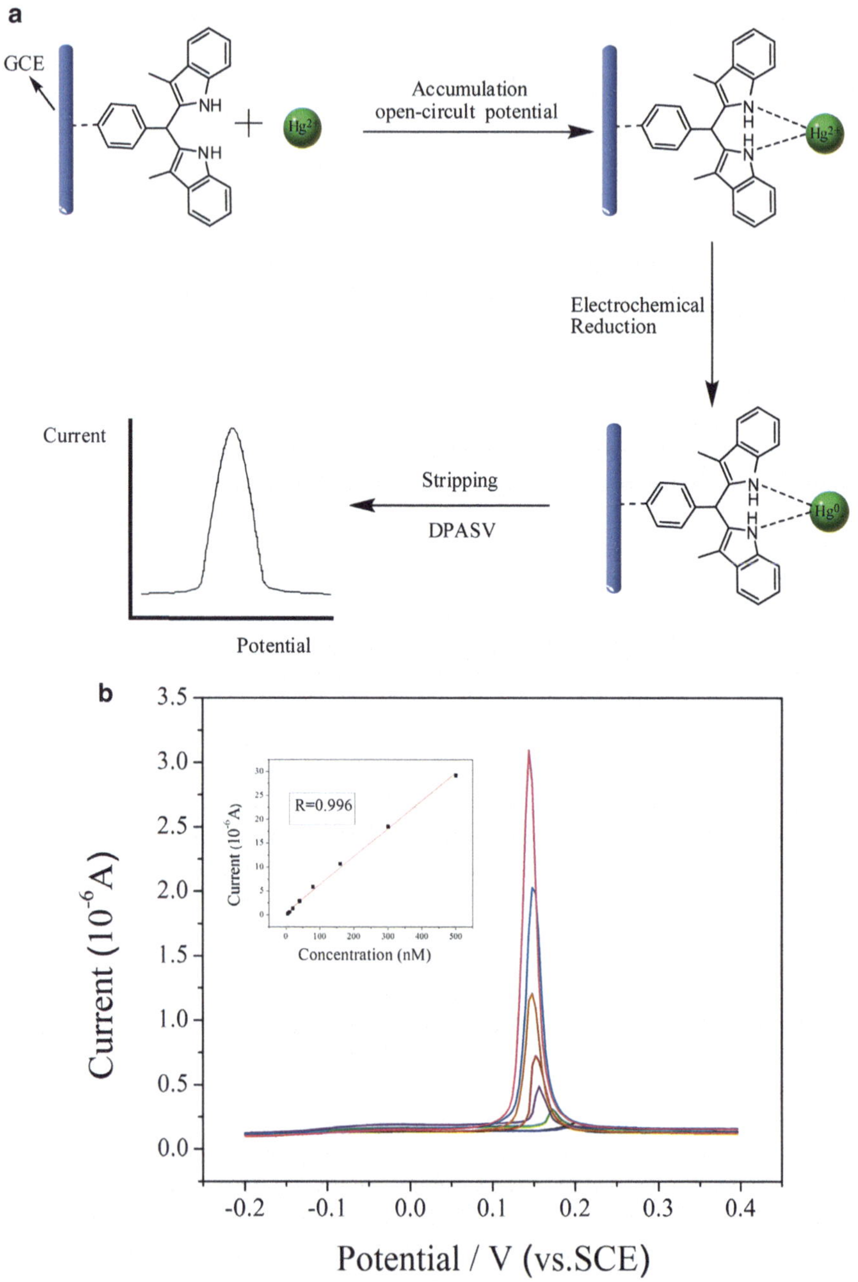

Fig. 10 (**a**) Schematics of the electrochemical sensor for detection of heavy metal in yellow river water and (**b**) Shows the electrochemical performance by depicting the relation between current and potential of the developed material using CNF [84]

6.2 Water Treatment and Purification

Like heavy metal detection, carbon fibres are suitable material for treatment and purification of pollutants present in water. For instance, novel PVA-based CNFs for water purification were synthesized via electrospinning, incorporating green tea solutions and TEOS and followed by calcination at 280 °C. This process reduced nanofiber diameter to ~250 nm (from ~500 nm) that enhanced the filtration potential of the material. The resulting CNF filters effectively removed contaminants which were evident by SEM and EDX analysis revealing a trapped nanoparticle containing Mg, Na, Ti, S, Si, and Fe [85]. Further, a 3D hierarchical CNF was incorporated with graphite felt to obtain tailored properties for removing pb ions from water. The developed material had shown high adsorption for pb^{2+} ion, oil, and other organic solvents form water. They exhibited excellent performance in water treatment, revealing strong adsorption, recovery, and recyclability which makes them a promising material for practical applications [86].

Fascinatingly, as shown in Fig. 11a, Carbonized loofah (CL) enriched with the carbon nanofibers were developed via CVD process (CL-CNF) for an efficient solar vapour generator. Both the CL and CL-CNF that are used in solar steam generation device have demonstrated strong light absorption and conversion. The CL-CNF SSGD, featuring a 2D water pathway, achieved a high evaporation rate of 1.72 kg m^{-2} h^{-1} and a 92.5% photothermal conversion efficiency under 1 kW m^{-2} solar irradiation. Solar powered purification effectively eliminated nearly 99% of the primary ions from the sewage. The reducing concentration reaches below the recommended levels for drinking water by WHO. Notably, metal ions like Ba^{2+}, Cu^{2+}, Mn^{2+}, and Ni^{2+}, initially at 100 mg L^{-1}, were significantly reduced (Fig. 11) [86]. Similar to this another study also investigated solar-based water purification. A nanoscale multiple interface strategy was developed using structured nanopores in carbon nanofibers (HPCNFs) which are formed via a sacrificial framework method have significantly enhanced the light absorption. Also, the HPCNF films reject ions (<0.05 mg L^{-1} salt ions) well in desalination. This resulted in producing 11.18 kg m^{-2} day^{-1} of freshwater from lake water which is adequate for a family of four to five members [87].

Functionalizing the materials with heteroatom is another solid strategy for tailoring the properties for specific purposes. In this direction, a process was developed to alter 1,2-dichloroethane (DCE) into the nitrogen and oxygen-doped CNFs using a Ni-Pd catalyst. The catalyst facilitated the growth of functionalized CNFs with high surface area (up to 470 m^2 g^{-1}) as well as porosity (up to 0.9 cm^3 g^{-1}). The CNFs contained 1.0–1.7% nitrogen and 3.6% oxygen. Also, demonstrated effective dichlorobenzene adsorption from water [88]. Apart from this, for water treatment, electrospun CNF-CNT composites with enhanced sorption and strength were

Fig. 11 (**a**) Shows the solar heat utilization for water purification using CNF (**b**) process for the CNF development (**c–h**) microstructural appearance of CNF and CL-CNF and (**i–l**) compositional elements of CNF with EDS, (**m**) the reduction in concentration level of different metal in water and (**n**) relationship between the absorption and wavelength of purified water and Rhodamine B [86]

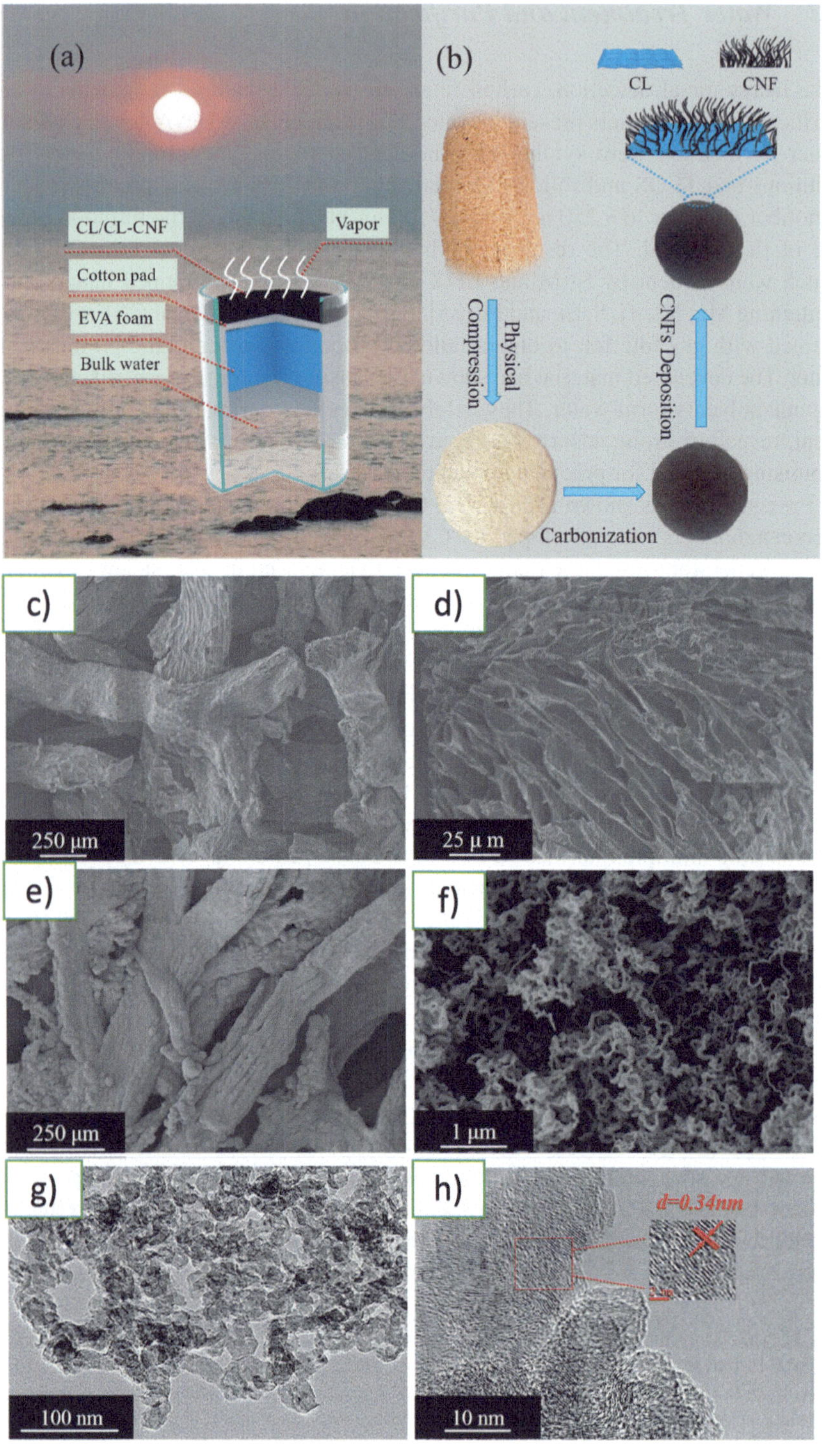
(a)
CL/CL-CNF
Vapor
Cotton pad
EVA foam
Bulk water
(b)
CL
CNF
Physical Compression
CNFs Deposition
Carbonization
c)
250 µm
d)
25 µ m
e)
250 µm
f)
1 µm
g)
100 nm
h)
d=0.34nm
10 nm

(continued)

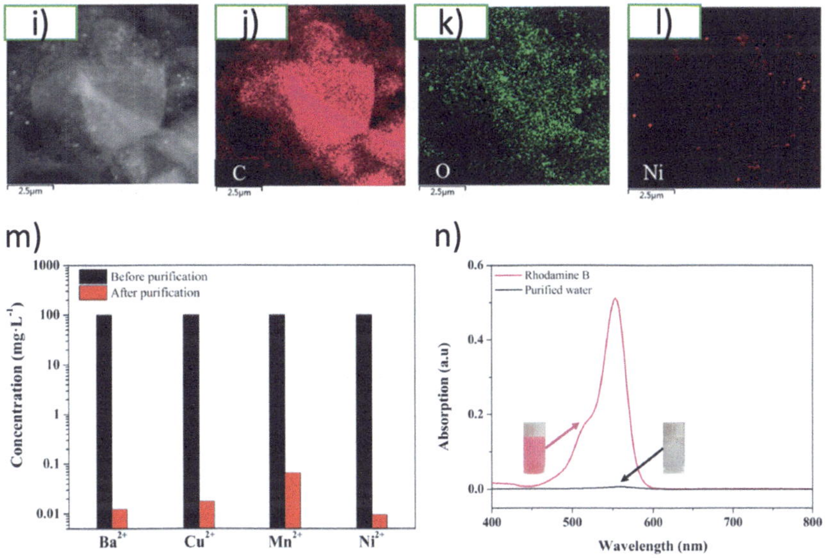

Fig. 6.11 (continued)

formed. Nanofiber characteristics were controlled by synthesis factors such as macro-porosity and CNT integration. Sulfamethoxazole and atrazine were successfully eliminated by the composites. Strength and performance were balanced by an optimized formulation that included 8% polyacrylonitrile, 2% CNT, 2.4% phthalic acid, 40% humidity, and 280 °C stabilization. In a flow-through system, this composite eliminated ten micropollutants, demonstrating both hydrophobic and particular binding interactions [89].

7 Sustainable Applications in Biomedical

The unique structural and functional characteristics of carbon nanofibers have made them an increasingly attractive material for various biomedical applications. Electroconductive CNFs derived from electrospun polyacrylonitrile were fabricated for a bone cell electrical stimulation. Mg-63 cells seeded on CNFs (SCNFs) were subjected to DC electrical fields. Cell growth significantly increased and reached to a peak proliferation (116.43%) at 100 μA. Increased alkaline phosphatase activity indicated the enhanced osteogenic activity. Conductive CNFs effectively promoted bone growth confirming their suitability for piezoelectric properties of bone for improved healing [90]. Similarly, electrospinning PVDF with addition of CNF addition and electrical poling enhanced the beta phase and piezoelectric effect. 0.5% CNF-PVDF nanofibers exhibited the highest beta phase content. After 30 V/μm poling, the nanofibers showed peak piezoelectric performance (5.8 V, 1.2 μA).

This clearly indicates the huge potential for such an affordable, non-toxic energy converter in wearable biomedical sensors [91].

Further, hybrid hydrogels of alginate and gelatin, enhanced with homogeneously dispersed CNFs were developed for 3D-printable and electroconductive tissue engineering scaffolds (Fig. 12a). As shown in Fig. 12b–d, the CNF incorporation improved the structure, and mechanical and electrical properties (Young's modulus 534.7 and conductivity 4.1 × 10–4 S/cm) and shear-thinning behaviour. This enabled efficient 3D printing. These conductive hydrogels therefore show possible viability for cardiac and neuronal tissue engineering applications [92].

8 Current Limitations and Challenges

Trade-off is one of the major limitations of carbon nanofiber-based materials that have detrimental effect on their performance for diverse applications. High conductivity, superior electrochemical kinetics, vast surface area and mechanical flexibility are few favourable characteristics of CNFs particularly for use in energy storage and

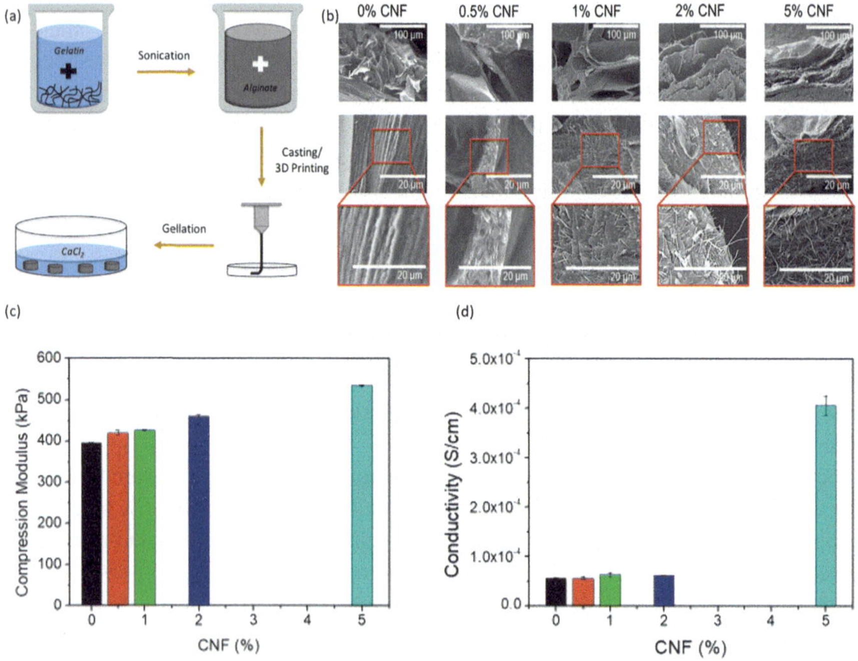

Fig. 12 (**a**) Procedure of developing electroconductive tissue scaffold using CNF, (**b**) SEM images of the developed material with varying CNF concentration in the solution, (**c**) the effect of CNF concentration on the compression modulus of the developed material and (**d**) electrical conductivity of the CNF-based material [92]

environment remediation. Nonetheless, it is still a massive challenge to balance all these desired traits in CNFs as they usually show trade-off. For example, while the micropores and mesopores are known to enhance the energy storage, they reduce the strength and flexibility of CNFs [93, 94].

Also, the material addition for characteristic activation techniques can have multiple effect on the structure and performance of the CNFs. Undesired pores can form by the addition of metal salts along with graphitization. The intricate interaction of distinctive materials makes it challenging to distinguish the effects of certain variable like conductivity, porosity, or heteroatom content which makes it more difficult to form the CNFs without deficiencies. Additionally, the current CNF research focus is on the performance enhancement that often leads to high production costs which is inapt for industry. Cost-effective scalability of CNFs requires addressing expensive precursors like PAN and toxic materials. The development of cheap, high-quality polymers and industry-scale electrospinning equipment is crucial for meeting future energy application demands [95].

9 Conclusion

The increasing demand for green technology is accelerating the development of carbonized biomass/polymer nanofibers as an environmentally and sustainably favourable materials. Several synthesis methods, including as chemical vapour deposition, electrospinning, and green synthesis, have substantial effect on the characteristics of the CNF. To maximize the performance of CNFs broad applications, the impact of critical process variables on the microstructure and behaviour of CNFs has been examined. Characterization techniques such as SEM, XRD, and Raman spectroscopy play significant role in evaluating the structural and compositional, behavioural attributes of CNFs. They have demonstrated remarkable potential in energy storage systems, water purification, and biomedical applications due to their superior conductivity, adsorption capacity, and biocompatibility. Despite of the associated advantages, CNFs confronts several challenges for large-scale production, material stability, and cost which require further research and innovation. Addressing the current problems and to enable the full potential of CNFs in distinct domains, needs multidisciplinary investigation.

References

1. Azwar, E., et al. (2018). Transformation of biomass into carbon nanofiber for supercapacitor application—A review. *International Journal of Hydrogen Energy, 43*(45), 20811–20821. https://doi.org/10.1016/j.ijhydene.2018.09.111
2. Popa, V. I. (2018). Biomass for fuels and biomaterials. In *Biomass as renewable raw material to obtain bioproducts of high-tech value* (pp. 1–37). Elsevier. https://doi.org/10.1016/B978-0-444-63774-1.00001-6

3. Cao, Q., et al. (2020). Electrospun biomass based carbon nanofibers as high-performance supercapacitors. *Industrial Crops and Products, 148*, 112181. https://doi.org/10.1016/j.indcrop.2020.112181
4. Yao, M., Bi, X., Wang, Z., Yu, P., Dufresne, A., & Jiang, C. (2022). Recent advances in lignin-based carbon materials and their applications: A review. *International Journal of Biological Macromolecules, 223*, 980–1014. https://doi.org/10.1016/j.ijbiomac.2022.11.070
5. Rakhi, T., Pragati, M., Prerna, B., & Jyoti, S. (2023). Multifaceted approach for nanofiber fabrication. In *Fiber and textile engineering in drug delivery systems* (pp. 253–283). Elsevier. https://doi.org/10.1016/B978-0-323-96117-2.00012-1
6. Kenry, & Lim, C. T. (2017). Nanofiber technology: Current status and emerging developments. *Progress in Polymer Science, 70*, 1–17. https://doi.org/10.1016/j.progpolymsci.2017.03.002
7. Rane, L. R., Savadekar, N. R., Kadam, P. G., & Mhaske, S. T. (2014). Preparation and characterization of K-carrageenan/Nanosilica biocomposite film. *Journal of Materials, 2014*, 1–8. https://doi.org/10.1155/2014/736271
8. Abdulhussain, R., Adebisi, A., Conway, B. R., & Asare-Addo, K. (2023). Electrospun nanofibers: Exploring process parameters, polymer selection, and recent applications in pharmaceuticals and drug delivery. *Journal of Drug Delivery Science and Technology, 90*, 105156. https://doi.org/10.1016/j.jddst.2023.105156
9. Deng, J., Li, M., & Wang, Y. (2016). Biomass-derived carbon: Synthesis and applications in energy storage and conversion. *Green Chemistry, 18*(18), 4824–4854. https://doi.org/10.1039/C6GC01172A
10. Zhang, F., Si, Y., Yu, J., & Ding, B. (2023). Electrospun porous engineered nanofiber materials: A versatile medium for energy and environmental applications. *Chemical Engineering Journal, 456*, 140989. https://doi.org/10.1016/j.cej.2022.140989
11. Yang, G., Li, X., He, Y., Ma, J., Ni, G., & Zhou, S. (2018). From nano to micro to macro: Electrospun hierarchically structured polymeric fibers for biomedical applications. *Progress in Polymer Science, 81*, 80–113. https://doi.org/10.1016/j.progpolymsci.2017.12.003
12. Arif, Z. U., Khalid, M. Y., Sheikh, M. F., Zolfagharian, A., & Bodaghi, M. (2022). Biopolymeric sustainable materials and their emerging applications. *Journal of Environmental Chemical Engineering, 10*(4), 108159. https://doi.org/10.1016/j.jece.2022.108159
13. Rahman Khan, M. M., & Chakraborty, N. (2024). Conducting polymer-based gel materials: Synthesis, morphology, thermal properties, and applications in supercapacitors. *Gels, 10*(9), 553. https://doi.org/10.3390/gels10090553
14. Yadav, M. D., Dasgupta, K., Patwardhan, A. W., & Joshi, J. B. (2017). High performance Fibers from carbon nanotubes: Synthesis, characterization, and applications in composites—A review. *Industrial and Engineering Chemistry Research, 56*(44), 12407–12437. https://doi.org/10.1021/acs.iecr.7b02269
15. Feng, L., Xie, N., & Zhong, J. (2014). Carbon nanofibers and their composites: A review of synthesizing, properties and applications. *Materials, 7*(5), 3919–3945. https://doi.org/10.3390/ma7053919
16. Kubley, A., et al. (2019). Smart textiles and wearable technology innovation with carbon nanotube technology. In *Nanotube Superfiber materials* (pp. 263–311). Elsevier. https://doi.org/10.1016/B978-0-12-812667-7.00012-4
17. Zhao, J., et al. (2020). Graphene quantum dot reinforced electrospun carbon nanofiber fabrics with high surface area for ultrahigh rate Supercapacitors. *ACS Applied Materials & Interfaces, 12*(10), 11669–11678. https://doi.org/10.1021/acsami.9b22408
18. Singh, K., Khanna, V., Chaudhary, V., Jasrotia, R., Prakash, C., & Al-Kahtani, A. A. (2024). Investigating the potential of powder metallurgy for fabricating graphene nanoplatelets reinforced copper nanocomposites. *Journal of Materials Research and Technology, 31*, 1258–1269. https://doi.org/10.1016/j.jmrt.2024.06.138
19. Deng, L., Young, R. J., Kinloch, I. A., Zhu, Y., & Eichhorn, S. J. (2013). Carbon nanofibres produced from electrospun cellulose nanofibres. *Carbon N Y, 58*, 66–75. https://doi.org/10.1016/j.carbon.2013.02.032

20. Qu, W., Hu, P., Liu, J., Jin, H., & Wang, K. (2022). Lignin-based carbon fiber: A renewable and low-cost substitute towards featured fiber-shaped pseudocapacitor electrodes. *Journal of Cleaner Production, 343*, 131030. https://doi.org/10.1016/j.jclepro.2022.131030
21. Wang, P., Wang, Y., & Tong, L. (2013). Functionalized polymer nanofibers: A versatile platform for manipulating light at the nanoscale. *Light: Science & Applications, 2*(10), e102. https://doi.org/10.1038/lsa.2013.58
22. Yang, J., et al. (2010). A mini review: Electrospun hierarchical nanofibers. *Journal of Dispersion Science and Technology, 31*(6), 760–769. https://doi.org/10.1080/01932690903333127
23. Anstey, A., et al. (2021). Nanofibrillated polymer systems: Design, application, and current state of the art. *Progress in Polymer Science, 113*, 101346. https://doi.org/10.1016/j.progpolymsci.2020.101346
24. Lv, X., Zhang, Y., Wang, Y., Zhang, G., Zhao, Y., & Liu, J. (2021). Formation of carbon nanofibers/nanotubes by chemical vapor deposition using Al_2O_3/KOH. *Diamond and Related Materials, 113*, 108265. https://doi.org/10.1016/j.diamond.2021.108265
25. Kairi, M. I., Khavarian, M., Bakar, S. A., Vigolo, B., & Mohamed, A. R. (2018). Recent trends in graphene materials synthesized by CVD with various carbon precursors. *Journal of Materials Science, 53*(2), 851–879. https://doi.org/10.1007/s10853-017-1694-1
26. Papadopoulou, F.-M., Soulis, S., Trompeta, A.-F. A., & Charitidis, C. A. (2024). Thin and flexible PANI/PMMA/CNF forest films produced via a two-step floating catalyst chemical vapor deposition. *Materials, 17*(23), 5812. https://doi.org/10.3390/ma17235812
27. Cui, R., et al. (2017). Phosphorus-doped helical carbon nanofibers as enhanced sensing platform for electrochemical detection of carbendazim. *Food Chemistry, 221*, 457–463. https://doi.org/10.1016/j.foodchem.2016.10.094
28. Chen, L., et al. (2020). A facile self-catalyzed CVD method to synthesize Fe3C/N-doped carbon nanofibers as lithium storage anode with improved rate capability and cyclability. *Journal of Materials Science and Technology, 44*, 229–236. https://doi.org/10.1016/j.jmst.2019.11.013
29. Abioye, A. M., & Ani, F. N. (2015). Recent development in the production of activated carbon electrodes from agricultural waste biomass for supercapacitors: A review. *Renewable and Sustainable Energy Reviews, 52*, 1282–1293. https://doi.org/10.1016/j.rser.2015.07.129
30. Kshirsagar, D. E., Dubey, H., Puri, V., Jadhav, V., & Sharon, M. (2022). CNFs from linseeds oil: A potential source of X-band microwave absorbers. *Materials Advances, 3*(9), 3989–3993. https://doi.org/10.1039/D1MA01252E
31. Wang, T., et al. (2020). Carbon nanofibers prepared from solar pyrolysis of pinewood as binder-free electrodes for flexible Supercapacitors. *Cell Reports Physical Science, 1*(6), 100079. https://doi.org/10.1016/j.xcrp.2020.100079
32. Huang, W., et al. (2020). A review on electrospinning nanofibers in the field of microwave absorption. *Ceramics International, 46*(17), 26441–26453. https://doi.org/10.1016/j.ceramint.2020.07.193
33. Sabantina, L., et al. (2018). Stabilization of electrospun PAN/gelatin nanofiber mats for carbonization. *Journal of Nanomaterials, 2018*, 1–12. https://doi.org/10.1155/2018/6131085
34. Moulefera, I., Trabelsi, M., Mamun, A., & Sabantina, L. (2021). Electrospun carbon nanofibers from biomass and biomass blends—Current trends. *Polymers (Basel), 13*(7), 1071. https://doi.org/10.3390/polym13071071
35. Nasir, M., Subhan, A., Prihandoko, B., & Lestariningsih, T. (2017). Nanostructure and property of electrospun SiO_2-cellulose acetate nanofiber composite by electrospinning. *Energy Procedia, 107*, 227–231. https://doi.org/10.1016/j.egypro.2016.12.133
36. Wang, J., et al. (2016). Interface miscibility induced double-capillary carbon nanofibers for flexible electric double layer capacitors. *Nano Energy, 28*, 232–240. https://doi.org/10.1016/j.nanoen.2016.08.043
37. Ghosh, S., Yong, W. D., Jin, E. M., Polaki, S. R., Jeong, S. M., & Jun, H. (2019). Mesoporous carbon nanofiber engineered for improved supercapacitor performance. *Korean Journal of Chemical Engineering, 36*(2), 312–320. https://doi.org/10.1007/s11814-018-0199-1

38. Xiong, J., et al. (2024). Electrospun biomass carbon-based porous nanofibers modified by green-dry cold plasma for gaseous formaldehyde adsorption. *Industrial Crops and Products, 222*, 119769. https://doi.org/10.1016/j.indcrop.2024.119769
39. Jiang, H., Yao, M., Wang, J., & Hong, W. (2024). High-performance biomass tar-based nanofibers with CoNi nanosheets prepared by electrospinning for supercapacitor electrode material. *Journal of Energy Storage, 98*, 113147. https://doi.org/10.1016/j.est.2024.113147
40. Chen, W., Wang, H., Lan, W., Li, D., Zhang, A., & Liu, C. (2021). Construction of sugarcane bagasse-derived porous and flexible carbon nanofibers by electrospinning for supercapacitors. *Industrial Crops and Products, 170*, 113700. https://doi.org/10.1016/j.indcrop.2021.113700
41. Rani, S. D., et al. (2020). NiMoO4 nanoparticles decorated carbon nanofiber membranes for the flexible and high performance glucose sensors. *Sensors and Actuators B: Chemical, 312*, 127886. https://doi.org/10.1016/j.snb.2020.127886
42. Mohammadpour-Haratbar, A., Mosallanejad, B., Zare, Y., Rhee, K. Y., & Park, S.-J. (2022). CO_3O_4 nanoparticles embedded in electrospun carbon nanofibers as free-standing nanocomposite electrodes as highly sensitive enzyme-free glucose biosensors. *Reviews on Advanced Materials Science, 61*(1), 744–755. https://doi.org/10.1515/rams-2022-0251
43. Qasim, M., Clarkson, A. N., & Hinkley, S. F. R. (2023). Green synthesis of carbon nanoparticles (CNPs) from biomass for biomedical applications. *International Journal of Molecular Sciences, 24*(2), 1023. https://doi.org/10.3390/ijms24021023
44. Omoriyekomwan, J. E., Tahmasebi, A., Dou, J., Wang, R., & Yu, J. (2021). A review on the recent advances in the production of carbon nanotubes and carbon nanofibers via microwave-assisted pyrolysis of biomass. *Fuel Processing Technology, 214*, 106686. https://doi.org/10.1016/j.fuproc.2020.106686
45. Bilge, S., Karadurmus, L., Sınağ, A., & Ozkan, S. A. (2021). Green synthesis and characterization of carbon-based materials for sensitive detection of heavy metal ions. *TrAC Trends in Analytical Chemistry, 145*, 116473. https://doi.org/10.1016/j.trac.2021.116473
46. Ng, C. H., et al. (2024). The roles of carbonaceous wastes for catalysis, energy, and environmental remediation. *Catalysis Communications, 187*, 106845. https://doi.org/10.1016/j.catcom.2024.106845
47. Flores–Oña, D., & Fullana, A. (2020). Carbon nanoparticles production using solvent assisted hydrothermal carbonization. *Diamond and Related Materials, 108*, 107960. https://doi.org/10.1016/j.diamond.2020.107960
48. Al-Rumaihi, A., Shahbaz, M., Mckay, G., Mackey, H., & Al-Ansari, T. (2022). A review of pyrolysis technologies and feedstock: A blending approach for plastic and biomass towards optimum biochar yield. *Renewable and Sustainable Energy Reviews, 167*, 112715. https://doi.org/10.1016/j.rser.2022.112715
49. Yogalakshmi, K. N., et al. (2022). Lignocellulosic biomass-based pyrolysis: A comprehensive review. *Chemosphere, 286*, 131824. https://doi.org/10.1016/j.chemosphere.2021.131824
50. Jiao, Y., Xu, K., Xiao, H., Mei, C., & Li, J. (2023). Biomass-derived carbon aerogels for ORR/OER Bifunctional oxygen electrodes. *Nanomaterials, 13*(17), 2397. https://doi.org/10.3390/nano13172397
51. Bokov, D., et al. (2021). Nanomaterial by sol-gel method: Synthesis and application. *Advances in Materials Science and Engineering, 2021*, 1. https://doi.org/10.1155/2021/5102014
52. Sam, D. K., Sam, E. K., Durairaj, A., Lv, X., Zhou, Z., & Liu, J. (2020). Synthesis of biomass-based carbon aerogels in energy and sustainability. *Carbohydrate Research, 491*, 107986. https://doi.org/10.1016/j.carres.2020.107986
53. Wang, Q., et al. (2023). Production of biomass-based carbon materials in hydrothermal media: A review of process parameters, activation treatments and practical applications. *Journal of the Energy Institute, 110*, 101357. https://doi.org/10.1016/j.joei.2023.101357
54. Wang, Y., et al. (2021). Biomass-derived carbon materials: Controllable preparation and versatile applications. *Small, 17*, 40. https://doi.org/10.1002/smll.202008079
55. Wu, Z., Liang, H., Hu, B., & Yu, S. (2018). Emerging carbon-nanofiber aerogels: Chemosynthesis versus biosynthesis. *Angewandte Chemie International Edition, 57*(48), 15646–15662. https://doi.org/10.1002/anie.201802663

56. Gao, Y., Huang, Y., Xin, X., Jiang, T., Zong, M., & Xu, J. (2024). Biomass modified carbon nanofibers use for flexible supercapacitor electrodes with thermal conductive performances. *Journal of Energy Storage, 99*, 113382. https://doi.org/10.1016/j.est.2024.113382
57. Kim, Y. K., Il Cha, S., Hong, S. H., & Jeong, Y. J. (2012). A new hybrid architecture consisting of highly mesoporous CNT/carbon nanofibers from starch. *Journal of Materials Chemistry, 22*(38), 20554. https://doi.org/10.1039/c2jm32631k
58. Lamsal, B., & Adhikari, R. (2024). Ways of tuning the morphology of electrospun carbon nanofibres for flexible supercapacitors. *Nano-Structures & Nano-Objects, 38*, 101137. https://doi.org/10.1016/j.nanoso.2024.101137
59. Kim, J., Heo, Y.-J., Hong, J.-Y., & Kim, S.-K. (2020). Preparation of porous carbon nanofibers with tailored porosity for electrochemical capacitor electrodes. *Materials, 13*(3), 729. https://doi.org/10.3390/ma13030729
60. Klankowski, S. A., et al. (2015). Anomalous capacity increase at high-rates in lithium-ion battery anodes based on silicon-coated vertically aligned carbon nanofibers. *Journal of Power Sources, 276*, 73–79. https://doi.org/10.1016/j.jpowsour.2014.11.094
61. Huang, C., et al. (2024). Biomass-derived multifunctional nanoscale carbon fibers toward fire warning sensors, supercapacitors and moist-electric generators. *International Journal of Biological Macromolecules, 256*, 127878. https://doi.org/10.1016/j.ijbiomac.2023.127878
62. Liu, C., et al. (2019). Preparation and electrochemical studies of electrospun phosphorus doped porous carbon nanofibers. *RSC Advances, 9*(12), 6898–6906. https://doi.org/10.1039/C8RA10193K
63. Zhu, J., et al. (2015). Nitrogen-doped carbon nanofibers derived from polyacrylonitrile for use as anode material in sodium-ion batteries. *Carbon N Y, 94*, 189–195. https://doi.org/10.1016/j.carbon.2015.06.076
64. Liu, S., & Ma, M.-G. (2020). Lignin-derived nitrogen-doped polyacrylonitrile/polyaniline carbon nanofibers by electrospun method for energy storage. *Ionics (Kiel), 26*(9), 4651–4660. https://doi.org/10.1007/s11581-020-03603-8
65. Tao, L., et al. (2019). Porous carbon nanofiber derived from a waste biomass as anode material in lithium-ion batteries. *Journal of the Taiwan Institute of Chemical Engineers, 95*, 217–226. https://doi.org/10.1016/j.jtice.2018.07.005
66. Singh, K., Trabelsi, Y., Mahajan, P., & Khanna, V. (2024). Role of graphene in next-generation electrochemical energy storage systems: Insights and innovations. *Journal of the Electrochemical Society, 171*(10), 100501. https://doi.org/10.1149/1945-7111/ad7bee
67. Mumtaz, A., Iqbal, J., & Oneeb, M. (2023). Iron oxide induced effects on the electrochemical stability and specific capacity of hafnium oxide electrode for supercapacitor applications. *Journal of Energy Storage, 74*, 109320. https://doi.org/10.1016/j.est.2023.109320
68. Beknalkar, S. A., et al. (2021). Fabrication of high energy density supercapacitor device based on hollow iridium oxide nanofibers by single nozzle electrospinning. *Applied Surface Science, 546*, 149102. https://doi.org/10.1016/j.apsusc.2021.149102
69. Muddasar, M., Beaucamp, A., Culebras, M., & Collins, M. N. (2024). High performance all lignin derived supercapacitors for energy storage applications. *Materials Today Sustainability, 26*, 100767. https://doi.org/10.1016/j.mtsust.2024.100767
70. Wang, B., Xu, J., Lu, F., Zhou, J., Wang, X., & Tan, F. (2024). Efficient utilization of lignin heterogeneity: Employing a simple approach to modulate the structure-effect relationship of lignin microspheres for its high-quality applications in supercapacitors. *Industrial Crops and Products, 221*, 119360. https://doi.org/10.1016/j.indcrop.2024.119360
71. Supchocksoonthorn, P., Hrimchum, N., Budsrirak, T., Intaraprasit, S., Thongsai, N., & Aussawasathien, D. (2023). Lignin based carbon fiber fabrics with hybrid doping approach as self-standing electrodes for supercapacitors. *Electrochimica Acta, 437*, 141523. https://doi.org/10.1016/j.electacta.2022.141523
72. Cao, Q., Zhu, M., Chen, J., Song, Y., Li, Y., & Zhou, J. (2020). Novel lignin-cellulose-based carbon nanofibers as high-performance Supercapacitors. *ACS Applied Materials & Interfaces, 12*(1), 1210–1221. https://doi.org/10.1021/acsami.9b14727

73. Tan, F., et al. (2025). Biomass-derived carbon nanofibers and electrolytes: Toward high-performance lignin-based supercapacitors with enhanced electrochemical performance. *Industrial Crops and Products, 224*, 120381. https://doi.org/10.1016/j.indcrop.2024.120381
74. Wang, H., et al. (2023). Preparation of hierarchical micro-meso porous carbon and carbon nanofiber from polyacrylonitrile/polysulfone polymer via one-step carbonization for supercapacitor electrodes. *Electrochimica Acta, 441*, 141827. https://doi.org/10.1016/j.electacta.2023.141827
75. Wang, H., Wang, H., Ruan, F., Feng, Q., Wei, Y., & Fang, J. (2023). High-porosity carbon nanofibers prepared from polyacrylonitrile blended with amylose starch for application in supercapacitors. *Materials Chemistry and Physics, 293*, 126896. https://doi.org/10.1016/j.matchemphys.2022.126896
76. Zheng, C., et al. (2023). Biomass-based carbon nanofibers modified with polyaniline for supercapacitor applications. *ACS Applied Nano Materials, 6*(15), 14136–14150. https://doi.org/10.1021/acsanm.3c02028
77. Zhu, J., Zhang, Q., Zhao, Y., Zhang, R., Liu, L., & Yu, J. (2023). Robust N-doping porous carbon nanofiber membranes with inter-fiber cross-linked structures for supercapacitors. *Carbon N Y, 202*, 13–25. https://doi.org/10.1016/j.carbon.2022.11.021
78. Aruchamy, K., Dharmalingam, K., Lee, C. W., Mondal, D., & Sanna Kotrappanavar, N. (2022). Creating ultrahigh surface area functional carbon from biomass for high performance supercapacitor and facile removal of emerging pollutants. *Chemical Engineering Journal, 427*, 131477. https://doi.org/10.1016/j.cej.2021.131477
79. Mahajan, P., Khanna, V., Singh, A., & Singh, K. (2024). Review—Advances in nanomaterial-based biosensors for heavy metal detection and remediation in soil. *Journal of the Electrochemical Society, 171*(11), 117527. https://doi.org/10.1149/1945-7111/ad9413
80. Wang, L., Peng, X., Fu, H., Huang, C., Li, Y., & Liu, Z. (2020). Recent advances in the development of electrochemical aptasensors for detection of heavy metals in food. *Biosensors & Bioelectronics, 147*, 111777. https://doi.org/10.1016/j.bios.2019.111777
81. Yang, Q., et al. (2023). Metal-free cysteamine-functionalized graphene alleviates mutual interferences in heavy metal electrochemical detection. *Green Chemistry, 25*(4), 1647–1657. https://doi.org/10.1039/D2GC02978B
82. Mishra, S., Budania, Y., Tyagi, A., Pratap Singh, S., Kumar, P., & Singh, S. (2024). Copper oxide anchored carbon nanofibers: A versatile platform for multiplex detection of antibiotics, heavy metals and pesticides. *Chemistry, an Asian Journal, 19*, 10. https://doi.org/10.1002/asia.202400241
83. Zhao, D., Wang, T., Han, D., Rusinek, C., Steckl, A. J., & Heineman, W. R. (2015). Electrospun carbon nanofiber modified electrodes for stripping voltammetry. *Analytical Chemistry, 87*(18), 9315–9321. https://doi.org/10.1021/acs.analchem.5b02017
84. Liao, Y., Li, Q., Wang, N., & Shao, S. (2015). Development of a new electrochemical sensor for determination of Hg(II) based on Bis(indolyl)methane/mesoporous carbon nanofiber/Nafion/glassy carbon electrode. *Sensors and Actuators B: Chemical, 215*, 592–597. https://doi.org/10.1016/j.snb.2015.04.006
85. Jamil, T., Munir, S., Wali, Q., Shah, G. J., Khan, M. E., & Jose, R. (2021). Water purification through a novel electrospun carbon nanofiber membrane. *ACS Omega, 6*(50), 34744–34751. https://doi.org/10.1021/acsomega.1c05197
86. Shen, Y., Li, L., Xiao, K., & Xi, J. (2016). Constructing three-dimensional hierarchical architectures by integrating carbon nanofibers into graphite felts for water purification. *ACS Sustainable Chemistry & Engineering, 4*(4), 2351–2358. https://doi.org/10.1021/acssuschemeng.6b00030
87. Luo, Q., et al. (2023). Hierarchical porous carbon nanofibers for highly efficient solar-driven water purification. *Science China Materials, 66*(8), 3310–3318. https://doi.org/10.1007/s40843-023-2431-3
88. Mishakov, I. V., et al. (2022). Water purification from chlorobenzenes using heteroatom-functionalized carbon nanofibers produced on self-organizing Ni-Pd catalyst. *Journal of Environmental Chemical Engineering, 10*(3), 107873. https://doi.org/10.1016/j.jece.2022.107873

89. Peter, K. T., et al. (2016). Synthesis, optimization, and performance demonstration of electrospun carbon nanofiber–carbon nanotube composite sorbents for point-of-use water treatment. *ACS Applied Materials & Interfaces, 8*(18), 11431–11440. https://doi.org/10.1021/acsami.6b01253
90. Samadian, H., Mobasheri, H., Hasanpour, S., Ai, J., Azamie, M., & Faridi-Majidi, R. (2020). Electro-conductive carbon nanofibers as the promising interfacial biomaterials for bone tissue engineering. *Journal of Molecular Liquids, 298*, 112021. https://doi.org/10.1016/j.molliq.2019.112021
91. Senthil Kumar, R., Sarathi, T., Venkataraman, K. K., & Bhattacharyya, A. (2019). Enhanced piezoelectric properties of polyvinylidene fluoride nanofibers using carbon nanofiber and electrical poling. *Materials Letters, 255*, 126515. https://doi.org/10.1016/j.matlet.2019.126515
92. Serafin, A., Murphy, C., Rubio, M. C., & Collins, M. N. (2021). Printable alginate/gelatin hydrogel reinforced with carbon nanofibers as electrically conductive scaffolds for tissue engineering. *Materials Science and Engineering: C, 122*, 111927. https://doi.org/10.1016/j.msec.2021.111927
93. Zhang, J., et al. (2019). Electrospun flexible nanofibrous membranes for oil/water separation. *Journal of Materials Chemistry A, 7*(35), 20075–20102. https://doi.org/10.1039/C9TA07296A
94. Liu, H., et al. (2014). Flexible macroporous carbon nanofiber film with high oil adsorption capacity. *Journal of Materials Chemistry A, 2*(10), 3557. https://doi.org/10.1039/c3ta14468b
95. Nie, G., Zhao, X., Luan, Y., Jiang, J., Kou, Z., & Wang, J. (2020). Key issues facing electrospun carbon nanofibers in energy applications: On-going approaches and challenges. *Nanoscale, 12*(25), 13225–13248. https://doi.org/10.1039/D0NR03425H

Carbon-Based Nanocomposites for Lightweight Composites

Satnam Singh, Akanksha Mathur, and Michele Forzan

1 Introduction

The innovations and rapid development in the materials science and technology domain led to the growing demand for performance-oriented materials which has completely transformed the applications and landscape of materials science. All the major engineering sectors such as aerospace, automotive, construction, electronics, sporting goods and equipment, and machinery parts [1–3] require materials that are not only strong but also lightweight and durable in long term. This is required to meet the current challenges such as energy efficiencies, environmental sustainability, and operational effectiveness and to keep the competition at par with the global players. Along with the development and research on the various materials in recent years; carbon-based nanocomposites have emerged as a groundbreaking class of materials that addresses diverse industrial needs and are potential candidates for future engineering classes of materials [4, 5]. Over the period various carbon-based materials have emerged which have redefined the standard of performance and highlighted their potential in lightweight superstrong materials across various engineering sectors [6, 7]. These materials are often used as composite materials for advanced applications, which consist of a matrix, typically a polymer, metal, or ceramic, reinforced with carbon-based nanomaterials such as carbon nanotubes (CNTs), graphene, SiC, WC, carbon nanofibers, and nano-graphite [8–10]. Such materials at the nanoscale due to their distinctive structures and exceptional intrinsic

S. Singh (✉) · A. Mathur
Multidisciplinary Engineering Department, The NorthCap University, Gurugram, Haryana, India
e-mail: satnamsingh@ncuindia.edu

M. Forzan
Department of Industrial Engineering, University of Padova, Padova, PD, Italy

V. Khanna (ed.), *Carbon-Based Nanocomposites for Sustainable Applications, Volume I*, Lecture Notes in Nanoscale Science and Technology 38,
https://doi.org/10.1007/978-3-031-95306-4_7

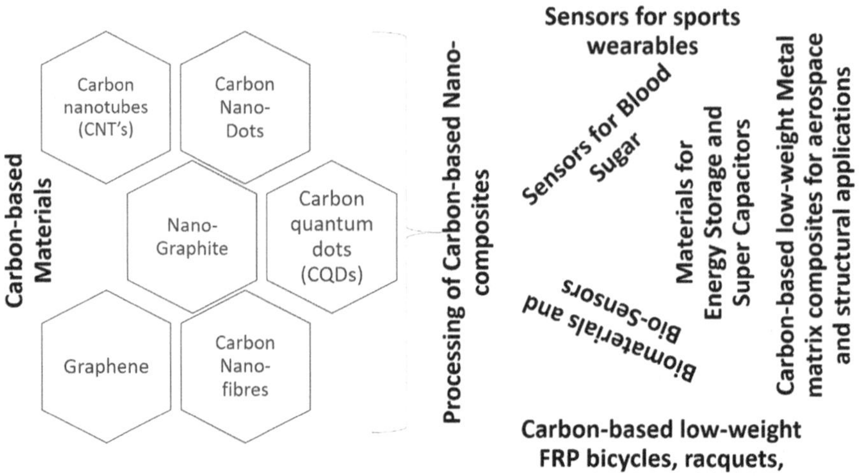

Fig. 1 Carbon-based materials and their potential application domains

properties qualify them to exhibit superior mechanical, thermal, and electrical characteristics, in comparison to conventional materials thus making them a prime choice for lightweight composites [11]. Figure 1 presents some of the advanced applications of carbon-based materials across the multidisciplinary domain including mechanical, sports, wearable electronics, biomaterials, sensors, cancer therapy, tissue engineering, gene delivery, etc. [12, 13].

The present work focus on the mechanical properties of carbon-based composites where the concept of lightweight coupled with strength and durability to enhances the material's usability and adaptability in advanced engineering applications. For decades, the emphasis has been on developing materials that combine strength, durability, and reduced weight to achieve higher efficiency and sustainability. Lightweight materials are particularly significant in sectors where weight reduction directly correlates with performance improvement. Some important real-life examples are:

- **Aerospace Sector:** In this sector, every kilogram of weight reduction in an aircraft results in substantial fuel savings, reduced emissions, and increased payload capacity. Advancements in materials science especially for high-strength and low-weight materials play a crucial role. The research by Dannenberg et al. [14] estimated that more than 1.3 billion litres of extra fuel were used due to excess weight in the decade around 1994 in the USA. Along similar lines, a study by Tom et al. [15] reported that the excess passenger weight in domestic transport systems for the USA between 1970 and 2010 revealed that 95.2 billion litres of extra fuel was required.
- **Automotive:** The automotive sector is undergoing many transformations, with a focus on improving specific strength, i.e. strength per unit weight. Lightweight, high-strength materials in vehicles can improve fuel economy, reduce emissions,

improve safety during impacts, and improve performance. It has been reported that using one ton of aluminium instead of cast iron or conventional steel in car manufacturing may reduce an average of 18 tons of greenhouse gas emissions during the life cycle of a middle-size sedan [16].

- **Sporting Goods and equipment:** The application of lightweight and durable materials is increasing in the sporting goods industry. Such materials can enhance the athlete's performance by reducing the equipment's weight while maintaining its robustness. Carbon-based materials such as graphene, carbon fibre, and graphite-based nanocomposites find extensive applications in sporting and wearable materials [17].

In pursuit of low-weight and high-strength materials, researchers have developed many alloys such as aluminium-based alloys, titanium alloys, and magnesium alloys [11]. However, these materials have provided the results in terms of weight reduction but limitations in balancing out the properties of impact resistance, high toughness, and specific strengths coupled with the cost and manufacturing processes involved still exists. Out of these materials, aluminium alloys are widely used in many engineering applications due to ease of manufacturing but still, other low-weight materials require a lot of research for sustainable solutions. These limitations of traditional materials and the gaps mentioned have paved the way for the further exploration of materials science in the domain of nanotechnology-driven solutions, particularly carbon-based nanocomposites in the advanced field of engineering and technology [18]. The manufacturing of these carbon-based composites further plays an important role in their utilization. Initially, carbon nanocomposites were quite costly however, over the years new manufacturing techniques led to sustainable manufacturing solutions and increased use of such materials in advanced applications such as wearable electronics, sensors, aerospace, sports industries, and mechanical parts.

In this chapter, a detailed study of carbon-based nanocomposites for lightweight composites is carried out. Many material combinations with carbon-based materials are presented with potential applications. Studies on the various manufacturing and fabrication processes involved in developing these materials are reported. The carbon-based materials focusing on the weight reduction of materials with improvement in properties are presented in the present work.

2 Carbon-Based Nanomaterials: A Game Changer

The research and development in the carbon-based materials domain over the years have reported the unique properties of these materials which are much advanced and have potential applications in the areas where conventional materials fail to deliver such as biosensors, wearable electronics, automotive sectors, and aerospace applications [19]. The carbon-based nanomaterials have gathered significant attention and worthiness due to their unique properties at the nanoscale levels. The discovery

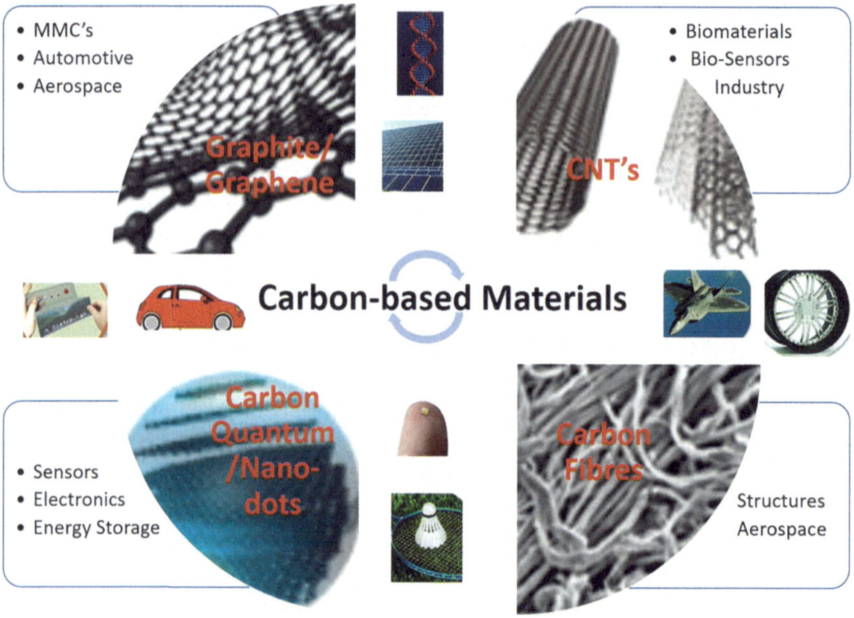

Fig. 2 Various carbon-based materials used in advanced applications

of helical microtubules in 1991 by Iijima [20] now commonly known as Carbon Nano-Tubes (CNT) and the report of graphene in 2004 by Geim and Novoselov (Nobel Prize winners in Physics, 2010) revolutionized the material science and opened new frontiers for many advanced applications. Figure 2 presents the carbon-based materials with their unique properties which are used in various advanced applications.

These carbon-based materials have gained popularity due to their unique properties which overcome the limitations posed by the conventional composite materials. Some of the advantages of carbon-based materials offered are:

- **Superior Strength-to-Weight Ratio:** Carbon nanomaterials significantly enhance the mechanical properties of composites, including tensile strength, modulus, and toughness, without adding excessive weight. This allows higher specific strength for carbon-based composite materials and enhanced applications in the aerospace domain.
- **Enhanced Thermal Conductivity:** The carbon-based materials can tailor the mechanical and thermal properties without adding weight restrictions. The heat dissipation and thermal stability of such materials are significantly high which are critical for many applications in electronics and aerospace domains.
- **Electrical Conductivity:** Carbon nanomaterials, particularly CNTs and graphene, can help in improving the electrical conductivity of materials, which allows such materials to be extensively used in the field of sensors, energy storage devices, and conductive coatings.

- **Multi-Functionality/Tailored Properties:** Carbon-based nanocomposites provide a unique combination and tailoring of mechanical, thermal, and electrical properties, which makes them suitable for a wide range of advanced applications.

2.1 Applications of Carbon-Based Nanomaterials/Composites

The versatile use and application of carbon-based materials in advanced applications are due to the unique combination of properties offered by these materials. Due to the unmatched qualities, the versatility of carbon-based nanocomposite materials is evident in their use across various industries [4]. Some of the industrial applications of these materials are presented in Fig. 3.

In aerospace applications, the lightweight carbon-based composites are used in the airframe components, wing structures, and fuselage of aircrafts. The low-weight composites offer improved fuel efficiency, reduced structural weight, and better thermal stability. Graphene, Carbon Fibre Reinforced Plastics (CRPF), and Titanium-based composites are used in next-generation aircrafts for high performances and efficient operational cost. The automotive parts such as engine components, body panels, and crash resistance structures are manufactured using carbon-based composites, which allow lower weight and better strength. The reduced weight allows better mileage and helps in reducing the greenhouse gas

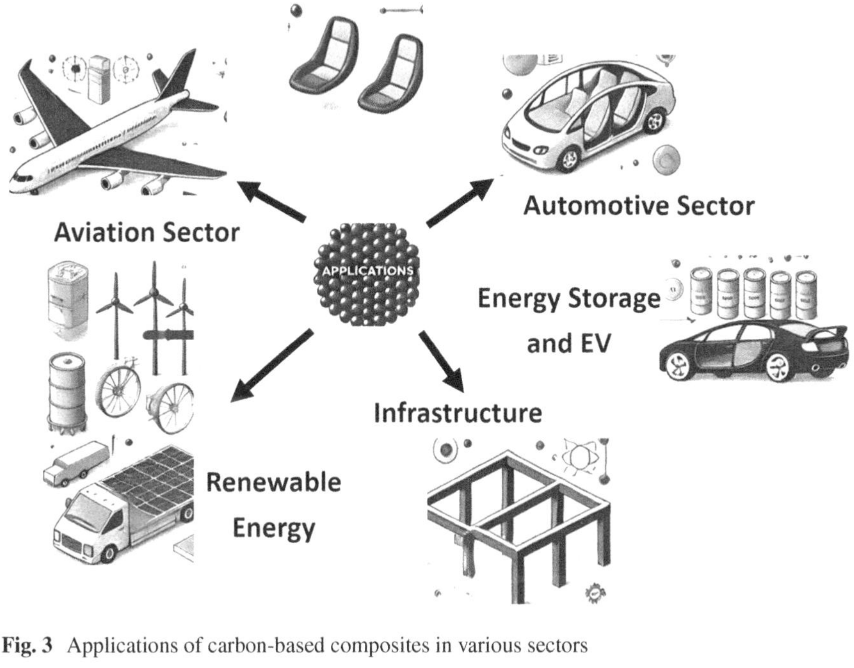

Fig. 3 Applications of carbon-based composites in various sectors

emissions. The presence of CNT-reinforced polymer composites improves the crash worthiness of automobiles by better impact absorption capabilities.

The improved properties offered by carbon-based nanocomposites such as high electrical conductivities, better sensitivity, thermal stability, and durability allows these materials to be used in conductive coatings, energy storage devices, wearable sensors, electrodes, etc. The application of graphene-based composites is widely used in biosensors for detecting biological molecules with high precision. The carbon-based materials such as CNT, graphene, carbon nanofibers, carbon nanodots, and carbon quantum dots are widely used in the efficient glucose sensors due to more surface area, porosity, and excellent catalytic properties [21]. The properties of high strength-to-weight ratios, durability, and strength allows carbon-based composites to be extensively used in sports good industry such as manufacturing of tennis racquets, golf clubs, bicycles, and skate boards. CNT-based composites and CRPF are widely used for manufacturing of sporting goods such as tennis racquets and bicycle parts. The advantage of lightweight improves the agility and durability of sporting goods.

2.2 Carbon-Based Materials Systems with Applications

Material systems play an important role in the advanced applications in the cut throat engineering environment, particularly as industries demand innovative solutions to meet various challenges in terms of performance, efficiency, and sustainability. A material system mainly comprises of a matrix phase, such as polymers, metals, ceramics, or concrete, which is reinforced with materials that enhance specific/functional properties. In recent years, carbon-based reinforcements like carbon nanotubes (CNTs), graphene, carbon nanofibers (CNFs), and nano-graphite have revolutionized the properties of advanced composites. These systems are tailored to deliver superior strength-to-weight ratios, excellent thermal and electrical conductivity, and enhanced durability. For example, the polymer matrix integrated with graphene is widely used in aerospace for lightweight structures and in electronics for efficient heat dissipation. Similarly, CNT-reinforced metals and ceramics offer promising solutions in high-temperature environments like turbine engines and aerospace applications.

Some of the materials systems for carbon-based materials with their potential applications in the industries are presented in Table 1.

The combination of materials depends on the requirement and tailoring of properties for specific applications. The presence of carbon-based reinforcements can significantly affect the functional properties of the materials. It requires a lot of optimizations for materials combinations, percentages, and processing for obtaining tailored properties. The versatility of these material systems enables their application in diverse fields such as automotive, aerospace, construction, biosensing, and renewable energy. As researchers are progressing and exploring more materials types; new class of hybrid and multi-functional composites are emerging. These

Table 1 Representation of carbon-based material systems and their applications in various industries

Materials system	Matrix material	Carbon-based reinforcement	Applications	Properties/ advantages
CNT/polymer composites [22]	Epoxy	Carbon nanotubes (CNTs)	Aerospace components, automotive body parts, and sports equipment	Lightweight, high strength, good conductivity
Graphene/ polymer composites [23]	Polypropylene/ nylon/HDPE	Graphene nanoplatelets/ powder particles	Heat sinks, EMI shielding, and flexible electronics	Excellent thermal conductivity, improved electrical properties, and low weight
CNT/metal composites [8]	Aluminium, magnesium,and titanium	Carbon nanotubes (CNT'S)	Lightweight automobile parts, aerospace structural components, and mechanical components	Enhanced strength-to-weight ratio, increased thermal conductivity
Graphene/ ceramic composites [24]	Silicon carbide	Graphene	High-temperature aerospace parts, heat shields, and turbine blades	High thermal stability, improved toughness, and high-strength modulus
CNT/ concrete composites [25]	Cement	Carbon nanotubes (CNT'S)	Construction, self-sensing materials, crack monitoring, and repair	Improved strength and toughness, improved electrical conductivity for structural health monitoring
Graphene/ polyurethane composite [26]	Polyurethane	Graphene nanoplatelets	Flexible sensors, wearable electronics devices	High elasticity, conductivity, and thermal stability. Improved strength and life
CNT/rubber composites [27]	Natural rubber	Carbon nanotubes	Tyre manufacturing, vibration damping materials	Improved mechanical properties, higher wear resistance
Nano-graphite/ polymer composites [28]	Polystyrene	Nano-graphite	Automotive dashboards, electronic housings	High stiffness, low density, and good thermal stability
Graphene/ Metal Composites [29]	Copper	Graphene	Heat exchangers, power electronics	Improved thermal and electrical conductivity, lightweight, and high strength with wear resistance

materials blend properties like self-healing, adaptability, and recyclability to address the growing demand for sustainability and smart materials in advanced sectors. However, the cost of such materials still remains the major factor for their enhanced utilization in the commercialized applications.

3 Processing/Manufacturing of Carbon-Based Nanomaterials

The processing and manufacturing techniques of carbon-based materials plays an important role in the effective utilization of such materials in the advanced applications. The sustainable solutions will lead to the cost-effective solutions which in turn will enhance the utilization rate of such materials. The processing and manufacturing of carbon-based materials involves a variety of techniques which allows the users to achieve desired combinations of properties for specific applications. Different methods use different approaches to ensure the effective mixing of carbon-based reinforcements into the various matrix materials. This allows optimal dispersion of reinforcements to achieve uniformity and to enhance the overall strength and performance of the composites. There are two aspects of carbon-based materials manufacturing; one is the synthesis of these carbon-based materials such as CNT, graphene, and carbon nanodots through the processes such as Chemical Vapour Deposition (CVD), Arc Discharge, Laser Ablation, Chemical Oxidation, and Chemical Synthesis [30]. The second aspect is to use the carbon-based materials as reinforcements and processing of composites through these materials by using a suitable matrix material through processes such as Solution Mixing, Melt Compounding, Hot Pressing Additive Manufacturing, and Microwave heating [31–33]. In the present chapter, the focus is on the second aspect, i.e. on the manufacturing of carbon-based materials composites/nanocomposites through various materials processing techniques. Table 2 presents some of the important manufacturing routes for carbon-based composite systems used for various applications.

All the processes have their own advantages and limitations. Some of the processes have the issue of scalability whereas some has issue of higher energy consumptions. A lot of work is going in this field where sustainable solutions are researched and focused. One of the promising solutions is novel microwave heating whish is quite useful for carbon-based materials as carbon-based materials absorbs microwaves easily and microwave processing is highly efficient which consumes less energy and lower processing times. The structure-properties correlations are also excellent in comparison to conventional processes. Still a lot of research is going on in the field of materials processing where sustainable solutions are researched which can be applicable to different materials systems. In recent years, work has been reported on sustainability index of manufacturing processes which allows multiple factors to be considered for sustainability [40] and is the need for future processes.

Table 2 Common processing techniques used for producing carbon-based composites

Techniques	Process	Advantages	Limitations	Applications
Solution Mixing Processing [31]	Used for dispersing carbon nanomaterials in a solvent and mixing with the polymer matrix	Process allows uniform dispersion of nanoparticles in the matrix suitable for thermosetting plastics	Process requires effective dispersion agents and is a time-consuming process	Process is suitable for processing epoxy composites for aerospace and automobile components
Melt Compounding [32]	Process involves mixing of carbon nanomaterials into a polymer melt during extrusion or injection moulding process	The process is scalable at large scale and is a cost-effective solution for plastics composites	The process has limited dispersion of reinforcements as compared to solution mixing process	Process is extensively used for automotive parts like bumpers and dashboards
In Situ Polymerization [34]	Process involves adding of carbon nanomaterials during monomer polymerization to form a composite	It provides strong interfacial bonding and excellent mechanical and electric properties	The process requires precise control of polymerization conditions for uniform distribution and excellent properties	The process is used for processing high-performance composites in electronics industry
Chemical vapour deposition (CVD) [35]	Process involves depositing carbon-based materials (e.g. graphene) onto a substrate in a controlled environment	The process produces high-quality graphene and CNTs, scalable for thin films. It can be used to produce CNT/cu nanocomposites	The process is expensive, time-consuming, and difficult to use for bulk applications	The process is used for fabricating transparent electrodes, flexible displays, and other electronics parts
Electrospinning [36]	Process involves creating carbon nanofibers/ nanocomposite fibres by spinning a polymer solution with carbon precursors	The process produces uniform nanofibers, scalable for industrial applications	The process is limited to specific materials and geometries	Used for nanofiber-based filtration systems for air/water, etc. nanofibers are used for solar cells, supercapacitor, radiation shielding, etc.

(continued)

Table 2 (continued)

Techniques	Process	Advantages	Limitations	Applications
Hot Pressing [32]	Process uses heat and pressure to densify and integrate carbon nanomaterials into a matrix material	The process produces excellent composites with good mechanical properties, suitable for metal and ceramic matrices	Process consumes high energy, and used for limited size parts. It has limited scalability	Used for fabricating aerospace structural components, turbine blades, and automobile components
Additive manufacturing (3D printing) [37]	Process involves layer-by-layer deposition of carbon-based composite material	Process can be used for fabricating complex geometries	Process requires specially formulated materials. Process is slower than conventional methods	Used for prototyping, complex shapes and custom medical implants
Microwave Sintering and Casting [38]	The process involves microwave energy to process the premixed powders to obtain nano/microcomposites	Process can be used for processing variety of composites with lower energy consumptions and lesser processing times	Difficult to process metallic-based materials and limited size. Difficult to scalability for larger components	Used for producing high-quality composites, with better structure-properties correlations. Low cost and sustainable solution for small components
Stir Casting [39]	The process involves melting of metallic matrix composites and adding reinforcements to obtain composites	Used for bulk metallic-based composites with a variety of reinforcements and matrix mainly aluminium based	The uniform dispersion of reinforcement is difficult and wettability of some reinforcements is an issue	The process is low-cost solution for bulk metallic composites

A lot of processes are available for processing carbon-based composites but still a lot of challenges exist for such composites. The major problem is with the dispersion and uniformity behaviour of carbon-based nanomaterials such as CNTs and graphene particles; which tends to agglomerate and tends to deteriorate the mechanical and electrical properties. Another important aspect is the poor wettability and poor bonding of nanoparticles with the matrix phase leading to interfaces with some defects. The high cost involved with the production of carbon-based nanoparticles and processing of nanocomposites limits the scalability of these materials in large-scale applications. Further, the production of carbon-based materials has a significant effect on environment and thus requires more recycling and sustainable fabrication methods for cost-effective and environment friendly solutions.

4 Future Directions in the Domain of Carbon-Based Composites

A lot of research is going on in the direction of sustainable manufacturing and processing of carbon-based lightweight composites for expanding their applications horizons. The future of carbon-based nanocomposites is focused in overcoming the challenges as explained in the previous paragraph and further expanding their application horizons towards the novel applications. The future trends are mainly in the following thrust area:

- Emerging trends shows that the researchers are working towards the hybrid composites where different nanomaterials are combined to harness new properties and applications. Researchers are using hybrid CNT-graphene composites for enhanced mechanical and electrical performance of the hybrid composite systems.
- Development of carbon-based smart materials which have the properties of self-healing, adaptive behaviour, and shape-memory for advanced engineering applications. The major advantage of these materials is combination of excellent properties with reduced weight. Such characteristics have opened a new era of smart materials for space exploration.
- The future trends are focused on sustainable production and fabrication of such composite material to lower the cost and enhanced their potential applications. A lot of work is now focused on recycling and reuse of such composites.
- The future trends show the application of carbon-based nanocomposites in medical implants, wearable sensors and energy storage devices. A lot of research is going on in these domains to harness the potential application of carbon-based materials again focusing on low-weight and high-strength characteristics.
- Work is going in the field of integration of AI and machine learning to optimize composite design, predict properties, and enhance the capabilities of such materials. A lot of work is in progress where carbon-based lightweight composites are explored for renewable energy systems, including wind turbines and solar panel support structures. Such materials are contributing to electric vehicle development by improving battery pack design and structural efficiency. Now the focus is on use of such materials in smart construction materials for self-monitoring and tracking health of infrastructures such as bridges and buildings which can produce warnings related to failure and can prevent fatal accidents.

5 Conclusions

This chapter has covered the potential applications, manufacturing aspects, and various materials systems that can be used for lightweight applications by using carbon-based nanocomposites. The major conclusions of the chapter are as follows:

1. Carbon-based nanocomposite materials such as carbon nanotubes (CNTs), graphene, and carbon nanodots (CNDs) have significantly contributed towards the lightweight properties with exceptional strength, thermal conductivity, and durability in many applications such as aerospace, structures, and machine components.
2. Carbon-based nanocomposite materials can be available with a variety of materials combination systems and exhibit diverse properties for potential applications across aerospace, automotive, biosensors, and electronics due to their lightweight and multi-functional properties. Research on hybrid systems combining different nanomaterials with different matrix can open new paths for enhancing performance and introducing novel functionalities.
3. There is a need for sustainable synthesis of such carbon-based nanomaterials and focus is now on green and sustainable processing of carbon-based nanocomposites for diverse applications. Researchers are working towards the recycling and reuse of such materials for improving the environmental aspects and introducing the circular economy among carbon-based materials.
4. Lightweight carbon-based materials and nanocomposites have immense potential as next-generation materials systems to be used for advanced applications. The continued research, development and innovation in synthesis, processing, and application will ensure the novel role of these materials in shaping the future of material science and industry. By addressing challenges and leveraging the emerging technologies such as AI and machine learning techniques, these composites will drive progress in creating sustainable, high-performance solutions.

References

1. Harussani, M. M., Sapuan, S. M., Nadeem, G., Rafin, T., & Kirubaanand, W. (2022). Recent applications of carbon-based composites in defence industry: A review. *Defence Technology, 18*(8), 1281–1300.
2. Patel, M., Pardhi, B., Chopara, S., & Pal, M. (2018). Lightweight composite materials for automotive—A review. *Carbon, 1*(2500), 151.
3. Wang, H., Li, S., Lu, H., Zhu, M., Liang, H., Wu, X., & Zhang, Y. (2023). Carbon-based flexible devices for comprehensive health monitoring. *Small Methods, 7*(2), 2201340.
4. Egbedina, A. O., Bolade, O. P., Ewuzie, U., & Lima, E. C. (2022). Emerging trends in the application of carbon-based materials: A review. *Journal of Environmental Chemical Engineering, 10*(2), 107260.
5. Zhang, J., Lin, G., Vaidya, U., & Wang, H. (2023). Past, present and future prospective of global carbon fibre composite developments and applications. *Composites Part B: Engineering, 250*, 110463.
6. Jung, Y., Cho, Y. S., Park, J. H., Cheon, J. Y., Lee, J. W., Kim, J. H., et al. (2023). Selective interbundle cross-linking for lightweight and superstrong carbon nanotube yarns. *Nano Letters, 23*(8), 3128–3136.
7. Li, L., Zhou, L., Zhu, J., Jian, M., & Zhang, J. (2024). Superstrong and tough silk fibers cross-linked with functionalized graphene. *Carbon, 226*, 119227.
8. Singh, K., Khanna, V., Singh, S., Bansal, S. A., Chaudhary, V., & Khosla, A. (2023). Paradigm of state-of-the-art CNT reinforced copper metal matrix composites: Processing, characterizations, and applications. *Journal of Materials Research and Technology, 24*, 8572–8605.

9. Singh, S., Gupta, D., & Kaushal, S. (2020). Dry sliding wear performance of Ni–SiC composites developed through an in situ microwave casting process. *Journal of Tribology, 142*(10), 101704.
10. Khanna, V., & Bakshi, B. R. (2009). Carbon nanofiber polymer composites: Evaluation of life cycle energy use. *Environmental Science & Technology, 43*(6), 2078–2084.
11. Nagaraju, S. B., Priya, H. C., Girijappa, Y. G. T., & Puttegowda, M. (2023). Lightweight and sustainable materials for aerospace applications. In *Lightweight and sustainable composite materials* (pp. 157–178). Woodhead Publishing.
12. Laraba, S. R., Luo, W., Rezzoug, A., Zahra, Q. U. A., Zhang, S., Wu, B., et al. (2022). Graphene-based composites for biomedical applications. *Green Chemistry Letters and Reviews, 15*(3), 724–748.
13. Juzenas, P., Kleinauskas, A., George Luo, P., & Sun, Y. P. (2013). Photoactivatable carbon nanodots for cancer therapy. *Applied Physics Letters, 103*(6), 063701.
14. Dannenberg, A. L., Burton, D. C., & Jackson, R. J. (2004). Economic and environmental costs of obesity: The impact on airlines. *American Journal of Preventive Medicine, 27*(3), 264.
15. Tom, M., Fischbeck, P., & Hendrickson, C. (2014). Excess passenger weight impacts on US transportation systems fuel use (1970–2010). *Journal of Transport & Health, 1*(3), 153–164.
16. Zhang, W., & Xu, J. (2022). Advanced lightweight materials for automobiles: A review. *Materials & Design, 221*, 110994.
17. Fan, Q., Duan, H., & Xing, X. (2024). A review of composite materials for enhancing support, flexibility and strength in exercise. *Alexandria Engineering Journal, 94*, 90–103.
18. Siwal, S. S., Zhang, Q., Devi, N., & Thakur, V. K. (2020). Carbon-based polymer nanocomposite for high-performance energy storage applications. *Polymers, 12*(3), 505.
19. Sudharsana, C., Anvarsha, N., & Kalyani, P. (2024). *Carbon-based nanocomposites: A comprehensive review of their multifunctional applications*. IntechOpen.
20. Iijima, S. (1991). Helical microtubules of graphitic carbon. *Nature, 354*(6348), 56–58.
21. Khan, M. E. (2021). State-of-the-art developments in carbon-based metal nanocomposites as a catalyst: Photocatalysis. *Nanoscale Advances, 3*(7), 1887–1900.
22. Byrne, M. T., & Gun'ko, Y. K. (2010). Recent advances in research on carbon nanotube–polymer composites. *Advanced Materials, 22*(15), 1672–1688.
23. Du, J., & Cheng, H. M. (2012). The fabrication, properties, and uses of graphene/polymer composites. *Macromolecular Chemistry and Physics, 213*(10–11), 1060–1077.
24. Markandan, K., Chin, J. K., & Tan, M. T. (2017). Recent progress in graphene based ceramic composites: A review. *Journal of Materials Research, 32*(1), 84–106.
25. Zhang, P., Su, J., Guo, J., & Hu, S. (2023). Influence of carbon nanotube on properties of concrete: A review. *Construction and Building Materials, 369*, 130388.
26. Huang, L., Yi, N., Wu, Y., Zhang, Y., Zhang, Q., Huang, Y., et al. (2013). Multichannel and repeatable self-healing of mechanical enhanced graphene-thermoplastic polyurethane composites. *Advanced Materials, 25*(15), 2224–2228.
27. Kumar, V., Alam, M. N., & Park, S. S. (2024). Review on functionalized CNTs reinforced silicone rubber composites for potential wearable applications. *Polymer Composites, 45*(14), 12503–12529.
28. Nugraha, A. D., Kumar, V. V., Gautama, J. P., Wiranata, A., Mangunkusumo, K. G. H., Rasyid, M. I., & Muflikhun, M. A. (2024). Investigating the characteristics of nano-graphite composites additively manufactured using stereolithography. *Polymers, 16*(8), 1021.
29. Nekahi, S., Vaferi, K., Nekahi, S., Vajdi, M., Sadegh Moghanlou, F., Baydogan, N., & Asl, M. S. (2023). Finned heat exchangers made of TiB2–SiC–graphene composites with enhanced heat transfer performance. *Journal of the Brazilian Society of Mechanical Sciences and Engineering, 45*(9), 497.
30. Rao, N., Singh, R., & Bashambu, L. (2021). Carbon-based nanomaterials: Synthesis and prospective applications. *Materials Today: Proceedings, 44*, 608–614.
31. Zhang, W. D., Shen, L., Phang, I. Y., & Liu, T. (2004). Carbon nanotubes reinforced nylon-6 composite prepared by simple melt-compounding. *Macromolecules, 37*(2), 256–259.

32. Ullbrand, J. M., Córdoba, J. M., Tamayo-Ariztondo, J., Elizalde, M. R., Nygren, M., Molina-Aldareguia, J. M., & Odén, M. (2010). Thermomechanical properties of copper–carbon nanofibre composites prepared by spark plasma sintering and hot pressing. *Composites Science and Technology, 70*(16), 2263–2268.
33. Singh, S., Gupta, D., Jain, V., & Sharma, A. K. (2015). Microwave processing of materials and applications in manufacturing industries: A review. *Materials and Manufacturing Processes, 30*(1), 1–29.
34. Mao, H. N., & Wang, X. G. (2020). Use of in-situ polymerization in the preparation of graphene/polymer nanocomposites. *New Carbon Materials, 35*(4), 336–343.
35. Faggio, G., Messina, G., Lofaro, C., Lisi, N., & Capasso, A. (2020). Recent advancements on the CVD of graphene on copper from ethanol vapor. *C, 6*(1), 14.
36. Lee, J. K. Y., Chen, N., Peng, S., Li, L., Tian, L., Thakor, N., & Ramakrishna, S. (2018). Polymer-based composites by electrospinning: Preparation & functionalization with nanocarbons. *Progress in Polymer Science, 86*, 40–84.
37. Hisham, M., Dileep, C., Jacob, L., & Butt, H. (2024). Additive manufacturing of carbon nanocomposites for structural applications. *Journal of Materials Research and Technology, 28*, 4674–4693.
38. Singh, S., Gupta, D., & Jain, V. (2016). Novel microwave composite casting process: Theory, feasibility and characterization. *Materials & Design, 111*, 51–59.
39. Kurt, H. I., Ergul, E., Basyigit, A. B., Oduncuoglu, M., & Yilmaz, N. F. (2023). Development and characterization of Al2024/magnesium oxide/CNTs/hybrid composites via stir casting method. *Proceedings of the Institution of Mechanical Engineers, Part E: Journal of Process Mechanical Engineering, 237*(5), 1854–1865.
40. Singh, S. (2022). Introduction to sustainability of manufacturing and materials processing technologies. In *Sustainable advanced manufacturing and materials processing* (pp. 1–8). CRC Press.

Carbon Nanocomposites for Stealth Technology: Innovations and Perspectives

Vijayeta Pal and Akhilesh Kumar Singh

1 Stealth Technology

Stealth technology represents a cutting-edge approach to reducing the detectability of objects, such as aircraft, ships, and vehicles, by radar, infrared, or other detection systems. They are also called as low-observable (LO) technology. It involves a variety of designing techniques and materials to be employed to produce vehicles or objects less visible by the waves like infrared, radar, sonar, and other exposure techniques. It's primarily used in military applications to enhance the survivability of aircraft, ships, and missiles by minimizing their visibility. The fundamentals of this field centered on developing advanced materials that effectively absorb or redirect electromagnetic waves, sound, and heat. Among these, carbon nanocomposites have emerged as transformative materials, offering excellent properties such as lightweight design, large electromagnetic absorption, thermally stable phenomenon, and mechanical strength. These innovations not only improve stealth performance but also enable multifunctional applications, paving the way for the next generation of adaptive and efficient stealth solutions.

V. Pal (✉) · A. K. Singh
School of Materials Science and Technology, Indian Institute of Technology, Banaras Hindu University, Varanasi, Uttar Pradesh, India
e-mail: vijayeta.idapt@itbhu.ac.in; aksingh.mst@iitbhu.ac.in

V. Khanna (ed.), *Carbon-Based Nanocomposites for Sustainable Applications, Volume I*, Lecture Notes in Nanoscale Science and Technology 38,
https://doi.org/10.1007/978-3-031-95306-4_8

2 Carbon Nanocomposites in Stealth Technology

These materials combine two or more distinct substances that remain separate within the final structure, resulting in enhanced properties. This engineered combination leads to materials with superior characteristics compared to their individual components. Composites are a combination of two or more dissimilar materials that stay in a unique form inside the final structure. They are better to traditional material structure due to their superior characteristics like as lightweight construction, elevated strength-to-weight ratio, enhanced corrosion resistance, durability, and the flexibility to alter thermal and electrical characteristics. Their personalized unique attributes and designing flexibility render them best suitable for innovative applications across many sectors [1, 2]. Moreover, carbon composites are ideal for stealth due to its better strength, lightweight quality, and effective radar absorption properties. They are vital for stealth technology in ships, cars, and aircraft because they usually made to offer thermal stability, structural integrity, and multifunctionality [3]. The number of present layers or walls, the size of the carbon material, the kind and density of flaws, and the chemical nature of functional groups on the carbon material's surface all greatly influence the structure and special qualities of these developed materials [4]. Therefore, they are having exceptional microwave absorption materials due to their superior dielectric qualities and logical structural design. They achieve good dielectric loss and produce remarkable microwave characteristics. Additionally, the carbon nanocomposite's low density contributes to the absorber's capabilities. Also, the surface of nanostructured carbon materials has a large number of crystal defects as well as unsaturated and dangling bonds, which greatly improves their interface polarization and multiple scattering [5]. In the past decades, numerous reports on carbon nanocomposite materials have highlighted their significance in stealth technology (Fig. 1).

3 Fundamentals of Carbon Nanocomposites

The word 'nanocomposite' specifically refers to a combination of compounds that are on the nano-meter size. The definition of nanomaterials has covered a wide range of systems, including one-dimensional, two-dimensional, three-dimensional, and amorphous materials. These material systems are made by combining distinct components mixed at the nano-meter scale. The properties of nanocomposite materials depend not only on the features of their individual components but also on their structure, morphology, and the spaces between them. Carbon nanocomposites are advanced materials formed by integrating carbon-derived nanostructures, like graphene, carbon nanotubes (CNT), or carbon fibres (CF), into a matrix material (metal, polymer, or ceramic). Their unique characteristics arise from the nanoscale dispersion of carbon components, which primarily enhance their mechanical

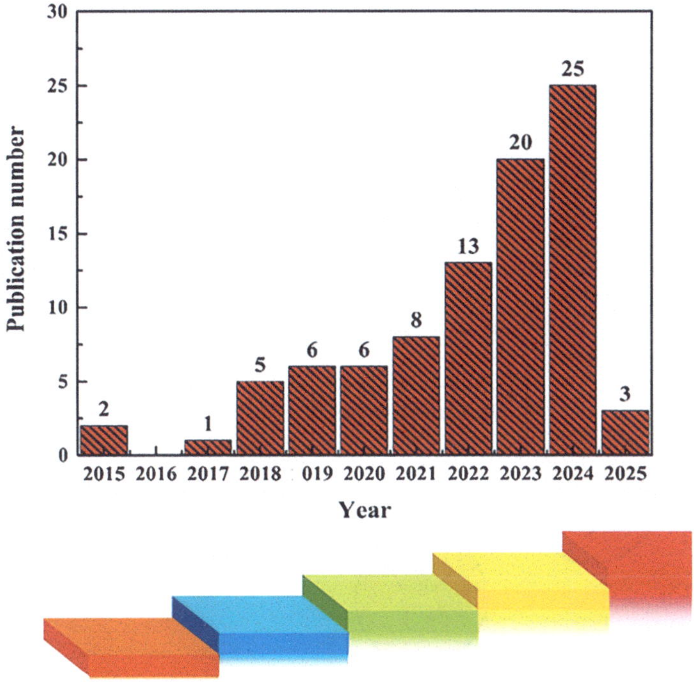

Fig. 1 Publication data of carbon nanocomposite materials for stealth technology. (From web of science (Jan 2025))

strength, chemical resistance electrical conductivity, and thermal stability features. The performance of these kinds of composites depends on factors like the type of carbon nanostructure, its interaction with the matrix, and the overall structure and distribution at the nanoscale. Because of their remarkable multifunctional qualities, they are broadly used in electronics, aerospace, energy storage, defence, and biomedical applications [1, 6].

3.1 The Categorization of Nanomaterials Based on Carbon

Carbon-based nanomaterials categorized into distinct types such as graphene and its derivatives, fullerenes, and single- and multi-walled carbon nanotubes, carbon nanofibers, amorphous carbon, carbon quantum dots, and diamond-like carbon, are briefly described in Fig. 2. Each type exhibiting unique structural and functional properties is briefly illustrated in the following paragraphs. They are showing

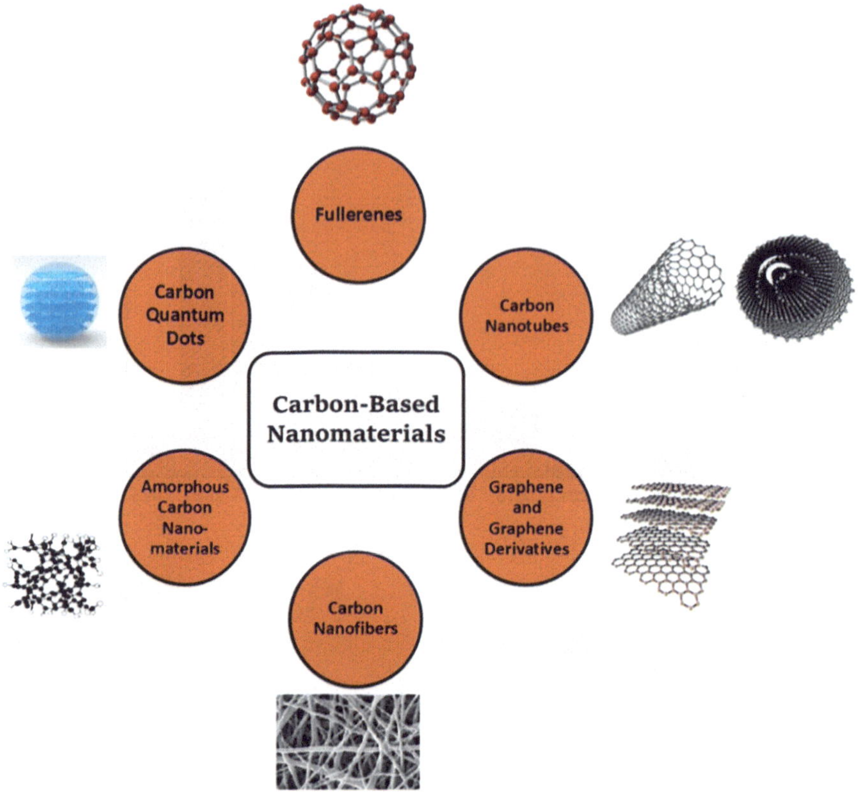

Fig. 2 Categorization of nanomaterials based on carbon

abundant applications across electronics, energy, medicine, and environmental technologies [6, 7].

3.1.1 Fullerenes

In general, fullerenes are a family of carbon-based nanomaterials made up completely by the carbon atoms systematically organized in the pentagonal and hexagonal structural patterns. They contain a distinctive spherical or cage-like structure. C60 is the most well-recognized fullerene, showing a soccer ball like configuration. Therefore, these zero-dimensional materials have outstanding electrical, optical, and chemical stability, which makes them suitable for use in the lubricants, photovoltaics, drug delivery, and superconductors. They are also beneficial in nanotechnology and medicinal field because of their special hollow nature, which allows them to enclose additional atoms or molecules.

3.1.2 Carbon Nanotubes (CNTs)

A novel category of carbon-based absorbent materials, carbon nanotubes (CNTs) have excellent mechanical strength, large conductivity, large aspect ratio, better high-temperature oxidation resistance, and adjustable dielectric properties. With average values of the elastic modulus in the order of 1 TPa and tensile strength in the range of 50–500 GPa, they show a number of advantages due to their capacity to alter the dielectric properties and their lightweight structures without losing mechanical qualities. More significantly, according to Gupta et al. [8], CNTs can be produced to yield good semiconducting characteristics with a current-carrying capacity of more than 109 A/cm and charge carrier mobility as high as 105 cm^{2}/V.s. This is particularly appealing for creating cost-effective and lightweight absorbers. Since carbon nanotubes are non-magnetic, their microwave absorbing properties comes from polarization, Ohmic losses, or multiple scattering due to their large specific surface area [9].

3.1.3 Graphene and Graphene Derivatives

Graphene and its derivatives have received a great attention among the technical organizations because of its excellent characteristics, for example, lightweight nature, zero band gap, large surface area, high electrical conductivity, large optical transmission, good mechanical stability, and large thermal conductivity, which catalysts rigorous interests in several fields. The first two-dimensional atomic allotrope of the element carbon, graphene, is composed of a single sheet of sp^2-bonded carbon atoms. It is a two-dimensional substance made up of a single layer of hexagonally latticed carbon atoms. Their adaptability for multifunctional usage (composites, sensors, energy storage, and biomedical) is increased by derivatives like graphene oxide (GO) and reduced graphene oxide (rGO), which add functional groups. They are essential to the development of electronics, nanotechnology, and renewable energy technologies [4, 6].

3.1.4 Nanofibers of Carbon (CNFs)

Graphene layers are usually layered in numerous orientations, with herringbone, platelet, and tubular, to formulated the cylindrical shaped nanostructures known as carbon nanofibers. Because of their special mechanical strength, electrical conductivity, and thermal stability, they are perfectly aligning for utilization in the multiple sectors like energy storage, composites, filtration systems, and catalysis. They show distinct shape and adjustable characteristics. Therefore, they may be customized into cutting-edge technologies similar high-performance electrodes and lightweight structures [4].

3.1.5 Amorphous Carbon Nanomaterials

They are made up of randomly distributed carbon atoms and lack a distinct crystalline shaped structure. They show a combination of sp^2 and sp^3 hybridized bonds, which provides them qualities like poor electrical conductivity, hardness, and good chemical stability. Carbon black and activated carbon are the great examples of them. They are frequently utilized in processes including energy storage, adsorption, catalysis, and composite reinforcement. They are suitable in a variety of sectors, including electronics and environmental technology, due to their adaptability [3].

3.1.6 Quantum Dots of Carbon

Generally, nanoscale carbon-based materials are called carbon quantum dots (CQDs). They often distinguished by their individual optical properties and intense fluorescence signature. Their diameters are typically less than 10 nm. Unusual features of these dots include biocompatibility, water solubility, high chemical stability, adjustable photoluminescence, and effective electron transport. Amorphous or somewhat crystalline in structure, CQDs are regularly functionalized with nitrogen (N) or oxygen (O) groups to rise their reactivity. They may be used in diverse applications due to their emission, which usually spans the visible to near-infrared spectrum range. The usage of CQDs in bioimaging field as fluorescent probes, sensing to identify contaminants or biomolecules, and to store energy in supercapacitors and batteries is extensive. They assist as carriers for targeted treatments in the distribution of medications, and they recover the procedures of water splitting and solar energy conversion in photocatalysis. They are also used in environmental remediation for wastewater treatment and optoelectronics in LEDs and displays. Because of their unique quantum-scale properties and the intrinsic nature of adaptability of carbon, CQDs are important materials for developing skill in a variety of domains. They have a low dielectric loss due to their outstanding dielectric qualities. Mainly, high graphitization levels in carbon-based nanomaterials result in better electrical conductivity, which raises the imaginary portion of the relative complex permittivity and produces excellent conduction loss. This naturally causes the dielectric loss of the carbon-based nanomaterials and enhances their electromagnetic microwave characteristics [4]. Moreover, the creation of carbon-based nanomaterials can inherit high specific surface area, porous, heterogeneous, and good interfacial polarization [6]. In the long run, carbon-based nanomaterials could develop comparatively powerful EM wave absorption capabilities. However, successful impedance matching is challenging because carbon-based nanomaterials have a low magnetic loss. Thus, it is usually challenging to achieve remarkable EM wave absorption efficiency for comparable absorber systems.

3.2 *Matrix Materials for Nanocomposites*

Matrix materials are the continuous phase in nanocomposites that hold the dispersed nanostructures, providing mechanical support, processability, and synergy with the functional properties of the fillers. For stealth applications, matrix materials are tailored to enhance electromagnetic wave absorption, mechanical durability, and environmental stability [10]. Figure 3 displayed a detailed classification on polymer, ceramic, and metal matrices, used in stealth technology.

3.2.1 Polymer Matrices

Polymer matrices are widely used because they are flexible, lightweight, and simple to process. They are particularly effective in applications requiring lightweight and conformal coatings. Low density, flexibility, electrical and dielectric tunability and corrosion resistance are some unique characteristics. A few examples of polymer matrices with their properties are given as below:

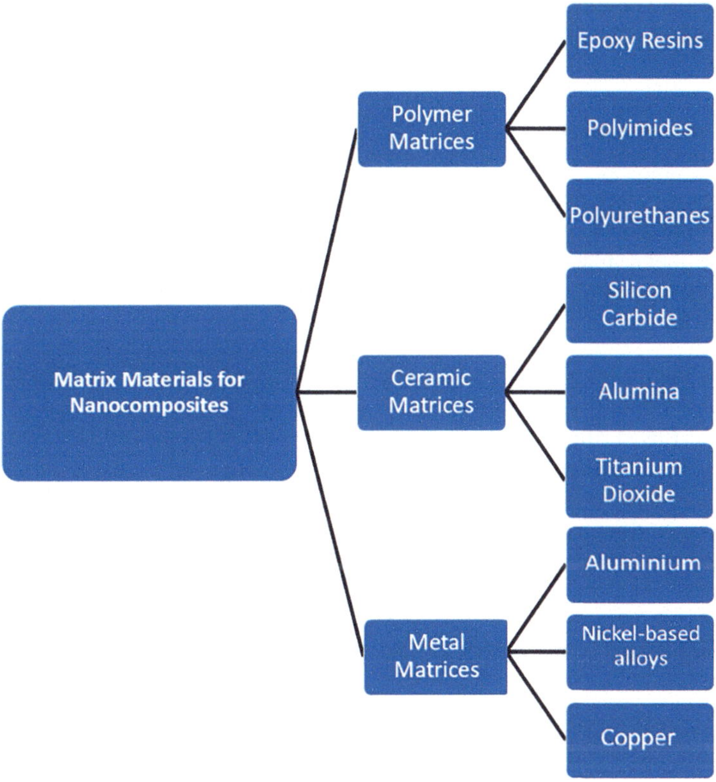

Fig. 3 Classification of nanocomposites using matrix materials

1. Epoxy Resins: High thermal and mechanical strength which is commonly used in radar-absorbing coatings.
2. Polyimides: Excellent thermal stability makes it suitable for high-temperature stealth applications.
3. Polyurethanes: Flexible and durable which often used for protective stealth coatings.

3.2.2 Ceramic Matrices

Ceramic matrices are used for their high-temperature resistance, mechanical hardness, and durability, making them ideal for extreme operational environments. The superior characteristics of ceramics matrices include thermal stability, high hardness, dielectric properties, and chemical stability. Some common examples of ceramic matrices are listed as below:

1. Silicon Carbide (SiC): It combines thermal conductivity with high mechanical strength which often used in high-performance stealth coatings.
2. Alumina (Al_2O_3): It offers superior hardness and dielectric properties.
3. Titanium Dioxide (TiO_2): They are effective as a dielectric material for radar wave absorption.

3.2.3 Metal Matrices

Metal matrices are primarily used in stealth applications requiring high electrical and thermal-specific conductivity, and mechanical robustness. The basic characteristics of metal matrices contain electrical conductivity, thermal conductivity, mechanical strength, and corrosion resistance. Most examples of metal matrices are shown as below:

1. Aluminium: They are lightweight and corrosion-resistant which makes them suitable for aerospace applications.
2. Nickel-Based Alloys: They are high-temperature resistance and electromagnetic properties, usually ideal for stealth coatings.
3. Copper: They contain excellent electrical and thermal conductivity which are frequently used in EMI shielding.

3.3 Key Properties Relevant to Stealth Applications

The main goal of stealth technology is to reduce detectability using a variety of techniques. The use of unique forms and materials to reduce the radar cross-section (RCS), absorbing radar signals with radar-absorbent materials (RAM) and suppressing infrared emissions to reduce heat visibility are the fundamental characteristics relevant to stealth applications. While visual stealth uses camouflage to blend in with the environment, acoustic stealth focuses on reducing noise. Sonar stealth employs anechoic coatings to evade detection underwater, whereas electromagnetic stealth seeks to minimize electronic transmissions. Each of these methods improves the capacity to work secretly and avoid detection.

A low radar cross-section (RCS) to limit wave reflection and high electromagnetic wave absorption to efficiently dissipate energy and reduce detectability are important characteristics pertinent to stealth applications. Covering a variety of frequency ranges, such as radar and infrared, requires broadband absorption. Particularly in aircraft applications, lightweight materials with excellent strength-to-weight ratios are crucial, and longevity guarantees performance in challenging environmental circumstances. Stealth performance is further improved by thermal management features like effective heat dissipation and suppression of infrared signatures. Optimized impedance matching is made possible by tunable permittivity and permeability, and consistent absorption from a variety of angles is guaranteed by wide angular performance. Eco-friendly, sustainable solutions are becoming more and more important in contemporary stealth technology, and multipurpose materials that combine stealth with structural or thermal features are becoming more and more popular (shown in Fig. 4).

Materials used in stealth technology must have particular qualities designed to lessen electromagnetic, acoustic, infrared, and visual signatures in order to successfully limit detectability. The following qualities are crucial for stealth applications.

3.3.1 Electromagnetic Properties: (Microwave Absorption Mechanism)

When microwaves hit blocking medium during propagation, they will reflect and transmit at the incident surface, as shown in Fig. 5a. If the impedance of the material is less than that of the air, the electromagnetic microwaves of the absorber system will be bounced outward [3, 11]. Microwaves are reflected more when the impedance mismatch is larger. Microwave radiation can effectively enter an absorber material system only when the electromagnetic impedance of the absorber matches that of the original propagation medium. Heat energy is produced as a result of interactions between the material and the microwaves travelling within it. Thus, impedance matching qualities, which define if microwaves may pass through a material's inside, and attenuation characteristics, which establish how well a substance can dissipate microwaves inside, define the MA performance of materials.

3.3.2 Strong Electromagnetic Wave Absorption

A strong electromagnetic wave absorption requires materials to efficiently absorb incoming electromagnetic waves across a broad frequency spectrum and convert the energy into heat or other non-reflective states. This process is mainly driven by several mechanisms including dielectric loss, magnetic loss, and impedance

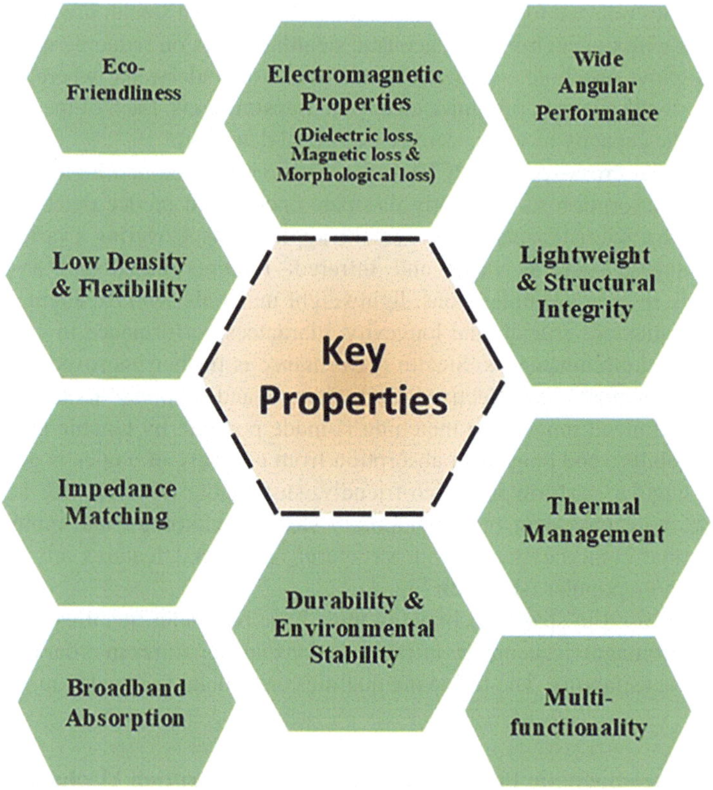

Fig. 4 Essential material characteristics boosting advanced stealth technology

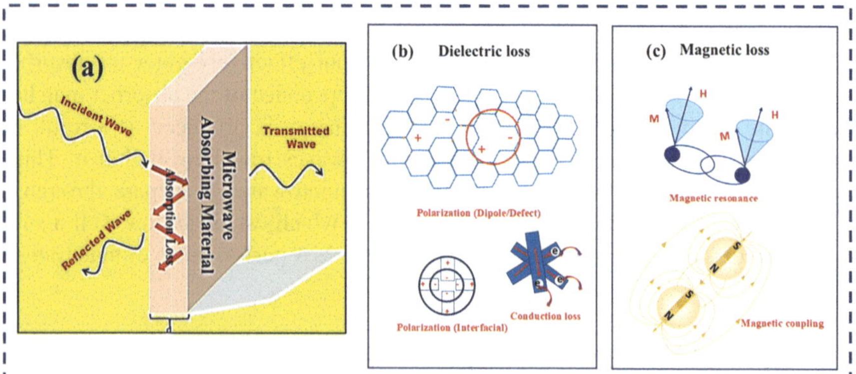

Fig. 5 (**a**) Schematic of EMW propagation on absorbing materials. (**b**) Dielectric loss mechanisms: interfacial polarization, dipole polarization, and charge carrier migration via conductive networks. (**c**) Magnetic loss mechanisms: natural resonance, magnetic exchange resonance, and eddy current loss

matching that significantly increase electromagnetic attenuation and absorption efficiency. The collective effect of magnetic and dielectric losses is the key for enhancing the material's ability to efficiently absorb microwave radiation. The capacity of a material to reduce electromagnetic wave reflections is generally determined by the interaction of many loss processes, which also affects the material's absorption efficiency. The material's electromagnetic properties, represented by complex permeability ($\mu_r = \mu' - j\mu'''$) and complex permittivity ($\varepsilon_r = \varepsilon' - j\varepsilon''$), generally define this performance. The real components (ε', μ') reflect the material's capacity to store electromagnetic energy, while the imaginary components (ε'', μ'') highlight its energy dissipation potential. Achieving precise balance in these parameters is the key for ensuring matching of impedance and maximizing the attenuation of incident.

3.3.3 Mechanism of Dielectric Loss

Since there is absence of magnetic particles in a single carbon-based microwave absorber (MA) material system, the primary type of loss that usually occurs within the material is dielectric loss, which includes conduction loss, interface polarization, dipole polarization, and defect-induced polarization. In general, conductive loss and polarization-relaxation loss are prime factor that make up the dielectric loss (Fig. 5b). Generally, the dielectric loss tangent ($\tan\delta = \varepsilon''/\varepsilon'$) is used to quantify a material's dielectric loss capability. A higher dielectric loss tangent indicates greater energy dissipation, making the material more effective for applications requiring significant microwave absorption. In conductive materials, the movement of carriers of charge (holes and electrons) generates due of its active electronic field's resistance, 'microcurrent', which is formed by the passage of charge carriers, such as electrons and holes in the conductive materials, which is then converted into heat. Electrical conductivity loss or conductive loss are terms used to describe this phenomenon. The electric energy is then consumed, and EMW absorption is accomplished. The imaginary part of permittivity is directly proportional to the electrical conductivity of the material, as per the free-electron theory (σ) of materials. The conductivity (σ) can be translated into the following formula: [12] $\varepsilon' = \sigma/2\pi\varepsilon_0 f$, where ε_0 is permittivity of vacuum. The material needs to be conductive in order to absorb electromagnetic waves (EMWs) effectively. In EMW absorption, there are mainly four types of loss of polarization-relaxation, i.e. (i) interfacial, (ii) dipole, (iii) ionic, and (iv) electronic [14, 13]. Interfacial polarization-relaxation takes place at the points of interaction of complex systems and further, at these interfaces, space charges build up or disperse unevenly in the presence of an electric field. The kind of polarization-relaxation shows a major part in EMW absorption and is common in composite materials or multiphase systems. The polarization resulting from the rotation (deflection or displacement) of the dipole moment (polar molecules, functional groups, etc.) in the direction of the alternating electronic field is responsible for the dipole polarization-relaxation process. The relative displacement of cations and anions in a high-frequency electric field is caused by ionic polarization-relaxation. It usually occurs in the frequency region of 10^3 to 10^6 GHz, when the ionic reaction is unable to keep up with the electric field's fast changes. However, electronic polarization-relaxation has to do with how an electric field causes electrons to move in relation to atomic nuclei. Like ionic polarization-relaxation, it takes place at extremely high frequencies and makes a very small contribution to the normal processes of EMW absorption. Ionic and electronic polarization-relaxation are frequently insignificant in our real-world EMW absorption since they only happen at very high frequencies. Thus, dipole type polarization-relaxation and its interfacial processes are the key causes of EMW absorption.

Debye theory states that energy dissipation results from polarization-relaxation losses, which happen when the polarization process cannot match the frequency of the alternating electric field. These losses are essential for improving the absorbing materials' EMW attenuation capacities [14, 15]. It can be represented as: $\varepsilon' = \varepsilon_\infty + \frac{(\varepsilon_s - \epsilon_\infty)}{1+(2\pi f)^2 \tau^2}$ and $\varepsilon'' = \frac{2\pi f(\epsilon_s - \varepsilon_\infty)}{1+(2\pi f)^2 \tau^2}$; where ε_s is static permittivity, ε_∞ is the permittivity at large frequency and τ indicates polarization-relaxation time. The modified connection associated between the real and imaginary part of permittivity of materials can be written as $\left(\varepsilon' - \frac{\varepsilon_s + \varepsilon_\infty}{2}\right)^2 + (\varepsilon'')^2 = \left(\frac{\varepsilon_s - \varepsilon_\infty}{2}\right)^2$, where a semicircle occurs with the centre of $[(\varepsilon_s + \varepsilon_\infty)/2, 0]$ and the radius of $(\varepsilon_s - \varepsilon_\infty)/2$ is usually called as Cole-Cole semicircle of the material describing the material's polarization-relaxation procedure in EMW absorber materials [16–18]. Especially, each point on the Cole-Cole semicircle signifies both the real part and imaginary part of the permittivity at exact frequency range. In practical investigation, a closer match between these permittivity components often results in a higher radius of the Cole-Cole semicircle of absorber material system.

3.3.4 Magnetic Loss Mechanism

Theoretical explanations for magnetic loss in material systems include domain wall resonance, eddy current loss, magnetic hysteresis loss, natural resonance, and exchange resonance; however, magnetic hysteresis loss is negligible, and the resonance related to domain wall only happens at low frequencies, well below the GHz spectrum that is relevant for microwaves [19]. In the microwave frequency range, natural resonance is the key phenomenon that contributes significantly to the total loss associated in magnetic material which generally responsible to absorb electromagnetic waves. Exchange resonance is the result of interactions between adjacent magnetic moments. Eddy current loss is caused by circulating currents caused by the changing magnetic field. Of them, magnetic attenuation is mostly caused by natural resonance, which can be mathematically modelled for further study.

3.3.5 Morphological Loss Mechanism

The capacity of material substances that absorb microwaves from the electromagnetic waves is significantly influenced by the shape and particle size. By altering the synthesis techniques, temperature, pressure, or doping materials, simple structures like spheres, rods, or flakes can develop into complex morphologies like flower-like or hierarchical porous structures. By changing the characteristics of the material, these sophisticated morphologies improve EMW absorption. Another important consideration is particle size. EMW interaction is enhanced when particle size is reduced to the nanoscale because of increased surface area, tunnelling effects, and special electrical and magnetic characteristics. Additionally, nanoscale particles increase active centres and decrease material density, which increases polarization-relaxation loss. If the size of the particles is below the depth of the skin, they decrease eddy current losses at high frequencies, improving absorption. EMW attenuation is further enhanced by the formation of monodomains by particles smaller than their critical size, which increase hysteresis loss because of restricted domain wall movement [20–22].

3.3.6 Impedance Matching

Impedance matching between the material and surrounding medium minimizes wave reflection at the material's surface, ensuring maximum energy penetration for absorption. The complex permeability and permittivity have a close relationship with the impedance matching of EMWs. The circumstance wherein: (i) the absorber's impedance value is necessary for optimal impedance matching when (i) the complex permittivity equals that of the complex permeability and (ii) it equals that of the open space. In such case, the incident EMWs' reflection is almost zero.

3.3.7 Tunable Permeability and Permittivity

For particular frequencies, the ability to adjust magnetic permeability (μ) and electrical permittivity (ε) assurances ideal impedance matching.

3.3.8 Low Radar Cross-Section (RCS)

The material should reduce the object's radar detectability by scattering as few electromagnetic waves back to the source as possible.

3.3.9 Electromagnetic Interference (EMI) Shielding

In order to avoid detection or signal disruption, the material should attenuate or block electromagnetic interference. Electromagnetic interference (EMI) shielding is the process of protecting electronic gadgets from unwanted electromagnetic radiation, which can cause malfunctions or data loss. It contains building EMI-preventing or EMI-absorbing barriers, typically made of conductive or magnetic materials. This process of shielding is essential for sensitive electronic devices to operate reliably and avoid interference from nearby devices. Metal screens, foils, and conductive coatings are frequently used materials.

3.3.10 Robustness and Ecological Reliability

For maritime applications, the developed materials must be able to withstand deterioration from corrosive environments including seawater and high humidity, temperature fluctuations, and UV radiation.

3.3.11 Broadband Absorption

A complete stealth coverage mostly requires the capability to absorb strong electromagnetic radiation in a variety of frequency spectrum (bands), such as radar (microwave) and infrared ranges.

3.3.12 Integrity of Structure and Lightweight

The ability of stealth materials is to tolerate external stressors including vibration, temperature changes, and mechanical loads, is often fixed by their structural integrity. Greater strength-to-weight ratios are necessary parameter for materials, especially in the sector of automotive and aerospace, where weight is a crucial feature.

3.3.13 Broad Angular Characteristics

To provide stealth from a variety of detection angles, the material must retain absorption effectiveness during a broad-spectrum range of incidence angles.

3.3.14 Thermal Control (Suppression of Infrared Signatures)

Materials must decrease heat emission to make them less visible by infrared sensors and other sensing devices. Low surface temperatures can be sustained with the use of effective thermal conductivity and dissipation procedures.

3.3.15 Multiple Functions

There is a cumulative demand for stealth materials that help numerous purposes, including electrical conductivity, thermal insulation, and structural support.

3.3.16 Microwave Absorbing (MA) Performance

In real-world applications, the reflection loss value (RL) or return loss is commonly used to evaluate the MA phenomenon of the material. Transmission line theory allows for the calculation of RL using the following formula: reflection los (RL) = 20 log 10$|Z_{in} - Z_0|$ $|Z_{in} + Z_0|$, where Z_{in} is the normalized input data of impedance of the MA material, while Z_0 is a constant magnitude that represents the input impedance parameter of any material in free space (377 Ω). Effective absorption bandwidth (EAB), as opposed to RL_{min}, is more convincing for practical applications, where MA materials must be able to minimize microwaves with a larger frequency spectrum as much as feasible. In general, it may be said that they efficiently convert more than 90% of the microwaves in a given frequency band when the magnitude of RL of any absorber materials fall below 10 dB; the range is known as EAB [20–23].

4 Advanced Innovations in Carbon Nanocomposites for Stealth

Carbon nanocomposites are at the forefront of stealth technology, offering groundbreaking solutions to absorb, deflect, or neutralize electromagnetic waves while ensuring lightweight, durable, and efficient materials. The blending of carbon-related nanostructures like graphene, CNTs, and quantum dots into polymer, ceramic, or metal matrices has significantly enhanced stealth capabilities by optimizing electromagnetic wave attenuation, thermal management, and mechanical integrity. It is not necessary that every carbon-based substance can be included into carbon-based nanocomposites for the absorption of electromagnetic waves. The exclusive electrical and structural characteristics that possibly contribute to the progress of carbon materials (graphene, carbon nanotubes, and porous carbon) are absent from many other carbon-based compounds. The prime features that make certain carbon materials suitable for producing superior nanocomposites for stealth applications are surface area, electrical conductivity, and structural qualities [3, 24]. Certain carbon compounds having unique special properties make them perfect for developing nanocomposites. The strong electrical conductivity of graphene and carbon nanotubes makes them excellent at absorbing electromagnetic radiation. By offering more surfaces for electromagnetic waves to interact with, materials with a high surface area, such as porous carbon, significantly increase absorption. Moreover, materials such as carbon nanotubes and graphene have special nanostructures that allow efficient energy dissipation, making them suitable for cutting-edge technology. Consequently, only materials with the required electrical, structural, and absorptive qualities are suitable for use in electromagnetic wave-absorbing nanocomposites, eventhough many carbon-based materials are useful for a variety of applications.

4.1 Advanced Materials Design

Advanced materials for stealth technology concentrate on multipurpose composites that control mechanical, thermal, and electromagnetic characteristics. When added to designed matrices, carbon nanostructures such as graphene, carbon nanotubes, and carbon quantum dots (CQDs) improve radar absorption, particularly in the 2–18 GHz range. In addition to lowering platform weight, these lightweight composites also improve mechanical durability under harsh environmental conditions, increase flexibility, and dissipate heat to reduce infrared fingerprints. Innovations such as eco-friendly components, adaptive systems for multi-spectral performance, and gradient materials for frequency optimization are making future defence technologies more efficient and environmentally sustainable.

4.2 Development of Composites Using Carbon

The fabrication of carbon composites is a new field of study that attempts to improve material qualities for a range of uses, such as environmental sustainability, energy storage, and aerospace. Researchers studied numerous recent progresses based on developing different hybrid composites, which combine graphene and carbon nanotubes (CNTs) in diverse matrices to form nanocomposites [1, 3, 23]. These developed composites are effective in applications like shielding against electromagnetic interference and structural components in stealth technologies due to their excellent electrical conductivity and mechanical strength. Together, graphene and CNTs expand the progress of high-performance materials for stealth technologies by enabling the customization of composite characteristics to resolve specific engineering challenges.

4.3 Carbon Nanomaterial Functionalization for Property Regulation

Functionalization is a necessary step to alter the properties of carbon nanomaterials to meet specific application requirements, especially in stealth applications. By tuning the surface chemistry of materials like graphene and carbon nanotubes, researchers may boost their radar-absorbing capabilities, making them essential for military and defence applications. One recent progress is the development of hybrid nanocomposites, which blend carbon-based materials with polymers and ceramics to optimize their absorption capability of incoming electromagnetic waves while keeping lightweight characteristics. Other significant improvement is the functionalization of carbon nanotubes with some magnetic materials, which greatly improve their capacity to absorb radar [4, 25]. This combination not only improves electromagnetic interference (EMI) shielding efficiency but also permits the development of flexible, lightweight coatings for stealth technology.

4.4 *Microwave Absorption Characteristics of Different Carbon-Based Composites*

4.4.1 Utilizing Biomass and Its Derivatives to Produce Carbon as Microwave Absorbers

According to S.T.A. Naqvi et al. [26], the structure and characteristics of carbon-derived biomass make it suitable for valuable absorber applications. They are allowing it to serve as an effective absorbing material candidate. As previously mentioned, a variety of loss processes are necessary for the suggested microwave absorption mechanism to function. Absorption bandwidth and intensity are two main metrics used to assess absorption performance. The parameter reflection loss (RL) is employed to calculate absorption intensity in the developed materials for stealth. Nine out of ten incident waves are absorbed when the reflection loss is −10 dB, and 99% of the incident waves are absorbed when the reflection loss is more than −20 dB. On the other hand, the absorbed bandwidth is determined by the frequency spectrum over which important reflection loss takes place. The absorption bandwidth increases with the range of reflection loss. Materials formed from biomass are further divided into subgroups as per their concept of loss mechanisms, which will be discussed in this section. The absorption performance metrics of carbon materials and its derivatives are assembled in Table 1.

Table 1 Absorption characteristics of carbon compounds obtained from biomass

Biomass	Absorption characteristics RL (dB); EAB (GHz); Thickness (mm)	References
Almond shell	−45.17; 5.5; 2.8	Fang et al. [27]
Spinach stem	−41.2; 1.5; 4.6	Wu et al. [28]
Pumpkin seeds	−29; 2.5; 7.4	Zhang et al. [29]
Corn stover	−57.1; 13.5; 3.25	Zhang et al. [30]
Watermelon	−37.2; 5.72; 2	Wang et al. [31]
Chicken feather	−44.6; 2.2; 1.68	Singh et al. [32]
Walnut	−31.8; 3.84; 1.5	Qiu et al. [33]

When biomass is transformed into carbon, its various porous structures don't change, resulting in various losses. When it comes to biomass carbon compounds' absorption capabilities, porosity is crucial. The material's porosity causes polarization when the incident wave strikes it, storing a significant amount of charge inside the material and reducing the applied wave's intensity. The signal is also attenuated by the numerous reflections, caused by the porous structure. Because of this, carbon compounds formed from biomass have greater reflection loss values due to their porosity. Fang et al. [27] created a specific biomass directly originated by porous carbon employing almond shells. They created carbon composites with various tiny shapes and achieved an RL of 45.17 dB at 5.5 GHz. Figure 6 displays the RL acquired for several materials. Using a direct carbonization process, Wu et al. [28] created porous carbons from the stems of spinach. These were extracted from spinach stems at different temperature such as 500 °C, 600 °C, 700 °C, and 800 °C. The porous carbon, known as HPC-S, was carbonized at 600 °C and had a 62.2 dB RL at 12.5 GHz. These results were obtained with an effective absorption bandwidth (EAB) of 7.3 GHz and a thickness of 2.71 mm.

The low dielectric loss in biomass derived from single material component results in poor absorption properties. Wide bandwidth and high absorption are extremely hard to achieve with a single system component, so loss mechanism medium is generally combined to create dielectric composites based on biomass. Multiple loss mechanisms, including conductive losses, cross-polarization, and polarization-relaxation, are provided by the use of such material, and these improve the absorption performance. Dielectric loss medium is mostly composed of carbon nanomaterials, silicon carbides, and conductive polymers. Additionally, carbon nanomaterials including MXene, reduced graphene, carbon nanotubes, and graphene have been utilized to create composites based on biomass. These substances enhance the performance of microwave absorption when mixed with biomass to provide superior impedance matching. Wang et al. [20–22] formed a CNT/cellulose aerogel at 550 °C, and the resultant composite had a reflection loss of 43.6 dB. Jiang et al. [34] designed MXene/cellulose aerogel in a similar way. Because MXene has a high conductivity, its inclusion enhanced the nature of conductive loss and dipolar polarization loss. The polarization of the interface between MXene and cellulose increased the dielectric loss. A reflection loss of 43.4 dB was attained, having a thickness of 2 mm and a frequency of 11.2 GHz.

4.4.2 Microwave Absorption Properties of Graphene and Its Composites

Graphene material systems mostly show high magnitude of dielectric constant. Among the carbon family, graphene is the thinnest material. It has great potential and considered as good microwave absorber materials. In general, graphene contains of a 2-D single layer of carbon atoms organized in a structure, hexagonal. The carbon-carbon (C–C) bonds show sp^2 hybridization, where the in-plane C–C bonds are amid the strongest connection in the specimen. Furthermore, the out-of-plane π bonds form a delocalized electronic network, which not only provides electronic

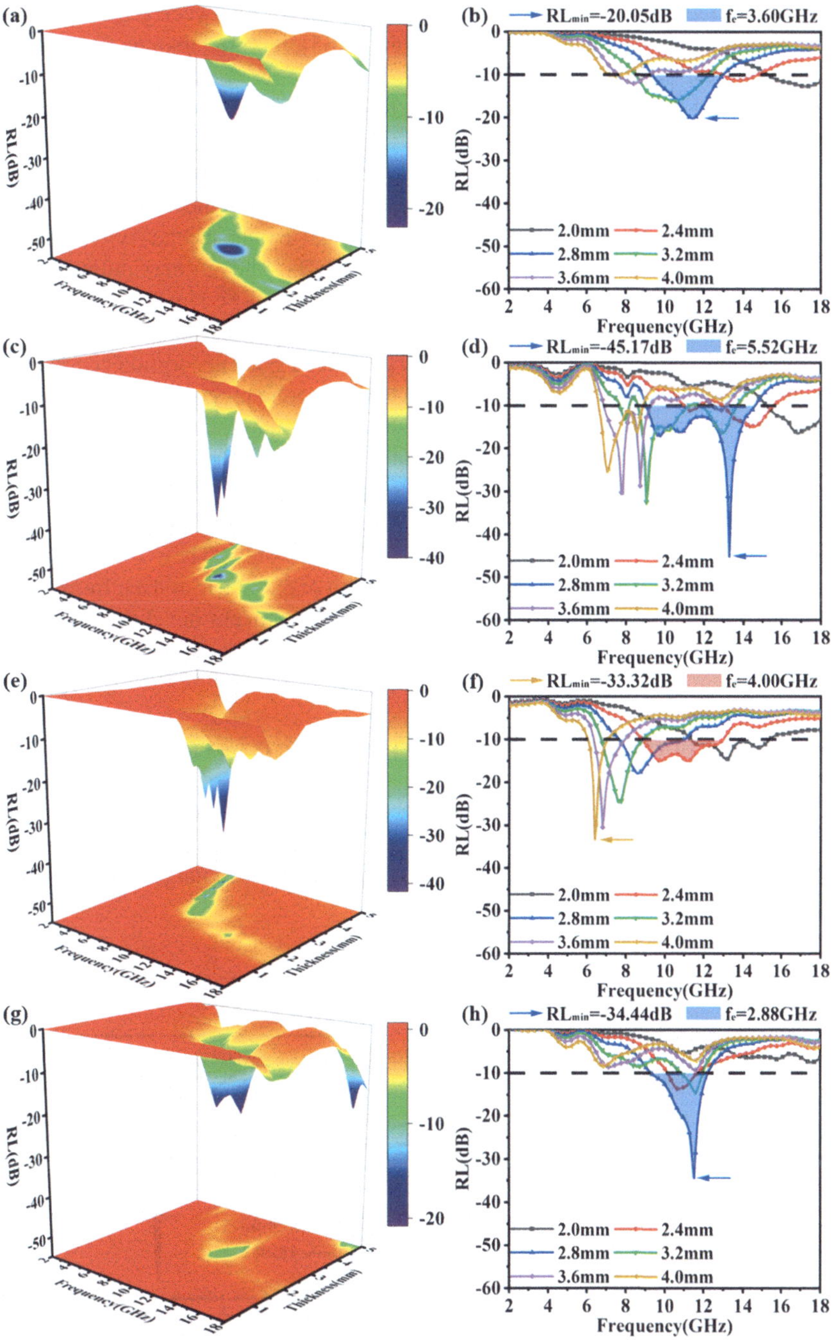

Fig. 6 Sample RL: (**a**) and (**b**) for PC-700, (**c**) and (**d**) for PC-800, (**e**) and (**f**) for CP-800, and (**g**) and (**h**) for PC-900. (Reproduced with approval from [27], Elsevier)

conduction in graphene but also permits weak interactions between graphene layers or between a substrate and graphene. Balci et al. [35], reported that the foam of GO showed inferior conductivity, then they showed a better shielding characteristics against the GO-film. This is attributed due to the microwaves experienced various reflection in the pores of the GO-foam.

Furthermore, graphene and magnetic components together had confirmed a possible candidate for superior microwave absorption performance. Ding et al. [36] discovered a unique combination of composites, i.e. Co nanosheets@rGO, showing an ideal bandwidth for absorption of 7.14 GHz and a maximum RL of −45.15 dB. Similar to this, Cao et al. [9] created Co_3O_4-rGO composites with an RL of −61 dB, claiming their performance was due to dipole and interfacial polarizations that could be adjusted. Wang et al. [37] achieved good absorption characteristics by designing a porous flower-like NiO@graphene composite. Improved impedance matching, magnetic loss, relaxation loss, and electrical loss were verified by a few graphene-ferrite-based composite materials, such as $NiFe_2O_4$-rGO [38], $CoFe_2O_4$/rGO [17, 18], $ZnFe_2O_4$-rGO [39], and $Ni_{0.4}Zn_{0.4}Co_{0.2}Fe_2O_4$-rGO [40]. To enhance the attenuation and impedance matching even further, Zhu et al. [41] discovered CoS_2@MoS_2@rGO ternary composites. They showed better absorption capabilities with an effective bandwidth of 6.24 GHz and a minimum RL of −58 dB (Fig. 7). These expansions validate the significant developments in the capability to absorb microwave radiation for graphene-based composite materials.

4.4.3 Properties of CNTs and Their Blends for Microwave Absorption

For improved microwave absorption (MA), carbon nanotubes (CNTs) have shown great promise due to their exclusive designs. The synthesis of CNTs involves several preparation techniques such as laser ablation, arc discharge, and chemical vapour deposition (CVD).

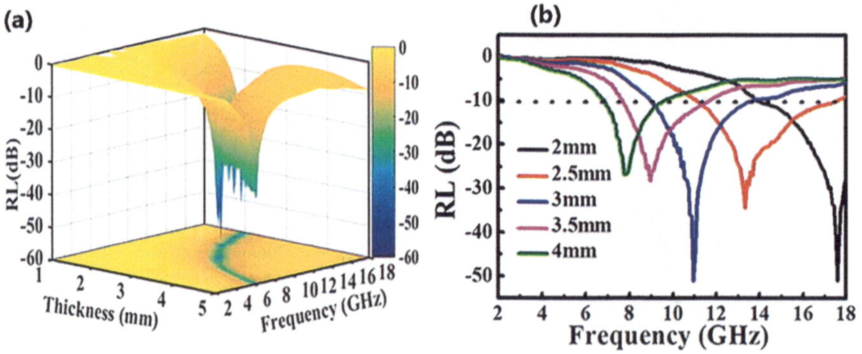

Fig. 7 (**a**) 3-D illustration and (**b**) reflection loss (RL) of the composite CoS_2@MoS_2/rGO with a 20% loading across the frequency spectrum of 2–18 GHz. (Reprinted with consent from [41])

CNTs are generally used as filler contents or go through metal catalysis derivation in MA. Non-magnetic CNTs serve as templates for functional materials and are typically dependent on dielectric loss. However, single MA loss shape and insufficient dispersion limit their efficiency. The performance of CNTs can be upgraded and the MA loss mechanism expanded by nanoparticle development and hydroxylation. The high-performance microwave absorption (MA) characteristics of virgin carbon nanotubes (CNTs) are limited due to short impedance matching. Impedance matching layers, which are usually made of magnetic nanoparticles or dielectrics like conductive polymers and transition metal oxides, are typically used to MA in order to further enhance it.

These layers improve MA characteristics by enhancing dipolar and interfacial polarization. For example, Tian and colleagues synthesized helical carbon nanotubes (HCNTs) with graphene capsules (GCs) on their exteriors. Resonance loss and a decrease in density of materials result from the adding of chiral factors, which greatly induce asymmetric polarization and boost cross-polarization. Figure 8a revealed that HCNTs at a 10 wt% mass ratio, showed an EAB of 6 GHz and an RL_{min} of −51.7 dB [42].

Further, it is quite hard to increase the effective absorption bandwidth (EAB) of carbon nanotubes (CNTs) as microwave absorber systems owing to their single loss phenomenon. In 2019, Ling et al. [44] formed a type of composite with a combination of CNTs with metal alloys, such as a three-layered type structure of carbonyl iron/CNTs/$Ce_2Fe_{17}N_{3-\delta}$, and attained a broad EAB of 12.6 GHz. Moreover, growing magnetic nanoparticles was attached with the surface of CNTs greatly expands the performance of MA due to magnetic losses like eddy current and hysteresis loss. Chen et al. [45] fabricated a ternary-based MA absorber material system combined of HWCNTs/Fe@Fe_3O_4 that was pretreated with the chemical compound H_2O_2 to boost dipolar polarization and enable the growth of Fe@Fe_3O_4 and EAB of 5.4 GHz is obtained. However, challenges like large conductivity and lack of magnetism could lead to impedance mismatch. Moreover, exact regulation over CNT diameter, length, and wall layers is crucial for modification efficiency and achieving stable, reproducible MA performance. Controlling the morphology of materials encapsulated within CNTs also impacts their absorption performance. By using melamine as the carbon source and Co particles as the catalyst, the generation of gas during the annealing process facilitates the formation of CNTs. The finally obtained product of cobalt-doped CNTs absorber composite material expands MA properties because of the large conduction loss and interface polarization, presented in CNTs. While the matching thickness of the developed materials was 2.1 mm, RL_{min} was −52.2 dB, and EAB was 4.56 GHz [46].

The basic phenomenon such as physical and chemical stability of CNTs catalyzed by MOFs and it remains intact during processing. Further, CNTs system resulting from annealing procedure ZIF-67 withstand some secondary sulfurization and creating a nitrogen-doped network in CNT and a layered type framework having a hollow shaped of CoS_2 particles. In reducing and inert atmospheres, secondary pyrolysis encourages the catalytic growth of Co-influenced CNTs, forming a networked structure on the framework's surface. This hollow structure helps to enhance

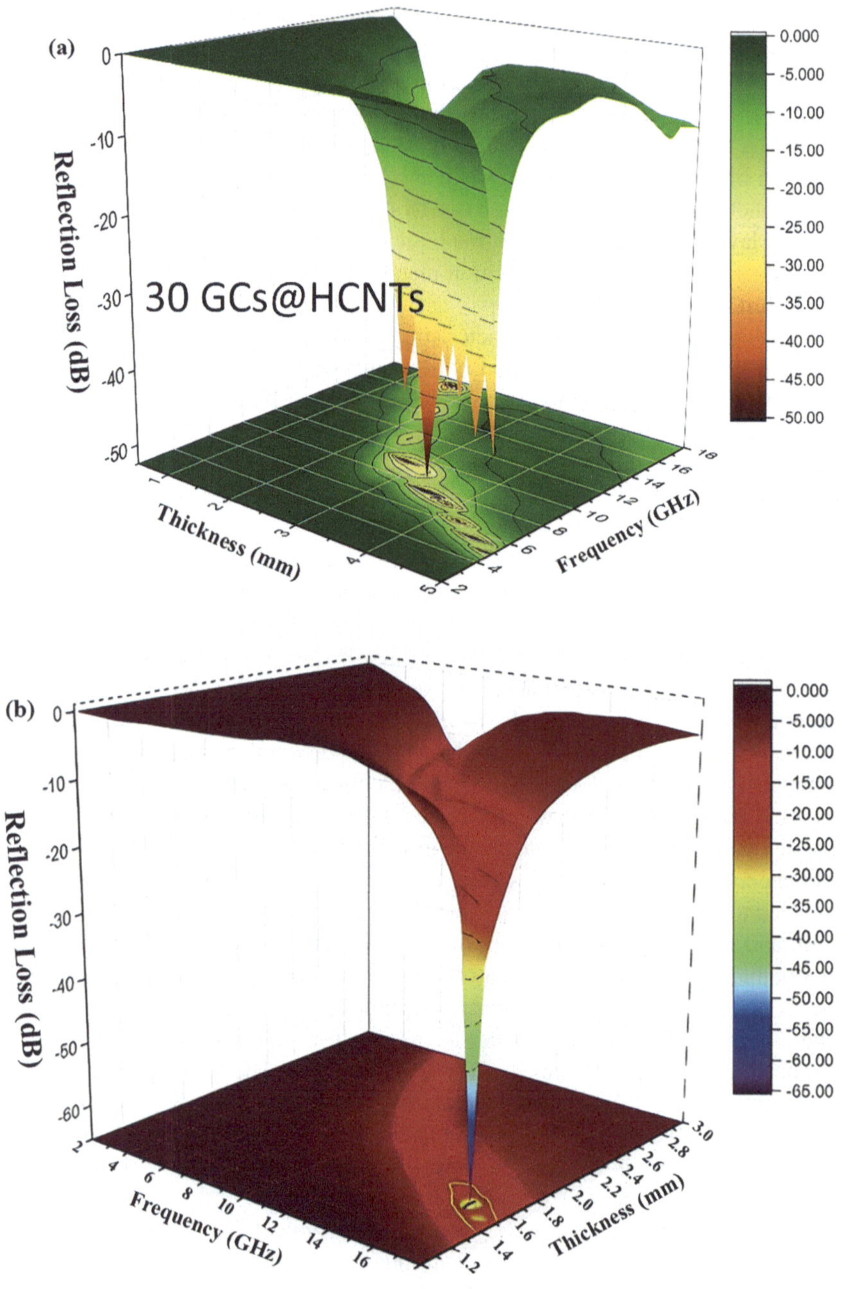

Fig. 8 (**a**) Shows 3-D reflection loss representations for GCs@HCNTs (reprinted from [42]) and CoS_2/NCNTs with thickness and frequency variation (**b**). (Reprinted with consent from [43])

impedance matching, boosting MA properties. At a particular thickness of 1.6 mm, the CoS_2/N-CNT composite accomplished an RL_{min} of −65 dB and an EAB of 6.2 GHz, depicts in Fig. 8b [43].

Investigation on CNT-based microwave absorption (MA) composite materials focuses on tuning the morphology and compositional design. With introducing defects in the materials and heteroatoms, they enhanced polarization and dielectric loss simultaneously, while structures like aerogels and microspheres improve dispersion and impedance matching, boosting EAB performance. However, different challenges remain, including complex compositions, unclear multi-component coordination mechanisms, and the need for advanced in situ characterization techniques. Moreover, hypothetical models are required for creating desired microstructures, such as core-shell and cyclic porous structures, in order to improve EAB functionality.

4.4.4 Microwave Absorption Characteristics of MOF-Derived Carbon Composites

Carbon materials formed from structures made of metal and organic, i.e. MOFs have received great attraction for MA because of their tunable nanostructures, which allow for control over defect polarization, magnetic losses, and interfacial polarization. MOFs are perfect starting materials for pyrolysis, which discovered carbon-based compounds by precisely adjusting their crystallographic, morphological, and chemical properties. The metal ions and ligands in the MOF, such as MIL (Cr-based) and Prussian Blue (Fe-based), affect the final composition. MOF derivatives expand its interface polarization, conduction losses, and impedance matching with their high conductivity, dipolar polarization, tunable graphitization, and superior dispersibility. These characteristics make them very promising and effective in microwave absorption applications.

Typically, monoatomic MOFs serve as templates for the manufacture of porous carbon composites comprising metal nanoparticles or oxides. Their synergistic outcomes between carbon shells and ferromagnetic cores, as well as their numerous magnetic/dielectric loss mechanisms, increase microwave absorption (MA). For enhanced MA characteristics, their structural and impedance matching properties are further tuned by annealing under inert gas and regulating various synthesis variables such as ligands and solvents. Using the condition of high-temperature pyrolysis of Ni-MOF in an inert atmosphere, Qiu et al. [47] were able to effectively produce a hollow nickel/carbon (Ni/C) microsphere. With an EAB of 5.1 GHz and an RL_{min} of −57.25 dB at a matched thickness of 1.8 mm, they confirmed remarkable MA characteristics.

The hollow shaped structure of carbon-nickel nanoparticles and their synergistic effect are key the point for the exceptional MA characteristics. Additionally, Cu-MOF@Co-MOF system precursors were thermally decomposed to produce core-shell and bimetallic type Cu/nitrogen-doped carbon@Co/nitrogen-doped carbon (Cu/NC@Co/NC) nanocomposites. The porous carbon skeleton boosted the criteria of impedance matching, and the dielectric loss was greatly reduced by changing the Co-to-Cu ratio. With an effective absorption bandwidth (EAB) of 5.19 GHz and an RL_{min} of −54.13 dB at 9.84 GHz, the developed composite accomplished notable microwave absorption features [48]. Further, a novel preparation method was employed by Shu et al. [49] to dope different multivariate MOF derivatives, such as FeCoNiMOFs, on to substrates like altered graphene nanocomposites using annealing process. The outstanding MA characteristics is attributed to the hollow structure of carbon-nickel nanoparticles and their synergistic mechanism. Further, composites were fabricated via thermal decomposition process of Cu-MOF@Co-MOF precursors. They showed a unique structure, having a core-shell and bimetallic Cu/nitrogen-doped carbon @ Co/nitrogen-doped carbon (Cu/NC@Co/NC) characteristics. By adjusting the Co-to-Cu ratio, the dielectric loss was greatly improved, and the porous skeleton of carbon boosted impedance matching. The composite reached excellent microwave absorption characteristics having an RL_{min} of −54.13 dB at 9.84 GHz and an effective absorption bandwidth (EAB) of 5.19 GHz [48]. Shu et al. [49] followed a unique approach in which annealing can be used to dope multivariate MOF derivatives, like FeCoNiMOFs, onto substrates like modified graphene composites. In graphene layers, the FeCoNi/C particles are evenly distributed as octahedra, and it becomes possible to tune the loss mechanism and MA functionality by adjusting the content of graphene oxide. They reported exceptional MA characteristics RL_{min} of −66 dB for 15.6 GHz with thickness of 1.53 mm and an EAB of 4.8 GHz.

4.4.5 Microwave Absorption Properties of Porous Type Carbon and Its Composites

In MA, porous carbon materials are highly valued for their outstanding qualities, movable apertures, and availability. In order to create porous biomass carbon, which is characterized by interfacial polarization, biomass materials are annealed in the inert gas. Because of its structure, it is possible that impedance matching is simply optimized and microwave reflectance is reduced. Zhou et al. [50] made porous carbon from fish skin, which is functionalized by nitrogen and oxygen. They reported an outstanding MA performance and achieved an EAB of 8.6 GHz (9.4–18 GHz) with 3 mm thickness and RL_{min} of −52.6 dB, as depicted in Fig. 9a, b.

Moreover, manganese oxide-added porous type carbon ((Mn_xO_y@C) absorber composites were synthesized by using two-step process. With notable dielectric characteristics, they reported low RL_{min} of −76.0 dB, having 5.2 GHz of EAB at a matching 2.0 mm thickness [51]. By maximizing impedance matching and enhancing microwave loss, biomass carbon's distinct structure and heteroatom doping improve MA performance. In addition to enhancing impedance matching properties, porous structures can increase the material's internal microwave transmission channel and withstand microwave losses. Designing particular gradient designs for high-performance applications and comprehending how pore size affects MA performance are still difficult challenges.

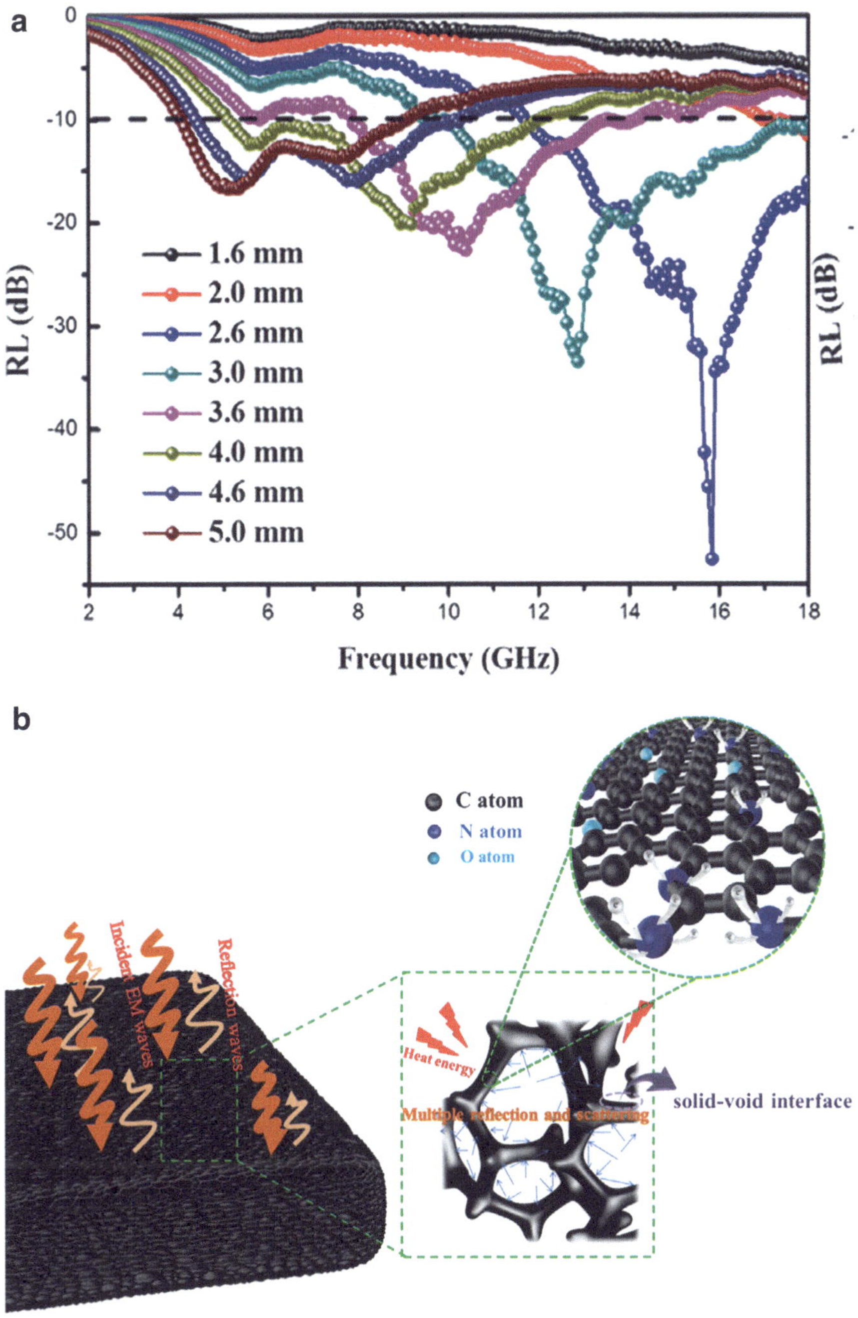

Fig. 9 (**a**) The magnitude of RL of CFs specimens at different frequency and thicknesses (**b**) schematic diagram of MA phenomenon of porous type carbon structures. (Reprinted with permission from [50])

4.4.6 Microwave Absorption Characteristics of Carbon Black and Its Composites

The imperfect pyrolysis or combustion process of hydrocarbons, which are mostly made up of carbons with trace quantities of hydrogen, oxygen, and sulphur, produces carbon black (CB), a black powder. Between 10 and 500 μm in size, CB particles are approximately spherical and have a propensity to combine in three dimensions. It is very promising for microwave absorption (MA) applications because of its key properties, which include small density, minor size, big surface part, low instability, and a well-designed morphology. To enhance the dielectric properties of matrix or fibres, CB can be used as a filler. After adjusting for Ni and CB content, CB/nickel hybrids on basalt fibres, achieved optimal MA functionality having EAB of 5.38 GHz and an RL_{min} of −35.29 dB at a thickness of 1.5 mm [52]. By regulating the content of flower-like Ni and CB, microwave absorbing performance was optimized in the composites. The accumulation of CB rises the variability of the surface and thereby improving both terms conductivity and polarization losses.

CB can also be used as a matrix to develop nanoparticles and materials with dielectric loss. Defect-rich carbon nanotubes (CNTs) can be formed on CB by chemical vapour deposition, which improves dielectric loss and produces a multilayer conductive network, hence disabling CB agglomeration and fabricating superior microwave absorbing performance [53]. CB showed better conductive nature. They easily control the matching characteristics of impedance of the composites and which makes it suitable for microwave absorbing applications.

5 Perspectives of Carbon Nanocomposites in Stealth Technology

The application of carbon nanocomposites in stealth technology has both great opportunities and challenges. Although these cutting-edge materials have better qualities, such as lighter weight and better absorption of electromagnetic waves, a number of problems need to be resolved before their full potential can be realized.

5.1 Current Challenges

For the fruitful integration of carbon-based nanocomposites into stealth application, a number of current problems must be addressed.

5.1.1 Performance and Cost-Effective Criteria

Todays, it is one of the major challenges in integrating carbon-based nanocomposites into stealth technology is striking an equilibrium between performance and cost. The production of good quality carbon nanomaterials, such as graphene and carbon nanotubes, is very costly and resource-intensive. For military and defence applications, where budgetary limitations are critical, the production of cost-effective manufacturing procedures that preserve the suitable performance attributes required for successful stealth capabilities is essential. Moreover, scalable production approaches and nanomaterial recycling strategies are being researched to lower these costs, but more research is necessary to make these technologies feasible for widespread applications [6, 54].

5.1.2 Health and Eco-Friendly Factors in Material Use

The use of carbon-based nanocomposites elevates serious environmental and health-related concerns. Due to the unique nature and properties of nanoparticles, their production and disposal can be hazardous and may have unforeseen adverse impacts on human health or the environment. Carbon nanotube inhalation, for instance, has been largely associated to respiratory concerns, which increases questions about safety protocols during manufacture, production, and use. Further, researchers [25] and other many coworkers are working harder in order to develop safer alternative options and conduct thorough assessments of these material's environmental effects, including life cycle analysis and potential harmfulness.

5.2 Trends for the Coming Years

In general, the field of stealth technologies stands to benefit significantly from the further progress of carbon-based nanocomposites. Several emerging trends have the possibility to alter military applications and increase the effectiveness of stealth technology. The current development of better radar-absorbing material systems, which combine many nanomaterials to enhance the absorption of electromagnetic waves across a wide frequency range, is one of major significant trend. Additionally, studies are being conducted on flexible and intelligent materials that could respond to environmental stimuli. In order to evade detection attempts, these developed materials might enable stealth devices to change their properties in real time. Also, continuous efforts are being made to deliver ecologically friendly alternatives that meet performance standards while lessening their adverse consequences, signifying the growth importance of sustainability in RAM systems [55].

In recent times, cutting-edge industrial technologies, such as three-dimensional (3-D) printing, may support the formation of complex new stealth-related structures while simultaneously optimizing material systems usage and improving the effectiveness of nanocomposites. Furthermore, the incorporation of artificial intelligence (AI) into the context of designing the material system is very similar to inspire revolution and enable the faster production of newly designed nanocomposite systems that require precise stealth characteristics. In the conclusion, they all suggest that stealth technology will certainly advance in the future to become more effective, adaptable, and environmentally conscientious. Industry assessments and research that have been released recently provide more details on these changes.

5.2.1 Sustainable and Bio-inspired Nanocomposites

In materials science and engineering, sustainable and bio-inspired nanocomposites are becoming more and more widespread in due to their ability to mimic natural processes while decreasing their adverse environmental significances. Additionally, nanomaterials like cellulose or chitin are frequently combined with renewable resources, such as plant biopolymers in order to improve the mechanical strength and promote biodegradability of these innovative materials. For instance, surfaces modelled after lotus leaves are impenetrable to water and dirt, while in the case of fibres especially natural fibres produce strong, lightweight materials that utilized in the packing and buildings. The prime goal of research is to design and optimize these materials for specific type applications, like antibacterial and thermal stability, while maintaining their environmental advantages. Future progresses will most likely focus on recycling waste and utilizing ecologically friendly manufacture approaches, which will surely lead to a greater use of sustainable and bio-inspired nanocomposite systems.

5.2.2 Materials with Multiple Usage for Multi-Spectrum Exposure

Materials with multiple usage are well known as multifunctional materials. They are key materials for multi-spectrum range stealth technology since they improve the concealment of military weapons against numerous detection approaches, including radar, infrared, and acoustic sensors. These innovative materials can effectively regulate heat signatures, lower radar cross-sections, and lower noise by integrating many other functions into a single composite complex structure. For the illustration, using elements that effectively absorb radar, such as conductive polymers and carbon-based nanomaterials, helps prevent hostile radar exposure. In thermal management, these materials reflect infrared radiation, which greatly helps to hide heat releases from vehicles and aircraft.

In the recent developments, they have included hybrid nanocomposites, which combine different form of carbon materials with polymers and ceramics to maximize performance across a range of spectrums although keeping the lightweight characteristics essential for aeronautical applications. The flexibility of stealth material systems is increased by the ability to exactly produce complex structures employing some latest manufacturing technologies, such as 3-D printing. Globally, the ongoing research and development of multifunctional material system has the potential to significantly enhance the effectiveness of stealth technology in defence applications.

5.2.3 Quantum Technology's Function in Stealth Breakthroughs

Quantum technologies have the potential to completely transform stealth capabilities in military applications. Quantum sensing, utilizing the ideas of quantum mechanics to identify the smallest variations in motion, temperature, and electromagnetic fields, provides unprecedented sensitivity. By providing exact situational awareness without relying on traditional tracking techniques, this enhanced detection capacity can significantly improve stealth systems and make it more difficult for adversaries to detect military operations.

Furthermore, the distribution of quantum keys, which is highly resistant to eavesdropping, enables secure information transfer in quantum communication systems. This feature is crucial for operational security since it protects important data from being intercepted during stealth missions.

Also, quantum computing has the capability of enhancing material design and simulations, which would facilitate the fabrication of more complicated nanomaterials with exceptional stealth qualities. These materials can enhance overall stealth performance by efficiently reducing radar signatures and regulating heat emissions. In conclusion, there are a number of advantages to using quantum technologies in stealth applications, including as enhanced sensing, secure communication, and better material design. These advantages will ultimately reshape the effectiveness of stealth operations in a military context.

5.3 Application Potential

Carbon nanocomposites have been employed in numerous applications in stealth technologies. Carbon nanocomposites have an extensive spectrum of potential for developing stealth technology across many platforms and applications, as shown in Fig. 10. In the aviation industry, they are integrated into airplane frames for radar and infrared stealth, enabling lightweight, efficient electromagnetic wave absorption as well as the temperature management. These materials' durability makes them ideal for underwater applications, and they reduce the radar and sonar signatures for military stealth. They are additionally employed for electromagnetic interference (EMI) shielding to protect delicate electronics, particularly for high-frequency applications. In multi-spectral circumstances, infrared camouflage boosts thermal stealth by obscuring heat signatures. These materials are employed as well to create flexible, lightweight, and adaptable protection in wearable military gear. Modern adaptive structures can effectively respond to their surrounding

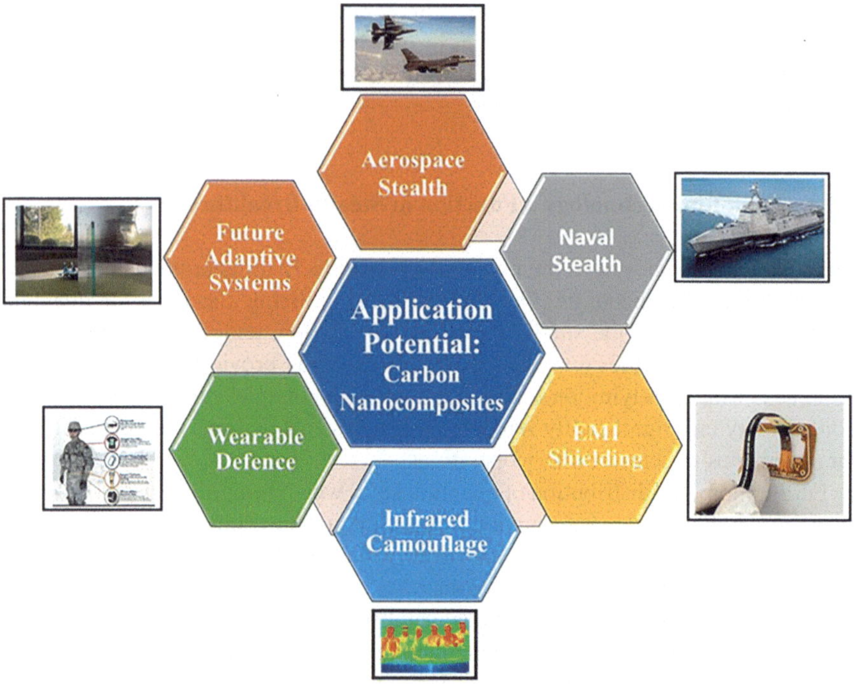

Fig. 10 Potential use of carbon nanocomposites in stealth technology applications

dynamical environment. They also offer a greater opportunity for application in both military and commercial technological fields in the future.

6 Overview of Developments Along with Potential Effects

Recent advancements in innovative materials, such as carbon-based nanocomposites, gradient structures, and adaptive systems, have completely transformed stealth and defence technologies. These materials allow for applications in aerospace, next-generation stealth platforms, and EMI shielding by offering exceptional electromagnetic wave absorption, lightweight properties, and multifunctional capabilities. Innovations like heat management systems and eco-friendly components ensure sustainability while enhancing performance. By offering broader frequency coverage, reducing environmental impacts, and enhancing durability in challenging conditions, these advancements significantly enhance both military and civilian technologies, paving the way for more effective and sustainable defence systems. By pushing these materials beyond current standards, innovative manufacturing methods are securing their role in the future of military technology. Additionally, their potential extends to commercial applications in telecommunications, aerospace, and automotive industries, creating avenues for groundbreaking solutions.

Nanotechnology and hybrid materials present lighter, more cost-effective, and efficient options for electromagnetic interference (EMI) shielding. The increasing need for compact shielding in smaller devices further underscores their significance. In general, carbon nanocomposites are redefining stealth and EMI shielding technologies. They provide outstanding performance and a broad array of applications, suggesting a bright future for both military and civilian sectors.

7 Conclusions and Future Outlook

Carbon-based nanocomposites, which provide unmatched electromagnetic properties, lightweight designs, and multifunctional capabilities, have made considerable strides in the field of microwave absorption in recent years. These substances have shown remarkable promise in stealth and electromagnetic interference shielding technologies. These materials include graphene, carbon nanotubes, and porous carbon. Performance has been further improved over a variety of frequency ranges while the environmental impact has been decreased that makes an important breakthrough such gradient structures for broadband absorption, adaptive systems for environmental responsiveness, and eco-friendly components. Multi-spectral stealth systems that can function across radar, infrared, and optical spectra will be the main focus of future advancements. Enhancing stealth performance will be largely dependent on dynamic and reconfigurable materials that adjust to shifting circumstances. The design of greener materials will be influenced by recyclable nanostructures and bio-based polymers, which will also prioritize sustainability. Utilizing developments in material science, computer modelling, and manufacturing technologies, a multidisciplinary approach will be necessary to address issues including improving impedance matching, increasing absorption bandwidth, and attaining scalable production. These innovations hold promise for revolutionary advancements in electronic devices, naval platforms, and aircraft, among other defence and civilian applications. Materials will transform current stealth systems and open the door to a safer, more sustainable technological future as they grow more effective, versatile, and environmentally friendly.

Acknowledgements The author (V. Pal) acknowledges the I-DAPT Hub Foundation for funding the DAPT Chanakya Postdoctoral Fellowship at IIT (BHU), Varanasi.

References

1. Ahmad, H., Tariq, A., Shehzad, A., et al. (2019). Stealth technology: Methods and composite materials—A review. *Polymer Composites, 40*(12), 4457–4472.
2. Pal, V., Singh, A. K., & Swain, P. K. (2023). New developments in piezoelectric polymeric composite materials for environmental energy usages. In *Novel polymeric materials for environmental applications* (pp. 411–456). World Scientific Publishing.

3. Zhang, F., Li, N., Shi, J.-F., et al. (2024). Recent progress on carbon-based microwave absorption materials for multifunctional applications: A review. *Composites Part B: Engineering, 283*, 111646.
4. Li, J., Zhou, D., Wang, P.-J., Du, C., et al. (2021). Recent progress in two-dimensional materials for microwave absorption applications. *Chemical Engineering Journal, 425*, 131558.
5. Wu, Y., Tan, S., Zhao, Y., et al. (2023). Broadband multispectral compatible absorbers for radar, infrared and visible stealth application. *Progress in Materials Science, 135*, 101088.
6. Kim, S.-H., Lee, S.-Y., et al. (2023). Carbon-based radar absorbing materials toward stealth technologies. *Advancement of Science, 10*, 2303104.
7. Devi, S., Pal, V., Yadav, N., et al. (2024). *Electrochemical exfoliation of graphene and its derivative*. Springer.
8. Gupta, A., Sakthivel, T., & Seal, S. (2015). Recent development in 2D materials beyond graphene. *Progress in Materials Science, 73*, 44–126.
9. Cao, M., Wang, X., Cao, W., et al. (2018). Thermally driven transport and relaxation switching self-powered electromagnetic energy conversion. *Small, 14*, 1800987.
10. Fu, Z., Pang, A., Luo, H., et al. (2022). Research progress of ceramic matrix composites for high temperature stealth technology based on multi-scale collaborative design. *Journal of Materials Science and Technology, 18*, 2770–2783.
11. Gunwant, D., & Vedrtnam, A. (2021). Microwave absorbing properties of carbon fiber based materials: A review and prospective. *Journal of Alloys and Compounds, 881*, 160572.
12. He, G., Duan, Y., & Pang, H. (2020). Microwave absorption of crystalline Fe/MnO@ C nanocapsules embedded in amorphous carbon. *Nano-Micro Letters, 12*, 1–16.
13. Liang, L., Li, Q., Yan, X., Feng, Y., Wang, Y., Zhang, H.-B., et al. (2021). Multifunctional magnetic Ti3C2T x MXene/graphene aerogel with superior electromagnetic wave absorption performance. *ACS Nano, 15*(4), 6622–6632.
14. Singh, P. P., De, A., & Khatua, B. B. (2024). Bio-inspired smart composite architecture for thermally tunable green EMI shielding. *Applied Surface Science, 643*, 158643.
15. Zhdanov, M. (2008). Generalized effective-medium theory of induced polarization. *Geophysics, 73*(5), F197–F211.
16. Lv, H., Guo, Y., Wu, G., et al. (2017). Interface polarization strategy to solve electromagnetic wave interference issue. *ACS Applied Materials & Interfaces, 9*(6), 5660–5668.
17. Liu, W., Tan, S., Yang, Z., & Ji, G. (2018b). Hollow graphite spheres embedded in porous amorphous carbon matrices as lightweight and low-frequency microwave absorbing material through modulating dielectric loss. *Carbon, 138*, 143–153.
18. Liu, Y., Chen, Z., Zhang, Y., et al. (2018a). Broadband and lightweight microwave absorber constructed by in situ growth of hierarchical $CoFe_2O_4$/reduced graphene oxide porous nanocomposites. *ACS Applied Materials & Interfaces, 10*, 13860–13868.
19. Pang, H., Duan, Y., Huang, L., et al. (2021). Research advances in composition, structure and mechanisms of microwave absorbing materials. *Composites Part B: Engineering, 224*, 109173.
20. Wang, Y. Y., Zhou, Z. H., Zhu, J. L., Sun, W. J., et al. (2021c). Low-temperature carbonized carbon nanotube/cellulose aerogel for efficient microwave absorption. *Composites Part B: Engineering, 220*, 108985.
21. Wang, Y.-Y., Sun, W.-J., Dai, K., et al. (2021a). Ultralight carbon nanotube/graphene/polyimide foam with heterogeneous interfaces for efficient electromagnetic interference shielding and electromagnetic wave absorption. *Carbon, 176*, 118–125.
22. Wang, B., Wu, Q., Fu, Y., & Liu, T. (2021b). A review on carbon/magnetic metal composites for microwave absorption. *Journal of Materials Science and Technology, 86*, 91–109.
23. Li, N., Shi, J.-F., Zhang, F., Jia, L.-C., Wang, Y.-Y., Yan, D.-X., et al. (2024). Peelable microwave absorption coating with reusable and anticorrosion merits. *ACS Applied Materials & Interfaces, 16*(4), 6462–6473.
24. Sharma, S., Parne, S. R., Panda, S. S., Sripada, G., & Suman. (2024). Progress in microwave absorbing materials: A critical review. *Advances in Colloid and Interface Science, 327*, 103143.
25. Jiang, B., Qi, C., Yang, H., Wu, X., Yang, W., et al. (2023). Recent advances of carbon-based electromagnetic wave absorption materials facing the actual situations. *Carbon, 208*, 390–409.

26. Naqvi, S. T. A., et al. (2023). Functionalization and synthesis of biomass and its composites as renewable, lightweight and eco-efficient microwave-absorbing materials: A review. *Journal of Alloys and Compounds, 968*, 171991.
27. Fang, X., Li, W., Chen, X., et al. (2022). Controlling the microstructure of biomass-derived porous carbon to assemble structural absorber for broadening bandwidth. *Carbon, 198*, 70–79.
28. Wu, Z., Tian, K., Huang, T., Hu, W., et al. (2018). Hierarchically porous carbons derived from biomasses with excellent microwave absorption performance. *ACS Applied Materials & Interfaces, 10*, 11108–11115.
29. Zhang, D., Liu, T., Cheng, J., et al. (2019a). Light-weight and low-cost electromagnetic wave absorbers with high performances based on biomass-derived reduced graphene oxides. *Nanotechnology, 30*, 445708.
30. Zhang, Z., Zhao, H., Gu, W., Yang, L., & Zhang, B. (2019b). A biomass derived porous carbon for broadband and lightweight microwave absorption. *Scientific Reports, 9*, 18617.
31. Wang, N., Wu, F., Xie, A. M., Dai, X., Sun, M., Qiu, Y., Wang, Y., Lv, X., & Wanga, M. (2015). One-pot synthesis of biomass-derived carbonaceous spheres for excellent microwave absorption at the Ku band. *RSC Advances, 5*, 40531–40535.
32. Singh, S. K., Prakash, H., Akhtar, M. J., & Kar, K. K. (2018). Lightweight and high-performance microwave absorbing heteroatom-doped carbon derived from chicken feather fibers, ACS sustain. *Chemical Engineer, 6*, 5381–5393.
33. Qiu, X., Wang, L., Zhu, H., Guan, Y., & Zhang, Q. (2017). Lightweight and efficient microwave absorbing materials based on walnut shell-derived nano-porous carbon. *Nanoscale, 9*, 7408–7418.
34. Jiang, Y., Xie, X., Chen, Y., Liu, Y. J., Yang, R., & Sui, G. X. (2018). Hierarchically structured cellulose aerogels with interconnected MXene networks and their enhanced microwave absorption properties. *Journal of Materials Chemistry C, 6*, 8679–8687.
35. Balci, O., Polat, E. O., Kakenov, N., & Kocabas, C. (2015). Graphene-enabled electrically switchable radar-absorbing surfaces. *Nature Communications, 6*, 6628.
36. Ding, Y., Zhang, Z., Luo, B., et al. (2016). Investigation on the broadband electromagnetic wave absorption properties and mechanism of Co_3O_4- nanosheets/reduced-graphene-oxide composite. *Nano Research, 10*, 980–990.
37. Wang, L., Xing, H., Gao, S., et al. (2017). Porous flower-like NiO@graphene composites with superior microwave absorption properties. *Journal of Materials Chemistry C, 5*, 2005–2014.
38. Zhang, Y., Wang, X., & Cao, M. (2018). Confinedly implanted $NiFe_2O_4$-rGO: Cluster tailoring and highly tunable electromagnetic properties for selective-frequency microwave absorption. *Nano Research, 11*, 1426–1436.
39. Yang, Z., Wan, Y., Xiong, G., Li, D., et al. (2015). Facile synthesis of $ZnFe_2O_4$/reduced graphene oxide nanohybrids for enhanced microwave absorption properties. *Materials Research Bulletin, 61*, 292–297.
40. Liu, P., Yao, Z., & Zhou, J. (2016). Fabrication and microwave absorption of reduced graphene oxide/$Ni_{0.4}Zn_{0.4}Co_{0.2}Fe_2O_4$ nanocomposites. *Ceramics International, 42*, 9241–9249.
41. Zhu, T., Shen, W., Wang, X., Song, Y., et al. (2019). Paramagnetic CoS_2@MoS_2 core-shell composites coated by reduced graphene oxide as broadband and tunable high-performance microwave absorbers. *Chemical Engineering Journal, 378*, 122159.
42. Tian, X., Wang, Y., Peng, F.-X., Huang, F., et al. (2021). Defect-enhanced electromagnetic wave absorption property of hierarchical graphite capsules@ helical carbon nanotube hybrid nanocomposites. *ACS Applied Materials & Interfaces, 13*, 28710–28720.
43. Yan, J., Huang, Y., Han, X.-P., Gao, X.-G., & Liu, P.-B. (2019). Metal organic framework (ZIF-67)- derived hollow CoS_2/N-doped carbon nanotube composites for extraordinary electromagnetic wave absorption. *Composites Part B: Engineering, 163*, 67–76.
44. Ling, A., Tan, G.-G., Man, Q.-K., Lou, Y.-X., et al. (2019). Broadband microwave absorbing materials based on MWCNTs' electromagnetic wave filtering effect. *Composites Part B: Engineering, 171*, 214–221.

45. Chen, C., Bao, S.-Z., Zhang, B.-S., Chen, Y., et al. (2019). Coupling Fe@Fe_3O_4 nanoparticles with multiple-walled carbon nanotubes with width band electromagnetic absorption performance. *Applied Surface Science, 467*, 836–843.
46. Wen, B., Zhang, J., Yang, G., Jing, D.-W., et al. (2022). Optimal electrical conductivity and interfacial polarization induced by loaded nanoparticles on carbon nanotubes for excellent electromagnetic wave absorption performance. *Journal of Colloid and Interface Science, 626*, 759–767.
47. Qiu, Y., Lin, Y., Yang, H.-B., et al. (2020). Hollow Ni/C microspheres derived from Ni-metal organic framework for electromagnetic wave absorption. *Chemical Engineering Journal, 383*, 123207.
48. Zhu, H.-H., Jiao, Q.-Z., Fu, R.-R., et al. (2022). Cu/NC@Co/NC composites derived from core-shell Cu-MOF@Co-MOF and their electromagnetic wave absorption properties. *Journal of Colloid and Interface Science, 613*, 182–193.
49. Shu, R.-W., Li, X.-H., Ge, C.-Q., & Wang, L.-Y. (2023). Synthesis of FeCoNi/C decorated graphene composites derived from trimetallic metal-organic framework as ultrathin and high-performance electromagnetic wave absorbers. *Journal of Colloid and Interface Science, 630*, 754–762.
50. Zhou, X.-F., Jia, Z.-R., Feng, A.-L., Wang, X.-X., et al. (2019). Synthesis of fish skin-derived 3D carbon foams with broadened bandwidth and excellent electromagnetic wave absorption performance. *Carbon, 152*, 827–836.
51. Liu, Y., Liu, X.-H., Xinyu, E., Wang, B.-B., et al. (2022). Synthesis of Mn_xO_y@C hybrid composites for optimal electromagnetic wave absorption capacity and wideband absorption. *Journal of Materials Science and Technology, 103*, 157–164.
52. Si, T.-T., Xie, S., Ji, Z.-J., Ma, C., et al. (2023). Modification of basalt fiber with Ni and carbon black formed synergistic microstructure for enhanced electromagnetic wave absorption. *Journal of Materials Science: Materials in Electronics, 34*, 1035.
53. Jiao, Y.-M., Song, Q., Yin, X.-M., Han, L.-Y., et al. (2022). Grow defect-rich bamboo-like carbon nanotubes on carbon black for enhanced microwave absorption properties in X band. *Journal of Materials Science and Technology, 119*, 200–208.
54. Zuo, D. Q., Jia, Y. Q., JianHua, X., & JiaJun, F. (2023). High-performance microwave absorption materials: Theory, fabrication, and functionalization. *Industrial and Engineering Chemistry Research, 62*, 14791–14817.
55. Zhang, S., Zhang, S., Zhu, P., et al. (2025). Recent achievements and performance of nanomaterials in microwave absorption and electromagnetic shielding. *Advances in Colloid and Interface Science, 335*, 103336.

Carbon-Based Nanocomposites for 3D Printing

Bancha Yingngam

Abbreviations

3D printing	Three-dimensional printing
ABS	Acrylonitrile butadiene styrene
CNFs	Carbon nanofibers
CNTs	Carbon nanotubes
DIW	Direct ink writing
DPL	Digital light processing
FDM	Fused deposition modeling
GO	Graphene oxide
MWCNTs	Multiwalled carbon nanotubes
PLA	Polylactic acid
SDGs	Sustainable development goals
SLA	Stereolithography
SLS	Selective laser sintering
SWCNTs	Single-walled carbon nanotubes
UV	Ultraviolet

B. Yingngam (✉)
Department of Pharmaceutical Chemistry and Technology, Faculty of Pharmaceutical Sciences, Ubon Ratchathani University, Ubon Ratchathani, Thailand
e-mail: bancha.y@ubu.ac.th

V. Khanna (ed.), *Carbon-Based Nanocomposites for Sustainable Applications, Volume I*, Lecture Notes in Nanoscale Science and Technology 38,
https://doi.org/10.1007/978-3-031-95306-4_9

1 Introduction

Additive manufacturing—or, more commonly, 3D printing—is a revolution in how things are designed and made, allowing the construction of pieces layer by layer [1]. Key techniques within this domain include fused deposition modeling (FDM), stereolithography (SLA), and selective laser sintering (SLS), each offering distinct advantages on the basis of material type. FDM, for example, relies on the extrusion of thermoplastic filaments, while SLA utilizes cured photopolymers, and SLS involves sintering-powdered materials [2]. The benefits of 3D printing are multifaceted, encompassing reduced prototyping costs and time, as well as enabling customization and sustainable manufacturing practices that minimize waste through on-demand production [3]. Raman et al. [4] highlighted the importance of this innovation in relation to the Sustainable Development Goals (SDGs), particularly SDG 9 (Industry, Innovation, and Infrastructure) and SDG 12 (Responsible Consumption and Production).

Even with these advances, traditional 3D printing materials typically lack mechanical properties, such as strength, stiffness, and durability, relative to materials created via traditional smart manufacturing methods [5]. In light of these hurdles, scientists are increasingly investigating carbon-based nanomaterials—including carbon nanotubes (CNTs), graphene, and carbon nanofibers (CNFs)—that can be incorporated into polymer matrices [6]. The incorporation of carbon nanomaterials with high aspect ratios, excellent mechanical strength, and unique electrical/thermal conductivity into polymer matrices resulted in excellent nanocomposites due to their extrusive features. These materials have a wide range of applications. This process allows for the fabrication of structural parts with a much better weight–strength ratio than traditional processes do, providing unique solutions for the aerospace and automotive industries [7]. In biomedical engineering, they enhance the performance of implants and tissue scaffolds, whereas in electronics, they contribute to the advancement of high-performance devices and sensors [8, 9]. This versatility is essential for promoting renewable energy initiatives, supporting sustainable cities (SDG 11), and facilitating access to clean energy (SDG 7) [10].

To successfully implement carbon-based nanocomposites in 3D printing processes, one should pay attention to material choice, dispersion methods, and compatibility with diverse printing techniques [11]. An important aspect of this integration is achieving a uniform distribution of nanomaterials within the polymer matrix, which is crucial for optimal printing performance and the desired characteristics of the final material. The development of scalable production methods, processing parameters, and sustainable materials across their entire life cycle continues to be imperative [3]. Future research directions may involve developing improved processes, understanding microstructure–property relationships, and diversifying applications. Future trends may focus on advanced nanocomposite synthesis, 4D printing materials, and self-healing polymer systems [12]. Simulation-based computational tools paired with artificial intelligence are also used for the optimization of material compositions as well as printing process parameters, in addition to increasing print quality and reducing material waste [13].

The above reports indicate the ongoing effort necessary to tailor carbon-based nanocomposites for 3D printing [14]. Improving further will require close collaboration among material scientists, engineers, and others interested in the success of the process, but this collaboration will benefit our understanding of material properties, processing protocols, and sustainable manufacturing pipelines. These carbon-based nanocomposites can be used as game changers to realize potential industrial applications, but only if efficient production is permitted can materials be used in reduced quantities, thus improving performance, paving the way for technological advancement, and contributing to the SDGs [15].

Thus, this chapter aims to provide a state-of-the-art overview of 3D additive manufacturing of carbon-based nanocomposites in terms of material synthesis, processing, and application potential, as well as pathways for further development. This chapter illustrates various carbon-based nanomaterials suitable for 3D printing and their unique properties and composites on the basis of a polymer matrix to foster the implementation of UN SDG 12. This chapter will also address 3D printing processes that are compatible with current carbon nanocomposites and characterize the properties and performance of printed parts for the enhancement of SDG 9. Finally, the application of carbon nanocomposites in various industrial sectors, including aerospace, automotive, healthcare, and electronics, is discussed, illustrating their potential role in the UN's SDGs 11 and 7. It will address points regarding scalability, standardization, and sustainability and potential solutions to some present-day limitations, provoking discussion on active research and future perspectives. This chapter will strengthen the future development of carbon-based nanocomposite dispersions for 3D printing, provide readers with an understanding of potential advancements, and better comprehend the intricacies and future developments of carbon-based nanocomposite dispersions for 3D printing.

2 Carbon-Based Nanomaterials for 3D Printing

Novel carbon-based nanomaterials have been extensively incorporated into 3D printing technology, as their exceptional properties make them a viable option for significantly improving the performance of printed components. This section discusses different generations of carbon nanomaterials (CNTs, graphene, CNFs, and other novel materials), their identification characteristics, and their utilization in 3D printing (Fig. 1).

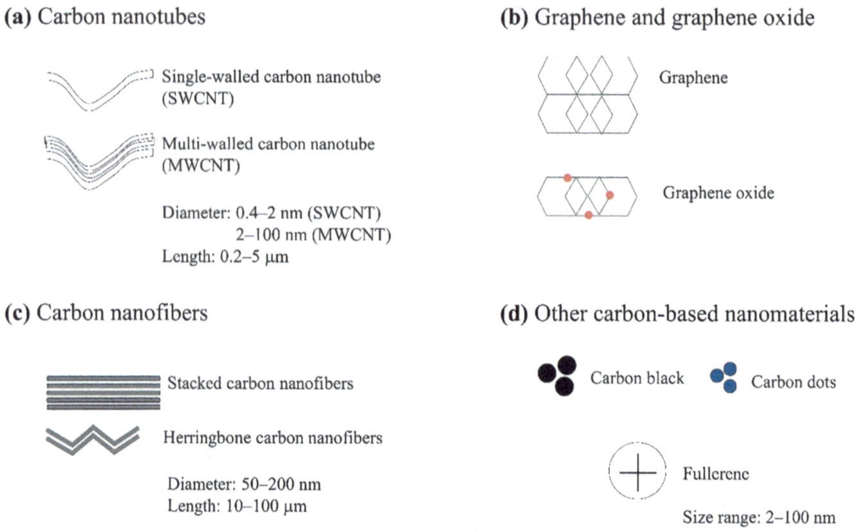

Fig. 1 Schematic displaying the main classes of carbon-based nanomaterials employed by means of 3D printing: (**a**) carbon nanotubes, (**b**) graphite and graphene oxide, (**c**) carbon nanofibers, and (**d**) other carbonaceous nanomaterial. (Created by the author with no permission needed)

2.1 Carbon Nanotubes

CNTs are cylindrical stacks of rolled sheets of the fundamental graphene framework [16] and are known as single-walled CNTs (SWCNTs) or multiwalled CNTs (MWCNTs) (Fig. 1a) [17]. These substances combine excellent mechanical performance with a Young's modulus of ~1 TPa and a tensile strength of ~100 GPa, making them the toughest materials on Earth [18]. In addition, the electrical conductivity of CNTs is 1000 times greater than that of copper [19], and the thermal conductivity can reach 3000 W/mK [20].

Research has highlighted the potential of CNTs as additive agents in polymer matrices for 3D printing (Fig. 2) [7]. Compared with standard polymer materials, their incorporation significantly enhances the mechanical properties of these matrices, resulting in improved toughness, tensile strength, and stiffness [8]. For example, reinforcing acrylonitrile butadiene styrene (ABS) filaments with MWCNTs has demonstrated enhancements in the mechanical characteristics of 20–30%, even at low load percentages [6]. Despite their benefits, challenges remain regarding the uniform dispersion and alignment of CNTs within polymer matrices, primarily due to their tendency to aggregate [21]. Strategies to address these challenges include modifying CNT surfaces and leveraging shear forces during the polymerization process [6].

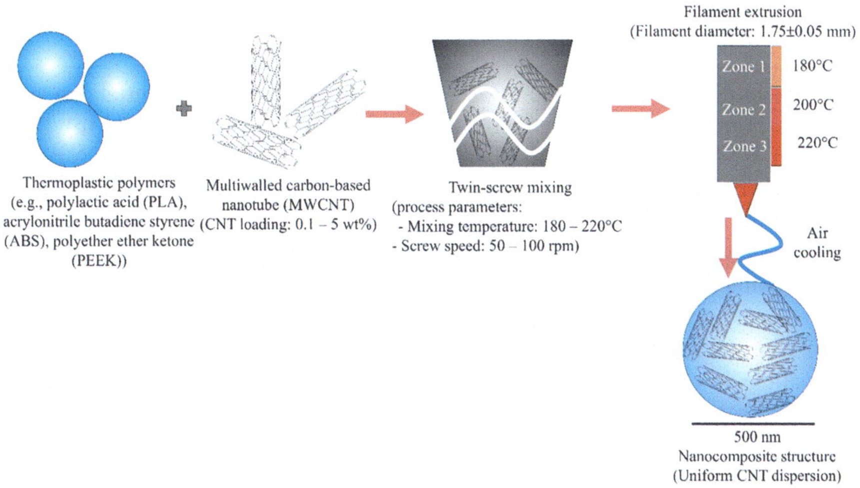

Fig. 2 Schematic illustration of the incorporation of carbon nanotubes into a polymer matrix to create a nanocomposite filament for 3D printing. (Created by the author with no permission required)

2.2 *Graphene and Graphene Oxide*

Graphene, a two-dimensional carbon allotrope, has emerged as a prominent nanomaterial in the 3D printing landscape (Fig. 1b) [1]. Characterized by a single layer of carbon atoms arranged in a hexagonal lattice, graphene has impressive mechanical, electrical, and thermal properties, including a Young's modulus of approximately 1 TPa and a tensile strength of approximately 130 GPa [22]. Its electrical conductivity is approximately 6000 S/cm, and its thermal conductivity can reach 5000 W/mK [23].

Graphene oxide (GO), a derivative of graphene with oxygen functionalities, offers enhanced solubility in polymer matrices, making it a valuable additive for improving the compatibility and stability of dispersions in 3D printing materials [8, 24]. Techniques such as solution processing, melt mixing, and in situ polymerization have been employed to incorporate both graphene and GO into various 3D printing filaments. Notably, adding less than 0.5 wt% graphene nanoplatelets to polylactic acid (PLA) filaments significantly increased the tensile strength and Young's modulus by 20–35% [25–27].

2.3 Carbon Nanofibers

Carbon nanofibers, another type of one-dimensional carbon nanomaterial, are synthesized primarily through catalytic chemical vapor deposition (Fig. 1c) [28]. Although CNFs exhibit lower mechanical and electrical properties than CNTs do, their cost-effectiveness and compatibility with polymer matrices make them advantageous for 3D printing applications. The addition of CNFs to polymer composites has been shown to enhance both tensile strength and thermal stability. For example, incorporating 3 wt% CNFs into PLA has resulted in a notable improvement in tensile strength of 23.3% and a reduction in elongation of 38.1% [29]. Additionally, CNFs contribute to the electrical properties of 3D-printed components, facilitating applications in electrostatic discharge protection and electromagnetic interference shielding [30].

2.4 Other Carbon-Based Nanomaterials

In addition to CNTs, graphene, and CNFs, a range of other carbon-based nanomaterials merit consideration in 3D printing (Fig. 1d):

- **Carbon Black:** Because of their incomplete burning of hydrocarbons, carbon black nanoparticles are used as competent fillers for enhancing the mechanical features, electrical conductivity, and ultraviolet (UV) resistance of polymer nanocomposites [31, 32].
- **Carbon Nanodots:** These zero-dimensional materials, with dimensions generally less than 10 nm, are recognized for their optical properties, such as photoluminescence, which render them suitable for use in sensing and bioimaging [33].
- **Fullerenes:** Three-dimensional carbon black (C60) with a unique mechanical value architecture, biointroductory studies, and energy applications [34, 35].
- **Carbon Nanohorns:** These nanomaterials possess a horn-like structure and present increased surface areas and absorption capacities in water treatment and gas sensor applications [34].

3 Synthesis and Processing of Carbon-Based Nanocomposites via 3D Printing

The integration of carbon-based nanomaterials into 3D-printed polymers is achieved through synthesis and processing methods that focus on enhancing the quality and properties of 3D-printed polymers. The key obstacles to overcome in this area include the effective dispersion of nanomaterials, effective interaction of the polymeric matrix and nanomaterials, and maintenance of the structural integrity of the incorporated nanomaterials throughout the processing steps. Herein, the author summarizes the relevant work related to lock-in methods of dispersion, functionalization strategies, incorporation, and filament manufacturing processes of carbon-based nanocomposites in 3D printing.

3.1 Dispersion Techniques

The dispersion of carbon-based nanomaterials in the polymer matrix directly impacts the efficiency of these materials as reinforcements. Although this feature is beneficial, it normally results in agglomerate formation and poor dispersion [36] owing to the high van der Waals forces associated with these materials [36]. To address these issues, several dispersion techniques have been developed:

- **Ultrasonication:** This method employs high-intensity ultrasonic waves to disrupt agglomerates and achieve uniform dispersion. The combination of localized heating and shear forces from cavitation helps in effectively dispersing the nanomaterials [3].
- **Shear Mixing:** By utilizing melt blending and solution plasticizers, shear mixing promotes better wetting and dispersion of nanomaterials within the polymer matrix. The high shear rates broke down the aggregate states, resulting in a uniform distribution of nanofillers [37].
- **In situ Polymerization:** Here, the nanomaterials are mixed with the monomer solution prior to polymerization, leading to the formation of nanocomposites with well-dispersed nanomaterials [38].
- **Surfactant-Assisted Dispersion:** Incorporating surfactants into a nanomaterial–polymer blend reduces the interfacial tension and prevents the formation of large aggregates. Surfactants stabilize nanomaterials through steric or electrostatic interactions, increasing their compatibility with the polymer matrix [39].

The choice of dispersion technique is contingent upon the characteristics of the nanomaterial–polymer system, the desired dispersion level, and the specific processing conditions. In some scenarios, a synergistic approach may be adopted to optimize dispersion and maximize the performance of the resulting nanocomposites [3].

3.2 Functionalization Methods

Functionalization serves as a strategic approach to enhance the dispersion and interfacial interactions of carbon-based nanomaterials within polymer matrices. The adhesion and compatibility of these nanomaterials can be significantly improved through various functionalization techniques:

- **Covalent Functionalization:** Functional groups (carboxyl, hydroxyl, or amine groups) are chemically attached to the surface of nanomaterials via oxidation or esterification methods [3].
- **Noncovalent Functionalization:** The noncovalent technique is different from the covalent approach, where molecules are adsorbed on nanomaterials through van der Waals forces or π–π interactions. This approach not only retains the original characteristics of the nanomaterials but also provides greater stability and dispersion in the polymer matrix [40].
- **Plasma Treatment:** A method for surface modification that employs ionized gases to deposit functional groups on the nanomaterial surface as needed for specific surface functionalities [41].
- **Grafting:** Grafting denotes the covalent bonding of polymer chains to the nanomaterial surface. This can be accomplished through a "grafting-from" strategy, in which polymer chains are initiated from the nanomaterials, or a "grafting-to" strategy, in which preformed polymers are attached [3]. Functionalization can result not only in better dispersion, adhesion, mechanical locking, and thermal stability but also in compatibility with applications.

3.3 Incorporation into Polymer Matrices

Completely dispersing carbon-based nanomaterials in polymer matrices is essential for producing high-quality nanocomposites for 3D printing. There are various ways this integration is achieved:

- **Solution Mixing:** When both the nanomaterials and the polymer are dissolved in a common solvent, they are mixed, and the solvent is evaporated. This approach facilitates their effective dispersion although their solubility and removal of the solvent can restrict their dispersion [42].
- **Melt Blending:** A nanomaterial is incorporated within a polymer melt through shear mixing processes such as extrusion or injection molding. In this process, heat and shear forces promote the dispersal of the nanomaterial into a polymer matrix and can be implemented for large-scale production [43].
- **In situ Polymerization:** As described earlier in this section, in situ polymerization is an approach in which nanomaterials are incorporated into a monomer solution [44].
- **Masterbatch Approach:** A concentrated nanomaterial and polymer mixture is prepared and diluted with extra polymer. Nanomaterials are added in this way to facilitate a more uniform distribution of nanomaterials and decrease the possibility of their clustering [37].

3.4 Filament Production for 3D Printing

The last part of working with carbon nanocomposites is related to the production of nanocomposite filaments dedicated for 3D printing. Filament production is performed mainly by extrusion, in which the nanocomposite material is melted and the paste is forced through a die to deliver a continuous filament usually with dimensions of ~1.75 mm or 2.85 mm in diameter [45, 46]. Some of the main aspects influencing filament production performance and quality are as follows:

- **Extrusion Temperature:** The temperature of the extruder must be optimal for the polymer to melt so that the nanocomposite can flow through the die. However, high temperatures can cause variations in polymer and/or nanomaterial properties to degenerate [47].
- **Screw Speed:** The screw speed in the extruder controls the shear rate and residence time of the nanocomposite material. It is important to maintain control over the screw speed so that dispersion occurs without allowing the material to agglomerate [48].
- **Die Design:** The design and arrangement of the extrusion die play a critical role in determining the filament properties—from diameter to surface finish. Defects, identified as gaps or nonuniform thicknesses, can be prevented with appropriate die designs [48].
- **Cooling and Spooling:** It is common for the temperature to decrease within the core before the filament is spooled on a spool for 3D printing, as this occurs through air or water systems after exiting extrusion [49].

These advances lead to 3D printing, improved performance, and applications through the efficient synthesis and processing of the material using carbon-based nanocomposites.

4 3D Printing Innovations in Carbon-Based Nanocomposites

The incorporation of carbon-based nanomaterials into 3D-printed feedstocks has made it possible to fabricate complex and functional parts. Nonetheless, for the success of these technologies, a careful approach to the selection and design of processes for carbon-based nanocomposites is essential for printing. The following section will shed light on some of the basic 3D printing techniques used these days, including FDM, SLA, and SLS, in addition to some advanced methods.

4.1 Fused Deposition Modeling

FDM, or fused filament fabrication (FFF), is a mature technique for carbon-based nanocomposite fabrication [50]. A schematic representation of the carbon-based nanocomposite FDM 3D printing process is presented in Fig. 3. FDM is mainly reliant on the melting and extrusion of a polymer filament (comprising nanomaterials) through the nozzle in stacks upon the substrate. Easiest and least expensive 3D printing technology (thus far)—investigators can print with a wide range of thermoplastics with FDM. FDM printing must provide a homogeneous mixture of nanofillers in the polymer matrix and excellent rheological properties that are critical for extrusion. Good layer bonding and dimensional accuracy require fine-tuning of printing parameters, such as the nozzle temperature, print speed, layer height, and infill density [51]. Several researchers have reported success in the FDM printing of individual carbon nanostructures with their respective polymers [51–53], resulting in the formation of different carbon-based polymer nanocomposites, such as CNT-ABS, graphene-PLA, and CNF-PEEK, with better mechanical, electrical, and thermal properties. Common challenges, such as the clogging of nozzles, poor surfaces, and insufficient surface finish, hinder the anisotropic behavior of layered deposition.

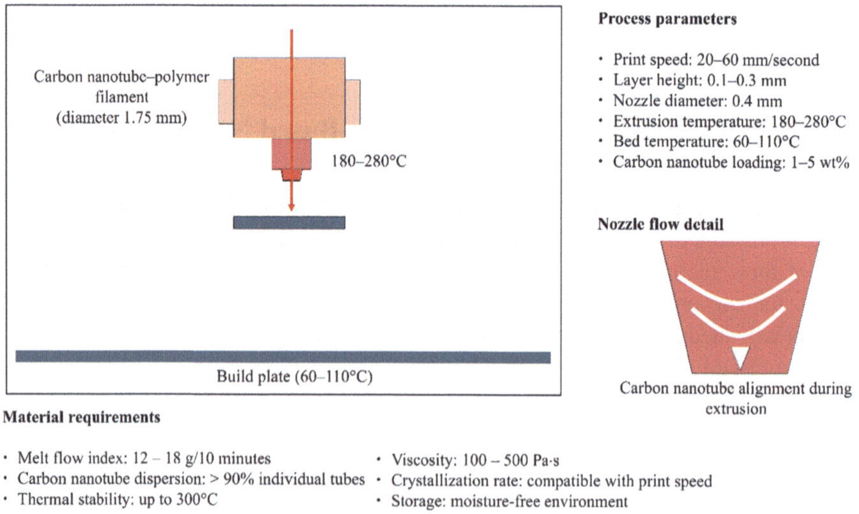

Fig. 3 Fused deposition modeling 3D printing process for carbon-based nanocomposites. (Created by the author with no permission)

4.2 *Stereolithography*

Stereolithography represents a vat photopolymerization technique that uses a laser to selectively solidify liquid photopolymer resins infused with carbon-based nanocomposites. In this method, a focused laser beam cures the resin in a precise, layer-by-layer manner, resulting in high-resolution parts with smooth surfaces and complex geometries [54]. The incorporation of carbon-based nanomaterials into SLA resins enhances the mechanical, electrical, and thermal performance of the printed objects. Nonetheless, achieving a uniform distribution of nanomaterials within the resin while preserving photopolymerization kinetics poses a significant challenge. This can be addressed through the functionalization of nanomaterials and the use of suitable dispersants and photoinitiators [55]. Various carbon-based nanocomposites, such as graphene-reinforced acrylate resins and CNT-modified epoxy resins, have shown improvements in strength, stiffness, and electrical conductivity [3]. However, the limited material options and the necessity of additional processing to eliminate uncured resin and support structures present constraints for SLA in nanocomposite manufacturing. Below Fig. 4 shows a schematic illustration of carbon-based nanocomposites obtained via a stereolithography 3D printing process.

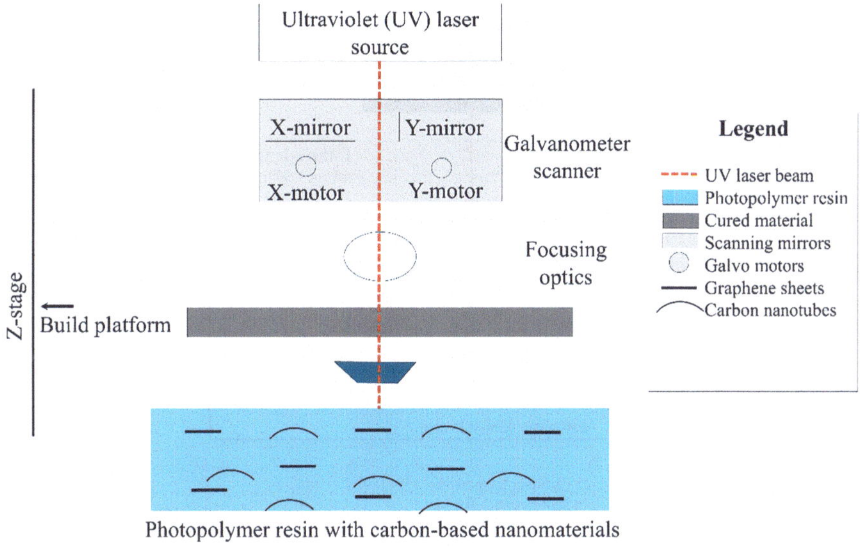

Fig. 4 Schematic illustration of carbon-based nanocomposites obtained via a stereolithography 3D printing process. (Created by the author with no permission)

4.3 Selective Laser Sintering

SLS is a form of powder bed fusion technology in which a laser beam is passed over layers of material to sinter the powder (including nanocomposites of carbon-based materials) into a solid structure. The SLS process employs a layer-by-layer approach, melting and binding particles until the final product is finished (Fig. 5). This method is characterized by high accuracy, acceptable mechanical properties, and the ability to use various materials (polymers, metals, ceramics) [56]. The mechanical, electrical, and thermal properties of the SLS powders were enhanced via the addition of carbon-based nanomaterials. However, a globally equity distribution of nanomaterials inside the powder bed, as well as the required powder packing and flow characteristics, is difficult, and problems with anisotropy, pore reproducibility, and uniformity can be solved by surface modification of nanomaterials, suitable binding agents, and flow additives [57]. SLS can print carbon-based nanocomposite-reinforced polyamides with improved mechanical strength, electrical conductivity, and thermal stability [2, 58]. Nonetheless, the cost of SLS devices and postprocessing steps to clean up unsintered powder and improve surface properties has made it unaffordable for nanocomposite fabrication.

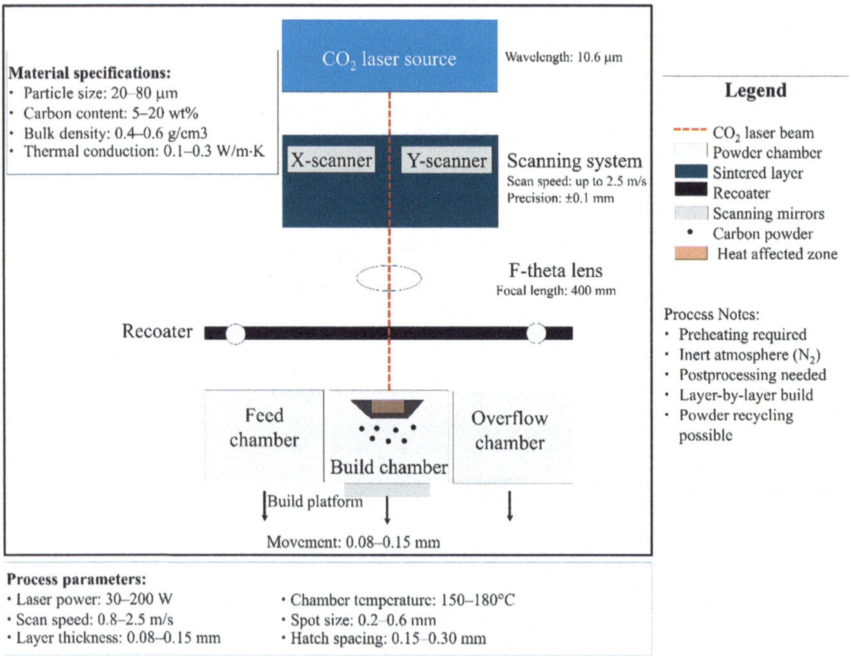

Fig. 5 Schematic illustration of carbon-based nanocomposites prepared via a selective laser sintering 3D printing process. (Created by the author with no permission)

4.4 Alternative 3D Printing Techniques

In addition to FDM, SLA, and SLS, several other noteworthy 3D printing methods for carbon-based nanocomposites exist:

- **Direct Ink Writing (DIW):** This technique uses a highly viscous ink containing dispersed carbon-based nanomaterials, allowing the creation of 3D micro- and nanostructures layer by layer. DIW is capable of achieving intricate geometries and fine detail [3].
- **Binder Jetting:** In this method, a liquid binder is selectively applied to a powder bed containing carbon-based nanocomposites, forming a "green part" that is subsequently sintered to achieve the desired strength and density [59].
- **Digital Light Processing (DLP):** DLP also uses a digital light projector to cure photopolymerizable resins with carbon-based nanomaterials via the same principle as SLA. Compared with traditional SLA, this method provides high resolution and fast print speeds [60].

New methods leading to improved performance make 3D printing of carbon-based nanocomposites a promising area to explore. Once the issues with the processes displayed above are overcome, the advanced materials and their functions will lead to new enabling capabilities and applications.

5 Properties and Performance of 3D-Printed Carbon-Based Nanocomposites

To a great degree, the incorporation of carbon-based nanomaterials in 3D-printed feedstocks has improved the functional properties of the printed parts. These improvements can be seen across a range of performance criteria, such as mechanical strength, electrical and thermal conductivity, and surface properties, thus creating opportunities for multifunctional designs. The next subsections focus on the main performance parameters of carbon nanocomposites in 3D printing.

5.1 Mechanical Properties

The utilization of CNTs, graphene, and CNFs greatly improves the mechanical properties of thermoplastics in 3D polymer composites [61]. According to the author's experience, for example, CNTs are added to acrylonitrile butadiene styrene (ABS), which increases the tensile strength by 30% and the Young's modulus by 40%. For materials with 0.58 wt% graphene added to polylactic acid (PLA), the increase in flexural strength is merely 25%, whereas the increase in flexural modulus is only 25%. The reasons for this improvement are mechanisms such as load transfer, crack bridging, and process deflection absorption [62]. However, the degree of performance enhancement depends on the usage of nanomaterials, along with their distribution in the polymer matrix, the interactions between the fillers and the polymer matrix, etc., which combine with each other [63]. This is also where the mechanical advantages can be enhanced depending on the printing parameters, including the layer height, infill density, and orientation [64].

5.2 Electrical and Thermal Conductivity

The employment of carbon-based nanomaterials allows for tailored conductive properties in 3D-printed components, depending on the specific application requirements [65]. Both CNTs and graphene possess excellent electrical conductivities of 12 S/m (CNTs) and 800 S/m (graphene), which form conductive composites with polymers that have numerous applications, including sensors, electromagnetic interference (EMI) shielding, and electrostatic discharge (ESD) protection [66, 67]. For example, PLA/CNT composites containing silver achieve a conductivity of 1.8×10^{-9} S/m, which is potentially suitable for a variety of electronic substrates [42]. Additionally, these nanomaterials increase the thermal conductivity, with CNTs able to transfer heat at a rate of 3000 W/mK and graphene able to transfer heat at 5000 W/mK [68]. These attributes generally promote their use in heat handling applications. The performance of composites mainly depends on the type and amount of nanomaterial, the distribution homogeneity of the additives in solution, and the establishment of an efficient interface [69]. Surfactant treatment and the application of hybrid fillers significantly improve the thermal conductivity [70].

5.3 Surface Properties

The surface properties of 3D-printed parts, such as their smoothness and wettability, are increased or improved when carbon-based nanomaterials are used. Studies of surface roughness have demonstrated the dependence on the type of nanomaterial used and the various printing conditions possible, which is beneficial for applications requiring high surface areas or cell attachment. The wettability can be well adjusted by adjusting the design of the surface chemistry [71]. For example, more oxidized CNTs offer hydrophilic surfaces, and oxidized/fluorinated graphene provides superhydrophobic surfaces. These properties support applications in self-cleaning technology and oil–water separation systems [72]. Moreover, the chemical reactivity of these nanocomposites can be enhanced by including catalytic groups or metal nanoparticles, allowing them to serve as support materials in chemical catalysis [73].

5.4 *Durability and Environmental Stability*

3D printing of carbon-based nanomaterials increases the strength of 3D-printed components because of the amazing mechanical, thermal, and chemical features of carbon materials. They increase the resistance to fatigue and creep because they can absorb and redistribute loads, thus stopping crack propagation. This leads to increased heat-transfer effectiveness and prevents polymer matrix degradation at elevated temperatures [74]. In addition, owing to their barrier properties and chemical inertness, these nanomaterials are highly resistant to corrosion, solvents, and chemicals [75]. These factors combined lead to a much greater stability of the as-prepared films. Even if functionalized or coated nanomaterials are utilized, these materials are stable to oxidative and UV radiation while also being moisture stable [60].

According to the author, the utilization of carbon-based nanomaterials in the 3D printing process paves the way for a host of benefits, improving the mechanical, electrical, thermal, and surface properties of the components that are printed while ensuring long-term durability and environmental stability. Thus, these materials will undoubtedly be fundamental drivers of further advancements in additive manufacturing.

6 Applications of 3D-Printed Carbon-Based Nanocomposites

The unique properties of carbon-based nanocomposites produced through 3D printing open a myriad of applications across diverse sectors. By leveraging complex geometries, these materials enable the design and manufacturing of functional components and structures that boast enhanced mechanical, electrical, thermal, and surface properties. This section explores the potential of 3D-printed carbon-based nanocomposites in various industries, including aerospace, automotive, biomedical, electronics, and energy storage (Fig. 6).

6.1 *Aerospace and Automotive Sectors*

3D-printed carbon-based nanocomposites are transforming component production in both the aerospace and automotive sectors. In aerospace, these materials can be found in structural components that offer both enhanced specific strength and vital electromagnetic interference shielding, as well as efficient thermal dissipation capability [76]. In the same vein, they likewise enable automotive engineers to construct improved parts expected for progressed accident display execution and for an expanded scope of features, for example, antistatic surfaces. Notable instances include CNT-PEI for parts of aircraft and graphene-ABS for parts of automobiles.

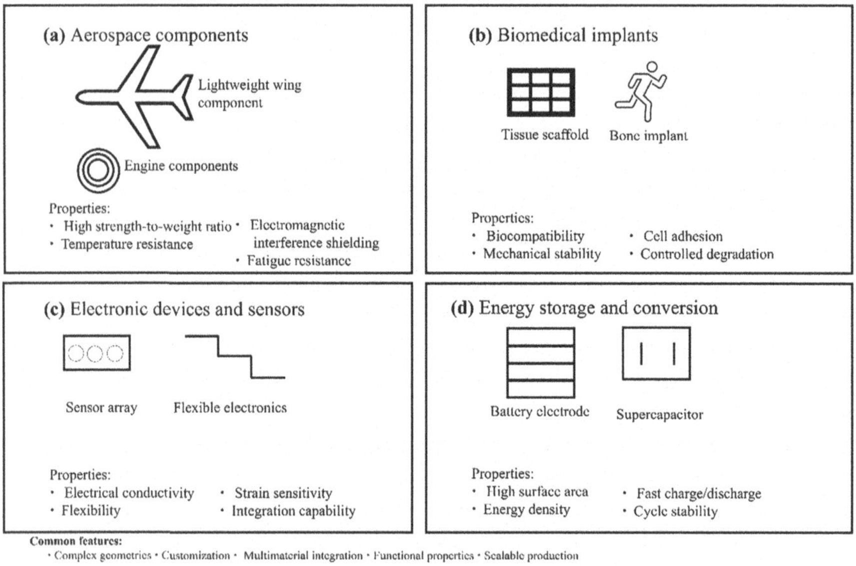

Fig. 6 Applications of 3D-printed carbon-based nanocomposites in various industries: (**a**) aerospace components, (**b**) biomedical implants and scaffolds, (**c**) electronic devices and sensors, and (**d**) energy storage and conversion devices. (Created by the author with no permission required)

These composites boast potentially superior mechanical and functional properties, yielding lighter, more performant designs that can be tailored with high resolution and lower cost [2, 77].

6.2 *Biomedical Engineering and Tissue Scaffolds*

Biomedical engineering, including tissue engineering, is an area where novelty is ever improving, with the use of 3D-printed carbon-based nanocomposites being an example. Biocompatible polymers such as polycaprolactone (PCL) or poly(lactic-co-glycolic acid) (PLGA) are being used to create composites with materials such as CNTs and graphene that improve mechanical and electrical performance while promoting tissue repair and regeneration [78]. Examples of applications include neural recording devices, in vivo biosensors, and drug delivery systems. For example, composites based on graphene oxide and PCL have already been studied for bone tissue engineering, and CNTs and alginate-based composites have shown increased capacity for cardiac tissue regeneration due to their increased conductivity [79, 80]. These enable the fabrication of complex, hierarchically arranged structures that replicate natural tissues, as well as the programmable release of active therapeutics and site-specific stimulation of cells [81].

6.3 Electronics and Sensor Technologies

Owing to the conductivity and mechanical strength of carbon-based nanocomposites, they have become important materials for printed and 3D-printed electronics and sensors. These materials may be processed to fabricate conductive, stretchable, and intelligent devices for use in wearable electronics, robotics, human–machine interfaces, and portable energy storage devices. These composites have a massive surface area and chemically active centers that are significant for enhancing the sensitivity and selectivity of sensor applications [11]. For example, CNT-thermoplastic polyurethane (TPU) strain sensors are used to detect motion, and graphene-PLA gas sensors are used to monitor volatile organic compounds [82, 83]. These advances pave the way for performance prototype applications involving specific electronic systems and environmental monitoring, safety, and health phenomena.

6.4 Energy Storage and Conversion

The synthesis strategy offered a basis for multiscale 3D-printed carbon-based nanocomposites at different levels, which are crucial for the miniaturization and development of next-generation energy storage and conversion devices according to the goals of sustainable development, including clean energy and climate change solutions. Taking advantage of their structural porosity and conductivity, high-power batteries and electrodes for supercapacitors are fabricated with highly optimized energy density structures to minimize their degradation [84]. Carbon-based nanocomposites can greatly enhance the performance of catalyst supports, gas diffusion layers, and proton exchange membranes in fuel cell applications [85]. Examples of such materials are graphene-PLA supercapacitor electrodes, which exhibit high capacitance, and flexible CNT-polyethylene glycol diacrylate (PEGDA) battery electrodes [86, 87]. 3D printing thus allows for innovative and environmentally conscious solutions across a range of sectors owing to its combination of natural material properties and design flexibility.

The author suggested that 3D-printed carbon-based nanocomposites have different industrial applications and can generate unusual combinations of material properties, the latter being possible only with the unique design space of 3D printing. They are discovering all sorts of new uses for them, along with investigating ways that they can become less used—the potential for revolutions in tech and sustainability unlocked.

7 Challenges and Future Prospects

Although substantial progress has been made in the exploration of carbon-based nanocomposites with 3D printing, several key challenges have been encountered. Solutions must, therefore, be framed for scalability versus cost efficiency, recycling versus standardization, quality assurance versus disposal methods, and how to integrate renewable resources. In this section, the author describes these bottlenecks sequentially and provides insight into future trends around these advanced materials.

7.1 Scalability and Cost-Effectiveness

One of the greatest hurdles is hindering the commercialization of carbon nanocomposites produced via 3D printing techniques because of their scalable and economic benefits. With 3D printing, there is considerable freedom of design, but mechanical 3D printing cannot compete with the mass production of metals. Faster printing technologies such as multinozzle systems and continuous liquid interface production (CLIP) have attracted some research interest. Together with adapted printing process parameters, these developments are intended to minimize production time and material expenditure. Furthermore, increased synergy between 3D printing and traditional manufacturing processes also offers a compelling path to lower cost production of complex parts [88].

7.2 Standardization and Quality Control

Control of the process parameters, standardization, and quality control of 3D-printed carbon-based nanocomposites are still daunting tasks. The properties of the final printed product depend significantly on factors such as the raw material quality, nanomaterial dispersion, printing, and postthermal treatment conditions. Without established standards, the reproducibility of results is an issue, and concerns regarding the safety of final products emerge. There are no existing standards for materials, methods, or testing protocols for comparative analysis across studies [89]. Hence, it is crucial to set standards in a holistic manner through the collaboration and consultation of industry players, academia, and authorities and regulators. Novel characterization techniques, including but not limited to X-ray computed tomography and in-process monitoring systems, could greatly improve quality control, enabling prompt detection of defects and control of the printing process in real time [90].

7.3 Sustainability and Recyclability

Another level of complexity in the evolution of 3D-printed carbon-based nanocomposites is the increasing sustainability constraints imposed by the global community (Fig. 7). Although these materials are known to have extreme properties and performance, the consumption of energy and greenhouse gas emissions during their preparation processes can be potential concerns, including the biodegradation and biological risks of nanomaterials. Owing to their complexity, which is characterized by strong interfacial adhesion, the recycling and disposal of these composites are complicated, adding to the waste management challenge [91]. By addressing sustainability and recyclability challenges in this industry, we can make strides toward the attainment of SDG 12 (Responsible Consumption and Production). This can be accomplished by encouraging sustainable manufacturing and minimizing waste to create a circular economy. Recently, scientific projects have concentrated on eco-friendly methods of green synthesis and on eco-friendly polymer matrices, such as PLA or polyhydroxyalkanoates (PHAs), which can contribute to global footprint reduction across a product's life cycle and promote biodegradability after disposal. The development of effective recycling techniques, such as solvent-based separation or thermal decomposition, may allow more carbon-based nanomaterials to be recovered to develop fully closed-loop systems with a circular economy. Sustainability benefits may also stem from improvements in the 3D printing process itself. This means reducing waste through support-free designs and recycling any excess print and minimizing energy consumption by optimizing printing parameters and using energy-efficient printers.

7.4 Emerging Trends and Potential Breakthroughs

Three-dimensional-printed carbon-based nanocomposites will soon be the next major focus. Figure 8 presents the patterns and potential breakthroughs in 3D-printed carbon-based nanocomposites according to the literature and the background of the author. Emerging trends include 3D printing with multiple materials and

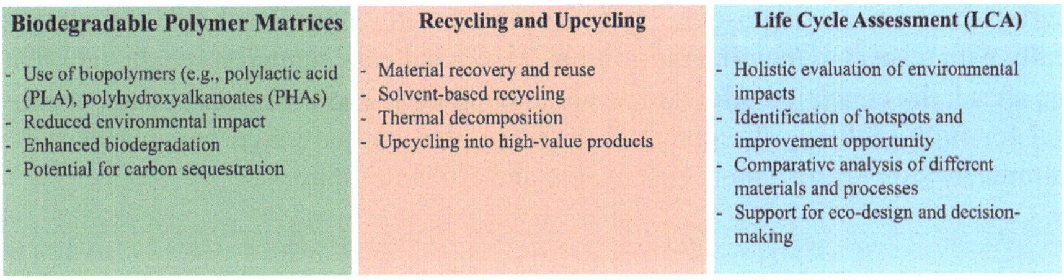

Fig. 7 Sustainability and recycling strategies for 3D-printed carbon-based nanocomposites. (Images created by the author; no permission is needed)

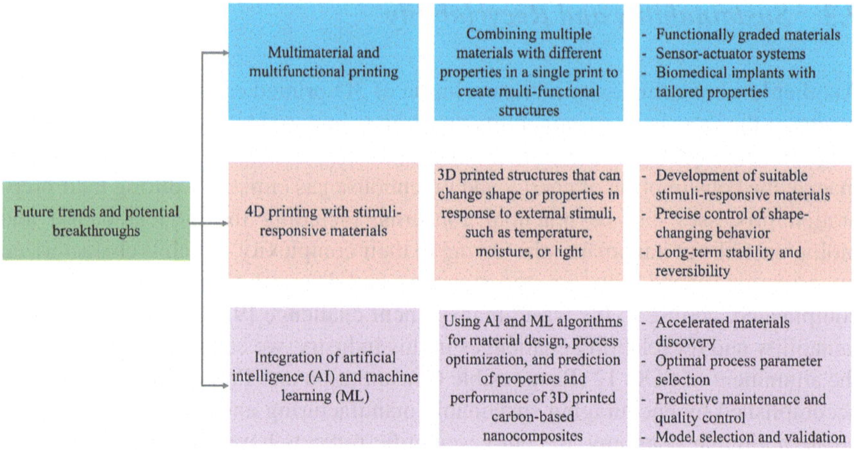

Fig. 8 Future trends and potential breakthroughs in 3D-printed carbon-based nanocomposites: multimaterial and multifunctional printing, 4D printing with stimuli-responsive materials, and integration of artificial intelligence and machine learning for material design and optimization. (Created by the author with no permission required)

functionally graded materials, enabling the fabrication of complex structures with diverse mechanical properties [92]. The addition of ceramics, metals, and biomaterials potentially enables the creation of smart structures that exhibit greater functionalities when we can improve their printability and processability [93]. In addition, 4D printing involves the ability to fabricate structures that change in response to environmental stimuli such as temperature, humidity, or light. The carbon-based nanomaterials employed in 4D printing have enabled the development of smart, responsive, and adaptive structures that target soft robotics and biomedical uses [94].

Artificial intelligence (AI) and machine learning in product design and manufacturing processes enhance efficiency through the value chain. The detailed knowledge of the cophilosophies of processing conditions and nanocomposite properties enables effective real-time enhancements in fulfilling the targeted processing conditions [13]. New carbon allotropes, such as graphyne, graphdiyne, and carbyne, along with other types of different graphene derivatives, nanoparticles, and metal–organic frameworks (MOFs), are expected to lead to the formation of advanced carbon-based nanocomposites. These innovations are aligned with SDG 9, specifically with respect to material innovation [95]. Despite the numerous challenges that lie ahead, the concerted efforts of academia, industry, and the government are critical for fully realizing the potential of 3D-printed carbon-based nanocomposites, ultimately supporting a more sustainable and efficient future.

8 Conclusion

In this chapter, the author explored the transformative potential of carbon-based nanocomposites within the realm of 3D printing. By utilizing advanced materials such as CNTs, graphene, and CNFs, these composites enhance the multifunctionality of printed components, setting a new standard for performance. The integration of carbon nanomaterials within polymer matrices significantly increases the mechanical, electrical, thermal, and surface properties of the final products. The author explained various preparation methods and processing techniques, emphasizing the critical roles of dispersion and interfacial adhesion in achieving optimal performance. However, despite promising advancements, challenges remain. Issues such as scalability, cost-effectiveness, standardization, quality assurance, economic viability, and recyclability continue to pose hurdles that must be addressed for widespread adoption. These innovative materials also align with multiple SDGs. For example, they contribute to SDG 9 (Industry, Innovation, and Infrastructure) through their applications in the aerospace and automotive sectors, enhance SDG 3 (Good Health and Well-being) via advancements in biomedical engineering, and support SDG 12 (Responsible Consumption and Production) by promoting sustainable manufacturing practices. Emerging trends such as multifunctional and multimaterial 3D printing, along with the advent of 4D printing, machine learning, and artificial intelligence, are paving the way for new possibilities. As research continues to evolve, the significance of carbon-based nanocomposites in fostering a sustainable and innovative economy cannot be overstated. Their role is crucial in achieving the objectives set out by the SDGs, heralding a new era in technology and manufacturing.

References

1. Shin, W., & Han, J. (2024). Multiscale analysis for characterization of tensile properties in extrusion-based 3D printed carbon nanocomposites. *Additive Manufacturing, 84*, 104085. https://doi.org/10.1016/j.addma.2024.104085
2. Hisham, M., Shebeeb, C. M., Dileep, C., Jacob, L., & Butt, H. (2024). Additive manufacturing of carbon nanocomposites for structural applications. *Journal of Materials Research and Technology, 28*, 4674–4693. https://doi.org/10.1016/j.jmrt.2024.01.049
3. Shahbazi, M., Jäger, H., Ettelaie, R., Chen, J., Kashi, P. A., & Mohammadi, A. (2024a). Dispersion strategies of nanomaterials in polymeric inks for efficient 3D printing of soft and smart 3D structures: A systematic review. *Advances in Colloid and Interface Science, 333*, 103285. https://doi.org/10.1016/j.cis.2024.103285
4. Raman, R., Lathabhai, H., Pattnaik, D., Kumar, C., & Nedungadi, P. (2024). Research contribution of bibliometric studies related to sustainable development goals and sustainability. *Discover Sustainability, 5*(1), 7. https://doi.org/10.1007/s43621-024-00182-w
5. Wang, J., Shi, Z., Gong, J., Zhou, X., Li, J., & Lyu, Z. (2024). 3D printing of graphene-based aerogels and their applications. *FlatChem, 47*, 100731. https://doi.org/10.1016/j.flatc.2024.100731

6. Fenta, E. W., & Mebratie, B. A. (2024). Advancements in carbon nanotube-polymer composites: Enhancing properties and applications through advanced manufacturing techniques. *Heliyon, 10*(16), e36490. https://doi.org/10.1016/j.heliyon.2024.e36490
7. Abedin, M. J., Hasan, I. U., & Rahman, M. Z. (2024). Carbon nanotube-polymer nanocomposites for 3D printing technology. In S. Moharana, L. Rout, & S. Sagadevan (Eds.), *Carbon nanotube-polymer nanocomposites: Synthesis, properties, and applications* (pp. 331–350). Springer Nature Singapore. https://doi.org/10.1007/978-981-97-6329-0_13
8. Gu, H., & Yao, F. (2024). An overview of three-dimensional fillers-polymer composites: Fabrications and applications for electronics and energy storage. *Polymer Reviews, 64*(4), 1322–1386. https://doi.org/10.1080/15583724.2024.2365782
9. Sharker, S. M. (2024). Emerging carbon-based nanomaterials for biomedical and bioengineering applications. In R. Javed, J.-T. Chen, & A. T. Khalil (Eds.), *Nanomaterials for Biomedical and Bioengineering Applications* (pp. 203–219). Springer Nature Singapore. https://doi.org/10.1007/978-981-97-0221-3_8
10. Sharifi, A., Allam, Z., Bibri, S. E., & Khavarian-Garmsir, A. R. (2024). Smart cities and sustainable development goals (SDGs): A systematic literature review of cobenefits and trade-offs. *Cities, 146*, 104659. https://doi.org/10.1016/j.cities.2023.104659
11. Shi, S., Jiang, Y., Ren, H., Deng, S., Sun, J., Cheng, F., et al. (2024). 3D-printed carbon-based conformal electromagnetic interference shielding module for integrated electronics. *Nano-Micro Letters, 16*(1), 85. https://doi.org/10.1007/s40820-023-01317-w
12. Zhang, J., Xu, M., Zhang, N., Wang, T., Zhang, Y., Yang, Z., & Wang, Q. (2024). Advances and applications of 4D-printed high-strength shape memory polymers. *Additive Manufacturing Frontiers, 3*(1), 200115. https://doi.org/10.1016/j.amf.2024.200115
13. Dananjaya, V., Marimuthu, S., Yang, R., Grace, A. N., & Abeykoon, C. (2024). Synthesis, properties, applications, 3D printing and machine learning of graphene quantum dots in polymer nanocomposites. *Progress in Materials Science, 144*, 101282. https://doi.org/10.1016/j.pmatsci.2024.101282
14. Kim, C.-H., Lee, S.-Y., Rhee, K. Y., & Park, S.-J. (2024). Carbon-based composites in biomedical applications: A comprehensive review of properties, applications, and future directions. *Advanced Composites and Hybrid Materials, 7*(2), 55. https://doi.org/10.1007/s42114-024-00846-1
15. Poonia, K., Singh, P., Ahamad, T., Le, Q. V., Phan Quang, H. H., Thakur, S., et al. (2024). Sustainability, performance, and production perspectives of waste-derived functional carbon nanomaterials toward a sustainable environment: A review. *Chemosphere, 352*, 141419. https://doi.org/10.1016/j.chemosphere.2024.141419
16. Dhanaselvam, P. S., Karthikeyan, B., & Vanitha, P. (2024). Principle and application of C-based nanomaterials: Graphite, graphene, and carbon nanotube (CNT). In Y. S. Song, L. R. Thoutam, S. Tayal, S. B. Rahi, & T. S. A. Samuel (Eds.), *Handbook of emerging materials for semiconductor industry* (pp. 49–64). Springer Nature Singapore. https://doi.org/10.1007/978-981-99-6649-3_4
17. Rudyak, V., Pryazhnikov, M., Minakov, A., & Shupik, A. (2024). Electrical conductivity of nanofluids with single- and multiwalled carbon nanotubes. Experimental study. *Nano-Structures & Nano-Objects, 38*, 101143. https://doi.org/10.1016/j.nanoso.2024.101143
18. Sun, J., Zhang, X., Wang, Y., Li, M., Wei, X., Liu, H., & Zhou, W. (2024). Experimental determination of the Young's modulus of individual single-walled carbon nanotubes with single chirality. *Nano Research, 17*(8), 7522–7532. https://doi.org/10.1007/s12274-024-6722-2
19. Parthasarathy, S., Moharana, S., & Behera, S. (2024). Carbon nanotube-polymer nanocomposites for wearable electronics. In S. Moharana, L. Rout, & S. Sagadevan (Eds.), *Carbon nanotube-polymer nanocomposites: Synthesis, properties, and applications* (pp. 461–487). Springer. https://doi.org/10.1007/978-981-97-6329-0_18
20. Feng, D., Zhao, Z., Li, P., Li, Y., Zha, J., Hu, J., et al. (2024). Multifunctional performance of carbon nanotubes in thermal energy storage materials. *Materials Today, 75*, 285–308. https://doi.org/10.1016/j.mattod.2024.04.005

21. Zhu, W., Zhao, J., & Tao, X. (2024). MoS_2-carbon based nanocomposites as anodes for lithium-ion batteries: A review. *Journal of Energy Storage, 84*, 110934. https://doi.org/10.1016/j.est.2024.110934
22. Matias, M. L., Pereira, C., Almeida, H. V., Jana, S., Panigrahi, S., Menda, U. D., et al. (2024). 3D printed MXene architectures for a plethora of smart applications. *Materials Today Advances, 23*, 100512. https://doi.org/10.1016/j.mtadv.2024.100512
23. Devi, N., Kumar, R., Singh, S., & Singh, R. K. (2024). Recent development of graphene-based composite for multifunctional applications: Energy, environmental and biomedical sciences. *Critical Reviews in Solid State and Materials Sciences, 49*(1), 72–140. https://doi.org/10.1080/10408436.2022.2132910
24. Yu, L.-H., Tao, X., Feng, S.-R., Liu, J.-T., Zhang, L.-L., Zhao, G.-Z., & Zhu, G. (2024). Recent development of three-dimension printed graphene oxide and MXene-based energy storage devices. *Tungsten, 6*(1), 196–211. https://doi.org/10.1007/s42864-022-00181-2
25. Giner-Grau, S., Lazaro-Hdez, C., Pascual, J., Fenollar, O., & Boronat, T. (2024). Enhancing polylactic acid properties with graphene nanoplatelets and carbon black nanoparticles: A study of the electrical and mechanical characterization of 3d-printed and injection-molded samples. *Polymers, 16*(17), 2449. https://doi.org/10.3390/polym16172449
26. Hou, Y., Wang, W., & Bartolo, P. (2024). In vitro investigations on the effects of graphene and graphene oxide on polycaprolactone bone tissue engineering scaffolds. *Bio-Design and Manufacturing, 7*(5), 651–669. https://doi.org/10.1007/s42242-024-00280-8
27. Iqbal, A. K. M. A., Harcen, C. S., & Haque, M. (2024). Graphene (GNP) reinforced 3D printing nanocomposites: An advanced structural perspective. *Heliyon, 10*(7), e28771. https://doi.org/10.1016/j.heliyon.2024.e28771
28. Li, X., Wang, X., Mei, D., Xu, C., & Wang, Y. (2024). Acoustic-assisted DLP 3D printing process for carbon nanofiber reinforced honeycomb structures. *Journal of Manufacturing Processes, 121*, 374–381. https://doi.org/10.1016/j.jmapro.2024.05.047
29. Kumar, N. H., Adarsha, H., Keshavamurthy, R., & Kapilan, N. (2024). Influence of carbon nano fiber addition on mechanical behavior of PLA based 3D printed polymer nano composites. *Journal of The Institution of Engineers (India): Series D, 105*(1), 33–47. https://doi.org/10.1007/s40033-023-00455-0
30. Wang, L., Men, J., Feng, J., Jiang, Y., Li, L., Hu, Y., & Feng, J. (2024). Extrusion 3D printing of carbon nanotube-assembled carbon aerogel nanocomposites with high electrical conductivity. *Nano Materials Science, 6*(3), 312–319. https://doi.org/10.1016/j.nanoms.2023.09.002
31. Mahmood, K., Sani, A., Sadiq, H., Tang, J., & Li, Q. (2024). Modification of silicone rubber by nanocomposites for enhancing physicochemical properties: A review. *Materials Science and Engineering: B, 310*, 117664. https://doi.org/10.1016/j.mseb.2024.117664
32. Yan, Y., Han, M., Jiang, Y., Ng, E. L. L., Zhang, Y., Owh, C., et al. (2024). Electrically conductive polymers for additive manufacturing. *ACS Applied Materials & Interfaces, 16*(5), 5337–5354. https://doi.org/10.1021/acsami.3c13258
33. Ravi, S. N., Rajendran, S., Madhumathi, G. S., Packirisamy, A. S. B., Vallinayagam, S., Khan, A. A., & Malik, A. (2024). Carbon nanomaterials: Pioneering innovations in bioimaging and biosensing technologies. *Journal of Molecular Structure, 1316*, 138987. https://doi.org/10.1016/j.molstruc.2024.138987
34. Paramasivam, G., Palem, V. V., Meenakshy, S., Suresh, L. K., Gangopadhyay, M., Antherjanam, S., & Sundramoorthy, A. K. (2024). Advances on carbon nanomaterials and their applications in medical diagnosis and drug delivery. *Colloids and Surfaces B: Biointerfaces, 241*, 114032. https://doi.org/10.1016/j.colsurfb.2024.114032
35. Siva, V., Kannan, S., Murugan, A., Shameem, A. S., Thangarasu, S., & Raja, A. (2024). Evaluation of zero-dimensional buckyball-based fullerene molecules in targeted drug delivery applications in dreaded diseases. In K. Biswas, Y. K. Mohanta, T. K. Mohanta, & M. Saravanan (Eds.), *Carbon-based nanomaterials in biosystems* (pp. 301–330). Academic Press. https://doi.org/10.1016/B978-0-443-15508-6.00018-X

36. Khan, S. A., Ghazi, S. M. U., Amjad, H., Imran, M., & Khushnood, R. A. (2024). Emerging horizons in 3D printed cement-based materials with nanomaterial integration: A review. *Construction and Building Materials, 411*, 134815. https://doi.org/10.1016/j.conbuildmat.2023.134815
37. Ribeiro dos Anjos, E. G., Passador, F. R., de Carvalho, A. B., Rezende, M. C., Sundararaj, U., & Pessan, L. A. (2024). Advanced ternary carbon-based hybrid nanocomposites for electromagnetic functional behavior in additive manufacturing. *Applied Materials Today, 40*, 102362. https://doi.org/10.1016/j.apmt.2024.102362
38. Rahman, M., Islam, K. S., Dip, T. M., Chowdhury, M. F. M., Debnath, S. R., Hasan, S. M. M., et al. (2024). A review on nanomaterial-based additive manufacturing: Dynamics in properties, prospects, and challenges. *Progress in Additive Manufacturing, 9*(4), 1197–1224. https://doi.org/10.1007/s40964-023-00514-8
39. Chaitra, V., Suma, N., & Uma, V. (2024). Electrical and electronic properties of cnt polymer nanocomposites. In S. Moharana, L. Rout, & S. Sagadevan (Eds.), *Carbon nanotube-polymer nanocomposites: Synthesis, properties, and applications* (pp. 47–81). Springer. https://doi.org/10.1007/978-981-97-6329-0_3
40. Kumar, A., Gautam, K., & Singh, N. (2024). State-of-the-art developments in surface functionalized carbon-based bio/nanocomposites for theranostic antibacterials, advanced bioimaging, and molecular bioelectronics inspired biosensing platforms. *Journal of Industrial and Engineering Chemistry, 140*, 103–145. https://doi.org/10.1016/j.jiec.2024.05.065
41. Shahbazi, M., Jäger, H., Ettelaie, R., Chen, J., Mohammadi, A., Kashi, P. A., & Ulbrich, M. (2024b). A smart thermoresponsive macroporous 4D structure by 4D printing of Pickering-high internal phase emulsions stabilized by plasma-functionalized starch nanomaterials for a possible delivery system. *Current Research in Food Science, 8*, 100686. https://doi.org/10.1016/j.crfs.2024.100686
42. Sharifi, J., Rizvi, G., & Fayazfar, H. (2024). Sustainable 3D printing of enhanced carbon nanotube-based polymeric nanocomposites: Green solvent-based casting for eco-friendly electrochemical sensing applications. *The International Journal of Advanced Manufacturing Technology, 131*(9), 4825–4837. https://doi.org/10.1007/s00170-024-13337-w
43. Fagundes, A. P., Cividini Neiva, E. G., Zimmermann, L. M., Padoin, N., Soares, C., & Gracher Riella, H. (2025). Melting and solution mixing in the production of photocatalytic filaments for 3D printing. *Chemical Engineering Science, 302*, 120862. https://doi.org/10.1016/j.ces.2024.120862
44. Shin, E. J., Son, J. H., Lee, H. S., & Lee, S. (2024). Manufacturing of conductive filaments for 3D/4D printing through in situ polymerization of biobased thermoplastic polyurethane/multiwall carbon nanotube composites. *ACS Applied Polymer Materials, 6*(3), 1751–1762. https://doi.org/10.1021/acsapm.3c02558
45. Doğru, A., Seydibeyoğlu, M. Ö., & Ayranci, C. (2025). The effect of interface enhancement on the mechanical properties of fiber-reinforced PA6 matrix composites in material extrusion-based additive manufacturing. *Progress in Additive Manufacturing, 10*, 361–374. https://doi.org/10.1007/s40964-024-00628-7
46. Rafiee, R., Amini, M. R., & Zehtabzadeh, H. (2024). Examining tensile properties in nanocomposite samples fabricated through extrusion-based additive manufacturing. *Arabian Journal for Science and Engineering, 49*(8), 10913–10926. https://doi.org/10.1007/s13369-023-08562-6
47. Park, S., Shi, B., Islam, M. M., He, J., Sung, D. H., Zhang, C., et al. (2024). Universal carbonizable filaments for 3D printing. *Advanced Functional Materials, 34*(51), 2410164. https://doi.org/10.1002/adfm.202410164
48. Tadi, S. P., Maddula, S. S., & Mamilla, R. S. (2024). Sustainability aspects of composite filament fabrication for 3D printing applications. *Renewable and Sustainable Energy Reviews, 189*, 113961. https://doi.org/10.1016/j.rser.2023.113961
49. Harb, S. V., Kolanthai, E., Pugazhendhi, A. S., Beatrice, C. A. G., Pinto, L. A., Neal, C. J., et al. (2024). 3D printed bioabsorbable composite scaffolds of poly (lactic acid)-tricalcium

phosphate-ceria with osteogenic property for bone regeneration. *Biomaterials and Biosystems, 13*, 100086. https://doi.org/10.1016/j.bbiosy.2023.100086

50. Li, H., Mu, Y., Wang, Q., Cai, D., Zhang, Y., Luo, Y., et al. (2024). Interlayer enhancement of 3D printed CF/PLA composites via localized microwave welding and annealing-induced crystallization. *Composites Part B: Engineering, 284*, 111737. https://doi.org/10.1016/j.compositesb.2024.111737
51. Lv, X., Pei, X., Yang, S., Zhang, Y., Wang, Q., & Wang, T. (2024). Tribological behavior of PEEK based composites with alternating layered structure fabricated via fused deposition modeling. *Tribology International, 199*, 109953. https://doi.org/10.1016/j.triboint.2024.109953
52. Kumar, S., Singh, H., Singh, I., Bharti, S., Kumar, D., Siebert, G., & Koloor, S. S. R. (2024). A comprehensive review of FDM printing in sensor applications: Advancements and future perspectives. *Journal of Manufacturing Processes, 113*, 152–170. https://doi.org/10.1016/j.jmapro.2024.01.030
53. Rafiee, R., & Amohaji, H. (2024). On the application of genetic algorithm for predicting the strength of CNT/ABS filaments using multiscale modeling. *Applied Physics A, 130*(7), 513. https://doi.org/10.1007/s00339-024-07657-7
54. Yue, X., Liu, M., Xia, Y., & Xiong, Z. (2024). Influences of two sizes of nano-Fe3O4 on stereolithography 3D printing properties of gel photoresist for photothermal-responsive hydrogel. *Materials Today Communications, 39*, 108881. https://doi.org/10.1016/j.mtcomm.2024.108881
55. Hossain, M. J., Tabatabaei, B. T., Kiki, M., & Choi, J.-W. (2025). Additive manufacturing of sensors: A comprehensive review. *International Journal of Precision Engineering and Manufacturing-Green Technology, 12*, 277–300. https://doi.org/10.1007/s40684-024-00629-5
56. Bose, S., Akdogan, E. K., Balla, V. K., Ciliveri, S., Colombo, P., Franchin, G., et al. (2024). 3D printing of ceramics: Advantages, challenges, applications, and perspectives. *Journal of the American Ceramic Society, 107*(12), 7879–7920. https://doi.org/10.1111/jace.20043
57. Ali, F., Kalva, S. N., & Koc, M. (2024). Advancements in 3D printing techniques for biomedical applications: A comprehensive review of materials consideration, postprocessing, applications, and challenges. *Discover Materials, 4*(1), 53. https://doi.org/10.1007/s43939-024-00115-4
58. Azam, M. U., Belyamani, I., Schiffer, A., Kumar, S., & Askar, K. (2024). Progress in selective laser sintering ofmultifunctional polymer composites for strain- and self-sensing applications. *Journal of Materials Research and Technology, 30*, 9625–9646. https://doi.org/10.1016/j.jmrt.2024.06.024
59. Tan, W. X., Chiang, P.-J., Tan, L. P., Ramanujan, R. V., Tan, M. J., Ong, A., et al. (2024). Adhesive and alloying properties of dual purpose polyfurfuryl alcohol binder for binder jet additive manufacturing of steel. *Additive Manufacturing, 86*, 104212. https://doi.org/10.1016/j.addma.2024.104212
60. Cortés, A., Bañón-Veracruz, M., Jiménez-Suárez, A., Campo, M., Prolongo, M. G., & Prolongo, S. G. (2024). Enhancing efficiency and sustainability of digital light processing 3D-printing by novel two-stage processing of carbon nanotube reinforced nanocomposites. *Journal of Materials Research and Technology, 29*, 3237–3248. https://doi.org/10.1016/j.jmrt.2024.02.018
61. Eivazzadeh-Keihan, R., Sadat, Z., Lalebeigi, F., Naderi, N., Panahi, L., Ganjali, F., et al. (2024). Effects of mechanical properties of carbon-based nanocomposites on scaffolds for tissue engineering applications: A comprehensive review. *Nanoscale Advances, 6*(2), 337–366. https://doi.org/10.1039/d3na00554b
62. de Rubeis, T., Ciccozzi, A., Giusti, L., & Ambrosini, D. (2024). On the use of 3D printing to enhance the thermal performance of building envelope—A review. *Journal of Building Engineering, 95*, 110284. https://doi.org/10.1016/j.jobe.2024.110284
63. Costabeber, G., Guerra, N. B., Brasil, G. S. A. P., Sasaki, J. C., Scontri, M., Burd, B. S., et al. (2024). Additive manufacturing of polylactic acid scaffolds dip-coated with polycaprolactone for bone tissue engineering. *Materials Today Communications, 40*, 109646. https://doi.org/10.1016/j.mtcomm.2024.109646

64. Skwarski, M., Kaczyński, P., Dmitruk, A., Makuła, P., & Ludwiczak, J. (2024). Optimizing mechanical properties and pioneering biodegradable polymer blends for superior energy-absorbing structures used in sport helmets. *Archives of Civil and Mechanical Engineering, 25*(1), 14. https://doi.org/10.1007/s43452-024-01075-5
65. Glogowsky, A., Korger, M., & Rabe, M. (2024). Influence of print settings on conductivity of 3D printed elastomers with carbon-based fillers. *Progress in Additive Manufacturing, 9*(4), 791–803. https://doi.org/10.1007/s40964-023-00483-y
66. Guadagno, L., Aliberti, F., Longo, R., Raimondo, M., Pantani, R., Sorrentino, A., et al. (2023). Electrical anisotropy controlled heating of acrylonitrile butadiene styrene 3D printed parts. *Materials & Design, 225*, 111507. https://doi.org/10.1016/j.matdes.2022.111507
67. Ponnamma, D., Yin, Y., Salim, N., Parameswaranpillai, J., Thomas, S., & Hameed, N. (2021). Recent progress and multifunctional applications of 3D printed graphene nanocomposites. *Composites Part B: Engineering, 204*, 108493. https://doi.org/10.1016/j.compositesb.2020.108493
68. Abd, G., Díaz, R. S., Gupta, A., Niepa, T. H. R., Mondal, K., Ramakrishna, S., et al. (2024). Carbon nanomaterials-based electrically conductive scaffolds for tissue engineering applications. *MedComm—Biomaterials and Applications, 3*(2), e76. https://doi.org/10.1002/mba2.76
69. Najafi, M., Forestier, E., Safarpour, M., Ceseracciu, L., Zych, A., Bagheri, A., et al. (2024). Biodegradable polylactic acid emulsion ink based on carbon nanotubes and silver for printed pressure sensors. *Scientific Reports, 14*(1), 10988. https://doi.org/10.1038/s41598-024-60315-z
70. Gackowski, B. M., Sharma, M., Koh, X. Q., Seng, D. H. L., Verma, D., Raveenkumar, V., & Idapalapati, S. (2024). Surface engineering of carbon nanotube-carbon fiber networks for enhanced strength in additive manufacturing of nylon composites. *Composites Part A: Applied Science and Manufacturing, 186*, 108383. https://doi.org/10.1016/j.compositesa.2024.108383
71. Oh, W.-J., Kim, D., Kim, D.-H., Chun, D.-M., Ha, J., & Kim, C.-S. (2024). Fabrication of CP-Ti structure with controllable wettability using powder bed fusion and eco-friendly post-process. *Scientific Reports, 14*(1), 4633. https://doi.org/10.1038/s41598-024-54958-1
72. Yu, S., Guo, J., Zhang, G., Chen, C., Zeng, S., Chu, C., et al. (2024). Improved broadband design of SiC/MWCNT absorbing materials through synergistic regulation of heterointerface structure and triple periodic minimal surface meta-structure. *Carbon, 226*, 119181. https://doi.org/10.1016/j.carbon.2024.119181
73. Hughes, K. J., Iyer, K. A., Bird, R. E., Ivanov, J., Banerjee, S., Georges, G., & Zhou, Q. A. (2024). Review of carbon nanotube research and development: Materials and emerging applications. *ACS Applied Nano Materials, 7*(16), 18695–18713. https://doi.org/10.1021/acsanm.4c02721
74. Ali, M. M., Nassrullah, G., Abu Al-Rub, R. K., El-Khasawneh, B., Ghaffar, S. H., & Kim, T.-Y. (2024). Influence of carbon nanotubes on printing quality and mechanical properties of 3D printed cementitious materials. *Developments in the Built Environment, 18*, 100415. https://doi.org/10.1016/j.dibe.2024.100415
75. Alshihabi, M., & Kayacan, M. Y. (2024). Effect of nanosized carbon nanotubes, titanium nitride and cubic boron nitride powders on mechanical and thermal properties of SLA 3D printed resin composites. *Polymer Composites, 45*(17), 15561–15573. https://doi.org/10.1002/pc.28850
76. Ramachandrarao, M., Khan, S. H., & Abdullah, K. (2025). Carbon nanotubes and nanofibers—reinforcement to carbon fiber composites—Synthesis, characterizations and applications: A review. *Composites Part C: Open Access, 16*, 100551. https://doi.org/10.1016/j.jcomc.2024.100551
77. Tuli, N. T., Khatun, S., & Rashid, A. B. (2024). Unlocking the future of precision manufacturing: A comprehensive exploration of 3D printing with fiber-reinforced composites in aerospace, automotive, medical, and consumer industries. *Heliyon, 10*(5), e27328. https://doi.org/10.1016/j.heliyon.2024.e27328
78. Kumar, R., Singh, R., Kumar, V., Ranjan, N., Gupta, J., & Bhura, N. (2024). On 3D printed thermoresponsive PCL-PLA nanofibers based architected smart nanoporous scaffolds for tissue

reconstruction. *Journal of Manufacturing Processes, 119*, 666–681. https://doi.org/10.1016/j.jmapro.2024.04.008
79. Chen, C. H., Dash, B. S., Ting, W. C., & Chen, J. P. (2024). Bone tissue engineering with adipose-derived stem cells in polycaprolactone/graphene oxide/dexamethasone 3D-printed scaffolds. *ACS Biomaterials Science and Engineering, 10*(10), 6425–6440. https://doi.org/10.1021/acsbiomaterials.4c00774
80. Dixit, M., Shrestha, L. K., Ariga, K., & Pati, F. (2024). Carbon-based nanoarchitectonics in advancing cardiac tissue bioprinting: A review. *ACS Applied Nano Materials, 7*(21), 24638–24652. https://doi.org/10.1021/acsanm.4c04441
81. Wang, Y., Liu, C., Song, T., Cao, Z., & Wang, T. (2024). 3D printed polycaprolactone/β--tricalcium phosphate/carbon nanotube composite—Physical properties and biocompatibility. *Heliyon, 10*(5), e26071. https://doi.org/10.1016/j.heliyon.2024.e26071
82. Raut, P., Kulkarni, D. S., & Kandasubramanian, B. (2024). The role of polyurethane and its composites in additive manufacturing of sensors. *Polymer-Plastics Technology and Materials, 64*(6), 850–862. https://doi.org/10.1080/25740881.2024.2439442
83. Triveni, B. V., Lalithamba, H. S., Bharath, H. S., Kumar, N. V., & Prashanth, G. K. (2024). Three-dimensional printed integrated electrochemical devices for various applications—A review. *ChemistrySelect, 9*(35), e202402463. https://doi.org/10.1002/slct.202402463
84. Putri, K. N. A., Intasanta, V., & Hoven, V. P. (2024). Current significance and future perspective of 3D-printed biobased polymers for applications in energy conversion and storage system. *Heliyon, 10*(4), e25873. https://doi.org/10.1016/j.heliyon.2024.e25873
85. Vanmathi, S., Awasthi, H., Pal, A., & Goel, S. (2024). IoT enabled carbon cloth-based 3D printed hydrogen fuel cell integrated with supercapacitor for low-power microelectronic devices. *Scientific Reports, 14*(1), 16935. https://doi.org/10.1038/s41598-024-67759-3
86. Munir, S., Ali, B., & Gul, S. (2024). The advancements of 3D-printed electrodes in electrochemistry. *Discover Electrochemistry, 1*(1), 14. https://doi.org/10.1007/s44373-024-00013-7
87. Yang, R., Chen, X., Zheng, Y., Chen, K., Zeng, W., & Wu, X. (2022). Recent advances in the 3D printing of electrically conductive hydrogels for flexible electronics. *Journal of Materials Chemistry C, 10*(14), 5380–5399. https://doi.org/10.1039/d1tc06162c
88. Banupriya, R., Jeevan, T. P., Divya, H. V., Yashas Gowda, T. G., & Manjunath, G. A. (2024). 3D-printed graphene-reinforced composites: Opportunities and challenges. *Polymer Composites, 46*(2), 1250–1266. https://doi.org/10.1002/pc.29068
89. Bin Abu Sofian, A. D. A., Lim, H. R., Chew, K. W., & Show, P. L. (2024). Advancing 3D printing through integration of machine learning with algae-based biopolymers. *ChemBioEng Reviews, 11*(2), 406–425. https://doi.org/10.1002/cben.202300054
90. Imran, M. M., Che Idris, A., De Silva, L. C., Kim, Y. B., & Abas, P. E. (2024). Advancements in 3D printing: Directed energy deposition techniques, defect analysis, and quality monitoring. *Technologies, 12*(6), 86. https://doi.org/10.3390/technologies12060086
91. Olawumi, M. A., Oladapo, B. I., & Olugbade, T. O. (2024). Evaluating the impact of recycling on polymer of 3D printing for energy and material sustainability. *Resources, Conservation and Recycling, 209*, 107769. https://doi.org/10.1016/j.resconrec.2024.107769
92. Rajendran, J., & Esfandyarpour, R. (2024). Revolutionizing personalized health: The frontier of wearable biomolecule sensors through 3d printing innovation. *Biomedical Materials and Devices*. https://doi.org/10.1007/s44174-024-00226-9
93. Hussan, K. S., Subramaniam, M. P., Kenz, K. T., Sreeram, P., Parvathi, S., Ps, S., et al. (2024). Fabrication and challenges of 3D printed sensors for biomedical applications-comprehensive review. *Results in Engineering, 21*, 101867. https://doi.org/10.1016/j.rineng.2024.101867
94. Agarwal, P., Mathur, V., Kasturi, M., Srinivasan, V., Seetharam, R. N., & S Vasanthan, K. (2024). A futuristic development in 3d printing technique using nanomaterials with a step toward 4D printing. *ACS Omega, 9*(36), 37445–37458. https://doi.org/10.1021/acsomega.4c04123
95. Ayanda, O. S., Mmuoegbulam, A. O., Okezie, O., Durumin Iya, N. I., Mohammed, S. A. E., James, P. H., et al. (2024). Recent progress in carbon-based nanomaterials: Critical review. *Journal of Nanoparticle Research, 26*(5), 106. https://doi.org/10.1007/s11051-024-06006-2

Applications and Implications of Exploring Carbon-Based Nanocomposites in Sustainable Cities Development

Abraham Barde, Victor Ugbetan Agbogo, Amina Hassan Bayero, Oryina M. Injor, Simeon Sunday Pama, Rifore Silas Belief, and Lawal Abdulshakur Isah

1 Introduction

Carbon-based nanocomposites (CNCs) refer to composites composed of advanced materials with carbon frameworks at the nanoscale, bridging carbon science and nanotechnology. [1]. To produce CNCs, carbon allotropes integrate with nanoparticles, metal oxides, or polymers, resulting in tailored nanomaterials [1–4].

Abraham Barde and Victor Ugbetan Agbogo contributed equally with all other contributors.

A. Barde (✉)
Department of Materials, University of Manchester, Manchester, UK

Air Force Institute of Technology, Kaduna, Nigeria
e-mail: abraham.barde@manchester.ac.uk

V. U. Agbogo · O. M. Injor
Department of Chemical, Metallurgical, and Materials Engineering (Polymer Division) & Institute of Nano-Engineering Research (INER), Tshwane University of Technology, Pretoria, South Africa

A. H. Bayero
Department of Polymer Technology, Nigerian Institute of Leather and Science Technology, Zaria, Nigeria

S. S. Pama
Department of Chemistry and Biochemistry, Ohio University, Athens, OH, USA

R. S. Belief
Department of Materials Science, Montana State University, Bozeman, MT, USA

L. A. Isah
Research Centre for Experimental Marine Biology and Biotechnology, University of the Basque Country, Leioa, Spain

V. Khanna (ed.), *Carbon-Based Nanocomposites for Sustainable Applications, Volume I*, Lecture Notes in Nanoscale Science and Technology 38,
https://doi.org/10.1007/978-3-031-95306-4_10

Carbon-based nanocomposites have enormous applications in techno-scientific domains with very exceptional and fascinating properties. Synthesis techniques including chemical vapor deposition, self-assembly, and sol-gel processes are employed to harness these properties, allowing precise control over their morphology and structure. The mechanical and electrical characteristics feature spanning diverse types of nanocomposite materials of different compositions of carbon nanotubes making it a potent sustainable city development. These nanocomposites are reckoned to be the man-made stiffest and strongest materials in the world today [5]. To attain these properties, chemical modifications have been employed over time. The excellent properties of their structure are in no doubt contingent on their minuscule size providing an opportunity to incorporate them into various matrices for enhanced performance [5].

Interestingly, carbonaceous nanofillers which include carbon nanotubes (CNTs) with graphene demonstrate a potential role in the development and design of nanocomposites owing to their improved functional, and structural, properties cum broad range of engineering across multiple engineering disciplines. Usually, the manipulation, distribution, and orientation of carbon nanotubes within nanocomposites seem to be an important phenomenon. For instance, the integration of graphene and carbonaceous nanotubes as fillers can lead to substantial transformation in the mechanical behavior and electrical behavior of composite matrices [6, 7].

The discovery and development of carbon nanotubes is an amazing achievement of human ingenuity. These light-in-pound substance features huge dimension ratios and are characterized by characteristics such as incredible mechanical properties. Their exceptional electrical properties and great adaptability have led to extensive research on them for over a decade or more in pertinence to multifunctional applications. However, the employment of CNTs in their unprocessed state could result in an inadequate representation of their innate behavioral characteristics, hence restricting technological advancement to macroscopic forms. According to the review by Sindhu et al. [8], it is essential to combine carbon nanotubes (CNTs) with synthesized engineering-grade polymers (such as thermoplastics, elastic materials, and conjugated polymeric materials) or bio-based polymers to make use of their inherent properties and create nanocomposites that have additional advantages. Developmental perspectives on the various tactics used for fabricating and designing nanocomposite components to attain functional structures are presented in the literature [8].

The physical features of synthesized CNT nanocomposites can be tailored with special target impetus on their chemical stability, and structural, heat-related, flow, electronic, and light-interactive properties. Key factors such as high thermal, mechanical strength, and high electronic conductance and expansive surface area, biocompatibility, distinct light-interactive characteristics, and chemical resilience make it a potential nanocomposite matrix for a revolution in the development of modern cities. Hence, the current work highlights select contemporary implementations and implications of carbon-based composites within the areas of sustainable city development.

2 Carbon-Based Nanocomposite in Environmental Monitoring

The growing concern over environmental pollution due to industrialization, fossil fuel use, forest fires, agricultural activities, inadequate waste management, urbanization, as well as the discharge of harmful contaminants into atmospheric and aquatic systems through cars, factories, and other sources has necessitated the need to explore the techno-innovative method of dealing with the challenges of which carbon-based nanocomposite is found to be a fascinating field, making it a widely researched scientific field in the recent times. Meanwhile, their distinctive characteristics, which cut across their large surface region, tunable reactivity, because their chemical resilience renders them versatile tools for wide-ranging uses in pollution control especially in the purification of polluted soil, water, and air.

2.1 Carbon-Based Nanocomposites in Water Treatment Applications

The most important natural resource in the universe is water, making it a significant material for sustainable development of cities and communities hence the need to devise a viable means of treating it from unwanted substances. Because of the rapid increase in urbanization and industrialization, water consumption has multiplied and met with diverse constraints. So many discharges and effluents like toxic heavy metals have caused enormous pollution in the water bodies in many environments [9–15]. Thus, swift and optimized extraction of those unwanted substances from water bodies is also a serious and paramount herculean task of a public health emergency. The incipient requirements for sanitized water have resulted in the creation of several carbonaceous nanotools for the quick cum effective pollutants remediation cum heavy elements from water, some of which include arsenic, copper, cadmium, lead, chromium, and so forth. Nanomaterials contain a very enhanced capacity to capture heavy metal contaminants from wastewater sources over conventional methods because of their elevated surface dimension, or area/volume ratio [16–18]. The capture of toxicity of heavy metals in aquatic environments is a critical matter and of utmost concern to Mother Earth. The promising vantage points of nanomaterials in sustainable environmental remediation future directions ought to be explored as a matter of urgency. Adsorption has been widely adopted to process heavy metal-laden industrial effluents owing to the adaptable, easy, and affordable nature of the method of remediation. The treatment of wastewater benefits from adsorbents like metal-oxide nanoparticles, carbonaceous particles, and nanocomposites, in addition to new advances in nanotechnology [19]. Runoffs from agricultural fields, residential zones, and industrial operations increase with the rate of growth in the world's population. Hazardous metallic materials are present in industrial activities, which typically end up in waterways. The inadequacies of traditional purification techniques in terms of removal efficiencies have necessitated the need for carbonaceous adsorbents. In place of the conventional method, nanotechnologies in recent times have been developed as innovative remediation strategies and are reckoned to offer advantages in water treatment [14, 20–22].

In a classic CNC synthesis utilizing the sol-gel technique, Wang et al. employed multi-wall CNT-based multi-wall CNT/TiO_2 nanocomposite materials for preparation, in result the authors reported that higher photocatalytic activity was noticed compared to MO solution breakdown using a single TiO_2 catalyst exposed to UV light. This outcome is undoubtedly due to the consistent nanoscale layering of TiO_2 particles coupled with an expansion of the light absorption spectrum. The elevated photoreactive efficiency and stability exhibited show great potential in wastewater treatment [20].

From history, various materials as displayed in Fig. 1 have been utilized as adsorbents for the treatment of water for use in cities and communities. Interestingly, the application of nanoscale technologies for chromium removal offers significant benefits over conventional adsorbents, rendering them a viable choice for optimized rapid kinetics, high interface area, acceptable regenerative rate, high purity, and exceptional contact charge-induced sorption capability, pore size distribution, and surface chemistry, making them a dependable method for purifying water [21, 23].

The up-and-coming perspective of nanomaterials in sustainable environmental applications cannot be overemphasized. Carbon nanotubes (CNTs) possess extensive surface-to-volume ratios, multilayered and tubular formations, plus enhanced

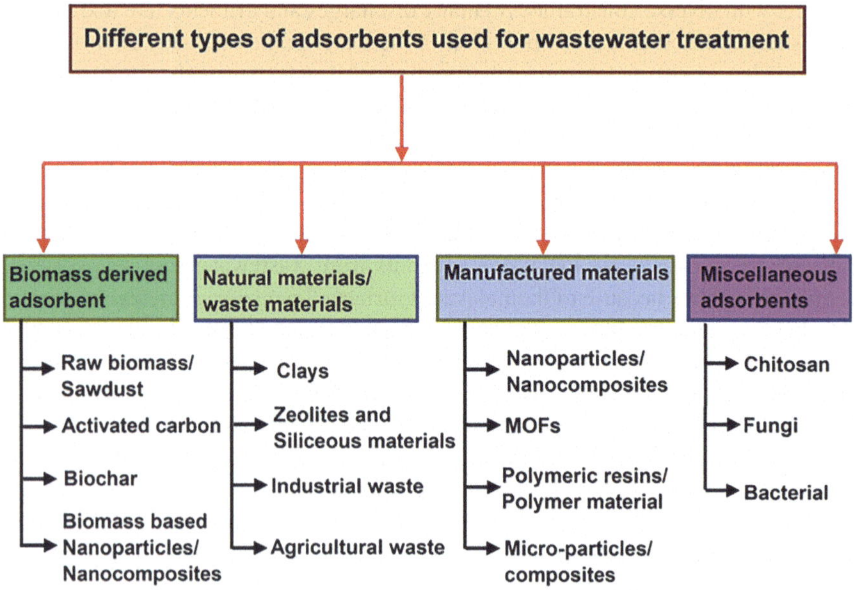

Fig. 1 Types of adsorbents for the treatment of water in cities and communities [23]. (Reproduced with copyright permission, Elsevier 2024)

adsorption potential compared to different synthetic organic compounds thus making them promising adsorbent materials for the treatment of water. The surface chemistry according to Zhang et al. of carbonaceous nanomaterials resembles characteristics of both activated carbon fibers (ACFs) and traditional granule-activated carbons (GACs) although they differ greatly in their pore topologies. GACs constitute a highly heterogeneous pore structure, with usually wide pore size distributions and slit-shaped while ACFs are those systems that are characteristically uniform, dominated by ellipsoidal micropores. Meanwhile, the fundamental configurations of CNTs are said to be comparatively more consistent than GACs and ACFs. CNTs tend to aggregate and are highly hydrophobic as susceptible to strong van der Waals forces extending longitudinally. This phenomenon usually causes CNTs to demonstrate a reduction in surface coverage despite expanded pore capacities via the development of microchannels and peripheral indentations within the clusters [24].

An environmentally friendly nanocomposite synthesized by Hassan et al., which was founded on nanoparticles ($CaCO_3$-NPs) and chitin (Cs) was merged to integrate the qualitative attributes of nanomaterials and bio-based polymers to defeat challenges of inadequate mechanical properties, biopolymers' fragility and poor removal capabilities. According to their reports, the as-prepared samples were investigated and characterized as well as employed in the adsorption of metallic contaminants from wastewater. Techniques such as utilizing XRD, TEM, and FTIR analytical techniques were employed to study the synthesized adsorbents. Based on their findings, the heavy metals uptake by the prepared and analyzed samples under various circumstances along with the utilizing XRD, TEM, and FTIR analytical techniques kinetics and after a thorough investigation of isothermal parameterization of the adsorption step, the findings showed that the composite removed 98.0, 94.8, 99.0, 97.9, 97.4, and 98.3% of Cu, Cd, Pb, Cr, Zn, and Ni, accordingly. In the meantime, both kinetic and thermodynamic investigation examinations demonstrated that the composites' uptake of these metal ions gave rise to the Langmuir isotherm and the pseudo-second order, respectively. Furthermore, it was discovered that the produced nanocomposite had the maximum metal absorption capabilities for Cd, Cu, Pb, Zn, Cr, and Ni ions, which were 83.33, 47.84, 98.03, 89.28, 62.11, and 63.69 mg/g, respectively [25]. Graphene oxide, carbon dots, fullerene, nanotubes, and derivatives are examples of carbon-derived nanoscale compounds that are used in the treatment and elimination of emerging contaminants from sewage bodies and have shown great promise as an underwater treatment method.

Based on the foregoing points, these nanomaterials demonstrate superior adsorption efficiency and specificity properties given their customizable chemistry, extensive surface area, and special structural behaviors that include elements such as π-π relationships, specific surface functionality, and electrostatic interactions. A notable experiment according to the reviewed work of Thakur et al. demonstrated the efficiency and Ismate violet 2R dye may be effectively treated and removed from infected water using made from carbon nanostructures as adsorbents, demonstrating an impressive highest elimination efficiency of roughly 91.71% at a dose of 10 mg/L in a short 120-min interval [26].

2.2 Carbon-Based Nanocomposites for Pollutant Remediation

Because functionalized CNCs have high-quality absorption capabilities for gases and volatile organic compounds (VOCs), they are added to air filtration systems to catch and eliminate contaminants, improving air quality. These substances can also be used as catalytic supports for the creation and design of advanced oxidation procedures (AOP), which will speed up the breakdown of airborne contaminants [1].

The carbon-derived materials (CBMs) in remediating heavy metal-contaminated soils have attracted increasing attention for their multifunctional properties and high efficiency. CBMs like CNTs, graphene derivatives, activated carbon forms, biochar, and carbon black particles are utilized for the heavy metal-polluted soil. For example, the processes involved in heavy metal removal from soil utilizing CBMs were outlined in the work of Abdullah et al. and listed as follows: Adsorption by matter, cationic-π coordination, complexation, precipitation, and electrostatic contact. Soil pH, minerals, organic matter, microbes, moisture, interacting ions, and material size are the main factors affecting the cleanup's impact. Concerns over safety hazards, elevated costs, and implementation challenges in attaining anions and cations co-passivating are the disadvantages that may hinder the use of CBMs. CPolymer-integrated carbon nanocomposites, such as CNFs, CNTs, graphene, and its oxidized forms can be used as filler components in the nanocomposites to eliminate pigments and dyes. The critical review of carbonaceous nanocomposites' performance by Abdullah et al. for extracting dyes and pigments via membrane filtration, adsorption, and photoreactive techniques indicated that CNCS can be a promising ingredient in remediating pollution in cities [4]. Considering the above characteristic properties of carbon-based nanocomposites, serious efforts have been made by Rivas-Sanchez et al. to spotlight the synergies existing between graphene oxide, carbon-based nanotubes, and mesoporous nanocomposites blended with algae, as well as their contribution to reducing emerging pollutant impacts. According to their reports and reviews, it was suggested that algae is also a viable material that possesses excellent potential for pollutant removal. However, the authors put forward suggested further evaluations across varied domains and material types that exploit its wider remedial objectives of the environment [27]. This implies that the material resources by linking their active functionality could culminate in end products that can be utilized for pigments, dyes, and many other forms of unwanted substances in the environs.

Meanwhile, addressing emerging contaminants (EC) has become an urgent matter of public emergency. Biochar has been applied in the work of Zhang et al., for EC treatment in wastewater owing to its satisfactory efficiency and low cost, but it also revealed that practically, its restricted adsorption capability is what makes its practical use so important. Multiple techniques, such as acid-base relationships alterations, steam activation, metal compound modification, hetero atoms doping, magnetization, and pellet milling, are utilized to change charcoal and steep it using a few physical properties to improve the treatment efficiency of this carbon-based material called biochar on ECs. This results in unique adsorption outcomes, mechanisms, and behaviors [28]. Intriguingly, modified biochar increased the removal capacity for newly discovered pollutants, demonstrating that the removal mechanisms largely and crucially followed two paths (catalytic degradation and adsorption). However, to progress, future attempts should standardize evaluation criteria for cost, environmental risk, and efficacy. They should also take advantage of the numerous prospects in robotics and expand the use of fortified biochar.

Again, the discharging of environmental pollutants owing to industrialization, urbanization, and other human activities constitutes a serious threat to environmental sustainability and the living beings in every city and community. To address this environmental challenge, the usual traditional methods have not been able to provide satisfactory results in treatment processes, hence the need for attention given to the modification of carbonaceous nanomaterials. The interest in carbon-based nanocomposites is in no doubt contingent upon their multifunctional features and excellent tunable properties [29–31]. Figure 2 displays features that agree with the above statements. Numerous efforts in developing and designing sensing technologies are evident in the works of literature over the past few years by several study writers, indicative that compact, cost-effective, and eco-friendly sensors are criteria [30–32]. Advanced biosensors employing innovative nanofabrication and eco-friendly synthesis methods are provided by nanotechnology research. Aguilar-Perez et al. [30] in their reviews, asserted that nanoscale biosensors applied for pollutant detection demonstrate high sensitivity with rapid detection in real-time monitoring. Aguilar-Perez's works narrowed down on progress in green nanotechnology-based chemical sensors for environmental applications, concentrating on detecting impurities in aquatic and terrestrial environments [30].

Strong evidence from the study to date suggests that nanomaterials are useful tools for industrial applications and could remove pollutants effectively and efficiently. To fully realize this potential, however, more study and advancement are required. Examples of these include investigating novel synthetic pathways and

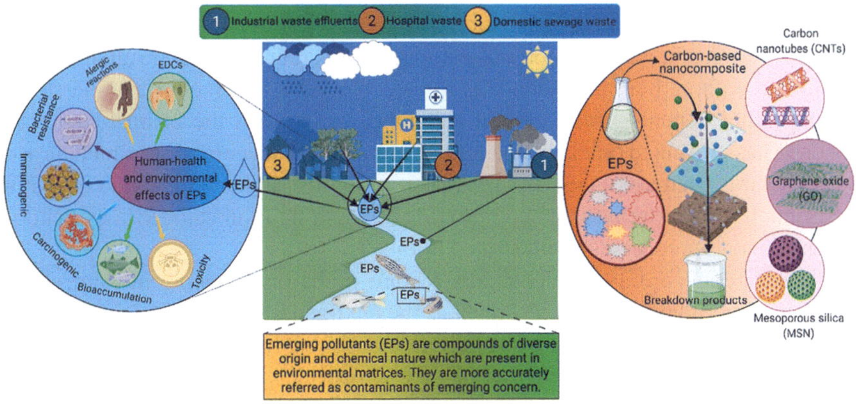

Fig. 2 Plate showing the sources and effects of pollution in the environment in cities [27]. (Reproduced with the copyright permission, Elsevier 2024)

improving the environmental dispersion and uptake of nanoparticles. Comparing the effectiveness of various NM kinds for cleaning up contaminants is also necessary. All things considered, this review emphasizes the enormous potential of nanomaterials to mitigate environmental contaminants and urges further study in this area [33].

3 The Use of Carbon-Based Nanocomposites in Concrete Structure

The duo characteristic features of CNCs, including lightweight and mechanical characteristics, make them a viable and potential material for concrete structures instead of steel reinforcing. CNTs are believed to be a hundred times stronger than steel and yet lighter. This is a highly desired property in the engineering science of emerging technology. This is suggestive of the fact that CNCs possess the ability to specify and revolutionize material construction in terms of high-strength plus lightweight properties. Regenerating structural materials with this carbon-based nanocomposite can bring a facelift to infrastructures in cities among other things. The lightweight constitution of CNT cum graphene, plus their extraordinary power, may help as it were to the design and innovation of materials that improve fuel economy in vehicles culminating in the design of durable and resilient infrastructures. More so, the research advancement around material miniaturization and increasing demands for smaller and more efficient substances coupled with the global campaign for eco-friendliness make CNCs a component of interest in sustainable and green technology.

Nanotechnology creates a creative approach to revolutionary civil engineering in self-purifying, dampness that senses and vibrates itself, self-treatment, self-healing, cement-reinforcing, and structural health monitoring in concrete processes. The best conditions for creating resilient structures like towers, bridges, and tall buildings can be created by adding nanomaterials to building materials, which can increase their flexural and compressive strength. Concrete qualities are altered when reinforced with carbon nanotubes due to their mechanical behavior and elasticity. Among other benefits, using carbon nanotubes to improve the characteristics of concrete has prevented cracking and protect structures from corrosion, which turned out to notably increase its use in this industry [34, 35]. Functionalization of nanotubes can improve strength and compromise strength flaws in cementitious materials and concrete processes as illustrated in Fig. 3.

Nanomaterials are a game-changing feature in the fast-growing modern world and nanotechnology is highly reckoned as the next-generation industrial revolution owing to the extraordinary characteristics demonstrated by them in their minuscule form. Graphene and graphene are among the most widely utilized nanomaterials in cement-based composites. Their incorporation has been shown to significantly enhance mechanical strength and durability, helping to address key limitations of

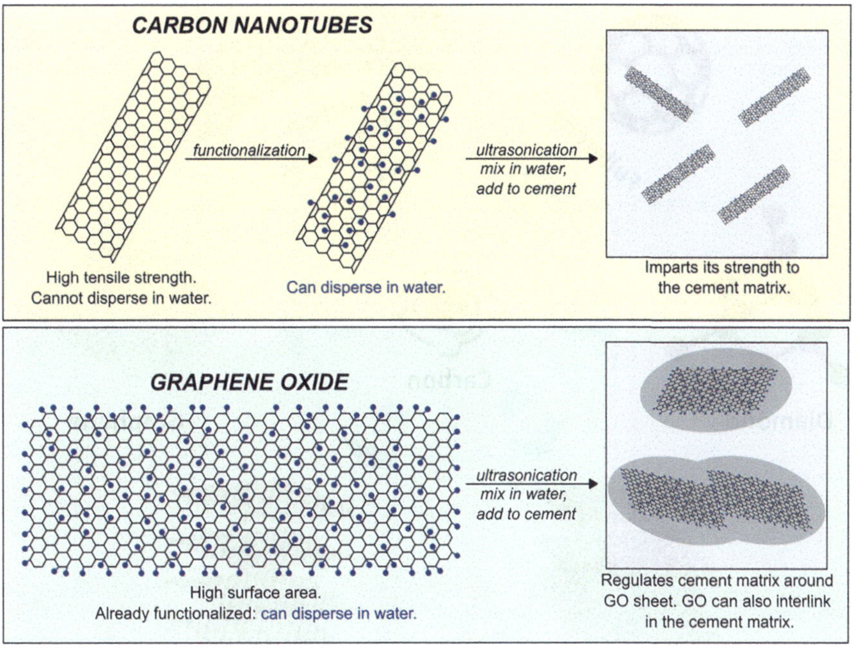

Fig. 3 Functionalization of carbon nanotube [36]. (Reproduced with the copyright permission of Elsevier. Elsevier, 2021)

conventional concrete and other cementitious materials. Different types of carbonaceous nanomaterials are illustrated in Fig. 4. Bautista-Gutierrez et al. conducted a research test on hardened, and durability characteristic features of concrete reinforced with graphene-based nanomaterials. According to their report review, graphene-based nanomaterials are found to overcome the various distresses associated with concrete based on the statistical interferences [38].

When it comes to environmental cleanup, nanoparticles have gained great interest. In light of their high reactivity and enormous surface area, nanoparticles operate better in environmental cleanup than other conventional approaches. They can be changed to provide applications with special features. Nanoscale substances exhibit high reactivity because of their higher number of sites that react and a large surface-area-to-volume proportion. Higher interaction with contaminants is made possible by these features, which causes the concentration of contaminants to drop quickly [26]. The creation of novel, multipurpose smart materials based on MWCNT and CNF-based composite structures) is paramount in sustainable city development and design in the modern age. CNF/MWCNT's material qualities are marginally inferior to those of SWCNT, or SWCNT variants. However, owing to their reduced expense, the CNF/MWCNT offers broader potential for practical implementations [25].

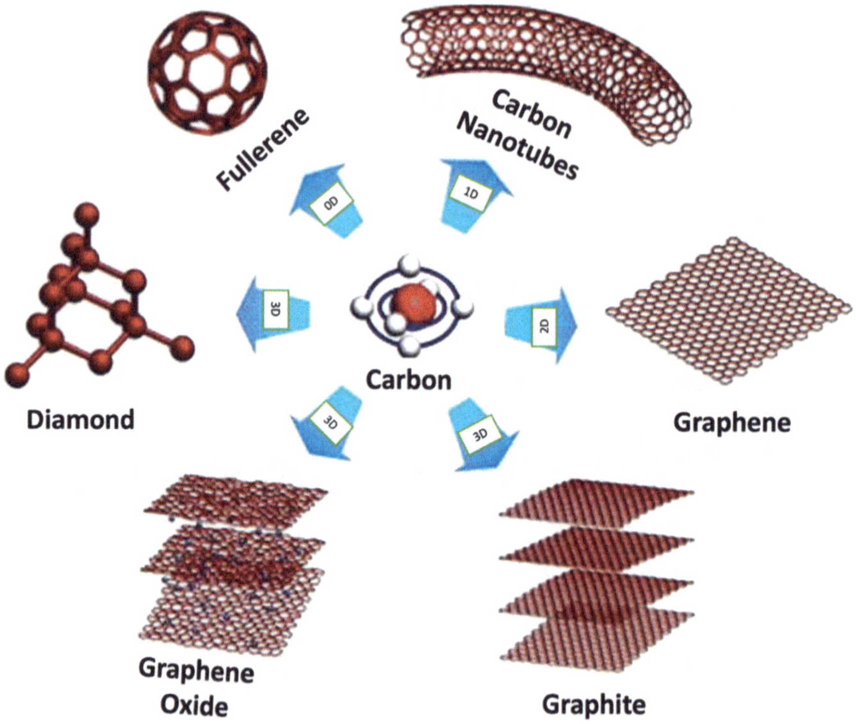

Fig. 4 Example of nanocarbon materials based on their structures [28, 37]. (Reproduced with the copyright permission of the publisher)

Nanotubes of carbon with several walls can enhance the thermal cement's mechanical characteristics when added to the cement matrix while polymethyl methacrylate (PMMA) is present. According to Liu et al., as shown in Fig. 5, the employment of unfunctionalized multi-walled CNTs in the reinforcement of cement did not show a notable impact on cement performance due to particle aggregation issues. However, using functionalized multi-wall CNT + in the present liquid monomer for the fabrication of composites is found to be more effective in creating a bridge effect and in improving cement properties.

Carbon-based nanocomposites when functionalized and integrated into construction materials will improve energy efficiency and thermal insulation in buildings; the key factor for their promising utilization in defense infrastructure, roadway projects, and sustainable urban initiatives [39–41].

Our expectations, perceptions, and capacity to manipulate the material world have all been altered by nanotechnology and will continue to be so. There is little doubt that these advancements are expected to significantly impact the construction sector and building materials hence our cities and communities. Carbon-based nanomaterials are known to be among the most effective and beneficial nanostructured reinforcements. Given their significant proportion of length to diameter, tiny

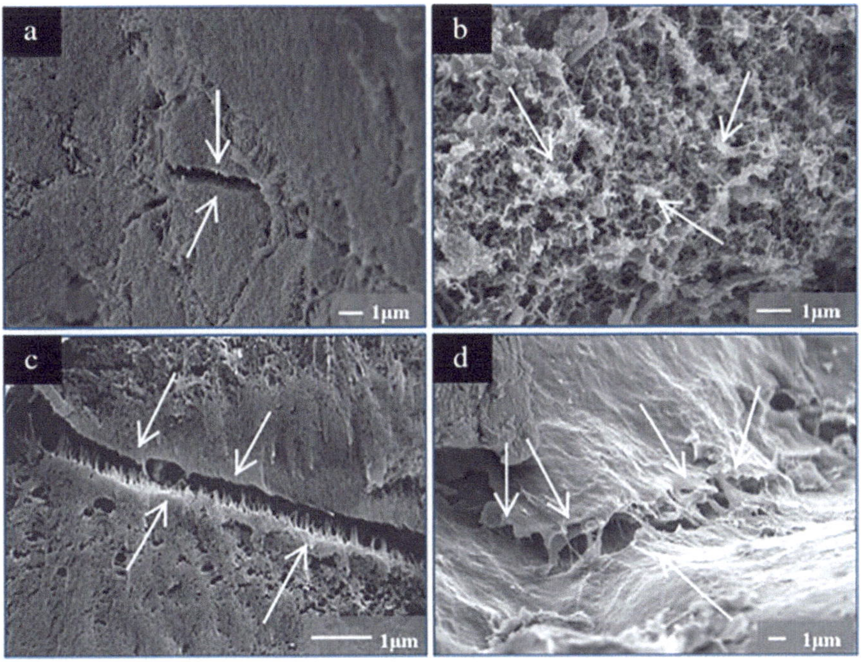

Fig. 5 SEM imaging depicting crack morphologies (**a**), crack infused with unmodified polymer and MWCNTs (**b**), functionalized MWCNT-polymer composite processed via high-intensity ultra-sonication in synthetic process × 5000 (**c**), and functionalized multi-wall CNT + polymer in a manufacturing process with ultra-sonication × 15,000 [34]. (Reproduced with the copyright permission of Elsevier)

size, low density, and distinct physical and chemical characteristics, they are ideal resources for reinforcements in multipurpose and intelligent cement-based substances. In this chapter, we also examine the latest developments in cement-based materials reinforced with carbon nanotubes, paying particular emphasis to their manufacturing processes, mechanical characteristics, electrical and piezoresistive characteristics, heat conductivity and damping characteristics, and their structural uses. Future obstacles to the creation and use of advanced CNT-integrated cementitious composites and structures are also addressed in the later section [42, 43].

Due to their exceptional qualities, CNTs are recognized as promising options for future high-efficiency and multifunctional cement composites in the twenty-first century. A uniform CNT suspension was produced by Wang et al., using a method that utilized ultrasonic treatment, mechanical agitation, electric field integration, and surfactant-assisted modification. According to their findings, the mechanical characteristics of CNTs/cement composites were practically enhanced by the low weight percentage of CNTs, suggesting that this resulted in improved compressive durability and structural resilience. The crafted CNTs/cement mixtures' impermeability and resistance to sulfate attack were enhanced by the additional CNTs. In the meantime, the produced mixtures' conductivity, electromagnetic absorption, and pressure-sensitive qualities were enhanced by the addition of CNTs [44], which established the framework and multipurpose concrete.

In the work of [45] MIP and SEM analyses were applied to characterize Portland cement–CNTs systems produced in their work. MIP analysis of the results, according to their report, indicated a reduction in total porosity with increasing CNT content. Thus, as seen in Fig. 6, the overall porosity of the CNT-containing mixtures dropped as the CNT content rose. Additionally, the number of mesopores decreased because of the significant effect of more CNTs, while the SEM approach revealed that CNTs were dispersed throughout the phases at which Portland cement pastes hydrate [45]. The incorporation of CNFs into cementitious materials is equally an effective method for the enhancement of the durability and ability to prevent cracks of the composite according to the work of Wang and Fan. However, the authors observed that the dispersion and influence mechanism of CNFs in concrete composites provide certain challenges and are controversial. The cement composites that were mixed with CNFs that had a reasonably excellent dispersion were made using ultrasonic treatment and a dispersant. Cement samples with various CNF doses and W/C ratios were also made to investigate the rheological characteristics and mechanical behavior. According to the author's report, the results showed that when the shear rate and doses of CNFs increased, the rheological variables, including consistency, shear strain, and stress due to yield, increased; however, the opposite pattern was observed when the W/C ratio increased from 0.28 to 0.32. In juxtaposition with the untreated sample, the cement composites containing 0.1 weight percent CNFs demonstrated rises in compressive power and tensile splitting force from 95.2 to 103.8 MPa and 6.60 to 7.38 MPa, correspondingly, with boosts of 9.0% and 11.8% per W/C of 0.30 [46]. Therefore, one could ascertain from these findings that the microscopic examination indicates that an optimal CNF dosage can significantly enhance pore architecture and interfacial properties in cementous composites given the CNFs have three effects: filling, bridging, and nano-core.

Meanwhile, the research that was carried out by Lavagna et al. involving the fabrication of multiple cementitious composites with diverse functionalized

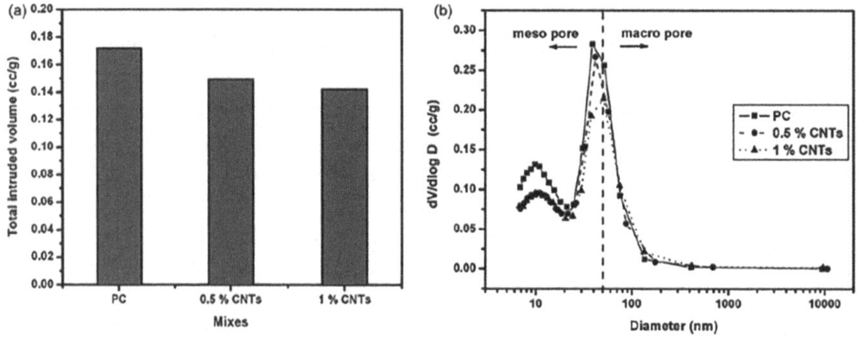

Fig. 6 The effect of CNTs in total intruded volume (cm^3/g) (**a**) and the arrangement of pore sizes (**b**) of Portland cement [34, 45]. (Reproduced with the copyright permission of Elsevier)

carbon-based additives as a result confirmed that surface modification through chemical functionalization or oxidative treatment improved the consistent dispersion and cohesive interaction of the reinforcements inside the cement matrices [47]. From the foregoing it is pointing that a synergistic effect between the carbon-based nanotube and carbon fiber was shown to yield a synergistic reinforcement effect, thus allowing achieving nearly twice the flexural strength of plain cement and exceeding sixfold fracture energy improvement [47–49].

Interestingly, Liu et al. observed in their research work that early-stage cracking of cement-based materials due to shrinkage deformation is a significant issue that can result in material failure under controlled settings. To address the material degradation brought on by cement-based material shrinkage, Liu et al. asserted that carbon nanotubes (CNTs) could be added to cement-based materials as a reinforcing agent. The new composites created via linear contraction and ring experiments are examined for autogenous shrinking and crack resistance. The findings demonstrated that the autogenous shrinkage of cement-based materials was considerably enhanced by CNTs, with a maximum decrease rate higher than 40%. Additionally, CNT greatly increased the cement-based products resilience to cracking. At 0.1 weight percent CNT concentration, the best results were observed [50]. Hence, the use of carbonaceous nanotubes prevents cement-based products from autogenously shrinking and prevents drying, and shrinking of materials made of cement to a certain degree. As a result, carbon nanotubes may be able to address the material degradation brought on by cement-based materials shrinking.

The nanotubes of carbon are expected to offer intriguing mechanical features, including significant stiffness and axial strength, due to their seamless cylindrical graphitic composition. However, due to their extremely small dimensions, these properties have not yet been directly measured. Treacy et al. did estimate Young's modulus of isolated carbon nanotubes by determining the amplitude of their intrinsic thermal vibrations using a TEM. In their findings, carbon nanotubes were discovered to have excellent elevated Young's modulus within the terapascal scale. Given their low density and high stiffness, they may be suitable as nanoscale reinforcements for durable, lightweight composites [51]. Going by the excellent electronic features and a thermal conductance surpassing that of diamond to mechanical behavior where the outperform modern materials in strength, rigidity, and durability, carbon-based nanotubes possess tremendous character for the design and development of fundamentally new material systems in sustainable city development. Specifically, carbon nanotubes' remarkable mechanical qualities and low-density present opportunities for the regeneration of composite materials blended with nanotubes [52]. In the twenty-first century, significant potential exists for the application of carbonaceous nanoparticles across a range of advanced material systems and devices. Because of the impressive stiffness and strength of carbon tube-reinforced nanocomposites, carbon nanotubes (CNTs) have the ability to provide both remarkable mechanical performance and versatile properties. As interest in nanotechnology and the use of carbon-based nanomaterials in various fields of endeavor has grown, efforts to improve dispersion and functionalization techniques have intensified. To use CNTs to strengthen polymer-based nanocomposites, it is necessary to ensure proper bonding within the CNTs and the polymer matrix as well as possible [52–54]. Thus, tendencies for nanocomposite materials strengthened using carbon-based tubes showing tremendous strength and stiffness indicating an opportunity for implementation and usage in modern times [55].

Carbonaceous nanotubes and nanofibers, graphene and graphene nanoplates, decreased graphene oxide, and graphene oxide are examples of carbon nanomaterials. These materials have excellent mechanical, chemical, and physical features. They can be effectively added to regular Portland cement to improve its mechanical qualities, resilience, and electrical/thermal permeability. They have also been used to change the functional attributes of cement-based goods, making them stronger, more intelligent, and more resilient composites. Du et al. conducted a comprehensive review on carbon-nanomaterials-enhanced cement-based composites (CN-CBCs) and discovered that the successful introduction of carbon nanocomposites into ordinary Portland cement resulted in improvements in durability, mechanical performance, and thermal-electrical conductivity, and altered the functional characteristics, creating composites that are stronger, more intelligent, and more resilient than traditional cement-based products [56].

4 Implications of Implementation of Carbon-Based Nanocomposites

A lot of potential and possible potential uses are identified for CNT-based technologies, such as energy conversion systems and storage technologies; semiconductors, probes, and links that are nanometer in size; sensors; hydrogen storage medium; conductive and strong composites; electron emission screens and radiative output systems. Nowadays, these applications are utilized in decorations and urban design solutions. Prototypes and sophisticated gadgets are among the other uses; hydrogen storage is controversial. However, CNT costs, variability in nanotube types, and constraints in fabrication techniques are critical hindrances for some important potential uses of SWCNTs [57–60].

Despite the unique characteristic properties of CNCs, it is pertinent to identify and discover long-term solutions to guarantee their sustainability, safety, scalability, and interoperability. The future and problems of carbon-based nanocomposites include integration and commercialization opportunities, toxicity and environmental concerns, scalability and cost-effectiveness, and new trends with possible uses.

4.1 Scalability and Cost-Effectiveness

Optimization of CNCs requires a balance between adaptability and efficiency/cost-effectiveness. Scalability is the shift from the scale of the lab to huge-scale production or industry. CNCs show impressive properties in the laboratory, but hard to retain them on a large scale. This has been found to hinder the implementation of these promising materials, perhaps due to the high demand for resourcefulness, and synthetic procedures.

Carbonaceous materials despite their key factors as possible adsorption agents for chromium removal from tap water, [23] in comprehensive review highlighted some practical challenges. These problems are as follows:

- *Selectivity*: CNC is characterized by low selectivity for some metals like chromium and arsenic due to their tendencies to reduce the efficacy and efficiency of addressing contaminants by adsorbing other compounds found in wastewater.
- *Reuse and Regeneration issue*: Application of CNC for reusability and regeneration of materials loaded with arsenic and chromium are challenging tasks that may result in residual contaminants.
- *Cost estimations*: The price and scalability of manufacturing carbonaceous compounds used for treating wastewater on a huge scale is relatively high compared to others.

4.2 The Effect of Catalysts on Carbon-Based Manufacturing

The goal of manufacturing carbon nanotubes at low cost requires the role of a catalyst. Usually, the impurities traceable in the product relate to the elemental catalyst employed during manufacturing. Thus, there is a need for purification. Those impurities can be removed chemically with acids or thermally. Thermal purification can be highly effective at removing impurities, but care must be taken as the structure of the nanotubes can suffer damage if the heat is excessive. Another of the main objectives and motivations for any prospective commercial use is the cost-effective, large-scale manufacture of CNTs in the field of nanotechnology [61]. As is popularly known, elevating a catalyst's activity enables the lowering of process temperature and pressure, which of course minimizes energy expenditures, which are vital considerations in large-scale industrial processes. Reducing the number of undesirable reagents and byproducts is another significant benefit of optimizing catalytic processes [62–65]. Lately, greater sensitivity has been the focus of catalyst-based manufacturing in the search for environmentally friendly methods for commercial catalysis that can result in a circular economy via minimizing byproducts, cutting down on reaction steps, and improving byproducts.

Reactor and separation apparatus capital expenditures and energy usage might be reduced with enhanced catalysts with higher selectivity, activity, and life. Carbon-based nanotube (CNT) catalysts are intriguing due to their significant catalyst efficiency, selectivity, stability, and metal substrate interactions [61, 66].

The development of efficient adsorbents through waste resource reutilization for the treatment of wastewater is a sustainable strategy for city development and environmental protection. Wang et al., in their research work, synthesized GNS-based adsorbent exhibiting a notably large surface area of about 1149.7 m^2/g utilizing a single-step CVD process utilizing catalysts sourced for the first time from sewage. It was discovered that the generated catalyst was primarily composed of MgO with trace amounts of Fe_2O_3, MnO_2, and CaO. Their results showed that the metal oxides in the catalyst increased MgO's catalytic function, which led to the production of GNS. Based on these procedures, the GNS yield is 144% greater than the graphene nanosheet on commercial MgO. The authors also went on to carefully investigate the methylene blue uptake by GNS, and GNS based on their reports showed a high adsorption capacity with rapid uptake rates for MB at various temperatures (293–313 K) and a broad pH spectrum (2–12). Wang et al. determined that GNS had their full potential for adsorption of 1092.1 mg/g and exhibited a high effectiveness of elimination of roughly 99.0% in 20 min. Additionally, they found that the fitted results utilizing the Langmuir isotherm and second-order kinetics well matched the adsorption data. According to Wang et al., there are several adsorption mechanisms involved in the accumulation of methylene blue upon GNS, including H-bonding, π-π stacking interaction, and electrostatic relationships [67]. The product showed good reusability based on this catalyst-based regeneration testing, retaining a 95% removal efficiency following five cycles. Furthermore, when GNS was used to treat simulated textile wastewater, substantial decolorization rates (98.5–99.9%) were obtained. Using inexpensive and effective adsorbents, this research study offered a sustainable wastewater treatment that benefited from the effective utilization of wastewater sources.

To enhance catalytic activity in polymer-based polymer electrolyte fuel cells (PEFCs), Du et al. incorporated groups of sulfonic acid to the outermost layer of platinum (Pt/CNT) catalysts reinforced with carbon nanotubes. This engraftment was achieved by both radical polymerizations in vivo of 4-styrene sulfonate with a thermal breakdown of ammonium sulfate. XRD, TGA, TEM, FTIR, and electrochemical methods were employed to characterize the resulting sulfonated Pt/CNT-catalysts. Because of the sulfonated Pt/CNT-catalysts' well-distributed distribution and easier proton access, the electrodes containing the in vivo radical polymeric process of Pt/CNT-catalysts sulfonated 4-styrene sulfonate based on their findings performed better than those with their unsulfonated counterparts. This suggests that sulfonation is an effective method to enhance efficiency and lower costs in Pt/CNT-based PEFCs. However, in electrodes containing Pt/CNT catalysts, the thermal decomposition of ammonium sulfate was observed to be less efficient than in those utilizing Pt/C catalysts supported on carbon black [68]. This is likely because of the substantial clustering of platinum particles on CNT surfaces under elevated temperatures, suggesting that, in contrast to Pt/C catalysts, Pt/CNT substrates are more responsive to temperature.

Maximizing the advantages of carbon nanotubes requires an in-depth understanding of their production and functionalization ought to be methodically developed and purified.

Considering the conventional wastewater treatment processes, it is evident that they are not designed to remove CECs (contaminants of emergent concern) including pharmaceutical wastes, pesticides, endocrine-disrupting chemicals, and personal care products. Given that they produce significant volumes of polluted sludge concurrently. According to Sarkar et al., modern wastewater remediation methods including membrane processes (NF, UF, RO), and although membrane-based bioreactors (MBRs), ozone, and catalytic oxidation can barely eliminate some toxins of growing worried, their use is hindered by its expensive cost. Designing and developing novel materials and techniques based on carbon-based nanotubes (CNTs) to remove CECs from urban wastewater may help the water supply chain find sustainable sludge management strategies [69]. Thus, this chapter is needed since sustainable city development demands methods for managing sludge in the water supply chain that are sustainable.

5 Future Perspectives and Challenges

Effective utilization of specific reactions combined with CNTs that are aligned or not provides a promising future for carbon-based nanocomposites. This application has created a rich and interesting field, understanding nanotube chemistry to address application-specific requirements such as interfacial strength for blending with polymers, and biocompatibility for nanotube biosensors surface modification of CNTs is essential. A uniform hexagonal lattice renders the sidewalls less reactive, meanwhile, CNT tips exhibit higher reactivity compared to their sidewalls as it were, hence, allowing various chemical compounds to be affixed to the tips of the nanotubes. Recently, some interesting processes for chemically altering both the interior and exterior walls of nanotubes have also been identified [55, 70]. Given the chemistry and exceptional and excellent properties of carbon-based nanotubes, they will continue to gain more ground in the emerging fields. Because of the high cost of synthesizing CNTs, the CNT products will continue to be expensive unless further research on the alternative or cheap method of production is intensified; otherwise, implementation will be impeded in the coming years.

The covalent bonding properties in carbon nanostructures have resulted in several designs and derivatives that widen and continue to expand their horizon as functional resources. However, the synthetic strategies depending on the findings of Salice et al. designed and optimized for functionalizing nanostructures could demand extended processing durations under severe conditions [71, 72]. By modifying current products and creating new ones, CNTs offer transformative potential for modern applications through their utilization as reinforcement materials for sustainable city development. Many professionals in the research and industrial sector today and in the future will likely come across numerous facets, particularly CNT science and technology, in their research adventures. Because of their fascinating chemistry and straightforward structural patterns, carbon nanotubes (CNTs) are suitable complementary materials for city development that will continue to garner much interest among architects, scientists, and engineers [71–75].

CNTs are projected as promising materials for 3D concrete printing (3DPC). According to Cui et al., adapting CNT in 3D concretes can significantly decrease the curing of 3DPC, thus enhancing efficiency, and even reducing 3DPC's curing shrinkage, enhancing printing excellence, and boosting 3DPC's initial strength. Therefore, it could occupy 3DPC pores, boost its toughness, and maintain shape stability post-extrusion [41]. However, research on CNT-based 3DPC remains limited in the literature, perhaps because the local 3D printing technology is in its early phase, and there is a lack of defined standards for the 3D printing procedure coupled with the elevated carbon material costs and intricate manufacturing processes. These factors are hindrances to CNT to improve cements in 3DPC. Meanwhile, 3D printing allows precise model-based designs when creating small designs, and CNT fortifying the 3DPC becomes feasible.

Nanocarbons like 1D nanotubes and 2D graphenes are comprehensively studied in the literature and reckoned to provide emerging flexible electronic platforms requiring high transparency, mobility, and flexibility [28, 37, 76–81]. Figure 7 shows diverse forms of carbon-derived materials that have potential for next-generation technology.

6 Conclusion

This review demonstrates that carbon-based nanocomposites serve as effective alternatives over traditional techniques for water purification, contaminant extraction, and concrete reinforcement. Nanotechnology may revolutionize civil

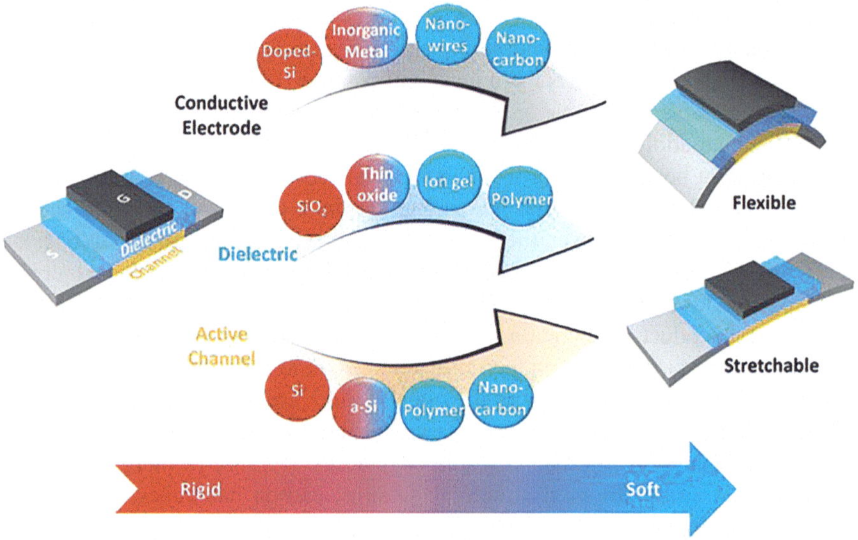

Fig. 7 Materials classification from rigid to soft as adapted from the work of Chae and Lee, indicating that carbon-based nanomaterials are suitable replacements for the conventional ones for next-generation soft applications [37]. (Reproduced with the permission of the copyright holder. Springer Nature, 2014)

engineering practices as many studies revealed that incorporating various nanoparticles into concrete significantly enhanced to improving concrete's mechanical properties, structural attributes, and longevity as well as introducing advanced functionalities, including self-cleaning, self-monitoring, fire resistance, etc. The utilization and the challenges involved in carbon-based nanocomposites for the sustainable development of cities have been brought to bear, highlighting notable applications, such as advancements in municipal wastewater processing, pollutants removal, and concrete structure reinforcement, among other things. Highlights of future perspectives plus the possible solutions and suggestions for the implications of CNTs were also discussed.

References

1. Wang, J., & Guo, X. (2020). Adsorption isotherm models: Classification, physical meaning, application and solving method. *Chemosphere, 258*, 127279. https://doi.org/10.1016/J.CHEMOSPHERE.2020.127279
2. Sindhu, M., et al. (2023). Characteristic studies of Ta_3N_5/BSC@PANI nanocomposites for hydrogen production via water-splitting under visible light irradiation. *Chemical Engineering Research and Design, 197*, 572–580. https://doi.org/10.1016/J.CHERD.2023.07.050
3. Moreau, N., et al. (2025). Physical methods for the preparation of cobalt nanoparticles for use in the synthesis of multiwalled carbon nanotubes. *Inorganics, 13*(1), 7. https://doi.org/10.3390/INORGANICS13010007
4. Morsy, M., Gomaa, I., Taha, W., Elhamid, A. E. M. A., Shoukry, H., & Mokhtar, M. M. (2023). Optimization and surface functionalization of carbon nanotubes using different hierarchical metal oxide nanoparticles. *HBRC Journal, 19*(1), 509–521. https://doi.org/10.1080/16874048.2023.2287768
5. de Luca, P., Siciliano, C., Nagy, J. B., & Macario, A. (2023). The role of carbon nanotubes in the reactions of heterogeneous catalysis. *Chemical Engineering Research and Design, 197*, 74–84. https://doi.org/10.1016/J.CHERD.2023.07.016
6. Sarkar, B., et al. (2019). Sustainable sludge management by removing emerging contaminants from urban wastewater using carbon nanotubes. In *Industrial and municipal sludge: Emerging concerns and scope for resource recovery* (pp. 553–571). Elsevier. https://doi.org/10.1016/B978-0-12-815907-1.00024-6
7. Du, C. Y., Zhao, T. S., & Liang, Z. X. (2008). Sulfonation of carbon-nanotube supported platinum catalysts for polymer electrolyte fuel cells. *Journal of Power Sources, 176*(1), 9–15. https://doi.org/10.1016/J.JPOWSOUR.2007.10.016
8. Wang, H., Yi, L., Huang, F., Huang, Q., & Zhou, T. (2024). Facile synthesis of graphene nanosheets on wastewater sediments for high efficient adsorption of methylene blue. *Separation and Purification Technology, 337*, 126366. https://doi.org/10.1016/j.seppur.2024.126366
9. Yadav, M. D., Joshi, H. M., Sawant, S. V., Dasgupta, K., Patwardhan, A. W., & Joshi, J. B. (2023). Advances in the application of carbon nanotubes as catalyst support for hydrogenation reactions. *Chemical Engineering Science, 272*, 118586. https://doi.org/10.1016/J.CES.2023.118586
10. Tahmasbi, B., Darabi, M., Moradi, P., Abbasi Tyula, Y., & Nikoorazm, M. (2024). Gadolinium Schiff-base complex on nanocomposite of graphene oxide magnetic nanoparticles as a robust, reusable and chemoselective nanocatalyst in the C–C coupling reactions. *Polyhedron, 258*, 117038. https://doi.org/10.1016/j.poly.2024.117038
11. Zheng, M., & Diner, B. A. (2004). Solution redox chemistry of carbon nanotubes. *Journal of the American Chemical Society, 126*(47), 15490–15494. https://doi.org/10.1021/ja0457967

12. Li, Q., Zhang, J., Yan, H., He, M., & Liu, Z. (2004). Thionine-mediated chemistry of carbon nanotubes. *Carbon, 42*(2), 287–291. https://doi.org/10.1016/j.carbon.2003.10.030
13. Basu-Dutt, S., Minus, M. L., Jain, R., Nepal, D., & Kumar, S. (2012). Chemistry of carbon nanotubes for everyone. *Journal of Chemical Education, 89*(2), 221–229. https://doi.org/10.1021/ed1005163
14. Wu, H. C., Chang, X., Liu, L., Zhao, F., & Zhao, Y. (2010). Chemistry of carbon nanotubes in biomedical applications. *Journal of Materials Chemistry, 20*(6), 1036–1052. https://doi.org/10.1039/b911099m
15. Salice, P., et al. (2014). Chemistry of carbon nanotubes in flow. *Journal of Flow Chemistry, 4*(2), 79–85. https://doi.org/10.1556/JFC-D-13-00031
16. Cui, Y., Wei, Q., Park, H., & Lieber, C. M. (2001). Nanowire nanosensors for highly sensitive and selective detection of biological and chemical species. *Science, 293*(5533), 1289–1292. https://doi.org/10.1126/science.1062711
17. Tans, S. J., Verschueren, A. R. M., & Dekker, C. (1998). Room-temperature transistor based on a single carbon nanotube. *Nature, 393*(6680), 49–52. https://doi.org/10.1038/29954
18. Kong, J., et al. (2000). Nanotube molecular wires as chemical sensors. *Science, 287*(5453), 622–625. https://doi.org/10.1126/science.287.5453.622
19. Baughman, R. H., Zakhidov, A. A., & de Heer, W. A. (2002). Carbon nanotubes—The route toward applications. *Science, 297*(5582), 787–792. https://doi.org/10.1126/science.1060928
20. Tasis, D., Tagmatarchis, N., Bianco, A., & Prato, M. (2006). Chemistry of carbon nanotubes. *Chemical Reviews, 106*(3), 1105–1136. https://doi.org/10.1021/cr050569o
21. Thostenson, E. T., Ren, Z., & Chou, T. W. (2001). Advances in the science and technology of carbon nanotubes and their composites: A review. *Composites Science and Technology, 61*(13), 1899–1912. https://doi.org/10.1016/S0266-3538(01)00094-X
22. Ma, P. C., Siddiqui, N. A., Marom, G., & Kim, J. K. (2010). Dispersion and functionalization of carbon nanotubes for polymer-based nanocomposites: A review. *Composites Part A: Applied Science and Manufacturing, 41*(10), 1345–1367. https://doi.org/10.1016/j.compositesa.2010.07.003
23. Treacy, M. M. J., Ebbesen, T. W., & Gibson, J. M. (1996). Exceptionally high young's modulus observed for individual carbon nanotubes. *Nature, 381*(6584), 678–680. https://doi.org/10.1038/381678a0
24. Liu, Y., et al. (2020). Autogenous shrinkage and crack resistance of carbon nanotubes reinforced cement-based materials. *International Journal of Concrete Structures and Materials, 14*(1), 43. https://doi.org/10.1186/s40069-020-00421-0
25. Hassan, S. S. M., El-Aziz, M. E. A., Fayez, A. E. S., Kamel, A. H., & Youssef, A. M. (2024). Synthesis and characterization of bio-nanocomposite based on chitosan and $CaCO_3$ nanoparticles for heavy metals removal. *International Journal of Biological Macromolecules, 255*, 128007. https://doi.org/10.1016/j.ijbiomac.2023.128007
26. Thakur, A., Kumar, A., & Singh, A. (2024). Adsorptive removal of heavy metals, dyes, and pharmaceuticals: Carbon-based nanomaterials in focus. *Carbon, 217*, 118621. https://doi.org/10.1016/j.carbon.2023.118621
27. Wang, B. M., Liu, S., & Han, Y. (2015). Research progress on carbon nanotubes reinforced cement-based materials. In *Key engineering materials* (pp. 487–493). Trans Tech Publications Ltd. https://doi.org/10.4028/www.scientific.net/KEM.629-630.487
28. Zhang, J., Chen, Z., Liu, Y., Wei, W., & Ni, B. J. (2024). Removal of emerging contaminants (ECs) from aqueous solutions by modified biochar: A review. *Chemical Engineering Journal, 479*, 147615. https://doi.org/10.1016/j.cej.2023.147615
29. Yun, Y.-H., et al. (2005). Multifunctional carbon nanofiber/nanotube smart materials. In *Smart structures and materials 2005: Smart electronics, MEMS, BioMEMS, and nanotechnology* (p. 184). SPIE. https://doi.org/10.1117/12.600276
30. Roy, A., Sharma, A., Yadav, S., Jule, L. T., & Krishnaraj, R. (2021). Nanomaterials for remediation of environmental pollutants. *Bioinorganic Chemistry and Applications, 2021*, 1764647. https://doi.org/10.1155/2021/1764647

31. Han, B., Yu, X., & Ou, J. (2011). Multifunctional and smart carbon nanotube reinforced cement-based materials. In *Nanotechnology in civil infrastructure* (pp. 1–47). Springer. https://doi.org/10.1007/978-3-642-16657-0_1
32. Asghar, N., et al. (2024). Advancement in nanomaterials for environmental pollutants remediation: A systematic review on bibliometrics analysis, material types, synthesis pathways, and related mechanisms. *Journal of Nanobiotechnology, 22*(1), 26. https://doi.org/10.1186/s12951-023-02151-3
33. Aguilar-Pérez, K. M., Avilés-Castrillo, J. I., & Ruiz-Pulido, G. (2020). Nano-sorbent materials for pharmaceutical-based wastewater effluents—An overview. *Case Studies in Chemical and Environmental Engineering, 2*, 100028. https://doi.org/10.1016/j.cscee.2020.100028
34. Aguilar-Pérez, K. M., Heya, M. S., Parra-Saldívar, R., & Iqbal, H. M. N. (2020). Nanobiomaterials in-focus as sensing/detection cues for environmental pollutants. *Case Studies in Chemical and Environmental Engineering, 2*, 100055. https://doi.org/10.1016/J.CSCEE.2020.100055
35. González-González, R. B., Parra-Saldívar, R., Alsanie, W. F., & Iqbal, H. M. N. (2022). Nanohybrid catalysts with porous structures for environmental remediation through photocatalytic degradation of emerging pollutants. *Environmental Research, 214*, 113955. https://doi.org/10.1016/j.envres.2022.113955
36. Šehović, E., et al. (2022). Thermodynamic valorisation of lignocellulosic biomass green sorbents for toxic pollutants removal. *Chemosphere, 307*, 135737. https://doi.org/10.1016/j.chemosphere.2022.135737
37. Rivas-Sanchez, A., Cruz-Cruz, A., Gallareta-Olivares, G., González-González, R. B., Parra-Saldívar, R., & Iqbal, H. M. N. (2022). Carbon-based nanocomposite materials with multifunctional attributes for environmental remediation of emerging pollutants. *Chemosphere, 303*, 135054. https://doi.org/10.1016/J.CHEMOSPHERE.2022.135054
38. Gan, J. S., Ashraf, S. S., Bilal, M., & Iqbal, H. M. N. (2022). Biodegradation of environmental pollutants using catalase-based biocatalytic systems. *Environmental Research, 214*, 113914. https://doi.org/10.1016/j.envres.2022.113914
39. Vidu, R., et al. (2020). Removal of heavy metals from wastewaters: A challenge from current treatment methods to nanotechnology applications. *Toxics, 8*(4), 101. https://doi.org/10.3390/toxics8040101
40. Soni, R., Pal, A. K., Tripathi, P., Lal, J. A., Kesari, K., & Tripathi, V. (2020). An overview of nanoscale materials on the removal of wastewater contaminants. *Applied Water Science, 10*(8), 189. https://doi.org/10.1007/s13201-020-01275-3
41. Wang, X. (2012). Nanomaterials as sorbents to remove heavy metal ions in wastewater treatment. *Journal of Environmental & Analytical Toxicology, 2*(7), 154. https://doi.org/10.4172/2161-0525.1000154
42. Palani, G., et al. (2021). Current trends in the application of nanomaterials for the removal of pollutants from industrial wastewater treatment—A review. *Molecules, 26*(9), 2799. https://doi.org/10.3390/molecules26092799
43. Nizamuddin, S., et al. (2019). Iron oxide nanomaterials for the removal of heavy metals and dyes from wastewater. In *Nanoscale materials in water purification* (pp. 447–472). https://doi.org/10.1016/B978-0-12-813926-4.00023-9
44. Mohanapriya, V., Sakthivel, R., Pham, N. D. K., Cheng, C. K., Le, H. S., & Dong, T. M. H. (2023). Nanotechnology—A ray of hope for heavy metals removal. *Chemosphere, 311*, 136989. https://doi.org/10.1016/j.chemosphere.2022.136989
45. Khaligh, N. G., & Johan, M. R. (2018). Recent application of the various nanomaterials and nanocatalysts for the heavy metals' removal from wastewater. *Nano, 13*(9), 1830006. https://doi.org/10.1142/S1793292018300062
46. de Beni, E., et al. (2022). Graphene-based nanomaterials in the electroplating industry: A suitable choice for heavy metal removal from wastewater. *Chemosphere, 292*, 133448. https://doi.org/10.1016/j.chemosphere.2021.133448

47. Jawed, A., Saxena, V., & Pandey, L. M. (2020). Engineered nanomaterials and their surface functionalization for the removal of heavy metals: A review. *Journal of Water Process Engineering, 33*, 101009. https://doi.org/10.1016/j.jwpe.2019.101009
48. Borji, H., Ayoub, G. M., Bilbeisi, R., Nassar, N., & Malaeb, L. (2020). How effective are nanomaterials for the removal of heavy metals from water and wastewater? *Water, Air, and Soil Pollution, 231*(7), 330. https://doi.org/10.1007/s11270-020-04681-0
49. Divate, S. B., Chavan, N. B., & Kadu, P. S. (2022). A review: Nanomaterials for removal of heavy metal from wastewater. *Journal of Pharmaceutical Negative Results, 13*(7), 7878–7881.
50. Parvin, F., Tareq, S. M., & Rikta, S. Y. (2019). Application of nanomaterials for the removal of heavy metal from wastewater. In *Nanotechnology in water and wastewater treatment: theory and applications* (pp. 137–158). Elsevier. https://doi.org/10.1016/B978-0-12-813902-8.00008-3
51. Yang, J., et al. (2019). Nanomaterials for the removal of heavy metals from wastewater. *Nanomaterials, 9*(3), 424. https://doi.org/10.3390/nano9030424
52. Zhang, S., Shao, T., & Karanfil, T. (2011). The effects of dissolved natural organic matter on the adsorption of synthetic organic chemicals by activated carbons and carbon nanotubes. *Water Research, 45*(3), 1378–1386. https://doi.org/10.1016/j.watres.2010.10.023
53. Dhand, V., Rhee, K. Y., Ju Kim, H., & Ho Jung, D. (2013). A comprehensive review of graphene nanocomposites: Research status and trends. *Journal of Nanomaterials, 2013*, 14. https://doi.org/10.1155/2013/763953
54. Katiyar, S., & Katiyar, R. (2024). A comprehensive review on synthesis and application of nanocomposites for adsorption of chromium: status and future prospective. *Applied Water Science, 14*(1), 11. https://doi.org/10.1007/s13201-023-02062-6
55. Veeman, D., et al. (2021). Sustainable development of carbon nanocomposites: Synthesis and classification for environmental remediation. *Journal of Nanomaterials, 2021*, 1–21. https://doi.org/10.1155/2021/5840645
56. Chiu, P.-W. (2003). *Towards carbon nanotube-based molecular electronics* (Doctoral dissertation). Freie Universität Berlin, Berlin, Germany.
57. Vargas-Bernal, R., Herrera-Pérez, G., & Tecpoyotl-Torres, M. (2017). The impact of carbon nanotubes and graphene on electronics industry. In *Encyclopedia of information science and technology* (4th ed.). IGI Global. https://doi.org/10.4018/978-1-5225-2255-3.ch253
58. Ito, A., & Tanaka, K. (2014). Applications of carbon nanotubes and graphene in spin electronics. In *Carbon nanotubes and graphene* (2nd ed., pp. 253–278). Elsevier. https://doi.org/10.1016/B978-0-08-098232-8.00011-5
59. Pinzón, J. R., Villalta-Cerdas, A., & Echegoyen, L. (2012). Fullerenes, carbon nanotubes, and graphene for molecular electronics. *Topics in Current Chemistry, 312*, 127–174. https://doi.org/10.1007/128_2011_176
60. Vargas-Bernal, R., Herrera-Pérez, G., & Tecpoyotl-Torres, M. (2018). *The impact of carbon nanotubes and graphene on electronics industry*. IGI Global. https://doi.org/10.4018/978-1-5225-7766-9.ch030
61. Boschetto, G., Carapezzi, S., & Todri-Sanial, A. (2021). Graphene and carbon nanotubes for electronics nanopackaging. *IEEE Open Journal of Nanotechnology, 2*, 120–128. https://doi.org/10.1109/OJNANO.2021.3127652
62. Pyo, S., Eun, Y., Sim, J., Kim, K., & Choi, J. (2022). Carbon nanotube-graphene hybrids for soft electronics, sensors, and actuators. *Micro and Nano Systems Letters, 10*(1), 9. https://doi.org/10.1186/s40486-022-00151-w
63. Chae, S. H., & Lee, Y. H. (2014). Carbon nanotubes and graphene towards soft electronics. *Nano Convergence, 1*(1), 15. https://doi.org/10.1186/s40580-014-0015-5
64. Wang, B., & Fan, C. (2018). Effect of carbon nanofibers on rheological and mechanical properties of cement composites. *Materials Research Express, 5*(6), 065058. https://doi.org/10.1088/2053-1591/aacdde
65. Li, Y. F., Zhou, J., & Gao, L. W. (2014). Effect of carbon nanotubes on mechanical properties of cement composites. *Advanced Materials Research, 915–916*, 768–771. https://doi.org/10.4028/www.scientific.net/AMR.915-916.768

66. Li, S., Shen, P., Zhou, H., Du, S., Zhang, Y., & Yan, J. (2022). Synergistic effects of CNTs/ SiO_2 composite fillers on mechanical properties of cement composites. *RSC Advances, 12*(42), 27253–27266. https://doi.org/10.1039/d2ra04127h
67. Lavagna, L., Suarez-Riera, D., & Pavese, M. (2023). Synergistic effect of carbon-based reinforcements on the mechanical properties of cement-based composites. *Journal of Composites Science, 7*(10), 430. https://doi.org/10.3390/jcs7100430
68. Du, M., Jing, H., Gao, Y., Su, H., & Fang, H. (2020). Carbon nanomaterials enhanced cement-based composites: Advances and challenges. *Nanotechnology Reviews, 9*(1), 115–135. https://doi.org/10.1515/ntrev-2020-0011
69. Manzur, T., Yazdani, N., & Emon, M. A. B. (2016). Potential of carbon nanotube reinforced cement composites as concrete repair material. *Journal of Nanomaterials, 2016*, 1421959. https://doi.org/10.1155/2016/1421959
70. Petrunin, S., Vaganov, V., & Sobolev, K. (2014). Cement composites reinforced with functionalized carbon nanotubes. *Materials Research Society Symposium Proceedings, 1611*, 133–138. https://doi.org/10.1557/opl.2014.769
71. Cui, K., Chang, J., Feo, L., Chow, C. L., & Lau, D. (2022). Developments and applications of carbon nanotube reinforced cement-based composites as functional building materials. *Frontiers in Materials, 9*, 861646. https://doi.org/10.3389/fmats.2022.861646
72. Sheikh, T. M., Anwar, M. P., Muthoosamy, K., Jaganathan, J., Chan, A., & Mohamed, A. A. (2021). The mechanics of carbon-based nanomaterials as cement reinforcement—A critical review. *Construction and Building Materials, 303*, 124441. https://doi.org/10.1016/J.CONBUILDMAT.2021.124441
73. Nochaiya, T., & Chaipanich, A. (2011). Behavior of multi-walled carbon nanotubes on the porosity and microstructure of cement-based materials. *Applied Surface Science, 257*(6), 1941–1945. https://doi.org/10.1016/J.APSUSC.2010.09.030
74. Bautista-Gutierrez, K. P., Herrera-May, A. L., Santamaría-López, J. M., Honorato-Moreno, A., & Zamora-Castro, S. A. (2019). Recent progress in nanomaterials for modern concrete infrastructure: Advantages and challenges. *Materials, 12*(21), 3548. https://doi.org/10.3390/MA12213548
75. Gjørv, O. E. (2011). Durability of concrete structures. *Arabian Journal for Science and Engineering, 36*(2), 151–172. https://doi.org/10.1007/S13369-010-0033-5
76. Liu, Y., Zhong, X., & Reza Mohammadian, H. (2023). Role carbon nanomaterials in reinforcement of concrete and cement; a new perspective in civil engineering. *Alexandria Engineering Journal, 72*, 649–656. https://doi.org/10.1016/J.AEJ.2023.04.025
77. Wang, Q., Wang, Y., Duan, B., & Zhang, M. (2016). Modified sol-gel synthesis of carbon nanotubes supported Titania composites with enhanced visible light induced photocatalytic activity. *Journal of Nanomaterials, 2016*, 1–6. https://doi.org/10.1155/2016/3967156
78. Fu, T., et al. (2023). Recent progresses, challenges, and opportunities of carbon-based materials applied in heavy metal polluted soil remediation. *Science of the Total Environment, 856*, 158810. https://doi.org/10.1016/j.scitotenv.2022.158810
79. Rao, N., Singh, R., & Bashambu, L. (2021). Carbon-based nanomaterials: Synthesis and prospective applications. *Materials Today: Proceedings, 44*, 608–614. https://doi.org/10.1016/J.MATPR.2020.10.593
80. Abdullah, N., Tajuddin, M. H., & Yusof, N. (2018). Carbon-based polymer nanocomposites for dye and pigment removal. In *Carbon-based polymer nanocomposites for environmental and energy applications* (pp. 305–329). Elsevier. https://doi.org/10.1016/B978-0-12-813574-7.00013-7
81. Gong, J.-L., et al. (2009). Removal of cationic dyes from aqueous solution using magnetic multi-wall carbon nanotube nanocomposite as adsorbent. *Journal of Hazardous Materials, 164*(2–3), 1517–1522. https://doi.org/10.1016/j.jhazmat.2008.09.072

Carbon-Based Nanocomposites for Wearable Electronics

Muhammad Shahab, Won Chun Oh, Sulaiman Al-Hashmi, and Kefayat Ullah

1 Introduction

Recently, remarkable advancement of flexible and wearable functional electronic gadgets has gained significant consideration because of their potential to revolutionize medical diagnosis methods [1]. These devices integrate features such as wearability, comfort, timely feedback, and remote operation, fostering the emergence of wearable health monitoring technologies, care systems, and personal health management. The exceptional properties of carbon materials can further drive the development of innovative wearable and flexible devices, enabling real-time health monitoring, effective disease prevention, and diverse functionalities [2]. Figure 1 illustrates the fabrication process of transparent wearable electronics from carbon materials for various electronic applications. Wearable and flexible electronics will be essential in personalized medicine by continuously monitoring a Person's physical activity and health condition. Flexible electronics provide the perfect foundation for creating customized, adaptable wearable technology because of its great flexibility and capacity to take different shapes [3].

In order to make wearable electronics that fit the curved and changing human body, extensive research has been done to create materials with properties similar to

M. Shahab · K. Ullah (✉)
Department of Applied Physical and Material Sciences, University of Swat, Swat, Pakistan
e-mail: kefsayed@uswat.edu.pk

W. C. Oh
Department of Advanced Materials Science and Engineering, Hanseo University, Seosan-si, Chungnam, Republic of Korea

S. Al-Hashmi
Center for Environmental Studies and Research, Sultan Qaboos University, Seeb, Oman

V. Khanna (ed.), *Carbon-Based Nanocomposites for Sustainable Applications, Volume I*, Lecture Notes in Nanoscale Science and Technology 38,
https://doi.org/10.1007/978-3-031-95306-4_11

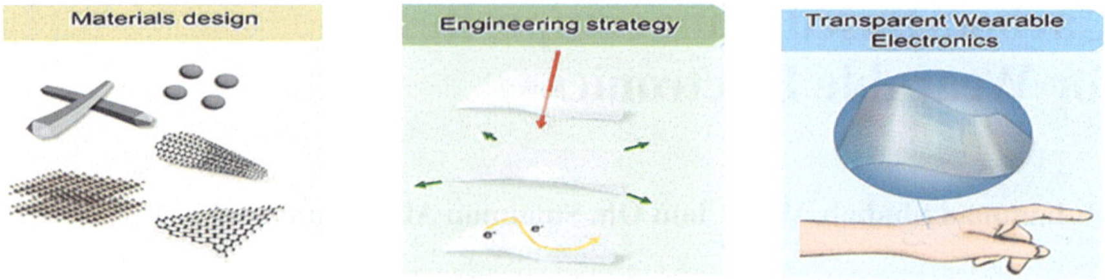

Fig. 1 Transparent electronics for wearable electronics application

skin, such as being highly stretchable and flexible. With advancements in material science, naturally stretchable and soft electronic devices have been developed [4].

Different types of carbon nanomaterials exhibit unique physical and mechanical properties enabling their application across various fields. Sensors and actuators are two key applications, functioning in opposite ways to facilitate the conversion of electrical signals and other forms of energy between the environment and the system's input/output units [5]. Also, because of its remarkable qualities, such as their excellent conductivity of electricity, clearly designed morphology, low weight, ease of chemical functionalization, tunable surface, mass fabrication scalability, and great thermal stability, carbon materials offer a possible option [6]. Particularly, graphene and carbon nanotubes (CNTs) can be organized into diverse multiscale macrostructures to create various types of flexible sensors [7].

Stretchable physical sensors are useful for creating a smooth and adaptable interface between electronics and health-related medical indicators such as body motion [8], pulse rate [9], muscle movement [10], body temperature [11], humidity [12], biochemical constituents (e.g., electrolytes and metabolites) [13], as well as electrophysiological signals (e.g., electrocardiogram (ECG), electro-encephalogram (EEG), electro-myogram (EMG) [14], facilitating the continuous collection of precise signals [15].

Consequently, numerous kinds of wearable and flexible electronic sensors have been established for detecting health-related indicators, as depicted in Fig. 2. These include strain/pressure sensors [16, 17], electrophysiological signals monitoring sensors, temperature/humidity sensors [2], electrochemical biosensors, and integrated sensing platforms [18, 19].

Flexible *strain/pressure sensors* have attracted significant research attention because of their potential uses in areas such as detection of human motion, interface systems between humans and machines, detection of electrophysiological signals, measurement of respiratory rate, artificial skin, and artificial intelligence. Within the realm of physical sensors, strain and pressure sensors have shown great promise for use in fields like machine-human interface, fitness tracking, and healthcare systems [20–22].

Flexible *electrophysiological sensors,* developed using flexible electronic materials are extremely important in medical industry, particularly because of the increasing occurrence of cardiac and brain vessels diseases. Research indicates that avoiding these disorders mostly depends on prompt diagnosis and ongoing

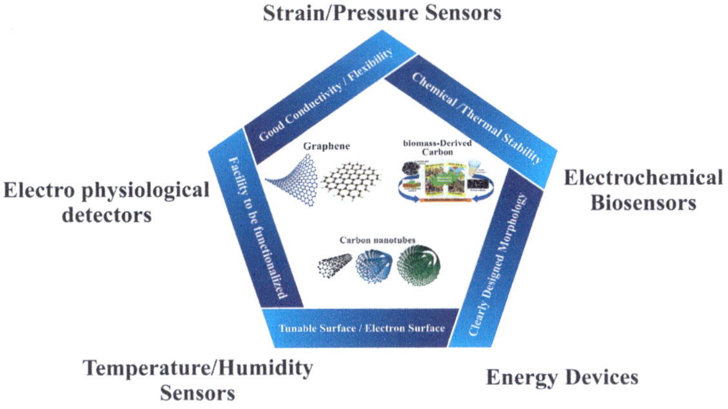

Fig. 2 Numerous wearable and flexible carbon-based electronic devices such as strain/pressure sensors, electrochemical biosensors, energy devices, temperature sensors, humidity sensors, and electrophysiological indicators

physiological signal monitoring. Even while conventional rigid sensors are still widely employed in medical monitoring, noise and motion distortions have a significant impact on the electrophysiological data (ECG, EEG, and EMG) they gather. Consequently, wearable smart monitoring devices utilizing flexible electronics are extremely important. Wearable devices have drawn a lot of attention as interest in personalized healthcare grows since it can monitor individual physiological indicators tailored to specific clinical needs. These unique biological parameters can be collected from various target organs, enabling the analysis of functions ranging from basic bodily movements to advanced clinical treatments [23, 24].

Drawing inspiration from the sensory capabilities of human skin, numerous bionic sensors have been designed to mimic the skin's ability to perceive various stimuli, finding extensive application in health monitoring. Considering the rising need for adaptable, high-performing sensors in wearable technology and health monitoring, this overview highlights recent advancements in *Temperature/Humidity*, and versatile sensors for monitoring of human health. Furthermore these flexible humidity sensors are capable of accurately monitoring simultanoues changes in temperature and humidity of a plant's microclimate, holds a great potential in precision agriculture. These sensors can deliver valuable insights for optimizing the crop growth environment and mitigating abiotic stresses, ultimately improving crop yields [25, 26].

The advancement of flexible and stretchable electronics has transformed biosensor technologies for studying biological systems. Flexible and stretchable *electrochemical biosensors* (FSECSs) allow for precise, rapid, and miniaturized detection of biochemical molecules in various biological contexts. Considerable advancements have been achieved in the production and functionality of FSECSs, particularly in electrode development, over the past decade. Moreover, electrochemical biosensors hold promise for offering quick and cost-effective diagnostic solutions, enabling the transition of clinical testing from centralized laboratories to point-of-care (POC) settings [27, 28].

Moreover, the latest generation of flexible and wearable power devices, along with conductive wires, has attracted significant attention in the field of energy storage. Supercapacitors (SCs) and rechargeable metal-ion batteries have emerged as two prominent options among various energy storage systems, evolving to meet the growing energy demands driven by the ongoing energy crisis. Wearable biomedical sensors are composed of sensors, data analyzers, data processors, and energy sources, integrated on stretchable natural fabrics or flexible synthetic polymers. This design ensures efficient and comfortable interaction with the human body [29–31]. Three members of the sp^2 family of carbon allotropes, each with unique dimensional properties, are included in the category of advanced carbon materials discussed in this review: One-dimensional carbon nanotubes, two-dimensional graphene, and three-dimensional graphite. In a crystal lattice structure resembling a honeycomb, each carbon atom in these materials makes covalent connections with three nearby carbon atoms [32]. As illustrated in Fig. 3, novel carbon materials have found widespread use because of their special properties as flexible and elastic sensors for a range of wearable and smart technologies as well as valuable components of the energy devices.

We start this chapter by outlining how carbon-based materials are used in stretchy and flexible electronics. Next, we present recent advancements in the controlled manufacturing and design of different carbon materials with tailored architectures for flexible electronics with performance, especially in smart wearables. There is a thorough review and discussion of the design, construction, operation, and performance of a variety of carbon-based flexible devices, such as conductive electrodes, power devices, strain and pressure sensors, temperature and humidity sensors, and electrochemical sensors. Furthermore, the potential of these technologies in healthcare monitoring systems are highlighted, and their usage in tracking human activity and other health-related indicators is investigated. Apart from their use in healthcare, energy conversion/storage devices are highlighted for their ability to facilitate multipurpose smart wearable systems. This study concludes by outlining the difficulties and exciting prospects for improved carbon materials to transform individualized health monitoring systems.

2 Carbon-Based Flexible Strain/Pressure Sensors

2.1 Materials and Types

In both academics and industry, strain and pressure sensors, which detect pressure and distortion by measuring pressure-induced electrical signals, have drawn a lot of interest because of their simple fabrication and well-established signal acquisition techniques, particularly for wearable electronics. These sensors are either applied to

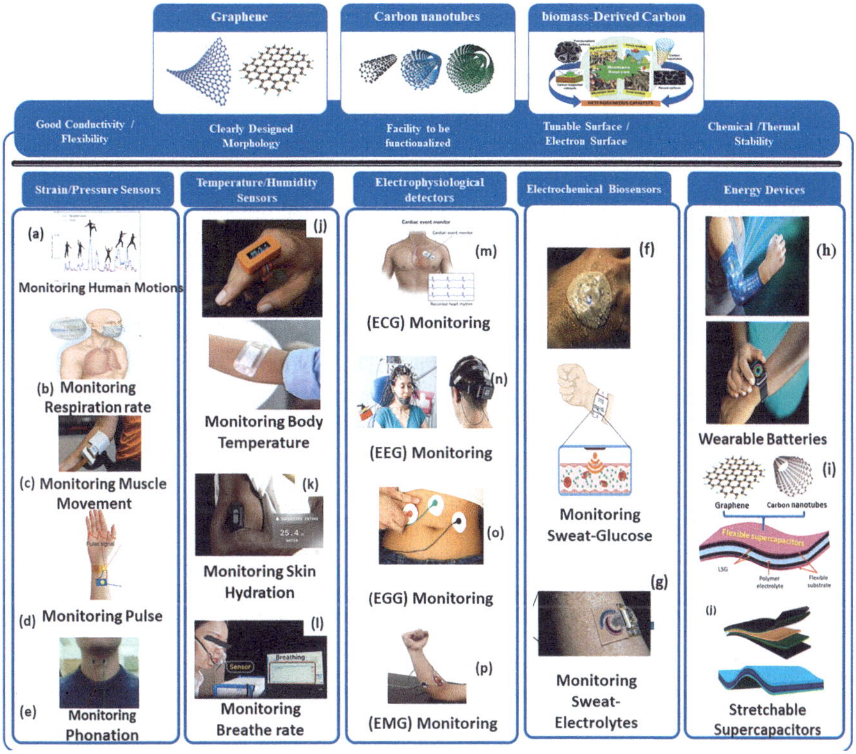

Fig. 3 Carbon-based energy devices and wearable sensors. Strain and pressure sensors: human gesture tracking, pulse checking, phonation, and muscle activity. Electrochemical biosensors: monitoring sweat-glucose and sweat-electrolytes. Energy devices: wearable batteries and flexible/ wearable supercapacitors for wearable power supplies. Temperature and humidity sensors: body temperature sensing, skin humidity. And, electrophysiological indicators: analyzing physiological signals like EMG, EEG, and ECG

the skin or incorporated into textiles, enabling real-time examining of health parameters like oxygen saturation, blood pressure, pulse, heart rate, respiration rate, and body motions. Different types of flexible and wearable strain/pressure sensors can be categorized based on their sensing mechanisms into piezoresistive, capacitive, electric, and triboelectric-type sensors [21, 33–35].

Throughout the various types of strain and pressure sensors, resistive sensors have been extensively studied by reason of their advantages, including rapid response, high adaptability, simplicity in fabrication, and a straightforward reading mechanism [36]. Strain/pressure sensors often utilize piezoresistive materials such as silicon and carbon-based compounds, which are integrated into the sensor's architecture to precisely measure mechanical strain or pressure and translate it into an electrical output. The creation of extremely sensitive sensors is made possible by piezoresistive effect, which makes them crucial for uses in robotics, medical devices, and motorized structures [6].

The capacitance-type strain/pressure sensor exhibits excellent performance, including highs sensitivity, fast response, and long-term stability. It accurately discriminates mechanical stimuli and meets essential criteria for gait monitoring, such as elasticity, antibacterial activity, wearable comfort, and durability, which has attracted increased research interest. The capacitive sensor has demonstrated its potential in wearable healthcare monitoring by effectively tracking physiological signals, including arterial pulse and muscle activation. While, several review papers have provided detailed insights into the transduction mechanisms and the development of flexible piezoelectric and triboelectric strain/pressure sensors [33, 37]. In this section, the main focus is a summary of the developments in carbon-based flexible resistive and capacitive strain and pressure sensors.

The benefits of biomass-derived carbons (BDCs), such as their vast availability, variety of forms, and distinguished electrical conductivity, make them highly promising materials for the progress of resistive strain and pressure sensors. Excellent conductivity, chemical stability, and superior mechanical properties are well-known attributes of carbon materials, such as graphene, carbon nanotubes (CNTs), carbon black (CB), and carbon fibers (CFs). Figure 4 shows how graphene and carbon nanotubes are used in flexible strain sensors. The field of smart wearable electronic devices has made sustainable use of these materials, especially in the construction of resistive strain/pressure sensors [16, 38]. On the other hand, an ultrathin porous

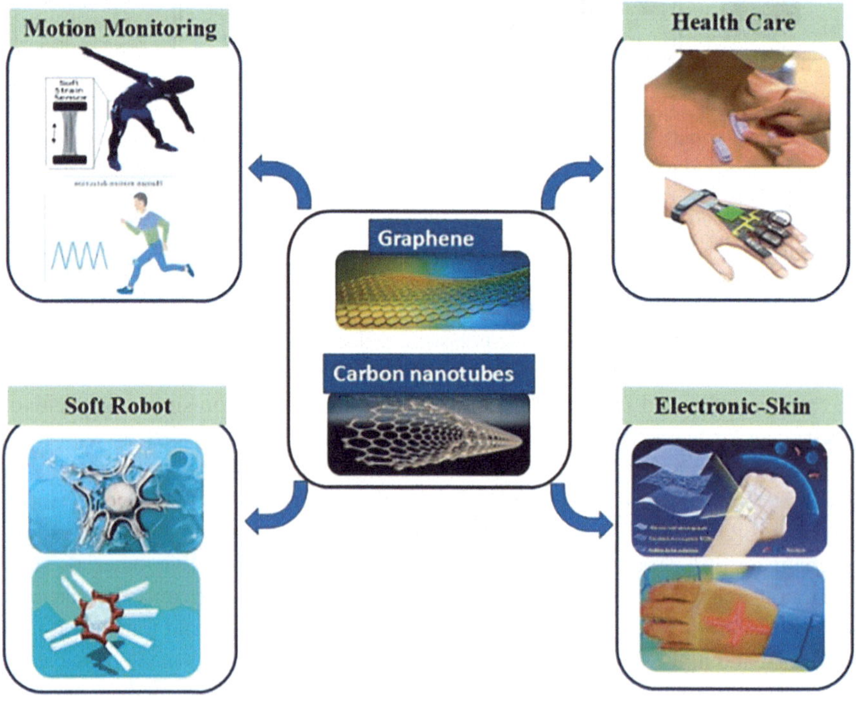

Fig. 4 Applications of graphene and carbon nanotubes in flexible strain sensors

polyvinylidene difluoride (PVDF) dielectric layer and a high-concentration carbon nanotube (CNT)-doped porous electrode are operated to design flexible capacitance-type strain and pressure sensors [37]. Furthermore, because of its potential for versatile sensing applications and its groundbreaking advancements in pressure-sensitive materials, MXene has drawn a lot of attention as a material with promise for capacitive sensors [39].

2.2 *Carbon-Based Flexible Strain Sensors*

2.2.1 Fabrication of Carbon Nanotubes (CNTs) Strain Sensors

Because of their exceptional electrical and mechanical characteristics, carbon nanotubes (CNTs) are frequently used as conductive materials in flexible strain sensors, which enhances the sensor's exceptional sensing capability [40]. High sensitivity (extreme gauge factor of about 1571), remarkable stretchable strength (stress greater than 24 MPa, strain greater than 400%), outstanding toughness (10,000 cycles at 10% strain), and a broad operating elongating range (0% to 400%) are just a few of the outstanding overall performance characteristics of the CNT-based strain sensor. The CNT strain sensor has been created for use in body activity examining, and electronic skins, by utilizing its high sensitive nature, wearability, and a wide range of applications [41]. CNT/polymer composite sensors are extensively operated in fields including food quality, human health, and structural safety monitoring because of their mechanical and functional qualities, which include piezo-resistance, stretchability, and flexibility. One common type of CNT is multi-walled carbon nanotubes (MWCNTs), which have diameters between 20 and 80 nm with characteristic ratios higher than 100. MWCNTs are frequently scattered across polymer matrices, such as epoxy resin, polydimethylsiloxane (PDMS), and polyvinylidene fluoride (PVDF) to generate these sensors [42]. In great-performance flexible strain sensors, carbon nanotubes (CNTs) are perfect for conducting nanofillers owing to their mechanical strength, flexibility, as well as biocompatibility. Strain sensor performance is assessed based on factors like sensitivity, strain range, durability, response time, and repeatability. However, a trade-off exists: sensors with large strain ranges generally have lower sensitivity, while those with high sensitivity have smaller strain ranges [43]. However, the disordered winding of CNTs creates structural defects and high contact resistance, limiting their electrical conductivity. Mono-dispersed ultra-long SWCNTs, with reduced agglomeration from π-π stacking interactions, facilitate the formation of efficient conductive networks. Additionally, directional filtration aligns aramid nanofiber (ANF)/SWCNT composite films into laminated structures, improving mechanical and electrical properties and achieving outstanding performance in electromagnetic interference (EMI) shielding, surpassing many advanced SWCNT-based materials [44].

2.2.2 Fabrication of Graphene-Based Strain Sensors

Two-dimensional carbon-based materials, such as graphene and its derivatives, like graphene oxide (GO) and reduced graphene oxides (rGO), offer a promising foundation for flexible strain sensors because of their special qualities, including exceptional flexibility, strong electrical and thermal conductivity, remarkable mechanical strength, and especially their high absorption throughout the visible and even near IR (NIR) [45, 46]. Currently, graphene preparation methods are primarily categorized into two approaches. The first is the bottom-up approach, which includes methods like chemical vapor deposition (CVD) [47], epitaxial growth [48]. The second is the top-down approach, encompassing methods like mechanical exfoliation, electrochemical exfoliation, and microwave exfoliation [49], which have been explored for wearable strain sensors. Here, we will briefly focus on CVD techniques and the production of graphene from exfoliated graphite. Building on the discussed methods of exfoliating graphene, one approach uses carbon nanosphere as exfoliants and water as a solvent; a microfluid homogenizer is used to attain a high shear rate. The resulting graphene, composed of two to five layers, is well-suited for producing flexible strain sensors via screen printing. These sensors demonstrate outstanding flexibility, maintaining stable performance even after 100,000 bending cycles or 1000 abrasion cycles. Furthermore, they can function as wearable flexible strain sensors, capable of monitoring resistance changes induced by limb movements [50]. Ruoff et al. utilized adhesive tape to separate graphene layers from graphite flakes. After exfoliation, the layers were dry-transferred onto a silicon wafer, a technique commonly known as the "Scotch tape" method. Subsequently, Novoselov and colleagues achieved the successful synthesis of high-quality monolayer graphene using this approach [51].

Chemical vapor deposition (CVD) encompasses techniques such as atmospheric pressure, low-pressure, thermal, and plasma-enhanced CVD. These techniques are thought to have great promise for producing huge quantities of monolayer graphene with outstanding quality. In addition to its biocompatibility, graphene is a very attractive platform for a variety of biosensing applications, including bioimaging and cancer diagnostics due to its vast surface area and adjustable structural morphology. While graphene synthesized through the CVD method has shown excellent performance in applications such as power conversion, its high production cost, complex process, and low yield remain significant challenges limiting its broader adoption [52]. An example of this is "laser-induced graphene (LIG)," which has attained significant attention in recent years as a promising material for developing flexible, high-performance, and cost-effective sensors [53].

2.2.3 Other Carbon-Based Strain Sensors

Apart from graphene and carbon nanotube, other carbon-based compounds such as carbon black (CB) [54], graphite [55], and nanodiamonds [56], which are comparatively more affordable, have also been utilized for the development of flexible/

wearable strain sensors. As an illustration of using carbon black materials in flexible strain sensors, in one work, carbon black/graphene (CB/Gr) conductive nanocomposites were transferred onto an Eco-flex substrate that had been sandpaper-pretreated to increase adhesion. The resultant sensor demonstrated remarkable sensitivity. It also exhibited impressive performance, including a quick response time of 60 (m/s) and exceptional durability [57]. Figure 5 lists several applications of wearable and flexible strain sensors.

2.3 *Carbon-Based Flexible Pressure Sensors*

The human body is sensitive to both internal pressures such as blood pressure, intracranial pressure, and intraocular pressure, as well as external forces. Excessive pressure can damage tissues and trigger adverse responses, including brain injuries [58], ocular hypertension [59], cardiovascular diseases [60], and skin ulcers [61]. Monitoring these pressures requires sensors that are compatible with the mechanical properties and curved shapes of soft biological tissues, ensuring accurate measurements without causing irritation [62].

Flexible pressure sensors are essential in wearable devices as they convert physical human motion signals into electrical signals. These sensors demonstrate excellent linearity across two pressure ranges: 0–20 and 20–70 kPa. Moreover, 2000 cyclic compression tests have verified their good cyclic stability. Additionally, the sensors efficiently track a variety of pressure applications like voice indicators, wrist and finger twisting, and other minor and major human gestures [63]. The detecting approaches of flexible micro-pressure sensors are essential to their operation and define their operating principles. Piezoresistive, capacitive, electric, and triboelectric sensors are the four primary categories of pressure sensors that can be distinguished based on these principles. Each kind uses a different technique to transform mechanical pressure into electrical signals, providing a range of features and uses for flexible and wearable sensing systems [64]. Flexible piezoresistive

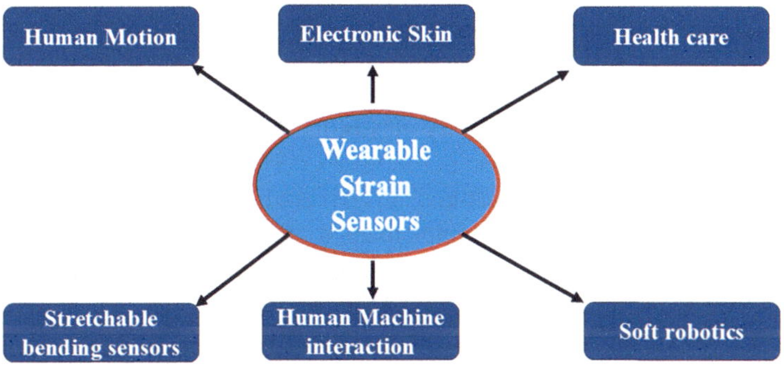

Fig. 5 Applications of wearable and flexible strain sensors

pressure sensors, in particular have drawn a lot of interest because of their quick signal conversion and ease of data collecting. Usually, a conductive material and a polymer material are used in the formation of these sensors. Numerous micro- and nanostructures have been created to improve the sensor's functionality and also to reduce the quantity of conductive materials. These include sandwich configurations, spherical microstructures, porous structures, biomimetic architectures, and interlocking designs, all of which helps to maximize the sensing properties [65]. Wearable pressure sensors frequently use flexible conductive polymer composites(CPCS), which are mostly composed of conductive fillers and elastomers [66, 67].Their outstanding sensing performance, wide range of material possibilities, and affordable manufacturing are the main reasons for their popularity. CPCs have remarkable flexibility and stretchability, thanks to elastomers as Ecoflex, thermoplastic polyurethane (TPU), and polydimethylsiloxane (PDMS) [68, 69]. Meanwhile, because of their better electrical and mechanical qualities, sophisticated nanomaterials such as metal nanowires, reduced graphene oxides (rGO), carbon nanotubes (CNTs), carbon black (CB), and nanoparticles are used as conductive fillers [70]. A piezoresistive pressure sensor was designed by combining carbon nanotubes (CNTs) with polyurethane sponge (PUS), building on the application of flexible conductive polymer composites (CPCs) in pressure sensors. By using a dipping drying technique to construct the sensor, CNTs were able to consistently stick to the PUS's skeleton. Quick reaction and regaining time, high sensitivity, and outstanding sustainable stability were among the remarkable performance outcomes of this design. Throughout 18,000 s of loading and unloading cycles at 0.1 Hz, the sensor operated reliably in the pressure range of 0 to 100 Kpa. Additionally, it was effectively used to track motion signals on different body areas, demonstrating its adaptability and wearable application potential [71]. Physiological signals such as blood pressure, heartbeat, and pulse rate are intimately associated with common illness like high blood pressure, and coronary heart diseases (CHD). In order to effectively monitor and treat cardiovascular disorders, wearable pressure sensors provide some useful tools for tracing these signals. The numerous uses of wearable and flexible pressure sensors are depicted in Fig. 6. Integrating wearable sensors to monitor both physiological and motion-related signals provides a comprehensive approach to health assessment and disease prevention [64, 72].

Besides, flexible capacitive pressure sensors have also gained considerable attention in fields such as wearable devices, electronic skin, medical diagnosis, physical health examining, and artificial intelligence. Their popularity stems from their notable features, including high sensitivity, flexibility, lightweight design, and ease of integration [73]. To improve capacitive sensor's performance, various microstructure designs, such as porous structures, air-gap configurations, wrinkled surfaces, and pyramid shapes, have been implemented. Ionic gel-based materials have recently used to replace the dielectric substance in the middle layer to these sensors, substantially enhancing their sensitivity. This modification forms electric double-layer (EDL) capacitance, resulting in super capacitive sensors with ultra-high sensitivity [74]. Typically, flexible capacitive sensors consist of two main components: a flexible dielectric layer positioned between two flexible electrodes. When pressure

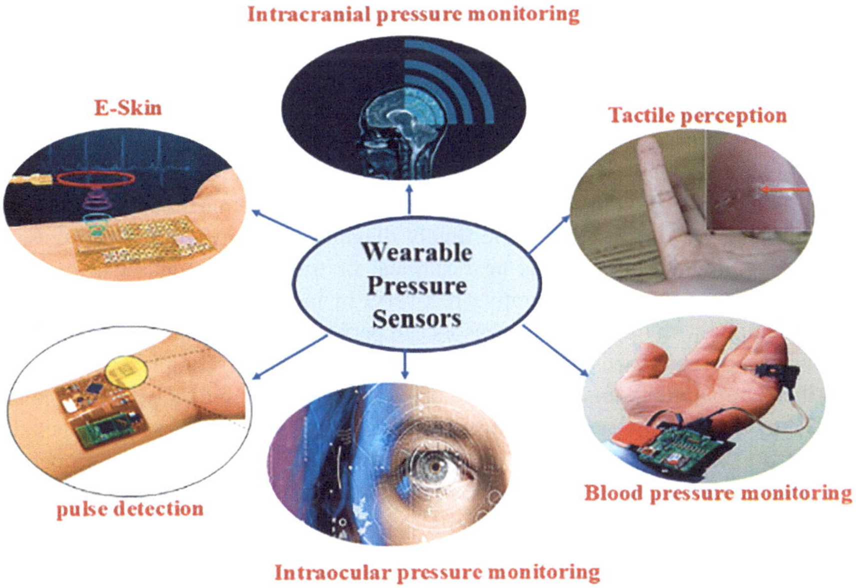

Fig. 6 Applications of wearable and flexible pressure sensors

is applied to the sensor, the dielectric layer compresses, causing an increase in capacitance. These sensors are capable of measuring a wide variety of pressures. As a result, flexible pressure sensors have proven to be highly useful in fields including electron-skin, human-machine interfaces, healthcare examining and motion recognition [75, 76]. Electronic skin (e-skin) is a wearable device designed to mimic the properties of human skin. It enables immediate and durable monitoring of various health indicators, including physical metrics like exercise, respiration, and blood pressure, as well as chemical markers found in saliva, sweat, and urine [77].

3 Carbon-Based Flexible Temperature/Humidity Sensors

Temperature of human body, skin hydration, and respiratory emission are essential health indicators that play a significant role in diagnosing medical conditions, driving the demand for wearable sensors capable of monitoring temperature and humidity. Recent developments have highlighted carbon-based materials as highly effective components for flexible sensors. These materials offer unique advantages, such as superior electrical conductivity, adaptability for modification and functionalization, a variety of scalable macroscopic structures, and inherent flexibility. The discussion that follows examined their potential in enhancing the performance of temperature and humidity sensors in greater detail.

3.1 Carbon-Based Flexible Temperature Sensors

3.1.1 Materials and Sensing Mechanisms

Temperature sensing is one of the key functionalities of flexible devices and is a crucial feature commonly needed in wearable technologies [78]. Attaching flexible temperature sensors to human skin enables continuous real-time monitoring of temperature changes, facilitating in the timely diagnosis and prevention of diseases. Tracking temperature variations near surgical or traumatic wounds can provide valuable insights into post-operative recovery. In industrial applications, robotic arms equipped with flexible force sensors can perform tasks like grasping, and integrating temperature sensors allows them to handle more complex operations. Flexible temperature sensors are also being incorporated into advanced robots, enhancing their tactile sensing abilities [79]. The mechanisms of sensing include thermistors [80], field effect transistor-based wearable biosensors [81], thermoelectric temperature sensor [82], etc. Flexible thermistor sensors (FTSs) are ideally suited for a wide range of applications requiring the monitoring of strain and dynamic temperature, owing to their remarkable thermosensitivity and mechanical flexibility [83]. For flexible temperature sensors based on thermistor, a range of carbon materials can be investigated, such as carbon nanotubes (CNTs) [84], and flexible temperature sensors that use two-dimensional materials like graphene [78]. The sensitivity of thermistors, which may be measured using the temperature coefficient of resistance (TCR), is the main topic in this section. The TCR illustrates how a thermistor's resistance varies in response to temperature variations. The mathematical expression for it is:

$$\propto = \frac{1}{R} \cdot \frac{\mathrm{d}R}{\mathrm{d}T}$$

where R is the thermistor's resistance, α is the temperature coefficient of resistance (TCR), and ($\mathrm{d}R_\mathrm{T}/\mathrm{d}T$) is the rate at which thermistor's resistance changes with temperature [85].

3.1.2 Carbon Nanotube (CNT) Thermistor-Based Temperature Sensors

Because of their remarkable properties, carbon nanomaterial-based flexible temperature sensors have shown a great deal of promise. These sensors have inexpensive manufacturing costs, great sensitivity, fast reaction times, and beneficial mechanical adaptability. Additionally, they display great cycling stability and reliability, making them perfect for diverse applications. The development of temperature sensing technology is aided by the special benefits of carbon nanoparticles [86]. Because of their exceptional thermal, electrical, and mechanical characteristics, carbon nanotubes (CNTs) have been used extensively as conductive fillers in flexible thermistors. For instance, flexible thermoplastic polyurethane

(TPU)/single-walled carbon nanotube (SWCNT) composites were created using a straightforward solution blending and thermal annealing process. The resulting temperature sensor exhibited remarkable accuracy (0.1 °C), excellent reproducibility, and high reliability within the temperature range of 30 to 100 °C [84]. A second innovation in CNT-based flexible thermistor sensors (FTSs) involves a hybrid design combining CNTs and MXene with a flexible substrate (paper/PDMS/Si_3N_4 nanoparticles). This sensor offers a broad temperature variety (−20 to 220 °C), high thermosensitivity (−0.52% $°C^{-1}$), fast response/recovery times (~300 m/s and 3 s), and durability over 2000 bending cycles. It also performs well in challenging conditions like compression, underwater, and on curved surfaces, with optimal performance achieved at a 1:2 CNT-to-MXene ratio [83].

3.1.3 Temperature Sensors Based on Graphene Thermistors

Graphene, a prominent two-dimensional (2D) material, is distinguished by its excellent heat conductivity, bulky surface area, and exceptional carrier mobility. Because of these characteristics, graphene is a leading candidate among carbon-based materials for the development of temperature sensors. Both theoretical and practice research indicate that thermal conductivity of monolayer graphene can reach up to 5300 Wm^{-1} K^{-1}, which is considerably higher than that of carbon nanotubes (3000 Wm^{-1} K^{-1}). This ultra-high thermal conductivity enables graphene-based temperature sensors to exhibit extremely fast thermal responses. Furthermore, graphene boasts a specific surface area of up to 2600 m^2/g, providing a much larger thermal contact area compared to other materials [87]. Wet spinning of graphene oxide liquid crystals (GO LCs) is a scalable method creating graphene fibers, making it perfect for flexible/wearable devices. Temperature sensors used the graphene fibers made from GO LCs as thermistors. These fiber thermistors demonstrated great repeatability, a low time constant of 0.33 ± 0.07 s, high linearity (correlation coefficient of 0.999), and a temperature coefficient of resistance (TCR ≈ −0.21 ± 0.01%/°C) [80]. Different sensing materials have been used to create a variety of wearable temperature sensors, including metals [88], carbon materials-based [89], and conductive polymers-based [90]. Of them, a derivative of graphene oxide (GO), that is reduced graphene oxide (rGO), stands out as a prominent material for temperature sensing within the carbon family. rGO exhibits remarkable electrical conductivity and is exceptionally conductive to heat. These characteristics make rGO the perfect material for temperature sensors, especially when combined with its good temperature coefficient of resistance (TCR) and sensitive thermal response [85].

3.1.4 Other Carbon Materials-Based Temperature Sensors

Other carbon-based materials, such as graphite and carbon black (CB), have been investigated for direct application or in conjunction with polymers to produce flexible temperature sensors in addition to graphene and carbon nanotubes [55, 91]. For instance, the production of a graphitic structure was a defining feature of the development of a flexible resistance-type temperature sensor using polyetherimide (PEI) and laser-induced graphene (LIG). The performance of the sensor was assessed for twisting cycles, humidity impacts, and temperature responsiveness. The findings showed that humidity has little effect on the PEI-LIG flexible thermistor's high sensitivity [92]. Additionally, carbon nanotubes (CNTS) and carbon black (CB) were added to polyvinyl alcohol/glycerol (PVA/Gly) matrix to create a conductive organohydrogel with a three-dimensional honeycomb structure. Hydrogen connections between PVA, glycerol, and water are essential to the development of the physical cross-linking network that forms the basis of the organohydrogel's structure. Because of their different aspects ratios and geometric arrangements, the CNTs and CB work in concert to improve the mechanical and electrical properties of PVA/Gly organohydrogel. The resultant organohydrogel has remarkable mechanical qualities, such as a strong tensile strength, and remarkable flexibility. Sensors made from these organohydrogels demonstrate high sensitivity to both tensile strain and temperature [93]. The applications of carbon-based flexible temperature sensors are illustrated in Fig. 7.

3.2 Carbon-Based Flexible Humidity Sensors

3.2.1 Materials and Sensing Mechanisms

Humidity is a crucial factor in environmental, ecological, industrial, and medical monitoring, both indoors and outdoors. Maintaining optimal humidity levels, (generally between 30% and 60%), is essential for supporting biological life, as variations outside this range can have a significant impact on quality of life [94]. Numerous humidity sensors have been created based on sensing mechanisms, including resistance-type [95], capacitance type, gravimetric, optical, also magnoelastic sensors. Because they can withstand significant water vapor condensation at high humidity levels, capacitance-type humidity sensors are typically more robust than resistance-type humidity sensors. Additionally, they have a wide capacitance range and can function efficiently at high temperature in response to variation in humidity [96]. Flexible humidity sensors frequently use carbon nanomaterials as active ingredients, including graphene (GO/rGO) and carbon nanotubes (CNTs) due to their exceptional chemical inertness, high electrical sensitivity, and customizable structural design [97]. A range of flexible humidity sensors employ carbon materials, as seen in Fig. 8.

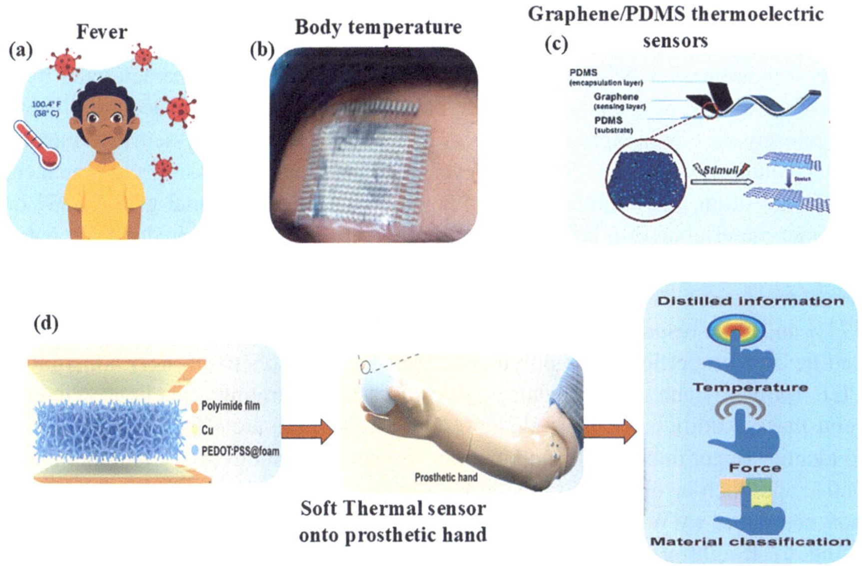

Fig. 7 Applications of carbon-based flexible temperature sensors

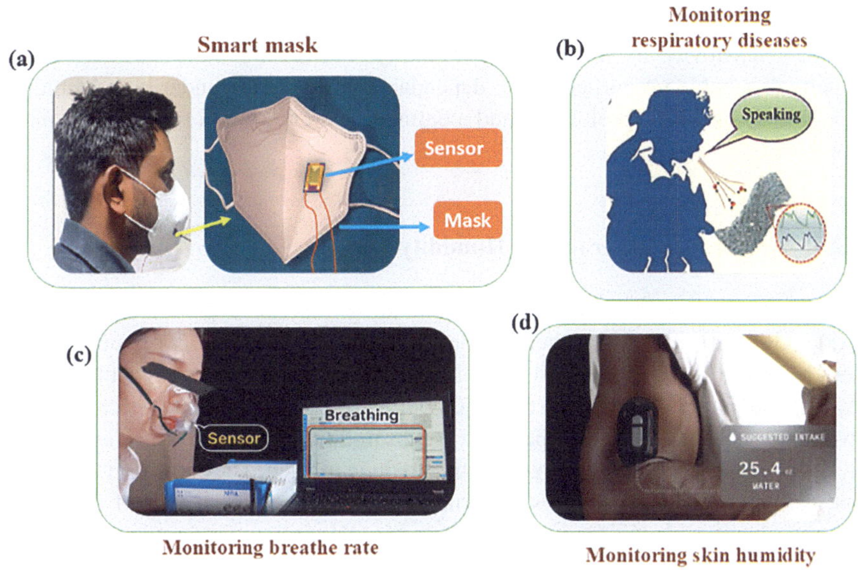

Fig. 8 Carbon materials-based flexible humidity sensors

3.2.2 Fabrication of Carbon Nanotube (CNTs) Humidity Sensors

Carbon nanotubes (CNTs) are one-dimensional nanomaterials characterized by a hollow assembly, outstanding electrical conductivity, and high surface area. These attributes make CNTs highly suitable for applications in humidity-sensitive materials and humidity sensors [98]. However, to enhance their sensing performance, CNTs are often functionalized with oxygen-containing functional groups [99] or polymer materials [100]. For example, a composite of CNTs and hydrophilic polymers was prepared, combining the hydrophilic nature of the polymers with the network structure of CNTs. This resulted in humidity sensors with a high response of 171% and rapid response/recovery times of 23 s and 10 s, respectively. The adhesive and flexible properties of the polymers allowed the sensors to strongly adhere to a PET (polyethylene terephthalate) substrate, demonstrating excellent winding robustness. Additionally, these flexible humidity sensors are effectively employed to identify finger movements, and monitor human respirations [101]. Another innovative approach involves the fabrication of humidity sensor using multi-walled carbon nanotubes (MWCNTs) affixed to the surface creases of a natural latex sheet using a straightforward swelling-self-assembly technique. The sensor's broad humidity detection range and quick humidity response time are made possible by the weak contact between MWCNTs and water molecules. This makes it appropriate for accurate 1 Hz human breathing detection. By increasing the area of contact between the MWCNTs and the airflow, the crumpled surface structure increases the sensor's sensitivity to variations in humidity. Furthermore, the surface's near-superhydrophobic properties assure dependable performance under high humidity conditions, enabling prolonged and continuous monitoring of human respiration [102].

3.2.3 Fabrication of Graphene Humidity Sensors

The presence of hydrophilic oxygen-containing functional groups in graphene oxide (GO) and reduced graphene oxide (rGO) makes them ideal for humidity sensors because they improve their capability to interact with water molecules [103]. Among 2D materials, graphene oxide (GO) has drawn a lot of interest because of its hydrophilic character and compatibility with solution processing, which are caused by the abundance of functional groups that include oxygen. The carboxyl (−COOH) and carbonyl (−C−OH) groups are primarily found along the edges of the sheet, whereas the epoxide (C−O−C) and hydroxyl (−OH) groups are found in the inside. A reduces graphene oxide (rGO) solution was drop-cast over a bimatted polyester substrate to create a flexible chemiresistive humidity sensor. Inject-printed silver integrated electrodes were used to pre-coat the substrate. The final sensor demonstrate great sensitivity in the 10–80% relative humidity (RH) range [104]. Capacitive humidity sensors usually use pristine graphene oxide (GO), whereas chemiresistive humidity sensors make considerable use of reduced graphene oxide (rGO), which is distinguished by its increased conductivity. The potential of graphene quantum dots

(GQDs) and 2D-layered graphene as humidity sensors has also been investigated because of its many adjustable characteristics. Comparing chemiresistive humidity sensors to other varieties reveals a number of benefit, such as economical production, simpler integration into complementary metal-oxide-semiconductor (CMOS) platforms, and effective operations [105]. Numerous reduction techniques, such as electrochemical, thermal, microwave, or laser-induced procedures, can transform pristine graphene oxide (GO) into reduced graphene oxide (rGO), which enhances its electrical conductivity [106]. Furthermore, the hydrophilic polymer known as polyvinyl alcohol (PVA), which is produced by hydrolyzing polyvinyl acetate, is distinguished by its exceptional hygroscopicity, affordability, and simplicity of film creation. Because it is biodegradable and non-toxic, it is a perfect material for wearable humidity sensors. This material must be hydrophilic to absorb water molecules, conductive for ion diffusion and migration, and flexible enough to meet the criteria for wearable applications [107]. Furthermore, graphene-based humidity sensors have emerged as a promising technology in agriculture to address the increasing demand for food, flexible and durable textile-based humidity sensors were developed and printed on four types of greenhouse fabrics using silver and carbon conductive inks. These sensors are designed to measure high humidity levels (80%, 90%, and 100% RH) and showed satisfactory dynamic responses in hysteresis tests, with response times of 15.8, 17.3, 24.8, and 25 s for G2S, W1S, B2C, and W1C fabrics, respectively. They also demonstrated excellent durability and repeatability, maintaining consistent performance even after extended UV exposure and rainfall. These features make them highly promising for smart agricultural applications [108].

4 Carbon Materials for Flexible Electrochemical Sensors

In addition to the tracking of physical factors related to health such as strain, pressure, and temperature, monitoring of chemical components related to health in actual time and without invasive methods is also very important since it can yield crucial new health insights, contributing to comprehensive and personalized healthcare. Electrochemical sensors operate by utilizing the electrochemical reactions of substance's electrode surface, transforming the detected signals into corresponding electrical signals. Among various biofluids, body surface sweat is rich in analytes and metabolites like glucose, lactic acid, urea, and inorganic salts that can provide valuable physiological insights about the body and are strongly correlated with blood levels, making it highly advantageous in wearable sensing applications [109]. Numerous techniques have been established to detect sweat analytes, including optical techniques like fluorescence sensing, surface-enhanced Raman spectroscopy methods, and colorimetric methods [110], as well as electrochemical sensing methods. Electrochemical sensing, in particular, is a popular and deep-rooted method for sweat analysis. Its popularity in wearable healthcare applications is attributed to its flexibility, biocompatibility, cost-effectiveness, lightweight nature, and high performance, which have garnered significant interest in the field [109,

111, 112]. Numerous uses of carbon-based electrochemical biosensors are illustrated in Fig. 9.

Various carbon materials, such as graphene oxide, reduced graphene oxide, graphene sheets, graphitic carbon nitride, carbon nanotubes (single-walled and multi-walled), porous carbon, quantum dots, fullerenes, diamonds, and three-dimensional porous carbons, have been widely applied as modifier materials in sensor fabrication [113]. Among these, carbon nanotubes (CNTs) are frequently employed as conductive electrode materials because of their excellent conductivity, remarkable stretchability, and ease of functionalization [114].

4.1 Fabrication of Carbon Nanotubes (CNTs) Electrochemical Sensor

Single-walled carbon nanotubes (SWCNTs) and multi-walled carbon nanotubes (MWCNTs), when arranged as aligned structures or network films on electrode surfaces, significantly enhance electrochemical performance. Compared to traditional rigid carbon-based materials like glassy carbon electrodes, CNTs offer superior advantages for the development of flexible electrodes [115].

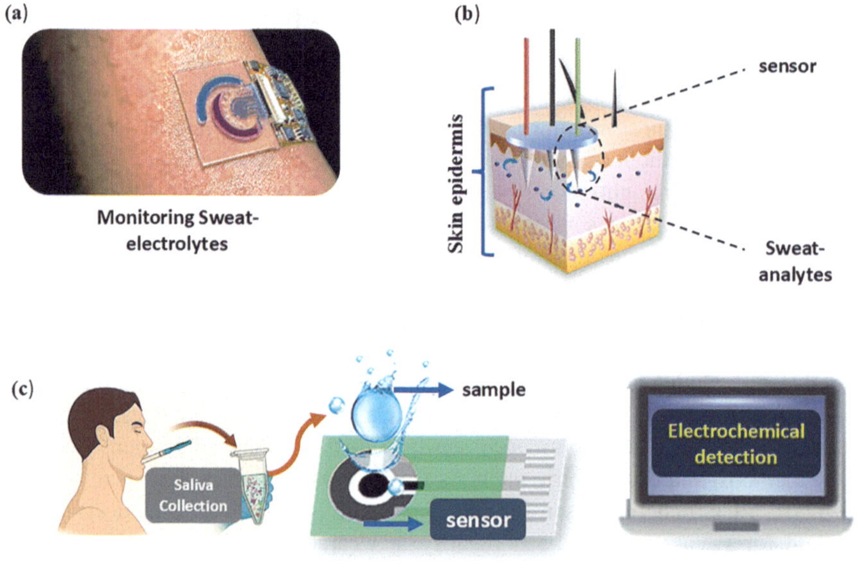

Fig. 9 Applications of carbon-based electrochemical biosensors

4.2 *Fabrication of Graphene Electrochemical Sensor*

Because of its exceptional qualities, graphene, which is made up of a single layer of sp^2-bonded carbon atoms arranged in a honeycomb lattice has attract a lot of interest in the field of electrochemical sensors. These include a large surface area, outstanding mechanical strength, greater electrical conductivity, ease of functionalization, and outstanding biocompatibility, making it an ideal material for improving the performance of electrochemical biosensors. Various graphene derivatives, such as graphene oxide (GO), reduced graphene oxide (rGO), graphene quantum dots, and three-dimensional graphene structures, have been extensively studied for electrochemical cancer biosensing [116].

5 Carbon-Based Flexible Power Systems

Many studies have been conducted to create wearable and flexible energy devices, such as flexible batteries (like lithium-ion batteries, high-energy metal batteries, sodium(Na) batteries) [117–119], supercapacitors [120], stretchable photovoltaics [121] and nanogenerators(e.g., triboelectric generators and piezoelectric generators) [122, 123], all designed to deliver reliable and efficient power sources for wearable electronics. Carbon materials, owing to their excellent electronic conductivity, large surface area, and simplification of chemical modification, particularly carbon nanotubes (CNTs), graphene, and fullerenes have been widely studied for various energy storage and conversion applications over the past few decades [124]. In this section, we will focus on flexible supercapacitors and metal-air batteries based on carbon nanomaterials due to their simple construction processes, excellent protection, and energy densities, and great potential for real-world use in wearable technology.

5.1 *Carbon Materials-Based Flexible Supercapacitors*

Because of their great power density, excellent rate capability, lengthy cycle life, and safe operation, supercapacitors (SCs) have drawn a lot of interest as a potential energy storage technology [125].Carbon materials including graphene [126], carbon nanotubes(CNTs) [127],and other carbon materials due to their unique properties have drawn a great deal of scientific attention for energy storage and conversion applications. Figure 10 shows numerous applications of flexible supercapacitors.

Supercapacitors are typically classified into two types: pseudocapacitors and electrochemical double-layer capacitors (EDLCs) [128, 129]. Pseudocapacitors involve redox reactions that occur at the surface of electrode, while EDLCs, which

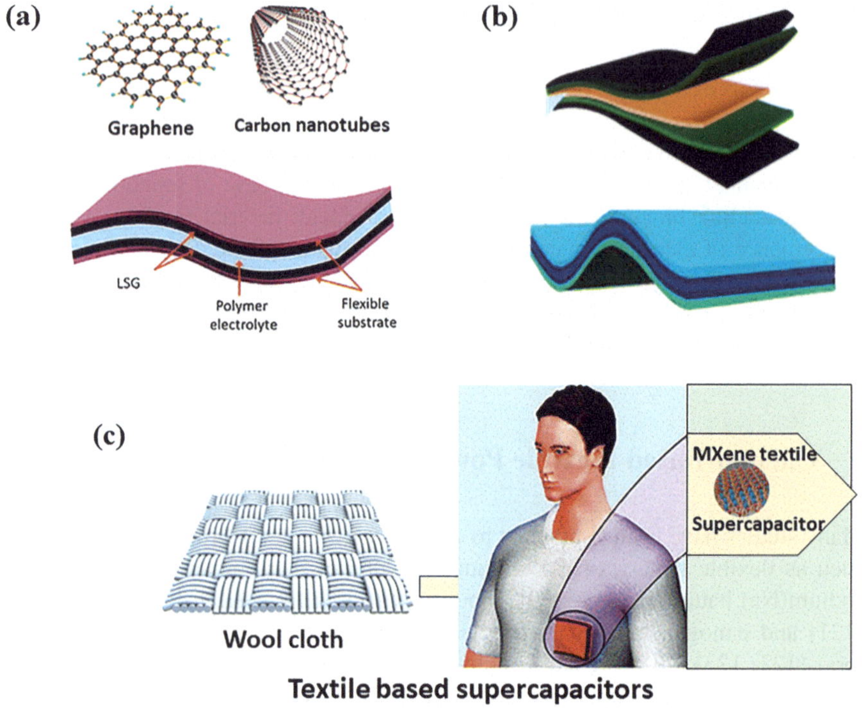

Fig. 10 Applications of applications of flexible supercapacitors

do not involve faradaic reactions, store charge by forming a double layer between the electrode and the electrolyte.

5.1.1 Carbon Nanotube (CNTs)-Based Flexible Supercapacitors

Carbon materials like carbon nanotubes (CNTs), graphene, and carbon black are particularly advantageous for EDLCs because of their porous structure, which provides a large interfacial area. Among these, CNTs stand out because of their high surface area, excellent thermal stability, high conductivity of electricity, and strong interaction with the electrolyte [130]. For example, a robust and ultralight MXene/carbon nanotube (CNT) aerogel is developed using a simple directional freezing and freeze-drying method. The strong π-π interactions between MXene nanosheets and CNTs result in a composite aerogel with excellent electrical conductivity (up to 1600 S m^{-1}), abundant pore structures, and remarkable mechanical properties. The all-solid-state flexible supercapacitor fabricated with this aerogel demonstrates impressive electrochemical performance, achieving 410.7 mF cm^{-2} at a current density of 0.8 mA cm^{-2}. Additionally, the flexible supercapacitor maintains stable cyclic voltammetry curves even when bent at angles up to 90°, and four devices connected in series are capable of powering an LED [131].

5.1.2 Graphene-Based Flexible Supercapacitors

Graphene, known for its exceptional electrical, optical, and mechanical properties, has found extensive applications in electrochemical energy storage devices, sensor development, and catalyst support. Owing to its unique characteristics, graphene is a highly promising material for use as an electrode in flexible supercapacitors [132]. For example, graphene oxide and greatly conductive graphene flakes are combined to create a graphene-based composite sheet. The graphene flakes contribute to the film's excellent electrical conductivity, while the graphene oxide enhances additional pseudocapacitance. This synergy results in a high specific capacitance, significantly surpassing that of pure graphene film. The flexible supercapacitors accumulated from this composite film demonstrate remarkable durability, maintaining its capacitance even when bent at various angles. After 4500 bending-releasing cycles (0–180°), the capacitance retention remains at 55.8%. With its outstanding mechanical and electrochemical properties, the reduced graphene oxide/graphene composite film holds great potential for use in flexible supercapacitors [133].

Despite significant advancements in carbon-based flexible/stretchable supercapacitors, several challenges remain for their practical implementation. In order to enhance large-scale manufacturing of these devices, future research in this area should concentrate on developing electrodes materials that combine exceptional mechanical qualities with superior electrochemical performance designing safe and effective flexible and stretchable solid-state electrolytes, and investigating scalable production techniques.

5.2 Flexible Metal-Air Batteries

Metal-air batteries, such as those using zinc (Zn), lithium (Li), magnesium (Mg), aluminum (Al), and iron (Fe), are gaining attention as energy storage options for flexible electronic devices. They offer high-energy density, low cost, abundant materials, eco-friendliness, and safe operation [134]. Wearable aqueous metal-air batteries, with high-energy density and intrinsic safety, hold significant potential as energy storage systems for low-power wearable devices. Aqueous electrolytes have stronger ionic conductivity, greater safety, lower cost, and better environmental compatibility, than non-aqueous electrolytes, making them ideal for wearable power sources. Mg-, Al-, Zn-, and Fe-air batteries are commonly studied as aqueous systems due to the compatibility of their metal anodes with water [135]. In contrast, Li-, Na-, and K-air batteries typically operate with non-aqueous electrolytes and pure oxygen due to their sensitivity to water and carbon dioxide in the air [136–138]. Flexible metal-air batteries have a variety of applications, as depicted in Fig. 11.

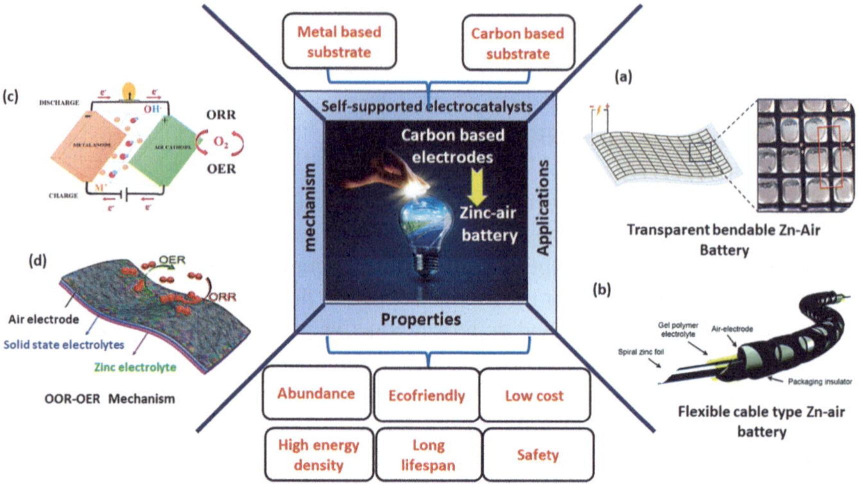

Fig. 11 Applications of flexible metal-air batteries

5.2.1 Zinc-Air (Zn-Air) Flexible Batteries

Among all aqueous metal-air batteries, zinc-air batteries stand out with a relatively high theoretical specific energy density of approximately 1218 Wh kg^{-1} and a working voltage of around 1.66 V. Zinc electrodes are more stable and safer in alkaline electrolytes, offering efficient charging capabilities. In contrast, aluminum-air and magnesium-air batteries face practical limitations due to lower working voltages and significant polarization issues in alkaline electrolytes despite their higher theoretical energy densities. As a result, zinc-air battery technology is considered the most cost-effective and viable option for future energy storage solutions [134]. Nanostructures carbon-based electrodes have shown great promise as oxygen electrode for zinc-air batteries in recent years. They offer exceptional advantages, including extensive surface areas, effective oxygen-catalytic sites, hierarchically absorbent structures, excellent roughness, and affordable [139]. Graphene enhances battery performance with its high specific surface area, excellent conductivity, and resilience under harsh conditions. Additionally, carbon nanotubes (CNTs) are regarded as ideal substrates for self-supported air electrodes due to their large surface area, exceptional mechanical strength, and efficient pathways for mass transport, resulting in superior overall battery performance [140]. To enhance the efficiency of rechargeable zinc-air batteries (ZABs), high-performance and environmentally friendly electrocatalysts for oxygen reduction (ORR) and oxygen evolution reactions (OER) are essential [141]. For example, to enhance the performance of Zn-air batteries, helical carbon nanotubes (HCNTs) encasing cobalt nanoparticles have been created as ORR/OER electrocatalysts. By exposing more C_0 N sites, offering superior electron conductivity, and promoting reactant mass transfer,

HCNTs increase ORR/OER activity. This study demonstrated the possibility of converting bamboo biomass into adaptable electrocatalysts, promoting the development of next-generation energy conversion and storage techniques [142]. Furthermore, the stretchable Zn-air battery can be incorporated into clothing and can tolerate human motion-induced bending and stretching, which makes it appropriate for use in wearable electronics. Clothing that has metal-air batteries attached to it can withstand a variety of deformations, including folding, winding, twisting, and extending, which is crucial for developing sophisticated equipment with electronic devices for contemporary society [143].

6 Carbon-Based Conductive Electrodes for Flexible Healthcare Procedures (Electrophysiological Sensing)

Conductive polymer-based flexible and stretchable electrodes are widely utilized across various applications, including wearable devices and sensors for the human body [144]. The use of carbon materials for flexible conductive electrodes is highlighted in this discussion, with particular attention paid to their uses in electrophysiological signal monitoring and as connectors for different wearable system component. Research on physiological signal sensors has focused on monitoring body and muscle motion [145], heart rate [146], pulse, body temperature, respiration, respiratory humidity [147], and electrophysiological signals like ECG, EMG, EEG, and EOG, as shown in Fig. 12. These advancements are crucial for developing advanced wearable healthcare technologies [18, 148]. Echocardiograms (ECGs) are

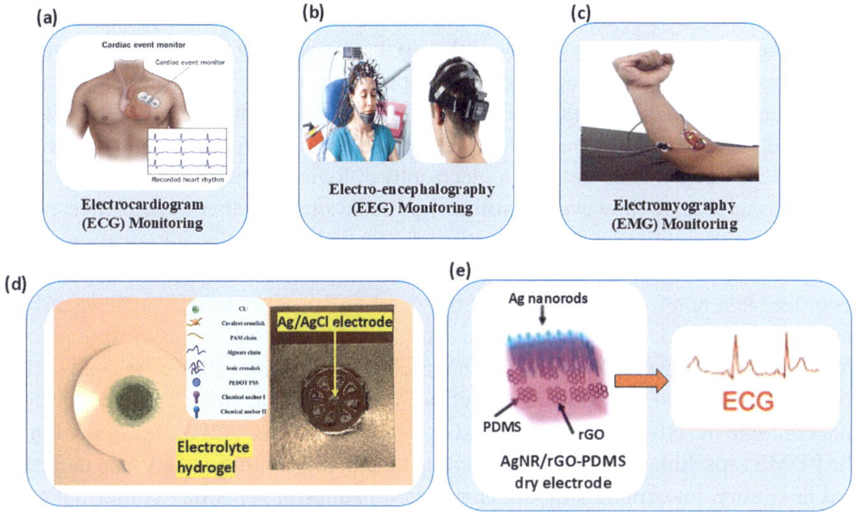

Fig. 12 Applications of carbon-based conductive electrodes for flexible healthcare systems

essential in assessing the structure and function of the heart, making them vital diagnostic tools. The largest cause of death worldwide is cardiovascular illnesses, which include failure of heart, hypertension, arrhythmias, and stroke. The World Health Organization (WHO) estimates that cardiac vascular diseases cause 17.7 million deaths a year, or almost 31% of the deaths worldwide [149].

To monitor ECG, EMG, EEG, and EOG signals using wearable devices, electrophysiological sensors need to be attached to the skin to capture electrical signals effectively. Minimizing resistance at the electrode-skin interface requires maintaining tight, conformal contact with the skin. The materials utilized in the manufacture of flexible electrodes have a significant impact on their performance. Both graphene oxide (GO), a two-dimensional carbon nanomaterial, and carbon nanotubes (CNTs), a one-dimensional carbon nanomaterial, are electrochemically functional, biocompatible, and favorably conductive. Because of these remarkable qualities, they are perfect candidates for a range of bioelectrodes, including neural electrodes, electrocardiography (ECG), electromyography (EMG), and electro-encephalogram (EEG). CNTs and GO are therefore commonly used in the formation of electrodes for the monitoring and analysis of bioelectrical signals [150].

6.1 Carbon Nanotubes (CNTs)-Based Conductive Electrodes for Electrophysiological Signals Detection

Because of their unique structure, which confers remarkable electrical, mechanical, and optical capabilities, carbon nanotubes (CNTs) hold great potential for a range of applications in biomedicine. Their applications in the construction of multipurpose electrode have made it possible to detect a range of electrophysiological signals, including pressure, ECG, EMG, and EEG, facilitating the early identification of disease and the tracking of physical activity. Because of their superior contact impedance, CNT-based electrodes are especially useful for electromyography (EMG) and electroencephalography (ECG). Further proving their adaptability in wearable systems, CNTs have also been effectively incorporated into electronic skin and clothing for the detection electrophysiological signals. The inherent propensity to aggregate, however, is still a significant obstacle that must be overcome in order to expand their application in sophisticated electrophysiological monitoring systems [151]. Developing dry electrodes for precise and reliable EEG signal recording is a noteworthy example of how carbon-based materials, such as carbon nanotubes (CNTs), can be used to build flexible conductors in biosensors. Polymethyl-siloxane (PDMS)/carbon nanotube (CNT) composites were used in this invention to construct flexible dry electrodes with isotropically crumpled exteriors that combine the advantages of wet and dry electrodes. Through the optimization of the PDMS crosslinker proportion, the composite demonstrated exceptional mechanical flexibility, low Young's modulus, and good adhesion, enabling conformal contact with the skin. By drastically lowering active artifacts during EEG signals recognition, this structure guarantees more accurate and dependable signal processing. The findings show the flexible, wavy CNT-based dry electrodes can perform better than conventional wet electrodes, providing enhanced functionality for clinical EEG monitoring and analysis in real time [150]. Additionally, a straightforward, inexpensive, solution-processing technique was used to create a multiwall carbon nanotubes/polydimethylsiloxane (MWCNTS/PDMS) composite-based bioelectrode. Over a frequency range of 20 Hz–1 KHz, the manufactured MWCNT/PDMS composite electrode showed electrode-skin resistance that was comparable to that traditional silver/silver chloride (Ag/AgCl) electrodes. Furthermore, there is no requirement for and intermediary gel layer or skin preparation because these electrodes can be put to dry skin. The consequences show that the dry electrodes based on MWCNT/PDMS are very effective in obtaining immediate EEG signals, outperforming commercially existing Ag/AgCl electrodes in both electromyography (EMG) and electrocardiography (EMG) studies [152].

6.2 Graphene-Based Flexible Conductive Electrodes for Electrophysiological Signals Detection

Graphene exhibits exceptional properties relevant to bioelectronics, including its sub-nanometer thickness, optical transparency (approximately 97% for visible light), high strength (130 GPa), stretchability (around 15%), excellent electrical conductivity, and biocompatibility. These remarkable characteristics make graphene an ideal material for the fabrication of cardio patches, flexible microelectrodes, and microprobes, which are widely used for monitoring and stimulating electrical activities in cardiac and neural cellular research [153]. Its high conformability makes it an ideal material for flexible electrodes, particularly when in contact with human skin. The ability of graphene-based electrodes to conform to the skin enhances the signal-to-noise ratio (SNR), reduces skin-electrode contact impedance, and improves signal quality. Because of these benefits, graphene is a very good material for flexible ECG electrodes, which could lead to long-term, high-quality ECG signal examining [154]. Weak bioelectrical signals, including electrocardiograms, electromyograms, can be captured and conducted by the rGO/PDMS dry electrodes with little distortion or attenuation. According to experimental data, eventhough the rGO/PDMS electrodes are dry electrodes, they produce better ECG signals with better signal-to-noise ratios (SNRs) than traditional wet Ag/AgCl electrodes while retaining comparable contact quality and electrode-skin resistance. Because of their outstanding performance, the rGO/PDMS electrodes are ideal for application in wearable, longstanding health monitoring systems [155].

7 Conclusions and Prospects

In conclusion, the performance and application of flexible/wearable electronics have been greatly improved by latest developments in the design and controlled manufacturing of carbon-based devices. Excellent electrical conductivity, great thermal, mechanical, and chemical strength, as well as the capacity to take on a variety of flexible shapes are just a few of the outstanding benefits that carbon materials like graphene, carbon nanotubes (CNTs), and other carbon allotropes have to offer. Their potential for a variety of uses is further increased by their simplicity in chemical functionalization. The development of flexible sensors, including strain/pressure sensors, temperature and humidity sensors, electrochemical sensors, as well as conductive electrodes and energy devices, has therefore made these materials extremely auspicious. Undoubtedly, their further development will be vital to the development of implantable and wearable electrical systems in the future.

The fabrication of cutting-edge carbon materials and technology for creating flexible electronics has enormous potential for next-generation medical treatment and healthcare systems. In the future, more attention should be paid to multifunctional wearable systems, which are currently rare. Numerous healthcare and activating monitoring domains, including ECG, EMG monitoring, glucose in perspirations or tears, body temperature, pulse, and breath monitoring are already using these systems. Moreover, intelligent technologies that can communicate with the human body hold great promises. We expect that next-generation healthcare and medical treatment systems will greatly benefit from the development of superior carbon materials and technology for the design and manufacturing of flexible electronics.

References

1. Khan, B., et al. (2024). Advancements in wearable sensors for cardiovascular disease detection for health monitoring. *Materials Science and Engineering: R: Reports, 159*, 100804.
2. Wang, H., et al. (2023). Carbon-based flexible devices for comprehensive health monitoring. *Small Methods, 7*(2), 2201340.
3. Gao, W., et al. (2019). Flexible electronics toward wearable sensing. *Accounts of Chemical Research, 52*(3), 523–533.
4. Won, D., et al. (2023). Transparent electronics for wearable electronics application. *Chemical Reviews, 123*(16), 9982–10078.
5. Zhou, X., & Cao, W. (2023). Flexible and stretchable carbon-based Sensors and actuators for soft robots. *Nanomaterials, 13*(2), 316.
6. Wang, X., et al. (2023). A review of carbon nanotubes, graphene and Nanodiamond based strain sensor in harsh environments. *C, 9*(4), 108.
7. Zhang, Y., et al. (2023). A review of wearable carbon-based sensors for strain detection: Fabrication methods, properties, and mechanisms. *Textile Research Journal, 93*(11–12), 2918–2940.
8. Ma, Z. (2024). Advances in graphene-assisted flexible substrate sensors for human motion monitoring. *International Journal of Electrochemical Science, 19*(9), 100760.

9. Xiao, W., et al. (2024). High-performance graphene flexible sensors for pulse monitoring and human–machine interaction. *ACS Applied Materials & Interfaces, 16*(25), 32445–32455.
10. Medagedara, M. H., et al. (2024). Advancements in textile-based sEMG Sensors for muscle fatigue detection: A journey from material evolution to technological integration. *ACS Sensors, 9*(9), 4380–4401.
11. Al-Qahtani, A. M., et al. (2023). Performance optimization of wearable printed human body temperature sensor based on silver interdigitated electrode and carbon-sensing film. *Sensors, 23*(4), 1869.
12. Liu, L., et al. (2024). Flexible humidity sensing Fiber with high sensitivity and stability for wearable weaving and physiological signal monitoring. *ACS Applied Nano Materials, 7*(12), 14458–14467.
13. Thakur, A., & Kumar, A. (2023). Recent trends in nanostructured carbon-based electrochemical sensors for the detection and remediation of persistent toxic substances in real-time analysis. *Materials Research Express, 10*(3), 034001.
14. Li, S., et al. (2020). Recent advances of carbon-based flexible strain Sensors in physiological signal monitoring. *ACS Applied Electronic Materials, 2*(8), 2282–2300.
15. Lin, Y., Qiu, W., & Kong, D. (2024). Stretchable and body-conformable physical sensors for emerging wearable technology. *Sensors & Diagnostics, 3*(9), 1442–1455.
16. Wang, H., et al. (2023). Advances in carbon-based resistance strain sensors. *ACS Applied Electronic Materials, 5*(2), 674–689.
17. Wang, G., et al. (2023). Multifunctional flexible pressure sensor based on a cellulose fiber-derived hierarchical carbon aerogel. *ACS Applied Electronic Materials, 5*(3), 1581–1591.
18. Shao, W., et al. (2023). Carbon-based textile Sensors for physiological-signal monitoring. *Materials, 16*(11), 3932.
19. Zhang, P., et al. (2024). Electrochemical and electrical biosensors for wearable and implantable electronics based on conducting polymers and carbon-based materials. *Chemical Reviews, 124*(3), 722–767.
20. Baloda, S. (2024). *Carbon nanomaterial based flexible pressure/strain sensors for wearable electronics*. BITS Pilani.
21. Raman, S., & Sankar, A. R. (2024). Recent progress in 2D textile-based piezoresistive strain and pressure sensors. *Journal of Micromechanics and Microengineering, 34*, 083001. https://doi.org/10.1088/1361-6439/ad5cfd
22. Xue, J., et al. (2024). New carbon materials for multifunctional soft electronics. *Advanced Materials, 37*, e2312596.
23. Zeng, T., Wu, Y., & Lei, M. (2024). Review: Developments and challenges of advanced flexible electronic materials for medical monitoring applications. *Advanced Composites and Hybrid Materials, 7*(5), 141.
24. Kim, M., et al. (2023). Emerging bio-interfacing wearable devices for signal monitoring: Overview of the mechanisms and diverse sensor designs to target distinct physiological bio-parameters. *Advanced Sensor Research, 2*(4), 2200049.
25. Li, J., et al. (2024). Flexible pressure, humidity, and temperature Sensors for human health monitoring. *Advanced Healthcare Materials, 13*(31), 2401532.
26. Ikram, M., et al. (2024). Flexible temperature and humidity sensors of plants for precision agriculture: Current challenges and future roadmap. *Computers and Electronics in Agriculture, 226*, 109449.
27. Zhao, Y., et al. (2023). Flexible and stretchable electrochemical sensors for biological monitoring. *Advanced Materials*. https://doi.org/10.1002/adma.202305917
28. Khan, A., DeVoe, E., & Andreescu, S. (2023). Carbon-based electrochemical biosensors as diagnostic platforms for connected decentralized healthcare. *Sensors & Diagnostics, 2*(3), 529–558.
29. Zeng, X., et al. (2023). Highly conductive carbon-based E-textile for gesture recognition. *IEEE Electron Device Letters, 44*(5), 825–828.

30. Thomas, S. A., Cherusseri, J., & Rajendran, D. N. (2024). Recent advancements on carbon fibers-based sustainable electrodes for flexible and wearable supercapacitors. *RSC Sustainability, 2*, 2403–2443.
31. Ehtesabi, H. (2023). Flexible and wearable sensors for biomedical applications. In *Flexible and wearable Sensors* (pp. 321–334). CRC Press.
32. Baydin, A., et al. (2022). Carbon nanotube devices for quantum technology. *Materials, 15*(4), 1535.
33. Li, X., et al. (2024). Capacitive pressure sensor combining dual dielectric layers with integrated composite electrode for wearable healthcare monitoring. *ACS Applied Materials & Interfaces, 16*(10), 12974–12985.
34. Yun, T., et al. (2023). Waterproof and ultrasensitive paper-based wearable strain/pressure sensor from carbon black/multilayer graphene/carboxymethyl cellulose composite. *Carbohydrate Polymers, 313*, 120898.
35. Sedighi, A., & Montazer, M. (2023). Nanomaterials for wearable, flexible, and stretchable strain/pressure sensors. In *Nanotechnology in electronics* (pp. 155–206). Wiley.
36. Xu, W., et al. (2024). Ultrasensitive and superelastic high-performance aerogel based on hollow carbon nanofiber/graphite composite with piezoresistive sensing applications. *Case Studies in Chemical and Environmental Engineering, 9*, 100629.
37. Zhong, Y., et al. (2023). Porous conductive electrode for highly sensitive flexible capacitive pressure sensor over a wide range. *Journal of Alloys and Compounds, 934*, 167919.
38. Wu, L., et al. (2023). Recent application progress and key challenges of biomass-derived carbons in resistive strain/pressure sensor. *Science China Materials, 66*(5), 1702–1718.
39. Qin, R., et al. (2024). Recent advances in flexible pressure Sensors based on MXene materials. *Advanced Materials, 36*(24), 2312761.
40. Wang, R., et al. (2023). Carbon nanotube-based strain Sensors: Structures, fabrication, and applications. *Advanced Materials Technologies, 8*(1), 2200855.
41. Wang, X., et al. (2023). High sensitive electrospun thermoplastic polyurethane/carbon nanotubes strain sensor fitting by a novel optimization empirical model. *Advanced Composites and Hybrid Materials, 6*(2), 63.
42. Huang, K., et al. (2023). Wireless strain sensing using carbon nanotube composite film. *Composites Part B: Engineering, 256*, 110650.
43. Chen, D., et al. (2024). Structure and function design of carbon nanotube-based flexible strain sensors and their application. *Measurement, 225*, 113992.
44. Zhang, D., et al. (2023). Mono-dispersion decorated ultra-long single-walled carbon nanotube/aramid nanofiber for high-strength electromagnetic interference shielding film with joule heating properties. *Carbon, 214*, 118315.
45. Chen, H., et al. (2023). Advances in graphene-based flexible and wearable strain sensors. *Chemical Engineering Journal, 464*, 142576.
46. Lee, B., et al. (2023). Automated approach to in vitro image-guided Photothermal therapy with top-down and bottom-up-synthesized graphene quantum dots. *Nanomaterials, 13*(5), 805.
47. Alwan, S. H., et al. (2023). A mini-review on graphene: Exploration of synthesis methods and multifaceted properties. *Engineering Proceedings, 59*(1), 226.
48. Chang, C. S., et al. (2023). Remote epitaxial interaction through graphene. *Science Advances, 9*(42), eadj5379.
49. Agrawal, A. (2025). Top-down strategies for achieving high-quality graphene: Recent advancements. *Journal of Industrial and Engineering Chemistry, 142*, 103–126.
50. Zhang, Y., et al. (2023). Mechanical exfoliation assisted with carbon nanospheres to prepare a few-layer graphene for flexible strain sensor. *Applied Surface Science, 611*, 155649.
51. Hosseinzadeh, A., et al. (2018). Graphene based strain sensors: A comparative study on graphene and its derivatives. *Applied Surface Science, 448*, 71–77.
52. Bahri, M., et al. (2023). Recent advances in chemical vapour deposition techniques for graphene-based nanoarchitectures: From synthesis to contemporary applications. *Coordination Chemistry Reviews, 475*, 214910.

53. Zou, Y., et al. (2023). Flexible wearable strain Sensors based on laser-induced graphene for monitoring human physiological signals. *Polymers, 15*(17), 3553.
54. Hong, W., et al. (2024). Flexible strain sensor based on nickel microparticles/carbon black microspheres/polydimethylsiloxane conductive composites for human motion detection. *ACS Applied Materials & Interfaces, 16*(25), 32702–32712.
55. Yang, C., et al. (2024). Graphite-doped flexible conductive hydrogel temperature-strain sensor. *Diamond and Related Materials, 142*, 110789.
56. Kausar, A. (2025). Nanodiamond reinforced shape memory nanocomposites—Versatile features and promises. *Polymer-Plastics Technology and Materials, 64*(1), 88–104.
57. Xia, J., et al. (2023). High performance strain sensor based on carbon black/graphene/Ecoflex for human health monitoring and vibration signal detection. *ACS Applied Nano Materials, 6*(20), 19279–19289.
58. Souza, D. A., et al. (2024). The neurological and Hemodynamics safety of an airway clearance technique in patients with acute brain injury: An analysis of intracranial pressure pulse morphology using a non-invasive sensor. *Sensors, 24*(21), 7066.
59. Wu, K. Y., et al. (2023). Advancements in wearable and implantable intraocular pressure biosensors for ophthalmology: A comprehensive review. *Micromachines, 14*(10), 1915.
60. Chowdhury, A. H., et al. (2023). Monitoring and analysis of cardiovascular pulse waveforms using flexible capacitive and piezoresistive pressure sensors and machine learning perspective. *Biosensors and Bioelectronics, 237*, 115449.
61. Lee, S., et al. (2023). A fabric-based wearable sensor for continuous monitoring of decubitus ulcer of subjects lying on a bed. *Scientific Reports, 13*(1), 5773.
62. Nie, Z., et al. (2024). Mechanically active materials and devices for bio-interfaced pressure Sensors—A review. *Advanced Materials, 36*(43), 2205609.
63. Zhang, T., et al. (2023). High-efficient flexible pressure sensor based on nanofibers and carbon nanotubes for artificial electronic skin and human motion monitoring. *Journal of Porous Materials, 30*(6), 1797–1806.
64. Hu, J., et al. (2023). Recent progress in flexible micro-pressure sensors for wearable health monitoring. *Nanoscale Advances, 5*(12), 3131–3145.
65. Lu, H., et al. (2024). Porous reduced graphene oxide@multi-walled carbon nanotubes/polydimethylsiloxane piezoresistive pressure sensor for human motion detection. *Materials Today Nano, 28*, 100512.
66. Du, H., et al. (2023). Electrical response, elastic property, and pressure sensing under bending of hybrid graphene/CNT/elastomer nanocomposites. *Composite Structures, 311*, 116838.
67. Allam, V., & Vandrangi, S. K. (2024). Carbon black and boron nitride conductive fillers dispersed porous conductive polymer composite-based piezoresistive sensor for biomedical applications. *Journal of Materials Science: Materials in Electronics, 35*(35), 2248.
68. Wang, G., et al. (2023). Thermoplastic polyurethane/carbon nanotube composites for stretchable flexible pressure Sensors. *ACS Applied Nano Materials, 6*(11), 9865–9873.
69. Luo, R., et al. (2023). Fragmented graphene aerogel/polydimethylsiloxane sponges for wearable Piezoresistive pressure Sensors. *ACS Applied Nano Materials, 6*(8), 7065–7076.
70. Zhao, Y., et al. (2023). Advancing the pressure sensing performance of conductive CNT/PDMS composite film by constructing a hierarchical-structured surface. *Nano Materials Science, 5*(4), 343–350.
71. Wang, F., et al. (2023). Wearable and cost-effective pressure sensor based on a carbon nanotube/polyurethane sponge for motion detection and gesture recognition. *ACS Applied Electronic Materials, 5*(12), 6704–6715.
72. Lopez, E. O., Ballard, B. D., & Jan, A. (2023). Cardiovascular disease. In *StatPearls*. StatPearls Publishing.
73. Dong, C., et al. (2024). Flexible capacitive pressure sensor: Material, structure, fabrication and application. *Nondestructive Testing and Evaluation, 39*(7), 1749–1790.
74. Cheng, A. J., et al. (2023). Recent advances of capacitive Sensors: Materials, microstructure designs, applications, and opportunities. *Advanced Materials Technologies, 8*(11), 2201959.

75. Basarir, F., et al. (2024). Edible and biodegradable wearable capacitive pressure Sensors: A paradigm shift toward sustainable electronics with bio-based materials. *Advanced Functional Materials, 34*(39), 2403268.
76. Xiong, X., Liang, J., & Wu, W. (2023). Principle and recent progress of triboelectric pressure sensors for wearable applications. *Nano Energy, 113*, 108542.
77. Yang, X., et al. (2024). Electronic skin for health monitoring systems: Properties, functions, and applications. *Advanced Materials, 36*(31), 2402542.
78. Lim, S., & Suk, J. W. (2023). Flexible temperature sensors based on two-dimensional materials for wearable devices. *Journal of Physics D: Applied Physics, 56*(6), 063001.
79. Liu, L., et al. (2024). Recent advances in flexible temperature Sensors: Materials, mechanism, fabrication, and applications. *Advanced Science, 11*(36), 2405003.
80. Jeong, M. H., et al. (2024). Rapid fabrication of graphene Fibers and Fiber-based thermistors for wearable devices. *ACS Applied Nano Materials, 7*(23), 26836–26842.
81. Nguyen, T. T.-H., et al. (2023). Field effect transistor based wearable biosensors for healthcare monitoring. *Journal of Nanobiotechnology, 21*(1), 411.
82. Gao, F.-L., et al. (2024). Multifunctional thermoelectric temperature sensor for noncontact information transfer and tactile sensing in human-machine interaction. *Advanced Functional Materials, 34*(1), 2309553.
83. Sun, B., et al. (2024). A strain-resistant flexible thermistor sensor array based on CNT/MXene hybrid materials for lithium-ion battery and human temperature monitoring. *Sensors and Actuators A: Physical, 368*, 115059.
84. Kong, C., et al. (2024). Preparation of ultra-sensitive flexible temperature sensors with thermal responsive waterborne polyurethane for enhanced temperature sensing performance. *Chemical Engineering Journal, 496*, 154246.
85. Li, X., et al. (2023). Wearable temperature Sensors based on reduced graphene oxide films. *Materials, 16*(17), 5952.
86. Chen, Z., et al. (2021). Flexible temperature sensors based on carbon nanomaterials. *Journal of Materials Chemistry B, 9*(8), 1941–1964.
87. Tang, C., et al. (2023). A review of graphene-based temperature sensors. *Microelectronic Engineering, 278*, 112015.
88. Soomro, A. M., et al. (2024). Textile-based flexible temperature Sensors for wearable and sports applications. *Physica Status Solidi (A), 221*(2), 2300523.
89. Hao, J.-M., et al. (2024). Positive temperature coefficient material based on silicone rubber/paraffin/graphite/carbon nanotubes for wearable thermal management devices. *Chemical Engineering Journal, 493*, 152427.
90. Zhao, D., et al. (2024). Flexible sensors based on conductive polymer composites. *Sensors, 24*(14), 4664.
91. Zhang, Y., et al. (2024). High-conductivity graphene/carbon black inks via interpenetrating networks for wearable fabric-based heaters and strain sensors. *Journal of Materials Chemistry C, 12*, 10083–10095.
92. Houeix, Y., et al. (2024). Flexible thermistors based on laser-induced graphene from polyetherimide. *Flexible and Printed Electronics, 9*(3), 035010.
93. Gu, J., et al. (2020). Multifunctional poly(vinyl alcohol) nanocomposite Organohydrogel for flexible strain and temperature sensor. *ACS Applied Materials & Interfaces, 12*(36), 40815–40827.
94. Beniwal, A., et al. (2023). Screen-printed graphene-carbon ink based disposable humidity sensor with wireless communication. *Sensors and Actuators B: Chemical, 374*, 132731.
95. Korotcenkov, G., et al. (2023). Paper-based humidity Sensors as promising flexible devices, state of the art, part 2: Humidity-sensor performances. *Nanomaterials, 13*(8), 1381.
96. Jude Iloabuchi, O. (2023). Humidity sensors, major types and applications. In C. M. T. Saeed, A. A. Mohammed, & K. S. Bahadar (Eds.), *Humidity sensors*. IntechOpen.
97. Meng, J., et al. (2023). Porous carbon nanofibres with humidity sensing potential. *Microporous and Mesoporous Materials, 359*, 112663.

98. Yu, X., et al. (2024). Carbon nanotubes for optimizing response stability and speed of acrylate hybrid copolymers humidity sensor. *Polymer Composites, 45*(11), 10516–10526.
99. Singh, S., et al. (2023). Functionalized multiwalled carbon nanotubes for highly stable room temperature and humidity-tolerant triethylamine sensing. *ACS Sensors, 8*(12), 4664–4675.
100. Jin, G., et al. (2023). A CNFs/MWCNTs aerogel film-based humidity sensor with directionally aligned porous structure for substantially enhancing both sensitivity and response speed. *Chemical Engineering Journal, 473*, 145304.
101. Ding, S., et al. (2023). Fast-speed, highly sensitive, flexible humidity Sensors based on a printable composite of carbon nanotubes and hydrophilic polymers. *Langmuir, 39*(4), 1474–1481.
102. Sun, Y., et al. (2023). Hydrophobic multifunctional flexible Sensors with a rapid humidity response for long-term respiratory monitoring. *ACS Sustainable Chemistry & Engineering, 11*(6), 2375–2386.
103. Liang, T., et al. (2023). Wrinkled reduced graphene oxide humidity sensor with fast response/recovery and flexibility for respiratory monitoring. *Sensors and Actuators A: Physical, 350*, 114104.
104. Laera, A. M., et al. (2024). Flexible humidity sensor based on chemically reduced graphene oxide. *Chem, 12*(12), 245.
105. Sharma, A., et al. (2024). *Graphene oxide (GO) and reduced graphene oxide (rGO) based humidity sensors, in nanotechnology: A quick guide to materials and technologies*. Bentham Science Publishers.
106. Noh, W., Go, Y., & An, H. (2023). Reduced graphene oxide/polyelectrolyte multilayers for fast resistive humidity sensing. *Sensors, 23*(4), 1977.
107. Guo, Y., et al. (2023). A self-powered PVA-based flexible humidity sensor with humidity-related voltage output for multifunctional applications. *Colloids and Surfaces A: Physicochemical and Engineering Aspects, 658*, 130700.
108. Tekcin, M., Tuzer Hamzaoglu, D. R., & Kursun, S. (2023). Flexible humidity sensor for smart agricultural applications. *Flexible and Printed Electronics, 8*(3), 035003.
109. Xu, L., et al. (2023). Advances in wearable flexible electrochemical sensors for sweat monitoring: A mini-review. *International Journal of Electrochemical Science, 18*(1), 13–19.
110. Wang, J., et al. (2024). Epidermal wearable optical sensors for sweat monitoring. *Communications Materials, 5*(1), 77.
111. Gao, F., et al. (2023). Wearable and flexible electrochemical sensors for sweat analysis: A review. *Microsystems & Nanoengineering, 9*(1), 1.
112. Tong, X., et al. (2023). Wearable electrochemical Sensors based on nanomaterials for healthcare applications. *Electroanalysis, 35*(2), e202200228.
113. Gibi, C., et al. (2024). Carbon materials for electrochemical sensing application—A mini review. *Journal of the Taiwan Institute of Chemical Engineers, 154*, 105071.
114. Xia, Y., et al. (2023). Wearable electrochemical sensor based on bimetallic MOF coated CNT/PDMS film electrode via a dual-stamping method for real-time sweat glucose analysis. *Analytica Chimica Acta, 1278*, 341754.
115. Gazzato, L., & Frasconi, M. (2025). Carbon nanotubes and their composites for flexible electrochemical biosensors. *Analysis & Sensing, 5*(1), e202400038.
116. Fu, L., et al. (2023). Strategies and applications of graphene and its derivatives-based electrochemical sensors in cancer diagnosis. *Molecules, 28*(18), 6719.
117. Xiang, F., et al. (2023). Recent advances in flexible batteries: From materials to applications. *Nano Research, 16*(4), 4821–4854.
118. Das, S., et al. (2023). Building sodium metal battery with Polyisoprene-based air-stable single-ion gel polymer electrolyte. *ACS Applied Energy Materials, 6*(10), 5113–5121.
119. Zaffar, A., et al. (2024). Carbon derivatives-based silicon as anode materials for lithium-ion batteries. In *Advanced hybrid nanomaterials for energy storage* (pp. 111–128). CRC Press.
120. Liu, S., et al. (2024). Flexible electrodes for aqueous hybrid Supercapacitors: Recent advances and future prospects. *Electrochemical Energy Reviews, 7*(1), 25.
121. Ali, I., et al. (2024). Advances in smart photovoltaic textiles. *ACS Nano, 18*(5), 3871–3915.

122. Sayam, A., et al. (2024). Carbon-based textile-structured triboelectric nanogenerators for smart wearables. *Advanced Energy and Sustainability Research, 5*(11), 2400127.
123. Chen, J., et al. (2024). Flexible BTO piezoelectric generator fabricated by hydrothermal method with the assistance of lignocellulose and carbon nanomaterials. *Chemical Engineering Journal, 497*, 154742.
124. Kothandam, G., et al. (2023). Recent advances in carbon-based electrodes for energy storage and conversion. *Advanced Science, 10*(18), 2301045.
125. Yang, Y., et al. (2023). Construction of three-dimensional carbon materials-based conductive bonding network in flexible supercapacitor electrodes. *Electrochimica Acta, 440*, 141751.
126. Shalini, S., et al. (2024). Progress in flexible supercapacitors for wearable electronics using graphene-based organic frameworks. *Journal of Energy Storage, 86*, 111260.
127. Kim, H., Cheong, J. Y., & Hwang, B. (2023). Mini review of technological trends of flexible Supercapacitors using carbon nanotubes. *Journal of Natural Fibers, 20*(2), 2204455.
128. Wang, C., et al. (2024). A simple route to functionalized porous carbon foams from carbon nanodots for metal-free pseudocapacitors. *Materials Horizons, 11*(3), 688–699.
129. Özçifçi, Z., et al. (2024). Production and characterization of activated carbon foams with various activation agents for electrochemical double layer capacitors (EDLCs) applications. *Colloids and Surfaces A: Physicochemical and Engineering Aspects, 690*, 133851.
130. Han, Y., et al. (2023). Review of flexible Supercapacitors using carbon nanotube-based electrodes. *Applied Sciences, 13*(5), 3290.
131. Xu, T., et al. (2023). Ultralight MXene/carbon nanotube composite aerogel for high-performance flexible supercapacitor. *Advanced Composites and Hybrid Materials, 6*(3), 108.
132. Zhang, Y., et al. (2023). Recent advances and perspectives on graphene-based gels for superior flexible all-solid-state supercapacitors. *Journal of Power Sources, 565*, 232916.
133. Chen, W., et al. (2023). A flexible composite film electrode and supercapacitor based on combined effect between graphene oxide and graphene. *Materials Science and Engineering: B, 297*, 116724.
134. Jiang, L., Luo, X., & Wang, D.-W. (2023). A review on system and materials for aqueous flexible metal–air batteries. *Carbon Energy, 5*(3), e284.
135. Li, H., Chen, J., & Fang, J. (2023). Recent advances in wearable aqueous metal-air batteries: From configuration design to materials fabrication. *Advanced Materials Technologies, 8*(8), 2201762.
136. Zhang, Z., et al. (2023). Addressing transport issues in non-aqueous Li–air batteries to achieving high electrochemical performance. *Electrochemical Energy Reviews, 6*(1), 18.
137. Sayahpour, B., et al. (2024). Quantitative analysis of sodium metal deposition and interphase in Na metal batteries. *Energy & Environmental Science, 17*(3), 1216–1228.
138. Ding, H., et al. (2023). Building electrode skins for ultra-stable potassium metal batteries. *Nature Communications, 14*(1), 2305.
139. Shao, W., et al. (2023). Carbon-based electrodes for advanced zinc-air batteries: Oxygen-catalytic site regulation and nanostructure design. *Electrochemical Energy Reviews, 6*(1), 11.
140. Wang, X., et al. (2023). Toward self-supported Bifunctional air electrodes for flexible solid-state Zn–air batteries. *Small Science, 3*(10), 2300066.
141. Kundu, A., et al. (2023). Metal–organic framework-derived advanced oxygen electrocatalysts as air-cathodes for Zn–air batteries: Recent trends and future perspectives. *Materials Horizons, 10*(3), 745–787.
142. Chen, Z., et al. (2024). Bamboo-modulated helical carbon nanotubes for rechargeable Zn-air battery. *Small, 20*(14), 2307776.
143. Agarwal, A., & Hussain, C. M. (2023). Metal-air batteries for wearable electronics: A case study for modern society. In *Concepts in smart societies* (pp. 298–312). CRC Press.
144. Lee, S.-J., & Kim, C.-L. (2023). Highly flexible, stretchable, durable conductive electrode for human-body-attachable wearable sensor application. *Polymer Testing, 122*, 108018.

145. Ali, M., et al. (2023). A flexible and biodegradable piezoelectric-based wearable sensor for non-invasive monitoring of dynamic human motions and physiological signals. *Advanced Materials Technologies, 8*(15), 2300347.
146. Han, L., et al. (2023). Health monitoring via heart, breath, and Korotkoff sounds by wearable Piezoelectret patches. *Advanced Science, 10*(28), 2301180.
147. Liu, E., et al. (2023). An overview of flexible Sensors: Development, application, and challenges. *Sensors, 23*(2), 817.
148. Zhang, Q., et al. (2023). Three-in-one portable electronic sensory system based on low-impedance laser-induced graphene on-skin electrode Sensors for electrophysiological signal monitoring. *Advanced Materials Interfaces, 10*(3), 2201735.
149. Zhang, T., et al. (2023). Flexible electronics for cardiovascular healthcare monitoring. *Innovation (Camb), 4*(5), 100485.
150. Oh, J., et al. (2024). Flexible dry electrode based on a wrinkled surface that uses carbon nanotube/polymer composites for recording electroencephalograms. *Materials, 17*(3), 668.
151. Zhang, Y., et al. (2024). Recent developments and trends of biosensors based on carbon nanotubes for biomedical diagnosis applications: A review. *Biosensors and Bioelectronics: X, 17*, 100424.
152. Sharma, P., et al. (2024). Multiwall carbon nanotube/polydimethylsiloxane composites-based dry electrodes for bio-signal detection. *IEEE Journal on Flexible Electronics, 3*(3), 108–114.
153. Lin, Z., et al. (2023). Graphene biointerface for cardiac arrhythmia diagnosis and treatment. *Advanced Materials, 35*(22), 2212190.
154. Dey, R., et al. (2023). Graphene-based electrodes for ECG signal monitoring: Fabrication methodologies, challenges and future directions. *Cogent Engineering, 10*(1), 2246750.
155. Baloda, S., et al. (2024). Reduced graphene oxide-polydimethylsiloxane based flexible dry electrodes for electrophysiological signal monitoring. *IEEE Transactions on Nanotechnology, 23*, 644–651.

The Future of Carbon-Based Nanocomposites in Sustainable Applications

Abhishek Attal, Bhuvnesh Kumar, and Mamta Dahiya

1 Introduction

The commercial utility of using carbon-based nanomaterials has been rising in numerous areas, such as modern technology, medicinal science, environmental science, and horticulture. These nanomaterials exhibit distinct characteristics that have caught the attention of researchers and industries that have thorough visualization of innovative ideas and approaches towards large-scale industrial production [1]. Carbon is an amazing element that can create a range of different setups, with different properties. These include the "hard" diamond and "soft" graphite carbon allotropes. In the past few decades, a new class of carbon-based materials, including carbon nanotubes (CNTs), fullerenes, and most recently graphene, have drawn considerable interest from science industries [2]. These "wonder materials" are known for their remarkable properties, rendering them very promising for a variety of applications. Fullerenes are special carbon allotropes, and the carbon structural evolution from graphite to fullerene to diamond shows a transitional phase. Fullerenes composed of sp^2-hybridized carbon atoms, connected by covalent bonds, shows brilliant symmetry. Fullerene molecule (C_{60}) made up of 60 carbon atoms forming a ball-like structure with the carbon atoms located at the vertices of 20 hexagons and 12 pentagons and has complex geometry [3, 4]. CNTs are another significant carbon nanomaterial with extraordinary mechanical properties. Those are rolled-up sheets of graphene and are tubular structures a few nanometres wide. These tubes may

A. Attal · M. Dahiya (✉)
Department of Mechanical Engineering, MAIT, Maharaja Agrasen University, Solan, Himachal Pradesh, India

B. Kumar
Department of Mechanical Engineering, SRM University, Delhi-NCR, Sonipat, Haryana, India

V. Khanna (ed.), *Carbon-Based Nanocomposites for Sustainable Applications, Volume I*, Lecture Notes in Nanoscale Science and Technology 38,
https://doi.org/10.1007/978-3-031-95306-4_12

have a different shape (chirality), sizes (diameter), and can be of different weight therefore can be used for a variety of applications [5]. The next associated form of allotrope is graphene which is a 2D or one single layer of hexagonally arranged carbon atoms. Carbon atoms are interlinked via sp^2 hybridization and have 0.142 nanometres bond lengths between two neighbouring atoms and a flat crystal lattice structure. CNTs and graphene are novel materials and very significant due to their unique structures and outstanding properties for advanced technology applications [6]. Figure 1 illustrates the different types of carbon nanocomposites (NCs). These nanostructures include various low-dimensional carbon allotropes, such as carbon black (CB), carbon fibres, CNTs, fullerenes, and graphene.

This review highlights the transformative role of carbon-based nanomaterials in addressing critical environmental challenges, particularly in wastewater, drinking water, and groundwater treatment, including industrial and pharmaceutical effluents. The exceptional electrical, thermal, and chemical properties of composite materials integrating CNTs and graphene bonds position them as promising solutions for these issues. Additionally, the article delves into recent developments and challenges in the growth of carbon and graphene quantum dot-derived nanomaterials for environmental remediation. It also uncovers the expanding applications of nano-adsorbents across diverse sectors such as aerospace, energy storage, chemical processing, sensor technology, and membrane science. By emphasizing the potential of carbon-based polymer nanocomposites in tackling environmental pollution,

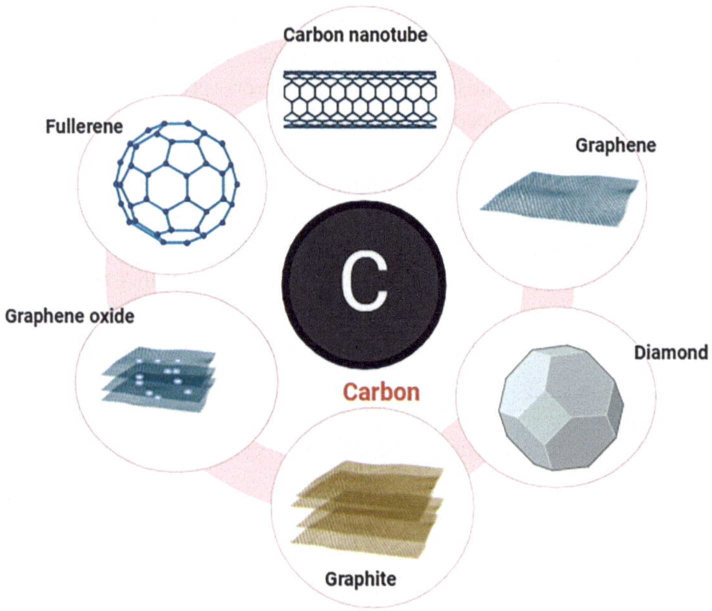

Fig. 1 Types of carbon-based nanocomposites

this article also highlights the role in fostering sustainable and innovative technological solutions for the future.

2 Carbon Nanomaterials for Environmental Applications

Due to their wide usage, carbon materials are effective in adsorbing modern contaminants from water. The most useful materials are nanometer size, with a large specific surface area, high-temperature resistance, pore size distribution, and various functional groups. They also facilitate the movement of pollutants through them, which decreases resistance. These properties make nanomaterials efficient in adsorbing gas pollutants. The number of papers published on carbon-based materials like graphene, GO, CNT, and carbon nanofillers (CNFs), that have been utilized in membrane separation to separate oil-water from 2005 to 2024 are shown in Fig. 2. Over the last 10 years, attention has steadily grown towards using graphene and CNF. It is observed from the literature review that CNT membrane technology has been used for a much longer duration.

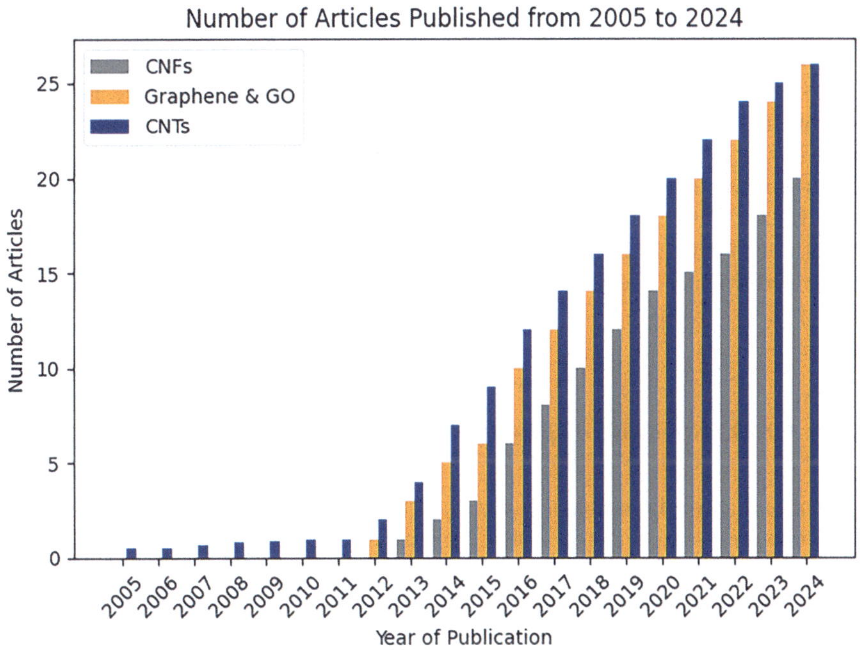

Fig. 2 Number of publications in indexed journals on carbon-based NCs for oily wastewater treatment from 2005 to 2024

2.1 Black Carbon

Black carbon particles are microscopic dark-coloured particulate matter consisting of pure carbon. It results from the partial burning of fossil fuels, biomass, and other organic matter. The physical and chemical properties of CB and soot are different. Carbon black is a nearly pure form of elemental carbon which has a structure similar to graphite. It contains very little organic compounds and inorganic materials (less than 1% for each). This makes it a lot cleaner and more refined than soot [7]. Polyaniline (PANI) is a conductive polymer which attracts attention due to its uniqueness in electrical behaviour, stability, and versatility. The structures of aniline and CB have been modified to arrive at this discovery. With these modifications, the resulting materials transitioned from basic spherical shapes to a combination of spherical and nanofiber-like elements, showcasing how changes in structure can affect the form of the end material [8]. PANI/CB NCs also have a higher electrical conductivity and higher stability to radiation in comparison to graphene-based potential nanotubes or NCs. Various reinforcing fillers like mica, carbon fibre, CB, melamine fibre, etc. have been used to improve the mechanical properties of these materials. From these, globally, CB is the most prominent [9, 10]. Compared to carbon nanotubes or any similar materials, the properties of PANI/CB nanocomposites are greatest. They are particularly more resistant to radiation damage and have greater electrical conductivity. Due to the lightweight, important numbers of PANI nanostructures also provide tiny, crystalline sites that help improve strength to prevent particle agglomeration and the effective distribution of nanoparticles [11]. Besides CB, different kinds of materials like carbon fibres, melamine fibres, and mica have also been used as reinforcement fillers to improve the mechanical properties of final products. Nevertheless, carbon black remains a vital reinforcement material worldwide [12]. Traditionally, micro fillers have been replaced with nanofillers such as nano clay in more recent years. These nanofillers are provided with a significantly higher surface area resulting in improved interaction and overall bonding at a molecular level for the surrounding material. This interaction allows for better performance and longevity of the composites than legacy filler technologies [13].

2.2 Graphene and Graphene Oxide

Graphene is a two-dimensional planar material with a hydrophobic nature, adding it to membranes is one way to enhance the mechanical, thermal, electrical, and functional properties of membranes. Graphene in particular improves mechanical strength (tensile strength approximately 130 GPa), thermal conductivity (up to 5000 W/mK), electrical conductivity (10^6 S/m), and improves membrane performance [14]. The development of carbon-based nanomaterials has progressed significantly, focusing on their superior characteristics, synthesis methods, and the role of carbon vapour in the production of these materials. The remarkable physical and chemical properties of carbon-based nanomaterials make them suitable for a wide range of applications, driving efforts to enhance their production.

Despite this, the hydrophobic nature of both graphene and graphite limits their use in developing hydrophilic membranes for oily wastewater treatment. On the other hand, graphene oxide (GO) exhibits aquaphilic properties due to its strong polar functional groups. The presence of epoxide, carboxylic, and hydroxyl groups on GO's surface significantly improves its compatibility with various polymers when in solution. As a result, GO has become a popular choice for designing hydrophilic membranes [15, 16]. Huang et al. developed ultrafiltration (UF) membranes with antifouling characteristics and enhanced stability by coating GO nanosheets onto a rough, porous substrate. They utilized a thin layer of nanoporous polyamide (200 nm) as a substrate for the GO coating via vacuum filtration, aiming to create micro/nano hierarchical roughness on the membrane surface. By optimizing surface roughness through cyclic oil-water separation experiments and contact angle measurements, they found that increasing the GO coating thickness reduces surface roughness. Membranes with polyamide support and a 5 nm GO coating have shown the strongest resistance to oil adhesion. This study concluded that a corrugated GO layer with nanoscale roughness on various commercially available porous supports resulted in membranes exhibiting low oil adhesion [17]. Chang et al. [18] demonstrated that carboxylation could enhance the hydrophilicity of graphene oxide membranes, leading to improved dye removal efficiency, which can be attributed to the higher surface charge density. Recently, GO membranes have also been useful for oil elimination in wastewater treatment [18]. Zhao et al. [19] incorporated Palygorskite nanorods between GO nanosheets to create laminated structures. These graphene oxide membranes have shown to be effective in separating water-containing oil emulsions, demonstrating their anti-oil properties [19].

Because graphene nanoparticles have significant, pertinent physicochemical properties so they have been utilized to remediate wastewater. Graphene NCs serve as nano-adsorbents in catalytic, photocatalytic, electrocatalytic and photoelectrocatalytic oxidation and reduction processes to decompose hazardous mixtures of toxicants and aid their catalytic degradation. The adsorption capacity is influenced by many key factors including the type and characteristics of the graphene material and metal ions, pH, temperature, contact time, acidic organic ligands, and co-ions [20]. The importance of O_2-containing functional groups is shown in GNS-500 and GNS-700 exhibited higher adsorption capacity for Pb_2þ than pristine graphene. As absorbents for a variety of materials, super-hydrophobic and superoleophobic porous graphene has been created [21]. It has a notable absorption capacity of over 90%, excellent selectivity, and appropriate recyclability. Furthermore, graphene-based nano-adsorbents have several advantages over CNT-based nanomaterials for the removal of toxic dyes from water. For example, single-layered graphene or GO nanosheets have two basal planes that are available for the adsorption of contaminants [22]. The established general mechanisms of adsorption and the benefits and drawbacks of employing graphene-based nanosorbents to remove metal pollutants are depicted in Fig. 3. Interestingly, there are no catalyst residues

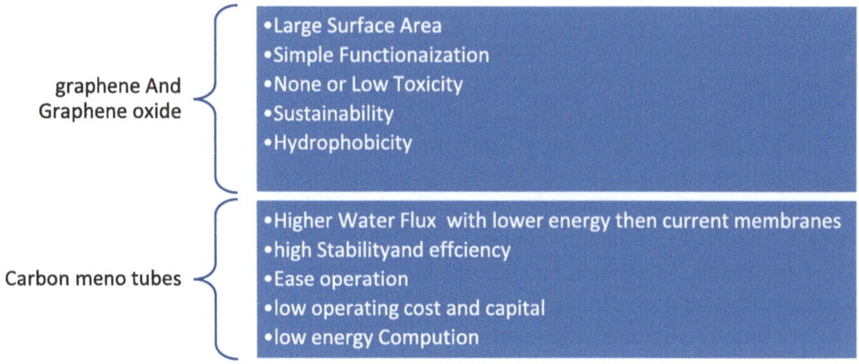

Fig. 3 Particular benefits of various carbon-based nanomaterials

at all in the final graphene-based nanomaterials. Commonly referred to as hydrophobic and electrostatic p–p interactions, it has positive interactions in ionic and/or molecular forms with a range of organic and inorganic pollutants [23].

One of the most efficient top-down methods to obtain graphene nanomaterials is electrochemical exfoliation [24]. On the other hand, advancements in synthesizing graphene-based nanomaterials from biomass offered a promising avenue for utilizing natural sources to convert bio-waste into graphene in an environmentally friendly manner. This method prioritizes the reuse of organic materials in nature (long-term sustainability). The benefits of biomass pyrolysis in graphene production are sustainable organic matter technology, environmentally friendly processing of bio-waste, and materials with high specific capacitance as well as high surface area. Nonetheless, drawbacks in this approach include limited scalability, laborious raw material preparation, and the highly disordered nature of produced graphene.

2.2.1 Carbon Nanotubes (CNTs)

CNTs are tubular nanostructures made of carbon with exceptional thermal, tensile, and electrical properties [25]. Their high specific surface area and oleophilic characteristics make them highly promising for applications such as high-permeation flux membranes for oil removal. Numerous studies have explored their potential in fabricating nanocomposite materials with advanced water-removal capabilities [26]. Gao et al. [27] demonstrated the development of ultrathin, super-hydrophilic membranes utilizing single-walled CNTs (SWCNTs) combined with titania (TiO_2) nanoparticles. Using the sol-gel method followed by UV light treatment, they successfully applied TiO_2 to SWCNT films, producing membranes with nanoscale pore sizes (20–60 nm) and ultrathin SWCNT layers (60 nm) that outperformed traditional mesh-based membranes [28, 29]. TiO_2 nanoparticles are activated by UV light to become photocatalytic and reactivate the super-hydrophilicity in oil-contaminated membranes. Accordingly, during several filtration experiments, the

NCs membranes exhibited self-cleaning and antifouling activities. Water permeation fluxes as high as 30,000 L m^{-2} h^{-1} bar^{-1} have been achieved. The membranes not only removed all the oil droplets with sizes below 3 μm but also achieved complete removal of greater than 100 nm oil droplets in water emulsions, demonstrating no droplet greater than 100 nm in the filtrate.

The significant aqueous solubility of drugs often leads to their accumulation in target tissues or cells in humans, animals, and plants. Annually, thousands of tons of active drugs and their components enter the environment, much of which comes from unused medications [31]. These substances frequently revert to active forms during metabolic processes, with urban sewage becoming heavily contaminated by diverse drug residues [32]. CNTs, owing to their cylindrical hollow structures and robust physicochemical interactions, have proven to be excellent adsorbents for such contaminants [30].

CNTs and CNT-based composites demonstrate exceptional dye absorption capabilities in aqueous systems. Both multi-walled nanotubes (MWCNTs) and SWCNTs are particularly effective due to their high surface area and the existence of versatile absorption sites on their inner and outer layers. These characteristics make CNTs valuable scaffolds for water purification applications, as illustrated in Fig. 4. Despite their advantages, the hydrophobic nature of CNT surfaces necessitates stabilization in aqueous suspensions to prevent aggregation, which could otherwise diminish their active surface area and overall effectiveness.

The irreversibility of raw CNTs has been in limited use due to their adsorption capacity being less than 1% for inorganic and organic compound parties due to low dispersion in the solution. These limitations have been mainly attributed to the aggregation of CNTs, which decreases the efficiency of contaminant adsorption [32]. Oxidation treatments, among others, are common functionalization techniques used to overcome this problem. Oxidation increased the dispersion of CNTs in media and their adsorption capacity to remove aquatic pollutants improves significantly.

On the other hand, oxidation is usually realized under extreme chemical conditions, which can destroy the structure of CNTs and result in secondary chemical waste that is hazardous to the environment. Therefore, new functionalization techniques that preserve structural integrity and minimize environmental impact are increasingly needed. The other restraint associated with CNTs is their high cost when compared to CB, carbon fibres, clay, and exfoliated graphite. This cost factor also makes it infeasible to implement CNTs as adsorbents to be used in large-scale water treatment applications [33]. Overcoming these hurdles demands inroads in low-cost synthesis and environmentally benign functionalization methods in order for CNTs to be scalable to industry.

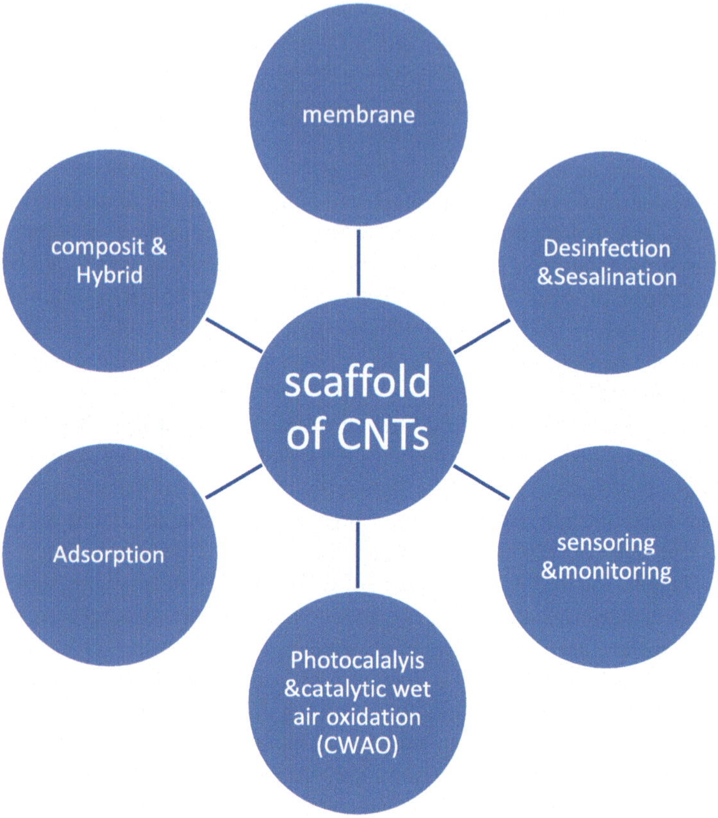

Fig. 4 CNTs are used as scaffolds in technologies for wastewater remediation [31]

2.2.2 Carbon Nanofibers (CNFs)

Along with CNTs, are fibrous cylindrical nanostructures known for their high oil adsorption capacity, particularly useful in oil-water parting processes. While CNFs and CNTs share comparable mechanical strength and electrical properties, CNFs possess a significantly higher functionalized surface area. The electrospinning technique is commonly employed for CNF production, resulting in chemically inert, highly porous, uniformly sized, and interconnected nanofibrous films [34].

Tai et al. explored methods to address the inherent low rigidity of CNF macrostructures in oil-water separation by incorporating silica (SiO_2) into the membrane design. By limiting the SiO_2 content to below 2.7 wt.%, they preserved the membranes' flexibility and toughness. The carbon-based electrospun nanofibrous membranes were further enhanced through crosslinking, improving their mechanical stiffness and enabling effective performance across diverse temperatures and pH levels [35].

CNFs exhibit intrinsic properties such as hydrophobicity, thermal stability, and chemical inertness, making them highly suitable for energy-efficient oil-water separation techniques like gravity-driven processes. Additionally, CNFs demonstrate excellent reinforcing capabilities, significantly enhancing the mechanical properties of NCs membranes.

The mechanical performance of CNF composites relies on factors such as diameter, aspect ratio, dispersion quality, polymer matrix adhesion, and the orientation of CNF fillers. Tensile strength improves as CNF diameters decrease, and proper dispersion further enhances composite properties. Chemical functionalization plays a crucial role in ensuring effective dispersion and adhesion of CNF fillers within the polymer matrix. Without proper compatibility, stress concentrations can form at the polymer/CNF interface, potentially lead to composite failure. Although CNFs' low rigidity and flexibility present challenges in membrane fabrication, studies suggest that blending CNFs with other materials can overcome these limitations [36].

Although CNT exhibits a high physical aspect ratio, CNF has shown a much higher electron transport across its sidewall than CNT so it shows higher chemical reactivity. This characteristic renders CNFs highly favourable for their functionalization and electrochemical utilization. Moreover, as discrete, free-standing forms, CNFs are extraordinarily durable. CNTs exhibit better mechanical and permeation properties; however, CNFs are more affordable this makes them much appealing for the commercial production of NCs membranes. CNTs, graphene, and GO are vital in addressing environmental challenges as these materials effectively adsorb heavy metals, dyes, and organic pollutants due to their large surface area and active sites. For instance, graphene-based materials have demonstrated high efficiency in removing arsenic and mercury from water, while CNTs can filter out organic dyes and hydrocarbons. These nanomaterials are also integrated into advanced oxidation processes, enhancing pollutant degradation through photocatalytic or Fenton reactions [37, 38]. Moreover, in air purification, CNTs and graphene are used to trap and neutralize toxic gases such as carbon monoxide and nitrogen oxides. They also support catalytic converters for vehicular emissions. Beyond pollutant removal, these nanomaterials play a role in environmental sensing by detecting low concentrations of contaminants, enabling early intervention [37]. While highly beneficial, their environmental and ecological impacts require careful assessment, as some studies indicate potential toxicity to aquatic organisms and ecosystems [38]. Ongoing research focuses on balancing applications with sustainability and safety.

3 Advancements in Carbon Nanocomposites for Remediation

Materials based on graphene as effective adsorbents for pharmaceuticals-contaminated wastewater treatment. Alteration of the experimental conditions, for example, enables the adjustment of their adsorption capacity [39]. Graphene materials are highly effective in addressing radionuclide and heavy metal contamination, with their adsorption efficiency impacted by various factors, including initial dosage and conditions such as pH, temperature, contact time, and the presence of organic ligands or other ions [40]. Studies have consistently demonstrated their success in eliminating toxic dyes and heavy metals from water [41, 42].

Gong et al. [43] analysed magnetic MWCNT nanocomposites as viable adsorbents for cationic dyes in aqueous environments. Composed of functionalized MWCNTs and iron oxide nanoparticles (IONPs), these nanocomposites effectively removed dyes like methylene blue (MB), neutral red, and brilliant cresyl blue. By studying adsorption kinetics, capacity, dosage, and pH, their results highlighted MMWCNT nanocomposites as a robust solution for dye adsorption [43]. Inyang et al. further investigated the use of biochar-CNT composites, evaluating the adsorption of methylene blue through batch experiments. Combining untreated hickory biochar (HC) and bagasse biochar (BC) with CNTs improved their physicochemical characteristics, such as thermal stability, surface area, and pore volume. Electrostatic attraction emerged as the primary mechanism for dye adsorption, confirming these composites as cost-effective solutions for removing dyes and organic pollutants [44]. Carbon nanocomposites (CNCs), comprising heterophasic materials with CNTs as structural elements, offer diverse applications. For instance, a hybrid material synthesized from poly(1,8-diaminonaphthalene) and carboxylated multi-wall carbon nanotubes (MWCNTs-COOH) demonstrated remarkable efficiency in extracting trace amounts of Cd^{2+} and Pb^{2+} ions from solutions [45].

Muneeb et al. [46] prepared a unique NC based on biomass that has used carbon for heavy metals (As, Cr, Cu, Pd, Zn) removal from wastewater. The maximum binding efficiency of the nanocomposite for these metals decreased at higher pH [46]. Tian et al. [47] developed a novel green and synergistic nanocomposite for an antibacterial application, where the surface of IONPs and AgNPs (IONP and AgNP) was attached to the surface of GO, and the whole nanocomposite was defined as GO-IONP-Ag nanocomposite. GO-IONP-Ag nanocomposite exhibited comparable better bactericidal activity against both Gram-negative and Gram-positive bacteria than pure AgNPs. Surprisingly, the composite showed strong bactericidal activity against Gram-positive bacteria at low concentrations [47]. In addition, the GO-IONP-Ag nanocomposite has the following advantages, such as easy recovery by magnetic separation, low cost, and environmentally friendly. These unique properties render it a viable multifunctional disinfectant in multiple domains [48]. CBN-based catalysts are very successful to degrade persistent environmental contaminants, including industrial chemicals and pesticides, 80% more effective than conventional chemical methods. Not only are these regenerative catalysts cost-effective but their use provides an environmentally friendly alternative to traditional pollution control methods.

The blend of carbon science and nanotechnology is reflected in the class of sophisticated materials known as CNCs, which are made of carbon-based structures with nanoscale dimensions [49]. The synthesis processes of chemical vapour deposition, sol-gel synthesis, and self-assembly are closely related to the definition and characteristics of CNCs, highlighting the need for precise control over composition, structure, and morphology for customized applications [50]. Excellent mechanical strength, strong electrical and thermal conductivity, a wide surface area, special optical properties, chemical stability, and biocompatibility are some of the primary characteristics of CNCs. As a result, these composites show promise for revolutionary applications and emphasize their importance as innovative materials with special properties that promote advancement in a range of scientific and technological fields [51].

CNCs are important and have great potential due to their special properties that make them materials that can be used in a variety of scientific and technological fields [52]. CNCs have remarkable mechanical strength, excellent electrical and thermal conductivity, a wide surface area and special optical properties [53]. By improving the charge storage capacity and cycling stability of lithium-ion batteries and supercapacitors, CNCs have shown great significance in the field of energy storage. This detailed study attempts to address issues in the field and outline future prospects for CNC research by incorporating insights from current developments and research investigations. We hope to highlight their revolutionary impact on five important areas—energy storage, sensing technology, catalysis, nano-medicine, and environmental remediation—by thoroughly examining the numerous uses of CNCs [54]. Nanocomposites are created through the process of sol-gel synthesis, which turns colloidal fluid (sol) into gel. Due to the uniform coating and the broadening of the absorption wavelength of TiO_2 nanoparticles, MWCNT-based MWCNT/TiO_2 NCs have been prepared in a typical CNC synthesis using the sol-gel method [55].

CNT composites in material innovation have demonstrated significant performance improvements, including a 60% increase in tensile strength and an order of magnitude increase in electrical conductivity. Yet issues such as ensuring homogeneous dispersion of CNTs in the polymer matrix and retention of nanoscale properties in macroscopic applications persist. By the same token, biodegradable CNT composites are a promising solution to support environmental sustainability; however, they also have drawbacks, including weak adhesive strength and complex fabrication methods, making them more complex to implement in practice. Future research perspectives involve investigating CBN applications across aerospace materials, next-generation energy storage systems, and advanced environmental sensors. CNTs and graphene: Nanostructured carbons have been integrated with existing nanobars to create a composite nanostructure which are characterized by significantly higher energy storage density and longevity when put to use in energy storage devices and also in terms of better sensitivity and reliability as environmental sensors. Building upon these advancements, the potential for CBNs to revolutionize sustainable applications is immense. As research continues to overcome challenges like the homogeneous dispersion of nanomaterials and scalable production techniques, the integration of these materials across diverse fields is expected to expand. Innovations in synthesis methods, such as 3D printing and green chemistry approaches, are likely to enhance the cost-effectiveness and environmental compatibility of CBN fabrication, making them more accessible for industrial-scale applications. Development of multifunctional nanocomposites that combine properties like high conductivity, mechanical strength, and catalytic activity will further broaden their usability in areas like renewable energy systems and environmental monitoring. For instance, hybrid materials integrating graphene and CNTs with other functional nanomaterials are emerging as highly efficient platforms for solar energy conversion, CO_2 capture, and wastewater treatment, addressing critical global challenges. The adaptability of CBNs to customized applications in aerospace, wearable electronics, and smart materials showcases their ability to meet the demands of rapidly advancing industries. Coupled with the growing emphasis on circular economy practices, the recyclability and biodegradability of certain CBN-based materials offer a path towards minimizing waste and resource depletion. As we move forward, interdisciplinary collaboration will be essential to unlock the full potential of carbon-based nanocomposites. By combining expertise in materials science, engineering, and environmental sciences, the future promises transformative applications that align innovation with sustainability, driving progress towards a greener and more efficient future.

4 Carbon-Based Polymer NCs

For the extensive development in materials science, polymer-based nanocomposites use nanotechnology to improve the characteristics and functionality of polymers. Compared to traditional polymer composites, these nanocomposites have better mechanical, thermal, electrical, and barrier properties because they are made of a polymer matrix (such as polyethylene, polypropylene, polystyrene, epoxy, polyurethane, and others) coated with nanoscale fillers or reinforcements (nanofillers can be inorganic nanoparticles, such as clay, silica, carbon nanotubes, graphene, or organic nanoparticles, such as dendrimers, CNTs, or hybrid nanoparticles) [56] The following are some benefits of polymer-based nanocomposites: Certain applications of CNTs in nanocomposite membranes make use of the material's physical characteristics to improve the membrane's mechanical stability or to disrupt the active layer's polymer packing in traditional reverse osmosis membranes [57].

Wang et al. have reported a hydrophilic NC surface coating on filtering membranes to enable high water flux. Likewise, with higher concentrations of oxidized MWNTs, the water permeability has been increased as well. This suggests that the introduction of MWNTs can effectively mitigate the restriction of water flux in the coating layer, which can be attributed to MWNTs that assist in improving permeability by disrupting the packing of polymer chains and forming void spaces [58]. Most recent composite membrane designs are based on coupling high-flow rates with the ability of carbon-based nanoparticles to adsorb, their corresponding antibacterial properties, and thermal stability. This could also be a safe and effective approach to benefit from the unique properties that nanoparticles possess for water and wastewater treatment if the stability of the incorporated nanomaterial over the whole membrane lifetime can be demonstrated [49]. Green NCs based on polymers and nanocarbons have been created for a variety of uses in fuel cells, solar cells, nanodevices, chemical sensors, and biosensors.

4.1 Cost-Efficiency and Scalability

One major obstacle still facing the manufacture of carbon-based nanocomposites is achieving scalability. When generated at scale, methods such as arc discharge for CNTs, and chemical vapour deposition (CVD) for graphene confront difficulties in preserving quality, consistency, and cost-effectiveness. Creating scalable procedures is essential to fulfilling the growing need for carbon-based nanocomposites in energy storage and electronics applications [59]. The fact that some high-quality graphene production techniques need resource-intensive techniques, which reduces their cost-effectiveness on a larger scale, highlights the scalability issue. To overcome scalability issues and make these nanocomposites economically feasible, new synthesis procedures, alternate precursor discovery, and production method innovations are continuously being developed. One important element affecting the broad acceptance of carbon-based energy is the cost of production [60].

Advancements in CNCs have revolutionized the field of environmental remediation, offering highly efficient solutions for removing pollutants from water, air, and soil. These nanocomposites combine carbon-based materials, such as CNTs and graphene, with other functional materials, enhancing their properties for targeted applications. In water treatment, graphene-metal oxide composites, such as graphene oxide-TiO_2, have demonstrated high efficiency in photocatalytic degradation of organic pollutants under UV light. These composites leverage the high surface area and electrical conductivity of graphene, coupled with the photocatalytic activity of metal oxides, to remove dyes, pharmaceuticals, and heavy metals [61]. Similarly, CNT-based composites have proven effective for adsorption and filtration processes. Multi-walled CNTs functionalized with iron oxides excel in removing arsenic and fluoride from water, making them viable for treating contaminated groundwater. In soil remediation, biochar-CNT hybrids have gained attention for adsorbing pesticides and hydrocarbons, enhancing soil quality while preventing pollutant leaching. These advancements underscore the potential of carbon nanocomposites in sustainable and efficient environmental clean-up while necessitating careful risk assessments for large-scale applications [38].

5 Sustainability in Industrial Applications

Carbon-based nanomaterials have gained significant attention to propel various sectors towards sustainable development. With high surface area, tunable morphology, electrical conductivity, and adsorption capabilities, they are considered excellent candidates for a prosperous field of industrial and environmental applications [1]. This section discusses the role of different carbon nanomaterials like activated carbon, CNTs, Gr, and GO towards sustainable development goals.

5.1 *Supercapacitors*

Supercapacitors are efficient electricity storage devices used in many consumer products and alternative energy systems. Their high energy density, speedy charge, and discharge rates, low heat generation, resilience, dependability, and the absence of removable components make them very favourable [62]. CNTs are valuable and desirable materials for high-performance supercapacitor electrodes. Because of unique properties such as high conductivity, high surface area, rapid charge storage, high mesoporosity, and good compatibility with electrolytes, they are suitable to be used for this application [63].

CNTs have been a great choice for storing energy in small or remote devices, where regular capacitors and electrodes need to be much larger because of their less efficient power-to-energy ratio [64]. These supercapacitors typically use materials such as carbon aerogel, black carbon, or carbon cloth as the active material [65]. A CNT supercapacitor has two principal components—the two electrodes (energy reserve component) and the electrolyte (the mobility carrier). The energy storage capacity of conventional carbon electrodes fades with a rise in the discharge current. However, if CNTs are used with other materials such as polypyrrole (PPy), the supercapacitor performs a lot better as it stores a lot more energy in a much more efficient manner, even at higher currents [66].

5.2 *Field Emission*

CNTs are by far the most familiar class of materials in this domain, and they are notable for their remarkable electrical properties and extremely sharp tips. Reducing the curvature radius of the tip leads to a more concentrated electric field and thus, higher field emission. This has an analogous effect to sharp lightning rods which enhance the emission because of their pointed geometry [67]. Adding CNTs may increase the electrical conductivity of the electrodes in a DSSC (Dye-Sensitized Solar Cell). What they were banking on is that with the CNTs the electrodes would conduct electricity much more effectively, allowing the solar cell to convert light into energy far better. The excessive aggregation of CNTs reduces the efficiency of dye-sensitized solar cells (DSSCs) by the dispersion of charge. The highly efficient energy conversion achieved at low energy and manufacturing costs is achieved by dye molecules, nano-crystalline metal oxides, and natural liquid electrolytes constituting DSSCs [68–70].

5.3 Water Treatment

Water contaminated by industrial effluents, agricultural runoff, and heavy metal contamination can significantly affect both human health and the environment. Carbonaceous nanomaterials, in particular CNTs and graphene-based materials, have shown excellent potential for removing pollutants and heavy metals in water, making them valuable for sustainable water treatment applications. Of these, activated carbon, a common adsorbent with high adsorption capacity, has found widespread use in filtration systems to eliminate organic pollutants from both drinking water and wastewater [71]. CNTs and graphene oxides have similar excellent adsorption performance, enabling them to remove heavy metals, organic dyes, and pharmaceuticals in wastewater [72]. With the plentiful active sites, which is caused by the rich oxygen functional groups on GO, GO have strong interactions with pollutants, therefore improving removal efficiency [73]. Because of their high surface-to-volume ratio, which provides a larger interface for interaction with contaminants, and the ease of functionalization of CBNs, which can enhance both selectivity and adsorption capability, they have been adopted for environmental applications [74].

5.4 Energy Storage

Beyond water purification, carbon nanomaterials are also pivotal in progressing towards energy sustainability, specifically in the areas of energy storage and energy conversion. Supercapacitors and batteries are commonly built on a polymer backbone incorporating graphene and CNTs in their structures due to their high electrical conductivity, mechanical strength, and high surface area. Example can be a supercapacitor, which when made with carbon nanotube shows better energy density and faster charge/discharge cycles compared with conventional capacitor materials, leading to a more effective energy storage alternative [75]. Due to the unique electronic properties of graphene, the potential of graphene-based materials has also been studied in the context of fuel cells, where high conversion efficiency can be achieved by boosting the electrochemical performance [76]. CBNs have been integrated into these systems to minimize reliance on non-renewable energy sources, leading to the creation of a sustainable energy system.

5.5 Pollution Control

Carbon-based nanomaterials also play an important role in green chemistry, which significantly contributes to sustainable development in another field of pollution control. CBNs have shown great success at targeting and neutralizing air pollutants such as particulate matter and nitrogen oxides (NO_2). Activated carbon is commonly employed in air filter systems because of its remarkable adsorption capacity, while CNTs and graphene oxide can be integrated into cutting-edge air purification devices, facilitating removal of a wider range of contaminants [77]. Carbon nanomaterial-based catalysts have also been researched for degradation of persistent contaminants in the environment, such as industrial chemicals and pesticides [78]. Such catalysts are efficient, cost-effective, and regenerative, offering a sustainable alternative to traditional pollution control techniques.

5.6 Biomedical Sensing and Diagnostic Devices

Because of their superior surface qualities and electrical conductivity, CNCs are utilized in the creation of biomedical sensors and diagnostic equipment [79]. Biosensors: The primary components for biomolecule detection in biosensors are functionalized carbon nanotubes and graphene-based materials. By detecting biomolecules, pathogens, and disease markers, these nanocomposites can be included into a variety of sensing devices, enabling early diagnosis and real-time monitoring [80]. Point-of-care devices: Quick and affordable diagnostic solutions are made possible by the integration of CNCs into point-of-care devices [81].

5.7 Applications of Wound Healing and Antimicrobials

CNCs' antimicrobial qualities and capacity to promote tissue regeneration make them promising for use in wound healing applications [82]. Potential options for coatings on medical equipment and wound dressings include materials like graphene oxide, which have demonstrated antibacterial action against a wide range of bacteria. By combining CNCs with graphene oxide, wound care products can hasten the healing process and help avoid infection [83].

5.8 Electronics and Smart Materials

Innovation in the creation of smart electronics and materials is being propelled by CNCs. The performance and usefulness of electronic devices could be completely transformed by incorporating these nanocomposites into common materials [49]. CNCs are unquestionably versatile materials with exceptional qualities that have a big impact on biomedical applications. Prolonged investigation and advancement in this field have the potential to convert these breakthroughs into workable treatments, enhancing patient outcomes and changing the healthcare industry [84]. The multifaceted properties of carbon-based nanocomposites, such as exceptional conductivity, mechanical strength, and chemical stability, have positioned them as pivotal materials for promoting sustainability across industrial sectors. These nanocomposites, enriched with carbon nanostructures like graphene and carbon nanotubes, are driving innovative solutions in energy, healthcare, environmental management, and advanced materials. In energy systems, their high surface area and electrical properties facilitate the design of efficient supercapacitors and energy storage devices. These technologies not only offer enhanced energy density and performance but also reduce dependency on non-renewable resources. Meanwhile, their use in field emission technologies advances energy-efficient displays and electronics, paving the way for eco-friendly innovations. Water treatment technologies leverage the catalytic and adsorptive capabilities of carbon nanocomposites to tackle contaminants, aligning with sustainable practices for clean water access. These materials are proving indispensable for addressing global challenges like water pollution and scarcity. Healthcare has seen transformative applications of these materials, particularly in biosensors and diagnostic devices, enabling rapid, precise disease detection. Their antimicrobial and regenerative properties also support wound healing solutions, fostering sustainable medical practices with minimal resource waste. In electronics, carbon-based nanocomposites are integrated into smart materials and lightweight, durable components, advancing sustainability in manufacturing and product use. Smart materials equipped with these nanostructures exhibit adaptive properties, enhancing efficiency and reducing resource consumption. Through these applications, carbon-based nanocomposites bridge technological advancement and environmental responsibility. Their versatile use across industries signifies a shift towards resource efficiency and greener practices. As research continues to refine their scalability and functionality, these materials are set to play an increasingly vital role in achieving long-term sustainability objectives across industrial domains.

6 Future Prospects for Carbon-Based Nanocomposites

CNT polymer composites have demonstrated dramatic enhancements in properties at the nanoscale; the transferability of these exceptional attributes to the macroscopic regime poses an important challenge. CNT composites have extraordinary properties, which is why ongoing research aims to address the current limitations in the upscaling of these materials without losing their excellent properties. A significant challenge is to achieve uniform dispersion of CNTs within the polymer matrix and to ensure the homogeneous performance of device over macroscopic structures. Despite a lot of motivation, little work has been done on the development of naturally reinforced biodegradable polymer composites with CNTs. Even though such "green composites" are environmentally friendly they exhibit a lot of issues including but not limited to poor adhesive strength between the short fibres and the polymer matrix, difficulty in maintaining the desired fibre orientation and nanoscale fabrication complexity. These problems limit the widespread use of biodegradable CNT composites and must be overcome through innovation.

Carbon nanomaterials have great potential in emerging industries (aerospace, battery technology, chemical industry, sensors, etc.). Future work is required to investigate new applications of CNT and graphene-based materials in these areas, especially in the specifications of next-generation energy storage systems and composite materials in light of aerospace needs, and sensors for environmental monitoring. If these challenges are overcome and the future directions pursued, carbon-based nanomaterials have the potential to advance sustainable development in diverse fields, minimize environmental pollution, maximize energy efficiency, and promote a circular economy. If the challenges of scale, dispersion, and fabrication can be met, CBNs could usher significant improvements in pollution control, energy efficiency, and sustainable industrial processes. Carbon-based nanocomposites have emerged as transformative materials with immense potential across diverse fields, including electronics, energy, healthcare, and environmental remediation. These materials, which incorporate carbon nanostructures such as graphene, carbon nanotubes (CNTs), and fullerenes, exhibit exceptional mechanical, thermal, and electrical properties. Future prospects for carbon-based nanocomposites lie in their capacity to address global challenges, such as sustainable energy production, environmental sustainability, and advanced manufacturing. In energy applications, carbon nanocomposites are critical in developing high-capacity energy storage systems like supercapacitors and lithium-ion batteries. Their superior conductivity and high surface area facilitate improved energy density and faster charge/discharge cycles, paving the way for more efficient renewable energy storage systems [85]. In healthcare, these nanocomposites are being explored for applications like drug delivery, biosensing, and tissue engineering due to their biocompatibility and functionability [86]. In environmental applications, carbon-based nanocomposites are promising for water purification, pollutant degradation, and CO_2 capture. Their catalytic and adsorptive properties enable efficient removal of contaminants, contributing to a cleaner environment[49]. Advances in scalable synthesis techniques and enhanced structural control will further drive the commercialization of these materials, making them integral to next-generation technologies.

7 Conclusion

Carbon-based nanomaterials represent an innovative approach towards sustainable development, proving to be effective agents in pollution mitigation and material innovation. They boast impressive abilities in seeking out and eliminating air pollutants like VOCs, particulate matter, and nitrogen oxides. The most ubiquitous adsorbent, activated carbon, has long been the most prominent of air filtration systems, due to its vast adsorption surface area, whereas novel adsorbent materials such as carbon CNTs and graphene oxide facilitate next-generation air purification devices. Furthermore, CBN-based catalysts demonstrated excellent performance in effectively degrading persistent environment pollutants, providing a cost-effective and reversible solution for industrial applications.

References

1. Julkapli, N. M., & Bagheri, S. (2015). Graphene supported heterogeneous catalysts: An overview. *International Journal of Hydrogen Energy, 40*(2), 948–979.
2. Zaytseva, O., & Neumann, G. (2016). Carbon nanomaterials: Production, impact on plant development, agricultural and environmental applications. *Chemical and Biological Technologies in Agriculture, 3*, 1–26.
3. Yadav, B. C., & Kumar, R. (2008). Structure, properties and applications of fullerenes. *International Journal of Nanotechnology and Applications, 2*(1), 15–24.
4. Taylor, R., & Walton, D. R. M. (1993). The chemistry of fullerenes. *Nature, 363*(6431), 685–693.
5. Zhou, H. W., Mishnaevsky, L., Yi, H. Y., et al. (2016). Carbon fiber/carbon nanotube reinforced hierarchical composites: Effect of CNT distribution on shearing strength. *Composites. Part B, Engineering, 88*, 201–211. https://doi.org/10.1016/j.compositesb.2015.10.035
6. Spanos, K. N., Georgantzinos, S. K., & Anifantis, N. K. (2015). Mechanical properties of graphene nanocomposites: A multiscale finite element prediction. *Composite Structures, 132*, 536–544. https://doi.org/10.1016/j.compstruct.2015.05.078
7. Cao, L., Yuan, D., Xu, C., & Chen, Y. (2017). Biobased, self-healable, high strength rubber with tunicate cellulose nanocrystals. *Nanoscale, 9*(40), 15696–15706.
8. Sharma, S., Verma, A., Rangappa, S. M., Siengchin, S., & Ogata, S. (2023). Recent progressive developments in conductive-fillers based polymer nanocomposites (CFPNC's) and conducting polymeric nanocomposites (CPNC's) for multifaceted sensing applications. *Journal of Materials Research and Technology, 26*, 5921–5974.
9. Ma, L., Dong, X., Chen, M., et al. (2017). Fabrication and water treatment application of carbon nanotubes (CNTs)-based composite membranes: A review. *Membranes (Basel), 7*(1), 16. https://doi.org/10.3390/membranes7010016
10. Su, X., Wang, R., Li, X., et al. (2022). A comparative study of polymer nanocomposites containing multi-walled carbon nanotubes and graphene nanoplatelets. *Nano Materials Science, 4*(3), 185–204.
11. Liu, T., Jia, S., Kowalewski, T., Matyjaszewski, K., Casado-Portilla, R., & Belmont, J. (2006). Water-dispersible carbon black nanocomposites prepared by surface-initiated atom transfer radical polymerization in protic media. *Macromolecules, 39*(2), 548–556. https://doi.org/10.1021/ma051659y
12. Zhu, X., Dudchenko, A. V., Khor, C. M., He, X., Ramon, G. Z., & Jassby, D. (2018). Field-induced redistribution of surfactants at the oil/water interface reduces membrane fouling on electrically conducting carbon nanotube UF membranes. *Environmental Science & Technology, 52*(20), 11591–11600. https://doi.org/10.1021/acs.est.8b02578
13. Wang, Z., Wang, X., Zheng, T., et al. (2022). High flux nanofiltration membranes with double-walled carbon nanotube (DWCNT) as the interlayer. *Membranes (Basel), 12*(10), 1011. https://doi.org/10.3390/membranes12101011
14. Wu, Y., Zhan, K., Yang, Z., Sun, W., Zhao, B., & Yan, Y. (2019). Graphene oxide/Al composites with enhanced mechanical properties fabricated by simple electrostatic interaction and powder metallurgy. *Journal of Alloys and Compounds, 775*, 233–240. https://doi.org/10.1016/j.jallcom.2018.10.158
15. Jayaramulu, K., Datta, K. K. R., Rösler, C., et al. (2016). Biomimetic superhydrophobic/superoleophilic highly fluorinated graphene oxide and ZIF-8 composites for oil–water separation. *Angewandte Chemie, International Edition, 55*(3), 1178–1182.
16. Li, J., Liu, L. Q., Xu, X. D., et al. (2015). Development of a measurement and control system for a 40l/h helium liquefier based on siemens PLC S7–300. *Physics Procedia, 67*, 1181–1186.
17. Huang, Y., Li, H., Wang, L., et al. (2015). Ultrafiltration membranes with structure-optimized graphene-oxide coatings for antifouling oil/water separation. *Advanced Materials Interfaces, 2*(2), 1400433.

18. Chang, Y., Shen, Y., Kong, D., et al. (2017). Fabrication of the reduced preoxidized graphene-based nanofiltration membranes with tunable porosity and good performance. *RSC Advances, 7*(5), 2544–2549. https://doi.org/10.1039/C6RA24746F
19. Zhao, X., Su, Y., Liu, Y., Li, Y., & Jiang, Z. (2016). Free-standing graphene oxide-palygorskite nanohybrid membrane for oil/water separation. *ACS Applied Materials & Interfaces, 8*(12), 8247–8256.
20. Khan, M., Assal, M. E., Tahir, M. N., et al. (2022). Graphene/inorganic nanocomposites: Evolving photocatalysts for solar energy conversion for environmental remediation. *Journal of Saudi Chemical Society, 26*(6), 101544.
21. Shi, M., Tang, C., Yang, X., et al. (2017). Superhydrophobic silica aerogels reinforced with polyacrylonitrile fibers for adsorbing oil from water and oil mixtures. *RSC Advances, 7*(7), 4039–4045.
22. Sheoran, K., Kaur, H., Siwal, S. S., Saini, A. K., Vo, D.-V. N., & Thakur, V. K. (2022). Recent advances of carbon-based nanomaterials (CBNMs) for wastewater treatment: Synthesis and application. *Chemosphere, 299*, 134364.
23. Li, G., Du, R., Cao, Z., et al. (2024). Research progress in graphene-based adsorbents for wastewater treatment: Preparation, adsorption properties and mechanisms for inorganic and organic pollutants. *C, 10*(3), 78.
24. Prekodravac, J. R., Kepić, D. P., Colmenares, J. C., Giannakoudakis, D. A., & Jovanović, S. P. (2021). A comprehensive review on selected graphene synthesis methods: From electrochemical exfoliation through rapid thermal annealing towards biomass pyrolysis. *Journal of Materials Chemistry C, 9*(21), 6722–6748.
25. Bansal, S. A., Khanna, V., Twinkle, Singh, A. P., & Kumar, S. (2022). Small percentage reinforcement of carbon nanotubes (CNTs) in epoxy(bisphenol-A) for enhanced mechanical performance. *Materials Today: Proceedings, 61*, 275–279. https://doi.org/10.1016/j.matpr.2021.09.225
26. Liu, Y., Su, Y., Cao, J., et al. (2017). Synergy of the mechanical, antifouling and permeation properties of a carbon nanotube nanohybrid membrane for efficient oil/water separation. *Nanoscale, 9*(22), 7508–7518.
27. Gao, S. J., Shi, Z., Zhang, W. B., Zhang, F., & Jin, J. (2014). Photoinduced superwetting single-walled carbon nanotube/TiO_2 ultrathin network films for ultrafast separation of oil-in-water emulsions. *ACS Nano, 8*(6), 6344–6352.
28. Gao, C., Sun, Z., Li, K., et al. (2013). Integrated oil separation and water purification by a double-layer TiO_2-based mesh. *Energy & Environmental Science, 6*(4), 1147–1151.
29. Xue, Z., Cao, Y., Liu, N., Feng, L., & Jiang, L. (2014). Special wettable materials for oil/water separation. *Journal of Materials Chemistry A, 2*(8), 2445–2460.
30. Ren, X., Chen, C., Nagatsu, M., & Wang, X. (2011). Carbon nanotubes as adsorbents in environmental pollution management: A review. *Chemical Engineering Journal, 170*(2–3), 395–410.
31. Mar, A. (2021). *Carbon-based nanomaterials*. RSC Publishing.
32. Ali, H., Khan, E., & Ilahi, I. (2019). Environmental chemistry and ecotoxicology of hazardous heavy metals: Environmental persistence, toxicity, and bioaccumulation. *Journal of Chemistry, 2019*(1), 6730305.
33. Ali, S., Rehman, S. A. U., Luan, H.-Y., Farid, M. U., & Huang, H. (2019). Challenges and opportunities in functional carbon nanotubes for membrane-based water treatment and desalination. *Science of the Total Environment, 646*, 1126–1139.
34. Liu, H., Cao, C.-Y., Wei, F.-F., et al. (2014). Flexible macroporous carbon nanofiber film with high oil adsorption capacity. *Journal of Materials Chemistry A, 2*(10), 3557–3562.
35. Tai, M. H., Gao, P., Tan, B. Y. L., Sun, D. D., & Leckie, J. O. (2014). Highly efficient and flexible electrospun carbon–silica nanofibrous membrane for ultrafast gravity-driven oil–water separation. *ACS Applied Materials & Interfaces, 6*(12), 9393–9401.

36. Al-Saleh, M. H., & Sundararaj, U. (2011). Review of the mechanical properties of carbon nanofiber/polymer composites. *Composites. Part A, Applied Science and Manufacturing, 42*(12), 2126–2142.
37. Mohapatra, L., Cheon, D., & Yoo, S. H. (2023). Carbon-based nanomaterials for catalytic wastewater treatment: A review. *Molecules, 28*(4), 1805.
38. Pikula, K., Johari, S. A., Santos-Oliveira, R., & Golokhvast, K. (2024). Joint toxicity and interaction of carbon-based nanomaterials with co-existing pollutants in aquatic environments: A review. *International Journal of Molecular Sciences, 25*(21), 11798.
39. Carmalin Sophia, A., Lima, E. C., Allaudeen, N., & Rajan, S. (2016). Application of graphene based materials for adsorption of pharmaceutical traces from water and wastewater—A review. *Desalination and Water Treatment, 57*(57), 27573–27586. https://doi.org/10.1080/19443994.2016.1172989
40. Xu, L., & Wang, J. (2017). The application of graphene-based materials for the removal of heavy metals and radionuclides from water and wastewater. *Critical Reviews in Environmental Science and Technology, 47*(12), 1042–1105. https://doi.org/10.1080/10643389.2017.1342514
41. Huang, Z.-H., Zheng, X., Lv, W., Wang, M., Yang, Q.-H., & Kang, F. (2011). Adsorption of lead(II) ions from aqueous solution on low-temperature exfoliated graphene nanosheets. *Langmuir, 27*(12), 7558–7562. https://doi.org/10.1021/la200606r
42. Sitko, R., Turek, E., Zawisza, B., et al. (2013). Adsorption of divalent metal ions from aqueous solutions using graphene oxide. *Dalton Transactions, 42*(16), 5682–5689. https://doi.org/10.1039/c3dt33097d
43. Gong, J.-L., Wang, B., Zeng, G.-M., et al. (2009). Removal of cationic dyes from aqueous solution using magnetic multi-wall carbon nanotube nanocomposite as adsorbent. *Journal of Hazardous Materials, 164*(2–3), 1517–1522.
44. Inyang, M., Gao, B., Zimmerman, A., Zhang, M., & Chen, H. (2014). Synthesis, characterization, and dye sorption ability of carbon nanotube–biochar nanocomposites. *Chemical Engineering Journal, 236*, 39–46.
45. Senusi, F., Shahadat, M., Ismail, S., & Hamid, S. A. (2018). Recent advancement in membrane technology for water purification. In *Modern age environmental problems and their remediation* (pp. 147–167). Springer.
46. Muneeb Ur Rahman Khattak, M., Zahoor, M., Muhammad, B., Khan, F. A., Ullah, R., & AbdEl-Salam, N. M. (2017). Removal of heavy metals from drinking water by magnetic carbon nanostructures prepared from biomass. *Journal of Nanomaterials, 2017*(1), 5670371.
47. Tian, T., Shi, X., Cheng, L., et al. (2014). Graphene-based nanocomposite as an effective, multifunctional, and recyclable antibacterial agent. *ACS Applied Materials & Interfaces, 6*(11), 8542–8548.
48. Zarei, M. (2017). Application of nanocomposite polymer hydrogels for ultra-sensitive fluorescence detection of proteins in gel electrophoresis. *TrAC Trends in Analytical Chemistry, 93*, 7–22.
49. Sudharsana, C., Anvarsha, N., & Kalyani, P. (2024). Carbon-based nanocomposites: A comprehensive review of their multifunctional applications. In *Nanocomposites—Properties, preparations and applications*. IntechOpen.
50. Feng, J., Su, B. L., Xia, H., et al. (2021). Printed aerogels: Chemistry, processing, and applications. *Chemical Society Reviews, 50*(6), 3842–3888. https://doi.org/10.1039/c9cs00757a
51. Singh, S., Bhardwaj, S., Tiwari, P., Dev, K., Ghosh, K., & Maji, P. K. (2024). Recent advances in cellulose nanocrystals-based sensors: A review. *Materials Advances, 5*, 2622–2654.
52. Domingues, R. M. A., Gomes, M. E., & Reis, R. L. (2014). The potential of cellulose nanocrystals in tissue engineering strategies. *Biomacromolecules, 15*(7), 2327–2346.
53. Kandhola, G., Park, S., Lim, J.-W., et al. (2023). Nanomaterial-based scaffolds for tissue engineering applications: A review on graphene, carbon nanotubes and nanocellulose. *Tissue Engineering and Regenerative Medicine, 20*(3), 411–433.

54. Manawi, Y. M., Ihsanullah, S. A., Al-Ansari, T., & Atieh, M. A. (2018). A review of carbon nanomaterials' synthesis via the chemical vapor deposition (CVD) method. *Materials (Basel), 11*(5), 822.
55. Sharma, A., Karn, R. K., & Pandiyan, S. K. (2014). Synthesis of TiO_2 nanoparticles by sol-gel method and their characterization. *Journal of Basic and Applied Engineering Research, 1*(9), 1–5.
56. Zhang, Z., Du, J., Li, J., et al. (2021). Polymer nanocomposites with aligned two-dimensional materials. *Progress in Polymer Science, 114*, 101360.
57. Arora, B., & Attri, P. (2020). Carbon nanotubes (CNTs): A potential nanomaterial for water purification. *Journal of Composites Science, 4*(3), 135.
58. Wang, X., Chen, X., Yoon, K., Fang, D., Hsiao, B. S., & Chu, B. (2005). High flux filtration medium based on nanofibrous substrate with hydrophilic nanocomposite coating. *Environmental Science & Technology, 39*(19), 7684–7691.
59. Adu, P. C. O., Aakyiir, M., Su, X., et al. (2024). Challenges and advancements in Elastomer/CNT nanocomposites with mechanochemical treatment, reinforcement mechanisms and applications. *Smart Materials in Manufacturing, 2*, 100053.
60. Haidar, A., Guimón, J., & Alon, I. (2024). Can graphene fuel a transformative change in energy storage technologies? A scenario analysis for the next two decades. *Technological Forecasting and Social Change, 202*, 123278.
61. Shalan, A. E., Makhlouf, A. S. H., & Lanceros-Méndez, S. (2022). Nanocomposites materials and their applications: Current and future trends. In *Advances in nanocomposite materials for environmental and energy harvesting applications* (pp. 3–14). Springer.
62. Zentel, R. (2020). Polymer coated semiconducting nanoparticles for hybrid materials. *Inorganics, 8*(3), 20.
63. Kausar, A. (2020). Polydimethylsiloxane-based nanocomposite: Present research scenario and emergent future trends. *Polymer-Plastics Technology and Materials, 59*(11), 1148–1166.
64. Penkova, A. V., Acquah, S. F. A., Piotrovskiy, L. B., Markelov, D. A., Semisalova, A. S., & Kroto, H. W. (2017). Fullerene derivatives as nano-additives in polymer composites. *Russian Chemical Reviews, 86*(6), 530.
65. Senthil Kumar, P., Yaashikaa, P. R., & Ramalingam, S. (2019). Efficient removal of nitrate and phosphate using graphene nanocomposites. In *A new generation material graphene: Applications in water technology* (pp. 287–307). Springer.
66. Mohamed, I. M. A., Dao, V.-D., Liu, C., Barakat, N. A. M., & Choi, H.-S. (2019). Synthesis of electrospun 1D-photoanode nanocomposite based on electrospinning followed by hydrothermal treatment for highly efficient liquid-junction photovoltaic devices. *Journal of Sol-Gel Science and Technology, 91*, 342–352.
67. Katseli, V., Economou, A., & Kokkinos, C. (2021). Smartphone-addressable 3D-printed electrochemical ring for nonenzymatic self-monitoring of glucose in human sweat. *Analytical Chemistry, 93*(7), 3331–3336. https://doi.org/10.1021/acs.analchem.0c05057
68. Vlachopoulos, N., Hagfeldt, A., Benesperi, I., et al. (2021). New approaches in component design for dye-sensitized solar cells. *Sustainable Energy & Fuels, 5*(2), 367–383.
69. Weerasinghe, J., Sen, S., Kumari, J., et al. (2021). Efficiency enhancement of low-cost metal free dye sensitized solar cells via non-thermal atmospheric pressure plasma surface treatment. *Solar Energy, 215*, 367–374.
70. Fu, G., Cho, E. J., Luo, X., et al. (2021). Enhanced light harvesting in panchromatic double dye-sensitized solar cells incorporated with bilayered TiO_2 thin film-based photoelectrodes. *Solar Energy, 218*, 346–353.
71. Bhatnagar, A., Hogland, W., Marques, M., & Sillanpää, M. (2013). An overview of the modification methods of activated carbon for its water treatment applications. *Chemical Engineering Journal, 219*, 499–511.
72. Fathy, N. A. (2021). Carbon-based nanomaterials for wastewater treatment. In *Aquananotechnology* (pp. 367–384). Elsevier.

73. Priyadharshini, S. D., Manikandan, S., Kiruthiga, R., et al. (2022). Graphene oxide-based nanomaterials for the treatment of pollutants in the aquatic environment: Recent trends and perspectives—A review. *Environmental Pollution, 306*, 119377.
74. Abdullah, T. A., Juzsakova, T., Hafad, S. A., et al. (2022). Functionalized multi-walled carbon nanotubes for oil spill cleanup from water. *Clean Technologies and Environmental Policy, 24*(2), 519–541.
75. Veeman, D., Shree, M. V., Sureshkumar, P., et al. (2021). Sustainable development of carbon nanocomposites: Synthesis and classification for environmental remediation. *Journal of Nanomaterials, 2021*(1), 5840645.
76. Su, H., & Hu, Y. H. (2021). Recent advances in graphene-based materials for fuel cell applications. *Energy Science & Engineering, 9*(7), 958–983.
77. Gulati, S., Kumar, S., Goyal, K., Arora, A., & Varma, R. S. (2022). Improving the air quality with functionalized carbon nanotubes: Sensing and remediation applications in the real world. *Chemosphere, 299*, 134468.
78. Ravi, S., & Vadukumpully, S. (2016). Sustainable carbon nanomaterials: Recent advances and its applications in energy and environmental remediation. *Journal of Environmental Chemical Engineering, 4*(1), 835–856.
79. Saguin, N. S. G., Maulik, G., Cao, X., et al. (2024). CNTs-based biosensors for enzyme detection. *Sensors and Actuators A: Physical, 377*, 115753.
80. Liang, X., Li, N., Zhang, R., et al. (2021). Carbon-based SERS biosensor: From substrate design to sensing and bioapplication. *NPG Asia Materials, 13*(1), 8.
81. Nayak, S., Blumenfeld, N. R., Laksanasopin, T., & Sia, S. K. (2017). Point-of-care diagnostics: Recent developments in a connected age. *Analytical Chemistry, 89*(1), 102–123.
82. Liang, Y., Liang, Y., Zhang, H., & Guo, B. (2022). Antibacterial biomaterials for skin wound dressing. *Asian Journal of Pharmaceutical Sciences, 17*(3), 353–384.
83. Dacrory, S., D'Amora, U., Longo, A., et al. (2024). Chitosan/cellulose nanocrystals/graphene oxide scaffolds as a potential pH-responsive wound dressing: Tuning physico-chemical, pro-regenerative and antimicrobial properties. *International Journal of Biological Macromolecules, 278*, 134643.
84. Mali, P., & Sherje, A. P. (2022). Cellulose nanocrystals: Fundamentals and biomedical applications. *Carbohydrate Polymers, 275*, 118668.
85. Siwal, S. S., Zhang, Q., Devi, N., & Thakur, V. K. (2020). Carbon-based polymer nanocomposite for high-performance energy storage applications. *Polymers (Basel), 12*(3), 505.
86. Murjani, B. O., Kadu, P. S., Bansod, M., Vaidya, S. S., & Yadav, M. D. (2022). Carbon nanotubes in biomedical applications: Current status, promises, and challenges. *Carbon Letters, 32*(5), 1207–1226.

Challenges and Limitations in the Application of Carbon-Based Nanocomposites

Ankoor Sura, Amanvir Singh, Arjun Singh, Vandana Sehrawat, Sudha Narwal, Bharti Dahiya, Lakshita Phor, and Sonia Nain

1 Introduction

Carbon-based material has been on a revolutionary path, from graphite to carbon nanotubes, fullerenes and rather recently graphene and graphitic carbon nitride. Such discoveries have transformed material science, and especially with the isolation of graphene in 2004, which brought about new properties such as high electrical conductivity, mechanical strength, and large surface area [1, 2]. These materials, engineered at the nanoscale, appear versatile candidates in tackling the world's most daunting problems: energy storage, environmental remediation, catalysis, and in biomedical sciences [3, 4].

However, such a rate of development generally hides really tough challenges and limitations tied to their synthesis, performance, and sustainability. Being carbon-based, it is known that nanocomposites have many unique properties [5, 6]. Practically speaking, however, they cause major problems for scaling, material stability under real conditions, environmental compatibility, and cost-effectiveness. Their optimization in performance or sustainable adoption needs overcoming these difficulties. This chapter fills that gap in knowledge with an in-depth look into "loop holes" within carbon-based nanocomposites with which future research and industrial improvements are guided. Knowing about them is not only some exercise in

A. Sura · A. Singh · A. Singh · V. Sehrawat · S. Narwal · B. Dahiya · S. Nain (✉)
Department of Chemistry, Deenbandhu Chhotu Ram University of Science and Technology, Murthal, Haryana, India
e-mail: sonianain.chem@dcrustm.org

L. Phor
University Centre for Research and Development, Chandigarh University, Mohali, Punjab, India

V. Khanna (ed.), *Carbon-Based Nanocomposites for Sustainable Applications, Volume I*, Lecture Notes in Nanoscale Science and Technology 38,
https://doi.org/10.1007/978-3-031-95306-4_13

academic rigor but is also essential toward the realization of the actual transformative potential [7–9].

In particular, carbon-based materials are frequently discovered as a new direction and solution for many different domains due to their outstanding advantages over conventional materials. The example of graphene is high conductance in contrast to other conventional metals, whereas graphitic carbon nitride possesses photocatalytic property, offering the greener paths for environmental remediation processes. Their vital importance for creating renewable energy technology seems clear from their utilization in solar cells, batteries, supercapacitors, and fuel cells. Their application in water purification and air pollution management and eventually, in the decomposition of microplastics gives great environmental value. However, alongside these benefits, issues such as poor dispersion, compatibility difficulties with matrices, and stable performance under operational conditions during a long period are to be addressed. Therefore, their limitations need to be studied in detail and possible alternatives for overcoming them should also be explored [10–12].

This chapter systematically discusses the issues and limitations of the carbon-based nanocomposites in its applications. From conversion to storage of energy sources, this book deals with matters that have arisen involving the solar cells, battery cells, supercapacitors, fuel cells, as well as hydrogen storage. Under the environmental applications section, aspects involve water purification, the management of air pollution and control, the degradation of pharmaceutical contaminants and micro-plastic. Further sections address challenges in thermoelectric and photothermal applications and biomedical and biosensing areas. This chapter, therefore, forms the basis of innovative strategies that can enhance performance, scalability, and sustainability in carbon-based nanocomposites through a holistic analysis of these limitations. Below is Fig. 1 that illustrates different carbon-based nanomaterials used in synthesis of nanocomposites.

2 Challenges and Limitations in Specific Applications

2.1 Energy Conversion and Storage

Energy conversion and storage technologies, the backbone of modern energy systems, have applications ranging from renewable energy harvesting to efficient storage and application in a huge number of electronic devices and vehicles. Among these, carbon-based nanomaterials have shown great promise due to their high electrical conductivity, large surface areas, and tunable properties. However, challenges persist in the field of multiple energy storage and conversion applications, including solar cells, chemical batteries, supercapacitors, fuel cells, and hydrogen storage [13–15].

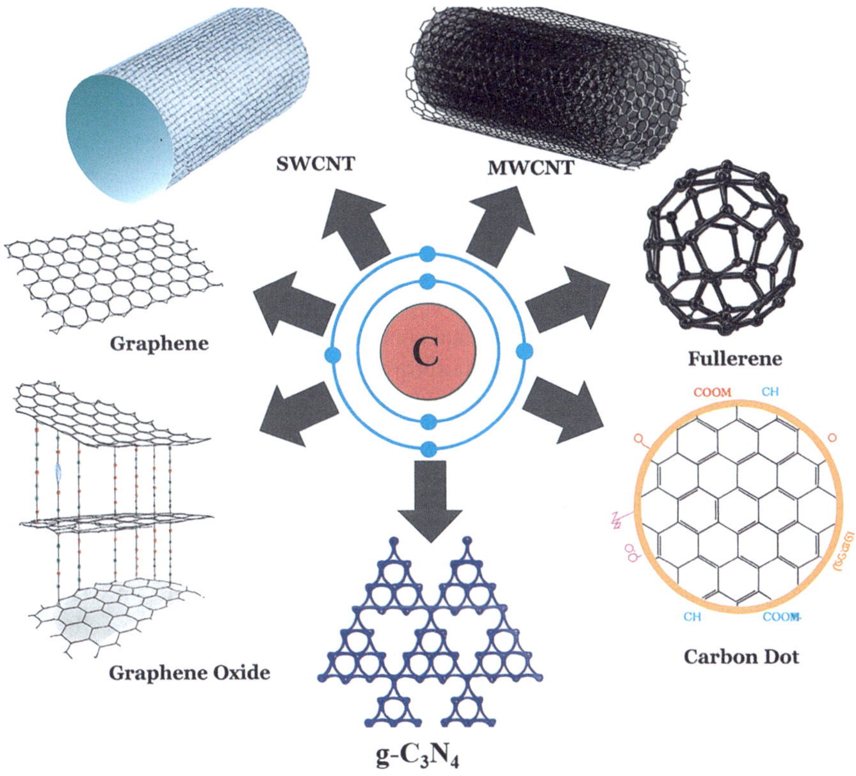

Fig. 1 Different carbon-based nanomaterials used in synthesis of nanocomposites

2.1.1 Solar Cells

Solar cells, especially organic and perovskite, hold great promise to meet the global challenges of energy and water. The challenges in their way for large-scale implementation are huge. Carbon nanocomposites in solar cells have shown promise in energy generation applications but have significant limitations in the areas of efficiency, stability, scalability, and operating conditions [16–18].

2.1.1.1 Efficiency and Stability in Carbon-Based Solar Cells

The carbon-based nanocomposite generally show lower efficiency compared to those based on silicon. For instance, one of the major issues is the non-optimal electron and hole transport across the device in these materials. The overall operational stability of these cells is also lower, specifically under real-world conditions of outdoor exposure, where it degrades faster than their silicon counterparts. For example, OPVs and PSCs are known to be subjected to degradation caused by moisture and oxygen and UV irradiation, which is abundant outdoor. Further development in the carbon-based nanocomposite area might be achieved with a carbon hybrid material like graphene-silicon hybrids or carbon composites combined with other stabilizing materials, for an improved stability and increased efficiency as well.

2.1.1.2 Scalability of Carbon Nanocomposite Solar Cells

The fabrication of carbon-based solar cells, especially in commercial size, is still challenging. The high synthesis cost processes and the restricted scalability of carbon nanomaterials limit their extensive application on a large scale. More importantly, despite their promises, carbon nanocomposites often face challenges in mass production due to issues concerning uniformity and reproducibility. Cost-effective techniques could help scale-up production and maintain high-quality material.

2.1.1.3 Light Absorption and Charge Separation

While carbon-based nanomaterials, especially graphene and CNTs, have high surface areas, their absorption of a very broad spectrum of light or efficient separation of charge carriers is still an issue. Thus, in organic and perovskite solar cells, poor absorption and suboptimal electron-hole pair separation lead to reduced performance from the device. This can be altered in principle by modifying the structural design in the material, that is, designing graphene CNT hybrid composites and incorporation of photoactive semiconductors, which might enhance light absorption and charge transport.

2.1.1.4 Lack of Standardized Performance Metrics

The challenge facing carbon nanocomposite-based solar cells is that their performance is measured by no standardized criteria. Temperature, humidity, and pressure variability as well as variations in device configuration may make the WER and PCE metrics highly variable, particularly in applications biased toward water purification. Maybe the standardization of testing protocols and benchmark performance metrics will further the understanding of what these devices can do in real-world conditions.

2.1.1.5 Durability and Environmental Conditions

Carbon nanocomposite-based photovoltaics pose serious drawback in terms of mechanical strengths which is the application concern to solar evaporator technology mainly used in wastewater, particularly industrial polluted industrial water as well as water pollution to seawater which comprises polluted aquatic ecosystems—all cases affecting mechanical degradation associated with adsorption of impurities on the materials of photovoltaic. Lastly, since the environmental agents would generally degrade the photovoltaic material strength at specific application duration, the resultant efficiencies and performance of most applications especially solar evaporator reduce. Such devices may be constructed to be more resilient with carbon nanocomposite materials, such as those mechanically resistant; graphene oxide-polymer composite composites; or graphene metal hybrids, such that it could support its decay while still remaining fully functional even in hostile operational environments.

2.1.1.6 Intermittent Nature of Solar Energy

The most important drawback is the intermittent nature of solar energy, dependent on the presence of sunlight. While carbon nanocomposite-based devices are very efficient under direct sunlight, their efficiency drops significantly at night or in cloudy and rainy weather conditions. This weakness can be overcome by combining the storage of energy systems or hybrid technologies that will ensure the constant functioning of these devices. For example, the incorporation of photothermal storage materials or battery systems inside a solar evaporator design could also increase efficiency in non-sunny times.

2.1.2 Chemical Batteries (Lithium-Ion, Sodium-Ion)

While carbon-based nanocomposite materials, including graphene, carbon nanotubes (CNTs), and fullerenes, offer exciting prospects for energy storage applications, particularly in lithium-ion (Li-ion) and sodium-ion (Na-ion) batteries, their widespread use is hindered by several key limitations [15, 19–21].

2.1.2.1 Energy Density

Despite the exceptional electrical conductivity and large surface areas of these composites, they typically face serious challenges to approach the level of energy density reached in standard battery chemistries, often lithium or sodium based significantly affecting such high-capacity applications as grid-scale energy storage and electric vehicles, for whom a much greater energy density forms a major metric. It also weakens competition to substances such as lithium cobalt oxide to this end since carbon-based materials are associated with relatively low energy density.

2.1.2.2 Cyclability and Stability

After many charging-discharging cycles, structure degradation causes severe capacity degradation in the carbon-based nanocomposite electrodes. In practice, for example, an available effective surface area to store ions gets lowered by the restacking layers of graphene and it raises electrical resistance that hinders ion transport. Overall cycle stability can be lowered down through this property. The structural defects caused by the low alignment also led to degradation in terms of conductivity for the composites by CNT's addition to electrode composites. Therefore, such materials have lower cycling stability and efficiency than the traditional electrode materials of a battery.

2.1.2.3 Integration with Other Battery Components

The challenge involves the successful incorporation of carbon-based nanocomposites with other constituents such as electrolytes and active materials. Although hybridization with alternative elements or compounds like metal oxides and sulfur might offer better performance, compatibility between these materials is difficult to ensure. Interfacial compatibility of carbon-based electrodes with electrolytes is a major requirement for eliminating possible problems of low ion conductivity or interfacial resistance that can otherwise affect the overall performance of the battery.

2.1.2.4 Scalability and Cost

High synthesis cost of carbon nanocomposites coupled with the manufacturing difficulty to scale-up for market use remains a significant deterring factor. The carbon nanomaterials involve complex processes currently that are not economically justified in terms of large-scale applications for the manufacturing of batteries. Since it costs too much to produce through high manufacturing cost, carbon-based nanocomposite makes for a less competitive although the established and considerably less expensive alternatives.

2.1.2.5 Electrocatalytic Activity

Carbon-based materials face great challenges to reach an efficient electrocatalytic performance in various energy storage and conversion applications even with their excellent electrical conductivity. For example, kinetics regarding sodium ion insertion or extraction from carbon-based electrodes, sodium-ion batteries suffer by relatively slow kinetics that results in relatively lower charge/discharge rates. Doping and surface functionalization to boost their electrocatalytic property is an active field of research, which remains very embryonic at this moment.

2.1.3 Supercapacitors

Carbon-based materials, especially in all-solid-state (ASC) supercapacitors with high mass loading, are plagued by several disadvantages that hinder their development for next-generation energy storage applications. Although they offer high conductivity, large surface area, and rapid charge/discharge rates, their integration into high-performance SCs involves several challenges [22–24].

2.1.3.1 Unclear Charge Storage Mechanisms

The mechanism of charge storage in carbon-based supercapacitors is still very unclear and results have varied from one study to another. This creates confusion on the appropriate interpretation of capacitance, among other electrochemical properties. Calculations of the performance metrics, such as energy density, power density, and capacitance (specific, areal, and volumetric), lack a standardized approach, making it impossible to compare different materials and devices in order to speed up research on the optimization of the performance of supercapacitors.

2.1.3.2 Fabrication Inconsistencies

Variation of mass loading, electrolytes types, nature of the material used in current collector, electrode preparation methods can very easily cause very large variations of device performance. Those variation modes affect uniformity and alter the total efficiency in supercapacitor device performance. Lack of standardization in preparation techniques for manufacturing supercapacitors becomes very challenging to compare in varied studies. Large-scale manufacturing becomes very difficult as a consequence.

2.1.3.3 Material Degradation

The carbon-based electrodes found in supercapacitors usually degrade with the repeated charge-discharge cycles, mainly under high-temperature or high-voltage conditions. Eventually, this causes degradation that leads to blockage in electroactive sites, causing the electrolyte to deplete and thus making the performance decrease in the overall supercapacitor. In addition, material dissolution in cycling negatively affects the long-term device stability and efficiency, especially in high-performance cases.

2.1.3.4 Voltage Limitations

Supercapacitors have a voltage window of 2.3 to 3.8 V, thus limiting its energy storage. This is one major limitation in cases that call for high energy storage, particularly for electric vehicles or grid storage, where other technologies are superseding the capabilities of supercapacitors. The voltage limitation, therefore, is the reason why carbon-based supercapacitors cannot gain wider use in energy-intense applications.

2.1.3.5 Electrode and Current Collector Challenges

For high mass loading all-solid-state supercapacitors, electrodes have to be ultrathin, lightweight, flexible, binder-free, and inexpensive. This is a great challenge to the manufacturing process. Ideal electrode design should have a high electroactive site number, electronic conductivity, and mechanical stability. The current collectors have to be ultralight and flexible but yet possess good electrochemical properties. All these factors increase the complexity in the synthesis of efficient, scalable, and stable electrodes and current collectors toward practical applications.

2.1.3.6 Aging and Thermal Management

Thermal management of supercapacitors is an important issue since prolonged cycling generates heat that may deteriorate the device. Aging phenomena such as capacitance reduction and internal resistance increase with time are normal and quite prominent at high mass loading or extreme cycling conditions. Thermal issues also cause more degradation and decrease the life of supercapacitors, making them unsuitable for long-term use in applications requiring high performance.

2.1.3.7 Scalability and Stability

Large-scale manufacturing of supercapacitors fails to achieve well-stabilized and uniformly material conditions. Supercapacitors should show good stability over many cycles in addition to providing long times of cycling and in extreme environmental parameters. There is, however, a potential issue with commercialization of massive manufacturing of carbon-based supercapacitors arising from these as material inconsistent, faults through manufacturing flaws, with such high-priced products.

2.1.4 Fuel Cells

Fuel cells, which transform chemical energy into electrical energy, show significant promise for renewable energy production. Carbon-based materials are crucial in increasing the performance of fuel cell electrodes and catalysts [25–27].

2.1.4.1 Electrocatalytic Activity

Carbon-based materials enhance the electrocatalytic activity of fuel cell electrodes. However, achieving effective electrocatalytic activity for reactions like the hydrogen evolution reaction (HER) remains a challenge. Carbon-based materials often require modifications to their structure or doping to achieve the necessary catalytic activity.

2.1.4.2 Durability and Stability

Fuel cells operating at high temperatures or under fluctuating conditions often face degradation of the electrode material over time, reducing efficiency. Carbon-based materials, while promising, often suffer from long-term instability when exposed to the harsh chemical and thermal conditions within fuel cells.

2.1.4.3 Cost and Scalability

The high cost of fuel cell technology, including carbon-based catalysts and electrodes, poses a major challenge. Additionally, the scalability of carbon-based materials for large-scale fuel cell production is still under research, with issues related to high production costs and the complexity of ensuring uniform quality across large batches.

2.1.5 Hydrogen Storage

Hydrogen storage is an integral part of the future of clean energy, mainly for fuel cell applications, and carbon-based materials are being researched for their potential in this respect because of their high surface area and the ability to adsorb considerable amounts of hydrogen. However, a few challenges remain before hydrogen storage employing carbon-based materials could become a feasible and economically feasible technology [28–30].

2.1.5.1 Limited Storage Capacity

The high surface area of carbon-based materials does not lead to high-density hydrogen storage. The storage capacity of hydrogen at moderate pressures and temperatures is necessary for any practical storage solutions. The inability to store sufficient hydrogen under reasonable conditions limits the material's application in real-world scenarios, especially for large-scale storage and use in fuel cells.

2.1.5.2 Thermodynamic and Kinetic Barriers

The adsorption of hydrogen on carbon materials is usually weak, and the thermodynamic conditions required for efficient hydrogen storage are harsh. In most cases, desorption and absorption kinetics of hydrogen are slow, which affects the efficiency of the hydrogen storage system. In order to make carbon-based materials more practical, these thermodynamic and kinetic barriers have to be overcome by modifying either the material or the storage systems.

2.1.5.3 Energy Density

Hydrogen has a lower energy density relative to the conventional fuel, thus making it harder to store large volumes efficiently. Even though carbon-based materials are developed to improve hydrogen storage capacity, it still remains a huge problem to obtain high volumetric and gravimetric energy densities without losing safety.

2.1.5.4 Material Compatibility

Hydrogen can induce embrittlement in certain materials, such as metals, which complicates the choice of materials for hydrogen storage vessels. Carbon-based materials need to be tested for their long-term resistance to hydrogen without degradation in order to ensure the structural integrity and reliability of storage containers over time.

2.1.5.5 Infrastructure Integration

The challenge of including hydrogen storage systems in the existing infrastructure is that large-scale investments and modifications are required. The current networks of distribution, storage, and pipelines are not suitable for hydrogen, both in terms of safety and scale. The upgrades to incorporate carbon-based hydrogen storage solutions require thoughtful planning and massive expense, making large-scale integration complicated.

2.1.5.6 Storage Cost

High costs for designing and maintaining hydrogen storage systems continue to be a primary hindrance to widespread applications. Innovations in the technology and materials involved need to be developed to significantly lower these costs, as only then will hydrogen storage become an economically viable proposition for industrial and large-scale applications.

2.1.5.7 Transportation

Transport of hydrogen is especially difficult because of its low volumetric energy density. Solutions being developed include high-pressure storage tanks, cryogenic containers, and dedicated pipelines. All these can be utilized for safe and cost-effective transport of hydrogen from the production site to storage or usage locations. These would place hydrogen in a suitable position as a practical and efficient energy carrier.

2.1.5.8 Cyclic Fatigue

Structural degradation over the period can occur when there are repeated cycles of filling and emptying in hydrogen storage systems. If it involves compressed hydrogen, it becomes critical for overcoming cyclic fatigue. Research should focus on developing materials of higher resilience along with advanced designs to improve the engineering approach and get over this deficiency.

2.2 Thermoelectric and Thermal Management Applications

The devices of carbon materials and nanomaterials for thermoelectric applications are highly limited by several challenges and issues that hamper the practical applications and commercialization of such devices [31–33]. The following are some of the important issues derived from the content given:

2.2.1 Low Thermoelectric Performance

Pristine carbon materials, including CNTs, graphene, and PEDOT:PSS, show low TE performance, generally with a ZT value less than 1. In order to achieve high TE efficiency close to that of inorganic semiconductors, considerable progress in material design by way of doping, nano structuring, and compositing is needed.

2.2.2 High Thermal Conductivity

The exceptionally high thermal conductivity of graphene and other carbon-based materials is advantageous in many, but detrimental in some applications, particularly concerning the thermoelectric efficiency of the devices. Ideally, for thermoelectric operation, the material must provide a path for charge flow but at the same time drastically limit the flow of heat. However, in most materials of high conductivity, there is too fast dissipation of heat, causing lowering of the thermoelectric efficiency. In tackling the problem, various approaches, including structural modification and chemical decoration, have been put forward. The latter is generally less explored or optimized and therefore leaves broad prospects of better solutions to enhancing TE performance.

2.2.3 Challenges in Bulk Network Formation

In order for thermoelectric devices to be efficient, carbon-based materials must be fabricated into well-ordered, three-dimensional (3D) bulk networks with good electrical conductivity and flexibility. It is challenging to achieve such aligned structures because it is difficult to organize the carbon materials at a macroscopic scale without losing their intrinsic properties. Disorganized networks impede the creation of effective temperature gradients, which are necessary for thermoelectric performance. Moreover, after forming such networks, it becomes an uphill task to sustain large temperature differences across the material, which will otherwise directly limit the overall performance of the device. The consequence of this fact necessitates research into the improved understanding and control of the fabrication of such networks toward better efficiency of the device.

2.2.4 Interface and Carrier Transport Issues

In the process of optimizing thermoelectric materials, a detailed understanding of the thermoelectric properties—such as the Seebeck coefficient (voltage generated in relation to the temperature difference measured) and the transport behavior of the charge carriers in the materials, like electrons or holes—is necessary. Yet these properties are less well understood for carbon-based materials, mostly at the nanoscale, where grain boundaries, defects, and impurities are dominant. The interfaces created between carbon materials and polymers or other constituents of thermoelectric devices are ill-characterized at the nanoscale although significant in influencing carrier transport, as well as the final efficiency, which is poorly accessible to analysis and subsequent optimization of performance in functional thermoelectric devices.

2.2.5 Scalability and Cost

Synthesizing high-performance TE materials at scale is economically and practically challenging. Methods like CVD, PVD, and additive manufacturing require optimization to become industrially viable, while maintaining quality and performance consistency.

2.2.6 Thermal and Structural Stability

Thermoelectric materials often are exposed to high temperature during their operational periods. These applications include waste heat recovery, for example. Prolonged exposure to such high temperatures results in the degradation of structure and performance over time. Thus, long-term reliability is the key concern in such situations. Carbon-based materials are typically stable at high temperatures, but they undergo structural changes over very long periods, reducing efficiency in such materials. For thermoelectric materials to be of commercial feasibility, these need to ensure stability in their structure and performance of numerous cycles of heating and cooling and therefore further research into this under harsh operating conditions involving their thermal and structural stability is indispensable.

2.2.7 Measurement Limitations

Accurately characterizing the thermoelectric properties—including electrical conductivity, Seebeck coefficient, and thermal conductivity—of materials is challenging, predominantly due to substrate effects and issues with experimental inconsistencies. Most standard measurement techniques are based on dubious assumptions and potentially introduce variability in the outcomes. More sophisticated methods of measuring thermal conductivity (such as the 3ω method) and thermal property measurements via time-domain thermoreflectance (TDTR) provide more reliable data but demand resources and specialized equipment; hence, they are not primarily used for routine testing but rather limit the potential scope of accurately assessing and comparing the performance of various kinds of thermoelectric materials under real-world conditions.

2.3 *Environmental Applications*

Environmental remediation has become an urgent priority due to the increasing levels of pollution-threatening ecosystems and public health. Carbon-based nanocomposites, with their exceptional properties such as high surface area, chemical tunability, and superior adsorption capacities, offer promising solutions for addressing various environmental challenges. These materials have shown potential in removing heavy metals and organic pollutants from water, controlling air pollution by capturing volatile organic compounds (VOCs) and particulate matter, degrading persistent pharmaceutical contaminants, and eliminating microplastics and other emerging pollutants. Despite their promise, the practical application of these nanocomposites faces several challenges, which must be addressed for effective large-scale deployment [34–36].

2.3.1 Water Purification (Heavy Metals, Organic Pollutants)

The purification of water is such an aspect of environmental sustainability that tackles the key issues of neutralizing water pollution and assuring access to clean and safe water resources. Carbon-based nanocomposites have become very prominent in this field due to characteristics such as a high surface area, excellent adsorption behavior, and tailorable functional groups. Although carbon-based nanocomposites possess potential, practical deployment in water purification faces many challenges that affect material stability, recovery, and cost-effectiveness. This section deliberates on these challenges and appraises the feasibility of carbon-based nanomaterials as a sustainable solution in the purification of water [37–40].

2.3.1.1 High Production Costs and Scalability

Advanced synthesis techniques, such as chemical vapor deposition, hydrothermal methods, and chemical exfoliation, are generally adopted for the production of carbon-based nanocomposites. The majority of the available methods require special equipment, controlled conditions, and costly reagents, raising the cost of production. Recent sophistication in the synthesis route has, to a certain extent, cut down the production cost, but the challenge remains with the scale-up for large industrial applications. Large-scale water treatment facilities need bulk nanocomposites, which are presently expensive and hence produced in inconsistent and irreproducible ways. It lowers the practicability of the real-world application of these materials.

2.3.1.2 Toxicity and Environmental Impact

Carbon-based nanocomposites may also have potential toxicity and an impact on the environment. Previous studies have indicated that SWCNTs and MWCNTs can show cytotoxic effects, produce reactive oxygen species, and tend to accumulate in biological and environmental systems. Long-term ecological effects of these materials are still poorly understood, therefore bringing into question their safety. Moreover, the disposal of used nanomaterials after water treatment may cause secondary pollution, which indicates the need for green synthesis procedures and sustainable methods of disposal.

2.3.1.3 Difficulty in Recovery from Aqueous Solutions

Recovery of carbon nanocomposites in powder or nanoscale forms from treated water is a difficult task because of their small size and high dispersibility. Traditional separation techniques, such as filtration or sedimentation, are generally inefficient, with a high risk of secondary contamination, where the residual nanomaterials can remain in the water or escape into the environment. To overcome this issue, magnetic carbon nanocomposites, which are a combination of carbon matrices with magnetic particles such as iron oxides, offer a highly effective solution. These materials can be easily and quickly recovered by using external magnetic fields. Such composites still maintain high adsorption and catalytic properties but eliminate the complexities of separation, reduce the risks of contamination, and enhance reusability for sustainable water purification practices.

2.3.1.4 Surface Functionalization and Solubility

Carbon-based nanomaterials, such as CNTs, have an intrinsic nature of hydrophobicity, and this makes them poorly dispersible and effective in an aqueous environment. Among the surface functionalization techniques, the attachment of polar functional groups improves their solubility to some extent and enhances the adsorption capacity. Nevertheless, such modifications increase the complexity of the synthesis, raise production costs, and may result in changes in the structure and chemistry of the materials. It always remains a great challenge to strike the optimal balance between functionalization and preserving the intrinsic properties of the nanocomposites.

2.3.1.5 Limited Adsorption Capacity in Complex Systems

Carbon nanocomposites show excellent performance in the removal of individual heavy metal cations because of their large surface area and numerous active sites. However, in realistic wastewater systems, these materials mostly face mixtures of contaminants, such as heavy metals, organic compounds, emerging pollutants, and biological impurities which are in competition for the limited available active adsorption sites. This competitive adsorption effectively lowers their overall efficacy and hence they become less effective for practical applications. Therefore, the development of advanced carbon nanocomposites with tailored selectivity for specific contaminants or designed for multi-pollutant environments is necessary. It has been observed that strategies involving functionalization, doping, or hierarchical structures can improve both the selectivity and the adsorption capacity of materials to enable efficient remediation of complex wastewater.

2.3.1.6 pH Dependency and Selective Adsorption

The removal of contaminants, however, highly depends on pH, which makes the application of carbon-based nanocomposites very challenging in real-life wastewater treatment systems. Wastewater generally contains a mixture of contaminants with different chemical compositions and pH. This means that in order to improve the removal efficiency, pretreatment steps, such as those stabilizing pH, should be performed before the treatment processes, increasing operational costs and complexity. In addition, selective adsorption of target contaminants is still challenging since the competing species in the wastewater will also bind on the active sites of nanomaterials, thus deteriorating their efficiency.

2.3.1.7 Regeneration and Longevity

The key to sustainable water purification is the ability of carbon-based nanocomposites to regenerate and be reused. However, this material is often subject to surface fouling by contaminants, blocking active sites; structural degradation due to repeated cycles of adsorption-desorption; and the gradual loss of functional groups, which are very important in pollutant removal. All these problems result in declining regeneration efficiency over time, therefore reducing their long-term applicability and increasing operational costs. To handle this, designing more robust nanocomposites with more durability and stable active sites is necessary. Several strategies, such as functional surface coatings, better structural integrity, and integration of self-cleaning mechanisms, can significantly enhance their regeneration capacity and assure cost-effectiveness and sustainability.

2.3.1.8 Limited Long-Term Stability

Long-term stability is an important concern for carbon-based nanocomposites since the performance may degrade under adverse operational conditions, including changing pH, high salinity, and organic matter. These can cause changes in the structure and chemistry of the materials that lead to a decrease in adsorption capacity and efficiency. Longer exposure could also lead to fouling, leaching of the active components, or structural degradation. This calls for advanced material design and durability assessments so as to ensure performance in real-world applications.

2.3.1.9 Production Consistency and Quality

The biggest challenge in synthesizing carbon-based nanocomposites is ensuring consistent quality, mainly because of the large-scale production. The variability in synthesis conditions results in impurities and defects that compromise the structural and functional integrity of the nanomaterials, which significantly affects their pollutant removal efficiency. In this regard, standardized and reproducible protocols of synthesis that ensure homogeneity in the properties of materials have to be developed. Only advanced quality control measures, scalable green synthesis methods, and rigorous testing can assure the industrial application performance consistency of the products.

2.3.2 Removal of Microplastics and Nanoplastics

The increasing levels of microplastics (MPs) and nanoplastics (NPs), along with other emerging contaminants in aquatic systems, present considerable ecological and health risks. Thus, innovative solutions for water treatment are urgently needed. Because of their high adsorption capacities, tunable functional groups, and the ability to be adapted to hybrid systems such as filtration and magnetic separation, carbon-based nanocomposites have appeared as promising materials for the efficient removal of MPs and NPs. These nanocomposites interact with MPs and NPs through adsorption, surface modification, and catalytic degradation, hence showing great potential for advanced technologies of water purification. However, several challenges and operational limitations have been found to be basic constraints in their practical applications [41–45].

2.3.2.1 High Coagulant Demand for MP/NP Removal

The negatively charged surface of MPs promotes their interaction with chemical coagulants, which increases the needed dosages of aluminum and iron-based coagulants for effective coagulation. The treatment cost is further increased, and the possibility of secondary pollution due to the residual coagulant ions becomes a concern. However, the efficiency in such coagulation tends to be very sensitive toward change in water properties like pH, presence of dissolved organic matter (DOM), hence complicating its practical implementation.

2.3.2.2 Membrane Fouling and Filtration Challenges

MPs and NPs contribute largely to membrane fouling in advanced filtration systems. The physical interaction of these foulants with membranes leads to pore blockages and the formation of cake layers that significantly reduce water permeability and increase the frequency of maintenance. During treatment, the conversion of MPs into smaller and more reactive NPs further escalates these fouling phenomena, which lead to high operational costs with regard to frequent membrane replacements.

2.3.2.3 Biofilm Formation and Biological Fouling

The surfaces of MPs and NPs serve as good substrates for biofilm formation, on which microorganisms can attach and proliferate. The formed biofilms reduce the adsorption efficiency of carbon-based nanocomposites, thereby promoting biological fouling in water treatment systems. The released microorganisms associated with biofilm from MPs risk secondary contamination in treated water, thereby undermining the safety and effectiveness of treatment processes.

2.3.2.4 Reduced Disinfection Efficiency

MPs and NPs can interfere with conventional methods of disinfection, including chlorination, ozonation, and UV irradiation. Biofilms on MPs can protect pathogens from the access of disinfectants; MPs can also physically block UV light and reduce its effectiveness. Furthermore, MPs interact with ozone and reduce its availability for microbial inactivation. This creates a need for more effort in disinfection, raising the complexity and cost of water treatment operations.

2.3.2.5 Challenges in Adsorbent Regeneration and Reuse

Regeneration and reuse of carbon-based nanocomposites are very essential for sustainable applications but are still technologically challenging. Most regeneration methods developed to date, such as thermal and chemical processes, are energy-consuming and often cannot renew the adsorption capacity fully. The incomplete desorption of MPs and NPs further limits the practicality of these materials, which makes it much more important to develop cost-effective and efficient techniques for regeneration.

2.3.2.6 Risk of Secondary Pollution from Nanocomposites

The potential for releasing harmful substances, including metal ions or nanoparticles into the water systems, makes the carbon-based nanocomposites an issue of concern in terms of the environment. The biggest challenge posed by these secondary pollutants in water treatment is in complex water matrices with high salinity and organic content, where the stability of nanocomposites may be threatened. This, therefore, calls for stringent material stability assessment and the development of more robust nanocomposites to mitigate such risks and ensure safe and effective use in water treatment systems.

2.3.3 Air Pollution Control (e.g., VOCs, Particulate Matter)

Carbon-based nanocomposites have shown immense potential in mitigating air pollution through applications such as adsorption, catalysis, and sensing of harmful pollutants. Their high surface area, tunable chemical properties, and superior conductivity make them suitable for capturing and neutralizing a range of airborne contaminants, including volatile organic compounds (VOCs) and particulate matter. However, despite their promising attributes, their practical application in air pollution control is hindered by several challenges and limitations [46–48].

2.3.3.1 Low Reactivity of Pollutants

Many air contaminants, such as VOCs, have chemically inert and nonpolar nature, leading to low interaction efficiency with the surface of carbon nanocomposites. This would further reduce their efficacy in capturing or neutralizing the pollutants. Higher reactivity by means of pre-concentration techniques or chemical modification of nanocomposites would further improve the performance but add complexity to increase the cost of air-purifying systems.

2.3.3.2 Aggregation of Nanomaterials

Carbon nanocomposites usually face aggregation resulting from van der Waals forces, thereby reducing the active surface area that is crucial for adsorption, catalysis, or sensing. This aggregation significantly reduces their efficiency and requires extra measures such as surfactants or oxidation processes to preserve the dispersion. However, such treatments can alter the intrinsic properties of the material itself, probably at the cost of their performance in applications related to air pollution.

2.3.3.3 Environmental and Health Concerns

The production and lifecycle management of carbon nanocomposites are associated with significant environmental and health risks. Many of them are synthesized using hazardous chemicals and energy-intensive processes that contribute to environmental degradation. In addition, their release into the ecosystems could lead to bioaccumulation and result in toxic effects on organisms. Long-term studies assessing these impacts are needed, as most of the current research is limited to short-term observations.

2.3.3.4 High Operating Temperatures

Many air pollution control applications of carbon nanocomposites, such as catalytic air purification and gas sensing, are required to operate at high temperatures (usually 400–550 °C) to reach optimal performance. Such high temperatures lead to increased energy input and operational costs, thus increasing the carbon footprint of these systems. In order to reduce this limitation, the development of low-temperature catalysts or alternative sensing mechanisms is one of the most important research tasks.

2.3.3.5 High Costs and Complex Fabrication

The synthesis of carbon nanocomposites usually requires expensive precursor materials and complicated synthesis methods, such as chemical vapor deposition and wet chemical synthesis. Additional performance-enhancing modifications, like doping or functionalization, also use special reagents and technologies that further increase the production costs. These economic challenges impede the scale-up of applications and industrial uses of carbon nanocomposites in air pollution control, especially for poor resource-challenged settings.

2.3.3.6 Lack of Standardization and Reproducibility

Inconsistent fabrication methods and testing protocols result in variable performance of carbon nanocomposites across different applications. For example, many gas sensors prepared from such materials face performance drift over time, either due to non-standardized calibration or material degradation. This lack of reproducibility is a key barrier to regulatory approval and industrial acceptance, and it clearly needs standardized manufacturing processes and rigorous quality control measures.

2.3.4 Pharmaceutical and Pollutant Removal

Carbon-based nanomaterials have emerged as promising materials for removing pharmaceutical pollutants and other contaminants from aqueous systems. Their large surface area, tunable functionalization, and excellent adsorption capacity make them effective in capturing diverse pollutants. However, their application in water treatment is limited by several technical and practical challenges that must be addressed to enhance their efficiency and sustainability [49, 50].

2.3.4.1 Low Reactivity of Carbon Materials

Although the pristine carbon materials, including graphene and CNTs, hold a high adsorption capacity, they intrinsically lack catalytic activity for the degradation of certain pollutants. Therefore, they have to be doped with active elements or hybridized with metal oxides to enhance their catalytic activity. This, however, increases the complexity and cost of synthesis.

2.3.4.2 Aggregation and Restacking

The tendency of carbon-based materials to aggregate or restack, arising from van der Waals forces, leads to a reduction in the effective surface area and accessibility of active sites. It dramatically compromises the efficiency of pollutant removal. In order to keep them in dispersion, advanced synthesis techniques, such as exfoliation or surface functionalization, have to be applied; however, these methods can be time-consuming and expensive.

2.3.4.3 Rapid Electron-Hole Recombination

High electron-hole pair recombination rates are common in carbon-based nanocomposites applied to photocatalytic applications, which significantly restrains the generation of reactive oxygen species for pollutant degradation. The construction of heterostructures, co-doping, or coupling with plasmonic materials can enhance charge separation and hence photocatalytic activity.

2.3.4.4 Photocorrosion and Stability Issues

Carbon-based photocatalysts, especially those combined with metal oxides, are prone to photocorrosion and structural degradation under long-term operation, which reduces their efficiency in successive cycles; hence, more robust materials or protecting coatings must be developed for long-term stability.

2.3.4.5 Difficult Recovery and Recycling

The small particle size and high dispersibility of carbon-based nanomaterials in water systems make their recovery challenging, often requiring advanced separation techniques. Inefficient recovery not only increases operational costs but also raises concerns about secondary contamination if residual nanomaterials are released into the environment.

2.3.4.6 Limited Light Absorption

Most carbon-based photocatalysts suffer from a drawback in efficiently utilizing visible light, even with tunable band gaps, which severely limits their effectiveness under natural sunlight and often requires the use of artificial UV light sources, further increasing the energy demands and costs of the treatment process.

2.3.4.7 High Synthesis Costs and Scalability

The preparation of high-quality carbon-based nanocomposites usually involves complex, energy-intensive multi-step synthesis processes using expensive reagents and advanced equipment. Such barriers strongly limit scalability, making it impossible to produce these materials on a large scale at an economically feasible cost. Consequently, their wide application to industrial-scale water treatment is still limited, rousing intensive research on more feasible and cost-effective production methods.

2.3.4.8 Incomplete Mineralization

Carbon-based photocatalysts tend to degrade pollutants to intermediate byproducts rather than fully mineralizing them to harmless compounds such as CO_2 and water. These intermediates can sometimes be more toxic or persistent in the environment than the original contaminants. Consequently, further treatment steps become necessary in order to annihilate these byproducts, which complicates the remediation process and increases both operational costs and environmental risks.

2.3.4.9 Environmental and Toxicity Concerns

The release of carbon-based nanomaterials into the environment during or after their application poses significant ecological and health risks. Long-term impacts are difficult to assess with limited toxicological data. These materials may persist in ecosystems and interfere with biological processes. Recovery methods and eco-friendly strategies for their disposal have been developed to reduce environmental pollution and assure sustainable applications.

3 Conclusion

Carbon-based nanocomposites have great potential to bring about new applications in energy storage, environmental remediation, and renewable energy technologies. However, challenges in scalability, durability, cost, and reusability impede their practical implementation. Innovative strategies should be developed to bridge the gap between research and real-world applications by fully addressing these limitations. Figure 2 illustrates challenges and limitations across various applications of carbon-based nanocomposites.

The first step toward overcoming the challenges in scalability is the development of green, energy-efficient, and economically viable synthesis methods. These may include the use of abundant and low-cost precursors, microwave-assisted techniques, or hydrothermal methods with low energy consumption. In addition, standardization of the synthesis protocols is necessary for reproducibility and consistency at an industrial scale. Integration of automated and high-throughput manufacturing techniques can further lower the costs and improve production efficiency, thus making the materials more accessible for large-scale applications. Another compelling issue is durability under operational conditions, especially for applications involving harsh environmental conditions such as fluctuating pH, high salinity, or organic pollutants. A new class of nanocomposites can be engineered to increase both structural and chemical stability in order to mitigate long-term degradation. In addition, the design of functionalization/doping strategies allows for tailoring materials to resist fouling while also improving selectivity to specific contaminants. In terms of their performance and stability, hybridizing carbon-based materials with metals, oxides, or polymers is another synergistic strategy. Recovery of nanomaterials by developing easily recoverable nanomaterials—like magnetic composites—is one

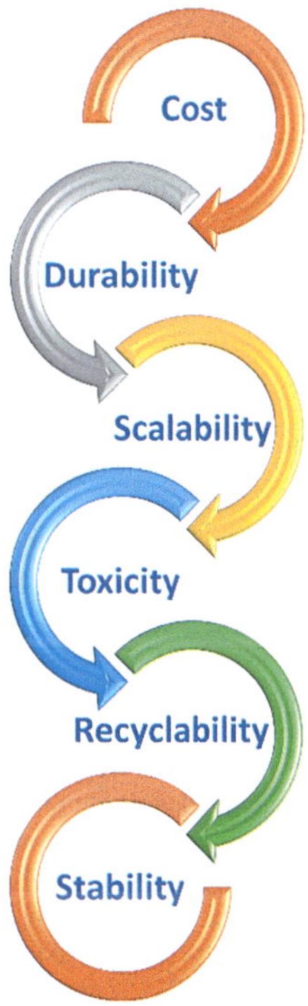

Fig. 2 Shared challenges and limitations across various applications of carbon-based nanocomposites

solution toward the challenges in reusability. Efficient separation from treated systems decreases secondary pollution and increases cost-effectiveness. To extend their life span and enhance their economic feasibility, advanced regeneration techniques by thermal or chemical treatment under mild conditions could be adapted for such materials. In this way, incorporating self-healing properties into nanocomposites enhances their durability and reusability through recovering minor structural damages. In order to have cost-effective applications, lifecycle assessments and recycling strategies are needed to reduce waste and environmental impact. The use of renewable or waste-derived carbon sources, such as biomass, for the synthesis of nanocomposites gives a sustainable alternative to conventional approaches. Moreover, incorporation of carbon-based materials into multifunctional devices can improve their usefulness—for example, integrating energy storage and pollutant removal in a single system—thereby offering the best cost-effectiveness. Innovative design can also help expand the scope of applications. For example, the design of nanomaterials that achieve complete mineralization of pollutants, rather than forming toxic intermediates, avoids the need for further treatment steps. In the same vein, the development of nanocomposites with hierarchical structures can overcome competitive adsorption in complex pollutant mixtures by enhancing adsorption capacities and catalytic activity. Lastly, interdisciplinary collaboration and public-private partnerships are important for fast-tracking the commercialization of these materials. Academia, industry stakeholders, and policymakers should work together to put in place enabling frameworks, including financial incentives for sustainable technologies and regulations related to safe disposal and recovery of nanomaterials.

With these challenges in mind, and the potential solutions presented, carbon-based nanocomposites can move from being just potential candidates to becoming transformative tools in solving global problems. Their effective and sustainable use holds the promise of advancing clean energy technologies, improving water and air quality, and ultimately contributing to a cleaner and more sustainable future.

References

1. Zhu, J. (2008). New solutions to a new problem. *Nature Nanotechnology, 3*, 528–529. https://doi.org/10.1038/nnano.2008.249
2. Song, X.-L., Chen, L., Gao, L.-J., Ren, J.-T., & Yuan, Z.-Y. (2024). Engineering g-C_3N_4 based materials for advanced photocatalysis: Recent advances. *Green Energy & Environment, 9*, 166–197. https://doi.org/10.1016/j.gee.2022.12.005
3. Liu, W., Li, Q., Yang, C., Shi, X., Wan, J., Jurado, M. J., Li, Y., Jiang, D., Chen, J., Qiao, W., Zhang, X., Fan, J., Peng, T., & He, Y. (2023). The role of underground salt caverns for large-scale energy storage: A review and prospects. *Energy Storage Materials, 63*, 103045. https://doi.org/10.1016/j.ensm.2023.103045
4. Liu, M., Xu, Y., Meng, Y., Wang, L., Wang, H., Huang, Y., Onishi, N., Wang, L., Fan, Z., & Himeda, Y. (2022). Heterogeneous catalysis for carbon dioxide mediated hydrogen storage technology based on formic acid. *Advanced Energy Materials, 12*, 2200817. https://doi.org/10.1002/aenm.202200817

5. Saleemi, M. A., Kong, Y. L., Yong, P. V. C., & Wong, E. H. (2022). An overview of antimicrobial properties of carbon nanotubes-based nanocomposites. *Advanced Pharmaceutical Bulletin, 12*, 449–465. https://doi.org/10.34172/apb.2022.049
6. Khan, M. E. (2021). State-of-the-art developments in carbon-based metal nanocomposites as a catalyst: Photocatalysis. *Nanoscale Advances, 3*, 1887–1900. https://doi.org/10.1039/D1NA00041A
7. Ambika, & Singh, P. P. (2021). Carbon-based nanocomposites: Preparation and application in environmental pollutants removal. In *Environmental remediation through carbon based nano composites* (Green energy and technology) (pp. 203–229). Springer. https://doi.org/10.1007/978-981-15-6699-8_10
8. Rivas-Sanchez, A., Cruz-Cruz, A., Gallareta-Olivares, G., González-González, R. B., Parra-Saldívar, R., & Iqbal, H. M. N. (2022). Carbon-based nanocomposite materials with multifunctional attributes for environmental remediation of emerging pollutants. *Chemosphere, 303*, 135054. https://doi.org/10.1016/j.chemosphere.2022.135054
9. Narwal, S., Sura, A., Singh, A., Azam, M., Alam, M., Duhan, S., & Nain, S. (2024). In-situ fabrication of N-doped carbon dots on novel blue-emitting Y_2TiWO_8: Ce^{3+} nanophosphor for thermal sensing and fingerprint detection applications. *Carbon N Y, 219*, 118768. https://doi.org/10.1016/j.carbon.2023.118768
10. Arun, J., Nirmala, N., Priyadharsini, P., Dawn, S. S., Santhosh, A., Gopinath, K. P., & Govarthanan, M. (2022). A mini-review on bioderived carbon and its nanocomposites for removal of organic pollutants from wastewater. *Materials Letters, 310*, 131476. https://doi.org/10.1016/j.matlet.2021.131476
11. Gul, A., Khaligh, N. G., & Julkapli, N. M. (2021). Surface modification of carbon-based nanoadsorbents for the advanced wastewater treatment. *Journal of Molecular Structure, 1235*, 130148. https://doi.org/10.1016/j.molstruc.2021.130148
12. Chen, T.-W., Kalimuthu, P., Veerakumar, P., Lin, K.-C., Chen, S.-M., Ramachandran, R., Mariyappan, V., & Chitra, S. (2022). Recent developments in carbon-based nanocomposites for fuel cell applications: A review. *Molecules, 27*, 761. https://doi.org/10.3390/molecules27030761
13. Chen, X., Cheng, P., Tang, Z., Xu, X., Gao, H., & Wang, G. (2021). Carbon-based composite phase change materials for thermal energy storage, transfer, and conversion. *Advanced Science, 8*, 2001274. https://doi.org/10.1002/advs.202001274
14. Zhang, T., & Ran, F. (2021). Design strategies of 3D carbon-based electrodes for charge/ion transport in lithium ion battery and sodium ion battery. *Advanced Functional Materials, 31*, 2010041. https://doi.org/10.1002/adfm.202010041
15. Kothandam, G., Singh, G., Guan, X., Lee, J. M., Ramadass, K., Joseph, S., Benzigar, M., Karakoti, A., Yi, J., Kumar, P., & Vinu, A. (2023). Recent advances in carbon-based electrodes for energy storage and conversion. *Advanced Science, 10*, e2301045. https://doi.org/10.1002/advs.202301045
16. Deshmukh, M. A., Park, S.-J., Hedau, B. S., & Ha, T.-J. (2021). Recent progress in solar cells based on carbon nanomaterials. *Solar Energy, 220*, 953–990. https://doi.org/10.1016/j.solener.2021.04.001
17. He, W., Zhou, L., Wang, M., Cao, Y., Chen, X., & Hou, X. (2021). Structure development of carbon-based solar-driven water evaporation systems. *Science Bulletin (Beijing), 66*, 1472–1483. https://doi.org/10.1016/j.scib.2021.02.014
18. Hadadian, M., Smått, J.-H., & Correa-Baena, J.-P. (2020). The role of carbon-based materials in enhancing the stability of perovskite solar cells. *Energy Environmental Sciences, 13*, 1377–1407. https://doi.org/10.1039/C9EE04030G
19. Poizot, P., Gaubicher, J., Renault, S., Dubois, L., Liang, Y., & Yao, Y. (2020). Opportunities and challenges for organic electrodes in electrochemical energy storage. *Chemical Reviews, 120*, 6490–6557. https://doi.org/10.1021/acs.chemrev.9b00482

20. Yuan, S., Lai, Q., Duan, X., & Wang, Q. (2023). Carbon-based materials as anode materials for lithium-ion batteries and lithium-ion capacitors: A review. *Journal of Energy Storage, 61*, 106716. https://doi.org/10.1016/j.est.2023.106716
21. Li, L., Zhang, D., Deng, J., Gou, Y., Fang, J., Cui, H., Zhao, Y., & Cao, M. (2021). Carbon-based materials for fast charging lithium-ion batteries. *Carbon N Y, 183*, 721–734. https://doi.org/10.1016/j.carbon.2021.07.053
22. Mevada, C., & Mukhopadhyay, M. (2021). Limitations and recent advances in high mass loading asymmetric supercapacitors based on pseudocapacitive materials. *Industrial and Engineering Chemistry Research, 60*, 1096–1111. https://doi.org/10.1021/acs.iecr.0c04811
23. Wang, Y., Zhang, L., Hou, H., Xu, W., Duan, G., He, S., Liu, K., & Jiang, S. (2021). Recent progress in carbon-based materials for supercapacitor electrodes: A review. *Journal of Materials Science, 56*, 173–200. https://doi.org/10.1007/s10853-020-05157-6
24. Dissanayake, K., & Kularatna-Abeywardana, D. (2024). A review of supercapacitors: Materials, technology, challenges, and renewable energy applications. *Journal of Energy Storage, 96*, 112563. https://doi.org/10.1016/j.est.2024.112563
25. Tellez-Cruz, M. M., Escorihuela, J., Solorza-Feria, O., & Compañ, V. (2021). Proton exchange membrane fuel cells (PEMFCs): Advances and challenges. *Polymers (Basel), 13*, 3064. https://doi.org/10.3390/polym13183064
26. Shaari, N., Kamarudin, S. K., & Bahru, R. (2021). Carbon and graphene quantum dots in fuel cell application: An overview. *International Journal of Energy Research, 45*, 1396–1424. https://doi.org/10.1002/er.5889
27. Hassan, M., Kanwal, S., Singh, R. S., Ali, M., Anwar, M., & Zhao, C. (2024). Current challenges and future perspectives associated with configuration of microbial fuel cell for simultaneous energy generation and wastewater treatment. *International Journal of Hydrogen Energy, 50*, 323–350. https://doi.org/10.1016/j.ijhydene.2023.08.134
28. Mulky, L., Srivastava, S., Lakshmi, T., Sandadi, E. R., Gour, S., Thomas, N. A., Shanmuga Priya, S., & Sudhakar, K. (2024). An overview of hydrogen storage technologies—Key challenges and opportunities. *Materials Chemistry and Physics, 325*, 129710. https://doi.org/10.1016/j.matchemphys.2024.129710
29. Rasul, M. G., Hazrat, M. A., Sattar, M. A., Jahirul, M. I., & Shearer, M. J. (2022). The future of hydrogen: Challenges on production, storage and applications. *Energy Conversion and Management, 272*, 116326. https://doi.org/10.1016/j.enconman.2022.116326
30. Hassan, Q., Algburi, S., Sameen, A. Z., Jaszczur, M., & Salman, H. M. (2024). Hydrogen as an energy carrier: Properties, storage methods, challenges, and future implications. *Environment Systems and Decisions, 44*, 327–350. https://doi.org/10.1007/s10669-023-09932-z
31. Zhang, Y., Zhang, Q., & Chen, G. (2020). Carbon and carbon composites for thermoelectric applications. *Carbon Energy, 2*, 408–436. https://doi.org/10.1002/cey2.68
32. Yun, J., Choi, S., & Im, S. H. (2021). Advances in carbon-based thermoelectric materials for high-performance, flexible thermoelectric devices. *Carbon Energy, 3*, 667–708. https://doi.org/10.1002/cey2.121
33. Dong, J., Suwardi, A., Tan, X. Y., Jia, N., Saglik, K., Ji, R., Wang, X., Zhu, Q., Xu, J., & Yan, Q. (2023). Challenges and opportunities in low-dimensional thermoelectric nanomaterials. *Materials Today, 66*, 137–157. https://doi.org/10.1016/j.mattod.2023.04.021
34. Gopinath, K. P., Vo, D.-V. N., Gnana Prakash, D., Adithya Joseph, A., Viswanathan, S., & Arun, J. (2021). Environmental applications of carbon-based materials: A review. *Environmental Chemistry Letters, 19*, 557–582. https://doi.org/10.1007/s10311-020-01084-9
35. Priya, A. K., Muruganandam, M., & Suresh, S. (2024). Bio-derived carbon-based materials for sustainable environmental remediation and wastewater treatment. *Chemosphere, 362*, 142731. https://doi.org/10.1016/j.chemosphere.2024.142731
36. Liu, D., Gu, W., Zhou, L., Wang, L., Zhang, J., Liu, Y., & Lei, J. (2022). Recent advances in MOF-derived carbon-based nanomaterials for environmental applications in adsorption and catalytic degradation. *Chemical Engineering Journal, 427*, 131503. https://doi.org/10.1016/j.cej.2021.131503

37. Sura, A., Singh, A., Singh, A., Narwal, S., Malik, P., Goyat, M. S., Mishra, Y. K., Nain, S., & Duhan, S. (2024). Efficient photo-oxidation of bisphenol a and tetracycline through sulfur-doped g-C_3N_4/CD heterojunctions. *Materials Advances, 5*, 5514–5526. https://doi.org/10.1039/D4MA00270A
38. Singh, A., Kaushik, V., Chahal, S., Goswami, A., & Nain, S. (2021). Efficient degradation of methylene blue dye and antibacterial performance of shape controlled RuO_2 nanocomposites. *ChemistrySelect, 6*, 10038–10050. https://doi.org/10.1002/slct.202102546
39. Chahal, S., Phor, L., Singh, S., Singh, A., Malik, J., Goel, P., Kumar, A., Kumar, S., Ankita, & Kumar, P. (2022). An efficient and unique method for the growth of spindle shaped Mg-doped cerium oxide nanorods for photodegradation of p-nitrophenol. *Ceramics International, 48*, 28961–28968. https://doi.org/10.1016/j.ceramint.2022.04.145
40. Singh, A., Goswami, A., & Nain, S. (2020). Enhanced antibacterial activity and photo-remediation of toxic dyes using Ag/SWCNT/PPy based nanocomposite with core–shell structure. *Applied Nanoscience, 10*, 2255–2268. https://doi.org/10.1007/s13204-020-01394-y
41. Padervand, M., Lichtfouse, E., Robert, D., & Wang, C. (2020). Removal of microplastics from the environment: A review. *Environmental Chemistry Letters, 18*, 807–828. https://doi.org/10.1007/s10311-020-00983-1
42. Shen, M., Song, B., Zhu, Y., Zeng, G., Zhang, Y., Yang, Y., Wen, X., Chen, M., & Yi, H. (2020). Removal of microplastics via drinking water treatment: Current knowledge and future directions. *Chemosphere, 251*, 126612. https://doi.org/10.1016/j.chemosphere.2020.126612
43. Zheng, H., Chen, Q., & Chen, Z. (2024). Carbon-based adsorbents for micro/nano-plastics removal: Current advances and perspectives. *Water Emerging Contaminants & Nanoplastics, 3*, 11. https://doi.org/10.20517/wecn.2023.74
44. Zhang, Y., Li, X., Xiao, M., Feng, Z., Yu, Y., & Yao, H. (2022). Effects of microplastics on soil carbon dioxide emissions and the microbial functional genes involved in organic carbon decomposition in agricultural soil. *Science of the Total Environment, 806*, 150714. https://doi.org/10.1016/j.scitotenv.2021.150714
45. Sura, A., & Nain, S. (2024). Visible light driven degradation of BPA and LDPE microplastic films using GO/SCN nanocomposite. *RSC Advances, 14*, 35336–35347. https://doi.org/10.1039/D4RA06055E
46. Gakhar, T., Basu, S., & Hazra, A. (2022). Carbon-metal oxide nanocomposites for selective detection of toxic and hazardous volatile organic compounds (VOC)—A review. *Green Analytical Chemistry, 1*, 100005. https://doi.org/10.1016/j.greeac.2022.100005
47. Han, B., Rupam, T. H., Chakraborty, A., & Saha, B. B. (2024). A comprehensive review on VOCs sensing using different functional materials: Mechanisms, modifications, challenges and opportunities. *Renewable and Sustainable Energy Reviews, 196*, 114365. https://doi.org/10.1016/j.rser.2024.114365
48. Saleem, H., Zaidi, S. J., Ismail, A. F., & Goh, P. S. (2022). Advances of nanomaterials for air pollution remediation and their impacts on the environment. *Chemosphere, 287*, 132083. https://doi.org/10.1016/j.chemosphere.2021.132083
49. Velempini, T., Prabakaran, E., & Pillay, K. (2021). Recent developments in the use of metal oxides for photocatalytic degradation of pharmaceutical pollutants in water—A review. *Materials Today Chemistry, 19*, 100380. https://doi.org/10.1016/j.mtchem.2020.100380
50. Oluwole, A. O., Omotola, E. O., & Olatunji, O. S. (2020). Pharmaceuticals and personal care products in water and wastewater: A review of treatment processes and use of photocatalyst immobilized on functionalized carbon in AOP degradation. *BMC Chemistry, 14*, 62. https://doi.org/10.1186/s13065-020-00714-1

Index

V. Khanna (ed.), *Carbon-Based Nanocomposites for Sustainable Applications, Volume I*, Lecture Notes in Nanoscale Science and Technology 38,
https://doi.org/10.1007/978-3-031-95306-4